T0389647

Emergence of Natural History

Series Editors

Aaron M. Bauer (*Villanova University, USA*)
Dominik Hünniger (*University of Hamburg, Germany*)
Andreas Weber (*University of Twente, The Netherlands*)

Editorial Board

Tom Baione (*AMNH, USA*)
Isabelle Charmantier (*Linnean Society, UK*)
Esther van Gelder (*Huygens ING, The Netherlands*)
Eric Jorink (*Huygens ING and Leiden University, The Netherlands*)
Sachiko Kusukawa (*Cambridge University, UK*)
Santiago Madriñan (*Universidad de Los Andes, Columbia*)
Dániel Margócsy (*University of Cambridge, UK*)
Henrietta McBurney Ryan (*independent scholar and art curator, UK*)
Staffan Müller-Wille (*University of Cambridge, UK*)
Florence Pieters (*University of Amsterdam, The Netherlands*)
Bert van de Roemer (*University of Amsterdam, The Netherlands*)
Kees Rookmaaker (*National University of Singapore*)
Paul Smith (*Leiden University, The Netherlands*)
Claudia Swan (*Washington University in St. Louis, USA*)
Mary Terrall (*UCLA, USA*)

VOLUME 6

The titles published in this series are listed at *brill.com/enh*

The Rhinoceros of South Asia

By

L.C. (Kees) Rookmaaker

With contributions by

Joachim K. Bautze
Kelly Enright

BRILL

LEIDEN | BOSTON

This is an open access title distributed under the terms of the CC BY-NC-ND 4.0 license, which permits any non-commercial use, distribution, and reproduction in any medium, provided no alterations are made and the original author(s) and source are credited. Further information and the complete license text can be found at https://creativecommons.org/licenses/by-nc-nd/4.0/

The terms of the CC license apply only to the original material. The use of material from other sources (indicated by a reference) such as diagrams, illustrations, photos and text samples may require further permission from the respective copyright holder.

The author acknowledges SOS RHINO for their generous support towards editorial assistance and picture management.

Every effort has been made to trace the copyright holders of the images in this volume. Any errors or omissions are regretted and should be brought to the notice of the publisher.

Cover illustrations: (Front) Raghogarh chiefs hunt rhinoceros. © Jagdish and Kamla Mittal Museum of Indian Art, Hyderabad. (Back) Watercolours by Joseph Wolf (1872) of *Rhinoceros unicornis, Rhinoceros sondaicus, Dicerorhinus sumatrensis* and *Dicerorhinus sumatrensis lasiotis.* © Archives, Zoological Society of London.

The Library of Congress Cataloging-in-Publication Data is available online at https://catalog.loc.gov

Typeface for the Latin, Greek, and Cyrillic scripts: "Brill". See and download: brill.com/brill-typeface.

ISSN 2452-3283
ISBN 978-90-04-54488-8 (hardback)
ISBN 978-90-04-69154-4 (e-book)
DOI 10.1163/9789004691544

Copyright 2024 by L.C. (Kees) Rookmaaker. Published by Koninklijke Brill BV, Leiden, The Netherlands.
Koninklijke Brill BV incorporates the imprints Brill, Brill Nijhoff, Brill Schöningh, Brill Fink, Brill mentis, Brill Wageningen Academic, Vandenhoeck & Ruprecht, Böhlau and V&R unipress.
Koninklijke Brill BV reserves the right to protect this publication against unauthorized use.

This book is printed on acid-free paper and produced in a sustainable manner.

In memoriam

My father

HENDERIK ROELOF ROOKMAAKER
27 February 1922–13 March 1977
Professor of Art History

My mother

ANNA MARIE ROOKMAAKER-HUITKER
10 August 1915–10 February 2003
Knight in the Order of Oranje-Nassau
Founder and Secretary of Stichting Redt een Kind

My wife

SANDRA MARY ANTHEA, BORN LAWRY
19 June 1947–17 September 2016
The loving companion of my life

∴

Contents

PART 2
The Javan Rhinoceros Rhinoceros sondaicus *in South Asia*

Acknowledgements

Whenever I watch the credits at the end of a movie acknowledging hundreds of crew and auxilliary staff, I tend to think that my research is done in 'splendid isolation' hidden away in an ivory tower. Fortunately that sentiment is far from reality. Although writing a book may sometimes feel like a monk's task, it is enlivened by truly remarkable set of people including family, friends, colleagues, fellow scientists, as well as the administrators and support staff in libraries, museums and other institutions.

How can we ever say thank you often enough? Definitely my sense of gratitude is real and immense to all those who have gone the extra mile to share their expertise or provide encouragement. It is a pleasant experience to be able to write a section of acknowledgements, yet it is done in the realization that so much more could be written about the involvement of each and every one of my correspondents. My rhino studies started over fifty years ago, and the preparation of this book has taken me many years. All persons who crossed my path in these long periods have contributed, each in their own special way. If any are omitted here, it is an oversight for which I apologize. At the start of this project to elucidate the range of the rhinoceros in South Asia, I told myself not to leave any stones unturned. Some hurdles could not be taken, yet many times the contents of emails or letters have brought exciting surprises, all contributing to an unprecedented historical perspective.

This book was first proposed as a report commissioned by A. Christy Williams on behalf of the WWF Asian Elephant and Rhino Program. It was envisaged to provide insights in the historical and current situation of all rhinos in South Asia in an account unhindered by a great scholarly apparatus of sources. The report was written in the course of 2013 and published in a greatly trimmed-down version in 2016. On my retirement from my university duties, the next step was to re-introduce the list of sources and expand this to include all the literature which had been added to the Rhino Resource Center over the years. My thanks are due to all colleagues who contributed data at the time.

My rhino research (or rhino passion) has always intruded every aspect of my personal life. It is not enough, yet I can do no more than list those who have made their special contributions. As they say, in no particular order, and therefore alphabetical for want of a better way. They are:

Pierre-Oliver Antoine; Reinoutje Artz; Joachim Bautze; Claudine Bautze-Picron; Wim Bergmans; Emanuel Billia; Shibani Bose; André Boshoff; Christopher Brack; Peter van Bree; CeCe Carter Sieffert; Timothy Howard and Elisabeth Clarke; Anwaruddin Choudhury; Paul Cooper; Cathy Dean; Helmut Denzau and Gertrud Neumann-Denzau; Divyabhanusingh; Joop van Doorn; Susie Ellis; Nina Fascione; Tom Foose; Denis Geraads; Colin Groves and Phyllis Dance Groves; Ajay Karthick; Graham Kerley; Andrew Laurie; Arjen Lenstra; Martyn Low; Esmond and Chryssee Bradley Martin; Anna Merz; Jim Monson and Isolde Monson-Baumgart; Pat Morris; Tanoj Mukherjee; Simon Muller; Erwin Neumayer; Peter Ng; Florence Pieters; Ellen Raven; Herman Reichenbach; Richard Reynolds; Richard Kees; Jan Robovskỳ; Nan Schaffer and Karen Dixon; Amit Sharma; Piet Smit; Russell Stebbings; Nico and Tineke van Strien; Willem van Strien; Marjolein van Trigt; Edward Turner; Rob Visser; Amirtharaj Christy Williams; John van Wyhe; Dan and Jacqueline Ziegler.

My family has been a constant support, starting with my parents, grandparents and aunts, closely followed by Hans and Aagje, Hinko, Carmina, Enric, Didac; Arjen, Marjolein, Eva, Orion; Marleen and Albert, Gerbert, Marieke, Dieke; Cathryn, Cameron, Lindsey, Meganne; Claire and Mark, Trevor, Justin; Simon and Angela, Alex, Christina. My wife Sandy who passed away too early and whose love, friendship and understanding was a continuous encouragement.

• • •

This book was published in the series *Emergence of Natural History* published by Brill in Leiden. The series editors were very encouraging, Kay Etheridge, Dominik Hünniger, Andreas Weber, and Aaron M. Bauer. The latter has been my regular mentor and liaison, proposed many suggestions to improve the text, and was a guide through the intricate publication process. The English was improved by Russell Stebbings. Two anonymous reviewers wrote their reports and added valuable comments. At Brill, Rosanna Woensdregt, Simona Casadio and Alessandra Giliberto were in charge of the production team.

In special cases, I have relied on the assistance of professionals and friends who have gone out of their way to answer all my often eclectic queries (according to continent):

EUROPE: Renate Angermann (Museum für Naturkunde, Berlin); Hans J. Baagøe (Museum of Zoology, Copenhagen); Guillaume Baron (Musée La Rochelle); Gennady Baryshnikov (Leningrad); Will Beharrell (Linnean Society of London); Natasha Illum Berg; Mieke Beumer; Bernhard Blaszkiewitz; Andrew Benton (LSE Library, London); Gert Bogels; Margherita Bolla (Museo Archeologico di Verona); Johannes Bornmann; Roger Bour; Sarah Broadhurst (ZSL); Helen and William Brock; Peter Burman (Hopetoun Papers Trust); Peter Byrne; Cécile Callou (Muséum national d'Histoire Naturelle, Paris); Alexander Cave; Simon Chaplin; Ann Charlton; Emmanuelle Choiseau (Muséum national d'Histoire Naturelle, Paris, Library); Marcus Clauss; Juliet Clutton-Brock; Wim Dekkers; Louise Detrez (Bibliothèque Nationale, Paris); Nélia Susana Dias; Stefan Dickers (Bishopsgate Institute, London); Gina Douglas (Linnean Society of London); Ruth Duncan (Gordon Highlanders Museum); John Edwards; Ingrid Faust; Arnaud Filoux; Christiane Funk (Museum für Naturkunde, Berlin); John Gannon; Klaus Gille (Hagenbeck Archives); Spartaco Gippoliti; Mara Grehl (Museum für Naturkunde Berlin); Frantz Grenet; Rupert Godfrey; Jos Gommans; Nicholas Gould (International Zoo News); Caroline Grigson; Peter Grubb; Helen Gwerfyl (Bangor University and Storiel Museum); Gijs van der Ham (Rijksmuseum, Amsterdam); Peter van Ham; Joanne Hatton (Horniman Museum); Gerard Heerebout; Eva Heidegger; Elizabeth Hensman (Horniman Museum); Sarah Hewitt; Jenny Hill (Paul Mellon Centre for Studies in British Art); Klaas van der Hoek (Allard Pierson, University of Amsterdam); Dana Holeckova (Dvur Kralove); Friederike von Houwald; Jiri Hruby; Dominik Hünniger; Gareth Hughes (Portland Collection, Worksop); Barney Jeffries; Annemarie Jordan Gschwend; Manfred Kaufmann (Weltmuseum Wien); Clin Keeling; Simon Keynes (Trinity College, University of Cambridge); James Kirwan (Trinity College Library, Cambridge); Ragnar Kinzelbach (Rostock); Heiner Klös; Heinz-Georg Klös (Zoo Berlin); Ursula Klös; Egge Knol (Museum Groningen); Martin Kohler and Thomas Kloeti (University Library of Bern, Ryhiner Collection); Izabela Korczyńska (Jagiellonian Library, Krakow); Richard Kraft; Regina Kramer (Schönbrunner Tiergarten); Petra Kretzschmar; Katrin Krohmann (Senckenberg, Frankfurt); Martin Krug (Hodonin Zoo); Chris Lavers; Ernst Lang; Karen Lawson (Royal Collection Trust); Nick Lindsay; Matthew Lowe (Museum of Zoology, University of Cambridge); Horst Lubnow; Arthur Lucas; Christoph Mallet; Paola Irene Galli Mastrodonato; Arthur MacGregor; Marie McInerney (Dublin City Library and Archive); Marie Meister (Musée Zoologique Strasbourg);

Stefan Merker (Staatliches Museum für Naturkunde, Stuttgart); Jim Middleton (Scarborough Museums Trust); Sandra Miehlbradt (Museum für Naturkunde Berlin); Emma Milnes (Zoological Society of London); Catherine Moorehead; Scott Morrison (Dunrobin Castle); Malgosia Nowak-Kemp; François Ory; Olivier Pagan (Zoo Basel); Roberto Portela Miguez (Natural History Museum); Felice Pozzo; Soleil Redwood (Horniman Museum); Maggie Reilly (Hunterian Library Glasgow); Simon Rhodes; Rachel Rowe (Smuts Librarian for South Asian and Commonwealth Studies); Michiel Roscam Abbing; Niki Russell (Hunterian Library Glasgow); James A. Secord (University of Cambridge); Willem van Schendel (University of Amsterdam); Lothar Schlawe; Franz Schwarzenberger; Elaine Shaughnessy; Kim Skalborg Simonsen (Givskud Zoo); Martin Sperlich; Beatrice Steck (Zoo Basel); Jan Stejskal; Klaus Stopp; Harro Strehlow; Ann Sylph (Zoological Society of London); Michelguglielmo Torri; Giuseppe Ugolini (Sapienza-Università di Roma); Claudia Valter (Germanisches Nationalmuseum); Hermann Walter; Peter Wandeler (Musée d'Histoire Naturelle Fribourg); Marie-Dominique Wandhammer; Mark Watson (Royal Botanic Garden, Edinburgh); Alwyne Wheeler; Peter Whitehead; Oscar Wilson; Samuel Wittwer (Stiftung Preußische Schlösser und Gärten Berlin-Brandenburg); Yassen Yankov (In & Out Club, London); Samuel Zschokke.

NORTH AMERICA: Walter Arader; Morex Arai (Huntington Library, San Marino); John Beusterien (Texas Tech University); Robert Campbell; Eric Dinerstein; James K. Galbraith; Peter Galbraith; Thomas Gnoske (Field Museum of Natural History Chicago); Annie Halliday (Mercer Museum, Doylestown); Bruce Hawley (Circus Historical Society); Julie Hughes; Kyle Jackson (Kwantlen Polytechnic University); Ken Kawata; Jane Kennedy; Erik L. Klee; Bill Konstant; Barry Landua (AMNH, Division of Anthropology); David M. Leslie; Barbara Maas; Kristen Mable (AMNH, Division of Anthropology); Ryan Mattke (John R. Borchert Map Library); Alison Petretti (Arader Galleries); Randy Rieches; Jane Ptolemy and Cheney Schopieray (William L. Clements Library, Ann Arbor); John Seyller (University of Vermont); Velizar Simeonovski; Sandra van der Sommen (Arader Galleries); Ingvar Svanberg; Leslie Squyres (Volkerding Study Center).

AFRICA: Suzannah Goss; Markus Hofmeyr; Peter Mundy; Lucy Vigne.

AUSTRALIA: Darrell Dorrington (ANU, Canberra); Barbara Nelson (ANU, Canberra); Karen Roberts (Museum Victoria, Melbourne).

ASIA: Hari Adhikari; Gopal Aggarwal; Jahan Ahmed; Nejib Ahmed; Abdul Mazid Ali (Manas); Jamir Ali (WWF

India); Tariq Aziz; N.C. Bahuguna (Chief Conservator of Forests, West Bengal); Rathin Barman (CWRC); Nirmala Reddy Barure (Bombay Natural History Society); Ashish Bashyal; Pranjal Bezbarua; Subhamoy Bhattacharjee (Wildlife Trust of India); Amal Bhattacharya (University of Rajgarh); Prem Bohara; Pranab Bora (Kaziranga); J. Bose (WWF-India); Anil Cherukupalli (WWF India); Sophie Bostock (Orientalist Museum); Swaraj Chitrakar and Christeena Chitrakar (Kiran Man Chitrakar Collection); Lisa Choegyal (Tiger Tops); Ashok Das (Pabitora); Asok Das; Bapkon Das; Kaushik Deuti (Zoological Survey of India, Kolkata); Divyabhanu Sinh Chavda; Tshering Dorji (Bhutan); Rahul Dutta; Deba Kumar Dutta (WWF India); Jack Edwards (Tiger Tops); Radhikaraje Gaekwad; Dipankar Ghose (WWF-India); Mudit Gupta (WWF-India); Ali Hassan Habib (WWF-Pakistan); Enamul Haque (International Centre for Study of Bengal Art); Buddhin Hazarika; Ravi Jung Karki; Subhash Kapoor; Nilutpal Kashyap; Raza Kazmi; Uzma Khan (WWF-Pakistan); Kamlesh Kumar (WWF-India); Sumanta Kundu (Corbett Foundation, Kaziranga); Satinder Lal; Suman Lama; Babu Ram Lamichhane; Sukianto Lusli; Jayanta Kumar Mallick; Varun Mani; Joseph Manuel; Vivek Menon and Amrit Menon; Dasrath Mishra; Jagdish Mittal (Hyderabad); Supriyo Nandi (Zoological Survey of India); John Payne; Nripen Nath (Pabitora Conservation Society); Dhipul Nath (WWF India); Abhinaya Pathak; Debabrata Phukon; Isma Azam Rezu; Shubham Rajak (Deccan College, Pune); Richard Kees (Jupiter Pipes); Uttam Saikia; Khari Chatrabhog Sangrahashala (Kashinagar); Vijay Sathe (Deccan College, Pune); Jayanta Sengupta (Victoria Memorial Hall, Kolkata); Atmaz Kumar Shrestha; Pankaj Sharma; Ruchir Sharma (WWF-India); Kushal Konwar Sharma (College of Veterinary Science, Guwahati); Balendu Singh; Bhupendra Singh Auwa (Maharana of Mewar Charitable Foundation); Dinkar Singh (Mussoorie); Manmohan and Mohit Singh; Rajeev Singh (Madan Puraskar Pustakalya, Kathmandu); Satya Priya Sinha; Naresh Subedi; Bibhab Kumar Talukdar (Asian Rhino Specialist Group); Mukul Tamuly; Rakesh Tewari; Kanchan Thapa; Ram Tharu; Santosh Tiwari (Valmiki Tiger Reserve); Sami Tornikoski; Arun Kumar Tripathi; Ian Williams (Spirare Books); Ahmed Sajjad Zaidi; Zainal Zahari Zainuddin.

Kelly Enright wishes to thank Kees Rookmaaker and Aaron Bauer for the opportunity to collaborate on this volume, and Nico Enright Imeson for letting me share my curious love of rhinos with him.

Joachim K. Bautze wishes to acknowledge the hospitality of the following people: In India: Maharao Bhim Singhji of Kotah and his son, Maharao Brijraj Singhji of Kotah, Devta Shridharlal (Kota), Madan Mohan Shastri (Kota), Capt. Ajay Singh Rathore and Shashi Palaitha (Kota), Maharawal Mahipal Singhji and Maharajkumar Harshvardhan Singhji of Dungarpur, Maharaj Daivat Singhji and Kirti Kumari (Mount Abu), Kumar Sangram Singhji of Nawalgarh (Jaipur), Shriji Arvind Singh Mewar (Udaipur), Rawat Nahar Singhji of Deogarh, Yaduendra Sahai (Jaipur), Om Prakash Tandon (Banaras). In the U.S.A.: Stuart Cary Welch (Cambridge and New Hampshire), Amy Poster and Bertram H. Schaffner (New York), Gursharan and Elvira Sidhu (Seattle). In Europe: Julian Sherrier and Gordon Howard Eliot Hodgkin (London), Konrad and Eva Seitz (Bonn), Ludwig Volker Habighorst (Koblenz), Pjotr and Gerlinde (Karlsruhe – Berlin).

The illustrations in this book were selected from books, manuscripts, maps, photographs, paintings, drawings and other artistic expressions. Preserved in libraries, museums and other collections, these treasures have become known through careful catalogues and recently on the expanding digital platforms. I have encountered often incredible generosity of the curators in providing the best possible images and to license them for publication. The current trend to make all cultural expressions freely available through dedicated open-access policies will definitely stimulate further research and allow better understanding of their significance.

SOS Rhino, a charitable organization based in the USA, is concerned with the conservation and management of the rhinoceros, with emphasis of the Sumatran Rhinoceros of South-East Asia. Through their directors, SOS Rhino has generously reached out and supported this project, providing the means to acquire and license the illustrations and maps, and also to edit any inconsistencies in the use of the English language. Without their assistance, this project would not have come to fruition in the current format.

TABLE 1.1   Institutions whose staff members have kindly helped to supply photographs of rhino-related
            images

Country, place	Institution name
Austria, Innsbruck	Universitäts- und Landesbibliothek Tirol
Austria, Vienna	Kunsthistorisches Museum
Austria, Vienna	Museum of Applied Art
Belgium, Leuven	University Library
Denmark, Copenhagen	Royal Library of Denmark
Denmark, Copenhagen	Zoological Museum
France, Paris	Bibliothèque Nationale
France, Paris	Cabinet des Medailles
France, Paris	Musée Carnavalet-Histoire
France, Paris	Muséum national d'Histoire Naturelle
Germany, Berlin	Museum für Naturkunde
Germany, Berlin	Schloß Charlottenburg, Sperlich collection
Germany, Berlin	Stadtmuseum
Germany, Darmstadt	Hessisches Landesmuseum
Germany, Hamburg	Archiv Carl Hagenbeck
Germany, Nuremberg	Germanisches Nationalmuseum
Germany, Stuttgart	Staatliches Museum für Naturkunde
Hungary, Budapest	Aquincum Museum
India, Bangalore	S.P. Lohia Foundation
India, Hyderabad	Jagdish and Kamla Mittal Museum of Indian Art
India, Jaipur	Maharaja Sawal Man Singh II Museum
India, Kashinagar	Khari Chatrabhog Sangrahashala
India, Kolkata	Victoria Memorial Hall
India, Kolkata	Zoological Survey of India
India, Kota	Collection of Maharao of Kotah
India, Mumbai	Bombay Natural History Society
India, Mussoorie	Kasmanda Hotel
India, Pune	American Institute of Indian Studies
India, Shyamnagar	G.M. Sekandar Hossain
Ireland, Dublin	City Library and Archive
Italy, Bologna	Biblioteca Universitaria
Italy, Florence	Museo dell'Opificio delle Pietre Dure
Italy, Florence	Uffizi Gallery
Italy, Milan	Musei del Castello Sforzesco
Italy, Napels	Museo Nazionale di Capodimonte
Italy, Rome	Sapienza-Università
Italy, Rome	Archivio Doria Pamphilj
Italy, Turin	Armeria Reale, Galleria Beaumont
Italy, Verona	Museo Archeologico di Verona
Nepal, Kathmandu	Madan Puraskar Pustakalya
Nepal, Kathmandu	Kiran Man Chitrakar Collection
Netherlands, Amsterdam	Allard Pierson collections
Netherlands, Amsterdam	Artis Library
Netherlands, Amsterdam	Rijksmuseum
Netherlands, Amsterdam	University Library (UBA)
Netherlands, Groningen	Groninger Museum

TABLE 1.1   Institutions whose staff members have helped supply photographs (*cont.*)

Country, place	Institution name
Netherlands, Leiden	Naturalis (RMNH)
Netherlands, Leiden	Rijksmuseum van Oudheden
Netherlands, Leiden	University Library (UBL)
Netherlands, The Hague	National Archives of the Netherlands (ARA)
Pakistan, Mohenjo Daro	Mohenjo Daro Archaeological Museum
Poland, Gdansk	Archiwum Państwowe w Gdańsku
Poland, Krakow	Jagiellonian Library
Qatar, Doha	Orientalist Museum
Switzerland, Bern	University Library of Bern, Ryhiner Collection
Switzerland, Zurich	ETH Library
UK, Aberdeen	Gordon Highlanders Museum
UK, Bangor	Bangor University, Storiel Museum
UK, Cambridge	University of Cambridge, Central Library
UK, Cambridge	University Museum of Zoology
UK, Cambridge	Royal Commonwealth Society
UK, Derbyshire	Kedleston Hall
UK, Edinburgh	University Library
UK, Glasgow	University of Glasgow, Hunterian Library
UK, Lincolnshire	Belton House
UK, Liverpool	Museum of Liverpool
UK, London	Bishopsgate Institute
UK, London	British Library (BL)
UK, London	British Museum (BM)
UK, London	British Pathé
UK, London	Horniman Museum and Gardens
UK, London	Hunterian Musem, Royal College of Surgeons of England
UK, London	In & Out, Naval and Military Club
UK, London	London Transport Museum
UK, London	National Army Museum
UK, London	National Maritime Museum, Greenwich
UK, London	Natural History Museum (NHM)
UK, London	Royal Collection Trust
UK, London	Victoria and Albert Museum
UK, London	Wellcome Library
UK, London	Zoological Society of London (ZSL)
UK, Oxford	Ashmolean Museum of Art and Archaeology
UK, Oxford	Bodleian Library
UK, Oxford	Pitt Rivers Museum
UK, Powys	Powis Castle
UK, Scarborough	Scarborough Museums Trust
USA, Ann Arbor, MI	William L. Clements Library
USA, Baltimore, MD	Walters Art Museum
USA, Chicago, IL	Art Institute of Chicago
USA, Chicago, IL	Field Museum of Natural History (FMNH)
USA, Gainsville, FL	Baldwin Library of Historical Children's Literature
USA, Iowa City, IA	Library of the University of Iowa
USA, Los Angeles, CA	J. Paul Getty Museum

TABLE 1.1   Institutions whose staff members have helped supply photographs (*cont.*)

Country, place	Institution name
USA, Los Angeles, CA	Los Angeles County Museum of Art
USA, Minneapolis, MN	University of Minnesota, John R. Borchert Map Library
USA, New Haven, CT	Yale Center for British Art
USA, New York, NY	American Museum of Natural History (AMNH)
USA, New York, NY	Brooklyn Museum
USA, New York, NY	New York Historical Society
USA, Philadelphia, MA	Arader Galleries
USA, Philadelphia, MA	University of Pennsylvania, Penn Museum
USA, San Marino, CA	Huntington Library
USA, Tucson, AZ	University of Arizona, Center for Creative Photography
USA, Washington, DC	National Gallery of Art

Colleagues and friends who have kindly contributed their photographs: Deepak Acharya; Gopal Aggarwal; Nejib Ahmed; Jamir Ali; Nishant Andrews; Hans J. Baagøe; N.C. Bahuguna; Rathin Barman; Ashish Bashyal; Joachim K. Bautze; Ramesh Bhatta; Subhamay Bhattacharyya; Lisa Choegyal; Anwaruddin Choudhury; Alain Compost; Asok Kumar Das; Helmut Denzau and Gertrud Neumann-Denzau; Kaushik Deuti; Deba Kumar Dutta; Jack Edwards; John Edwards; Andrew Gell; Frantz Grenet; Helen Gwerfyl; Sarah Hewitt; Julie Hughes; Sumanta Kundu; Andrew Laurie; Mathew Lowe; Esmond Bradley Martin; Marie Meister; Jim Monson; Pat Morris; Tanoy Mukherjee; Dhipul Nath; Ernst Neumayer; François Ory; John Payne; Debabrata Phukon; Shubham Rajak; Isma Azam Rezu; Richard Kees; Sudhir Risbud; Jan Robovský; Chandana Sarma; Amit Sharma; Dinkar Singh; Manmohan and Mohit Singh; Satya Sinha; Nico van Strien; Tineke van Strien; Ram Tharu; Arun Kumar Tripathi; Lucy Vigne; Yassen Yankov; Ahmed Sajjad Zaidi; Zainal Zahari Zainuddin.

Open Access

Open Access supporting Global Research and Nature Conservation through the Generosity of Sponsors and Subscribers.

THANK YOU.

Sponsors

Oak Foundation
oakfnd.org

Christy Williams and Kashmira Kakati   Erik L. Klee

International Rhino Foundation
rhinos.org

Tiergarten Schönbrunn – Vienna Zoo
www.zoovienna.at

Hodonin Zoo
www.zoo-hodonin.cz

Givskud Zoo – Zootopia
www.givskudzoo.dk

Aaranyak
aaranyak.org

Subscribers

Pierre-Olivier Antoine | Mieke Beumer | John Beusterien | Scott B. Citino | Cathy Dean | Wim Dekkers | John Edwards | Suzannah Goss | Marleen and Albert Hengelaar | Ken Kawata | Oliver Klohs | Martin Krug | Andrew Laurie | Nick Lindsay | Horst Lubnow | Sukianto Lusli | Claire and Mark Marshall | Francesco Nardelli | John Payne | Florence F.J.M. Pieters | Herman Reichenbach | Hans and Aagje Rookmaaker | Richard Kees | Oliver Ryder | Michael 't Sas-Rolfes | Franz Schwarzenberger | Bibhab K. Talukdar | Lucy Vigne | Dan and Jacqueline Ziegler

Figures

Maps

Tables

Datasets

Abbreviations

AD	Anno Domini, years after birth of Christ		km	kilometer(s), kilometre
ad.	adult		Lieut.	Lieutenant
ADC	Aide-de-Camp		m	meter, metre
am	morning before noon		mo.	month(s)
b.	born		MCB	Maharaja of Cooch Behar (Nripendra Narayan)
BC	Before Christ		Mt.	Mount
BCE	Before Common Era (= BC)		n.d.	no date(s)
BP	Before Present		NF	Neue Folge, new series
c.	circa		no.	number
cf.	confer, compare		NP	National Park
Capt.	Captain		n.s.	new series
CE	Common Era (= AD)		pers.com.	personal communication (usually by email)
cent.	century		pl.	plate
cm	centimeter		pm	afternoon, evening
Col.	Colonel		*R.*	genus *Rhinoceros*, in name of rhino
col.	coloured (plate, figure)		R.	River
d.	died		r.	reigned (for rulers)
D.	genus *Dicerorhinus*, in name of rhino		RF	Reserved Forest
Dept.	Department		RR	Rhino Region (table 1.2)
Div.	Division		rhino	rhinoceros, rhinoceroses
Dt.	District		rhinos	rhinoceroses
ed.	edition		Rs.	Rupee(s), currency
fig.	figure		sp.	species uncertain, after a generic name
fl.	*floruit*, active period		UK	United Kingdom
ft	feet		UP	Uttar Pradesh
ibidem	the same		WR	Wildlife Reserve
in.	inch(es)		WS	Wildlife Sanctuary
juv.	juvenile		yr	year(s)

Acronyms

AMNH	American Museum of Natural History, New York
ANU	Australian National University, Canberra
ASB	Asiatic Society of Bengal, Kolkata
BM	British Museum, London
BMNH	British Museum (Natural History), now NHM
BNHS	Bombay Natural History Society
DNPWC	Department of National Parks and Wildlife Conservation, Nepal
FAO	Food and Agriculture Organization
FFPS	Fauna and Flora Preservation Society
FMNH	Field Museum of Natural History, Chicago
ICZN	International Commission for Zoological Nomenclature
IM	Indian Museum, Kolkata
IUCN	International Union for the Conservation of Nature
MDM	Mohenjo Daro Museum, Pakistan
MNHN	Muséum National d'Histoire Naturelle, Paris
MPP	Madan Puraskar Pustakalaya, Library, Kathmandu
NHM	Natural History Museum, London; before 1992 BMNH
NMI	National Museum of India, New Delhi
NMP	National Museum of Pakistan, Karachi
NTNC	National Trust for Nature Conservation, Nepal
RCSE	Royal College of Surgeons of England
RCS	Royal Commonwealth Society
RCT	Royal Collections Trust
SSC	Species Survival Commission
UNDP	United Nations Development Program
V&A	Victoria and Albert Museum, London
VOC	Vereenigde Oost-Indische Compagnie, Dutch East India Company
WWF	World Wildlife Fund, World Wide Fund for Nature
ZSI	Zoological Survey of India, Kolkata
ZSL	Zoological Society of London

Introduction

1.1 The Rhinos of South Asia

The River Ganges embodies the life and culture of the Indian subcontinent as it flows steadily from the Himalayas for thousands of kilometers to the Bay of Bengal. Eternally paired with this mighty force was the rhinoceros. The English architect George Richardson portrayed the Ganges as "an old man of an austere aspect, crowned with palms, and pouring water out of a vase," with a rhinoceros as his companion "native of the country where this river glides" (fig. 1.1). This theme was repeated in a silver service of 1848 (fig. 28.1). The great megaherbivore continues to be found in the protected areas of Assam, West Bengal and Nepal and acts as an ambassador of the nations of South Asia.

The Greater One-horned Rhinoceros has been familiar since time immemorial, known from its heavy skin divided in conspicuous armour plates and the prominent single horn on the nose. This was the animal exported to Europe to become the quintessential vision of the rhinoceros outside its native haunts. The link between the River Ganges and the rhinoceros might have been a playful intervention, yet it is true that these animals are known predominantly from the valleys of the great rivers bounded in the north by the foothills of the Himalaya mountain range. The Indus, Narmada, Ganges and Brahmaputra once nourished populations of the rhinoceros across the northern half of the subcontinent. Over the centuries their range was reduced and in most of this territory no rhinos have been seen in the wild for at least one, sometimes even several centuries.

The primary aim of this study is to provide an accurate assessment of the distribution of the rhinoceros in South Asia in historical times. This region consists of the countries of Afghanistan, Pakistan, India, Nepal, Bhutan, Bangladesh, Sri Lanka and the Maldives. No remains of rhinoceros have ever been reported from the Maldives. Discussions in this book follow occurrences in a general west to east direction.

The checklist of South Asian mammals includes not one but three species of rhinoceros. Two of these have not survived in the region until the present day, and are among the rarest and most endangered animals on earth in their main habitats in South-East Asia. The Javan Rhinoceros is single-horned and armour-plated, often living in forested lowland coastal plains. The Sumatran Rhinoceros is double-horned, smaller and more hairy than other rhinos. It loves dense forests on higher mountain ranges to live its solitary life. When the three types are seen together, they can easily be distinguished from one another (figs. 1.2, 1.3). A few frequent visitors to zoological gardens in the past may have had this opportunity, while for contemporary naturalists such direct comparison is no more than a dream. In the wild only a handful of travelers and naturalists had the privilege to observe each of them in different locations. If direct comparison is impossible and opportunity for extended observation practically unique, it needs no further explanation that the three species have often been confused, even in the most respected handbooks.

The historical distribution ranges of the three species of rhinoceros known in South Asia can only be defined and delimited through a detailed study of the available literature and iconographic sources. Ever since I first became interested in the rhinoceros, over fifty years ago, the distribution of animal species has been my gateway to the past. Considering the elusiveness of rhinos in large sways of their habitat and the imperfection of the historical record, it is almost obvious that their respective ranges remain enigmatic in detail. Now in retirement it is an enquiry which I can pursue more vigorously. Maps of animal distribution are an interpretation of a series of events involving a particular species. Rhinos in South Asia have been observed, studied, measured, hunted, wounded, mounted as trophies, captured, exhibited, photographed, painted and sculpted by hundreds of people over hundreds of years in a wide variety of settings. Each of these individual events has a potential significance in the reconstruction of the historical distribution of the rhinoceros of South Asia. I hoped that to examine each and combine them, the larger picture would develop and allow a more accurate assessment.

It is only through the eyes of people that we might gain an understanding of past interactions with rhinos. The animals themselves were born, lived and died, in the company of their relatives, usually without leaving any physical trace. Historical enquiry with a view of understanding past ecologies and ranges of animal species depends in practice on the testimony of that small percentage of people who have shared their experiences in writing or have left other traces of their activities. Although a rhinoceros is a large and iconic animal which will get noticed, obviously many interactions with humans remain unrecorded.

© L.C. (KEES) ROOKMAAKER, 2024 | DOI:10.1163/9789004691544_002
This is an open access chapter distributed under the terms of the CC BY-NC-ND 4.0 license.

FIGURE 1.1 The River Ganges portrayed as an old man with an austere aspect, accompanied by a rhinoceros. The plate was designed by
George Richardson (1736–1817) for his book on icons of the world. It was "Published as the Act directs Nov. 23. 1776"
RICHARDSON, *ICONOLOGY*, 1779, VOL. 1, PL. 17 FIG. 62

FIGURE 1.2 Three species of rhinoceros in Asia: "*R. indicus* (*unicornis*), *R. sondaicus* and *R. sumatranus* (*bicornis*)." Anonymous engraving
LEISURE HOUR, 11 MAY 1872, NO. 1063, P. 297.
REPRODUCED BY KIND PERMISSION OF CAMBRIDGE UNIVERSITY LIBRARY

FIGURE 1.3 Five species of rhinoceros: Indian (twice), Sumatran (= Javan), Two-horned Sumatran, Two-horned African, Flat-nosed rhinoceros. This plate combines figures published by William Jardine (1784–1843) in his *Naturalist's Library*
JARDINE, *LEAVES FROM THE BOOK OF NATURE*, 1840, PL. 23

Even a specimen in an urban zoo is seen by thousands of visitors, most of whom leave absolutely no trace of the encounter. Therefore, a faithful reconstruction of past or present occurrences depends on traces in the widest variety of sources possible. In the case of rhinos, there are fossil remains, archaeological objects, rock paintings, art works on paper, canvas, and in sculptures, photographs, reports of encounters in the wild, specimens kept in captivity, trophies preserved by hunters and skeletons in scientific institutions, all part of a past which stretches through at least several centuries.

Peter Burges is a typical example of a person who is no longer recognized in any historical or scientific study. A respected gentleman, he was renowned in his hometown of Bristol as a big game hunter, as a photographer of wildlife and owner of a collection of trophies. Burges visited

India at least twice at the end of the 19th century, and was then enough of a celebrity to attract the American journalist Cleveland Moffett hoping for a revealing interview. Despite this local fame and the short-lived national exposure, Peter Burges (1853–1938) has disappeared from the record. The fate of his small museum is unknown, his photographs have disappeared, a diary or hunting notes may never have existed. Burges was one among hundreds, if not thousands, of people of different backgrounds who interacted with the rhinoceros. Nothing spectacular, nothing extraordinary, nothing of great scientific significance, yet a small element in the larger historical narrative. Many others have completely disappeared from the record, while some, like Burges, have made a small contribution to our general knowledge of the existence of the rhinoceros in past ages.

Writing as a zoologist, one of the great surprises during the years of research was the prolific depiction of the rhinoceros on wall paintings in the palaces and hunting lodges in Rajasthan. Particularly found in Kota, Bundi, Bijolia, Udaipur and Dungarpur, sporadically elsewhere, these paintings largely date from the 17th to 19th centuries and show rhinos in different settings, many of which are actual events rather than imaginary scenes. This iconography was studied in detail in the 1970s and 1980s by Joachim K. Bautze, now a retired German Professor of Indology, and analyzed in a seminal lecture appearing in 1985. He systematically catalogued the wall paintings found in the palaces of Rajasthan, apparently at a critical time because some are disappearing through vandalism and neglect. Realising that I could not do justice to these important iconographical treasures, I was very glad that Bautze agreed to list and explain the paintings of rhinos in a separate chapter under his own name (chapter 15). His study presents irrefutable evidence that these animals must have been observed and hunted in the vicinity of the centres of power in Rajasthan. This extends the historical distribution of the rhinoceros much further south than has been suggested in previous surveys.

It cannot be denied that this type of research sits in a grey area between science and humanities. Multidisciplinary one might say, but only in a kind of oblique way. The science of biology deals with organisms, like the rhinoceros, to understand nature in all its manifold facets as it is today. Researchers might read the odd "older" book or paper, while they are building on the knowledge of their predecessors. Taxonomists continue to improve our understanding of the world's biodiversity, and their discipline is one of the few that has to rely on past events when using specimens which were collected centuries ago, in order to decide upon the correct name to be used for any given animal following the prevalent rules to allow international discourse. The history of natural history should be part of every course of science, yet generally remains the domain of the interested amateur or the retired scientist looking at the work of their predecessors and past colleagues.

Historians explore ideas, cultures, personal and institutional biographies, focusing generally on the actions and works of men in the past. My work is historical in the sense of assessing how humans interacted with the rhinoceros in their territory. These data are needed to reconstruct the past range of these animals and to shape present understanding. There were rhinos in South Asia for a long time, hence the sources will necessarily span this long period. There is rarely a continuity or interrelation between events which might have been separated by many years, even decades, and would have been reported by unrelated types of observers. Each interaction must be investigated on its own merit. Despite this divergence in the sources, they all centre around the people who have shared their experiences. My enquiry has focused on these people, their lives and backgrounds, at the time when they had a chance to encounter a rhinoceros, each in their individual way.

The emphasis on past occurrences of rhinos dictates a geographical approach. Although this is more usual in a zoological context than in historical enquiry, this has proved the most acceptable way to organize the large amount of data used in this project. The series of chapters reflect this perspective, generally following a west to east directional path. This coincides to some extent with the historic age of the sources, some of the oldest known from

FIGURE 1.4
Scene of six rhinos enjoying their swim in the river. Detail of a painting by an unknown artist in the tradition of Kota in Rajasthan, undated
COLLECTION OF LATE MAHARAO BRIJRAJ SINGH OF KOTAH. PHOTO: JOACHIM K. BAUTZE

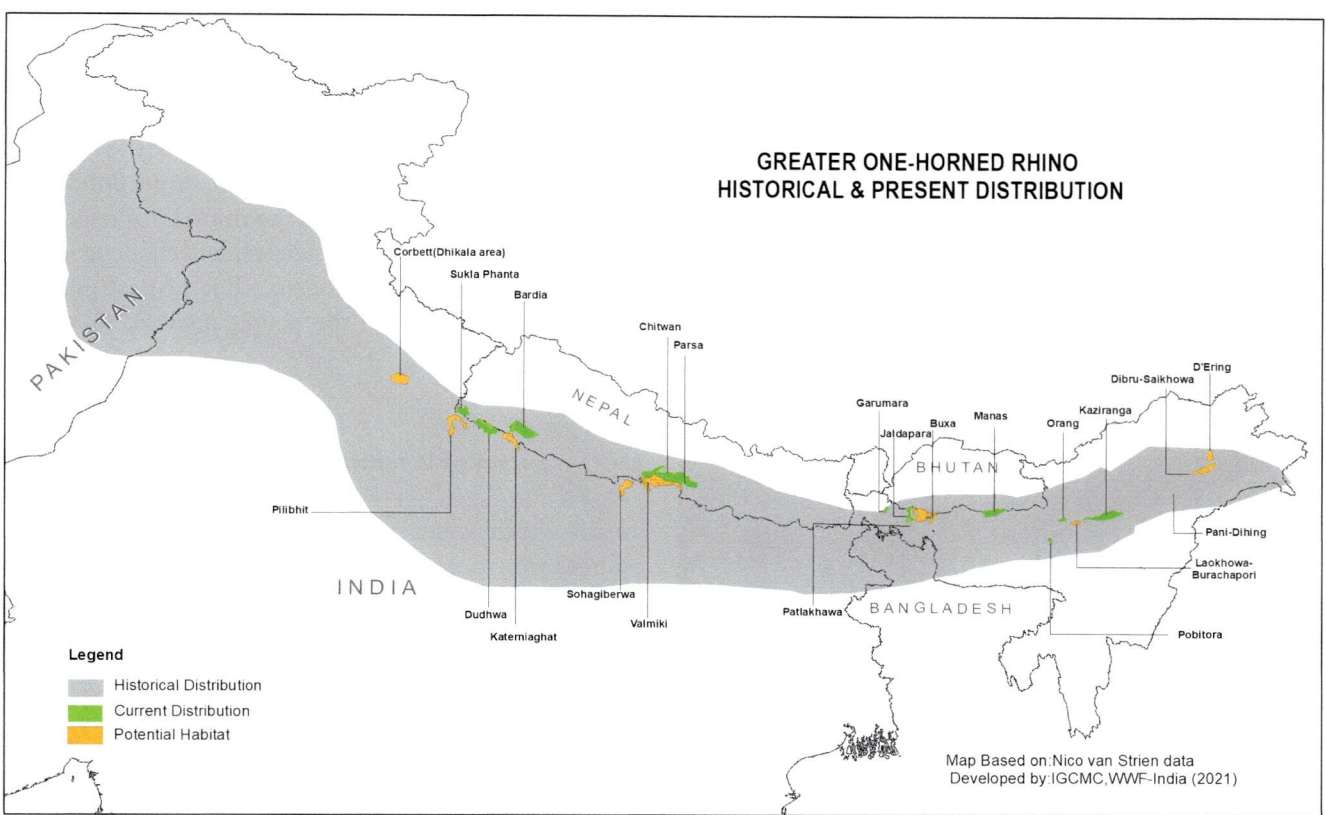

FIGURE 1.5   Map showing the current distribution of *R. unicornis* in protected areas (green) and potential habitats for expansion
COURTESY © WWF-INDIA 2021

Afghanistan and Pakistan, while the rhinoceros today still survives in Nepal and in the North-East of India. However, there are several sets of data which are best treated outside this geographical sequence, in order to combine the history of the fossil record, the rock art of ancient tribes, the artefacts of the Harappan civilisation, and also the cultural and educational significance of animals exhibited outside their original habitat in a captive setting.

Each recorded event involving a rhinoceros might give clues about its place in the cultural fabric of the Asian societies and about their changing impact on the natural habitat. The sources used are diverse and interdisciplinary, some more conventional than others. Written reports provide the bulk of the data, and in this research include passages from some very rare or unusual books, scientific papers as well as popular stories of hunting and adventure. Extensive use is made of illustrations of the rhinoceros, in rock art, in wall paintings, in sketches and drawings made in the field or in a studio, each evocatively telling its own part of the story. Photographs also provide glimpses of the past unparalleled for their ability to show events essentially unfiltered, which can be shared much better in this digital age than previously. All these sources help us to understand the role of the rhinoceros in culture and science.

In 1789 two British artists encountered a rhinoceros in Kotdwara, a small station in the foothills of the Himalayas in Uttarakhand, close to the western border of Jim Corbett National Park. I also traveled here some two centuries later: there were no longer any rhinos within hundreds of kilometers, none were seen, none were expected. This will not surprise anybody, because we are all aware that the spaces available for wildlife have reduced significantly over time. Without doubt, the current occurrence of the rhinoceros is a mere fraction of the wide swaths of countryside where once they were seen (fig. 1.5). And 'once' is not even all that long ago, although largely before our own generation. The data assembled here will provide a basis for further research into the mechanisms of extinction. Habitat loss to agriculture and urban expansion, climate change, excessive hunting and poaching, breakdown of law and order during armed conflict, all could be contributary factors. If we better understand what leads to extinction, we can take better decisions about our practices and methods of animal conservation. The causes must be more complex than the obvious human impacts on the natural world. Rhinos have ceased to exist in environments where some of the more common causes hardly seem to apply, being remote, less inhabited, rarely visited.

FIGURE 1.6 Rhinoceros knocking down a horseman. Drawing signed by Auguste-André Lançon (1836–1887) and Auguste Trichon (1814–1898). A one-horned rhino was transferred to an African story. Originally published in *La Mosaique* (Lançon 1876) with names of artists
GORDON STABLES, *WILD ADVENTURES IN WILD PLACES*, 1881, P. 108

The hunting of wildlife has taken its toll on the rhinoceros. Historical enquiries into the existence of animals must bring you face to face with a long series of dead animals, in personal accounts, in photographs and in museum collections. I can make no excuse for the sometimes gruesome pictures, which are part of human history and should be documented and not ignored. There is, to me at least, no sense to point fingers at people who lived in circumstances and worldviews sometimes different from ours. However, it is important to gain as much insight and understanding from these past events as possible. A dead rhino means a possibility to examine it, to measure it, to study its anatomy, to eat the meat, to preserve the hide, horn, skull, skeleton either as trophies or as specimens in a scientific institution, to draw it or photograph it to be shared with others, to record the situation in which it once lived, even to perform religious rites with it. If we can obtain at least some little items of information, then the death might have had at least a significance to us and future generations.

The rhinoceros was never a common animal in South Asia. Yet it was distinctive on account of its size and armour and horn, perhaps also for its stupidity and belligerence. Although hardly revered, the rhinoceros became part of the cultural fabric of South Asia as few other animals have. Today these animals are challenged by the changes in society and in the environment. They will still thrive in jungle areas as long as they can be protected from some outside influences, as is done in a spectacularly successful way in Nepal, in Bengal and in Assam. Protection will be a continuous battle as long as the rhino horn remains in demand and the wilderness areas remain challenged. This book contributes to an understanding of past practices and events and honours the people involved in science and in conservation.

1.2 Survey of Potential Sources

In the South Asian context, there is a wide variety of sources which can help to elucidate the range of the rhinoceros in the past, as well as the cultural significance of the animals. The research of this book has avoided any predetermined restrictions of date, language or discipline, although admitting that it was done by a zoologist in the western tradition. Some of the material adding to our knowledge of the interactions between rhinos and men is more obvious than others, which can here be elucidated. Most records date from the start of the 19th century towards the present, while the earlier ones generally are more challenging in nature. The localities associated with rhino events are defined in datasets and plotted on maps, using prefixes, numbers and symbols to differentiate between types of evidence.

In South Asia there are older records, like those of the Gupta era (5th century) and the Mughal period (16th century). There are further supporting types of evidence, comprised of fossil and subfossil skeletal remains, artefacts of the Harappan period, and depictions in rock art. This opens up a wide spectrum of cross-disciplinary data which depend on the interpretation and advice of experts in archaeology, indology, art history, palaeontology, geology and other disciplines rather beyond the reach of the usual zoological enquiry.

1.2.1 *Literature*
Written statements, generally published in books or journals, are the most common and accessible sources. Rhinos appear variously in careful scientific studies and surveys, in travel accounts and hunting stories, in gazetteers and reference works, even in fictional adventures. Some books are well-known, others have become rare and are seldom consulted. The digital age has revived some obscure works and provided easier searches in newspapers and popular magazines. Over 3000 items were found to include

information about the rhinoceros of South Asia, which form the backbone to the present enquiry.

1.2.2 *Gazetteers*

These governmental publication appeared in South Asia regularly with a wide breadth of data about a certain region. In case of fauna, the information is regularly repeated from one edition to the next, leading to potential false dates of some of the interesting events.

1.2.3 *Newspapers*

The digital age allows the use of newspapers far beyond the occasional and serendipitous access of former times. The scanning and text recognition process applied to national and local newspapers has opened up an entirely new field of research. Even if the search process has some inherent challenges, many relatively minor reports and events can be found in these pages. While there must still be many gems to be uncovered, several facts could already be verified or discovered especially (in the present case) in British and French newspapers.

1.2.4 *Photographs*

Published sources include photographs from the 1860s onwards. They provide a direct and unobstructed view of past events, with the advantage that only the skill of the photographers stands between us and the subject matter. There are several repositories of unpublished photographs, especially in the Nepalese context, providing otherwise inaccessible data.

1.2.5 *Iconography*

South Asian rhinos are depicted in many types of art work like paintings, drawings, engravings, book illustrations, wall paintings, photographs and maps. The wall paintings found in the palaces of Rajasthan have hitherto been poorly documented. When the subjects and ages of illustrations are known, these can provide incredible glimpses into past events and cultural fashions. The use of iconographical expressions is still underdeveloped in historical studies especially in a zoological context. A picture is worth a thousand words, and this remains true even if the reality is potentially distorted. Few artists were familiar enough with a rhinoceros to provide a perfect depiction of the skin folds, often horns were extended in size for dramatic purposes, yet they add much to our knowledge of historical events. Great pains have been made while researching this book to obtain images of some of the rarer illustrations, which add great details which otherwise would have remained obscure. The figures in this book are an integral part of the historical narrative and were selected with great care from the collections of museums and libraries on almost all continents.

1.2.6 *Artefacts*

There is a variety of ancient objects representing a rhinoceros found in South Asia. These include the archaeological remains of the Harappan period, coins of the Gupta era, and reliefs on the terracotta temples. These all have their own background and importance. They have been combined as artefacts and included in the datasets of localities (with the prefix M). In most cases, artefacts are poor indicators of rhino's occurrences because they were made by humans who of course could have migrated or obtained their knowledge far from the locality where they are found. Artefacts are important in the cultural dimension of rhino knowledge, hence their relevance should be explored.

1.2.7 *Rock Art*

The rhinoceros figures occasionally in the rock paintings or petroglyphs found in shelters across central parts of India, and more rarely elsewhere. Animals are identified as rhinoceros when they show a horn on their nose. The use of rock art in studies of mammal incidence has many potential pitfalls. In an African context, Boshoff & Kerley (2013: 25) discuss the symbolic nature of the depictions, the possibility of regional movements of humans, and the lack of certainty in species identifications, all leading them to the conclusion that rock art has very limited use in providing reliable supporting evidence. Concerning the rhinos of India, they can be identified only generally, as none of the paintings show enough definition of the characteristics of skin-folds or size which could be used to state it could be a specific kind. Age is another major obstacle. When first discovered, they were taken to be relatively recent. Currently experts allocate them to a much earlier period, maybe at least 5000 years before present. If this is correct, rock art has no relevance in a reconstruction of the historical distribution of *R. unicornis* or any other species of rhinoceros. However, it does testify to the presence of rhinos in earlier periods. For this reason, I have included the evidence to develop the protohistorical range (map 65.34).

1.2.8 *Excavated Skeletal Remains*

A few potentially more recent skeletal parts have been excavated in West Bengal and Bangladesh. These are unlikely to belong to extinct species and are therefore part of the modern records.

1.2.9   *Fossil Remains*

Generally consisting of skeletal material or skulls, often only recovered in fragments, fossil remains of rhinos have been found most commonly in the drier parts of Pakistan and India. Many of these have been allocated to species which have since become extinct as they date from the Pleistocene or earlier epochs. Much remains to be learned about the actual age of these remains and their relationships in systematic terms with the three living species. All records are combined in chapter 8 in an effort to check if the available studies can add to our understanding of the protohistorical and historical distribution of these species. Given the uncertainties about identifications and age, the records of these fossil remains have not been used in reconstructing these ranges.

1.2.10   *Toponyms*

The name of the rhinoceros sometimes appears embedded in the name of a geographical locality, known as a toponym. In South Asia, this is invariably a variant of the indigenous name of the animal (*ganda*). Toponyms might have historical significance, commemorating a certain event like killing a rhinoceros. However, there are many additional explanations. Toponyms have been recorded, but the localities have not been used in the reconstruction of the range of the rhinos.

1.2.11   *Specimens*

Rhinos that were killed provide potentially important evidence of these past events. In zoology, a specimen is a set of remains of an individual animal. This is primary important material because it can be examined and measured, and is mostly constant in nature. Specimens can include trophy heads, loose horns, skulls, or complete skeletons and complete hides, which can even be mounted for exhibition. Specimens are preserved in museums and scientific institutions, providing easy access, or sometimes in private collections of different sizes. The importance of any specimen increases with the accuracy of the accompanying data with locality, date and collector being of prime importance. These specimens provide avenues to past events unequalled in other sources as they continue to be available for taxonomic studies. Unfortunately, over time, many specimens have been lost through sale or neglect, many data have been lost, yet those remaining help to elucidate some historical dilemmas.

1.2.12   *Rhinos in Captivity*

Specimens that were captured alive in the wild and then kept in some kind of captive setting are important for

FIGURE 1.7   The 'Rhinocéros des Indes' shares the plate with the 'Rhinocéros bicorne d'Afrique.' The unknown artist has rather exaggerated the length of the horn. Throughout the 19th century, the original well-written text by Georg Leclerc de Buffon (1707–1788) was edited and adapted, like in the version aimed at young people by Pierre Blanchard (1772–1856), augmented by Jean-Charles Chenu (1808–1879), first published in 1849
BUFFON, *LE BUFFON DE LA JEUNESSE*, 1849, FACING P. 113

two reasons. First, the place of capture provides a geographic locality. More importantly, rhinos in captivity could potentially be seen by thousands while visiting menageries, circuses and zoological gardens. As encounters of rhinos in the wild have been a privilege for few people, to see one exported to urban centres made the animals known to a much wider public both in range countries and elsewhere.

1.3   Methods and Conventions

1.3.1   *Three Species of Rhinoceros*

There are three species of rhinoceros which lived in South Asia in historic times. Their common names in English have geographical elements, as Indian rhinoceros, Javan rhinoceros and Sumatran rhinoceros. As the Indian rhino is also resident in Nepal, nowadays many researchers prefer Greater one-horned rhinoceros, especially as it can conveniently be abbreviated to GOH (strangely leaving out the rhino part). The logical consequence to refer to the Javan rhino as Lesser one-horned rhino (LOH) has never become a preferred option. When specific identities are discussed, these common names become burdensome. Hence in this book the three species are consistently referred to by their scientific species names with the genus name abbreviated in these combinations:

R. unicornis	Greater one-horned rhino, Indian rhino – for *Rhinoceros unicornis*
R. sondaicus	Javan rhino, Lesser one-horned rhino – for *Rhinoceros sondaicus*
D. sumatrensis	Sumatran rhino, Hairy rhino – for *Dicerorhinus sumatrensis*

1.3.2   *Abbreviation of Rhinoceros*

In most cases, I am not in favour of the use of abbreviations. However, as this book is unashamedly all about the rhinoceros or about rhinoceroses, much space can be gained by using the short forms which have become customary in english: rhino and rhinos. Some interchange of words is allowed and quotes always follow the original.

1.3.3   *Chronological List*

The records of rhino events in this book are based on thousands of written sources. The details of all occurrences are provided in most chapters in the format of Datasets called the Chronological List, which include all the relevant data. All the exact references (author, date, page number) are provided here, and can be easily retrieved even if the descriptive sections discussing specific events do not include this information to avoid repetition. These chronological datasets are not meant to be light reading material, rather they provide a guide to sources and further literature on specific subjects.

1.3.4   *Dates*

Throughout this book, specific dates are given either in full or abbreviated in reverse sequence, following these examples: 21 February 1953 or 1953-02-21. When appropriate, usual abbreviations are followed for the twelve months of the year.

1.3.5   *Biographies and Life Dates*

The persons who encountered or discussed the rhinoceros are extremely important in the understanding of historical developments. I have been at great pains to elucidate the full names and life dates for each person. This necessitated research in a wide variety of sources, often retrievable on the internet, including gazetteers, lists of administrative and army personnel, lists of inhabitants, burial records, even newspaper announcements of births and deaths. The individual sources have not been identified in the present work. Many persons who interacted with rhinos, even those who wrote about them, are no longer generally known, yet their lives and contributions have actually made a difference to our understanding of the historical sequence. It was therefore important to rescue some of them from oblivion and add biographical details about these forgotten personalities.

Names are often confusing. I have signed this book as Kees Rookmaaker. Kees is a common name in Dutch, pronounced like the English 'case,' and it is how I have been known to everybody essentially from birth. However, it cannot be found in official records, where I can be found as Leendert Cornelis. Therefore, one person with two seemingly distinct sets of forenames.

TABLE 1.2   List of Rhino Regions employed in this book for easy reference and discussion

Number	Region	Map numbers	Protected areas
	Part 1		
Rhino Region 01	Afghanistan	9.2, 13.5	
Rhino Region 02	Pakistan	9.2, 10.3, 11.4, 13.5	
Rhino Region 03	Rajasthan	9.2, 10.3, 15.6	
Rhino Region 04	West India (Gujarat, Maharashtra)	9.2, 10.3	
Rhino Region 05	South India (Karnataka, Telangana, Tamil Nadu, Goa, Kerala)	Fossils only[a]	
Rhino Region 06	Sri Lanka	Fossils only[a]	
Rhino Region 07	Madhya Pradesh	9.2, 10.3, 16.7	
Rhino Region 08	Punjab, Haryana, Himachal Pradesh	10.3, 11.4, 17.8	
Rhino Region 09	Uttarakhand	17.8	
Rhino Region 10	Uttar Pradesh – Along Ganges	10.3, 18.9	
Rhino Region 11	Uttar Pradesh – North	18.9	Dudhwa, Katerniaghat

TABLE 1.2   List of Rhino Regions employed in this book for easy reference and discussion (*cont.*)

Number	Region	Map numbers	Protected areas
Rhino Region 12	Uttar Pradesh – East of Ghaghara River	18.09	
Rhino Region 13	Nepal – West	20.10	Suklaphanta, Bardia
Rhino Region 14	Nepal – Central	20.10	Chitwan
Rhino Region 15	Nepal – East	20.10	
Rhino Region 16	Bihar – South-West	28.11	
Rhino Region 17	Bihar – Ganges downstream from Patna to Munger	28.11	
Rhino Region 18	Bihar – Champaran and Darbhanga	28.11	Valmiki
Rhino Region 19	Bihar – Purnea east of Kosi River	28.11	
Rhino Region 20	Jharkhand (Rajmahal)	29.12, 30.13	
Rhino Region 21	West Bengal – The Middle of Bengal	30.13, 30.14	
Rhino Region 22	North Bangladesh – West of Jamuna-Brahmaputra	30.13, 30.14	
Rhino Region 23	North Bangladesh: East of Jamuna-Brahmaputra	30.13	
Rhino Region 24	North Bengal – Jalpaiguri	31.15, 32.16, 52.25	Gorumara
Rhino Region 25	North Bengal – Alipurduar	31.15, 32.16, 52.25, 64.34	Jaldapara
Rhino Region 26	North Bengal – Cooch Behar	31.15, 32.16, 32.17	
Rhino Region 27	Bhutan	38.19	Royal Manas
Rhino Region 28	Arunachal Pradesh	38.19, 38.20, 63.33	
Rhino Region 29	Meghalaya	37.18, 52.25	
Rhino Region 30	Assam (North-West) – Sankosh River	38.19, 38.20, 64.34	
Rhino Region 31	Assam (South-West) – Goalpara	30.17, 36.18, 37.20, 62.33	
Rhino Region 32	Assam (North) – Manas	38.20, 39.21	Manas
Rhino Region 33	Assam (North Central) – Darrang	38.20, 39.21	Sonai Rupai, Orang, Nameri
Rhino Region 34	Assam (South-Central) – From Guwahati to Golaghat	38.20, 39.21	Pabitora, Laokhowa, Kaziranga
Rhino Region 35	Assam – North Cachar	38.20	
Rhino Region 36	Assam (South-East) – Upper Assam from Golaghat to Tinsukia	38.20	
Rhino Region 37	Assam (North-East) – Upper Assam from Lakhimpur to Tinsukia	38.20	
	Part 2		
Rhino Region 38	Odisha (Orissa), Chhattisgarh	49.22	
Rhino Region 39	West Bengal – Western Sundarbans	50.23, 30.14	
Rhino Region 40	Bangladesh – Eastern Sundarbans	51.24	
	Part 3		
Rhino Region 41	Bangladesh – Chittagong	56.26	
Rhino Region 42	Mizoram	57.27	
Rhino Region 43	Tripura	58.28	
Rhino Region 44	Bangladesh – Sylhet	59.29	
Rhino Region 45	Lower Assam (Karimganj, Hailakandi, Cachar)	60.30	
Rhino Region 46	Manipur	61.31	
Rhino Region 47	Nagaland	62.32	

a   Map 8.1 (Fossil Rhinocerotidae) not included.

1.3.6 Rhino Regions

For the sake of clarity, the rhino habitats in South Asia have been divided in 47 Rhino Regions (table 1.2). They are based on the frequency of sightings and have no further zoogeographical relevance. These are used to reconstruct the final historical distribution of the three recent species of rhinoceros in the region. The current protected areas are shown on the applicable maps as a green circle, without providing information about the exact borders or extent.

1.3.7 Geographic Coordinates

In chapters with location data for rhino occurrences, the geographic coordinates of latitude and longitude are provided in decimal degrees. The general format used here is (for instance) 28.66N; 77.23E. These coordinates can often only be indications of the general area where rhinos were noticed. Encounters with wildlife were rarely described in sufficient detail for a particular cluster of trees besides a pond to be identified, and this type of detail would in any case be rather inconsequential as the animals tend to wander around a larger area. Therefore it is right to combine adjacent localities on the distribution maps.

1.3.8 Maps

The distribution maps were constructed starting from free software provided by QGIS. Design by Ajay Karthik and Richard Kees. For all maps, these are provided for research purposes only and make no claim to be politically or historically correct or to show actual borders.

1.3.9 Symbols of Localities on Maps

Place names associated with rhino events detailed in the Chronological Lists are expanded with their latitudes and longitudes in Datasets of Localities. These are again used, sometimes in combinations, in the construction of the maps of distribution in every given region. Each locality is shown with a symbol corresponding with the species and a number. Localities associated with petroglyphs and fossil occurrences are provided with separate symbols. The presence of artefacts, generally items connected with some kind of human activity, are shown with a special symbol in the case of archaeological sites, while their significance for the actual ranges of rhinos remains to be discussed for special cases.

Each locality has a number, which is unique throughout this work. Not every sequential number may be present. Each number in the datasets is preceded by a letter to help to separate different types of records (table 1.3).

All localities are numbered in a general west to east direction.

TABLE 1.3 Prefixes, numbers and symbols of records of various species and sets

	Prefix	Numbers
Records of *D. sumatrensis* in South Asia	D	11 to 99
D. sumatrensis – extralimital (Myanmar)	D	1 to 10
Records of *R. sondaicus*	S	101 to 199
Records of *R. unicornis*	A	201 to 499
Records of *R. unicornis* only after 1950	B	201 to 499
Records in Captivity (not shown on maps)	C	501 to 549
Records for re-introductions	E	551 to 599
Records of fossil remains of all species	W	601 to 699
Records of petroglyphs (rock art sites)	P	701 to 799
Records of all types of artefacts	M	801 to 899
Questionable records (not shown on maps)	Z	901 to 999

Symbols used on maps:

★ *R. unicornis*, ◆ *R. sondaicus*, ▲ *D. sumatrensis*, ⬟ Artefacts,
⬡ Rock art, petroglyphs, ◇ Fossils

1.3.10   *Orthography of Place Names*

The spelling of place names across the South Asian region tends to be fluid. This is to be expected since the original names came from indigenous languages. Many names exist in different formats, for example Guwahatti, Gowhatti, Gowhatty, Gauhati. This is a definite impediment to foreign researchers, especially in an age when internet search engines require some degree of accuracy and consistency. Most, not all, localities found in the literature of different epochs have been resolved.

1.3.11   *Changes in Place Names*

The names of many localities have evolved over time. Places are listed with the spelling found in the sources, followed by its name on a modern map in brackets. For the larger Indian cities, Mumbai has replaced Bombay, Chennai was Madras and Kolkata was Calcutta. In most cases, except in quotations, the modern names are used in discussions.

1.3.12   *References in Text*

This book has only few footnotes and all information has been accommodated in the text or in the datasets. It is standard practice in zoological literature to refer to sources of information or further reading in brackets within the text, where the author and date correspond with an entry in the bibliography or list of references at the end. Each reference is indicated by the author's surname, date, and (as far as possible) page range and figure number for the particular information. The format is: Author Year: Page – as in: Rookmaaker 1998: 212. If there were two authors, both are given (Rookmaaker & Groves), if more than two only the first followed by *et al.* (and others, *et alii*).

FIGURE 1.8   Logo of the Rhino Resource Center
DESIGNED BY WILLEM VAN STRIEN

1.3.13   *Pseudonyms of Authors*

In Victorian literature, especially in magazines describing hunting (*shikar*) experiences, it was customary to shield the names of persons involved by writing under pseudonyms or using initials instead of names. In retrospect, much valuable information was lost through this practice. A few recurrent pseudonyms have been resolved and the actual authors are used in the alphabetical listings.

1.3.14   *Literature Availability*

The rhinoceros of South Asia appears in an enormous variety of sources. This constitutes a major challenge in attempting a survey of literature and iconography like the present. Some books are excessively rare and may only exist as a single copy listed in the online library catalogues. I have been fortunate that librarians have made special efforts to check specific sources when asked. All references in this book have been consulted as either physical or digital copies and are genuine.

FIGURE 1.9
The young 'Saitaro' in Artis, the zoological garden of Amsterdam. This *R. unicornis* was born in the Tama Zoological Park in Tokyo, Japan on 20 December 1973 to parents imported from Assam. He died in the zoo on 24 May 1989. Drawing by the Dutch graphic artist Gert Bogels (b. 1959) of 1988
COLLECTION KEES ROOKMAAKER. ART WORK
© GERT BOGELS

1.3.15   *Bibliography*

This book is based on a set of 3061 sources listed at the end of this book. It is hoped that this will provide a gateway into further research. In case of doubt, it is recommended to consult the Rhino Resource Center (www.rhinoresourcecenter.com), which includes most of the items, often with more extensive information or digital access to the pages. Bauer & Lavilla (2022: lv) included a section explaining the methods to compile their bibliography, which is an example that I will gladly follow.

1.3.16   *The Rhino Resource Center*

This is an internet resource developed after recommendations during the International Elephant and Rhino Research Symposium held in Vienna in 2001 (Rookmaaker 2003a, 2003b, 2010). It was registered in The Netherlands on 1 August 2003 with Nico van Strien, Rob Visser and Esmond Bradley Martin as board members. Developed by the present author from material collected over several decades with the assistance of volunteer experts, it aims to list all publications about the rhinoceros in one single repository to streamline research and support nature conservation. The Rhino Resource Center (RRC) contains information of over 27,000 references to publications about rhinos and has an image gallery of over 6,000 items. The RRC is often used to develop conservation strategies and to enable innovative research as in Wilson et al. (2022a, 2022b). All references to rhinos since the start of the 21st century were combined in a single file (Rookmaaker 2020). The work of the RRC benefits from the long-term sponsorshop of SOS Rhino, the International Rhino Foundation and Save the Rhino International.

FIGURE 1.10
Rhino admiring a parrot, drawn by the German artist Theodor Gratz (1859–1947)
FLIEGENDE BLÄTTER FOR 1912, P. 152. DIGITAL COPY BY UNIVERSITÄTSBIBLIOTHEK HEIDELBERG

CHAPTER 2

Reading Rhinos through the Lens of Human-Animal Studies

Kelly Enright

Given the focus on rhinoceros horn today, it seems odd that Rudyard Kipling chose to write about rhino skin when creating a fable about the species in *Just So Stories* (1902). When his poor-mannered rhino steals a cake, its baker gets revenge by placing crumbs inside the rhino's skin. This makes him so itchy that he rubs vigorously against trees and rocks, seeking relief. As he scratches, his smooth skin folds and sags and wrinkles. The episode establishes in the species a "very bad temper" (Kipling 1902: 41). Kipling's tale is a sort-of colonial origin story that places upon the rhinoceros a peculiar explanation of perceived bully behavior and seemingly unexplainable rage. In many western tales of rhinos, both before and since, their physique is described as odd and ancient and their behavior as unfathomable.

In the Buddhist tradition, the traits of the rhinoceros are not reviled but admired. The Rhinoceros Sutra, or *Khaggavisana-sutta*, begins by describing rhinos as innately solitary animals whose isolation enables them to renounce violence and avoid pitfalls of society – both its temptations and its obligations. It lives "[w]ith no greed, no deceit, no thirst, no hypocrisy," according to the sutra, which concludes each verse: "wander alone like a rhinoceros" (Bhikkhu 2013). In this interpretation of rhino intentions, they are not grumpy hermits but tranquil monks. Their predilection to be left alone is not a sign of bad temper but a demonstration of the very temperament that can achieve enlightenment.

These representations illustrate the contrasting cultural constructions humans bring to their ideas about and interactions with non-human animals. Their observations are co-productions of human and non-human animal behaviors, shaped by both real and imagined encounters in specific times and places and passed down, creating traditions and legacies of species representation that have lasting implications for how a species is perceived and treated. Rhinoceros are not animals most humans are likely to encounter daily, making perceptions of them lean more on culture and tradition than on personal observations. Thus, the history of rhinoceros provides a lens into varied cultural imaginaries of a species that, by both of the above accounts, would rather be left alone.

The task of the rhinoceros historian is to find samples of stories, artworks, encounters, and specimens – and to read them as texts that demonstrate cultural values and perceptions. This investigation offers insight into how human-animal relationships are shaped, often providing context for the decline or abundance of a species, for if an animal is valued, its decline might be stopped. But value does not come in just one shape or size. Rhinoceros have been valued in ways that exterminate them (as trophies and pharmaceuticals) and in ways that save them (as endangered species). Sometimes, their value is confusing and contradictory, as in the practice of removing horns to stop poachers from killing. Is a rhinoceros truly a rhinoceros without this distinguishing feature – a protrusion that the artist Salvador Dali considered one of the most perfectly designed objects in all of nature?

While science defines animals biologically, anatomically, and ethologically, humanities scholarship investigates the role of animals in history and culture. Animal histories unwrap attitudes and conflicts that have led to current crises and reveal contrasting encounters and representations. The rhino portrayed by global conservation rhetoric today seems an altogether different creature than the agricultural pest who grazed indiscriminately for centuries. Unlike both Kipling's and Buddhism's rhinoceros, animals are not static, uniform objects. Essentializing their traits to explain canonized behaviors fails to consider their active agency within their specific environments, histories, and individual circumstances. Human-animal relationships at any moment or place in time are products of decades of negotiating needs and desires, as well as of individual experiences and attitudes.

In the past few decades, Human-Animal Studies has grown as an interdisciplinary field with contributions from history, literature, philosophy, anthropology, psychology, and art on the premise that animals have histories of their own and play a role in creating what we usually consider human history.[1] Such evidence-based scholarship is neither science nor activism but it complements both. The task of many animal scholars is to read into what humans

---

1 The literature is expansive, but some texts relevant to the ideas presented in this essay include Jørgenson (2019); Heise (2016); Dunlap (1988); Mighetto (1991); Herzog (2010); Fudge (2004); Montgomery (2015). Several books have been published that encompass an overview of the field, including Weil (2012); Taylor (2013); Marvin & McHugh (2014); Kalof (2017), and DeMello (2021).

© KELLY ENRIGHT, 2024 | DOI:10.1163/9789004691544_003
This is an open access chapter distributed under the terms of the CC BY-NC-ND 4.0 license.

FIGURE 2.1   The only living, Rhinoceros, or "Unicorn," in America. Now attracting such crowds of wonder loving people. At Barnum's Museum. Broadsheet issued as part of Barnum's Gallery of Wonders No. 9 by Nathaniel Currier (1813–1888) of 158 Nassau Street, New York. Probably issued in 1849, for one of the ventures of Phineas Taylor Barnum (1810–1891). Note the oversized horn
NEW-YORK HISTORICAL SOCIETY, PRINT ROOM: BROADSIDES PR-055-02-16-02

have said about animals and their actions to fill the silence in the archive. The voices of animals are not written, but if human intentions, activities, and expressions are examined critically, a more nuanced understanding of these shared experiences is possible (DeMello 2021: 27; Kalof & Montgomery 2011).

For example, writing the history of European colonization of North America cannot exclude the animals and plants they brought with them – their portmanteau biota, as historian Alfred Crosby called it in *Ecological Imperialism* (1986). Animals, plants, and diseases played an active role in reshaping landscapes, populations, and ecosystems in ways that contributed directly to the outcomes of human conflicts. Human intentions can be helped or hurt by animal bodies. In the region that is now Nepal's Chitwan National Park, the diminutive mosquito enacted its agency by preventing outsider intrusion through the spread of infectious diseases, keeping non-locals on the outskirts of the region for far longer

than they might have desired. As a consequence, rhino populations did not decline as rapidly here as elsewhere (Dinerstein 2016; Mishra 2009). Without this historical understanding, scholars might fail to understand that it is not human legislation, park creation, and conservation biology alone that preserves species. None of this happens in isolation of natural forces and ecological contributions. Rhino roam Chitwan, in part, because mosquitoes played defense.

Encroachments into rhinoceros territory in Africa and Asia happened both before and alongside colonialism, but the latter was a driving force in their extermination. Not only did outsiders introduce new agricultural methods that exploited more land that suppressed local and indigenous practices, they brought with them a culture that valued the heads and horns of rhinos as symbols of both individual and national pride. They killed, captured, and collected rhinos for monarchs, popular exhibits, and scientific inquiry. The field of natural history required physical evidence for investigation, seeking to catalog and categorize all life on earth. To both the Linnean project of classifying the world into nameable entities and natural theologians who investigated Biblical animal references, the rhinoceros' horn aligned with myths of the unicorn, or re'em, that sparked imaginations and interest. The Indian and Javan rhinoceros' single horn, in particular, made the animal seem to fit such descriptions. Its utter difference from the heraldic white horse with its long, slender horn made it a thing of curiosity, something to be considered and contemplated, not direct proof. Captive rhinos traveled Europe as audiences thirsted to see the real thing with their own eyes – both to witness the exotic and to decide for themselves what they thought of the creature. Unicorn associations continued even as citizens of empire became more familiar with rhinos. When the American showman P.T. Barnum exhibited one rhino in the mid-nineteenth century, he billed it as the "unicorn from scripture" (fig. 2.1). He beckoned crowds to witness the conflation of myth and reality embedded in the rhino's body (Enright 2008; Ridley 2004).

The rhetoric with which rhinoceros are most enmeshed is that of wilderness and wildness, created through human interpretation of rhinos' predilection for living apart from humans and their reputation for attacking without provocation. "Although not possessed of the ferocity of carnivorous animals, the rhinoceros is completely wild and untamable," wrote Georg Hartwig in a guide to "man and nature in the polar and equatorial regions of the globe" in 1875. Hartwig essentialized all rhinos, ignoring centuries-old Asian practices of training them to hunt and keeping them as amusements. In the eighteenth century, the Indian rhinoceros Clara toured Europe and was known to be so gentle and affectionate that she often licked her handler's face. A comic strip for Barnum's traveling menagerie just fifteen years before Hartwig's publication portrayed it as still, inactive, and lazily greeting visitors. While there are also reports of captive rhinos attacking their keepers, the practice of keeping and exhibiting rhinos continued to bring relatively calm creatures to public audiences without changing the type of essentializing rhetoric used by people like Hartwig. The wild and the tame rhino lived side by side in the western imagination and persist into the present day.

By the early twentieth century, the rhino's lousy temper had become rote in everything from scientific to children's literature and even shaped how scientists wrote about them. The legacy of Andrew Smith's declaration that rhinoceros "disposition is extremely *fierce*, and it universally *attacks* man if it sees him" (Smith 1838) was apparent nearly a century later, when Theodore Roosevelt theorized such attacks were contributing to their demise. "I do not see how the rhinoceros can be permanently preserved save in very out-of-the-way places," wrote Roosevelt, "the beast's stupidity, curiosity, and truculence make up a combination of qualities which inevitably tend to ensure its destruction" (Roosevelt 1910). Hollywood has cast rhinoceroses as raging and ravaging without explanation, threatening even the most resilient heroes. In *Hatari* (1962), rhinos face off with two icons of rugged masculinity – John Wayne and a Jeep. As he runs atop a train in *Indiana Jones and the Last Crusade* (1989), adventure-icon "Indie" is nearly emasculated by a circus rhino who pokes its horn through the roof right between his legs.

The juxtaposition of Jeeps and trains with rhinoceros is another trope seen often in popular culture's assessment of the species. Advertisements use rhinoceros as synonyms for toughness, as in a 1951 Armstrong Tires ad announcing the "Rhino-Flex" construction of their product has "None Tougher!" (Enright 2008: 115; 2012). A 2009 advert for Mitsubishi posed a rhinoceros over its tires while the body of a SUV was lowered onto its back. "It's more than technology. It's instinct," touted the ad (Mitsubishi 2009). A quick Google search reveals combinations of the words "rhino" and "tough" used to describe fishing rods, tools, knives, steel, water tanks, boots, glue, shelters, cell phone cases, and propane cylinders.

The rhino's image is wrapped up in the legacy of a specific type of ruggedness of character that is unafraid to confront even the largest and most formidable of beasts. In non-western cultures, the confrontation between rhino and human, and how the human fares in the interaction, has held associations with status and masculinity. In Redmond O'Hanlon's search for the Sumatran rhinoceros in Borneo, he finds the animal impossible to track. When he interviews an indigenous Ukit elder, he shares images of rhinos, and the man joyfully recounts hunting tales of his youth. O'Hanlon (1984) oddly ceases his search after talking to this man, saying he has found what he was looking for. This troubling ending implies that human knowledge is the endpoint of animal existence, that indigenous lives are as reflective of wildness as the lives of animals themselves, and that human connections to rhinoceros necessarily take the form of a hunt – whether with a gun, camera, or pen.

While scientific studies of rhinoceros bring excellent understandings of the species, historic ranges are still explained with phrases like "it may have also existed in" and "unconfirmed reports" (Khan 1989). Outside of representations and cultural capital, animal historians, contribute to the record of rhinoceros sightings. Such scientific uncertainty can benefit from historical investigation. While it may not all be recoverable, what is excavated can reconstruct historical habitats and lifeways that have impacted species for centuries, helping to explain species' vulnerability or resilience to specific impacts over time. Anthropology, too, has much to offer, as O'Hanlon's tale suggests. Taking indigenous stories, beliefs, and practices into account and weighing that knowledge equally with traditional science and scholarship provides a complete picture of global human-animal relations.[2]

Intensive examination of the cultural life of a species breaks apart the conflations of myth and science that have long associated rhinoceros with the idea of wildness. In rhino history, old ideas remain firmly planted even as attitudes about them shift. The homepage of the International Rhino Foundation (2022a) greets visitors with a photograph of a mother and baby rhinoceros with a headline stating their vision: "A World Where Rhinos Thrive in The Wild." The organization's mission and values reveal a closer connection to "science [and] political realities," but to grab attention and support, they choose to invoke "the wild" (International Rhino Foundation 2022b). Just as it was for Kipling and Roosevelt, the wild is a romantic place far from what they called civilization, by which they meant western culture. The problematic nature of wilderness is suggested further down IRF's homepage, where they state they are a U.S.-based organization with "on-the-ground programs in Africa and Asia where rhinos live in the wild." Conflating ideas of where wilderness exists with non-western nations is a legacy of colonialism and globalism that romanticizes landscapes, animals, and people.[3] This critique is not of IRF's exceptional work but of the language they, like many other conservation groups, feel they must use to generate empathy and support for endangered species. Rhinos are constantly placed in the wild, a place of the past and the imaginary, which does not offer the true vision of what it means to save a species in decline. It is ultimately a human-animal partnership. IRF states this more accurately when they share details of their work "supporting viable populations of the five remaining rhino species and the communities that coexist with them" (International Rhino Foundation 2022a).

If rhinos were to live in the wild, they would have to be time travelers. The rhetoric around species conservation looks backward and forward with its twinned goals of saving a species for future generations (ours and theirs) and

restoring ecosystems to a previous balance. In shorthand histories of endangered species, many assume the past was better, the present not good enough, and the future reason to hope. What is really meant by the vision that rhinos will once again live "in the wild" is that they will have lives independent of human protection. Saving a species involves scientific methods, social changes, shifts in cultural attitudes, political legislation, and in the case of rhinos, armed militia. Many a hunter from centuries past would be confused by the gun barrel facing away, not towards, the animal. Likewise, those who admired rhinos without needing a trophy would wonder where the wildness remains in a species that requires such diligent human protection.

Rhinos need bodyguards because humans place enormous value upon their horns. Looking at officially-established protections or conservation efforts is misleading in the case of rhinos. While habitat loss is also a danger that must be managed, poachers operate outside laws, national borders, and international trade systems. Thus, as protections for rhinos rose in the twentieth century, the value of its horn also increased to make them even more vulnerable in the twenty-first century. Nepal's rhino population dropped from 800 in the 1950s to less than 80 in the 1970s. They established Chitwan National Park to draw protective boundaries around the species and its habitat, but poaching continued. The region had seen declines before as horns, hooves, and hides were often used locally. This led to the designation of rhinos as a Royal Animal in 1846, protecting them from hunting by any other than the royal family of the Rana dynasty. This measure is credited with preserving a larger population than in other rhino habitats. How people have valued rhinos are directly tied to both their conservation and their decline (Tanghe 2017: 125–126; Martin 1985).

In 1976, World Wildlife Fund and the Nepalese government worked together to increase projects to protect rhinos near Chitwan and deployed guards from the forestry department before stationing the Royal Nepal Army members at the park. In 1982, the government worked with Smithsonian's National Zoological Park to establish the NGO now known as National Trust for Nature Conservation, but poaching still rose. As Nepal became a democracy in the 1990s, the community became more involved in conservation, creating buffer zones, but civil war would decrease these gains in the early 2000s in part because the military was unavailable to protect rhinos. Only when the community began to publicly pressure the government to enforce anti-poaching measures did the number of rhino deaths again decline. This timeline of conservation efforts reveals a few things essential to

---

2   West 2016; Ogden 2011; Govindrajan 2018; Sodikoff 2012; Rose et al. 2017.

3   Cronon 1995: 69–90; Nash 1967; Nelson & Calicott 1998, 2008; Oelschlaeger 1991.

interpreting environmental and animal histories. First, the establishment of boundaries, protections, and organizations are not the end of species preservation. Forces from outside written law who ignore borders still operate independently, wreaking havoc on even the best protective efforts. Second, conservation can be a top-down and a bottom-up process. Just as animals represent silences in the archive, so too are the voices of locals sometimes excluded. Not so in this case, as scholar Paul F. Tanghe points out. "Public authorities responded to social pressure, not vice-versa," he concluded, showing how the community began to value the presence and preservation of rhinoceros as part of their local identity. External tourism or market hunting were not instituted as regional economic drivers or incentives to preserve which, according to Tanghe, is a good thing. Tourism, he argues, "legitimiz[es] the economic exploitation and commodification of wildlife like rhino, and socailiz[es] community members and public authorities alike to see such wildlife through Market Pricing relational schemata, thereby suppressing moral mechanisms that could contribute to endangered species protection." In other words, how people think about rhinos is more important than their market value (Tanghe 2017: 165–170).

Such "moral mechanisms" and cultural incentives create strong bonds between communities and wildlife. When Tanghe asked locals why they wanted to protect rhinos, they replied: "I am Nepali!" Tanghe connects this to a sense of responsibility and belonging to a place, but it also belongs to heritage. Seeing rhinos as part of a local, long-standing identity creates a shared preservation of the environment and history. The stories people tell about their present are always informed by their past. Rhinos are a relic of a former way of being and worthy of preservation for this connection to cultural heritage. A cynic might say this is too anthropocentric a motive for conservation. Still, the idea that humans should preserve rhinos for their own sake is ethically inspirational but practically too ambitious. Culture will always make meaning of non-human others in ways that make sense to those inside the culture. Pretending that relationship does not exist is inaccurate and unproductive. Even the definition of wildness, wilderness, and who gets to be wild is a cultural construct. On a more direct note, the image of rhinos guarded by armed rangers plays directly into their status as culturally-important relics, as museum or palace guards stationed around important artworks or royal lineages.

Nepali attitudes shift the image of rhinos from heavily-guarded relics to members of living communities, from icons on the brink to animals with agency and allure. The commodification of wildlife is integral to

considerations of conservation and wildlife management. Understanding animals is not a one-way street; many roads are good, and examining historical examples, and cross-cultural comparisons in the work of humanities scholars can be instructive tools in making plans for the present and the future.

Rhinos have a long heritage of being exoticized. They have been symbols of empire, icons of the wild, and fulfillments of myths. They have traveled the globe and their form has contributed to dialogues greater than themselves. In Zocchi's painting *Allegory of the Continent of America* of 1760 (figure 10.12), a rhinoceros stands alongside American Indians and a Roman chariot in an amalgam of wild, ancient, and exotic elements, all out of place and time, the "others" of western culture. Like many animals, humans have found rhinos a resource for their own needs and desires. Throughout the centuries, their hide has been used as shield and armor; their hooves and horns intricately carved into chalices, knife handles, and sculptures; their heads displayed as trophies; their powdered horn ingested. Rhino parts remain in private collections and the storage rooms of museums, and though the trade is banned, it has not ended. Rhino horns now circulate in a global black market where poachers kill African rhinos to bring to Asia as pharmaceuticals and trophies.

Restricting hunting to an elite few is still a method used to help support rhinoceros conservation. In 2018, one hunter paid $400,000 for the opportunity to kill a black rhinoceros in Namibia's Mangetti National Park. One South African hunting lodge promotes hunts for white rhinoceros ranging from $55,000 to $129,000 depending on the size of the horn. They also offer a Green Hunt for $10,500, where the hunter only tranquilizes the animal for a photo shoot. In both cases, those providing hunts make sure to say the money is going toward the conservation of the species and that the rhinos killed are males who do not have breeding potential (Padilla 2019; Africa Hunt Lodge 2022). This method seems to devalue the wildness of the rhino, making sacrificial trophies of a few for the good of the many. Various other protection methods sacrifice wildness on behalf of conservation, such as poisoning or removing rhino horns to make them less appealing to poachers. Rigorous debate surrounded the practice of farming rhinos for their horn, as was allowed in South Africa, with some arguing it is a viable way to save the species. Each of these methods monetize rhino and shift their status from wild to domesticated. They remove something of what it means to be wild, belaying the idea that wildness is a past to which rhinos may return.

Through the lens of the rhino, complex human desires to protect a population reveal themselves. Both IRF and the owners of the Africa Hunt lodge value rhinoceros. Both wish to promote experiences with a wild animal and keep the species alive. Both might have some success in achieving this goal despite their differing ideas of what makes a rhinoceros valuable. They also both play upon the heritage of rhinos as legendarily wild. IRF wishes to separate the rhino from culture to appreciate a distant type of non-human nature that values the animal for its independent life, implying that the rhino is meant to live apart from human intervention. The hunting lodge, on the other hand, values the wildness of the rhino, informed by centuries of hunting tales as the most formidable of beasts. Both the park and the lodge have boundaries and breeding assistance, making these spaces elements of the human world. However, the idea that wildness still exists remains a profitable and psychologically desirable element of promoting interactions with rhinoceros.

The importance of conversing across cultures is vital for rhino conservation. Conservationists have pursued information campaigns, offering western scientific beliefs to those who believe in the healing ability of the horn, but the desire for it has not waned. The global market for rhino horns is banned, yet their numbers continue to decline. A 2015 publication, *The Costs of Illegal Wildlife Trade: Elephant and Rhino*, estimated that 29,000 rhinoceros survived worldwide, representing a 94% decline in the last century and a 60% decline in the previous 45 years (Tanghe 2017: 14; Smith & Porsch 2015: 16). That means that since the trade was banned in the 1970s, things have only gotten worse for rhinos and their horns. Conversely, IRF's "2021 State of the Rhino Report" stated that Nepals' greater one-horned rhino population had reached 752 – up from 107 in 2015. The report credits the success to the collaborative work of governments and non-profits and "the local communities that value their rhinos and other wildlife as national treasures" (International Rhino Foundation 2021). This cultural significance has developed a deep and abiding conservation ethic. There have been very few poachings in Chitwan in recent years. One that occurred in 2017 was followed by a "mourning rally" to protest, raise awareness, and demand action to find the criminals (Dudley 2017). This event reveals the importance of each and every rhino to the community and amplifies the importance of understanding animal representation and the role it plays in the lives of human and non-human animals alike.

The Greater One-Horned Rhinoceros Rhinoceros unicornis *in South Asia*

∵

The Indian or Greater One-Horned Rhinoceros *Rhinoceros unicornis*

The Greater One-Horned Rhinoceros with its distinctive skin folds is the quintessential rhinoceros of South Asia. The animal's iconic appearance and large size has made it a symbol of strength and vigour penetrating into many layers of the cultural tapestry of all people in the subcontinent. It is the ultimate pachyderm. This is the animal which has inspired a long series of eminent artists and scholars. It is the rhinoceros which for many centuries was the prototype of that family to the scientific community as well as to the general reading public. There has always been an understanding that there was an animal with prominent skin folds and a horn on the nose in South Asia. If that animal had been confined then, as now, to forests in Assam, Bengal and Nepal, news of its existence would not have reached the outside world as readily as it has done, speaking for a wider distribution across the region until a relatively recent age. Potentates and peoples admired the animal's power and physical strength, mirrored by an increasing popularity with travellers and traders.[1]

It did not take long before the rhinoceros was exported to the expanding European powers in search of trade in the valuable commodities of the east. Asian culture helped to shape the European cultures in the wake of the Middle Ages in what has been called the Century of Wonder by the American historian Donald Frederick Lach (1917–2000). Wonder and amazement indeed for creatures like the rhino embodying the splendour of the oriental courts. The first post-medieval rhino to reach Europe, brought by the Portuguese in 1515, was seen alive by thousands, but its fame spread quickly and decisively through the artistic superiority of the German artist Albrecht Dürer interpreting a sketch by an amateur artist (fig. 3.2). In western culture, no doubt his artistic representation epitomized the essence of the rhinoceros in the sixteenth century as much as today. Other artists, photographers, film makers and writers have tried to challenge this popular perception with more naturalistic portrayals, yet only slowly eroding the first impression of what a real rhino should be like.

The great naturalist Carl Linnaeus, whose genius introduced our current system of taxonomy and nomenclature, thought of this rhinoceros when inserting the species into his new classification encompassing the vastness of nature. At first it was the only type allowed, and it would take a remarkable amount of effort to displace the notion that all rhinos were essentially similar (fig. 3.3). Even Linnaeus was constantly working to improve his structure of the natural world. It was acknowledged in 1758, upon the publication of the 10th edition of the *Systema Naturae*, that he had reached the best foundation on which all subsequent work has been built. The rhino was then known to live in India, expressing the everlasting bond between the animal and the land, culture and people of South Asia.

The Greater One-Horned Rhinoceros is one of three species existing in Asia, and the only one endemic to South Asia (fig. 3.1). It occurred in India, Bangladesh, Bhutan, Nepal, Pakistan and Afghanistan, with current strongholds in North-East India and South Nepal. The exact historical and current distributions will be elaborated in the remainder of this book. The animal was known along the southern base of the Himalayas in the valleys of the major rivers, the Indus, Ganges and Brahmaputra. Today *R. unicornis* is listed in the IUCN's Red List as Vulnerable, which means that it is considered to be at high risk of unnatural extinction (Ellis & Talukdar 2019). Rhinos have been in decline in South Asia for many centuries and have disappeared from much of their former range before the nineteenth century. The latest status figures give a world population of *R. unicornis* of some 4,000 specimens. In the 15th century, before the expansion of human settlements and conversion of land to agriculture, it has been estimated that in the 35,800 km² of prime rhino habitat there might have lived 476,140 specimens (Dinerstein & MacCracken 1990). This is quite a daring estimate for a past situation for which we have little evidence, yet shows in stark terms how much the landscape has changed and what might happen if the decline of land suitable for wildlife is allowed to continue.

Outside South Asia, fossil remains have been recorded in the late Pleistocene from Thailand (Filoux & Suteethorn 2018) and China (Sun et al. 1998). However,

---

1 The biology and history of *R. unicornis* are introduced and explored by Guggisberg 1966; Laurie 1982; Laurie et al. 1983; Clarke 1986; Rookmaaker 1980, 1982, 1983a, 1984a, 1999c, 2002a; Zecchini 1998; Dinerstein 2003; Enright 2008; Bahuguna & Mallick 2010 and Divyabhanusinh et al. 2018, supplemented by a host of sources with often incredible insights of varied subject matter listed in the bibliography of this work.

© L.C. (KEES) ROOKMAAKER, 2024 | DOI:10.1163/9789004691544_004
This is an open access chapter distributed under the terms of the CC BY-NC-ND 4.0 license.

FIGURE 3.1 Watercolour by Joseph Wolf (1820–1899) of the great one-horned rhinoceros (*Rhinoceros unicornis*). Signed lower right "J. Wolf 1872."
This was later engraved for Sclater, *Transactions*, 1876, pl. 95
WITH PERMISSION © ZOOLOGICAL SOCIETY OF LONDON, ARCHIVES: DRAWINGS BY WOLF, VOL. III. MAMMALIA III.
UNGULATA I, NO. 70

FIGURE 3.2
Rhinoceros following the example of Albrecht Dürer (1471–1528),
with the characteristic twisted horn on the shoulder called the
"Dürer-hornlet." From a collection of watercolour drawings
assembled by the Italian polymath Ulysse Aldrovandi
(1522–1605) in Bologna
WITH PERMISSION © BIBLIOTECA UNIVERSITARIA DI
BOLOGNA: MS. 124, TAVOLE VOL. 1, PART 2 ANIMALI, F. 91

there is no evidence that the species during the last ten millennia ever crossed the mountains situated on the current border of Myanmar with either India or Bangladesh. This is a fact, which strangely and awkwardly continues to be ignored. Several range maps published in the last century show *R. unicornis* extending into South-East Asia, always without detailed analysis of data, and always without supporting evidence.

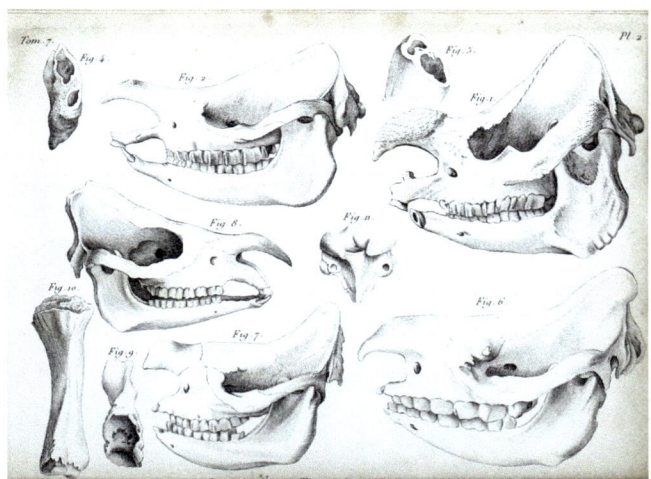

FIGURE 3.3 Skulls and bones of four recent species of rhinoceros published by Georges Cuvier (1769–1832) in 1806. The skulls represent 1. *R. unicornis*, 2. *R. sondaicus*, 6. *Diceros bicornis*, 7. *Diceros bicornis* (young), 8. *D. sumatrensis*. Plate drawn by the palaeontologist Charles Léopold Laurillard (1783–1858) and engraved by Henriette Couet (1792–1872)
ANNALES DU MUSÉUM D'HISTOIRE NATURELLE, PARIS, 1806, VOL. 7, PL. 2

FIGURE 3.4 Rhinoceros in Chitwan National Park, on an early morning, close to Meghauli Serai lodge
PHOTO: KEES ROOKMAAKER, 2017

FIGURE 3.5 Rhinoceros in a typical jungle of high grasses, taken by the Swedish photographer Bengt Berg (1885–1967) in North Bengal
BERG, *TIGER UND MENSCH*, 1935, P. 176. COURTESY: NATASHA ILLUM BERG

The Greater One-Horned Rhinoceros is found largely in the flat plains and swamps along watercourses, in riverine woodlands and areas with long grasses and reeds (fig. 3.4). This is the typical undulating *terai* area found along the base of the Himalayas. Access to water is essential for their mud baths and for the growth of their food plants. Rhinos create and maintain narrow paths through the tall reeds, which otherwise would be largely inaccessible. This typical habitat is well documented in photographs taken in Bengal by the Swedish naturalist Bengt Berg in the 1930s (fig. 3.5). Although rhinos are large animals which often gather in social groups, they can easily hide in the forests, where in the past they could only be approached on elephant back. There is very little evidence that *R. unicornis* was found in mountainous or even hilly terrain (Choudhury 2022).

This skin which looks as though it is divided into shields of armour has gained the species the appropriate name *Panzernashorn* in German, *Pantserneushoorn* in Dutch. In English this translates as Armour-plated Rhinoceros, which term, albeit descriptive, never gained momentum. When I started to read zoological works some fifty years ago, the species was commonly referred to as 'Indian Rhinoceros,' which compares well with the names based on geography of the other two Asian species, the 'Javan' and the 'Sumatran.' It is true, however, that the 'Indian' rhinoceros also occurs prominently at least in Nepal, hence the search continues for a proper and easy alternative. In zoos nowadays, especially in the USA, the name 'Greater One-horned Rhinoceros' has become popular, perhaps because it can be abbreviated to GOH.

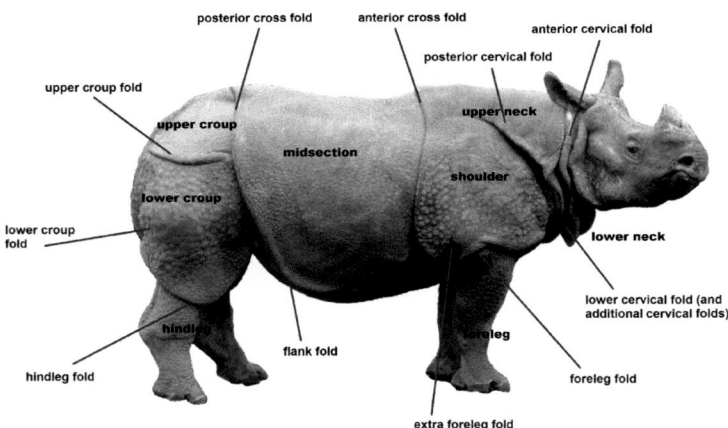

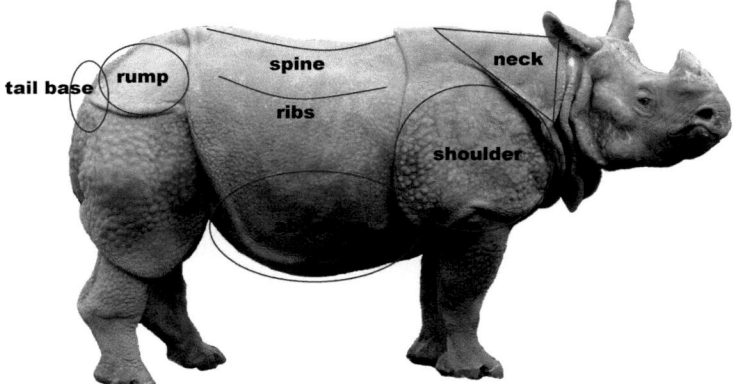

FIGURE 3.6
(A and B). Terminology for the folds and body regions of the rhinoceros
COURTESY © EVA HEIDEGGER AND MARCUS CLAUSS, ZURICH

For the present purpose, the name 'Indian' would be too confusing in a work with many geographic localities, and the unabbreviated GOH is too cumbersome. Therefore, to avoid any misunderstandings, I have used the scientific name *Rhinoceros unicornis*, throughout in its abbreviated form *R. unicornis*.

The folds in the skin of *R. unicornis* are the most prominent feature adding greatly to the animal's appeal and popularity. There are several folds in the neck region which appear to be constant in all ages and in both sexes. The best way to differentiate them in descriptions is shown in the diagrams of fig. 3.6. In older males there are lower cervical folds in the throat region forming a distinctive bib. The posterior cervical fold slopes backwards and upwards in the neck region, but never extends to the top of the shoulder (Neuville 1927). This sets *R. unicornis* clearly apart from *R. sondaicus*, in which the anterior and posterior cervical folds combine to form a triangular shield. With its large size, single horn and typical structure of the skin folds, *R. unicornis* is easily identified in almost all situations. The armoured hide is highlighted in Rudyard's Kipling's compelling story describing how the rhinoceros got his wrinkly skin, first published as a serial in 1898 with illustrations by Oliver Herford (fig. 3.7) and

then combined in the evocative *Just so stories for little children* of 1902.

R. unicornis is one of only two living species of rhinoceros with a single horn located on the anterior portion of the head above the nostrils. In most cases, these horns are relatively short when compared with the African species, with a horn length of over 40 cm being unusual and over 50 cm exceptional (fig. 3.10). The most comprehensive and authoritative source for the sizes of the largest rhino horns is the series of *Records of Big Game*, initiated and edited by Rowland Ward (1848–1912) as the director of this famous taxidermist firm in London (Morris 2003). The first edition appeared in 1892, and was regularly updated with six further editions during his life time, then continued throughout the 20th century. The books contained measurements of the largest trophies of all game animals. At first these were mostly confined to specimens which passed through the premises of Rowland Ward, and subsequently, with rising popularity owners were allowed to add their own data. The result is a veritable treasure trove of information, albeit with the passage of time increasingly difficult to use as only the names of the collectors (hunters) were recorded, combined with a general locality and sometimes a year. The books have no information

FIGURE 3.7 A page from the first publication in 1898 of the story by Rudyard Kipling (1865–1936) telling how the rhinoceros got his wrinkly skin. Illustrated by Oliver Herford (1863–1935)
ST. NICHOLAS, FEBRUARY 1898, VOL. 25, P. 275

regarding where each trophy was preserved, and as most were kept privately, it is a practically impossible task to retrieve them (Rookmaaker 2002c). According to the combined data in all editions of Rowland Ward's *Records of Big Game*, there were only three horns of *R. unicornis* larger than 50 cm (table 3.4). Another horn of 61.5 cm length and 2622 g weight from Assam was reported and illustrated from the Assam State Museum, Guwahati by Barua & Das (1969: 26), but the current depository is unknown (Rookmaaker 2020b). On World Rhino Day, 22 September 2021, the Government of Assam destroyed 2479 rhino horns from the State Treasury, which were collected from animals that died a natural death or were retrieved from poaching incidents (Yadava 2022). All horns were recorded and measured in advance and 94 were marked for preservation (figs. 3.8, 3.9). The longest *R. unicornis* horn was 57 cm along anterior curvature or 42.5 cm standing height, the heaviest horn was 3.05 kg. The average standing height of horns was 13.77 cm, average weight 560 gram, average basal circumference 43.47 cm.

In general size *R. unicornis* is among the larger living species in the family. It is important to establish the average or maximum sizes of the animals, especially to compare it with the sympatric *R. sondaicus*, which was in the past often said to be consistently smaller (chapter 46). This is not an easy task because rhinos can hardly be measured in the field while alive, and because different researchers might have used different methods. There are measurements for 17 trophy animals killed in Cooch Behar (Rookmaaker 2020b) which are here calculated in

TABLE 3.4 The greatest horn lengths recorded for wild specimens of *Rhinoceros unicornis*

	Locality	Length of horn	Source
1	Assam (Assam Forest Museum)	61.5 cm	Barua & Das 1969: 26
2	Assam (1909, Briscoe, in NHM)	61.0 cm = 24 in	Ward 1910: 464, Shikar 1909
3	India (T.C. Jerdon)	61.0 cm = 24 in	Ward 1892: 238
4	Assam (trophy horn)	57.2 cm	Laurie et al. 1983
5	Assam	57.0 cm	Assam Treasury measured 2021
6	Nepal (1901, Curzon at Morang)	54.0 cm = 21 ½ in	Ward 1903: 409; Ellison 1925: 33
7	Nepal (1933, Maharaja of Surguja)	49.5 cm = 19 ½ in	Ward et al. 1935: 337
8	Assam (1892, Ipswich Museum)	49.2 cm = 19 1/8 in	Ward 1892: 238; Ward 1907: 464
9	Assam, Singpho (1921, C.A. Elliot)	48.3 cm = 19 in	Ward 1899: 433; Ward 1922: 465
10	Assam (1913, D.H. Felce)	46.7 cm = 18 3/8 in	Ward 1914: 463
11	Assam (1902, M.H. Logan)	45.7 cm = 18 in	Ward 1903: 409
12	Assam (1870, Bengal hunters)	45.7 cm = 18 in	Pollok 1882b: 58
13	Cooch Behar (1898; largest in area)	41.3 cm = 16 ¼ in	Ward 1899: 433

The source refers only to the earliest edition of Rowland Ward's *Records* including this item.

FIGURE 3.8 Rhino horn burning in Assam on 22 September 2021. These horns marked for preservation were among the longest with lengths and weights of 36 cm (3.048 kg), 42.5 cm (2.568 kg), 38 cm (1.745 kg)
PHOTO © AMIT SHARMA

averages. Pollok (chapter 39) provides the size of only the two largest specimens with lengths of 381 and 406 cm, and height 188 cm. Other studies have been carried out in Nepal and on a limited basis in zoos. In a zoo environment, the average weight of adult males (n = 2) was 2373 kg, adult females (n = 2) 2043 kg (Holeckova 2009).

Size of *R. unicornis*
Length of body	maximum 430 cm
Height at shoulders	maximum 205 cm
Horn length	maximum 61 cm

The data of overall measurements combined in Table 3.5 present a rather equivocal picture. This is likely partly due to different methods used by the observers. It appears that females are smaller on average than males. A conclusion regarding differences between populations would be hazardous based on these limited data. In a comparison with *R. sondaicus*, the figures of body length do not present a clear result that the two species differ in size. Those of shoulder height show that the average males *R. unicornis* are a little taller (above 170 cm) than all specimens of *R. sondaicus*, but this is not clearly indicated for females. It is unlikely that the size can ever be fully diagnostic, especially in the field where only carcasses can be properly measured. In South Asia, therefore, *R. unicornis* is not consistently larger than *R. sondaicus*.

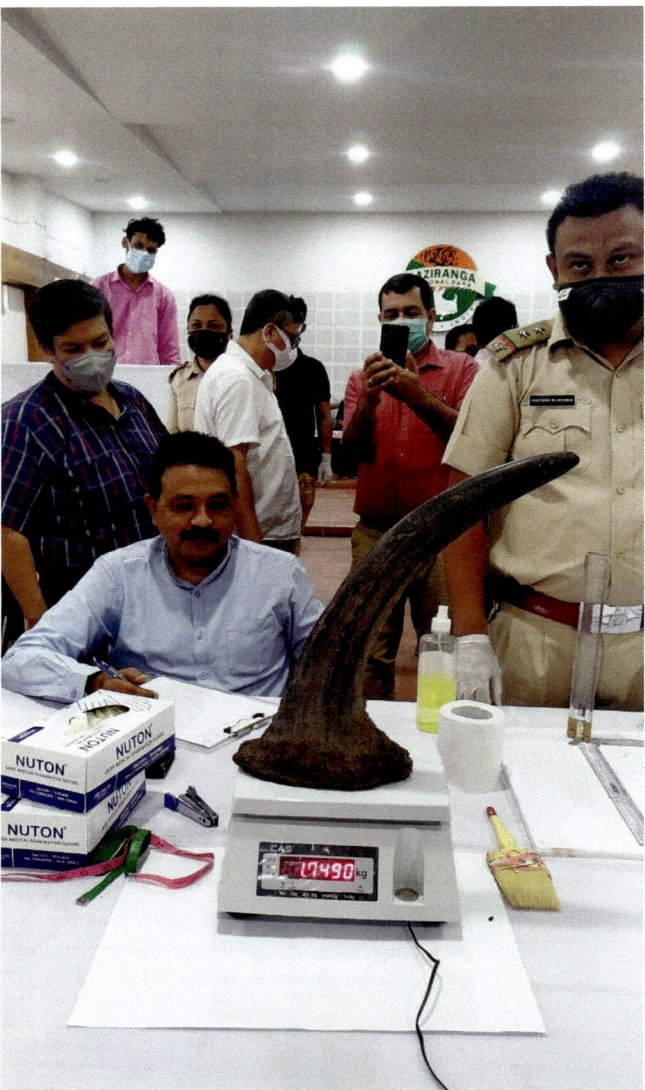

FIGURE 3.9 At the inspection of rhino horns in the Assam State Treasury before the burning exercise on 22 September 2021, this horn was the second longest found with 57 cm outer length, weight 1.749 kg
PHOTO © AMIT SHARMA

TABLE 3.5   Average length and height of adult males and females of *R. unicornis*

Locality	Body length		Shoulder height		Source
	Males	Females	Males	Females	
Cooch Behar	321 (n = 12)	307 (n = 1)	184 (n = 13)	174 (n = 4)	32.11
Assam	381, 406		188		39.5
Nepal	368–380 (n = 2)	310–341 (n = 4)	171–176 (n = 2)	148–164 (n = 4)	Laurie 1978
Nepal	346 ± 18.4 (n = 15)	335 ± 25.7 (n = 9)	172.3 ± 14.2 (n = 4)	149.3 ± 14.7 (n = 3)	Dinerstein 2003
Zoos			166–187 (n = 5)	154–166 (n = 7)	Houwald 2001

Expressed in cm. The number of animals measured is given in brackets in each case. Length is taken without the tail.

FIGURE 3.10   "The Indian rhinoceros R. unicornis Lin." Plate published in a zoological treatise by Edward Griffith, probably executed by Thomas Landseer (1794–1880). This shows that artists were innovative, because the animal looks full-grown and wild, although it was drawn in Polito's Menagerie at the Exeter 'Change' in London 1810–1814. The plate (hand coloured) was available earlier according to the imprint below the name: "London Published by G. & W.B. Whitttaker Septr 1824"
GRIFFITH, *ANIMAL KINGDOM*, 1826, VOL. 3, P. 424

Taxonomy and Nomenclature of *Rhinoceros unicornis* in South Asia

4.1 Taxonomy of the Greater One-Horned Rhinoceros

Rhinoceros unicornis is a monotypic species, which means that it is not divided into any subspecies or other geographic entities. Investigations of the possible differences between populations are complicated due to the small number of specimens with sufficient data in collections and to the possibilities of significant variation on a local level. The recent studies by Colin Groves (1982, 1993, 2003) have indicated that there is no reason to separate specimens from Nepal, North Bengal or Assam on any taxonomic level. The population in Chitwan NP shows a high genetic variability (Dinerstein & MacCracken 1990; MacCracken & Brennan 1993; P.Das 2014; Pathak et al. 2022). Starting from studies of captive animals, Zschokke et al. (2003, 2011) recommend that those originating from Nepal and those from India should be treated as separate Management Units and managers of captive facilities should avoid any crossing of individuals from different regions.

4.2 Principles of Nomenclature

The scientific names of animals are governed by a set of rules called the International Code of Zoological Nomenclature which has one fundamental aim: to provide the maximum universality and continuity in the names of animals compatible with the freedom of scientists to classify the animal kingdom according to their taxonomic judgements. The first Code was designed by the British geologist and ornithologist Hugh Edwin Strickland (1811–1853), with a first draft of 1837 being expanded to the publication of the Stricklandian Code in 1843 by the British Association for the Advancement of Science (Rookmaaker 2007b, 2010, 2011c). Strickland initiated the Code because he was aware that the same animal species was often described multiple times with different names leading to an increasingly chaotic taxonomic discussion. In order to stabilize the use of scientific names, he suggested that the only valid one would be the earliest available name. The starting point was arbitrarily fixed on 1 January 1758, the year in which the Swedish zoologist Carl Linnaeus (1707–1758) published the 10th edition of the pivotal *Systema Naturae*. Strickland's initiative has seen many changes over the centuries. The latest 4th edition of the current Code is governed by the International Commission of Zoological Nomenclature (ICZN 1999).

In the animal kingdom, the name of a species is a binomen consisting of a genus name (written with initial upper-case letter) and a species name (with initial lower-case letter). The name of a subspecies is a trinomen adding a subspecific name to the binomen (also with initial lower-case letter). Hence in *Rhinoceros sondaicus inermis*, *Rhinoceros* is the genus name, *sondaicus* the species name, and *inermis* the subspecific name. To avoid any confusion, It is advisable, but optional, in a taxonomic work to cite at least once the author and date of the species name, with punctuation as in *Rhinoceros sondaicus* Desmarest, 1822. A genus name can be abbreviated to its initial (R.). The author-date combination is enclosed in brackets when the species is transferred to a genus different from when it was first described.

Every genus, species and subspecies has potentially a name-bearing type designated by the author of the name or subsequently, usually called the holotype or type specimen. The geographical location where this specimen was captured, collected or observed is called the type locality. The type specimen and its associated type locality set the standard for that particular taxon. The International Code of Zoological Nomenclature is an important milestone which should be followed in every aspect. After the move of the Secretariat from London to Singapore, the present author was Secretary of the Commission from 2015–2016 with the assistance of Martyn Low, curtailed due to personal circumstances.

All living species of rhinoceros have been given multiple specific names and all except one have been divided into subspecies. Different names denoting the same taxon are called synonyms. A synonymy is a list of such synonyms, which is a tool used by taxonomists to establish which names are still valid and to help with an understanding of older literature.

In this chapter I provide an annotated synonymy for the genus *Rhinoceros* and for *R. unicornis*, with a discussion of subspecies, valid names, type specimens and type localities. The author and publication of each name follow the heading in each instance. All synonyms are listed, while only those used for taxa living in South Asia are placed in a wider historical context.

© L.C. (KEES) ROOKMAAKER, 2024 | DOI:10.1163/9789004691544_005
This is an open access chapter distributed under the terms of the CC BY-NC-ND 4.0 license.

4.3 The Genus *Rhinoceros* and Its Synonyms

The genus *Rhinoceros* is found in the *Systema Naturae* of Carl Linnaeus, of which the 10th edition published in 1758 is taken as the starting point of zoological nomenclature. This genus was subsequently split as new fossil and recent species of rhinoceros were discovered in Asia and elsewhere. *Rhinoceros* has remained in use for both one-horned species living in South and South-East Asia, *R. unicornis* and *R. sondaicus*. The generic name has five junior synonyms with either of these as type species: *Naricornis*, *Monoceros*, *Unicornus*, *Eurhinoceros* and *Monocerorhinus* (Rookmaaker 1983a: 28, 1983b). The genus *Ceratorhinus* used initially for the two-horned *sumatrensis* (54.2) is also found in combination with the single-horned species in a palaeontological context (Steinmann & Döderlein 1890: 774).

Rhinoceros Linnaeus, 1758
Carl Linnaeus. 1758. *Systema Naturae* [10th edition]. Holmiae, vol. 1, p. 56.

Type species: *R. unicornis* (selected by Pocock 1944a).

Historical context elucidated together with the species in 4.5.

Derivation of name *Rhinoceros*: Greek ῥινόκερως, from Greek ῥίς rhis (genitive ῥινός rhinos) (nose, nostril) and Greek κέρας keras (horn).[1]

Naricornis Frisch, 1775
Jodocus Leopold Frisch. 1775. *Das Natur-System der vierfüssigen Thiere*. Glogau, p. 31 (in Tabula Generalis).

Type species: *R. unicornis*.

Proposed by German naturalist Jodocus (Jost) Leopold Frisch (1714–1789) in a rare book reviewing the animal kingdom. The name is unavailable, because this publication was suppressed for nomenclatorial purposes being not consistently binominal.

Derivation of name *Naricornis*: Latin naris (nostril) or plural nares (nose) and Latin cornu (horn).

Monoceros Rafinesque, 1815
Constantine Samuel Rafinesque. 1815. *Analyse de la Nature*. Palerme, p. 56.

Type species: *R. unicornis*.

Name proposed without further explanation in a table of animal genera by the French-American polyglot and naturalist Constantine Samuel Rafinesque (1783–1840),

who is known for describing a plethora of zoological and botanical names (Ambrose 2010). The name *Monoceros* was used twice before as a generic name, in works which have both been suppressed for purposes of nomenclature. First the name was used for the legendary unicorn by Eberhard August Wilhelm von Zimmermann (1743–1815) in his *Geographische Geschichte* (1780, vol. 2, p. 157), second as a genus of Mollusca (Gastropoda) by Friedrich Christian Meuschen (1719–1811) in *Museum Geversianum* (1787, p. 290).

Derivation of name *Monoceros*: Greek μόνος monos (single) and Greek κέρας keras (horn).

Unicornus Rafinesque, 1815
Constantine Samuel Rafinesque. 1815. *Analyse de la Nature*. Palerme, p. 219.

Type species: *R. unicornis*.

Constantine Samuel Rafinesque (1783–1840) substituted this name for *Monoceros* found in the same work. The generic name *Unicornus* was earlier used by Pierre Denys de Montfort (1766–1820) in the combination *Unicornus typus* in Mollusca, Gastropoda, Muricidae (Montfort 1810, vol. 2: 454).

Derivation of name *Unicornus*: Latin unus (one) and Latin cornu (horn).

Eurhinoceros Gray, 1868
John Edward Gray. [1867] 1868. Observations on the preserved specimens and skeletons of the Rhinocerotidae. *Proceedings of the Zoological Society of London*, p. 1009.

Type species: *R. javanicus* (= *R. sondaicus*).

This name was proposed by John Edward Gray (1800–1875), zoologist at the British Museum in London, for rhino species with the "forehead and the nose behind the base of the horn flat, both in the living animal and skull." It was a subgenus which included the species *javanicus*, *unicornis*, *nasalis*, *floweri*, separated from *stenocephalus* in the subgenus *Rhinoceros*. Gray read his paper in the meeting of the Zoological Society of London on 12 December 1867, of which the *Proceedings* appeared in print in April 1868 (Duncan 1937).

Derivation of name *Eurhinoceros*: Greek εὖ (well) and Greek ῥινόκερως (rhinoceros).

Monocerorhinus Wüst, 1922
Ewald Wüst. 1922. Beiträge zur Kenntnis der diluvianen Nashörner Europas. *Centralblatt für Mineralogie, Geologie und Paläontologie*, p. 654.

Type species: *Rhinoceros (Monocerorhinus) sondaicus*.

Ewald Wüst (1875–1934), palaeontologist and botanist in Kiel, Germany, introduced 2 subgenera in the genus

---

1   The derivation of the scientific names of rhinoceros taxa (genera and species) was elucidated by Reinoutje Artz of Amsterdam, in this chapter as well as in chapters 8, 47 and 54.

Rhinoceros, named *Rhinoceros* and *Monocerorhinus*, the latter for a species characterized by females without horn.

Derivation of name *Monocerorhinus*: Greek μόνος monos (single), Greek κέρας keras (horn) and Greek ῥίς rhis (*genitive* ῥινός *rhinos*) (*nose*, nostril).

4.4 Names of the Living Species *Rhinoceros unicornis*

List of synonyms of *Rhinoceros unicornis* in chronological order. The summary is followed by historical and nomenclatorial annotations. In this case a full synonymy is provided because the species is only known to occur in South Asia. The taxonomic context of South Asian extinct species described in the genus *Rhinoceros* and allied taxa is investigated in chapter 8, with details of authorship, date, type specimen and type locality (8.6).

Rhinoceros unicornis Linnaeus, 1758
Carl Linnaeus. 1758. *Systema Naturae* [10th edition]. Holmiae, vol. 1, p. 56.
Type-specimen: not designated. Lectotype is the First Lisbon or Dürer Rhinoceros depicted on plate 38 in Jonston (1653, also 1657). Refer 4.5, 4.6.
Type-locality: "Africa, India" (Linnaeus 1758, vol. 1: 56), restricted to Assam (discussed in 4.5).
Derivation of name of *unicornis*: Latin unus (one) and Latin cornu (horn).

Rhinoceros rugosus Blumenbach, 1779
Johann Friedrich Blumenbach. 1779. *Handbuch der Naturgeschichte*, 1st edn. Göttingen, p. 134.
Type-specimen: not designated. Refer 4.7.
Type-locality: not stated, but obviously India.
Derivation of name *rugosus*: Latin adiective rugosus (wrinkled, shrivelled).

Rhinoceros asiaticus Blumenbach, 1797
Johann Friedrich Blumenbach. 1797. *Handbuch der Naturgeschichte*, 5th edn. Göttingen, p. 126.
Type-specimen: not designated. Refer 4.8.
Type-locality: East Indies ("Ost-Indien").
Derivation of name *asiaticus*: adjective form of the continent Asia.

Rhinoceros indicus Cuvier, 1816
Georges Cuvier. [1817] 1816. *Le Règne Animal*, Paris, vol. 1, p. 239.
Type-specimen: not designated. Refer 4.9.
Type-locality: Oriental Indies, especially beyond the Ganges ("Indes orientales, surtout au delà du Gange").

Derivation of name *indicus*: latinized adjective form of India.

Rhinoceros stenocephalus Gray, 1868
John Edward Gray. [1867] 1868. Observations on the preserved specimens and skeletons of the Rhinocerotidae. *Proceedings of the Zoological Society of London*, p. 1018, figs. 5, 6.
Type-specimen: Natural History Museum, London 1846.3.23.4 (no. 722e), received from the Museum of the Zoological Society of London. Refer 4.11.
Type-locality: Asia.
Derivation of name *stenocephalus*: Greek στενός stenos (narrow) and Greek κεφαλή kephalè (head).

Rhinoceros unicornis bengalensis Kourist, 1970
Werner Kourist. 1970. Die ersten einhornigen Nashörner. *Zoologische Beiträge*, p. 151.
Proposed conditionally, hence the name has no standing in zoological nomenclature. Refer 4.12.
Type specimens: There are two syntypes. First, the animal shown on plate 16 in *Ménagerie Royale* (1814), and another shown on plates 8 and 9 of Jardine's *Pachydermes* (1836). Refer 4.13.
Type-locality: Not stated, but from name implied to be Bengal, India.
Derivation of name *bengalensis*: latinized adjective form of Bengal.

4.5 History of the Specific Name *unicornis* in Genus *Rhinoceros*

The Swedish Professor of Medicine and Rector of the University of Uppsala, Carl Linnaeus (1707–1778) is known as the father of modern taxonomy because he first proposed the binominal system of nomenclature still in use today. This system was first suggested in 1735 when he published his *Systema Naturae*, which was updated in 12 editions during his lifetime. The botanical counterpart *Species Plantarum* became established in 1753. In the 10th edition of the *Systema Naturae*, Linnaeus recognized two species in the genus *Rhinoceros*: *R. unicornis* from "Africa, India," and *R. bicornis* from "India" (Rookmaaker 1998b). The second species is now considered to be the African Black Rhinoceros (*Diceros bicornis*) and does not need to be discussed here, following the general consensus that the locality "India" is incorrect.

The classification of the rhinoceros proposed by Linnaeus in the successive editions of the *Systema Naturae* is neither constant nor straightforward. It looks odd that the animal was first classed as a type of elephant (genus *Elephas*) and then twice in the order of *Glires*, among rodents rather than among ungulates (table 4.6). The combination of Africa and India for the localities is more or less correct when all types of rhino are combined in one heading. He may have copied this from the short paragraphs in Ray (1693), which is the only source quoted by Linnaeus in editions 2, 4, 5 and added to others in edition 12. There are five sources referring to *R. unicornis* in the 10th edition, which may have been selected because all these accounts were illustrated. In the 12th edition, these five are repeated, Ray is reinstated and one source is added to make a total of seven (Rookmaaker 1998b). The sources quoted for the rhinoceros by Linnaeus are the following.

(1) The *Synopsis* by the English naturalist John Ray (1627–1705) published in 1693 has one paragraph about the rhino found generally in Africa, Abassia, Bengala and Patane. There is no illustration (Ray 1693: 122–123).

(2) The Polish naturalist Johannes Jonston (1603–1675) in his *Historiae Naturalis de Quadrupedibus* of 1653 treated all information of the "rhinoceros" in one chapter, stating that the animal occurred in "remote regions of Africa, Abasia, many Asian places, the kingdoms of Bengala and Jacatra." Jonston (1653, pl. 38) illustrated his account with a copper engraving, showing a rendition of Dürer's Rhinoceros (fig. 4.1), again found in later editions (1657) and translations (Dutch 1660, English 1678).

(3) The Dutch physician Jacobus Bontius, or Jacob de Bondt (1592–1631) wrote about the rhino of the East Indies partly from personal knowledge obtained in Java, Indonesia. His notes were edited posthumously by Willem Piso (1611–1678), illustrated by a side-view and head of a rhino (Bontius 1658, part 3: 50–52; see 11.14).

(4) The German professor of anatomy Karl August Bergen (1704–1759) presented a public lecture on 16 October 1746 on the occasion of the visit of the Dutch Rhinoceros 'Clara' to Frankfurt an der Oder in July 1746 (5.4). Linnaeus owned two copies of this pamphlet (preserved in the Linnean Society of London, BL553C and BL545B), the latter of which has an added broadsheet sold at the time "Wahre Abbildung von einem lebendigen Rhinoceros oder Nashorn" (Bergen 1746; Rookmaaker & Monson 2000: 326).

(5) The German polymath Conrad Gessner (1516–1565) discussed everything then known about the rhinos in his *Historia Animalium*, published during his lifetime in Latin in 1551 and much abbreviated in German in 1563 and 1583. Linnaeus used the second

Latin edition of 1620 judging from the page number, which was illustrated with a copy of Dürer's Rhinoceros as in earlier editions (Gessner 1551: 553, 1563: cxxxi, 1620: 643).

(6) The Italian naturalist and collector Ulysse Aldrovandi (1522–1605) included an encyclopedic review on the rhino in his *Quadrupedum Omnium Bisulcorum Historia* first published in 1621 (p. 884), also illustrated by a Dürer Rhinoceros (fig. 3.2).

(7) The English ornithologist George Edwards (1694–1773) published a series of 210 bird paintings followed by another series with animals as well as plants. Edwards (1758: 22) has a short text accompanied by a coloured plate showing an elephant and a female rhino "drawn from the life in London, A.D. 1752." This may have been Clara, the Dutch Rhinoceros, as stated by Rookmaaker (1998a: 66), but Grigson (2015) suspects that it was actually a female imported on 1 October 1737 on the *Shaftesbury* and subsequently exhibited across the British Isles.

Most authors prior to Linnaeus made no attempt to differentiate a variety of types even when they were aware that not all rhinos were the same. In our modern classification, Bontius must have known *R. sondaicus* in Java. Bergen and Edwards described living specimens of *R. unicornis*. That species was definitely the one best known in Europe, because as far as we can establish all rhinos imported from 1500 to 1800 belonged to it (Rookmaaker 1973, 1998a).

For taxonomic purposes, the British zoologist Reginald Innes Pocock (1863–1947) fixed *R. unicornis* as the "genotype" (now type-species) of the genus *Rhinoceros* (Pocock 1944a). Although he does not state this clearly, he definitely implied that the specimen illustrated by Jonston (1653, pl. 38) is the type of *R. unicornis*. This is how I read his paper (cf. Rookmaaker 1998b: 78) and this restriction has been accepted by Groves (2003: 345) and Groves & Grubb (2011: 21). Jonston's plate is a copy of the famous woodcut issued by Albrecht Dürer in 1515, representing the rhino which arrived in Lisbon on 20 May 1515, now referred to as the First Lisbon Rhinoceros, alternatively Dürer Rhinoceros (4.6). This animal therefore is the lectotype (selected type-specimen) of *R. unicornis* Linnaeus, 1758.

An alternative view recently casually mentioned by Ridley (2022: 170) is an unfortunate digression and in my view mistaken. Ridley's essay highlights the life of Clara, the Dutch rhino, which toured through Europe 1741–1758, greatly enhancing the European understanding of the species in the 18th century (5.4). In a non-taxonomic context, she suggested that the depictions of Clara in the influential *Encyclopédie* by Diderot of 1768 and the *Histoire Naturelle* by Buffon of 1764 (4.9) had superseded Dürer's

TABLE 4.6   The classification of all species of rhinoceros proposed by Carl Linnaeus

Systema Naturae		Classis	Ordo	Genus	Species
Edition/town	Date, page				
1 Leiden	1735	Quadrupeda	Jumenta	Elephas	? Rhinoceros
2 Stockholm	1740, p. 40	Quadrupeda	4. Jumenta	23. Elephas	Elephas naso cornigero, Rhinoceros
3 Halle	1740, p. 46	Quadrupeda	iv. Jumenta	3. Elephas	Rhinoceros, das Nasen-Horn
4 Paris	1744, p. 69	Quadrupeda	4. Jumenta	23. Elephas	Elephas naso cornigero, Le Rhinoceros
5 Halle	1747, p. 48	Quadrupeda	4. Jumenta	23. Elephas	Elephas naso cornigero, Nasenhorn
6 Stockholm	1748, p. 11	Quadrupeda	5. Jumenta	25. Rhinoceros	2 species, no specific names
7 Leipzig	1748, p. 11	Quadrupeda	5. Jumenta	25. Rhinoceros	2 species, no specific names
8	1753	NIL			
9 Leiden	1756, p. 11	Quadrupeda	5. Jumenta	25. Rhinoceros	2 species, no specific names
10 Stockholm	1758, p. 56	Quadrupeda	v. Glires	22. Rhinoceros	unicornis, bicornis
11 Halle	1760, p. 56	Quadrupeda	v. Glires	23. Rhinoceros	unicornis, bicornis
12 Stockholm	1766, p. 104	Quadrupeda	vi. Belluae	36. Rhinoceros	unicornis, bicornis

FIGURE 4.1
Johannes Jonston (1603–1657), "Rhinoceros, Hornnase. Rhinocer." This copper-engraving printed vertical to the spine is distinguished by a left orientation with Tab. xxxviii in the upper right corner. Engraved by Matthäus Merian the Elder (1593–1650) working in Frankfurt am Main. The animal shows the small hornlet on the shoulders characteristic for the woodcut by Albrecht Dürer of 1515
JONSTON, *HISTORIAE NATURALIS*, 1653, PL. 38. COURTESY: JIM MONSON

representation and this "made Clara the lectotype of her kind." Although Linnaeus knew about Clara (from Bergen's *Oratio*), this lectotype designation is invalid according to the rules of nomenclature (ICZN 1999, art. 74.1).

According to the British zoologist Michael Rogers Oldfield Thomas (1858–1929), the type-locality of *R. unicornis* could again be found in the works by Jonston, referring to the unillustrated translation of 1657, which for our purpose is identical to the first edition of 1653 (Jonston 1657b: 321; Thomas 1911: 144). From the localities listed by Jonston, Thomas selected "Bengal" as the type locality of *R. unicornis*. Subsequently, Lydekker (1916: 48) changed this to "probably the sub-Himalayan terai of Assam." Because "Bengal" is rather uninformative, and could even be synonymous to the whole of India (South Asia) in the older literature, the locality Bengal was only favoured by Pocock

(1944b) and Laurie et al. (1983). Most taxonomic authorities have followed Lydekker to choose Assam.[2] According to the Rules of Zoological Nomenclature (ICZN 1999, article 76), the place of origin of the lectotype of a species becomes the type locality of the nominal species-group taxon, and cannot be fixed by a first selection. If we follow the early history of Dürer's Rhinoceros from its first known exhibition in Ahmedabad in 1514 (4.6), a capture locality in northern India, that is in Rajasthan or in Uttar Pradesh, between the Ganges and the Nepal border is far

2   Wroughton 1920: 311; Ellerman & Morrison-Scott 1951: 339, 1966: 339; Honacki et al. 1982: 311; Corbet & Hill 1992: 242; Rookmaaker 1980, 1998b; Grubb 1993: 372, 2005: 635; Groves 2003: 345; Alfred & De 2006; Choudhury 2016: 154.

more likely. The actual place of capture of this animal is of course pure speculation, so it is best for stability to maintain "Assam" as the type locality of *Rhinoceros unicornis* Linnaeus, 1758.

4.6 The Life of the First Lisbon Rhinoceros, or Dürer Rhinoceros, 1515–1516

This is the most famous rhinoceros in the history of the species. Given that it lived for less than a year in captivity in Europe, its impact is largely due to the enormous popularity of the woodcut of this animal issued in 1515 by the German artist Albrecht Dürer (1471–1528). Adorned with a characteristic small twisted horn on the shoulders, this image was the classic representation of the rhino for most people in the western world during three centuries. The history and iconography of the animal has been explored by Costa (1937).[3] There remains uncertainty if it was male or female. The animal's life can be traced in a series of 14 chronological steps.

(1) [1514] Capture date or locality unknown.
(2) To April 1514. Rhino kept in Champanel (50 km NE of Vadodara, Gujarat) owned by Sultan Shams-ud-Din Muzafar Shah II (Muzafar II) (1484–1526), the 9th ruler of the Muzaffarid dynasty (Barbosa 1518: 124).
(3) 16 April 1514. The animal was one of the presents of Muzafar II to the Portuguese Governor, Afonso de Albuquerque (1453–1515). Albuquerque sent a diplomatic mission to Muzafar II to negotiate approval to construct a fortress at Diu, with the ambassadors Diogo Fernandes de Beja (d. 1551) and Jaime Teixeira. The mission left Goa in February 1514 and arrived at Champanel in early April 1514, and then traveled to Champaner (Mondoval) to meet the Sultan. They departed 26 April 1514 and went to Surat (Surate) reaching 8 May. The rhino present had been kept elsewhere during this time.
(4) 16 May 1514. The rhino, accompanied by a native Oçem (mahout), arrived in Surat, and was handed over to the Portuguese ambassadors. They remained in Surat to wait for transport to Goa.
(5) 15 September 1514. Arrival in Goa after journey on board in one of three vessels.

(6) January 1515. Donated by Albuquerque to the Portuguese King Dom Manuel I 'the Fortunate' (1495–1521). It was shipped on board the *Nostra Senora da Ajuda*, captained by Francisco Pereira Coutinho, part of the fleet that left Cochin in early January 1515. The return journey of some four months around the Cape of Good Hope stopped only in three harbours, Mozambique, Saint Helena and Terceira in the Azores.
(7) 20 May 1515. Arrival on the river Tagus in Lisbon. The rhino was housed in King Manuel's menagerie at the Ribeira Palace in Lisbon, separate from his elephants and other large beasts at the Estaus Palace.
(8) 3 June 1515 (Trinity Sunday). Arena fight against an elephant in a courtyard enclosed by high walls between the Paço da Ribeira and the Casa da India. The rhino was declared victorious.
(9) June 1515. A sketch, now unknown, was sent together with a letter to Germany. The sketch and words of the letter were copied by the German artist Albrecht Dürer in a drawing and a woodcut. Another letter with information about the rhino was written by Valentim Fernandes de Moravia (d. 1519), a Moravian merchant and printer, of which only an Italian transcript survives (Valentim 1515; transcript in Gubernatis 1875: 389).
(10) December 1515. Rhino donated by the Portuguese King to Pope Leo X (1475–1521) in Rome, Italy. The previous year, the Pope had been very pleased with Manuel's gift of a white elephant, also from India, which had become known as Hanno. The rhino was transported on a ship (name unknown) commanded by João de Pina.
(11) January 1516. Transit stop at Marseilles, France, where the rhino was shown to the French King Francis I (1494–1574) on one of the islands in the bay on 22 January 1516.
(12) Late January or early February 1516. The Portuguese ship with the rhino was caught in heavy storm off the Italian coast at Porto Venere, and all perished. There were rumours that the animal tried to swim to shore. It did not survive, but possibly its remains washed ashore.
(13) 1516. There are numerous illustrations of the rhino with the hornlet on the shoulder copied from Dürer's woodcut in zoological and general literature almost throughout history since it was made in 1515. One such copy incorporated the motto: "Non buelvo sin vencer" ("I shall not return without victory") invented by Paolo Giovio (1483–1552), who also was one of the first to write about the life and death of

---

3 The work of Abel Fontoura da Costa (1869–1940) published in three languages in 1937 was followed by a series of detailed treatises by Coste 1946; Cole 1953; Lach 1965; Clarke 1973, 1986; Rookmaaker 1973, 1998: 80; Bedini 1997; Faust 2003, as well as a novel by Norfolk 1996.

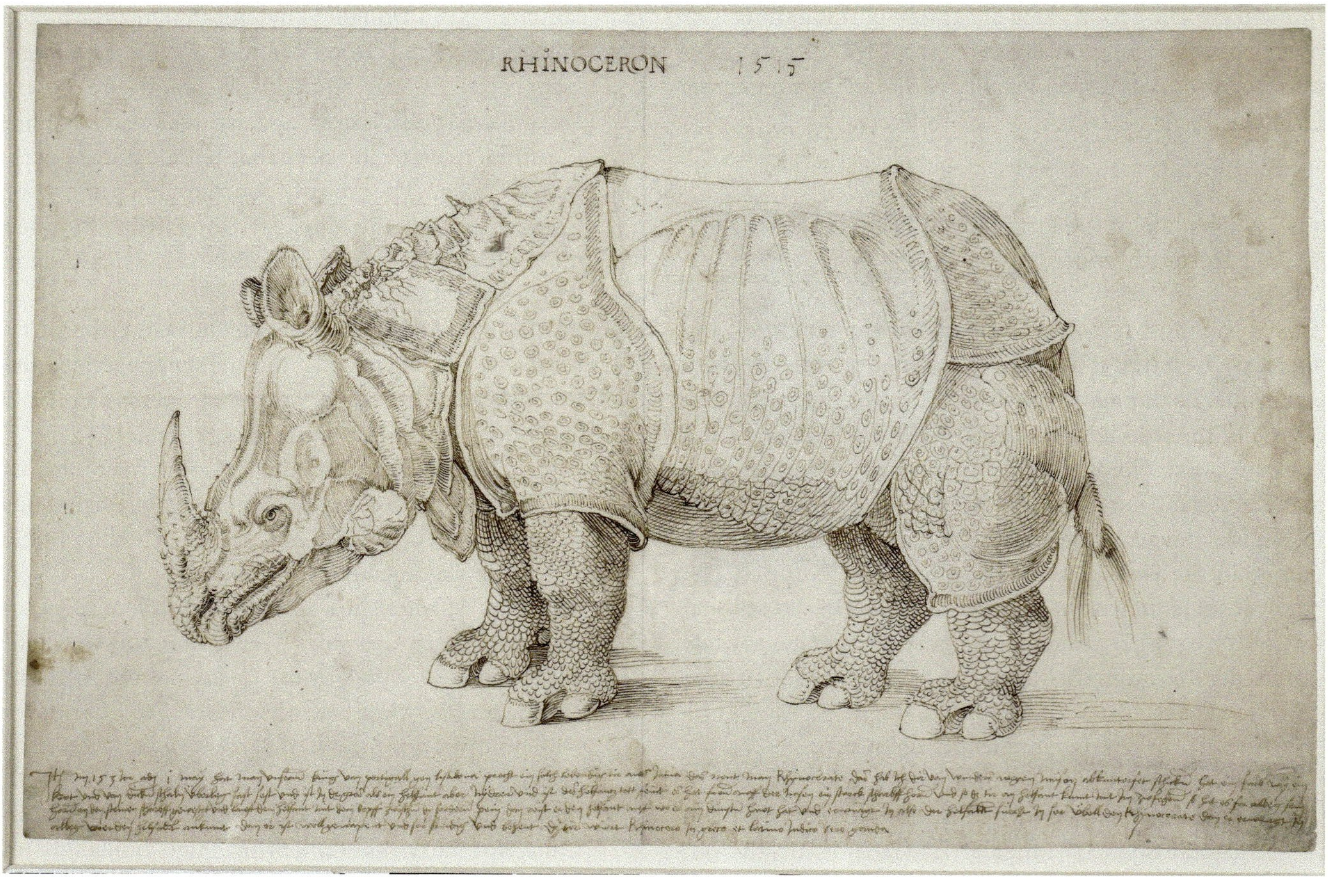

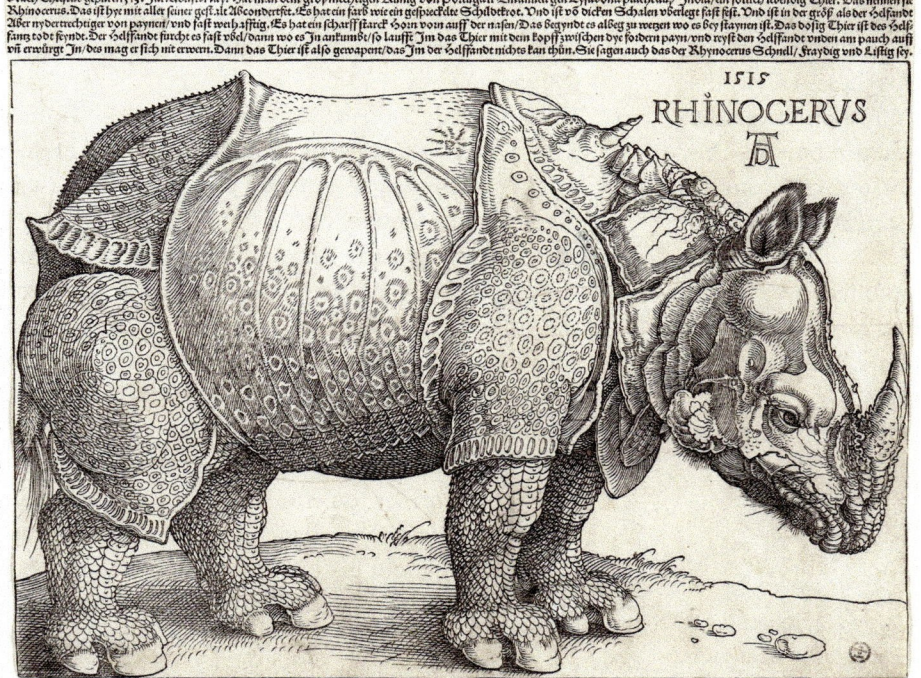

FIGURE 4.2   Drawing of "Rhinoceron" dated 1515 by the German artist Albrecht Dürer (1471–1528). It was sketched following an image forwarded
to Nuremberg from Portugal and shows the First Lisbon Rhinoceros. Size 27.4 × 45 cm
WITH PERMISSION © THE TRUSTEES OF THE BRITISH MUSEUM: SL,5218.161

FIGURE 4.3
Woodcut of the "Rhinocerus" dated 1515
by Albrecht Dürer. The first impression
was printed by Hieronymus Hölzel,
with text in five complete lines above
the image. Later German editions
always show the same text in 5 ½ lines.
Size 23.5 × 29.8 cm
NATIONAL GALLERY OF ART,
WASHINGTON: ROSENWALD
COLLECTION, 1964.8.697

FIGURE 4.4
Emblem of Alessandro de Medici with the motto "Non Buelvo sin Vincer" (I return not without victory)
PAOLO GIOVIO, *DIALOGO DELL'IMPRESE MILITARI ET AMOROSE*, 1559, P. 49

FIGURE 4.5 Francesco Granacci (1469–1543), "Joseph and his Brethren in Egypt." Painting, oil on wood, 1516. The First Lisbon Rhinoceros is introduced with its keeper in a small detail, on the right side of the central building in the background. Size 95 × 224 cm
WITH PERMISSION © UFFIZI GALLERY, FLORENCE, INV. 2152–1890

this rhino (first illustrated in Giovio 1559: 49). This emblem (fig. 4.4) was created for Alessandro de Medici (1510–1537), which opens a slight possibility that the remains of the animal made their way to Florence, Italy.

(14) February 1516. It is possible, but far from certain, that the skin of the dead rhino was mounted locally in Italy, or maybe back in Lisbon, and then forwarded to Pope Leo X. Despite intensive searches both in Rome and in Florence, there is no record that the rhino was ever displayed in either city (Bedini 1997). There are paintings by Francesco Granacci (fig. 4.5) and in the Palazzi Pontifica of the Vatican by Raphael (1483–1520) or by his assistant Giovanni da Udine (1487–1564), showing the animal as if alive, not as a stuffed or mounted animal (Clarke 1986: 27).

When Albrecht Dürer received the sketch and account of the rhino in Lisbon, he first made a drawing (fig. 4.2), followed by a woodcut dated 1515 (fig. 4.3). This depiction was popular and the woodcut exists in several states and versions (Cole 1953; Faust 2003). Besides the characteristic and rather exaggerated armour plating, the rhino has a good-sized single horn on the nose, as well as a smaller twisted horn on the shoulder. No known species of rhino shows a horn in this place, although additional horns of varying sizes have been found in captive specimens, perhaps caused by bruising (Hediger 1970). The influence of Dürer's Rhinoceros on zoological literature cannot be overstated, as it was copied hundreds of times in scientific and popular books as well as in countless different art forms.

4.7 History of the Specific Name *rugosus* in Genus *Rhinoceros*

The German zoologist Johann Friedrich Blumenbach (1752–1840) spent his academic career at the University of Göttingen, in addition to curating the zoological collections of the Königliches Akademische Museum established in 1773. He compiled a Handbook of Natural History (*Handbuch*) to serve as a guide for his course on natural history, first published in 1779 and then regularly updated in a series of 12 editions until 1830. In all editions he only recognized two varieties or species of rhinoceros, one Asian, the other African. Although initially he followed the example of Linnaeus (1758), his taxonomy had become outdated by the start of the 19th century when several new species had been described. Traditional in his classification, Blumenbach was very flexible in the use of scientific names, changing them according to any new insights he may have had. In the 1st and 2nd editions of the *Handbuch*, he maintained that the number of horns was variable, therefore the Asian single-horned and the African two-horned animals were merely varieties of the same species, which he called *Rhinoceros rugosus*. This name appears in Blumenbach (1779: 134) and (1782: 138), but it was never cited or referred to outside these two books, until it was rediscovered by Rookmaaker (2004). In the 3rd (1788) and 4th (1791) editions of the *Handbuch*,

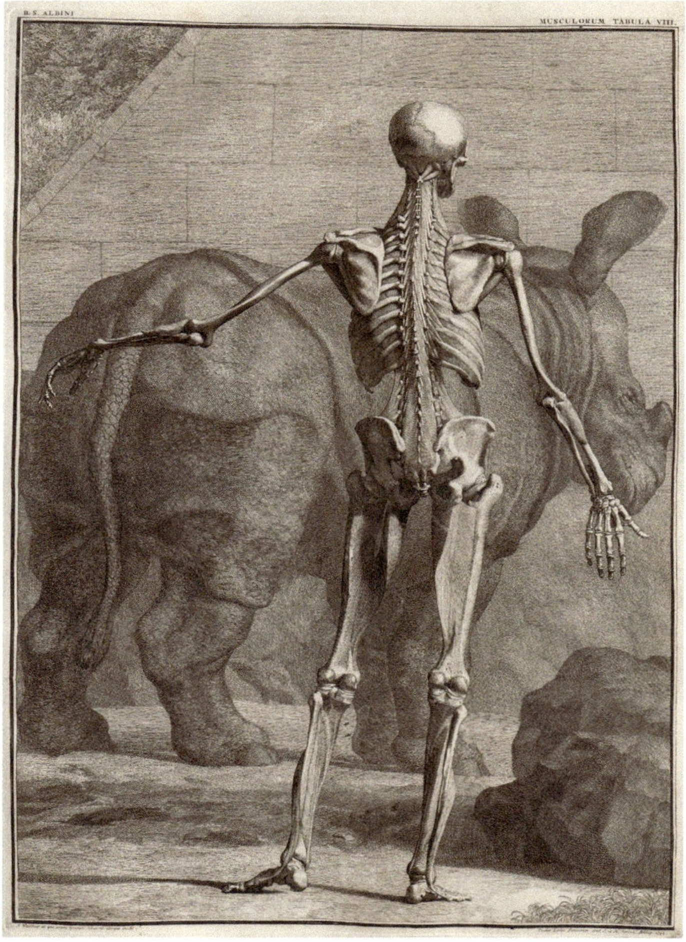

FIGURE 4.6
The Dutch Rhinoceros Clara seen behind a human skeleton. Drawn and engraved by Jan Wandelaar (1690–1759) after the living animal in Leiden or Amsterdam in 1741–1742, for the grand anatomical atlas by Bernard Siegfried Albinus (1697–1770) published in 1747. Size 56.8 × 40.4 cm. Inscription at the top: "B.S. Albini. Musculorum Tabula VIII"; lower left: "J. Wandelaar ad ipsa corpora hominum delineavit: idemque incidit" and lower right: "Prostat Leidae Batavorum apud J. & H. Verbeek, Bibliop. 1742." Published in Leiden by Johannes Verbeek (1696–1778) and Hermanus Verbeek (1698–1755). Identical plates are found in the English edition of 1749, with text changed to "C. Grignion sculp. – Impensis J. & P. Knapton, Londini, 1748" indicating the artist Charles Grignion (1721–1810) and the publishers John Knapton and Paul Knapton (1703–1755)
ALBINUS, *TABULAE SCELETI ET MUSCULORUM CORPORIS HUMANI*, 1747, PL. 8. COURTESY: ETH BIBLIOTHEK, ZÜRICH, ONLINE

FIGURE 4.7 Etching by Johann Elias Ridinger (1698–1767) of Clara, the Dutch Rhinoceros, when exhibited in Augsburg, Germany in May and June 1748. Size 34.5 × 27.8 cm
COLLECTION KEES ROOKMAAKER

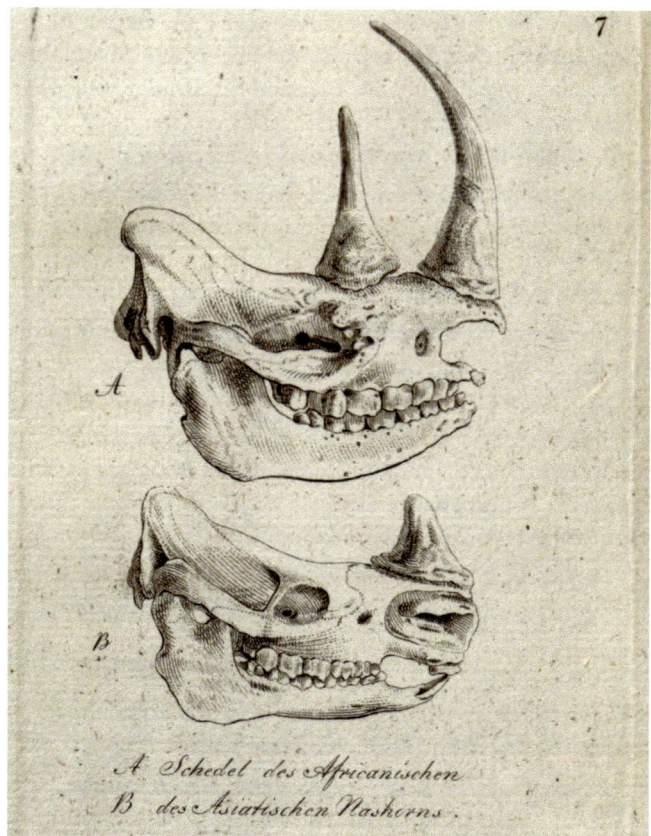

FIGURE 4.8 Skulls of the African (A) and the Asian (B) Rhinoceros
J.F. BLUMENBACH, *ABBILDUNGEN NATURHISTORISCHER GEGENSTÄNDE*, 1810, PL. 7

Blumenbach recognized two species, the Asian *Rhinoceros unicornis* and the African *R. bicornis*, using the names of Linnaeus. However, from the 5th (1797) edition onwards, he changed the names to *R. asiaticus* and *R. africanus* respectively. Blumenbach referred to two illustrations showing his *R. rugosus*, the plates by the Dutch artist Jan Wandelaar (1690–1759) published in the anatomical atlas by Albinus (1747, pls. 4, 8), and an engraving by Johann Elias Ridinger (1698–1767) of 1748, all three depictions of the Dutch Rhinoceros 'Clara' which toured the European continent from 1741 (5.4; figs. 4.6, 4.7). *Rhinoceros rugosus* Blumenbach, 1779 is a junior subjective synonym of *R. unicornis* Linnaeus, 1758.

4.8 History of the Specific Name *asiaticus* in Genus *Rhinoceros*

In the 5th edition of his Handbook of Natural History (*Handbuch*) of 1797, Blumenbach continued to distinguish one Asian and one African species of rhino, using here and in all subsequent editions the names *R. asiaticus* and *R. africanus* respectively. He also changed the diagnosis of the species. In the 3rd and 4th editions with

the name *R. unicornis*, he referred to two frontal teeth on each side ("dentibus primoribus utrinque binis"), changing this in the 5th and subsequent editions with the name *R. asiaticus* to be four frontal teeth on each side ("dentibus primoribus utrinque quaternis"). Blumenbach (1797) referred to Plate 7 of his own *Abbildungen naturhistorischer Gegenstände* of 1796 (fig. 4.8), depicting the skulls of an African and an Asian (Javan) rhino after the broadsheet issued privately by Petrus Camper (1722–1789) in 1787 (Rookmaaker & Visser 1982). Blumenbach lumped together in one species all rhinos known in Asia, which leads to a confused state in nomenclature. However, it is fair to say that *R. asiaticus* was in fact primarily meant to indicate the *R. unicornis* of Linnaeus (Rookmaaker 1983b). *Rhinoceros asiaticus* Blumenbach, 1797 is a junior subjective synonym of *R. unicornis* Linnaeus, 1758.

4.9 History of the Specific Name *indicus* in Genus *Rhinoceros*

The famous French naturalist Georges Cuvier (1769–1832) working at the Muséum National d'Histoire Naturelle in Paris took rhino systematics to a higher level (Rookmaaker 2015a). Like Linnaeus, Blumenbach and others before him, he summarized all animal species in his Animal Kingdom (*Règne Animal*) consisting of 4 volumes when it was first published. The first volume containing the rhino species is dated 1817 on the titlepage, but it has been established that it was available at least as early as December 1816, hence that year is in the citation (Roux 1976; Dickinson et al. 2011: 85). Cuvier gave no reason why he deviated from the earlier nomenclature, and while *R. indicus* must refer to the place where the species was known, the new name was essentially redundant. Cuvier's only reference is to the encyclopedic natural history by Georges-Louis Leclerc, Comte de Buffon (1707–1788), who treats the rhinoceros extensively as one species, purposely using only vernacular names (Buffon 1764, vol. 11: 174–203, pl. 7). *R. indicus* Cuvier, 1816 is a junior subjective synonym of *R. unicornis* Linnaeus, 1758.

4.10 Edward Blyth Recognizing Broad and Narrow Types of *Rhinoceros* Species

Edward Blyth (1810–1873) started researching ornithological subjects at a young age while he was running a druggist store in Tooting near London. From these early years he remained a prolific author in a wide variety of general and zoological journals. In 1841 he was appointed as Curator of the collections of the Asiatic Society of Bengal (ASB) in Kolkata (later integrated into the Indian Museum). Blyth persevered in his job until 1863 when he returned home due to ill health (Rainey 1874a; Grote 1875; Brandon-Jones 2006; Mathew 2015). Spending over twenty years in charge of the foremost Asian center for research and collecting of natural history, he was ideally placed to evaluate all new information which came in from every corner of South Asia. He maintained a constant correspondence with his colleagues in Britain, for instance with the geologist and ornithologist Hugh Edwin Strickland (1811–1853), well known for his endeavours to streamline zoological nomenclature. His letters were informative, but must have taken many hours to decipher (as they did for me), being written on both sides of thin paper, in a scrawly hand, sometimes text crisscrossing on the same page (Rookmaaker 2010: 189). Blyth described hundreds of new birds and contributed to the study of mammals.

Blyth wrote about the classification of rhinos in Asia (and beyond) in a pivotal paper published by the ASB in 1862, almost at the end of his curatorship. Even then, the ASB collection included only specimens of *R. sondaicus*

largely obtained in the Sundarbans, and Blyth examined only a few *R. unicornis* elsewhere. He confessed rather pointedly that he "only quite recently discriminated the two one-horned species; fancying, as a matter of course, that the numerous skulls of single-horned Rhinoceroses in the Society's museum, from the Bengal Sundarbans, &c., especially of the broad-faced type, were necessarily of the hitherto reputed sole Indian species" (Blyth 1862a: 159). If Blyth had been confused, the same keeps us on our toes with all identifications found in early literature about the rhinos of South Asia. It may be noted that John Edward Gray in London was not convinced that Blyth made the right interpretation of rhino classification and wrote to Charles Darwin that Blyth made some basic errors when identifying with mistaken confidence a number of skulls in the British Museum (Gray 1868b).

However, having settled the distinction to his own satisfaction, Blyth proceeded to separate the skulls into two types: "Of each there is a shorter and broader type, higher at the occiput, wider anterior to the orbits; and also a type the opposite of this, with every intermediate gradation. This amount of variation in the existing Asiatic species of the genus should intimate caution in the acceptance of all of the very numerous fossil forms that have been named by palaeontologists" (Blyth 1862a: 159). Remarkably, Blyth refrained from coining new names for each of these broader and narrower types, in fact he never mentioned these distinctions again in later publications. The two types were illustrated after early photographs taken by Thomas S. Isaac, until 1879 Superintending Engineer, Presidency Circle, and elected as ASB member in 1881 (figs. 4.9, 4.10). Both plates show a *R. unicornis* skull from Assam, and three *R. sondaicus* skulls from different Asian localities.

4.11 History of the Specific Name *stenocephalus* in Genus *Rhinoceros*

John Edward Gray (1800–1875) presented a new classification of the living species of rhino in a lecture delivered in London on 12 December 1867. One of the skulls in the collection of the Natural History Museum in London was a half-grown specimen which stood out on account of the length and slenderness of the nose and nasal bones, which he identified as a new species *Rhinoceros stenocephalus* (fig. 4.11). There were no details about origin or donor, merely that it had come in 1845 from the Museum of the Zoological Society of London. Founded in 1826 and terminated in 1855, this museum had a huge collection not

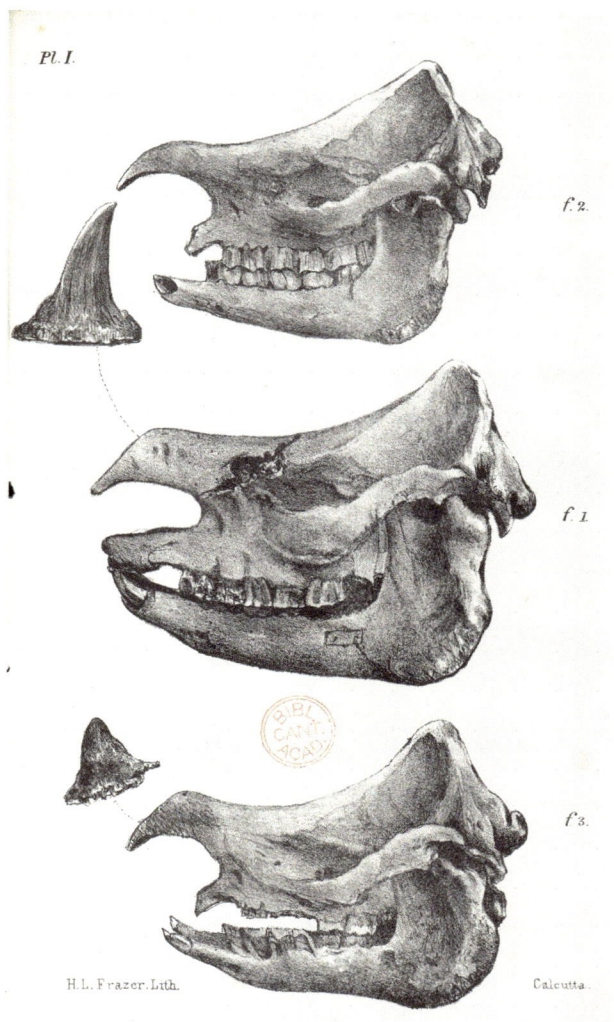

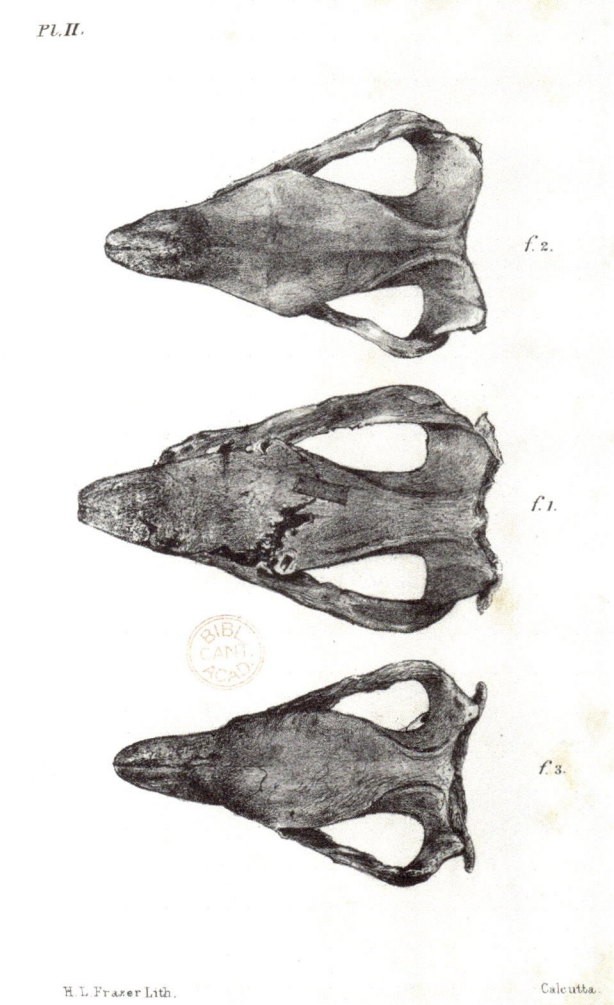

FIGURE 4.9 Three skulls in the collection of the Asiatic Society of Bengal in Kolkata, illustrating the paper by Edward Blyth in 1862. They represent (from top to bottom) the "narrow" type of *R. unicornis* from Assam, the "broad" type of *R. sondaicus* from the Sundarbans, and the "narrow" type of *R. sondaicus* from Java. The figures were engraved by H.L. Frazer after photographs by Thomas S. Isaac
BLYTH, *JOURNAL OF THE ASIATIC SOCIETY OF BENGAL*, 1862, VOL. 31, PL. 1. REPRODUCED BY KIND PERMISSION OF CAMBRIDGE UNIVERSITY LIBRARY

FIGURE 4.10 As Fig. 4.9, these skulls represent dorsal views of *R. unicornis* and *R. sondaicus*, with fig. 2 showing a specimen from Tenasserim (Myanmar) rather than the Sundarbans
BLYTH, *JOURNAL OF THE ASIATIC SOCIETY OF BENGAL*, 1862, VOL. 31, PL. 2. REPRODUCED BY KIND PERMISSION OF CAMBRIDGE UNIVERSITY LIBRARY

only from the Society's menagerie, but also from donations from across the world. A large part was incorporated in the last ten years of its existence into the British Museum, while other specimens were widely dispersed (Wheeler 1997). While the name *Rhinoceros stenocephalus* Gray, 1868 has been rarely used, it is a junior subjective synonym of *R. unicornis* Linnaeus, 1758.

4.12 History of the Specific Name *bengalensis* in Genus *Rhinoceros*

The German zoo-historian and collector Werner Kourist suggested that there have been a few problem-rhinos in captivity, which are difficult to identify as either *R. unicornis* or *R. sondaicus* from the available images. In his first comprehensive paper of 1970, Kourist pointed at two illustrations. The first in Kourist (1970, fig. 2) was the "Indian

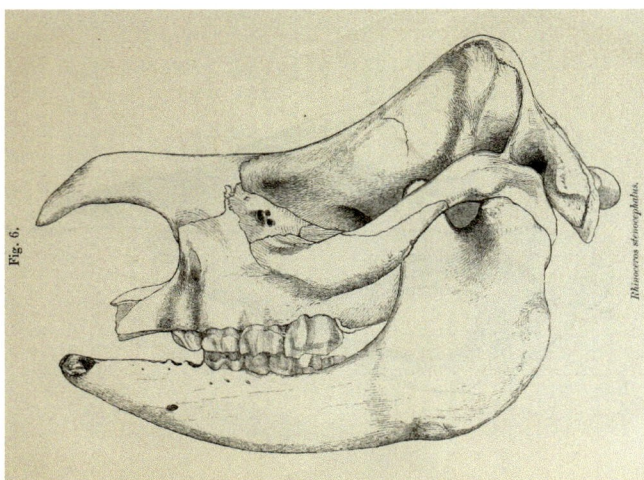

FIGURE 4.11 Skull of *Rhinoceros stenocephalus* in the collection of the Natural History Museum in London
GRAY, *PROCEEDINGS OF THE ZOOLOGICAL SOCIETY OF LONDON* FOR 12 DECEMBER 1867, PUBLISHED 1868, P. 1020, FIG. 6

FIGURE 4.12 Rhinoceros with a palm tree. Unsigned engraving of 1814, sometimes found in colour
MÉNAGERIE ROYALE, 1814, VOL. 1, PL. 16.
COURTESY: LIBRARY OF THE UNIVERSITY OF LEUVEN

FIGURE 4.13 "Le Rhinocéros" in the popular *Histoire Naturelle* by Georges-Louis Leclerc, Comte de Buffon (1707–1788). Plate VII, coloured in some copies, was drawn by Jacques De Sève (d. 1788), and engraved by Christian Friedrich Fritsch (1719–1774). After the original painting by Jean-Baptiste Oudry (1686–1755) made of the Dutch Rhinoceros Clara in Paris in January to April 1749
BUFFON, *HISTOIRE NATURELLE*, 1764, VOL. 11, PL. 7. COURTESY: JIM MONSON

Rhinoceros" published by Jardine (1836, pl. 8) depicting a male rhino owned by Thomas Atkins and mostly kept in the Liverpool Zoological Gardens from 1834 to 1842. The second in Kourist (1970, fig. 3) was the female rhino imported by William Jamrach living in the Berlin Zoo from 1874 as incorporated in an engraving by Gustav Mützel in Woldt (1882). Kourist (1974, fig. 1) added as a third example the rhino depicted on a plate in an anonymous *Ménagerie Royale* (1814) without further data.

Based on the depictions of these animals, Kourist (1970: 151) suggested that there might have been several forms ("mehrere Formen") of *R. unicornis*, or possibly individuals intermediary between *R. unicornis* and *R. sondaicus*, or finally two forms of *R. sondaicus*. After a short discussion, Kourist continued that perhaps a hitherto unrecognized form of the one-horned rhinoceros (*Rhinoceros unicornis bengalensis*) compared with the previously known (*Rhinoceros unicornis unicornis*) had a clear saddle formation of the neck shield, not unlike that of the Javan Rhino, connected to a smaller size of the body. When I commented (Rookmaaker 1983c: 45) that the new name "bengalensis" referred to the possible intermediary form mentioned in the preceding paragraph, this interpretation was not a correct representation of Kourist's intention. He wrote to me at the time, and while I had no chance then to remedy this misunderstanding on my part, with apologies hopefully the record can be set straight. Kourist proposed the name in case there was a form of *R. unicornis* displaying a more conspicuous saddle in the neck region (similar to *R. sondaicus*) combined with a smaller size than usual. As Kourist (1970: 151) proposed "*Rhinoceros unicornis bengalensis*" conditionally after 1960, this name is not available in zoological nomenclature (ICZN 1999, article 15.1).

Kourist was of course correct in stating that the animals in the three published plates mentioned were representations of problematic animals, which have caused him, myself, and other commentators to look closer at the evidence than otherwise would have been necessary. The history of the Liverpool Rhinoceros was reviewed by Rookmaaker (1993) identifying it as *R. unicornis* (4.13). Jamrach's Rhinoceros in Berlin is not easily sorted, but further investigations have led me to believe that it was in fact a specimen of *R. sondaicus* (47.12). The rather poor plate (fig. 4.12) in the *Ménagerie Royale* (Anon. 1814, pl. 16) looks to me like a copy based on Buffon (1764, pl. 7), hence showing the Dutch Rhinoceros 'Clara' which was sketched by Jean-Baptiste Oudry (1686–1755) when she was exhibited in Paris in January 1749 (fig. 4.13), and definitely a female *R. unicornis*.

4.13 The Life of the Liverpool Rhinoceros, 1838–1842

A male rhino was exhibited at the Liverpool Zoological Gardens from June 1834 to 1842 (Rookmaaker 1998: 82, 1993). In the popular series on natural history edited by the Scottish naturalist William Jardine (1800–1874) in the 1830s, the animal was depicted on two plates as "Indian rhinoceros, R. indicus" and discussed with a set of measurements obtained from a friend. The peculiar arrangement of the skin-folds on the plates and the relatively small length (390 cm) and shoulder height (142 cm) of the animal have led to suggestions that in fact it was *R. sondaicus*. In the hope of finding further clues to solve the dilemma, the life and remains of the animal are here reviewed.

The history can be pieced together using contemporary media reports, now becoming available on the British Newspaper Archive. The Liverpool Zoological Gardens off West Derby Road were opened to the public on Monday 27 May 1833 (*Liverpool Mercury* 1833-05-31). The owner was Thomas Atkins (1764–1848), who had previously toured with a small menagerie (Brown 2014). After his death on 6 June 1848, the zoo continued to be run by his sons Edwin (d. 1854) and John Atkins until it closed in 1864 (*Liverpool Standard and General Commercial Advertiser* 1848-06-13). The rhinoceros was brought to London on board the *Duke of Northumberland*, Captain William L. Pope, leaving Kolkata on 3 February and reaching London on 4 June 1834 (*Naval & Military Gazette* 1834-06-07). Anon. (1835) erroneously mentioned the rhino on the 'William Farleigh,' which is a journalistic mistake by this Dublin correspondent and pertained to Liverpool's elephant. This report also mentioned that the rhino was captured 1500 miles from Calcutta by some indigo planters. Indigo could be associated with Bihar, while 1500 miles would be in the region around Delhi, both of course then within the range of *R. unicornis*. Other rhinos were transported from Kolkata to London around the same time, with animals shown in the Surrey Zoological Gardens (from April 1834), the London Zoo (20 May 1834 to 19 September 1849), Wombwell's Menagerie (June to July 1836) and Manchester Zoo (1840 to 1842).

When the rhino arrived in London, Atkins purchased him later in the week for the grand sum of 1000 guineas (fig. 4.14). The arrival in Liverpool was noted on Thursday 26 June 1834 (*Gore's Liverpool General Advertiser* 1834-06-19 and 1834-06-26; *Liverpool Mercury* 1834-06-27). A *List of the Animals* exhibited at the zoo first published in 1834 mentions, without illustration, the "Indian rhinoceros (A male). Rhinoceros Indicus, Linn. – The present specimen is a remarkably fine lively animal, 4 years old" (Atkins 1834). While this rhino was usually kept in Liverpool, he was also taken for tours around the UK and Ireland, of which five have been identified. The exhibitions in London and Glasgow in 1834–1835 mentioned by Jardine (1836: 172) could not be verified.

(1) Dublin – 28 June to 27 July 1835. The rhino, said to be 3 years old, was transported from Liverpool to Dublin on a City of Dublin Steam-Packet and arrived on Sunday 28 June 1835. From the harbour it "was drawn in a caravan to the gardens by six of Guinness and Co.'s dray horses" (*Saunders's News-Letter* 1835-06-30). He stayed in the zoo for just a month, and was returned on Monday 27 July 1835 (*Saunders's News-Letter* 1835-07-25). The zoo paid £140 plus a young llama to Atkins (Rookmaaker 1998: 61). The rhino featured in the 29 August 1835 issue of the *Dublin Penny Journal*, with its likeness rather crudely drawn by the artist Horatio Nelson (fig. 4.15).

(2) Edinburgh – 16 January to 1 March 1836. The arrival of the rhino was announced in a local newspaper: "Arrival of the rhinoceros. The inhabitants of Edinburgh, and its vicinity, are respectfully informed

that the fine male Indian rhinoceros, from the Zoological Gardens, which cost the Proprietors One Thousand Guineas, has arrived, and is now Exhibiting at No. 13, South St David Street. Hours from 10 morning till 9 evening . Admittance, 1s. Trades-people and Children, 6d" (*The Scotsman* 1836-01-16, repeated 1836-01-23, 1836-01-30). After a few weeks, it was announced that the animal would return to Liverpool on 1 March 1836 (*The Scotsman* 1836-02-06, 1836-02-17, 1836-02-20). It must have been here that William Jardine's friend made his observations of the rhino, which was then said to be 6 years old.

(3) North England – 2 to 9 April 1836. A notice in the *Carlisle Journal* of 2 April 1836 announced the exhibition in the town on Saturday 2 April and 3 April 1836 only. It was on the way back to Liverpool and would be seen in Penrith on 5–6 April and in Kendal on 8–9 April 1836.

(4) Manchester – December 1836. An advertisement in the *Manchester Courier and Lancashire General Advertiser* of 3 December 1836 announced the arrival in the town, stating that it would be exhibited for a short time only.

(5) Cork, Ireland – 9 to 21 November 1842. Rookmaaker (2019e) was the first to draw attention to the visit of the rhino to Cork. The animal must have reached Cork on Wednesday 9 November 1842 and was advertised to be exhibited in the grounds next to the Theatre Royale by Richard C. Burke (*Cork Advertiser* 1842-11-09). The last known advertisement appeared on 21 November 1842, and the rhino may have been taken back to Liverpool at the time (fig. 4.19).

Before the Irish tour, the rhino was still mentioned as an exhibit of the Liverpool Zoological Gardens on 30 April 1842 (*Liverpool Mail*) and 6 May 1842 (*Liverpool Standard and General Commercial Advertiser*). The visit to Cork, if the same animal which seems likely, is then the last that is heard about him.

After death the hide was given to the British Museum, where it was mounted and catalogued by Gray (1843: 186) as *R. unicornis* "a. From Atkins' Menagerie." Sclater (1876a: 649) was told that it was no longer accessible, because when the *R. unicornis* from the London Zoo died in 1849, the skin was mounted on top of the existing specimen. Unexpectedly, the specimen still appeared in the catalogue later "88a. Stuffed. India. Atkins' Menagerie" (Gray 1873: 46), which remains the last dated entry.

Edward Blyth (1872e) recorded the presence of a skeleton in the Anatomical Museum of Guy's Hospital, Southwark, London. It was listed as *R. unicornis* by Pye-Smith (1874: 48). Blyth thought that it was in fact *R. sondaicus*, and most probably the remains of the Liverpool Rhinoceros (Gray 1868a: 1011). The skeleton is no longer in the hospital's museum. It may have been among specimens donated in 1926 to Brambell Natural History Museum, Bangor University, although not listed in the available records at the time (Helen Gwerfyl, email 2018). The skeleton and skull were examined by the rhino specialist Colin P. Groves in the 1980s when he identified it as *R. unicornis* (fig. 4.16). He found that the skull was still physically immature, which is at odds with the age of the Liverpool rhino being at least 10 years old when he died (Rookmaaker 1993).

FIGURE 4.14   Engraved vignette with a rhinoceros of the Liverpool Zoological Gardens
THE VISITOR'S HANDBOOK TO THE LIVERPOOL ZOOLOGICAL GARDENS, 1841. COURTESY: LIVERPOOL MUSEUM

FIGURE 4.15   The Rhinoceros during his visit to Dublin, Ireland in 1835. Drawn by the Dublin based artist Horatio Nelson (d. 1849), and engraved by Richard Clayton
DUBLIN PENNY JOURNAL, 29 AUGUST 1835, P. 68. COURTESY: DUBLIN CITY LIBRARY AND ARCHIVE

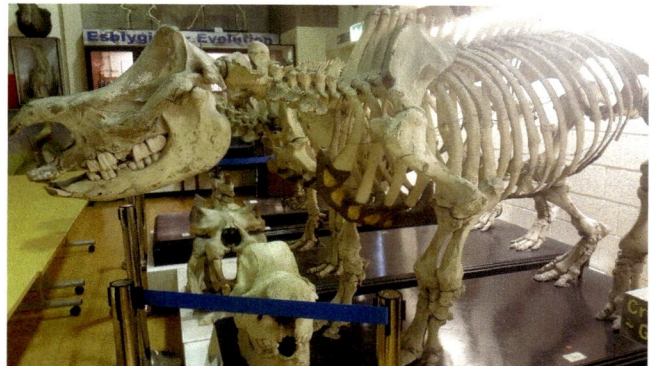

FIGURE 4.16   Skeleton of *R. unicornis* in the Brambell Natural History Museum, Bangor University, Wales. The specimen was probably transferred in 1926 from Guy's Hospital, Southwark, London
PHOTO: HELEN GWERFYL, BANGOR UNIVERSITY, 2018

The Liverpool rhinoceros was always advertised as an Indian *R. unicornis*. However, there were several claims that the animal actually was *R. sondaicus*, made, it must be said, by authors who presumably never saw him alive. Two main arguments were advanced in support of the claim, one regarding the arrangements of the skin folds, another regarding the animal's size. The general appearance in itself should be conclusive, because there are two well-presented plates of this rhino in Jardine (1834, pls. 8, 9), drawn by James Stewart (1791–1863) after life in Edinburgh in 1836 (figs. 4.17, 4.18). According to Sclater (1876a: 649), in this plate "the second fold of the skin across the back of the neck which distinguishes *R. sondaicus* from *R. unicornis* is plainly visible." Personally I do not find this entirely convincing, of course writing almost two centuries later. That second fold does not continue

FIGURE 4.17
"Indian Rhinoceros. Liverpool Zool. Gardens." Coloured engraving after drawings by James Stewart (1791–1863) made in Edinburgh in January to February 1836, engraved by William Home Lizars (1788–1859)
JARDINE, *THE NATURALISTS LIBRARY*, 1836, VOL. 13, PL. 8. COLLECTION KEES ROOKMAAKER

FIGURE 4.18
Frontal view of the rhinoceros in Edinburgh in 1836
JARDINE, *THE NATURALISTS LIBRARY*, 1836, VOL. 13, PL. 9. COLLECTION KEES ROOKMAAKER

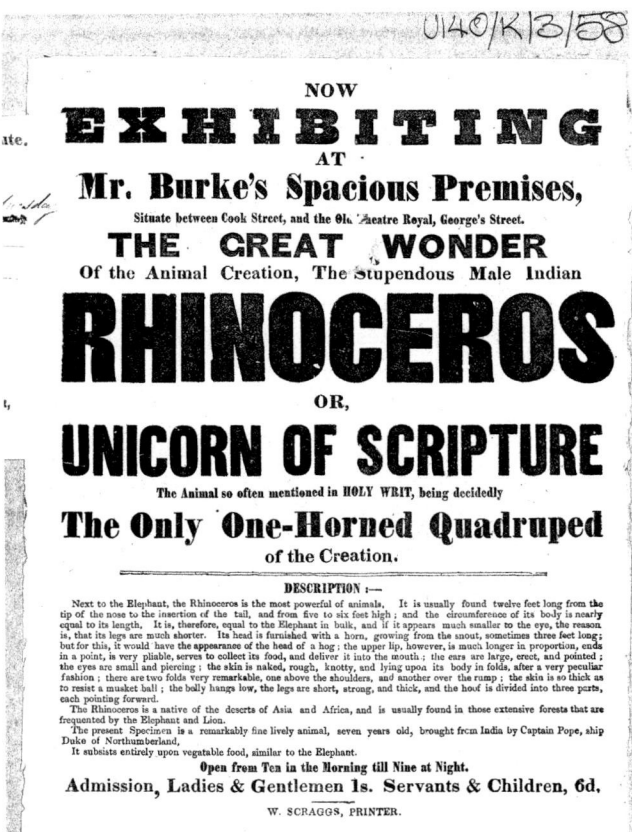

U140/K/3/58

NOW

EXHIBITING

AT

Mr. Burke's Spacious Premises,

Situate between Cook Street, and the Old Theatre Royal, George's Street.

THE GREAT WONDER

Of the Animal Creation, The Stupendous Male Indian

RHINOCEROS

OR,

UNICORN OF SCRIPTURE

The Animal so often mentioned in HOLY WRIT, being decidedly

The Only One-Horned Quadruped

of the Creation.

DESCRIPTION :—

Next to the Elephant, the Rhinoceros is the most powerful of animals. It is usually found twelve feet long from the tip of the nose to the insertion of the tail, and from five to six feet high ; and the circumference of its body is nearly equal to its length. It is, therefore, equal to the Elephant in bulk, and if it appears much smaller to the eye, the reason is, that its legs are much shorter. Its head is furnished with a horn, growing from the snout, sometimes three feet long ; but for this, it would have the appearance of the head of a hog ; the upper lip, however, is much longer in proportion, ends in a point, is very pliable, serves to collect its food, and deliver it into the mouth ; the ears are large, erect, and pointed ; the eyes are small and piercing ; the skin is naked, rough, knotty, and lying upon its body in folds, after a very peculiar fashion ; there are two folds very remarkable, one above the shoulders, and another over the rump ; the skin is so thick as to resist a musket ball ; the body hangs low, the legs are short, strong, and thick, and the hoof is divided into three parts, each pointing forward.

The Rhinoceros is a native of the deserts of Asia and Africa, and is usually found in those extensive forests that are frequented by the Elephant and Lion.

The present Specimen is a remarkably fine lively animal, seven years old, brought from India by Captain Pope, ship Duke of Northumberland.

It subsists entirely upon vegetable food, similar to the Elephant.

Open from Ten in the Morning till Nine at Night.

Admission, Ladies & Gentlemen 1s. Servants & Children, 6d.

W. SCRAGGS, PRINTER.

FIGURE 4.19   Handbill advertising the exhibition of a rhinoceros in November 1842 in Cork, Ireland

WITH PERMISSION © CORK CITY AND COUNTY ARCHIVES: RICHARD DOWDEN PAPERS, U140/K/3/58

across the back in the other illustrations of the Liverpool Rhinoceros, in the *Dublin Penny Journal* of 1835 and a vignette in a 1841 visitor's handbook. Also, I cannot see the the saddle in the neck region of *R. sondaicus*, which often is so eminently characteristic.

The small size of the rhino must have convinced Blyth (1872a) to identify it as *R. sondaicus*. Again this is inconclusive. Jardine (1836) gave a long list of measurements taken by a friend who had obviously spent time examining the rhino. For many of these, it is hard to find comparative figures, and size depends on the animal's age. If he was actually 6 years old, he would have been full-grown, but it was also described as young. An age of 4 years may also be possible, but again that is on the brink of adulthood. To focus on two measurements which are more widely available, Jardine gave the total length from the tip of the snout to the extremity of the tail as 12 ft 9 ¾ in (390.5 cm), and the height from the highest part of the back (shoulder height?) as 4 ft 8 in (142.4 cm). In adult males *R. unicornis*, total average length is 412 cm (Laurie et al. 1983), or 411.7 ± 20.6 cm (Dinerstein 1991); and the shoulder height is 163–193 cm (Laurie et al. 1983). In adult *R. sondaicus*, total length is 305–344 cm, shoulder height 120–170 cm (Groves & Leslie 2011). Therefore, in the Liverpool Rhinoceros the length is only within the range of *R. unicornis*, while the height is only within the range of *R. sondaicus*.

Due to the evidence and opinions of Blyth and Sclater, both expert naturalists, it has been surmised that the Liverpool Rhinoceros was an example of *R. sondaicus* by Reynolds (1961: 31), Guggisberg (1966: 117), Kourist (1970), Keeling (1984: 29), Reichenbach (2002), and probably other secondary sources. In my paper of 1993, I concluded that the evidence remained inconclusive, and that it would be better to be cautious and maintain the identification as *R. unicornis* (Rookmaaker 1993). The zoo historian Clin Keeling (1991: 16) as well as the taxonomists Groves & Leslie (2011) agreed. I still maintain that the Liverpool Rhinoceros must have been an example of *R. unicornis*.

The Greater One-Horned Rhinoceros (*Rhinoceros unicornis*) in Captivity in South Asia

5.1 Addition of 76 *Rhinoceros unicornis* in Captivity in South Asia

A total of 120 specimens of *R. unicornis* have been recorded in captivity in South Asia before 1950. This is an increase of 76 animals in comparison with my previous comprehensive list of rhinos in captivity (Rookmaaker 1998). Globally, there have been 757 specimens of *R. unicornis* in captivity until 2019.

R. unicornis is the Asian rhinoceros exhibited most widely both in South Asia and elsewhere. This chapter lists all wild-caught and captive-born specimens recorded in South Asia (Bangladesh, India, Nepal, Sri Lanka) from the earliest available records until the end of 1949. This listing therefore excludes (1) all animals arriving in a facility in 1950 or afterwards, and (2) all animals exported from South Asia and exhibited elsewhere in Asia, or in Australia, Africa, Europe and the Americas. There were 17 specimens which were exported after an initial (more or less extended) stay in India, and the local parts of their histories are included below, however they are not included in the total tabulation of South Asian specimens.

Most of the additional rhinos were kept by the local Indian rulers. The available records are generally no more than snapshots even for larger menageries like those in Lucknow (Oudh) or Vadodara (Baroda), and it is often impossible to say if a series of individual notices pertain to one or more animals. If any written registers do exist, they are likely to be in local languages and therefore beyond my capability. Many of the records are quite vague, none discuss which species it might be (therefore assuming *R. unicornis*), and only a few include illustrations. Some zoo-historians may feel that the number is inflated by the inclusion of such general and undefined reports, and while potential errors are unavoidable, it is also true that there is always some kind of evidence which should not be ignored.

5.2 Methods of Rhino Capture

Most rhinos were captured while they were still young, because adults were too large to be transported safely. It is likely that most were caught by men protected by the bulk of the elephants used to separate them from the mothers. Only a few actual stories of some wild-west like encounters have made it into print.

In March 1864, the adventurous and daring Major Charles Richard Cock (1838–1879) killed a mother rhino in the Garo Hills (Meghalaya), and then cursorily ordered his subordinates to catch the calf by hand:

> With B. shot a female rhino, which had a young one, which we determined to capture. As I had brought a strong cotton rope, we made the fellows get off the elephants and catch him. My mahout caught him by the ears first, and was immediately knocked over. We then all laid hold of him, and tied the cotton rope round his neck and middle, then fastened him up to the carcass of his mother, there being no tree available, and sending a pad elephant into Trickri killa, we summoned the whole population to drag him to the stockade. This was a work of great difficulty, as they had a mile to drag him through Assam grass, about 14 feet high, but at length they got him in all safe, and he soon took his milk kindly, drinking four quarts daily. (Cock 1865)

A similar scene was depicted by the Hungarian Count Andrássy de Mano Csikszentkiraly (1821–1891) in the 1859 edition of his book of his travels (fig. 48.1).

The hunters employed by Fitzwilliam Pollok in Assam (39.3) suggested catching a baby rhino using nets, just as they were used to catch a buffalo. Set in June 1867, he tells how the animal was subdued and soon got used to his human adversaries:

> Seetaram, an old hunter, and his nephew Sookur, a young phansic or nooser, volunteered, they went 3 pm, back midnight: In front were some six torch-bearers, and immediately behind them some twenty men, all but naked, staggered under some weight, which was attached to long poles and which was violently struggling and squealing most pitiously, enveloped in a net and tied with numerous ropes. This proved to be a young rhino. A shallow trench just long enough for the young beast to stand in was dug, and the animal itself, after having ropes

© L.C. (KEES) ROOKMAAKER, 2024 | DOI:10.1163/9789004691544_006

This is an open access chapter distributed under the terms of the CC BY-NC-ND 4.0 license.

FIGURE 5.1   Comic sequence by Ernest Henri Griset (1844–1907) illustrating the release of a rhinoceros from a pitfall by the adventurous
Cornelius Cracker, Esq. The story is in the captions. Cracker falls in the pit, followed by a lion and then a large rhinoceros. With
the pit now full, the rhino first helps the dangerous lion to escape, followed by a lift upwards to Cracker. Finally Cracker digs the
rhino to freedom: "Oh, does not a parting like this make amends?" Griset in Tom Hood (1835–1874), *The 5 Alls*, 1868, pp. 38–39
REPRODUCED BY KIND PERMISSION OF CAMBRIDGE UNIVERSITY LIBRARY

attached to its hind and front legs, round its body and neck, was relieved of its wraps, and, although it fought like a demon, deposited in its bed and secured to long pegs driven into the ground. I never saw a more savage monster, it went open mouthed at everybody, and had it got loose I have no doubt it would have floored many of us before it could have been recaptured. It was dawn before our captive was safe, and we retired to snatch another hour's rest. My work detained me about ten days in Burpettah, during which I had ample leisure to watch my new acquisition. It was no use attempting to go near it for the first 36 hours, our approach only irritated it and made it struggle, but on the second day the poor brute was much exhausted from want of sleep, food, and water, and also from the exertions it had made to escape. Sookur now with a jug of water approached

it, and poured some over its head; at first it threw its head about, but, a little water getting into its mouth, it became quieter, more water being poured over it, it opened its mouth and allowed the liquid to trickle down it. It was then fed with buffalo's milk, and then left undisturbed for another day. It had then mashed plaintains and milk given it, in three days it was untethered and allowed to roam about in a little stockade which had been erected round it, in a week it would follow its keeper about everywhere (Pollok 1871c: 469)

Although the stories give a completely wild impression, in reality the object was to keep the young animal alive and unscathed, otherwise it would have little value to an animal dealer. Due to difficulties in transporting such large animals by mechanical means, apparently it was common practice to make them walk long distances, of course accompanied by their long-suffering human keepers.

When rhinos were captured for zoological gardens in the the 20th century, invariably pit-traps proved to be the best and least offensive method. The conservationist E.P. Gee in Assam was definitely much in favour. The official pits used there were about 10 feet (3 m) wide and 6 deep (1.8 m) deep and would be dug in the middle of a rhino path, then camouflaged with sticks and grass. Once the rhino had fallen in, one of the sides would be dug out to form a gradient allowing the transfer of the animal to a cage, which could then be dragged to a stockade by an elephant (Gee 1964a: 159). Alternatively a rhino can be shot with a tranquillizer, a method possibly preferable in some conditions as the actual individual can be chosen in advance. The English illustrator Ernest Henri Griset responsible for numerous humorous phantasies has pictured the adventures of a rhino, in this case African, removed from an illegal pitfall by the elderly Cornelius Cracker, Esq. as part of a hunting story "not to say, fib" (fig. 5.1).

5.3   Total Number of Captive *R. unicornis* in South Asia to 1949

In the wild *R. unicornis* occurred and occurs only in the countries of South Asia. As this chapter provides details of the wild-caught animals kept in South Asia itself until 1949, additional data about the animals exported from the region, born in captivity, or shown in South Asia after 1950 are presented in summary. The extensive list for rhinos kept in any captive facility until 1994 by Rookmaaker (1998) gives details about 397 specimens of *R. unicornis*, of which 260 were wild-caught and 137 were born in captivity. The changes to the captive inventory after this date are included in the studbooks. The Indian studbook was initiated by the Kanpur Zoo (1994) and edited by Koliyal (2003). The latest (third) edition by the Central Zoo

Authority (2016) recorded data of 152 rhinos, of which 81 were wild-caught and 65 captive-born (and 6 unknown origin). At the end of 2016 there were 30 animals alive in 10 Indian zoos, 5 wild-caught and 25 captive-born. As these data are repeated in the international studbooks, the animal totals in this chapter are based on the latter.

The international studbook of *R. unicornis* was initiated by the director of the Basel Zoo Ernst Michael Lang (1913–2014), and is an example of careful and useful record keeping and publication (Lang 1994). Lang was an eminent zookeeper and was responsible for the major breeding program of the species (Lang 1956, 1961, 1967, 1968). The studbook editions carefully edited by the team at Basel Zoo for 2017 and 2019 provide information on 554 individuals, which include all animals in zoo facilities around the world from 1950 as well as some earlier specimens (Basel Zoo 2017, 2019). The latest update of data on 2021-12-31 gives a current captive population of 230 *R. unicornis* (117 males, 113 females) in 83 locations worldwide (Basel Zoo 2021).

In total, combining all sources, there have been 757 specimens of *R. unicornis* in captivity until the end of 2019 (table 5.7). Of these, 357 were captured in Nepal or India, and 400 were born in captivity. Besides the rhino born in Nepal in 1824 and two still-births, the first rhino birth occurred in Basel Zoo on 14 September 1956, starting a remarkable success story (table 5.8). In the 21st century (to 2020) only 12 rhinos were taken from the wild to bring new genetic material into the captive population.

The locality of capture is rarely known for any of the specimens taken into captivity before 1950. However, records much improved in the 20th century showing that 67 animals were obtained in Nepal and 115 in India (tables 5.9, 5.10, 5.11). In India, most animals were caught in Assam (mainly in Kaziranga NP), and only a relatively small number of these were exported. By contrast, most rhinos caught in Nepal (mainly in Chitwan NP) were exported (tables 5.9, 5.10).

• • •

Dataset 5.1: History of the Specimens of *R. unicornis* in Captivity

List of specimens of *R. unicornis* in captivity in South Asia up to 1949, in chronological order for each country. Each entry shows the first year in which the presence was recorded followed by an indication of the locality. The locality is followed by a year of death or transfer when available. Further information is found in the alphabetical listing below. This shows a total of 120 animals (118 wild-caught, 2 captive-born) in South Asia.

The list includes the 44 specimens previously presented in Rookmaaker (1998) identified by an asterix (*). There were 17 specimens which were exported after a short stay in the region, which are listed but not included in the South Asian totals.

Abbreviations: *   Also recorded in Rookmaaker (1998)
                 <   means: previous to the year stated
                 (nd)   means: no date known.

Bangladesh
*1780 Dhaka: Matthew Day (nd)
1861 Dinajpur fair (nd)

Nepal
1797 Kathmandu 1834
*1823 Kathmandu (nd)
*1823 Kathmandu (nd)
*1824 Kathmandu *birth* 1834 To Kolkata: unknown (nd)
1845 Kathmandu (nd)
1845 Kathmandu (nd)
1851 Kathmandu (nd)
*1907 Kathmandu (nd)
*1907 Kathmandu (nd)
*1907 Kathmandu (nd)
1932 Kathmandu 1939

Sri Lanka
1643 King of Bengale 1643 To Kandy (nd)

India
*1399 Delhi: Timur Bec (nd)
1600 Ajmer: Jahangir (nd)
*1615 Ajmer: Jahangir (nd)
*1615 Ajmer: Jahangir (nd)
1632 Etawah: Shah Jahan (nd)
1632 Etawah: Shah Jahan (nd)
*1659 Delhi: Aurangzeb (nd)
*1659 Delhi : Aurangzeb (nd)
1665 Agra: Aurangzeb (nd)
*1665 Kora Jahanabad (Tavernier) (nd)
*1671 Kassimbazar (nd)
1698 Udaipur: Palace (nd)
1698 Udaipur: Palace (nd)
1706 Kolkata Chandernagor 1716 Murshidabad: Nawab (nd)
1720 Manipur (nd)
1721 Bijolia: Maharana (nd)
1734 Udaipur : Palace (nd)
*1750 Patna: Young (nd)
1757 Murshidabad: Nawab (nd)
1760–1765 Lucknow: Nawab (nd)
1763 Patna: Vansittart 1763 Kolkata (transit) 1763
1765 Kolkata: Governor of Bengal (nd)
1769 Lucknow: Nawab (nd)
1775 Bundi: Umaid Singhji (nd)
*1780 Bhagalpur: Cleveland (nd)
1789 Lucknow: Nawab 1802
*1790 Gwalior 1790 Pune: Peshwa 1792
1790 Udaipur: Palace (nd)
1791 Kathmandu 1791 Kolkata: Governor of Bengal (nd)
1793 Gwalior 1793 Kolkata: Governor of Bengal (nd)
1808 Kolkata: Barrackpore (nd)
1810 Vadodara: Guikwar 1822
*1814 Lucknow: Nawab < 1856
*1814 Lucknow: Nawab < 1856
*1814 Lucknow: Nawab < 1856
*1819 Lucknow: Nawab < 1856
1819 Lucknow: Nawab < 1856
1819 Lucknow: Nawab < 1856
1819 Lucknow: Nawab < 1856

*1823 Vadodara: Guikwar 1848–1856
*1827 Kolkata: Barrackpore 1857
1830s Barpali: Raja of Bamra (nd)
1835 Fatigarh: Baiza Bai 1836 Allahabad: Baiza Bai (nd)
1835 Fatigarh: Baiza Bai 1836 Allahabad: Baiza Bai (nd)
1837 Kolkata: Barrackpore 1857
1837 Kolkata: Buddynath Roy (nd)
1838 Kolkata: Barrackpore (nd)
1840 Una: Baba Bikram 1857 59th Regiment (nd)
1841 Lucknow: Nawab < 1856
1841 Lucknow: Nawab < 1856
1841 Lucknow: Nawab < 1856
1841 Lucknow: Nawab < 1856
1841 Lucknow: Nawab < 1856
*1841 Lucknow: Nawab 1856 Kolkata: Barrackpore (Blyth) 1862
*1841 Lucknow: Nawab 1856 Kolkata: Barrackpore (Blyth) 1860
1845 Jaipur: Maharaja 1861
1848 Bettiah: Raja (nd)
*1848–1856 Vadodara: Guikwar < 1885
1850 Arki: Palace (nd)
1851 Mathura: Seth Chand (nd)
*1852 Allahabad: Rundheer Singh 1852 Lucknow: Nawab 1858
1855 Guwahati: Schlagintweit (nd)
*1856–1866 Vadodara: Guikwar < 1885
1858 Kolkata: Alipore Rajah (nd)
1858 Kolkata: Seven Tanks – 1870s
*1859 Kolkata: Maharaja of Burdwan 1861
*1859 Kolkata: Maharaja of Burdwan (nd)
1861 Udaipur: Maharaja (nd)
1862 Balrampur 1885
1864 Meghalaya: Cock (nd)
1865–1866 Vadodara: Guikwar (nd)
1866 Dibrugarh: Millett (nd)
*1867 Kolkata: Raja Mullick 1867
*1868 Bangalore: Lal Bagh 1870 Chennai (nd)
1868 Bangalore: Lal Bagh 1893
1869 Assam, Manas: Pollok (nd)
1870 Jeypore: Raja (nd)
1870 Cachar: Synteng (nd)
*1871 Kolkata: Raja Mullick 1871
1874 Alwar: Maharao (nd)
1875 Kolkata: Rutledge (nd)
1876 Junagadh: Nawab 1889
*1876 Gwalior: Maharaja 1903
*1877 Dumraon 1877 Kolkata Zoo 1880
*1878 Trivandrum Zoo 1900
1880–1885 Darbhanga: Maharaja (nd)
1885 Dungarpur: Maharawal (nd)
1886 Jaipur: Ram Newas (nd)
1886 Jaipur: Ram Newas (nd)
*1887 Kolkata: Wajid Ali Shah 1887 Kolkata Zoo (nd)
1888 Darbhanga: Maharaja (nd)
1888 Darbhanga: Maharaja (nd)
*1905 Kolkata Zoo 1932
*1909 Kolkata Zoo 1930
1909 Kolkata Zoo (nd)
1919 Dungarpur: Maharawal (nd)
1920s Assam: Wood's friend (nd)
*1925 Kolkata Zoo *birth* 1925
1925 Mayong, Assam (nd)
*1932 Kolkata Zoo 1970
*1932 Kolkata Zoo 1965

1936 Jaldapara: Forest Department 1936
1939 Dungarpur: Maharawal (nd)
*1941 Udaipur Zoo 1946 Jaipur: Ram Newas 1964
1941 Udaipur Zoo (nd)
*1944 Lucknow Zoo 1973

India – Exported outside Region after Longer Stay in India

*1514 Ahmedabad 1514 Surat 1514 Goa 1515 To Lisbon 1517 To Rome 1517
*1577 Goa 1577 To Lisbon and Madrid 1591
1620 Ajmer: Jahangir 1620 To Iran, then Turkey
*1683 Kolkata: Capt. Udall 1684 To London 1686

*1738 Patna: English Factory 1738 Kolkata (transit) 1739 London 1742/1744
*1738 Kolkata: VOC at Hooghly 1740 To Holland [Clara] 1758
*1769 Kolkata: Chandernagor 1769 To Versailles 1793
*1790 Lucknow: Nawab 1790 London (Clark's Rhino at The Lyceum) 1793
*1829 Kolkata: Buddynath Roy 1830 USA: Flatfoots 1835
*1831 Lucknow: Nawab 1835 London, died in transit 1835
*1833 Kolkata: Barrackpore 1834 Liverpool: Atkins 1841
1838 Kolkata: Barrackpore (nd) To Jakarta (nd)
*1864 Kolkata: Barrackpore 1864 London Zoo 1865 Paris 1874
1868 Mumbai: Aga Khan 1868 To King of Persia 1868
*1912 Kolkata Zoo 1912 London Zoo 1921
*1922 Mumbai Victoria Gardens 1922 London Zoo 1926
*1938 Kolkata Zoo 1939 Washington DC 1959

TABLE 5.7   Total numbers of *R. unicornis* in captivity through the past centuries

	1300–1799	1800–1849	1850–1899	1900–1949	1950–1999	2000–2019	Total
Wild-caught specimens of R. unicornis							
South Asia only	33	32	36	17	76	8	202
Export	9	17	48	26	51	4	155
Total	42	49	84	43	127	12	357
Captive-born specimens of R. unicornis							
South Asia only		1		1	42	33	77
Outside S. Asia				1	136	186	323
Total		1		2	178	219	400
Totals of wild-caught and captive-born specimens of R. unicornis							
TOTAL	42	50	84	45	305	231	757

This combines figures found in Rookmaaker (1998), new data in this chapter, and the records of the studbook (for data after 1950) kept by Basel Zoo (2017, 2019). South Asia only includes data for facilities in India, Nepal and Bangladesh (no rhinos recorded in Pakistan, Bhutan, Sri Lanka). Export figures of wild-caught animals and births outside South Asia include all specimens which left South Asia or were born outside the region.

TABLE 5.8   Numbers of births of *R. unicornis* according to continent

Births	1950s	1960s	1970s	1980s	1990s	2000s	2010s	Total
South Asia		4	10	16	12	16	17	75
Asia Rest		1	1		1	3	3	9
Africa								0
Australia							1	1
S. America					2			2
N. America		1	4	20	35	48	57	165
Europe	3	11	14	21	22	30	44	145
TOTAL	3	17	29	57	72	97	122	397

Three births recorded before 1950s are excluded. Data exported from the studbook of Basel Zoo (2017, 2019).

TABLE 5.9   Origin of wild-caught *R. unicornis* afterwards exhibited in South Asia from 1900 to 2019

Wild	1900	1910	1920	1930	1940	1950	1960	1970	1980	1990	2000	2010	Total
Nepal	6			3	3		2		3	1		2	20
Assam			2			6	12	19	22	8		3	72
Bengal				1								2	3
Bihar									1				1
UP											1		1
Unknown		1		1		2							4
Total	6	1	2	5	3	8	14	19	26	9	1	7	101

TABLE 5.10   Origin of wild-caught *R. unicornis* exported to countries outside South Asia from 1900 to 2019

Wild	1900	1910	1920	1930	1940	1950	1960	1970	1980	1990	2000	2010	Total
Nepal	5	2	6	4		1	4	2	11	8	4		47
Assam				3	5	9	9	6					32
Bengal													0
Bihar													0
UP													0
Unknown		1					1						2
Total	5	3	6	7	5	10	14	8	11	8	4	0	81

TABLE 5.11   Origin of all wild-caught *R. unicornis* from 1900 to 2019, combining data of tables 5.9 and 5.10

Wild	1900	1910	1920	1930	1940	1950	1960	1970	1980	1990	2000	2010	Total
Nepal	11	2	6	7	3	1	6	2	14	9	4	2	67
Assam			2	3	5	15	21	25	22	8		3	104
Bengal				1								2	3
Bihar									1				1
UP											1		1
Unknown		2		1		2	1						6
Total	11	4	8	12	8	18	28	27	37	17	5	7	182

5.4   List of Specimens of *R. unicornis* in South Asia up to 1949

Listed alphabetically according to locality of the collection, there are details of 120 specimens of *R. unicornis* in captivity in India, Bangladesh and Nepal. Localities are listed following current names of towns, states and countries. The list also includes 17 important specimens exported after a longer stay in India. Two rhinos were born in captivity in this set: Kathmandu in 1824 and Kolkata Alipore in 1925. All other specimens were presumably wild-caught.

Localities discussed here arranged by country or Indian states.

Bangladesh: Dhaka, Dinajpur.

Nepal: Kathmandu.

Sri Lanka: Kandy.

India:

Assam: Assam, Cachar, Dibrugarh, Guwahati.

Bihar: Bettiah, Bhagalpur, Darbhanga, Dumraon, Kassimbazar, Patna, Varanasi.

Delhi.

Gujarat: Ahmedabad, Junagadh, Vadodara.

Himachal Pradesh: Arki, Una.

India: Locality unknown.

Karnataka: Bangalore.

Kerala: Trivandrum.

Odisha: Barpali, Jeypore.

Madhya Pradesh: Gwalior.

Maharashtra: Mumbai, Pune.

Manipur: Manipur.

Rajasthan: Ajmer, Alwar, Bijolia, Bundi, Dungarpur, Jaipur, Udaipur.

Tamil Nadu: Chennai.

Uttar Pradesh: Agra, Allahabad, Balrampur, Etawah, Farrukhabad, Kora Jahanabad, Lucknow, Mathura.

West Bengal: Kolkata (14 facilities), Murshidabad.

Agra, Uttar Pradesh – Court of Aurangzeb

1. Period 1665–1682

In an album of animal drawings dated between 1665–1682, from the time of the Mughal Emperor Aurangzeb (1618–1707), there is a sketch of a rhino captured in the District of Patna (11.15, fig. 11.21).

Ahmedabad, Gujarat – Court of Muzaffar II

*1. 1514–1515 (to Goa, then exported to Lisbon, Portugal)

The first Lisbon Rhinoceros was made famous through the woodcut of Albrecht Dürer (4.6). It had been kept for an unknown period by Sultan Muzafar II (1484–1526) in Champanel (near Ahmedabad) before he presented it to the Portuguese Governor, Afonso de Albuquerque (1453–1515). The Portuguese ambassadors started from Champanel on 26 April 1514 and arrived in Surat on 16 May 1514. The animal was transported to Goa, where it stayed from 15 September 1514 to early January 1515 when it was taken on board to Lisbon, Portugal.

Ajmer, Rajasthan – Court of Jahangir

1. Period 1600–1604

Prince Salim is depicted in a miniature dated around 1600–1604 in a hunting scene, where five men are showing him a young rhino that has been captured (fig. 11.11). Prince Salim was the name of the Mughal Emperor Jahangir (1569–1627) before he accessioned the throne in 1605.

*2–3. Period 1615–1616

Two rhinos were in captivity in Ajmer when the English trader Thomas Coryat (1577–1617) visited between July 1615 and September 1616 (11.12).

4. Period 1620–1622 (Exported to Istanbul, Turkey)

Possibly obtained from the Mughal court during the time of Jahangir, the Shah of Iran presented a living rhino to Sultan Osman II (1603–1622) in Istanbul, Turkey about 1620–1622 (11.12).

Allahabad, Uttar Pradesh – Rundheer Singh

*1. 1852-02-12 (to Lucknow: Nawab)

Found in the camp of Rundheer (Randhir) Singh of Patti Saifabad, Uttar Pradesh. It was confiscated by Robert Henry Wallace Dunlop (1823–1887) and handed over to authorities in Lucknow (Anon. 1852a), where it was added to collection of the Nawab (6.10, fig. 6.14).

Allahabad, Uttar Pradesh – Camp of Baiza Bai

1–2. 1836 (from Farrukhabad)

1836-02-09. The Welsh traveler Fanny Parks (1794–1875) visited the camp of Baiza Bai (1784–1863), the exiled regent of Gwalior, who had two rhinos when she was in Farrukhabad in 1835. Parks again saw the two rhinos in Allahabad on 9 February 1836: "they fought one another rather fiercely; it was an amusement for the party" (Parks 1850, vol. 2: 44).

1836-10-21. Henry Edward Fane (1817–1868) witnessed a fight of 2 rhinos (Fane 1842, vol. 1: 50). Fane was Aide-de-Camp to Commander-in-Chief General Henry Fane (1778–1840), who during this short visit in Allahabad stayed with his brother William Fane (1789–1839). The animals belonged to an unnamed "Mahratta princess", undoubtedly Baiza Bai as stated by Fanny Parks earlier in the year.

Alwar, Rajasthan – Maharao Singh

1. 1874-09

When the railway between Delhi and Alwar (or Ulwur, 160 km south of Delhi) was first opened in 1874, some journalists visited the town. The ruler Maharao Sheodan Singh (1845–1875) showed them "a dwarf cow, a rhinoceros, and a white cow that filled me with a sense of admiration which years cannot obliterate" (TMWTEG 1874). The contemporary gazetteer of Alwar mentions a menagerie with "a good many birds, foreign and others, and a few wild beasts" without reference to a rhino (Powlett 1878: 118).

Arki, Himachal Pradesh – Raja Kishan Singh

1. Period 1840–1850

Raja Kishan Singh (1817–1877) of the Baghal Princely State made several improvements to the town of Arki, now in Himachal Pradesh. During the 1840s and 1850s he commissioned many paintings on the walls of the palace. There is a fresco of a captive one-horned rhino with two attendants

FIGURE 5.2   Rhinoceros with two attendants depicted on a fresco in the Palace of Arki, Himachal Pradesh. This photo taken in 1998 is unique
             and the mural probably no longer exists
             PHOTO © JOACHIM K. BAUTZE, 1998

(fig. 5.2). Not dated or signed, it is tentatively taken to represent a scene in the middle of the 19th century.

Assam, Manas area – F.T. Pollok

1. 1869-05-11
Fitzwilliam Thomas Pollok (1832–1909) captured a young rhino with a net at Busbaree (Manas area, Assam) after killing the mother (5.2). The animal was savage at first, but started to feed out of the keeper's hand in two days and follow him around after a week (Pollok 1871c: 469, 1898a: 175).

Pollok possibly captured other young rhinos of which no details are known (39.4). Captive rhinos could be sold for about Rs 600 or £60 (Pollok 1882a) and added to the budget for his excursions. Pollok also recorded three unsuccessful capture attempts, in January 1868 (trip 5), April 1870 (trip 16) and April 1872 (trip 26).

Assam – Locality Unknown

1. 1920s
A friend of the army officer Henry Stotesbury Wood (1865–1956) kept a rhino in an unknown locality, which used to go out every day with its keeper to fetch its fodder (Wood 1930: 63).

Balrampur, Uttar Pradesh – Maharaja of Balrampur

1. M 1862, 1885
A male rhino was present during a visit in 1885 of the Methodist missionary Samuel Knowles to Balrampur. He had also seen the same animal 23 years earlier, in 1862 (Knowles 1889: 49, 176). According to Hewett (1938: 48) it was donated together with two elephants by Swami

Harihar Anand in the time of Digbijai Singh, Maharaja of Balrampur (1818–1882).

Bangalore, Karnataka – Lal Bagh Park

Lal Bagh was a botanic garden and leisure park since the first half of the 19th century. In the 1860s the Superintendent was William New (d. 1873), followed by John Cameron until 1903 (Bowe 2012).

*1. F 1868 to 1893-03-08 †
The death of the female rhino was reported in the *Englishman's Overland Mail* (1893-03-08), stating that she arrived in 1868 together with a male. When Stanley Smyth Flower (1871–1946) visited Lal Bagh Park in April 1913, he found a curious empty cage formerly inhabited by a rhino (Flower 1914: 40).

2. M 1868 to 1870 (to Chennai)
The male which had arrived together with the female was incompatible, and "had to be deported to another State." It may have been the animal reported from 1870 in the Government Gardens of Madras (see Chennai).

Barpali, Odisha – Raja of Bamra

1. Early-19th century (1830s?)
When the Raja of Bamra married a daughter of one of the Rajas of Sambalpur, the dowry included a rhino which had been kept in Barpali (Ball 1877).

Bettiah, Bihar – Raja Kishore

1. 1848-01
The pseudonymous hunter Rohilla (1848) was invited to the palace of Kishore Singh Bahadur, 8th Raja of Bettiah (d. 1855). Their audience was interrupted by a "very tame and very stupid Rhinoceros."

Bhagalpur, Bihar – Augustus Cleveland

*1. No date (1780s?)

Augustus Cleveland (1754–1784), District Officer of Bhaugulpore (Bhagalpur) kept a rhino near his home "which I believe did not live very long" (Williamson 1807a: 45; 1807b, vol. 1: 168).

Bijolia, Rajasthan – Maharana Sangram Singh II

1. M Period 1721–1734

A male rhino is depicted in a mural associated with a ceremony which took place at the court in Bijolia during the reign (1710–1734) of Maharana Sangram Singh II (1690–1734) in Mewar (Udaipur) (fig. 15.46).

Bundi, Rajasthan – Rao Raja Umaid Singhji

1. Period 1775–1800

In a wall-painting dated to the last quarter of the 18th century, a rhino is shown engaged in a fight with an elephant (fig. 15.42). Both animals are mounted by riders. The rhino has no chains around feet or neck.

Cachar, South Assam – Synteng

1. M Period 1870s–1880s

The hunter-sportsman Synteng (1885), who remains pseudonymous (60.3), reported that rhinos were occasionally captured in Cachar, and he may have kept one for a short time: "the only youngster I ever secured I was fain to let loose again, as he steadily refused all food."

Chennai (Madras), Tamil Nadu – Government Gardens

The Government Gardens were located in Egmore on Anna Salai.

1. 1870-11 (from Bangalore) – alive in 1875

An anonymous correspondent saw a large rhino in the Government Gardens (*Preston Chronicle* 1871-03-18). This is likely to have been the male rhino which had proved incompatible with a female in Bangalore in 1868. The *Times of India* (1871-12-12) reported the temporary escape of the rhino. On a short visit to the People's Park, Benjamin Robbins Curtis (1855–1891) noted the rhino on 13 December 1875 (Curtis 1876: 225).

Darbhanga, Bihar – Maharaja of Darbhanga

The new palace of the Maharaja of Darbhanga was completed in 1883 designed by British architect Major Charles Outram Mant (1840–1881) with gardens laid out by the botanist Charles Maries (1851–1902). A photograph of 1888 shows "The Rhinoceros Park." Ehlers (1894, vol. 1: 357) comments on Darbhanga's zoological garden, without listing a rhino.

It is possible that the capture of one of these zoo animals was described in a report by Money (1893: 618), unfortunately undated (no. 1). The baby rhino, 3 feet (92 cm) high, was isolated in a patch of tree and grass jungle near the Nepal border. After several attempts, a noose was secured around his neck, but 20 men could not hold him. He was captured nevertheless, then brought to a stockade constructed in the hunting camp, where "after being in it some 12 hours, it became very quiet, and allowed milk to be poured down its throat. This was done constantly, and in little more than a day and a half it came out of its stockade for all intents and purposes a tame animal." After few weeks, the rhino reached Darbhanga where it was treated like a pet by the Maharaja. Money believed that this animal died after a few months because it was fed the wrong type of food.

1. Period 1880–1885

In the time of Maharaja Lakshmeshwar Singh Bahadur (1858–1898), the estate manager Major General Robert Cotton Money (1835–1903) mentioned that 1 baby rhino was caught in the Bhutan Duars during a hunting trip (Money 1893: 618). The report might date from the late 1870s or early 1880s (28.25).

2–3. 1888

There were 2 young rhinos in a "Rhinoceros Park" within the grounds of the Anand Bagh Palace or Lakshmeshwar Vilas Palace. In 1888 the palace was visited by photographers of the firm Messrs. Bourne and Shepherd of Calcutta, and several pictures were printed in *The Graphic* (Anon. 1888). One of the photographs shows two young-looking *R. unicornis* in an extensive enclosure (fig. 5.3).

Delhi – Timur Bec

*1. 1399-01

After Timur Bec or Timurlane (1336–1405) conquered Delhi, he organized a parade including elephants and rhinoceros (number unknown) as recorded in his reminiscences (13.7).

Delhi – Court of Aurangzeb

*1–2. 1659

François Bernier (1620–1688), who worked as a physician at Aurangzeb's court from 1658–1667, saw "rhinos" when he visited Delhi in 1659 (11.15). The number of animals is unclear, hence an estimate of possibly two.

Dhaka, Bangladesh – Matthew Day

*1. 1780s

Matthew Day (unidentified) had a rhino "kept in a park into which it was not very safe to venture" (Williamson 1807a: 45; 1807b, vol. 1: 168).

Dibrugarh, Assam – W. Millett

1. 1866-04-25
W. Millett of Tetellee Gooree (Teteliguri) had a young rhino for sale at Rs 350 (F. Buckland 1868: 217).

Dinajpur, NW Bangladesh – Annual Fair

1. 1861
The annual fair held in April at Nekh Murd, 50 km NW of Dinajpur (Dinagepore), usually sold cattle and other livestock. In 1861 also one rhino was offered, or maybe exhibited (Sherwill 1865: 28).

Dumraon, Bihar – Maharaja of Dumraon

*1. F 1877-02 (to Kolkata: Alipore Zoo)
A female rhino was donated to the Calcutta Zoo by Maharaja Maheshwar Bakhsh Singh (1803–1881), 9th Maharaja of Dumraon. The vexing question "now stirring the Calcutta intellect to its depth is, whether to train or to march the rhinoceros?" was decided in favour of a two-months overland journey (*Times of India* 1877-02-09 and 1877-04-12). It is possible that the animal had been kept by the Maharaja previous to his gift.

Dungarpur, Rajasthan – Court of Maharawal

There are three murals in the Juna Mahal in Dungarpur dating from different periods. The potential dates seem too far apart to refer twice to the same specimen. In all three cases, the rhino has a blanket over the back indicating life in captivity.

1. 1885-03
A rhino called 'Mohan Lal' was acquired by Maharawal Udai Singh II, depicted in a scene of a procession taking place in 1886 (fig. 15.51).

2. 1919
Adorned with ropes and chains, a rhino with a rider is shown in a mural dated 1919 (fig. 15.53).

3. 1939
A rhino mounted by a bearded man is depicted in a mural dating around 1939–1940 (fig. 15.52).

Etawah, Uttar Pradesh – Court of Shah Jahan

*1–2. 1632
1632-12-20. The English merchant Peter Mundy (1596–1667) stayed in the Agra and Surat areas from 1628–1634 (11.13). On one of his excursions on 20 December 1632, he had just passed Etaya (Etawah), when he heard that two rhinos had been sent to the town "to bee kept and fedd", and therefore obviously still alive (Mundy 1914, vol. 2: 171). Mundy had no time to turn back to get his first glimpse of a rhino. They were a gift of Saif Khan (d. 1640), Governor of Bihar 1628–1632 to the Mughal Emperor Shah Jahan.

Fatigarh (Farrukhabad), Uttar Pradesh – Baiza Bai

Baiza Bai (1784–1863) was the third wife of Shrimant Daulat Rao Sindhia (1779–1827), the ruler of Gwalior state, and after his death she acted as the Regent of the territory 1827–1833. After this she lived in exile until 1856 in various places.

1–2. 1835-04-15 (to Allahabad)
1835-04-15. While in Fatigarh, Baiza Bai was visited by the Welsh traveler Fanny Parks (1794–1875), who then saw "two fine rhinoceroses, which galloped about the grounds in their heavy style, and fought one another; the Ba'i gave five thousand rupees (£500) for the pair" (Parks 1850, vol. 2: 7). The two rhinos were still in the camp of Baiza Bai in Allahabad in February 1836.

Goa – Portuguese Embassy

*1. 1514-09-15 (from Ahmedabad via Surat) to 1515-01 (to Lisbon, Portugal)
The rhino donated by Muzafar II to the Portuguese King stayed in Goa from 15 September 1514 to early January 1515 before shipment to Portugal (see entry under Ahmedabad above and 4.6).

*2. F 1577 (to Lisbon, Portugal)
The Abada or Madrid Rhinoceros was shipped from Goa by the Portuguese Mission to Lisbon in 1577 (11.9).

Guwahati (Gauhati), Assam – Unknown Owner

1. 1855-12
Hermann Schlagintweit (1826–1882) traveled to India in 1854–1857 to make scientific investigations. He spent some months in Assam, visiting Guwahati and traveling up to Dibrugarh by steamer (38.17). He saw a single tame rhino in Gowahatti (Gauhati, Guwahati), where he stayed from 16 November to 21 December 1855 (Schlagintweit 1869, vol. 1: 433).

Gwalior, Madhya Pradesh – Maharaja of Gwalior

Baiza Bai, the exiled former Regent of Gwalior, is stated to have exhibited two rhinos while she was living in Fatigarh (Farrukhabad) in April 1835 and in Allahabad in 1836. There is no report of these animals in Gwalior.

1. 1790 (to Pune)
Mahadaji Scindia (1730–1794), the ruler of Gwalior 1768–1794, donated a rhino in 1790 to Peshwa Madhav Rao II (1774–1795) of Poona (Pune), Maharashtra.

2. 1793 (to Kolkata: Governor of Bengal)
Mahadaji Scindia sent a letter to Governor John Shore (1751–1834) in Kolkata informing that he had sent him a rhino (National Archives of India 1949, vol. 10: 247).

THE RHINOCEROS PARK

FIGURE 5.3   "The Rhinoceros Park" with two *R. unicornis* in the grounds of the Anand Bagh Palace in Darbhanga in 1888. Photograph by Messrs. Bourne and Shepherd of Calcutta
GRAPHIC, 1 DECEMBER 1888, P. 566

*3. 1876-02 to 1903
There are several references to rhinos during this period of 27 years.

1876-01. Albert Edward, Prince of Wales (1841–1910) visited Gwalior from 31 January to 3 February 1876 as guest of Maharaja Jayajirao Scindia (1843–1886). A fight between a rhino and a tiger was planned but was cancelled (*Homeward Mail from India* 1876-02-14).

1880s. Set in the Gwalior Fort before it was handed over to the ruling Scindia family in 1886, a story was doing the rounds of a dangerous game of football: "there is, or used to be, a rhinoceros at the capital at Gwalior whose homicidal instincts provided much sport for subalterns quartered at Gwalior before the fort was surrendered to Scindia. The game only required two players beside the rhinoceros, and was very simple. The beast had an exceedingly violent temper, and was confined in an enclosure half as large as Leicester Square, surrounded with pillars, between which a man could slip easily, but too close for the monster's bulk to pass. Somewhere inside the enclosure his food bucket lay, and the players, taking opposite sides of the enclosure, played football with the bucket. It differed from ordinary football, because only one player could play at a time – the rhinoceros was always looking after the other – and you could never get more than one kick, because the instant the rhinoceros heard the bang he was round and after you like a locomotive. That was the chance of the other player, who, dashing in full speed after the beast, was able to get one good kick while Rhino was narrowly missing his friend at the opposite barrier. Then it was his turn to flee, with the creature after him, and, for the

other to get a kick. Halftime was always called when the rhinoceros began to play cunning and lurked near the bucket" (*Times of India* 1900-03-06; E. Russell 1905: 630).

1890-03. It may have been the same animal seen in March 1890 when the Polish Count Josef Potocki (1862–1922) visited Maharaja Madho Rao Scindia (1876–1925). He saw a small menagerie with "a slow-walking rhinoceros" (Potocki 1896: 50). The red colour mentioned by Loisel (1912, vol. 3:18) is not clear in the original Polish account.

1903-01. During the preparations in January 1903 for the Delhi Durbar, one of the local rulers, probably the Maharaja of Gwalior, "carried away by his enthusiasm and his desire for effect at all cost, is actually seeking permission to bring a rhino" (*Homeward Mail from India* 1902-12-29).

India – Locality Unknown
The following four entries have no relation to each other. Due to uncertainties, these are not included in the totals in 5.3 or in Table 5.7.

1. 1830s
A young rhino captured in an unknown place, died on the way to Kolkata by being overfed (E. Roberts 1835, vol. 1: 129–130).

2. M Undated
A rhino in chains is depicted in a coloured drawing, without date or place (fig. 5.4). The item was sold at Sotheby's in the sale of the collection of the archaeologist Hagop Kevorkian (1872–1962) on 3 April 1978 (Sotheby's 1978). Rookmaaker (1997d) published the drawing, but further details remain elusive. The animal was not necessarily kept in South Asia.

3. Undated
A rhino tethered both at one front and one hind leg is shown in a drawing (fig. 5.5). Location and date are unknown, therefore the animal was not necessarily kept in South Asia.

4. Undated
A watercolour made in India in the 19th century (Freedman 2003) could depict a rhino in captivity (fig. 5.6).

Jaipur, Rajasthan – Maharaja of Jaipur
1. Period 1845 to 1861
A reference to a rhino in Jaipur in 1845 and a woodcut published in 1861 may indicate one or two specimens (here combined).

1845-12-04. After meeting Maharaja Ram Singh II (1835–1880) of Jaipur, the Russian aristocrat Prince Alexis Soltykoff (1806–1859) was taken on a tour through the town and saw a rhino roaming freely followed by

FIGURE 5.4   A male rhinoceros facing right chained by its front
feet to a low post. Size 23,7 × 37.2 cm. Probably early
19th century. Location unknown. Sold at Sotheby's,
3 April 1978, p. 110, lot 124
PHOTOGRAPH IN COLLECTION OF T.H. CLARKE

FIGURE 5.5   Rhinoceros in chains. The location and date are
unknown
PHOTOGRAPH IN COLLECTION OF T.H. CLARKE

FIGURE 5.6   Watercolour of the 19th century, without any
details. Ink and translucent watercolour on paper,
14 × 19.36 cm. Inscription in Hindi above animal:
"gaiṇḍa", see chapter 15 note 82
PRESENT WHEREABOUTS UNKNOWN; FROM
AARON M. FREEDMAN, *MALA KE MANKE*, 2003,
P. 198, NO. 102

FIGURE 5.7   Rhinoceros with a collar around the neck
embellished with three small bells, possibly drawn
in the late 19th century. Published by the Ceylonese
art historian Ananda Kentish Muthu Coomaraswamy
(1877–1947) from the collection of the painter
Gogonendranath Tagore (1868–1938)
COOMARASWAMY, *INDIAN DRAWINGS*, 1910, PL. 21
(LOWER FIGURE)

two people guiding the animal with sticks as if it was
a buffalo (Soltykoff 1848: 367, 1858: 435, see chapter 15
note 86).

[1861]. There is an Indian woodcut of a rhino in company
of its groom, published in 1861 as a frontispiece in
Pustak vidyāṅkur (Jaipur) (Bautze 1985: 427). The rhino
was paraded in front of the Maharaja Ram Singh II in
1861 or possibly earlier (chapter 15 note 87; fig. 15.67).

Jaipur, Rajasthan – Ram Newas Public Gardens
These gardens were opened in 1868 by order of Sawai Ram
Singh of Jaipur (chapter 15, note 98).
1–2. Period 1886 to 1903

There are records of one or two rhinos in Jaipur 1886–1903
on three separate occasions, which may all refer to the
same specimens. Three undated drawings could well
depict these animals, although details cannot be verified:
one sketch published by Coomaraswamy (1910, pl. 21)
shows a rhino with a thin horn carrying a collar with three
bells (fig. 5.7), and two possibly related watercolours by
a Rajasthani artist depict a rhino tied to a stick in the
ground (figs. 15.71, 15.72).
1886-01. Prince Louis Josef Jérôme Bonaparte (1864–1932)
during his tour of India (37.9) visited Jaipur where
he saw elephants dressed to fight one or more rhinos
(*Figaro* 1886-11-24).

1893-03-04. Franz Ferdinand (1863–1914), Archduke of Austria, meets Maharaja Madho Singh II (1862–1922) and describes how two rhinos were fighting in their enclosure in March 1893. The animals were painted with black paint (Franz Ferdinand 1895, vol. 1: 335).

1903-01-22. In the Ram Niwas Garden, a pair of rhinos was seen by Vyas Ramshankar Sharma, Secretary to Dhandevi Mahodaya from the United Provinces (Tillotson 2006: 173).

*3. F 1946 to 1964. (from Udaipur)
A female rhino was received from Udaipur, and was alive in 1964 (Rookmaaker 1998: 79).

Jaldapara, West Bengal – Forest Department

1. M 1936-09-18 to 1936-11-17 †
A male rhino calf wandering aimlessly near the Salkumar Forest of Chilapata Range was captured on 18 September 1936, and died after two months (Meiklejohn 1937: 2).

Jeypore, Odisha – Maharaja of Jeypore

1. 1870s
A rhino was marched down from Kolkata (ca. 1000 km) and sold to "Raja of Jaipur" (Jeypore), Maharaja Ramchandra Dev III (ruled 1860–1889) for Rs 16,000 (Ball 1880: 570).

Junagadh, Gujarat – Nawab of Kathiawar

There was also at least one R. sondaicus in Junagadh in 1900 (48.2).

1. 1876, 1889
1876. In early 1876, the english traveler Andrew Wilson (1831–1881) met Nawab Mohammad Mahabat Khanji II (1838–1882) who staged a fight with elephants for his visitor. There also was a rhino, but Wilson was told that the animal had killed either a man or an elephant every time when it was brought out for battle in the past, and therefore was no longer involved (Wilson 1876: 206).

1889-02-26. The astronomer John James Aubertin (1818–1900) was in Junagadh on 26 February 1889. Being shown around by Chhaganlal Harilal Pandya (1859–1935) on behalf of Nawab Mohammad Bahadur Khanji III (1856–1892), he saw an "enormous rhinoceros", presumably kept in an enclosure unconnected with the zoo where he visited later in the day (Aubertin 1892: 83).

Kandy, Sri Lanka – King Rajasinghe II

1. 1643
The German officer Johann Jacob Saar (1625–1664) recorded in his travel journal that in 1643 the King of Bengal sent an ambassador to the Emperor of Kandy (Candi) with a gift of a living rhinoceros hoping to receive elephants in return (Saar 1662: 89). Bengal was ruled by Shah Jahan's second son Prince Shah Shuja (1616–1661) from 1639 to 1647, while Kandy was ruled by King Rajasinghe II (1608–1687) from 1635–1687. As Saar only reached Sri Lanka in 1647, maybe the rhino was still alive at the time.

Kassimbazar (Cassamabasar), Bihar – Dutch Factory

*1. 1671
Local hunters presented a young rhino standing 5 feet (1.5 m) high, with a small horn, to Jacob Verburgh (d. 1680), Director of the VOC (Dutch East India Company) factory (Graaff 1701: 133).

Kathmandu, Nepal – King of Nepal

1. 1791 (to Kolkata: Governor of Bengal)
On 10 October 1791, Rana Bahadur Shah of Nepal (1778–1806) wrote to Major General Charles Stuart (1758–1828), representing the British Government in Kolkata, with the remark: "As requested by him the writer has been on the lookout for rhinoceros. His people caught one in the forest. As the animal is perfectly wild just now it is being tamed and will be sent down to Calcutta after the rains in the month of Kartik [November]" (National Archives of India 1949, vol. 9: 319). There are no further data.

2. Period 1797 to 1832
Hodgson (1832) mentioned a rhino kept at Kathmandu for 35 years "without exhibiting any symptoms of approaching decline." As it was probably still alive when Hodgson wrote, the animal could have arrived around 1797. The location of the Menagerie of King Rajendra Bikram Shah (1813–1881) is not known in detail.

*3. M 1823. Male rhino, father of calf born in 1824.

*4. F 1823. Female rhino, mother of calf born in 1824.

*5. M 1824-05 (birth) to 1834 (to Kolkata, but final destination not recorded)
Brian Houghton Hodgson (1800–1894) first went to work in Kathmandu in 1824 and stayed until 1844, since 1831 as British Resident. He had ample opportunity to study the local wildlife, then practically unknown (Cocker & Inskipp 1988; Lowther 2019). He was responsible for commissioning hundreds of watercolour drawings of mammals and birds, in preparation of a grand work on the region which never appeared. He also sent three rhino specimens to the British Museum, listed at the time of accession in 1845 as the skin of foetus, skull of just-born specimen, and a partial left side of lower jaw of R. unicornis (Gray 1846: 35; fig. 5.11). The specimens were transferred to

the Natural History Museum, where the skull is still present (no. 722b or NHM 1845.1.8.85).

Hodgson wrote about a male rhino born in Kathmandu in the menagerie of King Rajendra in May 1824 (Rookmaaker 1979; Datta 2004: 154). He followed its growth during the first year and provided several measurements subsequently (Hodgson 1825, 1826a, 1826b, 1827, 1832, 1834; see table 5.12). The animal was examined again in 1833 when it was sketched by a local artist.

There are two larger sets of drawings of animals made for Hodgson in Nepal, preserved in the library of the Zoological Society of London (ZSL, Hodgson Drawings) and in the Natural History Museum, London (NHM, Zoology Special Collections 88 f HOD). The ZSL set was donated to the Society by Hodgson in 1874. Both sets include drawings of *R. unicornis* showing the animal born in 1824 when it was nine years old, as well as parts of a skull.

The ZSL set contains 456 art originals of the mammals of Nepal, bound in 2 volumes, and has been digitized in 2015 accessible through the ZSL website. There is only one sheet with a rhinoceros (vol. 2, no. 50), with sketches and handwritten information on both sides (figs. 5.8, 5.9). The front has a complete coloured sketch showing a rhinoceros in lateral view, first published in Rookmaaker (1979). There are at least two other outlines of the rhinoceros sketched on the same paper, very faint, and invisible in the reproductions. The main title below the animal reads: "Rhinoceros 9 years old. male. March 1, 1833, Hab. Saul forest." Below the head: "Home Jany 42." There are some notes

written in pencil on the right side of the paper, now almost invisible but repeated on the reverse. On the reverse of the same sheet is another sketch of the rhino seen from the side, with below it "Rhinoceros 5 ½ years", therefore possibly made before the one on the front, but the reading of "5" is inconclusive. Across the rear part of the body are some longer notes, probably written by Hodgson:

Rhinoceros March 1, 1833. 9 yrs old.
[Illegible] to rump 7 – 4 – ½
Height afore 5 – 2 – 0
Length of head 2 – 5 – 0
Do [Length] of horn 0 – 6 – 0
Teeth perfect, penis erigible, scrotum empty. Book says teeth incisors 2-2/2-2, canines 0-0/0-0, molars 7-7/7-7. In this young animal the dental system is different & thus, incisors 1-1/1-1, canines none, molars 6-6/6-6 & so in a scull I have.

Incisors lateral separate, those of lower jaw elongated & projected forwards, convex before, flat & scarped behind. those of upper jaw flat & smooth crowned, low, long & narrow, grinding on inner scarped surface of lower molars remote from incisors – 6 inches. Close, increasing in size backwards. 1st nearly flat on crown, rest have crowns rather smooth between their serpentine edges.

This animal shewed first symptoms of puberty in [beginning of] his 10th year, when he went to Calcutta. He was born in the Durbar's menagerie as elsewhere recorded by me.

TABLE 5.12　　Growth of a male *R. unicornis* born in Kathmandu in May 1824

Age	Length of body	Shoulder height	Circumference	Length of head
3 days	3 ft 4 ½ in	2 ft	4 ft ¼ in	1 ft ½ in
(May 1824)	103 cm	61 cm	122.5 cm	31.7 cm
1 month	3 ft 10 in	2 ft 5 in	4 ft 5 in	1 ft 2 in
(June 1824)	117 cm	74 cm	134.6 cm	35.6 cm
14 months	5 ft 10 in	4 ft	7 ft	
(July 1825)	178 cm	122 cm	213.4 cm	
19 months	7 ft 3 in	4 ft 4 in	9 ft 5 in	
(December 1825)	221 cm	132 cm	287 cm	
8 years	7 ft 3 in	4 ft 10 in	10 ft 5 in	2 ft 4 in
(1832)	221 cm	147 cm	317.6 cm	71 cm
9 years	7 ft 4 ½ in	5 ft 2 in		2 ft 5 in
(1833)	224.8 cm	157.5 cm		74 cm

From the measurements provided by Hodgson (1825, 1826, 1832) and on the watercolour in ZSL (see Rookmaaker 1979). Recorded in feet (ft) and inches (in), the conversion to metric is added.

FIGURE 5.8
"Rhinoceros. 9 years old. Male. March 1, 1833. Hab. Saul forest." Drawing by anonymous artist made for Brian Houghton Hodgson (1800–1894) in the menagerie of the King of Nepal
WITH PERMISSION © ZOOLOGICAL SOCIETY OF LONDON: HODGSON PAPERS, VOL. 2, NO. 50

FIGURE 5.9
Reverse of the sheet on the rhinoceros with notes by Hodgson in the ZSL copy, showing a hitherto unpublished sketch of the same animal
WITH PERMISSION © ZOOLOGICAL SOCIETY OF LONDON: HODGSON PAPERS, VOL. 2, NO. 50

FIGURE 5.10
"Rhinoceros. 9 years old. mas. 1/15 nat. size. Hab. The Saul forest." Drawing by anonymous artist made for Brian Houghton Hodgson (1800–1894) in the menagerie of the King of Nepal
WITH PERMISSION © THE TRUSTEES OF THE NATURAL HISTORY MUSEUM: ZOOLOGY SPECIAL COLLECTIONS 88 F HOD

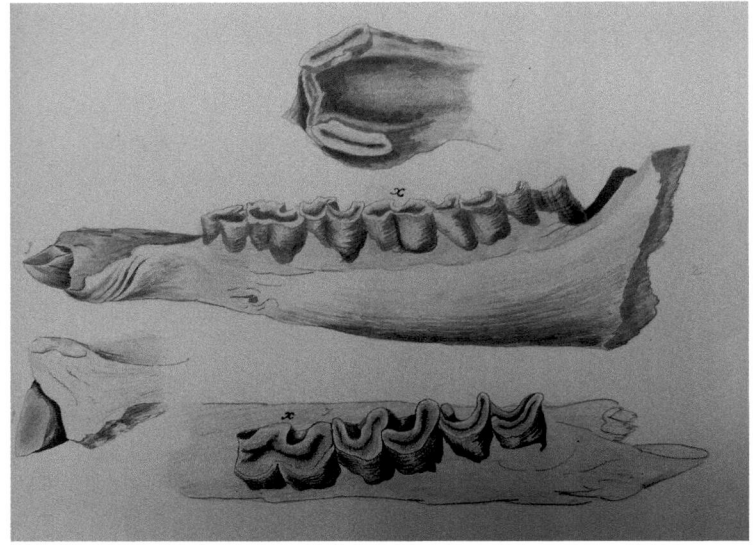

FIGURE 5.11
"Rhinoceros Unicornis of the Lower region of Nepal. 1. End
of upper jaw. 2, 3, 4. Lower jaw. Part VII. Pl. 1."
WITH PERMISSION © THE TRUSTEES OF THE
NATURAL HISTORY MUSEUM: ZOOLOGY SPECIAL
COLLECTIONS 88 F HOD

The transfer to Kolkata, presumably in 1834, mentioned here by Hodgson is not documented elsewhere and the fate of the animal remains obscure. This drawing is not signed, but a few others in this volume were made by "Raj Man Sing Chitrakar or artist", who must have been employed by Hodgson for this work and an early member of the well-known Chitrakar family of artists (24.14).

The NHM set was intended to illustrate a colour plate book on Nepalese mammals which never materialized. The drawings are more highly finished but lack the annotations. The drawing of the rhino at nine years seen from the side is a faithful copy of the main one in the ZSL set (fig. 5.10). Besides there is a sheet showing parts of the upper and lower jaws of a *R. unicornis* skull, possibly the specimen listed by Gray (1847: 72, 1862: 281) in the British Museum (fig. 5.11).

6–7. 1845-01
When Werner Hoffmeister (1819–1845) first entered Kathmandu during the visit of Prince Friedrich Wilhelm Waldemar (1817–1849) in January 1845, he crossed the town on elephant back. In an otherwise unidentified place, he passed "through a high, but narrow gate-way, into a court, where we saw several tame rhinoceroses, kept here on account of the custom of the country, which required that, on the death of the Rajah, one of these creatures should be slain, and imposes on the highest personages in the state the duty of devouring it" (Hoffmeister 1847: 148, 1848: 229). The actual number of rhinos kept in the Nepalese capital during the 19th century is irretrievable in the absence of eyewitness accounts.

8. 1851-01-12
On a diplomatic visit to Kathmandu in 1851, Francis Egerton, Earl of Ellesmere (1800–1857) "visited a rhinoceros, which they keep tied up in a yard; it was firmly secured to a post, by a chain round its neck, and by another one round its fore-legs; they said it was much inclined to be mischievous" (Egerton 1852, vol. 1: 217).

*9–11. 1907
According to Lydekker (1909), six male rhinos were captured alive during the annual hunt in 1907 (24.15). Three were kept by the Nepal Government to be tamed and trained for racing. The other three rhino were sold to the German animal dealer Carl Hagenbeck (1844–1913) and sold to zoos in Antwerp, Manchester and New York (S. Flower 1908: 12).

12. 1932, 1939
In an album of photographs taken in 1932 by the diplomat and explorer Philips Christiaan Visser (1882–1955), Dutch envoy to British India 1931–1937, there are two images of a young *R. unicornis* (figs. 5.12, 5.13). The animal is walking freely, supervised by a keeper, in front of Maharaja Kaiser Shumsher (1892–1964) and other dignitaries (Visser 1932).

In September 1939, Hassoldt Davis (1907–1959) in the company of Armand Georges Denis (1896–1972) was in Kathmandu (24.40), where he mentioned a private zoo in the premises of the Royal Palace, with a rhino "which must be killed at the exact moment of the Maharaja's death" (Davis 1940: 281). This is likely the specimen earlier reported for 1939 by Rookmaaker & Reynolds (1985: 152).

Kolkata (Calcutta), West Bengal
There have been rhinos in at least 14 different locations in or near the town of Kolkata, mostly known from reports in the 19th century. They are listed in a generally chronological order, as follows:

Captain Udall (from Gulkindall)	1683
French factory at Chandernagor	1706, 1769
Nawab of Bengal – see Murshidabad	1716

FIGURE 5.12
Rhinoceros shown by Maharaja Kaiser Shumsher to the envoy of The Netherlands, Philips Christiaan Visser (1882–1955) at a state visit in 1932
NATIONAL ARCHIVES OF THE NETHERLANDS, THE HAGUE: NL-HANA_2.21.284_37_001

FIGURE 5.13
Rhinoceros shown in 1932 to the Dutch envoy P.C. Visser by Maharaja Kaiser Shumsher (with white beard)
NATIONAL ARCHIVES OF THE NETHERLANDS, THE HAGUE: NL-HANA_2.21.284_37_001

Dutch VOC trading post at Hooghly	1738
British Governor of Bengal	1765, 1791, 1793
Barrackpore Park	1808, 1827, 1833, 1837, 1838, 1839, 1856, 1856, 1864
Maharaja Buddynath Roy	1829, 1837
Indian Rajah in Alipore	1858
'Seven Tanks' in Cossipore	1858
Raja of Burdwan	1859, 1859
Raja Rajendra Mullick	1867, 1871
Alipore Zoological Gardens	1877, 1887, 1905, 1909, 1909, 1912, 1925, 1932, 1932, 1938
W. Rutledge (dealer), Entally	1875
King Wajid Ali Shah of Oudh	1887
William Jamrach (dealer)	exports

Kolkata (Calcutta) – Captain Udall

1. F 1683 (to London, UK, where died 1686-09-26)
The mariner Edward Barlow (1642–1707) kept a unique illustrated journal during his travels around the globe. When he was in Kolkata at the end of 1683, he noted that "In the country are bred the great beasts called the 'rhinosarus', and many wild and cruel tigers, it being a very level country and full of woods and rivers" (Lubbock 1934: 361).

FIGURE 5.14 Rhinoceros depicted by Edward Barlow (1642–1707) in 1683
WITH PERMISSION © NATIONAL MARITIME MUSEUM, GREENWICH: JOD/4/1, BARLOW JOURNAL, P. 104

The journal has a plate of a rhino (Rookmaaker 2019a), to which Barlow later added that it was "The emblem of Risnosarus, that was brought from Bangall in ye yeare 1684 and sould at London for two thousand one hundred pound" (fig. 5.14). This reference identifies the animal as a female rhino transported on board the *Herbert* with Captain Henry Udall (Rookmaaker 1978b: 23, 1998: 82, 2007). The history of this animal in captivity in England is still only partly known, but it died on 21 September 1686 (figs. 5.15, 5.16). One of the sources about her life is the text on an engraving of a rhino modelled after Dürer, where the origin of the animal is said to be "the Court of the King of Gulkindall" (Rookmaaker 1978b, no. 4.1). Gulkindall

remains unidentified, and if it was Golconda, a kingdom around the present Hyderabad, this would tell us little about the actual provenance.

Kolkata (Calcutta) – French Factory at Chandernagor

1. 1706 to 1716 (to Nawab of Bengal)

Pierre Dulivier (1663–1722), Director of the French factory of Bengal, in 1701 received an order to buy a rhino for the menagerie of King Louis XIV of France. The animal was caught in 1706 and kept at Chandernagor. As all captains of commercial vessels refused to take the animal onboard, the animal soon became so large that it was impossible

FIGURE 5.15
"Exact Draught of that famous Beast the Rhinoserus that latley came." Broadsheet of the rhinoceros shown in London from August 1684
WITH PERMISSION © UNIVERSITY OF GLASGOW LIBRARY, HUNTERIAN LIBRARY (DOUGLAS COLLECTION): SP.COLL. AV.1.17, NO. 37

FIGURE 5.16
Anonymous drawing of the female rhinoceros in London from 1684
WITH PERMISSION © UNIVERSITY OF GLASGOW LIBRARY, HUNTERIAN LIBRARY (DOUGLAS COLLECTION): SP.COLL. AV.1.17, NO. 39

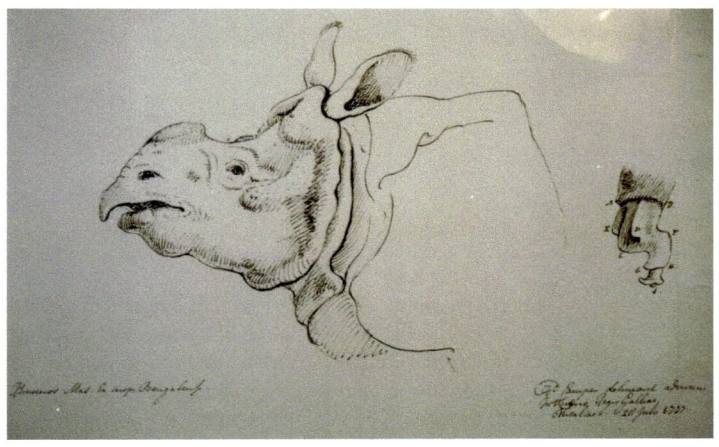

FIGURE 5.17
Sketch dated 28 July 1777 by Petrus Camper showing the Versailles Rhinoceros exhibited 1770 to 1793. The Latin text reads: "Rhinoceros Mas. Ex Bengalensi. P. Camper delineavit ad vivum. In Vendrio Regis Galliae. Versaliae 28 Julii 1777."
WITH PERMISSION © ALLARD PIERSON, UNIVERSITY OF AMSTERDAM: PETRUS CAMPER COLLECTION. PHOTO: KEES ROOKMAAKER

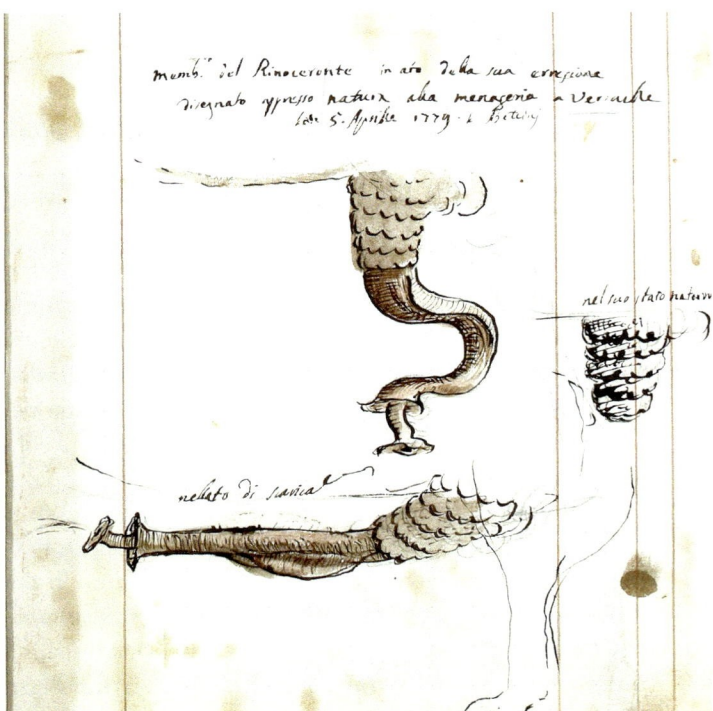

FIGURE 5.18
Detail of a page in a sketchbook by the Italian garden designer Francesco Bettini (1737–1815). He drew the penis of the rhino in Versailles in a state of erection at a visit on 5 April 1779. Text (upper figure): "mem^{br} del Rinoceronte in ato dela sua errezione, disegnato appresso natura ala menagerie a Versailles del 5. Aprila 1779 di Bettini"; (middle figure): "nel suo stato natural"; (lower figure) "in stato di scarica." Hence he drew the penis in a state of erection, in the natural state, and in state of discharge
WITH PERMISSION © ARCHIVIO DORIA-PAMPHILJ, ROMA : CAOS FARRAGGINE, VOL. 1, FOL. 218

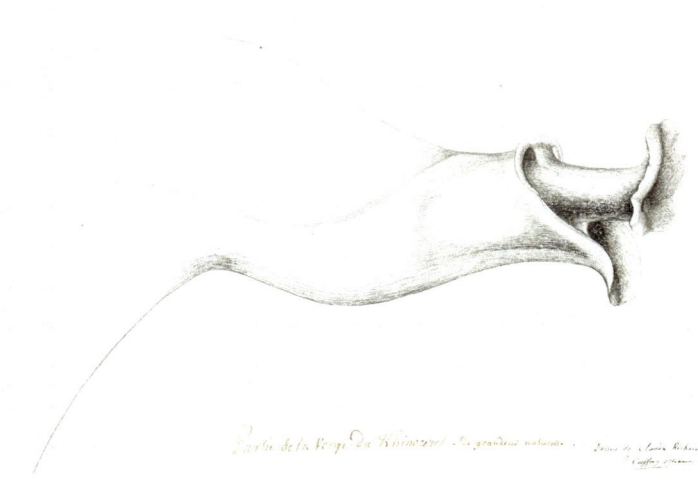

FIGURE 5.19
"La verge du rhinocéros, la dilatation inférieure de l'urètre ouverte" (Penis with the lower section of the urethra opened). One of the anatomical drawings made by Pierre Joseph Redouté (1759–1840) during the dissection of the Versailles Rhinoceros in 1793
WITH PERMISSION © MUSÉUM NATIONAL D'HISTOIRE NATURELLE, PARIS: VÉLINS, MNHN_VEL_PORTEFEUILLE_065_FOL048BIS

FIGURE 5.20
The mounted skin of the rhinoceros exhibited in the royal menagerie at Versailles from 1769 to 1791. During the 20th century a long horn belonging to a white rhino (*Ceratotherium simum*) was added to enhance the exhibit
POSTCARD ISSUED BY THE MUSEUM OF
NATURAL HISTORY IN PARIS. COLLECTION KEES
ROOKMAAKER

even to carry enough water for the journey. Therefore in 1716, the gentlemen of Chandernagor presented the rhino to the Nawab of Bengal. Dulivier had a bill of 1848 livres 7 sols 5 deniers for food of the rhino. At first the government refused to pay, but later it seems the amount was settled (Olagnier 1931: 526–527).

2. M 1769 (to Versailles, France, where it died 1793-09-23) A young male rhino was donated to King Louis xv by the French Governor Jean-Baptiste Chevalier (1729–1789), who had visited Assam territory in 1755 (38.8). The origin of the animal is unknown. He was shipped on the *Duc de Praslin* on 22 December 1769 and arrived at the port of Lorient on 11 June 1770, completing his journey on foot to the menagerie at Versailles (Rookmaaker 1983d, 1998: 95; Pequignot 2013). He was kept at the famous royal menagerie at Versailles (fig. 5.17). Having survived the French Revolution of 1789, he is said to have drowned in his pool on 23 September 1793. This was one of the first male rhinos on public display in Europe. This male was eclectically visualised by the Italian traveller Francesco Bettini (1737–1815) at the time of sexual discharge (Hays 2002: 96) (fig. 5.18).

After death he was dissected by the anatomist Felix Vicq d'Azyr (1748–1794), when Pierre Joseph Redouté (1759–1840) executed 36 accurate drawings of anatomical parts (Saban 1983).These included a sketch of the penis (fig. 5.19). The skeleton and hide were added to the collection of the Muséum d'Histoire Naturelle, exhibited for a long time with a 80 cm long horn belonging to an African white rhino (fig. 5.20), now restored to a more natural size but not from the animal itself.

Kolkata (Calcutta), West Bengal – Dutch Trading Post at Hooghly

*1. F 1738 to 1740-11-30 (to The Netherlands, died 1758-04-14)

Clara, or the Dutch rhinoceros, which was exhibited throughout Europe between 1741 and 1758, has become famous in recent literature and children's books. Before her arrival in Holland she had been kept as a pet in India, and her early history has to be pieced together from just a few clues found on the engravings or broadsheets sold during her travels, newspaper reports and a few books. The first steps towards unraveling her story were taken by Rookmaaker (1973, 1978a).[1] I take some pride in having first rediscovered the significance of this rhino in the modern era, as part of the research of my first academic paper published on 14 June 1973 while I was just finishing my secondary school education at the Vossius Gymnasium of Amsterdam.

Clara, the often used name of this female rhino, first appeared in print in 1748. Once in the handwritten text on a drawing by Anton Clemens Lünenschloss (1678–1763) made in the German town of Würzburg on 3 August 1748 where she is referred to as "Jungfer Clara" or Miss Clara (Brod 1958; Neubert 2007). Again, the same name spelled "Clar" is found in the cartoon text spoken by a Dutch sailor proposing a toast to her "Clar als gij belieft een mahl te

---

1 The story of Clara, the Dutch Rhinoceros, was first highlighted by Rookmaaker (1973) and Clarke (1974), and soon followed by a series of ever more detailed accounts of her history and iconography: Rookmaaker 1998, 2011b; Rookmaaker & Monson 2000; Clarke 1986; Verhey 1992; Faust 2003; Ridley 2004; Doornbos 2006; Wenley 2021; Ridley 2022; van der Ham 2022.

FIGURE 5.21
"Clar, als gij belieft een mahl te drincken" (Clar, what about a sip?). Small size broadsheet of 1747
WITH PERMISSION © JAGIELLONIAN LIBRARY, KRAKOW, POLAND: LIBRI PICTURATI A.38
MISCELLANEA CLEYERI, P. 51

FIGURE 5.22
Portrait of Jan Albert Sichterman (1692–1764), the owner of Clara, painted by Philip van Dijk (1683–1753) in 1745. He is depicted together with his youngest son Jan Albert (1733–1750) on the left and Gerrit Jan (1725–1796) on the right. Oil on canvas, 148.5 × 117 cm
WITH PERMISSION © GRONINGER MUSEUM: NO. 1993.02.12, PHOTO BY JOHN STOEL

drincken" (Clar would you like a sip?) on a small broadsheet (fig. 5.21; Faust 1976; Rookmaaker & Monson 2000, fig. 5).

The animal was born on 19 June 1738 in Assam. Her mother was shot using bow and arrows when she was still very young, after which she was caught in snares. She was then obtained by Jan Albert Sichterman (1692–1764), the Director of the Dutch Factory at Hooghly (Chinsurah),

just north of Kolkata (fig. 5.22). At first he kept her as a pet in the house (fig. 5.23), but when she grew too big, Sichterman arranged with the Dutch Captain Douwe Mout to take her to Holland. She was transported on board of the *Knappenhof*, which left Kolkata on 30 November 1740 and arrived in Holland on 20 July 1741 (van der Ham 2022: 46–53; fig. 5.24). Clara was then exhibited in Holland from August 1741 onwards, later touring in many European

FIGURE 5.23   Postcard produced by Sigmund Lebel (1851–1918) of
Vienna, printed by Philipp and Kramer (undated,
around 1900). The scene is not annotated but might
be reminiscent of Sichterman petting Clara when she
was still a baby
COLLECTION KEES ROOKMAAKER

countries. Her image depicted on a broadsheet sold by her owner was seen by the public and scientists across Europe (fig. 38.1), challenged by only few alternatives (figs. 5.25, 5.26), like one by Elias Baeck produced in Vienna in 1746 or another engraved at the end of her tour in 1754 by the German cartographer Johann Friedrich Endersch in the present Poland (Chodynski 2020). She allegedly died in London on 14 April 1758.

Many elements of the early history of this animal are probably less certain than the repetition in the literature would make us believe. The place of capture was stated in the broadsheets sold by Douwe Mout as "Asem" in the dominions of the "Great Mogol." Although Assam was never under direct Mughal dominion, it is still such an iconic rhino state that it is quite possible that a rhino was obtained there in the 1730s. It was certainly common practice, then as for centuries afterwards, to kill the mother rhino to obtain the young one. Clara's age is not very clear because the sources tend to repeat figures from previous years, largely based on uncertain estimates which Douwe Mout heard in India. The literature at the time often said that she was 3 years old on arrival in Holland, hence her birth in 1738. However, the earliest Dutch newspaper reports of 1741 stated that she was 2 years old (hence birth in 1739), while at the time of her death in 1758 she would have been 21 years old (hence birth in 1737). While the exact date will always remain unknown, a birthday on 19 June 1738 is a useful compromise (Rookmaaker & Monson 2000: 38). Future historians will wonder about the significance of the date, and may then recall that it is linked to my dear wife Sandra (Sandy) Mary Anthea, b. Lawry (1947-06-19 to 2016-09-17).

There are no depictions of Clara while she was in Bengal. After arrival in Europe, her head was sketched

FIGURE 5.24
Scene showing the *Knappenhof* ready to transport
Clara from Hooghly to Amsterdam in 1740. Detail of
broadsheet issued for him during his European tour with
Clara in 1747
WITH PERMISSION © RIJKSMUSEUM, AMSTERDAM:
FM 3786

Eigentliche und accurate Vorstellung
Des den 30. Octobr. Anno 1746. in der Kayserl. Residenz - Stadt Wien um 11. Uhr Vormittags auf
einem mit 8. Pferden bespannten Wagen, unter Begleitung 8. Cuirassiers, neu angekommenen Asiatischen Wunder-Thiers
Rhinoceros oder Nasen-Horn genannt, so in der Provinz Asem, unter dem Gebiet des Groß-Moguls gelegen,
gefangen worden.

FIGURE 5.25
The Dutch Rhinoceros Clara in Vienna in
October 1746 escorted by a court servant. Detail
of a broadsheet signed by Elias Baeck (1679–1747).
Size 39.7 × 30.5 cm
WITH PERMISSION © NÜRNBERG,
GERMANISCHES NATIONALMUSEUM,
GRAPHISCHE SAMMLUNG: INVENTAR-NR.
ST.N.10512, KAPSEL-NR. 1338

FIGURE 5.26
"Accurate Abbildung des Rhinoceros welcher
lebendig in Elbing Anno 1754 dn. 27 Novembr.
gesehen worden" (Accurate depiction of the
Rhinoceros which was seen alive in Elbing on
27 November 1754). Broadsheet issued by Johann
Friedrich Endersch (1705–1769), to show that
Clara's appearance had changed much during
the last years. He saw her in Elbing, now Elblag in
northern Poland
WITH PERMISSION © ARCHIWUM PAŃSTWOWE
W GDAŃSKU: APG 492/457, P. 121

by the visiting Swedish librarian Magnus von Celse (1709–1784) during the fair in Amsterdam in September 1741 (Broberg 2006; van der Ham 2022: 59, fig. 14). In 1742 she was sketched in Leiden by Jan Wandelaar (1690–1759), who was working on the grand anatomical atlas eventually published in 1747 by Bernard Siegfried Albinus (1697–1770), incorporating the animal as a front view and back view in the background of two plates (4.7, fig. 4.6). Clara was also sketched by the young Petrus Camper (1722–1789), who was an accomplished and prolific artist besides his illustrious career as a scientist. Three drawings (black and white chalk on blue-grey paper) were first rediscovered among his papers and illustrated by Rookmaaker (1978a), showing an oblique front view, two views of the head, and a full side view (also figured in colour by Verhey 1992:

31–34 and van der Ham 2022, figs. 15, 16, 17). An oil painting by Camper remains lost, after it was listed in the legacy of Laurentius Theodorus Gronovius (1730–1777): "no. 51. Rhinoceros or two depictions of the animal painted in oil on a canvas in a black frame, 12 inches high and 10 inches wide [31 × 26 cm]" (Gronovius 1778: 42). Two of these drawings include the date 1748 (figs. 5.27, 5.28), which is clearly repeated ("Rhinoceros, drawn and sculptured in the year 1748 in Leiden") and linked with the history of Clara in Camper's published lecture on the rhinoceros (Camper 1782: 139, 147). This date 1748 doesn't fit Clara's known history, neither do these drawings depict a full-grown rhinoceros. Camper's side-view with a scale along all sides seems to indicate an animal with length of about 280 cm and height of 140 cm. This is similar to the

FIGURE 5.27   Clara from the front, sketched by Petrus Camper
(1722–1789) in Leiden at the end of 1741 or start of
1742, despite the inscription "P. Camper f. 1748."
Size 21.7 × 16.4 cm
WITH PERMISSION © ALLARD PIERSON,
UNIVERSITY OF AMSTERDAM: PETRUS CAMPER
COLLECTION, PORTEFEUILLE A X

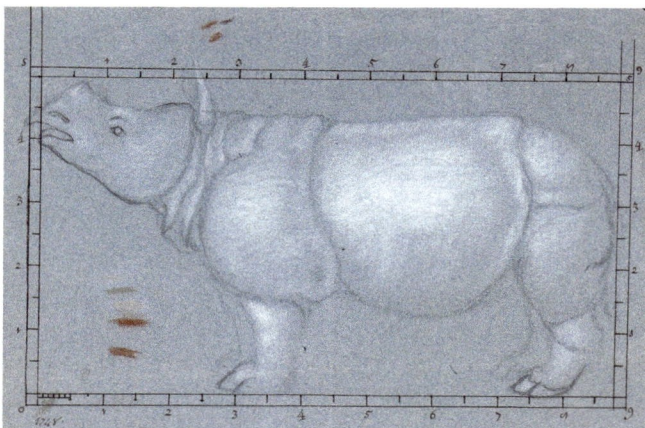

FIGURE 5.28   Side-view with measuring scales of Clara, the
Dutch rhinoceros, sketched by Petrus Camper in
Leiden in 1741 or 1742, despite the inscription "1748."
Size 28.2 × 19.9 cm
WITH PERMISSION © ALLARD PIERSON,
UNIVERSITY OF AMSTERDAM: PETRUS CAMPER
COLLECTION, PORTEFEUILLE A X

size of a male calf in Basel Zoo which at an age of 3 years
and 5 months measured 287 cm long and 131 cm high, with
a horn length of 16 cm (Lang 1961: 407). This shows that
Clara's assumed birth in the middle of 1738 and a proposed
date of Camper's drawings in late 1741 are the most likely
scenario.

In India, Clara was first kept as a pet by the Director
of the Dutch Factory located in Hooghly, now known as
Chinsurah, on the right bank of the Hooghly River. Jan
Albert Sichterman came to Bengal in 1725 and worked his
way through the ranks until he reached the top position
on 18 June 1734, which he continued until he left the country in 1744 (Feith 1915; Kühne 1995) (fig. 5.22). He married
Sibylla Volker Sadelijn (1699–1781) in 1721 and they had
4 sons and 4 daughters, who went for their education to
Holland. The life of Sichterman was the subject of an exhibition in the museum of his hometown Groningen (Jörg
et al. 2014). Sichterman probably received Clara as a political donation from Nawab Shuija-ud-din-Khan (1670–1739),
who represented the Mughal empire in Bengal. This was a
common friendly gesture in the course of trade negotiations and was usually reciprocated with other presents. As
Sichterman is thought to have received such a gift in 1738,
Clara's donation is ascribed to the Nawab first by Kühne
(1995: 44), unfortunately in the absence of the precise
circumstances. The rhino might have been just a month
old when she came to the household of Sichterman, small
enough to be allowed to walk in "the dining room when
ladies and gentlemen were eating, as a curiosity" (text of
1747 broadsheet). Such idyllic scenes could not last and
obviously another solution had to be found to keep this
ever-growing animal.

Sichterman negotiated with a visiting captain at the
end of 1740, suggesting that he could take the young rhino
to Holland, as she was still small enough to live on deck.
Douwe Mout (1705–1775), also called Douwe Jansz. Mout
van der Meer, was a Dutch captain born in Amsterdam,
who made regular journeys to the East for the Dutch East
India Co. (van der Ham 2022). On his third journey, he was
second in command as First Mate (Opperstuurman) on
the *Phoenix*, which left Texel 12 October 1739 and reached
Batavia 27 April 1740. Douwe Mout returned on the
Horssen first calling in Bengal, where he was promoted to
Captain (Schipper) on 18 September 1740 by Sichterman
due to the death of the original captain Jan Luders during the journey. While in Bengal, he was transferred to
another ship called the *Knappenhof* which was owned by
the VOC-Chamber of Rotterdam. The *Knappenhof* under
command of Douwe Mout and carrying the rhinoceros
started its return journey on 30 November 1740 in a small
fleet of 4 ships. After a transit stop at the Cape of Good

FIGURE 5.29   Clara in India, drawn by Sarah Hewitt for "The true story of the rhino who became a superstar!" first published in London in 2016
FROM *CLARA: THE TRUE STORY OF CLARA THE RHINO* © SARAH HEWITT 2016 WWW.BOOKSBYSARAH.CO.UK 'BOOKS BY SARAH PUBLISHING'

Hope from 2 to 20 March 1741, it reached the anchorage of Goeree (in Zeeland south of Rotterdam) on 20 July 1741.

Perhaps unexpectedly, Clara survived the journey very well, and soon after arrival in Holland she was taken to fairs in Amsterdam and Leiden to be shown as a complete novelty to the Dutch public. Douwe Mout became animal showman and as a true entrepreneur was able to take Clara around Europe in a waggon drawn by horses, showing her in an ever-increasing list of towns and countries, listed by Rookmaaker (1998: 61–67), and with new finds by Faust (2003: 54–59) and Van der Ham (2022: 183–185).

From humble beginnings, the story of her tour has become an inspiration to artists and authors in a variety of disciplines, and a welcome subject in stories for young children (fig. 5.29). She has become one of the most famous rhinos in the western world.

Kolkata (Calcutta) – British Governor of Bengal

1. 1765-12-18
The Governor Robert Clive, 1st Baron Clive (1725–1774) staged a fight between a rhino and an elephant, witnessed by the Dutch Governor George Lodewijk Vernet (1711–1775) and the French Governor Jean Law de Lauriston (1719–1797) in Fort William (Champion 1765; Curzon 1925: 257). Colonel Alexander Champion (d. 1823) noted that as the rhino "could not be moved there was no battle" (S.C. Ghosh 1970: 148).

2. 1791
The rhino offered by the King of Nepal was supposed to be sent to Kolkata in November 1791 (see Kathmandu). If it arrived in Kolkata still alive, it might have been kept by the Governor-General Charles Cornwallis (1738–1805).

3. 1793
The ruler of Gwalior Mahadaji Shinde (1730–1794) sent a letter to Governor John Shore (1751–1834) informing that he had sent him a rhinoceros (National Archives of India 1949, vol. 10: 247).

Kolkata (Calcutta) – Barrackpore Park (1800–1876)

When Richard Colley Wellesley, 1st Marquess Wellesley (1760–1842) was Governor of the Presidency of Fort William 1798–1805, he constructed a new mansion in Barrackpore, just north of the centre of Kolkata on the banks of the Hooghly River. He also started to keep animals in the surrounding park, which could be seen as the first zoological garden in India. Bishop Reginald Heber (1783–1826) visiting in 1823 never mentioned the presence of a rhino (Heber 1828). A single rhino is mentioned in 1827 (no. 2), and from 1837 to 1857 there must have been at least a pair (nos. 2 and 4). Sometimes animals in transit were deposited in the park while shipment was organized. Animals were no longer exhibited in Barrackpore Park after 1876 in favour of the new public zoo in Alipore.

The Curator of the Asiatic Society of Bengal, Edward Blyth (1810–1873) wrote in 1862 about a female rhino skull in the Calcutta Medical College, which was "one of a pair that lived about 45 years in captivity in Barrackpore Park. I have repeatedly seen the pair when alive, many years ago" (Blyth 1862a155). Blyth arrived in Calcutta in 1841, hence this is the first possible date of his observations.

There are no records of the first quarter of the century (1800–1825) which might tally with this remark. It is most likely that he saw the pair here listed as no. 2 (from 1827) and no. 4 (from 1837), but the longevity of even one of these as stated by Blyth cannot be verified. This is unfortunate because a stay of 45 years exceeds any other record in captivity (Rookmaaker 1998: 22). The female skeleton from the Medical College was transferred to the Indian Museum in 1879 and listed by Sclater (1891: 202, no. b).

1. 1808

Two unidentified hunters captured a rhino in the Rajmahal Hills in 1808 (29.1). It became tame after only two days, and was then taken to Kolkata, where it was expected in April (Sandeman 1868, vol. 4: 196, after *Calcutta Gazette* 1808-04-07). There is no record of its arrival.

2. 1827-12-12 to 1857

There are regular reports of rhinos in Barrackpore, a single animal from 1827, and a pair from 1837, until 1857. These instances are here combined as two animals (no. 2, no. 4), even though there is no certainty that the same animals were actually present and were not replaced after transfer or death.

1827-12-12. A pet rhino of Lord Amherst (1773–1857) attacked Jemmy G. (Shikarree 1830: 253).

1829. At the time of Governor Lord William Bentinck (1774–1839), the French naturalist Victor Jacquemont (1801–1832) saw a rhino in Barrackpore Park, kept chained on one foot to a tree on the edge of a meadow. It had been captured in the mountains on the other side of the Ganges, probably northern Bihar (Jacquemont 1841, vol. 1: 169; Geoffroy St. Hilaire 1842: 69).

1833. The English travel writer and poet Emma Roberts (1791–1840) visited Barrackpore where she saw one rhino content in a wallow (Roberts 1835, vol. 1: 129).

1837-02-11 The English novelist Emily Eden (1797–1869) heard that one of the rhinos in Barrackpore had a tendency to attack elderly gentlemen (Eleanor Eden 1872, vol. 1: 303). She expanded about this dangerous habit on 22 May 1837, when she also made it clear that there was a pair of rhinos in the park (Eden 1872, vol. 2: 23). Her brother, Governor Lord Auckland (1784–1849), she wrote, "has got an odd twist upon the subject of the rhinoceroses, and connives at their fence not being mended, so that they may roam about the park, whereby a respectable elderly gentleman, given to dining out at the cantonments, has been twice nearly frightened into fits. The story, now twice repeated, of the two beasts roaring as they pursue his buggy is very moving to hear; and his Excellency smiles complacently and says, 'Yes, they are fine beasts and not the least vicious.'"

1838. Adolphe Delessert (1809–1869) sees "des rhinocéros" (plural) in Barrackpore (Delessert 1843: 109).

1841-10-19. A male rhino broke loose and killed a sepoy of the 28th Regiment Native Infantry (*Reading Mercury* 1842-01-15, also without date in Grant 1860: 18).

1851. Frederick Fiebig took a photo of the rhino pool, showing one rhino (fig. 5.30). Symons (1935: 128) refers to a Rhino Tank, a name still used after the closure of the Barrackpore menagerie.

1851-01-12. Francis Egerton, Earl of Ellesmere (1800–1857) in the company of John Hunter Littler (1783–1856) sees 2 rhinos (F. Egerton 1852, vol. 1: 150).

1857. Colesworthey Grant (1813–1880) made a sketch of the rhino enclosure in Barrackpore showing 2 rhinos (Grant 1860: 18) (fig. 5.31).

3. M 1833 to 1834 (to Atkins in United Kingdom)

A 3-year old male rhino received in June 1834 by the English menagerie owner Thomas Atkins (1764–1848) in Liverpool had lived for some time in Barrackpore prior to shipment (4.12).

4. 1837-05-22 to 1857

This is the presumed second specimen referred to in the pair together with no. 2 of 1827. A pair was first mentioned by Emily Eden in 1837. Delessert saw more than one rhino in 1838. Francis Egerton saw a pair in 1851. The sketch of 1857 by Colesworthey Grant shows 2 specimens.

5. M 1838-04

A young male rhino was sent by the last Ahom King of Assam, Purander Singh (1807–1846), to Governor Lord Auckland (Lahiri 1954: 199).

6. F 1839 (to Jakarta, Indonesia)

Among papers relating to the work of the Natuurkundige Commissie voor Nederlands-Indië (Commission for Nature in the Dutch East Indies) there is a drawing of a young rhino (fig. 5.32). The written caption explains that it was a female "brought from Calcutta in 1839." This is most likely to refer to an arrival in Jakarta (Batavia), Indonesia, where the Commission was based since 1820.

7. M 1856-07 (from Lucknow) to 1862

An animal obtained in Lucknow at the sale of King of Oudh (6.11). The rhino was transported to Kolkata for Edward Blyth, but apparently transport to England delayed and the animal died in Kolkata. This and the next rhino might have been kept in Barrackpore.

8. F 1856-07 (from Lucknow) to 1860

From sale of King of Oudh, Lucknow, with previous male (6.11).

*9. M 1864-03-11 (to United Kingdom)

In July 1864, three *R. unicornis* arrived in England, one pair destined for the London Zoo, and a male for the Dublin

FIGURE 5.30
Photograph of rhino pool in Barrackpore Park in 1850 taken by Frederick Fiebig
WITH PERMISSION © BRITISH LIBRARY, FIEBIG COLLECTION

FIGURE 5.31
Sketch of rhino enclosure in Barrackpore Park by Colesworthey Grant (1813–1880) in 1857
GRANT, *RURAL LIFE IN BENGAL*, 1860, P. 18

FIGURE 5.32
"Rhinocéros vivant fem. âge moyen, apporté vivant de Calcutta en 1839" (Female rhinoceros alive, middle age, brought alive from Calcutta in 1839). Anonymous drawing preserved among papers of the Natuurkundige Commissie voor Nederlands-Indië
NATURALIS, LEIDEN, S2 I NC MAMMALIA 341-18, HTTPS://DH.BRILL.COM/NCO/VIEW/NCO _NNM001000392/MAKINGSENSE

Zoo. At least the adult male taken to London had been kept in transit for a while in Barrackpore Park prior to shipment (letter by Arthur Grote to ZSL, 1864-03-11).

Kolkata (Calcutta) – Maharaja Buddynath Roy

References to a rhino in a menagerie in either Chitpore or Cossipore in the first half of the 19th century both appear to indicate the mansion of Raja Buddynath (Baidya) Roy (d. 1860), third son of Maharaja Sukhmoy Roy (Chatterji 1929: 18). When passing through Chitpoor on 1 March 1824, Bishop Reginald Heber mentions that Baboo Budinâth Roy "has a menagerie of animals and birds only inferior to that of Barrackpoor" (Heber 1828, vol. 1: 90). The Zoological Society of London elected both the Raja and his son Cowar Roy Rajkissen on 16 March 1855 as members in recognition of their interest in zoology shown by maintaining a large menagerie (Chatterji 1929: 61).

*1. M 1829 (to U.S.A.: Flatfoots)

1829. The French naturalist Christoph-Augustin Lamare-Picquot (1785–1873), who shot two rhinos in the Sundarbans in 1829, stated that he had seen a living rhino many times at the house of the Nabab of Chittepour (Lamare-Picquot 1835: 61). Although he did not name the Nabab nor provided exact dates, the year 1829 and the menagerie of Raja Buddynath Roy are most likely.

1830. The first rhino to be exhibited in an American circus was a male arriving in Boston, Mass. on 9 May 1830 on the ship *Mary* (Reynolds 1968: 10; Thayer 1975; Rookmaaker 1998: 107). Advertised as the "Greatest Natural Living Curiosity Ever Exhibited in America", at first the animal was said to have been purchased in January 1830 from a "Rajah, or native prince of Calcutta" (eg. *Phenix Gazette* 1830-12-28), later changed in the media to a "Rajah of Benares" (eg. Anon. 1830: 620). It would have been captured near the Himalayan mountains in May 1829 aged 3 months.

In an archive associated with the businessman Marmaduke Burrough (1797–1844), who served as the American Consul in Kolkata 1828–1830 (William L. Clements Library, Ann Arbor, Michigan), there are two letters referring to this animal (38.13). Apparently Burrough had approached Buddynath Roy about the rhino, as seen from Roy's reply dated Calcutta 9 September 1829 that the animal would cost Rs 9500 only if a lion and lioness were bought at the same time. Sam W. Archer, one of Burrough's friends in Kolkata, wrote on 16 January 1830 that the rhino had been purchased by another party for Rs 7000 and was to be shipped on the *Mary* to America. When it arrived in Boston, ownership was claimed by the Flatfoots Association, a consortium of circus entrepreneurs, and

first assigned to the menagerie of June, Titus and Angevin. This rhino died in April 1835.

Having missed his chance to obtain this rhino, Burrough came into contact with Alexander Davidson, the Magistrate of Goalpara in Assam, probably when he visited the area. Davidson sent 2 rhinos to Kolkata, one direct to Burrough in March 1830 which was taken to Philadelphia arriving on 15 October 1830 (the second rhino in America) and another to a friend in December 1830 arriving in Boston on 10 May 1831 (the third rhino in America) (38.13, fig. 38.5).

2. 1837

Lady Honoria Marshall Lawrence (1808–1854) saw a pet rhino in the house of a Raja near Cossipore (Lawrence 1980: 45; Nair 1989: 613). Cossipore is adjacent to Chitpore, hence she might have meant the house of Raja Buddynath Roy.

Kolkata (Calcutta) – Indian Rajah in Alipore

1. 1858-05

The Swedish painter Egron Sellif Lundgren (1815–1875) visited a Rajah in Alipore, who had a menagerie showing a rhino with one horn (Lundgren 1872: 77).

Kolkata (Calcutta) – Mansion 'Seven Tanks' in Cossipore

Seven Tanks or Satpukur, now the Cossipore Club, is a mansion once owned by Ruplal Mullick and then by his son Sham Charan (or Shyamacharan) Mullick, the brother in law of Raja Rajendra Mullick.

1. 1858 or 1859 to 1870s

A rhino was seen at Seven Tanks during a visit by the pseudonymous Cadwalladar Cummerbund in the late 1850s (Cummerbund 1860: 164). On an unknown date (in the 1870s?), when Seven Tanks was owned by Shyamacharan Mullick, Swami Akhandananda (1858–1937) visited and saw a rhino in the garden next to a small pool (Akhandananda 1979).

Kolkata (Calcutta) – Burdwan (Bardhaman)

Maharaja Mahatab Chand Bahadur (1820–1879) had a residence in Burdwan, about 100 km NW of Kolkata, where there is now still a small zoological garden.

1. 1859 to 1861 †

Charles Buckland of the Bengal Civil Service had written (date unclear) to his cousin, Francis Trevelyan Buckland (1826–1880) about the Rajah in Burdwan "who keeps a really good menagerie. He has two rhinoceroses, who live in a large walled enclosure, in the centre of which is a reservoir of water" (F. Buckland 1861: 227, first printing of 27 June 1860; Hobhouse 1916: 203). In a "new edition" of the *Curiosities of Natural History* published in 1861, Buckland

(1861: 354) added in an addendum (therefore dating from the period after June 1860) that one of the rhinos had died.

2. 1859, 1863

The second rhino mentioned by Buckland (previous entry). On 21 February 1863, Russell Jeffrey (1807–1867), visiting missionary in Burdwan, saw one rhino (Jeffrey 1863: 114).

Kolkata (Calcutta) – Raja Rajendra Mullick

Raja Rajendra Mullick (1819–1887), also spelled Mallick or Mallik, built the Marble Palace in Muktaram Babu Street, Joresanko, adjacent to Chitpur road, completed in 1840. From 1854 onwards he kept animals in the gardens. The records of rhinos could refer to other houses owned by the Mullick family (see Seven Tanks), but the accession of a skull in July 1871 was donated by "Rai Rajendro Mullick, Bahadoor" (Indian Museum 1872).

*1. F 1867

There is a skull and mounted hide of an adult female *R. unicornis* in the Staatliche Museum für Naturkunde Stuttgart (no. 1218) obtained in 1867. The label states that it came from the Menagerie of Babu [illegible] Mullik, Calcutta (Rookmaaker 1998: 57). The museum bought the specimen from the taxidermy firm run by Edward Gerrard Jr. (1832–1927) in London. Although it must be one of their earlier transactions, Gerrard may have had an agent in Kolkata at the time, because other rhino specimens from India were registered at the Natural History Museum in London in 1870 (*R. unicornis*) and 1876 (*R. sondaicus*).

*2. F 1871

In the catalogue of the Indian Museum, William Sclater (1891: 202, no. l) listed a skull of *R. unicornis* from "Rajah R. Mullick, 1871." The accession was noted in July 1871 as a skull of *R. sondaicus* (Indian Museum 1872), which is presumably a wrong identification. The skull is now no. 19240 and still on display in the museum (Groves & Chakraborty 1983: 254).

Kolkata (Calcutta) – Zoological Gardens, Alipore

The Zoological Gardens in the suburbs of Alipore were inaugurated on 1 January 1876 by King Edward VII, then Prince of Wales. A rhino pit or enclosure was constructed but still empty later that year (B. 1876: 452). Besides the 10 specimens of *R. unicornis* listed here (2 on temporary exhibit in transit), there were also 3 *R. sondaicus* and 5 *D. sumatrensis*.

*1. F 1877-04-07 to 1880-05 † (from Dumraon)

Female rhino presented by Maharaja Maheshwar Bakhsh Singh (1803–1881), 9th Maharaja of Dumraon, Bihar (ruled 1844–1881). In the care of dealer William Rutledge, she walked from Dumraon to Kolkata. Died of tetanus (*Times of India* 1877-04-12; Sanyal 1892: 131; C. Buckland 1890a: 230).

*2. F 1887-12 (from Kolkata: Wajid Ali)

A female rhino was bought at the auction of the collection of Wajid Ali Shah (1822–1887), the exiled King of Oudh, in December 1887. The length of stay in the Zoo is not clear (6.12).

*3. F 1905-02 to 1932-01 †

Female rhino, donated by the Government of Nepal. She was transported from Raxaul to Kolkata by railway with the assistance of Charles Withers Ravenshaw (1851–1935), Resident to Nepal (Sanyal 1905). She was the mother of the calf born in the Zoo in 1925. The American naturalist Charles William Beebe (1877–1962) wrote a report of a visit to the zoo probably in early 1910, as the journal was issued in September (Beebe 1910). He took a photograph of the rhino enclosure measuring 2 acres and mentioned a pair of rhinos (nos. 3, 4) (fig. 5.33).

*4. M 1909-12-09 to 1930-06 †

Male, purchased together with another male (no. 5) for Rs. 15,000 from Nepal. The arrival in December 1910 mentioned by the Superintendent Bijay Krishna Basu (1910) is an error, which is confirmed by a notice of the addition of a pair of rhinos in the *Englishman's Overland Mail* (1909-12-09). Nuttall (1917: 632) published a photograph of one of the *R. unicornis* in the Calcutta Zoo (fig. 5.34). Fisher (1912) saw a (one?) "big one-horned rhinoceros" from Nepal.

5. M 1909-12-09 to (death unknown)

Male rhino purchased with previous male from Nepal (Basu 1910). This is not the animal sent to London Zoo in April 1912 (no. 6) because the arrival of that animal was recorded separately. Therefore it probably died soon, and remained unrecorded in the annual reports of the zoo.

*6. M 1912-01-25 to 1912-04-01 (Exported to London Zoo, UK)

Male rhino donated by Maharaja Chandra Shumsher of Nepal to King George V on 24 December 1911 (25.3). He lived in Alipore Zoo from 25 January to 31 March 1912, shipped on 1 April 1912 on the *ss Afghanistan*, reaching London on 21 May. He died in London Zoo on 27 April 1927.

*7. M 1925-10-09 (Birth)

Male rhino born dead, to the female of 1905 (no. 3) and male of 1909 (no. 4) (Ali 1927b).

*8. M 1932-03 to 1970-08-30 †

Male rhino donated by Marie Adelaide Freeman-Thomas, Marchioness of Willingdon (1875–1960), probably captured in Nepal with no. 9.

FIGURE 5.33
Enclosure for the rhinoceros, photographed by Charles
William Beebe (1877–1962) on a visit to the Alipore
Zoo in 1910. Two rhinos are visible
BEEBE, *NEW YORK ZOOLOGICAL SOCIETY BULLETIN*,
SEPTEMBER 1910, P. 693

FIGURE 5.34
"Indian rhinoceros, Zoological Gardens, Calcutta."
One of three specimens from Nepal in Alipore
Zoo 1909–1911
W.M. NUTTALL IN PLAYNE, *BENGAL AND ASSAM,
BEHAR AND ORISSA*, 1917, P. 632

*9. F 1932-10 to 1965-12-10 †
Female rhino donated by the Governor-General Freeman
Freeman-Thomas, first Marquess of Willingdon (1866–
1944), captured in Nepal during a visit of his son Lord
Ratendone (24.25).
*10. M 1938 to 1939-06 (Exported to Washington DC,
USA)
Male rhino captured in Kaziranga, temporarily exhibited until transfer to Washington DC, where it lived in
the National Zoological Gardens 1939-05-26 to 1959-01-09
known as 'Gunda' (Rookmaaker 1989: 112).

Kolkata (Calcutta) – William Rutledge (Dealer)
William Rutledge (1832–1905) was a dealer in birds and
wild animals at Entally in Kolkata (48.2).
1. around 1875
In a collection of watercolours and photographs related
to the work of T.C. Jerdon offered for sale at the Arader
Galleries in 2018 (48.2), there are photographs signed by

Captain William George Stretton (1836–1899). Because
Stretton first appeared in Kolkata references in 1875, these
items are unlikely to be earlier. There are three photographs of *R. unicornis*. One shows a young adult animal
with a chain around the neck attached to a tree trunk,
with the head facing a man in a suit with a topi, probably William Rutledge, signed by Stretton with the number
26/8 (fig. 5.35). The other two photographs are identical
and pasted to a single sheet, both showing a *R. unicornis*
from behind, with on the left the same person holding a
(measuring) stick, one of which is signed Stretton 70/2
(fig. 5.36). Probably these photographs were all taken at
the same time in the compound of Rutledge in Entally.

Kolkata (Calcutta) – King Wajid Ali Shah of Oudh
*1. F [unknown date] to 1887-12 (to Kolkata, Alipore Zoo)
While King Wajid Ali Shah of Oudh (1882–1887) was living
in Garden Reach, Metiabruz during his exile after 1856,

FIGURE 5.35
Young adult *R. unicornis*
photographed by Captain
William George Stretton (with
number 26/8), probably at the
premises of William Rutledge
at Entally, Calcutta, in 1875 or
afterwards. Size 14 × 10 cm.
From a miscellaneous collection
associated with T.C. Jerdon offered
for sale by the Arader Galleries
in 2018
WITH PERMISSION © ARADER
GALLERIES, PHILADELPHIA

FIGURE 5.36   Photograph of a young adult *R. unicornis* by
William G. Stretton (with number 70/2), probably at
the premises of William Rutledge at Entally, Calcutta.
Size 14 × 10 cm. From a collection sold in 2018 as
fig. 5.35
WITH PERMISSION © ARADER GALLERIES,
PHILADELPHIA

he had maintained a menagerie. When the animals were auctioned after his death, one female *R. unicornis* was obtained by the Zoological Gardens in Alipore (6.13).

Kolkata (Calcutta) – William Jamrach (Dealer)

Rhino exports 1855–1875 (table 5.13). These animals are not included in the totals as they did not stay long in India before shipment.

The animal dealership headed by Johann Christian Carl (Charles) Jamrach (1815–1891) based in London was responsible for the importation of many rhinos in Europe and America from the time of its establishment in 1840 (fig. 5.37). His son William Jamrach (1843–1891) often traveled to India to oversee operations, but there is no indication that he actually maintained an office. In Kolkata he probably worked together with William Rutledge who was established in Entally near the centre of the town. In Europe, Jamrach was associated with Carl Hagenbeck (1844–1913) based in Hamburg, Germany (in Sankt Pauli, first at the Spielbudenplatz and after 1874 at Neuer Pferdemarkt), and with Charles William Rice (1841–1879) based both in Hamburg and London (Rookmaaker 2014b).

In 1875, the German artist Heinrich Leutemann (1875) published a candid interview with William Jamrach, which can be summarized here. At that time Jamrach recognized that most larger zoological gardens in Europe had been supplied with at least one rhino resulting in reduced demand. Jamrach had already made twenty journeys to India, and probably would have added another twenty, as long as prospective owners were willing to pay a higher

FIGURE 5.37   "The Prince's menagerie on board the Serapis", giving
a view of animal transports in the 19th century
ILLUSTRATED LONDON NEWS, SATURDAY
13 MAY 1876, P. 61

and had to be brought to Kolkata, which was usually accomplished on foot, taking many months on the road, traveling by night and resting by day. Most of the young animals could be herded easily like sheep, but larger animals added much risk to the operation and several of these were lost on the onward sea voyage, which affected profits. The pair destined for Berlin in 1872, being some two years old and therefore getting too large, had to leave Kolkata in the rainy season. The ship could not approach the shore, hence Jamrach made a boat-bridge using 46 vessels strung together, anchored to the ground, and a walkway of planks 6 feet wide – it took only 3 minutes for the animals to walk from the shore to the ship. On other occasions, heavy weather or sickness resulted in the loss of the rhinos during the long voyage.

Despite these problems, Jamrach was quite successful in bringing the rhinos to Europe and America. He must have had agents in the field who are never identified by name. He also advertised in local newspapers, like in the *Pioneer* issued in Allahabad (1875-02-18 and 1876-01-08) for the purchase of tigers, elephants and rhinos, in those cases using a hotel in Calcutta as address to reply.

Kora Jahanabad, Uttar Pradesh – Unknown Locality

*1. 1665-12-02

The French diamond merchant Jean Baptiste Tavernier (1605–1689) encountered a rhino when he was traveling from Agra to Patna, close to Gianabad, now Kora Jahanabad, on 2 December 1665 (18.7).

Lucknow, Uttar Pradesh – Menagerie of Nawabs of Oudh

The reports of 19 rhinos kept by the Nawabs of Oudh are explored in chapter 6, summarized as follows (see

price. There were many hurdles to be negotiated, starting with the capture of the young animal, usually preceded by the death of the mother. Most of these animals were caught above the Ganges (meaning Bengal and Assam)

TABLE 5.13   List of specimens of *R. unicornis* exported by the animal dealer Jamrach from India 1850–1899

Sex	Arrival	Death	Disposal: town and facility
M	1855-06		Unknown: Auction Liverpool
F	1855-06		Unknown: Auction Liverpool
F	1856-05-24	1894-10-23	Vienna Schönbrunner Tiergarten
	1858-08-09		London, Jamrach
	1867	1867	Skulls of 2 animals died in transit
	1867	1867	Skulls of 2 animals died in transit
F	1871-09-21	1872-09-19	destination unknown
F	1872-09-19	1896-04-09	Berlin, Zoologischer Garten
M	1872-09-19	1909-10-26	Berlin, Zoologischer Garten
F	1872-04-26	1900-10-25	Cologne, Zoological Gardens
F	1875	1875	London, Rice (dealer)

tables 6.11, 6.12). Larger numbers were simultaneously recorded in 1819 (7 rhinos), 1824 (6 rhinos) and 1841 (12 rhinos). The total of 15 to 20 rhinos in the 1830s mentioned by Knighton (1855) is excluded because it is potentially inflated or fictitious.

1760–1765 Lucknow: Nawab (nd)

1769 Lucknow: Nawab (nd)

1789 Lucknow: Nawab 1802

1790 Exported from Laknaor to London (export)

*1814 Lucknow: Nawab < 1856

*1814 Lucknow: Nawab < 1856

*1814 Lucknow: Nawab < 1856

1819 Lucknow: Nawab 1824

1819 Lucknow: Nawab 1835

*1819 Lucknow: Nawab < 1856

1819 Lucknow: Nawab < 1856

1841 Lucknow: Nawab < 1856

1841 Lucknow: Nawab < 1856

1841 Lucknow: Nawab < 1856

1841 Lucknow: Nawab < 1856

1841 Lucknow: Nawab < 1856

*1841 Lucknow: Nawab 1856 Kolkata: Barrackpore Park (Blyth) 1862

*1841 Lucknow: Nawab 1856 Kolkata: Barrackpore Park (Blyth) 1860

*1852 Allahabad: Rundheer Singh 1852 Lucknow: Nawab 1858.

Lucknow, Uttar Pradesh – Nawab Wazid Ali Shah Zoo

*1. F 1944-02-02 to 1973-04-02 †
From Nepal, named Roxy (Rookmaaker 1998: 90).

Manipur – Emperor Meidingu

1. 1720
The *Ningthaurol* or Royal Chronicles of Manipur record that the Emperor Meidingu Pamheiba (1690–1751) was presented with a rhino caught by Tarao Palli Nagas, a tribe from the Chandel district in S.E. Manipur (Hodson 1912).

Mathura, Uttar Pradesh – Seth Lakhmi Chand

1. 1851-03-08
The merchant Zyn-ool-Addeen had brought a rhino to Agra to be sold to Seth Lakhmi Chand (d. 1866), a millionaire banker in Mathura, and the animal broke loose (*Agra Messenger* 1851-03-08).

Mayong, Assam – Local King

1. 1925 (assumed)
Local people in Mayong (Mayang) near Pabitora Wildlife Sanctuary recollect that during the tenure of "King Rohan Singha", a rhino calf was brought to the palace as a pet, but died from diarrhea after a few months (Talukdar 2000; Talukdar & Barua 2006). The sources are silent about the year, but according to a genealogy of 48 generations of Mayong kings, the fifth previous to the current incumbent was Bakat Singha Rahan, which might point to a date around 1925.

Meghalaya – Charles Richard Cock

1. 1864-03-19
Charles Richard Cock (1838–1879) shot a female rhino at Tickri-killa (Tikrikilla) and captured the young one (37.7). Cock (1865) sold the animal but the destination is unclear.

Mumbai (Bombay), Maharashtra – Aga Khan

1. 1868 (Exported, died in transit)
Aga Khan, Hasan Ali Shah (1804–1881) sent a rhino to Nasir al-Din Shah (1831–1896), King of Persia (Daftary 1990: 513). It is probably the same animal which reportedly drowned in the harbour when it was being loaded on British India Navigation Company's steamer *Java* (*Times of India* 1885-03-17).

Mumbai (Bombay), Maharashtra – Victoria Gardens

1. F 1922-01-10 to 1922-03-02 (Exported to London Zoo, UK)
Female rhino called Bessie donated by Maharaja Chandra Shumsher of Nepal to Edward, Prince of Wales during his visit on 18 December 1921. She was kept in Victoria Gardens from 10 January to 2 March 1922, leaving on the *ss. Perim* and arriving London 7 April 1922, where she lived in the Zoo to 28 April 1926 (25.4).

Murshidabad, West Bengal – Nawab of Bengal

1. 1716 (from Kolkata: Chandernagor)
Heading the Mughal administration of Bengal, the first Nawab was appointed in 1717 and Murshid Quli Khan (1660–1727) held the post until his death. The rhino kept at the French factory at Chandernagor just north of Kolkata was a diplomatic present to the Nawab, possibly as early as 1716. There are no data where the animal was kept or how long it survived.

2. 1757 to early 1770s
1757-05. A young rhino (the height of a donkey) was captured near Sakrigali in the Rajmahal Hills. It was seen by the French traveller Abraham Hyacinthe Anquetil-Duperron (1731–1805) in the first weeks of May 1757 (29.3). He was told that the animal was transported to the Nabab of Bengal, tied with ropes and guarded by 50 men (Anquetil-Duperron 1771: lii–liii).

Period around 1770. The account by the American traveler Bartholomew Burges (1740–1807) of his stay in India probably dates from the late 1760s. When he was at Muxadabag (Murshidabad), he saw a fight between elephants owned by the Nabob, likely Saif ud-Daulah (ruled 1766–1770). He also witnessed fights "of elephants, rhinoceroses, tygers, and wild buffaloes" (Burges 1790: 60), which could be the same occasion described in another part of his book: "a rhinoceros, and an elephant prepared for battle, were brought forth, who on the sight of each other, instantly approaching, begun with the utmost fury to engage. Provoked with his situation, he gave three sudden successive roars, followed by a violent jerk of his whole body backwards, which bringing the rhinoceros in front, with an incredible rapid pull of one of his teeth, he maimed him to such a degree, that unable to continue longer the fight, he turned tail and marched off and left him master of the field, amidst the shouts and acclamations of the crowd" (Burges 1790: 69).

Patna, Bihar – Mr. Young

*1. Period 1750–1780

Mr. Young of Patna (not identified) had a rhino which "occasionally walks the street" (Williamson 1807a: 45; 1807b, vol. 1: 168).

Patna, Bihar – English Factory

1. M 1738 (to London, where died c.1742–1744)

Procured by Humphrey Cole, the Chief of the English Factory in Patna, this 2-year old male rhino left Kolkata on 31 July 1738 and arrived in London on 2 June 1739. It was first exhibited in Eagle Street near Red Lion Square, where it was studied and drawn by the physician James Parsons. A detailed scientific description published in the *Philosophical Transactions of the Royal Society of London* in 1743, accompanied by two engraved plates (fig. 5.38), became the standard view of the rhinoceros for most of the 18th century (Parsons 1743). This animal might have been the one sketched by William Twiddy in Norwich in June 1744, but records after 1742 are unknown and maybe he died in that period. The history is further explored in 28.10.

2. 1763 (to London, probably died in transit)

Henry Vansittart (1732–1770), Governor of Bengal 1759–1764 obtained a rhino in Patna, Bihar in 1763 (28.11). It was forwarded to Kolkata and should have been shipped on the *Drake* in August 1763 as a gift to King George III, but there is no mention that it arrived in London.

Pune (Poona), Maharashtra – Menagerie of the Peshwa

*1. M 1790 to 1792

The menagerie of Peshwa Madhav Rao II (1774–1795) at the foot of Parvati Hill had a rhino donated (date unknown) by Mahadaji Scindia (1730–1794), the ruler of Gwalior 1768–1794. This male rhino was recorded five times between 1790 and 1792. There is no mention of the rhino after 1792.

1790-06-03. Edward Moor (1771–1848), officer of the East India Co., went to visit the Peshwa's menagerie, where he saw a rhino which was "a present from Scindia" (Moor 1794: 366; Parasnis 1921). He mentioned that clay models of all the animals were made at the request of the British Resident in the court of the Peshwa Mahrattas, Sir Charles Warre Malet (1753–1815).

1790–11. A watercolour drawing of the rhino was made in November 1790 by the Indian artist Gangaram Chintaman Tambat (Shaffer 2011: 7, 35; Hobhouse 1986, no. 8; Kincaid 1908: 88; Rookmaaker 1997b; Sardesai 1936: 223; Gode 1945; A. Das 2018b: 90). This representation was commissioned by C.W. Malet and remained in the family until it was offered for sale by Hobhouse in London in 1986. Malet added the following details, including some measurements, in his handwriting on the drawing: "This Drawing and Wax figure of a Rhinoceros belonging to Mudarao Narrian Peshwa was taken from the Life with great fidelity at Poona in November 1790 by Gangaram Chintaman Tambat, the Age of the Rhinoceros was supposed to be about 25 years of the following Dimensions viz. from the Outside of the Tail to the foremost wrinkle of the Neck ft 8 7 ½ ins [262.8 cm]. From the said wrinkle to the Top of the Nose ft 2 3 ½ [69.9 cm]. From the forefoot to the Shoulder Height ft. 5 11 [180 cm]. From the Hind foot to the Top of the Rump ft. 6. 3 [190.5 cm]. From the Root of the Horn nearest the nose to the point 1 [30.5 cm]. It was said by the Keepers that the Horn would grow. Poona 28 Oct. 1790. C.W.M. – This Animal is Retromingent but the genital parts when activated by concupiscence are projected as in the accompanying drawing in which the urinal line is also described. The colour of the animal is excellently conveyed in the wax Figure" (fig. 5.39).

1791-12-16. C.W. Malet wrote to the Governor-General Charles Cornwallis (1738–1805) that the Peshwa wished to reciprocate some donations and intended to send gifts including an elephant, a rhino, a horse, and some antelopes. He asks if it is best to send these by ship to Bombay (Sardesai 1936: 223).

FIGURE 5.38
Two views of the young male rhino sent from Patna in 1738 and studied by James Parsons (1705–1770) in London in 1739
PARSONS, *PHILOSOPHICAL TRANSACTIONS OF THE ROYAL SOCIETY OF LONDON*, 1743, PL. 2. COURTESY: JIM MONSON

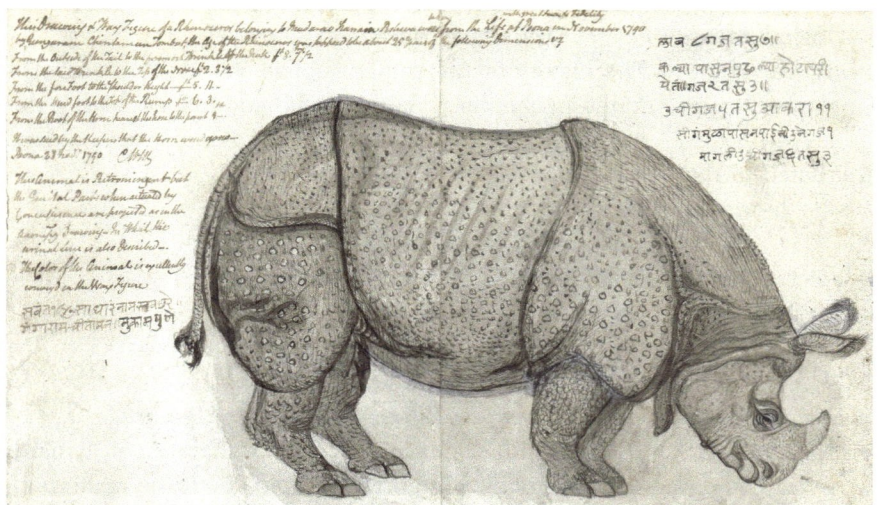

FIGURE 5.39
Gangaram Chintaman Tambat, A rhinoceros in the Menagerie of Peshwa Madhav Rao II (1774–1795) at Poona, November 1790. Watercolour and gouache, 12 × 36 cm
WITH PERMISSION © YALE CENTER FOR BRITISH ART, NEW HAVEN, CONN.: NO. B2006.14.33

1792-05-18. When David Price (1762–1835) lived in Poona, he went to the Peshwa's menagerie at the bottom of Pahrbutty Hill where he saw "the finest and most perfect model of a rhinoceros, that I have ever seen, either before or since. For, unlike the shapeless monster that we usually see exhibited, with his body enveloped in loose and flaccid folds of indurated hide, this stupendous animal was filled out to its utmost proportions" (Price 1836: 317). The keeper was able to get the animal to stand on its hind legs.

No date. Augustus Kincaid (1870–1954) was walking near Parvati Hill where he saw a grave of the Peshwa's rhino keeper who "one day ended his career with its horn through his body" (Kincaid 1908: 88).

Surat, Gujarat – See Ahmedabad, Gujarat, 1514

Trivandrum, Kerala – Zoological Gardens
*1. 1878-03 to 1900-06-16 †
Rhino exhibited (Rookmaaker 1998: 104). It was obtained in exchange from a native prince (Pettigrew 1882: 276).

Udaipur, Rajasthan – City Palace of the Maharanas
Rhinos in chains or otherwise in an obviously captive setting are depicted in art work of the successive Maharanas ruling in the first half of the 18th century. There must have been at least three rhinos, probably several more, which have here been combined conservatively as 2 specimens in the period prior to 1734 and 1 specimen in the period 1734 to 1751. A fourth and fifth rhino were painted in the 19th century.

1–2. Period 1698–1734

Period 1698–1710. A court painting on cloth shows animal fights in front of the palace of Maharana Amar Singh II (1672–1710) of Mewar, which include 2 rhinos (fig. 15.40).

Period 1710–1734. These animals could still be alive in the time of Maharana Sangram Singh II (ruled 1710–1734), when rhinos were kept at the palace (information from Bhupendra Singh Auwa, The City Palace, Udaipur).

3. Period 1734–1751

A painting shows Maharana Jagat Singh II of Mewar (ruled 1734–1751) hunting a rhino (fig. 15.48). The chains around the front feet indicate captivity. The inscription names the animal 'Fateh-Chand.'

4. Period 1790–1825

Period 1790–1810. A wall painting in the Chitram-ki-Burj in Udaipur shows a rhino in a staged fight with a large-eared (African?) elephant (fig. 15.41). The rhino has a belt around the neck and a rider on the back controls it with ropes.

1795. A Jain invitation letter, "vijñaptipatra" dated in the last two lines at the end of the scroll to May–June 1795 shows a tame rhino (fig. 15.49).

Period 1820s. A drawing by a Kota artist depicts a rhino kept at the entrance of the palace in Udaipur (fig. 15.50a, 15.50b).

5. Period 1861–1874

A mural made during the reign of Maharana Shambhu Singh in the Karjali Haveli of Udaipur shows a rhino with a rider fighting a stag (chapter 15).

Udaipur, Rajasthan – Zoological Gardens

Hughes (2013: 88) suggests that possibly Maharana Fateh Singh (1849–1930) acquired a rhino for the Zoological Gardens at Samore Bagh in Udaipur which had opened in 1897. Although early accessions included tigers, bears and lions from Gujarat, there is no evidence of a rhino at that time.

*1. F 1941 to 1946 (to Jaipur)

The female rhino was 3 years old on arrival. From Nepal (see next entry). Transferred to Jaipur Zoo (Rookmaaker 1998: 105).

2. M 1941

At the time of Maharana Bhupal Singh (ruled 1930–1955), a pair of rhino came from Nepal and were lodged at Samore Bagh (Bhupendra Singh Auwa, The City Palace, Udaipur, information 2018).

Una, Himachal Pradesh – Baba Bikram Singh Bedi

1. F 1840 to 1857

Baba Bikram Singh Bedi (d.1862) killed his brother Attar Singh on 25 November 1839. To atone for his crime, he washed his hands daily in the excrement of a rhinoceros (*Allen's Indian Mail* 1862-10-27). The animal was kept on his estates in Una (Oonah), east of Hoshiarpur, now in Himachal Pradesh. This female rhino was said to be ten years old in 1849, and was bought when it was one year old for Rs. 2500 (*Delhi Gazette* 1849: 56). The 59th Regiment of Bengal Native Infantry invaded Oonah in 1857. They confiscated the rhino of Bikram Singh. The animal was tame enough to be guided by a twig but got angry when this was applied too smartly (Girdlestone 1864: 170).

Vadodara (Baroda), Gujarat

The history of the rhinos kept in Baroda (Vadodara) is found in chapter 7. Between 1810 and 1886, at least 5 rhinos were kept by the Maharajas of Baroda, summarized as follows:

1810 Vadodara Guikwar 1822
*1823 Vadodara Guikwar 1856–1865
*1848–1856 Vadodara Guikwar < 1885
*1856–1866 Vadodara Guikwar < 1885
1865–1866 Vadodara Guikwar (death unknown).

Varanasi, Bihar – Maharaja of Banaras

There is no evidence that the Maharajas of Benares kept a rhino. The report circulated about the male rhino which arrived in Boston, Mass. on 9 May 1830 being kept previously by the "Rajah of Benares" in the media (like *Pittsburgh Weekly Gazette* 1830-05-28) is a mistake for a Rajah of Calcutta (Buddynath Roy) as found in other reports.

The French author Louis Jacolliot (1837–1890) visited India (38.12). His report of two rhinos kept by a Maharaja "below Benares", possibly in the 1860s or 1870s, may well be fictitious (Jacolliot 1875: 120).

Records of the Rhinoceros in Captivity in Independent Oudh

6.1 Captive rhinos in Oudh – Lucknow

An abundance of rhinos was kept in captivity in Lucknow, at least from 1775 when the Nawabs of Oudh (Oude, Awadh, Avadh) started to develop the town as their capital, until 1856 when the state was annexed by British forces. Reports allude even to a dozen rhinos seen simultaneously, and suggest that they might have been breeding (table 6.14). There are sadly no accounts or statistics which might help to identify individual specimens, no dates of arrivals, births or deaths. A very tentative list indicates the presence of 19 rhinos consecutively in Faizabad in the 1760s, in Lucknow from 1775 to 1856, and a further 2 in Kolkata during the exile of the last Nawab (table 6.15).

There is little to explain the definite fascination of the Nawabs with the rhinoceros. Of course they also kept quite large numbers of elephants, as well as tigers, cheetahs and other animals. Rhinos were found in the northern parts of the Oudh territory in the Lakhimpur and Bahraich regions. In the 18th century, the Nawabs enjoyed going on hunting expeditions, as seen in the sketches made for Jean-Baptiste Gentil (18.8). These were set up on an almost unimaginably large scale, with thousands of people in attendance, as narrated in 1793 by Captain Lewis Ferdinand Smith regarding a trip to Buckra Jeel (Bakhira Taal) near Gorakhpur: "Our party consisted of about 40,000 men, and 20,000 beasts, composed of 10,000 soldiers, 1000 cavalry, and near 150 pieces of cannon, 1500 elephants, 3000 carts or hackeries, and an innumerable train of camels, horses and bullocks. Carriages for women drawn by oxen. Also tigers, leopards, hawks, fighting cocks, fighting quails, and nightingales; pigeons, dancing women and boys, singers, players, buffoons, and mountebanks" (Smith 1806: 15). Rhinos had been expected but none were seen at this particular time, warned off by the noise and commotion.

It is rarely specified where rhinos were encountered in the town of Lucknow. There were stables, or enclosures, on the left (north) shore of the Gomti River in fields referred to as the Hazari Bagh (Hilton 1891: 122). So far no depictions or photographs are known showing the fields and stables where the rhinos could have been housed. The Hazari Bagh was connected with the town on the opposite shore by a ferry or pontoon bridge, and could be seen from the upper rooms in the Moti Mahal and from the 1810s from the verandah of the Shah Manzil constructed by Ghazi-ud-din (figs. 6.1, 6.2, 6.3). This viewing became popular in the time of Ghazi-ud-din after he initiated the performance of rhinos, elephants, tigers, bears and other animals in mutual fights (Walker 1865: 84; Alli 1874: 19; Fayrer 1900; Sharar 1975). Although many important visitors mentioned the presence of rhinos in Lucknow, their exhibition or mortal combats remained beyond the scope of artists, even though the Nawabs of Oudh were interested in art and poetry (Markel & Gude 2010; Llewellyn-Jones 2016). Much remains to be discovered about the Nawabs of Oudh and their passion for the rhinoceros.

FIGURE 6.1
A view of the Moti Mahal across the Gomti River in 1826. Part of an elongated diorama of Lucknow by an unknown artist. Watercolour and gouache, with gold; sheet 31 × 610 cm. Described when exhibited in Los Angeles and Paris by Markel & Gude 2010: 86, no. 56D
YALE CENTER FOR BRITISH ART, AVAILABLE DIGITALLY UNDER ORBIS:11747015

© L.C. (KEES) ROOKMAAKER, 2024 | DOI:10.1163/9789004691544_007
This is an open access chapter distributed under the terms of the CC BY-NC-ND 4.0 license.

FIGURE 6.2
View of the Hazari Bagh on the left shore of the
Gomti River where rhinos were kept in the fields
opposite the Moti Mahal and the Shah Manzil
in Lucknow. The river was bridged by a series
of pontoons connected by wooden planks. This
photograph was taken by Felice Beato (1832–1909),
who visited Lucknow in 1858 (further information
in Bautze 2007). The building in the background is
"The Martiniere" constructed in 1793 in a European
architectural style by the French Major-General
Claude Martin (1735–1800) as his own residence.
Silver albumen print 22,5 × 31,0 cm
WITH PERMISSION © P. & G. COLLECTION,
KARLSRUHE-BERLIN. COURTESY:
JOACHIM K. BAUTZE

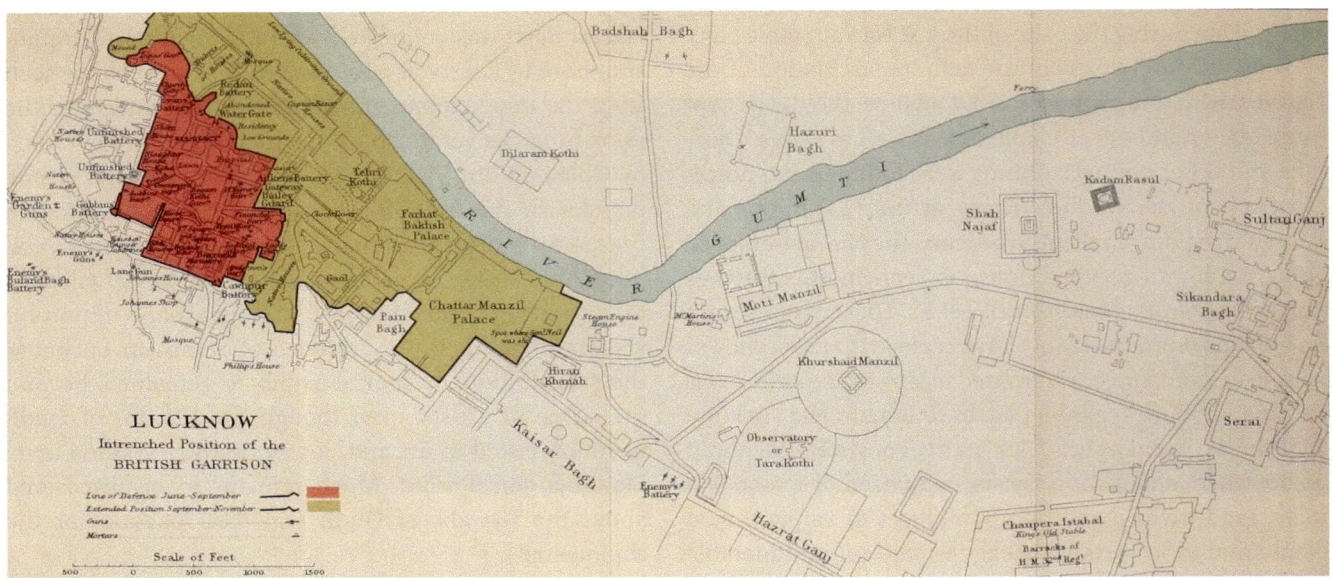

FIGURE 6.3   Map of Lucknow, entrenched position of the British garrison, engraved by Francis Sidney Weller (1849–1910). It shows the
position of the Moti Manzil and Hazuri Bagh on the River Gomti
MURRAY, *A HANDBOOK FOR TRAVELLERS IN INDIA, BURMA, AND CEYLON*, 1894, P. 232

• • •

Dataset 6.2: Chronological List of Events Indicating Rhinos in Captivity in Lucknow until the End of the 19th Century

Sequence: Date – Event – Reference – Specimen number according to table 6.15.

A. Period of Nawab Shuja-ud-Daula (1732–1775), Reigned 1759–1775

1760–1765 – Captive rhino guided by 4 men, as part of a larger painting of Nawab Shuja-ud-Daula on a lion hunt, probably dated early
1760s. The painter Mir Kalan Khan worked in Lucknow. Painting

oil on canvas, 72 × 84.5 cm. Clive Collection, Powis Castle, Powys,
Wales. National Trust NTPL Ref. No. 184225 – Losty 2002: 50, fig. 14;
Markel & Gude 2010: 2, no. 162; McInerney 2011: 618, fig. 12 – 6.2,
figs. 6.5, 6.6 – specimen 1.

1769 – 1 rhino captured. Jean-Baptiste-Joseph Gentil (1726–1799)
sketched hunting scenes including a captured rhino surrounded by
7 men – Gentil, Atlas in Victoria & Albert Museum, London (Atlas
IS25), p. 12 (scene 8) – 6.2, 18.8, fig. 18.4 – specimen 2.

B. Period of Nawab Asaf-ud-Daula (1748–1797), Reigned 1775–1797

1789 – Depiction of adult rhino with a chain and rope around the neck, once with inscription dated 1789 – Watercolour, anonymous (3 copies): (1) British Library, Wellesley Collection, India Office Library and Records NHD32 f. 4 (Archer 1962: 6; Rookmaaker 1998: 77, fig. 43); (2) Natural History Museum, Hardwicke Collection, Zoology 88 f HAR, size 292 × 427 mm (Magee 2013: 84); (2) Private collection Simon Keynes, Cambridge (pers.comm.) – 6.3, fig. 6.7 – specimen 3.

1790 – Young rhino exported from "Laknaor", shipped on the *Melville Castle*, Captain Philip Dundas (1763–1807), which left Calcutta 1790-01-18 and arrived in London 1790-06-02. Exhibited by Thomas Clark (1737–1816) in the Lyceum, The Strand, London, until July 1793 – Burt 1791; Bingley 1804: 487; Rookmaaker et al. 2015 – 6.4, figs. 6.8, 6.9 – specimen 4.

C. Period of Nawab Saadat Ali Khan (1752–1814), Reigned 1798–1814

1799-03-15 – William Tennant (1758–1813), army chaplain, sees a rhino of 12 years old in an apartment near the Palace – Tennant 1803, vol. 2: 410 – specimen 3.

1802-04-10 – George Annesley, Viscount Valentia (1770–1844), British politician, mentions 1 rhino in the menagerie – Valentia 1809, vol. 1: 117 – specimen 3.

D. Period of Nawab Ghazi-ud-din Haidar Khan (1769–1827), First King of Oudh 1814–1827

1814-10 – Amelia (Thackeray) Shakespear (1780–1824), British Civil Service, sees 2 young rhinos – Irvine 1938: 299 – 6.5 – specimens 5, 6 or 7.

1814-11-06 – 2 female rhinos "of middle growth" and 1 younger rhino were seen by Francis Edward Rawdon-Hastings, 1st Marquess of Hastings (1754–1826), Governor General of India 1812–1821 – Hastings 1858, vol. 1: 224 – 6.5 – specimens 5, 6, 7.

1819-10-21 – Colonel Thomas Lumsden (1789–1874), at Palace of Furreed (Farrad) Bakhsh, notes "near to the carriages I saw seven animals of the rhinoceros kind" – Lumsden 1822: 16 – 6.5 – specimens 5–11.

[1810s] – The gazetteer by Walter Hamilton (1774–1828) states that "the Nabob has a menagerie: the rhinoceros being the only remarkable animal in the collection" – W. Hamilton 1815: 498, 1820: 348 – specimens unclear (5–11).

1824-10-21 – Reginald Heber (1783–1826), Bishop of Calcutta 1823–1826, visits Lucknow and records 5 or 6 large rhinos "on the other side of the river Goomty, in a well-wooded park" – Heber 1828a, vol. 1: 381 (1st edn), 1828b, vol. 2: 58 (2nd edn) – 6.1, 6.5 – specimens included in 5–11.

1826-04-12 – A rhino-elephant fight was expected, but did not take place during visit of William Pitt Amherst, 1st Earl Amherst (1773–1857), Governor General 1823–1828 – Shouldham 1826: cli – 6.6 – specimens unclear (5–11).

1826-04-12 – Major Francis Gresley (1807–1880), Bengal Army, was present during Lord Amherst's visit. He wrote to his father describing his excitement at the prospect of watching "all kinds of fights: elephants together with tiger and buffalo and rhinoceros and bear with crocodile, which I dare say will be highly amusing and terrific." The rhino fight was abandoned – Shrestha 2009: 109 after Letter books of Francis Gresley (British Library, India Office, B120) – 6.6 – specimens unclear (5–11).

1820s [? 1827] – Abdul Halim Sharar (1860–1926) mentions rhino-tiger fight in time of Ghazi-ud-din – Sharar 1975: 117 – 6.6 – specimens unclear (5–7, 9–11).

E. Period of Nawab Nasir-ud-din Haidar Shah Jahan (1803–1837), Reigned 1827–1837

1827–12 – The intrepid lady traveller Fanny Parks, sometimes found as Parkes (1794–1875) visited Lucknow where she briefly mentioned a fight of "a rhinoceros against three wild buffaloes" – Parks 1850, vol. 1: 74 – 6.7 – specimen unclear (5–7, 9–11)

1827-12-14 – Rhino-tiger fight displayed during visit of Stapleton Cotton, 1st Viscount Combermere (1773–1865), Commander-in-Chief of India 1825–1830, reported by his Aides-de-Camp Edward Caulfield Archer (1771–1855) and Godfrey Charles Mundy (1804–1860) – Mundy 1832, vol. 1: 38; Archer 1833, vol. 1: 34 – 6.6 – specimen unclear (5–7, 9–11).

1830s – Presence of 15–20 rhinos recorded by William Knighton (1834–1900) after William Croupley. His account is at least partly fictitious – Knighton 1855: 253; Edwardes 1960 – 6.7 – specimens not included in table.

1831-01-18 – Fanny Parks (1794–1875) witnessed a fight between two rhinos which chased the spectators – Parks 1850, vol. 1: 177–178 – 6.7 – specimens unclear (5–7, 9–11).

1834 – Anonymous visitor saw several rhinos chained to trees – Anon. 1834b: 217; *Friend* 1835-03-14 – specimens unclear (5–7, 9–11).

1835-08 – Female rhino included in large gift of King Nasir-ud-din to King William IV; did not arrive in London, hence died in transit – Anon. 1837: 44; Brown 2001; Llewellyn-Jones 2016 – 6.8 – specimen 9.

F. Period of Nawab Muhammad Ali Shah (1777–1842), Reigned 1837–1842

1837-12-29 – Fight of 2 rhinos during visit of Prince Willem Frederik Hendrik (1820–1879) of The Netherlands, reported by Pieter Arriëns (1791–1860) – Arriëns 1853: 125; Pol 2016: 92 – 6.9 – specimens unclear (5–7, 9–11).

1837–12 – A rhino fight was scheduled, not witnessed by Emily Eden (1797–1869) and Fanny Eden (1801–1849), sisters of Governor-General George Eden, 1st Earl of Auckland (1784–1849) – Eden 1866, vol. 1: 88 – 6.9 – specimens unclear (5–7, 9–11).

1841-12-26 – Prince Aleksei Dmitrievich Soltykoff or Saltykoff (1806–1859), Russian diplomat, saw in the stables "Une douzaine de rhinocéros hideux et enormes étaient enchaines sous un long toit soutenu par des poudres" (12 rhinos chained under a long roof). From the description these stables were in or near the Chota Imambara of Hussainabad – Soltykoff 1848: 155, 157, 1858: 179 – 6.9, chapter 15 note 87 – specimens 5–7, 9–11, 12–16.

G. Period of Nawab Amjad Ali Shah (1801–1847), Reigned 1842–1847

1843-03-03 – The Prussian army officer Leopold von Orlich (1804–1860) visited the Menagerie on the left bank of the Goomty consisting of a large rectangular court surrounded by stables. There were 13 tigers, monkeys, rabbits, antelopes, quails and an ichneumon. However, 6 rhinos had been removed "to adorn various sepulchral monuments" and Orlich gave no indication that he had seen any of these – Orlich 1845a: 215, 1845b, vol. 2: 120, 1845c, vol. 2: 112 – specimens unclear (5–7, 9–16).

1843 – The German traveller Wolf Erich von Schönberg of Herzogswalde (1812–1887) visited the Iman-barah at Hussynabad (Hussainabad) noticing "some rhinoceroses and elephants, kept to amuse their masters by their combats in the arena" – Schönberg 1853, vol. 1: 134 – specimens unclear (5–7, 9–16).

1845-03-26 – Prince Friedrich Wilhelm Waldemar of Prussia (1817–1849) went to stables on the [north] bank of the Goomty, where he saw the enclosures for rhinos and elephants used for animal fights – Mahlmann 1853, part 2: 30; Kutzner 1857: 206; Burbach 1873, vol. 1: 429 – 6.9 – specimens unclear (5–7, 9–16).

[1840s?, date uncertain] – "At Lucknow may be seen many rhinoceros, some confined by a frail wooden railing, and others tethered to posts by ropes round a fore leg" – Old Shekarree 1860: 129 (the identity of the author is uncertain, now believed not to be Henry Astbury Leveson, who wrote under a similar pseudonym on hunting in Africa. The author gives no clue about the date of his observations).

H. Period of Nawab Wajid Ali Shah (1822–1887), King of Oudh 1847–1856 – Exiled to Calcutta in March 1856

1850s – Rhino fights mentioned by Joseph Fayrer (1824–1907), who was surgeon in Lucknow from July 1853 to 1858, and in charge of wild animals at the time of annexation in 1856 – Fayrer 1900: 92 – 6.9 – specimens unclear (5–7, 9–16).

pre-1855 – Account of "visit of a recent traveller to Lucknow", who saw in the menagerie "a rhinoceros tied to a tree", "I made my escape as quickly as possible, more especially as the rhinoceros began to shew signs of displeasure at our long stay" – Townsend 1855: 127 – specimen unclear (5–7, 9–16).

1855 – Major John Arthur Bayley (1831–1901) sees 3 rhinos in town driven along by horsemen armed with long spears – Bayley 1875: 140 – 6.9 – specimens 14–16.

1856-08 – Sale of King's menagerie (in Oudh). Media reports state that catalogue of auction included 3 rhinos – *Leeds Times* [and others] 1856-08-02 – 6.11 – specimens 17–19.

1856-08 – Sale of the King's menagerie. Edward Blyth (1810–1873) bought 2 rhinos at Rs 250 and transported them to Kolkata (further fate unknown) – Fayrer 1875b: 18; Grote 1875: x – 6.11 – specimens 18, 19.

[undated] – William Walker (1838–1908), writing as Tom Cringle in the *Times of India*, states that "the fights between elephants and rhinoceroses, which required to be viewed at a safe distance, took place across the Goomtee on the level ground in front of the Hazaree Bagh, the King and the court watching them from the verandah of the Shah Munzil" – Walker 1865: 84.

I. Rhino Confiscated from Rundheer Singh 1852

1852-02 – Rhino confiscated at raid of Rundheer Singh's camp on 12 February 1852, presented to Lucknow authorities – Anon. 1852a – 6.10 – specimen 17.

1858-03-27 – Rhino taken at Kaiserbagh by 53rd Regiment, commander Lt.-Gen. William Sutherland (1788–1862). Seen by Edmund Hope Verney (1838–1910) on 27 March 1858 – Jones 1859: 197; Verney 1862: 122 – 6.10 – specimen 17.

1858 – Soldier at time of Indian Mutiny was punished for shooting a tame rhino which had been captured by his regiment. His defence was that the hide should be too thick – Kinloch 1885: 62 – specimen 17.

J. King Wajid Exiled in Garden Reach, Calcutta 1856–1887

1876 – "Mr Routledge has just secured a splendid specimen of a rhinoceros standing only 2 ½ feet (76 cm) from the ground. It was captured for him at the Sunderbans" – *Times of India* 1876-09-09 – 6.11 – specimen 20.

1887-12-01 – 2 females – Sale of King's menagerie, catalogue commences with 2 female rhinos, priced at Rs 2000 each – *Amrita Bazar Patrika* (Calcutta) 1887-12-01 (from *Englishman*); Llewellyn-Jones 2016 – 6.11 – specimens 20, 21.

1887 – Female *R. sondaicus* arriving in Calcutta Zoo in 1887 had lived 10 years in Oudh's menagerie – Calcutta Zoo 1888; Noll 1889; Sanyal 1892: 131; Flower 1931: 203; Rookmaaker 1998: 122 – 6.11 – specimen 20.

1887 – Female *R. unicornis* acquired by Calcutta Zoo at sale of Oudh's menagerie – Calcutta Zoo 1888; Noll 1889; Anon. 1890: 234; Rookmaaker 1998: 55 – 6.11 – specimen 21.

∴

TABLE 6.14   Numbers of rhinos observed in Lucknow in respective years

1810		1820		1830	(15–20)	1840		1850	
1811		1821		1831	2	1841	12	1851	
1812		1822		1832		1842	several	1852	1
1813		1823		1833		1843	6	1853	
1814	2–3	1824	5–6	1834	several	1844		1854	
1815		1825		1835		1845	several	1855	3
1816		1826		1836		1846		1856	3
1817		1827	1–2	1837	2	1847		1857	
1818		1828		1838		1848		1858	
1819	7	1829		1839		1849		1859	

Details are found in the Chronological List. The numbers 15–20 in 1830 are from Knighton (1855) which may be fictitious in detail, hence not counted in totals.

TABLE 6.15   Tentative list of rhinoceros specimens kept by the Nawabs of Oudh

Faizabad

1. 1765. Captured in 1760–1765. No further data.
2. 1769. Captured in 1769. No further data.

Lucknow

3. 1789, 1799, 1802. Adult rhino. No further data.
4. 1790. Young rhino shipped to London, arriving 2 June 1790. Clark's Rhinoceros at The Lyceum.
5. 1814, 1819, 1824, 1841. female [1/7 in 1819, 1/6 in 1824, 1/12 in 1841], died before 1856.
6. 1814, 1819, 1824, 1841. younger female [1/7 in 1819, 1/6 in 1824, 1/12 in 1841], died before 1856.
7. 1814, 1819, 1824, 1841. young animal [1/7 in 1819, 1/6 in 1824, 1/12 in 1841], died before 1856.
8. 1819, one of seven rhinos in 1819, died before 1824 (when maximum 6 rhino were reported).
9. 1819, 1824, 1835. female [1/7 in 1819, 1/6 in 1824] to UK in 1835, died in transit.
10. 1819, 1824, 1841. 1 of 7 in 1819, 1/6 in 1824, 1/12 in 1841, died before 1856.
11. 1819, 1824, 1841. 1 of 7 in 1819, 1/6 in 1824, 1/12 in 1841, died before 1856.
12. 1841, 1855, 1 of 12 rhinos, died before 1856.
13. 1841, 1855, 1 of 12 rhinos, died before 1856.
14. 1841, 1855, 1 of 12 rhinos, died before 1856.
15. 1841, 1855, 1 of 12 rhinos, died before 1856.
16. 1841, 1855, 1 of 12 rhinos, died before 1856.
17. 1852, confiscated at raid of Rundheer Singh's camp. Not sold at auction of 1856. Died 1858.
18. 1841, 1855. 1 of 12 in 1841, sold at auction of menagerie in 1856, arrived Calcutta (Blyth).
19. 1841, 1855. 1 of 12 rhinos in 1841, sold at auction of menagerie in 1856, arrived Calcutta (Blyth).

Calcutta (Kolkata)

20. 1876. Female. Captured in Sundarbans (*R. sondaicus*) through Rutledge. Sold at auction in 1887. To Alipore Zoo in Calcutta.
21. 1887. Female. Arrival unknown. Second rhino included in Wajid's auction sale. To Alipore Zoo in Calcutta.

The potentially inflated numbers proposed by Knighton (1855) are not included. Larger numbers were simultaneously recorded in 1819 (7 rhinos), 1824 (6 rhinos) and 1841 (12 rhinos).

FIGURE 6.4
Two rhinos with feet chained, in a miniature painted in the late 18th century in India. The artist, locality and date are unknown. Image 6 × 15.2 cm
REPRODUCED IN BAUTZE (1985, FIG. 5) FROM THE COLLECTION OF KUMAR SANGRAM SINGH OF NAWALGARH

6.2   Capture of a Rhinoceros in 1769

During one of the regular hunting expeditions in the northern stretches of Oudh organised by Nawab Shuja-ud-Daula, a rhino was captured in 1769. This is the implication of a scene depicted in the Atlas assembled by the French army officer Jean-Baptiste-Joseph Gentil (1726–1799), who was at the court when it was still largely located in Faizabad (18.8). While the date 1769 is recorded in the Atlas, there is no further information about the fate of the little rhino in the painting (fig. 18.4). It could have been kept for a while either in Faizabad or in Lucknow where some of the official administration was located before the capital of Oudh moved there in 1775.

A miniature drawing of this period (fig. 6.4) showing two chained rhinos may represent a scene in Oudh, Jaipur or elsewhere (Bautze 1985: 423, fig. 5).

FIGURE 6.5   A royal lion hunt at Allahabad. Painting attributed to Mir Kalan Khan in Lucknow, 1760–1765. The central figure has
              been identified as Nawab Shuja-ud-daula. A captured rhino is escorted by three royal guards in the lower left corner.
              Watercolour and gouache, 72 × 84.5 cm
              CLIVE COLLECTION. POWIS CASTLE, POWYS, WALES © NATIONAL TRUST IMAGES/ERIK PELHAM

The painter Mir Kalan Khan was trained in the Mughal court and worked in Lucknow from about 1750 (Bautze 1993: 280). A painting called "A royal lion hunt at Allahabad" shows a rhino being guided by four men, and while it is not the main subject matter, there is no doubt that the animal had just been captured (figs. 6.5, 6.6). Although the location might have been in the region of Allahabad as suggested by the title, it is also thought that Mir Kalan Khan produced the painting in 1760–1765 at Lucknow (Losty 2002, McInerney 2011). There is no direct evidence that this event is linked to the capture in 1769 portrayed by Gentil. In view of the miniature of the two chained rhinos, it is not unlikely that there were two rhinos in captivity in Lucknow at the end of the 1760s.

6.3   Rhinoceros Exhibited in 1789

When Richard, 1st Marquess Wellesley (1760–1842) was Governor-General of India 1798–1805, he assembled a large collection of drawings (British Library, London). This includes a watercolour of a rhino with a rope around the neck, without inscription (Archer 1962). A very similar folio is found in the Hardwicke Collection (Natural History Museum, London), with the following inscription in pencil: "Rhinoceros unicornis. – drawn from a living animal in possession of Asoph-ud Dowlah – Vizir – Lucnow 1789" (Magee 2013: 84). A third copy without data is in the private collection of Simon Keynes, University of Cambridge (fig. 6.7). The animal has a chain and a rope

FIGURE 6.6
Detail of the hunt at Allahabad
showing a captured rhinoceros
CLIVE COLLECTION ©NATIONAL
TRUST IMAGES/ERIK PELHAM

around the neck, shows a horn of reasonable size, and looks adult. Blyth (1872c) noticed that head and lower limbs were quite smooth, which he thought to "represent a peculiar and undescribed species." However, it is clearly *R. unicornis*.

supposed to be this specimen (fig. 6.8). However a drawing by N. Burt of a young chained animal in 1791 appears to contradict this (fig. 6.9). The history of this rhino is clear, but the anomaly of the depictions has not been resolved (Rookmaaker et al. 2015).

6.4 Clark's Rhinoceros in London, 1790–1793

In 1790 a young rhino "from Laknaor, in the East Indies" was shipped to London on the *Melville Castle*, arriving on 2 June 1790. It was bought by the entrepreneur Thomas Clark (1737–1816) and exhibited at the Lyceum on The Strand, London until it died in July 1793. The painting of a large adult male rhino with a good-sized horn by the well-known animal artist George Stubbs (1724–1806) is

6.5 Groups of Rhinos during the Reign of Nawab Ghazi-Ud-Din

In October 1814, Amelia Shakespear took her children to see two young rhinos in Lucknow "which were perfectly tame and gentle and ate bread from the hands of several of the gentlemen." Just a few weeks later, the Governor-General Hastings was given an opportunity to see three rhinos: "Two were females of middle growth. The third was quite

FIGURE 6.7   Watercolour of a rhinoceros with ropes around the neck. An identical copy in the Natural History Museum, London (Hardwicke
             Collection) states that it was an animal alive in Lucknow in 1789
             WITH PERMISSION © PRIVATE COLLECTION OF SIMON KEYNES, CAMBRIDGE

FIGURE 6.8
George Stubbs (1724–1806), undated painting of a
male rhinoceros. Oil on canvas. Size 75 × 96 cm. From
the collection of the surgeon John Hunter (1728–1793)
WITH PERMISSION © HUNTERIAN MUSEUM,
ROYAL COLLEGE OF SURGEONS OF ENGLAND,
LONDON, RCSSC/P 267

young. They appeared tame and gentle." As the ages and sizes do not correspond, maybe there was already a larger group in the King's menagerie. In 1819 Colonel Thomas Lumsden saw seven rhinos when he arrived at the Farhad Bakhsh, within the Chattar Manzil Palace complex. It is likely that the rhinos, together with some other larger animals, were kept in a park on the north (left) bank of the Gomti River, where Bishop Reginald Heber in October 1824 described a peaceful scene, with "five or six very large rhinoceroses, the first animals of the kind I ever saw, and of which I found that prints and drawings had given me a very imperfect conception. They are more bulky animals,

FIGURE 6.9
Drawing of a rhinoceros in the Exeter 'Change Menagerie in London in 1791. Signed (lower left) N. Burt, Del.; (lower right) Record Sculpt. The draughtsman N. Burt was "of the Naval and Drawing Academy, Tottenham Court Road." Engraved by John Record

BURT, *DELINEATION OF CURIOUS FOREIGN BEASTS AND BIRDS*, 1791, PLATE OPPOSITE P. 9

and of a darker colour, than I had supposed, and the thickness of the folds of their impenetrable skin much surpasses all which I had expected. These at Lucknow are gentle and quiet animals, except that one of them has a feud with horses. They seem to propagate in captivity without reluctance, and I should conceive might be available to carry burthens as well as the elephant, except that, as their pace is still slower than his, their use could only be applicable to very great weights, and very gentle travelling. These have sometimes had howdahs on them, and were once fastened in a carriage, but only as an experiment which was never followed up" (Heber 1828a, vol. 1: 381). This must have been one of the very few instances where an attempt was made to let a rhino pull a carriage, obviously too difficult, unless it is in an imaginary setting like that painted by the Italian artist Guiseppi Zocchi in 1757 (fig. 12.18). Although Heber only mentions full-grown specimens, no younger ones, he implies that the rhinos were breeding in captivity, which was accepted as evidence by Darwin (1868, vol. 2: 165). Certainly, a group of 5 or 6 or 7 rhinos must have been quite a sight.

6.6   Rhinos Fighting other Rhinos and Tigers

Fights involving rhinos are first mentioned in the reign of Nawab Ghazi-ud-din, and were to become a more regular spectacle in later years. The first exact and reliable reference dates to the visit of the Governor-General Lord Amherst, who was informed about proposed fights on 12 April 1826. However, due to the ill health of the King, this did not actually take place (Shouldham 1826). A more successful attempt was made on 14 December 1827 in honour of the visit of Stapleton Cotton (1773–1865), the Commander-in-Chief of India when neither animal

was harmed: "A rhinoceros was next let loose in the open court-yard, and the attendants attempted to induce him to pick a quarrel with a tiger who was chained to a ring. The rhinoceros appeared, however, to consider a fettered foe as quite beneath his enmity; and having once approached the tiger and quietly surveyed him, as he writhed and growled, expecting the attack, turned suddenly round and trotted awkwardly off to the yard gate, where he capsized a palankeen which was carrying away a lady fatigued with the sight of these unfeminine sports" (Mundy 1832, vol. 1: 38).

Another fight between a rhino and tiger is described by the author and publisher Abdul Halim Sharar, which he sets in the time of Ghazi-ud-din, but his source is obscure. His description of the event is rather gruesome, and hopefully it was not a frequent occurrence: "The most interesting battles were between a tiger and a rhinoceros. Except for the underpart of its belly, the rhinoceros is brazen-bodied and the tiger's teeth and claws have no effect on it. With the assurance of this strength, the rhinoceros cared nothing for the strongest of foes and, lowering its head, was able to get under its adversary's belly and rip it open with the horn on its snout; all the entrails came tumbling out and the adversary's days were over. Only rarely did a tiger knock a rhinoceros flat on its back and then tear open Its belly with teeth and claws. Generally the rhinoceros managed to kill the tiger with a thrust of its horn" (Sharar 1975: 117). Tigers in the wild might sometimes attack young rhinos. Such scenes remain the realm of artistic interpretations, like one found in a drawing by the German artist Adolf Closs from the 1880s (fig. 6.10). A staged fight similar to those seen in Lucknow was depicted in dramatic fashion by the American artist Frederick Melville du Mond around 1900 (fig. 6.11). An equally fictitious scene of tiger fighting a rhino in the

FIGURE 6.10   "Un drame dans les jungles" (A jungle drama). Tiger
               attacking a young rhino which is rescued by the
               mother. Signed by the German painter and illustrator
               Gustav Adolf Closs (1864–1938) and engraved by
               Oskar Frenzel (1855–1915). Size 20 × 27 cm
               DECINTHEL, *LA CHASSE ILLUSTRÉE*, 1 OCTOBER
               1887, P. 317

F. Melville du Mond: *Kampf zwischen Rhinoceros und Tigern.*

FIGURE 6.11   Fight between rhino and a pair of tigers "Kampf
               zwischen Rhinoceros und Tigern." This is a fictitious
               scene created by the American artist Frederick
               Melville du Mond (1867–1927), who visited India
               in 1900. The artist signs (lower right corner) with
               the location "Bombay." He is known for drawings
               showing animal battles. This image (24 × 24 cm)
               was published in an unknown German language
               publication, probably soon after 1900
               COLLECTION KEES ROOKMAAKER

arena was drawn by John Alexander Harrington Bird published in a late posthumous reprint of a book by the Scottish novelist George John Whyte-Melville (1901) without further explanation (fig. 6.12).

6.7   The Time Of Nasir-Ud-Din in the 1820s

The conditions in Lucknow during the reign of Nawab Nasir-ud-din are the subject of the *Private life of an Eastern King* published in 1855 when the annexation of Oudh was being discussed. First published by "a member of the household" anonymously due to political and negative personal contents, the author William Knighton (1834–1900) had not been to Lucknow at the time and relied on information from the librarian Edward Croupley (Tait 1857). Despite being semi-fictitious and unillustrated, the book was quite popular, with two further printings in

1855, again in 1856 and 1857, followed by others later in the century. Bénédict-Henry Révoil (1816–1882) presented a French translation in 1858 and 1865, while Ludwig Thiele gave a German version in 1856. None of this allows one to dismiss the passages about the presence of rhinos or the animal fights. However, it does leave some room to doubt the veracity of the actual numbers, when Knighton writes about 15 to 20 rhinos "kept in the open park around Chaungunge, and allowed to roam about, at large, within certain limits" (Knighton 1855: 253). He writes more convincingly about the fights staged between two rhinos (preferably males) which would try to attack their opponent's belly with their horn (with an example from wild surroundings in fig. 6.13).

Fanny Parks, who lived in India 1822–1839 with her husband Charles Crawford Parks (1797–1854) working as a collector for the East India Company, seemed quite pleased by such entertainment which was perhaps not as harsh as

FIGURE 6.12 "With a short labouring trot he moves around the arena." Engraving signed by John Alexander Harrington Bird (1846–1936) showing a fictitious scene of a tiger fighting a rhino
WHYTE-MELVILLE, *THE GLADIATORS*, 1901, P. 151

FIGURE 6.13 Two rhinos arguing among themselves in the jungle. Imaginary scene drawn by the German artist Carl von Dombrowski (1872–1951)
SOURCE UNKNOWN, EARLY 20TH CENTURY.
COLLECTION KEES ROOKMAAKER

could be imagined: "When the elephant fights were over, two rhinoceros were brought before us, and an amusing fight took place between them; they fought like pigs. The plain was covered by natives in thousands, on foot or on horseback. When the rhinoceros grew fierce, they charged the crowd, and it was beautiful to see the mass of people flying before them" (Parks 1850, vol. 1: 177).

6.8 A Gift for King William IV in 1835

In the early 1830s, Nasir-ud-din felt compelled to send a splendid gift consisting of many items to the British King William IV (1765–1837) and his spouse Queen Adelaide of Saxe-Meiningen (Brown 2001; Llewellyn-Jones 2000: 11). The shipment left Lucknow on 20 January 1835, detailed in a list of 6 February 1835, signed by the Assistant to the Resident John Dowdeswell Shakespear (1806–1867), which included 2 elephants, 2 Arabian horses, 2 cows and "1 Female Rhinoceros, with Brocade Shool, or covering" (Anon. 1837: 44). Having reached Kolkata in March, the donation accompanied by Philip Friell representing the Nawab was forwarded on the *Duke of Argyll* reaching London 7 August 1835 (*London Evening Standard* 1835-08-08). The horses were dispatched to Windsor Castle, the elephants to the London Zoo and the Surrey Zoological Gardens. However, there was no rhino among the live arrivals, so she either died in Kolkata or during the voyage, without leaving further trace.

6.9 Rhinos in Lucknow before the Annexation in 1856

The Dutch Prince Willem Frederik Hendrik, third son of King Willem II (Wiliam II), made a tour on the Ganges as a teenager in 1837 (Pol 2016). He had the company of the

FIGURE 6.14
Capture on 12 February 1852 of Rundheer Singh
of Patti Syfabad. Sketch by Frank Slinger, Deputy
Commissioner at Allahabad
ILLUSTRATED LONDON NEWS, 15 MAY 1852,
P. 388

sisters of the Governor General, Emily and Fanny Eden.
He saw a fight of two rhinos which were still shackled at
the hind legs (Arriëns 1853).

The Russian aristocratic traveller and artist Prince
Alexis Soltykoff (1806–1859) was roaming around Lucknow
in December 1841 when he passed the stables situated in
a park near the mausoleum built for the Nawab's favourite
horse and saw 12 rhinos chained under a long roof supported by a beam (Soltykoff 1848).

Visiting the north bank of the Goomty River on
26 March 1845, the Prussian Prince Friedrich Wilhelm
Waldemar saw stables of an undefined number of rhinos, but apparently no fight was staged at that time
(Mahlmann 1853).

If there had been 12 rhinos in Lucknow in 1841, this
number must have dwindled quite considerably in the
next decade. The animal confiscated from Rundheer Singh
in 1852 was kept in Lucknow, most probably together with
those owned by the Nawabs. Joseph Fayrer still mentioned a fight including rhinos for an uncertain date in
the 1850s. The clearest report comes from Major John
Arthur Bayley for a visit in 1855: "One morning when riding into the town I was met by three rhinoceroses, driven
along by half-a-dozen horsemen armed with long spears"
(Bayley 1875: 140).

6.10   Rundheer's Rhinoceros 1852–1858

On 12 February 1852, Lieutenant-Colonel Robert Henry
Wallace Dunlop (1823–1887), joint magistrate at Jaunpur
in the south-east of Uttar Pradesh, took a small force of 34
native men to capture a local chief who was troubling the

government. Rundheer or Randhir Singh was the third son
of the rich zamindar Diwan Hirda Singh of Patti Saifabad
in Pratapgarh District, then on the outskirts of Oudh territory. When Dunlop entered the camp, Rundheer was
asleep and easily captured, later to be kept in confinement
in Lucknow until 1856 (Lethbridge 1893: 517; Anon. 1852b;
Anon. 1878a, vol. 3: 158). Dunlop found not only three hundred of Rundheer's followers, but also seven elephants
and an "unusually large and intractable rhinoceros." The
scene was the subject of a drawing made by Frank Slinger,
the Deputy Commissioner of Allahabad, for the *Illustrated
London News* of May 1852 (fig. 6.14). The rhino was handed
over to the authorities in Lucknow and then probably
became part of the Nawab's menagerie.

Or maybe it was the animal in the care of the 53rd regiment of the Bengal Native Infantry, recovered from the
town's Kaiserbagh during the troubles of the Mutiny of
1857 according to Oliver John Jones (1813–1878): "a tame
rhinoceros, who was reputed to be a hundred years old;
it certainly was nearly blind, and quite stupid, but very
good-natured, and would let one pull him by the horn or
rub his scaly coat of mail without showing the slightest
displeasure. He was taken up to their camp, and when I left
them, he was there in safety, and, if not in clover, certainly
in the midst of plenty" (Jones 1859: 197). On 27 March 1858,
the British officer Edmund Hope Verney (1838–1910)
observed "a wonderful rhinoceros in the camp, the property of the 53rd; he was found in Lucknow and is very
tame; every day he is driven to a well to drink, guided
by little taps from a twig which one would have thought
could hardly have been felt through his thick hide: if,
however, any one ventures to do more than touch him
very lightly with it, he at once gets angry. He is very old,

poor fellow, and suffers from some sort of ophthalmia, which has rendered him all but blind" (Verney 1862: 122). Apparently before long it was shot by mistake by one of the soldiers.

6.11 Sale of Oudh's Menagerie in 1856

The state of Awadh (Lucknow) was annexed by the British in February 1856, and King Wajid left the city to reach Kolkata on 13 May 1856. His menagerie in Lucknow could not be maintained and was put up for auction. An early list was circulating in April, with an unspecified number of rhinos (*Friends of India* 1856-04-03). A more detailed catalogue was mentioned in the British press in early August, pointing at an auction in July 1856. The list included 128 elephants, 2 giraffes, 18 tigers, 3 leopards, 4 lynxes, 60 hawks, 220 horses as well as 2,840 bullocks and 10,000 pigeons, not to forget 3 rhinos. Total receipts were Rs 1,50,000 (then about £15,000). It is likely that news of the auction had brought the Curator of the Indian Museum, Edward Blyth (1810–1873). In partnership with a German friend (probably the animal dealer Charles Jamrach), he bought many of the exotic animals at rock-bottom prices, like 16 tigers at Rs 10 each, a pair of rhinos at Rs 250, and a giraffe at Rs 500 (Fayrer 1875b; Grote 1875; Llewellyn-Jones 2000: 139). The third rhino, perhaps the one confiscated from Rundheer Singh, may not have been sold as it was too sick or old, and was cared for by the army. Blyth arranged for his lots to be transported to Kolkata, experiencing some casualties both on the journey and after arrival in town because onward shipment to Europe was difficult to obtain (Grote 1875; Rookmaaker 1997c; Brandon-Jones 1997). Blyth must have found accommodation for the two rhinos from Oudh somewhere in Kolkata. Presumably one of these animals was shipped on the *Sutledge* on 9 April 1858 to London, where it might have been cared for by Jamrach or another animal dealer. Its sale is unknown, as there are no records of any rhinos arriving in world zoos in the years 1856–1860 (Jamrach 1858; Rookmaaker 1998).

6.12 King Wajid Shah in Kolkata

Although the King was much devoted to his menagerie, he reached Kolkata in 1856 without it. However, he soon established an animal park at his new home called Garden Reach in Metiabruz, in Kolkata on the right side of the river (Llewellyn-Jones 2016). Besides 18,000 pigeons, there were to be 2000 other birds, animals and snakes (London *Times* 1874-10-27). The animal dealer William Rutledge (1839–1905) largely depended on the King for his business, supplying animals with an annual budget of 1 lakh (Rs 1,00,000) and the death of the King in 1887 precipitated its demise (Llewellyn-Jones 2016). The King also obtained at least two new rhinos while living in Metiabruz. One of these, standing 76 cm high, was captured in the Sundarbans for Rutledge (*Times of India* 1876-09-09) and must have been a specimen of *R. sondaicus*. The accession of the second specimen (*R. unicornis*) is not clear.

6.13 Auction in Kolkata in 1887

When Wajid Shah died on 1 September 1887, the animal collection was auctioned by Milton & Co., owners of a horse stable yard. An editorial in the *Amrita Bazar Patrika* (Calcutta) of 1 December 1887 talks of a catalogue "commencing with two female rhinoceroses, which are priced at Rs. 2000 each." A sale at half that price was arranged with Ram Bramha Sanyal (1850–1908), who was engaged as Superintendent of the Calcutta Zoo in Alipore since its inception in 1876 (Llewellyn-Jones 2016). The 1887–1888 annual report of the Calcutta Zoo (1888) lists the addition of an Indian and a Javan rhinoceros. The lifespan of this *R. unicornis* is unclear, but it may still have been alive in 1890 (Anon. 1890: 234). The *R. sondaicus* was alive when Sanyal (1892: 131) wrote his important treatise on the management of animals in captivity, but no record of its death has been found. In 1891 the zoo had a large income from the sale of rhino urine (Ehlers 1894, vol. 1: 372), remarkable if this would prove to have been the product of an animal as rare as a *R. sondaicus*.

CHAPTER 7

Records of the Rhinoceros in Captivity in Baroda

7.1 The Menagerie of the Gaekwads of Baroda

The Maharajas or Gaekwads of Baroda (Vadodara) in Gujarat kept rhinos during most of the 19th century. For any glimpses we are dependent on occasional reports by visitors and journalists. It is therefore hard to distinguish between individual specimens, when they arrived and when they died. There must have been at least 5 rhinos in Baroda (table 7.16). In 1875 there were two male rhinos in the arena (Barras 1885).

TABLE 7.16 Tentative list of rhinos kept by the Gaekwads of Baroda in the 19th century

A. 1810. Arrival not recorded. Died before 1822.

B. 1823 August. Arrival from Cooch Behar. Alive in 1848, 1856. Died 1856–1865.

C. 1848–1866. Arrival not recorded. Alive in 1866 (largest) and died before 1885, possibly much earlier.

D. 1856–1866. Arrival not recorded. Alive in 1866 (with large horn). Either no.C or no.D died before 1885.

E. 1865–1866. Arrived as immature animal. Death not recorded.

The first date or date range is the assumed year of arrival in Baroda.

• • •

Dataset 7.3: Chronological List of the Specimens of *R. unicornis* kept in the Menagerie of the Successive Gaekwads of Baroda (Vadodara)

Records are presented in this sequence: Date – Event – Sources – no., possible identity of specimen (with letters as in Table 7.16). They are arranged according to the reigns of the successive rulers.

A. Period of Maharaja Gaekwad Anandrao (1790–1819), Reigned 1800–1819

1810 – Lutfullah Khan (1802–1874) saw a rhino at a gate with the animal's name – Eastwick 1857: 42 – 7.2, specimen A.

B. Period of Maharaja Gaekwad Sayajirao II (1800–1847), Reigned 1819–1847

1822-05 – Bombay Government wants to obtain a rhino on the request of the Gaekwad – National Archives, Delhi 1822b (Public Department to C. Lushington 1822-05-16) – 7.3, specimen B.

1823-07 – Rhino captured in Cooch Behar is forwarded in charge of Gopaul Singh. Animal donated by the Governor General William Pitt Amherst, 1st Earl Amherst (1773–1857) – National Archives, Delhi 1823b (C. Lushington to W. Newnham 1823-07-17) – 7.3, specimen B.

1825-03-19 – Bishop Reginald Heber (1783–1826) sees a tame rhino donated by Lord Amherst – Heber 1828a, vol. 2: 125 (1st edn), 1828b, vol. 3: 5 (2nd edn) – 7.3, specimen B.

C. Period of Maharaja Gaekwad Ganpatrao (1816–1856), Reigned 1847–1856

1848-01-28 – Henry George Briggs (1824–1872) sees "the surviving rhinoceros presented by Lord Amherst" – Briggs 1849: 87 – 7.3, specimen B.

1856-02 – On the occasion of the arrival of the new Resident Cuthbert Davidson (1810–1862), a fight of two rhinos – *Bombay Times and Journal of Commerce* 1856-03-01 – specimens B, C.

D. Period of Maharaja Gaekwad Khanderao (1828–1870), Reigned 1856–1870

1865-06 – Louis-Théophile Marie Rousselet (1845–1929) witnesses a fight between two rhinos, described and illustrated – Rousselet 1871: 244 (*Tour du Monde*), 1875a: 124 (*L'Inde des Rajahs*), 1875b: 103 (*India and its Native Princes*), also found in new editions and translations of his book, as well as excerpts in popular magazine, enumerated in 7.5 – 7.5, fig. 7.1, specimens C, D.

1866-02 – Largest, the veteran, horn is a stump, 5 ft (150 cm) high – Baroda correspondent of *Bombay Gazette* 1866-02-14 – specimen C.

1866-02 – Second, nearly as large, very quiet. Horn 1 ½ ft (45 cm) long and 15 inch (38.1 cm) in circumference – *Bombay Gazette* 1866-02-14 – specimen D.

1866-02 – Third is a "bachcha" (young), not been here long. He is very good tempered, though stolid – *Bombay Gazette* 1866-02-14 – specimen E.

1866-02 – Rhino fight of an older and a younger rhino – *Bombay Gazette* 1866-02-14 (long and accurate description of the fight) repeated in *Bombay Gazette* 1866-03-17 – specimens D, E.

E. Period of Maharaja Gaekwad Malharrao (1831–1875), Reigned 1870–1875

1871-01 – Reception of Governor of Bombay, William Vesey-FitzGerald (1818–1885). He witnessed fight between an elephant and a rhino – *Englishman's Overland Mail* 1871-01-18 – specimen unclear (C, D, E).

© L.C. (KEES) ROOKMAAKER, 2024 | DOI:10.1163/9789004691544_008

This is an open access chapter distributed under the terms of the CC BY-NC-ND 4.0 license.

1871-01 – Menagerie includes a huge rhino – *Times of India* 1871-01-13 – specimen unclear (C, D, E).

1871 – Anonymous correspondent, witnesses fight between two rhinos – Anon. 1871b: 283 – specimens unclear (C, D/E).

F. Period of Maharaja Gaekwad Sayajirao III (1863–1939), Reigned 1875–1939

1875-06-09 – Witness of a fight between two rhinos – *Times of India* 1875-06-14 – specimens unclear (C, D/E).

1875-11-19 – Edward, Prince of Wales (1841–1910) witnesses a fight between two rhinos in the Old Palace. The animals refused to fight – Anon. 1875a, 1875b; *Examiner* 1875-12-18: 1409 – 7.7, specimens unclear (C, D/E).

1875-11-19 – Rhino fight during visit of Prince of Wales described and witnessed by William Howard Russell (1820–1907), by the journalist James Drew Gay (1846–1890), by Rupert Clement George, 4th Baron Carrington (1852–1929) – Gay 1876: 89; Wheeler 1876: 113; Russell 1877: 198; Goblet d'Alviella 1877: 60, 1892: 130; Penderel-Brodhurst 1911,

vol. 3: 98; Sergeant 1928: 30; Hibbert 2007: 311 (for Carrington) – 7.7, figs. 5.2 to 5.5, specimens unclear (C, D/E).

[1875] – Rhino fight between two large male rhinos witnessed by Colonel Julius Barras (d. 1894) without providing date – Barras 1885: 167–168 – specimens unclear (C, D/E).

1879-04-12 – Edward George Henry Montagu (Sandwich 1839–1916) witnesses rhino fight – Hinchingbrook 1879: 172.

1881-01-07 – Occasion of marriage of Gaekwad and Chimnabai I (1864–1884). Rhino fight – Anon. 1880: 411 – 7.8, specimens unclear (C, D/E).

1881-12-28 – Occasion of investiture of Gaekwad on 21st birthday. Rhino fight – Debans 1881; *Times of India* 1881-12-29; Anon. 1882: 149; Michel 1882: 74, 1893, vol. 3: 119, 127 – 7.8, specimens unclear (C, D/E).

1885-12-30 – One rhino exhibited in area – *Scotsman* 1886-01-30 – 7.8, specimen unclear (D/E).

1886-01 – This rhino had in former years been able to fight another rhino, but his old enemy is now dead – *Scotsman* 1886-01-30 – specimen C.

∴

7.2 The Rhinoceros of 1810

The earliest record of a rhino in Baroda was in 1810 when the Muslim scholar and teacher Lutfullah Khan (1802–1874) as a young boy spent many hours of leisure sitting with the keepers and staring at the animal (Eastwick 1857). It was then kept "at" (not near) one of the city gates bearing its name, still known as Gendi Gate or Gendi Darwaja (Desai 1916: 37). Originally the gate was constructed as part of the southern rampants of Kila-e-Daulatabad Fort, built by Muzaffar II (d. 1526) at the start of the 16th century. If we knew when the gate received its new name, we would probably know when the rhino arrived.

7.3 The Rhinoceros of 1823

From May 1822, a trail of documents shows that Maharaja Sayajirao II had requested the Bombay Government help him to obtain a rhino (National Archives of India 1822–1823). Charles Lushington (1785–1866), Secretary to the Bengal Government, wrote on 16 May 1822 to David Scott (1786–1830), Commissioner in Cooch Behar for assistance in this matter. Scott obliged on 7 August 1823 sending an animal in charge of Gopaul Singh: "The beast is rather unruly and as this man manages him better than any other who has tried it, I think he had better be retained in charge of the animal." Transported via Kolkata and Mumbai, the new rhino must have arrived in Baroda in late August 1823. In view of the Government's involvement,

it was taken as a donation from the Governor General William Pitt Amherst, 1st Earl Amherst (1773–1857).

It was seen a couple of years later, on 19 March 1825, by Bishop Reginald Heber (1783–1826), when it was "so tame as to be ridden by a mohout." Then it is quiet for many years, until the young English traveller Henry George Briggs (1824–1872) sees "the surviving rhinoceros presented by Lord Amherst" on 28 January 1848. Briggs would have talked to the keepers of the animal, and it is quite possible that this rhino was in Baroda for some 25 years. There is no mention of any animal fights for this solitary rhino.

7.4 Rhino Fights of the 1850s

The first time that a fight is mentioned is on the occasion of the arrival of Colonel Cuthbert Davidson (1810–1862) as the new Resident in Baroda, in February 1856. The event was reported by a correspondent of the *Bombay Times* and may be repeated here, as such fights in later years followed a very similar pattern:

> The arena itself disclosed a concourse of men, with their loins girt for active movements, bearing in their hands long spears; and along the side wall, opposite the Gaekwad's position, were small door ways, at each of which were stationed two or three soldiers – the whole scene was picturesque. The two animals destined to afford amusement to the barbarous tastes

of the court, were brought forward, well chained and tied, and a curtain was held before each till they were made free; when they were let at each other, they began most scientifically, each remaining on the defensive, their tusks being crossed – presently they became fiercer, and the fight was a succession of goring, biting, pushing, with an occasional lift from the ground of the weaker on the tusk of his antagonist; after a time, they were with difficulty separated, to the relief, it is to be hoped, of the English portion of the spectators, – oceans of water were poured over them, as well during the encounter as after; the vanquished one exhibited fearful wounds from gashes and bites (*Bombay Times* 1856-03-01).

The rhino of 1823 may have been joined by a second specimen between 1848 and 1856, or maybe the first one died and the Gaekwad obtained a new pair. All we are told is that ten years later, in February 1866, there were three rhinos, which for once are described in some detail (*Bombay Gazette* 1866-02-14). The first was the largest rhino, standing 5 ft (150 cm) high, with just a stump for a horn, called "the veteran" indicating that it had been in the collection for a while. The second was nearly as large, very quiet, and wielded a large horn, said to be 1 ½ ft (45 cm) long and 15 in (38 cm) in circumference, actually so large that the measurement is likely to be exaggerated. A third rhino was still young, and had recently arrived, being "very good tempered, though stolid." Despite this unusual profusion of detail, it is still difficult to be sure of individual specimens. It is unlikely that the 1823 rhino could have lived until 1866, given that the maximum longevity in captivity of *R. unicornis* is some 40 years (Rookmaaker 1998: 35). Maybe one of the 1856 rhinos was alive in 1866, the second was added in the intervening years, while the third young one probably arrived in 1865 or 1866.

7.5   The Rhino Fight Witnessed by Rousselet in 1865

The French writer Louis Rousselet (1845–1929) traveled in India from July 1863 to September 1868. He stayed in Baroda from 22 May to 3 December 1865. He took up photography in 1865, but maybe the living animals in the menagerie were too hard for the new techniques, as no rhino was photographed (Vignau 1992; Renié 1992; Bautze 2006). Rousselet witnessed a fight between two rhinos in June 1865, produced a written account and made a sketch. After return to France, his travel account was serialized in the well-read magazine *Le Tour du*

Monde, where the rhino fight in Baroda was illustrated in an engraving attributed to the French illustrator Émile Bayard (1837–1891). Rousselet then published *L'Inde des Rajahs*, a grand book with 317 engravings, first in 1875 and reprinted in 1877. An English translation edited by Charles Randolph Buckle (1835–1920) had four contemporary editions, of which those first available 1875-10-15 and in 1876 were published by Chapman and Hall in London, followed by those of 1878 and 1882 by Bickers and Son in London. An American edition was available in 1876 in New York from Scribner, Armstrong and Co. The illustration of the rhinos remained unchanged, and was included in excerpts and translations, at least in French, English, Dutch, German, Italian and Spanish works:

1. Rousselet 1871, p. 244 "Combat de rhinocéros, à Baroda" in the French magazine *Le Tour du Monde*.
2. Rousselet 1872a, p. 197 "Kampf zweier Rhinocerossen in Baroda" in the German magazine *Globus*.
3. Rousselet 1872b, p. 289 "Gevecht van rhinocerossen" in the Dutch magazine *Aarde en Haar Volken*.
4. Rousselet 1875a, p. 124 "Combat de rhinocéros, à Baroda" in his book *L'Inde des Rajahs*.
5. Rousselet 1875b, p. 103 "Rhinoceros fight at Baroda" in the English translation of his book, *India and its Native Princes* (fig. 7.1).
6. Rousselet 1876a, p. 103 "Rhinoceros fight at Baroda" in the second printing of *India and its Native Princes*.
7. Rousselet 1876b, p. 103 "Rhinoceros fight at Baroda" in the only American edition (New York, Scribner, Armstrong, and Co.) of *India and its Native Princes*.
8. Rousselet 1876c, p. 377 "Ein Rhinozeroskampf in Baroda vor dem Prinzen von Wales" in the German illustrated magazine *Über Land und Meer*.
9. Rousselet 1877a, p. 124 "Combat de rhinocéros, à Baroda" in the second edition of *L'Inde des Rajahs*.
10. Rousselet 1877b, p. 101 "Combattimento di rinoceronti" in the Italian translation of his book *L'India*.
11. Rousselet 1878, p. 25 "Lucha de rinocerontes en Baroda" in the Spanish translation of his book *La India de los Rajas*.
12. Rousselet 1878, p. 103 "Rhinoceros-fight at Baroda" (note dash in title) in the third printing of *India and its Native Princes*.
13. Rousselet 1882, p. 107 "Rhinoceros-fight at Baroda" (including dash) with the engraving vertical to the page, unlike all other French or English editions, in the fourth printing of *India and its Native Princes*.
14. Schlagintweit 1880, vol. 1, p. 229 (and in 2nd edition 1890, vol. 1, p. 229) "Rhinoceros-Zweikampf."
15. Feudge 1880, p. 323 "A rhinoceros fight."

FIGURE 7.1
"Rhinoceros fight at Baroda." Signed (lower left) Émile Bayard del. and (lower right) J. Cauchard sc. The plates were made after drawings by Louis Rousselet, not from photographs. It depicts a scene in June 1865. The two horns are likely to be imaginary ROUSSELET, *INDIA AND ITS NATIVE PRINCES*, 1875, P. 103. COLLECTION KEES ROOKMAAKER

16. Jeyes 1896, vol. 2, p. 12 "Rhinoceros fight at Baroda."
17. Ridpath 1912, vol. 2, p. 693 "Rhinoceros fight at Baroda."

It will be seen in this engraving that the two rhinos were fighting in front of a large audience, while the Gaekwad is not depicted here. There are two keepers ready with buckets to cool the animals. One could wish that this were a photograph and not a sketch, because awkwardly both animals have two horns. This is incongruous with all the other evidence, and absent from Rousselet's written account. Hence I believe that the addition of the second horns was artistic license (Rookmaaker 1998: 111). This is a real pity, because Rousselet's illustration was widely distributed at the time and therefore would have given a wrong impression to many of his informed readers. Rather pointedly, nobody in his era commented on this anomaly.

The scene set by Rousselet is faithfully mirrored by an account by the anonymous correspondent of the *Bombay Gazette*, who more than likely attended the same occasion. The audience first witnessed a fight between elephants, and he continued his account of the day:

> It was then announced that a rhinoceros fight would take place. Accordingly, two of these hideous animals were produced and set perfectly free. One of the animals was older and more stolid than the other, who, on being unchained, jumped and bucked all over the place, with an agility which no one would even have supposed such a creature to possess. The older one seemed to know what was coming, and kept himself cool for the encounter. Presently they came close together, and watched each other angrily with heads down to the ground, the younger one, however, did not want to fight, and turned away pursued by the other, who could not catch him. This happened several times, and we began to think that they would not engage; the spearmen began to prick the young one smartly in the rear, but this made him turn and charge them, when they scuttled off through the apertures in the walls. One man nearly came to grief; he was irritating the young one from behind, when the old animal, who was pursuing the other, came up, and without seeing the man apparently, bowled him over with his shoulder. Presently both animals, being worked up to the requisite pitch of ferocity, engaged in earnest. Their mode of fighting is very peculiar; they put their heads down as low as possible, each trying to get his horned nose under the throat of his antagonist, and working away to get a good hold, the one whose head is uppermost tries all sorts of jerks and shifts to get free and sometimes succeeds. There is a great deal of very skillful and quick wrestling in this struggle. As soon as one of them gets a firm hold he heaves the other slowly up, pressing forwards as he lifts. When one was well lifted into the air he generally came down unhurt, and instantly renewed the struggle; it was indeed a wonderful sight to see these huge creatures alternately lifting one another up, and exerting their marvelous strength. The fight lasted about half an hour, the combatants being liberally doused with water, to lay the dust which would otherwise have obscured their maneuvers. Both were very much exhausted, and grunted terribly, but their fury was unabated. It was now, however, considered

that they had had enough of it, and they were separated by letting off fireworks close to their noses; and they were then chained and led slowly off (*Bombay Gazette* 1866-02-14).

The journalist had already introduced the actors, clearly describing 3 animals, old, adult and young (specimens C,D,E in table 7.16).

7.6   Rhino Fights of the 1870s

The Baroda rhinos were made to perform their fights regularly, probably at least once or twice per year, whenever there were special occasions to impress visitors or maybe just to entertain the local people. In the time of Maharaja Malharrao this was reported twice in 1871, and in the time of Savijirao II on 9 June 1875, 19 November 1875, 12 April 1879, 7 January 1881, 28 December 1881 and 30 December 1885. Loisel (1912, vol. 3: 18) commented on the fact that the rhinos were painted just before these fights. In 1865 one was red, the other black, while in 1875 they were both blotched with red patches. These colours are lost in the monochrome illustrations.

7.7   Rhino Fights Performed for the Prince of Wales in 1875

The most significant occasion, certainly the most widely reported in the British press, was the event staged during the visit to Baroda of Albert Edward, Prince of Wales (1841–1910), future King Edward VII, in November 1875. The Prince was accompanied by a large number of prominent persons (Russell 1877). The Prince's Honorary Private Secretary, William Howard Russell (1820–1907) reported on the voyage for the *Times* (episodes later combined in a book), and the artist Sydney Prior Hall (1842–1922) made the official sketches. The general public was kept informed about the daily events in the *Illustrated London News* and the *Graphic*. The Prince's visit to Baroda lasted three days. On Friday 19 November 1875, after paying respects to the young Gaekwad, the visitors all went to see a series of animal and gladiator fights in the arena, probably on the outskirts of the city. One of the correspondents, George Pearson Wheeler (1847–1902) described the rhino fight in succinct terms: "The most amusing conflict of the day was between two rhinosceri. The beasts, who were made still more hideous than they naturally are, by being blotched over with red paint, were loth to fight at all. The greater coward of the two elicited roars of laughter by

sneaking and dodging all round the ring, apparently with a desire to preserve some dignity in the midst of his panic" (Wheeler 1876: 113).

Four sets of images immortalized the scene. It will be seen that the rhinos on these plates (unlike those of Rousselet) are distinctly single-horned and *R. unicornis*. The *Illustrated London News* had engaged William Simpson (1823–1899) to provide illustrations of the Prince's Indian tour. His engraving showed the two rhinos in close-up, both animals with large horns, and front feet shackled. All keepers and spectators are behind the animals, and the Prince's party can be seen in the top left (fig. 7.2). This is found twice in the newspaper, and once in a German translation (nos. 18–20):

18. *Illustrated London News*, Saturday 25 December 1875, p. 621 "A rhinoceros fight at Baroda before the Prince of Wales. From a sketch by one of our special artists." The engraving is printed vertical to the page in large size (34 × 24 cm).

19. *Illustrated London News*, Saturday 13 May 1876, p. [5] "Exhibition of rhinoceros-fighting at Baroda." In a special "Welcome Home number" with the engraving horizontal on the lower half of the page.

20. *Über Land und Meer, Allgemeine Illustrirte Zeitung* (Stuttgart), no. 19, 7 May 1876, p. 377 "Ein Rhinozeroskampf in Baroda vor dem Prinzen von Wales (S. 383)."

The second plate appeared in *The Graphic*, another London-based illustrated weekly. They had employed Sydney Prior Hall as their artist on the Prince's tour. Two single-horned rhinos, both shackled, are seen fighting (fig. 7.3). The background shows a garden scene with spectators, while next to the animals on the right are two men with spears, on the left one man, and a further two attendants behind the animals' heads. It was reprinted several times in London and in Paris (nos. 21–25):

21. *The Graphic* (London), 25 December 1875, p. 620 (lower half) "The rhinoceros fight at Baroda, before the Prince of Wales. From a sketch by one of our special artists."

22. Gay 1876, frontispiece, "A rhinoceros fight" (lower right) "From the Graphic."

23. Herbaut 1876, p. 172 "Voyage de S.A.R. Le Prince de Galles aux Indes – Un combat de rhinocéros à Baroda. – Voir page 167" in the French magazine *l'Univers Illustré*, 11 March 1876.

24. Roe 1878, p. 285 "Combats d'animaux dans les Indes. – Combat de rhinocéros. (Page 284)" in the French *Journal des Voyages*, 12 May 1878.

Very soon after the first publication in *the Graphic*, the same illustration was reworked (and signed) by

Paul-Adolphe Kauffmann (1849–1940). This third image is not only reversed, showing the two men with spears on the left, but also some parts of the background are slightly altered (fig. 7.4):

25. P.L. 1876, p. 4 "Le voyage du Prince de Galles. – Un combat de rhinocéros à Baroda" in the French magazine *l'Illustration, Journal Universel* for 1 January 1876.

FIGURE 7.2
"A rhinoceros fight at Baroda before the Prince of Wales."
Drawing by William Simpson (1823–1899)
ILLUSTRATED LONDON NEWS, 25 DECEMBER 1875,
P. 621

FIGURE 7.3
"The rhinoceros fight at Baroda, before the Prince of Wales" first published in *The Graphic* of 25 December 1875 drawn by Sydney Prior Hall (1842–1922)
FROM THE FRENCH TRANSLATION IN *L'UNIVERS ILLUSTRÉ*, 11 MARCH 1876, P. 172

FIGURE 7.4
"Un combat de rhinocéros à Baroda" drawn by Paul-Adolphe Kauffmann (1849–1940). Reversed from *The Graphic* of 25 December 1875
L'ILLUSTRATION, JOURNAL UNIVERSEL, 1 JANUARY 1876,
P. 4

FIGURE 7.5   Detail of "Le voyage du Prince de Galles dans l'Inde"
               drawn by François Avenet (1850–1888)
               LE MONDE ILLUSTRÉ, 15 JANUARY 1876, P. 44

Finally, there is a fourth illustration associated with the Prince's visit to Baroda in 1875. This was signed by the continental artist Alexandre Fernandus, a pseudonym of François Avenet (1850–1888). It is likely that this was copied from an earlier sketch in an English magazine, but I have not been able to identify it. The rhinos are a small part of a page with different Indian scenes (fig. 7.5):

26. *Monde Illustré* 1876-01-15, p. 44. Part of a larger plate entitled: "Le voyage du Prince de Galles dans l'Inde. – Baroda, Goa et Ceylan. – (Dessin de M. Ferdinandus)."

7.8   Rhino Fights of the 1880s

The Gaekwad of Baroda continued to care for the two rhinos. They were again brought out to fight each other on the occasion of his first marriage on 7 January 1881, and at the end of the same year on 28 December 1881 when he reached his 21st birthday and was officially appointed to rule the city. There is just one later report for 28 December 1885, at the second wedding of the Gaekwad. The newspaper correspondent only saw one rhino in the menagerie, being told that another animal "his old enemy" is now dead. There is no record when this last Baroda rhino of the 19th century passed away. When the Viceroy, Rufus Daniel Isaacs (1860–1935) visited Baroda in 1921 (Panemanglor 1926: 109), the rhinoceros was no more than a distant memory.

The Discovery of the Fossil Record of Rhinoceros in South Asia

8.1 Discoveries of Siwalik Fossils[1]

The study of palaeontology in India started in the 1830s with the discovery of large deposits of fossil bones in a region of North India referred to as the Siwaliks. The young Scottish surgeon Hugh Falconer (1808–1865) was posted as Superintendent of the Botanic Garden at Saharanpur a year after his arrival in 1831. With a general interest in natural history, he made excursions to the Dehradun valley, where he found fragments of a fossil tortoise in the winter of 1831 and again in 1834, establishing a tertiary age for the area. He shared these interests with Proby Thomas Cautley (1802–1871), an engineer in the public works department who would later be responsible for the construction of the Ganges Canal from Haridwar and Roorkee to Aligarh completed in 1854 (J. Brown 1980). Cautley in turn talked with his fellow engineers in the canal department, the senior John Colvin (1794–1871) and his junior assistants William Erskine Baker (1808–1881) and Henry Marion Durand (1812–1871). In October 1834, the latter two were hammering at rocks near Nahan or Sirmor (now in Himachal Pradesh), when Fateh Prakash (1809–1850), the Rajah of Sirmur, showed them a large mastodon molar weighing 12 pounds (Yule & Maclagan 1882: 15, H. Durand 1883, vol. 1: 36, Rolfe-Smith 2013). Baker and Durand were soon joined by Colvin, Cautley and Falconer, each spending time in the hills and employing workmen to dig the earth. Their zeal over the next few years resulted in the recovery of many tons of fossil remains. They preserved the specimens and tried to identify them as best as possible, writing several papers about these discoveries to the astonishment of the scientific community. Colvin returned home in 1836, Baker took over his position and moved away from the area, Durand left the canal department in 1837. Their collections were donated to the Indian Museum in Kolkata, the University of Edinburgh and Ludlow Museum (W. Baker 1850), as well as to the British Museum (now Natural History Museum, London) and the museum in the headquarters of the East India Company in Leadenhall Street, London (MacGregor 2018).

Falconer and his colleagues had no exclusive rights. The British officer Lieutenant-Colonel Lewis Robert Stacy (1787–1848) sent a collector to Nahan in 1836 (Stacy 1836).

He donated a large set of Siwalik fossils to the University of Oxford in 1836, where they were received by the palaeontologist William Buckland (1784–1856). Although Buckland (1837) devoted an evening lecture to this donation at the Ashmolean Society in Oxford on 5 June 1837, there was no attempt to catalogue or classify the specimens systematically. The university received another smaller donation from Colonel Charles Pratt Kennedy (1789–1875) in the same period. Later incorporated in the Oxford University Museum of Natural History, there were only a few jaw fragments with worn teeth referable to unspecified rhino species (Stimpson et al. 2022).

Although Cautley continued his work on the canal until his departure in 1854, he had little time to pursue his palaeontological interests. His collections were too large for the Geological Society in London and were transported in 214 chests to the British Museum arriving in 1842. This formed the basis for their important Siwalik collection to be used by many generations of palaeontologists. Falconer's collections also ended up in the British Museum, with some specimens incorporated in the Falconer Museum in Forres, Scotland opened in 1871. The amount of material was clearly overwhelming, and while the specimens were duly catalogued (below), there appears to be no information about the exact locality where individual items were discovered or their place in the geological strata.

They were supposed to belong to the Siwalik fauna, now considered to be late Pliocene to early Pleistocene. Apparently these finds of the 1830s came from a triangular area with border points of Chandigarh, Shimla and Paonta Sahib (currently in Haryana and Himachal Pradesh), centered around Nahan (HP), as shown by Badam (1979, fig. 10).

8.2 The *Fauna Antiqua Sivalensis*

It is likely that Falconer's cooperation with Cautley became less efficient when he was appointed Director of the Botanic Gardens of Calcutta in 1847 followed by his return home in 1855. Their last publication together under the names 'Falconer and Cautley' was the *Fauna Antiqua Sivalensis*, which was set up to be the grand opus on the Siwalik fossils. Advertisements appeared as early as 1844 as seen on the back cover of one of the works by Charles Darwin (1844). The first part with 64 pages of text and 14

---

1 This chapter has benefited from reviews of earlier drafts by Denis Geraads and Pierre-Oliver Antoine.

© L.C. (KEES) ROOKMAAKER, 2024 | DOI:10.1163/9789004691544_009
This is an open access chapter distributed under the terms of the CC BY-NC-ND 4.0 license.

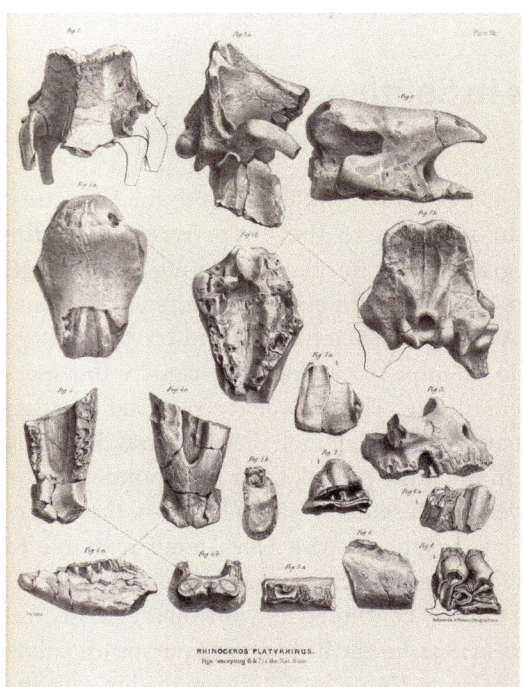

FIGURE 8.1   Skeletal remains and engraved new name of *Rhinoceros platyrhinus* attributed to Hugh Falconer (1808–1865) and Proby Thomas Cautley (1802–1871) dated 1847 *FAUNA ANTIQUA SIVALENSIS*, 1847, PL. 72. REPRODUCED BY KIND PERMISSION OF CAMBRIDGE UNIVERSITY LIBRARY

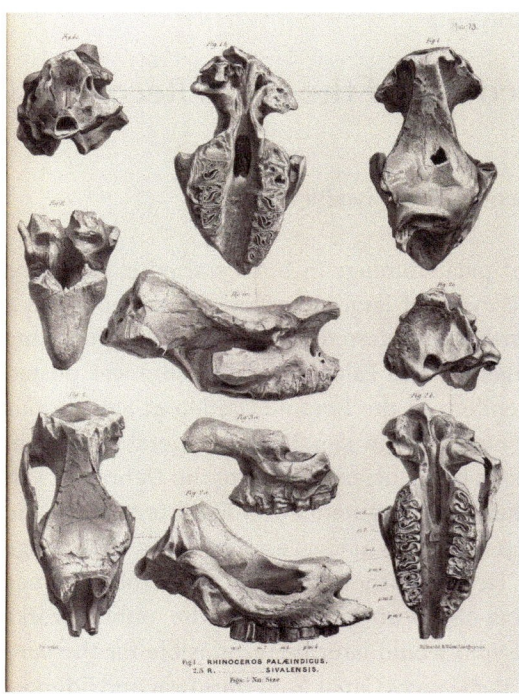

FIGURE 8.2   Skeletal remains associated with *Rhinoceros palaeindicus* and *R. sivalensis* attributed to Falconer and Cautley *FAUNA ANTIQUA SIVALENSIS*, 1847, PL. 73. REPRODUCED BY KIND PERMISSION OF CAMBRIDGE UNIVERSITY LIBRARY

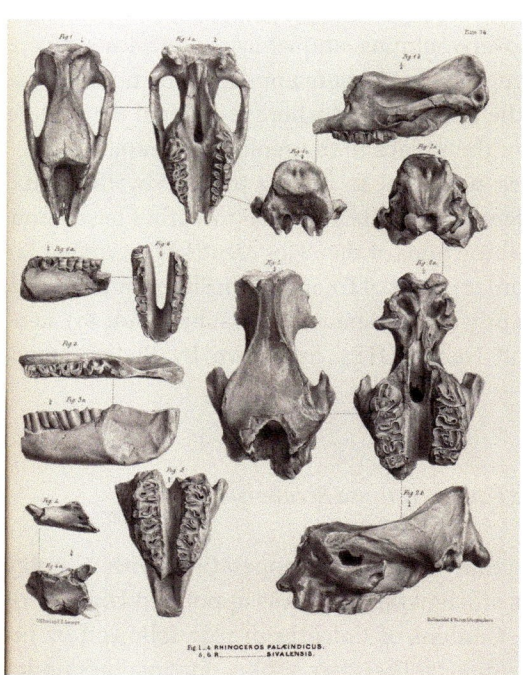

FIGURE 8.3   Skeletal remains associated with *Rhinoceros palaeindicus* and *R. sivalensis* attributed to Falconer and Cautley *FAUNA ANTIQUA SIVALENSIS*, 1847, PL. 74. REPRODUCED BY KIND PERMISSION OF CAMBRIDGE UNIVERSITY LIBRARY

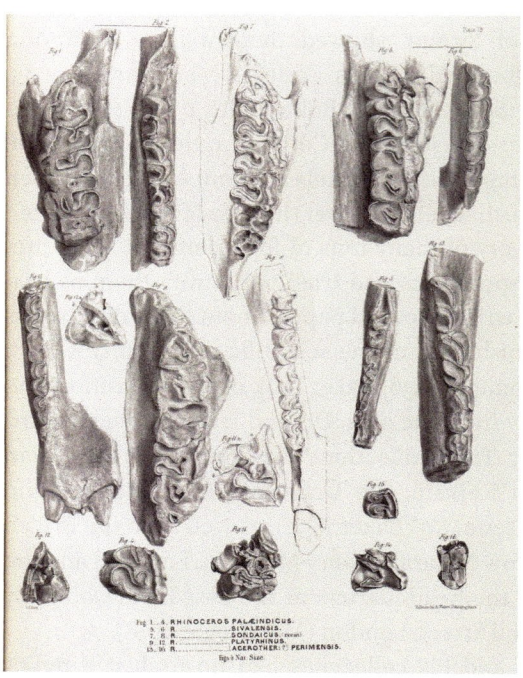

FIGURE 8.4   Skeletal remains of fossil Rhinocerotidae with engraved names *Rhinoceros palaeindicus, R. sivalensis, fossil R. sondaicus, R. platyrhinus, R. (Acerother.?) perimensis,* all attributed to Falconer and Cautley with date 1847 *FAUNA ANTIQUA SIVALENSIS*, 1847, PL. 75. REPRODUCED BY KIND PERMISSION OF CAMBRIDGE UNIVERSITY LIBRARY

plates on Proboscidea appeared in London in 1846, and the advertisements of the publishers Smith, Elder & Co. in London promised a total of 12 parts at intervals of four months (*The Atlas* 1847-01-02). However, all parts from the second onwards consisted of plates only without any explanatory text despite the price of a guinea per part and advertised "with descriptive letter-press." Part 9 of 1849 was the final one. Part 8 dated 1847 on the wrapper dealing with Suidae and Rhinocerotidae came with plates 69 to 80. Remains of different species of rhinoceros were shown on plates 72 to 79, each with multiple figures briefly explained in the engraved captions listing the species (figs. 8.1 to 8.4). These eight plates were drawn by Joseph Dinkel (1806–1891), George Henry Ford (1808–1876), B. George and Clara Cawse (fl. 1841–1867), and engraved by the lithographic firm of Charles Joseph Hullmandel (1789–1850) and Joseph Fowell Walton (1812–after 1863). The species names *Rhinoceros palaeindicus*, *R. platyrhinus*, *R. sivalensis*, *R. perimensis* were used here for the first time, and connected to the figures of specimens on the same plates, these are valid names attributed to Falconer and Cautley. Falconer was unable to contribute the promised text, and apparently there is no sign that any of it was ever written.

After Falconer died on 31 January 1865, his friend the surgeon Charles Murchison (1830–1879) edited all his published scientific papers, adding unpublished material found in notebooks or written on specimens. These *Palaeontological Memoirs and Notes* were collected in two illustrated volumes available in January 1868, including a chapter with descriptions of the plates in the *Fauna Antiqua Sivalensis*. The text of this chapter also appeared separately as a "reprint" (with different pagination) but awkwardly showing the date 1867 on the titlepage (unresolved). The books were a valiant effort to honour the name of a great scientist, while suffering from a lack of clear organization.

8.3 Perim Island Fossils

A second set of fossils from the Siwaliks age was discovered in 1835 on the small uninhabited Perim (Piram) Island in the Gulf of Cambay (Khambat), Gujarat, now considered to be from the late Miocene (Heissig 1972). The Scottish surgeon and botanist Charles Lush (1797–1845) had shown some of the finds to the Austrian traveler Baron Charles von Hügel (1795–1870), who sent samples to the Asiatic Society of Bengal in Kolkata (Hügel 1836). This caused quite a stir among the naturalists in the region, and further specimens, including a few rhinos, were in subsequent catalogues attributed to Major George

Fulljames (1809–1853) of the Bombay Native Infantry, Mrs. Leach, Albemarle Bettington (1812–1892) of the Bombay Civil Service, Walter Ewer (1787–1863) of the East India Company and Captain John Pepper (1794–1848), Commodore at Surat. Some of these remains were added to the collection of the Geological Society of London (Falconer 1845), and others are still preserved in the Indian Museum, Kolkata and the Natural History Museum, London.

8.4 Collections at the Asiatic Society in Kolkata

At the Asiatic Society of Bengal on Park Street, the fossil remains from the Siwalik hills in northern India and Perim were enthusiastically received by the Secretary James Prinsep (1799–1840). Apparently his enthusiasm was not matched by the ability of subsequent curators to deal properly with this enormous influx of large and heavy material. When Falconer was directing the Calcutta Botanic Gardens in the 1850s, he decided to catalogue the Siwaliks fossils at the Asiatic Society. This was done, assisted by Henry Walker (1804–1857), Professor of Anatomy and Physiology in the Medical College of Bengal, but not without considerable effort due to the state of the collections:

> "The main object of the present Catalogue was to identify, and determine the extent of the Indian Collections; and no language can exceed the appalling confusion, disorder and dilapidation in which they were found. Fossil bones from the Lias of England, from the Cape of Good Hope, Ava, Perim Island, the valley of the Nerbudda and the Sewalik Hills, huddled together in heaps, distributed over various rooms on the ground floor, and in ninety-nine cases out of the hundred, without a label or mark of any kind whatsoever to indicate whence they came! Many valuable specimens that had been presented to the Society were lost; others were found broken and the missing pieces either never recovered, or only after a search that extended over weeks. The only exception was in the case of Col. Colvin's second presentation, the specimens belonging to which were in glazed cases, in fair preservation and marked with distinctive numbers." And, "But this was not all. There is a pile of rubbish in the yard, known by a whimsical name, consisting of ejected materials, such as discarded rock specimens, broken corals, &c. It was the practice apparently, to pitch or sweep out fossil bones upon this heap" (Falconer & Walker 1859: 2,4).

Among the species of rhino, only one was identified to species level (*R. sivalensis*, p. 104), all others were left as "*Rhinoceros* sp."

8.5   Review by Richard Lydekker

These early efforts were continued by the English naturalist Richard Lydekker (1849–1915) in a series of memoirs on Siwalik animals. Educated at Cambridge, Lydekker in 1874 joined the Geological Survey of India, remaining in India until the death of his father Gerard Wolfe Lydekker (1811–1881). He then settled at Harpenden Lodge outside London (some 60 km from the NHM), where he married Lucy Davis and raised five children, Helen (1883–1956), Beatrice (1884–1970), Hilda (1886–1987), Gerard (1887–1917) and Cyril (1889–1915) (Clutterbuck 1983). Probably of independent means, he attended the Natural History Museum as a volunteer, producing scores of scientific and popular works on palaeontology, mammalogy and Indian natural history (Stearn 1981: 184). He was the author of the first name for the Central African white rhinoceros, *Rhinoceros simus cottoni* Lydekker, 1908 (now in genus *Ceratotherium*). Given his prolific output, influencing generations of scholars and amateur naturalists, a biography and scientific assessment is long overdue. While working in Kolkata in the 1870s, he examined the available fossil remains of rhinos and many other animal groups, reporting his finds largely in the *Palaeontologia Indica* "being figures and descriptions of the organic remains procured during the progress of the Geological Survey of India", a series of memoirs issued rather confusingly in 16 series of different topics, each with 1–4 volumes, each again divided into several parts. The Siwalik material was combined in Series 10 on "Indian Tertiary and Post-Tertiary Vertebrata" (1876–1881). Lydekker also catalogued the specimens of Siwalik mammals in the Indian Museum after their transfer from the Asiatic Society, followed in London by those in the British Museum (Lydekker 1885, 1886a).

8.6   The Fossil Species of Rhinoceros from South Asia

There is no indication that Falconer and his co-discoverers ever contemplated that the rhinos of the Siwaliks beds could be referable to any of the living species, *R. unicornis* or *R. sondaicus*. Lydekker (1880c: xii) stated quite clearly that it was unlikely that any of the recent species would have been found in the Pliocene or even in the Pleistocene. The Siwalik specimens therefore belonged to

extinct species, which naturally had never been named. This resulted in the description of a series of new species, which unfortunately were left poorly diagnosed through the failure to complete the text for the *Fauna Antiqua Sivalensis*. Lydekker's work was thorough, but suffered from a quick succession of papers in which new finds often made him change his mind about affinities and synonymies. However, this generosity in the establishment of new animal species was a common trend in natural history in the second half of the 19th century, when for instance John Edward Gray (1868) listed 10 species of recent rhinos (and 4 fossil ones) in his revision of the family (Rookmaaker 2015). The species described in the course of the 19th century from the Siwalik fauna even now still need a new reliable taxonomic revision to understand their status and validity (Antoine 2012, 2022; Turvey et al. 2020). While that is beyond the scope of this work, here are references to the 16 names proposed to fossil taxa in the genus *Rhinoceros* (as perceived by the describers) from South Asia, in chronological order.

Rhinoceros indicus fossilis Baker and Durand, 1836 (p. 593, pl. 15). Moginund, Morni Hills, Haryana. The same specimens were named *R. sivalensis* in 1847 and the name *fossilis* was only rarely mentioned as a synonym as in Lydekker (1881: 28). Maarel (1932: 68) showed that their material consisted of a mixture of various species. Although *fossilis* is senior to other names, it is likely that it should now be classed as a *nomen oblitum*, hence no longer to be used in nomenclature.

Rhinoceros angustirictus Cautley and Falconer, 1835 (p. 705). From tertiary strata in the Siwalik hills. Derivation of name *angustirictus*: Latin angustus (narrow) and Latin rictus (gaping, distended jaws). *Nomen nudum*, unavailable for purposes of nomenclature.

Rhinoceros palaeindicus Falconer and Cautley, 1847 (only on plates 73, 75, see figs. 8.2, 8.4). Nahan region. Upper Siwaliks. Late Pliocene to early Pleistocene. Derivation of name *palaeindicus*: Greek παλαιός palaios (ancient) and latinized adjective form of India. Remains obtained at Harwadi are in the Palaeontology Laboratory, Deccan College, Pune (Turvey et al. 2020: 6). This species belongs to the genus *Rhinoceros*, but its exact affinity remains unclear. It is taken as synonymous with *R. unicornis* by Laurie et al. (1983) and Antoine (2012).

Rhinoceros platyrhinus Falconer and Cautley, 1847 (only on plates 72, 75, see figs. 8.1, 8.4). Nahan region. Upper Siwaliks. Late Pliocene to early Pleistocene. Derivation of name *platyrhinus*: Greek πλατύρρινος (broad-nosed). The species was possibly two-horned (Antoine 2012). Because the name was attached to two figures which might in fact illustrate specimens belonging to two different species, the typology is uncertain. Matthew (1929: 534) selected NHM 36662 as type but also mentions NHM 36661 as neotype. The taxon was provided with a new diagnosis, based on the neotype NHM 36661, by Pandolfi & Maiorino (2016) as a valid species.

Rhinoceros sivalensis Falconer and Cautley, 1847 (only on plates 73, 74, 75, see figs. 8.2, 8.3, 8.4). Nahan region. Upper Siwaliks. Late Pliocene to early Pleistocene. Derivation of name *sivalensis*: latinized adjective form of Siwaliks (locality of fossil animal bones in northern India and Pakistan). This species is part of the genus *Rhinoceros*. It is regarded as synonymous with *R. sondaicus* by Groves & Leslie (2011) or with *R. unicornis* by Laurie et al. (1983) and Antoine (2012).

Rhinoceros (Acerotherium) perimensis Falconer and Cautley, 1847 (only on plate 75, see fig. 8.4). Perim (Piram) Island, Gujarat. Siwaliks, Late Miocene. Derivation of name *perimensis*: latinized adjective form of Perim. Specimens of the same species from the Miocene of the Bugti Hills, Baluchistan Province, Pakistan show that this is a valid species in a different genus (and different subfamily) known as *Brachypotherium perimense* (Antoine et al. 2013; Antoine 2023).

Rhinoceros namadicus Falconer, 1868 (p. 21). Nerbudda (Narmada) valley, Madhya Pradesh. Pleistocene. Derivation of name *namadicus*: latinized adjective form of Narmada as Namada. The name only was mentioned with locality: "a fourth [species of fossil rhinoceros] from the valley of the Nerbudda, *R. Namadicus*" (Falconer 1868a). *Nomen nudum*, unavailable for purposes of nomenclature.

Rhinoceros deccanensis Foote, 1874 (p. 4). Gokak, Belgaum Dt., Karnataka. Pleistocene. Derivation of name *deccanensis*: latinized adjective form of Deccan. The type specimen was found by Robert Bruce Foote (1834–1912) and described in 1874. A valid name which is rarely discussed. It is taken as a junior synonym of *R. unicornis* (Antoine 2012) or considered a *nomen dubium* being based on very incomplete material (Turvey et al. 2020: 6).

Rhinoceros namadicus Lydekker, 1876 (p. 14). Narbada (Narmada) valley, Madhya Pradesh. Richard Lydekker presented the first valid description of this species using an earlier name mentioned by Falconer in 1868, of which it is therefore a junior homonym. However, both in the explanation of the plates (Lydekker 1876, pl. 4 f. 5, 6) and in the later preface (Lydekker 1880c: viii), these specimens were referred to *R. unicornis*. This synonymy is later generally upheld (Hooijer 1946: 83; Laurie et al. 1983).

Rhinoceros planidens Lydekker, 1876 (p. 23). Siwaliks. Derivation of name *planidens*: Latin planus (flat) and Latin dens (tooth). The specimens were collected in the Siwaliks by William Theobald (1829–1908) of the Geological Survey of India. First described and figured as a new species by Lydekker (1876: 23, pl. 4 f. 7, 9), he (1879: 46, 1881a: 1) quickly changed his mind and synonymized it with *R. perimensis*, now *Brachypotherium perimense*. This is generally accepted (Hooijer 1946: 115).

Rhinoceros sivalensis var. *gajensis* Lydekker, 1881a (p. 40). Gaj beds. Lower Siwaliks. Late Oligocene and early Miocene. Derivation of name *gajensis*: latinized adjective form of Gaj. This early species has been referred to various genera, most recently to *Brachypotherium gajense* by Antoine et al. (2010) and Antoine et al. (2013).

Rhinoceros sivalensis var. *intermedius* Lydekker, 1884 (p. 5, pl. 1 f. 3). Lower Siwaliks. Late Oligocene and early Miocene. This species from the Chinji and Nagri formations has long been puzzling. Phylogenetic analysis (Antoine et al. 2003, fig. 4) and study of remains from the Potwar Plateau in Pakistan have led to the proposal of a new binomen *Diaceratherium intermedium* (Lydekker, 1884) by Antoine (2023).

Rhinoceros karnuliensis Lydekker, 1886c (p. 120). Karnool Dt., Andhra Pradesh (fig. 8.5). Late Pleistocene (Murty 1979). Derivation of name *karnuliensis*: latinized adjective form of Karnul. According to Hooijer (1946: 112), the species is probably indistinguishable from *R. sondaicus*, and has been taken as its junior synonym (Groves & Leslie 2011). Others were less certain and assigned it to *Rhinoceros* sp. (P. Roberts et al. 2014; Turvey et al. 2020). *Rhinoceros karnuliensish* of Lydekker (1886c: 121) and *Rhinoceros karnooliensis* Sathe, 2010 both are either incorrect subsequent spellings or inadvertent errors.

Rhinoceros sinhaleyus Deraniyagala, 1938 (p. 234). Sri Lanka, Ratnapura. Upper Pleistocene. This was first recognized by the zoologist and palaeontologist Paulus Edward Pieris Deraniyagala (1900–1976), who was director of the National Museum of Ceylon from 1939 to 1963. Details of locality in Dataset 8.4.

Rhinoceros sondaicus simplisinus Deraniyagala, 1946 (p. 162). Sri Lanka. Middle Pleistocene. Details in Dataset 8.4.

Rhinoceros kagavena Deraniyagala, 1958 (p. 122). Sri Lanka. Prehistoric age unknown. Details in Dataset 8.4.

8.7 Fossils from South and Central India

There were three early discoveries of fossil rhino specimens south of the Siwaliks. Robert Bruce Foote (1834–1912) surveyed fossil sites as part of his work for the Geological Survey of India from 1858 to 1891 (Pappu 2008). In May 1871, he was working in a riverbed near Chickdowlee (Chikdauli), Gokak, Belgaum Dt., Karnataka, and found a mandible and parts of a skull in good state of preservation.

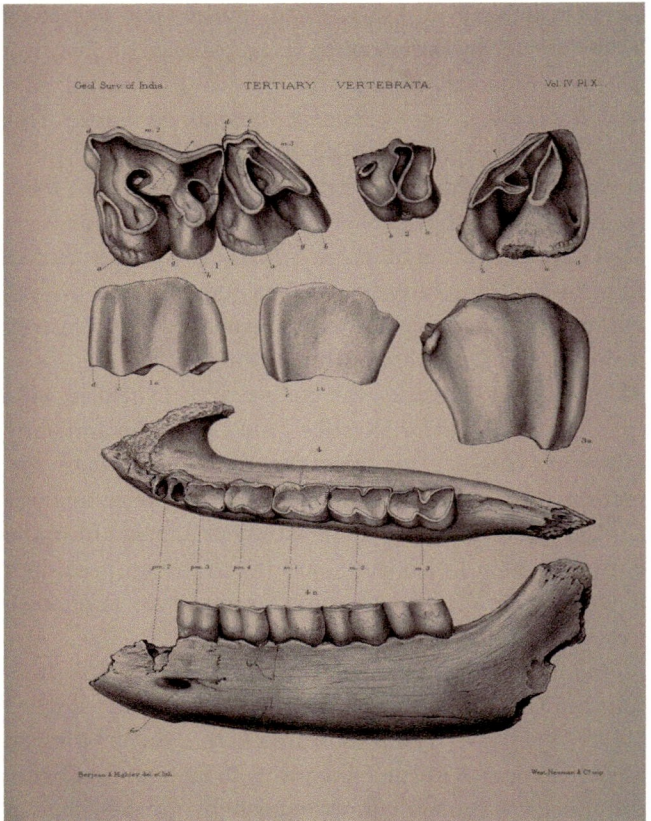

FIGURE 8.5   Specimens from the Kurnool Caves in Andhra
Pradesh. These include the type of *Rhinoceros
karnuliensis* (figs. 1, 2, 4) and a right upper true
molar assigned to *R. unicornis* found in a turbary
near Madras by Robert Bruce Foote (1834–1912) in
1871 (fig. 4). The plate was drawn and engraved by
Philibert Charles Berjeau (1845–1927) and Percy
Highley (1856–1929), published by the company of
William West, Newman & Co. of Hatton Gardens,
London
LYDEKKER, *MEMOIRS OF THE GEOLOGICAL
SURVEY OF INDIA, PALAEONTOLOGIA INDICA*, 1886,
SERIES 10, VOL. 4 PART 2, PL. 10

After comparing it with other known species, he described
it as *Rhinoceros deccanensis* in Foote (1874), stating that it
was smaller than *R. unicornis*.

In May 1884, the army officer Henry Bruce Foote
(1863–1932) helped his father in the exploration of the
Billa Surgam caves in Betamcherla, Kurnool Dt., Andhra
Pradesh (Foote 1884; K.N. Prasad 1998). He found teeth
and a mandible in the Cathedral and Charnel-House
caves. These were provisionally assigned to *R. sondaicus*,
but when further examined by Lydekker, he described
this Pleistocene species as *Rhinoceros karnuliensis*
(Lydekker 1886c: 120) and also figured the better specimens in the Indian Museum (Lydekker 1886b, pl. 10;
fig. 8.5). This same plate includes the representation of a

right upper true molar assigned to the recent *R. unicornis*
found in a turbary near Madras (Chennai), presumably
also found by Foote (Lydekker 1886b: 42).

A few fossil remains from the Narmada valley in central India were described as *R. namadicus* by Lydekker
(1876: 14), but the distinction of the species remained
uncertain due to lack of material. In the same general
area in July 1881, John Cockburn, the discoverer of the
first petroglyph (9.3), found a semi-fossilized rhino skull
in the valley of the Ken River, Uttar Pradesh, which he
thought would prove to be the same species as *R. unicornis*
(Cockburn 1883).

This paper by Cockburn was one of the first to suggest that there might be fossilized remains of rhinos in
India belonging to a recent species. Since his time, rhino
remains have been found in an increasing number of localities across the South Asian subcontinent (dataset 8.4).
These date from early Pliocene (Siwaliks) times especially
in Pakistan, but also from the more recent Pleistocene
(Mesolithic, Neolithic and Harappan). The literature on
these latter findings is extensively reviewed by Shibani
Bose (2014, 2018a, 2020) collating the base data for further
research. Following Bose's analysis, it is generally assumed
that the remains dating from the Late Pleistocene onwards
belong to *R. unicornis*, or otherwise a very closely allied
species. All palaeontological evidence has been incorporated by Geraads et al. (2021) in a comprehensive database
of Neogene and Quaternary rhino-bearing localities facilitating a more detailed understanding of the taxonomic
implications.

8.8   *Rhinoceros sondaicus* in Pakistan

Two sets of more ancient fossils from the Siwaliks
fauna in Pakistan have been referred to *R. sondaicus*
(A. Khan 2009). The palaeontologist Kurt Heissig (b. 1941),
the former Curator at the Bayerischen Staatssammlung
für Paläontologie und Geologie in Munich, recorded two
upper premolars assigned to *Eurhinoceros* aff. *sondaicus* among material collected in Pakistan in 1955–1956
(Heissig 1972: 29). More recently, both Abdul et al. (2014)
and Siddiq et al. (2016) identified *R. sondaicus* from two
premolars and a mandible found at Jari Khas, Mirpur
District and the Sardhok area, Gujrat District. The gap
between these finds and other occurrences of the species
may point at misidentifications attributed to the incom-
pleteness of the specimens. This should be the subject
of further investigations once other material becomes
available.

TABLE 8.17   List of sites where fossil rhinoceros remains have been found

Country or state	Palaeontological age	
	Mesolithic – Neolithic	Harappan or similar
Pakistan		Amri, Nausharo, Harappa
Gujarat	Langhnaj, Kanewal, Valotri, Khaksa	Surkotada, Shikarpur, Kuntasi, Lothal, Khanpur, Oriyo Timbo
Rajasthan		Kalibangan, Karanpura
Haryana		Madina
Madhya Pradesh	Narmada (Narbada)	
Uttar Pradesh	Damdama, Sarai Nahar Rai, Mahadaha	Banda
Bihar	Chirand	
Maharashtra	Inamgaon, Manjra Valley	
Andhra Pradesh	Kurnool	
Karnataka	Gokak	
Tamil Nadu	Payampalli, Sathankulam	

8.9   Subfossil Remains Excavated in India

In India the Pleistocene and Holocene rhinoceros remains were excavated from Mesolithic sites (ca. 7000 BCE) in Bihar, Uttar Pradesh, Madhya Pradesh and Gujarat, and in Harappan sites (ca. 2000 BCE) in Pakistan, Rajasthan, Haryana, Uttar Pradesh, and Gujarat (Badam 2013; Bose 2020: 303–305). A few remnants were found in the southern states of Maharashtra, Andhra Pradesh and Tamil Nadu (table 8.17). Younger remains which are sometimes dug up by chance are called subfossil here, and are treated like records of the modern species.

The number of sites with fossilized rhino remains is quite small. Bose (2020) recorded just 8 Mesolithic or Neolithic localities and 12 Harappan localities. Excavations outside her study area add no more than six further localities. Only five of these localities are within the known historical range of *R. unicornis*, therefore most fossils or subfossils were found outside the present range of the species, dictated by a climate more suited to preservation of skeletal material, but maybe less suited to the living rhinos. This leads to discussions of possible changes in the climate from wet forests to dry deserts in the course of time (Chitwalwala 1990; Manuel 2005), which may be pursued in a broader context.

The rhino is never a dominant species in the number of specimens. In most sites the animal is represented by just a few bones, many partially preserved. This is a limiting factor in the proper assessment of the implications for zoogeography, taxonomy and evolutionary processes well known to palaeontologists. Researchers from Lydekker (1886a) to Badam & Jeyakaran (1993) have highlighted the problem of the absence of comparative material of recent rhino species in Indian collections. There are several skeletons in the Indian Museum in Kolkata, but very few elsewhere. While most taxonomies of recent mammals are based on skull morphology and size, palaeontologists need to work with parts of bones from the postcranial skeleton, which therefore makes comparisons a difficult proposition. Guérin (1980) is one source where every bone of the recent species of rhino is illustrated, described and measured, which must be combined with an examination of the fossil bones themselves. In the descriptions of fossil specimens found in India, large mammal bones sometimes seem to be assigned to *R. unicornis* on the basis of their size. It is rare to find a comparison with *R. sondaicus*, or with the potentially older species like *R. namadicus*, *R. sivalensis*, *R. deccanensis* or *R. karnuliensis*. Budgetary and institutional restraints are hard to circumvent. At the moment there is a healthy population of *R. unicornis* in Nepal and in India, where of course some animals will die a natural death. It would be advisable to try to supply complete skeletons of adult males and females with complete data to key institutions, zoological as well as palaeontological, to help with future research activities.

8.10   Presence in Southern India

The presence of *R. unicornis* in the southern states of Maharashtra, Andhra Pradesh, Karnataka and Tamil Nadu needs further investigation and explanation. The few fossil remains seem to indicate that this species must have lived there until at least the end of the Pleistocene. A full picture does not yet emerge on the basis of less than a handful of fossil finds.

8.11   The Fossil Record of South Asian Rhinos

Despite the sparsity of the fossil record, the rhino once clearly had a wider distribution in South Asia than today.

For the Upper Pleistocene (ending 11,700 years ago) and subsequent periods of the Holocene, the species indicated is *R. unicornis* (Lydekker 1892; Badam 1979; Laurie et al. 1983: 2; Antoine 2012). The slight possibility of a wider

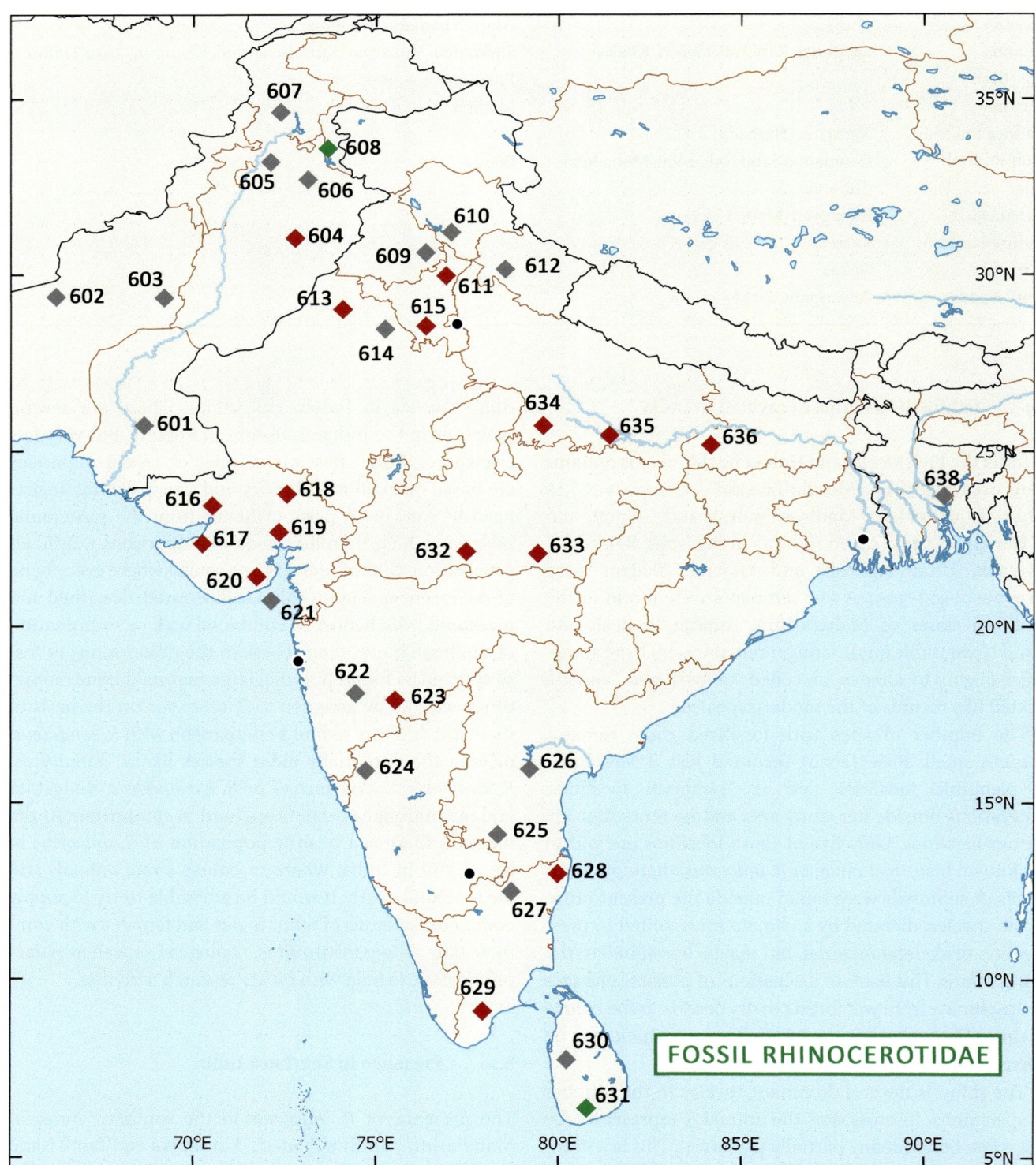

MAP 8.1   Localities where fossil remains of rhinoceros have been reported in South Asia. The numbers and places are explained in Dataset 8.4
◇   fossil records: grey = extinct or unknown species; red = *R. unicornis*; green = *R. sondaicus*
The species identifications follow the sources presented in Dataset 8.4.

former range of *R. sondaicus* and other allied (now extinct) species needs further investigation. There are no remains pointing at *D. sumatrensis*. Assuming it was *R. unicornis*, its distribution once extended to areas in western and southern India and Pakistan where it no longer occurs, and has not occurred in the historical period for which we have some written records as shown in the remainder of this book. However, the vestiges are slight and will need to be placed in a broader palaeontological context to resolve the multitude of current uncertainties.

The species identifications follow the sources presented in Dataset 8.4.

•••

Dataset 8.4: List of Localities Where Fossil Remains of Rhinoceros were Found in South Asia

The records are listed according to country and region (Northern India, Southern India, Sri Lanka, Pakistan) and for Indian localities according to states in alphabetical order. The numbers and places are shown on map 8.1.

The sequence of each record gives (first line) Locality – Coördinates – Map numbers; and (next line) Species – Palaeontological age (according to sources) – Material and Specimens – Sources.

Northern India

Bihar. Chirand, Saran Dt. – 25.74N; 84.82E – Map W 636
   R. unicornis – Neolithic, 1700 BCE – Fragments of ulna, left tibia, upper molar, complete left humerus (Zoological Survey of India). Excavated by Bindeshwari Prasad Sinha (1917–2002) of Archaeological Department, Bihar – Nath 1976; Nath & Biswas 1980; Manuel 2008; Bose 2020: 64

Gujarat. Kanewal Lake (Khaksar, Valotri), Kheda Dt. – 22.49N; 72.51E – Map W 619
   Rhinoceros sp. – Neolithic, 2000 BCE – 3 skeletal fragments – K. Momin et al. 1973; Bose 2020: 49

Gujarat. Kuntasi, Rajkot Dt. – 22.90N; 70.58E – Map W 617
   R. unicornis (assumed) – Harappan, 1900–1700 BCE – Bone fragments (unspecified). Found by Pappy Kuzuvelil Thomas (Deccan College) – Thomas et al. 1995, 1997; Bose 2020: 73

Gujarat. Khanpar, Morbi Dt. – 22.75N; 70.64E–Map W 617
   Rhinoceros sp. – Late Harappan, 1500–1000 BCE – 1 humerus. Found by Pappy K. Thomas in 1970s – Chitalwala 1990: 80; Bose 2020: 75

Gujarat. Langhnaj, Mehsana Dt. – 23.45N; 72.50E – Map W 618
   R. unicornis – Microlithic, 2000 BC – Left scapula, right humerus, talus, fragment of molar – Zeuner 1952; Clutton-Brock 1965; Bose 2018a: 37, 2020: 50

Gujarat. Lothal, Saragwala Dt. – 25.51N; 72.25E – Map W 619
   R. unicornis (assumed) – Harappan, 2000 BCE – Partial mandible, 2 terracotta figures – Rao 1962: 28, 1985: 485; Nath 1968, Nath & Rao 1985; Bose 2018a: 47, 2020

Gujarat. Oriyo Timbo, Bhavnagar Dt. – 21.80N; 72.10E – Map W 620
   R. unicornis (assumed) – 2980–2525 BCE – First phalanx. Evidence by Paul Charles Rissman – Rissman 1985: 160

Gujarat. Perim (Piram) Island, Gulf of Cambay – 21.60N; 72.35E – Map W 621
   R. perimensis – Upper Miocene – Pictured by Hugh Falconer (1808–1865) and Proby Thomas Cautley (1802–1871) – Falconer & Cautley 1847, pl. 75; Lydekker 1881a: 27; Hooijer 1946: 115

Gujarat. Perim (Piram) Island, Gulf of Cambay – 21.60N; 72.35E – Map W 621
   R. planidens – Upper Miocene – Described in 1876, referred to *R. perimensis* in 1880 – Lydekker 1876: 23, 1880c: xiii

Gujarat. Shikarpur (Valamiya Timbo), Kutch Dt. – 23.25N; 70.70E – Map W 616
   R. unicornis (assumed) – Harappan 1900–1700 BCE – Bone fragments (unspecified). Found by Pappy K. Thomas – Thomas et al. 1995, 1997; Bose 2020: 73

Gujarat. Surkotada, Kutch Dt. – 23.63N; 70.84E – Map W 616
   R. unicornis – Harappan 1790–1660 BCE – Few bones (unspecified) – Bose 2020: 73

Haryana. Pinjore valley, Siwaliks – 30.80N; 76.92E – Map W 611
   Rhinoceros sp. (fossil) – age unspecified – Cautley 1835

Haryana. Madina (near Delhi) – 28.95N; 76.42E – Map W 615
   R. unicornis – Harappan 1400–1200 BCE – Material: astralagus – Joglekar & Sharada 2016; Bose 2020: 70

Haryana. Moginund, Morni Hills – 30.73N; 77.00E – Map W 611
   R. unicornis fossilis – Pleistocene – Remains unspecified – Baker & Durand 1836; Hooijer 1946: 53; Laurie et al. 1983

Haryana. 6 miles east of Chandigarh – 30.72N; 76.97E – Map W 611
   R. sivalensis – Pleistocene (Upper Siwaliks) – Molar, specimen in AMNH New York. The fossil remains from the Siwaliks were collected by Barnum Brown (1873–1963) in 1924 – B. Brown 1925; Colbert 1935: 180

Himachal Pradesh. Nahan Dt. – 30.55N; 77.30E – Map W 610
   R. palaeindicus – Pleistocene (Upper Siwaliks) – Various skeletal elements – Falconer & Cautley 1847, pl. 73 f. 1, pl. 74 f. 1–4, pl. 75 f. 1–4; Pandolfi &Maiorino 2016

Himachal Pradesh. Nahan Dt. – 30.55N; 77.30E – Map W 610
   R. platyrhinus – Pleistocene (Upper Siwaliks) – Various skeletal elements – Falconer & Cautley 1847, pl. 72 f. 1–7, pl. 75 f. 9–11; Falconer 1868b, vol. 1, pl. 14

Himachal Pradesh. Nahan Dt. – 30.55N; 77.30E – Map W 610
   R. sivalensis – Pleistocene (Upper Siwaliks) – Various skeletal elements – Falconer & Cautley 1847, pl. 73 f. 2–3, pl. 74 f. 6, pl. 75 f. 5–6; Falconer 1868b, vol. 1, pl. 14

Madhya Pradesh. Sher River, Central Narmada valley – 22.65N; 78.70E – Map W 633
   R. unicornis – Pleistocene, 40.000 BCE – 3 horn cores – Badam & Sankhyan 2009; Bose 2020: 40

Madhya Pradesh. Central Narmada valley – 22.70N; 78.00E – Map W 632
   R. unicornis – Late Pleistocene, 11.000 BCE – Skeletal parts. Molars collected by William Theobald (1829–1908) in 1850s. – Lydekker 1880b: 33, 1886a: 12, 13, 1900: 22; Blanford 1891: 473; Hooijer 1946: 83; Badam 1979: 225; Manuel 2006

Madhya Pradesh. Central Narmada valley – 22.70N; 78.00E – Map W 632
   R. unicornis – Pleistocene – Left astragalus (no. 39686) presented by Charles Fraser (1799–1868), Bengal Civil Service – Falconer & Cautley 1847, pl. 76 fig. 18; Lydekker 1886d: 138

Madhya Pradesh. Narbada (Narmada) – 22.70N; 78.00E – Map W 632
  R. deccanensis – Pleistocene – Various skeletal remains – Lydekker 1880b: 33
Madhya Pradesh. Narbada (Narmada) – 22.70N; 78.00E – Map W 632
  R. namadicus – Pleistocene – Various skeletal remains – Falconer 1868a: 21; Lydekker 1876: 14, 1880b: 33, 1886a: 13; Trouessart 1898: 754
Madhya Pradesh. Narbada (Narmada) – 22.70N; 78.00E – Map W 632
  Rhinoceros sp. – Pleistocene – Bones collected by Charles A. Hacket of the Geological Survey – Lydekker 1886a: 13; Chauhan 2008
Punjab. Bari Nagal, Tehsil Kharar, Rupar Dt. – 30.75N; 76.60E – Map W 609
  R. barinagalensis Srivastava and Verma, 1972 (new species) – Lower Pleistocene, Pinjor stage – Mandible – Srivastava & Verma 1972; Laurie et al. 1983
Punjab. Gurha village, near Chandigarh – 30.85N; 76.70E – Map W 609
  Punjabitherium platyrhinum – Upper Siwaliks (Pinjor stage) – Specimen unspecified – E. Khan 1971
Punjab. Mirzapur and surroundings – 30.90N; 76.73E – Map W 609
  Coelodonta platyrhinus – Lower Pleistocene – Fragments of 2 maxillae in AMNH, New York – Pandolfi & Maiorino 2016
Punjab. Mirzapur and surroundings – 30.90N; 76.73E – Map W 609
  R. platyrhinus – Lower Pleistocene – Various bones and teeth. Locality taken from Colbert (1935) for the species – Pandolfi & Maiorino 2016
Punjab. Mirzapur Dt. – 30.90N; 76.73E – Map W 609
  Rhinoceros (unspecified) – Late Pleistocene – Specimen unspecified – Misra 1984: 45; Manuel 2008: 33
Punjab. Siswan – 30.86N; 76.74E – Map W 609
  Coelodonta platyrhinus – Lower Pleistocene – Fragments of maxillae in AMNH, New York – Colbert 1935: 178
Rajasthan. Kalibangan, Ganganagar Dt. – 29.43N; 74.15E – Map W 613
  R. unicornis – 2550–2000 BCE – Fragments of left tibia, right humerus, 4th metatarsal, 3rd metatarsal – Banerjee & Chakraborty 1973; Bose 2020: 72
Rajasthan. Karanpura, Hanumangarh Dt. – 29.12N; 75.05E – Map W 614
  Rhinoceros (unspecified) – Shoulderbone, pelvis, radius, ulna – Bose 2020: 72 after V.N. Prabhakar (unpublished)
Uttar Pradesh. Damdama, Pratapgarh Dt. – 25.52N; 79.70E – Map W 634
  R. unicornis – Mesolithic – 26 fossil bones – Thomas et al. 1995; Bose 2014: 67, 2020: 44
Uttar Pradesh. Mahadaba, Pratapgarh Dt. – 25.90N; 82.10E – Map W 635
  R. unicornis – Mesolithic – Molar teeth and a cervical vertebra. One bone identified as *R. unicornis* – Pandey 1990: 313; Bose 2014: 67, 2020: 47
Uttar Pradesh. Sarai Nahar Rai, Pratapgarh Dt. – 25.82N; 81.83E – Map W 635
  Rhinoceros sp. – Mesolithic, 8000 BCE – Bone – G. Sharma 1975; Manuel 2007: 233, 2008: 33; Bose 2014: 67, 2020: 46
Uttar Pradesh. Ken River, Banda – 25.40N; 80.30E – Map W 634
  R. unicornis – Semi-fossilized – Skull fragments and teeth. Collected by John Cockburn in July 1881. (Indian Museum, Calcutta, F.125) – Cockburn 1883: 56; Lydekker 1886a: 13, 1900: 22; Blanford 1891: 473
Uttarakhand. Hardwar. Low hills eastwards – 29.85N; 78.35E – Map W 612
  Rhinoceros sp. – Siwaliks – Molars – Falconer 1837: 233

Southern India

Andhra Pradesh. Kurnool (Karnul) caves, Betamcherla, Kurnool Dt. – 15.48N; 78.15E – Map W 626
  R. karnuliensis – Pleistocene – Third right upper premolar (Geological Survey of India F.236) collected by Henry Bruce Foote (1863–1932) in 1884 – Lydekker 1880c: 120, 1886b: 121, 1900: 22; Hooijer 1946: 112; Badam 1979: 211; Patnaik et al. 2008; Sathe 2010; Groves & Leslie 2011
Andhra Pradesh. Billasurgam cave complex, Kurnool Dt. – 15.43N; 78.18E – Map W 626
  Rhinoceros (unspecified) – Pleistocene – Material unspecified – Roberts et al. 2014
Andhra Pradesh. Palavoy, Anantapur Dt., 8 km SE of Kalyandurg – 14.50N; 77.15E – Map W 625
  Rhinoceros (unspecified) – South Indian Neolithic, 2278–1680 BCE – Material unspecified – Sathe 2010
Karnataka. Chickdowlee (Chikdauli), Gokak, Belgaum Dt. – 16.15N; 74.80E – Map W 624
  R. deccanensis – Late Pleistocene – Mandible, maxilla, few other fragments. Collected by Robert Bruce Foote (1834–1912) in 1871 – Foote 1874: 4, 1876; Lydekker 1886a (F.168); Krishne Gowda 1975: 309; Chauhan 2008
Maharashtra. Manjra Valley, Harwadi, Latur Dt. – 18.50N; 76.60E – Map W 623
  R. cf. unicornis – Age unspecified – 3 mandibles, 1 metacarpal, 1 radius distal fragment – Sathe 2010, 2015
Maharashtra. Inamgaon, Ghod River – 18.60N; 74.50E – Map W 622
  Rhinoceros (possibly *R. unicornis*) – Harappan, 1600–700 BCE – Bone fragments. In her pivotal work, Anneke Trientje Clason (1932–2008), Dutch archaeozoologist, was unsure if these large bones belonged to elephant or rhinoceros, and she gave both *R. unicornis* and *D. sumatrensis* as possibilities – Clason 1977: 19; Badam 1985: 414
Tamil Nadu. Madras (Chennai) – 13.57N; 79.75E – Map W 628
  R. unicornis – Pleistocene (?) – Right upper true molar found in a turbary by R.B. Foote – Lydekker 1886b: 42, pl. 10 fig. 3
Tamil Nadu. Payampalli, Vellore Dt. – 12.60N; 78.65E – Map W 627
  Rhinoceros (unspecified) – Neolithic, 1390 BCE – Material unspecified – Rao 1968: 28; Badam & Jayakaran 1993: 250
Tamil Nadu. Sathankulum, Tirunelveli Dt. – 08.45N; 77.90E – Map W 629
  R. unicornis – Late Pleistocene – Anterior portion of skull. Reconstructed for Chennai Museum by Gyani Lal Badam (b. 1940) – Badam & Jayakaran 1993

Bangladesh

Dhaka Division. Kapasia – 24.12N; 90.57E – Map W 638
  Rhinoceros (unspecified) – Remains unspecified, excavated in 1982 – K.Z. Hussain 1985

Sri Lanka

Kuruvita Gem Pit, Hiriliyadda – 07.48N; 79.90E – Map W 630
  R. kagavena Deraniyagala, 1958 (new species) – Age unspecified – Teeth – Deraniyagala 1958: 122
Pothu kola Deniya, Nivitigala, Sabaragamura Province – 07.00N; 80.45E – Map W 631
  R. sondaicus simplisinus – Middle Pleistocene – Teeth – Deraniyagala 1946: 162
Ratnapura – 06.50N; 80.50E – Map W 631
  R. sinhaleyus – Upper Pleistocene – Teeth – Deraniyagala 1938: 234

Pakistan

Amri, Dadu Dt., Sindh – 26.15N; 68.00E – Map W 601
  Rhinoceros (unspecified) – 3660–3020 BC – Remains unspecified – Bose 2020: 70

Bugti Hills, Balochistan – 29.00N; 69.10E – Map W 603

Rhinoceros sp. [*sivalensis, palaeindicus, unicornis*] – Miocene – Presence of several forms in the Siwalik strata – Antoine et al. 2010, 2013; Antoine 2012

Bugti Hills, Balochistan – 29.00N; 69.10E – Map W 603

Brachypotherium perimense – Miocene – Various skeletal remains – Antoine et al. 2013; Antoine 2022

Chakwal Dt., Punjab – 32.95N; 72.85E – Map W 606

Brachypotherium perimense – Miocene – Several premolars (University of the Punjab, Lahore) – Khan et al. 2012; Antoine 2022

Chinji Zone, Attock Dt., Punjab – 33.77N; 72.37E – Map W 605

Gaindatherium browni – Upper Tertiary – Near-complete skull (type AMNH 19409) – Colbert 1934

Harappa, Sahiwal Dt., Punjab – 30.65N; 72.00E – Map W 604

R. unicornis – 2500–1500 BC – Fragment of right scapula, length 47 cm. Mound F, Trench VI – Prashad 1936; Nath 1968; Bose 2020: 70

Jari Kas, Mirpur Dt., Kashmir – 33.13N; 73.75E – Map W 608

R. sondaicus, Punjabitherium platyrhinus – Pleistocene (Pinjor) – Remains unspecified – Abdul et al. 2014

Jari Kas, Mirpur Dt., Kashmir – 33.13N; 73.75E – Map W 608

R. kendengensis – Right upper third premolar – Sarwar 1971

Malhur, Haripur Dt., Khyber – 34.04N; 72.93E – Map W 607

R. perimensis – Miocene – Remains unspecified – Hooijer 1946: 114–117

Nausharo, near Mehrgarh, Kachi Dt., Balochistan – 29.46N; 67.57E – Map W 602

Rhinoceros (unspecified) – Harappan – Remains unspecified – de Geer 2008: 381; Bose 2020: 70

Potwar Plateau – 33.00N; 72.40E – Map W 606

Brachypotherium perimense, Rhinoceros sp., Rhinoceros aff. sondaicus, Rhinoceros aff. sivalensis, Dicerorhinus aff. sumatrensis – Siwaliks (Miocene) – Various skeletal remains – Antoine 2023

Sar Dhok, Jhelum Dt., Punjab – 32.96N; 73.68 – Map W 608

R. sondaicus – Pleistocene (Pinjor) – 2 premolars and a mandible – Khan 2009: 40; Abdul et al. 2014; Siddiq et al. 2016

Tatrot, Jhelum Dt., Punjab – 32.90N; 73.50E – Map W 606

R. sivalensis – Upper Siwalik – Remains unspecified – Abdul et al. 2014

⁛

The Rhinoceros in the Rock Art of Central India

9.1 Rock Art Depicting the Rhinoceros

The rhinoceros has the distinction of being included in the first rock painting or petroglyph discovered in India. The Ghormangur rock-shelter was found in 1883 by the amateur archaeologist John Cockburn in the Mirzapur District of Uttar Pradesh (9.2). Cockburn soon published a tracing and an extensive description starting a chain of many further discoveries in central India.

Most rock art shelters are concentrated in the Vindhya Hills around Bhopal in Madhya Pradesh and in Uttar Pradesh, while single sites are scattered across Odisha, Maharashtra, Rajasthan and Gujarat (9.4). These have been intensively studied over the years through the efforts and dedication of Vishnu Shridhar Wakankar (1919–1988) and the Austrian archaeologist and ethnologist Erwin Neumayer (b. 1949). My synthesis has benefited from the discussions and especially the illustrations of many examples in Neumayer (2011). The original drawings are usually painted with colours and are often in rock shelters in inaccessible hill areas. The study of their subject matter is facilitated by the tracings, which show the shape of animals like the rhinoceros clearly identifiable by the horn.

A selection of tracings depicting rhinos is sufficient to show that this animal played a role, even if minor, in the lives of the people of the ancient civilisation responsible for the art work. Based on the illustrations and a limited review of the literature, it is here investigated how these depictions help us to understand the distribution of the rhino in previous ages, discussing their age (9.5), the localities (9.6), the specific identity of the rhinos (9.7), ending with conclusions concerning the animal's range (9.8).

...

Dataset 9.5: List of Localities with Rock Paintings or Petroglyphs including One or More Representations of Rhinoceros

Sites (only found in India) arranged according to state and region, arranged from west to east. Rhino stands for a representations showing at least one rhinoceros. The numbers and places are shown in map 9.2.

General sequence of entries: Locality – Map coordinates – Figures – Map number; (Second line): Detail of petroglyph – References.

Rhino Region 04: Gujarat
Tarsang, Gujarat – 22.75N; 73.60E – Map P 701
Rhino, identification uncertain – Manuel 2005, 2007: 233, 2008: 34; Bose 2018a: 43, 2020: 63

Rhino Region 03: Rajasthan
Bilas River, Kota Dt., Rajasthan – 25.00N; 75.60E – Map P 704
Paintings in monochrome or red colour, including rhino – Vyas & Saran 1982; Manuel 2008: 34; Bose 2020: 63
Kanyadehe (Kanjadei), Chambal valley, Kota Dt., Rajasthan – 25.00N; 75.60E – Map P 704
Rhino – Manuel 2007: 233; Manuel 2008: 34

Rhino Region 04: Maharashtra
Chafe Dewood, Konkan, Ratnagiri, Mahrashtra – 17.20N; 73.35E – fig. 9.13 – Map P 702
Rhino 3 × 4 m – 9.5 – Sudhir Risbud, Ratnagiri, information 2019; Kevin Standage 2020
Vidarbha, Maharashtra – 19.21N; 76.80E – Map P 703
Rhino, identification uncertain – Pawar 2015

Rhino Region 07: Madhya Pradesh (North-West) – Mandsaur, Rajgarh
Dekan-Matschi Kalla, Mandsaur, Madhya Pradesh – 24.00N; 75.25E – fig. 9.8 – Map P 705
Rhino – Neumayer 2011 (M224)
Chaturbhoj Nath Nulla, Mandsaur, Madhya Pradesh – 24.00N; 75.25E – Map P 705
Rhino – Manuel 2005, 2007: 233, 2008: 34; Neumayer 2011 (M221, M222)
Narsinghgarh, Rajgarh, Madhya Pradesh – 24.00N; 76.70E – Map P 706
Rhino – Manuel 2008: 34

Rhino Region 07: Madhya Pradesh (North): Sehore, Hoshangabad, Raisen, Vidisha
Firengi (Firangi), Sehore, Madhya Pradesh – 23.15N; 77.10E – fig. 9.5 – Map P 707
Rhino – Neumayer 1993, fig. 139; 2011 (M213)
Adamgarh (Adamgadh), Hoshangabad, Madhya Pradesh – 22.70N; 77.75E – Map P 708
Rhino – Manuel 2008: 34
Kathothiya (Kathotia), Hoshangabad, Madhya Pradesh – 22.65N; 77.80E – fig. 9.2 – Map P 708
2 rhino in single composition – Manuel 2007: 233, 2008: 34; Neumayer 2011 (M201)
Marodeo shelter, 11 km from Pachmarhi, Hoshangabad, Madhya Pradesh – 22.41N; 78.42E – Map P 708
Large, aggressive rhino associated with smaller hunter carrying bow and arrows – Dubey 1992; Manuel 2005, 2007: 233, 2008: 34

© L.C. (KEES) ROOKMAAKER, 2024 | DOI:10.1163/9789004691544_010
This is an open access chapter distributed under the terms of the CC BY-NC-ND 4.0 license.

Bhimbetka, Raisen, Madhya Pradesh – 22.90N; 77.60E – fig. 9.6 – Map P 709

Rhino dashing one of its attackers into the air – Wakanker 1985; Manuel 2005, 2008: 34; Neumayer 1993 f. 131, 2011: 84 (M214); de Geer 2008: 381

Bhimbetka: Harni Harna – 22.90N; 77.60E – Map P 709

Rhino – Manuel 2005, 2007: 233

Bhimbetka: Lakhajuar – 22.90N; 77.60E – Map P 709

Rhino – Mathpal 1987: 105; Manuel 2007: 233, 2008: 34

Bhimbetka: Roup Village – 22.90N; 77.60E – Map P 709

Rhino – Manuel 2007: 233

Bhimbetka: Zoo rock – 22.90N; 77.60E – Map P 709

Rhino surrounded by 6 hunters with long spears tipped with microliths. Painted in transparent dark red colour. – Mathpal 1987: 105; Manuel 2006 fig. 1, 2007: 233

Gelpur (Dehgaon range), Raisen, Madhya Pradesh – 23.35N; 78.45E – fig. 9.4 – Map P 709

Rhino – Neumayer 2011 (M212)

Jaora (near Mandideep), Raisen, Madhya Pradesh – 23.10N; 77.50E – fig. 9.7 – Map P 709

A man is whirled into the air by a rhino's horn – Manuel 2007: 233; Neumayer 1993 f. 132, 2011: 85 (M217)

Kharwal (Kharbai), Raisen, Madhya Pradesh – 23.25N; 77.65E – Map P 709

Rhino – Manuel 2008: 34

Bairagarh, Raisen, Madhya Pradesh – 22.65N; 77.50E – Map P 710

Rhino – Manuel 2008: 34

Ghatla (Ghatala), Raisen, Madhya Pradesh – 23.35N; 77.75E – Map P 710

Rhino – Manuel 2008: 34

Putlikarar (near Narwar), Raisen, Madhya Pradesh – 23.35N; 77.95E – Map P 710

Animal with boar and rhino characteristics – Mathpal 1987: 17; Manuel 2008: 34

Ramchhajja, Raisen, Madhya Pradesh – 23.35N; 77.80E – Map P 710

Rhino – Manuel 2007: 233; 2008: 34; Neumayer 2011 (M211)

Urden, Raisen, Madhya Pradesh – 23.22N; 77.68E – fig. 9.3 – Map P 710

Rhino – Neumayer 1993, fig. 142, 196, 197; 2011 (M210)

Urden, Raisen, Madhya Pradesh – 23.22N; 77.68E – figs. 9.10, 9.11, 9.12 – Map P 710

Rhino with two horns – Cockburn 1899; Manuel 2007: 233; Neumayer 2011: 86 (M203, M208); Mondal & Chakraborty 2021

Hathitol, Raisen, Madhya Pradesh – 23.35N; 78.45 – Map P 711

Rhino – Manuel 2008: 3403

Gupha Mandir, Vidisha, Madhya Pradesh – 23.52N; 77.80E – Map P 712

Rhino – Neumayer 1993, f. 124

Rhino Region 07: Madhya Pradesh (North-East) – Jabalpur, Rewa

Katni, Jabalpur, Madhya Pradesh – 24.00N; 80.15E – Map P 713

In one composition a wounded rhino is chasing its hunter – Mathpal 1987: 17

Jhiriya: rock shelter no. 9, Rewa, Madhya Pradesh – 24.25N; 81.40E – Map P 714

Rhino – Manuel 2008: 34

Deorkothar (Deur Kothar), Itar Pahar, Rewa, Madhya Pradesh – 24.95N; 81.65E – Map P 714

Rhino – Manuel 2007: 233; Manuel 2005, 2008: 34

Morhana Nala (near Hanumana), Rewa, Madhya Pradesh – 24.80N; 82.16E – Map P 714

Rhino – Manuel 2008: 34

Rhino Region 10: Uttar Pradesh – Mirzapur

Lekhania (near Rajpur), Mirzapur, Uttar Pradesh – 24.61N; 82.96E – Map P 715

Rhino – Manuel 2008: 34

South Kaimur hills (= Mirzapur Dt.), Uttar Pradesh – 24.50N; 83.00E – Map P 715

Rhino – Manuel 2008: 34; Bose 2020: 60

Panchmukhi; Romp (near Robertsganj), Mirzapur, Uttar Pradesh – 24.60N; 83.04E – fig. 9.9 – Map P 715

Rhino – Manuel 2008: 34; Neumayer 2011 (M230); Bose 2020: 57, fig. 2.6

Ghormangur rock-shelter near Bidjeygurh, Mirzapur, Uttar Pradesh – 24.54N; 83.10E – fig. 9.1 – Map P 715

Discovered 1883-03-17. Rhino surrounded by hunters – 9.2 – Cockburn 1883: 58, pl. VII; Crooke 1899: 231; Drake-Brockman 1911: 200; Tewari 1987, 1992; P. Yule 1997: 29; Tiwari 2005: 215; Manuel 2005, 2007: 233, 2008: 34; Bose 2014: 67, 2020: 59

Kaimur range: Kerwa Ghat, Mirzapur, Uttar Pradesh – 24.53N; 83.10E – Map P 715

Rhino – Neumayer 2011 (M225); Bose 2020: 60, fig. 2.8

Kaimur range: Kerwa Ghat, Mirzapur, Uttar Pradesh – 24.53N; 83.10E – Map P 715

Rhino, butchering of animal – Neumayer 2011: 122 (M227)

Matahawa, Mirzapur, Uttar Pradesh – 24.50N; 83.10E – Map P 715

Rhino – Bose 2020: 57, fig.2.5

Rhino Region 16: Bihar – Kaimur

Kaimur range: Badki Goriya (Bhagwanpur), Bihar – 24.93N; 83.57E – Map P 716

Rhino – Bose 2020: 61; Tiwary 2014: 813

Kaimur range: Mithaiya Mand (not found, possibly near Bhagwanpur), Bihar – Map P 716

Rhino adult and baby – Bose 2020: 61

Rhino Region 20: Jharkhand – Hazaribagh

Isko (Isco) rock shelter, Hazaribagh, Jharkhand – 23.80N; 85.30E – fig. 9.14 – Map P 718

Rhino with single horn – Bose 2020, fig. 2.9; Saha & Rajak 2019: 502, fig. 11B; Rajak et al. 2020: 1048, fig. 6

Rhino Region 37: Odisha

Vikramkhol, Odisha – 21.81N; 83.77E – Map P 717

Rhino – Manuel 2008: 34

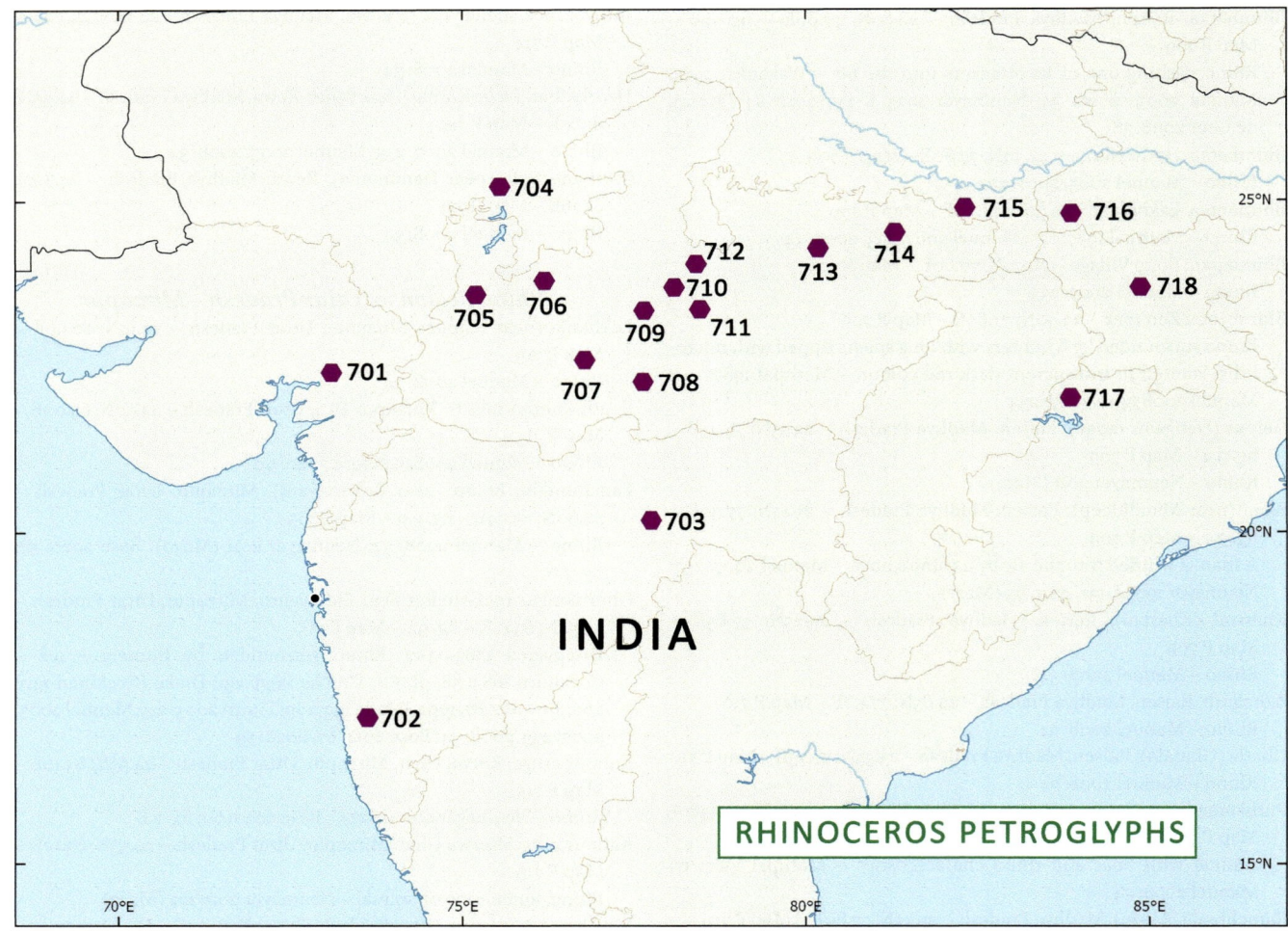

MAP 9.2   Localities of petroglyphs (rock art) showing the rhinoceros in India. The numbers and places are explained in Dataset 9.5
⬡   ROCK ART

9.2   Ghormangur Rock-Shelter

On 17 March 1883, John Cockburn was investigating the caves in the Ghormangur rock-shelter, situated a few miles from the ancient Vijaygarh Fort between Robertsganj and the Sone River in Uttar Pradesh. He must have realized the archaeological significance of his finds, because he took time to make tracings on the spot, to investigate their history and then to present his findings to the Asiatic Society of Bengal (Cockburn 1883). He could not ascertain the actual age of these rock paintings but thought that they could be no more than 700 years old. As this first well-documented example of rock art depicted a rhinoceros, the animal came to share in Cockburn's fame (Neumayer 2011).

Cockburn read this first paper on prehistoric rock art in India to the Asiatic Society of Bengal in Calcutta on 1 August 1883, and subsequently it was published in their journal on 24 October 1883. Signed by "J. Cockburn, Esq.",

it described this representation of a rhino hunt. Cockburn had transferred the image on the rock to tracing paper, showing an animal which is clearly a rhino with a single horn surrounded by men, one thrown in the air by the rhino, the others wielding spears with arrow-like heads. Despite the significance of Cockburn in the early history of rock art investigations, he himself has left only a few traces of his work and interests in natural history, archaeology and anthropology (9.3).

Cockburn explored the Vindhya Hills in Mirzapur district, Uttar Pradesh. He signed his name on a rock face near a Mesolithic painting in 1881 (Neumayer 2011: 3). He discovered the painting of the rhino and hunters illustrated in 1883. The shelter was locally called Ghormangur cave, and Cockburn stated that "its exact position is two miles due south of Mow Kullan bridge, and within three miles of the celebrated fortress of Bidjeygurh, and five of the river Sone. This rock shelter has the appearance of a

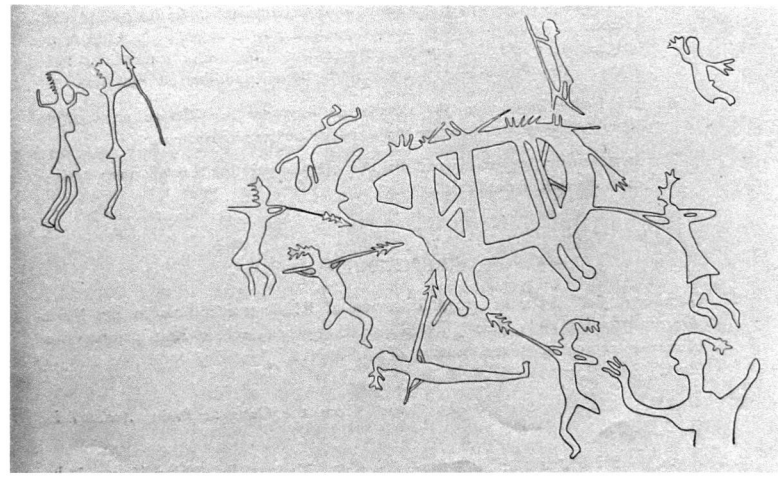

FIGURE 9.1
Tracing of the rock painting in Ghora Mangar rock
shelter, near Vijaygarh Fort, executed by Rakesh Tewari in
1979–1980
BULLETIN OF MUSEUMS AND ARCHAEOLOGY, 1987,
FIG. 2. COURTESY © RAKESH TEWARI

huge mushroom. It is a gigantic boulder, the remnant of some rocky ridge" (Cockburn 1883: 58). The main landmark, the fortress of Bidjeygurh is found with various spellings: Bidjegur, Beejaghar, Bidzegur, Bijaigarh, and more recently as Vijaygarh Fort.

The bridge below the Fort first constructed in the early 16th century across the Gaghar River at the village of Mow or Mau was painted by the English artist William Daniell (1769–1837), published as 'Bridge and Fort of Bidzegur' in the *Oriental Annual* of 1834, p. 177 (Archer 1980: 100). The shelter is 28 km south of Robertsganj and just north of the valley of the Sone River.

The Ghora Mangar rock shelter was again visited in 1979 by Rakesh Tewari, then an Exploration Assistant in the Uttar Pradesh State Archaeological Organisation. His detailed examination and new tracing of the painting show that there are some important elements absent from Cockburn's plate. Tewari (1987) published a new tracing made at the shelter (fig. 9.1). He found that the scene of 93 × 60 cm painted in dull ochre depicts 11 hunters attacking the rhino with pierced harpoons, five of which had become lodged in the animal's back, hip and lower leg. The tip of the horn is rounded, the mouth is more open, the harpoons look like they are made of wood with stone implements. Tewari suggested that the drawing was made during the Stone Age in the Mesolithic period (12,000 to 4,000 BP).

It may be noted here for the first time that this locality is not more than 150 km west of Rohtas, where on 26 December 1766 the surveyor Lewis Felix DeGloss saw a track of a rhino in the dense jungle (28.12).

9.3 The Life of John Cockburn

John Cockburn (1852–1902) had a short, eventful life. He mentioned twice that he had been Curator of the Allahabad Museum (Cockburn 1879, 1883: 63). A note in the *Times of India* (1872-07-12) suggests that Cockburn's application for Sub-Curator in Allahabad was initially unsuccessful, but possibly this decision was reversed. He then moved to Kolkata, as he listed a position of Officiating Assistant Osteologist, Indian Museum in the by-line of his 1879 paper. On 17 October 1881 he was temporarily promoted to Officiating 2nd Assistant to the Superintendent, John Anderson (*Englishman's Overland Mail* 1881-10-31), continuing until the start of 1883 (Indian Museum 1883: 4). This period of employment at the Indian Museum 1879–1883 was his most publicly productive, reading papers at the prestigious meetings of the Asiatic Society of Bengal, and publishing papers on various subjects (Cockburn 1879 to 1899). In 1883, he found employment in the Opium Department, first on probation, and with effect of 22 December 1884 confirmed as Assistant Sub-Deputy Opium Agent of the fourth grade (*Englishman's Overland Mail* 1885-03-24).

Cockburn remained active in his explorations of prehistoric relics, such as found during a three-months survey of the Singrauli basin in 1888. He was also active as a naturalist, contributing some bats to the Indian Museum collected in Alipore in 1877–1878, and fossil shells found at the Ken River near Banda in 1881 (Cockburn 1883: 57). Requested by Robert Armitage Sterndale (1839–1902), he contributed a chapter on the rhinoceros to his work on mammals (Cockburn 1884c). The amateur ornithologist and later Principal of Meerut College (1904–1923), William Jesse recalled meeting Cockburn while traveling in 1899: "Most fortunately, when I got into the Fatehgarh train, I found in the same carriage Mr. John Cockburn, of the Opium Department, who at one time used to collect for Mr. Hume [Allan Octavian Hume (1829–1912)], and had been a personal friend of the late Major Cock and others of that brilliant band of ornithologists who did so much in India in the seventies and early eighties, and who have now nearly all left the 'land of regrets' for ever, causing a void which it will be hard to fill up. One of Mr. Cockburn's hobbies is the study of the Serpent-Eagles, and we passed away the time very pleasantly, and, in my case, most profitably, as we discussed various things connected with Indian bird-life" (Jesse 1899).

A curious note in the *Englishman's Overland Mail* of 6 October 1898 states that Cockburn was selected to take the place of Alois Anton Fuhrer (1853–1930) as the head of the Archaeological Survey of India, saying that Cockburn "is probably well known to a good many of your readers, having worked in the Calcutta Museum, before he joined the Opium Department. Not only is Mr. Cockburn versed in folk-lore, in antiquarian research of every kind, but he also possesses an intimate acquaintance with natural history and can wield a very pretty pen. His 'Autobiography of a Muggur' and 'History of a Brahminy Duck' were clever sketches." The titles of those epistles cannot be retrieved, and possibly were serialized in local papers at the time. I have found no trace of Cockburn in the history of the Archaeological Survey, hence he must have continued in his previous post at the Opium Agency. John Cockburn, then "Sub-Deputy Opium Agent, Partabgarh" died on 14 October 1902 at Dehradun "of diabetic coma and carbuncle", aged 50 years (*Madras Weekly Mail*, 1902-10-30).

9.4   Rock art in Central India

The paintings in rock shelters were found mainly in Mirzapur near Varanasi (Uttar Pradesh) and in the Vindhya Hills near Bhopal (Madhya Pradesh). At least 42 sites have been documented with potential figures of a rhino (Dataset 9.6, map 9.2). Largely the sites were far from places with written evidence of the rhino, except for the single record in 1766 from Rohtas in SW Bihar. The animal is by no means common compared to other large mammals (Neumayer 2011). It is too hypothetical to suggest that Mesolithic men traveled some hundreds of kilometers northwards to go and hunt the rhino. If that means that the rhino in their age occurred in central India, then the animals were relatively rare, or otherwise they were only pursued occasionally because they were dangerous and hard to kill. It may be noted that the current landscape and climate is unlike the rhino's usual habitat. The climate may have changed over the years, but the rocky landscape must have remained largely similar, unsuitable for rhinos now as much as in the past.

9.5   Rock Art in Ratnagiri District, Maharashtra

Among the large number of petroglyphs discovered in the Ratnagiri District is one that shows the clear outline of a single-horned rhino (fig. 9.11). The existence of the ancient rock art was first revealed in the 1980s by Sudhir Risbud near the village of Niwali, 15 km from Ratnagiri. Since then, many examples without rhinos have been found on the Konkan coast and have been tentatively dated to the Mesolithic period (10,000 BCE).

9.6   Age of the Rock Paintings

The question about the chronology remains a matter of discussion among experts, because there are multiple details to be considered. The current consensus suggests that most of the rhino paintings were from the Mesolithic period, which was about 12,000 to 4,000 years before the present. The fossil remains of rhinos known from India are older (Pleistocene in age, over 11,700 years before present), the written records of course much more recent (six centuries at most). Although some rock paintings may be younger, it seems that most scholars now accept that the majority were made at least four millenia ago.

9.7   Types of Rhinoceros in Rock Art

There is generally little discussion about the species of rhino depicted in the rock art of central India (Tiwari 2000: 215). A detailed analysis is unnecessary, because the Mesolithic artists depicted a large animal with a horn, which is never realistic enough to be compared with pictures in a modern checklist. While *R. unicornis* is definitely the most likely candidate, there is no reason to dismiss *R. sondaicus* off-hand. If there were separate species of rhino in the late Pleistocene, like *R. namadicus* or *R. karnuliensis*, perhaps they only went extinct closer to the age when the rock paintings were made. Despite the uncertainties and possible intricacies, it is suggested, on the basis of our current zoogeographical and archaeological knowledge, that these petroglyphs contain representations of the single-horned *R. unicornis*.

One of the paintings in Urden shows a rhino with two horns (figs. 9.10, 9.11, 9.12). This should point at the double-horned *D. sumatrensis*, a species of course not associated with the Indian subcontinent except on the eastern fringes. It might be that the draughtsman had pre-existing knowledge of the animal handed down from his ancestors who might have migrated from regions where the two-horned rhinoceros existed (Mondal & Chakraborty 2021). Maybe the second horn was a flight of fancy of this early picture maker. While there is no immediate satisfactory explanation for this isolated example, in my view this evidence alone is inadequate to accept the presence of *D. sumatrensis* in this part of the continent.

9.8 The Rhinoceros in Mesolithic Times

In the context of the present investigation, it is an important question if the depictions of a rhino in the petroglyphs supports the conclusion that *R. unicornis* lived in parts of central India from Gujarat in the west along the Narmada River to Odisha in the east, at least until about 4000 years before the present.[1]

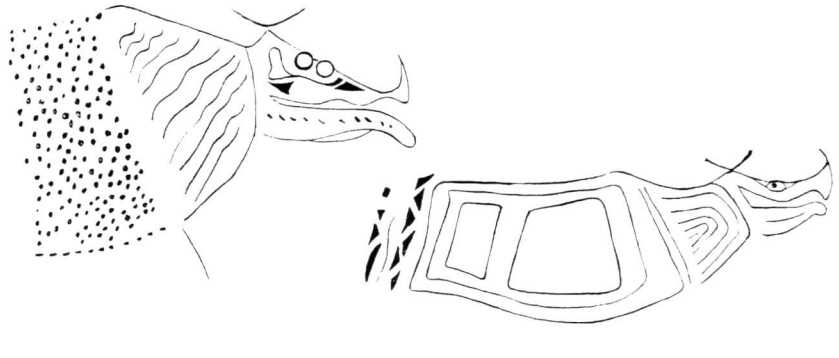

FIGURE 9.2
Mesolithic rock painting at Kathotia (Hoshangabad), Madhya Pradesh. Red, length of rhino 20 cm. Tracing by Erwin Neumayer
ROCK ART OF INDIA, 2011, NO. M201. © ERWIN NEUMAYER

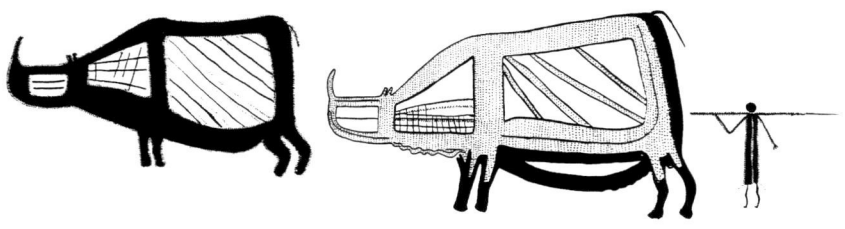

FIGURE 9.3
Mesolithic rock painting at Urden (Raisen), Madhya Pradesh. Red and yellow, height of man 27 cm. Tracing by Erwin Neumayer
ROCK ART OF INDIA, 2011, NO. M210. © ERWIN NEUMAYER

FIGURE 9.4
Mesolithic rock painting at Gelpur (Dehgaon range), Madhya Pradesh. Length of rhino 90 cm. Tracing by Erwin Neumayer
ROCK ART OF INDIA, 2011, NO. M212. © ERWIN NEUMAYER

1 The Mesolithic occurrence of rhinos in Central India was proposed by Chitalwala (1990), Tiwari (2000), Manuel (2006, 2007, 2008), Bose (2014, 2020) and Neumayer (1993, 2011).

Flaked

FIGURE 9.5
Mesolithic rock painting at Firengi (Sehore), Madhya Pradesh.
Length 100 cm. Tracing by Erwin Neumayer
ROCK ART OF INDIA, 2011, NO. M213. © ERWIN NEUMAYER

FIGURE 9.6
Mesolithic rock painting at Bhimbetka (Raisen), Madhya
Pradesh. Length of rhino 40 cm. Tracing by Erwin
Neumayer
ROCK ART OF INDIA, 2011, NO. M214. © ERWIN
NEUMAYER

FIGURE 9.7
Mesolithic rock painting at Jaora (Raisen), Madhya
Pradesh. Length of rhino 80 cm. Tracing by Erwin
Neumayer
ROCK ART OF INDIA, 2011, NO. M217. © ERWIN
NEUMAYER

FIGURE 9.8
Mesolithic rock painting at Dekan-Matschi (Mandsaur),
Madhya Pradesh. Length of rhino 24 cm. Tracing by Erwin
Neumayer
ROCK ART OF INDIA, 2011, NO. M224. © ERWIN
NEUMAYER

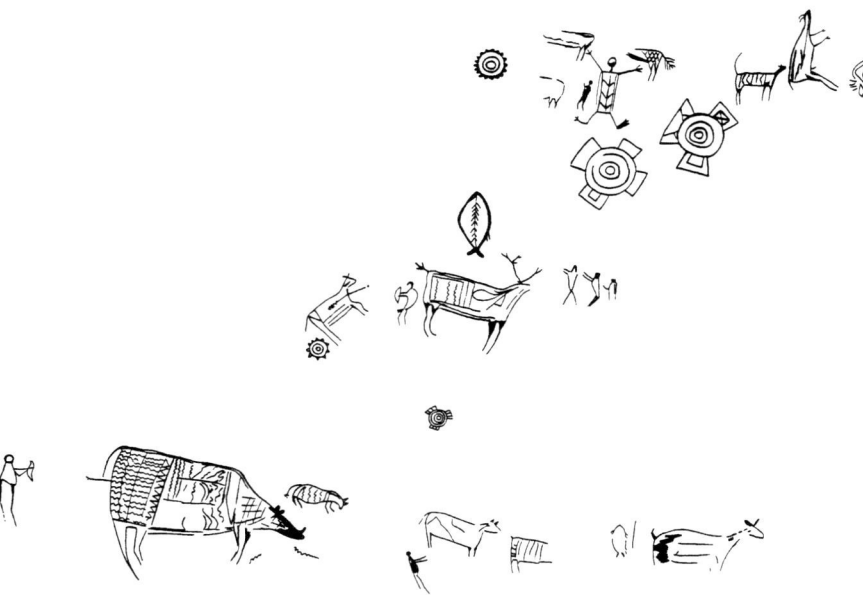

FIGURE 9.9
Mesolithic rock painting at Panchmukhi
(Mirzapur), Uttar Pradesh. Length of rhino
28 cm. Tracing by Erwin Neumayer
ROCK ART OF INDIA, 2011, NO. M230.
© ERWIN NEUMAYER

FIGURE 9.10
Mesolithic rock painting at Urden (Raisen),
Madhya Pradesh. Length of rhino 30 cm.
This is the only painting of a rhinoceros
with two horns. Tracing by Erwin Neumayer
ROCK ART OF INDIA, 2011, NO. M208.
© ERWIN NEUMAYER

FIGURE 9.11
Detail of the double-horned rhino at Urden.
Tracing by Erwin Neumayer
ROCK ART OF INDIA, 2011, NO. M208. © ERWIN
NEUMAYER

FIGURE 9.12
Enhanced tracing of the original rock painting of
the double-horned rhinoceros at Urden. Drawn by
Erwin Neumayer
ROCK ART OF INDIA, 2011, NO. M208. © ERWIN
NEUMAYER

FIGURE 9.13
Petroglyph at Chafe Dewood, Konkan,
Ratnagiri Dt., Maharashtra, found at the end of the
20th century
PHOTO: © SUDHIR RISBUD, RATNAGIRI, 2020

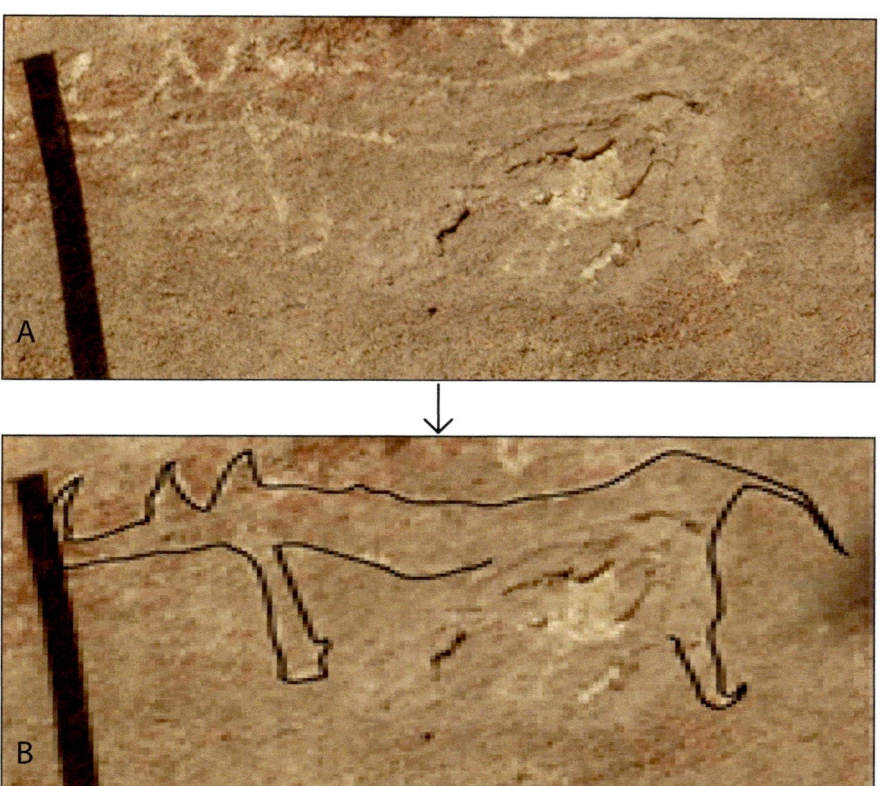

FIGURE 9.14
Petroglyph at Isko rock shelter, section A,
Hazaribagh, Jharkhand, found at a height of
2.45 m from the floor
PHOTO AND SUGGESTED OUTLINE:
© SHUBHAM RAJAK, PUNE, 2022

Zoologists have not yet incorporated possible rhino populations in central India in their view of the former range of *R. unicornis* (Rookmaaker 1984; Rookmaaker et al. 2017; Dinerstein 2015). Their hesitation, which reflects my own, is at least partly due to a lack of interdisciplinary exchange of evidence. It could be argued that the rock art is too early to be integrated in reconstructions of the historical range of a species, which in most instances should represent the situation from the 16th century onwards. The possibility that the rock artists did not depict the local environment must be too miniscule to be taken into account.

The data presented in chapters 14–15 show that rhinos were known in Rajasthan and adjoining areas until relatively recently. This will strengthen the possibility that the historical range of the rhino in India is in need of a new interpretation. The evidence of the rock paintings allows the working hypothesis that *R. unicornis* indeed inhabited parts of Gujarat, Rajasthan, Madhya Pradesh and Uttar Pradesh south of the Ganges until about 2500 BC or 5000 years before present. This is discussed in the Epilogue (65.1), where these records are incorporated in the Protohistorical Distribution of *R. unicornis* (map 65.36).

CHAPTER 10

Historical Records of the Rhinoceros in Harappan Settlements

10.1 First Excavations of Mohenjo Daro

The excavations in Mohenjo-Daro near the Indus River in the southern part of Pakistan from the 1920s revealed the existence of an important early civilisation which is thought to have flourished from 2600 to 1900 BCE. The objects retrieved by archaeologists like Rakhal Das Banerji (1885–1930) and John Hubert Marshall (1876–1958) included numerous steatite (soapstone) seals and terracotta objects (Jansen 1985). Later explorations showed that this Harappan or Indus Valley civilisation once extended through much of Pakistan and in the adjoining Indian states of Punjab, Rajasthan and Gujarat, even possibly southward to Maharashtra, as recently reviewed by Manuel (2007, 2008) and Bose (2018a, 2020). The Indus civilisation disappeared without leaving clues as to how to read its script.

10.2 List of Harappan Artefacts Depicting a Rhinoceros

The rhinoceros is found depicted on seals, modelled in small figurines, as well as in terracotta masks, copper tablets, a glazed cylindrical seal and a bronze image (figs. 10.1, 10.2). The seals sculpted in soapstone are typically small and square, about 3.5 × 3.5 cm and are the most common objects. The occurrences of Harappan object are listed according to excavation site with reference to their current depositories in dataset 10.6 and the localities are shown on map 10.3.[1]

10.3 The Rhino Species Is Only Tentatively Known

There is no doubt that in many objects there was a rhino showing a clearly defined single horn on the nose, besides characteristic skin folds and skin tubercles. Initially,

Marshall (1931, vol. 1: 387) was careful to identify these animals as "probably" *R. unicornis* and this has been followed in subsequent discussions. Bose (2020: 78) expresses this sentiment and asserts that there is no ambiguity about this fact. This confidence is likely to be well-founded. The animal is definitely a rhino, and being single-horned narrows the options to just two recent species. However, the figures are not naturalistic and clear enough to fully abandon the possibility that it might have been *R. sondaicus*, as this species was also represented in fossil remains found in Pakistan (8.8). Following the evidence discussed in my own paper on the distribution of rhinos in western India and Pakistan (Rookmaaker 2000) as well as the discussion by Brentjes (1978), *R. unicornis* is certainly the more likely candidate.

10.4 Rhinoceros above a Manger

In some seals the rhino is represented with its head above a trough or manger, sometimes with a head posture indicating that the animal is feeding from it (Bose 2020: 79). This could indicate that the rhino was kept in captivity (Conrad 1968; Bautze 1985: 406). Alternatively, Marshall (1931) identified the trough as some kind of cult object, and Atre (1990) saw the structure as a sign of dominance of the goddess over the largest animals (fig. 10.1).

10.5 Harrappan Bronze Sculpture from Maharashtra

Daimabad on the left bank of the Pravara River is a Harappan settlement excavated first in 1958–1959 and subsequently in the 1970s. Generally it is thought that it dates from the late-Harapppan period (1800–1600 BCE). In 1974 a local villager Chhabu Laxman Bhil unearthed four bronze sculptures, which he handed over to Shikaripura Ranganatha Rao (1922–2013) for the Archaeological Survey of India (Rao 1978). One of these is a rhino, single-horned, standing on two horizontal bars over two sets of wheels, 24.7 cm long, 29 cm high with a distance of 15.1 cm between the two sets of wheels (fig. 10.3), extensively described and figured by Yule (1985: 30, pl. 3 fig. 38). Debate on the religious or ritualistic significance of the finds is likely to continue.

---

1 The history of the rhinoceros in the Harappan civilisation is discussed by Marshall (1931, vol. 1: 387); Prashad (1936); Krumbiegel (1960: 15); Nath (1961: 361; 1969); Conrad (1968); Brentjes (1978: 159); Choudhury (1985a); Atre (1985); R.S. Sharma (2005); de Geer (2008: 380); Manuel (2008); Velmurugan (2017); Bose (2018a, 2020) and Rookmaaker (2000).

© L.C. (KEES) ROOKMAAKER, 2024 | DOI:10.1163/9789004691544_011
This is an open access chapter distributed under the terms of the CC BY-NC-ND 4.0 license.

FIGURE 10.1
Indus stamp-seal, carved from grey steatite, depicting a rhinoceros with head above the manger
WITH PERMISSION © THE TRUSTEES OF THE BRITISH MUSEUM: 1947,0416.4

FIGURE 10.2
Terracotta figure of a rhino found in Mohenjo Daro
WITH PERMISSION © MOHENJO DARO MUSEUM, PAKISTAN: ACC.NO. 50.273, DK 5462

FIGURE 10.3
Bronze sculpture of a rhino with one horn found in Daimabad, Maharashtra
ARCHAEOLOGICAL SURVEY OF INDIA, NEW DELHI.
PHOTO: JOACHIM K. BAUTZE

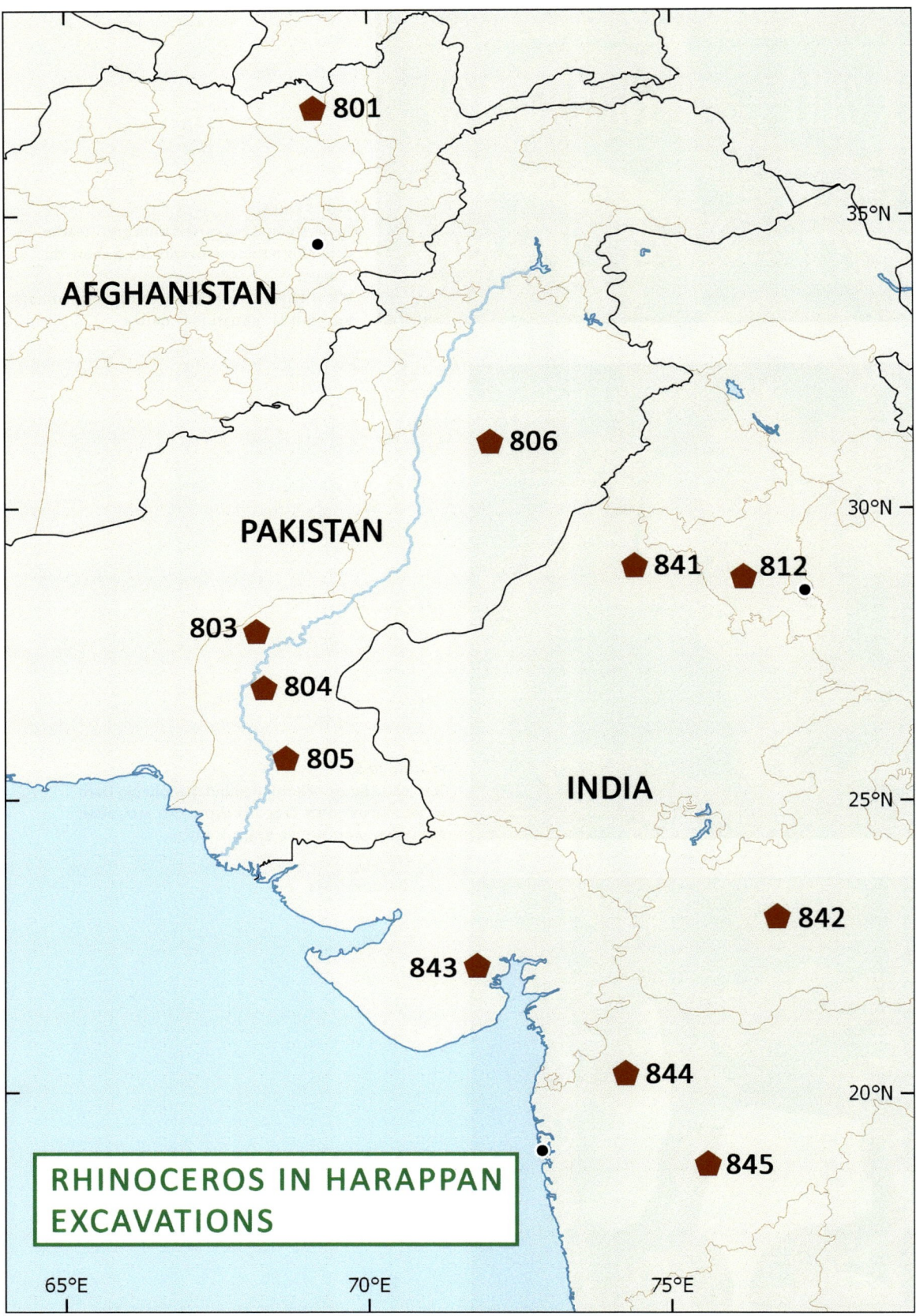

MAP 10.3   Localities of artefacts of rhinoceros from the Harappan period. The numbers and places are explained in
            Dataset 10.6

10.6 The Existence of the Rhinoceros Near the Indus River

The existence of such clear representations of a rhino in the objects of the Harappan culture is difficult to explain except by suggesting that the species occurred within the realm of the settlements. As the climate now seems too hot and dry to favour a rhino, this is a dilemma. Of course it is possible to reverse the argument to state that the presence of an animal like the rhino points to a change in the climate and hence in the environment over the last five millennia. The rhinos can only play a minor role in this much broader investigation conducted by the experts of the region.

It might be possible that rhinos lived along the banks of the Indus River and its tributaries where there would have been sufficient water resources. If there were some unspecified populations of rhinos in the region, it would still be unlikely that many inhabitants actually ever saw one in the wild, which is very much mirrored by the urban human settlements of today. There are no scenes that the rhinos were hunted, slaughtered or eaten. A rhino was still reported on the lower Indus River by Ibn Battuta in 1333 within 100 km from Mohenjo-Daro (13.6).

In the present context, the tentative conclusion is that *R. unicornis* occurred in the Indus River valley until about 2000 BCE (or 4000 BP). Earlier, this led me to believe that the historical distribution included a large part of the Indus Valley (Rookmaaker 2000, fig. 5), while it is more prudent to separate a southern and northern range in Pakistan (map 65.35). However, most zoologists adhere to the map advocated in the IUCN status survey published prior to my research (Foose & van Strien 1997: 9). This map shows a much more limited historical distribution extending westwards into the northern part of Pakistan only (eg. Amin et al. 2006; Schenkel et al. 2007), due to differences in the interpretation of the meaning of "historical" in these reconstructions of the past (65.1). This now traditional view is untenable in the light of the increasing evidence of rhino occurrence in the western parts of South Asia.

• • •

Dataset 10.6: List of Harappan Artefacts according to Locality

The numbers and places are shown on map 10.3. Museum depositories are abbreviated as NMI = National Museum of India, New Delhi; NMP = National Museum of Pakistan, Karachi; MDM = Mohenjo Daro Museum, Pakistan.

Afghanistan
Rhino Region 01: Afghanistan
Shortughai, Darqad Province – 37.20N; 69.30E – map M 801
1. Stamp with rhino – Franckfort 1983 – 13.3

Pakistan
Rhino Region 02: Pakistan
Mohenjo-Daro, Larkana Dt., Sindh – 27.32N; 68.13E – map M 803
2. Figurine in terracotta of rhino used as a toy – MDM, no. 50.273, DK 5462 – Marshall 1931, vol. 1: 387; Rookmaaker 2000, fig. 2 (colour).
3. Figurine in clay of rhino – NMI – Bautze 1985, figs. 1, 2.
4. Tablet depicting rhino – MDM no. 984 – Shah & Parpola 1991: 206 (M-1481)
5. Tablet depicting rhino – NMI no. 124 – Joshi & Parpola 1987: 110 (M-446)
6. Tablet depicting rhino – Archaeological Survey of India, nos. 63.10.231; 63.10.205 – Joshi & Parpola 1987: 110 (M-445, M-447)

7. Copper plate with rhino facing right, no manger. Weight 113.6 gr, size 57.9 × 52.1 mm, thickness 5.2 mm. These copper plates may have been used for printing, as even now images can be produced from them – Shinde & Willis 2014, colour fig. 2 (private collection)
8. Seals (steatite, soapstone) with rhino – Krumbiegel 1960: 15; Bautze 1985: 406; Geer 2008: 384
9. Seals with rhino – NMI nos. 119; 126 – Joshi & Parpola 1987: 67 (M-274, M-277); Bose 2018b: 48 (figure)
10. Seal with rhino 3.5 × 3.5 cm – NMI no. HR5992 – Lang 1961: 369
11. Seals with rhino – Archaeological Survey of India, nos. 63.10.140; 63.10.149 – Joshi & Parpola 1987: 67 (M-275, M-276); Bose 2018b: 46 (figure)
12. Seals with rhino – MDM nos. 501; 586; 50275; 703,55.64; 687; 437,55.53; 896 – Shah & Parpola 1991: 124–126 (M-1131; M-1132; M-1133; M-1136; M-1137; M-1139; M-1140); Bose 2018b: 46 (figured)
13. Seal with rhino facing left without manger – MDM no. 50.765 – Rookmaaker 2000, fig. 3 (colour)
14. Seals with rhino – NMP nos. 50.273; 50.272 – Shah & Parpola 1991: 124 (M-1134; M-1135)
15. Seal with rhino – Lahore, Dept. Archaeology, no. P-917B – Shah & Parpola 1991: 124 (M-1138)
16. Seals with rhino facing right, rectangular 3.8 × 1.5 cm – Siudmak 2016
17. Seal with rhino facing left, no. 341, lower part absent – Marshall 1931, vol. 1: 387

18. Seal with rhino facing right, no. 342, without trough –
     Marshall 1931, vol. 1: 387
19. Seals with rhino facing right, nos. 343 to 357, with small trough –
     Marshall 1931, vol. 1: 387; Nagar 1998: 78

Chanhu-Daro, Benazirabad Dt., Sindh – 26.17N; 68.32E – map M 804
20. Figurine in clay of animal composite of humped bull and rhino.
     People might not have seen a life rhino (excavation 1935–36) –
     Mackay 1938; MacKay 1943: 142; Bose 2020: 86

Alahdino, Matiari Dt., Sindh – 25.50N; 68.50E – map M 805
21. Seals with rhino – Fairservis 1982: 111

Harappa, Sahiwal Dt., Punjab – 30.64N; 79.90E – map M 806
22. Figurines of rhino, size is 3.1 × 6.0 × 8.5 cm – Dept. Archaeology
     and Museums, Pakistan – Bose 2018b: 47 (figure); https:
     //www.harappa.com/figurines/42.html
23. Figurine of rhino with a collar, size 3.3 × 8.7 × 4.5 cm – Dept.
     Archaeology and Museums, Pakistan – Bose 2018b: 47, figure;
     https://www.harappa.com/figurines/43.html
24. Figurines of rhino, total 34 recorded – Dales 1993
25. Seals with rhino, size 3.5 × 3.5 cm – Vats 1940: 307–308;
     Bautze 1985: 406
26. Seal with rhino showing trough – NMI no. Hr 5992/119 –
     Bose 2018b: 46 (figured)

India
Rhino Region 08: North-West India
Rakhigarh, Haryana – 29.30N; 76.10E – map M 812
27. Seal with rhino – A. Nath 1998

Rhino Region 03: Rajasthan
Kalibangan, Rajasthan – 29.50N; 74.15E – map M 841
28. Seal with rhino – Joshi & Parpola 1987: 305

Rhino Region 07: Madhya Pradesh
Dangwada, Ujjain Dt., Madhya Pradesh – 23.10N; 75.50E – map M 842
29. Image in terracotta of rhino – Manuel 2008: 35

Rhino Region 04: West India
Lothal, Gujarat – 22.51N; 72.25E – map M 843
30. Figurine of rhino, terracotta head, 4.5 × 6.2 cm – NMI –
     Bose 2018b: 47 (figured)

Rhino Region 05: South India
Varsus, Dhule Dt., Maharashtra – 21.00N; 74.20E – map M 844
31. Mask in terracotta – Manuel 2008: 35; Bose 2014: 72

Daimabad, Maharashtra – 19.70N; 75.70E – map M 845
32. Figurine of rhino on wheels – Rao 1978; Dhavalikar 1982: 363,
     1993; Bautze 1985: 409; Sali 1986; Yule 1985: 30, pl. 3 fig. 38 –
     fig. 10.3

∵

The Rhinoceros in the Arts and Sciences of the Mughal Period

11.1   The Mughal Empire

The Mughals ruled large parts of South Asia from the start of the 16th century until their decline two centuries afterwards. The successive Emperors were warriors, statesmen, politicians as well as patrons of sciences and arts. Four of the first six rulers are known to have interacted with the rhinoceros. They hunted them, they captured them and kept them in captivity, they described them and had them immortalized in paintings.

There is no immediate evidence that rhinos declined in the wild during this period. Babur and Humayun saw the animals in Pakistan. Babur and Jahangir also went to hunt them in the forests closer to the new administrative centres in northern India. These three, as well as Shah Jahan, have been depicted in connection with a rhinoceros. The well-known miniatures were first conceived during the time of Akbar at the end of the 16th century.

•••

Dataset 11.7: Chronological List of Records of Rhinoceros during the Mughal Dynasty

General sequence: Date – Locality (as in original source) – Event – Source – § number – Map reference.

The map numbers are explained in dataset 11.8 and shown on map 11.4. Records divided according to the reigns of the consecutive Mughal Emperors.

A. Period of Emperor Babur (Reign 1526–1530)

1513 – India, no locality – Babur had a detachment of 80 rhinos in a battle with King Cacander – Correia 1862, vol. 3: 573; Yule & Burnell 1886: 363 – 11.5

1519-02-16 – Karak-Khaneh or Karg-khana, on side of Sawati (Swabi) River – Babur hunt. Mother rhino and calf got away. A second calf wounded in bush fire was killed – Babur 1826: 253, 1922: 378 – 11.2, fig. 11.2 – map A 204

1525-12-10 – Bigram (Peshawar) near Siah-Ab (Black River) – Babur hunt. One rhino killed by Humayun. Two other rhinos shot – Babur 1826: 292, 1922: 450 – 11.2, figs. 11.3, 11.6 – map A 204

1520s – Peshawar and Hashnagar jungles – Babur: "there are masses of it [rhino] in the Pershawer (Parashawar) and Hashnagar jungles" – Babur 1826: 316, 1922: 490 (fol. 275b) – 11.2 – map A 203

1520s – Between Sind River and Bhira (Behreh) country (i.e. from Dinkot (Kalabagh) on Indus River east to Bhera on Jhalum River) – Rhino found – Babur 1826: 316, 1922: 490 (fol. 275b) – 11.2 – map A 205

1520s – India, no locality – Babur had a drinking vessel made of rhino horn. Possibly same as drinking cup in British Museum, London, Sloane Collection (SLMisc.1713) – Babur 1826: 316, 1922: 490; Clarke 1987: 348; Chapman 1999: 277 – 11.2, fig. 11.1

1525 – Saru or Sirwu (Ghaghara) River, Uttar Pradesh (near Bahraich) – Babur saw "masses there are also on the banks of the Saru-river in Hindustan" – Babur 1826: 316, 1922: 490 (fol. 275b) – 11.2 – map A 248

1529-03-25 – Chunar, Uttar Pradesh – Babur hears about a tiger and rhino by the river bank near his camp, but his hunt is unsuccessful – Babur 1826: 407, 1922: 657 – 11.2 – map A 239

B. Period of Emperor Humayun (Reign 1530–1540, 1555–1556)

1525-12-10 – Bigram – Babur hunt. One rhino killed by Humayun – Babur 1826: 292, 1922: 450 – 11.2, figs. 11.3, 11.6 – map A 204

C. Period of Emperor Akbar (Reign 1556–1605)

16th cent. – Sarkar of Sambal (District of Sambhal), Uttar Pradesh – Rhino numerous in time of Akbar – Jarrett 1891, vol. 2: 281 (Ain-i-Akbari) – 11.7 – map A 242

1570s – Akbar kept "rhinoceroses" in Agra Fort [no contemporary source quoted] – Grewal 2007: 60 – 11.7

1575 – Rhino exported and seen in Aleppo, Syria – Leonhard Rauwolf (1540–1596) saw a rhino from the East – Rauwolf 1583: 287; Rookmaaker 1998: 31 – 11.7

1577 – Rhino exported to Lisbon and known as Abada or Madrid Rhinoceros – The second living rhino in Europe since Roman times – Clarke 1986: 28–34; Rookmaaker 1973: 44, 1998: 91; Staudinger 1996; Jordan Gschwend 2015a, 2015b, 2018; Kuster 2015; Scarisbrick et al. 2016: 120–122 (no. 125); Beusterien 2019, 2020 – 11.9, fig. 11.9

1580 – India, no locality – Antonio Monserrate (1536–1600) saw war-elephant with coverings of rhino hide – Hosten 1912: 212 – 11.7

D. Period of Emperor Jahangir (Reign 1605–1627)

1580 to 1619 – India – Jahangir killed a total of 64 rhino in this period – Beveridge 1909, vol. 1: 369 – 11.10

1600 to 1604 – India – One rhino captured and shown to Prince Salim during a hunting expedition – Miniature in Los Angeles Country Museum, online https://collections.lacma.org/node/247632; Sotheby's 1977, lot 28; Okada 1992; A.K. Das 2018a: 77; – 11.10, fig. 11.11

1615 – Ajmer, Rajasthan – Two rhinos, from Bengal. Seen in menagerie by Thomas Coryat (1577–1617) – Coryat 1616: 246 – 5.4, 11.12 – map C 510

© L.C. (KEES) ROOKMAAKER, 2024 | DOI:10.1163/9789004691544_012

This is an open access chapter distributed under the terms of the CC BY-NC-ND 4.0 license.

1620 – Nurmahal, Punjab – Caravanserai built for Nur Jahan (1577–1645) – Cunningham 1882b: 62; photograph taken by Joseph David Beglar (1845–1907) in 1876 (Beglar 1876); A.K. Das 2018b, figure p. 91 – 11.12, fig. 11.12 – map M 811

1624 – Kul Nuh Ban (Nuh Forest) near Aligarh – Jahangir mentions killing a rhino with one bullet – Beveridge 1914, vol. 2: 270 – 11.12 – map A 241

E. Period of Emperor Shah Jahan (Reign 1627–1658)

1632–12 – Etawa (Etawah) – Peter Mundy (1600–1667) was told that King Shah Jahan had sent to Etawa "two great Rynocerosses to bee kept and fedd" – Mundy 1914: 186 – 5.4, 11.12 – map C 512

1665–1670 – India – Rhino-elephant fight witnessed by Shah Jahan and his four sons – Willem Schellinks (1623–1678) painting "A hawking party" – Orientalist Museum, Doha, no. OM.672; Bostock 2021 – 11.14, figs. 11.13, 11.14

F. Period of Emperor Aurangzeb (Reign 1658–1707)

1659 – Dehli (Delhi) – François Bernier (1620–1688) sees rhinos in a procession of animals – Bernier 1671: 75, 1891: 262 – 5.4 – map C 511

1665–1672 – District of Patna – Image of rhino captured in Patna is present in an album compiled at Aurangzeb's court in 1665–1672 – A.K. Das 2018a: 78 – 11.15, fig. 11.21

1696 – Captive, no place – King Christian V of Denmark had double horn of rhino killed at court of Indian Mogul – Jacobaeus 1696: 4, pl. 3 fig. 4; Jacobaeus & Laürentzen 1710, pl. 4 no. 31 – 11.15, fig. 11.22

...

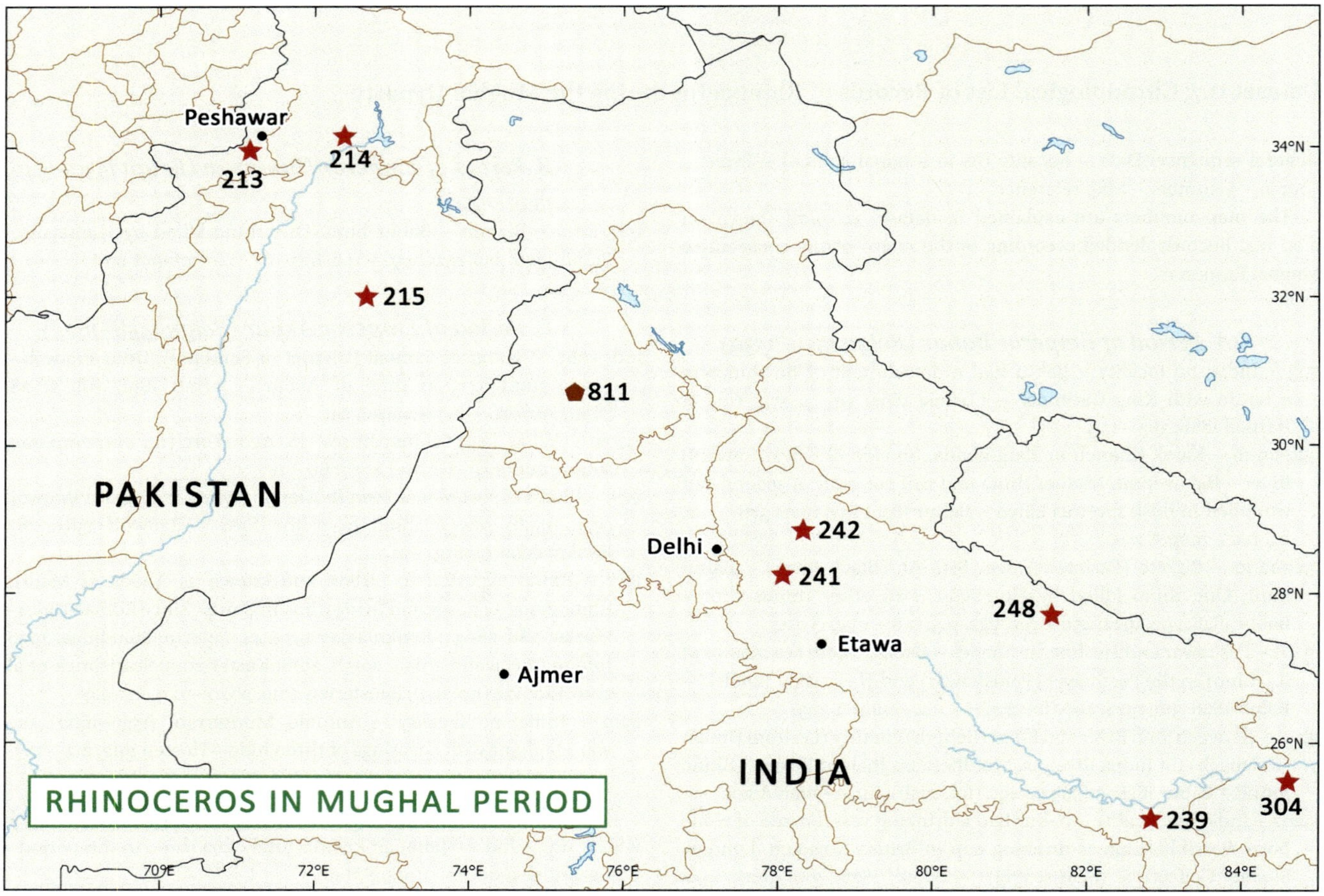

MAP 11.4   Records of rhinoceros during the Mughal period in South Asia. The numbers and places are explained in Dataset 11.8
★   Presence of *R. unicornis*
⬟   Artefacts

Dataset 11.8: Localities of Records of Rhinoceros during the Mughal Dynasty

All numbers and places are shown in map 11.4.

Rhino Region 02: Pakistan

A 213   Peshawar and Hashnagar jungles – 34.10N; 71.00E – 1520s

A 214   Bigram (Peshawar) near Siah-Ab (Black River) – 34.00N; 71.57E – 1525

A 214   Karak-Khaneh or Karg-Khana, on Sawati side (Swabi) – 34.09N; 72.53E – 1519

A 215   Between Indus River and Bhira (Behreh) – 32.28N; 72.58E – 1520s

Rhino Region 03: Rajasthan

*C 510   Ajmer – 26.44N; 74.63E – 1615 – captivity record

Rhino Region 06: North-West India

M 811   Nurmahal, Punjab – 31.09N; 75.59E – 1620

Rhino Region 07: Uttar Pradesh, Delhi

*C 511   Dehli (Delhi) – 28.63N; 77.34E–1659 – captivity record

A 241   Kul Nuh Ban near Aligarh – 28.20N; 78.30E – 1624

A 242   Sarkar of Sambal (Sambhal Dt.) – 28.70N; 78.30E – 16th cent.

*C 512   Etawa (Etawah) – 26.80N; 79.03E–1632 – captivity record

A 248   Saru or Sirwu (Ghaghara) River (near Bahraich) – 27.67N; 81.10E – 1525

A 239   Chunar – 25.11N; 82.86E – 1529

Rhino Region 11: Bihar

A 304   District of Patna – 25.80N; 84.92E – 1665–1672

* Localities not on map.

∴

11.2   The Times of Babur

Zahiruddin Mohamed Babur (1483–1530), born in current Uzbekistan, was the first Emperor of the Mughal dynasty. He wrote about his life in the *Baburnama*, composed in his Chagatai Turki language, and subsequently in 1589 in the time of his grandson Akbar translated into Farsi (Persian) by Abdur Rahim Khan-iKhanan. There are several manuscript copies, translated into English in 1826 by John Leyden (1775–1811) and William Erskine (1773–1852), into French in 1871 by Abel Pavet de Courteille (1821–1889), and again into English in 1922 by Annette Susannah Beveridge (1842–1929). The *Baburnama* is a chronological autobiography, of which large sections have been lost.

During the years in which Babur was expanding his realm into India from his base in Kabul and later Peshawar, he came to know about the existence of the rhinoceros. As far as we know, no wild rhino in Asia lived anywhere west of Kabul in human history. Babur saw rhinos in four areas, two now located in northern Pakistan and possibly adjoining Afghanistan, the other two in eastern Uttar Pradesh. The presence in Pakistan is only attested in very few sources, including the *Baburnama*, and needs serious consideration. The word which Babur used for the animal seen and hunted near Peshawar in the original version as well as in Persian could be read, in the absence of diacritical marks, either as 'gurg' meaning wolf or as 'karg' meaning rhinoceros (Ettinghausen 1950: 37). Assuming that the text has remained unaltered over the ages, the context and description eliminates the possibility that Babur intended to write about wolves, so the translation as rhinoceros must be correct.

On 16 February 1519, Babur was near the Sawati (Swabi) River, a tributary of the Indus River about 100 km east of Peshawar. Here was a place called *Karak-Khaneh* (Erskine) or *Karg-khana* (Beveridge), the home of the rhino. If this was an existing name, it indicates that it was already well-known for its rhino population. Babur's party first found a female rhino with a young one which both escaped. The jungle was set on fire, by which action a young rhino was injured and killed, and the remains were divided (Babur 1826: 253, 1922: 378):

> After starting off the camp for the river, I went to hunt rhinoceros on the Sawati side which place people call also Karg-khana (Rhino-home). A few were discovered but the jungle was dense and they did not come out of it. When one with a calf came into the open and betook itself to flight, many arrows were shot at it and it rushed into the near jungle; the jungle was fired but that same rhino was not had. Another calf was killed as it lay, scorched by the fire, writhing and palpitating. Each person took a share of the spoil. After leaving Sawati, we wandered about a good deal; it was the Bed-time Prayer when we got to camp.

On 10 December 1525, Babur was at a place called *Karg-awi* on the Siah-Ab or Siyah-ab (Black River) near Bigram (Peshawar). The party disturbed one rhino, which escaped and was pursued by Babur's son Humayun, who had never seen a rhino before. This animal was eventually killed, while two others were also killed by Babur's party (Babur 1826: 292, 1922: 450):

> Today I rode out before dawn. We dismounted near Bigram; and next morning, the camp remained on that same ground, rode to Karg-awi. We crossed the Siyah-ab in front of Bigram, and formed our hunting circle looking down-stream. After a little, a person brought word that there was a rhino in a bit of jungle near Bigram, and that people had been stationed near-about it. We betook ourselves, loose rein, to the place, formed a ring round the jungle, made a noise, and brought the rhino out, when it took its way across the plain. Humayun and those come with him from that side [Tajikistan, Afghanistan], who had never seen one before, were much entertained. It was pursued for two miles; many arrows were shot at it; it was brought down without having made a good set at man or horse. Two others were killed. I had often wondered how a rhino and an elephant would behave if brought face to face; this time one came out right in front of some elephants the mahauts were bringing along; it did not face them when the mahauts drove them towards it, but got off in another direction.

The legend of the enmity between elephant and rhino recurs from ancient times both in the Asian and Roman traditions (Ettinghausen 1950: 29). As in this instance the rhino left without fight, such a scene never became a standard ingredient in the Mughal miniatures.

On 23 March 1529, Babur was camped much further east near Chunar in present Uttar Pradesh, when a soldier reported to have seen a rhino nearby on the banks of the river. Babur and his party went to check this next day, but could not find the animal (Babur 1826: 407, 1922: 657).

In the section about the 'Fauna of Hindustan', Babur clearly described the rhino, showing that he was aware of the characteristics of the species. The translation by Leyden and Erskine seems to capture the essence best (Babur 1826: 316, cf. 1922: 489):

> The rhinoceros is another. This also is a huge animal. Its bulk is equal to that of three buffaloes. The opinion prevalent in our countries (Tramontana), that a rhinoceros can lift an elephant on its horn, is probably a mistake. It has a single horn over its nose, upwards of a span [23 cm] in length, but I never saw one of two spans. Out of one of the largest of these horns I had a drinking-vessel made, and a dice-box, and about three or four fingers' bulk of it might be left. Its hide is very thick. If it be shot at with a powerful bow, drawn up to the armpit with much force, and if the arrow pierces at all, it penetrates only three or four fingers [10 cm]. They say, however, that there are parts of his skin that may be pierced, and the arrows enter deep. On the sides of its two shoulder-blades, and of its two thighs, there are folds that hang loose, and appear at a distance like cloth housings dangling over it. It bears more resemblance to the horse than to any other animal. As the horse has a large stomach, so has this; as the pastern of the horse is composed of a single bone, so also is that of the rhinoceros; as there is a hoof in the horse's fore leg, so is there in that of the rhinoceros. It is more ferocious than the elephant, and cannot be rendered so tame or obedient. There are numbers of them in the jungles of Pershawer and Hashnaghar, as well as between the river Sind and Behreh in the jungles. In Hindustan too, they abound on the banks of the river Sirwu [Saru]. In the course of my expeditions into Hindustan, in the jungles of Pershawer and Hashnaghar, I frequently killed the rhinoceros. It strikes powerfully with its horn, with which, in the course of these hunts, many men and many horses were gored. In one hunt, it tossed with its horn, a full spear's length, the horse of a young man named Maksud, whence he got the nickname Maqsud-i-karg, or Rhinoceros Maksud.

The description probably suffers from the attempt at an accurate translation, and some parts like the comparison to the horse are not particularly clear. Nevertheless, an animal with a single horn, a thick hide, and heavy skin folds can only be a rhinoceros. The difficulty of piercing the skin with an arrow or even a bullet is a theme repeated by Jahangir, and found regularly in more recent hunting stories. An English soldier in the 1850s was punished when he put the legend to the test, killing a captive rhino in the process (Kinloch 1885: 62). The dangers attached to rhino hunting are realistic and the fame of Maqsud-i-karg is well-deserved. The legend that a rhino lifts an elephant until it is impaled on its horn is recorded by Zakarīyā ibn Muḥammad Qazwīnī (1203–1283), the 13th-century Persian geographer (Ettinghausen 1950: 22).

The drinking vessel made of rhino horn mentioned in Babur's Fauna of Hindustan has no reference to any possible anti-poisonous or medicinal properties of the material. It is suggested that this vessel is actually the boat-shaped cup dated around 1525, once in the collection of Hans Sloane (1660–1753) and in 1753 transferred to the newly founded British Museum (fig. 11.1).

FIGURE 11.1 Drinking cup of rhino horn. Length 15.2 cm, width 10.5 cm, height 5.5 cm

WITH PERMISSION © THE TRUSTEES OF THE BRITISH MUSEUM: SLOANE COLLECTION, SLMISC.1713

11.3 Interpreting Babur's Records

In the general description of the rhino quoted above, Babur gives two localities in Pakistan (Rookmaaker 2000). First he said he had killed several rhinos in the Parashawar and Hashnagar jungles, indicating the general region around Peshawar. It is not clear why the localities were repeated twice in the text. Second in the jungles between the river Sind (Indus) and Bhira (Behreh), or between the upper Indus River and Punjab. Babur also provided two localities in Indian territory, much further east. One is again quite general, when he said that the rhinos abound on the banks of the Saru River, now the Ghaghara River in Uttar Pradesh. This may have been based on information received, unlike the final record.

Given that Babur stated that the rhino was one-horned, the animals which he encountered could have been either *R. sondaicus* or *R. unicornis*. The zoologist Edward Blyth, possibly the first to quote Babur's references in relation to the rhino, suggested that the actual species could not be ascertained (Blyth 1862b, 1862d). Later he asserted that the species actually was *R. sondaicus* (Blyth 1875), followed by Kinloch (1903). Maybe this was his interpretation of a remark by Jerdon (1867) about rhinos on the banks of the Indus, inserted in a paragraph on *R. sondaicus*. Cockburn (1883) inclined towards *R. unicornis*. With reference only to the four passages in the *Baburnama*, no conclusion can ever be recommended as the characteristics are not detailed enough to separate the two species which are similar in general appearance.

11.4 Illustrations of the *Baburnama*

Five illuminated manuscripts of the *Baburnama* are known, some incomplete (Smart 1977; A.K. Das 2018a). They are preserved in (A) the Victoria & Albert Museum (V&A), London (with one expunged folio with rhino in the Museum Rietberg, Zurich; and another auctioned by Forge and Lynch in March 2023); (B) the British Library, London (Or.3714); (C) the State Museum of Oriental Art, Moscow, supplemented by folios in the Walters Art Museum, Baltimore, Maryland; (D) the National Museum, Delhi; and (E) the Alwar Museum, Rajasthan. The episode of Babur hunting at the Karg-khana in 1519 is depicted in 4 folios; Babur hunting with Humayun at Bigram in 1525 in 5 folios (2 double-spreads); and the description of the rhino in the Fauna of Hindustan in 2 folios (table 11.18). These illuminated manuscripts were all produced at the court of Akbar after he commissioned the Persian translation of 1589. The folio recently auctioned by Oliver Forge

TABLE 11.18 Rhinoceros depictions found in the five copies of the *Baburnama*

	Location of copy	Kargkhana 1519	Bigram 1525	Fauna of Hindustan	Chunar 1529	Comment
A	V&A		A37r			Rietberg folio is a missing
A(1)	Rietberg		A37l			page from V&A
A(2)	Auction 2023	no. 1				Forge and Lynch 2023
B	British Library	B43 r, f. 305v	B49r f. 351v, B49l 352r	B52 f. 379		Fig. 11.3, 11.6
C(1)	Moscow			C55		These 2 copies are part of
C(2)	Walters Art	W596 f. 21v				each other – fig. 11.2
D	New Delhi	D56r f. 215v	D62r f. 246v			
E	Alwar					no rhino images

This follows the classification introduced in Smart (1977) together with the folio (f.) number where available.

FIGURE 11.2   Babur hunting at the *Karg-khana* in 1519 showing several rhinos on the run and one rhino hide being prepared. Size 32 × 21 cm
WITH PERMISSION © WALTERS ART MUSEUM, BALTIMORE: MD. W596 FOL. 21R

FIGURE 11.3   Babur riding a chestnut horse at a rhino hunt near Bigram in December 1525. An elephant was taken close to a rhino which then ran away. Painting by Shivdas, 25.2 × 14.6 cm
WITH PERMISSION © BRITISH LIBRARY: OR.3714, FOLIO 352R

and Brendan Lynch in New York is signed by the artists L'al and Sarwan (Forge and Lynch 2023, no. 1). It is clear that none of the artists were present at the time, while it is unlikely that during Babur's period any rhino sketches were made. The rhinos depicted in the *Baburnama* were therefore based on material available to artists at the end of the sixteenth century.

11.5   The Rhinoceros in Battle

Babur may have had to contend with rhinos in warfare. The Portuguese historian Gaspar Correia (1492–1563) wrote in his *Lendas da Índia* (Legends of India) about a battle between Babur and King Cacandar, who probably

was Sikandar Lodi (d. 1517), the Sultan of Delhi between 1489 and 1517. He included a passage about Cacandar's army, which had "five well-arrayed battalions, consisting of 140,000 men on horseback and 280,000 on foot, and in front of them a battalion of 800 elephants, which fought with swords upon their tusks, and howdahs with archers and musketeers on their backs. And, in front of the elephants, 80 rhinoceroses (gandas), like the one that went to Portugal, and which they called 'bicha' (beast) fought strongly, carrying three-pronged iron weapons on the horn of their snout ... The Mogors took the advantage by shooting arrows, wounding many of the rhinos and elephants, which, as the arrows pierced them, turned and fled" (Correia 1862, vol. 3: 573, translation edited by John Beusterien from Yule & Burnell 1886: 363). The word *ganda*

FIGURE 11.4   A duel with a warrior seated on the back of
a rhinoceros. Folio from *Hamzanama*, *c.*1570.
Size 70 × 55 cm
MAK – MUSEUM OF APPLIED ARTS, VIENNA:
B.I. 8770/17. PHOTO: © MAK/GEORG MAYER

FIGURE 11.5   Jahangir on horseback in combat with the evil
champion Hizabr mounted on a rhinoceros. From a
manuscript of the *Shahnama*, *c.* 1750
WITH PERMISSION © VICTORIA & ALBERT
MUSEUM, LONDON: IS.256: 2 – 1952

and the description referring to the Lisbon Rhinoceros of
1515 (4.6) appear to leave no doubt that Correia meant the
rhino. Although it is not intrinsically impossible to have
a detachment of 80 rhinos, it does stretch the imagination. There are two images of a rhino taken to battle, in
both cases mounted by a warrior fighting an opponent
on foot or on horseback. One is found in a manuscript of
the *Hamzanama* dated 1570 in the Austrian Museum of
Applied Art (illustrated in A.K. Das 2018a: 69), the other
is in the *Shahnama* dated 1750 in the Victoria & Albert
Museum (figs. 11.4, 11.5).

11.6   The Times of Humayun

The only indication that the second Mughal Emperor
Humayun (1508–1556) ever saw a rhinoceros is found
in the recollections of his father Babur. Hunting in
December 1525, when he was still a teenager, Humayun
pursued and killed a rhino near Peshawar (Babur 1826:

292, 1922: 450). This is again mentioned by his grandson
Jahangir, who added that his father Akbar told him that
he witnessed such a hunt a few times in the company
of his father (Beveridge 1909, vol. 1: 102). The scene near
Peshawar given prominence in the *Baburnama* is shown in
several miniatures related to this description. Humayun is
depicted while inspecting the rhino killed by many spears
in the copy in the British Museum (OR 3714, folio 351v)
dated 1590–1593 (fig. 11.6).

During Humayun's reign, the rhino was again seen
near Peshawar by Seydi Ali Reis (13.8). Similar in date,
yet otherwise probably quite unrelated, there is a drawing
of a young single-horned rhino in a manuscript prepared
in 1547 at the court of Ibrahim Adil Shah I (1534–1558),
Sultan of Bijapur in South India (fig. 11.7). The *ʿAjaʾib
al-makhluqat* by Qazwīnī (1547) was translated from
Arabic to Persian and illuminated with 461 small paintings
(preserved in the British Library, Or.1621). The precursor
of this original illustration is not clear from the available
data.

FIGURE 11.6   Prince Humayan had joined the rhino hunt of his
father near Bigram in December 1525, and pursued
one animal which had run away. Painting by
Jagannath, 25.2 × 14.6 cm
WITH PERMISSION © BRITISH LIBRARY, OR.3714,
FOLIO 351V

FIGURE 11.7   Young rhinoceros drawn in the Deccan in the middle
of the 18th century in a translation of Qazwīnī at the
court of Bijapur
BRITISH LIBRARY, OR.1621, FOLIO 131R IN QATAR
DIGITAL LIBRARY: SEE QAZWĪNĪ 1547

11.7   The Times of Akbar

The third Mughal Emperor Akbar (1542–1605) reigned
from 1556 to 1605. Abū al-Fazl ibn Mubārak (1551–1602)
was the author of the administrative history of Akbar's
reign, known as the *Ain-i-Akbari*. It is unknown if Akbar
ever hunted or saw a rhino in the wild. Crooke (1899: 231)
must have misinterpreted his source to mention Akbar
hunting near Chunar. In the *Ain-i-Akbari* rhinos are only
mentioned once, stating that they were numerous in the
"Sarkar of Sambal", the region of Sambhal just east of
Delhi (Jarrett 1891). It was noted that shields were made
from the rhino skin, and finger-guards for bowstrings from

the horn. Examples of this latter production are now no
longer recognized, but the shields appear in Monserrate's
account of 1582.

The Portuguese Jesuit Antonio Monserrate (1536–1600)
was part of a mission from Goa to the court of Akbar in
1580–1582. In the Portuguese abstract of his experiences
"*Relaçam do Equebar, Rei dos Mogores*" known from
manuscripts of 1582, translated by the orientalist Henri
Hosten (1873–1935), he merely stated that the foreheads
of the war-elephants were covered either with metal
plates or ("de couro dãta") with rhino hide (Hosten 1912:
212). Monserrate also compiled a longer work in Latin,
Mongolicae Legationis Commentarius, finished in 1590
which again remained unpublished in his time. Here
he inserts a vivid impression of the elephant's fears of a
rhino: "It is hard to say what a dread elephants have of the
rhinoceros, although it is a much smaller beast. Moreover
the rhinoceros has an insolent contempt for the elephant.
When an elephant sees a rhinoceros, he trembles, cringes,

bends down, hides his trunk in his mouth, and humbly retreats until the rhinoceros has passed by. For the rhinoceros attacks from beneath, and buries the sharp horn on his nose in his opponent's belly. Meanwhile the elephant can do him no harm" (Hosten 1914: 584; Monserrate 1822: 87). Babur had already alluded to a similar belief. Monserrate probably found his inspiration in European works (12.6), but maybe he would not have expanded on the theme if he had not either witnessed such an occasion or seen elephants and rhinos in close proximity.

Akbar was fond of elephants, keeping multitudes of them in his stables in Agra Fort, while details about any rhinos have remained elusive (Grewal 2007: 60). Two rhinos are known to have been exported alive from India in Akbar's period. One was sent from Goa to Lisbon in 1577 and is now known as the *Abada* or Madrid Rhinoceros (11.9). In 1575, the German botanist Leonhard Rauwolf (1540–1596) saw a young "Rhinocerot" in the town of Aleppo, Syria, on its way to Constantinople, Turkey. He heard that the animal came from the East and was so ferocious that it had killed some 20 people when it was captured (Rauwolf 1583). It is unknown if it arrived at its Turkish destination.

11.8 Mughal Miniatures Showing the Rhinoceros

Akbar commissioned a translation of his grandfather's *Baburnama* from Turkish to Persian in 1589. He was a great patron of the arts and established a formal studio where Iranian master artists were in charge of over a hundred painters. The rhino is found in the miniature paintings from the 1560s in the *Hamzanama* and the *Shahnama*, and from the 1590s in the *Baburnama*, then until the late 18th century in a wide variety of mythological, historical and scientific texts (Beach 1981; Bautze 1985; A.K. Das 2018a). It is unlikely that any of the artists actually saw a rhino in the wild, and if they did, definitely not in Babur's retinue or in the same localities. Given that the miniatures were produced maybe sixty years after the events narrated in the *Baburnama*, they are artistic impressions rather than naturalistic or photographic portrayals. There are no paintings of Akbar associated with a rhino. Rhinos could still be found in the Mughal dominions in the 16th century, and were reported by Jahangir in the eastern part of Uttar Pradesh. Some rhinos may have been captured and taken to a menagerie. There are very few definite records of hunts, sightings or captures of rhinos as will be seen from the entries in this chapter.

The rhinos depicted in the Mughal miniatures are painted well enough that their general identity is irrefutable: they are large animals with folds in their skin, hooves on their feet, and always depicted with a single horn of a fair size. However, from a zoologist's perspective, none of the rhinos can be said to be shown photographically correct. The ears and the tail are often rather large, and the folds in the neck and shoulder region are shown as a set of rings rather than in the precise order found in the living animal. Despite such shortcomings, it cannot be denied that the depictions could not have been painted in this fashion without a good example, like a sketch made in the field, or from an animal seen in captivity, or of course an older painting. It is a little frustrating that there is no information where the first impressions of a rhino could have been obtained, especially in the periods of Akbar and Jahangir.

The question remains if these Mughal miniatures can tell us more about the rhino species then found in the western part of India and in Pakistan. From other texts we know that *R. sondaicus* occurred in the lower Ganges delta, and *R. unicornis* in the valleys of the Brahmaputra and Ganges along the southern fringes of the Himalayas. Specimens of either species could have been captured and transported to a Mughal menagerie. In case of wild animals, the paintings could provide a perception absent from a written description. Commentaries on the *Baburnama* by Blyth and Jerdon (11.3) leave the answer open, but among more recent authors there is definitely a preference for *R. unicornis*. However, the German archaeologist Burchard Brentjes (1929–2012) examined a painting of Prince Salim (Jahangir) made around 1600 (fig. 11.8). The small rhino seen running in the foreground has all the characteristics of *R. sondaicus*: smaller size, a curious arrangement of the folds, and the shape of the skin fold in the neck region (Brentjes 1969). There is no definitive answer. When comparing a set of 27 images of Mughal miniatures from different periods, my analysis (assisted by Andrew Laurie) is that 4 (15%) could possibly portray *R. sondaicus*, 7 (25%) were almost certainly *R. unicornis*, and in 16 (60%) the characteristics of folds and posture were indefinite. The conclusion must be that the Mughal painters had no intention to portray these animals in a naturalistic way, they wanted to show their ferocity and the courage required to subdue them.

11.9 The Life of Abada, the Madrid Rhinoceros, 1577–1591

The second post-medieval rhinoceros seen alive in Europe was shipped by the Portuguese mission from Goa in 1577.

FIGURE 11.8   Prince Salim hunting the rhinoceros, *c.*1600–1604
PRESENT LOCATION UNKNOWN, POSSIBLY A
PRIVATE COLLECTION IN NEW YORK

This female, known as the Madrid Rhinoceros, was exhibited first in Portugal and subsequently in Spain. New research has recently added considerably to the known history and iconography of this animal, from arrival in September 1577 to her death in 1591. It is likely that she was part of a set of gifts from the court of Akbar to the Portuguese in Goa. Unfortunately there is no evidence regarding the provenance of the animal. Beusterien (2020) suggested that the rhino was captured in the wild when about two years old after the mother was shot by a Mughal hunting party, rather than chosen from a set of rhinos in an otherwise unknown Mughal menagerie. Like all Mughal rulers, Akbar kept scores of animals. A Portuguese ambassador Petrus Tavares visiting in 1578 enumerates 300,000 horsemen and 20,000 elephants, besides 16,000 horses in the stable, 14,000 deer out of which 4000 are brought up in the house and 700 domesticated panthers and 10,000

oxen to draw carts, also 500,000 birds (Wicki 1970). A few rhinos would hardly have inconvenienced the keepers. In any case, it remains unclear where the animal or her mother would have lived in the jungle, in the region of Peshawar where Babur hunted, or near Sambhal close to Agra and Delhi as recorded in the *Ain-i-Akbari*, or indeed elsewhere in the *terai* of Uttar Pradesh and Bihar. The history of this specimen proves that rhinos were available to be hunted or captured in Akbar's realms in the course of the 16th century. In Europe, the Madrid Rhinoceros had relatively little impact on zoological history, although she must have been seen by large crowds in the Iberian peninsula. Her likeness was painted a few times, but definitely the image created by Philippe Galle in Antwerp in 1586 is the most iconic and the most copied in contemporary publications, characterized by the peculiar shape of the lower edge of the middle shield on the belly (fig. 11.9). Several depictions of the Abada in Madrid have come to light, like the drawing made by Adam Hochreiter of 1584 (fig. 11.10).

11.10   The Times of Jahangir

The fourth Mughal Emperor Jahangir (1569–1627), also known as Nur-ud-din Muhammad Salim or Prince Salim, ruled 1605–1627. He was a great hunter. In his memoirs, the *Tuzuk-i-Jahangiri*, his reference to the rhino is not straight-forward, because the word used to name the animal without diacritical marks can also mean wolf (as seen in the *Baburnama*). In the commonly used English translation by Alexander Rogers (1825–1910) and Henry Beveridge (1837–1929) published in 1909–1914, the word is translated as 'wolf', but others like Ali (1927a) suggest 'rhinoceros', and this is followed here. From the paintings produced during the reign of Jahangir (11.11), there appears to be no doubt that the Emperor and his retinue killed several rhinos. In the memoirs for the summer of 1624, Jahangir wrote:

> One of the strange things that happened was that one day I was on an elephant and was hunting rhinoceros in Aligarh in the Nuh forest (Kuh-i-Kul). A rhinoceros appeared, and I struck it with a bullet on its face, near the lobe of the ear. The bullet penetrated for about a span. From that bullet it fell and gave up its life. It has often happened in my presence that powerful men, good shots with the bow, have shot twenty or thirty arrows at them, and not killed. As it is not right to write about oneself, I must restrain the tongue of my pen from saying more (Beveridge 1914, vol. 2: 270).

FIGURE 11.9
The Madrid Rhinoceros engraved in 1586 by
Philippe Galle (1537–1612), working in Antwerp.
Broadsheet 17.5 × 27.0 cm. From the collection
of prints and drawings assembled Ulysse
Aldrovandi (1522–1605) in Bologna
WITH PERMISSION © BIBLIOTECA
UNIVERSITARIA DI BOLOGNA: MS. 124, TAVOLE
VOL. 6, PART 2, F. 63

FIGURE 11.10 "Zu Madrid l'anno 1584. Rinoceros o. L'Abada." Drawing by Adam Hochreiter (1550–1595), Chamberlain of
Archduke Ferdinand II of Austria. He saw the second rhino brought alive to Europe when the animal was in
Madrid, Spain. Watercolour on paper, 18 × 25 cm
WITH PERMISSION © UNIVERSITÄTS- UND LANDESBIBLIOTHEK TIROL, INNSBRUCK: COD. SERV. I B 42,
DRAWING PASTED INTO THE DIARY OF HOCHREITER

Apparently it was not the first time he had hunted the rhino, because he states that his companions had sometimes failed to kill the animal in his presence. The reason for this passage must therefore be that he was proud to shoot it with only one bullet. In an earlier part of the memoirs, Jahangir gives a list of all the game shot by him and his troops from his age 12 to age 50, or from 1580 to 1619, which included 86 tigers, 36 wild buffaloes and 64 rhinos (Beveridge 1909, vol. 1: 369). Beveridge translates wolves, Leach (1995, vol. 2: 1040) changes this to rhinoceros. A total of 64 rhinos is not impossible, but indeed high enough to cast some doubt regarding the correctness of the text or the translation.

The localities of Jahangir's hunts are not known. The 1624 hunt took place in the district of Kul, Kol, Koil, which is the current Aligarh (Beveridge 1914; Jarrett & Sarkar 1949, vol. 2: 104). There is no indication where the other 64 rhinos were killed, except of course that it must have been within the territories visited by Jahangir. It would be nice to trace any of the horns, shields or other trophies, but nothing is left beyond the paintings made during the reign of Jahangir.

11.11   Miniatures of Prince Salim with a Rhinoceros

There are five miniatures showing Prince Salim before his anointment as Emperor hunting rhinos from the back of elephants or horses, well described by Asok Kumar Das (2018a: 72), listed here with selected secondary literature:

1. Prince Salim hunting rhino. Private collection, New York City. Previously in the private collection of Otto Sohn-Rethel, Düsseldorf, Germany. Illustrated in A.K. Das 2018a: 73; Ettinghausen 1950, pl. 33; Brentjes 1969, fig. 1; Kühnel 1941, fig. 2, 1962, pl. 71; Störk 1977: 449; Divyabhanusinh 1999, fig. 9. Fig. 11.8.
2. Prince Salim visiting the hunting fields. Chester Beatty Library, Dublin. Illustrated in A.K. Das 2018a: 74; Leach 1995, vol. 2: 1040.
3. Prince Salim at a hunt. Los Angeles County Museum of Art (M83.137). Illustrated: Sotheby's 1977, lot 28; Okada 1992, fig. 4. Fig. 11.11.
4. Prince Salim in the hunting fields. Art Gallery of New South Wales, Sydney (EP 1.1969). Illustrated in A.K. Das 2018a: 75.
5. A prince on horseback hunting rhino. Attributed to Narsingh. Private collection. See Bautze 1985: 422, n. 43. Illustrated in A.K. Das 2018a: 76; Sotheby's 2015, lot 16.

Prince Salim's hunting trips depicted in these miniatures were made during the five years 1600 to 1604 when he

FIGURE 11.11   Prince Salim, later Jahangir, seated on an elephant with a gun, hunting deer and rhinoceros. On the left five men show him a rhino that was captured. From Salim's studio, Allahabad, 1600–1604.
Size 19.7 × 11.7 cm
LOS ANGELES COUNTY MUSEUM OF ART: M83.137

lived in Allahabad. Many rhinos were killed and their bodies are painted lying in heaps of up to four together, to show the great prowess of the future Emperor.

11.12   Menageries with Rhinos in Jahangir's Times

Jahangir's wife Nur Jahan (1577–1645) was responsible for building the Nurmahal, a caravanserai near Jalandhar, Punjab completed in 1620. The western gateway has several sculptured panels including two of a rhino next to an elephant (fig. 11.12).

In a miniature dated around 1600–1604, Prince Salim is depicted seated on an elephant holding a gun engaged in a general hunt (fig. 11.11). It shows one dead rhino, while next to it, a living rhino kept at bay by four men is offered

FIGURE 11.12
Nurmahal near Jalandhar built for Nur Jahan, ca. 1620
PHOTO: © GOPAL AGGARWAL, DECEMBER 2017

to Salim. This is just a glimpse, and the records keep us in the dark about the future fate of this particular specimen, which may well have lived for some time in one of the imperial parks. Two rhinos were in captivity in Ajmer when the English trader Thomas Coryat (1577–1617) visited between July 1615 and September 1616: "Hee keepeth abundance of wilde beasts … [including] unicornes; whereof two I have seene at his court, the strangest beasts of the world. They were brought hither out of the countrie of Bengala, which is a kingdom of most singular fertilitie within the compasse of his dominion" (Coryat 1616: 246). If brought from Bengala, further to the east, these animals cannot be associated with the hunting trips of Prince Salim.

Another living rhino made it all the way to Istanbul in Turkey, where one was recorded at the court of the Turkish sultan Osman II (1603–1622) It was donated together with four elephants by ambassadors representing the fifth Safavid king of Persia or Iran Shah Abbas I (1587–1629), who had good contacts with the Mughal rulers. This event, which probably took place in 1620–1622, is recorded in a manuscript drawing showing a stylized rhino with one horn (Stchoukine 1966: 104; Störk 1977: 480). A rhino was seen by the Venetian merchant Tommaso Alberti in Constantinople in November 1620 (Alberti 1889: 57).

11.13   The Times of Shah Jahan

The fifth Mughal Emperor Shah Jahan (1592–1666) reigned from 1627 to 1658. There are few references to any rhino during the reign of Shah Jahan in the middle of the 17th century. Whether he ever hunted these animals is unclear. He is depicted observing a fight between a rhino and an elephant by an artist in The Netherlands who could not have painted from personal experience (11.14). During his reign, his son Shah Shuja (1616–1661) was Governor of Bengal and in 1643 sent an ambassador with a living rhino to King Rajasinghe II (1608–1687) of Kandy, Sri Lanka (5.4).

11.14   The Hawking Party by Willem Schellinks

The miniatures produced by the Mughal artists became increasingly popular in Europe as a source of enjoyment and as objects to collect in the 17th century. They were also copied or adapted to local taste by a number of artists, many of them based in The Netherlands. One such painting by the Dutch artist Willem Schellinks (1623–1678) shows the Emperor Shah Jahan in the company of his four sons out for a hawking party (Gommans & de Hond 2015; Forberg 2017). The oil painting "A hawking Party" was sold twice at Sotheby's London (8 July 1981, lot 79, and 5 December 2007, lot 43) and is currently in the Orientalist Museum, Doha, Qatar (figs. 11.13, 11.14). It is signed but not dated, and therefore tentatively assumed to be completed 1665–1670. Schellinks never traveled to India and his painting is presumably based on an earlier unknown Mughal miniature (Lunsingh Scheurleer 1996). While acknowledging the importance of such rare portraits of Mughal rulers, the group is shown watching a fight between an elephant and a rhino. A preparatory drawing in the British Museum (reg. 1923,0113.20) excludes the fight, while a copy of the miniature in the Millionenzimmer of Schönbrunn Palace is damaged where the animal fight would have been.

The animal fight may have been Schellink's own invention, maybe to add an exotic element to the composition. Despite Babur's reference to the antagonism of rhino and elephant, which he tried to induce in 1519, the fight between the two megamammals was never part of

FIGURE 11.13   Willem Schellinks (1623–1678). Hunting scene with Shah Jahan and his sons, 1665–1670. Oil on canvas, 51 × 61.5 cm
WITH PERMISSION © ORIENTALIST MUSEUM, DOHA, QATAR: OM.672

FIGURE 11.14
Detail of the hunting scene watched by Shah Jahan,
painted by Schellinks
WITH PERMISSION © ORIENTALIST MUSEUM, DOHA,
QATAR: OM.672

FIGURE 11.15
Rhinoceros fighting an elephant as perceived by André
Thevet (1516–1590)
THEVET, *LES SINGULARITEZ DE LA FRANCE
ANTARCTIQUE*, 1558, P. 41

the Mughal art tradition. The theme was introduced in
European book illustrations around the middle of the
16th century but remained relatively unusual (Clarke 1986:
155). An early example is a small engraving illustrating a
geographical treatise by the French cosmographer André
Thevet in 1558 (fig. 11.15). Two centuries later, the subject
culminated in an etching by the well-known animal artist
Johann Elias Ridinger of 1760, where an elephant almost
crushes the rhino, which in turn is piercing his opponent's
belly with the horn, leaving an open question who would
have been victorious (fig. 11.16).

The animal fight painted by Schellinks has two remarkable features. First, the rhino has opened his mouth in an
attempt to bite the front leg of the elephant. Most examples of the rhino-elephant fight show the two animals
facing each other rather timidly without signs of aggression. A similar posture was drawn by the Florentine artist
Antonio Tempesta as part of a series of etchings of 1608
(fig. 11.17), although the rhino hardly touches the elephant
(Clarke 1986, fig. 123).

The second feature of the rhino as painted by
Schellinks is the animal's appearance. The majority of rhinos depicted in art from the 16th to 18th centuries were
based on the woodcut by Dürer of 1515 with its heavy body
folds and the small second horn on the shoulder (fig. 4.3).
Schellinks's rhino lacks this Dürer hornlet, and shows ears
turned backwards, a skin devoid of tubercles, an upturned
tail and hind feet wide apart. This type bears some
resemblance in some of these features to an anonymous
engraving first published in 1658 illustrating the account
by Jacobus Bontius on the rhino of Java (fig. 11.18). It was
added by the editor of the volume Willem Piso (1611–1678)
after a drawing made from a living animal in the (East)

FIGURE 11.16   Fight of elephant and rhinoceros dated 1760 from the
series "Kämpfe reissender Thiere" by Johann Elias
Ridinger (1698–1767). Size 35 × 29 cm
COURTESY: JIM MONSON

Indies ("in Indiis") sent to Johannes Wtenbogaert
(1608–1680), Receiver-General of Holland (Bontius 1658:
51; Coste 1946). According to Müllenmeister (1978), this
original was an undated watercolour signed by "P. Angel",
showing a rhino drawn from life in unstated locality
(fig. 11.19). There were two contemporary artists with the

FIGURE 11.17
Engraving by Florentine artist Antonio Tempesta
(1555–1630) from his series "Battling Animals"
dated ca. 1608. Size 9.1 × 12.9 cm
COURTESY: JIM MONSON

same name, Philips Angel of Middelburg (1616–1683) and Philips Angel of Leiden (1618–*c*.1664), the second of which traveled to India and the East Indies (Bol 1949). Assuming that Angel of Middelburg was the artist of this watercolour, Dittrich (1997) suggested that the rhino had been drawn alive in this Dutch town in the early 17th century. It is much more likely, however, that it was in fact Angel of Leiden, who might have drawn the animal in Indonesia.

Both Angel and Schellinks were known to the great painter Rembrandt van Rijn (1606–1669), with whom they shared a fleeting passion for Mughal miniatures and paintings (Schrader 2018). Schellinks could therefore have used Angel's watercolour as his inspiration for the rhino-elephant fight, or have seen only the published plate in Bontius (1658).

11.15 The Times of Aurangzeb

There are no indications that the sixth Mughal Emperor Aurangzeb (1618–1707) was ever able to hunt a rhino. However, during his time a few specimens were seen in his menageries, of which so far only three glimpses have been discovered.

François Bernier (1620–1688) was a French physician, who worked at Aurangzeb's court from 1658–1667. During one of his visits to Delhi in 1659, he mentioned the presence of rhinos ("des Rinocéros" in original French) without further elaboration (Bernier 1671). His silence could mean that he was so used to seeing these large animals that they needed no further introduction, although the

opposite could be expected, especially as he never mentioned them elsewhere during his wide-ranging travels. Surely based on Bernier's description, a Dutch compilation asserted that the Emperor would daily (except Sunday) enter a park surrounded by high walls to watch animal fights (Berkenmeyer 1729, vol. 3: 839). A similar scene of an animal park, this time in Lahor (Lahore), without a rhino, is depicted on a plate published by Pieter van der Aa (1659–1733) in Holland (fig. 11.20).

In 1718 Maharaja Sawai Jai Singh II of Jaipur (1700–1743) bought an Album containing 48 paintings, including 25 of various animals, unsigned but mostly annotated and a few dated from the period 1665–1682 (A.K. Das 1995). There are seven foreign animals: Grevy's zebra, African elephant, Barbary goat, Tibetan yak, Turkey, Ostrich and Cassowary. The Album contains one painting of a one-horned rhino on a green field portrayed in a naturalistic style (fig. 11.21), with an inscription stating that the animal was captured in the Suba (District) of Patna (A. Das 2018: 78). The compilation of such an Album seems to indicate that the animals were kept, exhibited or used in shows in one of the imperial facilities.

There was a rhino horn in the cabinet of curiosities of Christian V (1646–1699), the King of Denmark-Norway. It was listed in the illustrated inventory by Oligerus Jacobaeus (1650–1701) published in 1696 and again edited by Johannes Laürentzen (1648–1719) in 1710. The information in the second edition is more elaborate than in the first, in translation: "Double horn of a rhinoceros, of which the longer branch is two feet, and the other one foot long. The animal which carried the horn, died in the menagerie of the Great Mogol of India, from where this

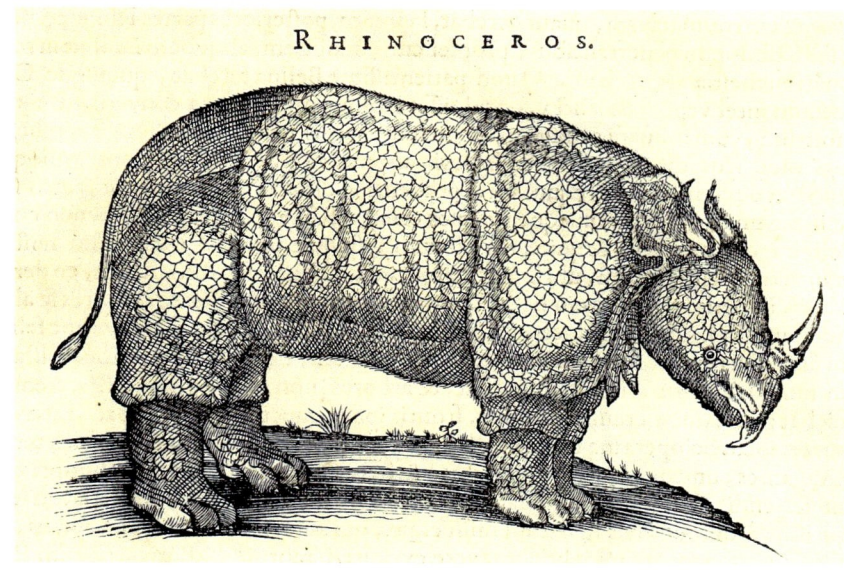

RHINOCEROS.

FIGURE 11.18
Rhinoceros illustrating text on the Javan
rhinoceros by Jacobus Bontius (1592–1631)
BONTIUS, *DE INDIAE UTRIUSQUE RE NATURALI
ET MEDICA*, 1658, P. 51

FIGURE 11.19
Undated watercolour of rhinoceros, signed
"P. Angel fecit" indicating either Philips Angel
of Leiden (1618–*c*.1664) or Philips Angel of
Middelburg (1616–1683). It shows a Javan
rhinoceros, similar to the one published by
Bontius in 1658.
DITTRICH, *BIJDRAGEN TOT DE DIERKUNDE*,
1997, P. 152. CURRENT LOCATION UNKNOWN

rare horn has been brought to Copenhagen by the traders of the Indies who are in our company, and presented to our King Christian V" (Jacobaeus & Laürentzen 1710: 4). The Danish Company had been formed in 1671 and had its center in Tranquebar in South India. It is likely that the horn was obtained by trade and was found to be unusual enough to be presented to the King. However intriguing this record is, no further details have come to light, either about the menagerie or about the transfer to Copenhagen. The specimen was illustrated in both editions of the catalogue (fig. 11.22), but it is unknown if it is still preserved. Because the reference to Jacobaeus was included by Linnaeus in his first description of *Rhinoceros unicornis* in 1758, this is a syntype of the species (Rookmaaker 1998b,

1999d; 4.5). The anterior and posterior horns were said to measure 60 cm and 30 cm respectively. These sizes are unusual for the double horned species *D. sumatrensis*, hence maybe the animal which had lived in a menagerie of Aurengzeb may be suspected to have been of African origin. A black rhino (*Diceros bicornis*) in an Indian menagerie in the 17th century has not been documented elsewhere, and was certainly unusual. However, as we can see in the Album of the Aurangzeb period, several species of African animals reached India in that period, like the Grevy's zebra and the African elephant. It was recorded that there were embassies from Abyssinia (Ethiopia) in 1665 and 1671 bringing gifts including live animals (A.K. Das 1995, 2012: 87).

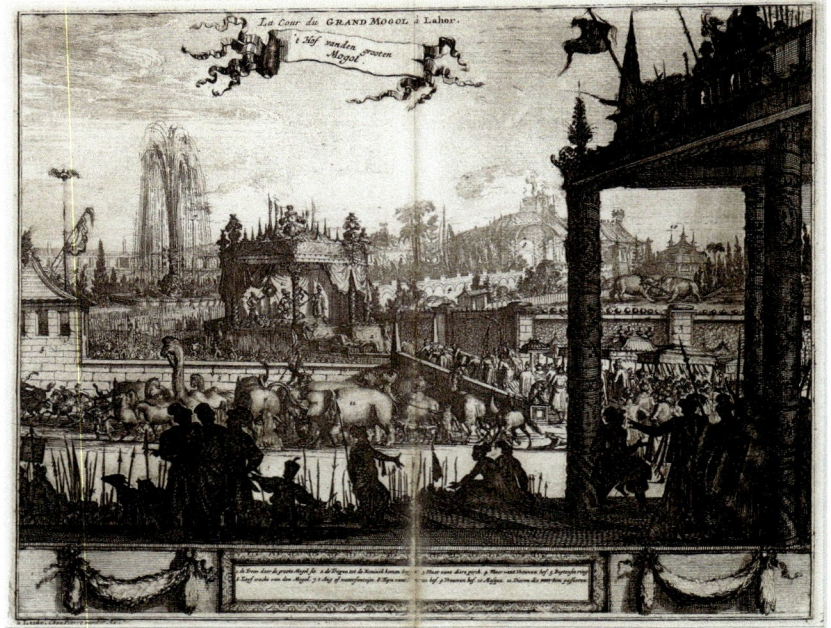

FIGURE 11.20
Court of the Great Mogol in Lahor, showing
the animal procession as described by François
Bernier. The structure on the right is labelled
"wall of the animal park", showing a fight of two
elephants behind it.
PIETER VAN DER AA, *LA GALERIE AGRÉABLE DU
MONDE*, 1729, VOL. 2, PL. 28. COURTESY: JIM
MONSON

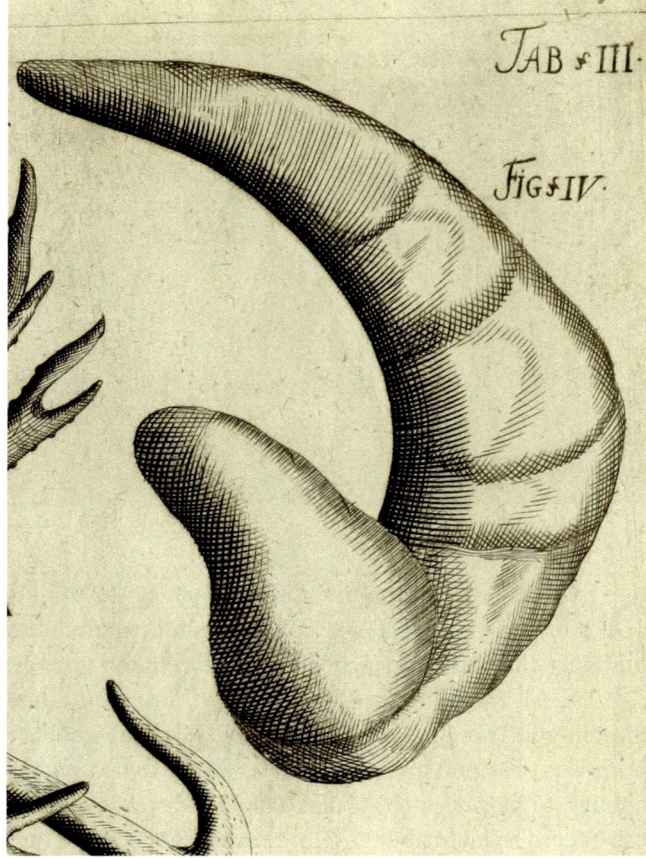

FIGURE 11.21   Rhinoceros in an album of natural history drawings
compiled at the court of Aurangzeb in the period
1665–1672. It is likely that the posterior cervical fold
is exaggerated because the general posture of the
animal resembles *R. unicornis*.
WITH PERMISSION © MAHARAJA SAWAI MAN
SINGH II MUSEUM, JAIPUR: NO. AG969

FIGURE 11.22   Double horn of rhinoceros, obtained in India from
a specimen that had lived in the menagerie of the
Great Mogol. The figure in the second edition of 1710
(pl. 4 no. 31) is slightly different.
JACOBAEUS, *MUSEUM REGIUM*, 1696, PL. 3 NO. 4

11.16 Gifts of Rhinos to Asian Rulers

Another indication of trade in live rhinos is found in specimens donated or sold to rulers in western Asian countries. During the time of Aurangzeb two or three specimens were recorded in Uzbekistan and Iran (Persia), known in all cases from short glimpses by foreign visitors:

1669. At the court of King Khan Abdul Aziz (ruled 1645–1680) of Bukhara (Bukoro), Uzbekistan, the Russian ambassador Boris Pazuchin was shown a rhinoceros (*karka*) among other animals (Unbegaun 1956: 549 following a Russian source; Rookmaaker 1998: 31).

1675–1683. During the reign (1666–1694) of Suleiman I, the eighth Safavid Shah of Iran, a rhino was recorded in the royal stables in Esfahan three times. In 1675, the French traveler Jean Chardin (1643–1713) saw and sketched a single-horned rhino (fig. 11.23), but incongruously said that it had been presented by an Ambassador from Ethiopia (Chardin 1711, vol. 3: 45, pl. 40). In 1676, the English physician John Fryer (1650–1733) described a rhino with one horn, said to have come from 'Bengala' (Fryer 1698: 287). On 30 July 1683, the German traveler Engelbert Kämpfer (1651–1716) obtained a sketch of a single-horned rhino (*kerqden*) in Esfahan, which is preserved in an album in

FIGURE 11.23
Rhinoceros seen in Esfahan, Iran in 1675
CHARDIN, *TRAVELS*, 1711, VOL. 3, PL. 40

FIGURE 11.24
Rhinoceros in an album compiled by Engelbert Kämpfer in Iran in 1683. Size 21 × 29.2 cm
WITH PERMISSION © THE TRUSTEES OF THE BRITISH MUSEUM: DEPT. ORIENTAL DRAWINGS, 1974,0617,0.1.29

the British Museum (fig. 11.24, see Rookmaaker 1978b: 33 no. 8.4). Given the shape of the lower outline of the middle shield, it is suggested that the drawing was in fact copied from the engraving of the Madrid Rhinoceros produced by Philippe Galle in 1586 (Clarke 1986: 166).

1699. In a list of rhino specimens in the King's Cabinet (Cabinet du Roi) in Paris, Louis-Jean-Marie Daubenton (1716–1800) included a bezoar stone obtained from an animal sent from India to the Shah of Persia, which had died on the way in 1699 (Buffon 1764, vol. 11: 210, no. 1056).

Exporting the Rhinoceros of India to East and West

12.1 Exporting the Image of the Rhinoceros

The existence of a rhinoceros on the Indian peninsula was known around the world from early times even though the range of these animals in the wild hardly crossed its borders. There are regular reports describing a rhino with a single horn in the annals of the Greek and Roman period, and at least one specimen reached Rome alive. The horn and skin were exported both as gifts and as items of trade, of which a few remarkable examples influencing western culture will serve as a guide to their significance. Scholars in the Renaissance and early modern period were well aware of the presence of a rhino in India itself, setting a benchmark for their discussions about the appearance of these pachyderms and the value of their horns. The rhinoceros was in fact often chosen in depictions of the continent of Asia as one of its most iconic inhabitants. In more modern days, the Indian jungle haunted by wild beasts and imaginary dacoit thugs, with the ubiquitous hazards of diseases, became a backdrop to stories of adventure and exploration with a wide appeal to the younger generation. When the dangerous threatening rhinos were included, these were depicted by artists working for magazines with popular appeal (examples in 50.11, 50.12) This chapter provides a few insights into this wide field of scholarship to show some remarkable examples of how the rhinos left the borders of their homeland to entertain and enlighten people across the world.

12.2 The Wild Ass of Ctesias in the 5th Century BCE

One of the earliest known references to the existence of the rhinoceros is found in a geographical treatise of the Greek physician and historian Ctesias (or Ktesias) living in the 5th century BCE. His work *Indica* is known from extensive fragments and remains authoritative for the period. Like all other sources of the Greek and Roman period, his writings on the rhino are discussed by the German orientalist Lothar Störk (b. 1941) in his dissertation of 1977. His guidance allows a better understanding of the ancient texts of the Greco-Roman period. Ctesias writes about the rhino in India, or at least, so it is assumed, because he never used the actual name of the animal. He wrote about "wild asses" described as being at least as large as horses, with heads of a dark red colour, blue eyes and white bodies, with a horn on their forehead, a cubit in length. As discussed by numerous commentators (Ball 1888; Karttunen 1989: 168; Lavers 2009; Bose 2020: 107), it is clear that Ctesias had no personal knowledge of any animal resembling the rhino, as his description has many elements which require a stretch of imagination. As his wild ass cannot be a unicorn, only the rhino qualifies as a single-horned animal in India. It would take further explorations to introduce the species to European scholars. The name rhinoceros or rinoceros, in Greek ρινόκερως, was used by Agatharchides (*De Mare Erythraeo*, 71) written in the second half of the 2nd century BCE in a work in an African context (Störk 1977: 314; Vendries 2016: 295).

12.3 The Rhinoceros of India in ancient Rome

The rhinoceros was well-known to the people of Rome, where between 55 BCE and 248 AD maybe a dozen of them were shown to the emperors and the public at large (Jennison 1937; Gowers 1950; Buttrey 2007; Rookmaaker 1998: 27). No doubt the majority of these animals came from Africa, because most evidence refers to the double horns of either the black rhino (*Diceros bicornis*) or the white rhino (*Ceratotherium simum*). A rhino seen at the games of Emperor Pompeius Magnus (Pompey the Great, 106–48 BCE) in 55 BCE was described by Pliny (*Historia Naturalis*, VIII.71) as single-horned ("unius in nare cornus"). This would point at *R. unicornis*, but because this text largely reflects an older work not related to rhinos in captivity, it is now generally thought that the animal was in fact African (Kinzelbach 2012a: 32; Vendries 2016: 301).

During the reign of Caesar Augustus (63 BCE–14 CE), there were three (maybe two) rhinos in the Italian capital. One was shown in 29 BCE at the celebrations of Octavian's triumph over Cleopatra. Another was seen on an unknown date in the *Saepta* completed in 26 BCE. A third animal was reported during the games of 8 CE as part of a fight with an elephant. The German zoologist Ragnar Kinzelbach (b. 1941) discussed the Artemidor-Papyrus from Alexandria of the 1st century BCE (Museo Egizio, Turin), where a small sketch is identified as part of the tail of a rhino. The authenticity of this Papyrus is debated among experts, and Kinzelbach's reliance on a small anatomical detail is not entirely convincing (Vendries 2016:

© L.C. (KEES) ROOKMAAKER, 2024 | DOI:10.1163/9789004691544_013
This is an open access chapter distributed under the terms of the CC BY-NC-ND 4.0 license.

FIGURE 12.1   Roman oil lamp made from terracotta, found in
             Mena, South-East of Verona, Italy, 1st century CE
             WITH PERMISSION © MUSEO ARCHEOLOGICO DI
             VERONA: INV. N. 24068

FIGURE 12.2   Roman oil lamp made of terracotta, start of
             1st century. Provenance unknown. Length 10.5 cm
             WITH PERMISSION © RIJKSMUSEUM VAN
             OUDHEDEN, LEIDEN: INV. BR 168

302). Kinzelbach suggests that the tail belonged to a rhino which was part of an embassy sent by the ruler of Bargosa, now Bharuch in Gujarat, to the Emperor of the Roman Empire. According to his reconstruction, the animal was transported on foot all the way on a journey which took about four years. Just five years old, it reached Samos in Turkey in 19 BCE and was exhibited in Rome in 11 BCE, and survived at least until 8 CE (Kinzelbach 2011a, 2011b, 2012a: 62, 2012b). This reconstruction of the life of a single animal unfortunately is built on a weak foundation. However, there were rhinos in Rome in the Augustinian era, and this could well have included one with an Indian origin.

If the textual references remain ambiguous, the iconographic evidence appears stronger. There are at least four oil lamps showing a single-horned rhino from this period (Vendries 2016, figs. 20–22). One of these was found in the vicinity of Verona, Italy and dates from the 1st century CE (Larese & Sgreva 1996; fig. 12.1). Others preserved in Leiden (fig. 12.2), Paris (fig. 12.3) and the Metropolitan Museum of New York (inv. 74.51.2162) may not have a clear provenance, but all of these clearly show a rhino with skin folds and a single horn. The same attributes are found in a small pottery object found in Aquincum (Budapest, Hungary) also dated 50–150 CE (Vendries 2016, fig. 29; fig. 12.4).

Given this evidence there is no doubt that there was at least one living specimen of *R. unicornis* in Rome between 29 BCE and 8 AD.

12.4   Rhino Horn Exported to China

In the second century BCE, the Chinese Emperor Qin Shihuang of the Zhou Dynasty sent out an enormous army to open up trade routes to acquire rhinoceros horns and elephant tusks needed in his frantic quest for immortality (Chapman 1999: 26). Much later, during the Ming (1368–1643) and Qing (1644–1911) dynasties, rhino horns became increasingly popular for the manufacture of exquisite carvings. The best introduction to the rhino horn carvings in China is the well-produced and well-researched volume by Jan Chapman, the former curator of the Chester Beatty Library in Dublin, who investigated the chronology and the specialities of the artists (Chapman 1999).

Horns from the African continent were preferred as they were generally much larger in size (Laufer 1914;

FIGURE 12.4   Ceramic object showing rhinoceros with a single
horn from the Roman period. Size 8 × 7.9 × 0.9 cm
WITH PERMISSION © MUSEUM OF AQUINCUM,
BUDAPEST

FIGURE 12.3   Oil lamp made of terracotta, of presumed Roman age
WITH PERMISSION © CABINET DES MEDAILLES,
PARIS: INV. 1937

FIGURE 12.5
Cup carved in China from an Asian rhino
horn in the early 17th century. It is mounted
in gold filigree work in the shape of a
magnolia flower, manufactured in Goa in
the second half of the 17th century. From
the collection of Rudolf II (1522–1612).
Size 15.5 cm high, 17.6 cm wide
WITH PERMISSION ©
KUNSTHISTORISCHES MUSEUM, VIENNA:
KUNSTKAMMER 3757

Jenyns 1954: 43). A trade from the Indian subcontinent in this commodity is poorly documented, and is hard to distinguish from horns originating in South-East Asia. It is likely that during the Ming dynasty, the cups were mostly carved from Asian horns, which changed to African in the 18th century (Jenyns 1999: 39).

Cups and carvings made by Chinese craftsmen from rhino horn were again exported to Europe, where they were coveted additions to the *Wunderkammer* for their exotic beauty (fig. 12.5). Recent scholarship has high-lighted some examples from the 16th and 17th centuries, in the Kunsthistorisches Museum in Vienna, Austria, and in Lugano, Italy (Jordan Gschwend 2000, figs. 14, 15; Chapman 1999: 234, fig. 338; Förschler 2017; Sass 2021).

12.5   Rhino Horns for the European *Wunderkammer*

There was a great taste of the exotic among collectors of natural curiosities in Europe. They might wait at the quay-side to take their first pick of rare or previously unseen objects brought by sailors returning from their travels. The actual origin was often lost and 'India' became syn-onymous to a fabled land of marvels and riches, with the consequence that some unusual or exotic objects proudly displayed in the *Wunderkammer* as 'Indian' could have originated from any region of the world (Keating & Markey 2011).

A survey of horns and other rhino parts revealed their presence in about 42 royal and private collections across most of the nations in Europe until the end of the 18th century (Rookmaaker 1999d). Although impressive, the list definitely only scratches the surface and many more examples will continue to be discovered. The rela-tive low number seems to indicate that rhino horns were imported through incidental sales by sailors and travellers and were never a main focus of dedicated trade.

Examples of the trade of rhino horns to The Netherlands in the 17th century were recorded by the Dutch mayor of Harderwijk, Ernst Brinck (1582–1649) in a series of unpub-lished notes on a variety of subjects, including the rhi-noceros on 2 pages (Roscam Abbing 2022). Rhino horns were then rare objects and therefore much prized for col-lections of natural objects. Brinck himself had seen one horn in the cabinet of Bernard Paludanus (1550–1633) in Enkhuizen in 1610, and three in the West India House in Amsterdam. In 1648 he received a horn himself from Simon Jacobs Domcekens (*c.*1613–1652), who had hunted a rhino with the King of Champa (Vietnam) in 1644 (Brinck [undated]; Swan 2021: 263). In Italy he noticed a horn in the Vatican as well as those of a male and female in the

FIGURE 12.6   Rhino horn cup commissioned in Goa or Sri Lanka in the mid-16th century by Constantino di Braganza (1528–1575). Height 21 cm
WITH PERMISSION © MUSEO NAZIONALE DI CAPODIMONTE, NAPELS: FARNESE COLLECTION, INVENTORY 10350-1870AM

collection of the Grand Duke in Florence. When Count Johan Maurits van Nassau-Siegen (1604–1679) returned from Brazil in 1644, he brought no less than 17 rhino horns, said to have originated from 'Angola' in the southern part of Africa.

The Portuguese community in Goa was responsible for the export of rhino horns, probably mostly unworked, to their home country. Philippe de la Très Sainte Trinité (1603–1671) recorded the sale of the commodity (Philippe 1649: 285). Two live rhinos passed through Goa in 1515 and 1576 when Indian rulers donated these animals to their counterparts in Lisbon (4.6, 11.9). The Portuguese nobleman Constantino di Braganza (1528–1575) commis-sioned a mounted horn cup when traveling either in Ceylon (Sri Lanka) or in Goa, to be presented to Princess Maria of Portugal (1538–1577) and now preserved in Naples (Jordan Gschwend 2000) (fig. 12.6). Raw horns imported from Asia were later crafted in workshops in Spain or Germany. Their provenance is a mute point, and a late edition of the *Pharmacopoeia* by Johann Schröder (1600–1664) is one of

FIGURE 12.7   Rhinoceros horn which was presented to
Pope Gregory XIV in 1590. The top section of the horn
is missing, and believed to have been cut off to be
administered to the Pope
WITH PERMISSION © AMERICAN MUSEUM OF
NATURAL HISTORY, NEW YORK, DIVISION OF
ANTHROPOLOGY: CAT.NO. A/741 A-C

FIGURE 12.8   Cover made to contain the rhino horn presented to
Pope Gregory XIV
WITH PERMISSION © AMERICAN MUSEUM OF
NATURAL HISTORY, NEW YORK

FIGURE 12.9   Bottom section of the cover inscribed "Gregorio XIIII
Pontifici Maximo"
WITH PERMISSION © AMERICAN MUSEUM OF
NATURAL HISTORY, NEW YORK

the few early sources to provide this origin clearly, when
he wrote that these horns are brought to Germany from
India ("aus Indien") where they are esteemed for their
great powers against poison (Schröder 1748: 1761).

12.6   European Beliefs in Medicinal Properties

Rhino horns and body parts were universally believed
to detect poisons and to have medicinal powers. Such
properties were more commonly associated with the
mythical unicorn, and transferred to rhino horns for a
variety of reasons. An introduction to the legends and
practices relating to unicorns through history is provided in the wonderfully erudite and wide-ranging treatise by the American scholar of English language Odell
Shepard (1884–1967) using a host of intricate early works
(Shepard 1967). Some horns reaching Europe, labelled
as unicorns, became legendary and valuable treasures

of even the most powerful rulers and leaders. One horn was donated to Pope Gregory XIV (1535–1591) by the Prior and Brothers of the Monastery of St. Mary of Guadalupe in Spain in 1590, the tip of which was cut off and ground to be administered to the dying clergyman (figs. 12.7, 12.8, 12.9). Maybe this exact specimen was seen by Ernst Brinck in the possession of Pope Paul V (1550–1621) in Rome in 1614 (Brinck [undated]). Still preserved, in the American Museum of Natural History, and identified by an inscription, it is believed to be the horn of an African rather than Asian rhinoceros (Lucas 1920).

The beliefs in curative powers of horns might well have existed in India as well and then influenced similar ideas in other parts of the world. This was mentioned by the Dutch traveller Jan Huygen van Linschoten (1563–1611), who lived in Goa as the Archbishop's Secretary 1583–1588, in his *Itinerario* of 1596 (English in 1598): "Their hornes in India are much esteemed and used against all venime, poyson, and many other diseases: likewise his teeth, clawes, flesh, skin and blood, and his very dung and water and all whatsoever is about him, is much esteemed in India, and used for the curing of many diseases and sicknesses, which is very good and most true, as I my selfe by experience have found, but it is to be understood, that all Rhinocerotes are not a like good" (Linschoten 1598: 70, chapter 47). The practices might have been more widespread than the silence in later sources appears to indicate.

12.7   Protective Armour of Rhinoceros Skin

Horns were not the only part of the rhinoceros exported. The skin of the animals could be used to manufacture shields and parts of protective armour. Such shields were expensive and were mainly decorative, kept among the treasures of the Indian rulers (14.4), and sometimes donated to visiting dignitaries. So far only two examples of protective armour partly made from rhino skin have come to light in European collections. There is a harness made to protect a horse in battle dating from the 15th century, where the breast plate is modelled from rhino skin, from unknown origin, now in the Armoury in Turin, Italy (fig. 12.10). An elaborate suit of armour modelled from rhino skin (fig. 12.11) said to be from Mandavie (Mandvi) in Gujarat, possibly dating from the 18th century or earlier, was bought from an Indian Rajah by Samuel Rush Meyrick (1783–1848) for his collection in Goodrich House, Herefordshire, UK (Skelton 1854, vol. 2, pl. 141; Egerton 1896: 68). This piece is now lost, not being included when the contents of Goodrich were bought by Sir Richard Wallace, 1st Baronet (1818–1890) to form the nucleus of The Wallace Collection, London (Arthur Bijl, email 2020).

FIGURE 12.10   Armour owned by the Nuremberg patrician Wilhelm Rieter von Boxberg (d. 1541). The horse is protected by a breastplate ("barda") made of rhinoceros skin. Origin unknown, manufactured in the 15th century
WITH PERMISSION © ARMERIA REALE, GALLERIA BEAUMONT, TURIN: DISPLAYED AT THE SECOND WINDOW

12.8   The Rhinoceros in European Literature

It is fair to say that most scholars in Europe from the 15th century onwards had a sense that the rhinoceros was part of the fauna of India. This is because the rhino was mentioned in several classical texts of Greek and Roman authors, including widely known examples by Aristotle and Pliny (table 12.19). The name was not always used consistently, and the Rhinoceros sometimes came under the guise of Abada, Monoceros or even Indian Ass. India as a locality might have been just a general indication of an exotic land, but nevertheless it reinforced the link between India and the rhinoceros.

The early rhinos which arrived in Europe alive from 1515 onwards were all shipped from India, which again contributed to the general understanding. Maybe for this reason the link between the rhino and India was not always very clearly stated. The list of books published between 1500 and 1750 in which it was explicitly mentioned that the rhino lived in 'India' or 'Bengal' (besides other

FIGURE 12.11   Body armour manufactured of rhino skin, said to
be from Mandvi, Gujarat, possibly 18th century.
Acquired by Henry Benedict Stuart, Cardinal
Duke of York (1725–1807) and sold to Samuel
Rush Meyrick (1783–1848) and his son Llewelyn
Meyrick (1804–1837), preserved in Goodrich House,
Herefordshire. Drawing by Joseph Skelton (1783–1871).
Current location unknown
SKELTON, *ENGRAVED ILLUSTRATIONS*, 1854, VOL. 2,
PL. 141

12.9   Rhinoceros as Allegory of America and Asia

It was fashionable in Europe during the Renaissance to personify the continents in elaborate allegories showing particular animals as characteristic attributes. This aspect of the history of art forms an important part of the iconographic investigations by Timothy Howard Clarke (1913–1993) published on 8 September 1986 in his indispensable *The rhinoceros from Dürer to Stubbs 1515–1799*. Tim Clarke and I were introduced to each other after the publication of my own first paper in a scientific journal on the early captive rhinos (Rookmaaker 1973) coincided with his well-illustrated treatment of the same subject in the *Connoisseur* (Clarke 1973, 1974). Through his love and expertise in ceramics, he discovered many examples which otherwise would have remained unexplained (Clarke 1976). Whenever we could meet in the following years in his house near Regent's Park in London, he would kindly bombard me with his latest discoveries of rhinos found in the most unexpected and elusive books, museums, churches and palaces. As a director of Sotheby's, specialising in European ceramics, he was able to rally support from a large range of colleagues and experts. The stories behind the depictions were often baffling to me, as a young student of biology, yet utterly exciting. I was developing a greater familiarity with the zoological and historical works, which I was reading in the quiet reading room of the Artis Library in Amsterdam. We had complimentary skills which made for mutual benefits while probing and discussing the historical record of the rhino. His masterful book of 1986 is full of small details which will continue to amaze and attract researchers in all kinds of disciplines. The hospitality of Tim and his wife Elisabeth and the hours of pleasant discussions on historical rhinos continue to fill me with gratitude.

countries) is therefore not particularly long (table 12.20). The animals which arrived alive in London, Amsterdam and Paris from the 1730s introduced the rhino of India to a wider public.

TABLE 12.19   List of Greek and Latin authors who mentioned the Rhinoceros living in India

Ctesias (5th century BCE): India, 25–26
Aristotle (384–322): Parts of animals, book 3, part 2, 663a
Megasthenes (350–290): De bestiis Indiae, fragment 15
Gaius Plinius Secundus (23–79): Naturalis historia, VIII. xxxi. 76
Quintus Curtius (1st century): Historiarum Alexandri, VIII. ix. 16
Claudius Aelianus (175–235): Historia Animalium, XVI. 20
Philostratus (190–230): Apollonius von Tyana, III. 2
Timotheus Gazaeus (5th century): On animals, 45
Cosmas Indicopleustes (6th century): Christiana Topographia, XI.

Clarke's notes on the rhinoceros associated with allegories of the continents are best read in their original context to savour the minutiae (Clarke 1986: 145–154). There are a few remarkable examples.

(1) In 1583 the Portuguese community in Antwerp erected a triumphal arch (*Arcus Lusitanorum*) which included India shown as a female figure with child holding a coconut, seated on a rhino (figs. 12.12,

12.13). A similar figure who must represent Asia is found, incongruously, on the frontispiece of a herbal called the *Theater of Plants* by the influential English herbalist and apothecary John Parkinson in 1640 (fig. 12.14). Although the animal feels out of place among the plants, the artist modelled the animal after the broadsheet issued by Philippe Galle in 1586 representing the Madrid Rhinoceros (fig. 11.9).

TABLE 12.20   Written references to the presence of rhinoceros in India, in books published from 1500 to 1750

Conrad Gessner (1516–1565): *Quadrupedibus viviparis*, 1551, p. 952

Sebastian Münster (1489–1552): *Cosmographiae universalis*, 1552, p. 1086

João de Barros (1496–1570): *Segunda decada da Asia*, 1553, p. 131

Garcia De Orta (1540–1599): *Aromatum historia*, 1567, p. 66

Johann Wilhelm Stuck (1524–1607): *Maris Erythri periplus*, 1577, p. 21

Joseph Boillot (b. 1560): *Nouveaux pourtraitz et figures de termes*, 1592

Jan Huyghen van Linschoten (1583–1611): *Itinerario*, 1596, p. 70

Antoine Colin (b. 1560): *Histoire des drogues*, 1602, p. 77

François Pyrard de Laval (1570–1621): *Voyage des François aux Indes Orientales*, 1611, p. 123

Wolfgang Franz (1564–1628): *Historia animalium sacra*, 1612, p. 109

Thomas Coryat (1577–1617): *Travailler for the English Wits*, 1616, p. 246

Ulysse Aldrovandi (1522–1605): *De quadrupedibus solidipedibus*, 1616, p. 382

Ulysse Aldrovandi (1522–1605): *Quadrupedum omnium bisulcorum historia*, 1621, p. 878

Edward Terry (1590–1660): *A voyage to East-India*, 1625, p. 109

William Bruton (n.d.): *Newes from the East-Indies*, 1638, p. 32

Thomas Heywood (d. 1641): *Porta pietatis*, 1638, p. 10

Vincent Le Blanc (1554–1640): *Les voyages fameux*, 1648, p. 242

Philippe de la Très Sainte Trinité (1603–1671): *Itinerarium Orientale*, 1649, p. 285

John Jonston (1603–1675): *Historiae naturalis de quadrupedibus*, 1653, p. 98

Ludovico Moscardo (1611–1681): *Memorie del museo*, 1656, p. 243

Samuel Bochart (1599–1667): *Hierozoicon*, 1663, p. 930

François Bernier (1620–1688): *Suite des mémoires sur l'Empire du Grand Mogol*, 1671, p. 75

Wouter Schouten (1638–1704): *Oost-Indische voyagie*, 1676, p. 59

Jean Baptiste Tavernier (1605–1689): *Les six voyages*, 1676, p. 71

Adam Boussingault (n.d.): *Le nouveau theatre du monde*, 1677, p. 181

Frederik Andersen Bolling (d. 1685): *Oost-Indiske Reise-bog*, 1678, p. 23

Nehemiah Grew (1641–1712): *Musaeum Regalis Societatis*, 1681, p. 29

Melchisedech Thevenot (1620–1692): *Relation de l'Indostan*, 1684, p. 129

Johann Christoph Wagner (1640–1703): *Interiora orientis detecta*, 1686, p. 101

Urbain Souchu de Rennefort (1630–1690): *Histoire des Indes orientales*, 1688, p. 341

John Ray (1627–1705): *Synopsis methodica animalium quadrupedum*, 1693, p. 122

Oliger Jacobaeus (1650–1701): *Museum regium*, 1693, p. 4

John Fryer (1650–1733): *New account of East India*, 1698, p. 288

François Bernier (1620–1688): *Voyages*, 1699, vol. 1, p. 42

Nicolaas de Graaff (1619–1688): *Reisen na de vier gedeeltens des Werelds*, 1701, p. 133

Jean Chardin (1643–1713): *Voyages en Perse*, 1711, vol. 3, p. 45

Jean Albert de Mandelslo (1616–1644): *Voyages célèbres et remarquables*, 1727, p. 376

John Carwitham (n.d.): *Natural history of four-footed animals*, 1739, p. 4

James Parsons (1705–1770): *Natural history of the rhinoceros*, 1743, p. 528

TABLE 12.20   Written references to the presence of rhinoceros in India, in books published from 1500 to 1750 (*cont.*)

John Harris (1667–1719): *Navigantium atque itinerantium bibliotheca*, 1744, p. 463

Carolus Augustus von Bergen (1704–1759): *Oratio de rhinocerote*, 1746, p. 26

Friedrich Gotthilf Freytag (1687–1761): *Rhinoceros e veterum scriptorum descriptus*, 1747, p. 3

Johann Schröder (1600–1664): *Pharmacopoeia universalis*, 1748, vol. 3, p. 1761

Jean Baptiste Ladvocat (1709–1765): *Lettre sur le Rhinocéros*, 1749, p. 10

The list provides the author's names, the short title of the book, date and page. This includes only first editions actually printed and available during the period.

FIGURE 12.12   "Arcus Lusitanorum." Portuguese triumphal arch erected for Solemn Entry at Antwerp of 18 July 1593. Engraved by Pieter van der Borcht (1545–1608) after designs by Maerten de Vos (1532–1603). The rhinoceros lacks a Dürer-hornlet
JEAN BOCH (1555–1609), *DESCRIPTIO PUBLICAE GRATULATIONIS*, 1595, P. 75. REPRODUCED BY KIND PERMISSION OF CAMBRIDGE UNIVERSITY LIBRARY

FIGURE 12.13   Allegory of India riding a rhinoceros
DETAIL FROM BOCH, *DESCRIPTIO*, 1595, P. 75

(2)   The Dutch cartographer Petrus Plancius (1552–1622) painted Asia as a lady with traditional attributes sitting on a rhino with Düreresque armour plating but without the horn on the shoulder, on his large map of 1594 (fig. 12.15). A very similar scene is found in a small vignette of Asia on the frontispiece of Beauvau (1615) designed by Jean Appier Hanzelet (1576–1647).

(3)   Asia is represented by the head of a rhino, with single horn, in the central part of the frontispiece of

a book of topographical scenes edited in 1599 by the German cleric and cartographer Georg Braun (fig. 12.16).

(4) The rhino is also known to accompany the allegory of America. An early example is found in a small detail of the frontispiece of a grand work of city views executed by the German artist Daniel Meisner (1638) with rhino and elephant in a peaceful coexistence (fig. 12.17).

(5) The grandest depiction in this genre depicts two rhinos brought into service pulling the chariot carrying the allegory of America. This scene is found on a colourful painting created by the Florentine artist Giuseppe Zocchi around 1760 (fig. 12.18), also copied in the same period on a plaque in pietra dura (Clarke 1986: 153, pl. 31).

The rhinoceros continues to be associated as an icon and mascot with Assam, India, South Asia and Asia until the present day.

FIGURE 12.14
Allegory of Asia riding a rhinoceros. Engraved by William Marshall for John Parkinson (1567–1650)
PARKINSON, *THEATRUM BOTANICUM: THE THEATRE OF PLANTS*, 1640, FRONTISPIECE

FIGURE 12.15   Asia seated on a rhinoceros. Detail of the top-right corner of the *Orbis Terrarum Typus De Integro Multis In Locis Emendatus* by Petrus Plancius (1552–1622) dated 1594. This same map was inserted in some copies of the *Itinerario* by Jan Huygen van Linschoten first published in 1596
COURTESY: DIGITAL COLLECTION OF THE ROYAL LIBRARY OF DENMARK, COPENHAGEN

FIGURE 12.16   The head of a rhinoceros appears to represent Asia in the central window of the title-page of *Urbium Praecipuarum Mundi Theatrum* by Georg Braun (1542–1622) dated 1599

FIGURE 12.17
Rhinoceros and elephant combined with an allegory of America by Daniel Meisner (1585–1625). The artist included this detail in the frontispiece of his *Sciographia Cosmica*, first published posthumously in 1634 and repeated in an identical second edition of 1678

FIGURE 12.18   Allegory of the continent America with her chariot being drawn by two rhinos. Oil on canvas by Giuseppe Zocchi (1711–1767) around 1760 as an example for a work in pietra dura. Size 40 × 54 cm
SU CONCESSIONE DEL MINISTERO DELLA CULTURA © MUSEO DELL'OPIFICIO DELLE PIETRE DURE DI FIRENZE

CHAPTER 13

Historical Records of the Rhinoceros in Afghanistan, Pakistan and Punjab

First record: 1333 – Last record: 1556
Species: *Rhinoceros unicornis*
Rhino Region 01 (Afghanistan), Rhino Region 02 (Pakistan)

13.1 Early Records from Afghanistan and Pakistan

Afghanistan and Pakistan are the western extremity of the historical range of the rhinoceros. The few available records are associated with famous names like Ibn Battuta, Timur Bec, Seydi Ali Reis and Babur. The discovery of a rock relief showing how two rhinos were hunted by a Sasanian King at Rag-I Bibi adds a further western extension. Interpretation of these data remains challenging requiring specialist knowledge of history, culture and linguistics. However, if accepted at face value, it is necessary to explain the presence of rhinos in unusually dry and mountainous environments. The Khyber Pass today hardly matches places where rhinos are found elsewhere, close to rivers or lakes in flat or slightly hilly country. Maybe such conditions existed then or even now in small areas along the Indus River, as perhaps they did where the Harappan civilisation flourished. The rhino records of the Harappan civilisation are found in chapter 10. The writings of the Mughal Emperor Babur are discussed in a wider context in chapter 11.

13.2 Absence of Rhinos in Punjab and North-Western India

There are no records of anyone encountering a rhinoceros in the large stretch of land between the Indus River in central Pakistan in the west and the Sirmaur mountains of Himachal Pradesh in the east, a distance of some 600 km. This major gap is rather obliterated in modern distribution maps of *R. unicornis* which show a continuous range from Uttar Pradesh westwards to Peshawar in northern Pakistan. These maps reflect the hope that a report in a medieval manuscript may have been overlooked. On the other hand, the gap may be real, begging the question if the rhino was exterminated, or never existed, in which case the populations in Afghanistan-Pakistan may have been isolated from early times, or connected eastward through Gujarat and Rajasthan. The fossils found in the Siwaliks of this region may one day provide an answer which to me at the moment remains elusive.

...

Dataset 13.9: Chronological List of Records of Rhinoceros in Afghanistan and Pakistan

General sequence: Date – Locality (as in original source) – Event – Source – § number, figure – Map number.
The map numbers are explained in dataset 13.10 and shown on map 13.5.

Rhino Region 01: Afghanistan

2000 BCE – Shortughai – Stamp with rhino in Harappan settlement – Franckfort 1983 – 13.3 – map M 801

1st cent. – Begram, the ancient capital Kapici – Rhino depiction within ivory medallion – Hackin 1954: 202, fig. 186; Bautze 1985: 414 – 13.4, fig. 13.1 – map M 802

3rd cent. – Rag-I Bibi – Rock relief showing Sasanian King Shapur I (reigned 240–270) hunting 2 rhino. Size 4.9 × 6.5 m – Cassar 2004; Grenet 2005; Grenet et al. 2007; Levine & Plekhov 2019 – 13.5, figs. 13.2, 13.3, 13.4, 13.5 – map M 802

1556 – Khyber Pass – Seydi Ali Reis (1498–1563) sees 2 rhinos when traveling from Peshawar westwards – Vambéry 1899: 63; Moris 1826: 94; Yule & Burnell 1886: 700; Rookmaaker 2000 – 13.8 – map A 211

Rhino Region 02: Pakistan

1333 – Indus River near Janani (Khairpur) – Ibn Battuta sees two rhinos, both ran away – Ibn Battuta 1855: 100, 1971, vol. 3: 596 – 13.6 – map A 212

1399-03 – Gebhan, lower Kashmir – Hunted by Timur Bec (1336–1405) – Petis de la Croix 1723, vol. 2: 94; Yule & Burnell 1886: 762; Ettinghausen 1950: 45; Gee 1964a: 151; Rookmaaker 2000; Geer 2008: 381 – 13.7 – map A 216

1519-02-16 – Karak-Khaneh on Sawati (Swabi) – Babur hunt with Humayun – 11.2

1525-12-10 – Bigram (Peshawar) on Siah-Ab – Babur hunted rhino – 11.2

1520s – Peshawar and Hashnagar jungles – Babur saw "masses of it" – 11.2

1520s – Between Indus River and Bhera – Babur records presence of rhino – 11.2

© L.C. (KEES) ROOKMAAKER, 2024 | DOI:10.1163/9789004691544_014
This is an open access chapter distributed under the terms of the CC BY-NC-ND 4.0 license.

1982 – Lal Suhanra National Park – *R. unicornis* male and female donated by King Bir Bikram Shah of Nepal (1944–2001). Both died in 2019 – Nawaz 1982; Walker 2003 – 13.9 – map D 551

Sources stating the historical presence of rhinoceros in Pakistan (including Mughal evidence, chapter 11): Blyth 1862a: 162, 1875: 50; Schlagintweit 1880, vol. 1: 229; Cockburn 1883: 60; Sterndale 1884: 407; Blanford 1892: 534; Finn 1929: 186; T.J. Roberts 1977: 159; Khan 1989: 2

• • •

Dataset 13.10: Localities of Records of Rhinoceros in Afghanistan and Pakistan

The numbers and places are shown in map 13.5.

Rhino Region 01: Afghanistan

M 801 Shortughai, Darqad Province – 37.20N; 69.30E – See Chapter 10
M 802 Rag-i-Bibi, Baghlan Province – 35.80N; 68.75E – 3rd cent.
M 802 Begram, Parwan Province – 34.98N; 69.20E – 1st cent.
A 211 Khyber Pass – 34.20N; 71.00E – 1556

Rhino Region 02: Pakistan

M 803 Mohenjo-Daro, Larkana Dt., Sindh – 27.32N; 68.13E – See Chapter 10
M 804 Chanhu-Daro, Benazirabad Dt., Sindh – 26.17N; 68.32E – See Chapter 10

M 805 Alahdino, Matiari Dt., Sindh – 25.50N; 68.50E – See Chapter 10
A 212 Janani, Khairpur Dt., Sindh – 27.70N; 68.90E – 1333
A 213 Peshawar and Hashnagar jungles – 34.10N; 71.00E – 1520s – See Chapter 11
A 214 Bigram on Siah-Ab (Peshawar) – 34.00N; 71.57E – 1525 – See Chapter 11
A 214 Karak-Khaneh on Sawati (Swabi) – 34.09N; 72.53E – 1519 – See Chapter 11
A 215 Between Indus River and Bhira – 32.28N; 72.58E – 1520s – See Chapter 11
M 806 Harappa, Sahiwal Dt., Punjab – 30.64N; 72.90E – See Chapter 10
A 216 Gebhan, Gujrat Dt., Punjab – 32.80N; 74.13E – 1399
D 551 Lal Suhanra NP – 29.50N; 72.00E–1982 – introduced

∴

13.3 Shortughai: Harappan Settlement in Northern Afghanistan

Excavations at the Harappan site of Shortughai, at the meeting point of the Kokcha and Amu Darya rivers in NE Afghanistan, have yielded objects dated between 2200–2000 BCE. There is one stamp depicting a rhino, which is very similar to types found further to the south (Franckfort 1983). Considering that this involves a unique example of a rhino depiction and that Shortughai was a known trading post, it is not inconceivable that this stamp had been transported from a town closer to the Indus valley.

FIGURE 13.1
Rhinoceros covered with a blanket found at the ancient capital Kapici in Afghanistan, dating from the first century CE
HACKIN, *NOUVELLES RECHERCHES ARCHÉOLOGIQUES À BEGRAM*, 1954. FIG. 186 U5

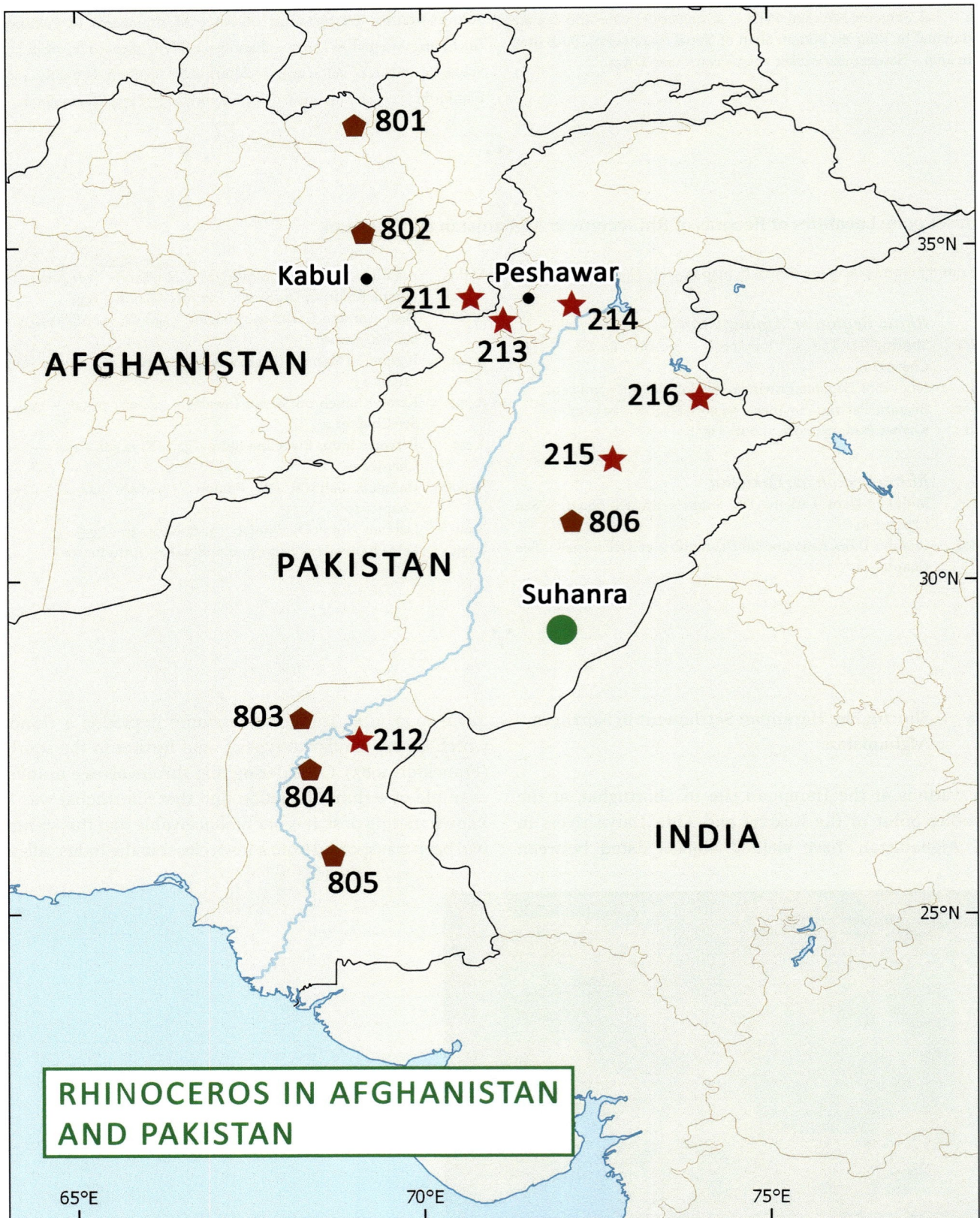

MAP 13.5   Records of rhinoceros in Afghanistan and Pakistan. The numbers and places are explained in Dataset 13.10. Suhanra National
Park has an introduced population
⭐ Presence of *R. unicornis*
⬟ Artefacts

FIGURE 13.2
Rock relief at Rag-i Bibi, near Pul-i Khumri, Baghlan Province, Afghanistan, discovered in 2001. A Sasanian King is hunting two large animals seen on the right, both identified as a rhinoceros
COURTESY © FRANTZ GRENET AND FRANÇOIS ORY, PARIS

FIGURE 13.3   Detail of the lower rhino figure carved in the rock at Rag-i Bibi, Afghanistan
COURTESY © FRANTZ GRENET AND FRANÇOIS ORY, PARIS

13.4   Ivory Object from the Ancient Capital Kapici

A rhinoceros stands within an oval-shaped medallion made of elephant ivory found in Begram, probably dating from the 1st century (Hackin 1954). Begram was the ancient capital of the Kushan Empire situated about 60 km north of Kabul. The animal has a short horn on the snout and appears to be covered by a blanket with a rhombic pattern (fig. 13.1).

13.5   Rag-i Bibi, Afghanistan: Rock Relief of 3rd Century

The rock relief at Rag-i Bibi in northern Afghanistan first became known to western archaeologists in 2002. Jonathan Lee visited the site in December 2003, followed by a mission led by Frantz Grenet in May 2004, together with François Ory and Philippe Martinez (Grenet et al. 2007). The relief shows a rider on horse-back with several attendants, inspecting a dead animal beneath the front leg of the horse, with a second animal trying to escape. The animals are identified as rhinos despite the absence of a horn and ears in the damaged sculpture, through the scales of the skin and their general posture. The relief is found about 10 km south of Pul-I Khumri, Baghlan Province, on a rock face 105 m above the valley floor, with a size of 4.9 m in height and a maximum width of 6.5 m. The relief shows Shapur I, a 3rd century Sasanian King of Iran, highlighting his feat of slaying a rhinoceros. The reason for this sculpture at this precise spot remains

FIGURE 13.4   Detail of the upper rhino figure carved in the rock at Rag-i Bibi, Afghanistan

COURTESY © FRANTZ GRENET AND FRANÇOIS ORY, PARIS

FIGURE 13.5   Artist's impression of the rhinoceros at Rag-i Bibi, Afghanistan

WITH PERMISSION © FROM GRENET ET AL. 2007

enigmatic as it was probably never easily seen from the caravan road that always passed in the distant valley. The rhinos are sculpted in such a way that it is likely that the sculptor might have seen the animals (figs. 13.2 to 13.5).

13.6   Ibn Battuta: Travels in 1333

On his well-known extensive travels, the Muslim scholar Ibn Battuta (1304–1369) left Kabul towards the valley of the Indus, reaching the river at Janani. This town, which no longer exists, must have been located around Khairpur, some 100 km north of Sehwan, which was his next destination (Husain 1976). Here he records his encounter with two rhinos in his journal:

> Description of the Rhinoceros (Carcaddan). After crossing the river of Sind called Banj Ab [Indus River], we entered a forest of reeds, following the track which led through the midst of it, when we were confronted by a rhinoceros. In appearance it is a black animal with a huge body and a disproportionately large head. For this reason it has become the subject of a proverb, as the saying goes *Al-karkaddan ras bila badan* (rhinoceros, head and no torso). It is smaller than an elephant but its head is many times larger than an elephant's. It has a single horn between its eyes, about three cubits in length and about a span in breadth. When it came out against us one of the horsemen got in its way; it struck the horse which he was riding with its horn, pierced his thigh and knocked him down, then went back into the thicket and we could not get at it. I saw a rhinoceros a second time on this road after the hour of afternoon prayer. It was feeding on plants but when we approached it, it ran away (Ibn Battuta 1971: 596).

It is unfortunate that the locality is uncertain, which is partly due to the absence of other sources from this period. Ibn Battuta is the only person to record the rhino this far to the south (Rookmaaker 2000). The description points to the rhino, but a horn of 3 cubits length (135 cm, at 1 cubit equaling 45 cm) is impossible for *R. unicornis*. The size of the head is equally not as we would put it today, and one wonders why these elements appear so much exaggerated. It might nevertheless be best to take the account on face value to admit the presence of the rhino in the lower Indus valley in that period.

FIGURE 13.6
Detail showing Timur supervising a rhino hunt in lower Kashmir in an illuminated copy of the *Zafarnama* dated 1533
WITH PERMISSION © BRITISH LIBRARY: IO ISLAMIC 137, FOLIO 307V

13.7   Timur Bec, 14th Century

Timur Bec or Tamerlane (1336–1405), the Iranian military leader and founder of the Timurid dynasty, captured Delhi in 1398. The history of his conquests and reign was written at the end of his life by Sharaf al-Dīn Yazdī as the *Zafarnama* or *Tārīkh-i jahāngushā'ī-yi Taymūr*, translated in 1723 by the French orientalist François Petis de la Croix (1653–1713).

Delhi. After the defeat of Delhi, Timur celebrated his victory on 4 January 1399 and "all the elephants and rhinoceroses were brought to Timur" (Petis de la Croix 1723, vol. 2: 63). There were 120 elephants of war, but we learn nothing more about the number of rhinos. These animals must have been captured elsewhere in the Indian region.

Lower Kashmir. In March 1399, Timur was within a few miles of Gebhan, a town on the southern border of 'Cachmir' (Kashmir), in the region of the upper reaches of the Indus River in northern Pakistan. Here Timur found a plain abounding with a great quantity of game, including rhinos. He spent some time hunting: "The soldiers took a great deal of game, and slew several rhinoceroses with their sabres and lances, though this animal is so strong, that it will beat down a horse and horseman with a single blow of its horn; and has so thick a skin, that it cannot be pierced but by an extraordinary force" (Petis de la Croix 1723, vol. 2: 94).

The hunt in Kashmir is occasionally found in illustrated copies of the *Zafarnama* (Ettinghausen 1950: 46). The rhino is shown in a manuscript dated 1533 where one of the officers cuts the animal's neck with a sword (fig. 13.6).

13.8   Seydi Ali Reis at the Khyber Pass

The Ottoman admiral Seydi Ali Reis (1498–1563), or Sidi Ali, had to make his way back overland from Gujarat to his home in Turkey in 1556. Once back, he wrote the *Mir'ât ül Memâlik* (The Mirror of Countries, 1557) describing the countries visited during the long journey. After he reached Peshawar, he continued westward to Kabul: "we crossed the Khyber Pass, and reached Djushai. In the mountains we saw two rhinoceroses (kerkedans) each the size of an elephant; they have a horn on their nose about two inches long" (Vambéry 1899: 63). Rhinoceros is a translation of the Turkish word *Kerkedan*. According to this recollection, he saw these animals near the Kotal or Khyber Pass, which is likely to be quite hilly if not mountainous country, unlikely to be suitable to *R. unicornis*. Maybe the rhinos were close to the Kabul River, maybe they were from the same population as seen by Babur near Peshawar in 1519 (Choudhury 2022). Large animals with short horns on the nose can hardly be anything but the rhinoceros.

13.9   Lal Suhanra National Park, Pakistan

A pair of rhinos was donated to the people of Pakistan by the King of Nepal, Birendra Bir Bikram Shah and arrived on 23 March 1982. On orders of Ghulam Jilani Khan (1925–1999), Governor of Punjab Province, the animals were housed in the Lal Suhanra National Park, 36 km to the east of Bahawalpur, established on 26 October 1972.

FIGURE 13.7
Two *R. unicornis* translocated from Nepal in
the Lal Suhanra National Park in 2014
PHOTO: AHMED SAJJAD ZAIDI

The rhinos were not allowed to roam free, and were kept in a newly constructed large paddock with a lake, where visitors could view them standing behind a barrier (fig. 13.7). The pair never reproduced. Media reports confirm that the female died on 12 January 2019 and the male on 21 February 2019, after 36 years 9 months 20 days (13,444 days) and 36 years 10 months 29 days (13,484 days) in captivity respectively. Despite being kept in an enclosure, these two animals have generally been included in counts of the wild population (Talukdar et al. 2011).

CHAPTER 14

Historical Records of the Rhinoceros in Gujarat

First Record: Pleistocene fossils – Last Record: Mesolithic (2000 BCE)
Species: *Rhinoceros unicornis*
Rhino Region 04

14.1   Rhinos in Gujarat

The rhinoceros has not been recorded in Gujarat for at least 4000 years. The arid conditions and the environment cannot sustain a permanent rhino population. This may point at a change in climate because once the animals were found in the eastern part of the state (Chitalwala 1990). Fossil remains of rhino skeletons have been reported from two localities dating from the Miocene (*Rhinoceros perimensis*, *R. planidens*) and nine localities from the Pleistocene or early Holocene (chapter 8). The more recent finds have been assumed to relate to *R. unicornis* although the fragments rarely allow positive identification. Among the Harappan settlements one terracotta head was recorded (chapter 10), and there is one tentatively identified image in a rock painting (chapter 9).

In more recent times, the first rhino which arrived in Europe after the Middle Ages was a donation of the ruler of Ahmedabad in 1514 (4.6). The capture site is unknown and is likely to lie beyond the borders of Gujarat. Nevertheless, allusions to the presence of the rhino in Cambaia must be based on the history of this specimen (14.2).

Modern Gujarat has three remaining connections with the rhinoceros. A large number of rhinos were exhibited by the Maharajas of Baroda (7.1) and the Nawab of Kathiawar (48.2) in the course of the 19th century. In an unfortunate twist, the usage of rhino horn in aphrodisiacs has become associated with the state (14.3). Gujarat has also remained one of the centers for the manufacture of shields from rhino hide until recently (14.4). The story of a rhino hunt in jungles south of Baroda published by the pseudonymous Peregrine Herne is evidently pure fiction (Herne 1858: 209).

R. unicornis, or any other rhino species, has not been known to live in Gujarat since 2000 BCE.

14.2   Rhinoceros in Cambaia

The Lisbon Rhinoceros imported in 1515 was associated with the King of Cambaia from the start. In a letter of June or July 1515, the German merchant Valentim Fernandes (d. 1519) wrote that the *ganda* was a present of the powerful King of the city of Combaia or Cambaia (Costa 1937: 33). Likewise, the Italian scholar Sigismondo Tizio (1458–1528) mentioned Cambaia as the origin of the rhino in a manuscript of 1516 (Garfagnini 1992: 13). In 1551 it appeared in the description of the rhino in the widely consulted natural history by Conrad Gessner (1551: 952). Later allusions to Cambaia or Cambay in relation to the occurrence of rhinos would be based on these sources.

The English chaplain Edward Terry (1590–1660) was part of an embassy to the Mughal kingdom and remained in Gujarat and Bihar for three years in 1616–1619 (Foster 1921: 288). Among the animals encountered during his visit, there were "some rhynocerots, which are large beasts as bigge as the fayrest oxen England affords; their skins lye platted, or as it were in wrinkles upon their backs" (Terry 1625: 471, 1655: 109). The link between these animals and Gujarat remains tenuous.

14.3   Rhino Horn as Aphrodisiac

For most of the 20th century, it was generally accepted that rhino horn was used as an aphrodisiac in many Asian countries, especially in China and in India. Unsurprisingly, there is little direct evidence or documentation, almost certainly none in the earlier works. Still it was a common and constantly repeated theme in the rhino literature to link the decline of animals in the wild with the use of the horn for its alleged hidden powers. In more recent years, the emphasis has shifted to the medicinal properties of rhino horn and other rhino parts, which remains a common practice in Chinese traditional medicine. Among the entrepreneurs of countries including Vietnam, rhino horn has more recently become a status symbol reflecting increasing wealth.

In the course of his essential surveys monitoring trade in rhino horn, Esmond Bradley Martin (34.6) made enquiries about the use of rhino horn as an aphrodisiac in India. He found that a belief in such powers was far from widespread and actually extremely rare (Martin & Martin 1982: 76; Martin 1983). There was no evidence at all from the main rhino areas in Uttar Pradesh, Bihar, West Bengal and Assam. Martin did manage to find a few medical practitioners in Gujarat who said that they had

© L.C. (KEES) ROOKMAAKER, 2024 | DOI:10.1163/9789004691544_015
This is an open access chapter distributed under the terms of the CC BY-NC-ND 4.0 license.

FIGURE 14.1   A rhino horn found in a store in Gujarat ready to be
pulverized and sold as medicine
PHOTO: ESMOND BRADLEY MARTIN IN 1980–1981.
COURTESY: LUCY VIGNE

suggested rhino horn to their patients (fig. 14.1). Indian researchers like Menon (1995: 82) are reluctant to accept that such practices continue. Probably there are fewer believers now, as possession of rhino horn is illegal, trade is very restricted, the product is extremely expensive, and the efficacy questionable.

Gujarat was perhaps unfairly singled out as the last vestige of an ancient belief. Many conservationists in Bengal like Shebbeare (1935: 1229) and Bahuguna & Mallick (2010: 274) refer to the use of rhino parts as aphrodisiac, and Lohani (2011a) even found the practice in a recent survey in Nepal, besides the more widespread application in local medicines (fig. 14.2).

In modern conservation literature it is close to an anathema to refer to any potential aphrodisiacal powers in rhino horn. It is stated with increasing confidence that rhino horn is not an aphrodisiac, while of course we can read about the practice in almost every book with a greater age. A myth or half-truth once established is almost impossible to eradicate as long as popular media continue to rely on tenuous stories which are repeatedly debunked in the more serious sources. I once contemplated adding to the research (Rookmaaker 2011d), which might need to be revived. In short, rhino horn is not an aphrodisiac and rhino horn has no medicinal virtues.

14.4   Manufacture of Shields from Rhino Hide

Shields made from rhino hide were valued for the strength and durability of the material. Found among the treasures of the Indian rulers and aristocracy, they largely date from the 16th to 19th century (Chapter 15, p. 196). Pant (1982: 71) suggests that "in almost all the museums of India having arms collections the shields made of rhino hide can be seen." Several shields are not recognized as being made of rhino hide, because "the shield of rhinoceros hide was occasionally left with a natural surface on the outside which was given a coat of black or brownish lacquer on both the inside and outside were smoothed completely and given a highly polished brown or black lacquered surface" (Pant 1982: 125, note 25). Among others, Pant (1982: 171–178, pls. LI, XLIII) presents "a typical example of shields used by the Rajput princes. It is made

FIGURE 14.2
Zoo keepers at the Yangon (Rangoon) Zoo of Myanmar in the process of collecting urine from their resident rhinos. The urine is routinely sold and used for medicinal purposes
PHOTO: ESMOND BRADLEY MARTIN IN 1980–1981. COURTESY: LUCY VIGNE

FIGURE 14.3
Shield made of rhinoceros hide inlaid with gold, rubies and diamonds (Meghani 2017: 173). Size 40.4 × 8.3 cm, made in Western India in late 18th or early 19th century. Donation by Vibhaji II Ranmalji, Jam Sahib of Nawanagar (1827–1895) to Albert Edward, Prince of Wales (1841–1910) in November 1875 in Bombay, during the Indian tour
ROYAL COLLECTION TRUST: RCIN 11458 © HIS MAJESTY KING CHARLES III 2023

FIGURE 14.4
Shield (*dhal*) of untanned rhinoceros hide with a finish of an unusual colour. The gold central medaillon is surrounded by four gilt bosses. Size 51.7 × 8.7 cm. This example was presented in 1875 to Albert Edward, Prince of Wales by Prithvi Singh (1838–1879), Maharaja of Kishangarh
ROYAL COLLECTION TRUST: RCIN 38404 © HIS MAJESTY KING CHARLES III 2023

of rhinoceros hide" and also illustrates several painted shields of this material.

The main centers of manufacture were in Gujarat (Ahmedabad, Surat, Vadodara, Kutch, Mandvi) and Rajasthan (Kota, Udaipur, Jodhpur, Bikaner). In the process of manufacture, the rhino hide is boiled to make the material transparent, after which paint and gilt are applied according to the wishes of the owner. The actual workshop is never recorded on the shield, hence their actual origin and date are difficult to establish.

The examples preserved are ornamental shields made for the rulers in Rajasthan, Hyderabad and other princely states (fig. 14.3). One shield is said to have been used in warfare, Rajput King of Mewar, Rana Pratap Singh I (1540–1597) in the battle of Haldighat in 1576 (Indian Museum, Kolkata, AT/95/1586). Another, once in the Lahore armoury, was owned by Govind Singh (1666–1708), the Sikh Guru (Spens 2014). There is one record of a full body armour of rhino hide once owned by a Maharaja of Gujarat (12.4). All these shields have one thing in common: they are artistically painted so that the actual surface of a rhino's skin is covered by a coat of dark lacquer on which the paint was applied (fig. 14.4). One of the few exceptions is a shield made of rhino skin in the collection of the armoury (*sileh khana*) of the Sawai Man Singh II Museum, City Palace Jaipur (figs. 15.61, 15.62). The caption informs: "Shield made of Rhino Hide, Jaipur, 18th century." The wrinkled surface of the skin is clearly visible in both images, as is the thickness of the hide.

Shields made from rhino hide are catalogued in a large number of museums or exhibitions, including the India Museum, London (Egerton 1880), the Indian Exhibition of 1895 (Earls Court 1895), the Purshotam Vishwam Mawjee Museum, Malabar Hill, Bombay (Anon. 1911b: 29) and the Muséum d'Histoire, Bern (Kalus 1974). This is just a small selection, as indicated by the large numbers of shields listed and figured in the recent digital lists with examples from the National Museum of New Delhi and the Salar Jung Museum of Hyderabad (museumsofindia.gov.in) and the Victoria & Albert Museum in London (collections.vam.ac.uk).

When Albert Edward, Prince of Wales (1841–1910) visited India in 1875–1876, many Indian rulers presented him a shield of rhino leather. These shields were catalogued, described and reproduced in two superbly illustrated volumes (C.P. Clarke 1898, 1910). The British royal collection still preserves many splendid examples inlaid with precious metals (Meghani 2017; rct.uk).

This shows that the use of rhino hide in the manufacture of shields was a regular practice spanning several centuries. The hides themselves will not provide further clues of provenance. It is thought that a large percentage of the material was imported from East Africa rather than procured in India itself (Martin & Martin 1982: 75). Much remains to be learned about the shields and other items used in India made from rhino hide.

A Pictorial Survey of the Rhinoceros in the Art of Rajasthan

Joachim K. Bautze

15.1   The Rhinoceros in Rajasthan

Today, Rajasthan, a state in North-Western India, is known for its arid climate. Ironically, a painting of the Indian rhinoceros which prefers a more humid climate for its habitat, is preserved in the probably driest region of the state, in Jaisalmer (district Jaisalmer). The rhinoceros is part of a painted wooden book-cover formerly protecting a text sacred to the Jains on palm leaf. It is datable to the end of the twelfth century and was published from 1951 onwards.[1] The presence of a rhinoceros in connection with the holy scriptures of the Jains should in no way be a surprise, as the eleventh of the 24 Jain *Tīrthaṅkaras* (saviour, spiritual teacher; literally: ford-maker), Śreyāṅsanātha, has the rhinoceros as his *cihna* (cognizance, emblem) which is usually shown at the pedestal of the *Tīrthaṅkara* (see 28.6).[2]

15.2   The Hunted Rhinoceros[3]

To date the earliest Rajasthani painting showing a hunt for a rhinoceros among other animals dates from the first half of the 16th century and illustrates a scene from the tenth book of the *Bhāgavata Purāṇa*, an ancient Indian text describing the childhood and youth of god Kṛṣṇa. The often-published painting[4] illustrates the following passage from the said text which is written on the reverse of this painting:

> One day, equipping himself with the Gāṇḍīva bow, two quivers with inexhaustible stock of arrows and putting on his armour, Arjuna, the victorious, mounted his chariot distinguished by the flag bearing a monkey-emblem, prepared himself for hunting. Arjuna, the destroyer of hostile warriors, accompanied by Śrī Kṛṣṇa, entered a dense forest infested with a number of tigers and wild beasts. He hunted down with his shafts several tigers, boars, buffaloes, antelopes (called *ruru*), *śarabhas* (a fabulous eight-legged animal, capable of killing lions), bisons, rhinoceroses, hares and porcupines.[5]

Earlier in the story as well as the manuscript, the birth of Kṛṣṇa is celebrated by the inhabitants of the village, in which he was discovered after his father Vasudeva brought him there by night. The painting in the collection of Eva and Konrad Seitz, Bonn, is inscribed on the back with the following text: "On the great festive occasion of the coming of Lord Kṛṣṇa, the Infinite, the Supreme Lord of the Universe, a variety of musical instruments were played on. The merry cowherds joyously sprayed and besmeared one another with curds, milk, ghee and water and they threw (balls of) butter at each other."[6]

Although a rhinoceros is neither shown in the painting nor is it mentioned in the text, it appears as sketch on the back of the painted manuscript folio (figure 15.1). The animal with head bent down and raised ears seems to be in the movement of running from right to left. The posture reminds of the picture with Arjuna's hunt and a scene from an illustrated Jain manuscript, the *Mahāpurāṇa* from 1540 showing the Vaijayantī forest.[7]

Paintings depicting a rhinoceros being hunted were found in Bundi, Kota, Bikaner, Udaipur, and Samod. The localities are shown in map 15.8 and explained in Dataset 15.11.

---

1   Puṇyavijayajī 1951, figure 35; Nawab 1959, p. 4[−5], plate v, figure 35; Nawab 1980, p. 11, col.plate 82.

2   Bautze 1985, p. 417, figure 3; Shah 1987, p. 146f; Glasenapp 1925, p. 491, plates 22–23.

3   If not mentioned otherwise, all photographs as well as copy-photographs were made by Joachim K. Bautze.

4   Hutchins 1980, col.plate p. 59 and descriptive text, p. 120f; Ehnbom 1985, p. 24f; Ehnbom 1988, p. 24f; Sotheby's NY, 21st and 22nd March 1990, lot 23; Das 2018, col.plate p. 83.

5   Bhāgavata Purāṇa, book X, chapter 58, verses 13–15. English translation quoted from Ganesh Vasudeo Tagare in Bhāgavata Purāṇa 1978, p. 1622. The Sanskrit word for rhinoceros in this part of the text is *khaḍga*.

6   Bhāgavata Purāṇa, book X, chapter 5, verses 13–14. English translation quoted from Ganesh Vasudeo Tagare in Bhāgavata Purāṇa 1978, p. 1284f.

7   Khandalavala / Moti Chandra 1969, p. 76, plate 19(b). In the description of the painting the rhinoceros is described as "goring a tree", cf. *ibidem*, p. 73.

© L.C. (KEES) ROOKMAAKER, 2024 | DOI:10.1163/9789004691544_016
This is an open access chapter distributed under the terms of the CC BY-NC-ND 4.0 license.

FIGURE 15.1
Sketch of a Rhinoceros on the back of a
painting which is part of the so-called
"Dispersed Bhagavata." Red ink on paper.
Rajasthan. Size not recorded
WITH PERMISSION © EVA AND KONRAD
SEITZ COLLECTION, BONN

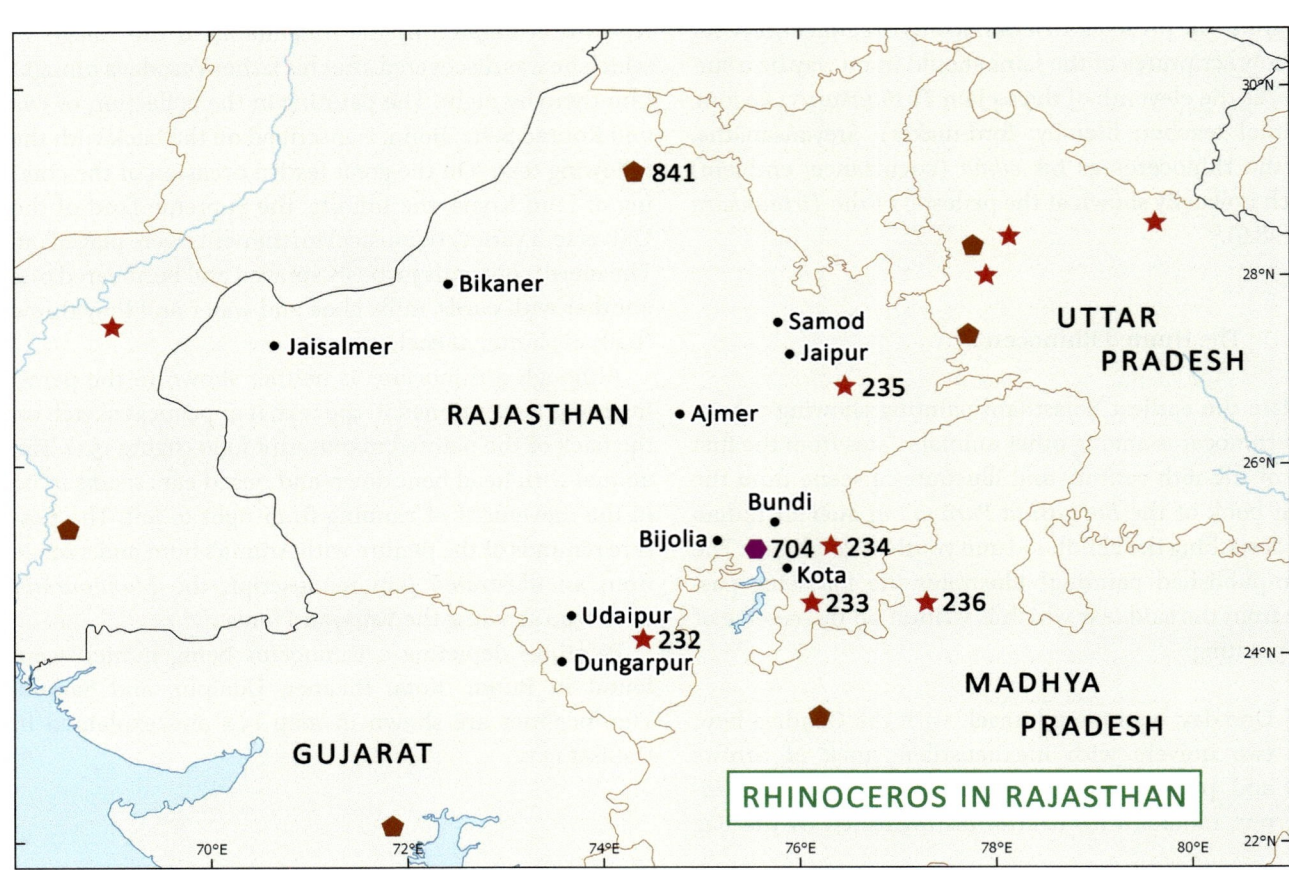

MAP 15.6   Records of rhinoceros in Rajasthan, with place names mentioned in the text. Locality 236 is just across the border in Madhya
Pradesh. The numbers and places are explained in Dataset 15.11

★   Presence of *R. unicornis*
⬠   Artefacts
⬡   Rock Art

Dataset 15.11: Localities of Records of Rhinoceros in Rajasthan and North-West Madhya Pradesh

The numbers and places are shown on map 15.8.
ref – these are localities mentioned in text where presence of rhino needs verification.

Rhino Region 03: Rajasthan

ref	Jaisalmer – 26.88N; 70.86E
ref	Bikaner – 27.94N; 73.33E
ref	Dungarpur – 23.84N; 73.70E
ref	Udaipur – 24.57N; 73.71E
A 232	Udaipur region – 24.37N; 74.35E
ref	Bijolia – 25.17N; 75.31E

ref	Bundi – 25.43N; 75.64E
P 704	Kota Dt. Bilas River – 25.00N; 75.60E
ref	Kota – 25.21N; 77.88E
A 233	Kota District – 24.65N; 76.00E
A 234	East of Kota – 25.60N; 76.30E
ref	Ajmer – 26.47N; 74.60E
ref	Jaipur – 26.88N; 75.79E
A 235	Jaipur region – 26.60N; 76.10E
ref	Samod – 27.20N; 75.80E
A 236	Raghogarh, Madhya Pradesh – 24.35N; 77.10E – 1690
M 841	Kalibangan, Rajasthan – 29.50N; 74.15E – fossil W613

FIGURE 15.2
Mahout and Hunter on an Elephant surprised by a Rhinoceros. Opaque watercolour on dry plaster. Mural on the western end of the southern wall within the Badal Mahal, Palace in Bundi. Height: 56 cm

Bundi

The earliest localizable hunting scene involving a rhinoceros is situated on the western end of the southern wall within the Badal Mahal,[8] the ancient fort and palace of Bundi (district Bundi, Rajasthan) (figure 15.2).[9] It is part of several hunting scenes and measures 56 centimetres in height. It is situated 2,45 metres above ground and dates from about the first quarter of the 17th century.

In an uneven landscape, interspersed with rocks but rich in vegetation, a mounted elephant running from right to left suddenly faces a standing female rhinoceros. The elephant has curled up its trunk and, in its excitement, has folded its ears forward, where attention is drawn to the swollen veins of the ears. The speed of the elephant causes the bells, which are attached to ropes, not to hang down but to be pulled upwards. The mahout (elephant rider) sitting in the neck of the animal raises both arms and the *ankuśa* (elephant goad) in amazement. The hunter on the elephant's back behind the mahout has taken up his bow and is drawing an arrow from his quiver. The rhinoceros has its tail and ears erect and seems to pause in its tracks with its mouth open. The artist leaves it to the viewer's imagination whether the hunter will be able to shoot or whether the rhinoceros will attack the elephant, which is rushing along resolutely. The painter

---

8 Also spelled "Baddala Mahal." It was built under Rao Bhoj (r. 1585–1607) of Bundi, father of Rao Ratan, cf. Bautze 1997, p. 40.

9 For a ground plan of the Badal Mahal in Bundi see Bautze 1987a, p. 341. For an exterior view of the Badal Mahal see *ibid.*, figure1. For general interior views cf. Bautze 1986b, p. 71 (western half); Bautze 1987a, figure 2 (eastern half); Bautze 1991, p. 15, figure 3 (western part); Bautze 2000, p. 16, plate 5 (northern wall). For an earlier published description of the rhino scene see Bautze 1987a, p. 121.

FIGURE 15.3
The Rhinoceros, detail from Figure 15.2

has skilfully heightened the underlying tension between the rhinoceros and the elephant. The male of a small wild boar family looks up at the edge of a stream, invisible to the elephant. Monkeys playing among the rocks in the background suddenly turn towards the elephant. A mountain goat, two gazelles, a deer and a hind, even a tiger, watch this encounter, but are only discovered by the viewer after some time among the rocks, as are the inconspicuous flowers with their red blossoms blooming in a large part of the landscape, and the birds of various kinds flying around. A pair of foxes that have just jumped over a stream at the bottom of the picture, separated from the rhinoceros only by another narrow stream, points to the next scene – seen from right to left – as the second fox looks around in the direction of a horse rearing up at some distance behind him.

The dark brownish skin of the rhinoceros shows numerous deep skin folds with a pattern of irregular squares distributed all over the body. One may get the impression that the back of the animal is covered by a kind of saddle (figure 15.3). The open mouth shows the teeth of the lower jaw. The horn is shown as outline only and apparently uncoloured. It somewhat resembles the dorsal fin of a shark.

The identity of the hunter on the elephant is uncertain. Only two rulers in this sequence of events can be identified with certainty: Rao Ratan of Bundi (r. 1608–1631)[10]

and his second son, the future ruler of Kota, Madho Singh (born: 18 May 1599, died: 1648).[11]

Kota

The Chhattar Mahal of Kota

Madho Singh became the first ruler of Kota (district Kota, Rajasthan), formerly a part of the Hara-state of Bundi, in 1631.[12] The earliest unretouched wall paintings available today within the palace of Kota were created either under the grandson of Madho Singh, Jagat Singh (r. 1658–1683), or another grandson of the said Madho Singh, Kishor Singh (r. 1684–1696), if not by the latter's son, Rao Ram Singh (c.1696/7–June 1707). Of these Kota-rulers, Jagat Singh can be identified in a mural of the eastern room of the Chhattar Mahal.[13] Portraits of Maharao Bhim Singh (r. 1707–June 1720) were inserted at a later date.[14]

Stylistically, the unretouched wall-paintings within the Chhattar Mahal of Kota can be dated to the end of the last quarter of the 17th century.

So far the largest known mural showing hunting scenes, including those of a rhinoceros, is situated on the northern wall of the eastern room of the Chhattar Mahal within the old fort of Kota. It measures 2,42 m high × 4,14 m wide.

---

10   For details based on the Indian calendar see Bautze 1986a, p. 88. For his comprehensive biography cf. Bautze 1997, p. 41f.

11   For details according to the Indian calendar see again Bautze 1986a, p. 95. For the miniature painting based on the mural within the Badal Mahal of Bundi see also Welch 1997, p. 141, cat. no. 32; Sangram Singh 1965, no. 128; Beach/Topsfield 1991, p. 90f, no. 35; Beach 1992, p. 183, figure 137; Hodgkin/Topsfield/Filippi 1997, p. 137, no. 79.

12   For details see Bautze 1997, p. 41f.

13   Bautze 1986c, figure No. 4.

14   Bautze 1989, plates 64–66.

FIGURE 15.4A   A dozen Rhinoceroses and their Hunters. Opaque watercolour and occasionally gold on dry plaster. Detail from the northern wall of the eastern room of the Chhattar Mahal within the *garh* of Kota. Size of the wall: 2,42 m × 4,14 m. The numbers 1–12 added to this painting are explained in the text

FIGURE 15.4B   Painting in the Chhattar Mahal of Kota, as figure 15.4 A, without the added numbers

The right hand part of this wall-painting includes 12 rhinoceroses, marked here by numbers (figure 15.4). This does not necessarily mean, that the artist intended to show 12 individual rhinos, especially since in India, this animal is known as being a more solitary animal. In one of the earliest Buddhist texts, the *Rhinoceros Sutra* (Sanskrit: *Khaḍgaviṣāṇa-gāthā*, the *Khaggavisāṇa-sutta* as part of the Sutta Nipāta of the Pāli Canon) it is repeated at the end of forty stanzas: "eko care khaggavisāṇakappo (wander alone like a rhinoceros)." Rather, the artist wanted to depict events of a continuous plot with some rhinos, an artistic practice known in India since more than two millenia.

Figure 15.5 shows the rhinoceros numbered "5." It has apparently pierced the left flank of the elephant with its horn. This attack probably causes the elephant to stumble, if not fall. The onslaught happened with such force that the hunter fell down from the animal's back while the mahout shoots arrows at the attacking rhino which is already hit by a few of them. Two running hunters also aim arrows at the rhino, trying to save the said hunter who is lying on the ground with dishevelled turban. The top right corner of this illustration reveals the hind part of the rhino numbered 3.

The rhino numbered 6 is of a somewhat more whitish complexion (figure 15.6). Although it was already hit by many arrows, with its lowered head it still managed to bring down two hunters on foot. Both unfortunate men have lost their turbans, in Rajasthani painting a sign of mortal danger. Another hunter on foot tries to run away, while at the same time he turns round and aims an arrow at the angry rhinoceros. Another hunter comes from

behind and also aims at the enraged animal. Two horsemen with round black bucklers approach at full speed. At the rhino's back, six elephants mounted with hunters rush to the scene to finish off the enraged rhino.

Rhino number 7 also ran down a rider and his horse. Being hit by a spear and several arrows it tries to escape two horsemen, one of which wounded the animal with another spear (figure 15.7). Two hunters on foot try to make their escape while aiming at the animal with their bows.

The rhinoceros marked with number 8 (figure 15.8) is trying to escape from an elephant that has put its trunk around the animal's neck. The tusks of the elephant bore into the young rhino's back, as does the spear of the hunter, who is also the pachyderm's mahout.[15] Either a preparatory drawing or a drawing after rhino number 8 was in the collection of Stuart Cary Welch (1928–2008) (figure 15.9), it corresponds to figure 15.8 in almost every detail.[16] Interesting is, what the Maharao of Kota, Shri Brijraj Singh, had to say about it: "In the seventeenth and eighteenth centuries the jungles of Kotah were vast, contiguous to the much larger and richer forests of central India, which stretched uninterrupted for hundreds of miles to the forests of Assam. It was common to find lions, wild buffaloes, and at times even rhinos along with tigers there. Wild elephants were found within two hundred

---

15   For earlier discussions and reproductions of this detail see Beach 1974, p. 43 and plate LXX, figure 74; Brijraj Singh 1985, p. 21, left column and figures 47–48.

16   For earlier publications of this drawing see Beach 1974, plate LXIX, figure 73; Brijraj Singh 2004, p. 146f, no. 47; Das 2018, p. 86, top.

FIGURE 15.5　　Rhinoceros numbered 5. Detail from Figure 15.4

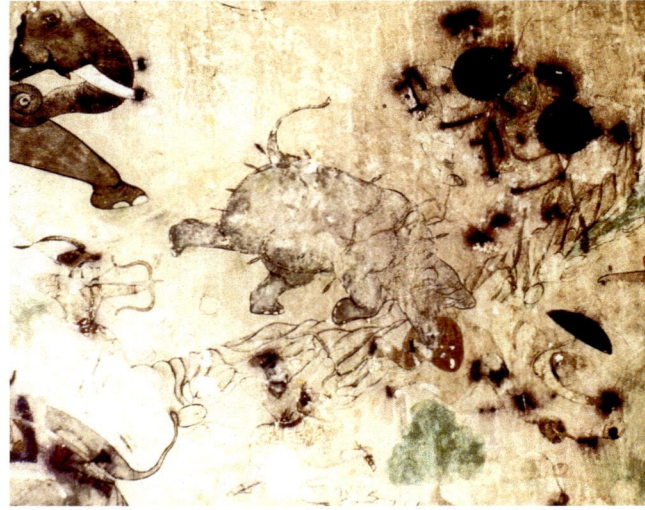

FIGURE 15.6　　Rhinoceros numbered 6. Detail from Figure 15.4

FIGURE 15.7　　Rhinoceros numbered 7. Detail from Figure 15.4

FIGURE 15.8　　Rhinoceros numbered 8. Detail from Figure 15.4

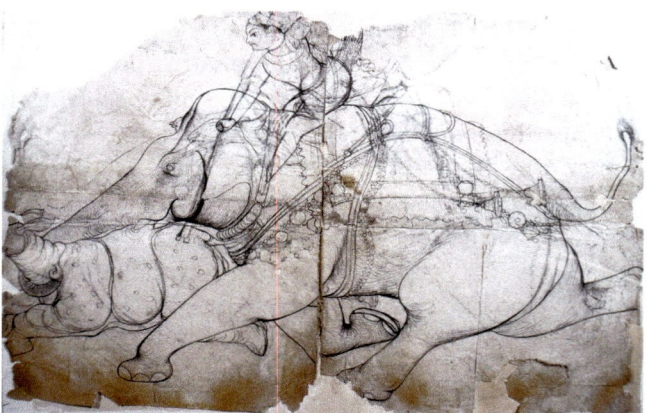

FIGURE 15.9　　"A Prince slays a Rhinoceros from an Elephant." Black ink over charcoal underdrawing on off-white laid paper. Size 30,0 × 46,3 cm. Promised Gift of Stuart Cary Welch (1928–2008) to the Harvard University Art Museums, Cambridge, Massachusetts, 78.1983

FIGURE 15.10　　Rao Bhao Singh of Bundi (r. 1658–1681) pursuing a Rhinoceros. Ink, water colour and gold on paper. Size 32,1 × 47,6 cm. Formerly in the collection of the late Stuart Cary Welch

miles of Kotah's border. So this scene may not be based on an artist's imagination; it is quite realistic."[17]

—————
17　　Brijraj Singh 2004, p. 146.

Another closely related drawing with some touches of colour, also formerly in the Cary Welch collection, shows a very similar scene with a mahout spearing the rhino in addition to a second man who had just shot an arrow at the fleeing animal (figure 15.10). In all previous discussions of this drawing the person spearing the rhino is identified with the said Rao Ram Singh and the drawing is dated accordingly, often even earlier.[18] The person in question, however, is not Rao Ram Singh of Kota but clearly Rao Bhao Singh of Bundi as was shown earlier.[19]

A kind of third variant of figure 15.8 is offered by figure 15.11, a detail of a large painting on cloth showing hunting scenes.[20] As with the previous examples, the elephant is lassoing the rhinoceros with its trunk while the rhino which had already been hit by several arrows which could only penetrate at the folds of the skin (figure 15.12). Most remarkable are the long ears of the rhino which resemble the horns of a water buffalo. These ears were in fact "elongated" by a later hand in order to fit with the other six water buffaloes roaming around in this painting on cotton cloth. As will be shown in the sequel, this painted fragment of cloth formerly belonged to a different hunting scene which was added here to a larger hunting scene illustrating basically the hunt on tigers and water buffaloes. When this fragment was added to the tiger and water buffalo hunt it was already incomplete: the elephant has lost its forelegs (figure 15.11).

A fourth variant is back-to-front (figure 15.13). It is painted on the left-hand part of the wooden lintel of the painted wooden door that leads from the western room to the eastern room of the Chhattar Mahal. It is obviously based on figure 15.8 and, while figures 15.9 to 15.11 were created during the first third of the 18th century, this was either painted or re-painted in the 20th century.

Also another detail from the northern wall of the eastern room of the Chhattar Mahal exists in different versions (figure 15.14). The mahout thrusts his spear into the rhino's head while the elephant pushes the fleeing animal to the ground with his left forefoot. The young animal with the small horn seems to roar in pain. Naturally, such dramatic scenes had to be practised by the artists with several preliminary drawings, such as figure 15.15, which reveals an earlier underdrawing in charcoal as well as a partly erased drawing in black ink on paper.

A variant of figure 15.14 exists also on cloth. It was also formerly in the collection of Stuart Cary Welch (figure 15.16). Here again, the "elephant is goring the side of the rhinoceros with its tusks while the rider stabs down

FIGURE 15.11   Maharao Durjan Sal (r. 1723–1756), son of Maharao Bhim Singh (r. 1707–1720) according to the inscription above the head of the mahout about to spear a wounded Rhinoceros. Opaque watercolour with gold on cotton fabric, a fragment from another painting on cloth showing the hunt on rhinoceroses. Formerly in the collection of Sir Howard Hodgkin (1932–2017)

FIGURE 15.12   Detail from Figure 15.11 clearly showing the horn of the Rhino with the later added horns of a water buffalo

FIGURE 15.13   A mounted Elephant "grabbing" a Rhinoceros with its trunk. Varnished watercolours on wood. Left-hand part of the wooden lintel of the painted wooden door that leads from the western room to the eastern room of the Chhattar Mahal, *garh* of Kota. Size not recorded

18   For earlier publications see Montgomery/Lee 1960, front cover and p. 45, cat. no. 36; Beach/Galbraith/Welch 1965, cat. no. 28; Welch/Beach 1965, p. 120, no. 27; Anonymous 1966, p. 54, top; Beach 1974, plates LXVII and LXVIII; Welch 1983, p. 80, figure 4; Beach 1992, p. 167, illustration 127; Welch 1997a, p. 16, figure 1 and detail, p. 29, figure 18; Welch 2004, p. 5, figure 5.

19   Bautze 1985b.

20   Topsfield/Beach 1991, p. 94f, cat. no. 37; Hodgkin/Topsfield/ Filippi 1997, p. 130f, cat. no. 74, with one more reference.

with a lance."[21] That this fragment of painted cloth along with the fragment shown in figure 15.11 belonged to a larger composition on fabric showing rhino hunts is corroborated by figure 15.17, another piece of painted cloth laid down on raw silk. Here, the rhino somehow managed to hurt the elephant with its horn after it was hit and wounded by numerous arrows which could only penetrate along the folds of the rhino's "armour", exactly as in Figures 15.12 and 15.16.

A mural in the manner of figure 15.14 was painted on the lowermost part of a wall as part of a courtyard within the Bara Devtaji-ki-haveli in Kota (figure 15.18). This *haveli*, a multi storeyed mansion situated not far from the *garh* of Kota contains numerous premises embellished with wall paintings from the late 18th century to the mid-19th century. Figure 15.18 belongs to the earlier phase, i.e., the last quarter of the 18th century.

Still, a further variant of figure 15.14 is offered by another painting on the lintel of the door leading to the eastern from the western room within the Chhattar Mahal (figure 15.19). Like figure 15.13 it is mirror reversed when compared to the presumed template and similar to the rhino numbered "9" in figure 15.4, where a horseman thrusts his spear into the rhino's neck.

The rhinoceros numbered "10" (figure 15.20) shows how the mounted elephant got hold of the rhino's right hind leg by twisting its trunk around it. Also here, the mahout thrusts his spear into the rhino's body which was hit by numerous arrows already. A horseman attacks the rhino frontally by hitting the animal on the head with his sword. Before that, the rhino must have brought down a hunter on foot.

Also, this scene must have inspired artists some one and a half centuries later. A painting in the collection of Shri Brijraj Singhji of Kotah shows three hunters, each of which on an elephant. Hunted, from top to bottom, are a rhinoceros, a tiger with a lion's mane and a water buffalo. Figure 15.21 shows a detail of this painting on paper. The hunter, marked with a golden nimbus, is identified as "Maharao Bhim Singh" on the front and back of the painting.[22] As in some previous examples, the scene is also reproduced laterally reversed here: The elephant grabs the left hind leg of the rhinoceros and tries to lift it.

The bull rhino, marked with number 12 and hit by several arrows, tries to escape from a rider who strikes the animal's rump with his sword when it encounters a mounted elephant whose mahout is unarmed. Figure 15.22

21   Reproduced: Sotheby's London, 31 May 2011, p. 54f, lot 29.
22   The inscription in Devanāgarī on verso reads: "mhārājedhīrāje mhārāvjī śrī bhīv sīghjī k[ī] chabī."

FIGURE 15.14   A mounted Bull Elephant imposing its left front leg on the shoulder of a fleeing Rhinoceros. Detail from Figure 15.4

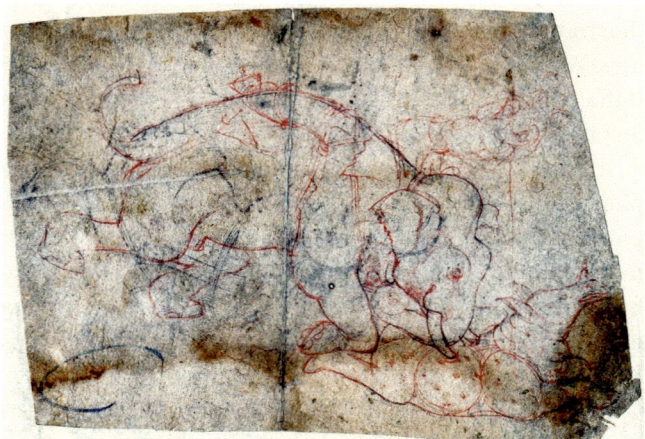

FIGURE 15.15   A Mahout on an Elephant spearing a Rhinoceros while the Elephant tries to thrust its tusks into the animal. Preparatory drawing on charcoal underdrawing and red ink on paper. Size 21,0 × 29,3 cm
WITH PERMISSION © P. & G. COLLECTION, KARLSRUHE-BERLIN

shows the horseman and the rhino lifting its head when it suddenly confronts the elephant which is, however, not shown in this detail.

Rhinoceros "12" was copied at some later date next to the left door leaf of the door wing on the eastern side of the door that opens to the western room of the Chhattar Mahal (figure 15.23). Also, the mounted elephant that faces the rhino in the earlier composition was copied, here, however, above respectively behind the rhino. The rhinoceros appears again in a panel on the right-hand wing of the door that separates the eastern from the western room within the Chhattar Mahal (figure 15.24). The four different representations of the rhinoceros on the

FIGURE 15.16   An Elephant and Rider hunting Rhinoceros. Fragment from a larger composition. Opaque watercolour and gold on cloth, laid down on raw silk. Kota. Size 21,0 × 36,0 cm. Formerly in the collection of Stuart Cary Welch

FIGURE 15.18   A rider on an Elephant spearing a Rhinoceros. Wall painting in the Bara Devtaji-ki-Haveli, Kota

FIGURE 15.19   Varnished painting on the wooden lintel of the door leading to the eastern from the western room within the Chhattar Mahal, *garh* of Kota. Size not recorded

FIGURE 15.20   Rhinoceros numbered 10. Detail from Figure 15.4

FIGURE 15.17   A wounded Rhinoceros attacking a mounted Elephant. Fragment of a larger composition, probably from the same work as Figures 11 and 16. Opaque watercolour and gold on cloth, laid down on raw silk. Kota. Size not recorded. Formerly in the collection of Stuart Cary Welch

FIGURE 15.21   Maharao Bhim Singh of Kotah (r. 1707–1720) and
friends out hunting. Opaque watercolours on paper.
Detail from a painting in the collection of Maharao
Brijraj Singh of Kotah. Size not recorded

FIGURE 15.22   Rhinoceros numbered 12. Detail from Figure 15.4

FIGURE 15.23   Varnished painting on wood next to the left door leaf
of the door wing on the eastern side of the door that
opens to the western room of the Chhattar Mahal,
garh of Kota. Size not recorded

FIGURE 15.24   Varnished painting on wood on a panel on the
right-hand wing of the door that separates the
eastern from the western room within the Chhattar
Mahal, *garh* of Kota. Size not recorded

FIGURE 15.25   Rhinoceros numbered 11. Detail from Figure 15.4

door that separates the eastern from the western room
of the Chhattar Mahal were definitely inspired or copied
from the older image template on the northern wall of the
eastern room of the Chhattar Mahal.

No hunter seems to aim his weapon at the rhinoceros
numbered 11 in Figure 15.4. It enters the large hunting
scene from right to left with a quick stride and does not
appear to have a hunter in front or behind it who has shot
his arrows at the animal, which has already been hit by at
least five arrows (figure 15.25).

FIGURE 15.26   Lower part of a large hunting scene within the western room of the Chhattar Mahal, *garh* of Kota. Varnished colours on dry plaster

FIGURE 15.27   Hunting scene involving two rhinos within the western room of the Chhattar Mahal, *garh* of Kota. Varnished colours on dry plaster

FIGURE 15.28   Retouched mural on the eastern wall of the eastern room of the Chhattar Mahal, *garh* of Kota

Also, the western room of the Chhattar Mahal offers some hunted rhinos. All murals within that room were repainted – or painted over – in the 20th century. Figure 15.26 shows the lower part of a larger hunting scene. The depiction was most likely again inspired by the hunting scene numbered 8 above. The mahout just shot at the rhino with his hunting rifle and hit it in the head. The hunter behind him aims an arrow at the animal, whose

neck is embraced by the elephant's trunk. The tips of the tusks were provided with blank push daggers (katar).[23]

Another hunting scene within the western room of the Chhattar Mahal opens the view to a double rhino hunt (figure 15.27). Two running rhinos, partly hit by arrows, are pursued by two mounted elephants. The mahout of the pachyderm in front of them lays an arrow while the mahout of the second elephant spurs his animal on with the *ankusha*.

The last hunted rhino is again in the eastern room, at the lower border of the painted eastern wall which has suffered considerably from water ingress. The horseman that thrusts his spear into the rhino's back as well as the rhinoceros itself were repainted at some later date, perhaps as late as the 20th century (figure 15.28).

The Jhala-ki-Haveli of Kota

The fortress of Kota contains not only palaces built under the respective rulers but also the palace of a minister who did not belong to the royal clan, the Jhala ki Haveli named after Rajrana Jhala Zalim Singh (1740–1823), who was for many years the chief minister, some say the regent, of the Kota state.[24] One part of the Jhala ki Haveli consists of an open rectangular court with an adjacent verandah followed by an almost windowless room on its eastern as well as western side. Both the walls of the verandahs as well as the almost windowless rooms behind are – or rather were – covered with murals dating from the last quarter of the 18th century.[25] The upper part of the eastern wall of the western room depicts at a length of 6,55 m different scenes showing Maharao Umed Singh (r. 1771–1819) and Rajrana Jhala Zalim Singh hunting. They are shooting and stabbing lions, tigers, water buffaloes, wild boars and rhinos from elephants, horses, and high seats. The four rhinos can only be approached by elephant, Figure 15.29 shows three of these rhinos. The main character in all cases is Maharao Umed Singh, who can easily be recognised by his radiant nimbus surrounding only his face.

If seen from south to north or right to left the southernmost rhino emerges from a thicket and seems to have surprised the Maharao and his companion on the back of the elephant (figure 15.30). While the Maharao is aiming an arrow at the rhinoceros on the elephant trying to escape, his companion swings a spear in the direction of that animal.

---

23   For a more recent essay on this kind of stabbing weapon see Nordlunde 2016, pp. 7–15 and pp. 83–183.
24   For details of his career see Bautze 1994.
25   For the fate of these murals see Bautze 1994 and Bautze 1996.

FIGURE 15.29   Three Scenes showing the young Maharao Umed Singh of Kotah (r. 1771–1819) hunting a Rhinoceros. Watercolour and gold on dry plaster on the upper part of the eastern wall of the western room on top of the Jhala-ki-Haveli within the *garh* of Kota

FIGURE 15.30   Maharao Umed Singh of Kotah (r. 1771–1819) on an Elephant surprised by a Rhinoceros. Detail of a mural on the upper part of the eastern wall of the western room on top of the Jhala-ki-Haveli within the *garh* of Kota

FIGURE 15.31   Maharao Umed Singh of Kotah (r. 1771–1819) spearing a Rhinoceros from the back of an Elephant. Detail from Figure 15.29

FIGURE 15.32   Maharao Umed Singh of Kotah (r. 1771–1819) aims an arrow at a Rhinoceros while his partner is about to shoot a Tiger. Detail from Figure 15.29

FIGURE 15.33   Maharao Umed Singh of Kotah (r. 1771–1819) chases a Rhino on an Elephant. Detail from Figure 15.29

In the next depiction, Umed Singh stabs the animal with a spear, which has already been pushed down by the elephant's trunk and tusks (figure 15.31). His companion, who may be the chief-minister, is about to send an arrow at the rhinoceros the skin of which shows a kind of large round pustules.

In the next scene, the rhino was apparently hit by an arrow shot by a hunter in the direction of the animal's run. This hunter on the elephant half hidden by a hill is shown in figure 15.29 at the top. Figure 15.32 illustrates how the rhinoceros suddenly looks around at the elephant on which Umed Singh is quickly approaching. The latter has put up an arrow to shoot at the frightened animal. Meanwhile, the archer behind him is aiming at a tiger whose head appears between two hills.

The northernmost and at the same time last depiction shows Umed Singh on an elephant pursuing a rhinoceros trying to escape, shooting, along with his companion behind him, arrows at the animal which has raised its tail (figure 15.33). It is striking that the horn of the rhinoceros

FIGURE 15.34   A Rhino Hunt. Opaque watercolour and gold on
                 paper. Bikaner. Size 20,3 × 30,0 cm. Collection of
                 Sir Cowasji Jehangir (1879–1962)

FIGURE 15.35   A Prince spearing a Rhinoceros from horseback.
                 Opaque watercolours and gold on paper. Bikaner.
                 Size 16,6 × 27,2
                 COLLECTION OF THE FONDATION CUSTODIA,
                 COLLECTION FRITS LUGT, PARIS

in the murals of this *haveli* always seems to point in the
wrong direction.

Bikaner

Bikaner (district Bikaner, Rajasthan) also produced several hunting scenes in which the rhino is the target. In the
hunting scenes that have become known, the rhinoceros
always attacks a hunter's horse by seeming to tear open
the horse's underside with its horn. Probably the earliest
depiction from about 1680, a miniature painting on paper,
was in the collection of the Bombay Parsee Sir Cowasji
Jehangir (1879–1962) (figure 15.34).[26]

The rhinoceros is pursued by seven horsemen and hit
by their arrows. The horse of an eighth rider catches it
with its horn between the hind legs, whereupon horse and
rider fall to the ground. A rider on a white horse who has
approached the rhinoceros from behind is about to stab it
in the anus with his spear.

A comparable depiction can be seen in a miniature
painting made in the first quarter of the 18th century,
which was in the possession of Spink & Son in London
in 1989 and is now in the collection of the "Fondation
Custodia Collection Frits Lugt" in Paris (figure 15.35). The
rhinoceros attacks a rider's horse at the same place and is
simultaneously hit in the anus by another rider with his
spear. Four more rhinos are visible beyond a range of hills
in the background, and the size of the ears of all five animals is remarkable. An inscription in the Nāgarī script on

FIGURE 15.36   A Rhinoceros attacking a horseman from behind.
                 Detail of a mural in the Chandra Mahal within the
                 Junagarh palace complex of Bikaner

the reverse of the painting apparently identifies the rider
on the rearing horse with the spear as the young Prithviraj
Chauhan (r. *c.*1178–1192),[27] but this is probably incorrect.

A mural painted towards the end of the 18th century
in the Chandra Mahal within the Junagarh palace complex of Bikaner shows a similar scene (figure 15.36). The
rhinoceros attacks a rider's horse at the said spot and is
simultaneously hit on the root of the tail by the spear
of a pursuing rider. Another rider with a lance ready to
throw approaches the rhinoceros from the front and from
behind. A kneeling marksman also aims a rifle at the rhinoceros. A beater on foot claps his hands behind him.

---

26   Reproduced: Khandalavala/Moti Chandra 1965, plate H;
       Das 2018, p. 87. This is probably identical with "A Rhino hunt,
       17th century" in *International Cultural Exhibition* 1945, p. 12,
       no. 12.

---

27   "rājā prathīrāj cohāṇ bālak."

The way the rhinoceros attacks a mounted horse was probably inspired by a 17th century Mughal painting in which two rhinos appear. One of these two animals stabs a mounted horse between both hind legs with its horn.[28]

Udaipur

An often-published miniature painting shows Maharana Amar Singh II of Udaipur (district Udaipur, Rajasthan; r. 1698–1710) spearing a rhinoceros from his elephant while a servant behind him waves a *chauri* with his right hand and holds a bow with his left.[29] As most of the authors have already remarked, the scene is strongly reminiscent of comparable situations in Kota paintings.

Chitram-ki-Burj

Just between cupola and vertical wall of the Chitram-ki-Burj[30] within the Mardana Mahal of the City Palace Museum of Udaipur the artists working for Maharana Bhim Singh (r. 1778–1828) painted a long series of several hunting scenes. On the bank of a body of water, two hunters fire their shotguns from a tree hideout at a tiger that has jumped at the head of a rhinoceros (figure 15.37). The attacked rhino is apparently shown three times to illustrate the movement of the animal as it leaves the water. Three more rhinos roam around in an open landscape.

Samod

The palace of Samod *aka* Samode (district Jaipur, *tehsil* Chomu)[31] is decorated with murals from the period of Maharaja Sawai Jai Singh III (r. 1819–1835) and Maharaja Sawai Ram Singh (r. 1835–1880) of Jaipur. A mural within the so-called "living room" presents a highly decorative axisymmetric composition at the bottom of the wall.[32] A section of this wall painting (figure 15.38) shows a duplicated lion, reminiscent of a tiger in its fur, jumping on the back of a rhinoceros and appearing to bite into the rhino's body. The body of the rhinoceros, which has a tail that is far too long, is covered with spots that resemble

FIGURE 15.37   A Tiger is being shot while frontally attacking a Rhinoceros. Detail of a wall painting done for Maharana Bhim Singh of Mewar (r. 1778–1828) within the Chitram-ki-Burj as part of the City Palace Museum, Udaipur

FIGURE 15.38   A Lion with the marks of a Tiger jumps on the back of a Rhinoceros. Detail of a mural showing an axisymmetric composition at the bottom of a wall within the so-called "living room" within the palace of Samod

measles. More depictions of rhinos as can be found within the murals of the palace at Samod are introduced in the sequel.

15.3   The Fighting Rhinoceros

Paintings depicting a fighting rhinoceros were found in Dungarpur, Udaipur, Bundi, Kota and Samod. The localities are shown in map 15.8 and explained in Dataset 15.11.

Dungarpur

The probably earliest representation within Rajasthani mural painting of a fighting rhinoceros is to be seen in the Juna Mahal, Dungarpur (district Dungarpur, Rajasthan).[33] Within the Juna Mahal it is situated in the northern part of the western wing of the ground floor above a door.[34]

28   *Coronation Durbar* 1911, p. 164, plate LXXIV(a). C.338. "Hunting Party disturbed by Rhinoceros. Lent by Bulāki Das, of Delhi."

29   Christie's 21 July, 1971, lot 146, plate 10; Dahmen-Dallapiccola 1976, p. 63, full-page coloured plate 23; Lunsingh Scheurleer 1978, p. 48f, cat.no.69, plate 32; Glynn 2011, p. 517, no. 23 and p. 522, figure 5.

30   For this monument see Cimino 2011, pp. 99–118 and colour plates pp. 238–244.

31   Cf. Cimino 2001, pp. 163–179; Sugich 1992, pp. 35–37 + coloured plates opposite p. 29; Georges 1996, p. 6, pp. 84–90; Martinelli/Michell 2004, pp. 8–9 and pp. 70–77.

32   For a full view see Georges 1996, p. 6; Cimino 2001, p. 169, figure 234.

33   For an excellent monograph on the architecture of this palace see Imig/Mahesh Purohit 2006.

34   For another mural within the same room see Bautze 2005, p. 511, figure 4; see also Imig/Mahesh Purohit 2006, p. 118, figures 131–132.

FIGURE 15.39   A white-coloured Rhinoceros attacking a black
Elephant. Detail of a mural in the northern part of
the western wing of the ground floor above a door
within the Juna Mahal, Dungarpur. Colours on wet
plaster

FIGURE 15.40   Detail from a painting on cloth showing Maharana
Amar Singh II of Mewar and his Court attending an
animal fight in front of the Palace at Udaipur
WITH PERMISSION © CITY PALACE MUSEUM,
UDAIPUR. ACC.NO. 2012.200012_R

The scene, set against a rocky reddish mountain shows
a dark elephant in fight with a fair-skinned rhinoceros
(figure 15.39). The almost black elephant braces itself
against the rhinoceros, which in turn tries to injure the
elephant between its front legs with its horn. As suggested
earlier, this mural dates from the late 17th century.[35]

Udaipur

A large-format painting on fabric in the collection of the
City Palace Museum, Udaipur, depicts the preparation of
a staged fight between a rhinoceros and an elephant at the
time of Maharana Amar Singh II of Mewar (r. 1698–1710)
(figure 15.40).[36] On the left side of the wall that usually
separates two fighting elephants stands a large elephant
stamping the ground with his right foreleg. A chain is put
around his left hind leg as well as his right foreleg. Both
chains, however, are not fastened to anything which allows
the animal to move freely. A rope or chain surrounds its
body as well as its neck. Two rhinos stand on the right
side of the wall, both are facing left. The one which faces
the elephant also has a chain or rope round its body and
neck. The chains around the foreleg and hind legs are free.
A somewhat smaller rhinoceros stands behind.

Chitram-ki-Burj

Another fight between a rhinoceros and an elephant can
be seen between cupola and vertical wall of the afore mentioned Chitram-ki-Burj within the Mardana Mahal of the
City Palace Museum of Udaipur, datable to the late 18th to
early 19th century (figure 15.41). A mounted elephant with
disproportionately large ears runs with its head up against
a rhinoceros, which also runs with its head up. Ropes are
tied around both opponents so that the riders can hold on
to them. A round white area between the eyes and ears of
both animals is framed in red at the edge except for a kind
of pie slice. A black liquid, a kind of rutting season juice,
runs out of this 'pie slice', which in the case of a bull elephant indicates his readiness to mate or fight. The painter
gave the rhinoceros a similar feature. Unlike in the afore-
mentioned painting on fabric, both animals are not separated from each other by a wall, nor do either have chains
on their feet. The large ears of the elephant seem to indicate that it is of African origin.[37] Both riders appear to be
sitting on some kind of cushion, with the rider of the rhinoceros apparently struggling to hold on to the ropes. Five
drivers on foot take part in the fight in their own way by
driving the animals at each other. One of them has fallen
and lost his turban; he is in danger of being trampled by
one of the animals. A hare seeks salvation in flight.

Karjari Haveli

The Karjari Haveli in Udaipur[38] has murals from the reign
of Maharana Shambhu Singh (r. 1861–1874). One of them
shows a fight of a mounted rhinoceros against a mounted
stag.[39] The rhino has lowered its head as has the stag. The
two bearded riders each seem to be encouraging their
animal to fight, which seems to be indicated by the hand
raised in each case, holding a club on the rhinoceros.

---

35   See again Bautze 2005.
36   Reproduced after Topsfield 2001, p. 133, figure 110b. For the full
painting see: *ibidem*, figure 110 and Das 2018, p. 88.

---

37   For a Mughal painting of an African elephant see e.g.,
Ehnbom 1985, p. 60f, cat.no. 22. For another example from
Udaipur see Cimino 2011, p. 149, figure 194 and col. plate 62,
p. 257; also p. 179, figure 255 and col. plate 75, p. 264.
38   For which see Cimino 2011, pp. 161–166.
39   Reproduced: Cimino 2011, p. 165, figure 226.

FIGURE 15.41   A mounted Rhinoceros fighting a mounted African
                Elephant. Mural between the cupola and vertical wall
                of the Chitram-ki-Burj within the Mardana Mahal of
                the City Palace Museum of Udaipur

Bundi

In the bazaar within the older part of the town of Bundi
stands a palatial mansion, the Govardhan Singhji ki Haveli,
with an inner courtyard the veranda of which is decorated
with wall-paintings from the last quarter of the 18th century. The lower part of the wall of this verandah shows different animals in white set against a red background. In
one scene (figure 15.42), a mounted rhinoceros runs into
a mounted elephant with its head held high. While the
mahout is goading the elephant, a second rider is about
to shoot an arrow at the rhino. The rider on the rhino's
back raises a whip with his right, his left-hand clings to
the rope which is fastened around the rhino's body. A running fourth man on the ground raises his left arm as if to
say something to the rider on the rhinoceros, while his
right hand holds a staff at the lower end, the upper part of
which seems to be wrapped with a wire.

Kota

The lower part of the verandah of the Bara Mahal[40] in the
Kota palace complex is decorated with a white marble (or
alabaster?) relief frieze with flower vases between which
various scenes are captured, including several animal
fights. A pair of fighting Indian antelopes suggests that
also the facing rhinos in the lower part of this relief are
about to engage in a fight (figure 15.43).[41]

What seems to be a friendly fight of two rhinos within
a painted cartouche (figure 15.44), is a mural in the inner

FIGURE 15.42   A mounted Rhinoceros being attacked by a mounted
                Elephant and its rider. Mural at the bottom of a wall
                as part of a verandah of the Govardhan Singhji ki
                Haveli within the old city of Bundi

FIGURE 15.43   Two Rhinos preparing for a fight. Relief at the
                bottom of the wall belonging to the relief frieze of
                the verandah of the Bara Mahal within the palace
                complex within the *garh* of Kota

room of the Bara Mahal.[42] Most unusual is the blue colour of the animal to our right, while the horn of both
animals resembles a snail shell. The "pox" on both rhinos
are evenly distributed and all of the same size. Based on
several inscriptions the murals in the Bara Mahal can be
dated between 1826 and 1829.

Samod

A detail of a mural within the so-called "living room"
within the palace of Samod reveals another friendly fight,
this time of a rhino and an elephant (figure 15.45). The

40   For a more general view of this particular part of the Bara Mahal
      see Gaekwad 1980, p. 75. Please note: the photograph does not
      show "walls and lamp-niches of the Durbar Hall ..." as stated in
      the caption, but the first room, a kind of verandah, of the Bara
      Mahal. The reliefs become apparent in the lower part of this
      reproduction.
41   Reproduced: Bautze 1985, p. 425, figure 7.

42   For a more general view of this particular part of the Bara Mahal
      see Michell 1994, p. 141.

FIGURE 15.44   A friendly fight of two Rhinos in a painted cartouche as part of the inner room of the Bara Mahal, *garh* of Kota

FIGURE 15.45   An Elephant "embraces" a Rhinoceros. Detail of a mural in the so-called "living-room" within the palace of Samod

FIGURE 15.46   A Rhinoceros witnessing a *tika*-ceremony. Detail of a mural on the eastern wall of the Mardana Mahal within the palace of Bijolia. The complete scene measures 90 × 118 cm

trunk with which the elephant seems to clutch the rhino's neck gives the impression of a friendly embrace rather than a life-threatening attack. This wall painting dates from the mid-19th century.

15.4   The Rhinoceros as Part of the Royal Court

Depictions of a rhinoceros in a royal court were found in Bijolia, Udaipur, Dungarpur. The localities are shown in map 15.8 and explained in Dataset 15.11.

Bijolia
Bijolia *aka* Bijoliya (district Bhilwara, Rajasthan) is mainly known for its ancient inscriptions and temples.[43] But there is also a rarely documented palace of the rulers of that place. Within the palace complex of Bijolia exists the Mardana Mahal, an approximately square room measuring about 3,24 × 3,17 m. The east wall of this room with a painted dome has a painted inscription

with the date 1701 CE.[44] The murals, however, show at least one event which must have happened in the 1720s if not 1730s. This wall painting on the east wall measures 90 × 118 cm (height precedes width) and shows Maharana Sangram Singh II of Mewar ("rāṇā sa[ṃ]grām / saṃgh jī" r. 1710–1734) applying *tika* to the forehead of apparently the local ruler, a certain Rao Mandhata ("rāv mānadhātājī"),[45] who appears in several murals of this room. Behind this local ruler, about whom almost nothing is known, sits Rao Bakhat Singh of Bedla ("rāv ba / ṣat sī / ghjī be / dalā rā"). Bedla (in Bargaon tehsil, district Udaipur) was a fief of the former Mewar state, as was Bijolia. Since Rao Bakhat Singh of Bedla ruled from 1721 to 1738, the ceremony, a kind of coronation or engagement ceremony, must have happened between 1721 and 1734. Be that as it may, this ceremony is attended by several courtiers, many of which are identified by an inscription. There are three mounted elephants, a white stallion, an elephant with large ears as seen in figure 15.41, and a solitary male rhinoceros with a red crescent of the moon painted upside down between its right eye and ear (figure 15.46).[46]

43   Tod 1920, Volume III, pp. 1796–1800.

44   The inscription starts with "śrī rāmjī" and ends, line 9 and 10: "vaisāk sudī 13 guré / saṃ° 1758" (Thursday, 13th of April/ May, 1701).

45   Rao Mandhata Singhji was the 11th Rao of Bijolia, his rule, however, started considerably later.

46   Reproduced: Bautze 1985, p. 424, figure 6.

Udaipur

During the reign of Maharana Jagat Singh II of Mewar (r. 1734–1751) there were at least two rhinoceroses at his court. By 1955 the Baroda Museum & Picture Gallery in Vadodara (Gujarat) acquired a Mewar painting showing a rhinoceros called "chhammī" according to an inscription on the top red border of the painting (figure 15.47).[47] Vadodara is the place, where on 19 November 1875 Albert Edward as the Prince of Wales (later Edward VII, r. 1901–1910) witnessed a rhinoceros fight (7.7).[48] The rhinoceros shown in Figure 15.47 is described as follows:

> The huge animal with its grotesque built, its folds of armour and the scaly hide looks like a monster of some bygone age.
>
> The heavy folds of the skin before and behind the shoulders and in front of the thighs are shown in a characteristic manner. The skin is also shown studded with masses of rounded tubercle. Our rhino is probably a female which is shown by its sharper and longer horn.
>
> There is, however, a mistake in the study because the short stumpy legs of the animal are each furnished with three toes and not two as shown in the painting.
>
> The forelegs of the animal are tied with a pair of silver coloured metallic chains. The vermilion marks on the forehead and the temple of the animal show the high esteem in which it is held by the Hindus. They set a great value on the blood and flesh of the animal and provide the main reason for its persecution. It appears that the rhinoceros is kept in captivity by the Mahārāṇā and is chained for reasons of safety.[49]

Another painting with a rhino and a very similar inscription[50] surfaced in the London art market in 1986 (figure 15.48). The name of the rhino ("ge[ṇ]ḍo") is Fateh-Chand. Like the one in Vadodara, it has only two toes per feet. The forelegs show a chain around each ankle. One of the three overseers holds a stick with a device atop the pole which is generally used for the separation of

7.   Mahārāṇā Jagatsingh accosting a rhinoceros, Rājput painting, Mewār, middle 18th century A. D.

FIGURE 15.47   Rhinoceros at the court of Udaipur. Opaque watercolour on paper. Size 22,9 × 44,5 cm
MUSEUM AND PICTURE GALLERY, BARODA, REG. NO. P.G. 5A. 420

FIGURE 15.48   The Rhino "Fateh-Chand" at the court of Maharana Jagat Singh II of Mewar (r. 1734–1751). Opaque watercolour on paper. Seen in 1986 in the London art market, present whereabouts unknown

fighting elephants. It is hence not clear if this rhino was kept for a pageant or a combat with another rhino or elephant. It is noticeable that the horn points forward and that the vermilion mark in the form of an inverted crescent moon next to the ear resembles again the same mark on the head of elephants kept at the Udaipur court.

A Jain invitation letter, "vijñaptipatra" dated in the last two lines at the end of the scroll to May-June 1795 CE[51] shows a tamed rhinoceros being pushed by the left hand of a gentleman holding a spear with his right hand. The animal wears a golden chain around its neck and there are silver chains around the ankles of the forelegs. It has a red dot encircled by white dots at its cheek, while the forehead is decorated with red streaks (figure 15.49). The

---

47   *Acc. No. P.G. 5A. 429.* I have not actually seen this painting. For the reading of the inscription, I rely on the transliteration offered by Gangoly in Gangoly 1960, p. 92, no. 40, where the size of the painting is given as 16 ½″ × 8″.

48   For a full description of this fight see Russel 1877, p. 198f.

49   Devkar 1957, p. 24.

50   "mhārāṇa: śrī jagat sīghjī [rest illegible]" and, by a different hand in bold letters: "ge[ṇ]ḍo phatecaṅd."

---

51   "s[amvat] 1852 / varaṣe matī cet vīd 5."

FIGURE 15.49   A tame rhinoceros in the city of Udaipur. Detail of a Vijñaptipatra (Jain letter of invitation) dated V.S. 1852 on the fifth of the dark half of [the month] chet (May–June 1795 CE). Opaque watercolour, ink, and gold on paper, formerly in the Collection of Ludwig Volker Habighorst, Koblenz, Germany

FIGURE 15.50B   Detail from Figure 15.50a showing the rhinoceros near the gate to the Palace

scroll measures 15,37 meter in length and illustrates plenty of scenes of everyday life in Udaipur.[52]

In the mid nineteenth century a rhinoceros was kept at the *tripoliya*, the gateway to the royal palace in Udaipur. This is shown by a duly inscribed[53] drawing by an artist from Kota, datable to the 1820s,[54] now in the Brooklyn Museum, New York (figs. 15.50a, 15.50b).[55] The artist first intended to depict the animal with a raised head but later decided to show it with its head down. The deer behind the rhinoceros seems to indicate that there was a kind of menagerie.

Dungarpur

One part of the "Aam Khas"[56] on the second floor within the Juna Mahal in Dungarpur is known as "Shish ka Kamra" or "room of mirrors" *aka* "room of glass."[57] It

FIGURE 15.50A   The Palace at Udaipur with a Rhinoceros. Sepia ink and colour on paper. Size 37,5 × 26,7 cm
BROOKLYN MUSEUM, NEW YORK. GIFT OF
MARILYN W. GROUNDS, 80.261.39

52   For more details about this scroll see Bonhams 2015, lot 112, pp. 116–117.

53   The Nāgarī-inscription in comparatively large letters at the bottom reads: "udapur kā ma[ha]l (Udaipur palace)." The inscription at the right hand border, below the rhinoceros, reads: "tarapol ... (Tripoliya, i.e. a gate with three thoroughfares)." For a 19th century photograph of this gate by Lala Deen Dayal (1844–1905) see Topsfield 1990, p. 9, figure 2.

54   A Kota drawing formerly in the collection of Christian Humann (1929–1981) shows Maharana Bhim Singh (r. 1778–1828) at Udaipur on an elephant during the Gangaur-festival, cf. Sotheby's 20 June 1983, p. 62f, lot 119 or Maggs Bulletin no. 40, 1986, p. 98, no. 99.

55   37,5 × 26,8 cm. Gift of Marilyn W. Grounds, acc.no. 80.261.39. Published: Cummins 1994, p. 317, D47.

56   This is the "Audienzsaal" in Imig/Mahesh Purohit 2006, pp. 127–133. For a good general overview of this palace see Sethi 1999, pp. 156–165; For the Aam Khas as part of the Juna Mahal see the photograph in Dwivedi 2002, pp. 72–73; Imig/Mahesh Purohit 2006, figures 149–157.

57   Lyons 2004, p. 40. Imig/Mahesh Purohit 2006, figures 155–156, here also called "Spiegelraum (room of mirrors)."

FIGURE 15.51   The Rhinoceros called "Mohan Lal" at the Court of Dungarpur leading a royal Gangaur Procession. Mirror glass, cut inlay of coloured glass and painting on dry plaster. Decoration done by the artists Nathu and Chagan for Maharawal Udai Singh II (r. 1846–1898) in the so-called "Shish ka Kamra" or "room of mirrors" aka "room of glass" on the second floor within the Juna Mahal, Dungarpur

FIGURE 15.52   A mounted Rhinoceros followed by a mythical Bird. Mural at the bottom of a wall as part of the so-called "Aam Khas", second floor of the Juna Mahal, Dungarpur

was decorated under and for Maharawal Udai Singh II (r. 1846–1898) who is shown, possibly in life size, on one of its walls[58] which are decorated with mirror glass, cut glass inlay and wall paintings. The bottom of the eastern wall of this room shows a royal procession, a so-called *savārī*, on the occasion of the Gangaur-festival,[59] which is led by a caparisoned rhinoceros (figure 50.51). This festive procession, a kind of pageant, has a panel with an inscription,[60] the translation of which informs: The honourable Maharawal Udai Singh with the honourable Keshri Singh [on the elephant] with the honourable crown-prince, Sir Khuman Singh ahead [on] the fifteenth of the bright half of [the month] Magh, V.S. 1943 [= 1886 CE]. Painted by the artists Nathu and his brother, Chagan.

Khuman Singh did not succeed his father to the throne of Dungarpur because he died on 30 October 1893, aged 37. Both the artists, Nathu and Chagan, the latter at times spelled Chhagan, came from Udaipur and were responsible for the paintings within the "room of mirrors", as all other inscribed paintings in there are also attributed to them. This Nathu is probably not identical with his namesake, from whom a miniature painting is dated as early as 1835.[61]

It seems by the time this *savārī* took place the rhinoceros did not yet belong to the said Maharawal Udai Singh. According to documents preserved in "The Maharawal Bijay Singh Research Archive" at Dungarpur,[62] Maharawal

---

58   See Sethi 1999, p. 160, bottom, where the rhino of figure 47 can be detected in the lower left corner; see also Coleridge 1997, pp. 136–137, a photograph by Deidi von Schaewen. For the portrait of the ruler see also Lyons 2004, p. 40, figure 14 and Imig/Mahesh Purohit 2006, p. 131, figure 155.

59   For this festival cf. Sharma 1978, p. 76 and figures 59–62. Two women carry the figures of Gauri and Shiva on their head towards the end of the depiction of this procession in the "room of mirrors" within the Juna Mahal.

60   "śrī māhārāval-jī śrī udai sīgh-jī / koṭārī-jī kecarī sīgh-jī āge śrī / māhārāval-kuṅvorā sāaib-jī śrī ṣuṅmaṅ / ṇ sīgh-jī savaṃt 1943 māhā sudī 15 / sītārā nāthu | bhāī cagan kalamī / da ... [rest not clear]." A diagonal labelling in a different hand adds: "acayārī gāṅṇa/gorānī." For the position of this label as part of the wall decoration see Imig/Mahesh Purohit 2006, p. 130, figure 156.

61   Cf. Topsfield 1980, p. 152, no. 218.

62   D.O. 62, House Hold, Expenditure of the Palace, Jan. 1884–Oct. 1887 CE, July 1885, for further details see Sahlström/Mahesh Purohit 1985.

FIGURE 15.53   A large, mounted Rhinoceros facing right. Mural in the lowest part of the wall in a side room of the so-called "Virat Sarup ki Odi", also known as "Virat Rup ka Kamra", Juna Mahal, Dungarpur

Udai Singh II acquired it only in March 1885 CE. The name of the rhinoceros was Mohan Lal.[63]

Also, at the bottom of the eastern wall in the northeastern corner of the Aam Khas a second rhino captures the attention of the viewer (figure 15.52). Although the rhinoceros has neither bridle nor harness, it is mounted by a bearded man with a turban and a kind of shillelagh in his left hand. The ropes fastened around the body of the animal keep the cushion on its back, on which the rider sits, in position. Chains are visible around all its feet. The rhino is followed by a mythical bird, probably Garuda, which holds several elephants with its claws and its beak.[64] This mural is of a considerable later date and was possibly painted as late as 1939–1940.[65] The difficulty in dating these murals is that the painters, Premchand and Kanhaiyalal,[66] gave their names along with their exact address in Nathdwara (district Rajsamand, Rajasthan), but they never mentioned a date.[67]

A third mural with a rhinoceros is in the lowest part of the wall in a side room of the so-called "Virat Sarup ki Odi", also known as "Virat Rup ka Kamra",[68] decorated with murals by both Premchand and Kanhaiyalal, dated 1919 according to Tryna Lyons (figure 15.53). A stag seems to escape from this disproportionately large rhinoceros, the rider of which holds a shillelagh in his right hand. The rhinoceros has no bridle, the ropes stretched around its body serve to hold the cushion on its back. Remarkable is a chain with a bell around the neck next to the chains placed around the forelegs.

15.5 The Unmolested Rhinoceros

A few murals show the rhinoceros peacefully alone or in a landscape among other wild animals. Rajasthani miniatures illustrating the meeting of Layla and Majnun in the wilderness as based on the famous old story told by the Persian poet Nizami (c.1141–1209) were often copied and are hence not considered here.[69]

Paintings depicting the unmolested rhinoceros were found in Bundi, Kota, Samod, Devgarh and Udaipur. The localities are shown in map 15.8 and explained in Dataset 15.11.

Bundi

A painting on paper in the collection of Gursharan and Elvira Sidhu (Seattle, WA, U.S.A) shows the veneration of a meditating saint seated on the skin of a black buck in front of his thatched hut by Maharao Ajit Singh of Bundi (r. 1770–1773) surrounded by numerous mammals and birds (figure 15.54a).[70] A detail (figure 15.54b), reveals two rhinos which seem to consider taking a dip in the water at their feet. The "horn" has only been mildly indicated.

Kota

In another painting on paper, presumably from Kota, several species of animals found refuge on an island, where they seem to be safe from the raging fire around (figure 15.55a). Among them a rhinoceros stands facing left (figure 15.55b). Its horn seems to echo the upper lip of its snout while between the eye and the neck a kind of gland seems to produce a liquid which strongly reminds of a similar gland between eye and ear of the elephant behind. With the male elephant, that liquid indicates that the animal is in a mating mood.

63   Mahesh Purohit undated, Appendix A, page 4, footnote 4.

64   For further examples in the murals of Udaipur see Cimino 2011, p. 179, figure 255 (= col. plate 77, p. 265) and p. 188, figure 266.

65   As suggested by Tryna Lyons on information supplied by Mahesh Purohit (1931–2021), cf. Lyons 2004, pp. 41–45.

66   His full name was apparently Kanhaiyalal Vitthaldas Sharma (1902–1998), cf. Ghose 2015, p. 19.

67   "|| kalamī, citrakār, rāmlāl, premcand, mukam śrīnāthjīdvarā || / || ṭhīkāṇā naihavelī mẽ jīla udepur mevāḍ rāṇākā mulak rājputānā ||"

68   Lyons 2004, p. 41f, figure 15 and figure 17. Imig/Mahesh Purohit 2006, p. 132, figures 158–159.

69   Cf. Kala 1961, p. 8f and plate 8 (= Das 2018, p. 88, top left); Pratap 1996, p. 108, figure 41; Beach/Nahar Singh 2005, p. 85, figure 101A. See also Topsfield 2008, no. 73, pp. 154–155. For a Mughal example see Topsfield 2008, cat.no. 30, pp. 68–69.

70   Bautze 1990, p. 103, figure 22.

FIGURE 15.54A   Maharao Ajit Singh of Bundi (r. 1770–1773) greeting a
Saint in a landscape. Opaque watercolours on paper
COLLECTION OF ELVIRA & GURSHARAN SIDHU,
SEATTLE, WA, U.S.A.

FIGURE 15.54B   Detail from Figure 15.54a showing two Rhinos

Devtaji-ki-Haveli of Kota

Several rooms within the Devtaji-ki-Haveli in Kota were
embellished with wall paintings between 1820 and 1840.[71]
One of these murals illustrates the well-known story of

<hr />

71   See for example Bautze 1987b or Varmā 1989, plates [phalak] 13
and 14.

FIGURE 15.55A   Different animals are protected on an island from the
fire around them. Opaque watercolours and gold on
paper. Banaras, Bharat Kala Bhavan

FIGURE 15.55B   Detail from Figure 15.55a showing a Rhinoceros

FIGURE 15.56   Two running Rhinos. Detail of a mural illustrating
the story of the salvation of the elephant king
(Gajendramoksha) in the Devtaji-ki-Haveli, Kota

FIGURE 15.57 Two Rhinos witnessing the Salvation of the King of Elephants. Mural at the Lakshmi-Narayan temple within the *garh* of Kota

FIGURE 15.58 A pair of Rhinos at the Salvation of the King of Elephants. Mural within the Bara Mahal, *garh* of Kota

Gajendramoksha or salvation of the king of elephants,[72] in which god Vishnu saves his devotee, an elephant, from the attack of a crocodile. In the company of nine elephants, two lions, two water buffaloes, two bears and one stag a pair of rhinos runs, like all the other animals, from right to left at the bottom of the mural (figure 15.56).

Lakshmi-Narayan Temple of Kota
The murals of the Lakshmi-Narayan temple within the *garh* of Kota are completely undocumented.

One of the numerous wall-paintings also illustrates the story of the salvation of the king of elephants. In the lower right corner of the composition datable to the late 1820s, two rhinos are facing left. They seem to follow a pair of running water buffaloes and a white lioness (figure 15.57).

Bara Mahal of Kota
The inner room of the Bara Mahal includes a large format mural of the Gajendramoksha. Two rhinoceroses are painted in the lower part of the composition, right above a miniature painting illustrating a particular festival of Shri Krishna inserted into the wall (figure 15.58).

Just near the painted cartouche with the two rhinos (figure 15.44) a kind of vignette contains a blue complexioned rhino resembling an inflatable rubber animal for the beach during holidays with young children (figure 15.59).

A miniature painting in the collection of late His Highness of Kotah, Maharao Brijraj Singhji (r. 1991–2022) illustrates how god Shiva accompanied by an orchestra formed by gods and saints like Brahma, Vishnu, Ganesha and Narada dances at the feet of Mount Kailash on top of which sits goddess Parvati on a throne (figure 15.60a).

FIGURE 15.59 A happy Rhino. Mural of a vignette in the inner room of the Bara Mahal within the *garh* of Kota

A detail (figure 15.60b), introduces half a dozen rhinos partly taking a bath in the river.

Samod
Palace of Samod
A detail of a mural within the so-called "living room" shows a lonely rhinoceros in the company of bears, boars, a tiger, antelopes and aquatic turtles (figure 15.61). The somewhat purple complexion is evenly covered with darker purple spots which are even apparent on the horn. This rhino is part of the same mural as is Figure 15.45.

---

72 For a summary of this legend see Mani 1975, p. 328f under "Indradyumna I."

FIGURE 15.60A　Mount Kailash, goddess Parvati and her dancing husband Mahadeva
MINIATURE PAINTING IN THE COLLECTION OF LATE HH KOTAH

FIGURE 15.61　A lonely Rhinoceros. Detail of a mural within the so-called "living room" in the Palace of Samod

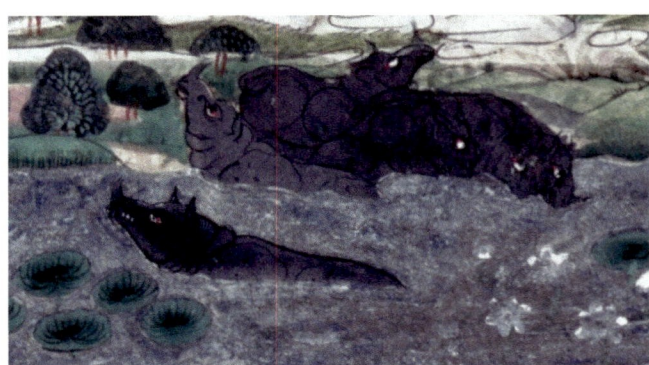

FIGURE 15.60B　Detail from Figure 15.60a showing 6 Rhinos at and in a river

FIGURE 15.62　Two Rhinos facing left. Detail of a mural within the so-called "reception room" in the Palace of Samod

A detail of another mural within the so-called "reception room"[73] reveals two rhinos standing side by side with large eyes looking upwards (figure 15.62). They are part of a tiger shoot and the capture of a blackbuck which happens, however, far away from them.[74]

Devgarh

Palace of Devgarh

The murals in the Moti Mahal within the palace of Devgarh (district Rajsamand, Rajasthan) include a white rhinoceros painted at the lower part of the wall (figure 15.63). This rhinoceros follows a tiger with an elephant's head. The murals on this part of the walls are ascribed to the artist Baijnath and are dated around 1845 CE.[75]

Udaipur

Nahar Odi, "a small shooting box just south of Lake Pichola that Maharana Fateh Singh restored in 1888 or 1889"[76] was also embellished with wall paintings.[77] Figure 15.64 shows two rhinoceroses facing each other in a landscape.

15.6　The Rhinoceros as Shown in a *Ragamala* (Rāgamālā)

A *ragamala*, literally a garland of *ragas*, is a sequence of mostly 36, but also 42 up to 251 illustrations. While the shorter sequences of 36 paintings are also known from

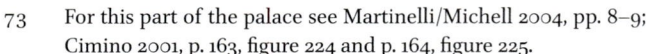

73　For this part of the palace see Martinelli/Michell 2004, pp. 8–9; Cimino 2001, p. 163, figure 224 and p. 164, figure 225.

74　For the full painting see Mehra 1993, pp. 36–37.

75　Beach/Nahar Singh 2005, p. 94, figure 115.

76　Purohit 1938, p. 26; Hughes 2009, p. 76f.; Hughes 2013: 88.

77　Hughes 2009, p. 77: "Most of the paintings serve to detail local sporting techniques and game, rather than recording specific happenings. Some seem to depict actual events."

FIGURE 15.63   Mural showing a Rhinoceros facing left in the Moti Mahal, a part of the Palace of Devgarh, ca. 1845 CE

FIGURE 15.64   Two Rhinoceroses facing each other in a landscape. Mural within the shooting box called "Nahar Odi" near Udaipur
PHOTOGRAPH COURTESY JULIE ELAINE HUGHES THROUGH KEES ROOKMAAKER

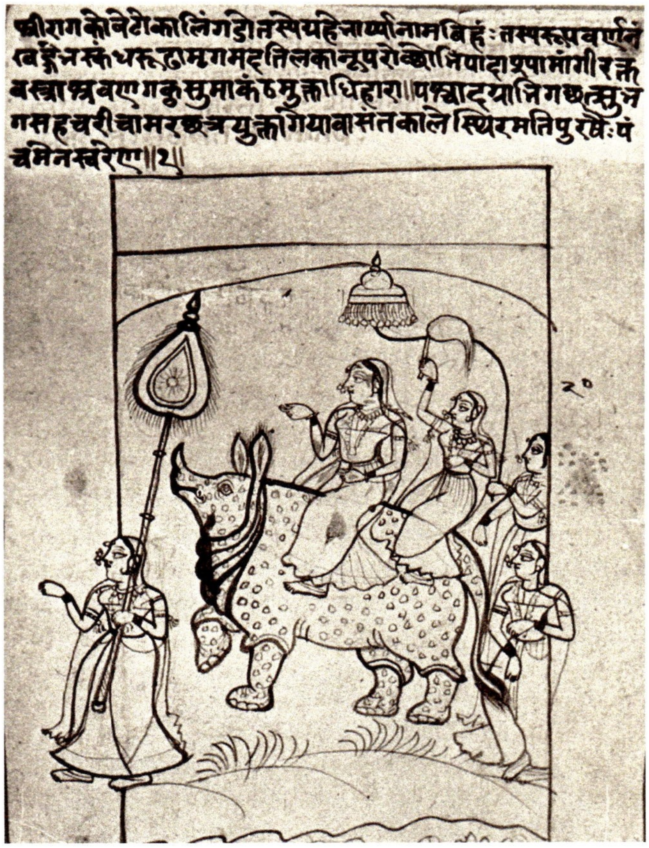

FIGURE 15.65   A *khākhā* (a sketch, a preparatory drawing) of a wife of Kāliṅga, the son of Rāga Śrī. From a Rāgamālā. Black ink on paper. Rao Madho Singh Museum Trust, City Palace, Kota. Size not recorded
AFTER BRIJRAJ SINGH, *THE KINGDOM THAT WAS KOTAH*, 1985, FIGURE 7

several wall paintings,[78] the larger series only exist on paper. A *raga* is also a sequence of musical notes. *Raga*, from the Sanskrit root rañj, meaning to dye, to colour, to affect somebody emotionally. A *raga* is hence a means to affect somebody's emotion which can be done through music, poetry, or painting or a combination of all three. "A definition of *rāga*, the remarkable and prominent feature of Indian music, cannot be offered in one or two sentences."[79] The painted *ragamala* tradition is restricted to South Asia.

The present illustration (figure 15.65),[80] a preparatory drawing for a painting, visualises according to the five-lined inscription on top, *bihaṃ* (?), wife of Kalinga,

the son of *raga* Shri. This means that this drawing belongs to a *ragamala* of which 251 folios were recorded. The fully painted *ragamala* is the work of the artist Ḍālu. It was completed during the reign of Maharao Guman Singh of Kotah (r. 1764–1771) in "nandgrām", Kota, on Tuesday, the second day of the bright half of [the month] Jyeshta, V.S. 1825 (May–June 1768 CE).[81] The description of this wife ("bhāryyā") informs that she sits on the shoulder of a rhinoceros ("khaḍge"), with a musk-mark on her forehead, wears anklets, a red garment, blossom earrings [and] an exquisite pearl necklace around her neck. Behind her walk attractive [female] companions with a "cāmara" (a kind of flywhisk made of the hair of the tail of a white yak) and an umbrella ("cchattra"). It should be sung during spring ("geyā vāsaṅtkāle") with some remarks on the musical notes of this *raga*.

78   Cf. Bautze 1987a.
79   Kaufmann 1968, p.v, for the detailed description see *ibidem*, pp. 1–25. See also Bautze 1987a, pp. 27–29.
80   Published first: Brijraj Singh 1985, figure 7.

81   Ebeling 1973, pp. 217–220.

15.7 The City of Jaipur and the Rhinoceros

It is probably true that "In almost all the museums of India having arms collection the shields made of rhinoceros hide can be seen."[82] The problem only is that usually they are often not recognised as being made of rhinoceros hide, because "The shield of rhinoceros hide was occasionally left with a natural surface on the outside which was given a coat of black or brownish lacquer or both the inside and outside were smoothed completely and given a highly polished brown or black lacquered surface."[83] Pant, together with other authors, presents "a typical example of shields used by the Rajput princes. It is made of rhinoceros hide [...]."[84] When Edward VII as Prince of Wales visited India in 1875–1876, many Indian rulers presented him a shield of rhinoceros' leather. These shields were catalogued, described, and reproduced in two superbly illustrated volumes.[85] All these shields have one thing in

FIGURE 15.66B Another view of the Shield made from the skin of a Rhinoceros in the Armoury of the Sawai Man Singh II Museum, City Palace Jaipur showing the thickness of the skin

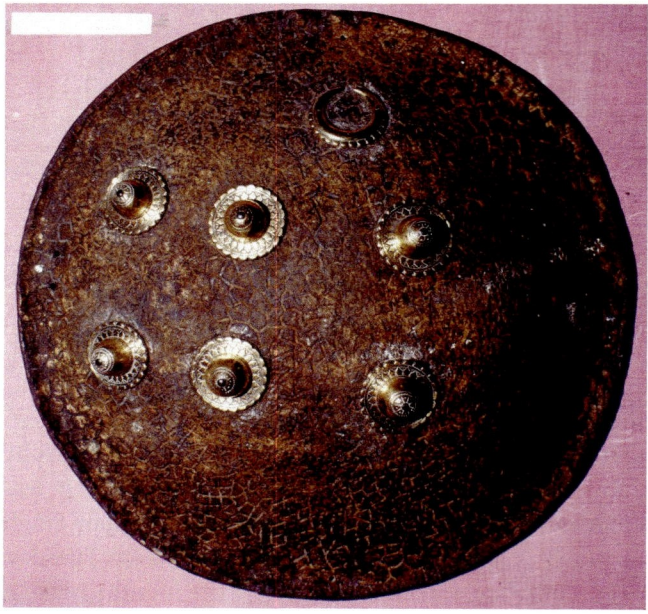

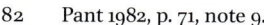

FIGURE 15.66A A Shield made from the skin of a Rhinoceros in the Armoury of the Sawai Man Singh II Museum, City Palace Jaipur, dated to the 18th century

---

82 Pant 1982, p. 71, note 9.
83 Pant 1982, p. 125, note 25.
84 Pant 1982, Appendix V, Chemical Analysis of a Shield, pp. 171–178; p. 171. The shield is illustrated on the front cover of the dust-wrapper or on p. 124, plate LI. Plate XLIII shows G.N. Pant holding this shield. Colour plate II also illustrates a painted rhinoceros shield, which Pant dates to the late 17th century CE. For more artistically painted shields made of rhino hide see Pant 1982, plates XXVIII, XXIX, XXXVI, LIII.
85 Clarke 1898, catalogue numbers 34, 206, 245 and 286. Clarke 1910, catalogue numbers 35, 38, 148, 252, 287, 460 and 551.

common: they are artistically painted so that the actual surface of a rhino's skin is covered by a coat of dark lacquer on which the paint was applied. Such shields were in use until the late 19th century, as can be glimpsed from the description of a young horseman in Udaipur by Edwin Lord Weeks (1849–1903) in 1893: "Other accessories were the sword-belt, crossing his breast and encircling his waist, of dark green velvet, richly worked with unalloyed gold, and thickly studded with emeralds, rubies, and brilliants; a transparent yellow shield of rhinoceros hide, with knobs of black and gold enamel; a sash of stiff gold lace,

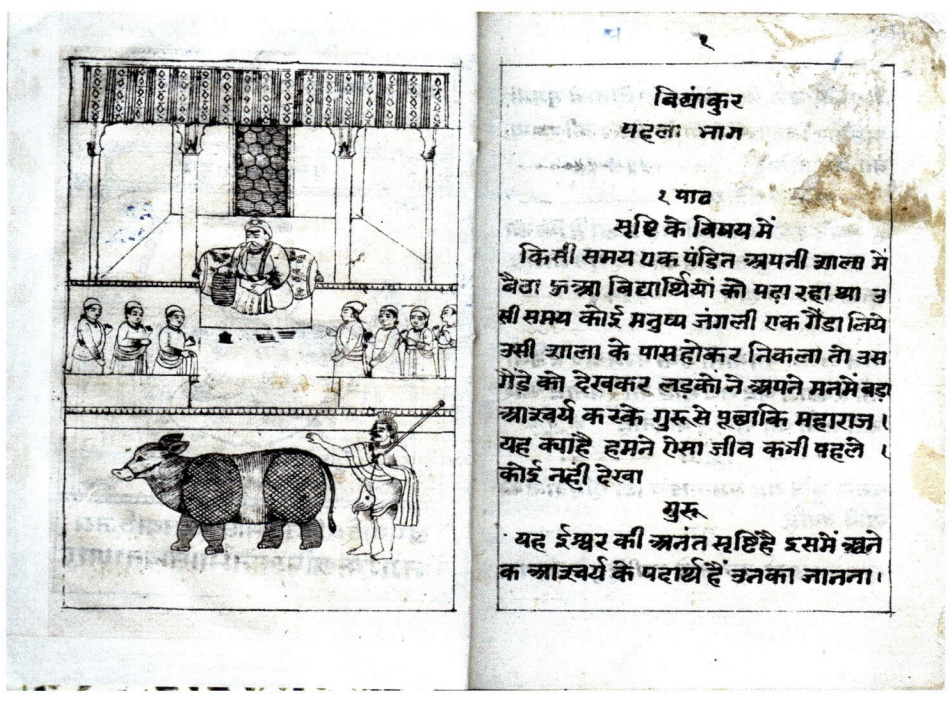

FIGURE 15.67
Back of the front page and page 1
from *pustak vidyāṅkur*, savāī jaynagar,
V.S. 1918. Page size: 21,2 × 14,7 cm
PRIVATE COLLECTION

with a crimson thread running through the gold; bracelets of the dainty workmanship known as Jeypore enamel thickly jeweled, which he wore on his wrists and arms; and there were strings of dull, uncut stones about his neck."[86]

One of the few exceptions is a shield made of rhino skin in the collection of the armoury (*sileh khana*) of the Sawai Man Singh II Museum, City Palace Jaipur (figs. 15.66a, 15.66b). The caption informs: "Shield made of Rhino Hide, Jaipur, 18th century." The wrinkled surface of the skin is clearly visible in both images, as is the thickness of the hide in Figure 15.66b.

In November 1845, the Russian artist and traveller Aleksei Dmitrievich Soltykov (1806–1859) remarked on a rhinoceros at Jaipur: "After that we passed over a verandah overlooking a rather large courtyard, and where velvet seats were placed; a small, very Indian carriage, harnessed to four gazelles, arrived, and was given a ride. Then came a rhinoceros at large, which two individuals were pursuing with sticks to direct it, as if it had been a buffalo."[87]

On the instructions of Maharaja Sawai Ram Singh of Jaipur (r. 1835–1880), a book titled "Germ of Knowledge (pustak vidyāṅkur)" was printed in Jaipur in 1861 at the printing press of Pandit Bansidhar.[88] This book has 170 numbered pages in which a teacher ("guru") explains to his disciple ("śiṣya") some principles of physics and astronomy. On page one the book commences with a first part, lesson one, about natural science. The text starts: Once upon a time, when a learned man (paṇḍit) sat in a school to teach, a wild man ("manuṣya jaṅglī") arrived with a rhinoceros ("gaiṇḍa").[89] This moment is captured by a full-page illustration facing page 1: a man holds a rhino by a rope, the animal is rather short legged and shows almost no horn (figure 15.67).

In 1867 Maharaja Sawai Ram Singh established the "Jaipur School of [Industrial] Arts", also known as "School of Arts and Crafts":[90] "For the purpose of teaching such new forms of Arts and Industry as the advancing civilization of the State, the intelligence and general prosperity of the people, have rendered imperative, and for the improvement of such branches of industry as already exist."[91] The idea of founding such an institution was reportedly the result of the Maharaja's earlier conversation with Sir Charles Edward Trevelyan (1807–1886) in Calcutta, the

---

86 Weeks 1896, p. 286.
87 Soltykoff 1858, p. 435: "Après cela nous passâmes sur un verandah donnant sur une cour assez vaste, et où des siéges de velours étaient placés; une petite voiture, très-indienne, attelée de quatre gazelles, arriva, et on lui fit faire un tour. Vint ensuite un rhinocéros en liberté, que deux individus poursuivaient avec des bâtons pour le diriger, comme si c'eût été un buffle."

88 "|| śrī savāī rām siṃhjī kī ājñā seṅ || / pustak vidyāṅkur / chapvāī paṇḍit baṃsīdhar ne savāī jay / nagar kachāpe-khāne meṅ || sambat 1918."
89 For another Rajasthani rhinoceros thus titled see Freedman 2003, p. 198f, no. 102. The inscription above the animal does not read "ghanda", as mentioned in the description, but clearly "gaiṇḍa."
90 Harnath Singh 1970, p. 70.
91 Hendley 1876, p. 125.

FIGURE 15.68A Brass Bronze of a left-facing Rhinoceros of the School of Art, Jaipur. Height, including base: 3,8 cm. Width, over all: 6,4 cm
WITH PERMISSION © P. & G. COLLECTION, KARLSRUHE-BERLIN

FIGURE 15.68B Brass Bronze of the Rhinoceros shown in Figure 15.68a, facing right

FIGURE 15.68C Underside of the Base of the Brass Bronze of the Rhinoceros shown in Figures 15.68a, 15.68b

FIGURE 15.69 Brass Bronze of a Rhinoceros facing right from the Jaipur School of Art. Height, including base: 4,8 cm. Width, over all: 7,5 cm
WITH PERMISSION © P. & G. COLLECTION, KARLSRUHE-BERLIN

first principal of the "Jeypore School of Arts"[92] was the Scot Colin Strachan Valentine (1834–1905) while the most influential principal of this institution from 1869–1873 was William Frederick de Fabeck (1834–1906). The first Indian principal, from 1875 to 1907, was Babu Upendra Nath Sen.[93] Products of this School of Art, in particular those made of brass, were soon exhibited in the "Albert Hall", i.e., the Jeypore Museum,[94] founded by Maharaja Sawai Madho Singh of Jaipur (r. 1880–1922),[95] the foundation-stone was laid earlier by the Prince of Wales on 6 February 1876.[96]

Page 23 of the *Illustrated Catalogue, School of Art, Jaipur, Rajputana*, lists "Brass Animals." The second column contains two rhinoceroses. The smaller one measures one and a half inches in height and costs 12 Annas (figs. 15.68a, 15.68b, 15.68c). The bottom of the stand clearly reads: "School of Art / Jaipur" (figure 15.68c).

The slightly larger model measures two inches and costs one Rupee and four Annas (figure 15.69). While the rhino of figures 15.68a to 15.68c shows traces of dark red paint, the rhinoceros model of Figure 15.69 shows no paint at all. Both models show a furrow-like line from the corner of the mouth to the base of the head.

The *Illustrated Catalogue, School of Art, Jaipur, Rajputana* does not list woodcarvings. At the end of the catalogue, it is pointed out that: "Besides the articles in the Catalogue, finest specimens of [...] Lacquered wood work &c., are kept in the School Show Rooms for sale." One of these specimens must have been a painted wooden model of a rhino (figure 15.70).[97] The large eyes and what appears to be an elongated mouth are striking. The short horn is painted white, the tail is black.

92   Also: "Jeypoor School of Arts", cf. Russel 1877, p. 460f.
93   Also spelled: Upendro Nath Sen or Opedro Nath Sen, cf. Showers 1909, p. 32; Benn 1916, p. 32, Harnath Singh 1970, p. 70.
94   Hendley 1895, plates X and XI.
95   Hendley 1895, p. 1.
96   Fayrer 1879, p. 109.

97   This example, in a private collection, measures 13 centimetres in height including the base.

FIGURE 15.70 Wooden model of a painted Rhinoceros presumably from the Jaipur School of Art. Height, including base: 13,0 cm. Width, over all: 19,8 cm
COLLECTION: © J.K. BAUTZE

FIGURE 15.72 Painting on paper of a rear view of a captive urinating Bull Rhinoceros facing left, presumably the same Animal as shown in Figure 15.71 and most probably by the same Artist. Size not recorded. Formerly in the collection of Kumar Sangram Singh of Nawalgarh (1926–1994) in Jaipur

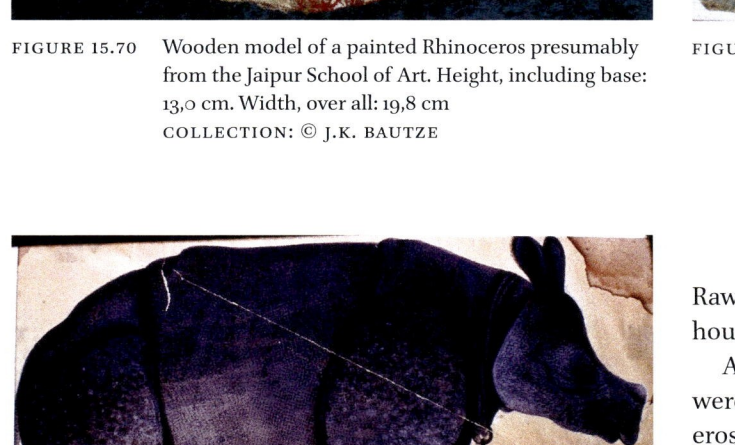

FIGURE 15.71 Painting on paper of a captive Rhinoceros facing right by a Rajasthani artist who had some training by a "company artist." Size not recorded. Formerly in the collection of Kumar Sangram Singh of Nawalgarh (1926–1994) in Jaipur

Rawal Madan Singh (r. 1928–1992), but often also at his house in Sansar Chand Marg in Jaipur.

Among the numerous Indian pictures in his collection were two that probably show one and the same rhinoceros in captivity (figs. 15.71 and 15.72). Technically, both paintings are by a Rajasthani artist, as can be shown, for example, by the correction of the outline, the purple colouring, and the stippled surface. Of particular note is the foreshortening in perspective in Figure 15.72. The artist probably painted one of the two rhinos kept in the "Zoological Section" of the "Ram Newas Public Gardens", begun in 1868 by order of Sawai Ram Singh of Jaipur.[99] On 4 March 1893 these rhinos appeared to have been painted a shiny black colour.[100] Also part of Sangram Singh's collection was a strange drawing with two rhinos facing each other (figure 6.4).[101]

Kumar Sangram Singh of Nawalgarh (Jhunjhunu district, Rajasthan, 1926–1994) was one of the greatest collectors of Indian paintings.[98] I was lucky to be his guest not only in Nawalgarh, where he introduced me to his father,

---

98 Cf. Sangram Singh 1965.

99 Showers 1909, p. 76f; Benn 1916, p. 83f.
100 Franz Ferdinand 1895, p. 335.
101 Bautze 1985, p. 423, figure 5. For another Rajasthani drawing see Gorakshkar 1979, p. 47, no. 379: "On an off white paper a beautiful and very realistic drawing of a rhinoceros has been drawn."

CHAPTER 16

Historical Records of the Rhinoceros in Madhya Pradesh

First Record: early period – Last Record: 1695
Species: *Rhinoceros unicornis*
Rhino Region 07

16.1   Rhinos in Madhya Pradesh

There is only one allusion to a rhino sighting in Madhya Pradesh during the historical period, in a painting made in Raghogarh close to the northern Rajasthan border at the end of the 17th century. This fits well with the presence of rhinos in historical times in the Ganges valley. They occurred at a small distance to the north, also to the east in the valley of the Sone River in south-west Bihar and to the west hunted with some regularity in parts of Rajasthan. Their presence in more ancient times may be inferred from fossils found in the Narmada valley (chapter 8), and particularly from the relatively large numbers of rhino images in the rock art shelters found across the province (details in chapter 9). The rhinoceros appears convincingly on the Sanchi Stupa (16.2). This depiction, and that on a coin, combined with a small number of references to the rhino in ancient texts all indicate the tentative possibility of the wider presence of these animals in former times.

∴

Dataset 16.12: Chronological List of Records of Rhinoceros in Madhya Pradesh

General sequence: Date – Locality (as in original source) – Event – Source – § number – Map reference
The map numbers are explained in dataset 16.13 and shown on map 16.7.

Rhino Region 07: Madhya Pradesh

Pleistocene – Narmada Valley – Various skeletal remains attributed to *Rhinoceros* sp., *R. deccanensis*, *R. namadicus* or *R. unicornis* – see chapter 8 (fossils)

Mesolithic (10,000–2,000 BCE) – Region in northern part of state, districts Jabalpur, Vidisha, Raisen, Hoshangabad, Sehore – Rhino depictions in rock art in various shelters – see chapter 9 (Rock Art)

Mesolithic (10,000–2,000 BCE) – Region in north-western part of state, districts Rajgarh, Mandsaur – Rhino depictions in rock art in various shelters – see chapter 9 (Rock Art)

Chalcolithic – Dangwada, Ujjain Dt. – Terracotta image of rhino – Manuel 2008 – see chapter 10 (Harappan)

undated – South of Ganges valley – Rhino may not have been found – Williamson 1807: 167; Fisher et al. 1969: 111

undated – Vindhya – Revakhanda of the *Skanda Purana* (V.iii 2, 16–24) states presence of rhino in Vindhya range – Awasthi 1978: 30 – map no. M 824

6th cent. BCE – Malwa region – Avanti coin. No. 28. Two-horned rhino stands facing right. The example is much worn – V. Smith 1906a: 154 (small sketch). The two 'horns' could be two ears – map no. M 821

321 BCE (Maurya period) – Central India – *Arthashastra*, the ancient Sanskrit treatise attributed to Kautilya states that the duties of the Forest Superintendent included collection of the skins of dead animals, including khadga (rhino) – Kangle 1963: 149, section 2.17.13 (English translation of the Arthashastra); Sagreiya 1969: 717

2nd cent. BCE (Maurya period) – Central India – Ringstone, with frieze of 7 animals including a rhino facing right. Patna Museum: PM/Arch.10747 – S. Gupta 1980: 58, pl. 25a

2nd cent. BCE – Bharhut Stupa – Animal identified as rhino in half-medaillon at the top of a north-east pillar, with bulky legs and long tail. [This is unlikely to actually represent a rhino] – Cunningham 1879: 43, pl. 36; Cockburn 1883: 57 – map no. Z 915

2nd cent. BCE – Sanchi – Relief on Stupa no. 2. Rhino identifiable by presence of horn – Marshall & Foucher 1902, vol. 1: 96, 173; Bautze 1985: 413, 2013; Manuel 2006, 2007: 235, 2008: 36; Karlsson 1999: 90, Bose 2018b: 53 (figure) – 16.2, fig. 16.1 – map no. M 823

1690–1695 – Raghogarh – Miniature depicting Raghogarh chiefs hunting rhinos – Mittal 2007, no. 24; Seyller & Mittal 2019, no. 28 – 16.3, fig. 16.2 – map no. A 236

∵

© L.C. (KEES) ROOKMAAKER, 2024 | DOI:10.1163/9789004691544_017
This is an open access chapter distributed under the terms of the CC BY-NC-ND 4.0 license.

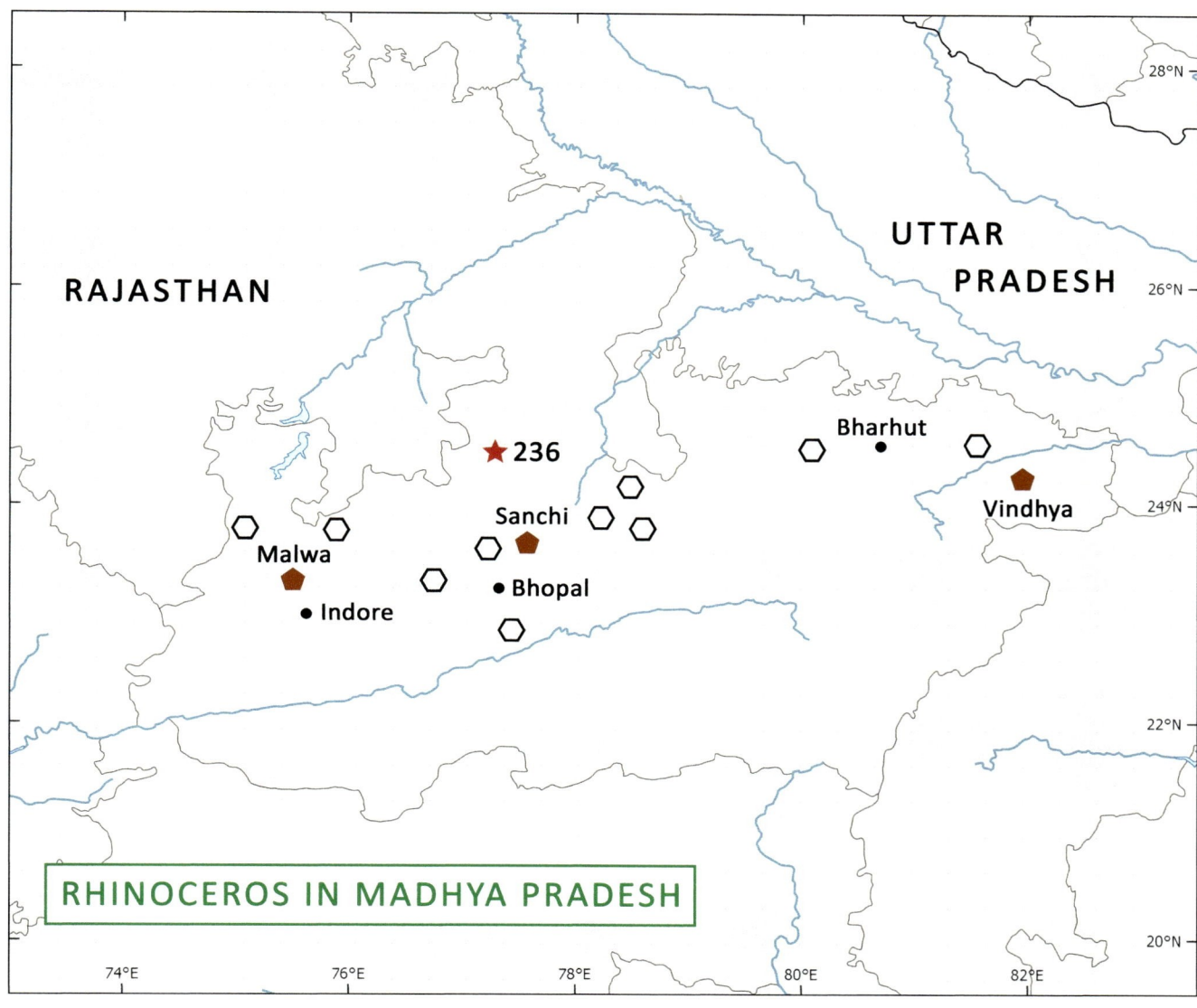

MAP 16.7   Records of rhinoceros in Madhya Pradesh. The numbers and places are explained in Dataset 16.13. The localities for Rock Art occurrences are explained in chapter 9

★ Presence of *R. unicornis*
⬠ Artefacts
⬡ Rock Art

• • •

Dataset 16.13: Localities of Records of Rhinoceros in Madhya Pradesh

The numbers and places are shown on map 16.7.

Rhino Region 07: Madhya Pradesh

M 821   Malwa region – 23.00N; 75.50E – 6th cent. BCE

A 236   Raghogarh – 24.35N; 77.10E – 1690

Z 915   Bharhut Stupa – 24.45N; 80.88E – unlikely record

M 823   Sanchi – 23.45N; 77.74E – stupa

M 824   Vindhya Range – 24.60N; 82.00E – early undated record

∴

16.2   A Rhino Depicted at the Stupa of Sanchi

There is one stone relief of a rhinoceros in the complex of buildings of Stupa no. 2 at Sanchi, Raisen District, dating from the 2nd century BCE. The animal is found in a half medaillon on the northernmost vedika pillar, standing on its own facing left in front of a bud of the lotus flower (fig. 16.1). Without the armour-plated skin, it is really only the clear indication of a horn which allows for the identification of a rhinoceros.

16.3   Rhinoceros Hunt in Raghogarh around 1695

There is one isolated miniature with a rhino hunt from the State of Raghogarh, now in the north-western part of Madhya Pradesh. It dates from the time of Raja Lal Singh, who ruled 1673–1697. The anonymous painter shows how one rhino is attacked by a dozen hunters on horseback, each armed with a shield and using spears, swords and arrows in their pursuit (fig. 16.2). The animal has wounded at least four warriors. In the background a further five rhinos are hiding in a forest trying to escape from an onslaught by two hunters with guns (Seyller & Mittal 2019). Although there is no inscription on either side, the painting is dated 1690–1695 on artistic grounds.

FIGURE 16.1
Relief of a rhinoceros with padma bud on Stupa no. 2 at Sanchi, Raisen District, Madhya Pradesh. Vedika, southeast quadrant, east entrance, left pillar, east face, detail, lower half medallion, 2nd century BCE
PHOTO: JOACHIM BAUTZE

FIGURE 16.2
Raghogarh chiefs hunt rhinoceroses, c.1690–1695. Painting 30.3 × 39.8 cm with red border
WITH PERMISSION © JAGDISH AND KAMLA MITTAL MUSEUM OF INDIAN ART, HYDERABAD

CHAPTER 17

Historical Records of the Rhinoceros from Himachal Pradesh to Uttarakhand

First Record: 1387 – Last Record: 1878
Species: *Rhinoceros unicornis*
Rhino Region 8 (North-West India), Rhino Region 9 (Uttarakhand)

17.1 The New State of Uttarakhand

The record of the rhinoceros at Kotdwara in 1789 is one of the more prominent encounters from this entire region, because the animal was seen, described and sketched. Hence its significance cannot be underestimated. The foothills of the state of Uttarakhand (constituted in 2000) continue the rhino habitat of Uttar Pradesh and Nepal in a westward direction. Rhinos were most regularly reported in Pilibhit District (formerly Rohilcund) along the Uttarakhand-Uttar Pradesh border in the 1860s, with the last specimen shot not later than 1873. The sighting in Kotdwara was slightly further west. Although none of the written accounts refer to actual encounters, the rhino might have occurred in the Dehradun valley. There is one reference from the 14th century in the extreme south-east of Himachal Pradesh in the Sirmaur Mountains.

• • •

Dataset 17.14: Chronological List of Records of Rhinoceros in Himachal Pradesh and Uttarakhand

General sequence: Date – Locality (as in original source) – Event – Source – § number, figures – Map number.
The persons interacting with a rhinoceros in various events are combined in Table 17.21. The map numbers are explained in dataset 17.15 and shown on map 17.8.

Rhino Region 08: Punjab, Haryana, Himachal Pradesh

1387 – Sirmor (Sirmaur) mountains – Prince Muhammad Khan hunted rhino – Elliot 1872, vol. 4: 16; Yule & Burnell 1886: 762; Ettinghausen 1950: 45 note 29 – 17.2 – map A 221

Rhino Region 09: Uttarakhand

Mesolithic – Hardwar (Haridwar) – Fossil bones of *Rhinoceros* sp. – See Chapter 8

13th cent. – Kumaon [Pilibhit region] – Rudradeva of Kurmacala gives method to hunt rhino from horseback – Shastri 1910: 11; Karttunen 1989: 171 – 17.5 – map A 224

Undated – Gharwal [area north of Haridwar] – Pandav Lila is a ritual play, which features the rhino – Briggs 1938: 8; Sax 1997, 2002; Channa 2005: 77 – 17.4 – map A 224

1755 – No locality – Engraving of rhino in rocky landscape used by Charles-Antoine Jombert (1712–1784) in a book to teach drawing – Jombert 1755, pl. 94 – fig. 17.1

1768 – Koumahouns (Kumaon) mountains – Jean-Baptiste-Joseph Gentil (1726–1799) mentions presence of rhino – Gentil 1822: 266 – 18.8 – map A 224

1789-04 – Coaduwar Gaut, Rohilcund (Kotdwara) – Rhino seen and sketched by the party of Thomas Daniell (1749–1840) and his nephew William Daniell (1769–1837), The text about the encounter in the *Oriental Annual* for 1834 was written by John Hobart Caunter (1792–1851) – Daniell & Caunter 1834a: 4–6; Archer 1980, nos. 44, 45; Rookmaaker 1999a, 1999b – 17.4, figs. 17.2 to 17.6 – map A 222

1789-04 – Coaduwar (Kotdwara) – First pencil drawing, first copy. No caption or signature. Rhino facing left, with depiction very close to the edge. Exhibited at the Helene C. Seiferheld Gallery in New York in December 1962 (current whereabouts unknown) – Seiferheld 1962; Rookmaaker 1999a, fig. 1 – 17.4, fig. 17.2 – map A 222

1789-04 – Coaduwar (Kotdwara) – First pencil drawing, second copy. No caption or signature. Rhino facing left. This example is since 1835 in the Hardwicke Collection in the archives of the Natural History Museum, London – Magee 2013: 83 – 17.4, fig. 17.3 – map A 222

1789-04 – Coaduwar (Kotdwara) – Second pencil drawing, undated, signed "W. Daniell." Rhino facing left, in a tropical landscape. Preserved in Paul Mellon Collection, Yale Center for British Art, New Haven, Connecticut, USA, no. B2001.2.775 – Rookmaaker 1999a, fig. 2 – 17.4, fig. 17.4 – map A 222

1789-04 – Coaduwar (Kotdwara) – Oil painting by Thomas Daniell, with title "A forest scene in the northern part of Hindoostan, with a rhinoceros." First exhibited at the Royal Academy of Arts, London, in May 1799. Preserved in Paul Mellon Collection, Yale Center for British Art, New Haven, Connecticut, USA, no. B2006.14.3 – Royal Academy 1799: 13, no. 299; Shaffer 2011: 4; Das 2018b: 89 – 17.4, fig. 17.5 – map A 222

1789-04 – Coaduwar (Kotdwara) – Engraved plate "The Rhinoceros." First published in the *Oriental Annual*, vol. 2, dated 1835 on the title-page but available in October 1834. The inscription identifies the artist William Daniell, engraver James Redaway (1697–1858), printer James Yates, and publisher Bull & Co., 26 Holles Street, Cavendish Square, London with proprietors Edward Bull (1798–1843) and Edward Churton (1812–1888). The date 1 October 1834 is printed on the plate – Daniell & Caunter 1834a, plate between pp. 4 and 5; Rookmaaker 1999a, fig. 3; 1999b, fig. 1 – fig. 17.6

1789-04 – Coaduwar (Kotdwara) – The text on the Kotdwara Rhinoceros in Daniell & Caunter 1834a: 4–6 is again found in French translation in Daniell & Caunter 1835b: 4–8, repeated in Urbain & Daniell 1840: 4. Extract in German in Daniell & Caunter 1834b, in

© L.C. (KEES) ROOKMAAKER, 2024 | DOI:10.1163/9789004691544_018
This is an open access chapter distributed under the terms of the CC BY-NC-ND 4.0 license.

English in Burton 1839: 313; Downing 1844: 385; Stead 1907: 41 – Daniell & Caunter 1834a: 4–6; Henderson 1834

1789-04 – Coaduwar (Kotdwara) – Engraving of 1834 copied (1) as aquatint resembling the original. Rhino is coloured brown, with only "Rhinoceros" as title – source unidentified

1789-04 – Coaduwar (Kotdwara) – Engraving of 1834 copied (2) as engraving, uncoloured. No title, but inscribed in English: (lower left) Drawn by W. Daniell R.A. (lower right) Engraved by J. Redaway. In: *Tableaux Pittoresques de l'Inde*, the French edition of the *Oriental Annual*, translated by P.J. Auguste Urbain, dated 1835 – Daniell & Caunter 1835b, plate following p. 4

1789-04 – Coaduwar (Kotdwara) – Engraving of 1834 copied (3) as engraving, uncoloured. Identical to the French edition in Daniell & Caunter 1835b. In: *L'Inde pittoresque: Calcutta*, by P.J. Auguste Urbain, 1840 – Urbain & Daniell 1840, plate following p. 4

1789-04 – Coaduwar (Kotdwara) – Engraving of 1834 copied (4) as steel engraving, with title "The rhinoceros in its native wilds." In: *Burton's Gentleman's Magazine and American Monthly Review*, Philadelphia, June 1839 – Burton 1839, plate facing p. 313

1789-04 – Coaduwar (Kotdwara) – Engraving of 1834 copied (5) as engraving, with title "The rhinoceros in its native wilds." In: *The Rover, a Weekly Magazine of Tales, Poetry and Engravings*, New York, vol. 2, no. 25, 1844 – Downing 1844, plate facing p. 385

1789-04 – Coaduwar (Kotdwara) – Engraving of 1834 copied (6) as engraving in a German translation, with title "Das Nashorn." In: Jacob Eberhard Gailer, *Wunderbuch für die reifere Jugend*. Stuttgart – Gailer 1842, plate 31. Also in subsequent editions of this work.

1789-04 – Coaduwar (Kotdwara) – Engraving of 1834 copied (7) as engraving, with title "147. Das Nashorn und seine Jagd." In: *Archiv für Natur, Kunst, Wissenschaft und Leben*, Braunschweig, vol. 8, 1850. This is a composite figure, with Daniell's rhino on the left in reverse, and other additions by the anonymous engraver. The number 147 refers to an anonymous article, the text of which does not mention

the animal seen by Daniell – Anon. 1850b, plate following p. 92 – fig. 17.7

1789-04 – Coaduwar (Kotdwara) – Engraving of 1834 copied (8) as engraving similar to Anon. 1850b. It is a similar scene in a rectangular frame with the same German title (without the number) published in an unverified source.

1835 – Deyra Dhoon (Dehradun) – Rhino and buffalo present in "the Deyra Dhoon, the Terraie" – Anon. 1834c: 306 – 17.3 – map Z 901

1843 – Bhogpore (20 km south of Haridwar) – Rhino in forest to the east – Davidson 1843, vol. 1: 71 – 17.3 – map Z 902

1850 – Rohilcund (Rohilkhand) – Rhino as far west as Rohilcund up to 1850s – Blanford 1891: 473; Trouessart 1898, vol. 2: 754, Lydekker 1900: 23 – map A 224

1860 – Dooeewalla – Rhino suspected, but proved to be elephant – Dunlop 1860: 37 – 17.3 – map Z 903

1860s – District of Philibeet (Pilibhit) – Mark Thornhill (1822–1900) suggested presence of rhino – Thornhill 1899: 305 – map A 224

1866 – Rohilcund and the Dhoon (Rohilkhand and Dehradun) – Rhino found – R.H.D. 1866: 485 – 17.3 – map Z 901

1867 – Rohilcund (Rohilkhand) – Rhino rare, as far west as Rohilcund "heard from sportsmen" – Jerdon 1867: 233 – 17.4 – map A 224

1870s – Near the boundary of Pilibhit District – Rhino killed by Robert Andrew John Drummond (1820–1887), commissioner of Rohilcund – Hewett 1938: 167 – 17.5 – map A 224

1871 – Rohilcund *terai* – Rhino known from deep jungle. Author was stationed at Shahjehanpore (Shahjahanpur); did not encounter rhino himself – Anon. 1871a: 374 – 17.5 – map A 224

1878 – Near Pilibhit – Last rhino in Uttar Pradesh [now Uttarakhand] in 1878 – Date uncertain. Laurie 1978: 9; Martin & Martin 1982: 29; Choudhury 1985a; Menon 1995: 56 – 17.5 – map A 224

1962 – Corbett National Park – Plan to re-introduce rhino – Gee 1962: 473 – map D 223

2007 – Corbett National Park – Plan to re-introduce rhino – Johnsingh et al. 2007 – map D 223

• • •

Dataset 17.15: Localities of Records of Rhinoceros in North-West India

The numbers and places are shown on map 17.8.

Rhino Region 08: Punjab, Haryana, Himachal Pradesh

M 811   Nurmahal, Punjab – 31.09N; 75.59E – 1620 – See Chapter 11
M 812   Rakhigarh, Haryana – 29.30N; 76.10E – See Chapter 10
A 221   Sirmor (Sirmaur), Himachal Pradesh – 31.00N; 77.35E – 1387

Rhino Region 09: Uttarakhand

*Z 901   Deyra Dhoon (Dehradun) – 30.34N; 77.90E – 1835, 1866 (unconfirmed reports)

*Z 902   Bhogpore (20 km south of Haridwar) – 29.65N; 78.10E – 1843 (unconfirmed report)

*Z 903   Dooeewalla – 30.20N; 78.11E – 1860 (unconfirmed report)

A 222   Coaduwar Gaut (Kotdwara), Rohilcund – 29.76N; 78.53E – 1789

*D 223   Corbett NP – 29.45N; 79.10E – 1962, 2007 – re-introduction

A 224   Rohilcund (Rohilkhand) and the Dhoon – 29.95N; 78.40E–1850, 1866, 1867, 1871

A 224   Kumaon (Pilibhit) – 29.10N; 79.90E – 13th cent., 1768

A 224   Philibeet (Pilibhit) Dt. – 29.10N; 79.90E – 1860s, 1870s, 1878

• •

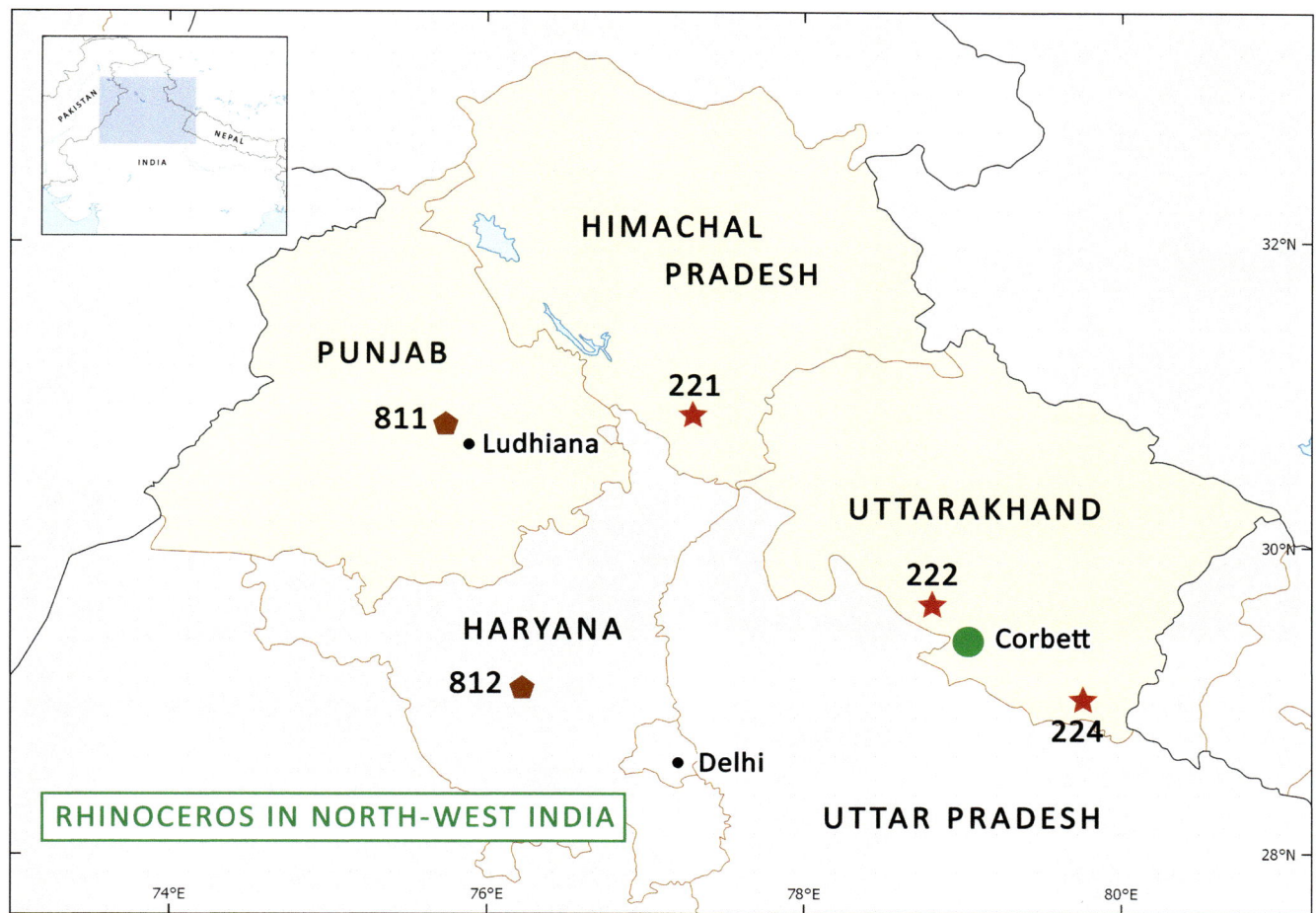

MAP 17.8 Records of rhinoceros in the Indian states of Punjab, Haryana, Himachal Pradesh, and Uttarakhand. The numbers and places are explained in Dataset 17.15. Corbett National Park had plans to re-introduce rhinos

★ Presence of *R. unicornis*

⬠ Artefacts

TABLE 17.21 List of persons who were involved in shooting or capturing rhino in Himachal Pradesh and Uttarakhand

Person	Date	Locality. Type of interaction
Prince Muhammad Khan	1387	Sirmor. Hunted rhino
Jean-Baptiste-Joseph Gentil (1726–1799)	1768	Kumaon. Presence of rhino
Thomas Daniell (1749–1840), William Daniell (1769–1837)	1789	Kotdwara. Observe and paint 1 rhino
Robert Andrew John Drummond (1820–1887)	1870s	Pilibhit. Kills 1

17.2 Prince Muhammed Shah in the Sirmaur Mountains in 1387

An early report about a rhino refers to the Sirmor mountains, now Sirmaur, located in the eastern part of Himachal Pradesh close to the Uttarakhand border and just over 50 km west of Dehradun. In the annals of the Sayyid dynasty of Delhi, the *Tarikh-i-Mubarak-Shahih* written around 1420, it is recorded that Prince Muhammed went to the mountains of Sirmor in the summer of 1387 to hunt rhino and sambar deer. Prince Muhammad Shah was the son of Sultan Feroze Shah Tughluq (1309–1388). There are no further details about this particular hunting trip. Sambar deer *Rusa unicolor* are found in many habitats

FIGURE 17.1
Rhino in rocky landscape. Drawn by François Boucher (1703–1770) and engraved by Quentin Pierre Chedel (1705–1762). From a book by Charles-Antoine Jombert (1712–1784) to teach drawing
JOMBERT, *MÉTHODE POUR APPRENDRE LE DESSEIN*, 1755, PL. 94. COLLECTION KEES ROOKMAAKER

excluding high mountains. Rhinos are not expected in higher elevations either, therefore this old record should refer to the lower lying valleys.

17.3   Valley of Dehradun

There are a few references which appear to indicate the presence of rhinos in the lower valleys near Dehradun. On close examination, these are actually not too promising. The anonymous lady writer in the *Asiatic Journal* of 1834 mentions the Deyra Doon (Dehradun) in such a general sentence including other species that there is too little substance to suspect that she actually believed that rhinos lived there (Anon. 1834a). Lieutenant-Colonel Charles James Collie Davidson (1794–1852) of the Bengal Engineers stated that rhinos lived in the hills towards the east from a small town Bhogpore on the banks of the Ganges, some 20 km south of Haridwar (Davidson 1843). Further north towards Dehradun at Dooeewalla (Doiwala), Robert Henry Wallace Dunlop (1823–1887) thought to have seen a rhino but on investigation it proved to be an elephant (Dunlop 1860). The fact that these references exist at all could mean that there were rumours in the region about the presence of rhinos until the middle of the nineteenth century. In more general terms, this is certainly an area where deer and other wildlife were found even in relatively recent times, as can be seen from the hunting reminiscences of the American missionary doctor John Calvin Taylor (1915–2013), who lived for many years in the area (Taylor 1980).

17.4   Daniell Observing a Rhino at Kotdwara in 1789

The Kotdwara Rhinoceros, observed in 1789, must rank high in the history of the species because it gives an almost unequalled insight into the rhino range supported by a detailed written account as well as sketches made on the spot. The British artist Thomas Daniell (1749–1840) had applied to the East India Company for permission to work in India as an engraver and arrived in Kolkata early in 1786, accompanied by his young nephew William Daniell (1769–1837). They were to stay for eight years visiting a large part of the country (Archer 1980; Cotton 1923; Foster 1951; Sutton 1955). From 29 August 1788 to November 1791, they went on a journey exploring the regions along the Ganges, reaching as far as Srinagar in the Garhwal of Uttarakhand. Their paintings during this time were combined in the first three volumes of *Oriental Scenery* published in London 1795, 1797 and 1801. They saw tracks of a rhino near Rajmahal in October 1788 (29.3). The next encounter was a great stroke of luck, definitely unsolicited because the Daniells rarely tried to depict any wildlife.

The event happened on Monday 20 April 1789 at "Coaduwar Gaut, Rohilcund", now Kotdwara in Uttarakhand, at the edge of the Himalayan foothills (Rookmaaker 1999a, b). The Daniells had reached here on 18 April after traveling overland from Delhi past Ampshahr, Amroha and Najibabad. They were aiming to reach Srinagar in the lower part of the Himalayas, well beyond the influence of the English administration, hence they had to wait for permission from Maharaja Pradyumna

Shah, 54th ruler of the Garhwal dynasty 1785–1804. While camped near Coaduwar, they walked in the area, and came upon the rhino. William Daniell kept a journal at the time (now British Library, MS Eur1268, vol. 1), which however makes no mention of the rhino. Therefore our knowledge of the sighting is entirely based on recollections consisting of an account by William Daniell published as late as 1835, two undated pencil drawings, and a painting of the scene by Thomas Daniell first exhibited in 1799. This lapse of time is somewhat disconcerting because obviously memories fade and in the case of the artwork you can wonder if the artists refreshed their impressions by checking another specimen. There were rhinos in various collections in London 1799–1800, 1810–1814, 1816–1820 and from 1834, all of which they could have seen, although most of those would have been younger than the Kotdwara rhinoceros.

Thomas Daniell made the rhino the subject of a medium-sized oil painting showing the animal from the side in a mountain landscape within a gorge in front of a large boulder (fig. 17.5). This painting was first exhibited at the Royal Academy of Arts in London in May 1799 with the title "A forest scene in the Northern part of Hindoostan, with a rhinoceros" (Royal Academy 1799, no. 299).

William Daniell (in Daniell & Caunter 1834a: 4) referred to a perfect sketch of the rhino in Kotdwara. There are actually two pencil drawings known (details in Dataset 17.14). The first shows only the rhino in lateral view, facing left. It is a powerful adult animal, with a sizeable horn (although the tip could be a later addition) and a slightly dropping left ear. There are two copies known of this first drawing, both without signature or inscription. One was exhibited (and presumably subsequently sold) at the Helene C. Seiferheld Gallery in New York in December 1962, attributed to Thomas Daniell (fig. 17.2). The other copy, with a slightly larger space around the animal, is preserved without attribution in the large collection of Indian drawings collected by Thomas Hardwicke (1755–1835) and bequeathed to the British Museum in 1835 (fig. 17.3). The second pencil drawing shows the rhino in a simple undefined landscape, facing left, with the same characteristics of horn and ear as the first drawing, although the position of the body and head are slightly changed (fig. 17.4). The only known copy is signed "W. Daniell", and is undated, hence could have been executed much later when William Daniell was working on the plates for the *Oriental Annual* of 1835.

In the early 1830s, William Daniell joined forces with John Hobart Caunter (1792–1851) to produce the *Oriental Annual*, a series of illustrated volumes about his Indian travels. Annuals were then a fashionable way of publishing, probably allowing publishers to test the market. The *Oriental Annual* was meant for the larger public, printed in a small size (20 × 13 cm), and each volume containing 22 or 25 engraved plates in monochrome. The expertise of Daniell provided significant quality. All the text was written by Caunter, who was a fashionable preacher in London as well as a prolific author. Son of George Caunter (1758–1812), the Governor of Prince of Wales Island in Malaysia, he trained for a military career and traveled as a Cadet of the 34th Foot through India from 1809 to 1814. Unable to settle, he went on to study theology at the University of Cambridge obtaining his B.D. in 1828. While the text in the *Oriental Annual* would have been based on the recollections of William (and maybe Thomas) Daniell, those were made accessible by Caunter. The collaboration of Daniell and Caunter continued in the three annual volumes for 1834, 1835 and 1836, the topic was changed in 1837 and work stopped apparently when William Daniell died in 1837. It may be noted that volume 2 for 1835 was available for sale as early as September 1834 from the London publishers Edward Bull and Edward Churton of Holles Street, Cavendish Square. It was reviewed and extensively extracted in the *Monthly Review* for November 1834 (Henderson 1834).

Caunter's text in the first three volumes of the *Oriental Annual* followed the travels of Daniell chronologically, but did not provide dates in the text. The encounter with the rhino after reaching Coaduwar Gaut is described as follows:

> Of the latter animal [rhinoceros] we were fortunate enough to obtain a view, which is by no means a usual thing, as it is not gregarious like the elephant, and therefore much more rarely met with. We had turned the angle of a hill that abutted upon a narrow stream, when, on the opposite side of the rivulet, we saw a fine male rhinoceros; it was standing near the edge of the water with its head slightly bent, as if it had been just slaking its thirst in the cooling stream. It stood, apparently with great composure, about two hundred yards above us, in an open vista of the wood. Mr. Daniell, under the protection of a lofty intervening bank, was able to approach sufficiently near to make a perfect sketch of it; after which, upon a gun being fired, it deliberately walked off into the jungle. It did not appear in the least intimidated at the sight of our party, which remained at some distance, nor at all excited by the discharge of the gun (Daniell & Caunter 1834a: 4).

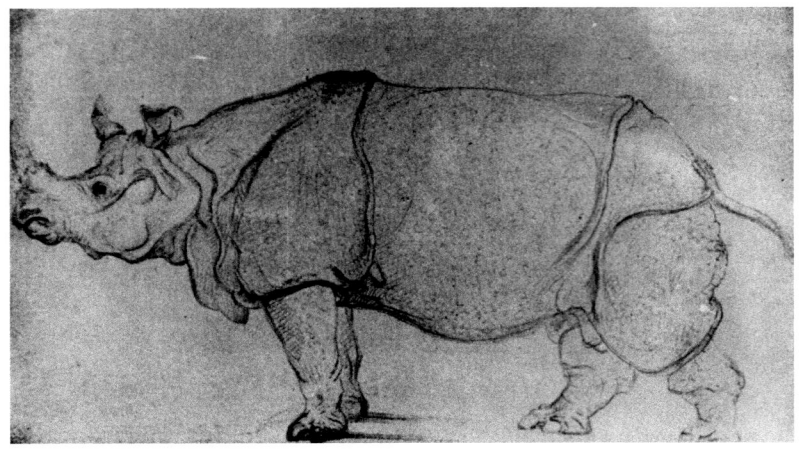

FIGURE 17.2
Pencil drawing of the rhinoceros seen at Kotdwara in
April 1789 attributed to Thomas Daniell. No signature,
size 21 × 35.5 cm
HELENE C. SEIFERHELD GALLERY, *ANIMAL
DRAWINGS*, 1962

FIGURE 17.3
Pencil drawing of the rhinoceros seen at Kotdwara in
April 1789. Unsigned, size 25.4 × 38.3 cm
NHM: ZOOLOGY 88 F HAR NO. 1.71 © THE TRUSTEES
OF THE NATURAL HISTORY MUSEUM, LONDON

FIGURE 17.4
Pencil drawing of the rhinoceros seen at
Kotdwara in April 1789. Signed by William Daniell.
Size 22.4 × 29.7 cm
WITH PERMISSION © YALE CENTER FOR BRITISH
ART, NEW HAVEN, CONN.: PAUL MELLON
COLLECTION, NO. B2001.2.775

The text continues with a general account of the rhino, which Caunter probably added from his general reading, without further reference to the animal seen at Kotdwara. This passage was repeated in subsequent reprints and translations. The date of the encounter is taken from excerpts of Daniell's unpublished Journal with details of his movements on 20 April 1789 while waiting to proceed to Srinagar, further into the mountains (Archer 1980, nos. 44, 45). The sketches and engravings confirm that the animal seen was a specimen of *R. unicornis*.

The only printed example of the Kotdwara rhino is an engraving found in the second volume of the *Oriental Annual* for 1835, bound lengthwise with the legend facing the spine (fig. 17.6). Just titled "The Rhinoceros", the additional printed information below the lower edge identified William Daniell as the draughtsman, James Redaway as the engraver, James Yates as the printer, and Bull & Co. as the publisher, with the date 1 October 1834. The rhino is shown in lateral view, facing right, joined by a second rhino in the background, in a mountainous landscape overlooking a lake with three deer and two storks. Apparently all examples of this first print of 1834 are in monochrome.

This image of the *Oriental Annual* was copied in the French translation of the series as well as in a variety of popular magazines, remarkably all published outside Great Britain (dataset 17.14). In the known examples, the drawing is left unchanged although looking less crisp, with

FIGURE 17.5   A forest scene in the Northern part of Hindoostan, with a rhinoceros, by Thomas Daniell. Oil on canvas, 81.3 × 68.6 cm. First exhibited at the Royal Academy of Arts, London, in 1799
WITH PERMISSION © YALE CENTER FOR BRITISH ART, NEW HAVEN, CONN.: PAUL MELLON COLLECTION, NO. B2006.14.3

FIGURE 17.6
"The Rhinoceros." Inscribed (lower left) Drawn by W. Daniell, R.A.; (lower right) Engraved by J. Redaway; (far right) Printed by J. Yates; (below title) London. Published Octr 1, 1834 by Bull & Co. 26 Holles Street, Cavendish Square
PLATE WITHOUT NUMBER IN THE *ORIENTAL ANNUAL* FOR 1835, PUBLISHED IN OCTOBER 1834, AFTER P. 4

Das Nashorn und seine Jagd.

FIGURE 17.7
"Das Nashorn und seine Jagd" (The rhinoceros and
its hunt). Half-page illustration
*ARCHIV FÜR NATUR, KUNST, WISSENSCHAFT UND
LEBEN*, BRAUNSCHWEIG, 1850, VOL. 8, AFTER
P. 92

the rhino facing to the right. However, the title is omitted or changed, and the lettering only includes the references (always in English as the original) to Daniell as draughtsman and Redaway as engraver. This version first appeared in the *Tableaux Pittoresques de l'Inde*, the French edition of the *Oriental Annual*, translated by P.J. Auguste Urbain, dated 1835 (Daniell & Caunter 1835b). Several examples in American magazines and German books for young people have been identified (fig. 17.7).

17.5 The Last Rhino in Pilibhit

There are several records from Pilibhit, previously also referred to as Rohilcund, Rohilkhand or Kumaun. An early allusion to the presence of rhino is found in a textbook on hunting written in the 13th to 16th century. Raja Rudradeva of Kurmacala (Kumaon) has the following advice for best results: "Five or six horsemen are quite enough for hunting rhinoceros (khadga). The horses should be quiet and well-trained in their notion. A horseman should strike the rhinoceros with small darts in quick succession on the back. If it turns back, then the horseman relying on the dexterity of his horse should at once run in its front; others should hit it from behind or skilled archers should pierce it with arrows on the sides" (*Syainika Sastra*, chapter 3 para. 37–40, after Karttunen 1989: 171).

The Garhwal region is also known for an ancient ritual involving the portrayal of a rhino in a dramatization of the Sanskrit versions of the *Mahabharata* (Briggs 1938; Sax 1997, 2002). The rhino would not have been chosen for this role if it had been unknown to live in the area at the time when the basic ritual was conceived.

The last rhino in Pilibhit District was killed in the early 1870s. Blanford (1891) was clearly mistaken to state that rhinos were not seen in Rohilcund after the 1850s, because there are several later records. Modern literature suggests a final date of 1878, more accurately read as 'before 1878.' In reminiscences of a shooting trip to western Nepal, the civil servant John Prescott Hewett (1854–1911) stated that "the last rhinoceros obtained in these parts was shot by the Hon. R. Drummond, I believe, near the boundary of the Pilibhit district in the early seventies" (Hewett 1938: 167). Hewett himself took no part in that particular hunt because he first reached India in 1878. The reference is to Robert Andrew John Drummond (1820–1887), son of the 6th Viscount Strathallan, who was with the Bengal Civil Service 1842–1878, and for some time Commissioner at Rohilcund. The locality and exact date cannot be verified. Drummond was known to have hunted, as he was responsible for shooting a tiger of 11 ft 9 in (358 cm), again without locality or date (Hewett 1938: 69; Fayrer 1875b).

CHAPTER 18

Historical Records of the Rhinoceros in Uttar Pradesh

First Record: 1017 – Last Record: 1933 and recent re-introductions
Species: *Rhinoceros unicornis*
Rhino Regions 10, 11, 12

18.1 Rhinos in Uttar Pradesh

The rhinoceros was mostly seen in the northern part of the province in the foothills of the higher mountain ranges along the northern borders with Uttarakhand and Nepal. The animals were reported in Bareilly and Pilibhit eastwards past the Gaghara River of the former region of Oudh to the Gorakhpur District. There are further records in the south between the Ganges and Yamuna Rivers, extending towards Agra and Aligarh, where apparently rhinos were confined to areas with the right environmental conditions. They were never particularly abundant in these southern areas and were extinct by the end of the 16th century. They survived much longer in the Himalayan foothills, until the end of the 19th century or locally even until the 1930s.

The powerful rulers of the native state of Oudh kept many rhinos in Lucknow and the immediate surroundings until the last Nawab was exiled to Kolkata in 1856 (6.1). Some of these were illustrated by contemporary artists and show the particular features of *R. unicornis*. There is no information where these specimens were captured, but it would not be unreasonable to suggest that most originated from the forested stretches of Oudh, now the northern part of Uttar Pradesh.

After 1870 rhinos were confined to Gorakhpur where the population would have been in contact with animals from Valmiki and Champaran in Bihar further east and from Nepal further north. Rhinos were reintroduced into Dudhwa National Park in 1984–1985 and the population is increasing slowly (19.1). Other national parks could become further second homes for the rhinos as long as protection can be guaranteed.

...

Dataset 18.16. Chronological List of Records of Rhinoceros in Uttar Pradesh

General sequence: Date – Locality (as in original source) – Event – Source – § number, figures – Map number.
The persons interacting with a rhinoceros in various events are combined in Table 18.22.

There are a few fossil remains found in Uttar Pradesh (chapter 8) and depictions in rock paintings in Mirzapur and Rewa Districts (chapter 9). The records are divided in two sections, first for the archaeological and cultural expressions, second for events involving rhinos in the wild. The events are listed according to the three Rhino Regions in Uttar Pradesh, Rhino Regions 10, 11, 12. The map numbers are explained in dataset 18.17 and shown on map 18.9.

Uttar Pradesh: Archaeology, Culture and Art

3rd cent. BCE – Bhita, Allahabad Dt. – Seal with figure of galloping rhino. Red and grey soapstone, 6.6 × 6.3 cm. Allahabad Museum – P. Chandra 1970: 36; M. Chandra 1973: 33; Bautze 1985: 413; Bose 2018b: 52 (figure), 2020: 89–90 (fig.2.16) – 18.2 – map M 854

2nd cent. BCE – Kausambi Archaeological site – Terracotta pottery figurine of a rhino in deep red ochre paint – G. Sharma 1969: 70, pl. 44B2; Bose 2018b: 51, 2020: 89. This might be same as the figure (possibly) of a rhino with broken back legs and horn, wide projections at ears (size 7.1 × 11 cm) in British Museum, London, no. 1942,0214.14 – 18.2 – map M 853

2nd cent. – Mathura: Bhutesar mound – Buddhist stupa excavated in 1870s. Three pillars with sculptures in relief known as the Bhutesvara Yakshis were transferred to the Indian Museum in Kolkata. The freeze above the pillars shows on the right a figure of a rhino facing left – 18.2, fig. 18.1 – map M 851

3rd cent. – Bulandshahr Dt.: Indor, or Indrapura, ancient mount near Anupshahar – Small sculpture in black slaty limestone of rhino. Also sculpted head of animal, possibly rhino. – Carlleyle 1879: 53; J. Menon et al. 2008 (about site). There is no indication of current location of these artefacts – map M 852

Medieval (6th to 16th cent.) – Bhita, Allahabad Dt. – Terracotta sculpture of rhino (no horn). Size 8 × 4 × 4 cm. Lucknow Museum no. 47.54 – Kumar 2020: 1795, fig. 1808 – 18.2 – map M 854

19th cent. – Uttar Pradesh – Plaque of a rhino inscribed on its body with Urdu letters meaning Muhamed and Shah, possibly referring to one of the Mughal rulers in the area – Divyabhanusinh 2018c: 105 (illustrated, courtesy M.K. Ranjitsinh; no provenance)

1900 – Varanasi – Howdah of Banaras kings in Vidya Mandir Trust Museum, Ramnagar Fort – Divyabhanusinh 2018b: 107 (illustrated) – 18.10

1966 – Varanasi – Rhino sculpture in Shri Vishwanath Mandir, within campus of Banaras Hindu University (constructed 1931–1966) – Joachim Bautze, pers.comm. 2020

1990 – Shakteshgarh, Mirzapur Dt. (19 km from Chunar) – Rhino statue outside the Fort, probably modern (late 20th century) – https://indowaves.wordpress.com/tag/chunar/ – map M 856

© L.C. (KEES) ROOKMAAKER, 2024 | DOI:10.1163/9789004691544_019
This is an open access chapter distributed under the terms of the CC BY-NC-ND 4.0 license.

Rhino Region 10: Uttar Pradesh – along Ganges

5th–6th cent. – Plains of the North Western Province (Uttar Pradesh and Uttarakhand) – Rhino found throughout – Cockburn 1883: 57

1017 – India about the Ganges – Al-Biruni (973–1050) states that "ganda exists in large numbers" – Sachau 1910, vol. 1: 203; Störk 1977: 421; Rookmaaker 1984a: 558 – 18.3

1529-03-25 – Chunar – Babur heard about a tiger and rhino by the river bank near his camp, but his hunt is unsuccessful – Babur 1922: 657 – 11.2 – map A 239

1608 – Oudh (Lucknow) – William Finch (d. 1613) found plentiful trade in rhino horns – Finch 1625: 436 – 18.6

1790 – Laknaor (Lucknow) – Clark's Rhino in the Lyceum, London 1790–1793 came from Laknaor (Lucknow) – 6.4

Rhino Region 11: Uttar Pradesh – North

1341 – Bahraich, on Saru (Ghaghara) River – Ibn Battuta (1304–1369) was attacked by a rhino, which was then killed – Ibn Battuta 1855, vol. 3: 100,356; Ibn Battuta 1971: 596,727; Rookmaaker 1984a: 559, 2000 – 18.4 – map A 247

1519 – Northern part of India – Depiction of rhino shown on Homem-Reinel Atlas or Miller Atlas of 1519 (Bibliothèque Nationale, Paris, Registre C; 28836, folio 3). The Atlas is attributed to the Portuguese cartographers Lopo Homem (1497–1572), Pedro Reinel (1462–1542) and Jorge Reinel (1502–1572) – George 1969: 128; Homem et al. 1519 – 18.5

16th cent. – Sarkar of Sambal (Sambhal) – Rhinos were numerous in time of Mughal Emperor Akbar (1542–1605) – Jarrett 1891, vol. 2: 281 (after Abul Fazl Allami) – 11.7 – map A 242

1622 – Kul Nuh Ban near Aligarh – Mughal Emperor Jahangir (1569–1627) mentions killing a rhino with one bullet – Beveridge 1914, vol. 2: 270 – 11.12 – map A 241

1665-12 – Gianabad (Kora Jahanabad) – Jean Baptiste Tavernier (1605–1689) saw rhino feeding upon millet-canes fed by boy – Tavernier 1676: 71, 1889: 114; Rookmaaker 1998: 78 – 18.7 – map C 514

1768 – Bahraich region – Shuja-ud-Daula (1732–1775), Nawab of Oudh, killed 2 rhino during a large hunt – Gentil Atlas IS25, p. 13, size 53.5 × 37 cm (Victoria & Albert Museum, London) – 18.8, fig. 18.5 – map A 247

1859-01 – Chauka River at Tillia (Tilleah) – Rhino on the banks of the river – Gordon-Alexander 1898: 348 – map A 246

1870 – Sitapur region – Rhino seen here but more rarely – Fayrer 1873b: 111 – map A 246

1876 – Bariney (Bareilly), in the *terai* of Rohilcund – Presence of rhino suspected during visit of Albert Edward, Prince of Wales (1841–1910) – Russell 1876: 282; Baldwin 1876: 145 – map A 244

n.d. – Kasmanda (north of Lucknow) – Female specimen of *R. unicornis* – Groves 1982: 253. No name of museum given. This locality not recorded elsewhere for a specimen and possibly refers to the rhino head in Oxford, see 1897 in RR12 – map Z 906

1984-03 – Dudhwa NP – Arrival of 2 male 3 female rhino from Pabitora, Assam – chapter 19 – map A 245

1985-04 – Dudhwa NP – Arrival of 4 female rhino from Sauraha, Nepal – chapter 19 – map A 245

2005-01 – Muradabad region – Stray animal from Nepal via Lagga Bagga. Animal darted and taken to Kanpur Zoo – B. Choudhury et al. 2009 – map A 243

2011 – Near Laggabagga (Lagabhaga) and Haripur, east of Pilibhit – Occasional strays, possibly wandering from Shuklaphanta NP in Nepal – Bista 2011 – map A 245

2014 – Kishanpur WS outside Dudhwa – Rhino observed – G. Menon 2014 – map A 245

Rhino Region 12: Uttar Pradesh – East of Ghaghara River

1525 – Saru or Sirwu (Ghaghara) River (near Bahraich) – Babur (1483–1530) saw "masses there are also on the banks of the Saru-river in Hindustan" – Babur 1922: 490 (fol. 275b) – 9.2 – map A 248

1750–1775 – While in Lucknow (Oudh) in the 1770s and 1780s, the Swiss adventurer Antoine-Louis Henri de Polier (1741–1795) obtained several works of art. One is a miniature with a standing portrait of Mirza Rustam signed by the artist Mihr Chand, and dated to the third quarter of the 18th century – Kheiri 1921: 17, fig. 45; Hickmann & Enderlein 1993: 17, from collection of Museum für Islamische Kunst der Staatlichen Museen zu Berlin – 18.9, fig. 18.7

1750–1775 – A second miniature obtained by De Polier in Oudh in the 18th century (see previous entry) shows a rhino together with ten other mammals and three birds. The depiction suggests wild animals rather than any in captivity – Kühnel & Ettinghausen 1933: 11, fig. 22; Kühnel 1937: 10, fig. 19 – 18.9, fig. 18.8

1769 – Gorakhpur (in Oudh) – Shuja-ud-Daula (1732–1775), Nawab of Oudh, killed at least 2 rhino and 1 is captured alive – Gentil Atlas IS25, p. 12 (Victoria & Albert Museum, London) – 18.8, figs. 18.3 to 18.6 – map A 250

1785 – Seoura (Gorakhpur Dt.) – Rhino (mother and calf) shown on map prepared for Jean-Baptiste-Joseph Gentil (1726–1799). British Library of London, India Office Library and Records, Prints and Drawings section Add.MS. Or. 4039, folio 19, size 34 × 22 cm – Gole 1988; Richard 1996; Dadlani 2005; Lahiri 2012: 87; Rookmaaker 2014a – 18.8, fig. 18.6 – map A 249

1789 – Buckarah Lake near Gurruckpore (Bakhira Taal) – Anonymous officer out with a party of seven (using 30 elephants and 1000 followers) expected to see a rhino – *Bath Chronicle* 1789-08-13 – map A 250

1793 – Buckra Jeel, near Gorakhpur (Bakhira Taal) – Nawab of Oudh hunting party, rhino suspected but apparently none killed – Smith 1806: 15, 1816: 541 – 18.9 – map A 250

1832 – Goruckpore *terai* (Gorakhpur) – Rhino present, not seen – Tiger 1832: 388 – map A 250

1839 – Saul forest along bases of Himalayan mountains – Rhino numerous – Royle 1839: lxx

1866 – Goruckpore (Gorakhpur) – Rhino scarce – RHD 1866: 485 – map A 250

1870s – Badi Tal, Sitapur region – H. shot two rhinos, some years ago – Fayrer 1875a: 277 – map A 250

1876 – Goruckpore (Gorakhpur) – Rhino now extinct, or nearly so – Baldwin 1876: 145 – map A 250

1877 – North-east part of Audh (Oudh) (Gorakhpur) – English officer witnessed fight between rhino and elephant – Anon. 1877a – map A 250

1897-12-27 – Gorakhpur District – Natural History Museum, Oxford, no. 21791. Mounted head. Donated by Harry George Champion (1891–1971), Imperial Forestry Institute (Oxford), 1951 – Oxford University Museum of Natural History, collections website [accessed 2015] – map A 250

c.1905 – Gorakhpur – A rhino which had wandered south from Nepal was the last killed in the district by Peter Henry Clutterbuck (1868–1951) when he was in charge. The Conservator of Forests Sainthill Eardley-Wilmot (1852–1929) recorded in 1910 that he brought part of the skin with him to England, where he sold it to a golfer who liked the material for his drivers "and attributed lightness and elasticity to the material" – Eardley-Wilmot 1910; Stebbing 1926, vol. 3: 190 – map A 250

1908 – Gorakhpur – A single rhino shot within the last few years – Hunter 1908b, vol. 12: 332 – map A 250

1930s – Oudh forest (Gorakhpur) – Strays from Nepal, formerly fairly numerous – Champion 1933: 43 – 18.11 – map A 250

1930s – Gorakhpur – Yezdezard Dinshaw Gundevia (1908–1986) visits place where a 'Bara Hakim' from Nepal and party had killed the last rhino – Gundevia 1984: 2 – 18.11 – map A 250

1987 and later – Katerniaghat ws – Strays seen occasionally from Bardia, Nepal – Martin 1996b; Thapa et al. 2013; Sinha et al. 2011b; Talukdar & Sinha 2013 – map A 248

2016 – Hastinapur (NE of Meerut) – Plan to re-introduce rhino – Rai 2016

2016 – Katerniaghat ws – Plan to re-introduce rhino – Rookmaaker et al. 2017: 25 – map A 248

2016 – Pilibhit Tiger Reserve – Plan to re-introduce rhino, across border of Uttar Pradesh and Uttarakhand – Rookmaaker et al. 2017: 25

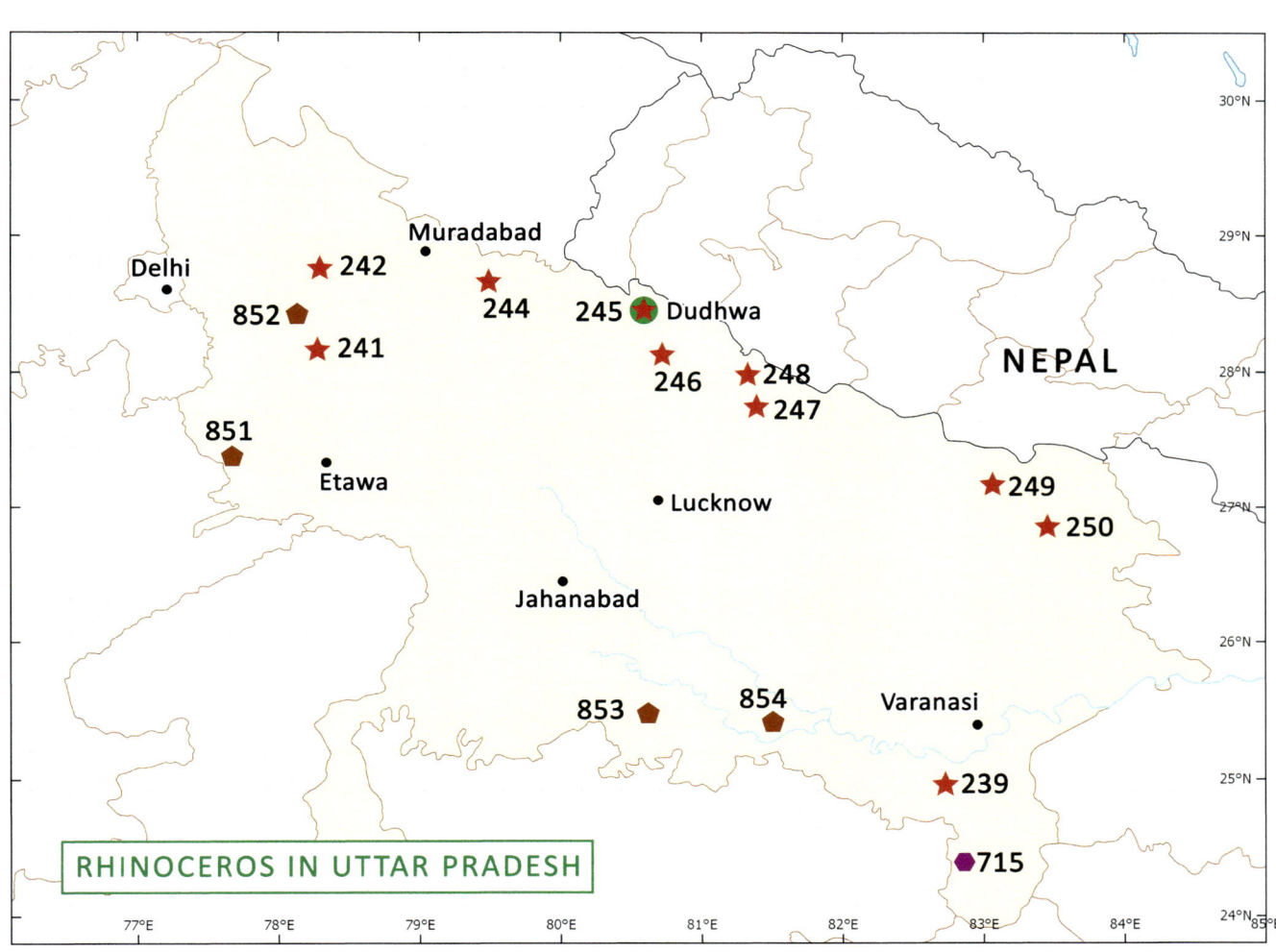

MAP 18.9   Records of rhinoceros in Uttar Pradesh. The numbers and places are explained in Dataset 18.17

★ Presence of *R. unicornis*

⬠ Artefacts

⬡ Rock Art

∴

Dataset 18.17: Localities of Records of Rhinoceros in Uttar Pradesh

All numbers and places are shown in map 18.9.

Rhino Region 10: Uttar Pradesh – along Ganges

*C 511	Dehli (Delhi) – 28.63N; 77.34E – 1659 – Chapter 11
M 851	Mathura – 27.45N; 77.70E – 2nd century BCE
M 852	Indor, or Indrapura, ancient mount near Anupshahar – 28.37N; 78.30E – 3rd cent.
*C 512	Etawa (Etawah) – 26.80N; 79.03E – 1632 – Chapter 11
*C 513	Laknaor (Lucknow) – 26.80N; 80.95E – 1790
*C 513	Oudh (Lucknow) – 26.80N; 80.95E – 1866
M 853	Kausambi Archaeological site – 25.37N; 81.40E – 3rd cent. BCE
M 854	Bhita, Allahabad Dt. – 25.30N; 81.80E – 3rd cent. BCE
*	Varanasi – 25.30N; 83.00E – 1900, 1966
A 239	Chunar – 25.11N; 82.86E – 1529 – Chapter 11
*M 856	Shakteshgarh, Mirzapur Dt. (19 km from Chunar) – 25.00N; 82.83E – modern art
P 715	Mirzapur Dt.: Ghormangur, Kaimur, Lekhania – 25.54N; 83.10E – Mesolithic – Chapter 9
P 715	Mirzapur Dt.: Matahawa, Morhana, Romp – 25.54N; 83.10E – Mesolithic – Chapter 9

Rhino Region 11: Uttar Pradesh – North

A 241	Kul Nuh Ban near Aligarh – 28.20N; 78.30E – 1622 – Chapter 11
A 242	Sarkar of Sambal (Sambhal) – 28.60N; 78.60E – 16th cent. – Chapter 11

A 243	Muradabad region – 29.00N; 78.75E – 2005 – stray from Nepal
A 244	Bariney (Bareilly) – 28.44N; 79.40E – 1876
*C 514	Gianabad (Kora Jahanabad) – 26.10N; 80.35E – 1665 – captive, not on map
A 245	Kishanpur WS – 28.38N; 80.40E – 2014
A 245	Dudhwa NP – 28.50N; 80.50E – 1984, 1985
A 245	Laggabagga (Lagabhaga) and Haripur – 28.44N; 80.55E – 2011
A 246	Badi Tal, Sitapur – 28.45N; 80.70E – 1870
*Z 906	Kasmanda (north of Lucknow) – 27.40N; 80.85E – unlikely record
A 246	Chauka River (Tillia) – 28.15N; 81.05E – 1859
A 247	Bahraich region – 27.60N; 81.60E – 1768
M 855	Sahet-Mahet, Gonda Dt. – 27.13N; 81.95E – c.1800?

Rhino Region 12: Uttar Pradesh – East of Ghaghara River

A 248	Saru or Sirwu (Ghaghara) River – 27.67N; 81.10E – 1341, 1525 – Chapter 11
A 248	Katerniaghat WS – 28.25N; 81.20E – 1987 ff., 2016
A 249	Bansi (Seoura) – 27.40N; 83.10E – 1785
A 250	Buckra Jeel, near Gorakhpur (Bakhira Taal) – 26.88N; 83.15E – 1789, 1793
A 250	Goruckpore (Gorakhpur) – 27.15N; 83.40E – 1608, 1769, 1832, 1866, 1876, 1877, 1897, 1905, 1908, 1930s, 1933

* not shown on map

∴

TABLE 18.22   List of persons who were involved in shooting or capturing rhino in Uttar Pradesh

Person	Date	Locality. Type of interaction
Ibn Battuta (1304–1369)	1341	Ghagara River. Attack by rhino which is killed
Babur (1483–1530)	1525	Ghagara River. Observes many rhino
Jahangir (1569–1627)	1622	Aligarh. Kills 1
Jean Baptiste Tavernier (1605–1689)	1665	Kora Jahanabad. Observes 1
Shuja-ud-Daula (1732–1775)	1768	Bahraich. Kills 2
Shuja-ud-Daula (1732–1775)	1769	Gorakhpur. Kills 2, captures 1 alive
Anonymous	1790	Lucknow region. One sent to UK
H.	1870s	Badi Tal. Killed 2
English officer	1877	Gorakhpur. Witnessed elephant-rhino fight
Anonymous, via Harry George Champion (1891–1971)	1897	Gorakhpur. Killed 1
Peter Henry Clutterbuck (1868–1951)	1905	Gorakhpur. Killed 1
"Bara Hakim" from Nepal	1930s	Gorakpur. Killed 1

Listed in chronological order for the main event.

FIGURE 18.1
Rhinoceros on a coping stone above the large pillars
known as the Bhutesvara Yakshis, dated around
2nd century CE. Indian Museum, Kolkata
PHOTO: JOACHIM K. BAUTZE

18.2 Archaeological Evidence

There are a few pottery objects and a seal found in archaeological excavations in Uttar Pradesh, especially in Bhita, Kausambi, Mathura and Indor. These range in dates from the 3rd century BCE to the medieval period ending in the 16th century CE. They attest to knowledge about the rhino, yet their small number might indicate the relative scarcity of the animals if not limited significance in the cultural expressions.

From the Bhutesar mound in Mathura, a set of three large pillars of a Buddhist stupa were excavated known as the Bhutesvara Yakshis. Exhibited since the 1870s in the Indian Museum, Kolkata, the coping stone (ushnisha) connecting the top of the pillars shows the figure of a rhino on the top right (fig. 18.1). The animal is well executed, with a single horn, but with few skin folds. It dates from the 2nd century.

18.3 Al-Biruni's History of India of 1017

The Iranian scholar al-Biruni (973–1050) spent a year in India in 1017 and wrote a History of India on his return. Besides discussing the difference between *ganda* and *karkadann* or *karg* from African sources, his description of the rhino is relatively clear: "The ganda exists in large numbers in India, more particularly about the Ganges. It is of the build of a buffalo, has a black scaly skin, a dewlap hanging down under the chin. It has three yellow hooves on each foot, the biggest one forward, the others on both sides. The tail is not long. The eyes lie low, father down the cheek than is the case with all other animals. On the top

of the nose there is a single horn which is bent upwards. The Brahmins have the privilege of eating the flesh of the ganda. I have myself witnessed how an elephant coming across a young ganda was attacked by it. The ganda wounded with its horn a forefoot of the elephant, and threw it down on its face" (Sachau 1910: 203). The locality "about the Ganges" is no more than a general indication.

18.4 Ibn Battuta Attacked by Rhino Near Bahraich

The famous early Muslim scholar and traveller Ibn Battuta (1304–1369) spent some years in Delhi during the reign of Sultan Muhammad bin Tughluq (d. 1351). In 1341, he went to the new town of Bahraich on the Saru (Ghaghara) river to see the tomb of the legendary warrior Ghazi Miyan. While passing through a forest of reeds, his party was attacked by a rhino. The animal was killed and its head was brought to the Sultan. Ibn Battuta noticed that the head was far larger than that of an elephant even though the animal itself was smaller, hence the saying "karkadann, head without body" (Ettinghausen 1950: 13; Rookmaaker 2000).

18.5 The Rhinoceros on Maps of the 16th Century

A depiction of a rhinoceros appears in the Atlas Miller of the early 16th century, preserved in the Bibliothèque Nationale, Paris. The animal is shown next to elephants in "Bengala Regio", located at the current state of Uttar Pradesh. There is no explanation where the designers of the map might have obtained the original figure, although a living rhino was shown in Lisbon in 1515 to 1516 (4.6).

FIGURE 18.2
A single-horned rhinoceros as depicted by European cartographers of the 16th century. This example is from the Portolan Atlas (or Vallard Atlas) attributed to Nicholas Vallard of Dieppe, Belgium, of 1547.
THE HUNTINGTON LIBRARY, SAN MARINO, CA: HM29, MAP 7

It was a remarkably common practice among the early cartographers to include a rhinoceros on their maps. Most of these depictions were located in the western part of the African continent, and invariably the animals were single-horned. Examples are found on the maps of Martin Waldseemüller in 1516, Ptolemaeus Argentorata in 1525, Peter Apian in 1530, Ulpius in 1542 and Nicholas Vallard in 1547 (George 1969: 147–150; Rookmaaker 2004c: 2, fig. 2). The last of these, now called Portolan Atlas, was produced in Dieppe, Belgium, in 1547, and provides a good example of a single-horned rhinoceros as depicted by the 16th century mapmakers (fig. 18.2).

18.6 Horn Trade Reported by William Finch in 1608

The English merchant William Finch (d. 1613) traveled in India in the time of the Mughal Emperor Jahangir from 1608 to 1611. When he visited Oude (Oudh), he noticed that there was an abundance of horns of the rhino (which he called Indian Ass). These were made into small buckler shields and drinking cups, some of which were said to be very valuable (Finch 1625: 436).

18.7 Tavernier Feeding Millet to a Rhino in 1665

The French diamond merchant Jean Baptiste Tavernier (1605–1689) had an encounter with a rhino during his sixth and last journey to India, when he was traveling from Agra to Patna. He was close to Gianabad, now Kora Jahanabad, on 2 December 1665: "you pass a field of millet, where I saw a rhinoceros eating stalks of this millet, which a small boy of nine or ten years presented to him. On my approaching the boy, he gave me some stalks of millet,

and immediately the rhinoceros came to me, opening his mouth four or five times; I placed some in it, and when he had eaten them, he continued to open his mouth so that I might give him more" (Tavernier 1676: 71). The passage seems sincere enough, and gave chance to Valentine Ball (1843–1895), the editor of the English translation, to elaborate on tame rhinos kept by Indian rajahs. However, should Tavernier not have been more surprised when he saw a boy feeding a rhino, an animal which arguably he saw for the first time.

18.8 Jean Gentil in Oudh, 1759–1777

The French army officer Jean-Baptiste-Joseph Gentil (1726–1799) reached Pondicherry in 1752 and was posted to the court of Nawab Shuja-ud-Daula of Oudh from 1759–1777. Resident in Faizabad during most of this period, he collected many historical items and worked on a new Atlas of the Mughal Empire. Commissioning local artists, Gentil assembled a series of watercolours depicting life at the court, which he annotated with French captions, now kept in the Victoria & Albert Museum, London (Atlas IS25). There are two pages illustrating royal hunts, which took place in 1768 and 1769. The first of these (p. 12) gives the year 1769 in the caption, referring to a hunt of ten days in the Province of Avad (Oudh). The page is illustrated with 11 small scenes of different events during these days, three of which include a rhino (fig. 18.3). Scene 5 has a dead rhino with a red belly amidst several other animals killed or captured; Scene 7 has another dead rhino among other slain animals; and Scene 8 shows not only a dead rhino in a pose similar to that in scene 5, but also a second rhino, obviously still alive, tied with ropes held by seven men (5.3).

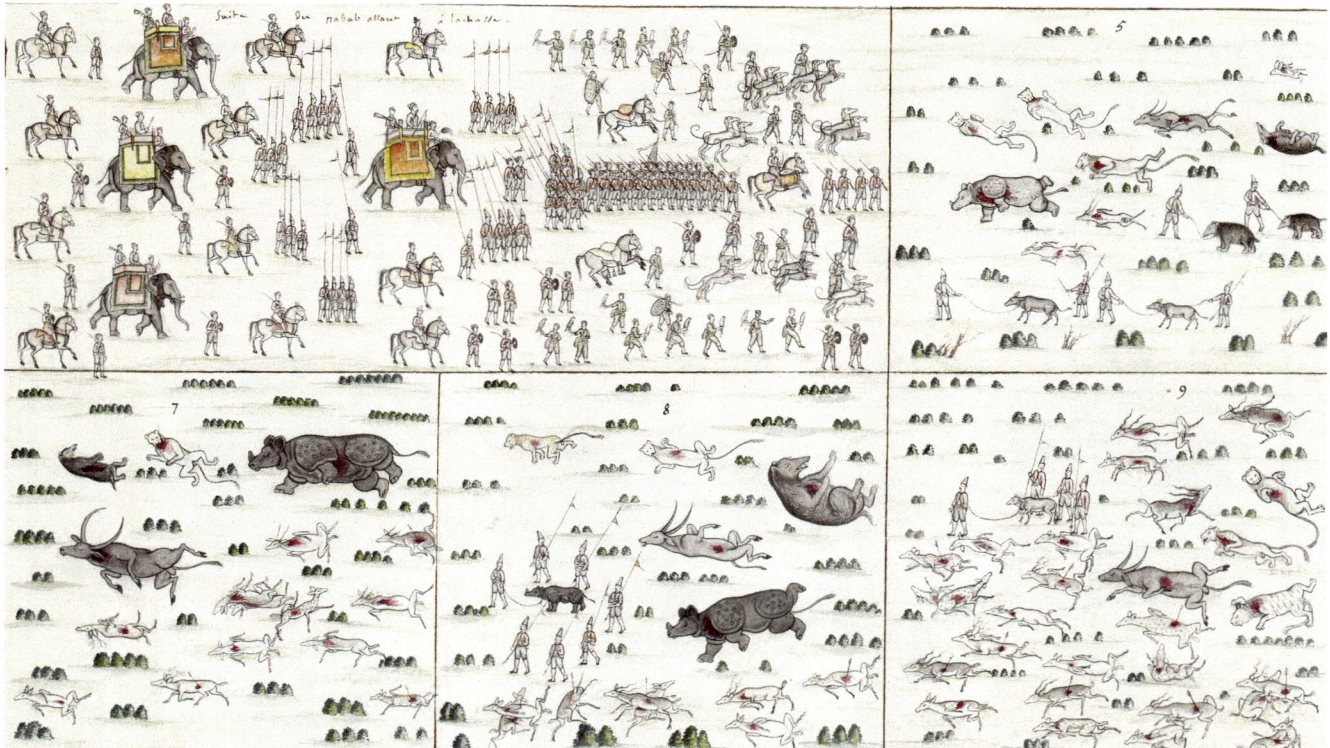

FIGURE 18.3   Watercolour drawing of an event during a hunt by the Nawab of Oudh in 1769. In this part of the sheet there are three
scenes showing the rhinoceros. Scene 5 (lower left): a dead rhino with a bleeding belly. Scene 7 (upper right): another dead
rhino. Scene 8 (lower middle): a dead rhino, as well as a young animal captured alive and surrounded by 7 soldiers
WITH PERMISSION © VICTORIA & ALBERT MUSEUM, LONDON: GENTIL ATLAS IS25, P. 12

On the next page (p. 13) of this Atlas, the Nawab of Oudh is depicted hunting with a force of 600 men commanded by Gentil (fig. 18.4). Dated 1768 in the caption, the hunt took place in the forests of Bahraich, some 130 km NW of Faizabad. The Nawab killed two rhinos after these animals had injured ten elephants. This scene is graphically depicted in one of the watercolours made by a local artist, showing two dead rhinos, two rhinos attacking elephants, as well as 15 of them on the run. The impression should be that rhinos were plentiful, and so forceful that only two of them could be slain by this mighty army.

In his memoirs published posthumously, Gentil (1822: 266) mentions the presence of rhinos in the Koumahouns mountains separating India from the great and small Thibet. This is shown on the map included in the same book as the hill region north of Lucknow, now just across the state border in Uttarakhand.

Another Atlas compiled by Gentil in 1785 presented the geography of the Mughal Empire. There are two copies of this manuscript, in the Bibliothèque Nationale of Paris and in the British Library of London. All maps in the London copy were embellished with little drawings of scenery, people, plants and several animals, both within the cartographic part and around the borders (Gole 1988).

Gentil employed a number of Indian artists, whose identities have been lost in time, but may have included Nevasi Lal and Mohan Singh. Three maps include images of the rhinoceros, in the regions of "Avad" (Oudh), "Bear" (Bihar) and "Bengale" (Bengal):

The map of "Avad" (Oudh) shows an adult rhino followed by a calf, in the region north-east of Bansi (figs. 18.5, 18.6). On Gentil's map the locality, maybe near a lake, is labelled Seoura, located just north of Bansi towards the mountains.

The map of "Bear" (Bihar) has one figure of a rhino north of Patna, between Muzaffarpur and Bettiah (figs. 28.13, 28.14).

The map of "Bengale" (Bengal) has one figure of a rhino in the general region of North Bengal (figs. 31.6, 31.7).

18.9   Hunts by Nawab Asaf-Ud-Daula of Oudh

Nawab Asaf-Ud-Daula (1748–1797) often went on hunting expeditions, very likely in forests of the northern part of the province towards the foothills of the Himalayas. Rhinos were rarely seen because they lived in dense forests, where they could not be hunted from the back of

FIGURE 18.4   A highly evocative watercolour showing an abundance of rhinos being hunted by the Nawab of Oudh in 1768
WITH PERMISSION © VICTORIA & ALBERT MUSEUM: GENTIL ATLAS IS25, P. 13

less well-trained elephants. The Nawab had over 1,200 elephants, however, and it would be unlikely that he did not
occasionally shoot a rhino, or maybe some juveniles were
captured.

There are portrayals of a rhino in two miniatures
obtained by the Swiss adventurer Antoine-Louis Henri
de Polier (1741–1795) in Lucknow in the 1770s or 1780s. The
first example shows a rhino roaming in the forest together
with other animals. This scene is the upper border of a portrait of Mirza Rustam signed by Mihr Chand, and dated to
the third quarter of the 18th century (fig. 18.7). The second
rhino is part of a miniature depicting ten different mammals and three birds. In this case, there is no information
about the artist who wanted to show a range of wild animals found in the surroundings of Oudh (fig. 18.8).

18.10   Rajahs of Benares (Varanasi)

There is no direct evidence in the form of written
accounts or trophies that the rulers of Benares State
hunted rhinos. As the animals were present in parts of

their territories around Gorakhpur and elsewhere, it is
likely that they occasionally pursued this activity. In the
Vidya Mandir Trust Museum, Ramnagar Fort, there is a
howdah with an artful depiction of a rhino in a landscape
(Divyabhanusinh 2018b).

18.11   The Last Rhinos of Gorakhpur

There are relatively few records of the presence of the
rhinos in the Himalayan *terai* from Pilibhit eastwards
to Gorakhpur from the mid-19th century onwards.
The region was close enough to urban centers to have
attracted hunters in case the wildlife would have warranted this. Rhinos were therefore rare, or only present
in a small number of localities. The Indian diplomat
Yezdezard Dinshaw Gundevia (1984), working at the time
of Jawaharlal Nehru, visited a place in Gorakhpur District
where the last rhino was killed, probably in the 1930s,
although he is sparse with the details when or exactly
where this took place. This may be the same animal
mentioned as early as 1910 by the Conservator of Forests

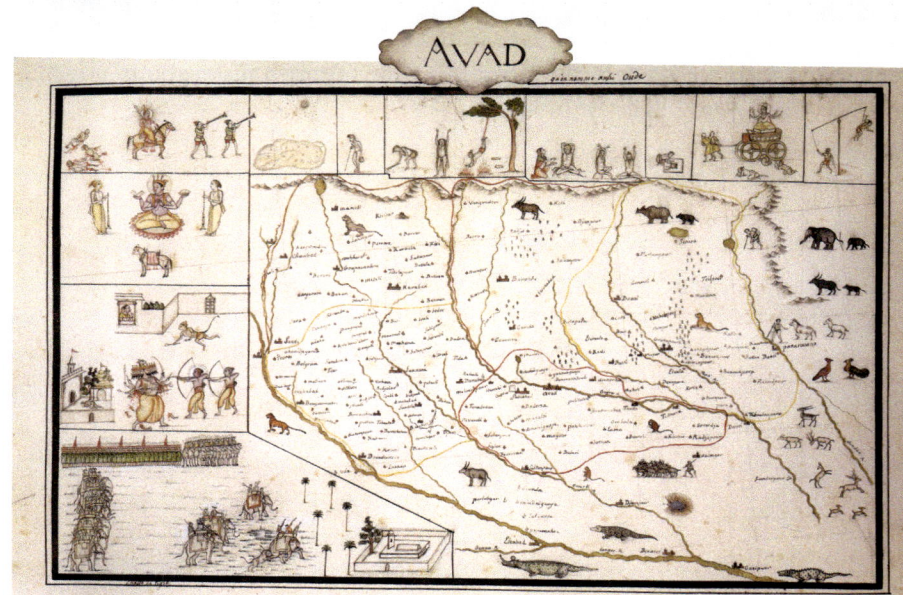

FIGURE 18.5
Map of 'Avad' (Oudh) by Jean Baptiste Joseph Gentil, dated 1785. It shows a mother rhino and her calf near a place called Seoura
WITH PERMISSION © BRITISH LIBRARY, INDIA OFFICE LIBRARY AND RECORDS: ADD.MS. OR. 4039, FOLIO 19

FIGURE 18.6
Rhino mother and calf near Seoura. Detail of Gentil's Map of Avad shown in fig. 18.5
WITH PERMISSION © BRITISH LIBRARY

FIGURE 18.7   Upper border of a miniature depicting Mirza Rustam, signed by Mihr Chand who worked in Lucknow in the third quarter of the 18th century
PHOTO: JOACHIM K. BAUTZE

FIGURE 18.8
Detail of a miniature obtained by Antoine-Louis Henri de Polier (1741–1795) in Lucknow in the 1770s or 1780s. The rhino is seen among other mammals and a few birds in a supposedly wild setting.
PHOTO: JOACHIM K. BAUTZE

Sainthill Eardley-Wilmot (1852–1929), who attributed the last killing to Lieutenant-Colonel Peter Henry Clutterbuck (1868–1951) around 1905. In more recent years, with an expanding rhino population in Nepal, stray rhinos are sometimes encountered in suitable habitats in the border regions. A new population was established in Dudhwa National Park in the 1980s, which has held its own and could be a blueprint for future translocations into this part of India.

Protecting the Rhinoceros in Dudhwa National Park, Uttar Pradesh

First Record: 1984 (re-introduction) – Last Record: current
Species: *Rhinoceros unicornis*
Rhino Region 10

19.1   The Establishment of Dudhwa National Park

Dudhwa Tiger Reserve in the Lakhimpur Kheri District, Uttar Pradesh, was established in its current administrative format in 2010. Some parts of the currently protected area were notified previously, starting with the Sonaripur Wildlife Sanctuary of 63 km² to protect Barasingha deer *Rucervus duvaucelii* in September 1958. The reserve was renamed Dudhwa in 1968 with an area of 212 km², followed by another addition of 200.2 km² in 1972. Thanks to the efforts of pioneer conservationist Kunwar "Billy" Arjan Singh (1917–2010), best known for his efforts to protect tigers and leopards, the area was declared a National Park on 1 February 1977 (Hart-Davis 2005). Dudhwa displays the typical features of a *terai* ecosystem and is surrounded by a buffer zone of reserved forests. At present, the national park covers an area of 490.3 km² and the buffer zone a further 190 km². In 1987, the park came under the purview of Project Tiger in conjunction with the Kishanpur Wildlife Sanctuary (established on 1 January 1973) and the Katerniaghat Wildlife Sanctuary (established on 31 May 1976).

• • •

Dataset 19.18: Chronological List of Administrative Events and Rhino Introductions in Dudhwa NP

General sequence: Date – Locality (as in original source) – Event – Sources

1958-09 – Sonaripur Wildlife Sanctuary established, 15,766 acres (63 km²) – Singh & Rao 1984: 10; Mathur & Midha 2008: 31; Sinha et al. 2011a: 43 (with date 1955)

1968 – Dudhwa Wildlife Sanctuary established with new name, new area 212 km² – Singh & Rao 1984: 10; Sinha et al. 2011a: 43

1972 – Dudhwa, addition of 200.2 km² – Sinha et al. 2011: 43

1973-10-07 – Kishanpur Wildlife Sanctuary established, Notification No. 1111/14-3-1/1971, area 203.41 km²

1976-05-31 – Katerniaghat Wildlife Sanctuary established, Notification No. 388/14-3-1, area 400.09 km²

1977-02-01 – Dudhwa declared a National Park. Area 490.3 km² and buffer zone of 190 km² – Sinha et al. 2011a: 43

1981-08-19 – Meeting of the Standing Committee of the Indian Board of Wild Life at New Delhi, chaired by Prime Minister Indira Gandhi (1917–1984). A rhino Sub-Committee was set up on 5 November 1979 under leadership of Samar Singh (b. 1940) and Kishore Rao. The recommendations in the report of the Rhino Sub-Committee were approved. Negotiations started with state governments of Uttar Pradesh and Assam – Singh & Rao 1984

1984-03-31 – Translocation of 5 rhinos (2 male, 3 female) from Assam – Singh & Rao 1984; Sale & Woodford 1984; Schenkel 1984; Hajra & Shukla 1984; Sale 1986; Sale & Singh 1987; Sinha 1991;

Suman 1994; Sinha & Sawarkar 2000; Oberai & Bonal 2002: 64–67 (with photographs)

1984-04-11 – Female Saheli aborted a calf on 7 April and died on 11 April 1984 – Gairola 1987; Menon 1995: 35

1984-04-20 – Rhinos released in Dudhwa core area – Singh & Rao 1984; Sale & Singh 1987: 82; Sinha 2005

1984-06-31 – Female Asha died when treating wounds – Sale & Singh 1987: 83

1985-04-01 – Translocation of 4 female rhinos from Nepal – Sale 1986; Sale & Singh 1987; Sinha 1991

1987 – Dudhwa NP under Project Tiger, combined with Kishanpur (1973) and Katerniaghat (1976), total area 884.9 km² – Sinha et al. 2011a: 43

1994-04-28 – A male (Lohit) born in Kanpur Zoo 1984-08-06 was transferred to Dudhwa. He was injured in a fight with male Bankey and returned (probably still in 1994) to Kanpur – Suman 1994; Sinha 2005

2010-06-09 – Dudhwa Tiger Reserve, Notification No. 1505/14-4-2010-872/2007 dated Lucknow 9 June 2010, with area 490.29 km² (without buffer zones)

2018-04 – Translocation within Dudhwa NP from Sonaripur Range to Belyara Range on 10–13 April 2018. Capture of 1 male (Napoleon), 3 females (Kalpana, Hemangini, Rohini) – Sharma et al. 2018

2021 – In the new population in Belyara Range, one female (Hemangini) gave birth in 2021, which calf died within a month. Two other babies were born in June 2021 – Ravi et al. 2022

•• ••

© L.C. (KEES) ROOKMAAKER, 2024 | DOI:10.1163/9789004691544_020

This is an open access chapter distributed under the terms of the CC BY-NC-ND 4.0 license.

TABLE 19.23   Status of the *R. unicornis* population in Dudhwa National Park 1984–2021

Date	Number	Natural death	Poached	Sources[a]
1984	3; 5			R164; R132
1985	7; 9			R164, R132
1987	7			R200
1989	6; 7; 9			R132; R211
1992	11			R198
1993	11; 12			R274, R275
1994	11; 13			R079; R211
1995	13			R098, R164
2002	17			R132
2004	21			R094
2005	21			R210, R213
2007	21; 23			R253; R210
2009	28; 29; 25–30			R210; R244
2010	30			R212
2011	30; 31			R210; R189
2012			1	R210
2013	32; 34	1	3	R210; R261
2015	30; 32			R251; R252
2016	30			R072
2017	32			R263
2019	39			news report
2020	42			news report
2021	42			R326

Different estimates in the same year are all provided.

a   References in Table: R072 Choudhury 2016: 154; R079 Dey 1994a; R094 Emslie et al. 2007; R132 Kalra & Sinha 2004; R164 Menon 1995: 15; R189 Prasad & Kumar 2011; R198 Roy 1993; R200 Sale 1986; R210 V. Singh 2015; R211 Sinha & Sawarkar 1993; R212 Sinha & Sinha 2010; R213 Sinha 2005; R244 Talukdar 2009c; R251 Talukdar 2015; R252 Talukdar 2018; R253 Talukdar et al 2011; R261 Thripathi 2013; R263 Tripathi 2018; R274 Vigne & Martin 1994; R326 Ravi et al. 2022.

19.2   Preparations for Reintroductions

In 1979, the Indian Board of Wildlife set up a sub-committee chaired by the British naturalist John B. Sale to make recommendations for suitable locations of new rhino populations. This Committee identified Dudhwa National Park as a possibility, which was confirmed after a visit in 1980 by the chairman of the IUCN/SSC Asian Rhino Specialist Group Rudolf Schenkel (1914–2003), himself an expert in rhino ecology (Schenkel 1984; Sale & Singh 1987: 82). Several feasibility studies were undertaken, including an experimental capture using drug immobilization in Assam in 1980, and a botanical survey in 1982 by Prabhat Kumar Hajra (b. 1940), Director of the Botanical Survey

of India, which showed that possibly 90 rhinos could eventually be accommodated in the reserve (Hajra & Shukla 1984). When preparations began, there was some opposition to translocation from Kaziranga to Dudhwa by students who believed that Assam would be stripped of one of its important resources (Baidya 1982, 1983, 1984).

In Dudhwa, a 27 km² core area was selected as the Rhino Reintroduction Area (RRA) in the South Sonaripur Range, comprising the entire Kakraha Block and a part of Chhota Palia Block. The RRA has nine permanent lakes. In order to prevent rhinos wandering into adjacent fields, the core area was enclosed by a two-strand electric fence, and a critical 9 km border section was further enhanced by the construction of a rhino-proof trench just outside the fence. Holding stockades were built, staff were sent to Assam for training, and trials for the capture operations were initiated (Sale & Singh 1987: 82).

19.3   Reintroduction from Assam in 1984

The first translocation from Assam to Dudhwa took place in 1984. The Assam Forest Department identified a group of 10 rhinos living just outside the Pabitora Wildlife Sanctuary as suitable candidates, because adequate protection could not be guaranteed for them. Between 11 and 21 March 1984, six animals were captured through immobilization with Immobilon, five close to Pabitora and one stray animal near Goalpara (Sale 1986). They were crated, revived and released in stockades close to the capture site, where they were examined by three veterinarians for any adverse effects of the capture operation. A large male escaped from the stockade during the night. The remaining five animals were first transported by truck to Guwahati, then flown from Guwahati to Delhi by specially chartered Russian Aeroflot IL-76 aircraft on 29 March (Singh & Rao 1984; Gairola 1987). The final leg of 430 km from Delhi to Dudhwa was achieved by road, arriving on 31 March 1984 (figs. 19.1, 19.2). These five animals were listed as follows, with ages as stated in 1984 (Sale 1986; Suman 1994; Sinha & Sawarkar 2000; Sharma & Gupta 2015):

Male	Bankey	age 7–8 years	died 2016-11-30 of old age
Male	Raju	age 25 years	died 1988-12-11 due to fighting
Fem	Pavitri	age 3–4 years	died 2013-01 after fight with tiger
Fem	Saheli	age 30 years	died 1984-04-11 after aborted pregnancy
Fem	Asha	age 16–17 years	died 1984-07-31 when being treating for a wound

FIGURE 19.1
Rhino in stockade at Dudhwa after the translocation from Assam in 1984. One of the few remaining photographs of this event published by Samar Singh and Kishore Rao
INDIA'S RHINO RE-INTRODUCTION PROGRAMME, 1984, PLATE FOLLOWING P. 12

FIGURE 19.2
Loading a crate at Guwahati Airport into the Russian Aeroflot Aircraft on 29 March 1984
SAMAR SINGH AND KISHORE RAO, *INDIA'S RHINO RE-INTRODUCTION PROGRAMME*, 1984, PL. FOLLOWING P. 12

The rhinos were initially kept in stockades where their health was monitored. The adult female Saheli was injured during the journey, and while undergoing treatment she aborted a baby of 7 April 1984. She was very weak and died from exhaustion and toxaemia on 11 April 1984 (Gairola 1987). A second adult female Asha had an accident when she was being immobilised for medical care a few weeks after release and died on 31 July 1984 (Sale & Singh 1987: 83). Three animals were released in the Kakraha area on 20 April 1984, followed by the large male Raju on 9 May 1984. Bankey (whose name means 'handsome and stud like') became the dominant male and was very aggressive to other males (fig. 19.3, 19.4). There were continuous fights between the two males Bankey and Raju, until the latter died after breaking his horn in mid-1988 (Sinha 1991: 7). Bankey became the only and later the dominant male until he died in November 2016 (*Indian Express* 2016-12-01).

19.4 Reintroduction from Nepal in 1985

As two adult females transferred from Assam had died soon after arrival, it was decided to add additional female rhinos from a different population. Talks were initiated with the Government of Nepal, resulting in an agreement to exchange sixteen domesticated Indian elephants for four female rhinos. The decision to obtain only females from Nepal was deliberate as the two males from Assam were deemed sufficient to ensure maximum genetic diversity. The capture operation took place 28 to 31 March 1985 in Sauraha, just north of Chitwan National Park. All four animals, estimated to be between four and six years old, were immobilized, crated and driven the 720 km to Dudhwa, arriving on 1 April 1985.

The first animal to arrive broke out of the stockade during the night, but the remaining three females soon settled down. They were released into the wild after a week. The names of the people who took part in these translocations from Assam and Nepal have not been recorded, except for R.D. Singh, then Dudhwa Range Officer, who was tragically killed by a tiger in April 1985 while caring for the rhinos. The four females received from Nepal on 1 April 1985 are listed as follows (Suman 1994; Sharma & Gupta 2015):

Fem   Swamvara   age 4–5 years   (alive 2015)
Fem   Narayani   age 5 years   (alive 2015)
Fem   Hemrani   age 4 years   (alive 2015)
Fem   Rapti   age 5–6 years   died 1991-09-25

19.5   Development of the Population

A third male (Lohit) was introduced into the population on 28 April 1994 (Suman 1994). Born at the Kanpur Zoo on 6 August 1984, he was wounded soon after arrival in a fight with the dominant male Bankey. After receiving treatment he was returned to Kanpur, and subsequently he was transferred to Lucknow Zoo on 6 April 1995 (Sinha 2005).

The rhinos have been continuously monitored by patrol teams with elephants. The Rhino Monitoring Centre was based in Salukapur Forest Rest House, which was renovated by Ram Lakhan Singh, the first director of Dudhwa NP (Sinha 2011a). Independent monitoring studies were the basis of academic dissertations by Tariq Aziz (1990) and Arun Kumar Tripathi (2002). Among the earliest participants in the long-term studies in Dudhwa was the wildlife ecologist Satya Priya Sinha (b. 1954), who wrote about his earlier experiences (Sinha 2011) and has published many interesting studies based on the ecology and behaviour of the rhinos in Dudhwa (Sinha 1991 to 2011), summarized in Sinha et al. (2011a).

Two years after these initial introductions, the remains of a newborn calf were discovered in a patch of tall grass in August 1987 (Sinha 1991: 7). The first successful birth was recorded in February 1989. Since then, the population has increased significantly, and all five surviving females brought from Assam and Nepal have produced calves. As of 2010, a total of 40 young had been born, six to the Assamese females, and 18 to the Nepalese females, followed by several second-generation births (with animal numbers following Sharma & Gupta 2015, if available). The totals are calculated up to 2015:

07 Saheli   1994-01-11, 1997-09-17, 2002-11-01, 2005-06-27, 2009, 2013-01 (total 6)
00 Pavitri   1991-08-04, 1995-09-21, 1997-10-02, 2007-09-14, 2012-11-06 (total 5)
03 Swayamvara   1989-10-12, 1991-08-10, 1994-10-07, 1998-08-06, 2004-07-29 (total 5)
04 Narayani   1987-09, 1989-06-01, 1992-07-31, 1999-11-21, 2001-08-27, 2004-08-31, 2011 (total 7)
05 Hemrani   1989-02-02, 1992-08-05, 1997-10-19, 2002-08-06, 2007-09-13, 2013-11-23 (total 6)
00 Rapti   1989-05-19, 1991 (total 2)
Second generation
10 Rajshree   1999-06-12, 2007-09-14, 2011 (total 3)
11 Rajeshwari   2002-09-07, 2005-03-09, 2007-09-16, 2011 (total 4)
14 Vijayshree   2006-05-21, 2008, 2012-10-12 (total 3)
17 Sada   2008, 2012-09-19 (total 2)
nn Rajrani   1995, 1999-01-02, 2001-11-01, 2004-07-30, 2006-10-07 (total 5).

FIGURE 19.3
The dominant male Bankey in Dudhwa National Park
PHOTO: SATYA SINHA

FIGURE 19.4   The head of the dominant male Bankey with a well-
developed horn
PHOTO: ARUN KUMAR TRIPATHI

The rhino population in Dudhwa NP had grown to 42 in 2020 (table 19.23). Although the rhino area is still fenced, over the years some animals have strayed outside and may have settled in the semi-cultivated land around the reserve (Kalra & Sinha 2004; Sinha et al. 2005). Only one case of poaching was reported on 27 November 2011 (Talukdar & Sinha 2013). A few rhinos have strayed south from Nepal (probably from the Bardia population). In 1996 one adult female rhino reached Dudhwa through Basanta Forest and was attacked and killed by a resident rhino inside the rhino holding area. Another sub-adult male rhino reached Sitapur, where it was captured on 2 November 2004, released again and then darted on 2 January 2005 near Moradabad, from where he was taken to Kanpur Zoo (Talukdar & Sinha 2013). This male called Nakul is listed in the studbook (no. 121) in Kanpur Zoo from 12 January 2005 and in Indira Gandhi Zoological Park, Visakhapatnam from 6 April 2013 (Central Zoo Authority 2016).

19.6   Katerniaghat Wildlife Sanctuary

Katerniaghat (also Katarniaghat) is located in the Bahraich District in the state of Uttar Pradesh, about 25 km east of Dudhwa National Park across the Girwa-Ghaghara River and a similar distance south of Nepal's Bardia National Park. The Wildlife Sanctuary established in May 1976 is part of the Dudhwa Tiger Reserve, a component of Project Tiger since 1987.

There were no resident rhinos in Katerniaghat, but dispersal from Bardia NP has been noted. The population in Bardia was first established when 13 rhinos were translocated from Chitwan in 1986 (chapter 23). Apparently, three of these animals found their way southward through the Khata corridor along the Girwa River. A male, a female and a sub-adult settled in Katerniaghat in 1989 (Sinha et al. 2011b: 55; Talukdar & Sinha 2013). In 1994 there were four rhinos in the sanctuary, but the origin of the fourth animal remains unexplained. The latest reports indicate that these animals are still there (2020), while no breeding has been reported. As Dudhwa NP is nearby, some interchange between the populations is to be expected but has never been reported. The rhinos of Katerniaghat are chance visitors to the area.

19.7   Spreading the Population within Dudhwa

To avoid overcrowding in the original Rhino Ranging Area in the South Sonaripur area of Kakraha range, it was proposed to create a second habitat elsewhere in the national park (Sinha & Sinha 2010; Talukdar et al. 2012). This was created in 2018 in the Bhadi-Churella sector of Belrayan Range with an area of 14 km² (Talukdar 2018; Tripathi 2018). The translocation of 1 male and 3 females was accomplished successfully from 9 to 13 April 2018 (Sharma et al. 2018), as shown in figures 19.5, 19.6, 19.7. All females gave birth during 2021, but one calf was lost due to accidental drowning. The home range of each adult rhino was monitored and shown to be between 3.66 and 4.97 km² (Ravi et al. 2022).

19.8   Proposed Developments

While the population in Dudhwa NP has grown through careful protection and management, the program is not entirely without setbacks. The powerful dominance of the male Bankey from the start of the reintroduction to his death in 2016 has meant that all second-generation rhinos in the park are genetically related (Sinha & Sinha 2010). As the founder animals arrived both from Assam and Nepal, the offspring are potentially hybrids of two different genetic lineages. Zschokke et al. (2011) recommend that therefore rhinos from Dudhwa should not be exported or used for captive breeding. The knowledge available about the population over several decennia might be a good foundation for a careful genetic study to understand the effects of hybridisation and inbreeding.

FIGURE 19.5
Rhino sedated during the translocation
procedure in Dudhwa NP in April 2018
PHOTO: NISHANT ANDREWS

FIGURE 19.6
Rhino being loaded on a crate during the
translocation procedure in Dudhwa NP in
April 2018
PHOTO: NISHANT ANDREWS

FIGURE 19.7
One of the four rhinos exploring new terrain
after translocation in Dudhwa NP in April 2018
PHOTO: AMIT SHARMA

CHAPTER 20

The Rhinoceros in Nepal

20.1 The Rhinoceros of Nepal

Rhinoceros unicornis inhabits the *terai* region of Nepal stretching along the southern border with India. The *terai* has a flat and undulating landscape traversed by multiple rivers which meander from the higher Himalayan mountains down to the Ganges basin. Nepal remained closed to foreign visitors until 1951. There were of course exceptions as long as permission could be obtained from the Kings of the Shah dynasty and the Prime Ministers or Maharajas of the Rana family (1846–1951). Laurence Oliphant (1829–1888), the son of the Chief Justice in Colombo, had a chance to travel with Maharaja Jung Bahadur to Kathmandu. It did not take them long to travel from the plains to the higher valleys in 1851, finding that "the country was almost totally uncultivated, and nearly all traces of roads disappeared as we traversed the green sward of the Terai of Nepaul, scattered over which were large herds of cattle, grazing on short grass, which extended in all directions over the vast expanse of flat country" (Oliphant 1852: 37). This was the region where rhinos were found, then in large numbers, now in a more delicate situation.

The rhinos only occurred in the low valleys which are the northernmost extension of the Indian plains. There is no evidence that they ever lived in the mountains and higher grounds. There were few reasons for foreigners to visit these low areas, especially as the climate was often reported to be unhealthy with a range of diseases like malaria making stays uncomfortable. Some of the earliest reports came from members of an Italian mission traveling to the mountains further north towards Tibet. Their letters home, written in the middle of the 18th century, mentioned the existence of the rhino, and once even included a piece of rhino horn for the pleasure of the Pope (24.7).

As long as access to Nepal was restricted, details of the fauna remained fragmentary. There is some information about the region between the Indian border and the capital Kathmandu. Here the Rapti and Narayan Rivers provide favourable habitat for rhinos, which can now be viewed through safari companies operating in the Chitwan National Park. The western and eastern districts of Nepal along the border remained unexplored, at least very few people were able to write about their experiences.

In the following chapters the available data in a historical perspective are discussed from Shuklaphanta and

Bardia in the west, along the central part where most activity was recorded, to the eastern district of Morang (map 20.10). Even where the literature is silent, it may be expected that some rhinos were present in all suitable habitats in the southern *terai* (fig. 20.1).

A photograph taken in Nepal published in the popular *Illustrated London News* of 1906 is possibly the first one taken of a *R. unicornis* in the wild (Rookmaaker 2021b). It was one of four photographs of curiosities of big game hunting (Anon. 1906c). The photographer is not identified and the images were credited to "the Illustrations Bureau" of 12 Whitefriars Street, London. The photograph must

FIGURE 20.1 Three rhinos with long horns in a tropical landscape. Printed by François Chardon (d. 1862): "Imp. F. Chardon ainé, 30 Hautefeuille Paris"
ACHILLE COMTE, *MUSÉE D'HISTOIRE NATURELLE*, 1854, PL. AFTER P. 68

© L.C. (KEES) ROOKMAAKER, 2024 | DOI:10.1163/9789004691544_021
This is an open access chapter distributed under the terms of the CC BY-NC-ND 4.0 license.

have been taken prior to 1906, arguably in Nepal, and shows a rhino in profile with a large forward-sloping horn standing on the edge of a pond (fig. 20.2).

The people of Nepal have always had a special recognition of the importance of the rhinoceros as the largest pachyderm in the country next to the elephant. Most significantly, the Shradda-Tarpan ceremonies are performed by rulers and have a religious significance (24.2). In recent times King Birendra issued a gold coin and a bank note with a depiction of a rhino (figs. 20.3, 20.4). The rhino is part of the cultural fabric of all people in this mountainous region.

•••

Dataset 20.19: Localities of Records of Rhinoceros in Nepal

There are only a few localities in the western and eastern districts. The events in the central part, around Chitwan National Park, are confined to a small area. All localities are combined in map 20.10. The exact records are explained in chapters 21 to 26.

Rhino Region 13: Nepal West
*	Sharda River – 28.95N; 80.10E – rhino absent
A 251	Shuklaphanta NP – 28.85N; 80.20E
A 252	Bilauri – 28.72N; 80.37E
A 253	Bardia NP – 28.55N; 81.35E

Rhino Region 14: Nepal Central
A 254	Tribeni (Gandak River) – 27.46N; 83.91E
A 255	Meghauli – 27.55N; 84.20E
A 256	Kasara – 27.56N; 84.32E

A 256	Sauraha – 27.58N; 84.50E
A 257	Devghat – 27.80N; 84.40E
A 258	Kantipur – 27.42N; 84.40E
A 258	Thori, Bhikna Thori – 27.36N; 84.59E
A 259	Hetauda, Hewtora, Hitounda, Hetowda – 27.44N; 85.02E

Rhino Region 15: Nepal East
A 260	Bagmati River north of Muzaffarpur – 27.00N; 85.50E – 1909
A 261	Saptari Dt. – 26.60N; 86.70E
A 261	West bank of Kosi River, at Patharghatta – 26.70N; 87.00E – 1909
*Z 911	Chainpur – 27.25N; 87.25E – 1797 (trade only)
A 262	Morang, east of Kosi River – 26.50N; 87.40E – 1901
*Z 912	Koshi Tappu WR – 26.65N; 87.00E – 1995

•••

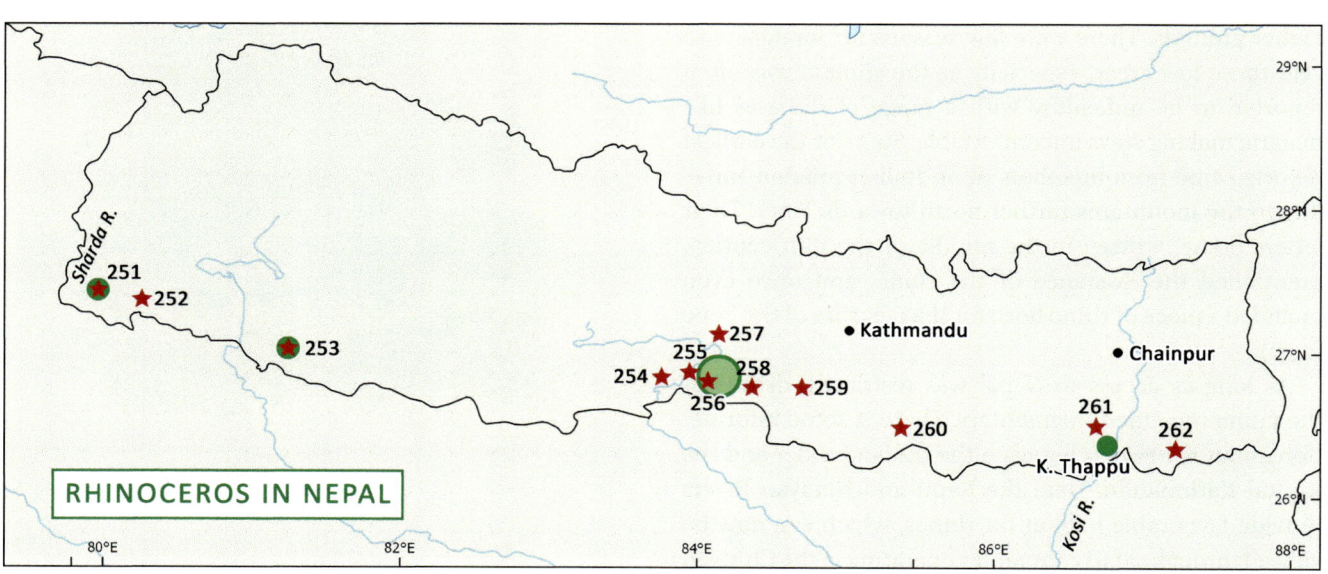

MAP 20.10   Records of rhinoceros in Nepal, located in the west (Rhino Region 13), centre (Rhino Region 14) and east (Rhino Region 15). The numbers and localities are explained in Dataset 20.19. Green circles indicate national parks.
★   Presence of *R. unicornis*

FIGURE 20.2
"Rhinoceros in the jungle Nepal." While
photographer and circumstances are
unknown, this image must have been taken
before 1906, hence it is likely to be the first
photograph of *R. unicornis* in the wild
ILLUSTRATED LONDON NEWS,
10 FEBRUARY 1906, P. 200

FIGURE 20.3
Gold coin with value Rs. 1000 of Nepal issued
in 1972 under King Birendra, 1972–2001.
Only 2176 pieces were struck

FIGURE 20.4
Banknote of Nepal for the value of
Rs. 100, first issued in 2012. One side shows
King Gyanendra Bir Bikram (b. 1947) wearing
a plumed crown
COLLECTION KEES ROOKMAAKER

CHAPTER 21

Historical Records of the Rhinoceros in Western Nepal

First Record: nil – Last Record: Strays in 1990s
Species: *Rhinoceros unicornis*
Rhino Region 13

21.1 The Western Reaches of Nepal

Given that Nepal remained closed to foreign visitors, the fauna of the *terai* west of Nepalgunj remained unknown. There is not a single definite record that a rhino lived in this region. It is unlikely that rhinos existed there when the Prince of Wales came to the Sharda River in 1876 or when Franz Ferdinand visited in 1893. Maharaja Juddha Shumsher came here in 1933 and 1936 to hunt, noting that no rhinos were expected (Smythies 1942).

In a survey of the mammal fauna of southern Nepal, the American zoologist David Lee Chesemore (b. 1939) visited western Nepal in 1965. He thought that rhinos might survive although their status was unclear. The Irish explorer Peter Byrne saw a rhino footprint in the Rani Tal area of Shuklaphanta in the 1990s and suggested that it was a stray from the Pilibhit area further south. He talked to local Tharus men, who believed rhinos might have been present "many years ago." Before the translocation of rhinos into Shuklaphanta National Park in 2000, there was already one male rhino in the sanctuary (Martin et al. 2009a).

To account for these rare rhino sightings in south-western Nepal, it is suggested that these were animals straying from the new populations in Pilibhit Tiger Reserve and Dudhwa National Park across Nepal's southern border. My conclusion is that there never was a resident rhino population in south-western Nepal in historic times.

∙∙∙

Dataset 21.20: Chronological List of Records of Rhinoceros in Western Nepal

General sequence: Date – Locality (as in original source) – Event – Source – § number, figures – Map number.
The localities are shown on map 20.10 and explained in Dataset 20.19. Records for Shuklaphanta NP after the translocation in 2000 are listed in Dataset 22.21, and those for Bardia NP after 1986 in Dataset 23.22.

1876 – Sarda (Sharda) River – No rhino found during shoot by Albert Edward, Prince of Wales (1841–1910) on invitation of Maharaja Jung Bahadur (1817–1877), Prime Minister and Commander-in-Chief of Nepal 1846 to 1877 – W. Russell 1877; Beresford 1914; Shaha & Mitchell 2001: 40 – 21.2

1876 – Sardah (Sharda) River – In former years it was to be met with in the forests bordering on the Sardah in Nepaul, but it is now extinct there or very nearly so – Baldwin 1876: 145

1893 – West Nepal – No rhino found during shooting expedition of Franz Ferdinand, Archduke of Austria (1863–1914) in western Nepal – Franz Ferdinand 1895; Höfer 2010 – 21.3

1933 – Naya Muluk, between Sarda (Sharda) and Rapti Rivers – No rhinos exist. Maharaja Juddha Shumsher (1875–1952) hunted here in 1933 and 1936 – Smythies 1942: 51

1965 – Bilauri, far S W of Nepal, 28.72N; 80.37E – Rhino may survive, status unknown – Chesemore 1970: 165 – 21.1, map 20.10

1990s – Shuklaphanta – Peter Byrne (b. 1925) showed a photograph of a rhino to local Tharus men, who said that "many years ago" there might have been some in the area – Peter Byrne, email 23.01.2018

1990s – Rani Tal in Shuklaphanta – Rhino track. Possible stray rhino from India – Byrne 2008: 6 – 21.1

2000 – Shuklaphanta – There were perhaps a handful of rhinos in the 20th century but by 2000 only one remained – Martin et al. 2009: 104

∙∙

21.2 The Prince of Wales on the Sharda River in 1876

Albert Edward, Prince of Wales was invited to visit the Kingdom by Maharaja Jung Bahadur in 1876. They met in the extreme west of Nepal, in the region of the Sharda River around Banbasa. The detailed itinerary provided by William Howard Russell (1820–1907), the Prince's Honorary Private Secretary during the trip, gives a description of the daily events in Nepal from 21 February to 6 March 1876. A total of 700 elephants were involved in

© L.C. (KEES) ROOKMAAKER, 2024 | DOI:10.1163/9789004691544_022
This is an open access chapter distributed under the terms of the CC BY-NC-ND 4.0 license.

FIGURE 21.1
"Indian hunting trophies, and zoological specimens, collected by H.R.H. The Prince of Wales." View of an exhibition in London following the tour of Albert Edward, Prince of Wales in western Nepal in 1876 engraved by Richard Hewitt Moore (d. 1910). The rhino skull (displayed to the right of the bear) was donated by the Nepal ruler and obtained in another part of Nepal
DETAIL FROM *ILLUSTRATED SPORTING AND DRAMATIC NEWS*, 7 APRIL 1877, P. 70

the chase of tigers and other game, but rhinos were not expected there.

The zoological collection assembled during the Prince's tour in India and Nepal in 1876 included three rhino skulls from Nepal, and must have been given to the party in February-March. The naturalist attached to the royal party was Clarence Bartlett (1849–1903), the son of the Superintendent of the London Zoo. After the Prince's return, the hunting trophies and zoological specimens were arranged for exhibition in the Picture Gallery of the Zoological Society's Gardens, opened to the public on Monday 6 February 1877 (*Cambridge Independent Press* 1877-02-10). The three rhino skulls were placed on a table adjacent to a huge skull of a Ceylonese elephant. Only one skull is depicted in an engraving by Richard Hewitt Moore in the *Illustrated Dramatic News* 1877-04-07 (fig. 21.1). The same newspaper reported that they were examined by William Henry Flower (1831–1899), because they were considered to be smaller than usual, who identified them as *R. unicornis*. It is not known how long the exhibition in the London Zoo lasted. All three skulls with nasal horns were added to the collection of the Royal College of Surgeons of England, curated by Flower, catalogued as nos. 2124 to 2126 (Flower 1884: 417). Another female rhino skull from the Nepal *terai* was donated in 1883 to the Natural History Museum, London, no. 1883.10.23.3 (Thomas 1906: 18). Elsewhere in London, at the Burlington Gallery, Piccadilly, there was an exhibition of over 200 drawings by the artist William Simpson (1823–1899) labelled as India "Special" (Simpson 1876). On show from 22 June to 30 September 1876, this exhibition had no image of a rhino, but several showing the Prince on a howdah with elephants, shooting bear and tiger.

21.3 The Visit of Franz Ferdinand of Austria in 1893

Franz Ferdinand, Archduke of Austria, visited India and Nepal in 1893. As shown on the map in his book, he traveled from Delhi, past Pilibhit, through the extreme south-western corner of Nepal from Dakna Bhagh to Sohela in March 1893 (Franz Ferdinand 1895). There is no mention of a rhino, nor did he bring home any rhino-related objects or photographs (Höfer 2010).

Protecting the Rhinoceros in Shuklaphanta National Park, Nepal

First Record: translocation in 2000 – Last Record: current
Species: *Rhinoceros unicornis*
Rhino Region 13

22.1 Rhinos in Shuklaphanta

Shuklaphanta in the south-western corner of Nepal was first gazetted as a Wildlife Sanctuary in 1976. The Maharajas of the Rana family have recorded shoots in this area in 1876, with the Prince of Wales, and in 1933 and 1936, without ever observing a rhino (21.1).

Four rhinos were translocated in March 2000 from Chitwan National Park. There have been at least six births up to 2013. A further 5 rhinos were transported from Chitwan in 2017. The status of the area was upgraded to National Park on 3 March 2017.

As Shuklaphanta borders on wildlife areas in Uttar Pradesh, like the Pilibhit Tiger Reserve and Dudhwa National Park, there is some dispersal across the border. One or two animals were already present in the 1990s, while the population figures can only be correct with such an exchange in mind (P. Adhikari 2003; Talukdar & Sinha 2013). In 2013 all nine resident rhinos were identified, numbered and photographed by a team led by Naresh Subedi for the National Trust for Nature Conservation (NTNC).

• • •

Dataset 22.21: Chronological List of Records of Rhinoceros in Shuklaphanta NP

General sequence: Date – Locality (as in original source) – Event – Sources – § number. Records in this region prior to 2000 are found in Dataset 21.20.

1963 – E.P. Gee visits and advises against translocation of rhinos to the area – Gee 1963c: 73

1976 – Gazetted as Royal Shuklaphanta Wildlife Reserve, 155 km² – Heinen & Kattel 1992

1994 – 1 rhino recorded – Nepal Government 1995: 12

2000 – One male rhino of original population remained, origin unknown (possibly Bardia) – Martin et al. 2009a: 104; K. Thapa et al. 2013: 347; Talukdar & Sinha 2013

2000-03 – Translocation of 1 adult male and 3 adult females from Chitwan to Shuklaphanta – Dinerstein 2003: 95; Byrne 2008: 7; Martin et al. 2009a: 104; Bhuju et al. 2009: 15; Thapa et al. 2013:347

2003 – 1 male from Dudhwa took up residence in Shuklaphanta – Dinerstein 2003: 95

2005 – 1 female was found dead close to the main east-west road west of Bauni bridge. She carried a 7-months old foetus; stomach full of worms – Byrne 2008: 7

2005, 2006 – Presence of female and calf, not part of translocation, possibly from Dudhwa through the Laljhadi corridor – Thapa et al. 2013: 349

2006 – 1 female found dead, probably from natural causes – Byrne 2008: 7

2008 – 3 rhinos strayed from Shuklaphanta into India – Talukdar 2008b

2017-04 – Translocation from Chitwan 1 male and 4 female – NTNC 2017: 50, 70

2017-03-03 – Gazetted as National Park, now 305 km² on World Wildlife Day

∴

22.2 Population Structure

The population of rhinos in Shuklaphanta is slowly increasing since the translocations from Chitwan NP in 2000 and 2017 (table 22.24). The population of individual rhinos is reconstructed based on the survey of Subedi et al. (2014), who gave the numbers 3001 to 3009 to the specimens observed (table 22.25). The identities of the earliest animals have not been established accurately.

© L.C. (KEES) ROOKMAAKER, 2024 | DOI:10.1163/9789004691544_023
This is an open access chapter distributed under the terms of the CC BY-NC-ND 4.0 license.

Kampf des Elephanten mit dem Rhinozeros.

FIGURE 22.1
Fight of the Elephant with the Rhinoceros ("Kampf des Elephanten mit dem Rhinozeros"). An unsigned plate of a gruesome event, included in a luxurious picture book for young people published by Christian Friedrich Grimmer 1834
GRIMMER, *NEUES DRESDENER BILDER-CABINET ZUR BELEHRENDEN UNTERHALTUNG FÜR DIE JUGEND*, 1834, PL. VII

TABLE 22.24   Status numbers for Shuklaphanta WR or NP 2000–2021

Year	Number	Reference[a]
2000	+ 4	Translocated from Chitwan
2000	1, 7	R159; R062
2003	6	R003
2007	6, 7	R253; R049, R229
2008	5, 6, 8–10	R131, R179, R243; R151; R062
2010	6	R151
2011	7	R013, R230, R260
2015	8	R215, R251
2017	+ 5	Translocated from Chitwan
2017	15	R308
2021	17	R327
2022	17	R331

a   References: R003 Adhikari 2003; R049 Bhatta 2008; R062 Byrne 2008: 7; R131 Jnawali et al. 2011: 33; R151 Martin & Martin 2010; R159 Martin 2001: 43; R163 Martin et al. 2009a: 107; R179 Nepal Government 2008; R215 Sireng 2015; R229 Strien & Talukdar 2007; R230 Subedi et al. 2013; R243 Talukdar 2008b; R251 Talukdar 2015; R253 Talukdar et al. 2011; R260 Thapa et al. 2013; R308 NTNC Nepal 2017; R327 Basel Zoo 2021; R331 Talukdar 2022.

TABLE 22.25   Details of rhino specimens found in Shuklaphanta from 2013

Sex[a]	Number and name	Data
M	3004 Chaudhar Bhale	Possibly stray animal previously in the area.
M	3002 Rani Pothi	Stray animal first recorded in 1995 near Rani Tal. Had 1 calf (3003).
M	3001 (Chitwan 01) Chitwan Bhale	From Chitwan in 2000-03. Usually in Taturgunj near border. Died 2014-01-10.
F	(Chitwan 02) Chitwan Pothi	From Chitwan in 2000-03. Had 1 calf (3005).
F	(Chitwan 03)	From Chitwan in 2000-03. No record in 2014, hence lost. Died 2005 (?).
F	(Chitwan 04)	From Chitwan in 2000-03. No record in 2014, hence lost. Died 2006 (?).
F	3003 Solu Pothi	Born 2001-11 to 3002. Had 2 calves (3006, 3008).
F	3005 Surya Pothi	Born 2006-11 to "Chitwan Pothi" and male 3001. Had 1 calf (3009).
U	3006 Bichuwa	Born 2007-10 to 3003 (father unknown).
U	3007 Bhadaure	Born 2009-10 to 3002 (father unknown).
U	3009 New Baby	Born 2011-11 to 3005 (father unknown).
U	3008 Kartike Baby	Born 2012-10 to 3003 (father unknown).
M	(Chitwan 05)	From Chitwan in 2017-04.
F	(Chitwan 06)	From Chitwan in 2017-04.
F	(Chitwan 07)	From Chitwan in 2017-04.
F	(Chitwan 08)	From Chitwan in 2017-04.
F	(Chitwan 09)	From Chitwan in 2017-04.

a   M = Male, F = Female, U = Unknown.

CHAPTER 23

Protecting the Rhinoceros in Bardia National Park, Nepal

First Record: translocation in 1986 – Last Record: current
Species: *Rhinoceros unicornis*
Rhino Region 13

23.1 Rhinos in Bardia

There are no historical records for the region in southern Nepal where Bardia National Park is situated. It is impossible to decide if this means that the species did not exist there in former times, or that it just went unnoticed due to the inability of foreigners to visit the area. Rhinos occurred around Katerniaghat (Uttar Pradesh) on the banks of the Ganghara River, therefore stray rhinos might well have wandered across the border without leaving particular records.

A total of 91 rhinos from Chitwan NP has been translocated to Bardia between 1986 and 2017 (23.2). Due to political instability in the area and dispersal of rhinos, the numbers have not increased significantly (table 23.27). There is adequate protection at this time in the park, providing hope for a stable population.

• • •

Dataset 23.22: Chronological List of Records of Rhinoceros in Bardia National Park

General sequence: Date – Event – Sources. Records in this region prior to 1986 are found in Dataset 21.20.

1969-07 – Area declared a Royal Shikar Reserve under the hunting regulations of 1966 – Bolton 1976

1976-03-08 – Establishment of Royal Karnali Wildlife Reserve – Bolton 1976; Heinen & Kattel 1992

1982 – Renamed Royal Bardia Wildlife Reserve – Amin et al. 2009

1984 – Area extended to 968 km² to include Babai River – Amin et al. 2009

1986-02 – Translocation of 4 (3 M, 1 F) from Chitwan to Babai Valley – Bauer 1988; Thapa et al. 2013

1986-12 – Translocation of 9 (5 M, 4 F) from Chitwan to Babai Valley – Hart-Davis 1986; Mishra & Dinerstein 1987; Bauer 1988; Dinerstein et al. 1990: 36; Jnawali & Wegge 1993; Dinerstein 2003: 94; Mishra 2008; Bhuju et al. 2009; Thapa et al. 2013

1988-12 – Gazetted as (Royal) Bardia National Park. Protected areas in Nepal are administered by the Department of National Parks and Wildlife Conservation (DNPWC) – Heinen & Kattel 1992

1990-01 – Translocation of 20 (6 M, 14 F) from Chitwan to Babai Valley – Dinerstein 2003: 94; Bhuju et al. 2009; Thapa et al. 2013

1990-02 – Translocation of 5 (2 M, 3 F) from Chitwan to Babai Valley – Thapa et al. 2013

1997 – Area of 327 km² surrounding the park was declared as a buffer zone – Amin et al. 2009

1999 – Translocation of 4 (2 M, 2 F) rhino from Chitwan to Babai Valley – Dinerstein 2003: 94; Bhuju et al. 2009

2000 – Translocation of 16 (8 M, 8 F) from Chitwan to Babai Valley – Dinerstein 2003: 94; Bhuju et al. 2009; Thapa et al. 2013

2001-03 – Translocation of 5 (2 M, 3 F) from Chitwan to Babai Valley – Pradhan 2001; Bhuju et al. 2009; Thapa et al. 2013

2002-03 – Translocation of 10 (5 M, 5 F) from Chitwan to Babai Valley – Bhuju et al. 2009; Thapa et al. 2013

2002 – Maoist insurgency leading to suspension of army guards in park – Martin et al. 2009a; Thapa et al. 2013

2003 – Translocation of 10 (3 M, 7 F) from Chitwan to Babai Valley – Bhuju et al. 2009; Thapa et al. 2013

2006 – Army posts re-established – Martin et al. 2009a

2016 – Translocation of 5 rhino from Chitwan to Babai Valley – NTNC 2017: 50, 61

2017-03 – Translocation of 3 (1 M, 2 F) rhino from Chitwan to Mulghat area and Babai Valley – NTNC 2017: 50, 61

2019-03 – Visit of author to Bardia NP in company of Prem Bohara and Suman Lama. No rhinos or tiger were observed.

© L.C. (KEES) ROOKMAAKER, 2024 | DOI:10.1163/9789004691544_024
This is an open access chapter distributed under the terms of the CC BY-NC-ND 4.0 license.

FIGURE 23.1
One-horned rhino in Bardia National Park,
Nepal
PHOTO: ANDREW GELL, NOVEMBER 2009

FIGURE 23.2
Rhino after sedation during the translocation
exercise from Chitwan NP to Bardia NP in 2016
PHOTO: WWF NEPAL

23.2   Translocations from Chitwan National Park

The area was first put under protection in 1976 as the Royal Karnali Wildlife Reserve, renamed after the extent was increased and a buffer zone was established to Bardia National Park in 1988. All rhinos in the park were originally translocated from Chitwan NP, about 450 km to the east. The first translocation took place in 1986 and has since been followed by several further movements (table 23.26). There are no reports of injuries or deaths resulting from the transport of the animals.

The first rhinos from Chitwan were released in the valley of the Karnali River on the western side of the park, while most subsequent releases were in the valley of the Babai River on the eastern side. The animals dispersed quite widely across the terrain, with several moving southward into Indian territory. The population settled relatively well and births were regularly reported.

TABLE 23.26   Translocations of rhinoceros from Chitwan to Bardia

Year	Males	Females	Total
1986	8	5	13
1990	8	17	25
1999	2	2	4
2000	8	8	16
2001	2	3	5
2002	5	5	10
2003	3	7	10
2016	2	3	5
2017	1	2	3
Total	39	52	91

FIGURE 23.3
Rhinoceros on the banks of the Babai River
in Bardia National Park
PHOTO: ASHISH BASHYAL

23.3   Period of Political Unrest

During the last years of the Nepalese Civil War, which was resolved in November 2006, there was much unrest in the Bardia region. In 2002 the army withdrew its posts from the Babai valley, and all patrolling was suspended (Martin et al. 2009a; Thapa et al. 2013). When the protection was restored, poaching continued. In 2008 probably around 20 rhinos remained, although estimates vary (Kock et al. 2009). The most comprehensive survey in 2009 showed the presence of 22 animals (15 adult, 1 subadult, 6 calves), all resident in the Karnali river valley (Amin et al. 2009; Martin & Martin 2010). An attempt to monitor individual specimens in 2011 showed that 6 (3 M, 3 F) animals translocated from Chitwan remained, including one male from the original 1986 set (Subedi et al. 2014). The remaining 26 (9 M, 12 F, 5 unknown) animals were all born in Bardia over the years.

A blind male named Nikunja Kane Bhale born in 1990 was cared for in an enclosure at the park headquarters in Thakurdwara (Subedi, no. 18). In 2003, Esmond Martin also photographed two rhinos in the camp cared for by army personnel and park staff (Martin 2004).

23.4   Protecting Rhinos in Bardia

The situation is stabilising. There are frequent patrols and studies in the park. There are limited facilities for both national and foreign visitors near the park headquarters. The density of rhinos is still quite low, and there are regular reports of animals moving southward to Katerniaghat, although park staff on both sides of the border cooperate to reduce such movements. With all historical incidents of poaching, insurgency and straying, the translocations to Bardia have not been an unmitigated success story. The tourist facilities need to be upgraded and the information available to international tourists should be streamlined to reduce false expectations of wildlife viewing. With sufficient budgets and continued efforts of army personnel and park staff, there is no reason why the rhino population in Bardia could not establish itself more firmly.

TABLE 23.27   Figures of total number estimates, natural deaths and animals poached for Bardia NP from 1986

Year	Number	Natural deaths	Poached	References
1986	translocated 13 (5/8)			R080, R130
1987	38	1		R013, R152
1989	13		1	R138, R211, R152
1990	translocated 25 (5/20)			
1992		4	2	R152
1993	40			R098

TABLE 23.27   Figures of total number estimates, natural deaths and animals poached for Bardia NP from 1986 (*cont.*)

Year	Number	Natural deaths	Poached	References
1993	39			R137
1993		1	4	R152
1994	10, 15, 39	2	1	R079, R130, R152
1995	40			R098
1997	40			R206, R253
1999	translocated 4			
2000	67			R013, R094, R159, R188, R229, R260
2000	translocated 16			
2001			2	R160
2001	translocated 5			
2002		2	3	R150, R160, R162
2002	translocated 10			
2003	73	1	9	R150, R162, R284
2003	translocated 10			
2004	100	1	2	R150, R162
2005	32			R013
2006	few		2	R094, R162
2007	30, 35		3	R049, R162, R229, R253
2008	17, 22, 31			R141, R013, R131, R179, R244
2009	22		0	R151
2011	24			R013, R180, R230
2015	21			R084
2015	29			R251
2017	translocated 3			
2019	22			media reports
2021	38			R331

(1/2 means 1 M, 2 F). M = Male, F = Female

References: R013 Anon. 2011; R049 Bhatta 2008; R079 Dey 1994a; R080 Dinerstein & Jnawali 1993; R094 Emslie et al. 2007; R098 Foose & van Strien 1997: 9; R130 Jnawali 1995; R131 Jnawali et al. 2011: 33; R137 Khan & Foose 1994: 6; R138 Khan 1989: 3; R141 Kock et al. 2009; R150 Martin & Martin 2006; R151 Martin & Martin 2010; R152 Martin & Vigne 1995; R159 Martin 2001: 43; R160 Martin 2004: 91; R162 Martin et al. 2009a; R179 Nepal Government 2008; R180 Nepal Government 2012; R188 Pradhan 2001; R206 Shrestha 1997: 24; R211 Sinha & Sawarkar 1993; R229 Strien & Talukdar 2007; R230 Subedi et al. 2013; R244 Talukdar 2009c; R251 Talukdar 2015; R253 Talukdar et al 2011; R260 Thapa et al. 2013; R284 Wegge et al. 2006: 700; R331 Talukdar 2022.

Historical Records of the Rhinoceros in Central Nepal

First Record: 1750 – Last Record: current
Species: *Rhinoceros unicornis*
Rhino Region 14

24.1 The Rhinoceros in Central Nepal

Nepal is one of the foremost rhino countries. Rhinos were concentrated in the low *terai* along the banks of the Rapti, Reu, Narayani and Gandak Rivers, nowadays protected as Chitwan National Park. This was where the rulers went to experience the jungle among abundant wildlife. The region was always associated with rhinos. Here they were observed, captured and hunted, more so than elsewhere in this mountainous country. Rhinos were royal game and could not be hunted without a special permit. This ban may have been instituted when the Rana family came to power in 1846. Even earlier, in the time of King Rana Bahadur Shah (1775–1806) there was talk of a general prohibition to kill a rhino when this custom was still widespread.

The diplomat Laurence Oliphant was told that Maharaja Jung Bahadur (1817–1877) was passionate about sport hunting. For most of the 19th century, there is scant written or pictorial evidence to link him or his successors Ranodip Singh Kunwar (1825–1885) and Bir Shumsher (1852–1901) to personal shoots involving rhinos. Chances are that these actually took place without leaving much trace. Nepal being closed to foreigners, only a few could have seen rhinos, a rare privilege accorded the British envoy Brian Hodgson in the 1820s and to occasional diplomatic missions later in the century. In the 20th century, during the reigns of Prime Ministers Maharaja Chandra Shumsher (1863–1929) from 1901 to 1929 and Maharaja Juddha Shumsher (1875–1952) from 1932 to 1945, regular shooting camps took place along the Rapti River in the *terai* of southern Nepal.

The rulers wanted to keep the rhino habitat safe because they enjoyed spending time in the jungle and being away from the constant administrative duties attached to their office. Perhaps more importantly, the animal possessed a religious significance.

24.2 The Importance of the Rhinoceros in Shradda Ceremony

In Hindu religion, the Shradda ceremony is performed by a male child to honour his dead ancestors. When conducted on the first annual death anniversary, it will lead to the everlasting peace of a departed soul and eventually liberate him from the eternal cycles of rebirth. In Nepal, the rhino is associated with the most effective type of Shradda. In the words of Maharaja Kaiser Shumsher, "The flesh and the blood of the rhino is considered pure and highly acceptable to the Manes, to whom high-caste Hindus and most Gurkhas offer libation of its blood after entering its disemboweled body. On ordinary Sradh days the libation of water and milk is poured from a cup carved from its horn. The urine is considered antiseptic, it is hung in a vessel at the principal door as a charm against ghosts, evil spirits, and disease" (Ellison 1925: 35). The most illustrious ceremony is the Shradda Tarpan, which in the Nepalese tradition should be performed once by every ruler. As soon as a rhino is killed, the stomach and intestines are cleaned out leaving a large empty cavity, quite big enough to take a human body. The person achieving the rite goes inside this cavity, drinks a cup of the rhino's blood and says the appropriate prayers (Briggs 1938: 130; Smythies 1974: 197; Mishra 2008).

The Shradda Tarpan in Nepal can only be performed by the Kings and Maharajas, and with their permission by family members, as they control all permissions to kill rhinos. Even the use of rhino horn is limited to a small number of wealthy or influential men. There are allusions that a rhino must be killed on the exact day one of the rulers dies (Oliphant 1852: 83; Davis 1940: 281). During the time of the Rana dynasty there have been 11 Prime Ministers or Maharajas, followed by 4 Kings, but only the ceremonies performed in 1933, 1936, 1938, 1953, 1979 have been documented. There is no published record about the killing of rhinos in captivity for such rites, which of course is no guarantee that this didn't happen.

Rhino horn is highly prized in Nepal not only for religious reasons. It is also used as medicine and aphrodisiac (Kennion 1932: 227; Smythies 1974; Mishra & Mierow 1976; Martin 1985; Lohani 2011), and by Yogis in worship when wearing rings of rhino horn around their finger as

© L.C. (KEES) ROOKMAAKER, 2024 | DOI:10.1163/9789004691544_025
This is an open access chapter distributed under the terms of the CC BY-NC-ND 4.0 license.

ordered by Lord Shiva (Briggs 1938). In the Newar culture of the Kathmandu valley, there is a legendary song about

Princess Mahavira who, disguised as a man, killed a rhino and received a reward (Lienhard 1974: 205).

•••

Dataset 24.23: Chronological List of Records of Rhinoceros in Central Nepal

General sequence: Date – Locality (as in original source) – Event – Sources – § no., figures.

From 1951 records are in chapter 26 (Chitwan). All localities are in Central Nepal (Chitwan, Rapti River area) unless specified. The localities are shown on map 20.10 and explained in Dataset 20.19. The persons known to have interacted with rhinos in Central Nepal are listed with details in Table 24.28. The records are divided in (A) early records of the 17th and 18th centuries; (B) records of the 19th century; (C) time of Maharaja Chandra Shumsher 1901–1929; (D) time of Maharaja Bhim Shumsher 1929–1932; (E) time of Maharaja Juddha Shumsher 1932–1945; (F) the modern period after 1945.

A. Early Records of the 17th and 18th Centuries

1681 – Kantipur (south of Chitwan) – Bidhata Indra, grandson of Hindupati, came to the Kantipur Palace escaping with a rhino. The cause of the flight is not given – Regmi 1961, vol. 1: 73

17th cent. – Lakshmi Temple, Bhaktapur – Two rhino sculptures on side of steps – Majupuria 1977: 128; Martin & Vigne 1995: 24; de Geer 2008: 35 – 24.6, fig. 24.2

1740 – Makwanpur – The Vatican missionary Cassiano de Macerata (1708–1791) states that dangers of the jungle include presence of tigers, elephants and rhino (rinoceronte) – Cassiano 1767: 9; Lunden 1994 – 24.7

c.1750 – Makwanpur – Antonio Agostino Giorgi (1711–1797) reports presence of rhino – Giorgi 1762: 431; Lévi 1905, vol. 1: 120; Regmi 1961, vol. 1: 82 – 24.7

c.1750 – Between Bettiah and Kathmandu – Marco della Tomba (1726–1803) reports presence of rhino – Tomba 1878: 46; Lévi 1905, vol. 1: 123 – 24.7

1756 – Devghat – Tranquillus of Apechio (1708–1768) mentions rhino – Vannini 1977: 78; Regmi 1961: 67 – 24.7

1756 – Devghat – King Tribikram Sen of Tanahun sends letter to Pope Benedict XIV (1675–1758) accompanied by rhino horn – Vannini 1977: 78; Regmi 1961, vol. 1: 67 – 24.7

1775 to 1806 – During reign of King Rana Bahadur Shah (1775–1806) there was talk of a general prohibition to kill a rhino – Regmi 1971: 121 – 24.1

1798 – Saptari terai – King Rana Bahadur Shah writes to Brahmananda Upadhyaya, Subha of Saptari to ban killing of rhino and request to capture baby rhinos – Regmi 1971: 121

B. Records of the 19th Century

1811 – Nepal terai – Forest greatly infested by rhinos and tigers – Kirkpatrick 1811, vol. 1: 19

1819 – Nepal terai – Rhino present – Hamilton 1819: 63

1834 – Rhino plentiful in low hills and forests – Hodgson 1832: 345, 1834: 98, 1844: 288 – 24.8

1844 – Hitounda (Hetauda) – Captain Thomas Smith killed 7 rhino accompanied by two Nepalese officers Sirdar Bowanee Singh and Sirdar Delhi Singh – Smith 1849: 342; Smith 1852, vol. 1: 86–87; Ruge 1854: 69–73; Rookmaaker 2004b. Smith's account was retold in Dutch by Kessel (1862: 49–61) "Ontmoetingen in Nepal" – 24.9

1845-01-06 – Hettaunda (Hetauda) – Prinz Friedrich Wilhelm Waldemar von Preussen (1817–1849) went out in the jungle "in order to see a rhinoceros" but he saw only "uncertain traces" (Hoffmeister). Waldemar shoots 2 tigers on 3 and 8 March 1845 elsewhere in the terai – Mahlman 1853, part 2: 14, text to plate XXI (Tigerjagd im Tarai); Kvaerne 1979: 37; Hoffmeister 1847: 138, 1848: 213. Mahlman 1853, part 2, plate VIII is a view of Hettaunda village (no animals)

1851-01-29 – Tirhaj (= terai) – The party of Francis Egerton Earl of Ellesmere (1800–1857) plans to hunt rhino – Egerton 1852, vol. 1: 204

1851-02-04 – Hewtora in terai (Hetauda) – An expected rhino hunt was aborted due to lack of riding elephants – Egerton 1852, vol. 1: 248

1851-12 – District around Chitaun is open country, "covered with long grass jungle rather than forest, and is very much infested with rhinoceros" – Oldfield 1880, vol. 1: 49

1851-12 – Maharaja Jung Bahadur (1817–1877) "killed several and wounded a large number of rhinoceros. Generally the elephants are afraid of them" – Oldfield 1880, vol. 1: 235

1851-12 – Chitaun (Chitwan) – Maharaja Jung Bahadur killed a rhino with one bullet. The skull and skin were preserved at Thappatalli, a residence of the Maharaja near Kathmandu – Oldfield 1880, vol. 1: 236

1851-12 – R. unicornis. Specimen donated by Henry Ambrose Oldfield (1822–1871), resident surgeon in Nepal 1850–1863. Natural History Museum, London, no. 1926.6.7.8 – Oldfield 1880; Pocock 1944b: 838

1852 – Maharaja Jung Bahadur is passionate in hunting tiger and rhino – Oliphant 1852: 26 – 24.1

1852 – Kathmandu, captive rhino. One rhino has to be killed when a Rajah dies – Oliphant 1852: 83 – 24.2

1857 – Maharaja Jung Bahadur no longer allowed any ladies to take part in rhino hunts after one was killed when the elephant was charged and knocked over – Kennion 1922: 603, 1932: 225

1861-11 – Thori – Maharaja Jung Bahadur shot 1 rhino, trophy carried to camp – Pudma Jung 1909: 255

1862-12 – Maharaja Jung Bahadur killed 4 rhino during 4 months hunt – Pudma Jung 1909: 266

1870-02-23 to 28 – Rhino present in terai. Duke of Edinburgh (King Alfred 1844–1900) hunts with Maharaja Jung Bahadur. No rhino killed – Fayrer 1870: 55, 184; Alfred Duke of Saxe-Coburg 1870

1873 – Nepal terai – Francis George Baring, 2nd Earl of Northbrook (1850–1929) killed 12 rhinos, while hunting with John Biddulph (1840–1920) of 19th Hussars and Colonel Hankey (not identified) – Homeward Mail from India 1873-04-07; Petite Gironde 1873-11-04 – 24.10

1875 – Nepal – R unicornis specimens donated by Evelyn Baring, 1st Earl of Cromer (1841–1917) to Indian Museum (nos. h, j), now in Zoological Survey of India nos. 2735 and 2736 (19243) or nos. GRM353, GRM354 – Sclater 1891, vol. 2: 202; Hinton & Fry 1923: 427; Groves & Chakraborty 1983: 253; Chakraborty 2004: 39

1876 – West Nepal – Albert Edward, Prince of Wales (1841–1910) hunts in West Nepal, without finding rhinos – 21.2

1880 – Nepal – R. unicornis horn and 3 nails donated by John Anderson (1833–1900) to Indian Museum (no. q) – Indian Museum 1881; Sclater 1891, vol. 2: 202; Hinton & Fry 1923: 427

1883-02 – Rapti River – Olly de Grey, in full Frederick Oliver Robinson, 2nd Marquess of Ripon, 4th Earl de Grey (1852–1923) and friends killed 6 rhino. His companions were Wilfrid Greenwood, William Duke of Portland, Charles Beresford, Baron Algernon Wenlock and Major Jackson (see Table 24.28) – Anon. 1884b; Portland 1937; Godfrey 2012. Photographs in photographic album dated 1883 preserved in J. Paul Getty Museum, Los Angeles (no. 84.XA.1170) (Bourne & Shepherd 1883) – 24.11, figs. 24.3 to 24.12

1883 – Nepal – *R. unicornis* female skull donated by Albert Edward, Prince of Wales, to Natural History Museum, London, no. 1883.10.23.3 – Thomas 1906: 18; Pocock 1944b

1885 – Rhino are preserved as Royal Game, only rulers are allowed to shoot – Kinloch 1885: 61, 1904: 61; Inglis 1888: 557; Clubman 1911: 325; Kennion 1922: 603, 1932: 225; Woodyatt 1922a: 150; Ali 1927a: 860; Smythies 1942: 45 – 24.1

1887 – Even with a pass to shoot in Nepal it is very difficult to get a rhino – R. Ward 1887: 17

1891 – Maharaja Bir Shumsher went to hunt tiger, without mention of rhino – Ehlers 1894, vol. 1: 293

1895 – Hetowda (Hatauda) – From May to October "the place is abandoned to the deadly orgies of the *terai* fever, to the loathsome leech and the filthy rhinoceros" – Ballantine 1895: 51

1898 – Maharaja Bir Shumsher donates to Gordon Highlanders regiment 2 kukri with handles made from horn of rhino which he killed. Preserved in Gordon Highlanders Museum (Aberdeen), number GH564 and GH608 – Shakespear 1912: 145 – 24.12, fig. 24.14

C. Period of Maharaja Chandra Shumsher (1863–1929), Reigned 1901–1929

1901 to 1929 – Maharaja Chandra Shumsher organized regular hunting camps in the Chitwan area, together with family members. No particulars of dates or events. One photograph taken by Dirgha Man Chitrakar on unknown date shows him sitting on top of a dead rhino – 24.13, fig. 24.15

1900s (?) – One Rana of Nepal is credited with having killed 97 rhinos in a month – A.K. Dutta 1991: 52 (not verifiable)

1904 – Charles Withers Ravenshaw (1851–1935), Resident to Nepal 1902–1905, killed 1 rhino. Ravenshaw was also instrumental in assisting the transport of a rhino from Nepal to the Calcutta Zoo in February 1905 (Sanyal 1905: 5) – *Times of India* 1904-03-21

1906 – Hunt organized for George, Prince of Wales (1865–1936) was canceled. One captive rhino was donated to London Zoo – 25.2

1907 – Nepal – *R. unicornis*, two skulls donated by the saddlery business of Watts & Co., to the India Museum, Calcutta. Still preserved in Zoological Survey of India, nos. 10437, 10438 – Groves & Chakraborty 1983: 253 – 24.15

1907-02 – Rapti River – Maharaja Chandra Shumsher killed 28 rhino and captured 6 calves – Manners-Smith 1909; Lydekker 1909 – 24.13

1908 – Naolpur Valley bordering on Chitwan – Maharaja Ganga Singh of Bikaner (1880–1943) and John Manners-Smith (1864–1920) killed 4 rhino and 104 tigers during 49 days shooting – Manners-Smith 1909; F.H. Brown 1920 – 24.16

1909 – Rapti Valley – Francis William Gordon-Canning (1854–1920), manager of indigo factory at Parsa, Champaran, Bihar, counted 20 rhino within a mile from camp when trying to take photographs (the images are now unknown) – Manners-Smith 1909; Rookmaaker 2021b

1909 – Tribeni on the Gandak River – The British Resident John Manners-Smith organized a hunt. Skin and bones of one rhino were presented to the Indian Museum, Kolkata. There is no later record of these specimens – Indian Museum 1911: 7 – 24.17

1909 – Tribeni on the Gandak River – *R. unicornis* specimen donated by Thomas George Longstaff (1875–1964) to Natural History Museum, London, nos. ZD.1937.3.1.1, 1938.6.28.8. Horn 15 in (38.1 cm) – Ward 1910: 465; Longstaff 1950: 149 – 24.17

1910 – *R. unicornis* horn 13 in (33 cm) length attributed to Maurice Loraine Pears (1872–1916), Scottish Rifles – Ward 1910: 465

1911 – Maharaja Chandra Shumsher, King George V and his party killed 18 rhino – 25.3

1911 – As part of King George's party, Charles George Francis Mercer Nairne Petty-Fitzmaurice (1874–1911) obtained horn 12 in (30.5 cm) length – Ward et al. 1914: 463 – 25.3

1911 – As part of King George's party, Adolphus, Prince of Teck (1868–1927) obtained horn 11 ¾ in (29.8 cm) length – Ward et al. 1914: 463 – 25.3

1911 – Last stronghold is in a small jungle valley near Kathmandu – Durand 1911: 56

1913 – Aubyn Bernard Rochfort Trevor-Battye (1855–1922) attends elephant capture operation. A rhino mother and calf were found; calf eaten by tiger – Trevor-Battye 1916

1919 – Nepal – Rhino said to be extinct – Hornaday 1919

1919 – British Envoy William Frederick Travers O'Connor (1870–1943) killed 1 rhino – O'Connor 1931: 346 – 24.18

1920 – Raja Bahadur Kirtyanand Sinha (1880–1938) killed 2 rhino and performed his Tarpan ceremony – K. Sinha 1934 – 24.19, figs. 24.28, 24.29

1921 – Edward, Prince of Wales (1894–1972) and party killed 10 rhino – 25.4

1921 – M. Maxwell, possibly shooting with Frederick O'Connor, obtained a horn of 15 in (38.1 cm) – Ward et al. 1922: 456 – 24.18

1922 – Maharaja Kaiser Shumsher killed 21 rhino, including the 2 mothers of calves taken alive. These were exported by animal dealer Frank Buck (1884–1950) to New York and Philadelphia (arrive 1923-05-24) – Buck & Anthony 1930: 48–89; Buck & Fraser 1942: 107–112; Buck 1931, 1936, 1953 – 24.20

1922 – Frederick O'Connor killed 13 rhino – O'Connor 1931: 346; Ellison 1925: 34 – 24.18, figs. 24.22, 24.23

1923-04-10 – *R. unicornis* female, skin, skeleton, skull and mandible, donated by Lorenz Hagenbeck (1882–1956) to American Museum of Natural History, New York (M-70445) – AMNH Database (online)

1923 – Gandak River – John Champion Faunthorpe (1871–1929) and Arthur Stannard Vernay (1877–1960) obtain 4 rhino for diorama in the American Museum of Natural History, New York – Vernay 1923a, b, c, 1931; Faunthorpe 1923, 1924a,b,c, 1930; Dyott 1923a, b, 1924; AMNH 1930; Rookmaaker 2021b – 24.21, figs. 24.30 to 24.35

1923-03-10 – Gandak River – George Miller Dyott (1877–1960), who was the photographer attached to the Vernay-Faunthorpe expedition, took photos of an adult female known as 'Lizzie' and her calf – Dyott 1923a (story only); Faunthorpe 1924a: 174, 1924c: 215; Ellison 1935, plate facing p. 34; Rookmaaker 2021b: 101 – 24.21, fig. 24.34

1923-03 – Gandak River – The photographer Dyott took cinematographic pictures of rhinos in the wild in Nepal, probably the first time that this was successfully attempted. The sequence was incorporated in a silent movie first shown in New York in November 1923 – Vernay 1923b, 1923c; Osborn 1922; Anon. 1923a – 24.21

1924-1935 – Undated photographs by Dirgha Man Chitrakar (1877–1951) showing members of the Rana family with rhino trophies in the field, depicting Juddha Shumsher (1875–1952); Kaiser Shumsher (1892–1964); Bahadur Shumsher (1892–1977); Shanker Shumsher (1909–1976); Mrigendra Shumsher (b. 1906); Surya Shumsher (d. 1945) – 24.14, table 24.29, figs. 24.16 to 24.21

1924–1935 – Undated photographs by Dirgha Man Chitrakar depicting King Tribhuvan Bir Bikram Shah (1906–1955) with rhino trophies in the field – 24.14, figs. 24.19, 24.21

1924 – Kaiser Shumsher is photographed showing a set of trophies including at least 46 rhino skulls in front of the Kaiser Mahal in Kathmandu. The caption states that these were the remains of one season's hunt. Published in the French illustrated magazine *L'Illustration* in January 1925, therefore possibly referring to the 1924–1925 winter season – Lévi 1925 – 24.40, fig. 24.72

1924 – Frederick O'Connor killed 5 rhino – O'Connor 1931: 346 – 24.18, figs. 24.25 to 24.27

1924 – Victor Bulwer-Lytton, 2nd Earl of Lytton (1876–1947) in hunt with O'Connor killed 2 rhino, one with horn of 13 ¼ in (34 cm). During the same trip 2 rhinos were killed by Capt. Harvey of the Gurkha Regiment – Lytton 1942: 144 – 24.18, fig. 24.23

1925 – Frederick O'Connor killed 1 rhino – O'Connor 1931: 346 – 24.18

1926 – John Prescott Hewett (1854–1941) and Thomas Sivewright Catto, 1st Baron Catto (1879–1959) killed 1 rhino, with horn 11 ½ in (29.2 cm) – Ward 1928: 438; Hewett 1938: 175; Catto 1944 – 24.23

1926 – Chitwan – Belle Willard Roosevelt (1892–1968) and Theodore Roosevelt Jr. (1887–1944) killed 2 rhino. The skins were mounted and displayed as part of a diorama in the William V. Kelley Hall of Asiatic Mammals of the Field Museum of Natural History in Chicago, Illinois. Specimens registered with catalogue numbers 25707 (collection no. 61) and 25708 (collection no. 62) – B. Roosevelt 1926; B.W. Roosevelt 1927: 241; E. Roosevelt 1959: 195; Chicago Field Museum 1929: 151, pl. 19; *Sunday Star*, Washington DC 1926-05-02, gravure section – 24.22, figs. 24.37, 24.38

1927 – Ian Cameron Grant (1891–1955), Cameron Highlanders, collecting for the "American National Collection" obtained horn 15 ½ in (39.4 cm) length – Ward et al. 1935: 335

1928 – Rhino still plentiful in the Chitawan district and along the Rapti river – Landon 1928, vol. 1: 292; Kennion 1928

D. Period of Maharaja Bhim Shumsher (1865–1932), Reigned 1929–1932

1930 – A rhino foot tobacco pot was manufactured by the taxidermy firm Van Ingen with brass plaque "Lt Col D.G. Bromilow, 20th Lcrs [Lancers], Nepal, 1930" for David George Bromilow (1884–1959) – Van Ingen website

E. Period of Maharaja Juddha Shumsher (1875–1952), Reigned 1932–1945

1932 – Lord Ratendone, later Inigo, 2nd Marquess of Willingdon (1898–1979) killed 2 rhino. A calf was captured and donated to Calcutta Zoo. Adrian Baillie (1898–1947) killed a third rhino. Ratendone was accompanied by the UK Ambassador to Nepal Clendon Turberville Daukes (1879–1947) and his wife Dorothy Maynard Daukes (1890–1967) – Daukes 1933; British Film Institute, London, amateur movie no. 444818 – 24.25

1932 – An English hunting party killed 7 rhinos. Possibly connected to Ratendone's hunt – Berg 1933: 123

1932 to 1945 – The shooting camps of Maharaja Juddha Shumsher were recorded in a sporting diary, later extracted by Evelyn Arthur Smythies (1885–1975) in 1942. In an album preserved in the Madan Puraskar Pustakalaya (MPP), Kathmandu there are 11 photographs including a rhino, all undated and unannotated (EAP166/2/1/27). Another album in MPP contains 2 photographs (EAP166/2/1/17) – Smythies 1942 – 24.27, figs. 24.40 to 24.48

1933 – Maharaja Juddha Shumsher killed 14 rhino – Smythies 1942: 83; Prasad 1996: 284 – 24.27, figs. 24.51, 24.52

1933-01-19 – Maharaja Juddha Shumsher performs the Shradda-Tarpan ceremony at Bardaha Camp – Smythies 1942: 90 – 24.27

1934 – Chitwan – Declaration of Rana hunting reserve – K. Thapa et al. 2013

1935 – Frederick O'Connor, the former Resident, killed 1 rhino with a horn of 12 in (30.5 cm). He was accompanied by George Granville Sutherland-Leveson-Gower, 5th Duke of Sutherland (1888–1963), his wife Eileen Sutherland (1891–1943) and Archibald Charles Edmonstone, 6th Baronet (1894–1954) – O'Connor 1935, 1940: 186 – 24.28, figs. 24.53, 24.54, 24.55

1936-02-23 – Maharaja Juddha Shumsher killed 1 rhino. He performs again the Shradda-Tarpan ceremony – Smythies 1942: 135, 164 – 24.29

1937 – Baron Carl Gustaf Emil Mannerheim (1867–1951) of Finland takes part in a hunt organized by the British Ambassador Frederick Marshman Bailey (1882–1967) and attended by Charles Ian Finch-Knightley, 11th Earl of Aylesford (1918–2008) and Wilton Lloyd-Smith (1898–1940). One rhino sighted – Shrestha et al. 2017: 18; movie "Tiger Hunt" preserved at British Film Institute, London, no. 12806 – 24.30

1938 – Viceroy 2nd Marquess of Linlithgow (1887–1952) killed 2 rhino near Bhikna Thori, and Colonel Cyril George Toogood (1894–1951) killed a 3rd rhino. Linlithgow traveled with his wife Doreen Maud Milner (1886–1965) and his three daughters Anne Adeline (1914–2007), Joan Isabella (1915–1989) and Doreen Hersey Winifred (1920–1997) – Smythies 1942: 100; *Sphere* 1939-07-22; Linlithgow 1938a (amateur movie in British Film Institute, London); Glendevon 1971: 108 – 24.31, figs. 24.56, 24.57

1938 – Kiran Shumsher (1916–1983) killed one rhino to perform the Shradda-Tarpan ceremony – Martin 1985, 2010: 1 – 24.32, figs. 24.58, 24.59

1939 – Maharaja Juddha Shumsher and party killed 35 rhino, captured 1 young rhino. Shooters included Roy Goodwin Kilburne (1892–1952), Umaid Singhji, 18th Maharao of Kotah (1873–1940), John Adrian Louis Hope, 1st Baron Glendevon (1912–1996), Mr. Grant, Geoffrey Lawrence Betham (1889–1963), also Min Shumsher (b. 1928), Bahadur Shumsher (1892–1977) and his grandson Nara Shumsher (1911–2006) – Smythies 1942: 108–116; Martin & Martin 1982: 31; Mishra 2008, pl. 9; Thapa 2013: 94; Shrestha et al. 2017: 18 – 24.33, figs. 24.60, 24.61

1939.01.28 – Min Shumsher (b. 1928), son of Maharaja Juddha Shumsher, killed his first rhino – Smythies 1942: 100 – 24.26, figs. 24.45, 24.46

1941 – Shooting party of members of the Rana family and guests at Sawari Hunting Camp – Anon. 1941 – 24.34, fig. 24.62

F. The Modern Period after 1945

1949 – Chitwan – Two Indian diplomats killed 2 rhino – Gee 1950a: 1732

1950s – Vijay Ananda Gajapathi Raju (1905–1965), Maharajkumar of Vizianagram, known as Vizzy, killed 2 or 3 rhino in Chitwan area – Anon. 1956; Islam & Islam 2004: 153, 157; Motilal 1970: 164 – 24.36

1950s – Kirpal Singh Majithia killed 1 rhino – Egen 2011 – 24.37

1951 – *R. unicornis* skull donated by Prince of Wales to NHM, no. 1951.11.30.2, probably belonging to earlier trip – Groves 1982

1952-04 – Raja of Kasmanda, Dinesh Pratap Singh (b. 1927) killed 1 rhino. Trophy preserved in Kasmanda Palace Hotel in Mussoorie, Uttarakhand – Dinkar Singh, pers.comm. – 24.38, fig. 24.63

1953 – King Mahendra killed 1 rhino and performed Shradda-Tarpan ceremony – Mishra 2008: 142 – 24.39

1956 – Two trophy heads inscribed "His Majesty the King Of Nepal 1956" in the Kaski Drawing Room of Narayanhiti Palace, Kathmandu, presumed killed by King Mahendra – Gurung 2014: 41; Rookmaaker pers.obs. 2018

1957 – The taxidermy firm of Van Ingen in Mysore received 6 rhino skins from Nepal – Ullrich 1964: 226 – 25.4 (about Van Ingen)

1958 – The taxidermy firm of Van Ingen received skins of mother and calf rhino – Ullrich 1964: 226

1958 – Wildlife Conservation Act of 1958 was instituted – Heinen & Kattel 1992; K. Thapa et al. 2013

1959-01 – Haranhari – An operation permitted by King Mahendra captured 3 rhino calves. No animals killed. Rhino plentiful – Gee 1959b: 73; K. Thapa 2013: 97

1960-04 – Ratification of Wild Life Conservation Act – Gee 1963c

1961 – Meghauli – Queen Elizabeth II (1926–2022) and Prince Philip, Duke of Edinburgh (1921–2021) are entertained by King Mahendra. Alec Douglas-Home (1903–1995) killed 1 rhino – M. Thapa 2013: 98; Shrestha et al. 2017: 22 – 25.4

1962 – Meghauli – *R. unicornis* 3 skeletons and skulls donated by Government of Nepal to Bayerische Staatssammlung, Munich:

1963/0159 juvenile, Narayanghar; 1963/0160 adult, Kush-Kush forest west of Meghauli; 1963/0160a juvenile, Kush-Kush forest – Bayerische Staatssammlung, Munich (pers.comm.)

1979 – King Birendra (1945–2001) killed 1 rhino for Shradda-Tarpan ceremony – Martin 1985: 15; Mishra 2008: 157 – 24.39

1983 – King Birendra obtained rhino horn of 11 in (28 cm) – Ward 1986: 564

1983 – Chitwan – *R. unicornis*, 2 skulls collected by Christen M. Wemmer (b. 1943) for Smithsonian Institution, Washington DC.: USNM 545847, on 1983-10-11 near Bridge No. 1; USNM 545848, on 1983-12-03 west of Dhuadhaura – National Museum of Natural History, Washington DC, website (2017)

1983 – Chitwan, west of Bhawanipur – *R. unicornis*, skull collected by Tiger project for Smithsonian, Washington DC. 1983-02-03 – National Museum of Natural History, Washington DC, website (2017)

TABLE 24.28   List of persons who were involved in rhino events in Nepal in chronological order

Name of person	Date	Locality. type of interaction
Captain Thomas Smith	1844	Hitounda. Killed 7
Sirdar Bowanee Singh	1844	Hitounda. Killed 7
Sirdar Delhi Singh	1844	Hitounda. Killed 7
Henry Ambrose Oldfield (1822–1871)	1851	Nepal. Obtained 1
Maharaja Jung Bahadur (1817–1877)	1851	Nepal. Killed several
Maharaja Jung Bahadur (1817–1877)	1861	Nepal. Killed 1
Maharaja Jung Bahadur (1817–1877)	1862	Nepal. Killed 4
Francis George Baring, 2nd Earl of Northbrook (1850–1929)	1873	Nepal. Killed 12
John Biddulph (1840–1920)	1873	Nepal. Killed 12
Colonel Hankey	1873	Nepal. Killed 12
Evelyn Baring, 1st Earl of Cromer (1841–1917)	1875	Nepal. Obtained 1
Maharaja Ranaudip Singh (1825–1885), Prime Minister	1883	Rapti. Killed 1
Frederick Oliver Robinson, 2nd Marquess of Ripon, 4th Earl de Grey (1852–1923)	1883	Rapti. Killed 2
Charles William de la Poer Beresford, 1st Baron Beresford (1846–1919)	1883	Rapti. Killed 1
William John Arthur Charles James Cavendish-Bentinck, 6th Duke of Portland (1857–1943)	1883	Rapti. Killed 2
Wilfrid Greenwood, later Edwin Wilfrid Stanyforth (1861–1939)	1883	Rapti. Killed 1
Algernon George Lawley (1857–1931), 5th Baron Wenlock	1883	Rapti. Killed 1
Maharaja Bir Shumsher (1852–1901)	1898	Nepal. Killed 1
Charles Withers Ravenshaw (1851–1935), Resident	1904	Nepal. Killed 1
Maharaja Chandra Shumsher (1863–1929)	1907	Rapti. Killed 28. Captured 6
John Manners-Smith (1864–1920), Resident	1907	Rapti. Present at hunt
Maharaja Chandra Shumsher (1863–1929)	1908	Naolpur. Killed 4 with party
Maharaja Ganga Singh of Bikaner (1880–1943)	1908	Naolpur. Killed 4
John Manners-Smith (1864–1920), Resident	1908	Naolpur. Killed 4
John Manners-Smith (1864–1920), Resident	1909	Gandak. Organized hunt
Thomas George Longstaff (1875–1964)	1909	Gandak. Killed 1
Francis William Gordon-Canning (1854–1920), indigo planter	1909	Rapti. Photographed 20
Maurice Loraine Pears (1872–1916), Scottish Rifles	1910	Nepal. Obtained horn

TABLE 24.28   List of persons who were involved in rhino events in Nepal in chronological order (*cont.*)

Name of person	Date	Locality. type of interaction
Maharaja Chandra Shumsher (1863–1929)	1911	Chitwan. His party killed 18
King George V (1865–1936)	1911	Chitwan. His party killed 18
Luke White, 3rd Baron Annaly (1857–1922)	1911	Chitwan. Killed 1 in King's party
John George Lambton, 3rd Earl of Durham (1855–1928)	1911	Chitwan. Killed 1 in King's party
Colin Keppel (1862–1947)	1911	Chitwan. Killed 1 in King's party
Horace Lockwood Smith-Dorrien (1858–1930)	1911	Chitwan. Killed 1 in King's party
Adolphus, Prince of Teck (1868–1927)	1911	Chitwan. Killed 2 in King's party
Bryan Godfrey-Faussett (1863–1945)	1911	Chitwan. Killed 2 in King's party
Charles G.F. Petty-Fitzmaurice (1874–1911)	1911	Chitwan. Killed 1 in King's party
John Manners-Smith (1864–1920), Resident	1911	Chitwan. Hosted King's party
Aubyn Bernard Rochfort Trevor-Battye (1855–1922)	1913	Nepal. Observed 2
William Frederick Travers O'Connor (1870–1943), Resident	1919	Nepal. Killed 1
Maharaja Ganga Singh of Bikaner (1880–1943)	1920	Nepal. Killed 4
Raja Bahadur Kirtyanand Sinha (1880–1938)	1920	Nepal. Killed 2
William, Prince of Wales (1894–1972)	1921	Chitwan. His party killed 10
Rowland Thomas Baring, 2nd Earl of Cromer (1877–1953)	1921	Chitwan. Killed 1 in King's party
Louis Francis Albert Victor Nicholas Mountbatten (1900–1979)	1921	Chitwan. Killed 1 in King's party
Alexander Charles William Newport (1874–1948)	1921	Chitwan. Killed 1 in King's party
Dudley Burton Napier North (1881–1961)	1921	Chitwan. Killed 2 in King's party
Bruce Arthur Ashley Ogilvy (1895–1876)	1921	Chitwan. Killed 1 in King's party
Perceval Landon (1869–1927)	1921	Chitwan. Killed 1 in King's party
Frederic Sinclair Poynder (1893–1943), 9th Gurkha Rifles	1921	Chitwan. Killed 1 in King's party
Rivers Berney Worgan (1881–1934), Military Secretary	1921	Chitwan. Killed 1 in King's party
M. Maxwell	1921	Nepal. Obtained horn
Frank Buck (1884–1950)	1922	Nepal. Exported 2
Frederick O'Connor (1870–1943), Resident	1922	Nepal. Killed 13
John Champion Faunthorpe (1871–1929)	1923	Chitwan. Collected 4
Arthur Stannard Vernay (1877–1960)	1923	Chitwan. Collected 4
Frederick O'Connor (1870–1943)	1924	Nepal. Killed 5
Victor Bulwer-Lytton, 2nd Earl of Lytton (1876–1947)	1924	Nepal. Killed 2
Captain Harvey	1924	Nepal. Wounded 2
John Manners-Smith (1864–1920), British Resident	1924	Nepal. Killed 2
John Prescott Hewett (1854–1941)	1926	Nepal. Killed 1
Thomas Sivewright Catto (1879–1959)	1926	Nepal. Killed 1
Theodore Roosevelt Jr. (1887–1944)	1926	Nepal. Killed 1
Belle Willard Roosevelt (1892–1968)	1926	Nepal. Killed 1
Kermit Roosevelt (1889–1943)	1926	Nepal. Visited rhino area
Hugh Wilkinson (1874–1939), British Envoy	1926	Nepal. Wounded 1
Ian Cameron Grant (1891–1955)	1927	Nepal. Obtained 1
David George Bromilow (1884–1959)	1930	Nepal. Obtained 1
Inigo, Lord Ratendone (1898–1979)	1932	Chitwan. Killed 2. Captured 1
Sir Adrian Baillie (1898–1947)	1932	Chitwan. Killed 1
Merton Lacey (1902–1996)	1932	Chitwan. Produces film
Maharaja Juddha Shumsher (1875–1952)	1933	Rapti. Killed 2
Nir Shumsher (1913–2013)	1933	Rapti. Killed 1
Baber Shumsher (1888–1960)	1933	Rapti. Killed 1
Surya Shumsher (d. 1945)	1933	Rapti. Killed 1

TABLE 24.28   List of persons who were involved in rhino events in Nepal in chronological order (*cont.*)

Name of person	Date	Locality. type of interaction
Narayan Shumsher	1933	Rapti. Killed 1
Surendra Shumsher	1933	Rapti. Killed 1
Brahma Shumsher (b. 1909)	1933	Rapti. Killed 1
Maharaja Ramanuj Saran Singh Deo of Surguja (1893–1965)	1933	Nepal. Killed 1
George, 5th Duke of Sutherland (1888–1963)	1935	Nepal. Killed 1
Frederick O'Connor (1870–1943)	1935	Nepal. Killed 1
Maharaja Juddha Shumsher (1875–1952)	1936	Chitwan. Killed 1
2nd Marquess of Linlithgow (1887–1952), Viceroy	1938	Chitwan. Killed 2
Cyril George Toogood (1894–1951)	1938	Chitwan. Killed 1
Walter Arthur George Burns (1911–1997)	1938	Chitwan. In Viceroy's party
Maharaja Juddha Shumsher (1875–1952)	1939	Chitwan. Killed at least 1
Bahadur Shumsher (1892–1977)	1939	Chitwan. Killed 2
Nara Shumsher (1911–2006)	1939	Chitwan. Killed 2
Min Shumsher (b. 1928)	1939	Chitwan. Killed 1
Roy Goodwin Kilburne (1892–1952)	1939	Chitwan. Killed 1
Maharao Umaid Singhji, 18th Maharao of Kotah (1873–1940)	1939	Chitwan. Killed 1
John Adrian Louis Hope, 1st Baron Glendevon (1912–1996)	1939	Chitwan. Killed 1
Mr. Grant (unidentified)	1939	Chitwan. Killed 1
Geoffrey Lawrence Betham (1889–1963)	1939	Chitwan. Killed 1
Maharaja Juddha Shumsher (1875–1952)	1941	Chitwan. Killed 1
King Tribhuvan (1906–1955)	1941	Chitwan. Killed 1
Prince Mahendra (1920–1972)	1941	Chitwan. Killed 1
Maharaja Vijay Ananda Gajapathi Raju of Vizianagram(1905–1965)	1950s	Nepal. Killed 2
Kirpal Singh Majithia	1950s	Nepal. Killed 1
Dinesh Pratap Singh (b. 1927), Raja of Kasmanda	1952	Nepal. Killed 1
Alec Douglas-Home (1903–1995)	1961	Chitwan. Killed 1
Christen M. Wemmer (b. 1943)	1983	Chitwan. Collected 2
Tiger Project (Smithsonian)	1983	Chitwan. Collected 1

The members of the Rana family and the royal family are included when they are identified in specific years.

24.3   Rhino Horn Products from Nepal

Rhino horn carvings from Nepal are relatively rare (Chapman 1999: 274). There is an undated libation cup of rhino horn showing the Hindu pantheon (fig. 24.1), and a carving of the "Green Tara" goddess said to be Nepalese (Chapman 1999, fig. 397). In the 1950s, Lee Merriam Talbot (1956b) photographed a large carved rhino horn (maybe African) on the steps of the royal palace. Three bowls dating from the 17th or 18th century were sold by Bonhams (2018), showing decorations relating to the god Vishnu.

Esmond Martin (1985) talked to members of the Silpakar family in Patan, who were famed for their expertise in carving. Their products were usually commissioned by the Rana family, hence few ever came on the market.

It would take two weeks to make a bowl decorated with motifs relating to the god Vishnu. There is no indication that horns or carvings were regularly exported.

24.4   The Annual Shooting Camps of the 20th Century

The Maharajas and Kings of Nepal had shooting camps almost every year from the start of the 20th century to the 1950s. Tigers could be shot everywhere, but rhinos were limited to districts in the south-east and in the south. With very few exceptions, these hunts took place in the present Chitwan area, along the banks of the Rapti, Reu, Narayani and Gandak Rivers. The most extensive preparations

FIGURE 24.1
Rhinoceros horn libation cup from Nepal. Bearing ten plaques outside and one inside, carved in relief and representing the Hindu pantheon from a Nepalese temple
PENN MUSEUM, UNIVERSITY OF PENNSYLVANIA
MUSEUM OF ARCHAEOLOGY AND ANTHROPOLOGY,
PHILADELPHIA, PA.: NO. A1131A

were made when King George V was invited in 1911 and Edward, Prince of Wales in 1921 (25.3, 25.4). The activities of Maharaja Juddha Shumsher in the 1930s were carefully recorded in a game book, which was edited and made public by Smythies (1942). This points to the possibility that similar game books were kept, maybe still in existence in official or private libraries. In this age of photography, many shooting camps were captured on film and the photographs were combined with those of other events in photo albums. Some of these have become available through recent digitization efforts, others have come on the auction markets, and more are likely still to be discovered.

These albums are powerful glimpses into the importance of the annual camps organized for the Nepalese rulers, for British ambassadors, and for others who could obtain a license. When the Rana family was deposed in 1951, there was a period of political vacuum, which gave chance to poachers to decimate the rhino population. The data for this more recent period are included in chapter 26 about Chitwan as a conservation area.

Without doubt rhinos were killed annually, probably as many as 20 to 30 in a season, sometimes fewer, sometimes more. A figure of 97 in one year found in recent literature cannot be verified, but 35 killed by Maharaja Juddha in 1938–1939 might not be too extraordinary. Most of these rhinos were preserved as trophies, generally as a head with horn to be mounted on the wall. Other parts of the animal, meat, skin, skull and skeleton, were kept or used by the guests and staff associated with the *shikar* camps, as well as by villagers. Despite the importance of the rhino as a trophy animal, the evidence about their exhibition, storage or disposal is particularly incomplete (24.40). Maybe they have all disappeared, maybe one day they will surface to bring a powerful testimony of a past age.

24.5 Estimates of Historical Rhino Numbers

Two centuries of public hunting ban has helped preserve a healthy rhino population. Their actual number is impossible to verify, in the absence of any type of scientific census, or written documentation. It is unlikely that an attempt to count was ever made. Best guesses have suggested a maximum number of 1000 rhinos in Chitwan. While the annual hunts would have had an impact, the number of rhinos killed did not result in any serious decline. The protection provided to rhinos by being royal game has ensured that there is still a thriving population in present-day Nepal. The population estimates for Nepal are mostly identical to those reported for Chitwan NP (26.6, table 26.32).

24.6 Rhino Statues Decorating the 17th Century Temple in Bhaktapur

The Nyata Poul or Siddhi Lakshmi Temple in the Durbar Square of Bhaktapur, on the east side of Kathmandu, dates from the end of the Malla dynasty in the 17th century. The steps leading up to the inner sanctum at the top of this seven-layered temple are embellished with animal statues including two rhinos with chains on their feet (fig. 24.2). Their presence does not appear to have any religious significance. A chained rhino might have been exhibited alive in the area at the time of building.

Werner Hoffmeister (1819–1845) accompanying Prince Wilhelm Waldemar in January 1845 mentioned depictions of rhino on an (unidentified) temple near the market.

24.7 Observations of Italian Missionaries in the 18th Century

In the course of the 18th century, the Italian Catholic Church sent several missions to the Kathmandu valley and Tibet (Lévi 1905; Vannini 1977). There are four accounts detailing the occurrence of rhinos in the *terai* region along Nepal's southern border.

Traveling in that period through the territory of the Rajah of Makwanpur, on the eastern side of Chitwan,

FIGURE 24.2   One of the rhinos on the steps of the Lakshmi Temple in Bhaktapur. Note the chains around the feet and around the neck
PHOTO: KEES ROOKMAAKER, 1982

the monk Antonio Agostino Giorgi and the missionary Cassiano de Macerata warn of the danger of wildlife including the rhino.

On the road from Bettiah (Bihar) to Kathmandu, also in the 1750s, the Capuchin friar Marco della Tomba expected to see a rhino.

The superintendent of the mission, Father Tranquillus of Apechio spent the summer of 1756 in the summer residence of the Raja of Tanahun at Devghat, on the north-western side of Chitwan. Abundant tigers, elephants, bears, rhinoceros and wild boars were found in the jungle. The Raja wrote to Pope Benedict XIV to offer him his homage and ask for priests, in a letter accompanied by a rhino horn as a gesture of good will.

24.8   Brian Hodgson Discovers the Fauna of Nepal

Brian Houghton Hodgson (1800–1894) first went to Nepal in 1824, later to become British Resident 1829–1831 and 1833–1844. He was a pioneer studying the region's fauna. He wrote a comprehensive introduction to the mammals of Nepal, which was published in two prominent journals of the Asiatic Society of Bengal in Calcutta and the Zoological Society of London. Hodgson found rhinos to be plentiful in the forests and hills of the lower regions of Nepal. The animals would stray into the rice fields of the *terai* after the rains to feed on the crops (Hodgson 1832, 1834, 1844). The private menagerie of King Rajendra Bikram Shah (1813–1881) in or near Kathmandu had (at least) a pair of rhinos, which produced a male calf in May 1824. Hodgson recorded his observations on these animals and commissioned a drawing (5.4).

24.9   The Observations of the Resident Thomas Smith in 1844

Captain Thomas Smith of the 15th Regiment of Bengal Native Infantry, once from Craig Avon near Neats, served as Assistant Resident in Kathmandu 1841–1845. Requested by King Rajendra Bikram Shah, he killed a rogue elephant. Although Smith (1849: 341) asserted that the head and tusks of this large elephant were preserved in the museum of the Earl of Derby in Knowsley, it is not listed in later inventories. Smith was accompanied by two officers Sirdar Bowanee Singh and Sirdar Delhi Singh. They hunted around Hetauda in March 1844, where they killed 7 rhinos (Smith 1852).

24.10   Francis Baring in Nepal in 1873

From 1872 to 1876, Thomas George Baring, 1st Earl of Northbrook (1826–1904) was Viceroy of India. His son Francis George Baring, 2nd Earl of Northbrook went for a hunting expedition in Nepal in the first months of 1873 with John Biddulph and Colonel Hankey. The party shot 2 elephants, 8 tigers, 12 rhinos, and a number of buffaloes.

His uncle Evelyn Baring, 1st Earl of Cromer served as the Viceroy's private secretary during his tenure. There is no record that he went on a hunting expedition in Nepal. However, he donated skulls of an adult male and of a juvenile male *R. unicornis* to the Indian Museum with date 1875. These specimens were catalogued by Sclater (1891), and still recorded recently in the collection of the Zoological Survey of India by Groves & Chakraborty (1983). The adult skull is on display, apparently previously with a label switched from another specimen.

24.11   Olly, Earl de Grey on the Rapti River in 1883

The Prime Minister of Nepal Ranaudip Singh Bahadur (1825–1885) had invited the Viceroy of India, George Robinson, 1st Marquess of Ripon (1827–1909), to participate in a special shoot in the Rapti Valley from 27 January to 18 February 1883. Preoccupied by political changes, the Viceroy arranged for his son Frederick Oliver Robinson, 2nd Marquess of Ripon, 4th Earl de Grey (1852–1923) to take his place. Known informally as Olly, he was famous in his days for the enormous number of birds which he shot in England every year. On this trip he was accompanied by Wilfrid Greenwood, William Duke of Portland, Charles Beresford, Baron Algernon Wenlock and Major Jackson. The Viceroy would have been welcomed by the British

FIGURE 24.3   Sketch of rhinoceros in a scrapbook by Frederick
Oliver Robinson, 4th Earl de Grey (1852–1923), made
during a hunting camp in the Rapti Valley in Nepal in
February 1883
WITH PERMISSION © PRIVATE COLLECTION

Resident of Nepal Charles Edward Ridgway Girdlestone
(1839–1889), who also joined the party together with at
least four other officers of his staff.

Olly de Grey kept notes and made a drawing (Godfrey
2012), Portland (1937) wrote about his reminiscences, and
the London magazine *The Graphic* included a two-page
spread in August 1884 (Anon. 1884b). According to the
caption in this presentation, the photographs were
taken on the spot by the staff of the Kolkata based photographic studio established in 1882 by Theodore Julius
Hoffmann (1855–1921) and Peter Arthur Johnston (d. 1885,
cf. *Aberdeen Free Press* 1885-07-08).

Sets of photographs of Olly's shooting trip were
combined in albums, one of which is preserved in the
J. Paul Getty Museum, Los Angeles with 57 albumen silver prints (Bourne & Shepherd 1883). This album is catalogued with the name Bourne & Shepherd, indicating
another photographic establishment in Kolkata, which
might have been involved in sales and distribution rather
than the actual photography in this instance. The album
contains eight prints showing rhino trophies (figs. 24.4 to
24.11).

Olly de Grey and party arrived at the camp on the
Rapti River on 27 January 1883 (*Times of India* 1883-02-06)
and on 18 February they were back in Kolkata (*Times
of India* 1883-02-19). During this period they killed in
total eight rhinos: two were killed by De Grey, one by

Greenwood, one by Wenlock, one by Beresford, two by
Portland, and another by Maharaja Ranaudip Singh.

In his manuscript notes (recently examined by Olly's
biographer Rupert Godfrey), Olly de Grey mentioned his
own experiences. First he killed a rhino, "an enormous
brute, a male with horn 10 inches in good condition." Then
Wenlock "hit a rhino in the head. I finished it with 4 bore
in the neck. Killed stone dead. I got another with a shot to
the neck." Olly de Grey made a rather simple sketch of a
rhino (fig. 24.3). He later only claimed 2 rhinos and must
have allowed the third one to be accredited to Algernon
George Lawley (1857–1931), 5th Baron Wenlock. The horn
of 10 in (25.4 cm) is reminiscent of a record horn listed
by Rowland Ward (1899: 433) without date belonging to
an alleged Cooch Behar trophy ascribed to the Duke of
Portland, possibly referring to this Nepal specimen.

Wilfrid Greenwood, later Edwin Wilfrid Stanyforth
(1861–1939) was a childhood friend of Olly de Grey. The
Album in the Getty Museum includes a photograph no. 1
showing three rhino heads with three hunters identified
as "Lord de Grey – M [Maharaja] – Wilfred Greenwood"
(fig. 24.4). The same rhino heads are in photograph 43
where the hunters are replaced by three Nepalese skinners (fig. 24.5).

Charles William de la Poer Beresford, 1st Baron
Beresford (1846–1919) shot one rhino, because he is shown
sitting on the head of the animal in a widely circulated
photograph (fig. 24.6).

William John Arthur Charles James Cavendish-Bentinck,
6th Duke of Portland (1857–1943) recalled the circumstances in his memoirs of 1937: "The rhino I killed charged
my elephant, and I was lucky enough to kill it with one
shot from my four-bore rifle. When it was lying dead on
the ground I noticed something struggling behind it, and
found that it was a baby rhino. Charlie Beresford and I
tried to catch it alive; but it was much too strong for us,
and our friends shouted to us to get back on to our elephants as there were two more rhinos close by, which had
been attracted by the squeals of the young one. Knowing
that the baby would die without its mother, we shot it"
(Portland 1937: 258). The head of this female rhino and her
young baby are shown in two photographs of the Album
(figs. 24.5, 24.9). Both trophies were set up and displayed
in Portland's residence at Welbeck Abbey, as shown in a
photograph (fig. 24.13). The natural history collections
in Welbeck were transferred to the Museum opened in
Wollaton Hall in Nottingham by the 6th Duke of Portland.
However, the rhino trophies did not survive the upheaval
of temporary storage during the world war and they
must now be considered lost (Gareth Hughes, Portland
Collection, pers.comm. 2019).

FIGURE 24.4
Hunting camp in the Rapti Valley in February 1883. Three rhino head trophies with "Lord de Grey – W [Maharaja Ranaudip Singh] – Wilfred Greenwood." Albumen silver print from album distributed by Bourne & Shepherd in 1883. Size of image 19.1 × 25.7 cm (7 ½ × 10 1/8 inch), size of mount 33.0 × 38.1 cm (13 × 15 inch)
J. PAUL GETTY MUSEUM, LOS ANGELES: NO. 84.XA.1170.1

FIGURE 24.5
Head of female rhino and body of her baby, both shot by the Duke of Portland in Nepal in February 1883. From album distributed by Bourne & Shepherd in 1883, 16.5 × 26.7 cm
J. PAUL GETTY MUSEUM, LOS ANGELES: NO. 84.XA.1170.5

Olly De Grey kept extensive game books of his shooting days on his estates, omitting his two Indian trips according to his biographer Rupert Godfrey, who studied these documents (Godfrey 2012). His proficiency as a hunter was widely circulated in advertisements by the gunmakers Holland & Holland in London, attesting to the efficacy of Schultze's Gunpowder in a "Record of Game killed by Earl de Grey from 1867 to 1900" showing a total of 370,728 individuals (Rapier 1901: 235; Menzies 1917: 154; Gladstone 1930: 177; King 1985: 110). The first column of this remarkable record is for rhino, and has a total of 2, both for the year 1882, which must refer to De Grey's Nepal excursion with an incorrect date (fig. 24.12).

24.12 Kukris Donated to Gordon Highlanders in 1898

In September 1898, Maharaja Bir Shumsher (1852–1901), Prime Minister from 1885 to 1901, donated two kukris to the 1st Battalion of the Gordon Highlanders (Shakespear 1912: 145). These ceremonial knives had handles made of the horn of a rhino which he had shot. On the battalion's return to Scotland, one of these kukris was presented to the Officers Mess, the other to the Sergeants Mess, in their Castlehill Barracks, Aberdeen (fig. 24.14). They are preserved in the Gordon Highlanders Museum in Aberdeen, described as "Gurkha Kukri with leather and silver scabbard, and handle of rhinoceros horn. Presented by King Edward VII's Own Gurkha Rifles."

Lord Charles Beresford

FIGURE 24.6
"Lord Charles Beresford" showing rhino trophy head and four feet of the animal killed in Nepal in February 1883. From album distributed by Bourne & Shepherd in 1883, 19.1 × 25.7 cm. This photo widely reproduced in Anon. 1884c, 1906b, Portland 1937: 261
J. PAUL GETTY MUSEUM, LOS ANGELES: NO. 84.XA.1170.8

Off with his head – so much for Buckingham.

FIGURE 24.7
"Off with his Head – so much for Buckingham." Head of rhino killed in Nepal in February 1883. From album distributed by Bourne & Shepherd in 1883, 9.8 × 14.6 cm
J. PAUL GETTY MUSEUM, LOS ANGELES: NO. 84.XA.1170.11

24.13   The Shoots during the Time of Maharaja Chandra Shumsher

Maharaja Chandra Shumsher (1863–1929) was Prime Minister of Nepal from 1901 to 1929 representing the powerful Rana family. From 1911 his reign was shadowed by King Tribhuvan (1906–1955) as the nominal head of state. Both the Maharaja and the King were enthusiastic hunters attending camps in the wilder parts of Nepal regularly in the first three decades of the 20th century. All evidence points to the fact that large numbers of their family members also spent time in the field, including Maharaja Chandra's sons Mohan (1885–1967), Baber (1888–1960), Kaiser (1892–1964) and Shanker (1909–1964). Although it is known that Chandra was present in Chitwan in 1911 and 1921 to welcome the guests of the British royal family (25.3,

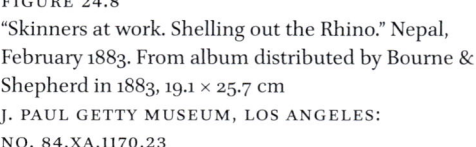

FIGURE 24.8
"Skinners at work. Shelling out the Rhino." Nepal, February 1883. From album distributed by Bourne & Shepherd in 1883, 19.1 × 25.7 cm
J. PAUL GETTY MUSEUM, LOS ANGELES:
NO. 84.XA.1170.23

FIGURE 24.9
"The Sport of Kings. In Nepaul the Rhinoceros is only killed by the Maharaja or his party." The specimens killed by the Duke of Portland in February 1883. From album distributed by Bourne & Shepherd in 1883, 19.1 × 25.7 cm. Also reproduced in the *Illustrated London News*, 10 February 1906
J. PAUL GETTY MUSEUM, LOS ANGELES:
NO. 84.XA.1170.31

25.4), there is only one early photograph which shows him after killing a rhino (fig. 24.15).

The written records for the period of Maharaja Chandra are particularly sparse. If a game book was kept by him or his officials, the details of this have remained hidden. Most narratives of this period relate to visits of foreign diplomats or expedition members who were given permission by the court to shoot a certain number of rhinos, all in the area of the Rapti River. Despite the lack of detailed information about dates, numbers of rhinos or tigers, or participants, there is ample evidence of a tradition among the ruling classes to stay in the wilderness, away from regular duties in the capital. This evidence usually derives from photographs, copies of which are found in different collections, with the majority taken by Dirgha Man Chitrakar (24.14).

The number of rhinos killed during these hunting camps is difficult to assess. There is one photograph showing a large series of tiger skins as well as 20 rhino skulls with horns (EAP838-1-10-1, no. 1, see 24.14). In the remaining photographs in the Chitrakar archive, excluding some obvious duplicates, there are a total of 46 individual

FIGURE 24.10
"Python and Rhinoceros." Nepal, February 1883. From album distributed by Bourne & Shepherd in 1883, 19.1 × 25.7 cm
J. PAUL GETTY MUSEUM, LOS ANGELES:
NO. 84.XA.1170.40

FIGURE 24.11
"Skinner preparing to skin." Hunting camp in Nepal in February 1883. From album distributed by Bourne & Shepherd in 1883, 19.1 × 25.7 cm. The right part of this photograph also found in *The Graphic* of 8 September 1884 (Anon. 1884c) and in *Navy and Army Illustrated* of 30 January 1904 (Anon. 1904b), where stated to be "from a photograph by Messrs. Johnston and Hoffmann, 31 Devonshire Street, W"
J. PAUL GETTY MUSEUM, LOS ANGELES:
NO. 84.XA.1170.43

rhinos. If this means a total of 66 rhinos over a period of ten years, this would give an annual average of some 7 rhinos killed during the camps. However, this may well disguise the reality. There is a photograph published in January 1925 (fig. 24.65) showing Kaiser Shumsher behind a set of at least 46 rhino skulls with horns, said to be the spoils of one season (Lévi 1925).

24.14 The Dirgha Man Chitrakar Collection of Photographs

Photography is the main source of information on the hunting expeditions of the Rana family and the King. Maharaja Chandra had engaged the artist Dirgha Man Chitrakar (1877–1951), first as Royal Painter, later also as Court Photographer. He was assisted by his son Ganesh

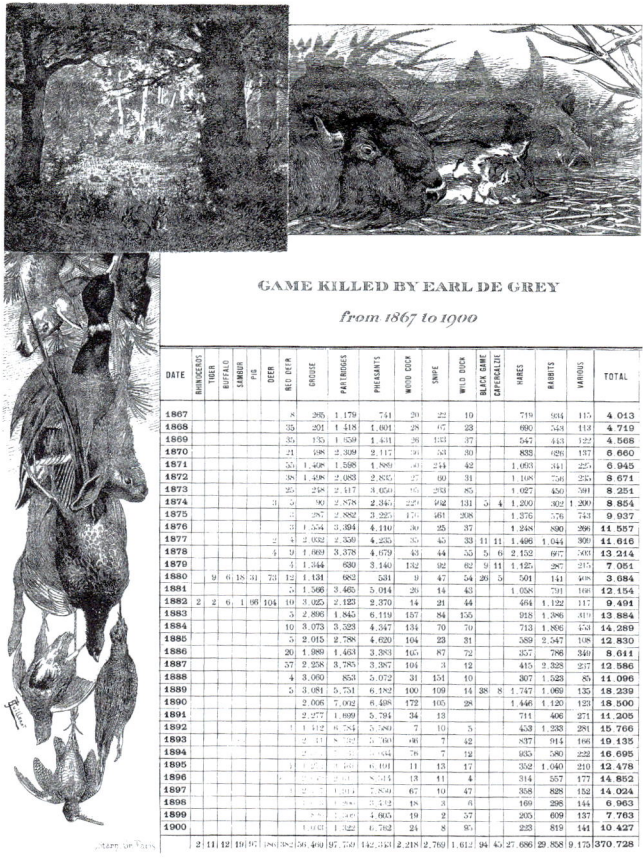

MAMA AND BABY

FIGURE 24.12 Game killed by Earl de Grey from 1867 to 1900. Reverse of a card from the business of Holland & Holland, 98 New Bond Street, London, advertising the efficiency of Schultze's Gunpowder

WITH PERMISSION © BODLEIAN LIBRARIES, UNIVERSITY OF OXFORD: JOHN JOHNSON COLLECTION, SPORT 10 (41)

FIGURE 24.13 "Mama and Baby." The head of a female rhino and the body of her baby killed by the 6th Duke of Portland in Nepal in February 1883. These trophies were once displayed in Welbeck Abbey

PORTLAND, *MEN, WOMEN AND THINGS*, 1937, P. 258

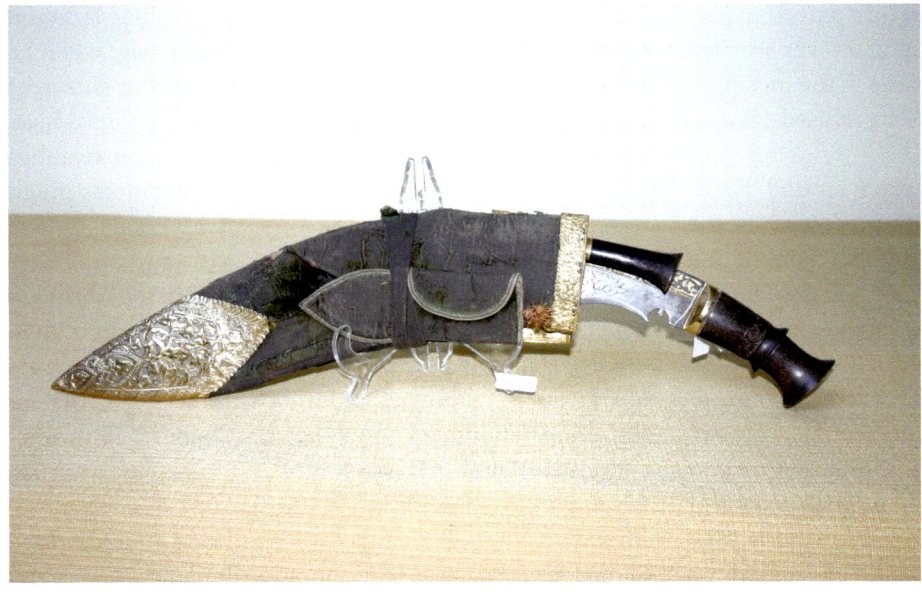

FIGURE 24.14
Kukri from Nepal with rhino horn handle, presented to the Gordon Highlanders by Maharaja Bir Shumsher in 1898

WITH PERMISSION © GORDON HIGHLANDERS MUSEUM, ABERDEEN: NO. GH608

FIGURE 24.15   Rhino killed by Maharaja Chandra Shumsher who is sitting on the back of the animal. The young
rhino might be alive. The other people, some of which will be members of the Rana family, are
unidentified. Undated photograph taken by Dirgha Man Chitrakar. Digitized from a glass plate
negative by the British Library Endangered Archives Program, no. EAP838-1-2-2, no. 9
PHOTO © KIRAN MAN CHITRAKAR COLLECTION, KATHMANDU

Man Chitrakar (1906–1985). Both spent several months each year in Chitwan to ensure that the main events could be captured on film. It is said that the Maharaja once asked Dirgha Man to send his son to Chitwan to be permanently stationed there and in return agreed to give him a large chunk of land in New Road. However, worried that he would lose his only son to malaria in the mosquito infested lands of the *terai*, Dirgha Man refused (Swaraj Chitrakar, pers.com. 2020).

The collection of photographs taken around Nepal during the 20th century are carefully preserved by the DirghaMan and GaneshMan Chitrakar Art Foundation in Kathmandu, run by Ganesh Man's grandchildren Swaraj and Christeena. It contains about 1800 glass plate negatives and 4000 acetate negatives. These were digitized through the Endangered Archives Program of the British Library and made available online (BL-EAP 2021e). There are scenes including a rhinoceros in four boxes with a total of 36 images, all on glass plates of 6.5 × 8.5 inch (16.5 × 21.6 cm), attributed to Dirgha Man Chitrakar.

The Chitrakar photographs are not annotated. There are no dates. Some persons can be recognized, sometimes tentatively, while others still remain unidentified. Apparently the images were never kept in a particular order hence any chronological sequence is hard to establish. There is reason to suggest that most were taken in the 1920s and 1930s. Maharaja Chandra Shumsher is found in one photograph which must date before 1929 when he died. His son Shanker Shumsher (1909–1976) is shown as a young man (teenager) giving a date range of 1925–1930. Likewise King Tribhuvan (1906–1955) was in his twenties in some photos taken in the field. Some examples might have been taken during the time of Maharaja Juddha Shumsher in the 1930s.

The list of 36 photographs with a rhino in the Chitrakar collection with tentative and sometimes uncertain identities of the persons involved is in Table 24.29. The names are given in abbreviated form without titles, including King Tribhuvan (1906–1955); Chandra Shumsher (1863–1929) and his sons Mohan (1885–1967), Kaiser (1892–1964), Shanker (1909–1976), and his grandson Mrigendra (b. 1906); Chandra Jung Thapa (n.d.); Juddha Shumsher (1875–1952) and his sons Bahadur (1892–1977) and Surya (d. 1945). Five plates were published by Davies (2005) and one by Mishra (2008).

TABLE 24.29   List of negatives in the Chitrakar Collection in Kathmandu with images of a rhinoceros

---

EAP838-1-2-2. A series possibly dating from the time of Chandra Shumsher.

| EAP838-1-2-2, no. 4 | Unknown man on rhino head with very long horn |
| EAP838-1-2-2, no. 9 | Chandra Shumsher, with many others. (fig. 24.15) |

EAP838-1-3-2. Different members of the Rana family

EAP838-1-3-2, no. 1	Chandra Jung Thapa (left) and 2 others
EAP838-1-3-2, no. 15	4 dead rhino in field; same occasion as 54 and EAP838-1-1-2, 51
EAP838-1-3-2, no. 18	Several unknown men
EAP838-1-3-2, no. 34	Mrigendra
EAP838-1-3-2, no. 38	Kaiser (fig. 24.16)
EAP838-1-3-2, no. 40	Several unknown men
EAP838-1-3-2, no. 46	Chandra Jung Thapa, Kaiser, one other with 7 rhino heads (Davies 2005)
EAP838-1-3-2, no. 51	Kaiser and Chandra Jung Thapa, and 2 others
EAP838-1-3-2, no. 52	One unknown man
EAP838-1-3-2, no. 53	One unknown man, 4 rhino heads, 6 tigers (Davies 2005)
EAP838-1-3-2, no. 54	same occasion as no.15, with many assistants (Mishra 2008)
EAP838-1-3-2, no. 59	Shanker and one other man, 4 rhino heads
EAP838-1-3-2, no. 60	Unknown man with 2 rhino heads, 2 tigers
EAP838-1-3-2, no. 62	Chandra Jung Thapa (Davies 2005)

EAP838-1-10-1. A box with a single image with results of a large hunt

| EAP838-1-10-1, no. 1 | Juddha, Bahadur, Mohan, many tigers, 18–20 rhino skulls (Davies 2005) |

EAP838-1-1-2. Series catalogued for Shanker Shumsher

EAP838-1-1-2, no. 7	Bahadur, Shanker, Tribhuvan
EAP838-1-1-2, no. 10	Shanker, Mrigendra, and 6 helpers (fig. 24.17)
EAP838-1-1-2, no. 20	Shanker, Mrigendra, and 6 helpers (like no. 10)
EAP838-1-1-2, no. 23	Tribhuvan
EAP838-1-1-2, no. 24	Tribhuvan (like no. 23) (fig. 24.19)
EAP838-1-1-2, no. 29	Kaiser, sitting on rhino head (Davies 2005)
EAP838-1-1-2, no. 36	Female rhino and dead calf, dead, no people
EAP838-1-1-2, no. 39	Rhino killed with elephants
EAP838-1-1-2, no. 42	Two rhino in grass, alive
EAP838-1-1-2, no. 51	Dead rhino with many helpers (together with EAP838-1-3-2, 15, 54)
EAP838-1-1-2, no. 60	Tribhuvan, Bahadur, Shanker, unknown (time as no. 7)
EAP838-1-1-2, no. 71	Bahadur, Kaiser, Shanker, Juddha (fig. 24.20)
EAP838-1-1-2, no. 75	Foreigners, also ladies, and many men
EAP838-1-1-2, no. 76	Shanker, Mrigendra, and unknown man
EAP838-1-1-2, no. 90	Shanker (same time as no. 76)
EAP838-1-1-2, no. 107	Shanker (fig. 24.18)
EAP838-1-1-2, no. 119	Surya, Shanker, Tribhuvan (fig. 24.21)
EAP838-1-1-2, no. 120	Surya, Shanker, Tribhuvan (same time as no. 119)
EAP838-1-1-2, no. 125	Surya, Shanker, Tribhuvan (same time as no. 119)

24.15   The Annual Shoot of 1907 with Manners-Smith

John Manners-Smith (1864–1920), Resident of Nepal 1905–1916, was present at a hunt in January and February 1907 organized by Maharaja Chandra Shumsher. This took place in the area on the Rapti River which had been prepared the previous year for the canceled visit of George, Prince of Wales (25.2). In total 14 male and 14 female rhinos were killed (Manners-Smith 1909). The account by Lydekker (1909) which refers to the same hunt has a slightly different total of 27 rhino. Six rhino calves were captured, ostensibly to restock the population in the

FIGURE 24.16   General Kaiser Shumsher (1892–1964) posing on a rhino in the Chitwan area of Nepal. Preserved
EAP838-1-3-2, no. 38
PHOTO © KIRAN MAN CHITRAKAR COLLECTION, KATHMANDU, GN_B_III_038

FIGURE 24.17   Shanker Shumsher (1909–1976) on right with his leg on rhino's head, and left of him Mrigendra
Shumsher (b. 1906), together with six unidentified helpers. Chitwan area, 1920's. Preserved
EAP838-1-1-2, no. 10
PHOTO © KIRAN MAN CHITRAKAR COLLECTION, KATHMANDU, GN_B_I_0010

FIGURE 24.18　Shanker Shumsher (1909–1976) sitting on a dead rhino in the Chitwan area. Preserved EAP838-1-1-2, no. 107
PHOTO © KIRAN MAN CHITRAKAR COLLECTION, KATHMANDU, GN_B_I_0107

FIGURE 24.19　King Tribhuvan of Nepal posing with a rhino after a hunt in *Terai*. Preserved EAP838-1-1-2, no. 24
PHOTO © KIRAN MAN CHITRAKAR COLLECTION, KATHMANDU, GN_B_I_0024

FIGURE 24.20
From left, Bahadur Shumsher (sitting on rhino head), Kaiser Shumsher, Shanker Shumsher and Juddha Shumsher (behind rhino with hunting cap on lap) with helpers during a hunting trip in Chitwan. Preserved EAP838-1-1-2, no. 71
PHOTO © KIRAN MAN CHITRAKAR COLLECTION, KATHMANDU, GN_B_I_0071

FIGURE 24.21
Surya Shumsher (left), Shanker Shumsher (center) and King Tribhuvan with hunted rhino in the Chitwan area of Nepal. Preserved EAP838-1-1-2, no. 119
PHOTO © KIRAN MAN CHITRAKAR COLLECTION, KATHMANDU, GN_B_I_0119

eastern part of Nepal. However, Lydekker heard that this was not pursued because all six animals were male, hence three were kept in Kathmandu, and three were sold to the German animal dealer Carl Hagenbeck, then exported to Antwerp, Manchester and New York.

24.16   The Annual Shoot of 1908 with the Maharaja of Bikaner

Invited by Maharaja Chandra Shumsher, Maharaja Ganga Singh of Bikaner (1880–1943) joined the shoot in the Naolpur valley near the Rapti River in February 1908. On this occasion 4 rhino were killed (Manners-Smith 1909).

24.17   Resident's Shoot of Christmas 1909

In a hunt organized by Manners-Smith for Christmas 1909 near Tribeni on the Gandak River, the explorer Thomas George Longstaff (1875–1964) shot one rhino. It had a horn of 38.1 cm listed by Rowland Ward (1910), later donated to the Natural History Museum in London (Longstaff 1950). The skin and bones of a rhino shot by Manners-Smith at the time was donated to the Indian Museum in Kolkata (Indian Museum 1911). Another two skulls from Nepal reached the Indian Museum in 1907, presented by the saddlery business of Watts & Co.

24.18   The British Envoy Frederick O'Connor
         1918–1925

Lieutenant-Colonel William Frederick Travers O'Connor (1870–1943) was the British Resident in Nepal 1918–1920, then British Envoy to Nepal 1921–1925. He returned on later occasions, shooting in 1926 and 1935. He was involved in the arrangements of the shoot organised for Edward, Prince of Wales in 1921 (25.4). O'Connor was an active hunter and went to the Chitwan area most seasons during his tenure. In his book of 1931, he presented a list of animals killed per year, without however providing any further details about locality, dates or companions (O'Connor 1931: 346).

Season of 1918–19: total rhino shot 1
Season of 1921–22: total rhino shot 13
Season of 1922–23: total rhino shot 0
Season of 1923–24: total rhino shot 5
Season of 1924–25: total rhino shot 1

On one of these occasions, O'Connor was photographed with the head of a rhino (fig. 24.22).

After completing the hunt with Edward, Prince of Wales on 18 December 1921, O'Connor stayed behind in the area (Ellison 1925). This would account for the 13 rhinos recorded for that season. With this date there is a horn of 38.1 cm obtained by M. Maxwell, who is not identified, but possibly part of O'Connell's group.

In 1924 O'Connor invited the Governor of Bengal Victor Bulwer-Lytton, 2nd Earl of Lytton (1876–1947) to his camp in the vicinity of Bhikna Thori. The party also included Lytton's wife, Pamela Chichele-Plowden (1874–1971), their two daughters Margaret (1905–2004) and Davidema (1909–1995), Lady Phyllis Windsor-Clive (1886–1971), Major Hugh Gordon Benton (1885–1931) and Captain Harvey of the Gurkha Regiment (Lytton 1942: 144). Lytton records that Harvey wounded two rhinos, while he personally killed two rhinos in the course of a week (fig. 24.23). One of these had a horn measuring 34 cm.

It is possibly this shoot in 1924 of which some photographs have been preserved considering the presence of two ladies in a few of them. These are found in an Album preserved in the Madan Puraskar

FIGURE 24.22   "Author and rhinoceros head" showing the British Envoy William Frederick Travers O'Connor in Nepal, on an unspecified date between 1918 and 1925
O'CONNOR, *ON THE FRONTIER AND BEYOND*, 1931, PL. AFTER P. 326

FIGURE 24.23   "My rhino" showing Victor, 2nd Earl of Lytton (1876–1947) in the Bhikna Thori area of Chitwan, Nepal, during a hunt in 1924
LYTTON, *PUNDITS AND ELEPHANTS*, 1942, P. 143

FIGURE 24.24
The British Envoy Frederick O'Connor
with his family and companions in front
of a dead rhino in Nepal. His hosts of the
Shumsher family accompany him. Undated,
possibly 1924
WITH PERMISSION © MADAN PURASKAR
PUSTAKALAYA: EAP166/2/1/8 NO. 177

FIGURE 24.25
Another photograph of the scene of 1924 in
fig. 24.24
WITH PERMISSION © MADAN PURASKAR
PUSTAKALAYA: EAP166/2/1/8 NO. 182

Pustakalaya library, Kathmandu and digitized through the Endangered Archives Program of the British Library (BL-EAP166/2/1/8). It contains 401 items, all undated and unannotated, half showing scenes in the jungle, others taken during the coronation of King Tribhuvan in 1913 (BL-EAP 2021a). However, the actual occasions remain uncertain in the absence of supporting data. Five photographs with a rhino have been identified:

EAP166/2/1/8 no. 143 – O'Connor sitting on rhino (not figured)

EAP166/2/1/8 no. 177 – O'Connor, Nepali team including Kaiser Shumsher with dead rhino (fig. 24.24).

EAP166/2/1/8 no. 182 – O'Connor and Nepali team with dead rhino (fig. 24.25).

EAP166/2/1/8 no. 199 – Loading the body of rhino, head of rhino lying in front (fig. 24.26).

EAP166/2/1/8 no. 206 – O'Connor and Nepali team with dead rhino (fig. 24.27).

24.19   Raja Kirtyanand Sinha at Harahia in 1920

Raja Bahadur Kirtyanand Sinha (1880–1938) of Banaili near Purnia, Bihar, was given permission by Maharaja Chandra

FIGURE 24.26
Loading the body of rhino, head of rhino lying
in front. Undated photograph in Nepal
WITH PERMISSION © MADAN PURASKAR
PUSTAKALAYA: EAP166/2/1/8 NO. 199

FIGURE 24.27
Dead rhino with Frederick O'Connor and
a Nepali team including Maharaja Kaiser
Shumsher. Undated, possibly 1924
WITH PERMISSION © MADAN PURASKAR
PUSTAKALAYA: EAP166/2/1/8 NO. 206

Shumsher to shoot two rhino. Hence in March 1920 he camped first at Baghahi, 19 km from Bhikna Thori railway station, and then at Harahia. He wounded one rhino at Baghahi, which could not be followed, but was later found dead at Sukhibar. Near Harahia (exact location unclear) he killed two male rhinos. The published account of this time in Chitwan is illustrated by two photographs showing the Raja and the two dead animals (figs. 24.28, 24.29).

On the spot of this first direct kill, the Raja performed the Tarpan ceremony with the blood of this animal. Then he "got rice soaked in the blood of the rhino and also got prepared a quantity of rhino meat, and having dried them

up, brought them home for use in Varshik Sharadhas" (K. Sinha 1934: 13).

24.20 Two Rhinos Captured for Frank Buck in 1922

The well-known American animal dealer and author Frank Buck was tasked to obtain Indian rhinos by William Temple Hornaday (1854–1937) and Charles Bingham Penrose (1862–1925), the Directors of the Bronx Zoo in New York and the Philadelphia Zoological Gardens. Staying in Kolkata later in 1922, he met Kaiser Shumsher and negotiated the capture of two young rhinos priced together at

FIGURE 24.28
"My first rhino" shot in Nepal in March 1920 by
Raja Kirtyanand Sinha of Banaili
SINHA, *SHIKAR IN HILLS AND JUNGLES*, 1934,
PL. 1

FIGURE 24.29
"My second rhino" shot in Nepal in March 1920
by Kirtyanand Sinha
SINHA, *SHIKAR IN HILLS AND JUNGLES*, 1934,
PL. 2

Rs 35,000 or $12,600. Kaiser headed the operation at the end of 1922, during which 21 rhinos were killed, including the mothers of the two calves (Buck & Anthony 1930: 60). There are no photographs in Buck's book, but some were used to illustrate short stories in magazines aimed at a young public (poorly printed and impossible to reproduce). Buck (1931) includes a photograph showing four dead rhinos surrounded by a large crowd, suggesting a 1922 date, but the same was recently published by Mishra (2008) dated to the 1930s (see Table 24.29, nos. 15 and 51). However, another illustration on the same page shows a young rhino in a wooden stockade (Buck 1931: 37). There is a further photograph showing how a crate with a rhino was loaded on the ship in the harbour of Kolkata (Buck 1936: 6). The whole episode was later also reworked as a cartoon reaching a new generation (Buck 1953). Early in 1923 Buck traveled to the railhead at Raxaul and

proceeded to the camp at Bilgange (Birganj) across the border into Nepal where the animals were ready to be transported. On 30 March 1923 in Kolkata, the shipment was loaded on the *Lake Gitano* to Hong Kong and then on the *President Wilson* to San Francisco. The female Bessie lived in the Bronx Zoo 1923-05-24 to 1962-01-25, and the other female Peggy in the Philadelphia Zoo 1923-05-24 to 1943-04-14.

24.21 The Vernay-Faunthorpe Expedition of 1923 in the Gandak Valley

Henry Fairfield Osborn (1857–1935), in 1908–1933 Director of the American Museum of Natural History, New York (AMNH), was first approached in 1921 by the former colonial administrator John Champion Faunthorpe (1871–1929)

FIGURE 24.30
John Champion Faunthorpe (1871–1929) on left
and Arthur Stannard Vernay (1877–1960) with a
rhinoceros killed in Nepal in March 1923. Photo
by George Miller Dyott (1877–1960)
VERNAY, *THE SPUR*, 1923. COURTESY:
UNIVERSITY OF IOWA, SPECIAL
COLLECTIONS AND UNIVERSITY ARCHIVES

FIGURE 24.31
One of the rhinos shot by Vernay in Nepal in
March 1923, photographed by George Dyott
FAUNTHORPE, *NATURAL HISTORY*, 1924,
P. 181; VERNAY, *THE SPUR*, 1923

about improving the displays of Asian mammals. This was put on a firmer basis once the antiques dealer and explorer Arthur Stannard Vernay (1877–1960) pledged his financial support, resulting in the Vernay-Faunthorpe expedition of 1923–1924 to Burma, Nepal and India.

Faunthorpe applied for permission to Maharaja Chandra Shumsher to collect rhino specimens in Nepal. The expedition members stayed for less than a week in the country as part of their larger tour of collection sites including Mysore and Khairigarh. They were advised to explore the valley of the Gandak River on the western boundary of Chitwan, which then was only accessible by a long march or a journey by boat. The camp was called Koalwa (Dyott 1923). Vernay was well-known as a big-game hunter, while Faunthorpe brought a wealth of local knowledge and contacts. They were accompanied

by the photographer George Miller Dyott (1877–1960), who captured the events in pictures as well as in a cinematograph film. There was also a taxidermist in the team, possibly Robert Henry Rockwell (1885–1973) who is said to have mounted the animals (Faunthorpe 1924a), assisted by the Indian skinner Pancham employed by the Ranee of Khairigarh.

Although there never was a full-scale written account of the expedition, there were several semi-popular articles by Vernay (1923a, b, c, 1931) including one in *The Spur*, by Faunthorpe (1923, 1924a, b, c, 1930) including those in *Natural History* and the *Wide World Magazine*, as well as by Dyott (1923a, 1923b, 1924) in the *Illustrated London News* and in the Spanish science magazine *Iberica*.

The expedition members were in their camp on the Gandak River from 10 to 14 March 1923. They observed

FIGURE 24.32
Rhinoceros shot in Nepal in 1923, photo by
George Dyott
DYOTT, *ILLUSTRATED LONDON NEWS*, 1923,
P. 433

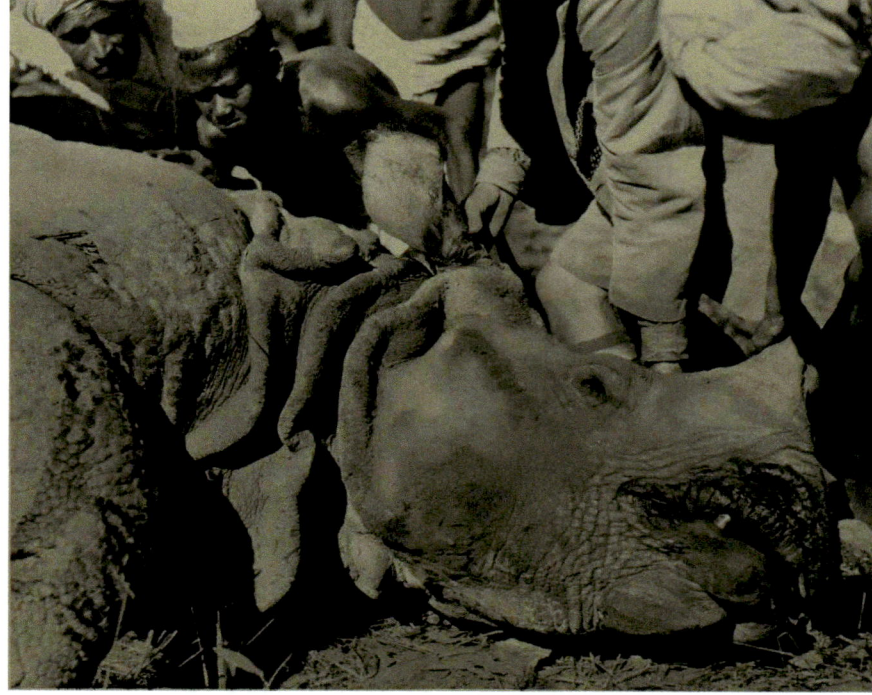

FIGURE 24.33
Head of rhino shot in Nepal in 1923, photo by
George Dyott
DYOTT, *ILLUSTRATED LONDON NEWS*, 1923,
P. 435

a female rhino which they called Lizzie with a calf. The object was to collect a male and female rhino, both with horns at least 30 cm long. They in fact killed 3 males and 1 (pregnant) female. This was their itinerary:

10 March 1923 (Saturday): first observation of female Lizzie and her calf. Animal left alive at end of expedition.

10 March 1923: 1 male rhino killed, hunter not stated. Horn measured 8 ½ in (21.6 cm).

11 March 1923 (Sunday): 1 male rhino killed by Faunthorpe, horn 12 ½ in (31.8 cm).

11 March 1923: 1 male rhino killed by Vernay, horn 12 in (30.5 cm).

14 March 1923 (Wednesday): 1 female rhino killed by Vernay, horn 8 in (20.3 cm). She was accompanied by a calf which was left. An embryo was found and preserved.

FIGURE 24.34   The female *R. unicornis* named 'Lizzie' in the Gandak Valley of Nepal followed by her half-grown calf. Photo by George Dyott
FAUNTHORPE, *NATURAL HISTORY*, 1924, P. 174

FIGURE 24.35   Advertisements for the movie "Jungle Life in India" shown in the Philharmonic Hall in London from 5 November 1923. This shows Trafalgar Square in London with two S-type buses fleet numbers S89 on route 3 and S454 on route 12
WITH PERMISSION © TFL FROM THE LONDON TRANSPORT MUSEUM: 1998/44789

The horn of the first rhino was worn and splintered down, and with a length of only 21 cm did not meet the requirements of the intended exhibition. The second male rhino shot by Faunthorpe had a much longer horn, while the skin showed several wounds and scratches. Another male and a female were then killed by Vernay, who was hunting on foot (Faunthorpe 1924a). Faunthorpe used a Jeffries .400 rifle and Vernay a Holland .465 (Vernay 1923a).

FIGURE 24.36
An early view of the diorama of the two
R. unicornis donated by Maharaja Chandra
Shumsher of Nepal, first opened to the
public on 17 November 1930 in the American
Museum of Natural History, New York
*AN ALBUM OF THE GROUPS IN THE
VERNAY-FAUNTHORPE HALL OF SOUTH
ASIATIC MAMMALS OF THE AMERICAN
MUSEUM OF NATURAL HISTORY, 1930*

The photographer George Dyott presented a comprehensive overview of his results in the *Illustrated London News* of 8 September 1923 (Dyott 1923a). Some other photographs were included in the papers by Vernay and Faunthorpe. In total rhinos appear in eight of the published photographs, probably representing three dead and two living specimens judging from the environment (figs. 24.30 to 24.34). Two images show the leaders Vernay and Faunthorpe standing behind a dead rhino in the first of which, they wear hats obscuring their faces (fig. 24.30). In the second image, they appear without hats on the other side of the animal (Dyott 1923a: 432, 1924: 230).

Dyott's proudest achievement must have been the photos of the female rhino which was named 'Lizzie' together with her half-grown calf. They saw them near the camp on the first day in the Gandak Valley, but decided that she did not have the requirements needed for the exhibition. He thought that it was the first photograph ever secured of a *R. unicornis* in the wild. Although this claim is incorrect, the achievement was remarkable (Rookmaaker 2021b).

Dyott also took cinematographic sequences with his 'Akeley' camera which has a simple and rapid elevating and traversing mechanism (Faunthorpe 1924a). This type of camera was developed in 1917 by Carl Akeley (1864–1926), curator of the American Museum of Natural History, and therefore was state-of-the-art when Dyott was in Nepal. These could be the first moving pictures showing rhinos alive in the wild. After completion of the expedition, Dyot compiled a silent 16 mm movie "Jungle Life in India" lasting 20 minutes: "third part is shot in Nepal. Rhino: we see their tracks, and then a cow and a calf appear on the left of the picture. They move about before us unsuspicious when all of a sudden they take

alarm, wheel and gallop away; the cow with the heavy and somewhat cumbersome gait of her kind and also of one who is conscious of the responsibilities of this life, while her care-free offspring dashes along more like some great ungainly puppy" (Vernay 1923c). The first private viewing of the movie was on 1 November 1923 at St. James's Picture Theatre in London, with a lecture by Dyott. It was shown to the public in the Philharmonic Hall for 4 weeks, beginning 5 November (Vernay 1923b, c). No program has survived, but the movie was advertised on London buses photographed on Trafalgar Square (fig. 24.35). The first reference to showing the movie in the USA is on 21 December 1923, when Faunthorpe presented a lecture in the American Museum of Natural History (Osborn 1922; Anon. 1923a). A copy is preserved in the archives (AMNH Special Collections, Film Collection no. 198, on ¾ inch U-matic videotape).

The expedition returned to the AMNH with 127 specimens of 42 mammal species (Osborn 1923). The rhinos are registered:

M-54454. Male *R. unicornis*, collected 11 March 1923. Complete skeleton (disarticulated), skull and mandible, whole skin. Mounted for exhibition. The horn of 12 in (30.5 cm) was listed in the records of Rowland Ward (1928: 438).

M-54455. Male *R. unicornis*, collected 11 March 1923. Skull, horn, skin (not mounted).

M-54456. Female *R. unicornis*, collected 14 March 1923. Skeleton, skin. Mounted for exhibition.

M-54457. Embryo. Male *R. unicornis*, found in female 54456.

M-245543. Skull and mandible. Found in the field on 14 March 1923.

There is no trace of the first male with the 8 inch horn killed on the first day. The mounted male and female were presented to the Museum by Maharaja Chandra Shumsher. They were part of a new diorama in the Vernay-Faunthorpe Hall of South Asiatic Mammals, first opened to the public on 17 November 1930 (fig. 24.36). In order to ensure the accuracy and natural feel of the background and structure of the diorama, Faunthorpe in 1928 went back to Nepal accompanied by the Assistant Chief of the preparation staff Albert E. Butler (1887–1969), and the artist Clarence Clark Rosenkranz (1871–1959) (Osborn 1928, 1930; Anthony 1930: 538). The director of the museum proudly stated that "the final result stands as the finest example of modern methods of Museum

FIGURE 24.37   One of the rhinos killed by Theodore Roosevelt's sister-in-law Belle Roosevelt in Nepal in January 1926
SUNDAY STAR, WASHINGTON DC 22 MAY 1926, GRAVURE SECTION. COURTESY: DAN ZIEGLER

FIGURE 24.38
An early view of the grand diorama of two Nepalese rhinos in the Field Museum of Natural History, Chicago
FIELD MUSEUM ANNUAL REPORT, 1929, PL. 19

exhibition. The finished South Asiatic Hall is superb. It surpasses our fondest dreams and expectations" (Osborn 1931: 39).

24.22   The Roosevelt Expedition Collects in Nepal in 1926

The brothers Theodore Roosevelt Jr. (1887–1944) and Kermit Roosevelt (1889–1943) undertook the James Simpson-Roosevelt Asiatic Expedition from May to November 1925. After they were joined by their wives on their return to Srinagar (Kashmir), the party was invited to Nepal by Maharaja Chandra Shumsher and the British envoy Hugh Wilkinson (1874–1939). They arrived at the station in Bhikna Thori on 9 January 1926, as told by Kermit's wife Belle Willard Roosevelt (1892–1968) in her recollections of 1927. On Tuesday 12 January 1926 Belle killed a female rhino with a single shot, and her calf was left to walk away. A second rhino, also female, was first wounded by

FIGURE 24.39   Lord Catto with a rhino trophy obtained in Nepal in 1926
TATLER, WEDNESDAY 26 APRIL 1944. REPRODUCED BY KIND PERMISSION OF CAMBRIDGE UNIVERSITY LIBRARY

Wilkinson and then killed by Theodore. A photograph was taken showing one of the dead rhinos together with Theodore, Kermit and Belle Roosevelt (fig. 24.37). The Roosevelts stayed only six days in Nepal.

The skins of the two rhinos were added to the collection of the Field Museum of Natural History in Chicago, Illinois (Chicago FMNH 1926: 74). In the course of 1927, the Museum developed a grand diorama, the William V. Kelley Hall of Asiatic Mammals in Hall 17 (fig. 24.38). It showed the two rhinos standing on the reedy bank of a river, with a background painted by Charles Abel Corwin (1857–1938). The mounted specimens were reproduced by the novel process of cellulose-acetate by the taxidermist Leon Louis Walters (1888–1956).

Belle Roosevelt wrote that both rhinos had been female, while the museum records list them as a male and female. The Roosevelt party had only been given permission to shoot two rhinos. Hence either an exchange was made in the field, or there was some other minor discrepancy in the records.

24.23   Hewett and Catto in 1926

In 1926, O'Connor invited John Prescott Hewett and Thomas Sivewright Catto, but may not have been present himself as he had left his official post. Shooting near Bhikna Thori, Hewett and Catto killed at least one rhino, with a horn of 29.2 cm listed by Rowland Ward (1928). Catto exhibited the rhino head trophy in the hall of his home in Holmbury St. Mary, Surrey (fig. 24.39).

24.24   The Shoots during the Time of Maharaja Bhim Shumsher

There is little evidence linking Maharaja Bhim Shumsher (1865–1932) directly with shooting expeditions during his short reign from 26 November 1929 to 1 September 1932. He might have performed the Tarpan ceremony in 1930 or 1931 which was usually timed within a year of accession of a new ruler. He was the host for the visit of Lord Ratendone in 1932 (24.25).

24.25   Lord Ratendone in Chitwan in 1932

On request of the Viceroy Freeman, 1st Marquess of Willingdon (1866–1941), Maharaja Bhim Shumsher gave permission to shoot a rhino to Freeman's son Lord Ratendone, 2nd Marquess of Willingdon (1898–1979) at the start of 1932. He was accompanied by the UK

Ambassador to Nepal Clendon Turberville Daukes and his wife Dorothy Maynard Daukes. The British Film Institute in London (no. 444818) preserves an amateur movie credited to Merton Lacey (1902–1996) as photographer, inaccurately dated to 1930. The credits of this movie add Maharaja Bhim Shumsher, Colonel Chandra Jung Thapa, Colonel Dilli Jung Thapa and Sir Adrian Baillie (1898–1947) as participants.

There were still "considerable numbers" of rhino in Chitwan. According to the description of the movie, Ratendone first shot a female rhino accompanied by a calf, Baillie shot a second rhino and Ratendone a third. The young rhino first ran away, but driven back to the mother it was captured alive. The Viceroy's wife Marie Adelaide Freeman (1875–1960) donated two rhinos to the Calcutta Zoo in March and October 1932, one of which would have been this young one captured in Nepal. At that time, Daukes (1934) estimated the value of rhino horn at £60 per item.

24.26 The Shoots during the Time of Maharaja Juddha Shumsher

Maharaja Juddha Shumsher (1875–1952) was Prime Minister from 1 September 1932 to 29 November 1945. A unique insight into his regular excursions in the jungle is provided through his sporting diary for the years 1933 to 1940. On his initiative, this was extracted and explained in a rare book published in 1942 by Evelyn Arthur Smythies (1885–1975). Together with his wife Olive Muriel (Cripps) Smythies (1890–1961), Smythies lived in Kathmandu 1940

to 1947 as Forest Advisor to the Government of Nepal (Smythies 1974; O. Smythies 1953, 1961; Anon. 1975). With light duties, he traveled across the length of Nepal. He first came to India in 1908 to work in the Forest Service, and retired to England in 1947. His book on *Big Game Shooting in Nepal* is a remarkable record from a country which then was still closed to casual visitors. Smythies does not elaborate on the participants of each hunt.

There is an Album of photographs in the Madan Puraskar Pustakalaya (MPP), Kathmandu from the time of Maharaja Juddha, preserved through the Endangered Archives Program of the British Library (BL-EAP 2021d). None of the images are individually annotated or dated. The photographs show that many members of the Rana family took part in the hunting camps, but details are elusive as Smythies gives only a few hints about the number of people present in each year. This album contains 11 photographs of rhinos and their hunters. These are reproduced here for the first time with the assistance of MPP (figs. 24.40 to 24.48), except no.16 of an indistinct living rhino disappearing in high grass, and no. 51 of Madan Shumsher near-identical to no. 31. There are two photographs of a young boy, who is likely to be Maharaja Juddha's son Min Shumsher (b. 1928), mentioned by Smythies (1942: 110) to have killed his first rhino on 28 January 1939. This may be the year for all these images, although this remains partly speculative.

Another album in MPP is connected with the times of Rani Jagadamba (1918–1988), the wife of Chandra's son Madan Shumsher (BL-EAP 2021c). This contains two photographs of rhinos in the Rapti River valley, as always without date. One is an indistinct image of a dead rhino

FIGURE 24.40
Maharaja Juddha Shumsher (middle), his brother Baber Shumsher (1888–1960) on the right with a foreign visitor on the back of a rhino in Chitwan. Date unknown
WITH PERMISSION © MADAN PURASKAR PUSTAKALAYA: EAP166/2/1/27/17

FIGURE 24.41
Maharaja Juddha Shumsher with members of his
family and officials behind a series of five rhino
heads
WITH PERMISSION © MADAN PURASKAR
PUSTAKALAYA: EAP166/2/1/27/30

FIGURE 24.42
Madan Shumsher (1909–1955), the youngest son
of Maharaja Chandra Shumsher, posing with his
foot on a rhino's head, possibly implying that it
is the one which he killed. A very similar photo is
available as no. 51 in the same album
WITH PERMISSION © MADAN PURASKAR
PUSTAKALAYA: EAP166/2/1/27/31

FIGURE 24.43
A member of the Shumsher family sitting on the head of
a rhino
WITH PERMISSION © MADAN PURASKAR PUSTAKALAYA:
EAP166/2/1/27/48

FIGURE 24.44 A member of the Rana family posing on the head of a rhino with his gun
WITH PERMISSION © MADAN PURASKAR PUSTAKALAYA: EAP166/2/1/27/54

FIGURE 24.45 One of the childen, possibly the Maharaja's youngest son Min Shumsher (b. 1928) sitting on a rhino's head
WITH PERMISSION © MADAN PURASKAR PUSTAKALAYA: EAP166/2/1/27/46

FIGURE 24.46 This could be Min Shumsher (b. 1928) also shown in no. 46. If the young animal in front of the dead rhino is still alive, he would be quite brave, and maybe the photograph was staged
WITH PERMISSION © MADAN PURASKAR PUSTAKALAYA: EAP166/2/1/27/57

FIGURE 24.47 An unidentified hunter sitting on the head of a rhino in Nepal
WITH PERMISSION © MADAN PURASKAR PUSTAKALAYA: EAP166/2/1/27/42

lying in high grass. A second shows two hunters on the back of a dead rhino (fig. 24.49).

A miscellaneous collection auctioned by Dominic Winter (2020) assembled by the reporter David Lomax (1938–2014) included a photograph of a hunting party with four dead tigers and one rhino head (fig. 24.50).

A note in Nepali script gives the name of Balakrishna Shumsher (1903–1981). The photograph deviates from the normal views in that it also shows ten women who must have joined their husbands in camp and wanted to see the results.

FIGURE 24.48
Young rhino brought to the camp
WITH PERMISSION © MADAN PURASKAR
PUSTAKALAYA: EAP166/2/1/27/53

FIGURE 24.49
Two Nepali men sitting on a dead rhino. The person
in front may be Maharaja Kaiser Shumsher, perhaps
in the 1920s or 1930s
WITH PERMISSION © MADAN PURASKAR
PUSTAKALAYA: EAP166/2/1/17/71

24.27 The Sporting Diary of Maharaja Juddha Shumsher for 1933

Following the extracts provided by Smythies (1942: 83–92), Maharaja Juddha Shumsher spent 23 days in Chitwan, from 7 to 29 January 1933. Rhinos were numerous enough to interfere with the usual tiger hunts. The first camp was at Patlahara on the north bank of the Rapti River, and the second camp called Bahara a few kilometers further west at the confluence with the Narayani River. No hunting had taken place in this area for over 70 years. This was the itinerary in 1933.

Patlihara Camp.

15 January 1933: 1 big male rhino killed by Maharaja Juddha after firing 19 shots from .465 rifle.

FIGURE 24.50   Group with 3 men and 10 women behind 4 tigers and a rhino head. The inscription in the lower right
corner refers to Balakrishna Shumsher (1903–1981)
WITH PERMISSION © DOMINIC WINTER AUCTIONEERS, CIRENCESTER, 16 DECEMBER 2020,
LOT 399

Bardaha Camp

18 January 1933: 3 rhinos were enclosed in a ring, but broke out before they could be shot.

19 January 1933: 1 rhino killed by Maharaja Juddha with several shots.

19 January 1933: 1 female rhino killed by Nir Shumsher (1913–2013), a son of Juddha. Her calf was captured alive and taken to camp.

19 January 1933: 2 rhino killed (shooter not identified). Maharaja Juddha performed the Khadga-rudhir Tarpan ceremony.

20 January 1933: 3 rhinos seen.

22 January 1933: 1 rhino killed by Juddha's brother Baber Shumsher (1888–1960).

22 January 1933: 3 rhinos killed by Juddha's sons, Surya Shumsher (d. 1945), Narayan Shumsher and Surendra Shumsher.

22 January 1933: 1 rhino killed by Baber's son Brahma Shumsher (b. 1909).

From Bardaha to Hetaura on road to Kathmandu

22 to 29 January 1933: 2 rhino killed (shooters not identified).

In total 14 rhinos were killed in January 1933, and a baby was captured (Smythies 1942: 92), with the shooter identified in eight cases. The photograph of the Maharaja inspecting a rhino is probably the scene on 15 January 1933 (fig. 24.51). The colour plate "Rhino prepares to charge" could be taken at any of the hunts mentioned in the book (fig. 24.52). The photographs published by Smythies to illustrate the events would have been taken from the game book, which is no longer known to exist. These were attributed to Samar Shumsher (1883–1958) and his son Balakrishna Shumsher (1903–1981), as well as the photographer B.D. Joshi (Smythies 1942: vi).

In the *Records* of Rowland Ward there are two horns from Nepal listed with date 1933, which may refer to the January hunt, or a later occasion. One horn of 15 in (38.1 cm) is attributed to Kaiser Shumsher (1892–1964), who was the Officer in charge of *shikar* in early 1933 (Smythies 1942: 163). Another horn of 19 ½ in (49.5 cm) is attributed to Ramanuj Saran Singh Deo (1893–1965), 115th Maharaja of Surguja, Madhya Pradesh.

On 19 January 1933 at Bardaha Camp, Maharaja Juddha performed the formal Khadga-rudhir Tarpan ceremony

FIGURE 24.51
"His Highness inspects his rhino", referring
to a rhino shot by Maharaja Juddha on
15 January 1933
SMYTHIES, *BIG GAME SHOOTING IN NEPAL*,
1942, PL. 24

FIGURE 24.52 "Rhino prepares to charge." It is not clear to which
of the hunts between 1933 and 1940 in Chitwan this
plate refers
SMYTHIES, *BIG GAME SHOOTING IN NEPAL*, 1942,
PL. 15

FIGURE 24.53 "The head of the rhino with twelve-inch horn" shot
by Frederick O'Connor in January 1935
COUNTRY LIFE, 15 JUNE 1935, P. 614. REPRODUCED
BY KIND PERMISSION OF CAMBRIDGE UNIVERSITY
LIBRARY

which should be an annual ritual at the anniversary of the father's death to help him on the road to Nirwana (24.2). The value of the ceremony in Nepalese tradition is increased when the vessel used to pour water is a rhino horn, again further when the liquid is rhino blood, and an ultimate merit freeing ancestors from reincarnation is accorded when the ceremony is made from inside the body of the rhino: "hence when a rhino is killed, the great mass of bowels and entrails are removed, leaving a vast cavity into which the man crawls to make the blood libation. But this last and rather unpleasant performance is not often done" (Smythies 1942: 90).

FIGURE 24.54
"The hunting party, including the Duke and Duchess of Sutherland, Sir Frederick O'Connor and Commander Edmonstone." A shoot in Nepal in January 1935
COUNTRY LIFE, 15 JUNE 1935, P. 614.
REPRODUCED BY KIND PERMISSION OF CAMBRIDGE UNIVERSITY LIBRARY

24.28 O'Connor Accompanies the Duke of Sutherland in 1935

Frederick O'Connor had arranged another visit in January 1935 when again he went to Bhikna Thori (O'Connor 1935, 1940: 186). He was accompanied by George, the Duke of Sutherland with his wife Eileen, and Archibald Charles Edmonstone. They stayed ten days and killed one rhino with a horn of 30.5 cm. The Duke of Sutherland lived at Dunrobin Castle in Scotland, where once they displayed three rhino heads (not necessarily from South Asia), which were sold off (Dunrobin Museum, pers.com. 2018-03). The two photographs illustrating the story in *Country Life* in 1935 are all that remains (figs. 24.53, 24.54). Probably referring to the same occasion, there is an unrelated photograph in the collection of Belton House, Lincolnshire, UK (fig. 24.55) with the inscription: "Jan 1935/To Dorothy the intrepid big game hunter/from Geordie." Geordie was the nickname of the 5th Duke of Sutherland, addressing this to Dorothy Kent (Cust) Power (1902–1966) of Belton House.

24.29 The Sporting Diary of Maharaja Juddha Shumsher for 1936

Maharaja Juddha undertook an extensive tour of the southern districts from 1 December 1935 starting in Mahottari and from there westwards through Sarlahi

FIGURE 24.55   Photo inscribed by the 5th Duke of Sutherland to Dorothy Cust (Power) in the collection of Belton House, Lincolnshire
WITH PERMISSION © NATIONAL TRUST: 1935 NT 436340

to Chitwan (Smythies 1942: 117). Only after reaching the Rapti Valley in February 1936, did they start to see rhinos (Smythies 1942: 135, 164). This was the itinerary for 1936.
18 February 1936: 1 rhino seen.
23 February 1936: 1 big male rhino killed by Maharaja Juddha Shumsher. There were at least ten or twelve other rhinos in this spot. The Khadga-blood Tarpan ceremony was performed.

24.30 Baron Mannerheim of Finland in Nepal in 1937

A national hero of Finland, Baron Carl Gustaf Emil Mannerheim (1867–1951) was part of a hunting party in Nepal in 1937 (Shrestha et al. 2017: 18). There is no rhino

FIGURE 24.56
"Rhinos break cover and charge" during
the 1938 shoot of the Viceroy, Victor
Linlithgow
SMYTHIES, *BIG GAME SHOOTING IN
NEPAL*, 1942, PL. 28

trophy in the Mannerheim Museum in Helsinki. The hunt was organized by the UK Ambassador Frederick Marshman Bailey, who is credited with a movie "Tiger Hunt" released in 1938 set in Nepal (British Film Institute, no. 12806). A rhino is sighted but left alone due to restrictions at the time.

24.31   The Viceroy Lord Linlithgow in Chitwan in 1938

In 1938 Maharaja Juddha Shumsher invited the Viceroy of India, Victor Alexander John Hope, 2nd Marquess of Linlithgow (1887–1952) to visit Nepal. He came with his wife Doreen and his three daughters. The party included the Viceregal staff, the British Envoy Geoffrey Lawrence Betham (1889–1963) with his wife and daughter, as well as the Legation Surgeon Colonel Rogers (Smythies 1942: 93; Glendevon 1971). The details of the shoot were set out by Smythies. This was the itinerary in December 1938.

8 December 1938 – Linlithgow killed one rhino at Kasra.

10 December 1938 – Linlithgow killed one rhino at Bhikna Thori.

10 December 1938 – C.G. Toogood killed one rhino at Bhikna Thori.

Linlithgow arrived on 3 December 1938 and was accommodated in a pleasant camp at Bhikna Thori, on the Indo-Nepalese border. On 8 December the hunt took place at Kasra, about 50 km towards the Rapti River. When the party arrived in their motorcars, two rhinos could be seen crossing the Rapti. That morning Lady Linlithgow killed a large tiger measuring 10 ft 8 in (325 cm), after which the Viceroy joined the Maharaja on his elephant Bhimgaj, staunch enough to remain standing when a rhino would

charge, and when this happened, the Viceroy aimed and "the rhino collapsed after two well-directed shots" (Smythies 1942: 97). On 10 December the party was again near Bhikna Thori. Linlithgow killed an old rhino "without teeth and a good horn", while a second rhino was shot by Colonel Cyril George Toogood (1894–1951), the Viceroy's Military Secretary (Smythies 1942: 100).

The official photographs include one showing two rhinos running away into the jungle (fig. 24.56) and another showing Linlithgow sitting on the shoulder of a dead rhino, with people on four elephants watching from behind (fig. 24.57). Linlithgow kept a Game Book which is still preserved in the archives of his residence Hopetoun House, near Edinburgh, Scotland, which has only few statistics of the hunt in Nepal and no rhino photographs (Peter Burman, Hopetoun Papers Trust, pers.comm. 2018). Neither is there any trace of the rhino trophies.

The Sphere (1939-07-22) published a series of pictures enlarged from a cinematic film showing some scenes of this hunt. The British Film Institute, London, still preserves a silent 16mm black and white movie with different activities associated with Linlithgow in India in 1938 and 1939, with detailed notes. This includes sequences described as "Men looking at a dead rhino (258); Lord Linlithgow sitting on the rhino (263); Members of the party standing around the body (270); Group: front row – Lord Linlithgow and his Nepalese hosts; back row – Lady Anne, Lady Doreen, Lady Linlithgow, Lady Joan and Captain W.A.G. Burns (277); The dead rhino (280)" (Linlithgow 1938). Major General Walter Arthur George Burns (1911–1997) of the Coldstream Guards was serving as the ADC to the Viceroy at the time.

FIGURE 24.57
Victor Alexander John Hope, 2nd Marquess of Linlithgow, the Viceroy of India, sitting on a rhino killed in December 1938 in the Chitwan area of Nepal
SMYTHIES, *BIG GAME SHOOTING IN NEPAL*, 1942, PL. 30

24.32 The Tarpan Ceremony of Kiran Shumsher in 1938

When Esmond Bradley Martin visited Nepal in 1982, he was able to meet Field Marshal Kiran Shumsher (1916–1983), a son of Maharaja Juddha. Kiran told how he killed a rhino in 1938 and used the animal to perform his Tarpan ceremony (Martin 1985). There is a photograph showing him standing with a gun behind a rhino, possibly taken at this time (Rana 2011). Esmond Martin took photographs of items made from the remains of the rhino by craftsmen in Patan, including an elaborate lamp designed by the General, a spice container from a foot, a flower pot, picture frames, two table lamps, a chandelier, a bowl and a jewel box (figs. 24.58, 24.59).

24.33 The Sporting Diary of Maharaja Juddha Shumsher for 1939

After the departure of the Viceroy on 12 December 1938, Maharaja Juddha Shumsher remained in his camp at Bhikna Thori. Some of the details of the events of the next 2 ½ months in the Chitwan area were told by Smythies (1942: 108–116) following the game books kept on the spot. First there was a period of rest and incidents involving wild elephants. As far as rhinos are concerned, the events resumed on 21 January, when there were obviously large numbers of the Maharaja's family and foreign visitors in camp. A total of 35 rhinos were killed during this time and a young one was captured, but Smythies only included details of the 18 rhinos killed in the first weeks. It

FIGURE 24.58 Field Marshal Kiran Shumsher in 1982 showing his treasures made of the skin of a rhino killed in 1938
PHOTO: ESMOND BRADLEY MARTIN

is possible that the album in Madan Puraskar Pustakalaya (MPP), Kathmandu from the time of Maharaja Juddha (EAP166/2/1/27) discussed in 24.26 refers to this shoot. This was the itinerary in 1939.

FIGURE 24.59
Items crafted in Patan from parts
of a rhino killed by General Kiran
Shumsher in 1938
PHOTO: ESMOND BRADLEY
MARTIN

FIGURE 24.60
Heads of three rhinos shot in
Chitwan in January 1939. The
hunters were Maharaja Juddha
Shumsher, his son Bahadur
Shumsher (left) and his grandson
Nara Shumsher (right)
THE SPHERE, SATURDAY
22 JULY 1939, P. 137

First camp at Bhikna Thori, 13 December 1938 to
26 January 1939

21 January 1939: 1 rhino killed by Roy Goodwin Kilburne
(1892–1952), Engineer in Nepal.

22 January 1939: 1 rhino killed by Maharao Umaid Singhji,
18th Maharao of Kotah (1873–1940). Possibly the horn
obtained at the time was made into a libation vessel
used by the rulers of Kota during their Tarpan ceremonies (Metzger & Brijraj Singh 2000: 99).

24–26 January 1939: 1 rhino killed by John Adrian Louis
Hope, 1st Baron Glendevon (1912–1996).

24–26 January 1839: 1 rhino killed by other sportsmen.

Second camp on the northern bank of the Narayani River,
27 January to 15 February 1939.

28 January 1939: 1 rhino killed by Mr. Grant (unidentified)
with Geoffrey Betham, the British Envoy.

28 January 1939: 1 rhino killed by the Maharaja's young son
Min Shumsher (b. 1928), then 10 years old.

29 January to 1 February 1939: 5 rhinos killed, 1 rhino calf
captured, shooters not identified.

3 February 1939: 1 huge old male rhino killed by Maharaja
Juddha Shumsher. Horn measured 20 in (50.8 cm), a
length not recorded in the Records of Rowland Ward.

3 February 1939: 4 additional rhino killed including by
Maharaja's son Bahadur Shumsher (1892–1977) and
his grandson Nara Shumsher (1911–2006). This day was
remarkable because three generations of the family
killed at least one rhino each. There are two photographs of this feat, one with three rhino heads and just
the three generations of shooters (fig. 24.60), another
with five rhino heads (the long-horned in the middle)
with seven family members (fig. 24.61).

5 February 1939: 1 rhino killed.

7 February 1939: 1 rhino killed.

Third camp at Jhawani Camp in heart of Chitwan,
16 February to 3 March 1939:

FIGURE 24.61
Five rhinos shot on 3 February 1939 in Chitwan. In the middle is Maharaja Juddha Shumsher. On his left his son Min Shumsher (then 10 yrs old), Bahadur Shumsher, and another family member. To the right of the Maharaja is Nara Shumsher and two others
SMYTHIES, *BIG GAME SHOOTING IN NEPAL*, 1942, PL. 33

FIGURE 24.62
Postcard issued in Nepal displaying the Sawari Hunting Camp in 1941. King Tribhuvan is on the left, Maharaja Juddha Shumsher on the right, with Prince Mahendra (1920–1972) next to him. The party stands behind four rhino heads
WITH PERMISSION © BILDER AUS NEPAL, COURTESY: JOHANNES BORNMANN

A further 12 rhinos killed to make a total of 38 rhinos in this 3 months period.

This shoot in the cold season of 1938–1939 was not unique but exceptional in the number of animals seen and shot, which also included 120 tigers and 38 leopards. There were an estimated 300–400 rhinos in the Chitwan area at the time (Smythies 1942: 49), and in some areas there were large concentrations, once a dozen were seen at once (p. 112). The correspondent of *The Sphere* (1939-07-22) heard that Maharaja Juddha was contemplating to give up rhino hunting for religious reasons. This may actually be the case, although he obviously did not forbid others to follow in his footsteps.

24.34 Sawari Hunting Camp in 1941

A late testimony of the continued interests of both the royal family and the Rana family is a postcard dating from

1941 (fig. 24.62). This was still a common scene even in the troubled years of war.

24.35 The Shoots after the Return of Monarchy from 1951

In 1951 the Rana dynasty lost power and the rule of Nepal reverted to the traditional monarchy, with King Tribhuvan (1906–1955) resuming this role until his death on 14 March 1955. The new royal family continued to hunt, especially in Chitwan. It is said that when Tribhuvan's son King Mahendra (1920–1972) died of a heart attack, he was spending time in the park (Gregson 2002: 109). His successor King Birendra (1945–2001) was less inclined to go hunting, but did not fully stop the tradition (Martin 1984; Bhatt 2002). The details of the regular hunts in Chitwan and elsewhere in the country of this period have not been published. There were some visitors to Nepal who were given permission to hunt. The three known examples all arrived during the time of King Tribhuvan and all came from India. Queen Elizabeth II of the United Kingdom honoured the kingdom with her visit in 1961, when she had a chance to spend time in the jungle (25.5).

24.36 The Cricketer Vijay of Vizianagram in Nepal in the 1950s

The Indian cricketer and politician Vijay Ananda Gajapathi Raju (1905–1965), Maharajkumar of Vizianagram, known as Vizzy, was an active hunter. During one of the seasons in the 1950s, he was in the *terai* of Nepal, where he killed 2 or 3 rhino. Islam & Islam (2004) published two photos of Vizzy with a dead rhino, neither of which are dated.

24.37 Kirpal Singh Majithia in the *terai*

Members of the prominent Indian landowner family Majithia made several hunting trips to the *terai* of Nepal when Surjit Singh Majithia (1912–1995) was the Indian Ambassador in 1947–1949 and following years. His brother Kirpal Singh Majithia was a well-known shikari at the time. The family used to send some of their own elephants to Nepal. Several family members would take part in these annual hunts, which privilege was extended to the estate's hungarian-born doctor Victor Egen (1910–1997) and his wife Amrita Damra Shergil (1913–1941). The reminiscences of their daughter Eva (Egen 2011) includes a photograph of a rhino killed by Kirpal Majithia, probably in the early 1950s.

24.38 Permit of the Raja of Kasmanda in 1952

The Raja of Kasmanda, Raja Dinesh Pratap Singh (b. 1927) obtained special permission in 1952 from King Tribhuvan to shoot one rhino in Chitwan. The trophy head is kept in the family and is now in Kasmanda Palace Hotel in Mussoorie, Uttarakhand (fig. 24.63). On the plaque of the trophy head is inscription: "Chitawan, Nepal, 1st April 1952, 10 1/8 inches" (25.6 cm), according to Dinkar Singh, the grandson of the Raja (pers.com. May 2018).

24.39 Shradda-Tarpan Ceremonies by Nepal Kings in 1953 and 1979

In 1953 King Mahendra is said to have performed the sacred Shradda-Tarpan ceremony (Mishra 2008: 142).

FIGURE 24.63   Trophy head of a rhino shot by Raja of Kasmanda, Raja Dinesh Pratap Singh in Chitwan in 1952, displayed in Kasmanda Palace Hotel in Mussoorie, Uttar Pradesh
PHOTO: © DINKAR SINGH, 2018

This was followed in December 1979 by King Birendra (1945–2001), which was a central theme of the autobiographical reminiscences of the conservationist Hemanta Mishra (b. 1945) published in 2008. Esmond Martin (1985) has a date 9 January 1981. A male rhino was killed near the Rapti River and taken to the Tarpan site in Kasra, where King Birendra was waiting with Queen Aishwarya (1949–2001), other members of the royal family and several Hindu priests. The King performed the Tarpan ceremony in the evening while priests recited their mantras and the Royal Nepalese Army Band played the national anthem. After this, other family members and officers were allowed to pay tribute to their ancestors in similar fashion (Mishra 2008: 183).

24.40 The Existence of Rhino Trophies

When we survey the evidence about all the persons hunting in Nepal, it is clear that many rhinos were among their victims (table 24.26). All counted together, there are records of 197 rhinos killed during the 20th century, while of course the real total may easily be double or triple that

FIGURE 24.64   Rhino skull with a large horn, without data, in the National Museum, Kathmandu
PHOTO: KEES ROOKMAAKER, 2019

number. Some rhinos were killed by foreign visitors, who may have taken the remains to their homes. However, the Maharajas of the Rana family and the Kings were responsible for at least a hundred rhinos. If the heads and horns of all these animals were set up as trophies, the question is what happened to those at the time and even, are they still in existence?

It is not really too difficult to find the odd rhino skull or trophy in the museums in Kathmandu. A survey of these examples in public exhibition (in 2019) shows that few have any type of history attached to them, although of course all originate from the Chitwan area. Some of these can be listed as follows:

Royal Hotel (since 1977 Yak & Yeti Hotel). When this hotel first opened by the well-known entrepreneur Boris Lissanevitch (1905–1985), a pair of rhino heads were displayed on either side of the hotel's main staircase (Peissel 1966: 25). No trophies seen in 2019.

Kaiser library. The library was badly damaged in the earthquake of 2015. The paintings and photographs of hunting scenes were no longer in the accessible galleries in 2019. There is one rhino skull in a glass case on display without label.

Natural History Museum, Swayambhunath. The exhibition includes a rhino skull and an embryo.

National Museum, Chhauni. The small natural history gallery contains a replica rhino head, a rhino skull, two feet, and a small mounted rhino (fig. 24.64).

Narayanhiti Palace Museum. There are two rhino trophy heads on each side of the staircase at back of the main entrance hall, with vignettes inscribed: "His Majesty the King of Nepal 1956." There are also rhino feet made into tables or stools: 2 in Bajura room, 1 in Jumla room, 2 in Dolpa room. A photograph showing a large carved rhino horn was taken in the 1950s "on the Palace steps at Kathmandu" (Talbot 1956).

These are only few glimpses what happened to the trophies assembled by the Rana family in the first half of the 20th century. Many of the heads or skulls with horns were preserved and stored in their palaces or sometimes exhibited.

One of the few indications that the Maharajas displayed some of their trophies in their palaces is found in the accounts of the Belgian filmmaker Armand Georges Denis (1896–1972) about the Singha Durbar. In September 1939, just before the outbreak of World War II, Denis was invited to visit Maharaja Juddha Shumsher. He was accompanied by his wife Leila, born Roosevelt (1906–1976) and the author Hassoldt Davis (1907–1959). Denis produced a documentary "Dangerous journeys" in 1944, which included footage from Nepal (no rhino). When Denis explained that he had come to Nepal to take pictures and movies of the wildlife, the Maharaja "led me off to another room. It was really a hall, some forty feet long, and I realised it was the Maharaja's own trophy room. An artist had been brought all the way from Paris to cover one whole wall with an immense mural depicting, in heroic proportions, the Maharaja hunting rhinoceroses in the Himalayas. Along the length of the hall ran a series of low marble pillars. There must have been at least twenty on each side of the room. On each, expertly mounted, rested the stuffed head of a rhino" (Denis 1963: 135; Martin & Martin 1982: 31). Davis (1940: 372) recalled the same occasion and told that they "drove to the 'museum', an abattoir of hunting trophies lined the walls. One in particular caught our attention, the head of a gigantic rhinoceros which sat upon a golden platter, for the back side of it had been sculptured and painted to depict all the gory innards which were seen when it was amputated. The Secretary said that in the cellar were heads and skins of animals piled four yards high and weighing 1640 pounds."

A similar occasion is found in recollections of the last Maharaja of Porbandar, Natwarsinh Bhavsinh (1901–1979) in the 1940s. Without specifying the location, he saw trophy skins of tiger, leopard, snow leopard, and rhino horns lining the walls, a setting fitting for a post-hunt dinner (Appel 2004: 289).

In a private zoo in the premises of the Royal Palace, there was a rhino "which must be killed at the exact moment of the Maharaja's death" (Davis 1940: 281) While driving in Durbar Square, Davis saw an enormous drum, 8 feet (2 m) in diameter, with a canvas of rhino skin. The drum was beaten in the morning to call for prayers.

FIGURE 24.65   Unattributed photograph published in January 1925 illustrating a paper by the French orientalist Sylvain Lévi (1863–1925).
The caption reads: The impressive outcome of one hunting season in Nepal. The remains of wild animals killed by the third son
of the Maharaja, Prince Kaiser, who is seen seated on the verandah in front of his working quarters ("L'impressionant tableau
d'une saison de chasse de Nepal. Les dépouilles des bêtes fauves abattues par le troisième fils du Maharaja, le Prince Kaiser, que
l'on vois accouché sur le veranda précédant son cabinet de travail"). The display included at least 46 rhino skulls with horns
L'ILLUSTRATION, JOURNAL UNIVERSAL, 17 JANUARY 1925, P. 59

From the account of Denis, it appears that the trophy hall was in the royal palace in Kathmandu, while Davis introduced a short drive between the palace and the so-called museum. There are no photographs, nor details about the French artist or the contents of the cellar filled with trophies. The palace was the Singha or Singa Durbar dating from 1905, which was the official residence of the Prime Ministers until 1950. The building was largely destroyed by fire in 1973 and again damaged after rebuilding in the earthquake of April 2015. Singha Durbar nowadays houses different government departments. Esmond Martin (2009a) wrote that in late 2007 it was not yet known how many horns were in the King's Palace in Kathmandu. A report about an enquiry of the wildlife trophies mentions a store on the third floor of the Singha Durbar containing 19 trophies which were shifted to the Home Ministry (Dahal 2019). Maybe none of this survived the past political turmoil and natural earthquakes. No photograph has yet emerged of the hall with forty rhino heads and a large painted mural. It would be a testimony of days which have now truly passed.

There is a disturbingly telling photograph, probably taken in 1924, showing at least 46 rhino skulls and horns – not heads with skins. The person in charge of the collection was Kaiser Shumsher seen on the verandah of the Kaiser Mahal in the center of Kathmandu (fig. 24.65).

In Chitwan NP with a population of hundreds of rhino, many die annually from natural causes. The park authorities remove the horn and hooves which are considered "royal trophies." Before 1951 these were sent to a store in the Hanuman Dhoka palace complex, and later to the King's Wildlife Office (Martin 1985). The records kept at the palace apparently have been transferred to the park headquarters in Kasara (Kunwar 2012: 287). Jha et al. (2015) examined 214 genuine rhino horn samples securely stored at the Armed Forest Protection Training Centre, Tikauli, Chitwan and at the Office of the Chitwan National Park, Kasara. The small museum maintained by the National Trust for Nature Conservation (NTNC) next to their headquarters in Sauraha also shows a few rhino specimens, including 2 embryos in alcohol, one small mounted baby rhino and 4 skulls without horns, which might be replicas (Shrestha et al. 2017: 150). These stores where rhino horns are kept must be guarded and do not benefit from further exposure.

Encounters with the Rhinoceros by the British Royal Family in Nepal

25.1 Three Royal Shoots in Central Nepal

Nepal has always been an independent country. The rulers have maintained a good relationship with India and beyond with the British royal family. To build and preserve the diplomatic standing there have been five occasions when the Nepalese King or Prime Minister invited the royal family of the United Kingdom to come and spend time in the jungle areas of Nepal. Albert, Prince of Wales, went to the western part of Nepal in 1876, where no rhinos were present (21.2). The planned visit of George, Prince of Wales in 1906 had to be cancelled due to an outbreak of cholera (25.2). King George V had another opportunity in 1911 (25.3), followed by Edward, Prince of Wales in 1921 (25.4) and Queen Elizabeth in 1961 (25.5). The three successful visits took place in the Chitwan area.

25.2 George, Prince of Wales in West Nepal Canceled in 1906

A hunting expedition in Nepal had been planned for the Prince of Wales (1865–1936), later King George V, on the invitation of Maharaja Chandra Shumsher in the course of his India tour in 1906. However, on 19 February 1906 an outbreak of cholera in the hunting camps was confirmed, effectively canceling the royal visit despite the preparations which had been made on a grand scale (Reed 1906: 401). The Maharaja donated a collection of live animals to the Prince of Wales, which was accepted and handed over to London Zoo.

The donation of this collection of Nepalese animals was announced in the meeting of the Zoological Society of London as early as 22 February 1906 (Anon. 1906b). From Nepal the animals were transported by train via the station of Mohameh (Mokama Ghat) on the Ganges. They arrived in Kolkata on 5 April 1906 (*Englishman's Overland Mail* 1906-04-05), where they were temporarily housed in the Alipore Zoo. Including a young male rhino, the collection left Kolkata on 6 May 1906 on the *Tactician* in charge of Arthur Thomson, and arrived at Tilbury Docks (London) on 8 June 1906 (Finn 1906a: 21). The expenses of the transport had been defrayed by Herbrand Arthur Russell, 11th Duke of Bedford (1858–1940). The animals were first exhibited on the north bank of Regent's Canal (Anon. 1906b). The rhino named 'Carlo' was transferred

to the Rhino House and lived in London Zoo until 4 January 1924 (Rookmaaker 1998: 86).

The rhino was attended by two keepers delegated by the Calcutta Zoo, who are included in a photograph by Walter Pfeffer Dando (1852–1944) taken soon after arrival (fig. 25.1). A different photograph by Dando of the rhino seen on his own was included in a small *Guide to the Prince of Wales Collection* with general information about the species compiled by Frank Finn (1868–1932), until 1903 Deputy Superintendent of the Indian Museum (Finn 1906a: 21). A photograph of the young rhino in London Zoo by Walter Sidney Berridge (1881–1950) was published by Finn (1906b). A stereographic image with text by Edward Kay Robinson (1855–1928), the editor of the popular nature magazine *The Country-Side*, was offered for sale to readers in February 1907 as no. 23 of The Country-Side Stereograph Zoo Series (Robinson 1907: 184, see fig. 25.2). Another unsigned photograph of the rhino

FIGURE 25.1   Young male rhino 'Carlo' donated by Maharaja Chandra of Nepal to George, Prince of Wales in 1906. Photographed by Walter Pfeffer Dando (1852–1944). This was first published by Robinson (1910: 95) DANDO, *WILD ANIMALS AND THE CAMERA*, 1911, P. 60. REPRODUCED BY KIND PERMISSION OF CAMBRIDGE UNIVERSITY LIBRARY

© L.C. (KEES) ROOKMAAKER, 2024 | DOI:10.1163/9789004691544_026
This is an open access chapter distributed under the terms of the CC BY-NC-ND 4.0 license.

FIGURE 25.2
The male *R. unicornis* from Nepal living in the London Zoo from 8 June 1906 to 4 January 1924. This stereographic image was available in February 1907 THE COUNTRY-SIDE STEREOGRAPH ZOO SERIES, NO. 23. COLLECTION KEES ROOKMAAKER

with one Indian attendant appeared in the *Illustrated Sporting and Dramatic News* (Anon. 1906b) and Lewis Medland (1845–1914) took one in the paddock in 1906 (Edwards 2012: 153).

25.3 King George V in Nepal in December 1911

King George V and Queen Mary (1867–1953) made a tour of India in connection with the grand Delhi Durbar of 7 to 16 December 1911. After this event, the King was invited by the Government of Nepal for an informal shoot in the *terai*. Traveling via the railway station at Bhikna Thori, the King reached the first camp at Sukhibar on 18 December. He was welcomed by Maharaja Chandra Shumsher (1863–1929) and three of his sons Mohan (1885–1967), Baber (1888–1960) and Kaiser (1892–1964), as well as the British Resident John Manners-Smith (1864–1920). He was accompanied by a large number of high-ranking officers and nobility. To maintain privacy, journalists were not allowed to join the camp, although there were several photographers. The official narrative of the King's tour was written by John William Fortescue (1859–1933), the Royal librarian and archivist, and published in 1912.[1] The King and his party left Nepal on 28 December 1911 after spending 11 days in two camps on the Rapti River.

Fortescue (1912: 201) calculated that in total the party bagged 39 tigers (*Panthera tigris*), 18 rhinos and 4 sloth bears (*Melursus ursinus*), of which the King personally accounted for 21 tigers, 8 rhinos and 1 bear. Awkwardly, when the daily diary is followed carefully, he only refers to 13 rhinos, of which 6 for the King. On first count, the press reported the shooting of 13 rhinos until 27 December (*Homeward Mail from India* 1911-12-30), and with only one extra day added, the total was changed to 18 rhino at the end of the trip (*Homeward Mail from India* 1912-01-06). Apparently the record keeping had been fragmentary, or maybe had not included animals added by the Nepalese contingent.

Following Fortescue (1912), Reed (1912) and Day (1935), the rhino events can be summarized. The shooters are identified in table 25.30.

18 December 1911 (Monday): By road from Bhikna Thori Railway Station to Sukhibar (21 km) in 40 cars.

First camp: Sukhibar (Sakhi Bar) on the south bank of the River Rapti (27.55N; 84.28E).

18 December 1911: 2 rhino killed by King George V.

18 December 1911: 1 rhino killed by Luke White, with John George Lambton.

19 December 1911 (Tuesday): 1 rhino killed by Colin Keppel.

19 December 1911: 1 rhino killed by Horace Lockwood Smith-Dorrien.

20 December 1911 (Wednesday): 2 rhinos seen which ran away.

20 December 1911: 1 large male rhino killed by King George V.

20 December 1911: 1 rhino killed by Adolphus, Prince of Teck.

20 December 1911: 1 rhino chased for an hour the elephant carrying Clive Wigram.

21 December 1911 (Thursday): 1 male rhino killed by Bryan Godfrey-Faussett and Charles Fitzmaurice. The latter obtained a horn 12 in (30.5 cm) long (Ward 1914: 463).

---

1 Most subsequent accounts of the shoot in Nepal in 1911 rely on Fortescue (1912): Brooke 1912; Reed 1912: 223; Anon. 1912b; Gabriel & Luard 1914; Brooks 1922; Day 1935; Rookmaaker et al. 2005; Kloska 2014; Rahi 2017.

22 December 1911 (Friday): 1 rhino killed by Bryan Godfrey-Faussett.

Second camp: Kasra (Kasara) Camp, 11 km from first camp, on Rapti River (27.56N; 84.41E).

23 December 1911 (Saturday): no rhino killed.

24 December 1911 (Sunday): presentation of collection of live animals (1 male rhino) by Kaiser Shumsher to King George V.

25 December 1911 (Monday): 1 female rhino killed by King George V. The full-grown calf ran away.

25 December 1911: 1 rhino killed by King George V.

25 December 1911: 1 rhino killed by Adolphus, Prince of Teck.

26 December 1911 (Tuesday): 1 rhino killed by King George V.

27 December 1911 (Wednesday): no rhino killed.

28 December 1911 (Thursday): no rhino killed.

28 December 1911: Return from Kasra to Bhikna Thori, the King proceeds by train to Kolkata.

The proceedings of camp life were captured both by artists and photographers. The royal artist George Percy Jacomb-Hood (1857–1929) and royal photographer Ernest Brooks (1875–1957) were in the party. Four sketches showing a rhinoceros by Jacomb-Hood are in the Royal Collection Trust, each inscribed with a description of the event and the date when the artist completed his work. These are here published for the first time:

RCIN 931118. The King on an elephant inspecting a rhino he had killed. 1911-12-18 (fig. 25.3).

RCIN 931121. Major Wigram on an elephant chased by a rhino. 1911-12-21 (fig. 25.4).

RCIN 931125. Captain Godfrey-Faussett on an elephant chased by a rhino. 1911-12-23 (fig. 25.5).

FIGURE 25.3
"Nepal Dec. 18, 1911." King George V on an elephant inspecting a rhinoceros he had killed. Drawing by George Percy Jacomb-Hood
ROYAL COLLECTION TRUST: RCIN 931118
© HIS MAJESTY KING CHARLES III 2023

TABLE 25.30   Shooters of rhinos in the Chitwan area of Nepal in 1911

Person	Dates of rhino shot
King George V (1865–1936)	1911-12-18, 1911-12-18, 1911-12-20, 1911-12-25, 1911-12-25, 1911-12-26
Luke White, 3rd Baron Annaly (1857–1922), Permanent Lord-in-Waiting to the King 1910–1921	1911-12-18 (with Lambton)
John George Lambton, 3rd Earl of Durham (1855–1928)	1911-12-18 (with White)
Colin Keppel (1862–1947), Admiral of Royal Navy	1911-12-19
Horace Lockwood Smith-Dorrien (1858–1930)	1911-12-19
Adolphus, 2nd Duke and Prince of Teck (1868–1927), Personal Aide-de-Camp to the King 1910–1914	1911-12-20, 1911-12-25
Bryan Godfrey-Faussett (1863–1945), Equerry to the King	1911-12-21 (with Fitzmaurice), 1911-12-22
Charles George Francis Mercer Nairne Petty-Fitzmaurice (1874–1911)	1911-12-21 (with Faussett)
Major Clive Wigram (1873–1960), Equerry to the King	nil (chased on 1911-12-20)

RCIN 931128. Infant rhinoceros, being one of the Christmas presents from the Maharaja to the King. 3 men stand beside the animal. 1911-12-24 (fig. 25.6).

The French press followed the royal tour with equal comprehensiveness as their British counterparts. An unidentified artist working for the *Petit Journal* in Paris produced a grand scene of the royal party seated on seven elephants, with King George pointing his gun at two large rhinos next to a small pond (fig. 25.7 from Anon. 1911a). Published on 31 December 1911 and depicting a life-like tableau, this is an indication of the speed of communication set up between Nepal and the home front.

All known photographs associated with the shoot organized for King George in Nepal in 1911 are part of larger albums. Two albums were assembled for the royal family, one album has an unknown history, while four known presentation albums were the work of the photographic firm of P.A. Herzog and P. Higgins of Mhow. These four might have been commissioned by the Maharaja of Nepal, because some copies include a letter from him to the recipient. There is no record which of the two partners might have been present in camp. Altogether there is a wealth of photographic evidence about this time in Nepal, even though there is a lack of certainty regarding what is shown.

FIGURE 25.4   "Major Wigram chased by a Rhino. Dec. 21, 1911." Drawing by George Percy Jacomb-Hood
ROYAL COLLECTION TRUST: RCIN 931121 © HIS MAJESTY KING CHARLES III 2023

FIGURE 25.5   "Captain Godfrey Faussett and his Rhinoceros. Dec 23, 1911." Drawing by George Percy
Jacomb-Hood
ROYAL COLLECTION TRUST: RCIN 931125 © HIS MAJESTY KING CHARLES III 2023

FIGURE 25.6   "Infant rhinoceros, one of the Xmas presents to His Majesty from the Maharajah. Nepal Dec. 1911."
Drawing by George Percy Jacomb-Hood
ROYAL COLLECTION TRUST: RCIN 931128 © HIS MAJESTY KING CHARLES III 2023

FIGURE 25.7   "Aux Indes Anglaises. Une grande chasse en
l'honneur de S.M. George v." A great hunt by King
George v. Unsigned coloured engracing, size
31 × 45 cm
PETIT JOURNAL, PARIS, 31 DECEMBER 1911, P. 424

The First Photo Album of Queen Mary

The most comprehensive photographic albums were
those owned by Queen Mary. These were part of a donation of books, scrapbooks and 29 photographic albums
to the Royal Commonwealth Society (RCS) in London in
1950, transferred to the University of Cambridge in 1992.
The individual items are usually without any explanatory
text, making it is impossible to correlate the photographs
with the actual events. Only a few of these photographs
have been more widely distributed in print. The two
albums from the collection of Queen Mary in the Royal
Commonwealth Society offer a unique insight into the
daily proceedings.

The first album in the RCS (QM 20) has the title "Indian
Tour 1911–1912" on the cover. This oblong album measuring
42 × 37 cm is bound in black leather. It contains 520 photographs taken during the whole tour, with those reflecting
the hunt in Nepal numbered 197 to 333. It has an informal

look with the photographs arranged in an artistic fashion
across the pages. There are 8 photographs with a rhino, all
different from those in QM 21.

QM 20-244: dead rhino with elephants behind (fig. 25.8).

QM 20-261: dead rhino (not illustrated).

QM 20-294: dead rhino lying on back being measured
(fig. 25.9).

QM 20-296: dead rhino lying on side, no people (fig. 25.10).

QM 20-299: dead rhino (same as 296) surrounded by elephants (fig. 25.11).

QM 20-301: dead rhino being cut (fig. 25.12).

QM 20-304: dead rhino being cut (fig. 25.13).

QM 20-330: dead rhino with hunters standing behind the
body, including Kaiser Shumsher (fig. 25.14).

The Second Photo Album of Queen Mary

The second album in the Royal Commonwealth Society
(QM 21) is more formal with all photographs neatly
arranged. This large oblong volume (52 × 37 cm), beauti-
fully leather-bound, is entitled: "H.I.M. The King-Emperor
of India's Shooting in the Nepalese Terai, December, 1911",
with the arms of Nepal embossed above it. It has 41
leaves with a total of 278 photographs, numbered, but
not otherwise annotated. I identified 13 photographs
which include a rhino. Four are again found in another
album: Presentation copy 1 (below) identified as ANU
(Rookmaaker et al. 2005, figs. 3.6).

QM 21-045: dead rhino with King George holding a gun
standing next to the animal (fig. 25.15).

QM 21-092: rhino head. Photo published by Reed (1912:
223) (fig. 25.21).

QM 21-125: dead rhino being cut (fig. 25.16).

QM 21-126: dead rhino being cut, with elephants behind
(fig. 25.17).

QM 21-127: dead rhino, probably same specimen as 126
(fig. 25.18).

QM 21-128: rhino with severed head surrounded by people
(fig. 25.19). Also in ANU 36.

QM 21-165: young rhino alive, donated on 24 December
1911 (fig. 25.30). Also in ANU 74.

QM 21-177: mounted young rhino trophy, an animal not
otherwise recorded (fig. 25.27).

QM 21-206: dead rhino from behind, with elephants
(fig. 25.20).

QM 21-219: rhino alive in grass, some elephants in distance
(fig. 25.21).

QM 21-235: same rhino as 206 (fig. 25.22). Also in
ANU 54.

QM 21-242: dead rhino from front, partial elephants behind
(fig. 25.23).

FIGURE 25.8
Dead rhino surrounded by elephants and people, Nepal 1911. King George is seen in the howdah on the tusked elephant
PHOTO © ROYAL COMMONWEALTH SOCIETY: ALBUM QM 20, NO. 244.
REPRODUCED BY KIND PERMISSION OF CAMBRIDGE UNIVERSITY LIBRARY

FIGURE 25.9
Dead rhino being measured, Nepal 1911
PHOTO © ROYAL COMMONWEALTH SOCIETY, ALBUM QM 20, NO. 294.
CAMBRIDGE UNIVERSITY LIBRARY

QM 21-243: rhino being cut (fig. 25.24). Also in ANU 30, and Woodyatt (1922a) (fig. 27.3).

In this series, no. 219 (fig. 25.21) may not be the best of photographs by today's standards, but it is one of the earliest known attempts to show a rhino alive in the jungle. So far, it is the second known and the first taken in Nepal (Rookmaaker 2021b).

Photographic Presentation Albums

The presentation albums show the name of the firm "Herzog & Higgins, Mhow" on the front cover, as well as a title embossed in gold "H.I.M. The King Emperor of India's Shooting in the Nepalese Terai. December 1911."

Four examples with this description are known. It is likely that each of the officers in the King's suite received one, at least 17 according to the enumeration in Fortescue (1912: 189). These may have been individually designed because the number of images and even the exact measurements (mostly 47 × 36 cm) are not consistent. Only the first album has been available for consultation.

Presentation copy 1. National Australian University, Canberra (signature: Menzies 2108458) consists of 16 pages and 50 photographs. It once belonged to Sajjan Singhji Bahadur, 14th Maharaja of Ratlam (1880–1947), who is not known to have participated in the Nepal hunt. The album was discovered by Dr. U.N. Bhati in a rural home

FIGURE 25.10
Dead rhino lying on its right side, Nepal 1911
PHOTO © ROYAL COMMONWEALTH SOCIETY, ALBUM
QM 20, NO. 296. CAMBRIDGE UNIVERSITY LIBRARY

FIGURE 25.11   Dead rhino (same as no. 296) surrounded by elephants and people, Nepal 1911
PHOTO © ROYAL COMMONWEALTH SOCIETY, ALBUM QM 20, NO. 299. CAMBRIDGE
UNIVERSITY LIBRARY

FIGURE 25.12
Dead rhino being cut, surrounded by people,
Nepal 1911
PHOTO © ROYAL COMMONWEALTH SOCIETY,
ALBUM QM 20, NO. 301. CAMBRIDGE
UNIVERSITY LIBRARY

FIGURE 25.13
Dead rhino being cut, Nepal 1911
PHOTO © ROYAL COMMONWEALTH SOCIETY,
ALBUM QM 20, NO. 304. CAMBRIDGE
UNIVERSITY LIBRARY

FIGURE 25.14
Dead rhino with unidentified hunters, Nepal 1911
PHOTO © ROYAL COMMONWEALTH SOCIETY,
ALBUM QM 20, NO. 330. CAMBRIDGE
UNIVERSITY LIBRARY

FIGURE 25.15
King George V with a dead rhino in
Nepal 1911
PHOTO © ROYAL COMMONWEALTH
SOCIETY, ALBUM QM 21, NO. 45.
CAMBRIDGE UNIVERSITY LIBRARY

FIGURE 25.16
Dead rhino being cut, Nepal 1911
PHOTO © ROYAL COMMONWEALTH
SOCIETY, ALBUM QM 21, NO. 125.
CAMBRIDGE UNIVERSITY LIBRARY

in Madhya Pradesh. It may be incomplete with the back cover missing. All images are available on the library's website (ANU 2021), and the photographs showing a rhino were discussed and reproduced in Rookmaaker et al. (2005). The four photos with rhinos are also found in RCS Album QM21.

Presentation copy 2. Album sold at Christie's London on 10 September 1992 (lot 19), with 160 photographs. It contains a letter dated 15 July 1912 from Maharaja Chandra

to the Royal Surgeon, Richard Henry Havelock Charles, 1st Baronet (1858–1934).

Presentation copy 3. Album sold at Bonhams London on 3 November 2015 (lot 64) and again at Sotheby's New York on 11 December 2017 (lot 97), with 179 photographs. It contains a letter from Maharaja Chandra to Admiral Colin Keppel.

Presentation copy 4. Asian Art Museum of San Francisco (no. 2005.64.160), obtained in 2005 from the

FIGURE 25.17   Dead rhino being cut with elephants behind, Nepal 1911
PHOTO © ROYAL COMMONWEALTH SOCIETY, ALBUM QM 21, NO. 126. CAMBRIDGE UNIVERSITY LIBRARY

FIGURE 25.18   Dead rhino, same as in no. 126, lying on its side, Nepal 1911
PHOTO © ROYAL COMMONWEALTH SOCIETY, ALBUM QM 21, NO. 127. CAMBRIDGE UNIVERSITY LIBRARY

FIGURE 25.19
Rhino head and body surrounded by people, Nepal 1911. Also found in ANU no. 36
PHOTO © ROYAL COMMONWEALTH SOCIETY, ALBUM QM 21, NO. 128. CAMBRIDGE UNIVERSITY LIBRARY

206   46

FIGURE 25.20
Dead rhino, probably same specimen as
no. 235
PHOTO © ROYAL COMMONWEALTH
SOCIETY, ALBUM QM 21, NO. 206.
CAMBRIDGE UNIVERSITY LIBRARY

219   8

FIGURE 25.21
A rare photograph of a living rhinoceros in
the jungles of Nepal in 1911, running away
from the elephants and hunters
PHOTO © ROYAL COMMONWEALTH
SOCIETY, ALBUM QM 21, NO. 219.
CAMBRIDGE UNIVERSITY LIBRARY

collection of the American surgeon and collector William Keve Ehrenfeld (1934–2005), while the original recipient is unclear. This includes a photograph showing trophies of 7 tigers, 2 bears and 2 rhino (fig. 25.25), which was reproduced in Reed (1912: 221) and Fabb (1986, no. 96).

Another type of album was offered for sale by Bates and Hindmarch, Cheltenham in 2019. It is oblong (44.5 × 31.5 cm) and contains 208 photographs, including 50 of the Nepal shoot. These were taken by the royal photographer Ernest Brooks. It is not known if any rhino depictions were included.

Trophies

The taxidermy of the shoot had been awarded to the Theobald Brothers, a firm which operated in Mysore until 1919. It was run by Albert G.R. Theobald (d. 1919), and his two sons Charles and William. Charles Theobald had been present in the field, and the King personally

FIGURE 25.22
Dead rhino lying on its left side with elephants
behind it, Nepal 1911. Also found in ANU 54
PHOTO © ROYAL COMMONWEALTH SOCIETY,
ALBUM QM 21, NO. 235. CAMBRIDGE UNIVERSITY
LIBRARY

FIGURE 25.23
Dead rhino lying on its left side, Nepal 1911
PHOTO © ROYAL COMMONWEALTH SOCIETY,
ALBUM QM 21, NO. 242. CAMBRIDGE UNIVERSITY
LIBRARY

FIGURE 25.24
Cutting the skin off a dead rhino, Nepal 1911.
Also ANU 30 and Woodyatt (1922, pl. facing p. 143,
see fig. 27.2)
PHOTO © ROYAL COMMONWEALTH SOCIETY,
ALBUM QM 21, NO. 243. CAMBRIDGE UNIVERSITY
LIBRARY

FIGURE 25.25
"A splendid bag: 7 tigers, 2 rhinos, 2 bears."
Photograph signed by Herzog and Higgins.
A copy is found in an album (Fabb 1986,
no. 96) in the Asian Art Museum of
San Francisco
STANLEY REED, *THE KING AND QUEEN IN
INDIA*, 1912, P. 221. COURTESY: UNIVERSITY
OF CAMBRIDGE, CENTRE OF SOUTH ASIAN
STUDIES

FIGURE 25.26
"A fine head." Photograph signed by Herzog and
Higgins
REED, *THE KING AND QUEEN IN INDIA*, 1912,
P. 223

presented him with a gold scarf pin set with diamonds. The trophies arrived in Mysore in January 1912 including 39 tigers, 4 bears, several rhinos (*Englishman's Overland Mail* 1912-01-25). This firm was well-known for high-quality mounts and is still remembered in Theobald Road in Nazarabad, just north of the zoo in Mysore.

A photograph of "a fine head" signed by Herzog and Higgins shows how the trophies left the field (fig. 25.26). A series of trophies assembled in Nepal and possibly donated to the King's party included one small mounted juvenile rhino and two loose horns (fig. 25.27). Probably all the rhino trophies were set up as heads and distributed to the various persons responsible for their shooting. Two horns entered the list of records of Rowland Ward, one

of 11 ¾ in (29.8 cm) for the Duke of Teck, and another of 13 ½ in (34.3 cm) for King George V. One trophy head was donated by the King to the Natural History Museum, where it used to hang on one of the walls in the exhibition space (C.N. 1913; Bryden 1914; Lydekker 1916, fig. 17) (fig. 25.28). None of the other specimens have been retrieved.

Donation of Living Animals (1911)

On Sunday 24 December 1911 the King received a large collection of living Nepalese animals destined for the London Zoo, including a young male rhino (Fortescue 1912: 202; Jacomb-Hood 1925: 61). A photograph of the young animal was published in *The Sphere* of 17 February 1912 (fig. 25.29), and another was included in QM21 no. 165 (fig. 25.30),

FIGURE 25.27
Trophies of the King's shoot in Nepal, including a mounted baby rhinoceros
PHOTO © ROYAL COMMONWEALTH SOCIETY, ALBUM QM 21, NO. 177. CAMBRIDGE UNIVERSITY LIBRARY

FIGURE 25.28
Mounted head of Great One-horned Rhinoceros in Natural History Museum in London, no. ZD.1913.10.19.1. This photograph by C.N.
THE SKETCH, 26 NOVEMBER 1913; AND *THE SPHERE*, 21 MARCH 1914

identical to ANU 74 (Rookmaaker et al. 2005, fig. 6). Another presentation was a bowl of rhino horn loaned to the Victoria & Albert Museum (Chapman 1999: 272). The living animals were transported to Kolkata, where they were exhibited from 25 January to 31 March 1912 (Anon. 1912a; Basu 1912). On 1 April 1912 all animals were loaded on board the ss *Afghanistan*, reaching London on 21 May, accompanied by the Superintendent Bijay Krishna Basu (Basu 1913). The rhino was photographed by Frederick Willam Bond (1887–1942) in the Zoo in 1912 (Edwards 2012: 153) and died on 27 April 1927 (Rookmaaker 1998: 87).

25.4 Edward, Prince of Wales, in Nepal in December 1921

Edward, Prince of Wales (1894–1972), the son of King George V and Queen Mary, visited Nepal from 14 to 21 December 1921 on the invitation of Maharaja Chandra Shumsher.[2] The Prince and others in the party killed a total of 10 rhinos on the Rapti River in Chitwan, according to the detailed account of the shoot prepared by the

2　The visit of William, Prince of Wales to Nepal in 1921 was described by Ellison 1922, 1925; Anon. 1922a, 1922b; *The Sphere* 1922-01-14; Phillips 1922; Walker 1922: 88–90; Windsor 1922; Times of India 1922, chapter III; O'Connor 1931: 344, 1940: 176; Brook Northey 1937: 118; Bhatt 1977: 102; Ziegler 1990.

FIGURE 25.29
An early photograph (unattributed) of a young
male rhinoceros donated by Maharaja Chandra
Shumsher to King George on 24 December 1911
THE SPHERE, 17 FEBRUARY 1912

FIGURE 25.30
Young male rhinoceros donated to King George on
24 December 1911. Also found in ANU 74
PHOTO © ROYAL COMMONWEALTH SOCIETY,
ALBUM QM 21, NO. 165. CAMBRIDGE UNIVERSITY
LIBRARY

official naturalist Bernard Cuthbert Ellison. His notes
and illustrations first appeared in 1922 in the *Journal* of
the Bombay Natural History Society, to which body he
was attached for a short time as curator (Ellison 1922:
691). It was again published in slightly revised format,
together with accounts of sport in India, in a book ready
on 23 April 1925 (Ellison 1925). The shooters are identified
in Table 25.31. The daily itinerary regarding rhino events
is here.

14 December 1921 (Wednesday): camp at Bhikna Thori. No
   rhino.

Camp at Dhoba, 32 km from Bhikna Thori

15 December 1921 (Thursday): 1 rhino wounded by Louis
   Mountbatten (1900–1979). The animal was later recovered, only skull and horn were saved (Ellison 1925: 13).

15 December 1921: 1 rhino wounded by Dudley Burton
   Napier North.

Camp at Kasra

16 December 1921 (Friday): 1 female rhino killed by Dudley
   North. Remains picked up after the party left Nepal
   (Ellison 1925: 14–15).

16 December 1921: 1 rhino killed by Dudley North using a
   double-barrelled .470 Gibbs rifle with a solid bullet.

17 December 1921 (Saturday): 1 female rhino killed by
   Frederic Sinclair Poynder of the 9th Gurkha Rifles. This
   rhino had a foetus, 4 ft 1 in (124.5 cm) long with a weight
   of 54 kg.

18 December 1921 (Sunday). Maharaja Chandra Shumsher
   donated a collection of live animals, including 1 young
   female rhino.

19 December 1921 (Monday): 1 male rhino killed by Prince of Wales, at Sarasoti Koli, together with General Kaiser Shumsher. Ellison (1922: 687) stated that the horn was broken, which is not visible in the photographs. Rowland Ward (1928: 438) credits the Prince with a horn of 12 in (30.5 cm).

19 December 1921: 1 male rhino killed by the surgeon Alexander Newport, at Sarasoti Koli.

19 December 1921: 1 male rhino killed by the journalist Perceval Landon, at Dhoba. The animal was 9 ft 10 in (300 cm) long and 6 ft (182 cm) high.

20 December 1921 (Tuesday): 1 male rhino shot by Louis Mountbatten, at Dhoba, together with the Military Secretary Rivers Berney Worgan.

20 December 1921: 1 female rhino shot by Bruce Ogilvy, at the 15th milestone.

20 December 1921: 1 female rhino killed by Rowland Thomas Baring, at the 15th milestone. The head of this animal was displayed in the show rooms of Gerrard and Sons at Camden Town, London (Ellison 1925: 174).

21 December 1921 (Wednesday): last day. No rhino listed.

Despite the apparent detail, the rhino count is unclear. For the first rhino of 20 December, Ellison attributes the animal to Worgan in the text, and to Mountbatten in the table. Mountbatten (1987: 221) doesn't mention shooting a rhino, but a trophy head in his possession was mounted by Rowland Ward in London (Morris 2003: 63; figure

Person	Dates of rhino shot 1921-MM-DD
William, Prince of Wales (1894–1972)	1921-12-19
Rowland Thomas Baring, 2nd Earl of Cromer (1877–1953)	1921-12-20
Bernard Cuthbert Ellison	naturalist, taxidermist
Perceval Landon (1869–1927), journalist	1921-12-19
Louis Francis Albert Victor Nicholas Mountbatten (1900–1979)	1921-12-15 (wounded), 1921-12-20
Alexander Charles William Newport (1874–1948), surgeon	1921-12-19
Dudley Burton Napier North (1881–1961)	1921-12-15 (wounded), 1921-12-16, 1921-12-16
Bruce Arthur Ashley Ogilvy (1895–1876), Equerry to Prince of Wales	1921-12-20
Frederic Sinclair Poynder (1893–1943), 9th Gurkha Rifles	1921-12-17, 1921-12-17 (foetus)
William Frederick Travers O'Connor (1870–1943)	envoy, photographer
Rivers Berney Worgan (1881–1934), Military Secretary	1921-12-20

TABLE 25.31   Shooters of rhinos in the Chitwan area of Nepal in 1921 and some other participants

FIGURE 25.31
A rhinoceros mounted by the taxidermy firm of Van Ingen in Mysore leaving the workshop. The actual specimen has not been identified
PHOTO © COURTESY PAT A. MORRIS

in Ellison 1925, facing 188). The two rhinos wounded on 15 December are not counted in the contemporary total, even though both animals were picked up after the Prince had left, maybe because the skins had deteriorated. There was also an attempt to ensure that the first rhino was credited to the main guest according to an unwritten tradition, made complicated by the fact that William really preferred playing polo and cricket. The Prince killed just one rhino on 19 December, although O'Connor (1931: 344) mentions another rhino wounded by the Prince of Wales and picked up after the end of the shoot. The best total count of the Prince's shoot in 1921 is 2 rhinos wounded and recovered later, 9 rhinos killed of which one had a foetus to make a total of 10 (Ellison 1925: 181). The Prince accounted for one of these. In addition a living young female rhino was donated to the party to be transported to London Zoo.

All rhinos were shot along a road which was made through the jungle from Bhikna Thori northwards, then along the Rapti River to Kasara and the confluence with the Narayani River. Ellison (1922, 1925) includes a map of the area with the route taken by the Prince's party.

Although the rhino head trophies must have been set up and exported, none are in the public domain. A head belonging to Mountbatten was mounted by Rowland Ward, another to Cromer by Gerrard, others possibly by Van Ingen or Theobald in Mysore, whose work is described by Ellison (1925: 179) without referring to a rhino. The taxidermy firm of Van Ingen was established in Mysore in 1912 by Eugene Melville van Ingen (1865–1928) and later run by his children: John de Wet (1902–1993), Henry Botha (1904–1996), James Kruger (1905–?), Edwin Joubert (1912–2013), Gretchen (1914–1916) and Rosamund (1916–2006). Many rhino trophies must have passed through his premises although sadly exact records are no longer accessible (Morris 2006: 95, 136) (fig. 25.31).

Photographic Evidence

There were a number of photographers who captured some events of the shoot. Ellison (1925) has 13 figures with a rhino (one taken by Faunthorpe, 24.21), of which four were previously used in his earlier paper (Ellison 1922). O'Connor (1931) has 2 rhino plates. An album of photographs held by the Madan Puraskar Pustakalaya library, Kathmandu and preserved through the Endangered Archives Program of the British Library (BL-EAP 2021e) has a series of photographs taken during the hunt, five including a rhino, of which four have never been previously published.

The rhino killed on 19 December by the Prince of Wales features in most of the photographs. Once the animal was killed, it was pictured from the front without any persons (fig. 25.32). Then the Prince of Wales posed in front of

FIGURE 25.32   "Rhinoceros shot by the Prince of Wales." A front view of the rhino killed on 19 December 1921
ELLISON, *H.R.H. THE PRINCE OF WALES'S SPORT IN INDIA*, 1925, FACING P. 19

the dead animal wearing a special Nepalese kukri in his belt. Maybe the photographer was the British Resident Lieutenant-Colonel William Frederick Travers O'Connor, who inserted this item with a signature and date as a frontispiece in his own book (fig. 25.33). The same rhino is shown with the Prince of Wales, Kaiser Shumsher, O'Connor and North (fig. 25.34), with a group of the Maharaja's family (fig. 25.35), and with Ellison and other staff (fig. 25.36).

The Prince of Wales took part in the skinning of this rhino, an activity rarely shown in other settings. Ellison himself published one photograph of this event (fig. 25.37), while four others were preserved in the archival records (figs. 25.38, 25.39, 25.40, 25.41).

FIGURE 25.33
Edward, Prince of Wales in front of a male rhino killed on the Rapti River in Nepal on 19 December 1921, with his signature. The same photograph was published without signature by Ellison (1922: 690, 1925: 20)
O'CONNOR, *ON THE FRONTIER AND BEYOND*, 1931, FRONTISPIECE. REPRODUCED BY KIND PERMISSION OF CAMBRIDGE UNIVERSITY LIBRARY

FIGURE 25.34
"H.R.H. The Prince of Wales and his Rhinoceros." The photo shows from left to right: William Frederick Travers O'Connor (1870–1943), Maharaja Kaiser Shumsher (1892–1964), Edward Prince of Wales (1894–1972) and Dudley Burton Napier North (1881–1961)
O'CONNOR, *ON THE FRONTIER AND BEYOND*, 1931, FACING P. 302. CAMBRIDGE UNIVERSITY LIBRARY

FIGURE 25.35
The rhinoceros killed by Edward, Prince of Wales on 19 December 1921. The Prince is seen standing behind the animal wearing his kukri. Also in the picture are Maharaja Kaiser with a kukri and his relatives and officials. The same photograph was published by Ellison (1925: 20 bottom)
PHOTO © MADAN PURASKAR PUSTAKALAYA: EAP166/2/1/12/180

FIGURE 25.36
Rhino shot by Prince of Wales on
19 December 1921, with Nepalese
people
ELLISON, *H.R.H. THE PRINCE OF
WALES'S SPORT IN INDIA*, 1925,
FACING P. 174, TOP FIGURE

FIGURE 25.37
"H.R.H. takes a hand with his Kukri
in decapitating a rhinoceros." Also in
Ellison (1922: 684)
ELLISON, *H.R.H. THE PRINCE OF
WALES'S SPORT IN INDIA*, 1925, P. 21
TOP FIGURE

FIGURE 25.38
"Prince interested at separating
Rhino body with axe." Also published
by Ellison (1922: 684, 1925: 21) as
"A further stage of the ritual"
PHOTO © MADAN PURASKAR
PUSTAKALAYA: EAP166/2/1/12/186

FIGURE 25.39
"Rhino shikar, Prince with Khukri." The Prince is in the centre, accompanied on the right by (assumed) Bernard Cuthbert Ellison, who was in charge of the taxidermy
PHOTO © MADAN PURASKAR PUSTAKALAYA: EAP166/2/1/12/187

FIGURE 25.40
"Prince looking the way to cut rhino"
PHOTO © MADAN PURASKAR PUSTAKALAYA: EAP166/2/1/12/188

FIGURE 25.41
"Prince cutting rhino body with khukri"
PHOTO © MADAN PURASKAR PUSTAKALAYA: EAP166/2/1/12/197

FIGURE 25.42
Head of rhino killed by Rowland Thomas
Baring, 2nd Earl of Cromer (1877–1953), at
the 15th milestone on 20 December 1921
ELLISON, *H.R.H. THE PRINCE OF WALES'S
SPORT IN INDIA*, 1925, FACING P. 174

FIGURE 25.43   Trophy of head of the rhino shot by Louis
Mountbatten on 20 December 1921
ELLISON, *H.R.H. THE PRINCE OF WALES'S SPORT IN
INDIA*, 1925, FACING P. 188 LOWER LEFT

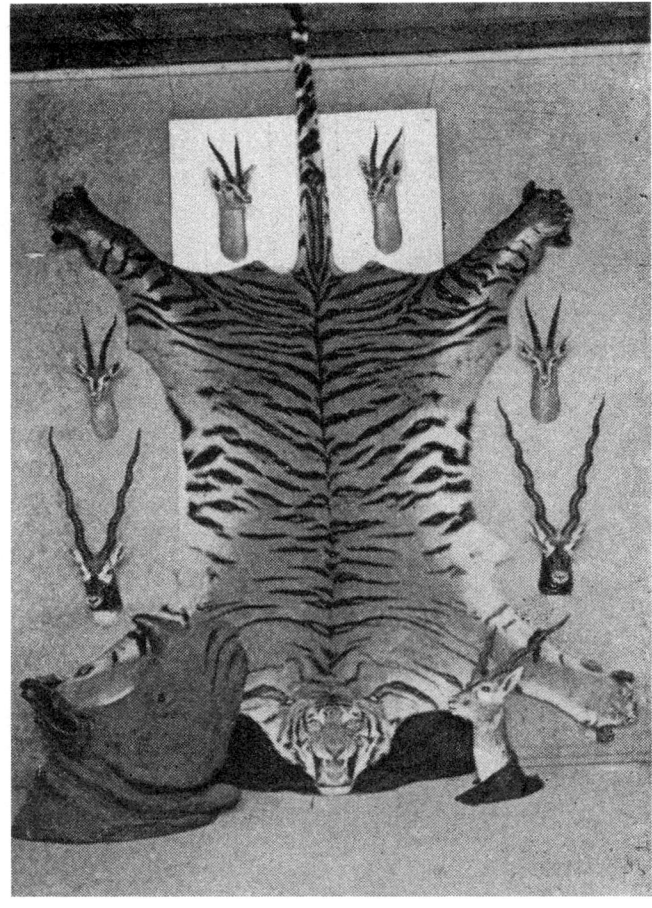

FIGURE 25.44   Trophy head of rhino obtained by Louis Francis
Albert Victor Nicholas Mountbatten (1900–1979)
on 20 December 1921 in Nepal, displayed in the
showroom of Rowland Ward
ELLISON, *H.R.H. THE PRINCE OF WALES'S SPORT IN
INDIA*, 1925, FACING P. 188 LOWER RIGHT

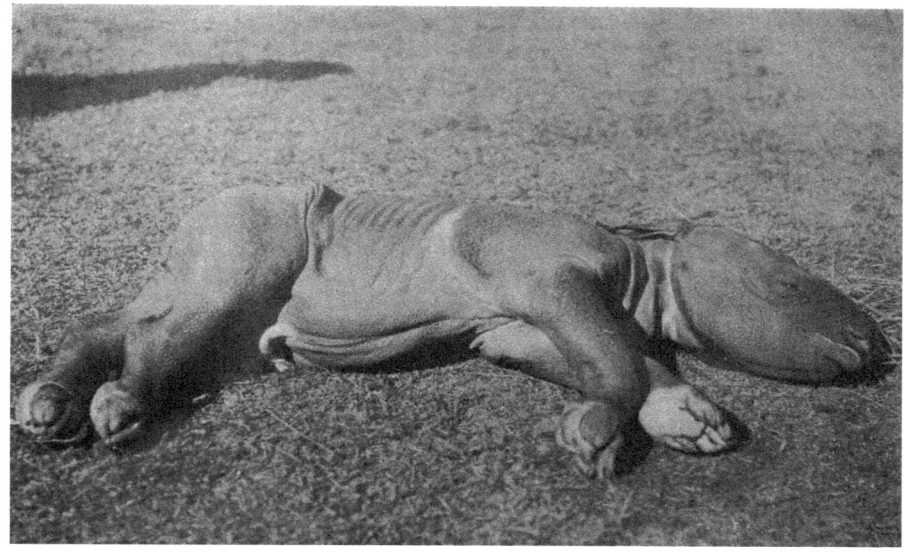

FIGURE 25.45
"An unborn rhinoceros calf" from a female
rhino shot by Frederic Sinclair Poynder
(1893–1943) on 17 December 1921
ELLISON, *H.R.H. THE PRINCE OF WALES'S
SPORT IN INDIA*, 1925, FACING P. 33

The head of the rhino killed by Cromer on 20 December was photographed (fig. 25.42). The animal killed by Mountbatten on 20 December is found on photographs contributed by him to Ellison's book (figs. 25.43, 25.44). There is a photograph of the foetus recovered from the female rhino of 17 December (fig. 25.45).

Donation of Living Animals (1921)

Maharaja Kaiser Shumsher on behalf of the Nepalese people donated a set of live animals to the Prince of Wales on 18 December 1921, including a young female rhino later called Bessie. It is assumed that she is the animal photographed while still in Nepal (fig. 25.46). Bernard Ellison was responsible for the transport from Nepal to Mumbai, arriving on 10 January 1922. The animal collection was temporarily housed in Victoria Gardens, where the rhino was pictured twice (figs. 25.47, 25.48). The collection was transported on the ss. *Perim* leaving Mumbai 2 March and arriving London 7 April 1922. Bessie lived in London Zoo until 28 April 1926 when she died of sarcoma of heart and lungs (Duncan 1922; Rookmaaker 1997e, 1998: 87).

25.5 Queen Elizabeth II in Nepal in February 1961

Queen Elizabeth II (1926–2022) and Prince Philip, Duke of Edinburgh (1921–2021) were invited to shoot in the Meghauli area of Chitwan by the Nepalese King Mahendra Bir Bikram Shah Dev (1920–1977). Staying from 26 February to 1 March 1961, they admired the wildlife and the landscape, while leaving any shooting to their retinue (Pearson 1986). The hospitality arrangements were made by Boris Nikolayevich Lissanevich (1905–1985), the foremost early hotelier of Nepal (Peissel 1966: 260; Mahato & Dahal 2021). The journalist Ruth Lynam and the photographer Hank Walker (1921–1996) followed the events in the field for a documentary in *Life Magazine* (Lynam 1961).

On Monday 27 February 1961 a female rhino with calf was surrounded by 305 elephants. She was killed through simultaneous shots fired by Alexander Frederick Douglas-Home (1903–1995), Foreign Secretary, and Michael Edward Adeane (1910–1984), Private Secretary to the Queen (Mishra 2010: 164; photo in *Daily Mirror* 1961-03-03). The trophy was shared as reported in a letter by Douglas-Home: "Sir Michael Adeane and I both hit the rhinoceros, and I am certain it was my shot that killed it. We are sharing it. I am having the horn and the front feet, and Sir Michael is having the back end. I am not certain what I will do with the feet. Probably make them into wastepaper baskets" (Pearson 1986). This event was filmed by the American filmmaker Ellis Dungan (1909–2001) for the British news. This footage is preserved by British Pathé as part of a newsreel entitled "Queen in Kathmandu." One of the frames shows a dead rhino lying in the grass surrounded by several unidentified men (figs. 25.49, 25.50).

FIGURE 25.46
A living specimen of the Great one-horned
Rhinoceros, photograph attributed to the office
of Maharaja Kaiser Shumsher. This may be the
female donated to the Prince of Wales in Nepal on
18 December 1921
ELLISON, *H.R.H. THE PRINCE OF WALES'S SPORT
IN INDIA*, 1925, FACING P. 35

FIGURE 25.47
Young female rhino presented by the King of Nepal
to the Prince of Wales in 1921, shown during a short
transit stop in the Victoria Gardens of Bombay in
February 1922. Also published in Ellison (1922: 691)
ELLISON, *H.R.H. THE PRINCE OF WALES'S SPORT
IN INDIA*, 1925, P. 45

FIGURE 25.48
The female rhinoceros in Mumbai, when in transit
from Nepal to London. A detail appeared in the
Daily Mirror of 15 February 1922, p. 16
WIDE WORLD MAGAZINE, MARCH 1922

FIGURE 25.49
Queen Elizabeth II in a howdah in Nepal, February 1961. From the news reel "Queen in Kathmandu"
WITH PERMISSION © BRITISH PATHÉ, ID: 1704.02

FIGURE 25.50
Rhinoceros shot in Nepal on 27 February 1961 by Alexander Frederick Douglas-Home and Michael Edward Adeane. From the news reel "Queen in Kathmandu"
WITH PERMISSION © BRITISH PATHÉ, ID: 1704.02

CHAPTER 26

Protecting the Rhinoceros in Chitwan National Park, Nepal

Established: Chitwan National Park in 1973
Species: *Rhinoceros unicornis*
First Record: 1800 – Last Record: current
Rhino Region 14

26.1 Chitwan in the Wake of Rana Rule

When the Rana rule came to an end in February 1951 and King Tribhuvan regained the traditional throne, Nepal went through a period of political turmoil. The hunting reserves of the Rana period, which had protected wildlife, were suddenly left without legal framework. Assam's Conservator of Forests, Patrick Donald Stracey (1906–1977) visited Chitwan in 1957 and found that there was a dedicated Rhino Protection Officer with a staff of 152, but only one elephant. Poaching appeared rampant and 20 to 30 rhino carcasses were found annually. Stracey (1957) estimated that there were a total of 500–600 rhinos in 1957.

IUCN's first staff ecologist, Lee Merriam Talbot (1930–2021), in his authoritative report *A Look at the Threatened Species* (1960), mentioned rumours in September 1958 that a band of Indian poachers had entered the Rapti Valley and slaughtered all rhinos, maybe as many as 500 animals. That set alarm bells ringing and IUCN requested E.P. Gee, then one of India's foremost rhino experts, to visit Nepal and report on the situation. In March-April 1959, Gee surveyed the rhino areas and prepared a report that was published in *Oryx*, the journal of the Fauna Preservation Society (now Fauna & Flora International). Gee (1963c) after a second visit in 1962 estimated that 185 rhinos lived in the valleys of the Narayani, Rapti and Reu Rivers. He confirmed widespread poaching resulting in the deaths of at least 70 rhinos each year since 1951.

•••

Dataset 26.24: Chronological List of Records of Rhinoceros in Chitwan National Park

General sequence: Date – Event – Sources – Paragraph number, figures.

1941–1946 – In the Rapti Valley, Maharaja Juddha Shumsher employed a Special Service of 70–80 game wardens, to prevent rhino poaching – Smythies 1974: 197 – 26.1

1945 – Strict preservation by the Nepal Government has saved rhino from extinction – Harper 1945: 378; Gee 1959b: 61

1950 – Malaria eradicated, settlement by people from hills – Nepal Government 2006: 5

1951 – Poaching increased after end of Rana rule – Gee 1959b; Spillett & Tamang 1966

1953 – Olive Muriel Cripps Smythies (1890–1961) and her husband Evelyn Arthur Smythies (1885–1975) observed rhino in Chitwan – O. Smythies 1953: 214–229, 1961

1954 – Poachers took 72 rhino, and several rhinos were killed in floods – Talbot 1960: 191

1957 – Publication of first wildlife law in Nepal offering legal protection to rhinos and their habitat – Heinen & Kattel 1992

1957-09 – Poaching gangs entered Rapti Valley and killed as many as 500 rhino – Talbot 1960: 192

1959-01 – Mahendra Mriga Kunja (Mahendra Deer Park) of 68 square miles (176 km²) was formally opened by King Mahendra – Gee 1959b: 61; Spillett & Tamang 1966; Bhatt 2002; Kandel 2012; Thapa et al. 2013

1959-03 – Edward Pritchard Gee (1904–1968) travels from Assam for intensive rhino survey in Chitwan area – Gee 1959a, 1959b

1960-04 – Formalization of Wild Life Conservation Act – Gee 1963c; Thapa et al. 2013

1962-01 – Forest Act came into operation – Gee 1963c

1962 – Scattered reports exist of rhino living along the Bagmati River at the eastern border of the Birganj Forest District (east of Chitwan). None seen by the American biologist David Lee Chesemore. James Mangalraj Johnson, one of the Indian foresters, found rhino tracks here in 1962 – Chesemore 1970

1962 – *R. unicornis*, 3 skulls in Munich from Kush-Kush forest near Meghauli, donated by Nepal Government – Bayerische Staatssammlung, Munich, Germany

1963 – Rhino live in the *terai*, particularly in valleys of Rapti and Sapt Gandak, north of Gorakhpur Dt. – Stracey 1963: 66

1963 – Rhinos are completely protected. Poachers kill about 20–30 animals annually – Stracey 1963: 66

1963-03 – E.P. Gee, second visit – Gee 1963c

1963-03 – Mahendra National Park seems to have discontinued following incursion of settlers occupying most of the low-lying rhino habitat – Gee 1963c

1964 – Royal decree declared a rhino sanctuary (now Chitwan). Creation of a special guard force, the Gainda Gasti (Rhino Patrol) comprising of 130 armed men, later responsible for protection outside park boundaries – Heinen & Kattel 1992; Thapa et al. 2013 (with date 1959); Shrestha et al. 2017: 7

1964 – Removal of 22,000 villagers (4400 families) from rhino sanctuary and relocated elsewhere – Willan 1965; Caughley 1970 – 26.02

© L.C. (KEES) ROOKMAAKER, 2024 | DOI:10.1163/9789004691544_027
This is an open access chapter distributed under the terms of the CC BY-NC-ND 4.0 license.

1966 – Census by James Juan Spillett (1932–2018) on 17–18 March 1966 showed a total of 366 rhinos: 172 non-sexed, 67 adult males, 83 adult females, 44 young. Most seen in Baguri block (157 adult, 23 calves). Total may be 400 – Spillett 1966a

1968 – Completion of Tiger Tops hotel within the rhino sanctuary – Bauer 1972: 303–310; MacDougal 1978 – 26.4

1968 – Start of Trisuli Watershed Project under Food and Agriculture Organization (FAO), and the United Nations Development Program (UNDP). The FAO biologist Graeme James Caughley (1937–1994) surveyed areas and made recommendations about the need for wildlife protection in the country – Caughley 1970; Heinen & Kattel 1992

1969 – Rhinos threatened mainly by habitat reduction and competition with domestic livestock. There are 200 armed guards in the reserve – Fleming 1969; Zschokke & Baur 2002

1970 – Survey to initiate wildlife conservation projects by the Scottish naturalist John Blower (1922–2020) – Blower 1973; Heinen & Kattel 1992

1970 – Lowell Jackson Thomas (1892–1981) filmed an episode of the CBS TV program series 'High Adventure' – Scott 1971

1971 – IUCN requested world's zoos to refrain from buying or accepting any rhino originating in Nepal to allow recovery – Anon. 1971: 157

1971 – Habitat being destroyed by thousands of cattle and buffalo, except 150 acres around Tiger Tops – Waller 1971, 1972a

1972–1976 – William Andrew Laurie (b. 1950) conducted early ecological study of rhinos in Chitwan, December 1972 to June 1976. His Ph.D. thesis was a landmark scientific study of the ecology of *R. unicornis* – Laurie 1977, 1978, 1978b, 1982, 1983, 1984, 1986 – 26.3

1973 – National Park and Wildlife Conservation Act 2029 (1973) – Heinen & Kattel 1992

1973 – Royal Chitwan National Park was established under the 1973 National Parks and Wildlife Protection Act. It has the rhino in its official logo. It covers an area along Rapti, Narayani and Reu rivers of 210 square miles (544 km^2) – Bolton 1977: 475; Martin 1985: 14; Nepal Government 2006: 5; Thapa et al. 2013 – fig. 26.3

1973 – Start of National Parks and Wildlife Conservation Project under FAO and UNDP. Goals included the effective management and conservation of wildlife and their habitats, the development of a national park and reserve system, and the development of the Department of National Parks and Wildlife Conservation. The project ended in July 1979 – Heinen & Kattel 1992

1975 – Protection of park handed to Nepal army – Martin 1992

1979 – Visit of Heinz-Georg Klös (1926–2014), Director of Berlin Zoo, with his wife Ursula – Klös & Klös 1980

1980 – Immobilization and capture of 8 rhinos for zoological gardens – Leyrat 1982: 45–52; Mishra et al. 1982: 466

1982 – Establishment of King Mahendra Trust for Nature Conservation – Thapa et al. 2013

1983 – Christen M. Wemmer (b. 1943) obtained 3 skulls for Smithsonian Institution, Washington DC – Website (2017)

1983 – Explorer Wilfred Patrick Thesiger (1910–2003) took photos – Pitt Rivers Museum, Oxford, online register

1984 – Chitwan declared World Heritage Site – Nepal Government 2006: 5

1984–1988 – Ecological studies by American conservationist Eric Dinerstein (b. 1952) adding immensely to the available data. His book *The return of the unicorns* (2003) provides an accessible review of old and new data on the natural history of *R. unicornis* – Mishra & Dinerstein 1987; Dinerstein & Jnawali 1993; Dinerstein &

MacCracken 1990; Dinerstein & Price 1991; Dinerstein et al. 1990; Dinerstein 2003; Dinerstein 2006; Dinerstein 2013; Dinerstein 2015

1985 – Inside park boundaries, rhinos are protected by 500 armed men of the Royal Nepalese Army who carry out foot patrols twice daily. Outside the Park's borders, an additional force of 200 men from the Royal Nepalese Forest Department's Rhino Patrol stand guard – Martin 1985: 14

1988 – Rhinos in Chitwan 1975–1988 increased by 3.7 % per annum – Laurie 1997

1993 – Gainda Gasti was absorbed in the existing cadre of armed forest guards – Thapa et al. 2013; Martin 1992

1993 – In the decade 1984–1993, an average of 4.7% of rhino population was poached annually – Vigne & Martin 1995

1995 – In Chitwan, impact of tourists on individual rhinos manageable – Lott & MacCoy 1995

1998 – Rhinos disappeared from Barandabhar forest, formerly the Mahendra Deer Park, after Padampur village relocated to Saguntole – Kandel 2012

1998-03-22 – Nepal government at Tikauli burnt many rhino trophies, except potentially valuable horns. Parts included skin pieces, 83 horns, nails, teeth, 9 skulls – Martin 1998

2000 to 2007 – Poaching increased because the Royal Nepalese Army based inside the two parks feared attacks from Maoist insurgents and withdrew from 30 guard posts to reinforce their remaining 14. Chitwan's communication repeater station broke and intelligence funding was cut – Khan et al. 2003; Martin 2004; Martin et al. 2009a

2001-11 – State of emergency. Troops taken away from guard duty in Chitwan. About 40 rhino killed by poachers this year – Wielandt 2002

2006-09-23 – In a helicopter accident in the Kanchenjunga Conservation Area all 24 passengers and crew were killed, including Tirtha Man Maskey (1948–2006), Narayan Poudel, Mingma Norbu Sherpa, Chandra Prasad Gurung, as well as Harka Gurung, Yeshi Choden Lama, Jennifer Headley, Jill Bowling Schlaepfer, Matthew Preece, Margaret Alexander – Dinerstein 2006; Mishra 2008: 213; Thapa 2013

2008 – New census found 408 rhinos: 277 adult, 51 sub-adult, 80 calves – Acharya 2008; Amin et al. 2009

2008 – King Mahendra Trust for Nature Conservation (1982) was renamed as National Trust for Nature Conservation (NTNC) – Thapa et al. 2013

2011 – Block count in Chitwan showed 503 rhino (1 per km^2) being 66% adult, 12% subadult, 22% calves – Subedi et al. 2013

2012 – Visit by Bernhard Blaszkiewitz (1954–2022), Director of the Tierpark in Berlin – Blaszkiewitz 2012

2012 – Release of movie 'Khaag', meaning rhino horn, about the conservation activities of Kamal Jung Kunwar, Chief Warden of Chitwan 2013–2015 – Kunwar 2012

2015 – Parsa Wildlife Reserve, 3 rhino present, strayed from Chitwan – Talukdar 2015

2016 – Rautahat (Parsa WR), Bara Dt. has 3–4 rhino permanent, others stray from Chitwan. One rhino poached in 2016 – Rimal et al. 2018 – fig. 26.2

2017 – Rhino habitat quality reduced through bittervine (*Mikania micrantha*) forcing rhinos to move outside protected areas, and reducing connectivity between populations resulting in inbreeding and reduced genetic diversity – Aryal et al. 2017

2017-03 – Visit by Kees Rookmaaker (b. 1953) staying at Meghauli, guided by naturalists Pradip, Suman Lama and Varun Mani. A repeat visit was made in 2019 – figs. 26.1, 26.7, 26.8

FIGURE 26.1   Rhino in Chitwan National Park near Sauraha
PHOTO: KEES ROOKMAAKER, MARCH 2019

FIGURE 26.2   Rhino killed by poachers in Chitwan National Park, Nepal
PHOTO: DEEPAK ACHARYA, 2007

26.2   New Human Settlers in Chitwan

Chitwan had traditionally been sparsely populated largely due to the prevalence of malaria. After the end of the second world war, Nepal with the help of foreign aid invested

heavily in a program to eradicate the disease using the insecticide DDT. This opened up the area for human settlement, deemed a political necessity, because many people had been affected by floods in the hills and were without land or livelihood. The Rapti Valley Multi-Purpose

FIGURE 26.3   Logo of Chitwan National Park displayed outside the
Park Headquarters in Sauraha
PHOTO: KEES ROOKMAAKER, 2019

Development Project settled 2,500 people per year in the western portion of Chitwan, and the population totaled at least 12,000 in March 1959 when Gee visited. About 70% of the forest and grassland habitat had been cleared for agriculture, causing the rhino population to plummet.

Despite the immense changes in the Chitwan region, King Mahendra (ruling from 14 March 1955) continued the tradition of royal hunts and invited Queen Elizabeth II and Prince Philip in February 1961 (25.5). Although the royal pair preferred the camera, one tiger and one rhino were killed by their party. The newspaper reports at the time created a public protest, and the Secretary of the Fauna Preservation Society, Charles Leofric Boyle (1899–1999), expressed his dismay in a letter to the *Times* (Boyle 1961) wondering why a highly endangered animal like the rhino had been hunted.

In 1962, the Forest Department petitioned the Nepal Government about the rising number of illegal settlements and encroachment in the forests and the wildlife sanctuary. The committee of enquiry, appointed in April 1963, found considerable disturbance, upon which in October 1963 the Government created a Settlement Commission under an Assistant Minister, with membership of Conservator of Forests, Govind Narsingh Raimajhi (Willan 1965). The Commission was provided with powers to remove illegal squatters and resettle them elsewhere. As a result, in 1964–1965, as many as 4,400 families or

about 22,000 people were moved to areas outside the park. An extension to the south of the Rapti River was proposed to incorporate the main rhino habitat. Rhinos were then more frequently spotted in the area of the King Mahendra National Park where they had not been seen in recent years, and 35 calves were counted.

26.3   Establishment of Royal Chitwan National Park

Following a helicopter survey in June 1968, the New Zealand ecologist Graeme James Caughley (1937–1994) estimated that there were 81 to 108 rhinos in Chitwan. Another survey in May 1972 suggested a population of 120–147 animals. These estimates may have been on the low side because visibility is poor in the riverine habitats preferred by the rhinos. The surveys had been part of a process necessary to establish the Royal Chitwan National Park, which was gazetted in 1973 covering 544 km², following approval by King Mahendra in December 1970. National Park Regulations were introduced on 4 March 1974. In 1977 the park was extended to a total of 932 km², with the new land mainly covering the hill forests, which are not prime rhino habitat.

In 1977, the park was enlarged to its present area of 932 km². A bufferzone of 766 km² was added in 1993 to the north and west of the Narayani-Rapti river system, and between the south-eastern boundary of the park and the international border with India. In the north and west of the protected area, the Narayani-Rapti river system forms a natural boundary to human settlements. Adjacent to the east of Chitwan National Park is Parsa Wildlife Reserve, and contiguous in the south is the Valmiki National Park, an Indian Tiger Reserve. This coherent protected area of 2,075 km² represents the Tiger Conservation Unit (TCU) Chitwan-Parsa-Valmiki, which covers a block of alluvial grasslands and subtropical moist deciduous forests.

26.4   Two Early Studies of Rhinoceros Ecology and Behaviour

The first detailed ecological and behavioural study of the greater one-horned rhinoceros in Nepal was undertaken by a student from the University of Cambridge, William Andrew Laurie (b. 1950). He studied the rhinos in Chitwan from December 1972 to December 1975 with short breaks of a month, with a return in June 1976, altogether 3 years solid field-work (fig. 26.4). The results of his observations were carefully set out in his dissertation (1978), which remains a principal source of authoritative and accurate information on the species. Laurie decided that the only

FIGURE 26.4   Andrew Laurie returning from his observations while studying the ecology and behaviour of the rhinoceros in Chitwan National Park in 1975. He is crossing the Rapti River at Sauraha on the elephant called Prem Prasad

PHOTO COURTESY: ANDREW LAURIE

FIGURE 26.5   Tiger Tops in the Meghauli section of Chitwan National Park was constructed with just four rooms in 1963

PHOTO: © RON RANSON 1964. COURTESY: LISA CHOEGYAL

FIGURE 26.6
Tiger Tops in 1971 when the number of rooms was doubled
PHOTO © COURTESY: JACK EDWARDS

FIGURE 26.7
Visitors to Tiger Tops of Chitwan National Park arriving at the Meghauli Airstrip in a Royal Nepal Air twin otter in the late 1970s
PHOTO © COURTESY: LISA CHOEGYAL

reliable method of understanding the rhino population was to prepare individual identification cards with photographs. It took him 2 ½ years to complete the inventory of 226 rhinos, ultimately estimating a population of 270–310 in 1976.

A second large-scale study of the natural history of *R. unicornis* was conducted by the American scientist Eric Dinerstein (b. 1952). At first he had gained experience with the Nepalese landscape in Bardia as a Peace Corps volunteer in 1975. His studies were focused on rhinos from 1984 to 1988 in Chitwan NP under the guidance of the Conservation and Research Center of the U.S. National Zoological Park, and returned regularly to Nepal as chief scientist of the World Wildlife Fund – United States. He obtained permission to capture, immobilize, and measure rhinos on a regular basis, which gave him a unique opportunity to study each individual in greater detail. His experience contributed to the success of the rhino translocations from Chitwan to Bardia. His book *The Return of the Unicorns* (2003) is a pivotal and detailed study of the behaviour, feeding ecology and reproduction among other facets of the natural history and conservation of the rhinos of Nepal.

26.5 Tiger Tops, the Pioneering Tourist Lodge

American big game hunters John Coapman (1927–2013) and Toddy Lee Wynne (1924–1987) came to Chitwan in the autumn of 1963 (Liechty 2017a: 74, 2017b). They decided to establish a tourist lodge where visitors could stalk and photograph tigers, rhinos and other wildlife (Islam & Islam 2004: 82). With the help of Boris Lissanevitch (1905–1985), the manager of The Royal Hotel in Kathmandu, they approached King Mahendra and his younger brother Prince Basundhera (1921–1977). They obtained a concession with a monopoly of 15 years to develop tourism in Chitwan, all for a token fee of $500 per annum. Coapman proceeded to construct an airstrip and build a lodge partly modelled on the famous Tree Tops in Kenya's Aberdare Mountains built in 1932, where Queen Elizabeth acceded to the British throne on 6 February 1952. Tiger Tops started with just four rooms built on stilts (fig. 26.5), expanded in the 1970s with further rooms and facilities (figs. 26.6, 26.7). The setting was idyllic: "His aim is to give his guests an opportunity to live right in the jungle, and to see and hear its wild inhabitants at night as well as in the day time, particularly tigers. From the hotel one looks northwards, across the river, to the tall grass swamps on Bimli island, where there are always several rhinos. Behind the hotel are the wooded Churia Hills which are the haunt of sambar. Further away to the north are the jungle-clad hills of the Rapti valley, and beyond them, on clear days in winter one can see some of the giant Himalayan peaks" (Willan 1965).

In 1972, Coapman was bought out by two entrepreneurs who loved Nepal and wanted tourists to have a chance to see the incredible wildlife in an original setting. John Raymond Edwards (1935–2009) and Charles (Chuck) McDougal (b. 1930) had started the first wildlife tourism company, Nepal Wildlife Adventure, operating jungle treks, fishing and hunting expeditions. The operation was afterwards run by Edwards' children, Kristjan, Anna Tara, Tim and Jack. Tiger Tops was closed and abandoned in 2012 when the Government banned all overnight stays within the national park boundaries. It was the end of an era, when tourists from across the world had been among many celebrities and world leaders, like Prince Philip Duke of Edinburgh (in 1986), Princess Anne (in 1981), Robert Redford (in 1981), Goldie Hawn, Mick Jagger (in 1990), Henry Kissinger (in 1986) and Jimmy Carter (Choegyal 2017). They all came to view the park's major attraction: the splendid rhinoceros.

26.6   Changes in the National Park

Chitwan was designated as a UNESCO World Heritage site in November 1984. In the original evaluation, it was noted that "Royal Chitwan meets three criteria for World Heritage natural properties. The park is an outstanding example of geological processes and biological evolution as the last major surviving example of the natural ecosystems of the Terai region criteria. The research on the natural history of the area has been an important contribution to man's knowledge of ecological systems in the Terai. The park also contains superlative natural features of exceptional natural beauty in terms of its scenic attractions of forested hills, grasslands, great rivers and views of the distant Himalayas. Additionally, the park provides critical and viable habitat for significant populations of several rare and endangered species, especially the one-horned Asian rhino and the gharial. The current management of the park is of a high standard and the Government of Nepal has clearly demonstrated that it recognizes the value of the park as an important part of Nepal's heritage" (UNESCO 1984).

When the National Park was gazetted in 1973, the Rhino Guards Unit (*Gainda Gasti*) was created. Its personnel were responsible for protection and intelligence obtained on the park borders and buffer zones, while the Nepalese Army patrolled the park's interior. In 1975, there were 200 armed soldiers and officers, increasing in 1987 to three companies and in 1988 to an entire battalion. The presence of the army ensured minimal poaching. In fact, from 1977 to 1983, not a single rhino was killed in the region.

The confused political situation around 1990 resulted in poor morale among the military personnel. In that year, 700 armed men were stationed in Chitwan to man 40 posts. There were shortages of transport and radios, as well as a breakdown of the intelligence gathering system. Gainda Gasti's original strength of 185 was reduced to 124 in January 1991, and they struggled due to a lack of vehicles and domesticated elephants. Consequently, the numbers of rhinos killed by poachers peaked at 17 in 1992. Poachers usually only remove the horn of the animal which, at the

time, sold for $8,000 to $10,000 to middlemen living in the Narayangadh and Tadi bazaar close to Chitwan. In 1993, five fake horns made of buffalo and cow horn by an entrepreneur in the Gorkha District were sold at $22 each, but everyone involved was arrested, as the sale of fake horns is illegal.

In December 1993 the Gainda Gasti was amalgamated with the Forest Guards. The name of the unit was changed to Armed Forest Guards and their main duty became the protection of trees outside the reserve. The tension between wildlife and villagers increased, as rhinos wandered out of the reserve. In 1993, one man was killed by a rhino outside the park, and five illegal grass cutters and a member of the park staff were injured. Villagers often entered the reserve to look for firewood or to graze cattle, despite fines of Rs 60 for trespassing and Rs 20 when a domestic animal is found. From March to December 1993, 11,000 people were arrested by the army inside the park and 25,000 cattle were impounded (Martin 2010).

In 1996, an area of 750 km² surrounding the park was declared a buffer zone to create a feeling of local ownership, to engage with the local communities in conservation, and to reduce their dependence on the park for natural resources. The Government of Nepal made a provision of 30–50% of the park revenue for community development and natural resource management in the buffer zone.

26.7   Rhino Numbers in Chitwan

On 1 June 2001, Crown Prince Dipendra killed King Birendra. Considering the political vacuum that followed and the increased power of Maoist insurgents, organized crime syndicates saw their chance. Poaching increased: 12 rhinos in 2000/01, 38 in 2001/02 and 28 in 2002/03. The poachers operated in gangs of several men, when an average horn could earn US $2,000–$3,500 for the group. Horns are generally transported by middlemen to Kathmandu, by which time the price has already doubled. The reason for the increase in poaching was the closure of several guard posts after the state of emergency was declared. The number of soldiers remained the same, but they were concentrated in eight posts, leaving large parts of the park unprotected. There was also less funding for intelligence and a general fear and breakdown of law and order, all working to the advantage of poachers and traders. Some of these effects were understood and reversed when the situation improved at the end of 2003. However, poaching remained quite heavy for the next few years.

In March 2008, a census was organised by the Department of National Parks and Wildlife Conservation

FIGURE 26.8
Rhino calf in Chitwan National Park
PHOTO: RAM THARU, 2019

FIGURE 26.9
Modern rhino statue on a quiet street crossing in
Sauraha, Chitwan, Nepal
PHOTO: KEES ROOKMAAKER, MARCH 2019

FIGURE 26.10
In Nepal, rhinos in a shop in Sauraha, Chitwan
PHOTO: KEES ROOKMAAKER, MARCH 2019

(DNPWC) in collaboration with National Trust for Nature Conservation, WWF and Zoological Society of London. A total area of 470.2 km² of potential rhino habitat, both inside the national park and in the surrounding buffer zone, was surveyed. It required 3,107.5 elephant hours to complete the census, which revealed a total of 408 rhinos in the park and surrounding areas. A similar census in 2011, with just as much effort and elephant time, indicated an increase to 503 animals in just three years.

The rhino population in Chitwan National Park has recovered well, increasing from a hundred animals in the 1950s to nearly 700 at present (table 26.32). The park is well managed and protection is adequate. Human-wildlife conflicts remain inevitable and are incorporated in overall conservation strategies. The park's wildlife is an important element in the tourist industry, which is now a major source of international revenue. There is a good partnership with conservation bodies, which enhances the role of the army, the government and community groups. Chitwan is the second most important stronghold for the one-horned rhino and, with continued vigilance, will continue to be a haven for *R. unicornis* (figs. 26.8, 26.9, 26.10).

TABLE 26.32   Population estimates of *R. unicornis* in Chitwan NP

Year	Number	Natural death	Poached	Sources
1900	1000			R083, R257
1910	1100			R124, R174
1928	hundreds			R136
1930	300			R039
1936	200			R124
1939	300–400			R217
1942	300–400			R194
1950	800, 1000			R013, R047, R083, R155, R148, R178, R188
1952	50			R106, R107
1953	200, 1000		20–30	R204, R109, R194, R225
1954	48		72, 75	R007, R174, R235, R325
1955	1000			R256
1957	400, 600, 500–600			R013, R083, R109, R194, R225, R226, R265
1958	48, 100		60, 500	R109, R194, R265, R108, R174, R235
1959	300		12	R013, R047, R083, R109, R117, R208, R265
1960	200–225		75	R111
1961	160			R111, R117, R208
1962	60–80, 185		18	R084, R148, R111
1963	185, 300			R207, R226
1964	40, 185			R266, R112
1965	150, 180			R268, R293
1966	100, 165, 362			R013, R117, R218, R222, R117, R208, R218, R219
1968	81–108, 165			R013, R063, R135, R155, R096
1970	70, 125–150			R279, R169
1971	80, 120–147			R117, R185
1972	60, 80, 147			R083, R280, R013
1973	200	6	5	R047, R117, R152
1974		8	2	R152
1975	270–310	2		R083, R152, R167, R188
1976	270–310	1	2	R146, R152
1977	270–300	5		R056, R152, R155
1978	270–310			R013, R145, R152
1979		6		R152
1980	400	8		R081, R152, R168
1981	300	3		R152, R292

TABLE 26.32   Population estimates of *R. unicornis* in Chitwan NP (*cont.*)

Year	Number	Natural death	Poached	Sources
1982		7		R152
1983		3		R152
1984		2	2	R152
1985	400	2		R152
1986	400	2	3	R125, R152, R167
1987		6		R152
1988	358–376	1	3	R013, R080, R082, R083, R085, R148, R152
1989	375	8	1	R152, R211
1990		3	3, 12	R152, R171
1991	400		1	R152, R171
1992	375–400		17	R152, R157
1993	375–400	11	7	R144, R152, R157
1994	440–466	4	1	R013, R083, R152, R157, R158, R206
1995	446–466	7	1	R157, R158
1996		7		R158
1997		2	1	R158
1998		9	3	R001, R087
1999		23	6	R001, R004, R159, R258
2000	544	28	16	R001, R004, R159, R188, R258
2001	600	12	12, 15, 37	R001, R004, R150, R160, R163, R258
2002	600	16	35–38	R001, R004, R135, R150, R160, R163, R258
2003		14–17	19–22	R001, R004, R150, R160, R163, R258
2004		17	9–11	R001, R004, R150, R163, R258
2005	372	11	15	R001, R004, R049, R150, R163, R178, R258
2006	372	8	10–19	R001, R004, R015, R084, R163, R258
2007		7	1–14	R001, R004, R015, R163, R258
2008	408	9	7	R004, R013, R015, R131, R151, R163, R179, R258
2009	408	8	9–12	R004, R015, R084, R151, R258
2010		13	11–13	R004, R015, R258
2011	503	14	1–4	R004, R013, R015, R180, R230, R258
2012		9–11	1	R004, R015, R258
2013		9	1	R015, R258
2014		10	0	R015, R258
2015	605		0	R015, R251
2021	694			R331, also 3 in Parsa NP

This includes official figures on rhino natural deaths and numbers poached, from 1900 to present.

References in Table. R001 Acharya 2008; R004 Adhikari 2015: 79; R007 Amritaghata 1955; R015 Aryal et al. 2017; R039 Bechhold 1930; R047 Bhatt 1977: 81; R049 Bhatta 2008; R056 Bolton 1977: 475; R063 Caughley 1970; R080 Dinerstein & Jnawali 1993; R081 Dinerstein & MacCracken 1990; R082 Dinerstein & Price 1988: 406; R083 Dinerstein 2003: 83; R084 Dinerstein 2015: 108; R085 Dinerstein et al. 1990: 36; R096 Fitter 1968: 59; R103 Gee 1950a: 1732; R106 Gee 1952a: 224; R107 Gee 1953a: 341; R108 Gee 1958: 353; R109 Gee 1959b: 67; R111 Gee 1963c; R112 Gee 1964a: 153; R117 Goodwin & Holloway 1972; R124 Harper 1945: 378; R125 Hart-Davis 1986; R131 Jnawali et al. 2011: 33; R135 Kemf & van Strien 2002: 6; R136 Kennion 1928; R144 Lahan 1993: 18; R145 Laurie 1978: 12; R146 Laurie 1982; R148 MacCracken & Brennan 1993; R150 Martin & Martin 2006; R151 Martin & Martin 2010; R152 Martin & Vigne 1995; R155 Martin 1985: 14; R157 Martin 1996b: 12; R158 Martin 1998: 88; R159 Martin 2001: 43; R160 Martin 2004: 91; R163 Martin et al. 2009a: 107; R167 Mishra & Dinerstein 1987; R168 Mishra et al. 1982: 420; R169 Mitchell 1977: 405; R171 Mountfort 1991; R174 Mukherjee 1963: 45; R178 Nepal Government 2006: 5; R179 Nepal Government 2008; R180 Nepal Government 2012; R185 Pelinck & Upreti 1972; R188 Pradhan 2001; R194 Rohr 1959: 955; R204 Shebbeare 1953: 143; R206 Shrestha 1997: 24; R207 Simon 1966; R208 Simon 1967; R211 Sinha & Sawarkar 1993; R217 Smythies 1942: 49; R218 Spillett & Tamang 1966; R219 Spillett 1966a: 492; R225 Stracey 1957: 764; R226 Stracey 1963: 243; R230 Subedi et al. 2013; R235 Talbot 1960: 191; R251 Talukdar 2015; R256 Thapa 2005: 3; R257 Thapa 2013: 95; R258 Thapa 2016; R265 Ullrich & Ullrich 1962: 187; R266 Ullrich 1964: 226; R268 Ullrich 1965b: 98; R279 Waller 1971: 12; R280 Waller 1972a: 5; R292 Whalley 1981; R293 Willan 1965; R325 Martin & Martin 1982: 31; R331 Talukdar 2022.

Historical Records of the Rhinoceros in Eastern Nepal

First Record: 1797 – Last Record: 1995 (?)
Species: *Rhinoceros unicornis*
Rhino Region 15

27.1   Rhinos in Eastern Nepal

Only one, maybe two, rhinos are known to have been killed on the Kosi River. Few early visitors to Nepal would have needed to pass through the eastern parts of the country. Even visits of the Kings and Prime Ministers were infrequent. Rhinos occurred in the *terai* close to the southern border, where populations were contiguous with those in the northern districts of Bihar (28.1).

The Koshi Tappu Wildlife Reserve on the banks of the Kosi River was established in 1976, covering 175 km² of wetlands. Rhinos were listed as part of the fauna in the Red List of 1995, but in truth this is unlikely to be correct.

• • •

Dataset 27.25: Chronological List of Records of Rhinoceros in Eastern Nepal

General sequence: Date – Locality (as in original source) – Event – Source – § number – map number.
The localities are shown on map 20.10 and explained in Dataset 20.19.

1797-02 – Chainpur – Royal order regarding supply of live rhinoceros calves, dated February 1797 – Regmi 1995: 31 (note 24) – 27.2

1856 – Kosi River – Rhinos occur, but only in places not visited during the 1854–1857 expedition of Hermann Schlagintweit (1826–1882) and his brothers – Schlagintweit 1880, vol. 1: 229 – 38.17 – map A 261

1878 – Nepal *terai* – James Inglis (1845–1908) shot a rhino in Nepal territory by mistake – Inglis 1878: 214, 1888: 557, 1892: 557

1880 – Eastern *terai* – Rhino present. Horns from this area are more valued than those from western *terai* "much longer, finer in texture, and better coloured, having little or no white in them" – Oldfield 1880, vol. 1: 237

1880 – Kuster received hunting pass without permit for rhinos or elephants – Kuster 1880 – 27.04, fig. 27.1

1880s – Bobia on Kosi River – John Joseph Shillingford (1822–1867) shot one and was imprisoned for this – Philippe d'Orléans 1892: 157 – 27.02

1888 – Kosi River – Prince Henri d'Orléans (1867–1899) failed to see any rhino for six weeks, despite permission to shoot four. He was accompanied by his nephew Philippe d'Orléans (1869–1926), Douglas Graham, 5th Duke of Montrose (1852–1925) and the indigo-planter Gwatkin Williams (1845–1891) – Henri d'Orléans 1889: 287; Philippe d'Orléans 1892: 157; Levesque 1888; Anon. 1889; Jacques d'Orléans 1999: 45 – 27.4, fig. 27.2

1900 – *Terai* from Bhootan to Nepaul – Rhino present. General area possibly including East Nepal – Russell 1900: 339

1901-04 – Morang, east of Kosi River – Viceroy George Curzon (1859–1925) killed 1 rhino. The Viceroy's team included the British Resident in Nepal Thomas Caldwell Pears (1851–1921), the Private Secretary Walter Roper Lawrence (1857–1940), the Military Secretary Everard Baring (1865–1932), the Surgeon Ernest Harold Fenn (1850–1916), four Aides-de-Camp Baron Clive Wigram (1873–1960), Claude Champion de Crespigny (1873–1910), Viscount Gerald Berkeley Portman (1875–1948), Henry Molyneux Paget Howard, 19th Earl of Suffolk (1877–1917), as well as Colonel Gordon and Major Armstrong attached to the Nepal Embassy – *Indian Daily News* 1901-04-04; *Times of India* 1901-04-18; *Homeward Mail from India* 1901-04-27, 1901-05-04; Ronaldshay 1928, vol. 2: 168; British Library, Curzon Collection (no. 430/77) – 27.5, figs. 27.4 to 27.10

[1901?] – [Morang] – According to recollections published by Nigel Gresley Woodyatt (1861–1936), the Viceroy George Curzon, his Aide-de-Camp Robert George Teesdale Baker-Carr (1867–1931) and the Military Secretary Everard Baring (1865–1932) killed 2 rhinos. As Curzon shot only 1 rhino in 1901 and Baker-Carr was in the UK at the time, locality and date are probably garbled. The accompanying photograph was taken in Chitwan in 1911 – Woodyatt 1922a: 149, plate facing p. 143 – 27.5, fig. 27.3

1909 – Morang – Rhino present – Manners-Smith 1909; Gee 1959b: 79

1909 – West bank of Kosi River, at Patharghatta – Rhino present – Manners-Smith 1909

1909 – Banks of the Bagmati River north of Muzaffarpur – Rhino present – Manners-Smith 1909

1916 – East Nepal – Rhinos stray into Purnea, but none recently shot – K. Sinha 1916: 6

1923 – Morang – John Champion Faunthorpe (1871–1929) hunted here. No rhino present – Faunthorpe 1924a

1927 – Nepal – J.B. Norman, manager of Pipra Indigo Estate in Bihar, credited with horn of 14 in (35.6 cm) length. Considering his usual residence, the animal could have been obtained in eastern Nepal – Ward et al. 1935: 335

1934 – Morang, bordering Purnea – Number of rhino is so small that the Government has strictly prohibited rhino-shooting – K. Sinha 1934: 8

1957 – Kosi area – Few scattered rhinos – Stracey 1957: 764

1963 – Valley of Kosi – Rhino present – Stracey 1963: 66

© L.C. (KEES) ROOKMAAKER, 2024 | DOI:10.1163/9789004691544_028

This is an open access chapter distributed under the terms of the CC BY-NC-ND 4.0 license.

1995 – Koshi Tappu WR – Rhino present – Nepal Government (Red List) 1995: 12

2010 – Koshi Tappu WR – Rhino not listed among 21 mammals observed – Chhetry & Pal 2010

∵

27.2 Sporadic Early Sightings

In 1797 there was a royal order for the District of Chainpur requesting the supply of rhino calves to Kathmandu. Chainpur is a mountainous regions, and was perhaps a larger administrative centre at that time. From the late 1870s onwards there are sporadic reports of rhinos from the banks of the Kosi River and further east in Morang District. One was killed without permission by John Joseph Shillingford in the 1880s. Other hunting parties often left without ever observing a rhino, like the Schlagintweit brothers in 1856 and Henri d'Orléans in 1888. The only well-documented record was the hunt by the Viceroy George Curzon in 1901 (27.5). He shot one rhino in Morang

with a horn of record length. All the subsequent statements are indefinite and probably written by authors who had no chance to visit this remote region. It is therefore impossible to be certain when the rhino became extinct in eastern Nepal.

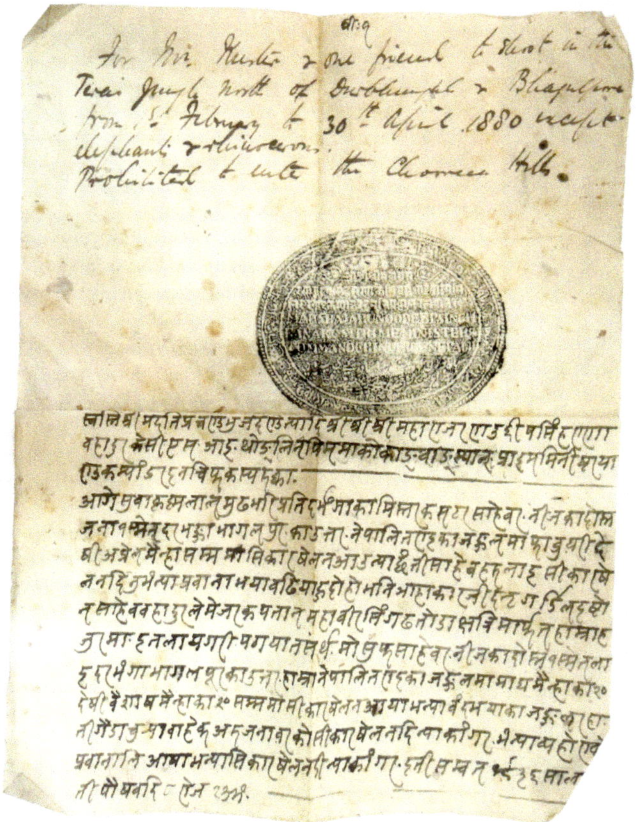

27.3 A Hunting Pass for Hunting in the Eastern *terai*

The hunting of wildlife was strictly regulated by the authorities in Nepal. Permits were needed for this purpose. As a rare example, one of these was preserved for a hunt in the area north of Dharbangah in 1880 (fig. 27.1). Written in Nagari script, a translation states: "For Mr. Kuster & one friend to shoot in the Terai jungle north of Durbhangah & Bhagulpore from 1st February to 30th April 1880 except elephant & rhinoceros. Prohibited to enter the Chomes Hills" (Kuster 1880). Kuster remains unidentified.

27.4 Prince Henri d'Orléans on the Kosi River in 1888

Prince Henri d'Orléans of France went to shoot near the Kosi River in 1888. He camped six weeks in the area together with his nephew Philippe d'Orléans and the 5th Duke of Montrose, under the supervision of an English indigo planter Gwatkin Williams. He had obtained permission to kill four rhinos, and although they visited jungles where the animals had been, the Nepalese attendants claimed that no elephant was fit for such a hunt. According to the published reports, the party never actually saw a rhino. The animal collections of Philippe d'Orléans were set up by Rowland Ward and displayed until 1912 in Wood Norton Hall, Evesham, then transferred to a mansion in Anjou, Belgium, until their final transfer to a special building next to the Natural History Museum in Paris in 1928 (Pichot 1908; Rode 1934). Although photos by Carl Vandyk taken in Wood Norton around 1908 show several rhino heads, including at least one single-horned of unknown provenance (fig. 27.2), there is no trace of these in later inventories (Dias 2012).

FIGURE 27.1   Hunting pass in Nagari Script allowing Kuster to hunt in 1880

WITH PERMISSION © HORNIMAN MUSEUM AND GARDENS, LONDON: MANUSCRIPT NN15460

FIGURE 27.2
The Cynegetic Museum at Wood Norton, the
residence of Philippe d'Orléans. Photo by Carl
Vandyk (1851–1931) in 1908 showing on the far wall
three heads of the African *Diceros bicornis*, and on
the left wall a single-horned *R. unicornis*
PICHOT, *LES SPORTS MODERNES*, MARCH 1908

FIGURE 27.3
"Nepalese shikaris skinning a dead rhinoceros and
separating the head with an axe." Photo: Herzog and
Higgins, Mhow, C.I. This was taken during the royal
hunt in Chitwan in 1911
WOODYATT, *MY SPORTING MEMORIES*, 1922,
PL. FACING P. 143

27.5   The Viceroy Curzon in Morang in 1901

The Viceroy Lord Curzon was invited in March 1901
by the Government of Nepal to shoot in the eastern part
of the country, on the Kosi River north of Forbesganj. On
behalf of Maharaja Chandra Shumsher, he was received
by General Jit Bahadur Khatri (d. 1913) and Colonel Hark
Jung Thapa, who was in charge of the 220 elephants. The
Viceroy's team included several members of his staff, as
well as the British Resident in Nepal Thomas Caldwell
Pears and others attached to the Nepal Embassy (*Indian
Daily News* 1901-04-04). Also in camp were the photogra-
pher P.A. Herzog and in charge of catering Henry Gottlieb
Wutzler (d. 1915), proprietor of Charleville Hotel in
Mussoorie (*Civil & Military Gazette* 1901-03-26).

For three weeks they were based in four different
camps. This was the itinerary.

Friday 29 March: arrive Forbesganj station at 1.48 PM.

29 March 1901: move to Sahibgunge camp (*Indian Daily
News* 1901-03-28).

29 March 1901: started shooting.

2 April 1901: camp moved 10 miles.

8 April 1901: camp again moved 8 miles.

9 April 1901: crossed River Kosi, one mile broad.

Sunday 14 April: camp moved 16 miles.

Sunday 14 April 1901: news of a rhino.

Monday 15 April 1901: 1 rhino shot and pursued for a long
time, not killed. Rhino passed close to camp.

Wednesday 17 April 1901: 3 rhinos found in heavy grass
jungle.

FIGURE 27.4
"V[iceroy] firing at the first Rhino" showing
Lord Curzon during a hunt in the Nepalese *terai*
east of the Kosi River on 17 April 1901. Photo by
P.A. Herzog
WITH PERMISSION © BRITISH LIBRARY:
CURZON COLLECTION, 430/77, NO. 116

FIGURE 27.5
"Round the dead Rhino." Lord Curzon in Nepal
on 17 April 1901, photographed by P.A. Herzog
WITH PERMISSION © BRITISH LIBRARY,
CURZON COLLECTION: 430/77, NO. 117

17 April 1901: Curzon killed 1 rhino, while 2 others ran away.
Thursday 18 April 1901: Curzon rides 20 miles to Forbesganj
station and departs.

Rhinos were only reported at the end of the shoot. On
the last day, 17 April 1901, the only rhino was killed by the
Viceroy. In the evening he wrote to his wife: "I saw the
great brute dimly standing in a sort of tunnel that he had
forced for himself through the bottom of the grass. He
turned and fled. I fired a shot that caught him in the neck
and sent him over like a rabbit. Then you never saw such
a commotion. He kicked and plunged, and we had to pour

at least a dozen shots into him before he was finished off"
(Ronaldshay 1928, vol. 2: 168). The bag of the shoot totaled
12 tigers, 1 rhino and 1 leopard (*Homeward Mail from
India* 1901-05-04).

In the reminiscences of the Gurkha officer Nigel Gresley
Woodyatt (1922: 149), Curzon is credited with two rhinos
while hunting with his Aide-de-Camp Robert Baker-Carr.
However, Baker-Carr could not have been present then,
because news reports state his journey to London from
February to September 1901 (*Gentlewoman* 1901-09-21).
Probably Woodyatt's recollections were garbled, although

FIGURE 27.6
"The dead rhino." Lord Curzon in Nepal on
17 April 1901, photographed by P.A. Herzog
WITH PERMISSION © BRITISH LIBRARY,
CURZON COLLECTION: 430/77, NO. 118

FIGURE 27.7
"The first Rhino." Lord Curzon in Nepal on
17 April 1901, photographed by P.A. Herzog.
The same item in a formal album of "Herzog
and Higgins" is available with handwritten text
by Subodh Rana, son of Kiran Shumsher. This
identifies the persons as the Viceroy (middle), left
of him Colonel Hark Jung Thapa, and right of him
Jit Bahadur Khatri
WITH PERMISSION © BRITISH LIBRARY:
CURZON COLLECTION, 430/77, NO. 119

I have been unable to identify any other hunt in which Curzon killed two rhinos. The photograph cursorily inserted by Woodyatt refers to the royal hunt in Nepal in 1911 (23.3, fig. 27.3).

The Curzon Collection in the British Library contains a green half-leather album (36.6 × 50.5 cm) with a title stamped in gold on the front cover "H.E. the Viceroy's Shooting Tour, Nepal, Tarai, April 1901." It has 121 photographs taken during this camp in Nepal taken by P.A. Herzog on behalf of the photographic firm of Herzog & Higgins based in Mhow (*Englishman's Overland Mail* 1901-04-25). A similar album was sold by the auction houses of Bonhams (2011-10-04, lot 302), Sotheby's London (2014-10-08, lot 361) and again Sotheby's New York (2017-12-11, lot 95). The copy

in the British Library from Curzon's personal library is the easiest to access. There are six photographs of Curzon shooting "the first rhino" and scenes taken near the animal's body (figs. 27.4 to 27.9).

Curzon is credited with an extraordinarily long rhino horn from Nepal with a length of 21 ½ in (54.6 cm) and a circumference of 24 ¾ in (62.8 cm). The specimen was first listed in the 4th edition of the *Records* of Rowland Ward (1903: 439) and shown in an illustration, again found in the 5th edition of 1907 (fig. 27.10). While this points to a connection with Curzon's camp in 1901, it is noteworthy that none of the contemporary reports mention that the rhino had a very long horn. The photographs in the Album do not appear to show a record horn, but they are taken

FIGURE 27.8 "The first Rhino." Lord Curzon in Nepal on 17 April 1901, photographed by P.A. Herzog
WITH PERMISSION © BRITISH LIBRARY: CURZON COLLECTION, 430/77, NO. 120

FIGURE 27.9 "Waiting for the second Rhino." Lord Curzon in Nepal on 17 April 1901, photographed by
P.A. Herzog
WITH PERMISSION © BRITISH LIBRARY: CURZON COLLECTION, 430/77, NO. 115

at such an angle that it is hard to be certain. Yet only one rhino was taken during this hunt. An alternative possibility might be that the record horn was a donation to the Viceroy from Maharaja Chandra Shumsher and possibly obtained elsewhere. The trophy head with the long horn similarly shaped as the one illustrated by Rowland Ward is preserved in Kedleston Hall, Derbyshire, the ancestral home of Curzon (fig. 27.11). The house also displays a stool made of a rhino lower leg with a wooden top.

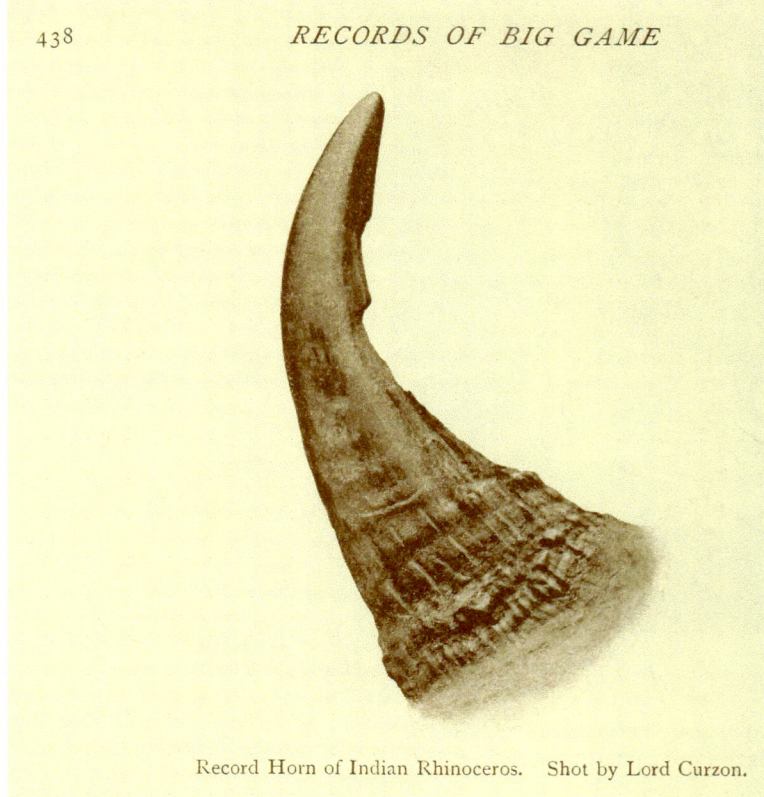

Record Horn of Indian Rhinoceros.   Shot by Lord Curzon.

FIGURE 27.10
Record horn of *R. unicornis*, shot by Lord Curzon,
measuring 21 ½ in (54.6 cm)
ROWLAND WARD, *RECORDS OF BIG GAME*, 4TH EDN,
1903, P. 438

FIGURE 27.11
Trophy head of a rhino killed
by Lord Curzon, probably the
specimen of 17 April 1901 in Nepal,
displayed in Kedleston Hall,
Derbyshire
WITH PERMISSION © NATIONAL
TRUST

Historical Records of the Rhinoceros in Bihar

First Record: 1632 – Last Record: current
Species: *Rhinoceros unicornis*
Rhino Regions 16 to 19

28.1 Rhinos in Bihar

The large northern state of Bihar was once prime rhinoceros habitat. Bihar is wedged between the Himalayas in the north and the Ganges River in the south. Rhinos occurred in the long stretch of *terai* along the base of the mountains across from the border with Nepal as well as near the Ganges, Kosi, Gandak and Sone rivers traversing the region. The Ganges River was the main thoroughfare for travelers going inland from Kolkata to Patna and beyond. The portrayal of the Ganges as a venerable old man accompanied by a rhino by the English architect George Richardson combines the river and the animal in an iconic emblem (fig. 1.1). This symbolism was perpetuated in an ornament of the testimonial silver service presented to Edward Law, 1st Earl of Ellenborough in 1848 (fig. 28.1). The accounts by Thomas Williamson in the *Oriental Field Sports* and *Foreign Field Sports* relating to animals seen close to the river are important because together with the illustrations by Samuel Howitt they were widely circulated for a long time after their first publication in 1807.

The single early record of 1766 in the south-western corner of Bihar, near Rohtas on the Sone River, has so far been overlooked. It opens up the interesting perspective that this area was still forested and uninhabited enough to allow a large animal species like the rhino to exist up to the middle of the eighteenth century. The locality is also close to areas where rock art sites depicting rhinos were discovered, belonging to a much earlier era, but perhaps rhinos persisted longer there than has generally been understood.

FIGURE 28.1   River Ganges accompanied by a rhinoceros. In 1848, Edward Law, 1st Earl of
Ellenborough (1790–1871), Governor-General of India 1842–1844, was presented
with a large testimonial silver service. Executed by the artists Frank Howard
(1805–1866) and Alfred Brown working for the jewellers firm of Hunt and Roskell,
it included two ornaments for the ends of the table representing the rivers Indus
and Ganges. The latter is accompanied by a rhino (left, partly obscured) said to
be "most admirably executed. It was, we believe, taken from a living model, and
such care was used to secure accuracy, that a cast of the skin was made to obtain
a correct resemblance of its texture" (Anon. 1848). Parts of the service were
exhibited in Crystal Palace during the Great Exhibition of 1851 (Anon. 1851b) and
illustrated in the guide-book (Anon. 1852c).
DETAIL FROM *THE CRYSTAL PALACE AND ITS CONTENTS*, 17 FEBRUARY 1852,
P. 296

© L.C. (KEES) ROOKMAAKER, 2024 | DOI:10.1163/9789004691544_029
This is an open access chapter distributed under the terms of the CC BY-NC-ND 4.0 license.

Further north, the Maharajas of Darbhanga hunted rhinos in their forests west of the Kosi River until the end of the 19th century. There were regular reports from the districts of Champaran near the Gandak River, joining good rhino habitat across the Nepalese border. As this was closest to the trading town of Patna, the rhinos captured and exported in the 18th century, would have been found here. It may well be true that the last specimen was shot in 1949. In more recent times, rhinos have started to resettle in Champaran straying into the area in search of fresh pastures. There may be a chance to consolidate this trend through trans-boundary cooperation.

The northern districts from the Kosi River eastwards to West Bengal were inhabited by rhinos only in suitable places, which became fewer as the human population and the demand for agricultural lands increased. Rhinos were observed here, in Purnea and near the confluence of the Kosi and Ganges Rivers, at least until the 1870s. The rhino killed during the 1871 hunt in Purnea by the well-known Shillingford family of hunters and indigo-farmers is important because it provided the first mounted specimen of *R. unicornis* ever exhibited in the Indian Museum in Kolkata.

It is hard to say how long the rhino survived in the region east of the Kosi River. One of the rulers of the area, Raja Bahadur Kirtyanand Sinha (of Banaili) shot two rhinos, but had to travel to Nepal to achieve this. Reading his book (K. Sinha 1916) about wildlife in Purnea, leaves the impression that the species could still be found in the early 20th century, although he mentions no definite records. Rhinos probably were gone by the end of the 1920s, or if any were left, they did not survive the changes to the landscape during the Kosi Barrage project of the 1960s.

• • •

Dataset 28.26: Chronological List of Records of Rhinoceros in Bihar

General sequence: Date – Locality (as in original source) – Event – Sources – § number, figures – Map number.

The persons interacting with a rhinoceros in various events are combined in Table 28.33. Places and numbers are explained in Dataset 28.27 and shown on map 28.11.

The records are listed according to the four Rhino Regions 16 to 19 in Bihar.

Rhino Region 16: Bihar – South-West

1st cent. – Gaya Dt. – Punch-marked coins – V. Smith 1906a: 132; Theobald 1890: 217 – 28.4 – map A 303

1st cent. – Magadha kingdom of Bihar – Punch-marked coins – Gupta & Hardaker 2013 (symbol 320) – 28.4 – map A 303

undated – Gaya Dt. – Four rhinos sculptured in black marble, once in the India House Museum in Calcutta (currently unknown) – Blyth 1872c – map A 303

6th cent. – Camundi (Cheyenpur, near Bhabua) – King Puskara, son of Lord Varuna, liked to hunt rhinos – S. Sharma 2002 – 28.5 – map A 302

1766-12-26 – Berealpour, near Rottasgur (Rohtas, on Sone River) – Lewis Felix DeGloss saw tracks of 'Rynosserus' – Buchanan-Hamilton 1926: 176; Phillimore 1945: 25 – 28.12 – map A 301

Rhino Region 17: Bihar – Ganges Downstream from Patna to Munger

1700 BC – Chirand, Saran Dt. – Skeletal fragments identified as *R. unicornis* – see chapter 8 – map W 636

3rd cent. BCE – Pataliputra, ancient capital north of Patna – During the reign of Chandragupta Maurya (321–298 BCE), animal fights including rhinos – V. Smith 1906b, vol. 2: 111; Bautze 1985: 413 – 28.2 – map A 304

100 BCE – Murtaziganj (near Patna) – Soapstone disc, 5.1 cm diameter, depicting rhino with prominent horn – Shere 1951: 184 pl. VII.3; M. Chandra 1973: 29; Gupta 1980, pl. 25a; Bautze 1985: 413; Bose 2020: 90, fig. 2.17 – map A 304

1632 – Puttana (Patna) – The English merchant Peter Mundy (1596–1667) was offered "3/4 of a hundred weight" (about 35 kg) of rhino horn for sale by Nundollol (Nanda Lal) – Mundy 1914, vol. 2: 171 – 28.7 – map A 304

1632 – Etaya (Etawah) – Peter Mundy missed to see a pair of rhinos that were taken to Emperor Shah Jahan as a gift of Saif Khan (d. 1640), Governor of Bihar 1628–1632 – Mundy 1914, vol. 2: 186 – 5.4

1649 – Bengal near Ganges – Philippe de la Très Sainte Trinité (1603–1671) mentioned presence of rhino – Philippe 1649: 285– 28.8

1659 – Muguer (Munger) – Niccolao Manucci (1638–1717) said Shah Shuja (1616–1661) heard about rhino in the hills behind Munger – Manucci 1907: 334; Mandala 2019: 47 – 28.8 – map A 307

1665–1672 – Patna – Depiction of rhino captured in Patna in an Album compiled at Aurangzeb's court in 1665–1672 – A. Das 2018a: 78 – 11.15, fig. 11.20 – map A 304

1670 – Pattana – Thomas Bowrey (1659–1713) mentioned "Rhinocerots" in this kingdom. The text was edited by Richard Carnac Temple (1850–1931): "the Woods in this Kingdome afford great Store of those deformed Annimals called Rhinocerots, and many of them are taken younge and tamed." and "Soe that soe farre as is reported of them to be Utter Enemies to the Elephant I doe confide in, for in all Kingdoms where are found the Rhinocerot the Elephant is not found wild there, nor dare the tame ones frequent the Woods, As for instance, Pattana, Bengala, and Java Major" – Bowrey 1905: 221,223, figure redrawn on p. 222 – 28.9, fig. 28.7 (new photo) – map A 304

1738 – Patna region – Young male rhino, obtained by Humphrey or Humffreys Cole of the Patna Factory, sent to London where exhibited from 1 June 1739 until at least 1742. Studied and drawn by James Parsons (1705–1770) in his capacity of assistant to the physician James Douglas (1675–1742) in London – Parsons 1743; Rookmaaker 1973, 1978b: 29, no. 5.19, 1989: 83; Clarke 1986: 41–46; Faust 2003: 40, no. 699; Hanson 2010; Grigson 2015 – 5.4, 28.10, figs. 28.9 to 28.12 – map A 304

1738 – Patna region – Broadsheet published on 10 October 1739 showing the male rhino in London sent from Patna. Engraved by Gerard van der Gucht (1696–1776) after drawings by James Parsons. There are multiple copies in the University of Glasgow Library (Hunterian Library) no. DL.1.32, and others listed by Rookmaaker (1978b: 29) and Faust (2003: 40); also in collection Kees Rookmaaker – 28.10, fig. 28.8

1738 – Patna region – James Parsons presented a lecture on the history and morphology of the rhinoceros at a meeting of the Royal Society of London on 9 June 1743. Based on his own drawings, the published paper was accompanied by two plates of the animal in side-view (tab. I) and in a frontal and rear view (tab. II), engraved by James Mynde (1702–1771) – Parsons 1743, translated Parsons 1747 (German) and 1760 (French)

1746 – Bengal on Ganges – The German naturalist Carolus Augustus von Bergen (1704–1759) mentions presence of rhino. This book was among the sources listed in Linnaeus (1758) for *R. unicornis* – Bergen 1746: 26 – 4.6

1763 – Bihar – Henry Vansittart (1732–1770) obtained a living rhino to be sent to King George III in London. It did not arrive in UK – Malcolm 1836: 256 – 5.4 (Patna), 28.11

1765 – Monghyr (Munger) – Surveyor Louis (Lewis) Felix DeGloss found rhino tracks close to town – Hirst 1917: 40 – 28.13 – map A 307

1767-10-26 – Between Piprah and Ganduk (Gandak) River – Lewis DeGloss saw tracks of 'Rynosseroces' – Phillimore 1945: 26 – 28.13 – map A 306

1779 – Ganges valley – Rhino accompanying 'Ganges' portrayed as a venerable man by George Richardson (1736–1817) – Richardson 1779, vol. 1, pl. 17 – 28.1, fig. 1.1

1780s – Western side of Ganges – Rhino now seldom found – Williamson 1807: 167 – 28.16

1787 – Monghyr (Munger) – Maharaja Ramchandra Singh Deo I of Patna (ruled 1765–1820) is attacked by rhino close to Monghyr. Dirk van Hogendorp (1761–1822) did not encounter the animal – Hogendorp 1887: 66 – 28.15 – map A 307

1788 – Derriapore, near Munger on south bank of Ganges – Rhino seen goring horses of soldiers. Scene depicted as "Anecdote of Hunters & Rhinoceros" drawn by Samuel Howitt (1756–1822) – Williamson 1807a: 172; Howitt 1814: 113, 1819: 113–114 – 28.17, figs. 28.15 to 28.18 – map A 305

1800 – Geyaspur – When an invalid officer obtained the land (30 years earlier?), there were rhino but these disappeared – Martin 1838a, vol. 1: 321 – map A 305

1829-09-11 – Patna region – "Rhinoceros hunting", drawing by Charles D'Oyly (1781–1845) – D'Oyly 1829 – 28.20, fig. 28.23 – map A 304

1836 – Bykontpore (Baikathpur), just east of Patna – A servant of H.C.'s friend M. was disemboweled by a rhino. He examined the wound immediately, and heard him say afterwards that it was a very clean cut (with incisors) – H.C. 1836 – map A 304

1838 – Shahabad (south of Patna) – Rhino absent – Martin 1838a, vol. 1: 502

1838 – Patna region – Rhino absent – Martin 1838a, vol. 1: 225 – map A 304

1859 – Borders of the Ganges – Rhino present – Goodrich 1859, vol. 1: 637

1893 – Sonepore – Rhinos traded in market. Noted by Godefroy Durand (1832–1896) – G. Durand 1893 – map A 304

Rhino Region 18: Bihar – Champaran and Darbhanga

5th cent. – Northern Bihar (Champaran) – Emperor Kumaragupta I could have hunted rhino in northern Bihar – Sohoni 1955, 1956; Prakash 1962, 1963; Sircar 1966; Chandra 1973: 29; Y.B. Singh 1981; A.N. Singh 1983; Bautze 1985: 416; Tandon 2006 – 28.3 – map A 309

5th cent. – Ganges valley – Kumaragupta I (414–455), Gupta Emperor. Gold coins of 'Rhinoceros-Slayer' type – Nagar 1949; Altekar 1954, 1955, 1957; J. Walker 1957; Raven 2012, 2019 – 28.3, fig. 28.2

1766 – Ganduk (Gandak) River – Lewis DeGloss was informed about capture of 2 rhinos – Phillimore 1945: 26 – 28.13 – map A 309

No date – Gandak River – *Gandak* means rhino – Tiwari 2000 – map A 309

1768 – Rhinoceros wilds (between Bettiah and Raxaul) – Locality on map produced by Thomas Jefferys (1719–1771), English cartographer – Jefferys 1768 – 31.4, fig. 31.3 – map A 310

1780s – Regional mountains [at Boggah (Bagaha)] – Rhino-elephant fight witnessed by Major William Lally (1761–1803). The scene depicted as "Rhinoceros bayed by Elephants" by the British artist Samuel Howitt (1756–1822) in plate XI of *Oriental Field Sports*. The text of this book was supplied by Thomas George Williamson (1759–1817) – Williamson 1807, pl. 11 – 28.16, figs. 28.15, 28.16, 28.17 – map A 308

1780s – [Boggah] – Williamson's text was repeated in *Select Reviews of Literature and Spirit of Foreign Magazines* (Williamson 1811). The scene on plate XI in Williamson (1807) was copied in a drawing owned by Ananda Kentish Muthu Coomaraswamy (1877–1947), suggested to have been produced in Jaipur in the 19th century (Coomaraswamy 1912, pl. 22). It was reworked in several illustrated popular works until the 1840s: Bertuch 1816, pl. 20; Howitt 1833: 237; Williamson 1835: 68–69; J.H.C. 1836; Howitt 1837; Knight 1848: 125 – 28.16, fig. 28.18 – map A 308

1785 – "Bear" (Bihar), north of Patna – Rhino image on map of "Bear" (Bihar) by Jean-Baptiste-Joseph Gentil (1726–1799) – Gole 1988; Rookmaaker 2014a – 28.14, figs. 28.13, 28.14 – map A 310

1815 – Hurdeen near Baraguree (south of Raxaul) – Group of hunters killed adult male and female, and one young rhino in one day – Anon. 1817; Anon. 1818b; Percy & Percy 1821: 126 – 28.19 – map A 310

1820 – Sarun Dt. (Bettiah and Champaran) – Rhino in the *terai* – W. Hamilton 1820: 275 – map A 310

1827 – Mountains on the other side of the Ganges – Origin of captive rhino in Barrackpore near Kolkata in 1827 – Jacquemont 1841: 169 – 25.4 – map A 310

1848-01 – Bettiah – In palace of Kishore Singh Bahadur, 8th Raja of Bettiah, 'Rohilla' saw a "very tame and very stupid Rhinoceros" – Rohilla 1848: 209 – 5.4 (Bettiah), 28.21 – map A 309

1848 – NW of Betteeah (Bettiah) – Rhino killed by a party from Gorruckpore – Rohilla 1848: 216 – 28.21 – map A 309

1860s – Champaran District – Found not uncommon 100 years back – Mukherjee 1963: 46 – map A 309

1870s – Darbhanga near Nepal border – William Gordon Gordon-Cumming (1843–1930) shot rhino – Money 1893: 616 – 28.25 – map A 312

1881 – Darbhanga near Nepal border – Rhino hunted by Maharaja of Darbhanga – *Times of India* 1881-04-02; *Homeward Mail from India* 1881-04-25 – 28.25 – map A 312

1883 – Darbhanga near Nepal border – Maharaja of Darbhanga on shooting trip killed 2 rhino – *Times of India* 1883-04-20 – 28.25 – map A 312

1890s – Tribeni ghat (Triveni) – Jungle used to attract rhino – P. Choudhury 1960: 592 – map A 308

1899 – Champaran – Rhino extinct before start of 20th century – Lahan 1993: 2 – map A 309

1909 – Banks of Bagmati River north of Muzaffarpur – Rhino present – Manners-Smith 1909 – map A 311

1930 – Bagaha – Occasional footprint of rhino strayed from Nepal – M. Martin 1949: 97, 144 – map A 308

1936 – N.W. Champaran – Occasional stray visitor from Nepal – Harper 1945: 377; Shebbeare 1953: 144 – map A 309

1939 – Champaran – Last rhino killed, according to Mrs. Jamal Ara (d. 1987) – Ara 1948: 285, 1954: 68; Sinha 2011 – 28.26 – map A 309

1949 – Ramnagar jungle – Occasional stray rhinos from Nepal – Houlton 1949: 120 – map A 308

1952 – Champaran – 2 rhinos, strays from Nepal – Gee 1952a: 224 – map A 309

1960 – Hathimalkhanta forest (Gonauli) – Strays from Nepal – P. Choudhury 1960: 34 – map A 308

1960 – Balgangwa forest – Strays from Nepal – P. Choudhury 1960: 34 – map A 308

1960 – Rampur-Madanpur range – Strays from Nepal – P. Choudhury 1960: 34; Sinha 2004: 47 – map A 309

1960 – Champaran – Rhino killed – Sinha 2011; Mazumdar & Mahanta 2016: 25 – map A 309

1963 – North Champaran – Strays from Nepal – Talbot 1960: 191; Mukherjee 1963: 46 – map A 309

1977 – Champaran – Strays from Nepal – Shahi 1977: 2 – map A 309

1978-05-10 – Valmiki ws – Protection as Wildlife Sanctuary for 462 km² under notification no. S.O. Wildlife-12/78-2336 – 28.27 – map A 308

1982 – Valmiki – Male rhino wandered from Chitwan (Nepal). Caught and transferred to Patna Zoo (studbook 157, Raju) – Rookmaaker 1998: 99; Sinha 2011; Central Zoo Authority 2016 – map A 308

1990-03-06 – Valmiki Wildlife Sanctuary – Addition of 419 km² under notification no. S.O. 136 – 28.27 – map A 308

1990-12-05 – Valmiki National Park – Established in part of Valmiki ws with an area of 335 km² (notification no. S.O. 989) – 28.27 – map A 308

1994-03-11 – Valmiki National Park – Restructured as part of Project Tiger reserves, covering 840.26 km² of forest (notification no. Vanyaprani 11/94-303-E/Va.Pa) – 28.27 – map A 308

2004 – Bihar – Rhino no longer in Bihar – Sinha 2004

2011-03 – Valmiki Tiger Reserve – Female strayed from Nepal – Sinha 2011; Kumar 2013 – 28.27 – map A 308

2012-08-07 – Valmiki Tiger Reserve – Reserve divided into a core area of 589 km² and a buffer zone of 310 km² (notification no. Vanyaprani 22/08–608(E)/Va. Pa of 26 November 2013) – 28.27 – map A 308

2016 – Champaran – Strays from Nepal – A. Choudhury 2016: 154 – 28.27 – map A 309

Rhino Region 19: Bihar – Purnea east of Kosi River

1632 – Poroonia (Purnea) – Rhino found in herds of 30–40 animals – Mundy 1914, vol. 2: 171 – 28.7 – map A 317

1768-01 – Rangamatty – Major James Rennell (1742–1830) saw tracks – Phillimore 1945: 32 – map A 314

1810 – Purnea, southern marshy tract – Rhino killed by indigo planter "into whose premises he had fortunately thrust himself" – W. Hamilton 1828: 431; O'Malley 1911: 13; Buchanan-Hamilton 1928: 284 – 28.18 – map A 317

1811 – Pir Pahar, north bank of Ganges above Bhagalpur – Rhinos found – Buchanan-Hamilton 1930: 93 – map A 315

1813 – Purneah Dt. (Purnea) – Rhino killed by Charles B-d (Charles Buckland) – *Calcutta Gazette* 1814-06-09: 9 – map A 317

1814-05 – Indigo Factory at Rampore Kolassee (Kolasi) – Rhino, which had first alarmed people at another indigo factory to the west of Kolassee, was followed on foot and alone by Charles B-d (Charles Buckland), who was wounded by the animal in head and neck. The rhino moved to Rampore Kolassee where it was killed by the factory's owner M-e (? Alex Mackenzie) – *Calcutta Gazette* 1814-06-09: 9 – map A 317

1814 – Poornea (Purnea) – Rhinos cross the Ganges "infesting the low lands in the southern part of the district of Poornea" – *Calcutta Gazette* 1814-06-09; Sandeman 1868: 362 – map A 317

1840 – Koasee (Kosi) in Purnea – Rhinos killed 40–50 years ago – Baker 1887: 251 – map A 315

1846, 1850 – Bhagalpur District – Walter Stanhope Sherwill (1815–1890) mentions rhino – Sherwill 1854: 5; Buchanan-Hamilton 1939: 253; Basu 1935: 148; Gardner 2016 – map A 316

1848 – Bhangulpore (Bhagalpur) – Rhino sometimes comes down to river – Anon. 1848: 144 – map A 316

1870-04 – Purneah – Rhino seen but not shot (with photo of hunting party led by John Joseph Shillingford) – Bruiser 1870 – 28.22, fig. 28.24 – map A 317

1871-04 – "D-pore jungle" in Purneah (Purnea) near Ganges; Coosey desert (Kosi) – Annual hunt of planter John Joseph Shillingford (1822–1867) of the Kolassy Factory, attended by his sons George William Shillingford (1844–1896) and Joseph Lay Shillingford (1845–1889), together with six other hunters. Male rhino killed by George W. Shillingford; remains donated to Indian Museum, no. GRM197 – Shillingford 1871; Bruiser 1871; Blyth 1872a; Sclater 1891, vol. 2: 202; P. Choudhury 1963: 14; Chakraborty 2004: 39 – 28.22, fig. 28.25 – map A 317

1871 – Coosey desert (Kosi) – Hunt securing 12 head of rhino [mistaken number; the list of other trophies same as those of Shillingford hunt of 1871] – Bradshaw 1878: 116 – map A 317

1874 – Koosey River (Kosi), Battabarree jungles – Rhino shot by Premnarain Singh, a young Zemindar, attended by James Inglis (1845–1908) – Inglis 1874, 1878, 1888: 556, 1892: 556 – 28.23, fig. 28.26 – map A 317

1870s – Korah (Korha) – One, maybe two, rhino shot by Alexander George Macdonald Wodschow (b. 1849) – P. Choudhury 1963: 97 – 28.24 – map A 317

1870s – Korah (Korha) – Korah known as "Gena-bari" after rhino shot [now Gerabari] – P. Choudhury 1963: 97 – 28.24 – map A 317

1878 – Koosee (Kosi) north of Bhaugulpore (Bhagalpur) – Occasional rhino in tracts with tall elephant grass – Inglis 1878: 104; Graham 1878, vol. 1: 268 – map A 315

1880 – Purnea – Rhinos used to be shot about 30 years ago – K. Sinha 1916: 6, 1917: 7 – map A 317

1880 – Kosi River – Rhinos used to occur 50–60 years ago – Harper 1945 after Govt. Bihar 1936; Houlton 1949; Shahi 1977 – map A 315

1880 – Kosi River – Rhinos occur – Schlagintweit 1880, vol. 1: 229; Trouessart 1898, vol. 2: 754 – map A 315

1880 – Purnea – Raja Bahadur Kirtyanand Sinha (1880–1938) knew of an elephant still living in 1916, once charged by rhino; its entrails came out, but was treated and recovered – K. Sinha 1916: 6 – map A 317

1880s – Cosi (Kosi) River – Rhino killed in annual shoot of Maharaja of Durbhanga – Money 1891 – 28.25 – map A 315

1880s – Bhagulpore (Bhagalpur) Dt. – Presence of rhino during shoot of Maharaja of Darbhanga – Money 1891 – 28.25 – map A 316

1886 – Purneah (Purnea) Dt. – Skull of male *R. unicornis* donated by Joseph Lay Shillingford (1845–1889) in Bombay Natural History Society – Bombay NHS 1886 – 28.22 – map A 317

1887 – Koasee (Kosi) in Purnea – Rhinos now absent – Baker 1887: 36,251; S. Sinha 2004: 47 – map A 315

1890 – Purnea – Hunt by Felice Scheibler (1856–1921); no rhino – Scheibler 1900 – map A 317

1892 – Purnea – Rhino present – Sanyal 1892: 132 – map A 317

1893 – Purnea country – Edward Braddon (1829–1904) went once on unsuccesful rhino hunt – Braddon 1893, 1895: 75 – map A 317

1900 – Saharsa District – Rhinos used to come from Nepal – S. Sinha 2004: 47 – map A 315

1917 – Purnea – Former occurrence. No longer present – K. Sinha 1917: 7 – map A 317

1948 – Purnea – Rhinos wander south from Nepal terai – Ara 1948 – map A 317

1954 – Kosi River – Rhino sanctuary should be set up in Bihar – Ara 1948, 1954 – map A 315

1959 – Kosi River – Last rhinos killed at start of Kosi River Barrage project, 1959–1963 – Seshadri 1969: 93 – map A 315

∴

TABLE 28.33   List of persons who were involved in shooting or capturing rhino in Bihar

Person	Date	Locality. Type of interaction
Kumaragupta I (414–455)	5th cent	Ganges valley. Hunts rhino
Humffreys Cole of the Patna Factory	1738	Patna region. Sends 1 rhino to UK
Henry Vansittart (1732–1770)	1763	Bihar. Sends 1 rhino to UK, not arrived
Louis (Lewis) Felix DeGloss	1765	Munger. Finds tracks
Louis (Lewis) Felix DeGloss	1766	Rohtas (Sone River). Finds tracks
Louis (Lewis) Felix DeGloss	1767	Gandak River. Finds tracks
James Rennell (1742–1830)	1768	Rangamatty. Finds tracks
William Lally (1761–1803)	1780s	Bagaha. Sees elephant-rhino fight
Maharaja Ramchandra Singh Deo I of Patna (ruled 1765–1820)	1787	Munger. Attacked by rhino
Thomas George Williamson (1759–1817)	1788	Derriapore. Rhino seen goring horses
Indigo planter (unnamed)	1810	Purnea. Kills 1
Charles Buckland	1813	Purnea. Kills 1
Alex Mackenzie, Charles Buckland	1814	Kolasi. Kill 1
Party of hunters	1815	Hurdeen. Kill male, female, and young
Charles D'Oyly (1781–1845)	1829	Patna. Hunt rhino
H.C.'s friend M.	1836	Baikathpur. Attacked by rhino
Rohilla	1848	Bettiah. Observes tame rhino in Palace
Party from Gorruckpore	1848	Bettiah. Kill 1
John Joseph Shillingford (1822–1867), Bruiser	1870	Purneah. Observe 1 rhino
George William Shillingford (1844–1896)	1871	Purneah. Kills 1
Premnarain Singh, James Inglis (1845–1908)	1874	Koosey River. Kills 1
Alexander George Macdonald Wodschow (b. 1849)	1870s	Korha. Kills 1 or 2
William Gordon Gordon-Cumming (1843–1930)	1870s	Darbhanga. Kills 1
Maharaja of Durbhanga	1880s	Kosi River. Kills 1
Maharaja of Durbhanga	1883	Darbhanga. Kills 2
Joseph Lay Shillingford (1845–1889)	1886	Purneah. Donates 1 skull of rhino
Anonymous	1939	Champaran. Kills 1
Anonymous	1960	Champaran. Kills 1
Patna Zoo	1982	Valmiki. Capture 1 stray from Nepal
Satya Sinha	2011	Valmiki. Observes 1 stray from Nepal

Listed in chronological order for the main event.

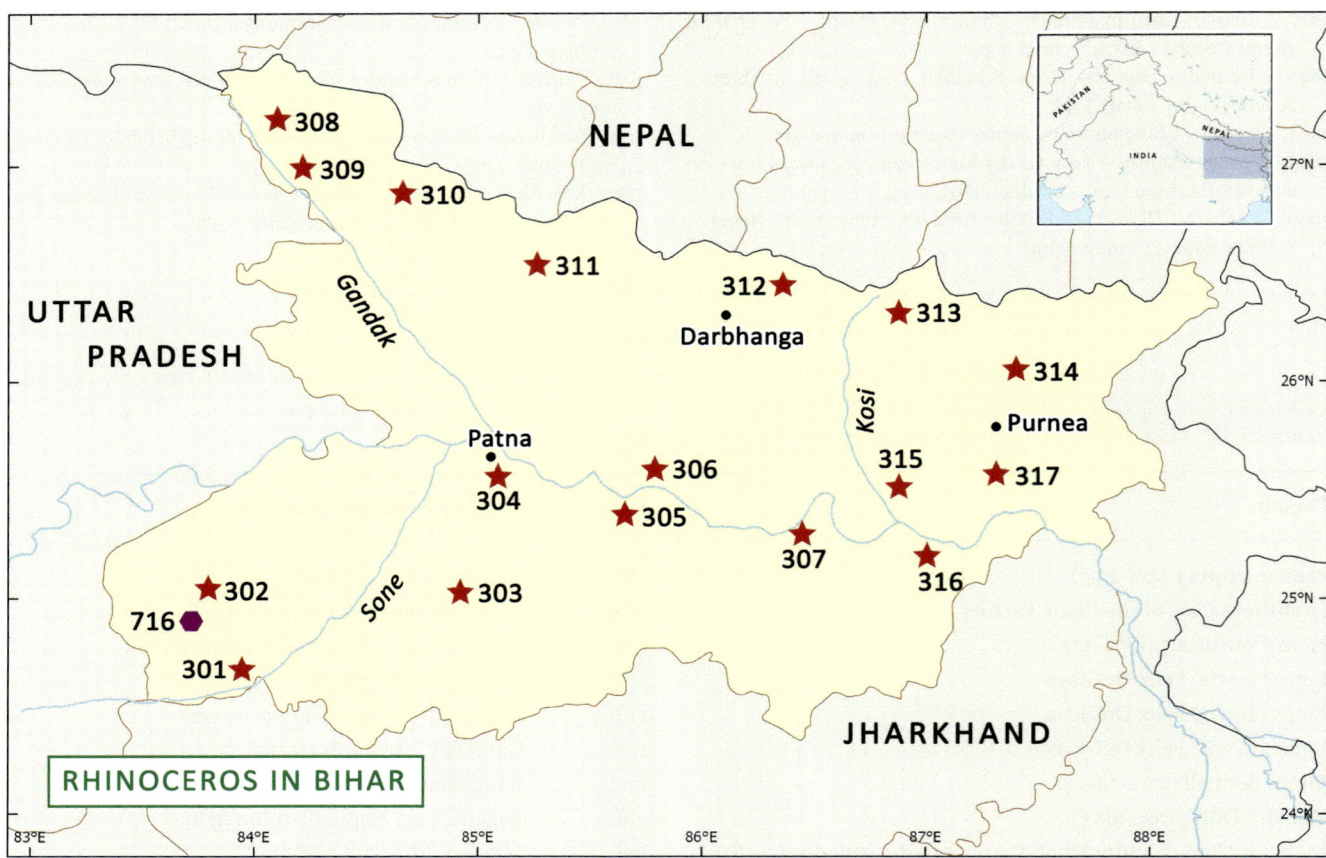

MAP 28.11   Records of rhinoceros in Bihar. The numbers and places are explained in Dataset 28.27
⭐ Presence of *R. unicornis*

...

Dataset 28.27: Localities of Records of Rhinoceros in Bihar

The numbers and places are shown on map 28.11.

Rhino Region 16: Bihar – South-West

A 301   Berealpour, near Rottasgur (Rohtas) – 24.58N; 83.90E – 1766-12-26

P 716   Kaimur. Kaimur range: Badki Goriya, Mithaiya – 24.65N; 83.60E

A 302   Camundi (near Bhabua) 25.05N; 83.60E – 6th cent.

A 303   Gaya Dist. – 24.80N; 85.00E – undated, 1st cent.

Rhino Region 17: Bihar – Ganges Downstream from Patna to Munger

A 304   Pataliputra - 25.75N; 85.00E–3rd cent.

A 304   Murtaziganj (Patna) – 25.60N; 85.05E – 100 BC

A 304   Pattana (Patna) – 25.60N; 85.05E – 1632, 1670, 1738, 1838

A 304   Magadha kingdom, Bihar – 1st cent.

A 304   Sonepore – 25.70N; 85.17E – 1893

A 304   Bykontpore (Baikathpur) – 25.50N; 85.38E – 1836

A 305   Geyaspur – 25.35N; 85.95E – 1800

A 305   Derriapore – 25.35N; 86.00E – 1788

A 306   Between Piprah and Ganduk River – 25.45N; 86.00E – 1767

A 307   Monghyr (Munger) – 25.37N; 86.47E – 1660, 1765, 1787

Rhino Region 18: Bihar – Champaran and Darbhanga

A 308   Balgangwa forest – 27.34N; 83.88E – 1960

A 308   Tribeni ghat (Triveni) – 27.44N; 83.93E – 1890s

A 308   Hathimalkhanta forest (Gonauli) – 27.34N; 83.98E – 1960

A 308   Valmiki Tiger Reserve – 27.30N; 84.10 – 1982, 2011

A 308   Ramnagar – 27.20N; 84.16E – 1949

A 308   Bagaha – 27.20N; 84.16E – 1780s, 1930

A 309   Rampur-Madanpur range – 23.20N; 84.00E – 1960

A 309   NW Champaran – 27.20N; 84.16E–1860s, 1899, 1936, 1939, 1952, 1960, 1977, 2016

A 309   Bettiah – 26.81N; 84.51E – 1848

A 310   Sarun Dt. (Bettiah and Champaran) – 1820

A 310   Rhinoceros Wilds (between Bettiah and Raxaul) – 26.80N; 84.80E – 1768

A 310   Hurdeen near Baraguree – 26.90N; 84.80E – 1815

A 310   Bihar north of Patna – 26.70N; 84.95E–1665–1672, 1785, 1829

A 311   Bagmati River north of Muzaffarpur – 26.65N; 85.30E – 1909

A 312   Darbhanga, Nepal border – 26.50N; 86.10E – 1870s, 1881, 1883, 1888

Rhino Region 19: Bihar – Purnea East of Kosi River

A 313   Kosi River (north) – 26.30N; 86.95E – 1877, 1948, 1954, 1959
A 314   Rangamatty – 26.00N; 87.73E – 1768
A 315   Saharsa District – 25.90N; 86.60 – 1900
A 315   Kosi River, Purnea – 25.50N; 86.80E – 1878, 1880, 1887

A 315   Pir Pahar – 25.50N; 86.80E – 1811
A 316   Bhagalpur District – 25.20N; 87.00E – 1846, 1848, 1850, 1880s
A 317   Purnea, Poroonia – 25.50N; 87.25E – 1632, 1814, 1880, 1886, 1890, 1892, 1893, 1916, 1948
A 317   Coosey desert, Purnea near Ganges – 25.50N; 87.25E – 1871
A 317   Purnea, southern marshy tract – 25.50N; 86.80E – 1810
A 317   Battabarree jungles (lower Koosey R.) – 25.50N; 86.80E – 1874
A 317   Korah (Korha) – 25.62N; 87.40 – 1875
A 317   Rampore Kolassy (Kolasi), near Katihar, 25.58N; 87.50E – 1814

28.2   The Ancient Capital of Pataliputra

During the reign of Chandragupta Maurya (321–298 BCE), animal combats were a favourite diversion and included fights of bulls, rams, elephants and rhinos. They took place in the ancient capital Pataliputra, then situated just north of Patna where the old course of the Ganges and the Son Rivers meet.

28.3   Rhino-Slayer Coins of Kumaragupta I

Many coins were produced during the Gupta Empire showing the kings engaged in various activities (Raven 2012, 2019). The gold coins known as the 'Rhinoceros-Slayer' type date from the reign of Kumaragupta I (414–455). The obverse (front) of these coins shows the king on horseback holding a sword in his right hand, attacking a rhino with its head turned back. The reverse shows the goddess Ganga with a lotus in her right hand. The legend reads "Bharta khadgatrata Kumaragupto jayatyanisam" (Ever victorious is the Lord Kumaragupta who is protector by sword and protector from rhino), using the Sanskrit word *khadga* meaning both 'sword' and 'rhinoceros.' The first four coins of this type were discovered in the Bayana hoard in 1946 in Rajasthan (Altekar 1954, 1955, 1957; Nagar 1949). A well-preserved coin is in the British Museum, no. 1955,0407.1 (Walker 1957; fig. 28.2). Examples

FIGURE 28.2   Gold coin of the 'Rhinoceros-Slayer' type from the period of Kumaragupta I in the 5th century. Diameter 18 mm, weight 7.91 gram

WITH PERMISSION © ASHMOLEAN MUSEUM OF ART AND ARCHAEOLOGY, OXFORD, NO. HCR6580 of this type are relatively rare, as only 16 have been documented so far, and are attributed to Mint Group 8 (Ellen Raven, pers.com. 2020-09). The rhino is clearly depicted with one horn and circular notches on the skin. Given the accuracy of the depiction, it has been suggested that Kumaragupta himself engaged in rhino hunting. As the rhino is best known from Assam, Mukherjee (1955) and Chinmulgund (1955) suggested that this coin commemorates the conquest and annexation of Kamarupa (Assam) during Kumaragupta's reign. However, the rhino also occurred across many parts of the Ganges valley within the borders of the Gupta empire in the 5th century. (Paragraph reviewed by Dr Ellen Raven.)

28.4 Punch-Marked Coins with a Rhinoceros

The rhino, or definitely an animal with a protrusion on the nose, is seen in the small punch-marked coins which belong to a period from 400 BCE. They have been found in several Indian localities, including the Gaya District of Bihar (Smith 1906a), in southern Bihar during the ancient Magadha kingdom (Gupta & Hardaker 2013) and in Odisha (Mishra 2003: 199). Theobald (1890) confirmed that the rhino was only rarely found on these coins and referred them to *R. sondaicus* rather than *R. unicornis*. This conclusion is unwarranted because the punch-marked coins are far too small to allow a depiction in enough detail to show the characteristics of different species of rhino.

28.5 Rhino Hunts of King Puskara

The 6th-century King Puskara, son of Lord Varuna, is said to have had a particular fondness for hunting rhinos (Sharma 2002). According to a passage in Bāṇabhaṭṭa's *Harṣacarita* of the 7th century, the King was killed while he was engaged in this particular pastime. Puskara's capital was Camundi, identified as Candamundi (Cheyenpur) near Bhabua in SW Bihar. It is not specified where his hunts took place.

28.6 The Jain Saviour Śreyāṅsanātha with His Rhinoceros Emblem

Born in Sarnath in Bihar in an early age, Śreyāṅsanātha, the 11th Tirthankara (saviour, spiritual teacher; literally: ford-maker) of the Jain religion was linked to the rhinoceros which was his *cihna* (emblem, cognizance) since at least the 5th century (15.2). The animal appears in representations both in statues and in miniatures only from the 14th century onwards (Bautze 1985: 416; Glasenapp 1925: 491, pls. 22–23). These were found in various locations across the north and west of India, especially in Uttar Pradesh, Rajasthan and Gujarat.

For a zoologist, it is unlikely that these representations of the rhino in Jainism are linked to the actual presence of rhinos in a particular location, while reflecting the continued significance of the animal in religious and cultural contexts (dataset 28.28).

• • •

Dataset 28.28: Some Examples of Artistic Expressions Linking Śreyāṅsanātha to a Rhinoceros

The following selection is mentioned in western literature.

5th cent. (Gupta period). – Udayagiri and Khandagiri caves, Odisha – Jain tradition. Rhino depicted as *cihna* on pedestals associated with Śreyāṅsanātha, the 11th Tirthankara – Illustrated: Mohapatra 1975: 360, pls. 89 f. 2, 102 f. 2, 111 f. 10.

6th cent. (Gupta period) – No locality – Jain tradition. Rhino as cognizance on pedestal of Śreyāṅsanātha, who stands naked with an attendant on either side. Image 1.3 × 1.1 ft (40 × 33 cm). Preserved in Sarnath Museum – Bautze 1985: 417; Sahni 1914: 328 no. G62.

6th to 8th cent. – No locality – Jain tradition. Rhino mentioned as *cihna* of Śreyāṅsanātha, in the Tiloyapannati by Yativrsabha – Bautze 1985: 417.

10th cent. – Jain stupa at Kankali Tila, Mathura, Uttar Pradesh – Jain tradition. Freeze with rhino as symbol of Śreyāṅsanātha – Smith 1901b, vol. 2: 40, pl. 80; Manuel 2007: 235, 2008: 36 – map M851 – 18.2, fig. 28.3.

10th–11th cent. – Sahet-Mahet (ancient Sravasti), district Gonda, Uttar Pradesh – Jain tradition. Rhino depicted in centre of pedestal to a Panca-tirthi sculpture of Śreyāṅsanātha in sandstone. Size 79 × 46 × 14 cm. Preserved in Lucknow Museum, Antiquity no. J-856. – Bautze 1985: 417; Shah 1987: 147; V. Kumar 2019: 777, fig. 734 – map M854 – 16.2.

11th cent. – Cedi area, Madhya Pradesh – Jain tradition. Rhino depicted as *cihna* of Śreyāṅsanātha. Preserved in Nagpur Museum – Bautze 1985: 417; Shah 1987; B.C. Bhattacharya 1974: 47, pl. 16 (illustrated).

11th cent. – Rajasthan – Jain tradition. A rhino is found on the wooden cover (*patli*) of a sacred manuscript of the 13th century. The animal faces left and shows a small horn and an unnatural indication of the skin-folds. Preserved in the Jaina Bhandara at Jaisalmer – Bautze 1985: 416. Illustrated: Nawab 1959, pl. v, 1980: 11, pl. 82; A. Ghosh 1975: 399, pl. 268A – fig. 28.6.

1228 – No locality – Jain tradition. Rhino mentioned as *cihna* of Śreyāṅsanātha. In: Digambara text *Pratisthasaroddhara* dated samvat 1285 – Bautze 1985: 417.

14th cent. – Rajasthan – Jain tradition. Rhino depicted as *cihna* of Śreyāṅsanātha. Three statues in white marble. Preserved in Jaina temple at Sanganer, Rajasthan – Illustrated: Bautze 1985: 417, fig. 3 – figs. 28.4, 28.5.

Undated – No locality – Jain tradition. Rhino depicted in pedestal of a Digambara statue of Śreyāṅsanātha – Illustrated: Bose 2018b: 54.

Undated – No locality – Jain tradition. Rhino depicted in pedestal of a Shivetambara statue of Śreyāṅsanātha – Illustrated: Bose 2018b: 54.

1516. – No locality – Jain tradition. Miniature, 3 cm in height. Part of an illustrated manuscript of the Aranyakaparvan, 3rd book of the Mahabharata. Preserved in Asiatic Society, Mumbai – Bautze 1985: 418; Das 2018a: 82.

1540 – Uttar Pradesh – Jain tradition. Rhino shown roaming in a forest with other animals . Folio of illustrated text, the *Mahapurana* of the Digambaras. Painted at Palam near Delhi. Preserved in Sri Digambara Jain Atisaya Kshetra, Jaipur – Bautze 1985: 419; Khandalavala & Chandra 1969: 76, col.pl. 19b.

16th cent. – No locality – Jain tradition. Rhino depicted on a folio of a dispersed set of the *Bhagavata Purana*, illustrating Krishna in a chariot shooting at animals including rhino, but not mentioned in text – Bautze 1985: 419; Khandalavala & Mittal 1974.

FIGURE 28.3
Rhino on a frieze from the Jain stupa at Kankali Tila, Mathura, Uttar Pradesh
VINCENT A. SMITH, *THE JAIN STUPA AND OTHER ANTIQUITIES OF MATHURA*, 1901, PL. 80 FIG. 3

28.7   Peter Mundy's Visit to Bihar in 1632

A rhinoceros was first reported in Purnea by the English merchant Peter Mundy (1596–1667), based upon information received at Patna in November 1632. According to hunters employed by the ruler, "rinoserosses, heere called Ghendas" were so common in Poroonia (Purnea) that herds of up to 30 or 40 could be found. The animals were hunted for their skins and horns, which at the time were used in the manufacture of bracelets worn by women. Apparently Mundy himself never saw a rhino, either captive or wild.

28.8   The Terrifying Woods behind Munger

The Venetian explorer Niccolao Manucci (1638–1717) spent much of his life at the Mughal court. Shah Shuja,

the second son of Emperor Shah Jahan and Empress Mumtaz Mahal, retreated to Muguer (Munger) after a battle with Mir Jumla (1591–1663) in 1659. This town was known then as "the Key of the Kingdom of Bengal because it is at the foot of hills, and near it are extensive jungles, called by the dwellers there Burianguel, that is 'Terrifying woods' because in them are many wild beasts, tigers, rhinoceroses, wild buffaloes and other animals" (Manucci 1907: 334).

28.9   Bowrey's Drawing at Patna in 1670

Thomas Bowrey (1659–1713) reached Pattana in 1670. The place was surrounded by forests where he heard of the existence of many "of those deformed animals called Rhinocerots" which were often tamed (Bowrey 1905). The manuscript left by this young English traveler includes a

FIGURE 28.4   Rhinoceros *cihna* on the pedestal of a depiction
of Śreyāṁsanātha in Sanganer (Rajasthan),
*c.*14th century
PHOTO: JOACHIM K. BAUTZE

drawing of one of these rhinos, although maybe just a bit too clumsy in showing the armour to suspect that it was a portrait of an animal which he might have seen there (fig. 28.7).

28.10   Rhinoceros Shipped to Europe in 1739

Patna had a factory or trading centre of the British East India Company from the 17th century. In 1738, Humffreyes or Humphrey Cole, the Chief of the factory 1732–1743, procured a young male rhino, most likely in the forest north or east of the town. The animal was taken to Kolkata, where it was sketched by Thomas Gregory-Warren, a gunner of Fort William (Carwitham 1739), which may well be the ink drawing now preserved in the 'Douglas Collection' (fig. 28.9). The animal was shipped to London on board the *Lyell*, captain James Acton, arriving on 1 June 1739. It was exhibited in Eagle Street near Red Lion Square in London, from 9 AM until 8 PM on payment of half-a-crown. A large broadsheet was issued on 10 October 1739 inscribed to Cole "for the Favour he has done the Curious in sending it over to England" (fig. 28.8). The unknown owner subsequently exhibited this male rhino at the Golden Cross on Charing Cross in November and December 1739, at the Bell Inn on Haymarket in November 1740, and then on tour in Stamford in March 1742 and Lincoln to Gainsborough in April 1742.

FIGURE 28.5
Detail of the rhinoceros *cihna* shown in fig. 28.4
PHOTO: JOACHIM K. BAUTZE

FIGURE 28.6   *Patli* or wooden cover of a manuscript sacred in the Jain tradition, dated to the 13th century, kept in Jaina
Bhandara at Jaisalmer
SARABHAI MANILAL NAWAB, *THE OLDEST RAJASTHANI PAINTINGS FROM JAIN BHANDARS*, 1959, PL. 5

FIGURE 28.7
"The Rhinocerot" drawn by the English traveler Thomas Bowrey in Pattana (Patna) in 1670
WITH PERMISSION © BRITISH LIBRARY: MSS EUR D782, P. 97

Probably the third rhino to reach London alive, it made scientific history because it was carefully studied by the physician James Douglas (1675–1742) and his assistant James Parsons (1705–1770). The latter read a detailed report to the Royal Society on 9 June 1743, accompanied by three of his own drawings (Parsons 1743), also available in German (1747) and French (1760). The description of the rhino by Parsons remained the standard for almost the rest of the century, extensively copied for instance in the influential animal encyclopedia by the French Count de Buffon (1764), which again was reprinted and translated for another century.

Parsons made a series of drawings while he examined the rhino in London in July 1739 (Rookmaaker 1978b). These were assembled by Douglas and when he died on 2 April 1742, his papers passed to the great surgeon and collector William Hunter (1718–1783) who then was tutoring his son. In total, 13 sheets with sketches were preserved in the Douglas Collection, now in the Hunterian Library, University of Glasgow (fig. 28.10). Two were done in pencil, the others are in red chalk. Often there is a rough sketch as well as a more finished drawing for each position, like side views, front views or hind views. Three further sketches signed "IPS MD del" are in the Sloane Collection, British Museum (5261, nos. 47–49) showing a side-view, the penis, and the hooves. Another sheet with drawings is in a private collection (fig. 28.11). These sketches by Parsons were used to produce three engraved plates illustrating his paper presented to the Royal Society and published in the *Philosophical Transactions*, showing a lateral view, two oblique front views and a series of seven miscellaneous rhino parts including the feet and the penis of the

1739 animal. Parsons (1743: 528) referred to two oil paintings of the rhino seen from the side (fig. 28.12). One was then owned by the physician Richard Mead (1673–1754) in Bloomsbury House, London, but cannot be identified in the catalogues of the sales of his collection in 1755. The other was in the hands of his widow Elizabeth Reynolds (d. 1786) and then also disappears (Nichols 1812: 487). I remember waiting in one of the rooms of the mammal department of the Natural History Museum on Cromwell Road in London in July 1975, when my eye fell on a painting which I then immediately knew to be one of those painted by Parsons, previously unrecognized and of unknown provenance (Thackray 1995: 57).

28.11 Capture of a Rhino by Vansittart in 1763

After the British Governor of the Bengal Presidency, Major-General Robert Clive (1725–1774) returned home from his second tour in India in 1760, he heard that King George III wanted to see an elephant alive. He must have written to his colleague, Henry Vansittart (1732–1770), English Governor of Bengal from 1759 to 1764, for appropriate action. In the first months of 1763, Vansittart traveled to Mongeer (Munger) and Patna in Bihar, and on his return he wrote to Clive, in a letter dated Fort William 25 February 1763:

> Upon what you mentioned to me the King's Desire to see an Elephant, I have provided two small ones, such as the Europe Captains told me they could manage to carry. One goes in the Clinton, the other

PINOKEPOC.

An Exact Figure of the
RHINOCEROS
That is now to be Seen in
LONDON.

Inscrib'd to Humffreyes COLE Esq.
Chief of The Hon.ble East India Company's Factory at PATNA in the
Empire of The Great MOGUL for
the Favour he has done the Curious
in Sending it over to England.

Publish'd October 10. 1739.

FIGURE 28.8
Broadsheet of a male rhinoceros in London in
October 1739. Engraved by Gerard van der Gucht
(1696–1776). The preliminary drawings were
sketched by the physician James Parsons
(1705–1770). There is a heading PINOKEROC in
quasi-greek style, and text below the drawing:
"An Exact Figure of the Rhinoceros That is now
to be seen in London. Inscrib'd to Humffreyes
Cole Esq. Chief of the Hon.ble East India
Company's Factory at Patna, in the Empire of
the Great Mogul, for the Favour he has done
the Curious in sending it over to England. –
Published October 10, 1739." Size of sheet
31.5 × 40 cm, of depiction 21.5 × 33 cm
COLLECTION KEES ROOKMAAKER

FIGURE 28.9
Ink-drawing possibly showing the rhino of
1739 when in Kolkata, drawn by Thomas
Gregory-Warren
WITH PERMISSION © UNIVERSITY OF
GLASGOW LIBRARY, HUNTERIAN LIBRARY
(DOUGLAS COLLECTION): SP.COLL. AV.1.17,
NO. 26

in the Hardwicke. I have also got two Houdas or
Castles [howdahs] making for them at Patna which,
I hope will be in time to be sent in the Ashburnam.
In that ship I send a Persian Mare which is said to be
of a very fine Blood, and all the Judges assure me she
will be much valued for a Breeder and to compleat

the Collection, I send a Rhinoceros in the Drake. –
All these I have directed to be delivered either to you
or my Brother, intending they shall be presented to
His Majesty by you both in such Form & Manner as
you shall judge most proper (British Library, Mss Eur
G37/30/1 ff. 8–10).

FIGURE 28.10
Preliminary drawing (rhino facing right) by James Parsons for a broadsheet showing a male rhinoceros in London in 1739
WITH PERMISSION © UNIVERSITY OF GLASGOW LIBRARY, HUNTERIAN LIBRARY (DOUGLAS COLLECTION): SP.COLL. AV.1.17, NO. 43

FIGURE 28.11
Sketches of the rhinoceros seen in London in 1739 by James Parsons, similar to the plates published in the *Philosophical Transactions of the Royal Society* in 1743
WITH PERMISSION © COLLECTION JIM MONSON

This is followed by another letter from Vansittart to Clive on 20 April 1763, with some further confirmation:

> In my letter of the 25th February by the Clinton, Drake & Hardwicke, I mentioned my sending two Elephants, one in the Clinton, and one in the Hardwicke, a Rhinoceros in the Drake, and a Persian Mare in the Ashburnam, to be presented through your hands and my Brothers to His Majesty if you thought them fit. – The two Elephants went down to Ingolee but that intended for the Clinton proved too large and she could not take it on board. Neither are the Castles yet arrived from Patna, they shall be sent next season (British Library, Mss Eur G37/30/1 ff. 48–49).

A Persian horse sent on the *Ashburnam*, an elephant on the *Hardwicke*, and a rhino on the *Drake*. When the *Hardwicke*, captain Brook Samson, arrived in London on 21 August 1763, the young male elephant was presented to the King by Henry's elder brother Arthur Vansittart (1767–1804) and Lord Clive (Malcolm 1836: 256), or, following a contemporary broadsheet (British Library 1914,0520.686), donated by Brook Samson on 27 September 1763. The animal was kept in Buckingham House Gate, and died on 24 July 1776 (Plumb 2010; Grigson 2016). The ship *Drake*, captain John Smith, also arrived together with the *Hardwicke* in August, but there is no mention in the newspapers or elsewhere that a rhino was on board. It is likely that the animal had died in Kolkata, or otherwise during the voyage. Nevertheless, from Vansittart's letters it seems clear

FIGURE 28.12   James Parsons (1705–1770), Rhinoceros in London, 1739. Oil on canvas, 122 × 147 cm. Found in a storeroom
                of the British Museum in 1878 and transferred to the Natural History Museum, then thought to be from the
                collection of Hans Sloane
                WITH PERMISSION © THE TRUSTEES OF THE NATURAL HISTORY MUSEUM, LONDON

that he obtained it in Patna, and therefore it must have been captured either in the forest north of Patna or in the Rajmahal hills to the east.

28.12   A Remarkable Record from the Sone River in 1766

This southernmost written record of a rhinoceros in central India dates from 1766. As part of the large-scale geographical survey of India led by Major James Rennell (1742–1830), the region south of Patna was traversed by the Polish Major Lewis Felix DeGloss. He left a journal of which excerpts were selected by Phillimore (1945). De Gloss left Patna on 25 November 1766 to survey the country further south, traveling along the River Zoane (Sone) to Doudnagar (Daudnagar) and Rottasgur (Rohtas). After halting at the Fort, he attempted to continue southwards, but he found that progress was almost impossible due to heavy jungle full of tigers, deer, and other game.

On 25 December 1766 he reached Berealpour, and on the next day he reported seeing the track of 'Rynosserus.' It is quite certain that he was in the area between Rohtas and the confluence of the Sone and Koel Rivers, probably on the left bank.

28.13   Two Animals Seen on the Gandak River in 1767

Lewis Felix DeGloss continued his contributions to the survey of India supervised by James Rennell. On 16 October 1767 he set out from Monghyr traveling westward along the north bank of the Ganges River towards Hajipur (Patna) and the confluence of the Gandak River. In his journal of 26 October, he recorded that after leaving the village of Piprah his progress was slow due to high reed jungles. But he heard that in the previous year, people had placed a trap and captured two rhinos in this vicinity. There is no information about the disposal of these animals.

FIGURE 28.13 Jean Baptiste Joseph Gentil, *c.*1785, map of 'Bear' (Bihar), including a rhinoceros in the Bettiah region
WITH PERMISSION © BRITISH LIBRARY, INDIA OFFICE LIBRARY AND RECORDS: ADD.MS. OR. 4039, FOLIO 17

FIGURE 28.14 Detail of the rhinoceros figure on the map of Bihar prepared for J.B.J. Gentil *c.*1785, from fig. 28.13
WITH PERMISSION © BRITISH LIBRARY

28.14 Rhinoceros on the Map of Bihar by Gentil Dated 1785

In the Atlas produced by the French army officer Jean-Baptiste-Joseph Gentil (18.8), the map of 'Bear' (Bihar) shows the image of a single-horned rhino in the region around Bettiah (figs. 28.13, 28.14). Another map drawn by English cartographer Thomas Jefferys (1719–1771) in 1768 has the "rhinoceros wilds" in a very similar general locality (fig. 31.3).

28.15 Hogendorp's Accident in 1787

Dirk van Hogendorp (1761–1822), Director of the Dutch factory in Patna, had an accident when traveling by boat from Kolkata towards Patna in 1787 (Hogendorp 1887). Stranded at Mongheer (Munger), his host Colonel Barrington warned him that the overland journey was hazardous on account of an enraged rhino which lately had attacked everything on this route. Hogendorp thought that the dangers were exaggerated and went on his way. After a few miles he encountered the Maharaja of Patna Ramchandra Singh Deo I proceeding by elephant to Kolkata, who stopped for a greeting and was told about the rhino. On his return to Patna, the Maharaja told Hogendorp that soon after their meeting his elephant was attacked by the rhino resulting in a terrible fight, fortunately without casualties.

28.16 The *Oriental Field Sports* of 1807

A series of anecdotes of life in India written by Thomas George Williamson (1759–1817) were combined with plates after his designs in the *Oriental Field Sports* of 1807. It appeared in several editions and both text and plates found their way into pirated popular magazines and books in the first half of the 19th century. Williamson had spent 20 years in the 3rd European Regiment of the East India Company army, starting as a cadet in 1778, promoted to Captain on 1 June 1796 and leaving in 1798, after his premature dismissal due to his criticism of the military establishment (Dodwell 1838: 270; VCPH 1935; Edwards 1980). The few sources about his life contain little detail about his postings in India, or his residence after his return to England, although he is known to have died in Paris in October 1817. Williamson had an expansive writing style, which is also evident in his notes about the rhino, which are entertaining to read but contain only limited historical detail.

Williamson included two observations of the rhino in one chapter. The first involving officers who witnessed an attack on their horses at Derriapore in 1788 was illustrated in the *Foreign Field Sports*, another companion volume of 1813 (28.17). It is unclear if Williamson ever saw a rhino himself, but I suspect that he may have been one of the officers involved in that incident. He also seems to speak with some authority about captive animals kept by three Englishmen in the 1780s: Augustus Cleveland (1754–1784) at Bhagalpur; Matthew Day at Dacca; and Mr. Young at Patna (5.4).

Williamson' second observation entitled "A Rhinoceros bayed by Elephants" was the subject of plate XI in the *Oriental Field Sports* (fig. 28.16). It illustrates how Major William Lally (1761–1803) was engaged in an elephant capture operation, probably in the 1780s, when he witnessed a fight between a rhino and a herd of elephants. The only locality provided in the text states that Lally had "arrived at the summit of a low range of hills." In another chapter, Williamson (1807b, vol. 1: 148, text to pl. 10) writes about "my worthy friend, the late Major Lally, when on command at Boggah, caught many [elephants] in this manner." Let me assume that Boggah, now Bagaha on the Gandak River in Champaran, is the locality of the fight depicted in plate XI.

The popularity of Williamson's anecdotes owes much to the decision of the publisher Edward Orme to combine them with prestigious colour plates executed by the English artist Samuel Howitt (1756–1822). Although the title of the work states that plates were executed after designs by Williamson, there is no evidence that this meant more than proposing the subject matter. The artist Samuel Howitt has become well-known for his depictions of natural history subjects in a variety of formats (Gilbey 1900, vol. 2: 36). His portrayal of the rhinoceros must have been based on specimens exhibited in London as Howitt never traveled to India. There is a sheet of sketches of a rhino's head signed by Howitt labelled "Studies from nature at Exeter 'Change" (fig. 28.15). The Exeter 'Change and the Lyceum were two adjacent buildings on the Strand in central London containing a menagerie owned by three successive proprietors: Thomas Clark (1737–1816), Gilbert Pidcock (1743–1810) and Stephen Polito (1763–1814). Three rhinos were kept here, known as Clark's Rhinoceros (shown 1790–1793), Pidcock's Rhinoceros (shown 1799–1800) and Polito's Rhinoceros (shown 1810–1814) after their owners (Rookmaaker et al. 2015). The rhino sketches most likely represent Pidcock's Rhinoceros, a young male imported between June and November 1799 and found dead in January or February 1800 (Bingley 1804).

The book combining the text by Williamson and the aquatinted plates by Howitt was first announced with the working title *The Indian Sportsman* as early as 1800 by the publisher John Debrett (1753–1822) located opposite

Burlington House, Piccadilly, London (Debrett 1801: 87). He advertised aquatint plates of 18 × 13 inches (45.7 × 33 cm), issued at 2 guineas per pair, with expected completion in April 1801. This was endorsed by a notice in February 1801 when Howitt (1801) listed 50 plates, 28 ready for inspection, and the remaining 22 "in a forward state, but not yet fit for exhibition." Among this second set, there were no. 49 'Rhinoceros attacking Horses at their Pickets', and no. 50 'Rhinoceros bayed by Elephants.' The latter preliminary aquatint may in fact have been the one owned by Leger Galleries (1976, no. 26; Rookmaaker et al. 2015, fig. 6): 'A rhinoceros hunted (or bayed) by elephants' (42.9 × 29.7 cm), signed by Howitt, originally from the collection of Henry John Reynolds-Moreton, 3rd Earl of Ducie (1827–1921). There is no evidence that Debrett actually published the work or that the plate with the rhino and elephants was available before 1806.

The production of this grand and expensive work of 40 plates was undertaken by Edward Orme (1775–1848) of New Bond Street, London with the new title *Oriental Field Sports*. It was first issued in twenty monthly installments with additional wrappers from June 1805 to December 1806. An early announcement was inserted in the *Morning Post* (London) on 1805-06-17, probably before any part was issued. Five installments were available in November 1805, together with a Prospectus describing the whole work (*Morning Chronicle* 1805-11-06). The earliest known copy of this Prospectus in the British Library dates from July or August 1806 when 15 installments had been issued and provides a detailed description of the proposed contents (Orme 1806). Each installment was priced at 1 guinea. Bound with this *Prospectus* is a printed page stating "January 1, 1807. This splendid work of Oriental Field Sports, is this day completed, in twenty numbers, at Twenty-one Shillings Each."

Lally's account of the fight between rhino and elephants was illustrated by Howitt on the 11th plate. As the total of 40 plates was divided over 20 installments, plate XI would have appeared in part 6. The date 1 January 1806 found on some of the early folio copies fits with this sequence. The folio-sized plate XI in this installment and in the first bound folio edition of 1807 was entitled "A Rhinoceros hunted by Elephants" (fig. 28.16), changed to "A rhinoceros bayed by Elephants" in the text chapters. Once included in the *Oriental Field Sports*, plate XI is found in different sizes according to the edition in which it was published. The second version of the plate found in the quarto and octavo editions has no title, but includes artist and publisher details, a plate number in the top-right corner and the date 1 August 1807 (fig. 28.17).

The *Oriental Field Sports* was sold in a variety of formats and editions, all with a long elaborate title. There were at least nine different editions, each with 40 plates, which are identified by date, publisher or size. There is reference to "undomesticated" animals in the title of the first folio edition, but corrected to "domesticated" afterwards. The artist's name is either spelled Howitt or Howett. Despite the apparent popularity of the plates and descriptions, no complete translation is known, and the book has remained rare in European continental libraries.

Editions of the Oriental Field Sports

1. First folio edition 1805–1806. Issued in 20 monthly instalments between June 1805 and December 1806 (or possibly January 1807). The titlepage would have been included in the final part (Orme 1806: 4). A copy consisting of 20 parts in original blue paper wrappers is in the Yale Center for British Art (Folio B 2010 5) and a sample is viewable on their website (in 2022). Plate XI may well have been sold separately as well, and is further described under no. 2 below. There are advertisements stating availability of 5 installments or parts (*Morning Chronicle* 1805-11-06), 10 parts (*British Press* 1806-04-18), and 18 parts "to be completed by Christmas" (*British Press* 1806-12-09).

2. First folio edition bound in one volume, dated 1807. Large oblong folio (46.7 × 58.4 cm), pp. i–ii, 1–150 (Williamson 1807a). Published in London, "Printed by William Bulmer and Co. Shakspeare Printing-Office, for Edward Orme, Printseller to His Majesty, Engraver and Publisher, Bond Street, the corner of Brook-Street." The title includes reference to "undomesticated animals" and "drawings by Samuel Howett." Plates show titles both in English and in French. Plate XI has the title "Rhinoceros hunted by Elephants" (left) and "Un Rhinoceros chassé par des Elephants" (right). The smaller inscriptions refer to Samuel Howitt as draughtsman, Henri Merke (1760–1814) as engraver, Edward Orme as publisher, with the date 1 January 1806 (fig. 28.16).

3. Edition dated 1807. Quarto (4to) format (33 cm), in 2 volumes, pp. xiv, 306 and iii, 239, [xi] (index) (Williamson 1807b). The plates are inscribed with names of artist and publisher but have no separate caption. Published in London, "Printed for Edward Orme, Printseller to His Majesty, New Bond Street." The title is shorter than in the first edition, refers to "domesticated animals" and "drawings by Samuel Howitt."

4. Edition dated 1808. Quarto (4to) format (28 cm), in 2 volumes, as no. 3 (Williamson 1808). Published in London, "Printed by W. Bulmer and Co. Cleveland-Row, St. James's; for Edward Orme, Bond-Street, the corner of Brook-Street, and B. Crosby and Co.

Stationers' Court." The title again refers to "domesticated animals", but "drawings by Samuel Howett." The added publisher is Benjamin Crosby (1768–1815). An advertisement in the *Oxford University and City Herald* (1808-12-24) refers to "a new edition, in royal quarto and imperial octavo, of the Oriental Field Sports." The *Norfolk Chronicle* (1809-02-11) repeats this with the date 6 February 1809, providing the price of the 4to edition as 5 guineas and the 8vo edition as 3 guineas. The text to plate XI in vol. 1, pp. 163–177 is headed "A Rhinoceros bayed by Elephants." The plate facing p. 163 does not repeat the title. In the List of Plates (vol. 1, p. xv) plate XI is referred to as "A Rhinoceros hunted by Elephants."

5. Edition dated 1808. Octavo (8vo) format, in 2 volumes, as no. 4, same page count but smaller page size (Williamson 1808). The plates are folded into the volume, with the same dimensions when opened as the quarto format. Issued simultaneously with the 4to edition (no. 4).

6. McLean edition dated 1819. Folio format (45 × 57 cm), in 1 volume, pp. viii, 146, iv (Williamson 1819c). Published in London, "Printed for Thomas McLean, Bookseller and Publisher, by Howlett, 50, Frith Street, Soho." Thomas McLean (1788–1875) was a publisher in Haymarket. Title as no. 4 with "Samuel Howett."

7. Second Orme edition, undated, probably 1819. Folio format (34.5 × 24 cm), in 1 volume, pp. xiii, 455, iii (Williamson 1819a). Published in London, "Published by Edward Orme, Printseller to His Majesty, New Bond Street." Printed by W. Lewis, 21 Finch-Lane Cornhill. Title as no. 3 with "Samuel Howitt."

8. Young edition dated 1819. Quarto (4to) format (33 cm), in 2 volumes, pp. xiv, 306, and iv, 239, xiv (Williamson 1819b). Published in London, "Printed for H.R. Young, 56, Paternoster-Row." Title as no. 3 with "Samuel Howett." Identified as "Second edition" below the volume number on the title-page.

9. Reprint of no. 7 dated 1828. Folio format (32.7 × 24 cm), in 1 volume, xii, 455, i pp (Williamson 1828). In London, "Published by Edward Orme, Printseller to His Majesty, New Bond Street." Plates are watermarked 1828.

Illustrations of Rhino in Oriental Field Sports

Williamson's anecdotes and Howitt's illustration were regularly pirated in popular magazines, often with little regard for the original features. The text on the rhino without illustration was reprinted (only missing two paragraphs) in an American magazine *Select Reviews of Literature* (Williamson 1811). A copy of plate XI is found as an uncoloured drawing, once in the collection of Ananda

Kentish Muthu Coomaraswamy (1912, pl. 22). There were at least eight published copies of the plate, probably quite a few more, mainly from the 1830s. These illustrations were often poor in quality, and the number of elephants is reduced from six to four and even to two, maybe because the animals were found to be too crowded in the original.

1. "Der Kampf des Rhinoceros mit dem Elephanten" or "Combat du Rhinoceros et de l'Eléphant" (Verm. Gegenstaende/Mélanges CCXXVI) in *Bilderbuch für Kinder* by Friedrich Justin Bertuch (1747–1822) – in German and French (Bertuch 1816, vol. 9, pl. 20), coloured plate like original with 6 elephants.

2. "Elephant attacked by a Rhinoceros" in *The Menageries: Quadrupeds* by Charles Knight (1791–1873) – in English (Knight 1833: 186), uncoloured plate with only 2 elephants (rhino faces left as in original). This figure is again found in his *Natural History* (Knight 1844: 195, uncoloured plate), also in Frost (1856: 179, small uncoloured plate).

3. "Combat entre un Rhinocéros et des Eléphants" in *Musée des Familles (Lectures du Soir)* for 1834 – in French (Howitt 1834: 297), uncoloured plate with 4 elephants (rhino faces left) (fig. 28.18).

4. "Combattimento tra il Rinoceronte e gli Elefanti" in *Album: Giornale Illustrata* 9 May 1835 – in Italian (Howitt 1835: 68–69), uncoloured plate with 4 elephants (rhino faces left).

5. "Rhinoceros attacked by Elephants" in *Saturday Magazine* 2 January 1836 – in English (J.H.C. 1836), uncoloured plate with 6 elephants, reversed (rhino faces right).

6. "O Rhinocerote atacado" in *O Panorama* 16 September 1837 – in Portuguese (Howitt 1837: 154), uncoloured plate with 4 elephants, reversed (rhino faces right).

7. "Das Nashorn im Kampf mit Elephanten" in *Heller-Blatt oder Magazin zur Verbreitung gemeinnütziger Kenntnisse* (Breslau) for 1840 – in German (Howitt 1840), uncoloured plate with 4 elephants (rhino faces left).

8. "Ein Rhinoceros von Elefanten angegriffen" in *Pfennig-Magazin für Belehrung und Unterhaltung* 30 August 1845 – in German (Howitt 1845: 273), uncoloured plate with 4 elephants, reversed (rhino faces right).

28.17 The *Foreign Field Sports* of 1813

When Thomas George Williamson wrote his *Oriental Field Sports* (28.16), he included in his text an account of an incident at Derriapore in 1788. The same is found

FIGURE 28.15   Samuel Howitt, "Studies from nature at Exeter
'Change", undated [?1799]. Sepia wash
WITH PERMISSION © PRIVATE COLLECTION

FIGURE 28.17   Rhinoceros attacked by elephants. Plate XI in
the quarto edition of Williamson's *Oriental Field
Sport*, 1808 (no. 3). There is no title, but inscribed
(lower left) Williamson & Howitt, (lower right)
J. Clark Etched, (middle) Published by Edwd
Orme, Bond Street, Augst 1st 1807 & by B. Crosby &
C^o Stationers Court
THOMAS WILLIAMSON, *ORIENTAL FIELD SPORTS*,
1808 (SECOND QUARTO EDITION), VOL. 1, PL. 11

FIGURE 28.16   "Rhinoceros hunted by Elephants" – "Un Rhinoceros chassé par des Elephants." Signed lower left: Saml Howitt del.
from the original design of Capt Thos Williamson, (lower right) H. Merke sculp., and (middle) Edwd Orme Excudit.
There is a round vignette between the English and French titles with number "N° XI." Below the title is text: Published
& sold Jany 1. 1806 by Edwd Orme, Printseller to His Majesty, 39 Bond Street, London
THOMAS WILLIAMSON, *ORIENTAL FIELD SPORTS*, 1807 (FIRST FOLIO EDITION ONLY), PL. 11

ATTAQUE D'UN RHINOCÉROS PAR DES ÉLÉPHANTS.

FIGURE 28.18
Illustration of a rhinoceros under attack by elephants, derived from the plate by Samuel Howitt in the *Oriental Field Sports* of 1807. The number of elephants is reduced to four only, and the landscape is drawn much simpler
MUSÉE DES FAMILLES: LECTURES DU SOIR, 1834, P. 297. COURTESY: JIM MONSON

ANECDOTE OF HUNTERS & RHINOCEROS.

Howitt del.   *Published & Sold Jan.t 1.st 1813, by Edw.d Orme, Bond Street, London.*   *Merke Sculpt*

FIGURE 28.19   Anecdote of Hunters & Rhinoceros. Howitt del. Merke sculpt. Published & sold Jan.1st, 1813 by Edw. Orme, Bond Street, London. Drawn by Samuel Howitt (1756–1822), engraver Henri Merke (1760–1814), publisher Edward Orme (1775–1848). The same plate is in the edition of 1819
SAMUEL HOWITT, *FOREIGN FIELD SPORT*, 1814

FIGURE 28.21 "Rhinoceros hunting. See Vaillant's second journey, vol. 3, page 40. Published by J. Wheble. Jany 1, 1799" Signed at lower left corner: Howitt. Vaillant is the French ornithologist François Levaillant
SPORTING MAGAZINE, OR MONTHLY CALENDAR, 1799, VOL. 13, FACING P. 161

FIGURE 28.20 "Shooting a Rhinoceros. Published May 31st. 1820 by J. Wheble & J. Pittman, 18 Warwick Square. London." Signed at lower left corner: Howitt. Illustrating the story by Thomas Williamson (unattributed) about a rhino on the south bank of the Ganges east of Patna
HOWITT, *SPORTING MAGAZINE, OR MONTHLY CALENDAR,* 1820, VOL. 6, P. 88

FIGURE 28.22 Watercolour of a rhinoceros by Samuel Howitt, ca. 1810–1812. Signed "Howitt" in lower left corner
WITH PERMISSION © COLLECTION MADAME MARIE-LOUISE ADAMS, PARIS

in abbreviated form in the *Foreign Field Sports* of 1813. Williamson stated that in 1788, two officers stationed at Dinapore (Danapur, near Patna) traveled along the Ganges downstream towards Monghyr (Munger). While camping at Derriapore on the south bank of the river, they saw a rhino attacking their horses. This was a ferocious animal standing 6 feet (183 cm) high at the shoulders which had been terrorizing travellers and villagers for some time. It may well have been the same animal which had troubled the Maharaja of Patna just a year earlier in 1787 (28.15). Eventually the animal was killed by a local hunter, reaping him a reward of £300 sterling offered by the authorities. Williamson must have known the officers, or maybe he was even one of them. The place Derryapour appears on a map of 1784 just west of Barh, on the south bank of the

Ganges. There was no illustration attached to the account in the *Oriental Field Sports.*

The *Foreign Field Sports* were published by Edward Orme in London in 11 monthly parts of 10 plates each in the course of 1813. Possibly included in the 6th part, there were two rhino anecdotes of "Hunters and a Rhinoceros" and "Rhinoceros hunting", the latter taken from the African adventures of the French traveller François Levaillant (1753–1828), best known for his splendid illustrated books about birds of the world (Rookmaaker 1989; Rookmaaker et al. 2003). The Derriapore incident described in the first anecdote is illustrated in an aquatint by Samuel Howitt

FIGURE 28.23   "Rhinoceros hunting", signed (lower left) C. D'Oyly delᵗ 11th Sept. 1829; (lower right) Behar
                Lithographic Press. Lithograph by Charles D'Oyly (1781–1845)
                D'OYLY, *INDIAN SPORTS NO. 2*

entitled "Anecdote of Hunters & Rhinoceros" (fig. 28.19), at first available both in coloured and uncoloured states. The complete volume has a title-page dated 1814, while the plates are dated 1 January 1813 (Howitt 1814). There was a second edition in 1819 published by H.R. Young (Howitt 1819: 113).

Another abbreviated version of Williamson's story was written to accompany a different illustration of a rhino by Samuel Howitt (1820) in the *Sporting Magazine* (fig. 28.20). The same journal had already published a much earlier sketch by Howitt (1799), illustrating an African adventure but definitely showing a pair of one-horned rhinos being attacked by dogs (fig. 28.21). The rhino in the plate of 1820 is almost identical to a sketch by Howitt, undated but most probably showing one of the rhinos in the Exeter 'Change Menagerie (fig. 28.22; Rookmaaker et al. 2015, fig. 7).

28.18   An Indigo Plantation in Purnea in 1810

Rhinos became rare enough in Purnea district that individual specimens are singled out in later reports. Walter Hamilton (1774–1828), the author of the *East-India Gazetteer*, heard in 1810 that a rhino was seen in the marshy woods of southern Purnea, but it was shot by an (unnamed) farmer when "fortunately" wandering onto an indigo estate (W. Hamilton 1828).

28.19   The *Percy Anecdotes* of a Hunt in 1815

The *Percy Anecdotes* were a series of 20 volumes originally issued 1821–1823, full of short stories to help readers in dinner-talk. They were purportedly written by Sholto and Reuben Percy, in fact pseudonyms for Joseph Clinton Robertson (1787–1852) and Thomas Byerley (1789–1826). Volume 19 includes a 'Letter from the North-East frontier of British India' about a rhino hunt written at the time of the Nepal or Gurkha war. The *Percy Anecdotes* do not claim originality, and in fact this letter had appeared previously in the *Asiatic Journal* of 1817, the *Literary Journal* of 1818 and probably elsewhere. The anonymous author writing from "Camp, N.E. Frontier, May 1815" was hunting with three fellow officers near Hurdeen, between Bettiah and Raxaul. On the banks of a small lake, they first found two young rhinos, killed one while the other escaped. This

G.S. A.P. W. DE C. J.S. J. H. G. H.C. DR. B.
              F.C.   H.W.S.

A Good Bag.

FIGURE 28.24 "A good bag." A faded photograph of the Shillingford family on a shooting tour in 1870, the year when the rhinoceros escaped. People in picture are identified underneath with their initials, from left to right, tentatively identified as: G.S. (George William Shillingford), A.P. (Picachi), F.C. (Cawes), J.S. (Joseph Lay Shillingford), H.W.S., H.C., (Henry Cave), Dr. B., J.H.G., W. de C. (perhaps Bruiser)
ORIENTAL SPORTING MAGAZINE, 1870, VOL. 3, P. 219

brought the furor of three large male rhinos and attacks on the elephants followed. In the end, the hunters killed a male and female rhino, "the largest of them was above 6 feet [183 cm] high" (Percy & Percy 1821: 126).

28.20 *Indian Sport* by D'Oyly of 1829

Charles D'Oyly (1781–1845), the Commercial Resident of Patna 1821–1831, was a prolific artist of Indian scenes. He established the Behar Amateur Lithographic Press to enable him to print his own sketches as well as others. In 1828–1829 he produced a 2 volume set with lithographs illustrating *Indian Sports*. The plate entitled "Rhinoceros hunting" is signed and dated 11 September 1829 (fig. 28.23). Given the exact date, it is likely that D'Oyly depicted a specific event, of which the details are yet unknown. The animal on the plate is obviously a rhino, but in reality the armour-plated skin and the general appearance are less than realistic.

28.21 A Rhino Hunt Near Bettiah in 1848

In January 1848, a writer using the pen name Rohilla and his party set out from Segwolee (Sagauli) and made their way to Bettiah. Here they were invited to the palace of Kishore Singh Bahadur, 8th Raja of Bettiah, where their audience was interrupted by a "very tame and very stupid Rhinoceros." The party went on to an area where they heard about a group of rhinos. They could not reach them, but "a party went in the following month from Gorruckpore to attack these beasts and upon arriving at the ground they fell in with a solitary Baboo, with a single elephant, who had just then and there killed a Rhinoceros with a single ball" (Rohilla 1848).

28.22 Hunt by Shillingford Family in 1871

On 8 April 1871 a rhino was shot on the lower Kosi River during a hunt organized by the sons of the indigo planter John Joseph Shillingford (1822–1867) of the Kolassy Factory. George William Shillingford and his brother Joseph Lay

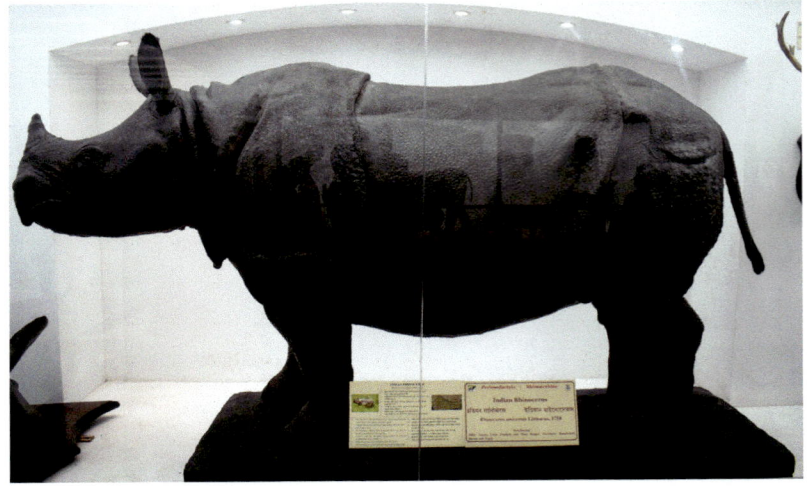

FIGURE 28.25

Mounted *R. unicornis* in the galleries of the Indian Museum, Kolkata. Male shot during the Shillingford hunt in 1871 in Purnea, Bihar

PHOTO: © TANOY MUKHERJEE, 2018

Bull rhinoceros. A glorious prize.

FIGURE 28.26

"Bull rhinoceros. A glorious prize", signed (lower right) "Vincent Brooks, Day & Son, Lith." The animal was shot in Purnea in 1874 as reported by the indigo planter James Inglis. Plate engraved in London by the lithographic company founded by Robert Alfred Brooks (1815–1885), possibly after a photograph

JAMES INGLIS, *TENT LIFE IN TIGERLAND*, 1888, FACING P. 298

Shillingford were joined by six other hunters, three of whom had come up from outside the region, probably Kolkata as they arrived on the steam ferry plying between Sahibganj to Carragola (Karagola). The family organized such hunts every year and spared no expense, as on this occasion 35 elephants and staff were present. A rhino had been spotted during the hunt of 1870 (Bruiser 1870), on which occasion the party was photographed (fig. 28.24). The report of the 1871 hunt was signed by 'Bruiser', who may be identified with 'W. de C.' as initials of his companions occur also in the text. Later I.J.W. (1909) of Redland, Bristol claimed to have taken part in the hunt. Others might have been members of the local Johnson, Cave and Picachi families. Choudhury (1963) states that the rhino was killed at Korha within a few miles from Purnea town, but the detailed report by Bruiser (1871) shows that the

event happened on the Kosi River near the confluence with the Ganges, in the unidentified "D-pore" jungle. The animal was shot by L.W. and G.W.S., the first unidentified, the second clearly George William Shillingford. No further rhinos were seen until the hunt ended on 20 April 1871.

The remains of the dead rhino were brought into camp on 9 April 1871. John Anderson (1833–1900), Curator of the Indian Museum, had been alerted in advance and had sent a taxidermist. The report of the Indian Museum (1871) included a list of taxidermists employed at the time, being C. Swaries with assistants J. Lopies, Shaik Gulloo, Sahik Hinghoo, Shaik Harroo, Peer Bux, Babu Khan, Kurreem, Islam Khan, Robiram, Abdool Suban, and 5 boys. It is unknown who came to Purnea. The hunters took measurements before skinning the animal: length from nose to tip of tail 14 ft (366 cm), height at shoulder 6 ft 5 in (206 cm), circumference of base of horn 1 ft 11 in (69.5 cm), and height of horn 9 in (23 cm). It took 20 men to load the skin on a cart and put it on the steamer at Karagola to be transported to Kolkata. The male rhino was officially donated by George W. Shillingford to the Indian Museum (1872). Once mounted, the specimen was added to the display in the gallery of the Indian Museum, where it is still seen (fig. 28.25). It was the first example of *R. unicornis* available to view in Kolkata. When William Sclater (1891) wrote his catalogue, he only listed bones of the feet, therefore possibly the skull and skeleton never reached the museum. In 1886, a skull in the collection of the Bombay Natural History Society was catalogued as a donation by J. [Joseph Lay] Shillingford.

28.23   James Inglis on the Kosi River in 1874

The indigo planter James Inglis (1845–1908) only saw two rhinos during his stay near the Kosi River (Inglis 1878, 1888). One night, maybe in 1874, he was alerted to the presence of a rhino in the Battabarree jungles (not located). Premnarain Singh, a young neighbouring Zemindar, had preceded him and killed the animal when stuck in quicksand. It measured 11 ft 3 in (342.8 cm) from snout to tail, 6 ft 9 in (205.8 cm) height, horn 6 ½ in (16.5 cm). An illustration of a dead rhino surrounded by hunters on four elephants was first included in Inglis (1888: 298) and again in the reprint of 1892 (fig. 28.26). The "glorious prize" is clearly *R. unicornis*.

28.24   Wodschow's Hunt in 1870s

The gazetteer of Purnea compiled by Pranab Chandra Roy Choudhury (b. 1903) in 1963 states that a rhino was killed by Alexander George Macdonald Wodschow (b. 1849), manager of the Korah Indigo Factory. This record may have been found when he researched the origin of the local name *genabari* or *gerabari* (P. Choudhury 1963). It is likely that the hunt occurred in the 1870s or a little later.

28.25   Hunting Parties of the Maharaja of Darbhanga, 1880s

Lakshmeshwar Singh Bahadur (1858–1898), Maharaja of Darbhanga, organized annual hunting parties near the Nepal border. The English manager of the large estate from 1876, Major General Robert Cotton Money (1835–1903), said that they usually went out at the end of March to shoot for about a month, and that it was an honour to be invited to one of these large parties, accompanied by as many as 200 elephants (Money 1891). The only glimpses involving a rhino were reported in the *Times of India* of 2 April 1881, when one was seen but not shot, and in the *Times of India* of 20 April 1883 reporting a bag of 1740 animals including 2 rhinos. Probably on another occasion in the 1870s, William Gordon Gordon-Cumming (1843–1930) shot a rhino near the Nepal border (Money 1893). In 1888, The Maharaja kept two young rhinos in his "Rhinoceros Park" in the gardens of his palace (5.4, fig. 5.3).

28.26   Last Rhino in Champaran in 1939

The naturalist Mrs. Jamal Ara, married to a forest officer in Bihar, has placed on record that the last rhino was shot in Champaran in 1939 by a man who obtained permission on the plea that the rhino had already been wounded by a poacher (Ara 1948, 1954).

28.27   Valmiki Tiger Reserve

The Valmiki forest in the West Champaran District provides a typical *terai* landscape. It is contiguous with Chitwan National Park and Parsa Wildlife Reserve in Nepal, as well as with the Sohagibarwa Wildlife Sanctuary in Uttar Pradesh. These forests were managed for timber production by the Zamindari estates of Ramnagar and Bettiah, until they came under the purview of the Forest Department in 1950 and 1953–1954. To provide protection to the habitat, Valmiki Wildlife Sanctuary was declared on 10 May 1978, with its status changing to National Park on 5 December 1990 and to Tiger Reserve on 11 March 1994 (Sinha 2011).

Valmiki is close to Nepal's Chitwan National Park, where the rhino population has been increasing in recent years. Some animals strayed southwards to the marshy land on the banks of the Gandak River in the Madanpur Forest Range. The first pair was spotted there in 2001–2002 by villagers who were not used to seeing such large mammals in their fields. In 2003, it was reported that a female rhino had delivered a calf. According to the Forest Department, an additional five rhinos settled in the Madanpur and Valmikinagar ranges in 2005. Two rhinos have been killed in railway accidents on the Bagaha to Chhitauni Railway line which passes through the forests south of the national park. The first was a female killed in April 2007, and the second a 10-year old animal killed in March 2013 (*Daily Pioneer*, Patna 2013-03-20). In January 2008, a male rhino was found drowned in a canal just west of the reserve and across the border in Uttar Pradesh. That would have left three rhinos in Valmiki in 2011 (Sinha 2011).

An additional three rhinos came to the park from Chitwan in 2010 and appear to have settled in Valmiki (Kumar 2013). Two instances of poaching have been recorded: a female rhino in the western portion of the reserve in March 2011, and a second animal in April 2013.

Reports obtained by the National Park staff appeared to show that two male rhinos were resident in the Madanpur forest area of Valmiki, but this could not be substantiated in August 2013. During that survey, a baby rhino was observed in a paddy field near the Madanpur range, which probably had traveled from Chitwan. Such movements across the international border will continue while the numbers of rhinos in Nepal are increasing. This can provide a good opportunity to explore possibilities of a trans-boundary approach to nature conservation.

CHAPTER 29

Historical Records of the Rhinoceros in Rajmahal Hills, Jharkhand

First Record: 1788 – Last Record: 1860
Species: *Rhinoceros unicornis*
Rhino Region 20

29.1 Rhinos in the Rajmahal Hills

The Rajmahal Hills are a stretch of low undulating forests on the southern bank of the Ganges River, where its course bends from its easterly direction towards the south, from Bhagalpur past Rajmahal to Farakka. Once part of Bihar, the eastern section is now in Jharkhand state. Rhinos may have lived in much of this territory, although most records refer to their presence in a restricted area including Sahibganj, Sikrigalli, Rajmahal and Farakka. Some of this land allows the animals access to the river and the jungle around it without entering the hills.

In the early literature there was no discussion about the species found in Rajmahal, which might mean that it was supposed to be *R. unicornis*, the usual one-horned type. Of course, few people actually stated to have seen one (table 29.34). Only Governor Hastings shot one or more in 1822, and Duvaucel was attacked by one in 1823. Working in Kolkata, Edward Blyth (1862a) was the first to make a judgment, and he imagined the animals were *R. sondaicus*, without stating the reason for his choice. He received some support, notably from Jerdon (1867), until Blanford (1891) said clearly that he thought that Blyth was mistaken and that these animals were *R. unicornis*. The latter species definitely gets the majority vote from that point onwards. Shebbeare (1953: 141) was uncertain. If the water-colour by Sita Ram (fig. 29.1) was in fact made in Rajmahal in 1822 and if the artist tried to depict an actual animal seen in the field, the rhino was definitely one-horned, but unfortunately the position allows no clear view of the shoulder-region where the triangular shield of *R. sondaicus* might have clinched the matter.

The rhino known in Purnia north of the Ganges was certainly *R. unicornis* (chapter 28). The region of Malda eastwards across the Ganges from Rajmahal has a small number of indeterminate records (chapter 30). In the absence of clear evidence, but given the small distance between Purnia to the north and Malda in the east, it is my view that the rhinoceros of Rajmahal belonged to *R. unicornis*.

...

Dataset 29.29: Chronological List of Records of Rhinoceros in Jharkhand

General sequence: Date – Locality (as in original source) – Event – Source – § number – Map number.

The persons interacting with a rhinoceros in various events are combined in Table 29.34. The map numbers are explained in dataset 29.30 and shown on map 29.12.

Rhino Region 20: Jharkhand

*c.*1650 – Ragiamahol (Rajmahal) – Palace of Shah Shujah (1616–1661), second son of Shah Jahan, contains figures of rhino – Valentijn 1726, vol. 5: 166 – 29.2 – map A 322

1757-05 – Sacrigali (Sakrigali) – The French traveller and indologist Abraham Hyacinthe Anquetil-Duperron (1731–1805) saw a rhino captured in the mountains. It was taken to the Nabab of Bengal by a group of over 50 men – Anquetil-Duperron 1771: lii–liii – 29.3 – map A 321

1788-10 – Mooteejerna (Moti Jharna) waterfall – Rhino tracks seen by William Daniell (1769–1837), in the company of Thomas Daniell (1749–1840) – Daniell & Caunter 1834a: 101; Archer 1980: 41; Rookmaaker 1999a – 29.4 – map A 321

1808 – Rajemahal Hills – Two hunters killed 6 rhinos – *Calcutta Gazette* 1808-04-07; Sandeman 1868, vol. 4: 196

1808 – Rajemahal Hills – One rhino captured, tame after 2 days, expected in Calcutta [no report of arrival] – *Calcutta Gazette* 1808-04-07; Sandeman 1868, vol. 4: 196

1810 – Hills between Rajmahal and Sakarigali (Sakrigali) – Some rhinos left, but rarely seen – Buchanan-Hamilton 1939: 292 – map A 321

1811-01 – Motijharna – The Bagicha at the foot of the hills is a favourite haunt of rhinos – Buchanan-Hamilton 1939: 40 – map A 321

1811-01 – Lake straight west from Mosaha – Large lake, the haunt of wild rhino – Buchanan-Hamilton 1930: 93 – map A 322

1814-05 – Between Siclugully (Sakrigali) and Rajemahal – Gentlemen from Poornea killed 6 or 7 rhinos – *Calcutta Gazette* 1814-06-09; Sandeman 1868, vol. 4: 196 – map A 321

1814-07 – Terriagully (or Teliagarhi) – Governor-General Francis Edward Rawdon-Hastings, 1st Marquess of Hastings, Earl of Moira (1754–1826) [hereafter: Governor Hastings] wounded a rhino – *Calcutta Gazette* 1814-08-11 – 29.5 – map A 321

1819 [= 1820] – Rajemal hills – Governor Hastings killed 3 rhinos – Mundy 1832, vol. 2: 181; S. Shrestha 2009: 260 – 29.5

© L.C. (KEES) ROOKMAAKER, 2024 | DOI:10.1163/9789004691544_030

This is an open access chapter distributed under the terms of the CC BY-NC-ND 4.0 license.

1820-12-14 – Maharajpoor in Rajmahal hills – One rhino killed by Captain Charles William Brooke (1784–1836) of Commissariat Department during the trip of Governor Hastings – Hastings 1821; *Bombay Gazette* 1821-01-17; Cockburn 1884: 140 after *Bengal Hurkara* 1820-12-14 – 29.5 – map A 321

1823-01-24 – Sakrigali – Alfred Duvaucel (1793–1824) severely wounded by rhino – Arago 1840: 232, 1856: 49; Coulmann 1862; Rookmaaker 2019d, 2021a; Low 2019; Dorai & Low 2021 – 29.6 – map A 321

1824 – Futhipore, south of Rajmahal – Lieutenant-Colonel Charles Ramus Forrest (1748–1827) saw tracks of a rhino, several of which are found on these hills – Forrest 1824: 133 – map A 323

1828-03-01 – Rajemal hills – Rhino found in swamps in thickest forest – Mundy 1832, vol. 2: 181

1828-08-10 – Rajmahal Hills at Boglipoor, Bhaugulpoor – Rhino found. Killed by local Puharrees with poisoned arrows – Heber 1828a, vol. 1: 214 (1st edn), 1828b, vol. 1: 282 (2nd edn)

1831 – Motee Jhurna waterfall – Captain Henry Charles Baskerville Tanner (1835–1898), Survey of India, found a large rhino-looking fossil skull, later found to be incorrectly identified – Sherwill 1854: 67 – 29.7

1834 – Rajmahal – Rhino found – Anon. 1834c

1836-03 – Rajmahal, Oodooah Nullah (Udhwa) – Group of 7 hunters saw rhino on 10 March. Killed one on 13 March 1836 – Royal Tiger 1836 – map A 323

1836-04 – Raj Mehal – Robin Hood and friend T-v-s killed 1 rhino with one ball – Robin Hood 1836 – map A 322

1837 – Rajmahal – Rhino found – E. Roberts 1835, vol. 1: 128

1838 – Rajmahal – Rhino occasionally but very rarely seen – Martin 1838b, vol. 2: 146

1843 – Rajmahal – Rhino abound – Davidson 1843, vol. 2: 52

1842 – Sukreegullee (Sakrigali) – Rhino found – Anon. 1842a: 35; Anon. 1851a: 274 – map A 321

1842 – Rajmahal – Smaller one-horned rhino found – Anon. 1872a

1843 – Blue mountains, or hills of Rajmahal – Rhino found – Bellew 1843, vol. 2: 108 – map A 321

1843 – Radjmahal (Rajmahal) – Erich von Schönberg (1812–1887) is told to expect rhinos in a hunt, but little wildlife encountered – Schönberg 1852, vol. 1: 62

1844-11-31 – Sickri-gali (Sakrigali), 18 miles above Rajmahal – Fanny Parks (1794–1875) stated that rhinos abound here – Parks 1850, vol. 2: 398, 1851: 32 – map A 321

1846, 1850 – Rajmahal Hills – Walter Stanhope Sherwill (1815–1890) says rhino have retreated to the Northern and North-eastern face of the Rajmahal hills, where they find cover in the dense forest – Sherwill 1854: 5, 20; Buchanan-Hamilton 1939: 253; Basu 1935: 148; Gardner 2016 – map A 321

1857 – Taljhari – Missionaries of Church Missionary Society living in a spot where rhino used to roam – Storrs 1877, 1895; Hobbes 1893, vol. 1: 98 – map A 322

1862 – Rajméhal hills – *R. sondaicus*, fast verging on extirpation – Blyth 1862a: 151, 1862b: 32; Murray 1866: 172

1862 – Rajmahal – *R. sondaicus* occurs. Thomas Caverhill Jerdon (1811–1872) visited Rajmahal in 1862 – Anon. 1862; Jerdon 1867: 234

1863 – Rajmahal – Rhino has become extinct during life of Henry Yule (1820–1889) – Yule 1863: 19

1867 – Land of the Sontals – Rhino quite extinct – Man 1867: 218

1878 – Raymahal Hills – *R. sondaicus* found – Brandt 1878: 36–37; Craig 1897, vol. 2: 473

1885 – Sonthal Parganas – *R. sondaicus* disappeared c.1885 – Thompson 1932: 154

1962 – Hazaribagh National Park – Plan to re-introduce rhino – Gee 1962: 473

Species in Rajmahal Hills stated as *R. sondaicus*: Blyth 1862a: 162; 1863a: 137; RHD 1866: 485; Murray 1866; Jerdon 1867; Blyth 1872b; Brandt 1878; Cockburn 1884; Sclater 1891, vol. 2: 203; Craig 1897, vol. 2: 473; Thompson 1932; Choudhury 2016: 155.

Species in Rajmahal Hills stated as *R. unicornis*: Hunter 1877, vol. 14: 43; Ball 1880: 228; Cockburn 1883: 60; Baker 1887: 251; Blanford 1891: 473; Trouessart 1898, vol. 2: 753; Lydekker 1900: 23; Harper 1945: 376; Mukherjee 1963: 46, 1966: 26, 1982: 52; Shahi 1977: 2; Martin & Martin 1982: 29; Rookmaaker 1984a; Menon 1995: 56; Sinha 2004: 47.

∴

TABLE 29.34   Records of rhinos killed, sighted or captured in the Rajmahal Hills of Jharkhand

Locality	Persons	Date	Killed	Sighted	Captured
Mooteejerna	Daniell	1788		tracks	
Rajemahal	Party of 2 hunters	1808	6		1 (fate unknown)
Sicligully	Poorneah party	1814	6–7		
Terriagully	Hastings	1814		1 wounded	
Rajemal hills	Hastings	1820	3		
Maharajpoor	Brooke	1820	1		
Sakrigali	Duvaucel	1823		2 in attack	
Futhipore	Forrest	1824		tracks	
Rajmahal Hills	Heber	1828		mentioned	
Oodooah Nullah	Party of 7 hunters	1836	1	seen	
Raj Mehal	Robin Hood	1836	1		

MAP 29.12   Records of rhinoceros in the Rajmahal Hills (Jharkhand). The numbers and places are explained in Dataset 29.30
★ Presence of *R. unicornis*

• • •

Dataset 29.30: Localities of Records of Rhinoceros in Jharkhand

The numbers and places are shown on map 29.12 and in a wider context on map 30.13.

Rhino Region 20: Jharkhand

A 321   Blue mountains, Hills of Rajmahal – 1843
A 321   Terriagully – 25.24N; 87.68E – 1814
A 321   Sakrigali, Sukreegullee, Sickri-gali – 25.25N; 87.70E – 1755, 1823, 1842, 1844
A 321   Maharajpoor east of Sahibganj – 25.24N; 87.70E – 1820

A 321   Moti Jharna, Mooteejerna, Motee Jhurna waterfall – 25.21N; 87.73E – 1788, 1811
A 322   Taljhari – 25.10N; 87.75E – 1857
A 322   Lake west from Mosaha – 25.11N; 87.75E – 1811
A 322   Ragiomahal (Rajmahal) – 25.05N; 87.82E – 1650
A 323   Oodoooah Nullah (Udhwa) – 24.97N; 87.80E – 1836
A 323   Futhipore, south of Rajmahal – 24.50N; 87.85E – 1824
*   Hazaribagh NP – 24.10N; 85.34E – 1962 – plan of re-introduction
*not on map

•
• •

29.2   Rajmahal as Hunting Country

The Rajmahal hills were known to be full of wild animals. The English naturalist Thomas Pennant (1726–1798)
describing an imaginary journey through India, assured
his readers that the governors settled in Rajmahal "on
account of the quantity of game of chace, which the
neighbourhood offered" (Pennant 1798, vol. 2: 286).
Rajmahal was the capital of the Mughal seat of Bengal, in
1595–1608 and again 1639–1660 at the time of Shah Shujah,
the second son of Shah Jahan. When the French traveler
Jean Baptiste Tavernier (1605–1689) passed the town of
Rajmahal on 4 January 1666, he said that it "was formerly
the residence of the Governors of Bengal, because it is a
splendid hunting country" (Tavernier 1889: 102).

The Dutch compiler François Valentijn (1666–1727)
mentioned the palaces of Ragiamahol (Rajmahal) in
his extensive history of the Dutch East India Company,
and described the gardens surrounding the buildings:
"Around every court, of which there are several together,
there are beautiful gardens, equal to any of the European
counterparts. Every garden has its own special fountains
and waterworks, which have been artfully constructed
from alabaster marble as well as blue and white stone,
and decorated with many fine copper effigies of lions,
dragons, rhinoceros, and other wild and tame animals"
(Valentijn 1726, vol. 5: 166).

29.3   Rhino Capture Near Sakrigali in 1757

The French scholar Abraham Hyacinthe Anquetil-
Duperron (1731–1805) was in India 1755–1761, mainly
around Surat, but traversing parts of Bengal in 1756–1757.
He gave an extensive account of his pioneering peregri-
nations, which includes a paragraph of an event in the
Rajmahal Hills in the first two weeks of May 1757: "Près
de Sacrigali, je rencontrai un Rinoceros nouvellement pris
dans les montagnes, que l'on menoit au Nabab de Bengale.
Il etoit à-peu-près de la hauteur d'un âne. On lui avoit lié
le corps en travers avec de grosses cordes; & deux grands
cables tenant à ces cordes par des noeuds, lui prolongeoient
le corps de chaque côté. Cinquante hommes dirigeoient
chacun de ces cables; de manière que quelqu'effort que fit
l'animal, il étoit obligé de céder au plus petit mouvement
de tous ces bras réunis" (Anquetil-Duperron 1771: lii–liii).
In summary, he stated that he saw a young rhino, newly
captured in the mountains at Sakrigali. It was transported
by an army of over 50 men to the Nabab of Bengal. The
animal might have survived, as animal fights including
a rhino were reported around 1770 at the court of Saif

ud-Daulah (ruled 1766–1770), who then had a residence
in Murshidabad, some 170 km south of Sakrigali (5.4
Murshidabad).

29.4   Daniell Passing the Rajmahal Hills in 1788

During their well-documented journey from Kolkata to
the source of the Ganges, the British artist Thomas Daniell
and his nephew William Daniell passed the Rajmahal
Hills, like so many travelers before them (17.4). According
to the unsurpassed narrative of their journey by Mildred
Agnes Archer (1911–2005), these hills were covered with
thick jungles where tigers, rhino and elephants lurked
(Archer 1980: 41). This has led myself to extrapolate that
they must have known the presence of these wild animals
from accounts of earlier travelers (Rookmaaker 1999a:
205). Although this is likely to be correct, there is only one
reference to a rhino at this point of their travels, and at
the same time they were among the first to notice the animal in print. The Moti Jharna or Fall of Pearls, just south
of Sakrigali, can be seen from the riverbed by all passing
ships. In their journal of 8 October 1788, hiking the few
kilometers from the Ganges to the waterfall, they recorded
that "on the way William was thrilled to see the footprint
of a rhinoceros" (Daniell & Caunter 1834a: 101).

29.5   Governor Hastings in Rajmahal

Francis Edward Rawdon-Hastings, Earl of Moira (1754–
1826) was Governor General of India 1812–1821, during
which time he embarked on two major tours of inspection. On the first of these, he started out from Barrackpore
on 28 June 1814 and followed the Ganges upstream to
Haridwar (Uttar Pradesh) reaching it in December 1814.
Hastings went on a tiger hunt at Terriagully, near Sakrigali,
on 21 July 1814. Although the day is skipped in his Journal
for this period (Hastings 1858, vol. 1: 90; Losty 2015: 42),
a newspaper reported how the Governor killed a tiger of
297 cm (9 ft 9 in) in length (*Calcutta Gazette* 1814-07-28).
Probably on the next day, Hastings also wounded a rhino
but it "contrived to escape" (*Calcutta Gazette* 1814-08-11).

Hastings again went to the Rajmahal Hills on a con-
valescent tour in the winter of 1820–1821. An anonymous
correspondent recorded at Maharajpoor (Maharajpur)
close to Sakrigali on 14 December 1820: "We came this
morning from Seerkunda [Surkundah] (about six or eight
miles), and having heard from the shikarees that there
were rhinoceroses on the way, we penetrated through
very thick jungles all along, intending to fire at nothing

FIGURE 29.1   Elephants hauling a dead rhinoceros in hill country. Watercolour by Sita Ram, artist for the Governor-General Francis Edward
Rawdon-Hastings. Size 17.2 × 27.6 cm. Possibly the animal shot by the party of Hastings at Maharajpur in the Rajmahal Hills,
Bihar on 14 December 1820. From an album assembled for Lady Flora Hastings
WITH PERMISSION © BRITISH LIBRARY: ADD OR 5006

but these animals" (Hastings 1821). A letter printed in contemporary newspapers tells us what happened next, stating that the party "encountered three Rhinoceroses. Col. Nichols wounded one of the first two seen, (an immense fellow) and knocked him down the first shot, but he rose again and got off; another was started, and passing Capt. Brooks was by him shot dead, the ball entering the folds of the neck. This one was dragged into the Camp by three Elephants" (*Bombay Gazette* 1821-01-17). The shooter may have been Captain Charles William Brooke (1784–1836) of the Bengal Presidency.

The army officer Godfrey Charles Mundy (1804–1860) mentions a similar incident, but with an (incorrect?) date 1819, of Hastings and a large party of friends killing 3 rhinos in these hills (Mundy 1833). The feat was remembered even when Edward Ward Walter Raleigh (1802–1865) of the Bengal Medical Service was traveling on 23 August 1827 and anchored his boat at the same spot where Hastings killed a rhino (S. Shrestha 2009: 260).

Sita Ram was an Indian artist engaged by Hastings to make watercolour drawings during a trip from Kolkata to Delhi 1814–1815 and also in the Rajmahal Hills in 1820

(Losty 2015). There are 34 paintings of natural history subjects assembled in two albums by Flora Campbell, Lady Hastings (1780–1840). One of these by Sita Ram, undated, shows "Elephants hauling a dead rhinoceros in hill country" (fig. 29.1). Although there is no contemporary title and hence no locality for this event, the scene corresponds with the events in December 1820 showing the rhino pulled along by three elephants. If so, this is the only known depiction of a rhino in the Rajmahal Hills. The fate of the animal remains obscure.

29.6   Attack on Alfred Duvaucel in 1823

The French collector Alfred Duvaucel (1793–1824) is best known for spending a few months with his friend Pierre-Médard Diard (1794–1863) in the employ of Thomas Stamford Raffles in Singapore and Sumatra (Rookmaaker 2021a). Duvaucel was the stepson of Georges Cuvier, the influential naturalist of the Natural History Museum in Paris, and he had gone out to increase the collections of that great institution. At first he stayed two years (1817–1818) in the French factory Chandernagor near Kolkata, where he returned in 1820 to continue his activities. He then traveled to Dacca and Sylhet, and in 1822 made a plan to explore the Himalayan mountains around Kathmandu on his way to Tibet (Dorai & Low 2021). Accordingly he went by boat up the Hooghly and Ganges to the vicinity of Rajmahal. Here he stopped for a few days as we know from letters he wrote to his mother Anne Marie Sophie Coquet du Trazail (1764–1849), which were preserved and edited by his childhood friend Jean-Jacques Coulmann (1796–1870). Duvaucel wrote four letters in January 1823 from Sielygalli near Bajemel, which resolves to Sakrigali near Rajmahal, followed by three letters from Boglipour, now Bhagalpur (Coulmann 1862). Here he had a near-fatal incident involving a rhino (Rookmaaker 2019d).

On the evening of Friday 24 January 1823, Duvaucel told his mother that earlier in the day his letter writing was interrupted when news reached him in his temporary residence in Sakrigali that a rhino had been wounded by a local hunter who had shot the animal with an arrow in the eye. He gathered his servants and seven or eight men armed with bows and arrows. He found the animal, approached up to ten steps ready to shoot it, presumably with a gun, and so to add an unexpected treasure to the museum's holdings. However, he was suddenly attacked by a second rhino which he had not noticed in the excitement, and was thrown ten paces through the air. The animal continued his attack and inflicted a deep wound in the right thigh, before disappearing in the bushes. Duvaucel tried to shoot both rhinos, but their fate is not recorded here. Duvaucel himself lost a lot of blood, felt unable to walk, although his leg was not broken and his muscles were not severely affected. Medical attention was only available in Bhagalpur, so his men put him in an oxcart to carry him the three miles to the banks of the Ganges. He was attended by a Scottish doctor, who kept him under observation until the second week of February. It is unclear what happened next, whether Duvaucel continued his journey upstream to Patna, or returned to Chandernagor. When his health did not improve, he decided to return home, but on reaching Madras in August 1824, he suddenly passed away. After the news reached Paris, his uncle Frédéric Cuvier stated that Duvaucel had been so much weakened by the rhino attack that he was unable to fight ensuing bouts of

Il voit le monstre venant de son côté. (Page 50.)

FIGURE 29.2
Alfred Duvaucel attacked by a rhino in the Rajmahal Hills. Impression signed by Jean Adolphe Beaucé (1818–1875)
JACQUES ARAGO, *OEUVRES ILLUSTRÉES*, 1856, P. 49

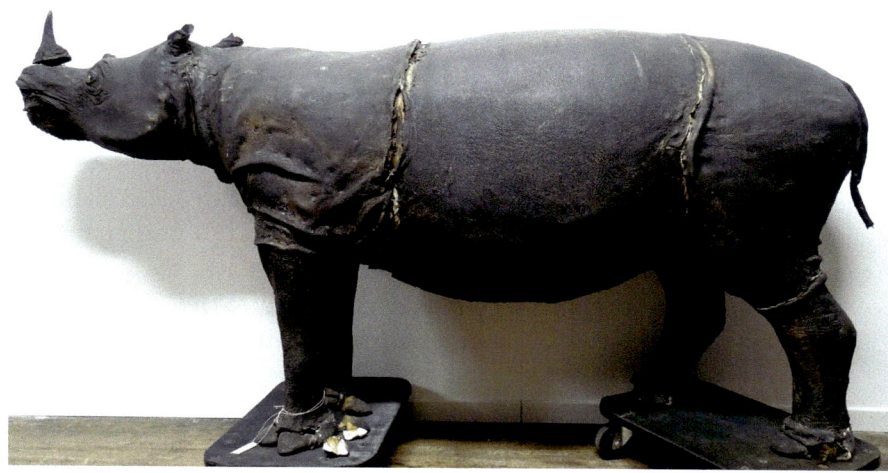

FIGURE 29.3
Mounted specimen of a two-horned Sumatran rhinoceros, *Dicerorhinus sumatrensis*, in the stores of the Musée Zoologique in Strasbourg (no. Mam-01505). Although unlikely to be the correct specimen, it has been said to have been the rhino which attacked Alfred Duvaucel in the Rajmahal Hills near Sakrigali in January 1823
PHOTO: MARIE MEISTER, JULY 2018

dysentery (Duvaucel 1824: 278; F. Cuvier 1825a). In subsequent obituaries, the dysentery was generally given as the cause of death (Eyries 1824; Mahul 1825), the rhino attack was forgotten, and the chronology of his last travels was confused.

An embellished account of Duvaucel's incident was written by the French explorer and artist Jacques Étienne Victor Arago (1790–1855), intent on literary acclaim rather than historical truth. In Arago's version, Duvaucel had gone out with twenty hunters, wounded a rhino and tried to hide from the furious animal behind a tree on the edge of a ravine. When he looked what was happening, he was hurled into the air and thrown into the valley. The rhino killed one hunter, wounded three others, then disappeared (Arago 1840: 232). No need to linger on this version, except that in a posthumous edition of Arago's works of 1856, the story was illustrated by the engraver Jean Adolphe Beaucé, with neither Duvaucel nor the rhino drawn from direct observation (fig. 29.2). The same drawing made its way into another fictional hunt by a wild huntsman in *Harper's Weekly* of 7 March 1857 (Anon. 1857).

It is unlikely that Duvaucel was able to salvage any of the rhinos encountered on that fateful day in the Rajmahal Hills. Coulmann (1862: 143) stated that the rhino responsible for Duvaucel's death could be seen in France, mounted in a museum in Strasbourg. Dominique Auguste Lereboullet (1804–1865), the second Director of the Musée Zoologique in Strasbourg from 1837, made the same suggestion about a rhino specimen obtained as a donation from Paris in 1829 (Lereboullet 1837). When I was pursuing my research into the death of Duvaucel, I had high hopes at last to find an actual rhino from the Rajmahal Hills. By good fortune, the specimen is in fact still present in the stores of the Strasbourg museum, standing on a pedestal,

but without a contemporary label or associated history (no. Mam-01505). There is always a chance with old specimens that data are confused in exchanges, and maybe this has happened somewhere in the historical sequence, because the mounted animal in Strasbourg is clearly a specimen of the two-horned *D. sumatrensis* (fig. 29.3; Rookmaaker 2019d, fig. 3). It is therefore highly unlikely to be Duvaucel's nemesis, because in my view this species has never been known or even suspected in the hills on the south bank of the Ganges.

29.7 Rumours in the Early 19th Century

Most travelers who passed Rajmahal on the Ganges, even those who spent time in the region, never had a chance to actually see a rhino in the wild. The surveyor Francis Buchanan (1762–1829), later known as Francis Hamilton, surveyed this region in early 1811 and was quite positive about the presence of the animals. In his description of the waterfall of Moti Jharna, he observed that "in the floods it finds its way to the Ganges by two routes. The first is nearly opposite to the cascade, the second forms a long channel that winds for along way south by the roots of the hills, is named Bagicha, and joins an old channel of the Ganges south from Masaha. The Bagicha at the foot of the hills is a favourite haunt of the Rhinoceros." Between Masaha and the hills there was a large lake, where the rhinos could be found.

When the author and artist Fanny Parks (1850, vol. 2: 398) passed Sakrigali in November 1844, she noted that there should be a lot of animals in the area, like "bears, tigers, rhinoceroses, leopards, hogs, deer of all kinds." However, she didn't see any of them herself (fig. 29.4).

Just a few years later, Walter Stanhope Sherwill (1854), surveying the Bhaugulpoor District in 1846–1850, already talked about the former presence of rhinos in Kankjole, south of Rajmahal. He thought that rhinos must have moved further inland, to the northern and north-eastern slopes of the Rajmahal Hills. At Moti Jharna he mentioned that his surveyor colleague Henry Charles Baskerville Tanner had found a fossil skull which resembled a rhino, but later such bones in the area were confirmed to be limestone constructions.

29.8 Disappearance from Rajmahal around 1860

The reports in the literature become increasingly vague from the 1860s. Sir Henry Yule (1820–1889) was the most specific in saying that the disappearance of the rhinos from Rajmahal had occurred during his lifetime, before 1863 (Yule 1863: 18). Cockburn (1884) also talked about the recent extinction of the rhino from this district, but did not cite any occurrence after 1822. Maybe the animals did not disappear into the more hilly areas inland, but in fact were persecuted to extinction in the wake of advancing civilisation on the banks of the river which were their natural habitat. A date around 1860 may be proposed.

FIGURE 29.4 Sikri-Gall (Sakrigali), sketched by Fanny Parks (1794–1875) in 1844
PARKS, *GRAND MOVING DIORAMA OF HINDOSTAN*, 1851, P. 32

Historical Records of the Rhinoceros in Former Undivided Bengal

First Record: 1609 – Last Record: 1886
Species: *Rhinoceros unicornis*
Rhino Regions 21 to 23

30.1   West Bangladesh and Adjoing West Bengal

The area covered by this chapter is administratively complex and only the redundant term 'undivided Bengal' covers its extent. It is bordered by the traditional rhino habitat in North Bengal in the north, by Assam in the east and by Jharkhand and Bihar in the west. It needs to be investigated how far southward the rhino might have occurred, and also which species was involved.

There are reports of only three sightings and four animals killed, mainly dating from the first half of the 19th century (table 30.35). Therefore the area cannot have included much habitat which attracted the animals, which is likely the result of settlements and agriculture. The records are discussed largely according to the districts of the state of West Bengal and the current divisions of Bangladesh. It cannot be decided how long the rhino persisted in parts of the region. Even for the accounts from the middle of the 19th century, the reporters implied that the animals must have strayed into the territory, either 100 km from North Bengal or 150 km from the Rajmahal Hills.

A rhinoceros is depicted on panels in five terracotta temples of which many examples were built across Bengal up to the 19th century (30.8). It needs to be examined to what extent this contributes to our knowledge of the range of the rhino in this part of India and Bangladesh.

Elsewhere in Bangladesh, in the south *R. sondaicus* was common in the Sundarbans (chapter 51), and in the east *D. sumatrensis* occurred in Chittagong (chapter 56) and Sylhet (chapter 59). However, for the current area it is more plausible to expect *R. unicornis*. Although this is generally implied or accepted in the literature, it is difficult to be completely certain in the absence of specimens in museums. The depictions found in the temples were never meant to show a zoological image. Nevertheless, in my view the rhino which occurred east of the Ganges and along the Padma River was *R. unicornis*. From the small number of definite accounts, it appears that *R. unicornis* reached its southernmost part of its range near Rajshahi on the Padma River.

The rhinos killed in Rajshahi in 1822 and in Rangpur in 1837 and 1850 must be the last animals in the region. They might even have been strays. The later references are generally contained in the district gazetteers, which are very useful documents but not usually updated with the latest information. When the presence of the rhino is mentioned in Murshidabad in 1857, this may refer to an earlier event. Rhinos were likely absent from the western parts of Bangladesh from around 1820.

∴

Dataset 30.31: Chronological List of Records of Rhinoceros in West Bengal, India (Dinajpur, Malda, Murshidabad) and Bangladesh (Dhaka, Mymensingh, Rajshahi, Rangpur)

General sequence: Date – Locality (as in original source) – Event – Sources – § no. – Map reference

Species identification follows original source (with names updated). The persons interacting with a rhinoceros in various events are combined in Table 30.35. The map numbers are explained in dataset 30.32 and shown on map 30.13.

Rhino Region 21: the Central Districts in West Bengal, India

9th cent. – [Temple] Jagjivanpur, Malda Dt. – Terracotta plaque depicting a rhino in right lateral profile, excavated in 1996–2005 by the team led by Amal Roy. State Archaeological Museum, Kolkata, No. 04.18, size 26.7 × 23.2 × 59 cm – Roy 2012, 2013: 213 – 30.8 – map M 863

14th cent. – [Mosque] Pandua, Malda Dt. – Adina Mosque, terracotta panel which contains portions of an elephant and rhino, according to Alexander Cunningham (1814–1893). The object is not seen now, as a 1931 photograph of the pulpit shows no sign of any structure depicting a rhino – Cunningham 1882a: 91; Khan & Stapleton 1931: 133, fig. 33; Bautze 1985; Rookmaaker & Edwards 2022: 78 – 30.8 – map M 862

1655 – [Temple] Bishnupur, Bakura Dt. – Krishnarai Temple, terracotta panel said to depict a rhino. A close examination of the photo taken by Asok Kumar Das shows the animals to be boars with large tusks – Das 2018b: 92

© L.C. (KEES) ROOKMAAKER, 2024 | DOI:10.1163/9789004691544_031
This is an open access chapter distributed under the terms of the CC BY-NC-ND 4.0 license.

1750s – [Temple] Baranagar, Murshidabad Dt. – Gangeshwar Siva Temple, terracotta panel with a rhino with hunters – Das 2018b: 91; Haque 2014, pl. 213; Michell 1983, no. 374; Bautze 1985: 415; Rookmaaker & Edwards 2022: 75 – 30.8, fig. 30.7 – map M 861

1806 – Forests of Peruya (Pandua, near Adina Mosque) – Rhino supposedly accompanied 2 elephants straying from Morung – Buchanan-Hamilton 1833: 133 – 30.6 – map A 331

1812 – Old capital Gur (Gaur) near Malda – Armand Magon de Clos-Doré (1780–1850), hunting with Capt. W., saw rhino – Magon de Clos Doré 1822: 82, translated in Guhathakurta & van Schendel 2013: 80–82; Rookmaaker & Edwards 2022: 71 – 30.6 – map A 332

1820 – Berhampur, Murshidabad Dt. – Rhino in chess set made of ivory and wood – Das 2018b: 92; National Army Museum, London, NAM 1962-03-46-5 – 30.7, fig. 30.2

1835 – Murshidabad – Gouache painting of party on elephant hunting rhino and other animals – Victoria & Albert Museum IS.33: 12–1961 – 30.7

1838 – Puraniya – Rhino lately seen. [Puraniya is south of English Bazar (Malda), but here mistake for Purnea, see 30.5] – R.M. Martin 1838c: 184 – 30.5

[1840s, not dated] – Pabna Division (giving localities near Shahsadpur in previous chapter on hog-hunting) – Rhino found in densest jungle. The author William Henry Florio Hutchisson (1773–1857) gives no indication if or where he had seen rhino himself – Hutchisson 1883: 233 (date uncertain, likely after 1835 when author arrived in Murshidabad)

1854 – Pergunnah Shikarpoor (Malda) – Rhino has been seen – Pemberton 1854: 5; Sengupta 1969: 10; Bahuguna & Mallick 2004 – map A 331

1857 – Murshidabad Dt., northern part – "Rhinoceros have been seen in the north of the District" – Hunter 1876b: 34; Agrawal et al. 1992: 113; Bahuguna & Mallick 2004 – 30.7 – map A 333

1876 – Malda Dt. – Rhino very rare – Hunter 1876b: 34; Banerjee 1964: 242 (as R. unicornis); Bahuguna & Mallick 2004; Agrawal et al. 1992: 113; Sengupta 1969: 10 – map A 331

1892 – Maldah (Malda) – Rhino present – Sanyal 1892: 130 – map A 331

Rhino Region 22: North Bangladesh – West of Jamuna-Brahmaputra

9th cent. – [Temple] Paharpur, Rajshahi Division – Somapura Mahavihara, terracotta panel showing a rhino with rider, possibly a demi-god or gandharva – Dikshit 1938: 56–72, pl. 46; Panchamukhi 1951; Das Gupta 1961: 31; Bautze 1985: 415; Uddin & Rezowana 2012: 195, fig. 15.24, 2015: 135, fig. 34; Rookmaaker & Edwards 2022: 71 – 30.8, fig. 30.2 – map M 866

1704–1722 – [Temple] Kantanagar, Rangpur Division – Kantaji Temple, two terracotta plaques with rhino surrounded by hunters – Das 2018b: 92; Haque 2014, pl. 167; Michell 1983, no. 617; Bautze 1985: 415; Rookmaaker & Edwards 2022: 74 – 30.8, figs. 30.5, 30.6, 30.7 – map M 864

1609 – Ghoraghat region, Rangpur Dt. – Rhino present in times of Abu'l-Hasan Asaf Khan (1569–1641), father of Mumtaz Mahal – Sarkar 1928: 146 – map A 342

1800 – Bangladesh – R. unicornis, previous occurrence – De Geer 2008: 380

1800 – Bangladesh outside Sundarbans – R. sondaicus, previous occurrence – Simon 1966

1820 – Rungpoor Dt. (Rangpur) – Rhino not uncommon, quite harmless – Hamilton 1820: 203 – map A 343

1820 – Dinagepoor Dt. (Dinajpur) – Rhino can scarcely be said to be known – Hamilton 1820: 223

1822 – Anarpore on Mohamady (Mahananda) River, Rajshahi – Rhino killed, stated to measure 7 ft 5 in (226 cm) high – B H B A M B T 1822 – 30.5 – map A 341

1837 – Gozgotto near "R-pore", shown on Rennell's map of 1794 as Guzgottah (now near Gajaghanta) between Rangpur and Teesta River – Rhino killed was male, with pretty large horn, measuring 4 ft 5 in (145 cm) high and 12 ft 2 in (370 cm) long – Gallovidian 1837: 276; animal identified as R. sondaicus by Blyth 1870: 146, 1872e: 3105 – 30.2, 52.2 – map A 343

1838 – Ronggopoor (Rangpur) – Rhino not uncommon – Martin 1838c: 574; Anon. 1841b: 269 – map A 343

1838 – Ronggopoor (Rangpur) – There are 60–70 hunters, who kill 1–2 rhino each year. Horns bought by merchants in Dhubri – Martin 1838c: 574 – map A 343

1850 – Goggut (Ghaghat) River, Rungpore (Rangpur) – Capt. G. killed large rhino "the largest brute of the genus ever seen in Rungpore" (without measurements) – Asmodeus 1850: 80 – 30.2 – map A 343

1850 – [Temple] Puthia, Rajshahi Dt. – Pancha Ratna Govinda Temple (mid 19th century), within the Puthia Rajbari. Terracotta plaque with rhino – Haque 2014, pl. 214; Das 2018b: 92; Bautze 1985; Rookmaaker & Edwards 2022: 78 – 30.8, fig.33.6 – map M 865

1853 – Unknown locality – Hunting party from Rampore Bauleah (old name for Rajshahi) killed 6 rhinos. The report names Mr. Knowles and a magistrate of Birbhum. The Bombay Times 1853-04-25 refers to this as a "Rungpoor party" – Bombay Times and Journal of Commerce 1853-04-22, p. 762: (from Bengal Hurkaru 1853-04-12) – 30.5

1886 – Rungpore (Rangpur) – Slopes north of Rungpore – Simson 1886: 188, 193 – map A 343

1886 – Dinagepore Dt. (Dinajpur) – Rhinos killed by friends, who preserved their skulls – Simson 1886: 193

1900 – Bangladesh – R. unicornis extinct – Sarker & Sarker 1984

1950 – Kaliagunj, West Dinajpur – Mask of rhino – Indian Museum, ethnographical object – map M 864

1982 – Bangladesh – R. unicornis perhaps survives – Honacki et al. 1982: 311

Rhino Region 23: North Bangladesh – East of Jamuna-Brahmaputra

1640 – Sahad, Mymensingh – Rhino killed by Mirza Nathan (Alauddin Isfahani), officer in the Mughal army during reign of Jahangir (early 17th century) – Borah 1936: 644 following Nathan's Baharistan-i-Ghaybi – map A 344

1868 – Dacca Division – Rhino present – Reynolds 1868: 324 – 30.4

1868 – NW part of Mymensingh – Rhino has been killed; very rarely seen – Reynolds 1868: 252; Sachse 1917: 11 – 30.3 – map A 344

1875 – Maimansingh Dt. (Mymensingh) – Rhino rarely seen but occasionally found – Hunter 1875b, vol. 5: 391 – 30.3 – map A 344

1884 – Dacca (place of residence of collector) – R. unicornis NHM skull, 1884.1.22.3, locality "Dacca" from George Peress Sanderson (1848–1892) – Pocock 1944b: 838 – 30.4

1886 – Region where Mymensingh joins Assam – R. unicornis found. Rhinos killed by friends, who preserved their skulls – Simson 1886: 188, 193 – 30.3 – map A 344

1989 – Northern Bangladesh – R. unicornis female strayed from Gorumara NP; had to be physically brought back – Choudhury 2016: 154

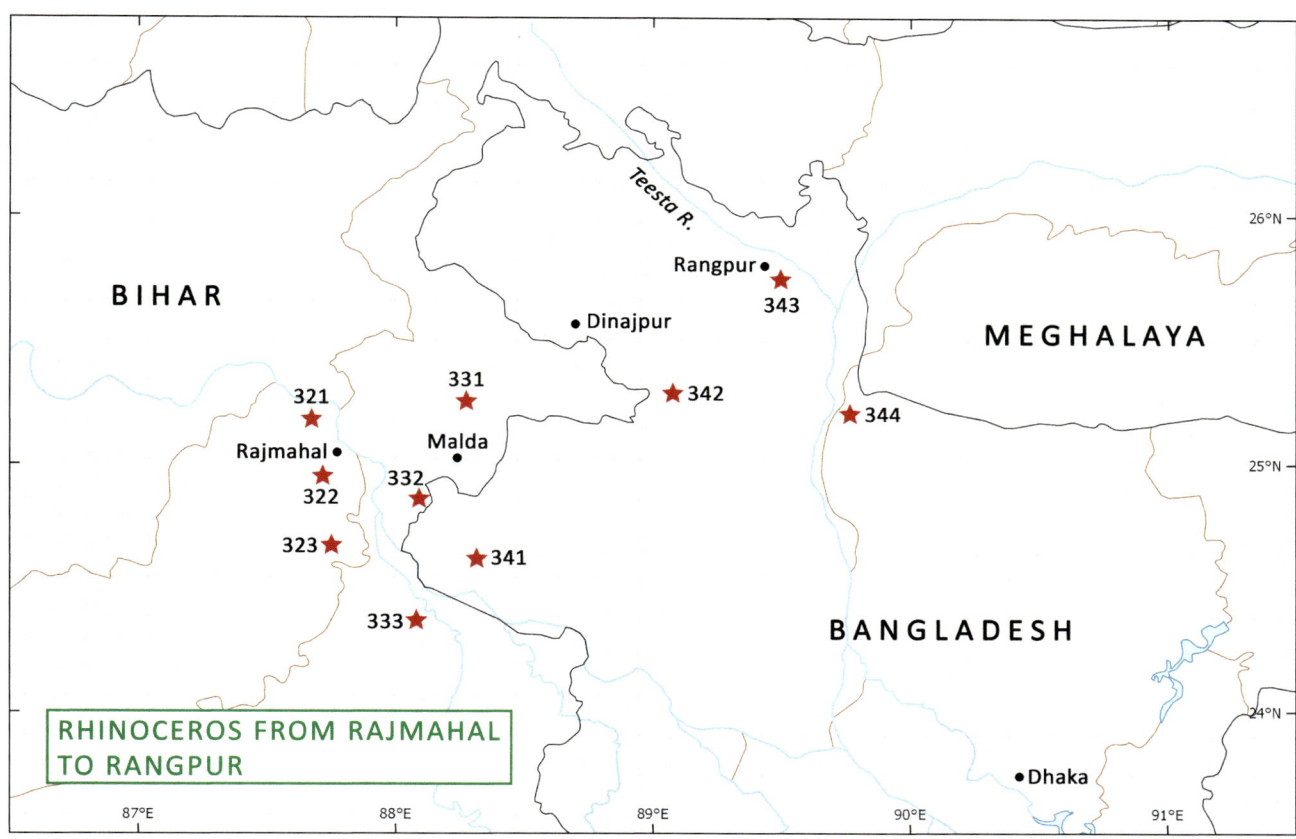

MAP 30.13 Records of rhinoceros from the Rajmahal hills of Jharkhand through West Bengal to Dinajpur, Rangur and Mymensingh. The numbers and places are explained in Dataset 30.32

⭐ Presence of *R. unicornis*

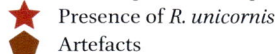 Artefacts

• • •

Dataset 30.32: Localities of Records of Rhinoceros in Middle West Bengal and NW Bangladesh

The numbers and places are shown on map 30.13.

Rhino Region 21: the Middle of Bengal

A 331 Forests of Peruya (near Pandua) – 25.15N; 88.15E – 1806
A 331 Pergunnah Shikarpoor (Malda) – 25.00N; 88.30E – 1854
A 331 Maldah (Malda) – 25.00N; 88.30E – 1876, 1892
A 332 Old capital Gur (Gaur) – 24.75N; 88.08E – 1812
A 333 Murshidabad – 24.15N; 88.13E – 1857
A 333 Berhampore (Berhampur) – 24.15N; 88.13E – 1820

Rhino Region 22: North Bangladesh – West of Jamuna-Brahmaputra

A 341 Anarpore on Mohamady (Mahananda) River, Rajshahi – 24.65N; 88.27E – 1822
*   North Bangladesh – 26.40N; 88.55E – 1989 (stray)

*   Dinagepoor Dt. (Dinajpur) – 25.63N; 88.70E – 1820 (absence?), 1886
*M 864 Kaliagunj, West Dinajpur – 25.63N; 88.70E – 1950 – [object only]
A 342 Ghoraghat, Rangpur – 25.26N; 89.25E – 1609
A 343 Gozgotto, Guzgottah, Goggut (Ghagat) River, Rangpur – 25.83N; 89.25E – 1837, 1850
A 343 Rungpoor (Rangpur) – 25.70N; 89.30E – 1820, 1822, 1838, 1886

Rhino Region 23: North Bangladesh: East of Jamuna-Brahmaputra

A 344 Sahad, Mymensingh – 25.25N; 89.80E – 1650
A 344 NW of Mymensingh Dt. – 25.25N; 89.80E – 1868, 1875, 1886
* not on map

∵

TABLE 30.35   Records of rhinos killed, observed or depicted in NW Bangladesh and middle West Bengal

District – division	Locality	Date	Killed	Sighted	Temple decorations
Bangladesh					
Rangpur	Guzgottah	1837	one, 145 cm		
Rangpur	Goggat	1850	one		
Rangpur	Kantanagar	18th cent.			yes
Mymensingh	Sahad	1640	one		
Rajshahi	Anarpore	1822	one, 225 cm		
Rajshahi	Paharpur	9th cent.			yes
Rajshahi	Puthia	1850			yes
India: West Bengal					
Malda	Peruya	1806		seen	
Malda	Gaur	1812		seen	
Malda	Jagjivanpur	9th cent.			yes
Murshidabad		1800–1850		seen	
Murshidabad	Baranagar	18th cent.			yes

FIGURE 30.1
Rhinoceros hitting a tree, drawn by Armand-Lucien
Clément (1848–1920)
LARIVE AND FLEURY, *DICTIONNAIRE FRANÇAIS
ILLUSTRÉ DES MOTS ET DES CHOSES*, 1889, VOL. 3,
P. 519

30.2   Rangpur Division, Bangladesh

There are just two instances when a rhino was killed near Rangpur, in 1837 and 1850. The habitat must have been marginal despite the proximity to known rhino habitats in Jalpaiguri and Cooch Behar across the Teesta River (fig. 30.1). In 1850, the officer Captain G. killed one on the Ghaghat River in the vicinity of Rangpur, said to be "the largest brute of the genus ever seen in Rungpore" (Asmodeus 1850). In a letter dated "R – pore 24th Feb. 1837" A Gallovidian (1837) told how he killed a rhino which had been terrorizing the neighbourhood of Gozgotto. The locality is found as Guzgottah about 8 km north of Rangpur on contemporary maps. The measurements of this male rhino with a pretty large horn, given a height of 145 cm and a length of 370 cm, are too small for a full-grown *R. unicornis*. It could have been a specimen of *R. sondaicus* as suggested by Blyth (1870), who unfortunately put the event erroneously in the Garo Hills of Meghalaya. Without further evidence, I suggest that this was a juvenile male *R. unicornis* (52.2).

Evidently some officials of the government suspected the presence of rhino in Rangpur, because in 1822 they wrote to the Collector of the town asking him to obtain a live rhino for the Gaekwar of Baroda. Having no hope to catch one, he wrote to a friend in Bijni, near Manas in Assam, the "Samindar of Pergunnah Bignee, on the subject of procuring a Rhinoceros, and as his people are constantly in the habits of taking them, I hope to be able to comply with the desire of Government" (National Archives of India, Home Department, Public, 30 May 1822, no. 27). This time, an animal was obtained from Cooch Behar (5.4).

The rhino in Rangpur, never common, was not found again after 1850. If Martin (1838c) is correct that in his days in the Rangpur area there were some 60 or 70 active hunters who would each shoot 1–2 rhinos every year, for the horns to be sold in Dhubri (Assam), that would account for the early demise.

30.3 Border of Mymensingh and Assam

The rhino was absent from Mymensingh in the north of Bangladesh. Only the 17th century *Baharistan-i-Ghaybi* mentions the killing of a rhino in Sahad, which the editor of the volume calculates to be on the banks of the Brahmaputra (Borah 1936). Two gazetteers by Reynolds (1868) and Hunter (1875b) mention that rhino were once killed, "although now rarely seen" in the north-west of Mymensingh. Just as generally, Francis Bruce Simson (1868) has *R. unicornis* where "Mymensingh joins Assam." Maybe there were some strays from Cooch Behar or Gorumara, but never a permanent population.

30.4 Dhaka Division, Bangladesh

There were no rhinos in this central part of Bangladesh. A skull of *R. unicornis* in the Natural History Museum, London (1884.1.22.3) was donated by George Peress Sanderson (1848–1892) from "Dacca", but as that was his place of residence, it is likely that the animal was shot elsewhere. Williamson (1807) said that Matthew Day (unidentified) had a captive rhino at his house in Dacca in 1807 (5.4).

30.5 Rajshahi Division, Bangladesh

There is only one report of a rhino in the western part of Bangladesh. In March 1822, a party of hunters from Rajshahi killed a rhino, in a place called Anarpore close to the banks of the Mahananda River. The signature B H B A M B T is probably a combination of all their initials. They provided a long list of measurements of this male rhino: length from nose to extremity of tail, 15 ft 10 in (482 cm); height 7 ft 5 in (226 cm); tail 15 in (38 cm); circumference of body 14 ft 8 in (446 cm) and horn 21 in (53 cm). The size looks quaintly inflated and must be treated with some suspicion. They said that a rhino was unusual in the area and thought that it had strayed from the Morung hills in Bihar.

There is no reason to discredit this isolated record while recognizing that this was never traditional rhino country. The author of a general history, Robert Montgomery Martin (1803–1868) mentioned the species in Puraniya district south of Malda, but his reference was probably a misreading for Purnea (Bihar), because that is where another rhino thrust himself on the premises of an indigo planter (Hamilton 1828). Equally improbable is a short notice from a correspondent of Rampore Bauleah (old name for Rajshahi) in a newspaper of April 1853 that Mr. Knowles and party had killed 6 rhinos and other game. The reporter must have forgotten to insert that these animals were shot in another part of the country.

30.6 Malda District, West Bengal

In Malda and Dinajpur districts, only stray rhinos were reported twice, in 1806 and 1812. The author of a historical survey of India, Francis Buchanan-Hamilton (1762–1829) had heard about two elephants who had strayed in 1806 from the north to the forests of Peruya, north of Malda and close to Adina Mosque. These animals were accompanied by a rhino according to the local reports.

In 1812, a French nobleman went to the ruins of the ancient city of Gaur, not to understand the history, but to look for tigers and other wildlife which had come to live in the bamboo forests. Armand Magon de Clos-Doré (1780–1850) teamed up with Capt. W. and seated on two domesticated elephants they looked for wild buffalo, hyena and tiger (Magon 1822). A passing glimpse of a rhino ended their pursuit, scared as they were that their elephants would take fright. Sixty years later, the Collector of Malda, John Henry Ravenshaw (1833–1874) photographed the ruins in relative safety, unhindered by any wildlife (Ravenshaw 1878).

FIGURE 30.2   Bishop with shape of rhinoceros in a chess set made of ivory and wood from Berhampur area, ca. 1820
WITH PERMISSION © NATIONAL ARMY MUSEUM, LONDON: NAM 1962-03-46-5

30.7   Murshidabad District, West Bengal

Among the multifaceted reports in one of the gazetteers edited by William Wilson Hunter (1840–1900) there is a remark that rhinos had been seen in the northern part of Murshidabad (Hunter 1876b). This evidence is too scant to accept that the animals were ever resident in the district.

Two portrayals of the rhino were manufactured in Murshidabad. One is a white Bishop in the form of a rhino, part of a chess set of ivory and wood made by a local craftsman in Berhampur, around 1820, for the European market (Das 2018b: 92) (fig. 30.2).

In the Victoria & Albert Museum in London there is a painting, gouache on mica, showing a hunting party on elephants with a rhino. Made in Murshidabad ca. 1835, it was acquired in 1948 through W.G. Archer from the collection of Ishwari Prasad (number: IS.33: 12–1961). The painting, too dark to be reproduced, shows a rhino with a sizeable horn standing quite tamely on the right. The animal in fact may have been modelled after the engravings by William Daniell (17.4).

∙∙∙

Dataset 30.33: Localities of Terracotta Temples

The numbers and places are found on map 30.14.

Rhino Region 21: the Middle of Bengal

M 861   Baranagar, Murshidabad Dt., West Bengal – 24.25N; 88.23E – 18th century

M 862   Pandua (Adina Mosque), Malda Dt., West Bengal – 25.15N; 88.16 – 14th cent.

M 863   Jagjivanpur, Malda Dt., West Bengal – 25.04N; 88.40E

Rhino Region 22: North Bangladesh – West of Jamuna-Brahmaputra

M 864   Kantanagar, Rangpur – 25.78N; 88.67E – 1722

M 865   Puthia, Rajshahi – 24.37N; 88.83E – 1850

M 866   Paharpur, Rajshahi – 25.05N; 88.97E – 9th cent.

∙∙
∙

30.8   Terracotta Temples in Undivided Bengal

An early terracotta plaque excavated at the archaeological site of Jagjivanpur in Malda Dt. dates from the start of the Buddhist Pala Kingdom in the 9th century A.D. (Roy 2012). Among the great wealth of items, there is one plaque depicting a single-horned rhino in profile (fig. 30.3). The find shows that there was some knowledge of the animal in the region.

A large number of temples largely made from terracotta tiles have been constructed across the regions of Bengal and Bangladesh from medieval times. Among all the scenes found in these structures, there are a few depictions of a rhino (Bautze 1985; Divyabhanusinh et al. 2018). The recent survey by Rookmaaker & Edwards (2022), completed with the help of Dr. Joachim Bautze, has identified four structures which include a portrayal of a rhino, while two others mentioned in the literature could not be verified.

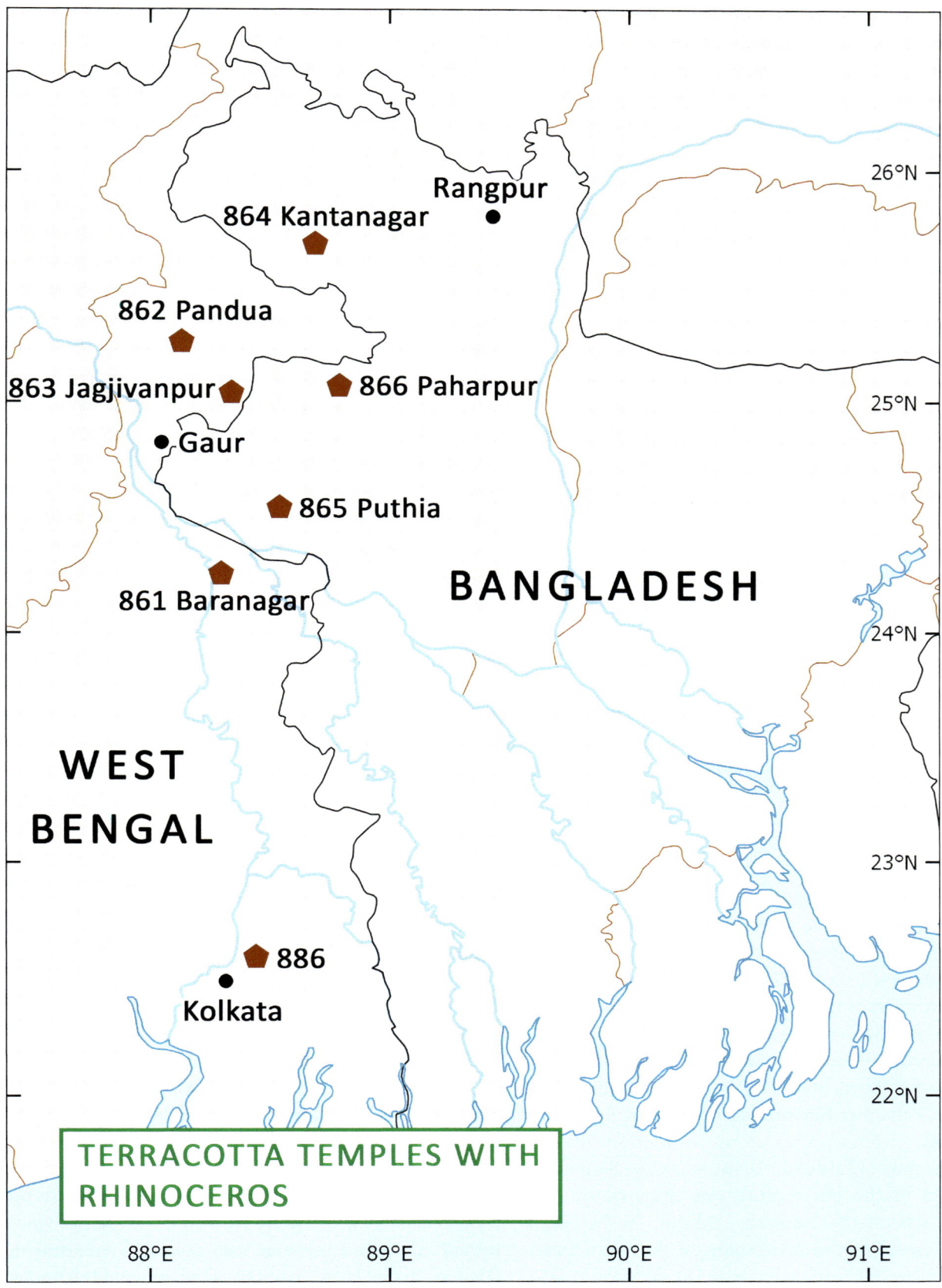

MAP 30.14   Localities of terracotta temples in Bengal with rhinoceros depictions. The numbers and places are explained in
Dataset 30.33. For record M886 (Chandraketugarh) see Dataset 50.59

FIGURE 30.3   Terracotta plaque depicting a rhino in lateral
profile, excavated in 1996–2005 at the archaeological
site of Jagjivanpur in Malda District. State
Archaeological Museum, Kolkata, No. 04.18,
size 26.7 × 23.2 × 59 cm
FROM AMAL ROY, *JAGJIVANPUR*, 2012, PL. 80B

FIGURE 30.4   Rhinoceros with rider on a 9th century terracotta
panel found at Paharpur, Rajshahi Division, now
in the Indian Museum, Kolkata. From the original
photograph taken by Kashinath Narayan Dikshit,
preserved in Archaeological Museum, Government
of Bangla Desh 2603–50. Information from Claudine
Bautze-Picron

The Adina Mosque in Pandua (Malda, West Bengal) built
in the 14th century might have contained a panel with a
rhino. It was mentioned in the report of an archaeological
tour made in 1880 by Alexander Cunningham (1814–1893),
founder of Archaeological Survey of India: "There are five
complete panels, and two half panels, which have been
cut through. These two contain portions of an elephant
and rhinoceros" (Cunningham 1882: 91). However, a more
recent photograph of the place indicated does not show
the animal (Khan & Stapleton 1931: 133, fig. 33).

The Krishnarai Temple built in 1655 in Bishnupur, about
150 km north-east of Kolkata, has a panel showing an animal with a horn like a rhino (Das 2018b: 92). A close examination identifies the animal as a boar with large tusks.

A rhino has been identified in panels in terracotta
temples in Paharpur, Kantanagar, Puthia and Baranagar
(map 30.14).

In Paharpur (Rajshahi Division) the Somapura
Mahavihara of the 9th century was surveyed by Rao
Bahadur Kashinath Narayan Dikshit (1889–1944),
Director-General of the Archaeological Survey of India
from 1937–1944. In his original report, Dikshit (1938, pl. 46)
stated that there were three examples of a rhino, but only
one of these was illustrated and can now be identified
(fig. 30.4). Although the horn is absent (maybe broken off)

and the skin folds are obscured, this must show a rhino. It
is depicted with a rider in a typical flying position, who
might have wielded a sword, in which case he might represent a *vidyadhara* or semi-god in a rare example of the
animal acting as a *vahana* or mount of a god.

In Kantanagar (Rangpur, Bangladesh) the Kantaji
Temple built from 1704 to 1722 contains two plaques with
a rhino. In both cases the animal has a body full of scales,
no skin folds, a smooth head with a tiny horn, and a long
upturned tail (figs. 30.5, 30.6, 30.7). A saddle or cloth
draped on the top of their back might point at some kind
of domestication or captivity. However, this is contradicted by their setting in scenes including hunters on foot
and on horseback.

The Govinda Temple in Puthia (Rajshahi, Bangladesh)
was built in the middle of the 19th century. There is one
terracotta panel with a rhino showing a scaly skin, divided
by curved horizontal lines into several compartments
(Haque 2014, pl. 214). The horn is indistinct in the photograph (fig. 30.8).

In Baranagar (Murshidabad, West Bengal), the
Gangeshwar Siva Temple was built around 1750 close to

FIGURE 30.5
Rhinoceros attacked by two hunters with guns. Terracotta panel in the Kantaji Temple, Kantanagar, Bangladesh
PHOTO © JOACHIM K. BAUTZE

FIGURE 30.6
Rhinoceros hunted by a rider on horseback as well as two persons with guns. Terracotta panel of the Kantaji Temple in Kantanagar, Bangladesh
PHOTO © JOACHIM K. BAUTZE

FIGURE 30.7
Detail of the second Kantanagar panel with a rhinoceros
PHOTO © JOACHIM K. BAUTZE

FIGURE 30.8
Rhinoceros in a terracotta panel of the Govinda Temple in Puthia, Rajshahi Division
FROM HAQUE, *TERRACOTTAS OF BENGAL*, 2014, PL. 214

FIGURE 30.9
Rhinoceros attacked by six hunters with gun. Terracotta panel in the Ganeshwar Shiva Temple in Baranagar, mid-18th century
PHOTO © ASOK KUMAR DAS

the banks of the Hooghly. There is one terracotta panel with a rhino. Six hunters are pointing their guns at the rear of the animal, which has a scaly skin, a short but clearly visible horn, and three horizontal skin folds on the side (fig. 30.9).

The rhino is a rare element in the terracotta temples. In two instances (Kantanagar and Baranagar) the animal is shown as a target for hunters with guns. Das (2018b) identifies the men as Europeans referring to their penchant for this kind of sport. Without further context it is hard to decide if there is any relationship between the scenes depicted on these panels and the actual locality where the temples were constructed. They may be based on accounts which had reached the local rulers and artists from elsewhere. In Puthia there is a human presence possibly unrelated to hunting. The cloth on the back of the rhino in one panel at Kantanagar combined with two

European hunters might indicate that these men captured the animal and possibly transported it to a place closer to the temple.

All rhinos in the terracotta panels show shell-like scales all over the body, which is reminiscent of the mosaic-like patterns on the skin of *R. sondaicus*. The skin folds are generally shown in an unnaturalistic portrayal. The single horn in each case is insignificant in size. There is not enough evidence to reach a clear conclusion which species was depicted. If *R. sondaicus*, the animal or a depiction of the animal must have come from the Sundarbans. If *R. unicornis* as the more likely alternative (Rookmaaker & Edwards 2022), the animals might have been seen in North Bengal or in the Rajmahal Hills. Unfortunately, all conclusions must remain tentative given the absence of documentary evidence and the portrayals of the rhino without showing the main characteristics.

Historical Records of the Rhinoceros in North Bengal

First Record: 1768 – Last Record: 1970s outside protected areas.
Current in Gorumara and Jaldapara. Strays throughout region.
Species: *Rhinoceros unicornis*. Few historical records of *R. sondaicus*
and *D. sumatrensis*
Rhino Regions 24, 25, 26

31.1 Rhinos in North Bengal

North Bengal is prime habitat for the rhinoceros. The animals are encountered frequently throughout this territory, from early days to the present. Although North Bengal is not an official term, it helps to refer to the range of rhinos in the West Bengal districts of Darjeeling, Jalpaiguri, Alipurduar and Cooch Behar. Four large rivers flow through the territories southward, the Teesta (or Tista) River and the Jaldhaka River in the western part, the Torsa (or Torsha) River in the center, and the Sankosh River in the east forming the border with the state of Assam. The usual species found here is *R. unicornis*, while there are indications of the sporadic occurrence of *R. sondaicus* and *D. sumatrensis* (31.2).

Rhinos lived in areas just below the Himalayas generally called the *duars* (also *dooars* or *dwars*). This land east of the Teesta River was governed by Bhutan until 1865 and then annexed by the British after the Anglo-Bhutan war (fig. 31.1). The name "duar" originates from a system of rent-collecting employed by the Bhutan Government, which divided the land in small districts, each with a name ending in 'duar.' The word 'duar' in Hindi means 'door', hence in this region the *duars* may have been the 'doors' to Sikkim and Bhutan in the north. Originally, the entire tract of foothills east of the Teesta River was known as the 'Bhutan duars', stretching for over 100 km into Assam. The 'Sikkim duars' or 'Darjeeling duars' referred to a region on both sides of the Teesta from Jalpaiguri towards the mountains. By the early 20th century, the term had become limited to the tea district that stretched between the Teesta and Sankosh Rivers, and is largely within the boundaries of Jalpaiguri and Alipurduar Districts. The northern boundary of the *duars* is either the foot of the first hills or the tops of the first ridge, which in Buxa may reach up to 2000 m. From the foothills southwards, a strip of land around 10 km wide forms a plateau intersected by rivers and streams, merging with the true plains that continue in a downward slope. The northern portion contained forests and thickets of dense vegetation, where elephants, deer, tigers, buffalo and rhino were found. The southern portion of the *duars* consists of rich black soil where historically rice, cotton and tobacco were grown.

The British army started to send troops to Jalpaiguri in the 1850s aiming to further their interests eastwards of the Teesta River. That is when reports started to appear about English officers hunting rhinos, often on the east bank of the river. The young Lieutenant Robert Cecil Beavan, for instance, was quite open about his illegal excursions in 1859 into Bhutan territory for sport and exploration. After the Bhutan War in 1865 there was more chance to travel eastwards from Jalpaiguri, resulting in frequent shooting parties by army officers, politicians and other travelers until the end of the century (table 31.36).

The regular hunting expeditions organised by the Maharaja of Cooch Behar from 1871 to 1909 largely took place in three areas of North Bengal (chapter 32). He usually shared these trips with an impressive line-up of high-standing persons, as told in the Maharaja's hunting book (Nripendra 1908). Although there are few formal statements regarding the legal status of the land, it is clear that hunting in Cooch Behar was off-limits to the general public. It was difficult enough in the early days to find and shoot a rhino, as the animals lived among the high grasses in muddy parts of the jungle. This was a real deterrent to the large contingent of sportsmen who were able to visit the area. Even a senior officer like Colonel Kinloch in the 1870s could only afford to hunt rhinos on rare occasions. He said that such elaborate arrangements had to be made to obtain the necessary elephants with mahouts, food and guides, that "no one but a millionaire could afford to organise an expedition without assistance."

The rhino was usually encountered between the Teesta and Jaldhaka rivers north of Jalpaiguri, which town was often the starting point for excursions into the interior (Rhino Region 26, see 31.3). Some 19th-century authors referred to rhinos in the Sikkim *terai* generally, which might mean the region west of the Teesta river, although no definite records from this territory were found and maybe the term was used rather loosely also for the hills between Sevoke on the Teesta and Nagrakata on the Jaldhaka. The single newspaper report from Darjeeling town in 1922 is likely to be fictional, as even a stray rhino would not climb the mountains up to a height of 2000 m. East of the Jaldhaka River there is quite a large stretch

© L.C. (KEES) ROOKMAAKER, 2024 | DOI:10.1163/9789004691544_032
This is an open access chapter distributed under the terms of the CC BY-NC-ND 4.0 license.

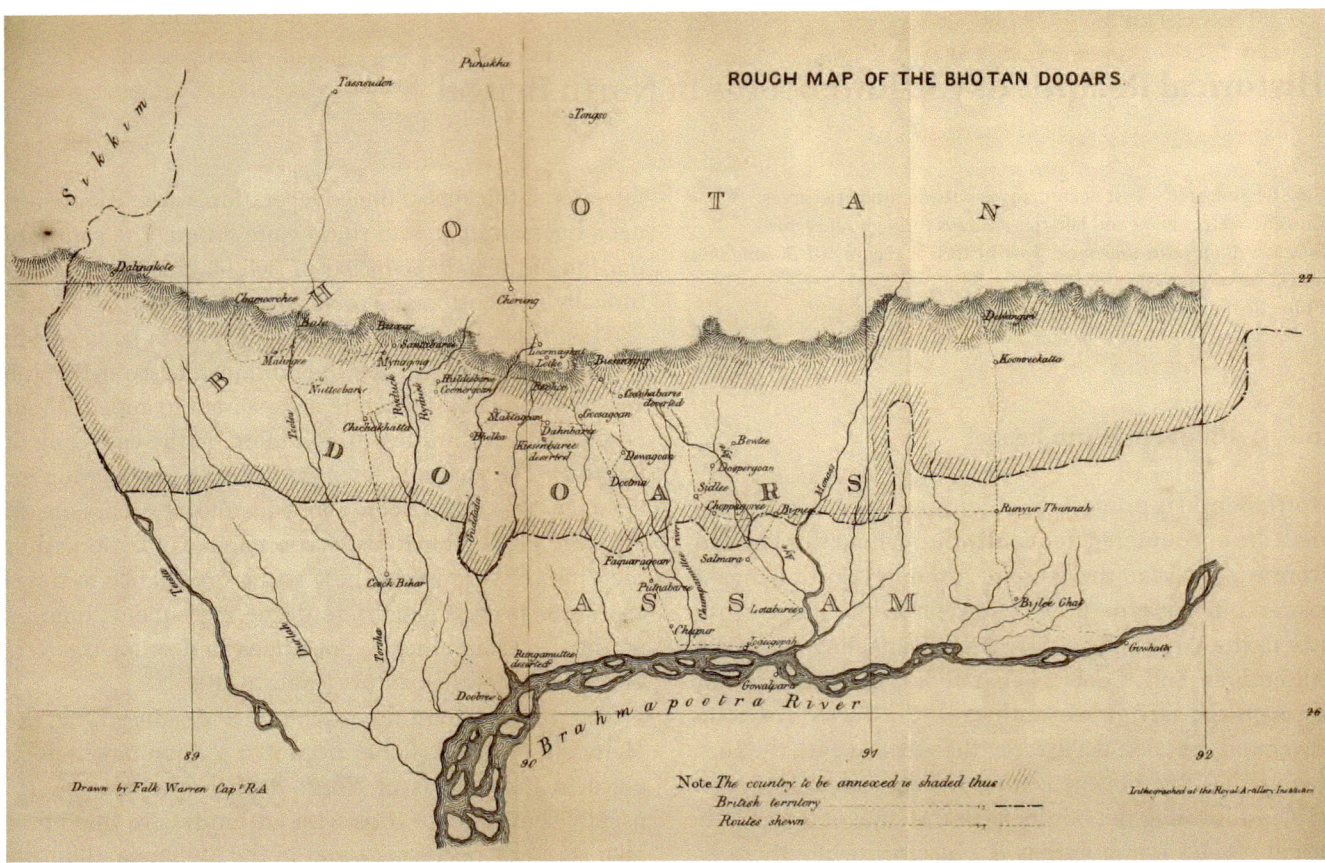

FIGURE 31.1   "Rough Map of the Bhotan Dooars" showing the terminology of this part of West Bengal and Assam, south of the current country of Bhutan, used in the middle of the 19th century. From a report by Falkland George Edgworth Warren (1834–1908) published in 1866

where rhinos were never recorded, which might just mean that it was less accessible to sportsmen and travelers. The animals were again found in the Jaldapara area and further east between the Torsha and Sankosh rivers.

Rhinos were once common in the Sankosh-Raidak region in the eastern part of the state. The Maharaja of Cooch Behar hunted here for most of the years until 1907. Bist (1994) is right to state that ironically this was the first region where rhinos went extinct. This is more likely the result of a lack of protection in the absence of gazetted reserved forests rather than the known hunting activities which would reduce the population by a few individuals per year. The report by Fawcus (1943) is quite open that the poaching in the early 1930s went largely undetected because the rhinos were killed in high grass areas which were never patrolled by the limited forest staff allocated to forests with little economic interest. The Assamese and Mech poachers accused of the slaughter in Jaldapara might have passed through the Raidak area quite unnoticed for many years. The Alipurduar District was greatly developed with tea plantations and other agricultural practices, leaving little room for any wildlife.

31.2   Three Rhino Species in North Bengal

North Bengal is a stronghold of *R. unicornis*. If no further distinction is made in an early report, it must be assumed that the encounters were with this larger species. There is one record for *D. sumatrensis*, from the Dalgaon Forest (chapter 64). There are five records for *R. sondaicus*, in locations ranging from the Teesta River in the west to the Buxa District in the east (chapter 52). It is remarkable that all events involving *R. sondaicus* occurred in places also inhabited by *R. unicornis*, placing the two species always in close proximity. This may call for an explanation which will need further research into the relationships between species. Both *D. sumatrensis* and *R. sondaicus* were extinct in North Bengal at the end of the 19th century.

31.3   Definition of Three Rhino Regions in North Bengal

Sampat Singh Bist, former Principal Chief Conservator of Forests in West Bengal, wrote an important paper on the history of the rhino in the state in 1994. He divided the range into three populations and five sub-populations, largely based on the locations of the shooting camps of the Maharaja of Cooch Behar. From east to west and indicated by Roman numerals, these were (I) the Sankosh-Rydak (Raidak) population stretching from the border with Assam to Buxa; (II) the Jaldapara and Patlakhawa sub-populations around the Torsa River, and (III) the Gorumara population near the Teesta River. He found that rhinos were extinct in area I in 1970, and in area II much earlier.

West of the Teesta River, there are no reports of rhino occurrence from Darjeeling District (West Bengal), Uttar Dinajpur (West Bengal), Rangpur (Bangladesh) and Kishanganj (Bihar). Although it is only a distance of about 100 km, this is remarkable. The habitat may have been unsuitable or the region may have seen early settlement and expansion of agriculture.

The divisions suggested by Bist are followed here, but re-arranged in a direction from west to east, hence Rhino Region 24 (Bist III), Rhino Region 25 (Bist II) and Rhino Region 26 (Bist I). The same terminology is used in 32. 9 for the locations of the Maharaja's shoots.

Rhino Region 24

North Bengal: Teesta River to Jaldhaka and Diana Rivers. This corresponds with Bist Area III. The Maharaja of Cooch Behar hunted here in 1884 (32.9). This region includes the present Gorumara National Park. Most records indicate *R. unicornis*, while there are four of *R. sondaicus*.

Rhino Region 25

North Bengal: Torsa River (in Jaldapara). Bist (1994) separates two sub-populations. Most records indicate *R. unicornis*, while there are one each for *R. sondaicus* and *D. sumatrensis*.

Sub-population B-1 (Bist II-A) is the northern stretch from the present Jaldapara NP and Chilapata reserves to Bhutri and the adjoining forests in Buxa and Cooch Behar Divisions. The Maharaja of Cooch Behar shot here in 1892, 1893, 1904 (32.9).

Sub-population B-2 (Bist II-B) is the southern stretch including the Patlakhawa reserve and extending to Pundibari near Cooch Behar. The Maharaja of Cooch Behar shot here in 1905 (32.9).

Rhino Region 26

North Bengal: Sankosh-Rydak (Raidak). Bist (1994) separates two sub-populations.

Sub-population C-1 (Bist I-A) includes the Rydak (Raidak) and Bhalka forests in the present Buxa Tiger Reserve. This is in Alipurduar District. The Maharaja of Cooch Behar shot here in 1878, 1881, 1884, 1891, 1893, 1895, 1896, 1904, 1905 (32.9).

Sub-population C-2 (Bist I-B) is located further south in Garodhat (Garadhat) between the town of Cooch behar and the Sankosh River, which is the border with Assam. This is in Cooch Behar District. The Maharaja of Cooch Behar shot here regularly, in 1883, 1884, 1885, 1886, 1887, 1889, 1890, 1891, 1892, 1893, 1895, 1896, 1897, 1898, 1900, 1901 (32.9).

• • •

Dataset 31.34: Chronological List of Records of Rhinoceros in North Bengal

General sequence: Date – Locality (as in original source) – Event – Source – § number, figures – Map number.

The map numbers are explained in dataset 31.35 and shown on map 31.15.

The hunting records of the Maharaja of Cooch Behar 1871–1909 are discussed in chapter 32. The localities are included in the list but the records themselves are not repeated here. After 1930 most rhino sightings were in Gorumara NP and Jaldapara NP with details in chapters 33 and 34.

The records are divided according to Rhino Regions (31.3) preceded by general comments without precise localities. The persons interacting with a rhinoceros in various events are combined in Table 31.36. Abbreviation used: MCB for the Maharaja of Cooch Behar, Nripendra Narayan (1862–1911).

North Bengal – General Records

1768 – "Rhinoceros Wilds" (second instance) indicated in region of Cooch Behar on map produced by Thomas Jefferys (1719–1771), English cartographer – Jefferys 1768 – 31.4, fig. 31.3

1781 – "Royaume du Rhinocéros" or Kingdom of the Rhinoceros – Indication in the region of North Bengal on a map prepared by Louis Brion de la Tour (1756–1823), Cartographer Royal to the King of France – Brion de la Tour, map dated 1781 (Ryhiner Collection, University of Bern); Bautze 1985: 415 n. 25 – 31.5, figs. 31.4, 31.5

1800 – Bengal – William Carey (1761–1834), the founder of Serampore College, states rhino presence – Carey 1800: 139

Undated, maybe 1840s – Without locality (Bengal ?) – Author recollects "a Bengal Nuwab who killed two rhino, and so proud was he of this, that he caused one to be lugged along from the jungle to his capital, in a decomposed state, and it was then very indifferently cleaned and stuffed with straw, and in a short time became so very offensive as to be a public nuisance. As the Nuwab would not remove it, some young men in the neighbourhood very quietly poked it into the river when unusually dark" – Old Shekarree 1860: 129

1849 – Bengal (no locality) – *R. unicornis*, juvenile female. Articulated skeleton, and skull. Obtained in 1849 or 1852 from Gustav Adolph Frank (1809–1880), animal dealer in Amsterdam. Specimen in Naturalis Museum, Leiden, old no. a, current Mam.17921 – Jentink 1887: 167, 1892: 197

1850 – Bengal – Rewards were sanctioned for the destruction of wild animals, at a rate of Rs 5 per rhino – G. Singh 2014: 35 (Proceedings of the Lieutenant Governor of Bengal during February 1870, Judicial Department, Proc. No. 179–180; West Bengal State Archive)

1851 – North Bengal – Pseudonymous author was present at killing of 7 rhino in unclear locality – Veritas 1856: 302

1852 – North Bengal (?) – Pseudonymous author. Killed 7 rhino – Veritas 1856: 302

1853 – North Bengal (?) – Pseudonymous author. Killed 6 rhino – Veritas 1856: 302

1855-03-17 – North Bengal (?) – Killed 3 rhino. Rhino chased elephant carrying the hunter – Veritas 1856: 302

1855-03-27 – North Bengal (?) – Killed 2 rhino. A young rhino captured. Survival of the captured animal not verifiable – Veritas 1856: 302

1855-03 – Bengal (no locality) – American businessman Charles Huffnagle (1808–1860) on a hunting expedition from elephant back shot 25 tigers and 5 rhinos. He had donated another rhino hide and skeleton to the Museum of the Asiatic Society of Bengal in 1850 – Blyth 1850: 88; Anderson 1916: 8 (quoting manuscript diary entry written by hand in *Bengal Almanac* for 1855; no longer retrievable in Mercer Museum and Library & Fonthill Castle, Doylesdtown, Pa.) – 51.11

1862 – Bhutan *terai* – Hermann Schlagintweit (1826–1882) offers skull to Smithsonian Institution, Washington, DC – Smithsonian 1862: 84

1867 – North Bengal (?) – Heinrich Friederich Lawaetz Melladew (1842–1925) of the Horse Guards finds 4 rhino, not shot – Melladew 1909: 11

1869 – North Bengal – Reward for shooting rhino is Rs 5 – Rainey 1869: 593

1873 – Bengal – Horn sold at 50–60 rupees per seer (about 1.25 kg). The hide when cut and polished makes capital whips – Wanderer 1873: 535

1896 – Bengal – Reward for shooting rhino is Rs 20 – Martin & Martin 1982: 29; Menon 1995: 56

1899 – By notification No. 843T.R. of 13 October 1899, the killing of rhino in the reserved forests of Darjeeling and Jalpaiguri districts was made penal. It is reported that no further measures for the preservation of this animal are at present needed – Wild 1900: 35; J.W.G. 1900

c.1900 – Bhutan duars (Shikarpur) – Fictional story set on fringe of Himalayas. Rhino presence reported, but none seen – Gouldsbury 1909

1909 – Bengal *terai* – Rhino scarce; shooting prohibited – White 1909: 322; Eardley-Wilmot 1910; Trafford 1911: 69; Menon 1995: 39

1914-11-10 – North Bengal – Notification No.10479 For. (10 November 1914): season closed for whole year for rhino – Muriel 1916: 63

1919 – Western Duars – Rhino strictly preserved for 15 years, slowly increasing in numbers – Milligan 1919: 12

1922 – Phobtshering Tea Estate, Darjeeling – Rhino on rampage in town. Shot by Mr. Ward Wilson and Mr. Pascal. – Roy 2003: 65 (after *Darjeeling Advertiser* 1922-05-10) – 31.17. The story is most likely fictional

1932-06-23 – Bengal Act VIII of 1932: The Bengal Rhinoceros Preservation Act, 1932. Passed by Abdelkerim Abu Ahmed Ghuznavi (1872–1939) – Bengal Government 1932; Berg 1933: 51; Ullrich 1971: 21

1954 – North Bengal – Shooting rights were leased to three Game Associations in Darjeeling, Jalpaiguri and Cooch Behar, which proved largely ineffective – Dutt-Mazundar 1954: 158

1963 – North Bengal – *R. unicornis* present in two small pockets – Stracey 1963: 66

1968–1972 – North Bengal – Period of heavy rhino poaching by families of Karjee, Baraik (from Mech community) and Tamang (from Nepal) – West Bengal Forest Dept. 1993: 6; Martin 1996a; Bist 1997

1982–1985 – North Bengal – Period of heavy rhino poaching – West Bengal Forest Dept. 1993: 6

1983–1997 – Siliguri – Rhino products exported through Siliguri. At least 18 rhino horns were seized – Martin 1999

Rhino Region 24: North Bengal – Jalpaiguri (Bist Area III)

1869 – Julpigoree (Jalpaiguri) – Charles, hunting with Raja of Julpigoree, did not find any rhino – Charles 1840: 114

1859 – Butan side of the river Teesta, not far from Jalpigéri (Jalpaiguri) – Captain Frederick Richard Norman Fortescue (1823–1867) shot adult male rhino. Skull with horn presumably taken to UK, examined in Calcutta by Blyth and figured – Blyth 1862a: 155, pl. 1 f. 1, pl. 2 f. 1; Jerdon 1867: 233 – 31.8, figs. 31.8, 31.9 – map A 354

1860 – Near Jalpaigori (Jalpaiguri) – Presence of large numbers of rhino – Kinloch 1903: 161 – map A 354

1862 – Bootan frontier (Jalpaiguri) – Ashley Eden (1831–1887) saw numerous fresh footprints, hence rhino must be numerous – Eden 1865: 192 – map A 354

1864 – Eastern bank of Teesta River across from Julpigoria (Jalpaigori) – Lieutenant Robert Cecil Beavan (1841–1870) saw herd of 7–8 rhinos and killed 2 – Beavan 1865, 1868: 388 – 31.9 – map A 354

1864-12 – Kranti (Apelchand Forest) – Thomas Andrew Donogh (b. 1818) finds rhino in abundance – Donogh 1872: 248 – 31.11 – map A 353

1864-12-26 – Maynagori – T.A. Donogh kills old rhino, 8 feet (2.4 m) high – Donogh 1873a: 65 – 31.11 – map A 354

1865-01-12 – Maynagori – T.A. Donogh finds rhino in abundance – Donogh 1870: 1567 – 31.11 – map A 354

1865-11 – Jangiri, Bhutan Dooars (imaginary name =? Jalpaiguri) – Rhino in pool, not full grown, small horn – Poins 1868 – map A 354

1865 – Jalpaiguri – The sepoys of a Gurkha regiment quartered at Jelpigori were very keen and successful rhino hunters, and had killed no less than seventy in the course of a year, which made them much money – Elwes 1930: 50 – 31.18 – map A 354

1866 – Sikhim *terai* – Rhino very common – R.H.D. 1866: 485 – map A 351

1867-12 – Kyranti (Kranti), 15 miles from Maynagori – Saw rhino which escaped – King Pippin 1870 – map A 353

1868 – Jalpaiguri Dt. – In 1868, rewards paid for destruction of 12 rhinos, recorded by John Colpoys Haughton (1817–1887), Commissioner of Cooch Behar 1865–1873 – Haughton 1879: 30 – map A 354

1868 – Julpigooree (Jalpaiguri) – Injuries of people inflicted by rhino are common, stated by David Boyes Smith (1833–1889), Sanitary Commissioner for Bengal – D.B. Smith 1868: 206 – map A 354

1869 – Julpigoree (Jalpaiguri) – James Howard Thornton (1834–1919) heard that Commissariat elephant was charged by rhino in a shooting party east and north of Julpigoree [may refer to incident on 1870-04-15] – Thornton 1895: 165 – map A 354

1869 – Jalpaiguri Dt. – In 1869, rewards paid for destruction of 20 rhinos – Haughton 1879: 30 – map A 354

1870-04-15 – Sivole (Sevoke), Teesta River – Henry John Elwes (1846–1922) visited the *terai* to shoot rhino, together with Colonel Francis Charles Bridgeman (1848–1924) and Field Marshall Francis Wallace, 1st Baron Grenfell (1841–1925). Sees rhino. One rhino charged an elephant – Elwes 1930: 48; Sharpe 1906: 345 – 31.12 – map A 351

1870-04-22 – Bhutan Duars – Elwes meets (at Titalya or Tentulia) the police superintendent Richard Percival Davis (1840–1915), "who had probably shot more rhinoceros in the Dooars than any man in India" – Elwes 1930: 48 – 31.12 – map A 354

1870-04 – Ramshaihat (Ramshai) – Elwes hunts with Field Marshall Francis Wallace Grenfell. They shoot several rhino – Grenfell 1925: 29; Elwes 1930: 51; Stebbing 1930: 159 – 31.12 – map A 355

1870-04-27 – Ramshaihat (Ramshai) – Lieutenant-Colonel William Osborne Barnard (1838–1920) wounded a rhino – Elwes 1930: 51 – 31.12 – map A 355

1871 – Jalpaiguri vicinity – Rhino present – Wanderer 1871: 26 – map A 354

1873-02 – Bhootan Duars near Jalpaiguri – Killed 4 rhino. One female, 11 ft (335 cm) long, 4 ft 8 in (142.2 cm) high – Wanderer 1873: 534 – map A 354

1873-03 – Bhootan Duars near Jalpaiguri – Kills 4 rhino in 3 days – Wanderer 1873: 537 – map A 354

1873-03-12 – Bhootan Duars near Jalpaiguri – Shot 2 rhino. A friend killed one with a magnificent horn – Wanderer 1873: 537 – map A 354

1873-03 end – Bhootan Duars near Jalpaiguri – Killed 5 rhino including one with horn of 15 inch (38 cm) – Wanderer 1873: 537 – map A 354

1875–1877 – Jalpaiguri – Joseph Carne MacDonell (1849–1932) saw female rhino with young in a small swamp – MacDonell 1929: 87 – map A 354

1875 – Bhootan *terai* – Young female shipped to London by the dealer William Jamrach (1843–1891). Identified as *R. unicornis* on arrival. It died in premises in London of Charles William Rice (1841–1879). Skeleton in Royal College of Surgeons of England, London, no. 2127; destroyed in war bombing – Sclater 1875a: 82, 1876a: 648; Rainey 1875a, 1877; Flower 1884: 418; Rookmaaker 1998: 88 – 31.13 – map A 354

1875 – Bhotan *terai* – Skull of *R. unicornis*. Royal College of Surgeons of England, London, no. 2128. Probably same animal as no. 2127. Donation by Alfred Henry Garrod (1846–1879) – Flower 1884: 418 – map A 354

1876 – Darjiling Dt. *terai* – Rhino pretty abundant – Hunter 1876c: 39 – map A 351

1877-03 – Ghorabhandah (Gorubathan) – Hunting with Maharaja of Darbhanga, Major General Robert Cotton Money (1835–1903) and Charles Theophilus Metcalfe (1837–1892) killed 7 rhinos, captured 1 – Money 1893 – 31.14 – map A 352

1877 – Ghorabhandah (Gorubathan) – *R. sondaicus*. Robert Cotton Money (1835–1903) killed one – 52.4 – map A 352, S 146

1878 – Bhutan Duars – Alexander Angus Airlie Kinloch (1838–1919) shot 2 rhino after his trip organized by the Maharaja of Cooch Behar in April 1878, on subsequent occasion in same district – Kinloch 1885: 63, 1892: 84–87 – 31.15 – map A 354

1878-05 – Left bank of the Ti'sta (Teesta) River – *R. sondaicus*. A friend of Kinloch killed one – 52.4 – map S 147

1881 – Moraghat, Bhutan Duars – *R. unicornis*. Frederik August Möller (1852–1915) shot one on 6 June 1881. Specimen in Zoological Museum of Copenhagen, Denmark, accession number CN-525 – 31.16, figs. 31.16, 31.17 – map A 358

1881 – Moraghat, Bhutan Duars – *R. sondaicus*. Frederik August Möller (1852–1915) killed a female – 52.6 – map A 358, S 148

1884 – Jalpaiguri – Rhino no longer roam – *Homeward Mail from India* 1884-06-30: 632 – map A 354

1884 – Jaldhaka River – MCB and party killed 4 rhino – 32.9 Shoot 18, 1884 (a) – map A 357

1886-02 – Bhutan Duars (Jalpaiguri) – Alexander Kinloch kills old male rhino, figured in book – Kinloch 1892: 88–90, 1903: 161–162 – 31.15, figs. 31.12, 31.13 – map A 354

1887 – Darjeeling *terai* – Rhino present – Ward 1887: 17 – map A 352

1887 – Julpigoree (Jalpaiguri) – Rhino shooting still possible, between December and July – Baker 1887: 36 – map A 354

1887-05 – Foot of Bhutan Hills – Robertson Pughe (1822–1906), Superintendent of Police, saw a group of 6 rhino – Baker 1887: 257 – map A 354

1889-11-25 – Teesta River, Kaliguri Guest House (?) – Rhino presence found by William Sproston Caine (1842–1903) – Caine 1890: 349 – map A 351

1890 – Sikkim Duars along Darjiling railway – Rhino present in this general area – Caine 1890: 347 – map A 351

1890s – Jalpaiguri – One fine rhino killed – J.W.G. 1900 – map A 354

1890s – Jaldhaka Region – Calculated 20 rhinos in 1890s – Bist 1994 (area III) – map A 354

1893 – Bhutan Dooars – Rhino used to be plentiful – Money 1893: 616 – map A 354

1894 – Julpiguri (Jalpaiguri) – Mr. Savi (unidentified), superintendent of khedda operations, secured 220 elephants, and killed one rhino – *Englishman's Overland Mail* 1894-04-04 – map A 354

1895 – Tendu (Chapramari) – Hunting party of Charles Alfred Elliott (1835–1911), Governor of Bengal. Rhino "charged into a beating elephant and was shot by Ghatru, the well-known elephant catcher" – *Englishman's Overland Mail* 1895-03-20 – map A 356

1895 – Jalpaiguri – *R. unicornis* "hati gera", found in the forests; also in swampy *khas* lands. Are becoming scarce. Meches, Garos and Rajhansis eat the flesh – Sunder 1895: 103 – 64.5 – map A 354

1895 – Jalpaiguri – *R. sondaicus*. Kuku gera exists in district – Sunder 1895: 103 – 52.7 – map A 354, S 147

1895 – Gorumara – Notification as Reserved Forest – chapter 33 – map A 355

c.1900 – Jalpaiguri Dt. – Charles Elphinstone Gouldsbury (1849–1920) mentions rhino along northern border – Gouldsbury 1913: 35 – map A 354

1900 – Jalpaiguri *terai* – Killed 1 "fine" rhino tracked by his mahout. Identity of author unknown – J.W.G. 1900 – map A 354

c.1900 – Bhutan [Duars] – *R. unicornis*. Skin of male. No details. Naturhistoriska Museet, Göteborg, Sweden. Catalogue no. M.987 – Göteborg Museum, Sweden – map A 354

c.1900 – Bhutan Duars – George Cecil (n.d.) shot a couple of rhinos – Cecil 1911: 1080 – map A 354

1900 – Jalpaiguri and Buxa – Permit issued to shoot a rhino to General George Luck (1840–1916), Commander-in-Chief, Bengal Command 1898–1902 – Revenue Department, Forests, Proceedings for Nov. 1900 (Summary p. 155) – map A 354

1900 – Eastwards of Tista (Teesta) valley – Rhino present – Lydekker 1900: 23, 1918, vol. 2: 117 – map A 354

1900 – Bhootan Dooars – Rhino abundant; frequenting swampy ground and dense jungles – Russell 1900: 339 – map A 354

1909 – Jalpaiguri – Rhino decreasing. Protected in reserved forests – Hunter 1909: 224 – map A 354

1919 – Jalpaiguri – Strict preservation. Not as common as formerly – Inglis et al. 1919: 825; Stebbing 1920: 128 – map A 354

1919 – Bhootan [Duars] – Rhinos in reedy and grassy swamps – Hornaday 1919 – map A 354

1922 – Darjeeling Dt. – Rhino is rare – Dozey 1922: 166 – map A 351

1929–1935 – Jalpaiguri – Heavy poaching by local people – Shebbeare 1935: 1229 – map A 354

1932 – Sevoke – *R. unicornis* first occurs about 51 miles south-east of Darjeeling – Hobley & Shebbeare 1932: 21 – map A 351

1936 – *Duars* (60 miles from Siliguri) – Diana Hill of Ranikhet saw a rare rhino. A local hunter said that there was a big herd in the region – Hill 1937 – map A 357

1946 – Nagrakata – Philip R.H. Longley (b. 1898) saw rhino on road near Nagrakata Club – Longley 1969: 132 – map A 357

1949 – Gorumara Wildlife Sanctuary – Notification of status – Chapter 33 – map A 355

1949–1951 – Hillajhora RF (Jaldhaka region) – Rhino present – Bist 1994 – map A 357

1969 – Maynaguri (Jaldhaka) – Rhino wandering from Gorumara – Bist 1994 – map A 354

1969 – Chapramari (Jaldhaka) – Rhino wandering from Gorumara. Chapramari was declared a Reserved Forest in 1895, and a Wildlife Sanctuary in 1998 – Bist 1994 – map A 356

1969 – North Diana, South Diana (Jaldhaka) – Rhino wandering from Gorumara – Bist 1994 – map A 355

1992-08 – Mahananda WS – Female rhino strayed from Gorumara into the forests of Apalchand range and moved as far west across the river Tapeta as Gulma in Mahananda sanctuary. Died in Apalchand forest in August 1992 – Bist 1994 – map A 351

1996 – Chapramari Wildlife Reserve – Straggler from south (Gorumara). Same animal wandered into Darjeeling District – Choudhury 2013: 225 – map A 356

2015 – Mahananda WS (on Teesta R) – A male rhino found wandering on 2015-12-11. Transferred to Bengal Safari, Siliguri (new zoo) – Anon. 2016; West Bengal Forest Dept. 2016 – map A 351

Rhino Region 25: North Bengal – Alipurduar (Bist Area II)

1800 – Buxadewar Hill, north of Chichacotta – Samuel Turner (1759–1802) hears about rhino, animal not seen – Turner 1800: 20 – map A 363

1865-02-09 – Falakata – Thomas Andrew Donogh (b. 1818) kills one rhino with sharp horn – Donogh 1873b: 221 – 31.11 – map A 359

1865-02-13 – Falakata – T.A. Donogh kills four rhino, one with large horn – Donogh 1869: 559, 561, 563 – 31.11 – map A 359

1865-02-21 – Falakata – T.A. Donogh kills one rhino – Donogh 1873c: 575 – 31.11 – map A 359

1865-03-07 – Falakata – T.A. Donogh kills three rhino including a pregnant female; embryo preserved – Donogh 1875: 216 – 31.11 – map A 359

1865 – Bala to Buxa – Captain John Henry Baldwin (1841–1908) saw huge 'gainda' which ran away – Baldwin 1876: 145 – map A 363

1865 – Buxa – Commanding officer Colonel S – d shot rhino, referring to William Joseph Fitzmaurice Stafford (1820–1887) – Baldwin 1876: 145; Ruggles 1906: 144 – 31.10 – map A 363

1890s – Torsa Region – Calculated 100 rhinos in 1890s – Bist 1994 (area II) – map A 362

1892 – Torsa River – MCB and party killed 4 rhino – 32.9 Shoot 29, 1892 (b) – map A 362

1893 – Torsa River – MCB and party killed 1 rhino – 32.9 Shoot 31, 1893 (b) – map A 362

1895 – Dalgaon Forest – *D. sumatrensis*. "Rhinoceros Malayan" exists in locality – 64.5 – map D 92

1900 – Chilapata, Buxa division – *R. sondaicus*. James Wyndham Alleyn Grieve (1872–1939) killed one – 52.8 – map A 362, S 149

1904 – Torsa River – MCB and party saw rhino – 32.9 Shoot 48, 1904 (c) – map A 362

1905 – Torsa River – MCB and party killed 1 rhino – 32.9 Shoot 51, 1905 (b) – map A 362

1908 – Buxa Duar – Gordon Casserly (1869–1947) saw rhino once. Prohibited to shoot – Casserly 1914: 98 – map A 363

1908 – Cooch Behar – Jitendra (1886–1922), second son of Maharaja Nripendra Narayan killed rhino close to the palace in town – Casserly 1914: 298 – 32.9 for 1895 – map A 364

1911 – Torsha River north of Silitorsa – John Frederick Gruning (1870–1922) sees over 20 tracks in area – Gruning 1911: 13 – map A 362

1911 – Cooch Behar – *R. unicornis* is common – Kauffmann 1911: 158 – map A 363

1914 – Alipur Duar – Gordon Casserly (1869–1947) mentions presence of rhino – Casserly 1914: 4 – map A 363

1920 – Buxa Duars – Prohibition to shoot rhinos – Stebbing 1920: 128 – map A 363

1920s – Cooch Behar – Joseph Veasy Collier (Indian Forest Service) killed rhino with long horn of 12 ¼ inch (31 cm) – Ward et al. 1928: 438 – map A 363

1920s – Cooch Behar – Major Henry Algernon Hildebrand (1883–1947), 34th Poona Horse Regiment, shot rhino with long horn of 12 ½ inch (31.8 cm) – Ward et al. 1928: 438 – map A 363

1921 – Buxa – Rhinos break down fences – Bengal Forest Dept. 1921: 6 – map A 363

1930 – Buxa – Rhino common – Mukherjee 1963: 46 – map A 363

1930–1949 – Patlakhawa (Torsa) – People in Cooch Behar vouch for the presence of "dozens" of rhinos in the region – Bist 1994 – map A 361

1932 – Kuch Behar – Indian rhino may survive – Thompson 1932: 154 – map A 363

1937 – Buxa division – Rhino present. Numbers unknown – Harper 1945: 377 (Senior Conservator of Forests, Bengal, 1937) – map A 363

1941-03-13 – Jaldapara notified as Game Sanctuary – Chapter 34 – map A 360

1940s–1950s – Torsa Region (area II) – Rhinos from Jaldapara wander to forests of Nilpara Range (Bhutri, Godamdabri, Bharnabari, Rangamati) – Bist 1994 – map A 362

1940s–1950s – Torsa Region (area II) – Rhinos from Jaldapara wander to forests of Madarihat Range (Khaliburi, Till and Dumchi) – Bist 1994 – map A 362

1944-03 – Cooch Behar – Trophy head inscribed "Cooch Behar, 14-3-44" in Ramgarh Lodge (Taj hotel), Jaipur – https://hiveminer.com/Tags/jamuwaramgarh – map A 363

1949 – Cooch Behar – Nancy Valentine (1928–2017) photographed with dead rhino – Stiles 2017 – 30.10, fig. 32.29 – map A 363

1950s (?) – Cooch Behar – Maharaj Colonel Kesri Singh (b. 1927) shooting with Maharaja of Jaipur states that male rhino attacked the elephant with the Maharaja – K. Singh 1979: 154 – map A 363

1954-01 – Chilapata (Torsa) – V.S. Rao recorded rhino in Khairbari, Bhutri, Salkumar, and Basti (Chilapata) – Bist 1994 – map A 362

1954-01 – Mairadanga (SW Jaldapara) – V.S. Rao recorded the rhino in a private Jute forests near Mairadanga – Bist 1994 – map A 360

1955–56 – Patlakhawa (on Torsa R.) – Presence of 10 rhino – Dutt-Mazundar 1954: 158; West Bengal Forest Dept. 1993: 1 – map A 361

1970 – Patlakhawa – Regular habitat of rhinos until about 1970 – Bist 1994 – map A 361

1974 – Patlakhawa – No rhino present in census of 1973–74 – Bist 1994 – map A 361

1982 – Jalpaiguri division – Rhino present – Mukherjee 1982: 49 – map A 363

1982 – Nimati Range, Buxa – Rhino wandered from Jaldapara to 1982 – Bist 1994 – map A 363

1983 – Buxa Tiger Reserve – Established in Feb. 1983, when no rhino present – Bist 2008 – map A 363

1984 – Torsa Region – Rhinos wandered to forests of Titi and Jalgaon until 1984 – Bist 1994; West Bengal Forest Dept. 1993: 1 – map A 362

1985 – Patlakhawa – Occasional visits of rhinos from Jaldapara. None recorded since 1985 – Bist 1994; West Bengal Forest Dept. 1993: 1 – map A 361

2018 – Patlakhawa – Recommendation to transfer rhino to this area – Indian Times 2018-05-27 – map A 361

Rhino Region 26: North Bengal – Cooch Behar (Bist Area I)

1785 – Rhino depicted in area of Cooch Behar on map prepared for Jean-Baptiste-Joseph Gentil (1726–1799) – Gole 1988; Rookmaaker 2014a; Gentil Atlas in British Library of London, India Office Library and Records, Prints and Drawings section Add.MS. Or. 4039, folio 14 – 18.8, 31.6, fig. 31.7

1801 – Cooch Behar – Maharaja Harendra Narayan (1783–1839) often went out to shoot rhino – Ghose 1874: 172; Chaudhuri 1903: 277; Ray 2015: 448 – 31.7

1877–1897 – Sankosh Region – Maharaja of Cooch Behar Nripendra Narayan (1862–1911) killed 104 rhino and wounded 18 here in 20 years – Bist 1994 – chapter 32

1871 – Raidak-Sankosh – MCB and party kill 2 rhino – 32.9 Shoot 1, 1871 (a)

1873 – Raidak-Sankosh – MCB and party kill 4 rhino – 32.9 Shoot 2, 1873 (a)

1877 – Raidak-Sankosh – MCB and party kill 15 rhino – 32.9 Shoot 4, 1877 (a); Shoot 5, 1877 (b); Shoot 6, 1877 (c)

1878 – Raidak-Sankosh – MCB and party kill 16 rhino – 32.9 Shoot 7, 1878 (a); Shoot 8, 1878 (b)

1879 – Raidak-Sankosh – MCB and party kill 11 rhino – 32.9 Shoot 9, 1879 (a)

1881 – Raidak-Sankosh – MCB and party kill 2 rhino – 32.9 Shoot 12, 1881 (b)

1882 – Raidak-Sankosh – MCB and party kill 9 rhino – 32.9 Shoot 13, 1882 (a); Shoot 14, 1882 (b); Shoot 15, 1882 (c)

1883 – Raidak-Sankosh – MCB and party kill 7 rhino – 32.9 Shoot 16, 1883 (a); Shoot 17, 1883 (b)

1884 – Raidak-Sankosh – MCB and party kill 3 rhino – 32.9 Shoot 19, 1884 (b); Shoot 20, 1884 (c)

1885 – Raidak-Sankosh – MCB and party kill 11 rhino – 32.9 Shoot 21, 1885 (a); Shoot 22, 1885 (b)

1886 – Raidak-Sankosh – MCB and party kill 19 rhino – 32.9 Shoot 23, 1886 (a)

1887 – Raidak-Sankosh – MCB and party kill 4 rhino – 32.9 Shoot 24, 1887 (a)

1889 – Raidak-Sankosh – MCB and party kill 5 rhino – 32.9 Shoot 25, 1889 (a)

1890s – Sankosh Region – Calculated 120 rhinos in 1890s – Bist 1994 (area I)

1890 – Raidak-Sankosh – MCB and party kill 6 rhino – 32.9 Shoot 26, 1890 (a)

1891 – Raidak-Sankosh – MCB and party kill 5 rhino – 32.9 Shoot 27, 1891 (a)

1893 – Raidak-Sankosh – MCB and party kill 7 rhino – 32.9 Shoot 30, 1893 (a)

1895 – Raidak-Sankosh – MCB and party kill 2 rhino – 32.9 Shoot 34, 1895 (b); Shoot 35, 1895 (c)

1896 – Raidak-Sankosh – MCB and party kill 7 rhino – 32.9 Shoot 36, 1896 (a); Shoot 37, 1896 (b)

1897 – Raidak-Sankosh – MCB and party kill 1 rhino – 32.9 Shoot 38, 1897 (a)

1898 – Raidak-Sankosh – MCB and party kill 6 rhino – 32.9 Shoot 39, 1898 (a)

1900 – Raidak-Sankosh – MCB and party kill 1 rhino – 32.9 Shoot 42, 1900 (a)

1901 – Raidak-Sankosh – MCB and party kill 5 rhino – 32.9 Shoot 43, 1901 (a)

1904 – Raidak-Sankosh – MCB and party saw rhino – 32.9 Shoot 46, 1904 (a); Shoot 47, 1904 (b); Shoot 49, 1904 (d)

1930 – Cooch Behar (Sankosh Region) – Rhino were common, but now exterminated – Wood 1930: 69; Bist 1994 – map A 373

1947–1968 – Bholka (Bhalka) on Raidak – Rhino present – Bist 1994 (Forest Dept. Annual reports) – map A 371

1953 – Bholka (Bhalka) range – Proposal to establish a game sanctuary was not pursued – Bist 2008 – map A 371

1955-56 – Bholka (Bhalka) (Raidak area) – Presence of 10 rhino, straying from Manas in Assam – West Bengal Forest Dept. 1993: 1 – map A 371

1970 – Sankosh-Rydak Region – Rhino extinct. Except strays from Jaldapara – Bist 1994

• • •

Dataset 31.35: Localities of Records of Rhinoceros in North Bengal

The numbers and places are shown on map 31.15.
(MCB), Records of the Maharaja of Cooch Behar (details in chapter 32).

Rhino Region 24: North Bengal – Jalpaiguri (Bist Area III)

A 351 Sivole (Sevoke), Teesta River – 26.90N; 88.50E – 1870, 1932

A 351 Sikhim terai (along Teesta River) – 26.90N; 88.50E – 1866, 1878

A 351 Darjiling Dt. terai (for Sikhim terai) – 26.90N; 88.50E – 1876, 1887, 1922

A 351 Mahananda WS – 26.90N; 88.50E – 1992, 2015

A 351 Kaliguri Guest House (Teesta River) – 26.70N; 88.50E – 1889

A 352 Ghorabhandah (Gorubathan) – 26.95N; 88.70E – 1877

S 146 Ghorabhandah – 26.95N; 88.70E–1877 – presence R. sondaicus

A 353 Kranti, Kyranti (Apalchand Forest) – 26.72N; 88.70E – 1864, 1867

A 354 Maynagori, Maynaguri (Jaldhaka) – 26.60N; 88.80E – 1864, 1865, 1969

A 354 East bank of Teesta River at Julpigoria (Jalpaiguri) – 26.70N; 88.85E – 1859

A 354 Jalpaiguri, also Julpigooree, Julpaiguri – 26.70N; 88.80E

A 354 Bhootan Duars (Jalpaiguri) – 26.70N; 88.80E – 1862, 1873, 1884, 1885

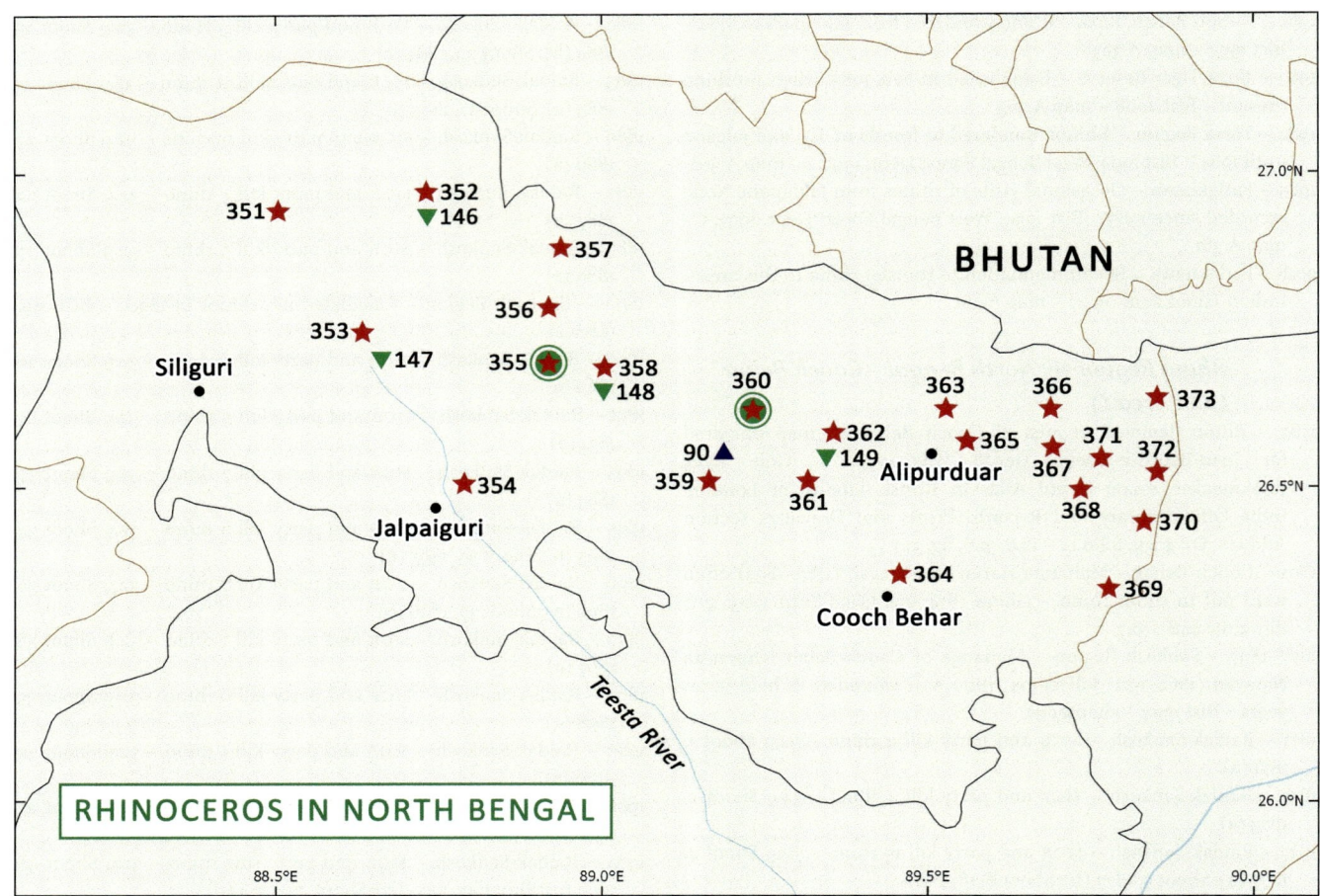

MAP 31.15   Records of rhinoceros in North Bengal. The numbers and places are explained in Dataset 31.35. Green circles indicate national parks
- ★ Presence of *R. unicornis*
- ⬟ Artefacts
- ▽ Presence of *R. sondaicus*
- ▲ Presence of *D. sumatrensis*
  Green circle: national park

S 147	Teesta River – 26.75N; 88.63E – 1878 – presence *R. sondaicus*
S 147	Jalpaiguri – 26.75N; 88.65E –1895, 1919 – presence *R. sondaicus*
Z	Jangiri, Bhutan Dooars – 1865 – fictional locality
Z	Shikarpur, Bhutan Duars – 1900 – fictional locality
A 355	Gorumara National Park – 26.75N; 88.80E – 1949
A 355	Ramshaihat (Ramshai) – 26.72N; 88.85E – 1870
A 356	Chapramari – 26.85N; 88.85E – 1969, 1996
A 356	Tendu (Chapramari) – 26.85N; 88.85E – 1895
A 357	Hillajhora (Hila) RF – 26.95N; 88.87E – 1949–1951
A 357	Nagrakata – 26.87N; 88.90E – 1946
A 357	Nagrakata, on Jaldhaka River – 26.87N; 88.90E – 1884 (MCB)
A 357	Lower Tondoo (Tendu) – 26.80N; 88.90E – 1884 (MCB)
A 357	Diana Hill (60 miles from Siliguri) – 26.85N; 89.00E – 1937
A 357	North Diana, South Diana (Jaldhaka) – 26.85N; 89.00E – 1969
A 357	Ambari (Upper Tondu) – 26.87N; 89.04E – 1884 (MCB)
A 358	Moraghat, Bhutan Duars – 26.77N; 89.02E – 1881
S 148	Moraghat – 26.77N; 89.02E–1881 – presence *R. sondaicus*

Rhino Region 25: North Bengal – Alipurduar (Bist Area II)

A 359	Falakata – 26.49N; 89.20E – 1865
D 092	Dalgaon Forest (Dhulagaon) – 26.65N; 89.15E – 1895 – presence *D. sumatrensis*
A 360	Mairadanga (SW Jaldapara) – 26.57N; 89.21E – 1954
A 360	Jaldapara National Park – 26.70N; 89.30E – chapter 34
A 361	Patlakhawa (on Torsa R.) – 26.51N; 89.34E – 1930, 1954, 1955, 1970, 1974, 1985, 2018
A 361	Patlakhowa (Patlakhawa) – 26.51N; 89.34E – 1905 (MCB)
A 362	Torsa Region – 1890s, 1940s, 1943, 1984
A 362	Torsha River north of Silitorsa – 26.60N; 89.35E – 1911
A 362	Chilapata (Torsa) – 26.60N; 89.35E – 1954
A 362	Chelapata (Chilapata) – 26.60N; 89.40E – 1892, 1893, 1904 (MCB)
S 149	Chilapata, Buxa Dt. – 26.55N; 89.40E – presence *R. sondaicus*
A 363	Nimati Range, Buxa – 26.65N; 89.43E – 1982
A 363	Alipur Duar – 26.60N; 89.55E – 1914
A 363	Buxadewar Hill, north of Chichacotta – 26.60N; 89.55E – 1800

A 363   Buxa Duars – 26.60N; 89.55E – 1865, 1908, 1920, 1921, 1930, 1937

A 363   Buxa Tiger Reserve – 1983

A 363   Bala – 26.70N; 89.56E – 1865

A 364   Cooch Behar town – 26.35N; 89.50E

Rhino Region 26: North Bengal – Cooch Behar (Bist Area 1)

A 365   Mahakalguri – 26.54N; 89.69E – 1896 (MCB)

A 366   Raidak River – 26.56N; 89.75E – 1886, 1889, 1890, 1891, 1893, 1895, 1896, 1897, 1904, 1905, 1907 (MCB)

A 367   Parokata – 26.49N; 89.71E – 1878, 1881, 1882, 1884 (MCB)

A 368   Rossik bheel, Rasikbil – 26.42N; 89.72E – 1885, 1886, 1889, 1891 (MCB)

A 368   Chengtimari, Dorko – 26.42N; 89.71E – 1886 (MCB)

A 368   Madhoobasha – 26.43N; 89.77E – 1884 (MCB)

A 368   Jorai River – 26.45N; 89.80E – 1895 (MCB)

A 369   Falimari on Sankosh River – 26.38N; 89.81E – 1884, 1889 (MCB)

A 370   Garud Haut – 26.44N; 89.84E – 1883, 1887, 1889, 1891, 1892, 1893, 1896 (MCB)

A 371   Bara Bhalka, Raidak River – 26.52N; 89.84E – 1891, 1897, 1904 (MCB)

A 371   Bholka (Bhalka) on Raidak River – 1947–1968, 1953, 1955–56

A 372   Dal Dalia, Daldal – 26.52N; 89.80E – 1885, 1890 (MCB)

A 373   Haldibari, Sankosh River – 26.60N; 89.85E – 1884, 1891, 1893, 1896 (MCB)

TABLE 31.36   List of persons who were involved in shooting or capturing rhino in North Bengal

Person	Date	Locality. type of interaction
Maharaja Harendra Narayan (1783–1839)	1801	Cooch Behar. Regular rhino hunts
Veritas (pseudonym)	1851–55	Bengal. Killed many rhino
Charles Huffnagle (1808–1860)	1855	Bengal. Killed 5
Robert Cecil Beavan (1841–1870)	1859	East of Jalpaigiri. Killed 2
Frederick Richard Norman Fortescue (1823–1867)	1859	East of Jalpaigiri. Killed 2
Hermann Schlagintweit (1826–1882)	1862	Bhutan *terai*. Obtained skull
Ashley Eden (1831–1887)	1862	Bootan. Observed tracks
Thomas Andrew Donogh (b. 1818)	1864–65	Maynaguri and Falakata. Killed several rhinos
Captain John Henry Baldwin (1841–1908)	1865	Buxa. Rhino seen
William Joseph Fitzmaurice Stafford (1820–1887)	1865	Buxa. Killed rhino
Sepoys of Gurkha Regiment	1865	Jalpaiguri. Killed many rhino
Heinrich Friederich Lawaetz Melladew (1842–1925)	1867	North Bengal. Observed 4
King Pippin (pseudonym)	1867	Mynagori. Observed 1
Henry John Elwes (1846–1922)	1870	Teesta River. Party charged by rhino
Richard Percival Davis (1840–1915)	1870	Said to have killed many rhino
William Osborne Barnard	1870	Gorumara. Wounded 1
Francis Wallace Grenfell (1841–1925)	1870	Gorumara. Killed 1
Wanderer (pseudonym)	1873	Bhootan Duars. Killed several rhino
Joseph Carne MacDonell (1849–1932)	1875	Jalpaiguri. Rhino seen
Maharaja of Darbhanga	1877	Bhutan Duars. Party killed 7
Major Robert Cotton Money (1835–1903)	1877	Bhutan Duars. Party killed 7
Charles Theophilus Metcalfe (1837–1892)	1877	Bhutan Duars. Party killed 7
Alexander Angus Airlie Kinloch (1838–1919)	1878	Bhutan Duars. Horn 30.5 cm recorded by Rowland Ward
Frederik August Möller (1852–1915)	1881	Moraghat. Killed 1
Robertson Pughe (1822–1906)	1887	Bhutan Hills. Observes group of 6 rhino
J.W.G.	1890s	Bengal. Killed 1
Savi	1894	Jalpaiguri. Killed 1
Charles Alfred Elliott (1835–1911)	1895	Tendu. Charged by rhino
George Cecil	1900	Bhutan Duars. Killed 2
General George Luck (1840–1916)	1900	Jalpaiguri. Received permit for 1 rhino

TABLE 31.36   List of persons who were involved in shooting or capturing rhino in North Bengal (*cont.*)

Person	Date	Locality. type of interaction
Gordon Casserly (1869–1947)	1908	Buxa. Observed 1
Jitendra Narayan (1886–1922)	1908	Cooch Behar town. Killed 1
John Frederick Gruning (1870–1922)	1911	Torsa River. Saw tracks
Joseph Veasy Collier	1920s	Cooch Behar. Horn 31.1 cm recorded by Rowland Ward
Henry Algernon Hildebrand (1883–1947)	1920s	Cooch Behar. Horn 31.8 cm recorded by Rowland Ward
Bengt Magnus Kristoffer Berg (1885–1967)	1932	Jaldapara. Took photographs
Diana Hill, of Rhaniket	1936	Jalpaiguri. Saw 1 rhino
Jaipur resident	1944	Cooch Behar. Killed 1
H.E. Tyndale	1940s	Jaldapara. Took photographs
Philip R.H. Longley (b. 1898)	1946	Nagrakata. Observed 1
Nancy Valentine (1928–2017)	1949	Cooch Behar. Killed 1
Maharaja of Jaipur	1950	Cooch Behar. Hunted rhino
Kesri Singh (b. 1927)	1950	Cooch Behar. Hunted rhino
Bengal Safari, Siliguri	2015	Captured 1 rhino found straying

Listed in chronological order for the main event. This excludes participants of the shoots organized by the Maharaja of Cooch Behar (found in 32.12).

31.4   "Rhinoceros Wilds" on Map of 1768

The English cartographer Thomas Jefferys (1719–1771), established at Charing Cross in London, produced a map of the East Indies in 1768. Curiously, this map contains the text "Rhinoceros Wilds" in two places, once north of Patna, and again north of regions called Rungpour and Beyhar, the latter where Cooch Behar is now (fig. 31.3). These indications would have been based on earlier maps or maybe on a yet to be revealed survey.

31.5   The Rhino Lands of Brion de la Tour in 1781

The rhino was recorded in North Bengal on a map prepared in 1781 by Louis Brion de la Tour, Cartographer Royal to the King of France (figs. 31.4, 31.5). He clearly marked a "Contrée de Rhinocéros" (land of the rhino) in the region of Bramsong, between Morong and Bootan Bisnee (Brion de la Tour 1781).

31.6   The Map of Bengal Produced by Gentil in 1785

When the French officer Jean-Baptiste-Joseph Gentil (1726–1799) produced his maps of India in 1785, he included some small drawings of the animals found in the region (Atlas in British Museum). His map of "Bengale" shows a rhino just outside the north-eastern border of the state, in the general area of Cooch Behar (figs. 31.6, 31.7). It is likely that there was no special intention to place the animal in this exact spot as the drawings were largely ornamental (see 18.8).

31.7   The Shoots of Harendra Narayan of Cooch Behar

The 18th Maharaja of Cooch Behar, Harendra Narayan (1783–1839), the ancestor of Nripendra, was a formidable hunter and wrestler. He went out to shoot rhinos in the region of Cooch Behar on a regular basis. Ghose (1874) mentions Harendra visiting Lukkhipore (Lakhipur, Assam) in 1801 where he hunted rhinos, tigers and buffaloes for a few days. He also issued a special order that he must be informed when anybody would see a rhino, in order that he could go out to shoot it.

31.8   Fortescue on the Teesta River

In his classification of the rhino specimens in the Asiatic Society of Bengal, Edward Blyth (1862) illustrated what he referred to as "a very fine example of the narrow type of skull of *Rhinoceros indicus*; a splendid adult male, with its horn" (figs. 31.8, 31.9) The skull had been obtained in 1859

FIGURE 31.2
A typical scene: *R. unicornis* in high grass jungle
sketched by Edward Percy Stebbing (1870–1960)
STEBBING, *THE DIARY OF A SPORTSMAN
NATURALIST IN INDIA*, 1920, P. 171

FIGURE 31.3   Detail of a map of the East Indies produced by the cartographer Thomas Jefferys in 1768. Note the "Rhinoceros Wilds", twice: in
Buttiah and north of Beyhar
JEFFERYS, MAP OF THE EAST INDIES, 1768

by Major Frederick Richard Norman Fortescue of the 73rd Regiment of the Bengal Native Infantry, "on the Butan side of the river Teesta, not far from Jalpigari." Fortescue died in Meerut in 1867, while Blyth (1862) said that he had taken the rhino trophy to England.

31.9   Beavan's Excursion from Jalpaigori in 1864

Robert Cecil Beavan (1841–1870) of the 62nd Native Infantry since 1858 was working with the Bengal Survey Department. He showed great aptitude for ornithology and his name is remembered as an alternative of the Grey-headed or Beavan's Bullfinch (*Pyrrhula erythaca* Blyth, 1862). He also wrote a paper on his encounters with the rhino (Beavan 1865). In 1864, he was stationed in Julpigorie (Jalpaiguri), then an outpost on the frontier

with Bhutan. The territory across the river was out of bounds to officers, but sometimes, with the silent agreement of the local chief, they made small forays to shoot for fun. One day Beavan, together with Colonel M. and an officer R., went to hunt rhino. After some time they saw a group of seven animals, of which they killed two. As nothing is said about trophies, perhaps they were not taken due to the surreptitious nature of the excursion.

31.10   A Rhinoceros on the Road to Buxa in 1865

Captain John Henry Baldwin, in his well-known book *The Large and Small Game of Bengal* (1877), recalled that "during the campaign of 1865, an advance guard, when marching very early one morning between Bala and Buxa, suddenly came upon a huge 'Gainda' standing in

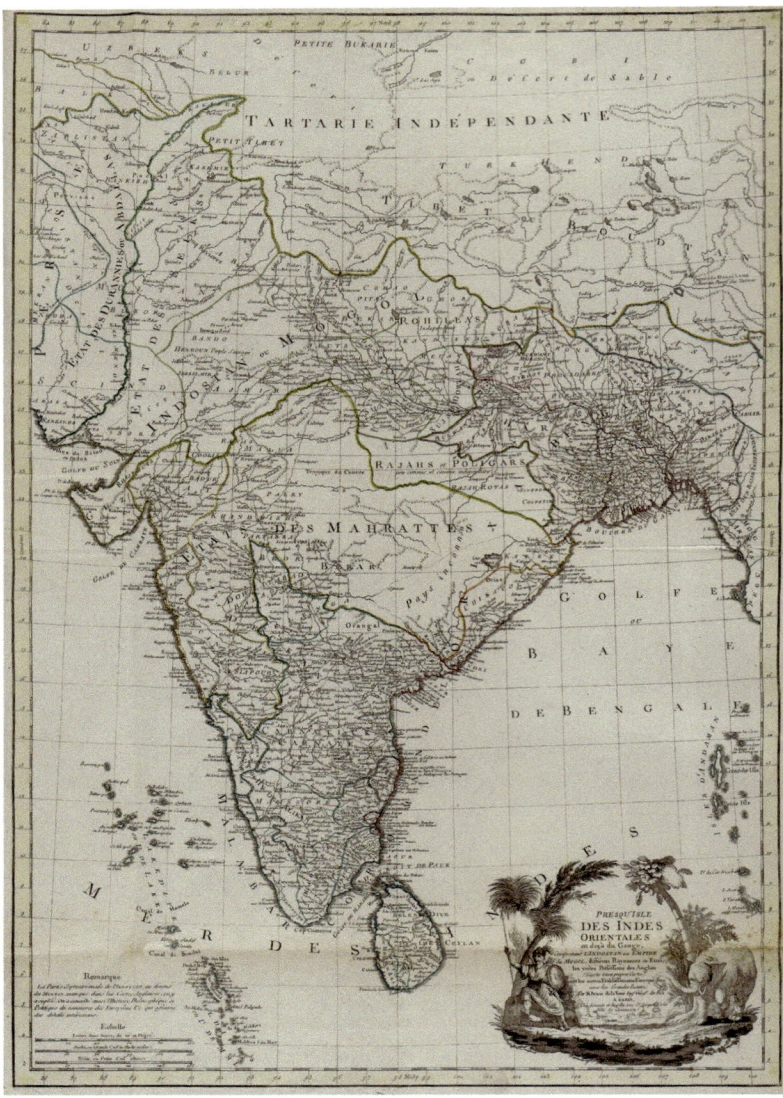

FIGURE 31.4
Map of India by Louis Brion de la Tour (1756–1823)
prepared in France in 1781: Presqu'isle des Indes
Orientales en deçà du Gange, comprenant l'Indostan
ou Empire du Mogol. Size 74 × 50 cm
WITH PERMISSION © UNIVERSITY LIBRARY OF
BERN: RYHINER COLLECTION, MUE RYH 7102: 6

FIGURE 31.5
"Contrée de Rhinocéros" (land of the rhino) indicated in North
Bengal on the map by Louis Brion de la Tour in fig. 31.4
WITH PERMISSION © UNIVERSITY LIBRARY OF BERN, MUE
RYH 7102: 6, DETAIL

FIGURE 31.6
Rhinoceros depicted on the map of
"Bengale", Bengal prepared by Colonel Jean
Baptiste Joseph Gentil in 1785
WITH PERMISSION © BRITISH LIBRARY,
INDIA OFFICE LIBRARY AND RECORDS:
ADD.MS. OR. 4039, FOLIO 14

FIGURE 31.7
Detail of rhinoceros and elephant on the
map of Bengal of Gentil in 1785
WITH PERMISSION © BRITISH LIBRARY,
SEE FIG. 31.6

the middle of, and completely blocking, the narrow path. The animal, however, quickly wheeled round, and disappeared in the jungle. Later, a very fine rhinoceros was shot by my commandant, Colonel S–d, in the neighbourhood of Buxa." The officer was William Joseph Fitzmaurice Stafford. Apparently, Baldwin was not part of the campaign, and he never actually shot a rhino.

31.11   Thomas Andrew Donogh in Jalpaiguri

The *Oriental Sporting Magazine* (OSM) from 1869 to 1877 included a series "Bhootan journal of tiger-shooting" in at least 28 installments. Relating hunting expeditions in 1864 and 1865, these were signed by T.A.D. From internal evidence, this hunter can be identified as Thomas Andrew Donogh (b.1818) who worked in the Bengal Civil Service in Punjab from 1835. The first instalment published in 1869 (OSM 2, no. 20) reproduced his chapters 12 and 13, while the remaining chapters 1 to 24 were published in sequential issues of the journal from vol. 5 no. 49 (1872) to vol. 10

no. 120 (1877). Donogh was in the region from the Teesta River eastwards to the Torsha River, and his encounters with the rhino were dated between December 1864 and March 1865.

In December 1864, marching in the Apalchand Forest north of Kranti, Donogh found a country where the rhino abounds, and he wonders if they can be used as beasts of burden because he has seen some which were tame and docile (Donogh 1872: 248). On his return near Maynaguri, 26 December 1864, he killed his first rhino, an old specimen that was close to 8 feet (2.4 m) high at the shoulders (Donogh 1873a). As there were many rhinos in this area, he actually only shot specimens with a good horn (Donogh 1870: 1567, 1873a: 65).

Donogh moved to the area of Falakata on the Mujnai River, where he killed nine rhinos: one on 9 February 1865 with a sharp well-formed horn (Donogh 1873b); four on 14 February 1865, including one with a large horn of a cubit length (45 cm) which he kept (Donogh 1869); one on 21 February 1865 (Donogh 1873c); and three on 7 March 1865, which included a pregnant female, of

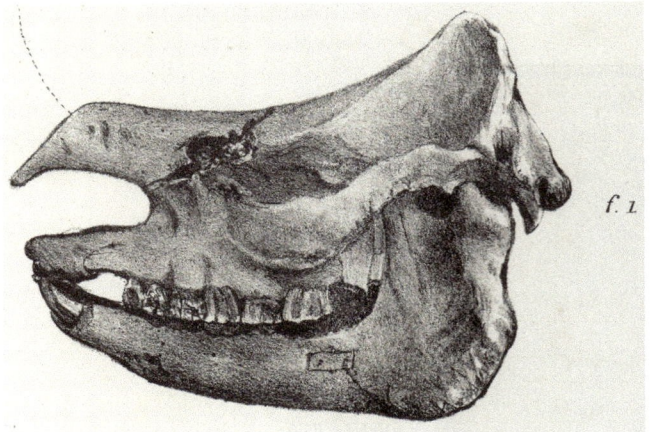

FIGURE 31.8   Lateral view of the skull of the "narrow type" of *R. unicornis* in the Asiatic Society of Bengal killed on the east bank of Teesta River by Capt. Frederick Richard Norman Fortescue (1823–1867), ca. 1859
BLYTH, *JOURNAL OF THE ASIATIC SOCIETY OF BENGAL*, 1862, PL. 1 FIG. 1

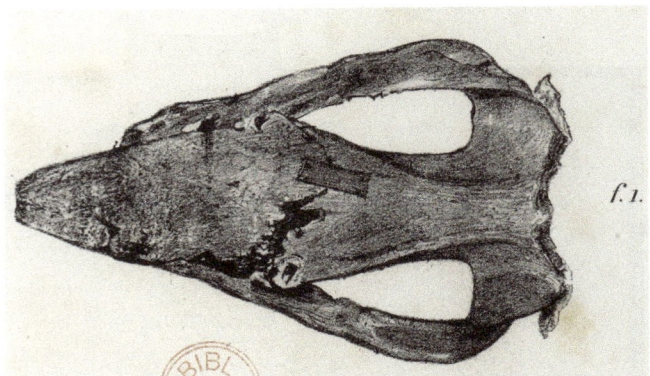

FIGURE 31.9   Dorsal view of the skull obtained by Fortescue near the Teesta River
BLYTH, *JOURNAL OF THE ASIATIC SOCIETY OF BENGAL*, 1862, PL. 2 FIG. 1

which he kept the hide of the embryo as a curiosity (Donogh 1875).

31.12   The Hunt by Henry John Elwes in 1870

The hunt by Elwes and his fellow officers, who all later attained high ranks in the army, only became public through their memoirs, and then only partly. If there were still hunting logs or diaries at that time, these are now lost, together with any trophies. I have followed Elwes (1930) and Stebbing (1930) for dates in April 1870, but strangely Grenfell (1925) refers to September 1870. Henry John Elwes (1846–1922) and Francis Charles Bridgeman arrived in Siliguri in the second week of April 1870, with an introduction to John Colpoys Haughton, Commissioner of Cooch-Behar 1865–1873. Here they found Francis Wallace Grenfell, 1st Baron Grenfell and William Osborne Barnard ready to join their hunt. For several days, they camped in the area around Sevoke on the Teesta River, where many rhinos moved around in the high grass, but apparently they just wounded one. On 22 April 1870, halting at Titalya close to Jalpaiguri, they met the police superintendent Richard Percival Davis, who was said to have "shot more rhinoceros in the Dooars than any man in India." A similar distinction was credited to the soldiers of the Gurkha regiment stationed in Jalpaiguri from the time of the Bhutan war, who "were very keen and successful rhinoceros hunters, going out on foot in small parties and creeping along the paths formed by them in the dense reed beds. They had killed no less than seventy in the course of a year or

so, and as the horns are highly valued in native medicine, they had made quite a lot of money by their hunting." In most literature, only the actions of officers are given in some detail, but quite possibly a lot more damage was done to the wildlife population by the rank and file. Elwes and his fellow officers continued their hunt for two weeks, moving northwards as far as the Jaldhaka river, near the present Gorumara NP, eventually killing a total of eight rhinos.

31.13   Rhino from the Bhutan *terai* Shipped to London in 1875

In January 1875, the animal dealer William Jamrach was shipping a young female rhino from Kolkata to London. He had asked a local artist Khaliludin Ahmed to prepare a drawing, because he suspected that this specimen might belong to an unknown species, obtained in the "Bhootan terai." Both Jamrach and H. James Rainey agreed that the surface of the skin was more rough and uneven than that of *R. unicornis*, and of a black colour throughout; ears broad, and devoid of the fold in the neck which is characteristic of *R. sondaicus* (Rainey 1875a, 1877). Philip Lutley Sclater exhibited the drawing in the meeting of the Zoological Society of London of 16 February 1875 (Sclater 1875b), and after arrival of the animal added the outcome of the event. Sclater, joined by Alfred Henry Garrod and Abraham Dee Bartlett examined the young rhino and could not differentiate her from the usual *R. unicornis* (Sclater 1876a: 648). The animal was kept in the premises of Jamrach's associate Charles William Rice in St. George's, London, where she soon died. The skin was discarded, while the skeleton was purchased by the museum of the Royal College of Surgeons of England, where it was catalogued as no. 2127 "skeleton of young female" *R. unicornis* "from an animal which died in captivity in this country" (Flower 1884: 416). The specimen no longer exists.

31.14 The Maharaja of Darbhanga in Jalpaiguri in 1877

In a popular magazine with adventures for boys, there is an unexpected contribution about a hunt organized in the Jalpaiguri District for Lakshmeshwar Singh Bahadur (1858–1898), the Maharaja of Darbhanga (1860 to 1898). Written by the Maharaja's estate manager Major General Robert Cotton Money, these events probably took place in March 1877 when Money was employed in the Bihar estate. While camping at Ghorabhandah (Gorubathan) they were surrounded by swampy land full of stretches of cardamum and wild plaintain. Money shot a rhino with a fine long horn, his friend the Commissioner Charles Theophilus Metcalfe killed another and wounded a third, while the Maharaja had shot a rhino before going in pursuit of a baby rhino, which eventually was caught. In total seven rhinos were killed that day, which included one which charged Money, being "one of the smaller species of rhinoceros, at all times the most pugnacious." This second species would have square angular tubercles and less fully developed folds (52.4).

31.15 The Six Rhinos of Alexander Kinloch

Alexander Angus Airlie Kinloch (1838–1919), officer of the 60th Royal Rifles, joined the army in 1855. From 1870 he was stationed in India as Regimental Musketry Instructor with the rank of Captain. As no musketry was performed in the hot weather, he enjoyed six months leave annually for at least 15 years, giving him chance to travel widely and pursue his love for big game shikar. From 1890 he served in the Royal Rifle Corps, before retiring with honorary rank of Major-General in 1895 (Hare 1929). At the outset he had hopes to bag every sort of large game found on the continent of India and furnish a complete guide for future sportsmen. Although he realized that he would not accomplish this personal goal, he wrote a book with details of his experiences. His *Large game shooting in Thibet, the Himalayas, and Northern India* was first published in 1869 with a reprint (or second edition) in 1875 without any mention of the rhino. The third edition "revised and enlarged" was issued as *Large game shooting in Thibet, the Himalayas, Northern and Central India* in

July 1885, and again updated in August 1892. The number of reprints shows that the book was popular in his days, providing guidance, hints and background to the natural history of the animals that could be hunted in India.

Kinloch had three chances after 1875 to visit areas of the *duars* where the rhinos were found, each time starting out from Jalpaiguri. He does not specify his exact destinations, but on his map he highlights his itineraries either around the Teesta River (his Sikhim *terai*) or further east in Cooch Behar (fig. 31.10). He was clear in his statement that in total he killed six rhinos personally, while others were credited to his friends (Kinloch 1903: 161, 1904: 61).

The first trip in April 1878 was on "invitation from a friend", who in fact was a disguise for the Maharaja of Cooch Behar. Kinloch was joined by S. and D., here identified as Thomas Smith and Godfrey John Bective Tuite-Dalton (1840–1911). Together they accounted for 11 rhino in just over a week (MCB shoot 8, see 32.9). While "in one of the best heavy-game shooting districts in Bengal", Kinloch shot 3 rhinos himself. His first rhino killed on the third day was a large male, which was followed by a small female on day 4 and a large animal with a massive horn at the end of the week (1885: 63–65, 1892: 85–88).

From Cooch Behar Kinloch proceeded to a locality on the left bank of the Teesta River, where in May 1878 his companion S. (probably the same Thomas Smith) killed a rhino identified by Kinloch as a specimen of *R. sondaicus* (52.5). Elsewhere (p. 102) he wrote about "a rhinoceros" shot during this trip.

Kinloch (1892: 88) states in one sentence that "on a subsequent visit to the same jungles I shot two more 'Rhino', one with a single bullet." As this follows the paragraphs about his time in Cooch Behar in April 1878, the same jungles must still refer to that region, and might have followed soon after the more formal shoot organized by the Maharaja. These were Kinloch's 4th and 5th rhinos. Rowland Ward (1899: 433) included one horn of 12 inches (30.5 cm) from the Bhutan Duars among the best specimens of *R. unicornis*.

On a further trip in February 1886, Kinloch went alone, using two elephants, in the Bhutan Duars. He killed a huge male rhino with a short horn (his 6th), said to be old because three teeth were missing (Kinloch 1892: 90). He took time to preserve the trophy and some parts of the skin as whips.

Kinloch was told that there had been large numbers of rhino in the *duars* in the 1860s. Interestingly, he mentioned that several sportsmen earned considerable sums by the sale of rhino horn. This was, he said, one of the reasons why rhino were less frequently found, combined with the clearance of large tracts for tea plantations. Kinloch

FIGURE 31.10   Detail of the map of North Bengal published by Alexander Kinloch in 1892
KINLOCH, *LARGE GAME SHOOTING IN THIBET, THE HIMALAYAS, NORTHERN AND CENTRAL INDIA*, INSERTED AT
END. COURTESY: THOMAS GNOSKE, CHICAGO

FIGURE 31.11   Group of trophies. The specimens obtained by
Alexander Kinloch include the skull of *R. unicornis*
with a short horn. Printed by Thacker, Spink, & Co.,
Calcutta
KINLOCH, *LARGE GAME SHOOTING*, 3RD ED.
SECOND PRINTING, 1892, FRONTISPIECE

reflected on the general opinion among sportsmen and naturalists that the number of animals and the extent of the forests were drastically reduced during the latter part of the 19th century, and that this was at least partly due to excessive hunting (Kinloch 1903: 161):

> Forty years ago rhinoceros were extremely numerous, and several might easily be killed in one day. Owing to indiscriminate slaughter of both sexes and all sizes, their numbers have been terribly reduced; but there are enough left to enable a well equipped sportsman to be pretty sure of obtaining one or two specimens. With these, I think that he ought to be content, although I must plead guilty to having shot six. Even if I had the chance, I would never shoot another, unless it had an extraordinarily good horn.

The 1885 edition of Kinloch's *Large Game Shooting* had no illustration of a rhino among its new photogravures. The 1892 edition had a new frontispiece with a group of trophies, including a rhino skull with a short horn (fig. 31.11). The mounted head of the old male with short horn killed in 1886 was prominently shown in a photogravure (fig. 31.12). Kinloch must have taken this trophy to Scotland, because a different photograph taken by William Mayor, photographer in Forfar, is found in a chapter contributed by Kinloch to a general book of sports

FIGURE 31.12 "The Great Indian Rhinoceros. Rhinoceros unicornis." Head of male rhino killed in February 1886. Printed by Thacker, Spink, & Co., Calcutta
KINLOCH, *LARGE GAME SHOOTING*, 3RD ED. SECOND PRINTING, 1892, FACING P. 83

FIGURE 31.13 Rhinoceros shot by Major-General Kinloch in Bhutan Duars, 1886. This photograph was taken after Kinloch's return to Scotland by William Mayor, photographer in Forfar
AFLALO, *THE SPORTS OF THE WORLD*, 1903, P. 161

(fig. 31.13). Edited by Frederick George Aflalo (1870–1918), the illustrated volume *Sports of the World* first appeared in fortnightly parts during 1903, with Kinloch's contribution in part 5 of 17 January 1903. The volume was available as bound copy without illustrations in 1904 and again with illustrations in 1905. Some illustrations were sourced by Aflalo, and include reproductions of two drawings by the British artist Lawson Wood (1878–1957) entitled "Making off" and "Close quarters", both of which portray general impressions of a rhino hunt (figs. 31.14, 31.15).

31.16 The Swedish Tea Planter Möller in Moraghat in 1881

The Swedish tea planter Frederik August Möller (1852–1915) shot a young *R. unicornis* on 6 June 1881 at Moraghat near Binnaguri. The skull and lower mandible were donated to the Zoological Museum of Copenhagen, Denmark in June 1887 (figs. 31.16, 31.17). Möller killed a specimen of *R. sondaicus* in the same locality a few months earlier (fuller discussion in 52.6).

31.17 A Fictional Rhino in Darjeeling Town in 1922

The hill station of Darjeeling would be too high in the mountains to expect even a stray rhino. There was a story recently uncovered by the journalist Barun Roy (b. 1977), which despite the overload in detail is likely to be fictitious. None of the people mentioned could be identified. According to the *Darjeeling Avertiser* of 10 May 1922, this happened: "7 May, Darjeeling. The entire town today was witness to the madness of an animal which caused many injuries and loss of property. Spotted at the very dawn among the dense foliage surrounding the St. Paul's School by native milkmen, the animal as big as a wild buffalo upon being hit by stones ran down towards the residential areas. In a short period it was at the very heart of Chowrasta where it drank from the fountain near the Bellevue Hotel causing great havoc among the populace. Mr. Price who had been leaving for Lebong miraculously escaped when the beast running down toward Lebong nearly ran over him. A poor native labourer was not that lucky as the beast hit him throwing him far away, perhaps as far as 20 feet. The residence of the Bustee

FIGURE 31.14
"Making off." View of a (virtual) rhinoceros hunt from elephant back by the British artist Lawson Wood (1878–1957). This was used to illustrate a paper by Kinloch in the compendium by Aflalo
F.G. AFLALO, *THE SPORTS OF THE WORLD*, 1903, P. 161

CLOSE QUARTERS.

FIGURE 31.15
"Close quarters" showing a rhino attacking an elephant with a hunter in a howdah. Drawn by Lawson Wood
F.G. AFLALO, *THE SPORTS OF THE WORLD*, 1903, P. 163

FIGURE 31.16
The skull of *R. unicornis* obtained in Moraghat in 1881 and donated to the Zoological Museum of Copenhagen in 1887, number CN-525
PHOTO © HANS J. BAAGØE, 2018

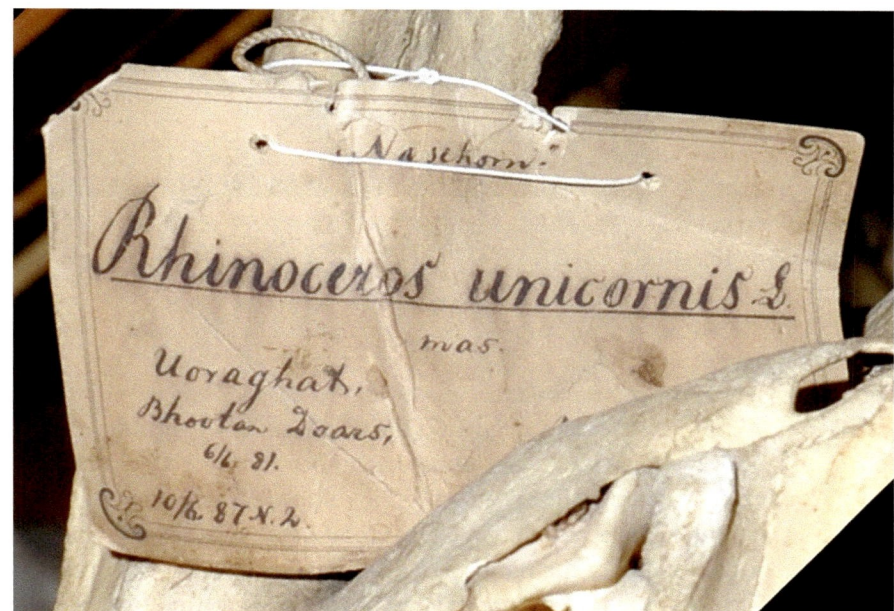

FIGURE 31.17
Label of the skull of *R. unicornis* collected by Möller in Moraghat in 1881
PHOTO © HANS J. BAAGØE, 2018

meanwhile ran to protect themselves. It was only at the foot of the hill when the beast disappeared towards the Phoobtshering Tea Estate that Mr. Price was able to take a good look at it. The beast was a massive rhinoceros that had appeared out of nowhere. Incidentally, two of the most well known hunters of the Doars valley, Mr. Ward Wilson and Mr. Pascal were present at the Damried Hotel in Darjeeling and upon being impressed into service shot the rampaging rhinoceros at the Phobtshering Tea Estate. The rampaging rhinoceros left fifteen natives wounded and numerous houses and sheds thoroughly destroyed." The Phubsering or Phoobsering Tea Garden is just north of Darjeeling, with tea plantations ranging from 1000 to 1800 meter in altitude.

31.18 Rewards Allocated for the Destruction of Rhinos in Bengal

The government of Bengal gave rewards for killing destructive wild animals, including rhino. Although the documentation of this official practice is elusive, the system was already well in place in the middle of the 19th century. As early as 1848, a reward of Rs 5 was paid for killing a rhino, in Nowgong which was then still part of Bengal. The same amount of Rs 5 was also the going rate in 1850 (G. Singh 2014: 35). There is a further glimpse of the procedures in a police report by John Colpoys Haughton (1817–1887), the Commissioner of Cooch Behar, who gives details about the number of rhinos for which rewards were paid. In District Julpigoree (Jalpaiguri), in 1868 for 12 rhino and 2 rhino cubs, 1869 for 20 rhino, and in District Goalpara 1868 for 1 rhino (Haughton 1879). In 1896 the amount had increased to Rs 20, which was continued by the Government of West Bengal until this program was abandoned in 1910. These rewards were instituted with the intention to protect people from an abundance of wild animals, recognising that annually quite a few deaths were reported after wildlife conflicts. Rhinos are dangerous animals, just through their sheer bulk and unpredictability, and attacks remain common both in the range states as in captivity (Kawata 2014).

31.19   Poaching and Protection in the 1930s

The early efforts to preserve rhinos in Bengal were spearheaded by Edward Oswald Shebbeare (1884–1964). He joined India's Forest Department in 1906 and had become quite a legend by the time he retired in 1938 as Senior Conservator of Forests in Bengal, when he went to Malaya to serve as Chief Game Warden (Gee 1964c; Seow 2010; Westaway 2018). After four years detained as a prisoner of war 1942–1945, he resumed his post, finally retiring in 1947 (fig. 31.18). Before 1930, Shebbeare's concerns about the state of wildlife in Bengal were largely ignored. It was well-known that the trade in rhino horn was lucrative, a single horn fetching Rs 2000 or 150 pounds, destined for the Chinese market. However, rhinos were largely protected by the remoteness of their strongholds and the fact that bullets fired by local 'gas-pipe' guns were unable to penetrate their thick skin. The situation was assessed in 1943 in a report commissioned by the Government of West Bengal and authored by Louis Reginald Fawcus of the Indian Civil Service on behalf of an impromptu Game and Game Fishes Preservation Committee. He found that poaching increased when muzzle-loading guns became more generally available to the local Mech tribesmen of the *duars* region. The first poachers came from Assam but, in a spate of poaching in the early 1930s, they were joined by local Meches to form gangs: "Their plan was to build a light bamboo staging about 8 ft. above the ground at strategic points, usually where two well-worn rhino tracks met, and lie up when the moon was nearly full. Sooner

FIGURE 31.18   Edward Oswald Shebbeare (1884–1964) photographed by Bengt Berg in 1932
BERG, *PÅ JAKT EFTER ENHÖRNINGEN*, 1932, P. 17.
COURTESY: NATASHA ILLUM BERG

or later a victim was bound to pass and received a heavy bullet at a range of a few feet" (Fawcus 1943). At first the poaching went undetected because there were only few forest officers active in the rhino areas and they could not penetrate the high grass jungles favoured by the animals. When the remains of a rhino killed by poachers was discovered outside the grass, Shebbeare requested the government for military backup. There was much opposition for half a year, shots were exchanged, arrests were made and prosecutions were successfully obtained.

It took the forest department several years to recognize the toll which this poaching activity was taking. Shebbeare, as one of the most senior forest officers in Bengal and one committed to preservation of wildlife, put in a lot of effort to get the attention of the politicians and senior administrators. His successful lobbying resulted in the passing of the Bengal Rhinoceros Preservation Act (Bengal Act VIII of 1932) by the Bengal Legislative Council on 1 April 1932. Jaldapara was subsequently established as a game sanctuary, followed by Gorumara. This stabilised the situation, providing more powers to apprehend poachers, but it is unlikely that poaching stopped altogether. Rhino numbers might have dipped to less than 25 in the 1930s, and only slowly recovered with subsequent periods of increased illegal poaching recorded in 1968–1972 and 1982–1985 (table 31.37).

31.20 Changes in Rhino Numbers in North Bengal

There were no estimates of rhino numbers in the forests of North Bengal published in the 19th century. Bist (1994) calculated the total retrospectively for 1890 as 120+ in Sankosh, 100+ in Torsha and 20+ in Jaldhaka. This can be accepted as a minimum number, but chances are that rhinos were actually far more abundant, because most hunters found rhinos to be plentiful. It was recognized that rhinos sometimes traveled widely to reach rice and corn fields.

While rhino numbers diminished through hunting and the clearing of forest areas to make way for tea plantations and agriculture, there is no evidence that the animals were close to extinction or even particularly threatened in the early 20th century, unlike the situation in Assam. For a long time, hunting was legal in Bengal if a permit could be obtained. The idea of a closed season for game hunting became prevalent at the turn of the century (J.W.G. 1900), which at first probably meant that forest officials refused to issue the necessary permit. However, the rhino range continued to be eroded and was concentrated in the forests of Gorumara and Jaldapara. When extensive poaching threats were uncovered around 1930 and brought under control, the total number of rhinos was estimated to be less than 50. A detailed census was not carried out at the time.

Rhinos had gone from Sankosh-Raidak around 1930. The rhinos were protected in West Bengal when Jaldapara was recognized as a Game Sanctuary in 1932 (chapter 34). Gorumara was always a Reserved Forest and received enhanced status in 1949 (chapter 33). These have remained the strongholds for rhinos and other wildlife with increasing numbers. Some rhino have strayed outside the reserves, where they are much at risk. North Bengal has always been a favourite abode for the rhinoceros as long as human disturbance is limited.

TABLE 31.37   Estimates of the status of rhinoceros in the northern part of West Bengal

Locality	Year	Numbers	References
Bengal	1888	scarcer than it used to be	R002
West Bengal	1890	240	R055, R156
Bengal	1890	240	R055
Bengal	1904	< 12	R227
West Bengal	1929	200	R306, R309
West Bengal	1932	24; 35–40; 40–50	R045; R044, R127; R309
Buxa region	1933	increasing	R124
Bengal – outside reserves	1935	12	R306
Jalpaiguri	1937	4–5	R043, R124
Torsa	1937	56	R043, R124
West Bengal	1943	200	R055
Buxa region	1948	couple	R164
Buxa region	1949	couple	R055, R164
Buxa region	1950	couple	R055
Bengal Duars	1950	30	R103
Buxa region	1952	fair number	R164
Bengal Duars	1952	30	R199
Bengal	1953	58	R107
Buxa region	1953	fair number	R055
Buxa region	1953–1958	exist	R055, R164
Bengal	1954	43	R007
West Bengal	1958	43; 48	R119; R108
Buxa region	1959	10	R055, R164

TABLE 31.37   Estimates of the status of rhinoceros in the northern part of West Bengal (*cont.*)

Locality	Year	Numbers	References
Bengal	1963	50	R226
Bengal	1964	65	R112
Buxa region	1966–1968	exist	R055, R164
West Bengal	1966	80	R037
West Bengal	1968	55	R096
Bengal – outside reserves	1989	32	R138, R211
Bengal – outside reserves	1993	25	R164

References in table: R002 Adams 1888: 251; R007 Amritaghata 1955; R037 Banerjee 1964: 241; R043 Meiklejohn 1937: 8; R044 Berg 1933: 118; R055 Bist 1994; R096 Fitter 1968: 59; R103 Gee 1950a: 1732; R106 Gee 1952a: 224; R107 Gee 1953a: 341; R108 Gee 1958: 353; R112 Gee 1964a: 153; R119 Grzimek 1958: 118; R124 Harper 1945: 378; R127 Hobley & Shebbeare 1932: 21; R138 Khan 1989: 3; R156 Martin 1996a; R164 Menon 1995: 15; R199 Ryhiner 1952b: 161; R204 Shebbeare 1953: 143; R211 Sinha & Sawarkar 1993; R226 Stracey 1963: 243; R227 Street 1963: 100; R306 Shebbeare 1935; R309 Fawcus 1943

The Hunting Camps of the Maharaja of Cooch Behar

32.1   The Rulers of Cooch Behar

It was a tradition of the rulers of Cooch Behar to go hunting in the forests of their territory, and sometimes beyond. From the late 1870s these outings developed into grand affairs with no expense spared in luxurious camps catering for all needs. Nripendra Narayan Bhupa (1862–1911), the Maharaja of Cooch Behar was a formidable hunter and sportsman from an early age. His annual shoots became a prized experience and were attended by an amazing list of aristocrats, high-ranking officers, wealthy businessmen, even royalty, mainly from the United Kingdom but also from other European countries and India itself. Their combined experiences have added to the appreciation of Indian wildlife, contributed to the knowledge of the rhinoceros, and of course added trophies both to private collections and to public museums. Most details of these shoots are found in the game book published by Maharaja Nripendra in 1908, supplemented by additional published and unpublished sources.

32.2   *Thirty-Seven Years of Big Game Shooting in Cooch Behar* (1908)

Nripendra Narayan Bhupa, Maharaja of Cooch Behar (hereafter MCB), published a major record of 37 years of big game shooting from 1871 to 1907. This book provides an unparalleled insight into his regular hunting expeditions in West Bengal and Assam. Entitled *Thirty-Seven Years of Big Game Shooting in Cooch Behar, the Duars, and Assam: A Rough Diary*, the book is known in two largely identical editions:

(1)   Published in Bombay: Printed at The Times Press, with date 1908.

(2)   Published in London: Rowland Ward, Limited, 167, Piccadilly [and] Bombay: Bennett, Coleman & Co., The Times Press, with date 1908.

Both are octavo size (23.5 × 18 cm, 9 ¼ × 7 ¼ inch), pp. [i–viii], ix–xxviii, 1–461 (plus 6 unnumbered pages). In the first Bombay edition, the title-page (p. [v]) is preceded by a blank page, frontispiece, a half-title and a blank page (pp. i–iv). In the London-Bombay edition, pp. i–iii are blank, followed by the frontispiece facing the title-page. There are 151 photographic illustrations, all but two listed on pp. xvi–xxii. There are six unnumbered pages (one

blank) with five illustrations, found after pp. 12, 96 and 410. The book is dedicated to King Edward VII.

The first Bombay edition is dated 1908 and may have been issued in a small print run, or maybe was meant to be presented to friends of the author. I have not found any advertisements or reviews. The second London-Bombay edition is more or less a reprint with just a few small changes in the order of the initial pages. Aimed at the British market, it was first advertised (priced at 21 shillings) in October 1909, for instance on 23 October 1909 in the *Field* (fig. 32.1), and reviewed on 6 November 1909 in the *Illustrated Sporting and Dramatic News*. Although dated 1908 on the title-page, I doubt that this second edition can have been available for sale before October 1909.

The contents of the Maharaja's book are said to be based on his game-book which was kept as a record of the shoots (MCB 1908: 1). Press reports in newspapers as early as 1891 refer to this game-book kept since 1876, which "would, if published, show some interesting records of Indian sport, ... during the last 15 years" (*Leicester Daily Post* 1891-09-14). Actually no notes were kept in 1872, 1874, 1875 and 1876. It may be remembered that the Maharaja was only seven years old in 1871 when he was first taken out into the jungle, and it is remarkable that anything was written down at the time. Although the text in the published book of 1908 is likely to be a faithful mirror of the game-book, it is certainly regrettable that the original is no longer known to exist. If it is ever found, maybe hidden among some old papers in a library or attick, it would be a remarkable record of a bygone era.

In this chapter, I will refer to Maharaja Nripendra as MCB, and to his book as (1908: xx), where xx represent the page numbers. As the two editions of the book are identical, references are found in both of these.

32.3   Manuscript Sources

Margaret Elphinstone Rhodes (1925–2016), a cousin of Queen Elizabeth II, was the daughter of Sidney Herbert Buller-Fullerton, 16th Lord Elphinstone (1869–1955) and Mary Frances Bowes-Lyon (1883–1961). She wrote in her memoirs that her father went on three big game hunting trips to India in the 1890s staying with the Maharaja of Cooch Behar: "I have his game book, in which he precisely recorded for the three visits a bag of 13 tigers,

© L.C. (KEES) ROOKMAAKER, 2024 | DOI:10.1163/9789004691544_033
This is an open access chapter distributed under the terms of the CC BY-NC-ND 4.0 license.

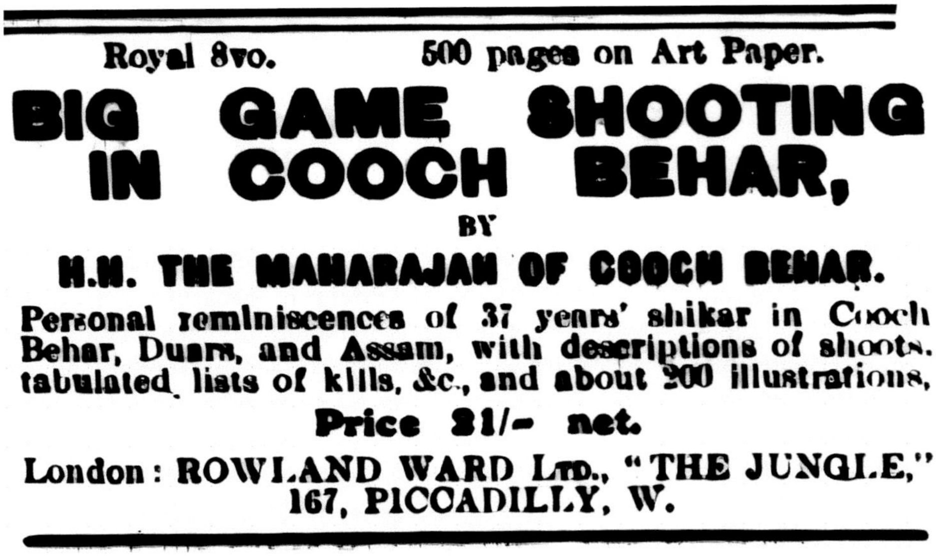

Royal 8vo.   500 pages on Art Paper.

BIG GAME SHOOTING IN COOCH BEHAR,

BY

H.H. THE MAHARAJAH OF COOCH BEHAR.

Personal reminiscences of 37 years' shikar in Cooch Behar, Duars, and Assam, with descriptions of shoots. tabulated lists of kills, &c., and about 200 illustrations,

Price 21/- net.

London : ROWLAND WARD Ltd., "THE JUNGLE," 167, PICCADILLY, W.

FIGURE 32.1
First advertisement of Nripendra's book in *The Field* (London) of 23 October 1909

3 leopards, 21 rhino, 39 buffalo, 10 python" (Rhodes 2012). Lord Elphinstone's Game Book of large size is still known, kept in private hands, and the contents have kindly been made available for the present investigation. It contains entries for shoots in 1894, 1895 and 1899, handwritten by Sidney Elphinstone and illustrated with contemporary photographs. The details can be found below combined with the entries from MCB (1908), referred to as the *Elphinstone Game Book*.

Such game books were important souvenirs for the people who went hunting in India or elsewhere. It is likely that MCB helped his guests to write these, providing information and photographs. Although I suspect that several are still preserved in private libraries, only very few have come to light. The same can be said about photo-albums with images of the shoots, sometimes preserved as part of longer travels in India.

Stephen Hungerford Pollen (1869–1935) compiled such a photo-album, which entitled "A hunting-party with the Maharaja of Cooch Behar, 1895–1896" was auctioned by Bonhams on 9 April 2009 (Pollen 2009). It contains 76 gelatin silver prints, 90 albumen prints, including a tiger shot by Pollen, scenes of the Raidak River and a rhino shot by the Maharaja. The current whereabouts of the album is unknown. I hope one day more of these important glimpses into the past will be found and made available for research.

There is every indication that the large annual shooting camps were attended by photographers, contracted for the occasion by MCB. There are 151 photographs illustrated in MCB (1908), none of them dated or accredited. The earliest photographs may date from 1885. In 1891 Ehlers (1894, vol. 1: 385) noted the presence of a photographer called Schirmer from Bourne & Shepherd of Calcutta. In 1894,

the *Englishman's Overland Mail* (1894-03-14) has McKenzie from Messrs. Johnston and Hoffman. Such photographers would supply their products to the Maharaja as well as to his guests, and these would have formed the basis for any game books or similar records.

32.4   Sources on Cooch Behar Hunts by Participants

While Nripendra (1908) remains the most extensive source about the shoots in Cooch Behar, there were many participants who published their own impressions. These are valuable first because they add details not found in the 1908 book, secondly because they helped to spread information about the rhinoceros and other animals in North Bengal. There were also reports in newspapers giving in some cases a daily account of the proceedings, of which further details are found in the list of shoots (32.9). It is likely that in some instances the journalists were allowed to follow the hunts and possibly even stay in the camps with other guests and support staff. The following accounts by participants have been identified, according to the hunts identified in each year (table 32.38).

TABLE 32.38   References to the hunts in Cooch Behar in works by participants

Shoot	Reference
1877 (b)	Greville 1917: 300
1878 (b)	Kinloch 1885: 63, 1892: 84–87
1880 (a)	Lawley 1933: 25
1884 (a)	Anon. 1884a (Lord Mayo)
1884 (a)	Pontevès de Sabran 1886: 101 (J.K. Fraser)

TABLE 32.38   References to the hunts in Cooch Behar (*cont.*)

Shoot	Reference
1886 (a)	Timler 1888
1886 (a)	Barclay 1932: 137
1887 (a)	Breteuil 2007: 109
1889 (a)	Hewett 1938: 90
1890 (a)	Scheibler 1900: 195, 211
1890 (a)	Devee 1916: 110 (Suniti Devi)
1891 (a)	F. Hamilton 1919: 151, 1921: 8
1891 (a)	Moss 2003: 217 (Marques Ailsa)
1891 (a)	Ehlers 1894: 384–397
1892 (a)	Warren 1903: 178 (Prince Victor)
1892 (a)	Harrison 1892: 118–121
1893 (a)	Jenkins 1893
1894 (a)	Hewett 1938: 91
1894 (a)	Scheibler 1900
1895 (a)	Seton-Karr 1897; Blathwayt 1898: 342
1895 (a)	Elphinstone Game Book (see 32.3)
1896 (a)	Elphinstone Game Book (see 32.3)
1896 (a)	Moffett 1898a, 1898b (Peter Burges)
1898 (a)	Lal Baloo 1902
1899 (a)	Mancini 1899
1899 (a)	Lonsdale 1899: 541
1899 (a)	Elphinstone Game Book (see 32.3)
1905 (b)	Buchan 1920: 63 (F. Grenfell)

32.5   Maharaja Nripendra and His Family

Nripendra Narayan succeeded his father as ruler of Cooch Behar in 1863 when he was just 10 months old. Until 1880, when the title of Maharaja was conferred on him, he had no legal powers. A family tree is explored by Moore (2004: xi–xii) and online by Buyers (2020). Nripendra was the son of Narendra Narayan (1841–1863) and Nishiyama. He had one sister Ananda Mayi Devi (1860–1887) and a half-brother Jotindra Narayan (1861–1920).

Nripendra married in 1878 with Suniti Devi (1864–1932), the daughter of Keshub Chandra Sen (1838–1884). They had four sons: Rajendra or Rajey (1882–1913), Jitendra (1886–1922), Victor (1887–1937), Hitendra (1890–1920); and three daughters: Sukriti, Girlie (1884–1958), Prativa, Pretty (1891–1923) and Sudhira, Baby (1894–1963). Moore (2004) has explored the life of this princely family. Playne (1917) has a chapter on the family of the state of Cooch Behar with some historical photographs. MCB (1908) has a photograph of the Maharaja and his four sons taken in 1907 (fig. 32.3).

The Maharaja and his wife regularly visited the United Kingdom (figs. 32.4, 32.5). Due to the Maharaja's ill-health in later life, they went to live in London, where they arrived on 22 May 1910. They attended the coronation of King George V and Queen Mary on 22 June 1911. In July, they took a bungalow in Marina Court Avenue, Bexhill-on-Sea, East Sussex (Burl 1998). Nripendra Narayan died there on 18 September 1911, just 48 years old.

FIGURE 32.2
The Victor Jubilee Palace of Cooch Behar, built in 1887, modelled after Buckingham Palace
PHOTO: KEES ROOKMAAKER, 2017

FIGURE 32.3
"Family group, 1907." Nripendra, the Maharaja of Cooch Behar in 1907 with his four sons Rajendra (1882–1913), Jitendra (1886–1922), Victor (1887–1937), Hitendra (1890–1920)
NRIPENDRA, *THIRTY-SEVEN YEARS OF BIG GAME SHOOTING IN COOCH BEHAR*, 1908, FACING P. 411

FIGURE 32.4   Nripendra, Maharaja of Cooch Behar, painted in 1887 by Sydney Prior Hall (1842–1922). Oil on panel, 40.7 × 29.5 inch (103 × 75 cm)
ROYAL COLLECTION TRUST: 403827 © HIS MAJESTY KING CHARLES III 2023

FIGURE 32.5   Suniti Devi, Maharani of Cooch Behar, painted in 1887 by Sydney Prior Hall. Oil on panel, 103 × 75 cm
ROYAL COLLECTION TRUST: 403601 © HIS MAJESTY KING CHARLES III 2023

On some shoots, MCB was accompanied by his family members (Devee 1916). Although the record is likely to be incomplete, the following dates have been found.

Brother Jotindra – joined 1871 (a), 1873 (a), 1884 (b), 1884 (c), 1892 (b), 1893 (a), 1898 (a).

Maharani Suniti Devi – joined 1891 (a), 1895 (c), 1898 (a), 1900 (a), 1903 (a), 1905 (b), 1907 (b).

Son Rajendra (Rajey) – joined 1898 (a), 1899 (a), 1903 (a), 1905 (b), 1906 (a), 1907 (a), 1907 (b).

Son Jitendra – joined 1898 (a), 1902 (a), 1904 (d), 1905 (b).

Son Victor – joined 1900 (a), 1903 (a), 1906 (a).

Daughter Sukriti – joined 1898 (a), 1900 (a).

32.6   The Practicalities of the Shoots

The Maharaja of Cooch Behar had exclusive rights to hunt rhinos in the region under his jurisdiction (Bist 1994). Anybody wishing to shoot in the area needed permission (Lydekker 1908). It is not unlikely that applicants might have been invited to join the large annual shoots which MCB described in his book.

The shooting trips were organized in great detail and no expense was spared. Some actual figures of expenditures were published in the *Annual Administration Report of the Cooch Behar State*, amounting in 1900-01 to Rs 16,361; 1901-02 Rs 20,037; 1902-03 Rs 32,375; 1903-04 Rs 31,251; and 1904-05 Rs 8,623 (J. Pal 2015).

The annual shoots lasting several weeks most of the years were definitely grand affairs. There were enough elephants, howdahs, trackers, shikaris, skinners and support personnel in camp (Ehlers 1894, vol. 1: 385; Peissel 1966: 148). Everything was provided to make life in camp comfortable, and there was well-prepared and plentiful food and drink, even ice brought in daily from Calcutta: "the hospitality of the Maharaja was unbounded, life in the tents the ideal of good feeling and comfort, and, most important of all, the sport was bound to be good" (Hewett 1938: 90). A special insight was written by the correspondent of the *Englishman's Overland Mail* (1891-03-18), stating that "the wonders of the camp itself are almost equally worthy of notice. It is seldom even in the best of Indian camps de luxe that one finds the arrangements so complete in all their details. Every day brought the mails and supplies of ice from Calcutta, and the tired sportsman could always count on an exquisite dinner, to be eaten to the strains of the Maharaja's band. Besides a doctor, the camp boasted a barber, a photographer, and a post-master, and was better supplied with all the comforts of civilised life than many

an Indian station. But with all its delights, this dainty camp world have been reft of its chief attraction had the Maharani not been present with the Maharaja, to do the honours and to preside over her little court with the grace, tact, and courtesy for which she is famed."

On return to camp, the team had high spirits: "All in camp are as jolly as sand boys. The ladies go out every day, being as keen as the men, though they properly hide their faces when blood is shed: and after a tiring day the strains of the Maharaja's band under the able leadership of Herr Konig, and the sweet melodies sung by Mrs. Gordon, are most soothing additions to the comfort of the Camp" (*Englishman's Overland Mail* 1882-03-07).

32.7   The Total Number of Rhinos Killed in Cooch Behar Shoots

MCB (1908: 449) listed a total of 207 rhinos in his table of "Total big game shot during the thirty-seven years 1871–1907." The annual totals are again found in tables throughout the book, and also total to 207 rhinos. In five years, the number of rhinos in the tables differ from the number mentioned in the text: 1878 mentions 16 rhinos in text (11 in table), 1882 (a) 3 in text (2 in table), 1885 (b) 4 in text (3 in table), 1893 (b) 1 in text (0 in table), and 1901 (a) 5 in text (3 in table). Hence 10 rhinos appear absent from the tables (Debnath & Sarkar 2021). The total number of rhinos killed during the expeditions in Cooch Behar in 1871–1907 should be adjusted to 217 in total (table 32.39). Besides MCB and his family, 33 persons who were credited with shooting a rhinoceros could be identified (table 32.40).

It is often said in more recent reviews that MCB's penchant for hunting has led to the extinction of rhinos in India (e.g. Chaudhuri 1903: 100; Menon 1995: 56; Bist 1994). Note that MCB rarely commented on the overall status of any of the animal species, although sometimes he seems to have realized that numbers were diminishing. Although the argument is not entirely without merit, it definitely needs to be remembered that in other parts of India rhinos were hunted more often and probably in far greater numbers by military personnel, administrative officers, teaplanters, in fact by anybody with access to riding elephants and guns. The available published reports were mainly written by visitors or foreign personnel, but beyond that, when firearms became more widely available, some local people definitely would have contributed their share.

TABLE 32.39   The total number of rhinos killed in Cooch Behar shoots, 1871–1907

		1880	7	1890	6	1900	1
1871	2	1881	8	1891	5	1901	6
1872		1882	9	1892	6	1902	6
1873	4	1883	7	1893	8	1903	0
1874	5	1884	19	1894	5	1904	0
1875		1885	11	1895	9	1905	1
1876		1886	18	1896	7	1906	0
1877	15	1887	4	1897	1	1907	0
1878	16	1888	0	1898	6		
1879	11	1889	5	1899	9		
Subtotal	53		88		62		14
TOTAL	217 rhinos						

TABLE 32.40   List of persons who were credited with the kill of at least one rhino in Cooch Behar

[1877] Francis Richard Charles Greville, 5th Earl of Warwick, 5th Earl Brooke (1853–1924)

[1878] Alexander Angus Airlie Kinloch (1838–1919)

[1880] Algernon George Lawley, 5th Baron Wenlock (1857–1931)

[1880] Geoffrey Nicolas Dawnay (1852–1941)

[1880] Henry St. John Kneller (1848–1930)

[1884] Dermot Robert Wynham Bourke, 7th Earl of Mayo (1851–1927)

[1884] Colonel James Keith Fraser (1832–1895)

[1884] Richard Augustus D'Oyly Bignell (1847–1906)

[1885] George Augustus Parker, Viscount Parker (1843–1895)

[1885] Field Marshal Frederick Sleigh Roberts, 1st Earl Roberts (1832–1914)

[1886] John Clavell Mansel-Pleydell (1817–1902)

[1886] Colonel Aleksandr Karlovich von Timler (Tümler)

[1887] Marie Elisabeth Laure Le Tonnelier de Breteuil (1868–1918)

[1889] George Irwin, Bengal Civil Service

[1889] Count Imre Széchényi of Sárvár-Felsővidék (1825–1898)

[1890] Alexander Evans Gordon (1845–1920)

[1891] Colonel Henry Streatfeild (1857–1938)

[1892] Prince Christian Victor of Schleswig-Holstein (1867–1900)

[1892] Charles Harbord, 5th Baron Suffield (1830–1940)

[1892] Lieutenant-Colonel Ernest Harold Fenn (1850–1916)

[1892] James Jonathan Harrison (1858–1923)

[1893] Major-General William Lambton (1863–1936)

[1894] Felice Scheibler (1856–1921)

[1895] Heywood Walter Seton-Karr (1859–1938)

[1895] Captain William Bertram (1859–1915)

[1896] Peter Burges (1853–1938)

[1896] Lieutenant-Colonel Stephen Hungerford Pollen (1869–1935)

[1896] Hans Heinrich xv von Hochberg, Prince of Pless (1861–1938)

[1898] Alfred 'Chips' Ezra (1872–1955)

[1899] Henry Alfred Joseph Doughty-Tichborne, 12th Baronet (1866–1910)

[1899] Prince Vittorio Emanuele of Savoy-Aosta, Count of Turin (1870–1946)

[1901] Lieutenant-Colonel Reginald Heber-Percy (1849–1922)

[1901] Adolf von Waldstein (1868–1930)

[1909] Gilbert John Elliot-Murray-Kynynmound, 4th Earl of Minto (1845–1914).

32.8 Trophies

MCB and the Maharani came to England for an extended visit from May to December 1887. He brought with him a large collection of animal specimens, which were taken to the workshop of the well-known taxidermist Rowland Ward (1848–1912) at 166 Piccadilly, London, between the Ritz Hotel and Fortnum & Mason (Morris 2003; Jackson 2018). On the occasion of the Colonial and Indian Exhibition in 1886, MCB had contributed skins of tigers and an elephant displayed by Rowland Ward in a dramatic exhibit, before being sent back to the Palace in Cooch Behar (Ward 1913: 75). Overshadowed by that exhibit and apparently not specifically listed in the catalogue, Ward mentioned that through MCB's courtesy, "overhead I have disposed several heads of the rhinoceri. One of these especially is a very fine specimen killed by the Rajah. The whole skin was sent to me, but I had not time to model the beast entire. When the heads reach me, they are just a shapeless mass" (Ward 1886: 76).

From June 1887 the window display at Rowland Ward's premises "The Jungle" included excellent models of two heads which the Maharaja pronounced "the finest he ever saw" (*Globe* 1887-06-02; *Times of India* 1887-06-21). According to these newspapers, Ward "has made the rhinoceros hide hard as timber, as delicate in colour yellow amber, and as smooth and transparent a thick sheet of hard ice." There is no record which specimens these might have been, especially as already 136 rhinos had been shot during MCB's shoots 1871–1887. While the two heads were shipped back to India, Ward also sketched them adding a handwritten caption "Indian Rhinoceros (Rhinoceros Unicornis) from Kuch Behar." This first appeared on the cover of the 4th edition of his *Sportsman's Handbook* which was dated 1887 (Ward 1887) but only available from 25 February 1888 at 3s 6d (Ward 1888). This drawing of the two rhino heads continued to appear in all subsequent editions of the *Handbook*, both on the cover and on p. 110, in the *Records of Big Game* (1899), and for many years as part of the stationary of the firm (figs. 32.6, 32.7, 32.8). Besides the two heads, Ward also manufactured some beautiful tables, whips, letter-racks, card-trays, inkstands, all from rhino hide (*Derby Daily Telegraph* 1887-06-03).

MCB was one of Rowland Ward's major clients. The taxidermist's indebtedness is shown by the inclusion of the drawing of rhino heads in many of his publications and on his letterhead, as well as by the fact that the largest specimens shot in Cooch Behar were highlighted in the influential *Records of Big Game*. MCB sent most of his trophies of all species to Rowland Ward in London, but records do not exist. Once prepared, Ward would have shipped them back to Cooch Behar, or maybe also to other participants of the shoots. It is to be expected that the trophies were put on display or attached to the walls of the new Victor Jubilee Palace constructed in 1887 in the center of Cooch Behar where it still stands today (fig. 32.2). Recent sources (Martin 1996a: 33; Ray 2015) place these specimens in the dining hall, however I am unable to find them on contemporary photographs (Playne 1917; Campbell 1907). Felice Scheibler (1899a: 50, 1899b, 1900: 188), who visited in 1890–1894, stated that innumerable trophies of the hunt were seen along the broad staircase and in the halls. James Harrison (1892: 116) visiting in February 1892 remarked that in the right wing "one of the halls is filled with trophies, some of them being immense – all, in fact, 'record' heads; there are tiger, bear, leopard, elephant, buffalo, and rhinoceros besides, no end of smaller heads and horns."

FIGURE 32.6   Two heads of rhinos from Cooch Behar, first published on the cover of Rowland Ward, *The Sportsman's Handbook* (from 4th edition of 1887 and onwards)

Indian Rhinoceros (Rhinoceros Unicornis) from Kuch Behar

FIGURE 32.7   "Indian Rhinoceros (Rhinoceros Unicornis) from Kuch Behar" sketched by Rowland Ward, now integrated in his *Records of Big Game* (3rd edition, 1899, p. 430)

FIGURE 32.8   The rhinos from Cooch Behar featured prominently on the letterhead of Rowland Ward
FROM P.A. MORRIS, *ROWLAND WARD*, 2003, P. 130.
COURTESY © PAT MORRIS

FIGURE 32.9
Skeleton of *R. unicornis* in the Zoological Survey of India (no. 19262), donated by the Maharajah of Cooch Behar in 1879
PHOTO: © TANOY MUKHERJEE, 2018

In his pictorial history of Bengal, Campbell (1907, vol. 2: 287) also mentioned "a fine collection of trophies of the chase in the Cooch Behar Palace." There may well have been a special trophy room, or even stores where some of the specimens were preserved. Sadly nothing is left of this past glory, and not only that, there appears to be absolutely no trace of what might have happened to any of these animal remains in the course of the 20th century. Like so many other testimonies of a past age, their potential historical and scientific value is lost.

In 1903, Ward (1913: 160) suggested to MCB that he might donate a rhino specimen to the Natural History Museum in London. Duly sanctioned, a male rhino set up by Rowland Ward was received in January 1903 (no. 1903.2.13.1) and at first temporarily exhibited in the central hall of the museum in South Kensington (*Globe* 1903-01-30; Lydekker 1903). A skull was also included (Pocock 1944b: 838). The animal's size is recorded as 10 ft 4 in (315 cm) long, 5 ft 5 in (165.7 cm) high, with a short horn. Although said to be an unusually fine specimen, it was surprisingly one of the smaller animals, suggesting that it was a young adult.

The only other rhino specimen known to have been donated by MCB is a complete skeleton and skull of

R. unicornis presented to the Indian Museum in Kolkata in 1879 (fig. 32.9). It was listed in the catalogue by Sclater (1891: 202, no. e), and is still available, no. ZSI 19262 or GRM 198 (Chakraborty 2004: 39).

A few trophies found their way to the collections of other participants (table 32.41). It must be said that it is unlikely that the list is complete. Many horns or trophy heads, maybe also items made from the feet, should have made their way to private homes, but few have been recorded in detail.

32.9 The Details of the Annual Shoots

All shoots organized by MCB are analysed below in chronological order. Dates and numbers of animals killed are given by MCB in tables at the end of the chapters. Not all dates are accurate, and the number of animals do not always tally with numbers in the text (32.7). MCB in most years went on a big shoot lasting over a month, usually starting in February during the dry season, when much of the grass had been burned. There were also regular short excursions to places not too far from the capital Cooch Behar. The list of all shoots is in Table 32.42. The outlines below follow a set format.

1. The first line provides the sequential number of the shoot, the date range, and the number of rhinos killed.
2. The second line shows the main region where the shoot took place.
3. This is followed by a list of localities as found in MCB (1908) with the original spelling and sequence, showing the general distance between places. These localities where rhinos were found are in 32.12 (datasets 32.36, 32.37) and shown on map 32.16.
4. Next is a list of participants, following the names and sequence used by MCB, except that family members have been grouped together. There were about 340 people from Europe and India, and a small number from elsewhere. While it has not been possible to identify them all, in 191 cases the full names were found. These are indicated by an asterisk * after the surname, referring to the list of participants with known identity in 32.12 and Table 32.44.
5. The dates on which rhinos were seen, wounded or killed are given, using all available sources. These are used to reconstruct the localities in 32.10.
6. The final section has any remarks or comments about the interactions with rhinos.

TABLE 32.41   Known trophies of rhinos shot in Cooch Behar, as recorded in the literature

Shoot 1877 (b). Francis Richard Charles Greville, 5th Earl Brooke (1853–1924). Horn in Easton Lodge in Essex, destroyed in fire in 1918.

Shoot 1887 (a). Marquise Marie Elisabeth Laure Le Tonnelier de Breteuil (1868–1918). Head with horn in Breteuil Castle in France, auctioned in 2012.

Shoot 1892 (a). James Jonathan Harrison (1858–1923). Trophy in Brandesburton, Yorkshire, UK.

Shoot 1892 (a). Prince Christian Victor of Schleswig-Holstein (1867–1900). Trophy, whereabouts unknown.

Shoot 1894 (a). Felice Scheibler (1856–1921). Trophies in Villa Scheibler located in Rho, near Milan, Italy.

Shoot 1896 (a). Peter Burges. Two rhino skulls in his home in Bristol, UK.

TABLE 32.42   List of shoots of the Maharaja of Cooch Behar 1871 to 1907

No.	Year	Dates	Number of days	People	Area	Rhinos killed
1	1871 (a)	no record		4	Probably Raidak area	2
	1872	no record				
2	1873 (a)	no record		4	Probably Raidak area	4
3	1874–1876	no record				5
4	1877 (a)	01–22 to 02–08	18	7	Probably Raidak area	1
5	1877 (b)	02–20 to 03–03	13	7	Probably Raidak area	10
6	1877 (c)	December		5	Raidak River	4
7	1878 (a)	March		9	Probably Raidak area	5

TABLE 32.42 List of shoots of the Maharaja of Cooch Behar 1871 to 1907 (*cont.*)

No.	Year	Dates	Number of days	People	Area	Rhinos killed
8	1878 (b)	April		4	Raidak – Sankosh area	11
9	1879 (a)	03–04 to 03–29	26	5	Sankosh River	11
	1880	01–22 to 01–27				
	1880	February				
10	1880 (a)	02–28 to 03–14	15	5	Assam, Khagrabari east of Sankosh	7
	1880	03–31				
	1880	December				
11	1881 (a)	02–14 to 02–26	13	11	Assam, Khagrabari east of Sankosh	4
	1881	03–03				
12	1881 (b)	03–24 to 03–26	3	5	Raidak – Sankosh area	4
	1882	02–06 to 02–07				
13	1882 (a)	02–25 to 03–04	8	13	Raidak – Sankosh area	3
14	1882 (b)	March	7	5	Raidak – Sankosh area	2
15	1882 (c)	04–04 to 04–15	12	5	Probably Raidak area	4
	1883	February				
16	1883 (a)	02–18 to 03–23	34	9	Raidak – Sankosh area	6
17	1883 (b)	04–03 to 04–06	4	4	Probably Raidak area	1
	1884	January				
18	1884 (a)	02–14 to 03–25	41	8	Jaldhaka and Raidak regions	16
	1884	03–31				
	1884	04–03 to 04–09				
19	1884 (b)	04–18 to 04–22	5	7	Raidak – Sankosh area	2
20	1884 (c)	05–19 to 05–21	3	4	Raidak – Sankosh area	1
	1884	June				
21	1885 (a)	02–28 to 03–21	22	17	Raidak – Sankosh area	7
22	1885 (b)	03–30 to 04–06	8	5	Raidak – Sankosh area	4
23	1886 (a)	02–12 to 03–12	29	13	Raidak – Sankosh area	18
	1886	December				
	1887	02–08				
24	1887 (a)	02–18 to 03–09	20	12	Raidak – Sankosh area	4
	1888	02–25				
	1888	04–23				
	1888	07–18				
25	1889 (a)	02–12 to 03–18	35	19	Raidak – Sankosh area	5
	1889	April				
	1889	June				
	1889	July				
	1890	01–07 to 01–23				
	1890	02–05				
26	1890 (a)	02–14 to 03–19	34	28	Raidak – Sankosh area	6
	1890	April				
	1890	May				
	1890	June				
27	1891 (a)	02–14 to 04–14	60	22	Raidak – Sankosh area	5
	1891	July to 07–26				
28	1892 (a)	02–22 to 03–12	19	27	Assam, east of Sankosh River	5
29	1892 (b)	04–25 to 05–04	10	7	Torsa River, Cooch Behar	1

TABLE 32.42   List of shoots of the Maharaja of Cooch Behar 1871 to 1907 *(cont.)*

No.	Year	Dates	Number of days	People	Area	Rhinos killed
30	1893 (a)	02–11 to 03–20	38	25	Raidak – Sankosh area	7
31	1893 (b)	04–10 to 04–27	18	14	Torsa River, Cooch Behar	1
32	1894 (a)	02–27 to 03–14	16	20	Dhubri and Goalpara, Assam	5
33	1895 (a)	02–20 to 03–21	30	15	Assam, Manas River	7
34	1895 (b)	04–14 to 04–28	15	16	Raidak – Sankosh area	seen
35	1895 (c)	12–01 to 12–15	14	17	Raidak – Sankosh area	2
36	1896 (a)	02–20 to 03–27	36	14	Raidak – Sankosh area	5
37	1896 (b)	April		4	Raidak – Sankosh area	2
38	1897 (a)	03–11 to 04–17	38	11	Raidak – Sankosh area	1
	1898	01–30 to 02–08				
39	1898 (a)	02–14 to 03–23	38	17	Raidak – Sankosh area	6
	1898	April				
	1898	May				
40	1899 (a)	02–23 to 03–23	29	14	Assam, Manas River	7
41	1899 (b)	03–24 to 03–29	6	2	Assam, Bansbari	2
	1899	04–10 to 04–19				
42	1900 (a)	02–10 to 03–23	42	20	Raidak – Sankosh area	1
43	1901 (a)	03–01 to 04–16	47	12	Raidak – Sankosh area	5
44	1901 (b)	April		2	Assam, Manas River	1
	1902	01–21 to 02–03				
45	1902 (a)	03–01 to 04–13	44	11	Assam, Manas River	6
	1902	12–13				
	1903	02–07 to 02–11				
	1903	02–16 to 04–04	51	25	Raidak – Sankosh area	nil
	1903	November				
	1903	December				
46	1904 (a)	01–17 to 02–16	22	10	Raidak – Sankosh area	seen
47	1904 (b)	03–15 to 03–26	12	10	Raidak – Sankosh area	seen
48	1904 (c)	04–04 to 04–09	6	8	Torsa River to Jaldhaka River	seen
49	1904 (d)	12–17 to 12–27	11	10	Raidak – Sankosh area	seen
50	1905 (a)	02–18 to 02–26	9	12	Assam, Khagrabari east of Sankosh	wounded
51	1905 (b)	04–02 to 04–23	22	13	Torsa River and Raidak River	1
	1905	December				
	1906	03–22 to 04–30	40	11	Assam, Manas River	
	1906	November	5	10	Chilapeta	
	1907	02–18 to 02–24	7	15	Assam, Khagrabari east of Sankosh	
	1907	03–13 to 04–01	20	13	Raidak – Sankosh area	
Shoots after the publication of MCB (1908)						
	1908	February			Torsa River	
	1908	March			Assam, Lakhipur	
52	1909	02–11 to 02–17	7	11	Assam, Manas River	1
	1910	February				
	1912	January				

All shoots found in MCB (1908) are included. Only those during which rhinos were mentioned are numbered and provided with a letter to identify them. Other shoots are only listed by date. Dates are in MM-DD format.

1871

Shoot 1. 1871 (a) – no dates – 2 rhino.

Participants (4): MCB, W.O.A. Beckett*, R.H. Renny*, brother Jotindra.

This was MCB's first recorded shoot, when he was 7 years old. It is likely that this occurred in the Raidak River area, where he would return often in subsequent years. The area is about 25 km from the town of Cooch Behar. It would be interesting to know if he was already responsible for shooting the two rhinos.

1872 – No Record

1873

Shoot 2. 1873 (a) – no dates – 4 rhino.

Participants (4): MCB, [Thomas] Smith*, Kneller*, brother Jotindra.

No data.

1874 – No Record

Shoot 3. Records for 1874–1876 were lost. MCB (1908: 4) thought that about 20 tigers and 5 rhino were shot.

1875 – No Record

1876 – No Record

1877

Shoot 4. 1877 (a) – 22 January to 8 February 1877 (18 days) – 1 rhino.

Possibly Raidak River area.

Participants (7): MCB, Dr. Benjamin Simpson*, Southby*, Wilson, E. Hewett, Kneller*, Dalton*.

Rhino recorded in table (1908: 12), no further details. MCB was now 13 years old.

Shoot 5. 1877 (b) – 20 February to 3 March 1877 (13 days) – 10 rhino.

Raidak River: Takuamari and Falimari.

Participants (7): MCB, Lord Brook [Greville]*, Sir Robert Abercrombie*, Dr. Simpson*, Dr. Roberts, Dalton*, Dr. Cummins.

No details of dates or localities for rhino. Francis Richard Charles Greville, 5th Earl Brooke (1853–1924) had left Calcutta on 15 February 1877 to reach Dhubri on 22 February. He states that he went shooting in the Bhootan Hills until 9 March together with "C", Assistant-Commissioner (Greville 1877b, 1878: 244). A newspaper report said that his party including Abercrombie killed 9 rhinos in that period (Greville 1877a). Greville (1917: 300) recalled: "I had

been warned that the track my rifle was to command was greatly favoured by the rhinos and so it proved, several came my way, and I had the good fortune to account for four. I kept only one trophy." The trophy horn stood in a cabinet in the hall of Easton Lodge in Essex, which was largely destroyed in a fire in 1918. The photograph of "A corner of the Entrance Hall, Eaton Lodge" in Greville (1917: 304) does not include a rhino trophy head on the wall.

Shoot 6. 1877 (c) – December 1877 – 4 rhino.

Raidak River area.

Participants (5): MCB, General Kinloch*, Col. Mant, Kneller*, Dalton*.

No details of dates or localities for rhino.

1878

Shoot 7. 1878 (a) – March 1878 – 5 rhino.

Possibly Raidak River area.

Participants (9): MCB, Duke Maliano (now Duke Grazzioli*), Don Julio*, and his brother Marquis Pizzardi*, also Sage, Dalton*, Southby*, Simpson*, Hewett.

Text states that 5 rhinos were killed, but MCB did not include them in the table (p. 12). No details of dates or localities for rhino. Three Italian noblemen took part in this shoot, after they had attended the marriage to the Maharaja to Suniti Devi on 6 March 1878 (Chaudhuri 1903: 422). The listing in MCB (1908: 6) is rather garbled and must refer to Mario Grazioli, his brother Giulio Magliano, and Carlo Alberto Pizzardi.

Shoot 8. 1878 (b) – April 1878 – 11 rhino.

Raidak – Sankosh area.

Participants (4): MCB, [Thomas] Smith*, Kinloch*, Dalton*.

No details of dates or localities for rhino. MCB (1908: 6) says that in April a total of 11 rhino were shot by the three participants. General Alexander Angus Airlie Kinloch (1838–1919) took part in this hunt which he described in some detail as to events without specifying localities (Kinloch 1885: 63–65, 1892: 85–88, see 31.15).

Day 1. One rhino wounded.

Day 3. First rhino killed by Kinloch. Old male, height 5 ft 9 in (175.2 cm), length to root of tail 10 ft 6 in (320.0 cm).

Day 4. Second rhino killed by Kinloch, small female.

Day 4. Third rhino killed by Smith, perfect horn 1 foot (30.5 cm) long.

Day 5–7. Killed several more rhino and lost others.

Day 7. Kinloch killed his third rhino, largest that was killed during the trip, massive horn, no measurements kept.

1879
Shoot 9. 1879 (a) – 4 to 29 March 1879 (26 days) – 11 rhino.
Raidak – Sankosh area.
Locality: Garud Haut on Sankosh River.
Participants (5): MCB, Simpson*, Baillie, Hewett, Kneller*.
No details of dates or localities for rhino.

1880
In March 1880, the Viceroy conferred the title of Maharaja Bahadur upon Nripendra of Cooch Behar, "in recognition of the high rank and dignity of the holder" (*Homeward Mail from India* 1880-03-25).

Shoot 10. 1880 (a) – 28 February to 14 March (15 days) – 7 rhino.
Assam, Khagrabari, east of Sankosh River.
Localities: Pargaon, Khagrabari.
Participants (5): MCB, Dawnay*, Lawley*, Cammell, Kneller*.
Rhinos shot:
1880-03-01, 1 killed by Henry Kneller.
1880-03-05, 2 killed by Algernon Lawley.
1880-03-07, 1 killed.
1880-03-14, 2 killed by Geoffrey Dawnay.
MCB (1908: 11) writes that 6 rhinos were killed in March. As the table (p. 12) records 7 in total for the year, the additional rhino could have been obtained elsewhere. There is an almost daily account of this shoot in the autobiography of Algernon George Lawley (1857–1931), providing dates and names of people shooting the rhinos and other animals (Lawley 1933: 25).

1881
Shoot 11. 1881 (a) – 14 to 26 February 1881 (13 days) – 4 rhino.
Assam, Khagrabari, east of Sankosh River.
Localities: Khagrabari.
Participants (11): MCB, Ashton*, Fraser* (= Lord Saltoun), Thomas, Turner, A. Apcar*, Alexander*, Kneller*, Major Jarrett*, Wilson, Dalton*.
Rhinos shot:
1881-02-15, 1 rhino killed, 1 wounded at Khagrabari.
1881-02-17, 1 killed at Khagrabari.
1881-02-21, 1 killed at Khagrabari.
1881-02-22, 1 killed at Khagrabari.

Shoot 12. 1881 (b) – 24 to 26 March 1881 (3 days) – 4 rhino.
Raidak – Sankosh area.
Localities: Parokata.
Participants (5): MCB, Fraser*, Kinloch*, Dalton*, Kneller*.
Rhinos shot:

1881-03, 4 killed in 3 days, at Parokata.
1881-03, 1 wounded at Parokata.

1882
This year, soon after his 18th birthday, MCB made several trips to the jungle. Rhinos were encountered on three of these excursions.

Shoot 13. 1882 (a) – 25 February to 4 March 1882 (8 days) – 3 rhino.
Raidak – Sankosh area.
Localities: Garud Haut, Malakata, Mahakulgoree, Parokata, Berbera.
Participants (13): MCB, Sir Ashley Eden*, Sir William Eden*, Lawler [Lawlor*], Schalch*, M. Eden*, Lord Ilchester*, Lord Durham*, Dr. Simpson* and Miss Simpson*, Henry*, Gordon [Evans Gordon*] and Mrs. Gordon*, Moore*. The *Englishman's Overland Mail* (1882-03-07) adds Jotindra Narayan, Kneller* (in charge of the camp), Mrs. C. Wilkins.
This hunt was organised in honour of the visit of Ashley Eden, the Lieutenant-Governor of Bengal. In the text, MCB (1908: 18) mentions 3 rhinos, but only 2 are included in the annual table (p. 23). A journalist of the *Englishman* (1882-03-07) mentioned 3 rhinos. This is the first time ladies were included in the hunting party.
Rhinos shot:
1882-02-25, 2 rhinos killed.
1882-02-27, a large rhino killed.

Shoot 14. 1882 (b) – March 1882 (7 days) – 2 rhino.
Raidak – Sankosh area.
Localities: Parokata.
Participants (5): MCB, Southby*, Gladstone, Browne*, Count Seeback*.
Rhinos shot:
1882-03, 2 killed near Parokota.
1882-03, 2 wounded near Parokota.

Shoot 15. 1882 (c) – 4 to 15 April 1882 (12 days) – 4 rhino.
Possibly Raidak – Sankosh area.
Participants (5): MCB, Garth*, Mansel*, Spragge, Gordon*.

1883
Shoot 16. 1883 (a) – 18 February to 23 March 1883 (34 days) – 6 rhino.
Raidak – Sankosh area. Main annual shoot.
Localities: Garud Haut, Deogong.
Participants (9): MCB, Cammell, Dawnay*, Davies Fitzgerald, Lucas, Peck, Thornhill, Kneller*, Dalton*.

The party killed 6 rhino, without details of dates or shooters. On 2 March a rhino attacked two elephants.

Shoot 17. 1883 (b) – 3 April – 6 April 1883 (4 days) – 1 rhino.
Locality: not recorded, probably Raidak – Sankosh area.
Participants (4): MCB, Munro, Dalton*, Scott.
During this trip 1 rhino was killed and 4 were wounded.

1884

Shoot 18. 1884 (a) – 14 Feb – 25 March 1884 (41 days) – 16 rhino.
Jaldhaka and Raidak regions. Main annual shoot.
Localities: Ramshahai Haut, Lower Tondoo, Dhoobjhora, Upper Tondoo, Ambari, Nagrakata, Haldibari Haut, Falakata, Theamari Gaut on Raidak River, Parokata.
Participants (8): MCB, Earl of Mayo*, Neville Chamberlain*, Bignell*, Shillingford*, Colonel Keith Fraser*, Colonel Money*, Moore.
The table (1908: 38) records 16 rhino, while the tally in the text is 15 rhino. The party shot rhino in both the Jaldhaka and Raidak areas. The following can be recorded with approximate dates and locality:
(In the Jaldhaka area)
1884-02-24, 1 killed at Lower Tondoo (Tandu).
1884-03-06, 1 killed at Ambari.
1884-03-07, 2 killed and 5 wounded at Ambari.
1884-03-10, 1 killed at Nagrakata, others seen in distance.
(In the Raidak area)
1884-03-19 and following days, 5 killed near Haldibari (Theamari Gaut). During the week, one large rhino was seen but not shot. Both the Maharaja and Bignell were charged by rhinos. At least one rhino each was attributed to MCB, Bignell and Mayo. Fraser also killed one on an unknown date.
1884-03-23, 5 killed at Parokata.
They started at Ramsai and stayed in the area of the Jaldhaka River, probably because an elephant capture operation was conducted by Joseph Lay Shillingford (1845–1889) from Purnea. MCB added 12 elephants to his stable, increasing the total number to 54 (Anon. 1885a: 56; *Scientific American* 1884-09-13). The party moved to different camps because they found the area poor in wildlife. In the middle of March they started to travel eastward to the more familiar grounds near the Raidak River.

Their arrival in Tondoo near the Jaldhaka River was recorded in the *Times of India* (1884-02-28), where the correspondent said that the elephants had migrated northwards due to the forest fires. News from Nagrakata stated that 4 rhinos had been killed between 6 and 10 March (*Times of India* 1884-03-20).

The most illustrious guest in 1884 was Dermot Robert Wynham Bourke (1851–1927), 7th Earl of Mayo, an Irish peer and son of the late Viceroy. Recalling his time in Cooch Behar, he claimed at least one rhino in the week of 19 March (Anon. 1884a), but it is unknown if any trophy was displayed in his home Palmerstown near Dublin, Ireland.

Another participant, Colonel James Keith Fraser (1832–1895) returned to Kolkata on 24 March 1884 after some weeks of "marvellous sport", stating that he killed 3 rhinos (Pontevès 1886: 101).

Shoot 19. 1884 (b) – 18 April to 22 April 1884 (5 days) – 2 rhino.
Locality: Madhoobhasha.
Participants (7): MCB, Khitindro [Jitindra], General Wilkinson, Major Cook*, Gordon*, Charley Briscoe*, Jackson.
There are no details about this shoot in MCB (1908). The locality Madhoobhasha is not otherwise recorded, possibly it was in the Raidak area where they often went to hunt.

Shoot 20. 1884 (c) – 19 May to 21 May 1884 (3 days) – 1 rhino.
Raidak – Sankosh area.
Locality: Falimari.
Participants (4): MCB, Khitindro [Jitindra], Bignell*, Briscoe*.
There are no further details.

1885

Shoot 21. 1885 (a) – 28 February to 21 March (22 days) – 7 rhino.
Raidak – Sankosh area. The main annual shoot.
Localities: Chota Bhalka, Deogong, Dal Dalia, Falimari, Bara Bhalka, Barbara, Parokata, Rasik Bhil, Mahakalgooree.
Participants (17): MCB, General Frederick Roberts*, General Stewart*, General Godfrey Clark*, Dr. Benjamin Simpson*, Colonel Pole Carew*, Neville Chamberlain*, Hume*, Gordon*, Major Garth, G. Garth, Bignell*, Viscount Parker*, Col. St. Quinton, General Paget*, Knyvett*, Jamieson.
A total of 7 rhino were killed, at least one was wounded.
There are details about two of these:
1885-03-01, 1 killed, no exact locality, at 150 yards with a .500 Express gun by Viscount Parker.
1885-03-18, 1 killed, at Dal Dalia.
The male rhino killed on 18 March was larger than most, 401 cm length, 279 cm girth. When General Roberts returned to Kolkata on 5 March, he claimed bagging one rhino (*Times of India* 1885-03-09).

FIGURE 32.10
"The Bag, 1885 Shoot" showing animals shot
during the annual hunt. This is the earliest
photograph in MCB's book with a rhino killed
during the shoots
NRIPENDRA, *THIRTY-SEVEN YEARS OF BIG
GAME SHOOTING IN COOCH BEHAR*, 1908,
P. 37. REPRODUCED BY KIND PERMISSION
OF CAMBRIDGE UNIVERSITY LIBRARY

A photograph shows the first tiger of this 1885 season with ten of the hunters behind the animal (MCB 1908: 34). Another shows "The Bag, 1885 Shoot" (fig. 32.10) probably taken after return, with 7 tiger skins and at least 3 rhino skulls with their horns attached (1908: 37).

Shoot 22. 1885 (b) – 30 March to 6 April 1885 (8 days) – 4 rhino.

Raidak – Sankosh area.

Localities: Guigaon, Jaldoa, Rossik Bheel.

Participants (5): MCB, C.H. Moore, C.R. Hills, A. [Alick] Evans-Gordon*, Charley Briscoe*.

A smaller party killed 4 rhino in these 8 days. The table (1908: 39) records 3 rhino, but in the text 4 rhino are clearly mentioned. The shooters of the individual rhinos are not recorded. The animals were found at Jaldoa and Rossik Bheel.

1886

The London Zoo received a male *R. unicornis* on 25 December 1886 donated by MCB through the intervention of the Surgeon-General Benjamin Simpson (1831–1923), who was one of the Maharaja's guardians when he studied in England in 1878–1879. There is no information about the rhino's age on arrival or when he was captured (Sclater 1887). Named Tom, the animal lived in the zoo until 30 December 1911 (Rookmaaker 1998: 86). He is seen in a postcard of 1905 (Edwards 1996: 130), in an unsigned photograph of 1900 standing in the pool (Edwards 2012: 148), and in a photograph by Henry Sandland taken around 1895 (Morley & Friederichs 1895: 3). When he died, his height at withers was 5 ft 10 ½ in (179 cm) and his weight was calculated at 3612 lbs or 1638 kg (Pocock 1912).

Shoot 23. 1886 (a) – 12 February to 12 March 1886 (29 days) – 18 rhino.

Raidak and Sankosh Rivers.

Localities: Garad Haut, Rossik Bheel and Chengtimari, Berbera jungle and Falimari, Chengtimari, Dorko, Sunkos River, Sal Forest Reserve, Tara Bagh.

Participants (13): MCB, Gojendro (junior) [Gajendra*], Capt. Harbord*, Capt. Bignell*, Prince Esterhazy*, Colonel Timler* and Prince Maslof* (Russian), Colonel Upperton*, Major Gordon*, Bignell*, Charley Briscoe*, Major Mansel*, Moore.

Rhinos shot:

1886-02-16, 5 killed before lunch at Rossik Bheel and Chengtimari. Mentioned by Barclay (1932: 137).

1886-02-19, 4 killed at Chengtimari.

1886-02-24, 3 killed near Dorko or Chengtimari.

1886-03-05, 1 killed near Sunkos River.

1886-03-09, 5 killed after lunch in Sal Forest Reserve (near Sunkos River).

Colonel Aleksandr Karlovich von Timler (Tümler), accompanying the Russian Prince Nikolai Maslov, wrote a diary of experiences in India, including the hunt of rhino in Cooch Behar in February 1886 (Timler 1888). It mentions a photograph taken at the time, now unknown.

Besides MCB, the shooters are not specified, but there is an undated photograph (1908: 44) of a dead rhino together with (presumably) John Clavell Mansel-Pleydell sitting on the animal, flanked by five other men (fig. 32.11).

1887

Shoot 24. 1887 (a) – 18 February to 9 March 1887 (20 days) – 4 rhino.

Lower Sankosh River.

FIGURE 32.11
"The Rhino shot by Mansel." The person sitting
on the rhino's back is probably John Clavell
Mansel-Pleydell (1817–1902). The five others are not
identified
NRIPENDRA, *THIRTY-SEVEN YEARS OF BIG GAME
SHOOTING IN COOCH BEHAR*, 1908, P. 44

FIGURE 32.12
"Our first Rhino, 1887 Shoot." The animal was killed
on 21 February 1887 at Guigaon by the Marquise de
Breteuil. Only support staff are shown in the picture
NRIPENDRA, *THIRTY-SEVEN YEARS OF BIG GAME
SHOOTING IN COOCH BEHAR*, 1908, P. 48

Localities: Garud Haut, Guigaon, Chota Bhalka, Deogong, Haldibari.

Participants (12): MCB, Earl of Annesley*, Count and Marquise de Breteuil*, A. Norman, Sir B. Simpson*, L. Daniell, Percy Simpson, Bignell*, Ridgway, Gordon*, Cameron.

MCB thought that the rhinos were particularly wild (hidden) this time, meaning that the country needed some rest. A total of 4 rhinos were shot:

1887-02-21, 1 killed, 1 wounded, at Guigaon.

1887-02-24, 3 killed, at Deogong.

One of the rhinos shot on 24 February was a male 354 cm long, 183 cm high. The first rhino killed on 21 February 1887 is shown in a photograph (fig. 32.12) with four elephants but without any of the guest shooters (1908: 48, reproduced Anon. 1909a). This animal was shot by Marquise Marie Elisabeth Laure Le Tonnelier de Breteuil (1868–1918), the only sister of Henri Charles Joseph Le Tonnelier de Breteuil, Marquis de Breteuil (1847–1916). The Marquis recorded the hunt in his diary stating that a total of 4 rhinos were killed (Breteuil 2007: 109). The trophy head was preserved in their castle in Yvellines, northern France and was auctioned in 2012 (Reyssat 2011, lot 34, with figure). It had a plaque with text: "Rhinoceros unicorne indicus, mâle, Gârad-hât Cooch-Behar, Indes Anglaises, 27 Février 1887, Miss de Breteuil." The more likely date was 21 February 1887, and Miss de Breteuil should be the Marquise.

1888

There were several shorter hunting parties this year. No rhino were killed to allow them to recover from previous activities in their habitat.

1889

Shoot 25. 1889 (a) – 12 February to 18 March 1889 (35 days) – 5 rhino.

Raidak – Sankosh River area.

Localities: Takuamari, Horseshoe jungle, Falimari, Haldibari on Sunkos River, Bamoner Shali, Bara Hamua, Moshamari, Deogong, Berbera, Khagribari, Rossik Bheel.

Participants (19): MCB, Earl of Scarborough*, Lord Galway*, Crawley, Dudley Leigh*, Sir Benjamin Simpson*, Sir Henry Lennard*, Currie, Bignell*, Gordon*, Hendley*, Colonel Luard, Colonel Boileau*, Count Hoyos*, Lord Ancram*, Eyre-Coote, Count Szechenyi*, J.P. Hewett*, George Irwin*.

A total of 5 rhino were killed this year.

1889-02-18, 1 killed, 2 wounded, near Falimari.

1889-03-03, 2 wounded, near Bara Hamua on Sunkos River.

1889-03-08, 1 fine male killed, at Moshamari.

1889-03-17, 3 killed, towards Rossik Bheel.

Sir John Prescott Hewett (1854–1941) recorded some events of this hunt in *Jungle Trails* (Hewett 1938: 90). He states that the biggest rhino (of 8 March?) was killed by George Irwin and the Hungarian Count Imre Széchényi, who are probably the men seen on a photograph (1908: 61) of the "big bull rhino" (fig. 32.13).

1890

Shoot 26. 1890 (a) – 14 February to 19 March 1890 (34 days) – 6 rhino.

Sankosh River area, partly in Assam.

Localities: Takuamari, Falimari, Deotakata, Turkani, Chota Balka, Baradogla, Dal Dalia, Haldibari, Sunkos Reserves.

Participants (28): MCB, Lord and Lady Claud Hamilton*, Hughes*, Pole-Carew*, Brasier-Creagh*, Pakenham, Gordon* and Mrs. Gordon*, Capt. and Lady Streatfeild*, Bignell* and Mrs. Bignell*, Lady Prinsep, Sant, Sir Benjamin Simpson*, Turner, Grove, Garth*. Also Sweet, Nirmal, Gojendro*, General Auchinlech*, G. Apcar, Hendley*, Chevalier Scheibler*, General Hills*, Charley Hills*.

The party this year included at least five ladies. Mrs. Helen (Evans) Gordon shot a tigress on 11 March, widely reported in the press at the time. The Maharani Suniti Devi might have joined this year, as she mentioned that in her memoirs (Devee 1916: 110).

Just 4 rhino are detailed in the text:

1890-02-14, 1 wounded by Gordon, at Takuamari. Later found dead, and probably included in the total. Also mentioned in *Englishman's Overland Mail* (1890-02-25).

1890-03-05, 2 killed in Dal Dalia (Garadhat area).

1890-03-12, 1 killed in Sunkos Reserves.

The table states the hunting of 6 rhinos, but the newspapers at the time give only 5 specimens (*Englishman's Overland Mail* 1890-03-18). When Felice Scheibler (1900: 195) joined the hunt in the course of March, the party had already shot 5 rhino. When he left (1900: 211) the total was 6 rhino, but it doesn't appear that he was among those responsible. Felice Scheibler returned to Cooch Behar in 1892 and 1894, and his photographs are discussed in 1894 (a).

1891

Shoot 27. 1891 (a) – 14 February to 14 April 1891 (60 days) – 5 rhino.

FIGURE 32.13
"Big bull Rhino. 1889 Shoot." The hunters are probably the Hungarian Count Imre Széchényi (1825–1898) and George Irwin
NRIPENDRA, *THIRTY-SEVEN YEARS OF BIG GAME SHOOTING IN COOCH BEHAR*, 1908, P. 61

Raidak – Sankosh River area.

Localities: Chukchuka jungles, Bara Bhalka, Horseshoe jungles, Chota Dal Dalia, Kookoor Chubi, Garad Haut, Guigaon, Rossik Bheel, Bara Hamua, Deotakata, Haldibari, Gordongunj.

Participants (22): MCB and Maharani, Harry Hungerford, Brasier-Creagh*, Lord Fred Hamilton*, Streatfeild* and Lady Florence Streatfeild*, L. Daniell, Gordon* and Mrs. Gordon*, Bignell* and Mrs. Bignell*, Mrs. Toomey*, Sir Edward Sassoon* and Lady Sassoon*, Lord Ailsa*, Charley Harbord*, Vere, Cumberlege, Maharaja of Mourbhanj*, Kiddell, Ehlers*. Possibly a few others joined, listed in *Englishman's Overland Mail* (Wednesday 1891-02-18) as Mr. Shaut and Mr. Sen.

The party included at least five ladies, among whom was Maharani Suniti Devi.

During this long hunt, 5 rhino were killed, of which 4 were mentioned in the text, all without identifying the actual shooter:

1891-02-18, 1 wounded near Bara Bhalka.

1891-02-19, 1 wounded in Horseshoe Jungles.

1891-03-08, 1 killed at Rossik Bheel.

1891-03-16, 1 male killed near Deotakata.

1891-04, 2 killed near Gordongunj.

According to Frederick Spencer Hamilton (1856–1928) in his memoirs (Hamilton 1919: 151, 1921: 8), one of the rhino was shot and killed by his nephew Henry Streatfeild (1857–1938), probably the one recorded in Rowland Ward (1892: 238) with horn length 5 ½ in (14 cm) and bulky circumference of 36.8 cm. Hamilton mentioned that the party drove from Cooch Behar to Assam on a road specially prepared for the occasion, in a coach with the Australian trainer Louis Joseph Oakley (1854–1955) at the wheel (Moore 2004: 108).

Archibald Kennedy, 3rd Marquess of Ailsa (1847–1938) was present at the killing of one of the rhinos, even if he did not fire the last shot (Moss 2003: 217).

The German traveler Otto Ehrenfried Ehlers (1855–1895) also participated this year, but a rhino is not recorded for him (Ehlers 1894, vol. 1: 384–397). It is noteworthy that he mentions that one tent was occupied by Schirmer, a photographer representing the firm of Bourne & Shepherd of Calcutta.

1892

Shoot 28. 1892 (a) – 22 February to 12 March 1892 (19 days) – 5 rhino.

Assam, Khalabari, east of Sankosh River.

Localities: Bara Dogla, Sapkata, Do-Mahana, Pachkoldoba, Bara Bhalka, Turkani, Sunkos River, Jamba Mech's village, Dal Dalia.

First Viceregal shoot. Participants (27): MCB, Viceroy Lord Lansdowne* and Lady Lansdowne*, Prince Christian Victor*, Lord de Vesci*, Harbord*, Valletort*, Scheibler*, Sir Benjamin Simpson*, Bignell* and Mrs. Bignell*, Miss Lowis, Lord William Beresford*, General Viscount Frankfort*, Powney, Harrison*, Lord Borthwick, Brown. *Homeward Mail* (1892-03-21) adds: Doctor Fenn*, Colonel Boileau, W. Grenfell*, Mrs. Grenfell*, Mr. Lowis* and Mrs. Lowis*. Harrison (1892: 118) also adds Mr. Powis with his wife and daughter.

The list of MCB includes at least three ladies, but possibly five. The Viceroy and his wife left on 28 February, together with his retinue.

MCB records that 5 rhino were killed without identifying the shooters (*Englishman's Overland Mail* 1892-03-02). However, some names were supplied by the correspondent recorded in the *Homeward Mail* (1892-03-07, 1892-04-14).

1892-02-22, 1 killed at Bara Dogla, shot by Prince Christian Victor.

1892-02-23, 1 killed at Sapkata, and 1 wounded by Charles Harbord.

1892-02-24, 2 killed near Do-Mahana. A male shot by Ernest Harold Fenn was 182 cm high with a 30.5 cm horn. A female was first wounded by Felice Scheibler, then dropped down by MCB.

1892-03-02, 1 killed at Turkani, by Captain Edward Bignell and James J. Harrison.

This was the first time that a Viceroy visited Cooch Behar. The Viceroy at the time, Henry Petty-Fitzmaurice, 5th Marquess of Lansdowne (1845–1927) was accompanied by his wife, Maud Evelyn Petty-Fitzmaurice, Marchioness of Lansdowne (1850–1932). They stayed for the first week, leaving 28 February. As usual, their presence also meant regular updates in the press. Lansdowne did not have a chance to shoot a rhino, even though he must have seen some in the area. When MCB published his book, a photograph of "The party, 1892" including the Viceroy was added afterwards (on an unnumbered page). At the time, there were 15 gentlemen, 5 ladies and an unidentified child.

The shoot was also attended by British royalty. Prince Christian Victor of Schleswig-Holstein (1867–1900) was the son of Prince Christian (1831–1917) and Princess Helena (1846–1923), a daughter of Queen Victoria. In his edited memoirs, he wrote "February 22. Went out shooting. I got a rhino" (Warren 1903: 178). In MCB, no rhino is killed on that day, but also the count exceeds those listed by one.

Rowland Ward (1903: 409) listed a horn of 12 ½ in (31.75 cm) from Cooch Behar attributed to James Jonathan Harrison (1857–1923), who was spending some months of sports hunting in India and Sri Lanka. MCB does not record his shot. Harrison (1892: 121) published his experiences in a very rare book, which confirms that he killed one rhino together with Bignell on 2 March, and that he took some photographs (figs. 32.14, 32.15). It is likely that the horn was displayed among the hunting trophies in his house in Brandesburton, Yorkshire, said to be one of the largest private collections until 1923 (photos in Harrison 1910, but Indian rhino not visible).

Shoot 29. 1892 (b) – 25 April to 4 May 1892 (10 days) – 1 rhino.
Chilipati area, in Buxa north of Cooch Behar, on Torsa River.
Localities: Chelapata, Forest reserves.
Participants (8): MCB, McLaughlin, Gojendro, Jotindro, Sant, Sujey, Suresh, Baxi.

One rhino was killed:
1892-04-26, 1 killed, 1 wounded, by MCB in the Forest Reserves near Chelapata.

1893

Shoot 30. 1893 (a) – 11 February to 20 March 1893 (38 days) – 7 rhino.
Raidak – Sankosh River area, partly in Assam.
Localities: Mahakalguri, Reserve Forests, Raidak River, Reserves north of Haldibari Road, Haldibari, Hamua, Garad Haut, Deogaon, Bhalka Balasee, Bhalka reserves, Guigaon, Bara Dogla, Majerdabri.
Participants (25): MCB and Maharani, Mr. Astley, Mrs. Astley, Mr. Harrison, Mrs. Harrison, Warneford, Lowis*, Miss Lowis, Mr. Bignell*, Mrs. Bignell*,

FIGURE 32.14
Photograph of the rhinoceros killed on 2 March 1892 during shoot 28 in Cooch Behar by James Jonathan Harrison (1858–1923). The record horn of 31.75 cm is hidden in the grass and cannot be seen
WITH PERMISSION © SCARBOROUGH MUSEUMS TRUST: COLLECTION OF HARRISON'S GLASS PLATES

FIGURE 32.15
Jungle scene with the rhinoceros killed by J.J. Harrison in 1892
WITH PERMISSION © SCARBOROUGH MUSEUMS TRUST: COLLECTION OF HARRISON'S GLASS PLATES

Miss Sen*, Col. Tillotson*, Daniell, Brown, Valletort*, Lambton*, Shields, Jotindro, Galeffi, Sujey. Also West, Burgess (Burges*), G. Burgess (Burges), Carnegy*.

Total 7 rhino in table (1908: 119), all recorded separately without shooters:

1893-02-17, rhino tracks seen in forest reserves near Raidak River.

1893-02-19, 1 killed, size given, near Raidak River.

1893-02-20, 2 killed, a male and female, near Raidak River.

1893-02-22, 1 killed, Reserves north of Haldibari Road.

1893-03-03, 1 wounded, at Hamua.

1893-03-05, 1 killed, at Hamua.

1893-03-09, 1 killed at Deaogaon near Garad Haut.

1893-03-17, 1 wounded at Bara Dogla.

1893-03-18, 1 killed at Bara Dogla.

Colonel Atherton Edward Jenkins (1859–1945) of the Rifle Brigade also attended this shoot until 26 February, but incongruously he is not listed in MCB. Jenkins (1893) published an account, in which he states that the first rhino killed on 19 February was shot by Major-General William Lambton (1863–1936). The horn of this animal was credited to Lambton in Rowland Ward (1896: 282) measuring 11 in (28 cm).

Shoot 31. 1893 (b) – 10 to 27 April 1893 (18 days) – 1 rhino.

Chilipati area, in Buxa north of Cooch Behar, on Torsa River.

Localities: Chelapata.

Participants (14): MCB, Sir Charles Elliott*, Nolan, Currie, General Lance*, Boileau, Elliott (again), Dyne, Lister, Lowis*, Forbes, Bignell*, McLaughlin, Baxi.

No rhino is recorded for this trip in the table (p. 119), but one is mentioned in the text (p. 118):

1893-04-10, the party saw 5 rhino, in Chelapata forest.

1893-04-11, rhino tracks seen.

1893-04-26, 1 killed and 1 wounded in Chelapata.

The presence of Charles Alfred Elliott (1835–1911), Lieutenant Governor of Bengal, was recorded in the *Englishman's Overland Mail* (Wednesday 1893-05-03) stating that rhinos were seen but not shot up to 23 April when he left the camp.

1894

Shoot 32. 1894 (a) – 27 February to 14 March 1894 (16 days) – 5 rhino.

Assam, between Dhubri and Goalpara.

Localities: Camps at Dudnath Pahar and Khalsamari.

Participants (20): MCB, Lord William Beresford*, Lord Wolverton*, Sir Benjamin Simpson*, Lord Dalrymple*, Count D'Harnoncour*, Count Scheibler*, Gordon*, Mackinnon*, Bignell*, Hewett*, Scott, Lowis* and

Mrs. Lowis*, Miss Lowis, Herbert, young Gouripur. *Englishman's Overland Mail* (1894-03-14) adds Countess Scheibler*, Mr. Wilson (Lord Beresford's trainer) and Mr. McKenzie (photographer from Messrs. Johnston and Hoffman).

Total 5 rhino:

1894-02-27, 1 killed and 1 wounded (lost) at Dudnath Pahar.

1894-03-02, 1 killed and 1 female wounded (retrieved 4 March), at Samerdanga, "the rhino jungles at Samerdanga were quite the most unbeatable I had ever seen, thick *khagra* interlaced with wild rose bushes."

1894-03-18, 1 killed and 1 wounded (later retrieved) at Kaee Mari near Khalsamari camp.

The first camp was at Dudnath Pahar "on the right bank of the Brahmaputra. It is a picturesque spot with hills all round and on both sides of the river" (*Englishman's Overland Mail* 1894-03-14). The second camp at Khalsamari was on the south bank of the river.

For Lord William Beresford (1847–1900) this was the last shoot before his return home (*Englishman's Overland Mail* 1894-01-31).

This was the second shoot, after 1889, attended by John Prescott Hewett (1854–1941) and recorded in his book (Hewett 1938: 91).

There were three ladies in the party. Cooch Behar's Superintendent Edmond Lowis had brought his wife Susan and his daughter. The Italian count Scheibler was accompanied by his wife Ernestina.

For Felice Scheibler (1856–1921), this was his third and last hunt with the Maharaja (earlier 1890, 1892). His book *Sette Anni di Caccia Grossa* was first published in 1900, reprinted in Italian in 1910 and 1928, but never translated. It contains two photographs of dead rhinos, both signed by V. Turati, who was probably related to his wife whose mother's maiden name was Turati (figs. 32.16, 32.17). It is not definitely recorded when these photographs were taken, maybe in 1892 and 1894 respectively if they were inserted in the text at the appropriate place. One record horn of 8 ¾ in (22.2 cm) was ascribed to "Countess Scheibler" (probably the Count) in Rowland Ward (1899: 433). In Villa Scheibler located in Rho, near Milan, Italy, there was a large hall full of trophies, including three heads of Indian single-horned rhinos, as seen in a photograph taken in 1900–1909 by Carlo Fumagalli (figs. 32.18).

1895

When the Maharaja's second son Jitendra (b. 1886) was still a school boy, an unusual event happened while his father was away in Assam, hence probably in 1895. As told by Casserly (1914: 28), a villager came to the palace in Cooch Behar with news about a rhino some 5 miles

FIGURE 32.16

"Morte del rinoceronte" (death of the rhinoceros). Photograph taken in 1892 (or perhaps 1893 or 1894) during a hunt in Cooch Behar. The four European hunters are not identified

FELIX SCHEIBLER, *SETTE ANNI DI CACCIA GROSSA*, 1900, P. 193. COURTESY: SPARTACO GIPPOLITI

FIGURE 32.17

"Rhinoceros unicornis." View how a dead rhinoceros is moved by a large number of local attendants. Photograph probably taken in 1894 during a hunt in Cooch Behar attended by Count Felix Scheibler

SCHEIBLER, *SETTE ANNI DI CACCIA GROSSA*, 1900, P. 218. COURTESY: SPARTACO GIPPOLITI

from town. Rhinos had not been seen so close to town for a long time. Jit sent the only available elephant to the spot and followed on his bicycle. A rhino appeared from a clump of bamboos and "apparently objecting to the presence of the cows, charged furiously at them. Up went their tails and off went the cows. Round and round the field they raced, the young heifers leaping and frisking like black buck, while the rhino lumbered heavily after them. The villagers scattered and fled. The scene was so comical that Jit, standing like a circus-master in the centre of the ring, could hardly stop laughing long enough to lift his rifle and take aim. At last he fired; and the rhinoceros checked, stumbled forward a few paces

and collapsed in an inert mass on the ground." A remarkable story.

Shoot 33. 1895 (a) – 20 February to 21 March 1895 (30 days) – 7 rhino.

Assam, Bhuiapara near Manas River.

Localities: Simlaguri, Bansbari, Bhuiapara.

Participants (15): MCB, Capt. Harbord*, Capt. Firman*, St. Clair, Lord Elphinstone*, Sydney-Parker*, O'Donnell, Seton-Karr*, Hugh Fraser, Lowis*, N. Sen*, Col. Allan Gardner*, McCabe*, Dick-Cunyngham*. Also: S.P. Saut.

In total 7 rhino were killed in this trip (*Englishman's Overland Mail* 1895-04-10; *Globe* 1895-04-27).

FIGURE 32.18   Trophies of three heads of *R. unicornis*, on the left wall behind the elephant. This photograph was taken by Carlo
Fumagalli in the 1900's in Villa Scheibler, the residence of Count Felix Emil Scheibler in Rho, near Milan, Italy
WITH PERMISSION © CIVICO ARCHIVIO FOTOGRAFICO, MUSEI DEL CASTELLO SFORZESCO, MILANO: PROT. N° CF
2018/54

1895-02-21, 1 wounded at Simlaguri.

1895-03-10, 1 male killed by Seton-Karr, at Bhuiapara.

1895-03-12, 2 killed at Bhuiapara, 1 wounded and found
dead later.

1895-03-13, 1 female killed at Bhuiapara, skull and horn
pictured (1908: 149).

1895-03-19, 1 male killed at Bhuiapara, largest ever shot by
MCB.

1895-03-20, 1 killed and 3 seen at Bhuiapara.

As Lord Sidney Elphinstone took part in this shoot, the
details with dates and game found were recorded in the
Elphinstone Game Book (32.3). His list of participants only
differs by the addition of S.P. Saut, while he has P. Sen
instead of N. Sen. He records the sizes of the male rhinos of
10 March and 19 March, which might be the skulls shown
on a photograph (fig. 32.19). Apparently Elphinstone did
not shoot any of the rhinos. Elphinstone and Firman continued on a separate excursion afterwards (38.24).

The first rhino on 10 March was ascribed to Heywood
Walter Seton-Karr (1859–1938), who is said to have fired
four times. Seton-Karr (1897; Blatwayt 1898: 342) describes
the hunt and says that when the animal was killed "the

body was photographed, and when we moved away the
Maharaja's trained skinners had already begun to remove
the shields of hide from the flanks and shoulders and the
head and feet." This animal had a good size, length 310 cm,
height 185 cm and horn of 34.3 cm.

The female rhino killed on 13 March had a large horn
measuring 16 ¼ in (41.3 cm) attributed in Rowland Ward
(1899: 433) to MCB but with date 1898 (possibly when the
specimen was registered by the firm). The skull with horn
is pictured in MCB (fig. 32.20).

A few days later on 19 March, the Maharaja killed a
rhino, measuring 335 cm length, 194 cm height and a
horn of 33.0 cm: "the biggest rhino I have seen or shot.
Magnificent beast in every way. I think he is a record as
regard height also" (1908: 454) – apparently not photographed at the time.

Shoot 34. 1895 (b) – 14 to 28 April 1895 (15 days) – rhinos
observed.

Raidak – Sankos River area.

Localities: Bhalka Balashi (Sunkos River), Forest Reserves,
Jorai River, Barbera.

FIGURE 32.19
Bag of 1895 from Elphinstone Game Book. There are two rhino skulls with horn in front of the series of tiger skins
WITH PERMISSION © *ELPHINSTONE GAME BOOK* FOR 1895; PRIVATE COLLECTION

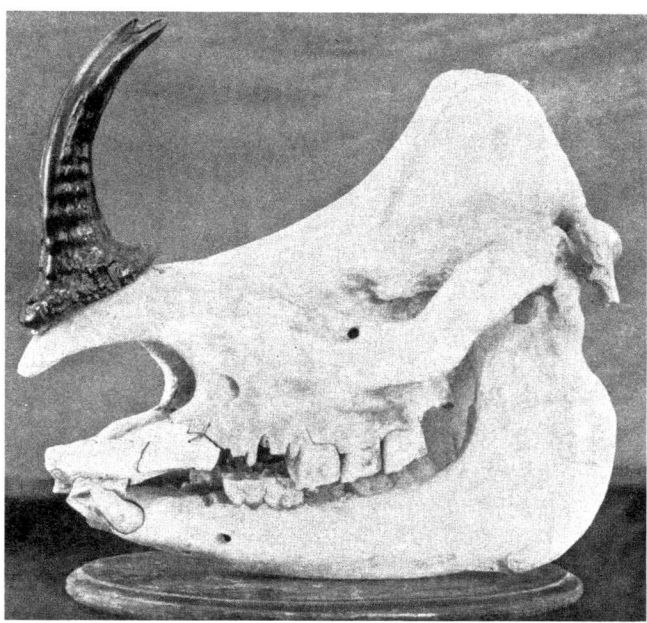

FIGURE 32.20 "Record Cow Rhino's Skull. Horn 16 ¼ inch." Skull of a female rhino killed on 13 March 1895, with a horn of 41.3 cm
NRIPENDRA, *THIRTY-SEVEN YEARS OF BIG GAME SHOOTING IN COOCH BEHAR*, 1908, P. 149

Participants (12): MCB, Firman*, Baron Massow, Apcar, Perree*, Lehmann, Seton-Karr*, P. Sen, E. Ezra, Alfred Ezra, Hawkins, Delafosse*.
No rhinos were killed during this trip, but there were tracks:
1895-04-28, tracks of rhino, near the Jorai River.

Shoot 35. 1895 (c) – 1 to 15 December 1895 (14 days) – 2 rhino.
Raidak River area.
Localities: Mahakalguri, Forest reserves, Raidak River.
Participants (17): MCB and Maharani, General Yeatman-Biggs*, H. Elliott, Lister, Bertram* and Mrs. Bertram*, Nolan, Lowis*, Lowis Junior*, Georges, Lane-Anderson*

and Mrs. Lane-Anderson*, Plowden*, Wintour, Peter Sen, Sujey.
Rhinos shot:
1895-12-02, 1 wounded in the Raidak River forest reserves.
1895-12-06, 2 killed near Raidak River. A male was attributed to Bertram, a female to Peter Sen.
1895-12-07, 1 wounded by Sujey, Wintour and Nolan.
Given that during the February hunt some large rhinos were shot, it is remarkable that on 6 December 1895 another one was found. This male was first shot by General Arthur Yeatman Biggs and Captain Lister, but the final shot came from Captain William Bertram (*Englishman's Overland Mail* 1895-12-18). Therefore it was called "Bertram's rhino" on the plate (1908: 165), reprinted in Campbell (1907, vol. 2: 288) and Playne (1917: 457) (fig. 32.21). It measured 391 cm in length (including tail) and 201 cm in height. This animal was said to be larger than the biggest rhino of 19 March 1895, and strangely is not included in MCB's table of large specimens. MCB did not see the animal in the field, but had sent Hatashu, one of his hunters, with the photographer to measure it. Maybe he did not fully trust the result, especially since the height "takes a good bit of swallowing, even now" (1908: 166).

1896

Shoot 36. 1896 (a) – 20 February–27 March 1896 (36 days) – 5 rhino.
Raidak – Sankosh River.
Localities: Mahakalguri, Guigaon, Baman Kote, Bakla River, Bara Dogla, Raymana, Haldibari, Bara Hamua, Jorai River, Theamari Reserves, Garad Haut.
Participants (14): MCB and Maharani, Prince and Princess Hans Henry of Pless*, P. Burgess (Burges*), Lowis*, Pollen*, N. Sen, Plowden, Count Hochberg*, P. Sen, J. Ezra, Aaron Ezra, Sir Ben Simpson*. Also, Elphinstone*, who is not in the initial list, but is mentioned bagging a bison (1908: 184).

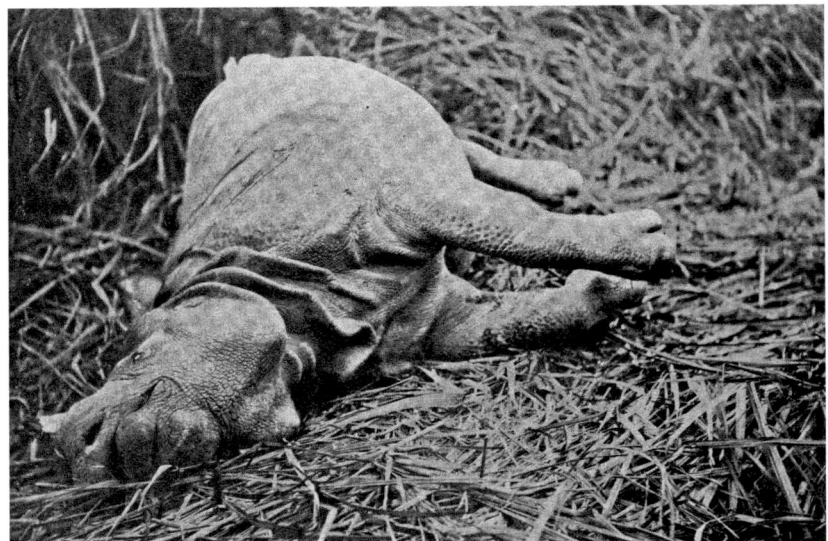

FIGURE 32.21
"Bertram's Rhino." Large male *R. unicornis* killed on
6 December 1895 by Captain William Bertram in
Cooch Behar
NRIPENDRA, *THIRTY-SEVEN YEARS OF BIG GAME
SHOOTING IN COOCH BEHAR*, 1908, P. 165

FIGURE 32.22   Rhinoceros killed during the Cooch Behar shoot (no. 36) of February–March 1896. The lady is Daisy
of Pless next to her husband the Prince of Pless. On the left is the Maharaja and his son Rajey, then
14 years old, but not recorded among participants
WITH PERMISSION © *ELPHINSTONE GAME BOOK* FOR 1896; PRIVATE COLLECTION

In total 5 rhinos shot (one by MCB), with details of 4
rhinos:
1896-02-21, 1 wounded at Guigaon.
1896-02-24, 1 male killed in Mahakalguri area by Burges
(date from Elphinstone).
1896-02-26, 1 female killed in Mahakalguri area by Pollen
(date from Elphinstone).

1896-02-28, 1 killed at Baman Kote in reserves.
1896-03-05, 1 killed and 4 others seen in Raidak reserves (2
killed in Elphinstone).
1896-03-13, rhino seen ar Raymana.
Lord Sidney Elphinstone was part of this shoot, strangely
not mentioned in the original party (1908: 175), but
referred to once later (1908: 184). The *Elphinstone Game*

FIGURE 32.23
Rhinoceros photographed by the
Conservator of Forests, Walter
Francis Perrée (1871–1950). It
might be the male rhino killed on
17 April 1897 near Garad-Haut, or
another one taken elsewhere
*BADMINTON MAGAZINE OF SPORTS
AND PASTIMES*, AUGUST 1906,
VOL. 23, P. 232

Book separates the period 22–27 February, and adds a second rhino on 5 March where MCB only has one (32.3). The total of 5 rhinos is correct. It includes a photograph of one of the rhinos (fig. 32.22), showing the Polish Prince of Pless, Hans Heinrich XV von Hochberg (1861–1938), who joined the hunt together with his wife Princess Daisy (1873–1943). She said nothing about rhinos, but she killed a tigress, the skin of which was displayed at their castle Fürstenstein, now Książ Castle in southwest Poland (Daisy of Pless 1931: 33).

A photo-album of this shoot was compiled by Stephen Hungerford Pollen, and is said to include a photograph of a rhino shot by MCB (32.3).

A photograph taken by Peter Burges labelled "A dead rhinoceros" might show one of the animals killed either on 24 February 1896 or in April 1896 (Pollok and Thom 1900: 48; Moffett 1898a: 508, 1898b: 145; fig. 39.2). Another photo in the same book (p. 448) shows a display of trophies including the skulls of at least three rhinos, hence this would suit this particular instance (also in Moffett 1898b: 136; fig. 39.3). The American journalist Cleveland Moffett (1863–1926) visited Peter Burges in his home in Bristol, introducing him as a quiet English gentleman renowned for his big game hunting (Moffett 1898a, b). There were two rhino skulls on the walls, obtained in Cooch Behar in 1896, which is the year mentioned for Burges's reminiscences.

Shoot 37. 1896 (b) – April 1896 – 2 rhino.
Raidak – Sankosh River.
The rhinos were shot by Michel, Burgess (Burges*), Ezra and Buxi (1908: 192).

1897
Shoot 38. 1897 (a) – 11 March to 17 April 1897 (38 days) – 1 rhino.
Raidak – Sankosh River.
Localities: Bagjhora, Raidak Balashi, Chota Dal Dalia, Bara Dogla, Baman Kote, Haldibari, Bara Bhalka, Parakota, Jorai River, Garad Haut, Horse-shoe jungle.
Participants (11): MCB, General Yeatman-Biggs*, Aylmer*, Allen, A. Ezra, Plowden, Burgess (Burges*), Smyth, Good, Ashton*, Sujey, Perree*.
Rhinos shot:
1897-03-15, 2 rhino seen at Raidak Balashi.
1897-04-17, 1 male killed in Horse-shoe jungle near Garad-Haut, by Smyth and others.
The male rhino killed on 17 April measured 373 cm in length and 178 cm in height. As this is the only rhino killed when Walter Perree was present, it could be this animal shown in a photograph "by Mr. W.F. Perree Dhubri, Assam, India" published in 1906 (Perree 1906; fig. 32.23).

Soon after this shoot, there was a severe earthquake in the region on 12 June 1897 (Luttman-Johnson 1988). The third storey of the new palace in Cooch Behar collapsed during the tremors. Even the jungle and waterways changed, as MCB recorded after a later visit: "The earthquake has altered and changed the whole face of the country, and many patches of good jungle that I got game out of in 1896 are now mere swamps full of fasan. The small nullahs especially are very bad, being crossable in only a few places" (1908: 246).

1898

Shoot 39. 1898 (a) – 14 February to 23 March 1898 (38 days) – 6 rhino.

Raidak – Sankosh River.

Localities: Raidak River, Cheepla River, Raidak Sapkata, Bara Dogla, Pentaguri, Guigaon, Hamua, Chengmari, Baman Kote, Longa River, Takum Takum Khas, Longa River, Pechadavri.

Participants (17): MCB and Maharani, Frewen, Jaucourt*, A. Ezra, Rajey, Adam, Neufville, Jotindro, Peter [Sen], Plowden*, Sujey, Lawson, Gauripur, Jitendra, Sukriti, Kebla.

Rhinos shot:

1898-03-16, rhino seen on Longa River.

1898-03-18, 3 killed at Takum Takum Khas, one by Ezra, two by MCB.

1898-03-20, 2 males, 1 female killed on Longa River.

The female killed by Ezra on 18 March was 163 cm high. A male killed on 20 March measured 361 cm length, 184 cm high.

The killing of one of the tigers is recounted by Lal Baloo (1902), unfortunately without date or place. He writes that he was joined by the Maharanee, a Calcutta merchant, two French nobility, a member of the Viceregal staff, a globe-trotting Australian and 2 others. His experiences most likely apply to this year, although his list of participants is difficult to match.

1899

Shoot 40. 1899 (a) – 23 February to 23 March 1899 (29 days) – 7 rhino.

Assam, Bansbari (Manas).

Localities: Simlaguri, Kamargaon, Bansbari, Dhowbheel (near Bansbari), Bhuiapara.

Participants (14): MCB, Count of Turin*, Prince Teano*, Count Carpenetto*, Lord Lonsdale (H.C. Lowther)*, Lord Elphinstone*, Sir Henry Tichborne*, Mr. Cecil Plowden, Prall, Rajey, Gurdon*, Vanderbyl*. Also Sir Benjamin Simpson and Percy Hall (Anon. 1899a).

A total of 7 rhinos were shot (Anon. 1899d):

1899-02-24, rhino tracks seen near Simalguri.

1899-03-07, 2 males, 3 females killed at Dhowbheel (near Bansbari).

1899-03-11, 1 seen near Bhuiapara.

1899-03-13, 1 killed near Bhuiapara.

1899-03-18, 1 killed in Bhuiaparea area by Tichborne.

Lord Sidney Elphinstone was part of this shoot, but in his *Elphinstone Game Book* just records that a total of 7 rhino were killed (32.3). He also mentions that he continued on a private hunt together with Philip Vanderbyl from 24 to 29 March 1899 (38.24).

On 7 March, no less than 13 head of big game were killed before twelve o'clock, which MCB called the "grandest day's shooting I have ever had" (1908: 236). The trophies for this day are carefully displayed in a photograph (fig. 32.24).

One of the rhinos was killed by the Italian Prince Vittorio Emanuele, Count of Turin, shown on a photograph printed in close-up in MCB (1908: 244) and with a wider angle by Mancini (1899) showing the Count on the howdah (figs. 32.25, 32.26). He may in fact have accounted for no less than three rhinos (Mancini 1899; Anon. 1899b; MacFarlane 1900: 527). A recent catalogue of the specimens and photographs in his collection does not list a

FIGURE 32.24 "A Record Day's Bag: 13 head of Big Game in 2 hours – 7 Buffalo, 5 Rhino and 1 Bison" showing the animals killed on 7 March 1899
NRIPENDRA, *THIRTY-SEVEN YEARS OF BIG GAME SHOOTING IN COOCH BEHAR*, 1908, P. 237

FIGURE 32.25 "Rhino dropped in his tracks." The animal was killed by Prince Vittorio Emanuele, Count of Turin in March 1899 in Assam
NRIPENDRA, *THIRTY-SEVEN YEARS OF BIG GAME SHOOTING IN COOCH BEHAR*, 1908, P. 244

FIGURE 32.26 "Rinoceronte ucciso nell'Assam del Conte di Torino. (Il Principe sta sull'elefante)." Prince Vittorio Emanuele, Count of Turin stands in the howdah. Photograph attributed to Onorato, Prince of Teano
MANCINI, *ILLUSTRAZIONE ITALIANO*, 1899, P. 69

rhino trophy (Finotello & Agnelli 2015). Lonsdale (1899: 541) also recalled his time in Assam.

Shoot 41. 1899 (b) 24 to 29 March 1899 (6 days) – 2 rhino. Assam, Bansbari.

Persons: Elphinstone*, Vanderbyl*.

After the first shoot (1899 (a)) and followed by a third one on 10–19 April (no rhino recorded), Elphinstone and Vanderbyl had a few days hunting on their own (Anon. 1899e). The *Elphinstone Game Book* (32.3) only states that on 25 March they moved to Jalagaon, on the east bank of the Manas River. During six days, the two hunters managed to shoot 2 rhinos, but no tiger. Although not mentioned by Elphinstone, Philip Vanderbyl killed a rhino with record horn and took a few photographs (38.24). The 2 rhinos obtained at this time were included in the annual table (1908: 248), hence this shoot is included here despite the absence of MCB.

1900

Shoot 42. 1900 (a) – 10 February to 23 March 1900 (42 days) – 1 rhino.
Raidak – Sankosh River – Patgaon (Assam). This was the main annual shoot.

Localities: Bara Dogla, Pimulaguri, Haldibari, Samuktola, Roymana, Patgaon, Aye River, Lanka River, Saralbhanga River, Kachugaon, Dhoombarjhar jungles, Chota Bhalka, Chukchuka, Haldibari, Bhalka.

Participants (20): MCB and Maharani, Lord Stavordale*, Lord Hyde*, Earl of Suffolk*, Mr. Colvin* and Mrs. Colvin*, E. Colvin*, Hare, Tichborne*, Baker-Carr*, James, Mr. Hawkins and Mrs. Hawkins, N. Sen, Sant, Sujet, Miss Maclean, Sukriti, Victor.

Rhinos shot:

1900-02-22, 1 seen near Roymana.

1901-03-03, 1 killed, 5 seen on the Lanka River.

1901-03-13, 2 seen in Dhoombarjhar jungles.

The participation of Lord Stavordale, Lord Hyde and Earl of Suffolk was mentioned in the *Times of India* (1900-03-13 and 1900-03-21). The list of outstation participants was also included in a short paragraph in *Homeward Mail from India, China and the East* (1900-03-12).

1901

Shoot 43. 1901 (a) – 1 March to 16 April 1901 (47 days) – 5 rhino.

Raidak – Sankosh River – Patgaon (Assam). This was the main annual shoot.

Localities: [Bengal] Towards Chukchuka, Jorai River, Falimari, Kumargaon, Garad Haut, [Assam] Patgaon, Saralbhanga River, Forest Reserves, Lanka River, Doragaon, Champamoti River, Patgaon, Roymana, Doombar Jhar Kas, [Bengal] Haldibari, Muktaigaon, Dhoompara, Raidak River, Salsarabari.

Participants (12): MCB, Grimston*, Count Waldstein*, Mr. Heber Percy*, Mrs. Heber Percy, James, Herbert, Tichborne*, Raja of Pudokota*, Beaumont*, J. Ezra, Plowden*.

Rhinos seen and shot:

1901-03-10, 1 killed, 2 seen in Forest Reserves near Sankosh River.

1901-03-16, 1 young killed (by mistake) in Forest Reserves near Lanka River.

1901-03-21, 1 seen in Forest Reserves.

1901-03-30, 1 seen near Patgaon.

1901-04-02, 1 seen, 1 killed at Doombar Jhar Kas by MCB and Heber Percy.

1901-04-02, 2 killed in same area by MCB and Waldstein.

The text gives details of 5 rhinos, which should be authoritative. In the two tables with numbers shot in 1901 (1908: 303, 315), rhino totals are listed as 1 in March and 2 in April. The newspaper (*Englishman's Overland Mail* 1901-04-25) also has a total of 3 rhinos. The female "exceptionally large for cow rhino" of 10 March stood 175 cm at shoulder and had a horn of 36.8 cm. One of the animals of 2 April stood 174 cm high.

Shoot 44. 1901 (b) – April 1901 – 1 rhino.

Assam, Manas region.

After the main annual shoot, Beaumont and Arthur (unidentified) continued in Assam "the place Pollok wrote about" (1908: 314) and shot 1 rhino.

1902

Shoot 45. 1902 (a) – 1 March to 13 April 1902 (44 days) – 6 rhino.

Manas River, Assam. This was the main annual shoot.

Localities: Gosiengaon, Bholkadota Holkadoba, Lahapara, Daimasi River, Dhun Bheel, Kalamabari, Halapakri, Kalibhanga, Bakie River.

Participants (11): MCB, Orr-Ewing, Sir B. Simpson*, A. Ezra, Lord Villiers*, James, Blackett, Richards, N. Sen, Sujey, Jit. Anon. (1905a) refers to the presence of Colonel E.W. Baird* and his wife*.

Rhinos seen and shot:

1902-03-06, 1 male rhino killed at Gosiengaon.

1902-03-23, 1 male and 2 females killed on Daimasi River near Lahapara.

1902-03-25, 2 killed and 1 wounded at Dhun Bheel near Lahapara.

A photograph (1908: 318) "1902 shoot" shows one of the rhinos (possibly the male of 6 March) in front of several elephants, three carrying a howdah (fig. 32.27). That male rhino was one of the largest on record with 389 cm length, 179 cm height and a horn of 31 cm. On 10 March MCB also bagged his longest tiger, 10 ft 5 in (317 cm). The total number of rhinos was mentioned in the *Homeward Mail from India, China and the East* (1902-04-19).

The medical officer Benjamin Simpson in total participated in 14 hunts between 1877 and 1902. There is no record that he ever killed a rhino at any time. He may have been happy to take photographs, of which one was published by Pollok and Thom (1900, p. 467) (fig. 39.4).

Rowland Ward (1903: 409) credits a horn from Cooch Behar of 12 inch (30.5 cm) to A. Ezra. MCB includes A. Ezra in 1895, 1896, 1898, 1902 but once (p. 179) he uses the full first name Aaron. Alternatively, maybe this was Alfred "Chips" Ezra (1872–1955), a son of a Kolkata merchant, who presented a lecture at the Zoological Society of London in 1917 on his experiences in India (Ezra 1917).

Game was plentiful in this part of Assam close to the Manas River, as suggested by MCB (1908: 331): "We had by this date completed exactly a fortnight's shooting at Lahapara, and though from 19th to the 22nd little or nothing was bagged, taking the whole 14 days together, thirty-three head of big game were killed. With such a bag one would naturally expect the country to be depleted of beasts worth shooting, but this was not so. So far as one could judge, the district remained full of big game, including a fair number of tiger."

1903

Following the Durbar in Delhi celebrating the succession of Edward VII as Emperor of India in the first weeks of January, the usual animal shoot was held in the Raidak-Sankosh area of Cooch Behar from 16 February to 4 April 1903 (51 days). About 25 people participated. No rhinos were recorded this year, which might be indicative of a diminishing number of animals. This seems to be confirmed by Chaudhuri (1903: 100) who said that due to the scarcity of game in Cooch Behar, two large tracts "in the north of Parganas Cooch Behar and Tufangunj which contained good jungle" had been constituted into Forest Reserves.

FIGURE 32.27
"1902 Shoot." A dead rhinoceros surrounded by trained elephants
NRIPENDRA, *THIRTY-SEVEN YEARS OF BIG GAME SHOOTING IN COOCH BEHAR*, 1908, P. 318

1904

As Lord Curzon, the Viceroy, was expected for a week's shoot in April, there were several shorter trips rather than one long one.

Shoot 46. 1904 (a) – 17 January to 16 February 1904 (22 days) – rhino seen.
Raidak – Sankosh area.
Localities: Cheeplah, Bara Dogla, Alipur Duars.
Participants (10): MCB, Mr. Bankier, Mrs. Bankier, Hammond*, French*, Beecher, James, Mr. Miller, Mrs. Miller, Sujey.
Rhinos:
1904-01-22, rhino present in reserves.

Shoot 47. 1904 (b) – 15 to 26 March 1904 (12 days) – rhino tracks.
Raidak – Sankosh area.
Localities: Bakla.
Participants (10): MCB, Hammond*, Jyotsna, Mr. Edwards*, Mrs. Edwards, Miss Porter, P. Sen, Ghose, Sujey, James.
Rhinos:
1904-03-19, rhino tracks in the reserves.

Shoot 48. 1904 (c) – 4 to 9 April 1904 (6 days) – rhino seen.
Torsa River to Jaldhaka River.
Localities: Madari Hat, Chilapata, Ramshahaihat, Tondu, Khateemaree, Dinah River.
Participants (8): MCB, Viceroy Lord Curzon*, Marindin*, Baring*, Bird, Adam [= Vernon] Keighley*, Farrington*, Armstrong.

Rhinos:
1904-04-04, rhino seen near Chilapata in the reserves.
MCB decided to travel a little further from home than usual to entertain the Viceroy, Lord Curzon. In the region of the Torsa River, they came across rhinos but apparently none were shot. However, a report in the *Illustrated Sporting and Dramatic News* mentioned that several rhinos were shot by the Maharaja, but none by the Viceroy (Anon. 1904a: 518). Lord Curzon did get at least a tiger, shown with him in a photograph "Lord Curzon's first Tiger" (1908: 376) including all the dignitaries in full costume. A total of 5 tigers were found this time.

Shoot 49. 1904 (d) – 17 to 27 December 1904 (11 days) – rhino seen.
Raidak – Sankosh area.
Localities: Chukchuka, Jorai River, Balashi, Garad Haut.
Participants (10): MCB, Mr. Edwards*, Mrs. Edwards, Miss Porter, Miss -Bainbridge, P. Sen, Jit, James, Ghose, Juggins.
Rhinos seen:
1904-12-20, 2 seen at Balashi, just above Bhalka.

1905

Shoot 50. 1905 (a) – 18 to 26 February 1905 (9 days) – rhino wounded.
Assam, Kachugaon.
Localities: Kalabari, Hali River, Kachugaon, Pechadabri.
Participants (12): MCB, Viceroy Curzon*, Perree*, Lord Lamington*, Baring*, Meynell, Akers-Douglas*, Ezechiel, Peter, Hammond*, Macnab, Howell.

Rhinos shot:

1905-02-20, rhino wounded at Kachugaon near Forest Bungalow.

The Viceroy Lord Curzon returned for a second shoot, this time just across the Sankosh River in Assam. A photograph shows 11 members of the party (1908: 388). Although there were rhinos in the area, none were killed. The party secured 10 tigers.

Shoot 51. 1905 (b) – 2 to 23 April 1905 (22 days) – 1 rhino.
Torsa River (West Cooch Behar) and Raidak River.
Localities: Patlakhowa.
Participants (13): MCB and Maharani, Count Quadt*, F. Grenfell*, R. Grenfell*, Rajey, Jit, Ghose, Sujey, Juggins, Peter Sen, Corbett*, Hammond*.
Rhinos shot:

1905-04-12, 3 rhino seen near Patlakhowa.

1905-04-13, 1 wounded in Raidak reserves.

1905-04-18, 1 killed and several seen, in reserves.

The rhino killed was a female with a height of 173 cm. The twins Francis and Riversdale Grenfell found tiger shooting a bit too civilised as it did not require special physical strength: "yet it is worth a very long journey to see the immense jungle, the elephants, and all the wild and delightful surroundings of the Indian forests" (Buchan 1920: 63).

1906

There were two shoots this year, in Assam and in Chilapeta. No rhinos were included in the reports.

1907

This year again there were two shoots, one in Assam and one in Cooch Behar. The first shoot in February 1907 was organised for the Viceroy, Earl de Minto who had come with his wife Lady Mary (Anon. 1909c). No rhino recorded. Their first day was the most exciting with 7 tigers killed in just a few hours. There is an illustrated account by an officer of the Highland Light Infantry (Anon. 1907). The 15 participants are shown in a photograph in MCB (1908: 412).

This was the last shoot included in the book of the Maharaja.

32.10   Shoots in Cooch Behar after 1908

MCB's book ends with the events of 1907. The usual shoots continued in the next three years, when the Maharaja became unwell and died on 18 September 1911 in Bexhill, East Sussex. Some details about the trips in 1908–1916 are found in other sources.

1908

There was a shoot in February 1908 in the Torsa River area, with a camp in Sili Torsa (Anon. 1908: 28). There is no mention of a rhino.

Again in March 1908, there was a shoot organised on the invitation of Babu Bhola Nath Chaudhuri, Zemindar of Lakhipur, Assam, with camps at Satasia and Kaladanga south of the Brahmaputra (Anon. 1908: 28). No rhino was recorded.

1909

Shoot 52. 1909 (a) 11 to 17 February 1909 (7 days) – 1 rhino.
Assam, Manas.
Locality: Lohapara, 20 km from Sorupeta Railway station, Manas area.
Participants (11): MCB, Viceroy Gilbert, 4th Earl de Minto*, Lady Mary de Minto*, their daughter Lady Eileen*, Victor Reginald Brooke*, James R.D. Smith*, Warren Crooke-Lawless*, Francis Scott*, Rudolf Jelf*, Arthur Dentith*, Rivett Carnac* (*Homeward Mail* 1909-02-27).
Rhino shot:

1909-02-12, a rhino attacked the elephant of Victor Brooke.

1909-02-13, 1 killed by Viceroy de Minto.

The Viceroy had been unsuccessful in killing a rhino on his previous trip in 1907, but this time large numbers of rhinos were seen in the chosen area. On Monday 13 February the elephants surrounded several rhinos, one of which (a male) was secured by Minto (*Times of India* 1909-02-17). On the previous day his military secretary Victor Reginald Brooke (1873–1914) was attacked by a two rhinos, resulting in one of the earliest photographs of a *R. unicornis* in the wild, certainly a rare record of this action (Rookmaaker 2021b, see fig. 40.2). The Viceroy's shoot took place within the North Kamrup Reserve Forest which had been created on 29 July 1907, or definitely very close to it. Together with the enormous expense, this brought several complaints in the media (40.3). Although it is said that Minto had been led to believe that there were great herds of rhino in the area, he took the criticism to heart and did not allow any of his staff to kill any of these animals (Milroy 1916).

1910

A report of a shooting trip is contained in an unpublished letter of 23 February 1910 written at the Maharaja's shooting camp near Dhubri by Auberon Claud Hegan Kennard (1870–1951)*, officer in the London Rifle Brigade (Kennard 1910). He told his mother-in-law that others in the party were General John Cowans*, Colonel Reginald Lynch-Staunton*, Captain Kitson and R.B. Railston. A rhino is not recorded during this shoot.

1912

Prince Wilhelm of Sweden (1884–1965) was invited to the annual hunt in Cooch Behar in early 1912 (William of Sweden 1915: 253). This must have been organised by the new Maharaja Rajendra. No rhinos were shot.

1916

After MCB's second son Jitendra Narayan became Maharaja, he was joined by his wife Indira Devi (1892–1962), his brothers Hitendra and Victor, and his sister Prativa (Pretty) on a shooting trip (*Englishman's Overland Mail* 1916-03-10). While the locality is not in the report, they were able to secure 4 leopards, 1 tiger and 2 rhinos. One of the rhinos stood 18 cubits high at the shoulder.

1950s

Few subsequent reports involving the family of Cooch Behar have been located. The days of the grand luxury camps of Nripendra were long gone. However, Nripendra's grandson, Jitendra's son Maharaja Jagaddipendra Narayan (1915–1970) was known as a great hunter and during his tours hosted other rajahs and high officials of the Indian Government. In 1949, Jagaddipendra went out shooting with his first wife, the American actress Nancy Valentine (1928–2017). They must have killed one rhino then, shown in an undated photograph of Nancy in the jungle (Stiles 2017).

Curiously, a small notice in *Sports Illustrated* (1956-02-27) advertised that one of the grandchildren of the Maharaja of Cooch Behar was carrying on the traditional tiger shoots from the backs of elephants, by inviting paying guests who apply to the Maharaja's Himalayan Shikar Syndicate. This was in the same period as a sequence described as a tiger hunt with the Maharaja of Cooch Behar, part of the movie *The Big Hunt*, released in 1959, directed by George Sherwood (45.8).

32.11 Sizes of Rhinos Connected with the Cooch Behar Shoots

MCB (1908: 454) has an appendix listing the largest specimens, including sizes of 10 males and 3 females. Measurements of another 8 rhinos are given in the course of the text. Three of the larger specimens (nos 14, 15, 17) are not included in his table, possibly because the Maharaja suspected that measurements were inaccurate. Rowland Ward (1910) also has a list of the larger specimens from Cooch Behar. All data from the Cooch Behar shoots are combined in Table 32.43 and discussed by Rookmaaker (2020b).

32.12 Localities for Rhinos in Cooch Behar

Although MCB (1908) mentions many localities, such names of villages, rivers, lakes have changed over the years, and are hard to trace on modern maps. MCB (1908) includes a foldout map, bound at the end, which shows the locations of hunting camps with an 'x' mark, but only some are named. Most names mentioned in the text are not found on the map. This is probably not as disastrous as it sounds because the places were quite close to each other. The full list of all localities where rhinos were found is below and provides approximate identifications and modern names where possible (dataset 32.37).

In his book, MCB often refers to "forest reserves" or "hunting reserves" without any explanation. Pal (2015) found mention of forest reserves in the *Annual Administration*

TABLE 32.43   Sizes of *R. unicornis* measured during the shoots of the Maharaja of Cooch Behar

No	Sex	Date	Locality	Total length between sticks	Length of body between sticks (excl.tail)	Girth	Shoulder height	Horn length	MCB 1908
			Large specimens included in the appendix in Nripendra 1908: 454						
1	M	1887-02-24	Deogong	354	307	277	183	29.8	p. 45. In table as 1886
2	M	1886-02-16	Rossik Bheel	381	328	295	184	26.7	p. 42 where girth 284, height 174
3	M	1889-03-08	Moshamari	361	312	284	185	33.3	
4	M	1891	Raidak River	371	330	295	178		
5	M	1892-02-24	Do-Mahana	373	320	315	182	30.5	

TABLE 32.43 Sizes of *R. unicornis* measured during the shoots of the Maharaja of Cooch Behar (*cont.*)

No	Sex	Date	Locality	Total length between sticks	Length of body between sticks (excl.tail)	Girth	Shoulder height	Horn length	MCB 1908
6	M	1893-02-19	Raidak Reserves	358	318	290	184	33.0	
7	M	1895-03-10	Bhuiapara, Assam	361	310	279	185	34.3	
8	M	1895-03-19	Bhuiapara, Assam	389	335	302	194	33.0	largest ever
9	M	1898-03-20	Longa River	361	310	284	184		
10	M	1902-03-06	Gosiengaon, Assam	389	319	300	179	31.0	
11	F	1890	Sankosh River	371	307	315	175	23.5	largest female
12	F	1895-03-13	Bhuiapara, Assam	nil	nil	nil	nil	41.3	picture p. 149
13	F	1901-03-10	Sankosh River	nil	nil	nil	175	36.8	height p. 287
		Other specimens, apparently not in table, mentioned in course of the book							
14	M	1885-03-18	Dal Dalia	401		279			p. 32
15	M	1886-03-09	Sal Forest Reserve	422	356	284			p. 45
16	F	1894-03-04	Samerdanga				182		p. 124
17	M	1895-12-06	Raidak Reserves	391			201		p. 166, picture p. 165
18	M	1897-04-17	Garad Haut	373	312		178		p. 210
19	F	1898-03-18	Takum Takum Khas				163		p. 226
20	M	1901-04-02	Doombar Jhar Kas				174		p. 308
21	F	1905-04-18	Raidak Reserves				173		p. 392
		Sizes of specimens included in Rowland Ward's *Records of Big Game*							
			Cooch Behar	429			194	41	Ward 1910: 464
			Cooch Behar	399			185	35	Ward 1910: 464
15	M	1886-03-09	Cooch Behar	422			184		Ward 1910: 464
			Cooch Behar – MCB					33.0	Ward 1896: 282
			Cooch Behar – MCB					34.7	Ward 1899: 433
12	F	(1898)	Cooch Behar – MCB				198	41.3	Ward 1899: 433
		[1892]	Cooch Behar – James J. Harrison					31.8	Ward 1903: 409
			Cooch Behar – A. Ezra					30.5	Ward 1903: 409
		1893-02-19	Cooch Behar – Lambton					28.0	Ward 1896: 282
		1899	Cooch Behar – P.B. Vanderbyl					22.9	Ward 1903: 409
		1894	Cooch Behar – Scheibler					22.2	Ward 1899: 433
		1891	Cooch Behar – Henry Streatfeild					14.0	Ward 1892: 238

Measurements are converted to metric cm, where 1 foot = 30.48 cm, 1 inch = 2.54 cm, 1 hand = 10.16 cm.

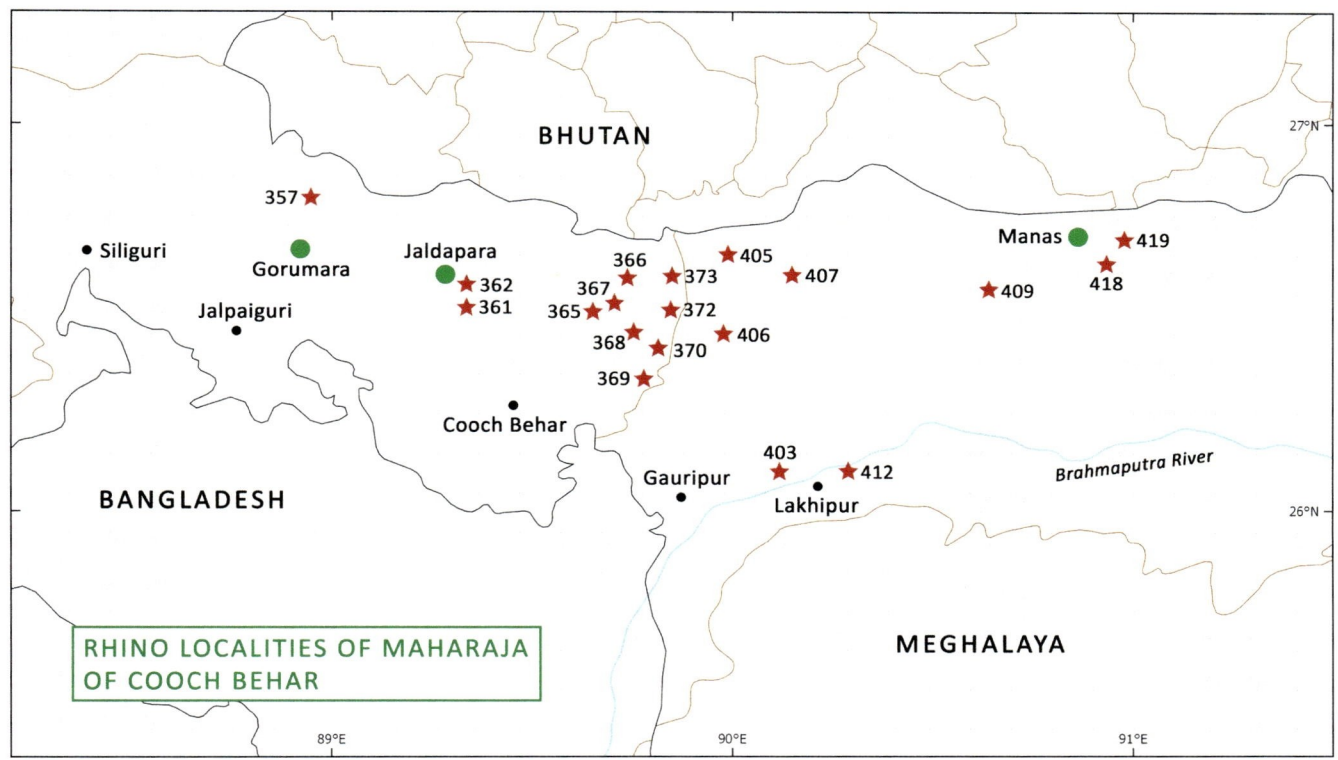

MAP 32.16   Hunting localities of the the Maharaja of Cooch Behar in North Bengal and Assam. The numbers and places are explained in
Dataset 32.37
★   Presence of *R. unicornis*

Reports of the Cooch Behar State. The report of 1882–1883 referred to reserves in Garodhat and Patlakhawa area. Similarly, reserves were listed in 1898–1899 aggregating 10,418 bighas in Taluk Bakshirbosh Putimari, and aggregating 19,980 bighas in Taluk Bara Shalbari, Chengtimari, Atea Mochar, Takoamari, Rampore-Garbhanga, Paglirkuti, Rashikbil, Khgribari, Jaldhoa and Pholimari, at Tufangunj. The term "hunting reserves" indicates that these must have been lands where hunting was forbidden to anybody without a license, but the authority and legislation is unclear.

32.13   Participants of the Shoots

For every shoot, the game book of the Maharaja (1908) includes a list of participants, reproduced in 32.9 and 32.10. In keeping with the nature of his records, MCB in most cases identifies the people by surname or title only. To gain insight into the social tapestry of life in the camps over time, and to understand the impact of organized shoots in the political climate of conservation and preservation, it is necessary to reconstruct the identities in more detail. The full list of persons visiting the shoots of Cooch Behar from 1871 to 1916 contains some 320 names. In total, 190 persons have been recognized, using a large variety of sources. For persons with more common surnames or

with different family members in the country the task of identification has often not led to clear conclusions and could not be incorporated.

It is difficult to know if the lists in the book are complete and accurate, or if they included only those who were active in the shoots. Subordinate staff like assistants or photographers were not likely to be noticed. In the autobiography of Maharani Suniti Devi, she mentions about 48 people who might have attended the shoots, or maybe visited the palace around the same time, referring to 14 persons who appear to be absent from the Maharaja's game book: Lady Bayley; Colonel Frank Dugdale and Lady Eva Dugdale; Mrs. Eldridge (nurse of Suniti); Lord and Lady Galloway; Sir R. Garth (father of Mrs. Evan Gordon); Hari Mohum Chatterjee (engineer); Lady Hewitt; Lord Jersey; Mrs. Pemberton (sister of Mrs. Evan Gordon); Lord Pembroke; Lady Prinsep; and Duke of Sutherland (Devee 1916). The Maharaja's children and other family members (32.5) were definitely happy to take part in the shoots or stay in the camps.

The list in Table 32.44 is therefore no more than a start to unravel the intricacies of the shoots in the jungles of Cooch Behar. Still, it shows that the shoots attracted many noblemen and high-ranking officers, mostly from Britain (or British stationed in India), but participants from other nations were definitely just as welcome.

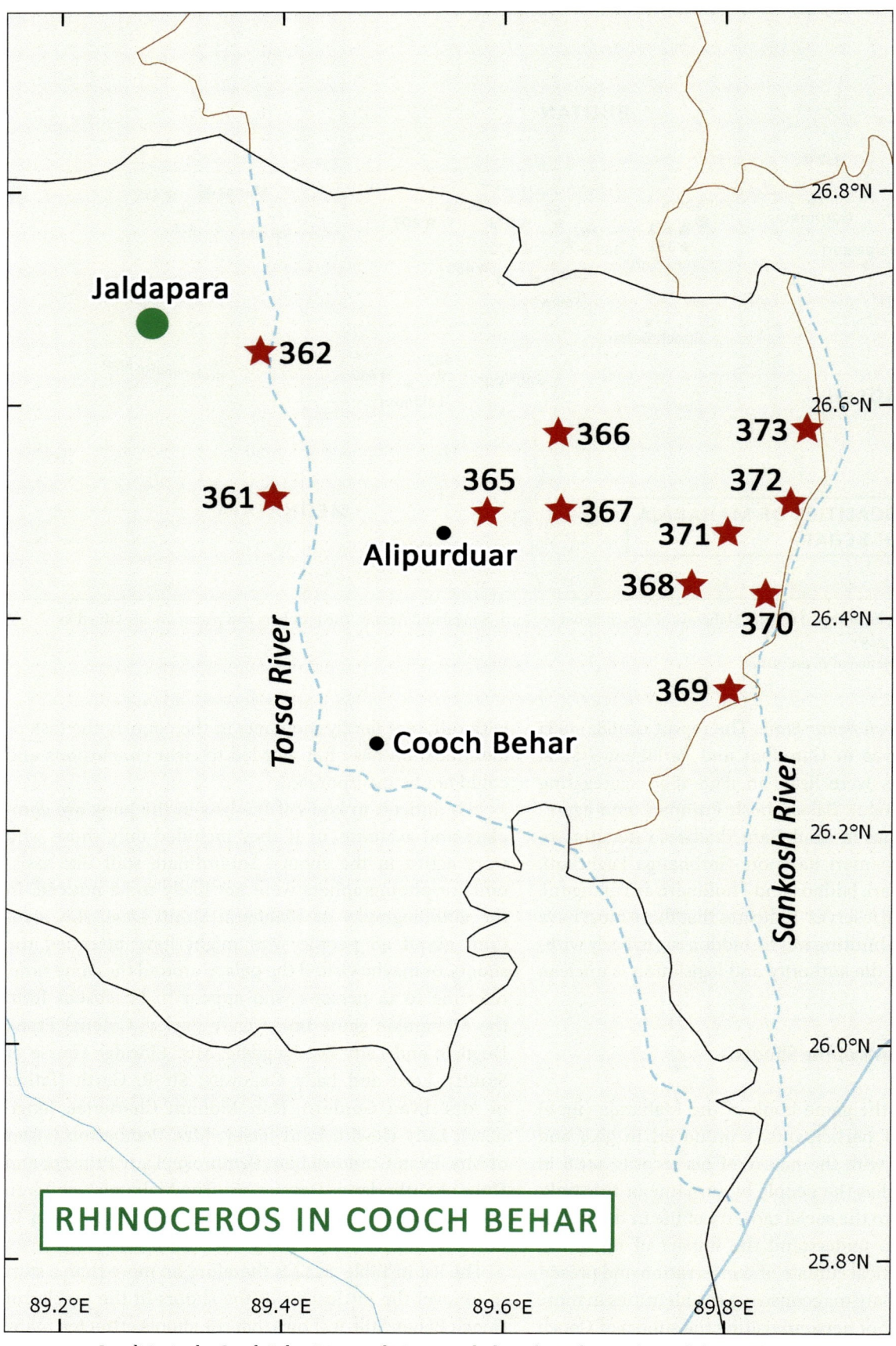

Jaldapara

⭐362

⭐366   373⭐

361⭐   365   372⭐
              ⭐   ⭐367
● Alipurduar   371⭐
                    368⭐
              ⭐ 370   26.4°N
                    369⭐

● Cooch Behar

Torsa River

Sankosh River

26.8°N

26.6°N

26.2°N

26.0°N

25.8°N

89.2°E   89.4°E   89.6°E   89.8°E

RHINOCEROS IN COOCH BEHAR

MAP 32.17   Localities in the Cooch Behar District of West Bengal where the Maharaja obtained rhinos. The numbers and places are explained in Dataset 32.37

•••

Dataset 32.36: Chronological List of Localities of Rhinoceros Recorded during the Hunts of the Maharaja of Cooch Behar in West Bengal and Assam

Localities are approximate. Bist (1994) discussed and named the areas in West Bengal, which is reflected in the following list. MCB also hunted in parts of Assam separated from Cooch Behar by the Sankosh River, also in the Manas area of Assam north of Barpeta, and in the Gauripur-Goalpara area on the Brahmaputra River.

The list includes all known and named localities where rhinos were killed according to MCB (1908) as found in 32.9. They are presented according to regions, referred to as Rhino Regions 24, 25, 26, 30, 31, 32 (where Rhino Regions 24, 25, 26 are similar to Bist's III, II and I). In each case, the entries include date (YYYY-MM-DD), locality as in original, current name or description, and the map reference (map nos. 32.16, 32.17 or 36.18). The available grid references are combined with the current locality.

Rhino Region 24. Jaldhaka River (*Jalpaiguri, West Bengal*)

This is Area III, the Gorumara population in Bist (1994: 46). The localities are shown in map 32.16.

Date	Original locality	Current locality	Map no.
1884-02-24	Lower Tondoo	Tandu (Tondu), on Jaldhaka River 26.80N; 88.90E	map 357
1884-03-06	Ambari	Ambari (Upper Tondu) 26.87N; 89.04E	map 357
1884-03-07	Ambari	Ambari	map 357
1884-03-10	Nagrakata	Nagrakata, on Jaldhaka River (right bank) 26.87N; 88.87E	map 357

Rhino Region 25. Torsa River (*Alipurduar, West Bengal*)

This is Area II, the Torsa population in Bist (1994) in the Patlakhawa Reserve. The localities are shown in map 32.17.

Date	Original locality	Current locality	Map no.
1892-04-26	Chelapata, Forest Reserves	Chilapata, in Buxa 26.60N; 89.40E	map 362
1893-04-26	Chelapata	Chilapata	map 362
1904-04-04	Chilapata	Chilapata	map 362
1905-04-12	Patlakhowa	Patlakhawa (on Torsa River) 26.51N; 89.34E	map 361

Rhino Region 26. Raidak – Sankosh River (*Central and Eastern Parts of Cooch Behar*)

This is Area I, Sankosh-Rydak population in Bist (1994). On his map Bist has the northern part as population IA, the southern part as IB. The localities are shown in map 32.17.

Sankosh-Rydak listed without details for areas where rhinos were found: 1871 (a), 1873 (a), 1877 (a), 1877 (b), 1877 (c), 1878 (a), 1878 (b), 1879 (a), 1882 (a), 1882 (b), 1882 (c), 1883 (a), 1883 (b), 1896 (b).

Date	Original locality	Current locality	Map no.
1878-04	Parokata	Parokata, 23 km east of Alipurduar 26.49N; 89.71E	map 367
1881-03-26	Parokata	Parokata	map 367
1882-03	Parokata	Parokata	map 367
1883-02	Garud Haut	Garadhat or Garodhat, on Sankosh River 26.44N; 89.77E	map 370
1884-03-19	Haldibari (Theamari Gaut)	Haldibari on Sankosh River 26.60N; 89.85E	map 373
1884-03-23	Parokata	Parokata	map 367
1884-04	Madhoobasha	Madhurbhasa 26.43N; 89.77E	map 368
1884-05	Falimari	Falimari on Sankosh River 26.38N; 89.81E	map 369
1885-03-18	Dal Dalia	Daldal 26.52N; 89.80E	map 372
1885-04	Jaldoa and Rossik bheel	Rasikbil, near Raidak River 26.42N; 89.72E	map 368
1886-02-16	Rossik Bheel and Chengtimari	Rasikbil	map 368
1886-02-19	Chengtimari	Chengtimari, near Rasikbil 26.42N; 89.71E	map 368
1886-02-24	Dorko or Chengtimari	Chengtimari 26.42N; 89.71E	map 368
1886-03-05	Sunkos River	Sunkos or Sankosh River 26.40N; 89.83E	map 370
1886-03-09	Sal Forest Reserve	Sankosh River	map 370

(cont.)

Date	Original locality	Current locality	Map no.
1887-02-21	Guigaon	not found but in Garadhat area	map 370
1887-02-24	Deogong	not found but in Garadhat area	map 370
1889-02-18	Falimari	Falimari on Sankosh River 26.38N; 89.81E	map 369
1889-03-03	Bara Hamua (Sankosh R.)	not found but in Garadhat area	map 370
1889-03-08	Moshamari	not found but in Garadhat area	map 370
1889-03-17	Rossik Bheel	Rasikbil	map 368
1890-02-14	Takuamari	Takoamari, near Rassikbil 26.41N; 89.75E	map 368
1890-03-05	Dal Dalia	Daldal 26.52N; 89.80E	map 372
1890-03-12	Sunkos Reserves	Sunkos or Sankosh River	map 370
1891-02-18	Bara Bhalka	Balka, on Raidak River 26.52N; 89.84E	map 371
1891-02-19	Horseshoe Jungles	not found but in Garadhat area	map 366
1891-03-08	Rossik Bheel	Rasikbil, near Raidak River	map 368
1891-03-16	Deotakata	not found but in Rasikbil area	map 368
1891-04	Gordongunj	not found but in Haldibari area	map 373
1893-02-19	Raidak River	Raidak River	map 366
1893-02-20	Raidak River	Raidak River	map 366
1893-02-22	Reserves north of Haldibari Rd	Haldibari area	map 373
1893-03-05	Hamua (Sankosh R.)	not found but in Garadhat area	map 370
1893-03-09	Deaogaon near Garad Haut	Garadhat, on Sankosh River	map 370
1895-04-28	Jorai River	Jorai 26.45N; 89.80E	map 368
1895-12-06	Raidak River	Raidak River	map 366
1896-02-21	Guigaon	not found but in Garadhat area	map 370
1896-02-24	Mahakalguri	Mahakalguri, near Samuktala 26.54N; 89.69E	map 365

(cont.)

Date	Original locality	Current locality	Map no.
1896-02-26	Mahakalguri	Mahakalguri, near Samuktala	map 365
1896-02-28	Baman Kote (reserves)	not found but in Garadhat area	map 366
1896-03-05	Raidak reserves	Raidak River	map 366
1896-03-13	Raymana	not found but in Haldibari area	map 373
1897-03-15	Raidak Balashi	Balashi, near Bhalka 26.50N; 89.85E	map 371
1897-04-17	Horse-shoe jungle	not found but in Garadhat area	map 366
1898-03-16	Longa River	not found but in Rasikbil area	map 368
1898-03-18	Takum Takum Khas	not found but in Rasikbil area	map 368
1898-03-20	Longa River	not found but in Rasikbil area	map 368
1904-01-22	Raidak reserves	Raidak River	map 366
1904-03-19	Raidak reserves	Raidak River	map 366
1904-12-20	Balashi	Balashi	map 371
1905-04-18	Raidak reserves	Raidak River	map 366

Rhino Region 30. Khalabari, East of Sankosh River (Assam, Western Part North of Brahmaputra)

Area listed without details where rhinos were found: 1880 (a). The localities are shown in map 36.18.

Date	Original locality	Current locality	Map no.
1880-03-01	Khagrabari	Khagrabari 26.73N; 90.44E (north of Sidli)	map 409
1880-03-05	Khagrabari	Khagrabari	map 409
1880-03-07	Khagrabari	Khagrabari	map 409
1880-03-14	Khagrabari	Khagrabari	map 409
1881-02-15	Khagrabari	Khagrabari	map 409
1881-02-17	Khagrabari	Khagrabari	map 409
1881-02-21	Khagrabari	Khagrabari	map 409
1881-02-22	Khagrabari	Khagrabari	map 409
1892-02-22	Bara Dogla	not found but in Sapkata area	map 407
1892-02-23	Sapkata	Sapkata near Patgaon 26.59N; 90.21E	map 407

(cont.)

Date	Original locality	Current locality	Map no.
1892-02-24	Do-Mahana	not found but in Sapkata area	map 407
1892-03-02	Turkani	not found but in Sapkata area	map 407
1893-03-18	Bara Dogla	not found but in Sapkata area	map 407
1894-02-27	Dudnath Pahar	Dudnath, north bank Brahmaputra 26.20N; 90.31E	map 403
1900-02-22	Roymana	Raimona 26.65N; 89.86E	map 405
1901-01-30	Patgaon	Patgaon 26.57N; 90.21E	map 407
1901-03-03	Lanka River	not found but in Patgaon area	map 407
1901-03-13	Dhoombarjhar jungles	Dumbazar 26.62N; 90.00E	map 406
1901-03-10	Forest reserves (Saralbhanga)	Saralbhanga River 26.63N; 90.21E	map 407
1901-03-16	Lanka River (Forest Reserves)	not found but in Patgaon area	map 407
1901-04-02	Doombar Jhar Kas	Dumbazar	map 406
1905-02-20	Kachugaon	Kachugaon 26.56N; 90.09E	map 406

Rhino Region 31. Assam South of Brahmaputra

The localities are shown in map 36.18.

Date	Original locality	Current locality	Map no.
1894-03-02	Samerdanga	not found but in Lakhipur area	map 412
1894-03-18	Kaee Mari near Khalsamari camp	Kaimari, south bank of Brahmaputra 26.32N; 90.22E	map 412

Rhino Region 32. Manas Area (Assam)

Area listed without details where rhinos were found: 1906 (a). The localities are shown in map 36.18.

Date	Original locality	Current locality	Map no.
1895-02-21	Simlaguri	Simlaguri, east of Manas River 26.48N; 90.96	map 418
1895-03-10	Bhuiapara	Bhuiapara, east of Manas River 26.70N; 90.00E	map 418
1895-03-12	Bhuiapara	Bhuiapara	map 418
1895-03-13	Bhuiapara	Bhuiapara	map 418
1895-03-19	Bhuiapara	Bhuiapara	map 418
1895-03-20	Bhuiapara	Bhuiapara	map 418
1899-02-24	Simlaguri	Simlaguri	map 418
1899-03-07	Dhowbheel (near Bansbari)	Bahbari 26.61N; 91.00E	map 418
1899-03-13	Bhuiapara	Bhuiapara	map 418
1899-03-18	Bhuiapara	Bhuiapara	map 418
1899-03	Bansbari	Bahbari	map 418
1901-04	Manas River	Manas River	map 419
1902-03-06	Gosiengaon	not found but in Manas area	map 419
1902-03-23	Daimasi River near Lahapara	Lahapara 26.75N; 91.10E	map 419
1902-03-25	Dhun Bheel near Lahapara	Lahapara	map 419

...

Dataset 32.37: Localities of Records of Rhinoceros in Cooch Behar

Rhino Region 24: Jalpaiguri, North Bengal

A 357   Nagrakata, on Jaldhaka River – 26.87N; 88.90E – 1884

A 357   Lower Tondoo (Tendu), on Jaldhaka River – 26.80N; 88.90E – 1884

A 357   Ambari (Upper Tondu) – 26.87N; 89.04E – 1884

Rhino Region 25: Alipurduar, North Bengal

A 361   Patlakhowa (Patlakhawa) 26.51N; 89.34E – 1905

A 362   Chelapata (Chilapata) Forest Reserves – 26.60N; 89.40E – 1892, 1893, 1904

Rhino Region 26: Cooch Behar, North Bengal

A 365   Mahakalguri, near Samuktala – 26.54N; 89.69E – 1896

A 366   Raidak River: Forest reserves – 26.56N; 89.75E – 1886, 1889, 1890, 1893, 1895, 1896, 1904, 1905, 1907

A 366   Baman Kote (reserves) – not found but in Raidak River area – 1896

A 366   Horse-shoe jungle (Raidak R.) – not found but in Raidak River area – 1891, 1897

A 367   Parokata, 23 km east of Alipurduar – 26.49N; 89.71E – 1878, 1881, 1882, 1884

A 368   Rossik bheel, Rasikbil, near Raidak River – 26.42N; 89.72E – 1885, 1886, 1889, 1891

A 368   Jaldoa (Rossik Bheel) – 26.42N; 89.72E – 1885

A 368   Chengtimari, Dorko – 26.42N; 89.71E – 1886

A 368   Takuamari, Takoamari, near Rassikbil – 26.41N; 89.75E – 1890

A 368   Deotakata – not found but in Rasikbil area – 1891

A 368   Longa River – not found but in Rasikbil area – 1898

A 368   Takum Takum Khas – not found but in Rasikbil area – 1898

A 368   Madhoobasha, Madhurbhasa – 26.43N; 89.77E – 1884

A 368   Jorai River – 26.45N; 89.80E – 1895

A 369   Falimari on Sankosh River – 26.38N; 89.81E – 1884, 1889

A 370   Garud Haut, Garadhat or Garodhat, on Sankosh River – 26.44N; 89.84E – 1883

A 370   Deaogaon (Deogong) on Sankosh River – 26.44N; 89.84E – 1887, 1893

A 370   Guigaon – not found but in Garadhat area – 1887, 1896

A 370   Bara Hamua (Sankosh R.) – not found but in Garadhat area – 1889, 1893

A 370   Moshamari – not found but in Garadhat area – 1889

A 371   Bara Bhalka, Balka, on Raidak River – 26.52N; 89.75E – 1891

A 371   Balashi, Raidak Balashi, near Bhalka – 26.50N; 89.75E – 1897, 1904

A 372   Dal Dalia, Daldal – 26.52N; 89.80E – 1885, 1890

A 373   Haldibari (Theamari Gaut) on Sankosh River – 26.60N; 89.85E – 1884

A 373   Gordongunj – not found but in Haldibari area – 1891

A 373   Raymana – not found but in Haldibari area – 1896

A 373   Reserves north of Haldibari Rd – Haldibari area – 26.68N; 89.86E – 1893

Rhino Region 30: Assam – Sankosh River, North of Brahmaputra

A 403   Dudnath, north bank Brahmaputra – 26.20N; 90.31E – 1894

A 405   Raimona – 26.65N; 89.86E – 1900

A 406   Dhoombarjhar jungles, Dumbazar – 26.62N; 90.00E – 1901

A 406   Kachugaon RF – 26.38N; 90.00E – 1905

A 407   Patgaon – 26.57N; 90.21E – 1901

A 407   Sapkata near Patgaon – 26.59N; 90.21E – 1892, 1893

A 407   Turkani – not found but in Sapkata area – 1892

A 407   Forest reserves, Saralbhanga River – 26.63N; 90.21E – 1901

A 409   Khagrabari (north of Sidli) – 26.73N; 90.44E – 1880, 1881

Rhino Region 31: Assam – Goalpara, South of Brahmaputra

A 412   Khalsamari, south bank of Brahmaputra – 26.09N; 90.32E – 1894

A 412   Samerdanga – not found but in Lakhipur area – 1894

Rhino Region 32: Assam – Manas, North of Brahmaputra

A 418   Bansbari (Bahbari) – 26.61N; 91.00E – 1899

A 418   Simlaguri – 26.48N; 90.96 – 1895

A 418   Bhuiapara, east of Manas River – 26.70N; 91.00E – 1895, 1899

A 419   Lahapara – 26.75N; 91.10E – 1902

A 419   Gosiengaon – not found but in Manas area – 1902

..

TABLE 32.44   List of persons participating in a shoot in Cooch Behar

Robert John Abercromby, 7th Baronet (1855–1895)

Archibald Kennedy, 3rd Marquess of Ailsa (1847–1938)

Aretas Akers-Douglas, 1st Viscount Chilston (1851–1926)

A.L. Alexander, resident engineer in Cooch Behar

Schomberg Henry Kerr, Lord Ancram, 9th Marquess of Lothian (1833–1900)

Hugh Annesley, 5th Earl Annesley (1831–1908)

Thomas Ashton, appointed Superintendent of the Philkhana (elephant stables) in Cooch Behar, 1890

Colonel John Claud Alexander Auchinleck (1836–1892)

Lieutenant-General Sir Fenton John Aylmer, 13th Baronet (1862–1935)

Colonel Edward William Baird (1864–1956)

Millicent Bessie (Clarke) Baird (1871–1936)

Major Robert George Teesdale Baker-Carr (1867–1931), ADC to Lord Curzon

Brigadier-General Everard Baring (1865–1932), Military Secretary to Lord Curzon

Capt. H.R. Beaumont

William Ostliffe Adams Beckett (1831–1871), Assistant Commissioner and Settlement Officer in Assam

Lieutenant-Colonel Lord William Leslie de la Poer Beresford (1847–1900), ADC to several British viceroys

Edward Duncan Frederick Bignell (1851–1914), Indian Staff Corps

Mrs. Katherine Harriet (Lowis) Bignell (1856–1900)

Richard Augustus D'Oyly Bignell (1847–1906), MCB private secretary

Colonel Thomas Smalley Boileau (1851–1933)

Lieutenant-Colonel George Washington Brazier-Creagh (b. 1860)

Henri Charles Joseph Le Tonnelier de Breteuil, Marquis de Breteuil (1847–1916)

Elisabeth Laure Le Tonnelier de Breteuil (1868–1918)

Charles B. Briscoe, Superintendent of Police in Cooch Behar

Francis Richard Charles Greville, 5th Earl of Warwick, 5th Earl Brooke (1853–1924)

Victor Reginald Brooke (1873–1914), military secretary to the Viceroy

Lord Henry Ulick Browne (1831–1913), British Commissioner Cooch Behar 1876–1883

Peter Burges (1857–1938), big game hunter and photographer settled in Bristol (Clifton)

Captain Rivett Carnac

Major Harry George Carnegy (1865–1905), Political officer

Emanuele Coardi di Bagnasco e Carpeneto (b. 1849), Italian nobleman

Colonel Sir Neville Francis Fitzgerald Chamberlain (1856–1944)

General Godfrey Lewis Clark (1855–1918)

Elliot James Dowell Colvin (1885–1950), son of JRC Colvin

Major John Russell Colquhoun Colvin (1858–1935), State Council, Cooch Behar, from 1899

Mrs. Jessie Rosaline (Webster) Colvin (1861–1949)

Major J. Neild Cook, Civil Surgeon

Edward James Corbett (1875–1955), army officer and naturalist

General Sir John Stephen Cowans (1862–1921)

Colonel Warren Roland Crooke-Lawless (1863–1931), surgeon

George Curzon, 1st Marquess Curzon of Kedleston (1859–1925), Viceroy 1899–1905

John Hew North Gustav Henry Hamilton-Dalrymple, 11th Earl of Stair (1848–1914)

Godfrey John Bective Tuite-Dalton (1840–1911), Commissioner of Cooch Behar

John Daniell (1878–1963), tutor of Maharaja's son Hitendra, cricket player

Geoffrey Nicolas Dawnay (1852–1941)

Claude Fraser Delafosse (1868–1950), Principal of Victoria College 1893–1896

Arthur William Dentith (1894–1959), Superintendent of Cooch Behar

TABLE 32.44   List of persons participating in a shoot in Cooch Behar (*cont.*)

Lieutenant Colonel William Henry Dick-Cunyngham (1851–1900)

John George Lambton, 3rd Earl of Durham (1855–1928)

Captain M. Eden, ADC to Ashley Eden

Sir Ashley Eden (1831–1887), Lieutenant-Governor of Bengal

Sir William Eden, 7th and 5th Baronet (1849–1915)

Otto Ehrenfried Ehlers (1855–1895), German traveller

Sir Charles Alfred Elliott (1835–1911), Lieutenant-Governor of Bengal

Sidney Herbert Buller-Fullerton-Elphinstone, 16th Lord Elphinstone (1869–1955)

Nikolaus III [Miklós Pál], Prince Esterhazy (1817–1894), Hungarian nobleman

Alexander Evans-Gordon (1845–1920), Superintendent of Cooch Behar 1883–1891

Mrs. Helen Frances (Garth) Evans-Gordon (1857–1939)

Lieutant-Colonel Alick Evans-Gordon, son of Alexander Evans-Gordon

Sir Henry Anthony Farrington, 6th Baronet of Blackheath (1871–1944)

Lieutant-Colonel Ernest Harold Fenn (1850–1916), Surgeon to the Viceroy

Lieutenant Humphrey Brooke Firman (1858–1916), 16th Queen's Lancers

Major-General Reymond Hervey de Montmorency, 3rd Viscount Frankfort de Montmorency (1835–1902)

Alexander William Frederick Fraser, 19th Lord Saltoun (1851–1933)

Colonel James Keith Fraser, CMG (1832–1895)

Sir Edward Lee French (1857–1916), Indian Police

Gajendra (Gojendro) Narayan, senior (1856–1919), cousin of MCB

Gajendra Narayan, junior (d. 1930), son of MCB's cousin

George Edmund Milnes Monckton-Arundell, 7th Viscount Galway (1844–1931)

Colonel Alan Coulston Gardner (1846–1907)

Sir Richard Garth (1820–1903), Chief Justice of Bengal

Gordon, see Evans-Gordon

Mario Grazioli, 3rd Duke of Santa Croce di Magliano (1848–1936),Italian nobleman

Francis Octavius Grenfell (1880–1915)

Lieutenant Riversdale Nonus Grenfell (1880–1914)

Lieutenant-Colonel Rollo Estouteville Grimston (b. 1861), Viceroy's bodyguard

Philip Richard Thornhagh Gurdon (1863–1942), Commissioner

Lord Claud John Hamilton (1843–1925)

Louisa Jane Hamilton, Duchess of Abercorn (1812–1905), wife of Lord Claud

Frederick Spencer Hamilton (1856–1928)

Sir Egbert Laurie Lucas Hammond (1854–1936), Superintendent at Cooch Behar 1903–1908

Charles Harbord, 5th Baron Suffield (1830–1940)

Count Hubert Karl Sigismund Joseph Franz de la Fontaine und d' Harnoncourt (1850–1920), German nobleman

James Jonathan Harrison (1858–1923), officer and big game hunter

Lieutenant-Colonel Reginald Heber-Percy (1849–1922), author of chapter on shikar in 1894

John Lupton Hendley (1858–1935), Civil Medical Officer of Jalpaiguri 1887–1892

Sir Edward Richard Henry (1850–1931), Assistant Magistrate Collector

Hans Heinrich XV von Hochberg, Prince of Pless (1861–1938), Polish aristocrat

Daisy, Princess of Pless (Mary Theresa Olivia, née Cornwallis-West) (1873–1943)

Sir John Prescott Hewett (1854–1941), author of *Jungle Trails* 1938

General James Hills-Johnes (1833–1919)

Charles Hills (1847–1935), younger brother of General Hills-Johnes

Count Fritz Hochberg (1868–1921), brother of Hans Henry of Pless

Count Ernst Karl Heinrich Hoyos-Sprinzenstein (1856–1940), German nobleman and adventurer

Major-General Thomas Elliott Hughes (1830–1886)

TABLE 32.44   List of persons participating in a shoot in Cooch Behar (*cont.*)

Allan Octavian Hume (1829–1912), zoologist in India

George Herbert Hyde Villiers, 6th Earl of Clarendon (1877–1955), Lord Hyde 1877–1914

Henry Edward Fox-Strangways, 5th Earl of Ilchester (1847–1905)

George Irwin, Bengal Civil Service

Colonel Henry Sullivan Jarrett (1839–1919)

François Lévisse De Montigny, Comte de Jaucourt (1825–1906)

Captain Rudolf George Jelf (1873–1958), ADC to Viceroy

Colonel Atherton Edward Jenkins (1859–1945), Rifle Brigade

Vernon Aubrey Scott Keighley (1874–1939), ADC to Viceroy

Auberon Claud Hegan Kennard (1870–1951)

Alexander Angus Airlie Kinloch (1838–1919)

Henry St. John Kneller (1848–1930), Guardian of the Maharaja's sons, to 1884

Alexander Vansittart Knyvett, Superintendent of Police

Major-General The Hon. Sir William Lambton (1863–1936)

Charles Wallace Alexander Napier Cochrane-Baillie, 2nd Baron Lamington (1860–1940), Governor of Bombay 1903–1907

General Fredrick Lance (1837–1913)

George Lane Anderson, businessman at Calcutta

Mrs. Mary Beatrice Anderson, wife of George Lane Anderson

Henry Petty-Fitzmaurice, 5th Marquess of Lansdowne (1845–1927), Viceroy 1888–1894

Maud Evelyn Petty-Fitzmaurice, Marchioness of Lansdowne (1850–1932), wife of the Viceroy

Lawlor, M.P. of Ireland

Algernon George Lawley (1857–1931), 5th Baron Wenlock

Francis Dudley Leigh, 3rd Baron Leigh (1855–1938)

Sir Henry Arthur Hallam Farnaby Lennard, 2nd Baronet (1859–1928) of Wickham Court

Hugh Cecil Lowther, 5th Earl of Lonsdale (1857–1944)

Susan Mary (Currie) Lowis (1844–1909)

Edmond Elliot Lowis (1836–1918), Superintendent at Cooch Behar 1891–1896

Frank Currie Lowis (1872–1963), son of E.E. Lowis

Colonel Reginald Kirkpatrick Lynch-Staunton (1880–1918)

Archibald Donald Mackinnon (1864–1937), medical officer

Giulio Magliano (1849–1933), Italian nobleman (brother of Grazioli)

John Clavell Mansel-Pleydell (1817–1902) of Smedmore House

Nikolai Nikolaevich, Prince Odoevsky-Maslov (1856–1929), Russian nobleman

Dermot Robert Wynham Bourke, 7th Earl of Mayo (1851–1927)

R. McCabe, Deputy Commissioner

Frederick Hubert McLaughlin (1840–1911), Assistant Commissioner Mymensingh

Charles Randal Marindin (1851–1935), Commissioner of Darjeeling

Gilbert John Elliot-Murray-Kynynmound, 4th Earl of Minto (1845–1914), Viceroy 1905–1910

Lady Mary Caroline (Grey) of Minto (1858–1940), wife of Viceroy Minto

Lady Eileen Nina Evelyn Sibell Elliot-Murray-Kynynmound (1884–1938), daughter of Viceroy Minto

Major-General Robert Cotton Money (1835–1903), Manager of the Durbhunga Estate

Charles H. Moore, businessman in Calcutta

Raja Sriram Chandra Bhanj Deo (1871–1912), Maharaja of Mourbanj (Mayurbhanj, Odisha)

General Arthur Henry Fitzroy Paget (1851–1928)

George Augustus Parker, Viscount Parker (1843–1895)

Walter Francis Perree (1871–1950), Conservator of Forests

Carlo Alberto Pizzardi (1850–1922), Italian nobleman

Cecil Ward Chicheley Plowden (1864–1944), MCB private secretary

TABLE 32.44   List of persons participating in a shoot in Cooch Behar (*cont.*)

Lieutenant-General Reginald Pole-Carew (1849–1924)

Lieutenant-Colonel Stephen Hungerford Pollen (1869–1935), ADC to Viceroy

Martanda Bhairava Tondaiman (1886–1928), Rajah of Pudukkottai (South India)

Bertram Fürst von Quadt zu Wykradt und Isny (1849–1927), German nobleman

Robert Home Renny, Deputy commissioner

Field Marshal Frederick Sleigh Roberts, 1st Earl Roberts (1832–1914)

Sir Edward Albert Sassoon, 2nd Baronet (1856–1912), British businessman

Lady Sassoon, Aline Caroline de Rothschild (1867–1909)

Aldred Frederick George Beresford Lumley, 10th Earl of Scarbrough (1857–1945)

Colonel Vernon Ansdell Schalch (1849–1935)

Countess Ernestina (Pullé) Scheibler (b. 1871)

Count Felix Emil Scheibler (1856–1921), Italian nobleman

Captain Francis George Montagu Douglas Scott (1879–1932), ADC to Viceroy

Nikolaus Graf von Seebach (1854–1930), German nobleman

Nirmul Sen, MCB's military secretary in Cooch Behar

Heywood Walter Seton-Karr (1859–1938), soldier and game hunter

Joseph Lay Shillingford (1845–1889), indigo planter in Purneah, in charge of kheddah operations

Sir Benjamin Simpson (1831–1923), Indian Medical Service Bengal 1853–1890 (once guardian to MCB)

Miss Simpson, daughter of Benjamin Simpson

Thomas Smith, Deputy Commissioner of Cooch Behar in 1876

Lieutenant-Colonel James Robert Dunlop Smith (1858–1921), private secretary to the Viceroy

Richard Southby, Manager of Selim Tea Estate, Darjeeling

Giles Stephen Holland Fox-Strangways, 6th Earl of Ilchester (1874–1959), Lord Stavordale until 1905

Field Marshal Sir Donald Martin Stewart, 1st Baronet (1824–1900)

Colonel Henry Streatfeild (1857–1938)

Lady Florence Beatrice (Anson) Streatfeild (1860–1946)

Henry Molyneux Paget Howard, 19th Earl of Suffolk (1877–1917)

Sidney Parker (1852–1897), son of 6th Earl of Macclesfield

Count Imre Széchényi of Sárvár-Felsővidék (1825–1898), Hungarian nobleman

Onorato Caetani, 14th Duke of Sermoneta, 4th Prince of Teano (1842–1917), Italian nobleman

Sir Henry Alfred Joseph Doughty-Tichborne, 12th Baronet (1866–1910)

Colonel Lionel Tillotson (1845–1921)

Colonel Aleksandr Karlovich von Timler, Russian officer

Mrs. Flora Toomey, wife of George Toomey, indigo plantation owners

Prince Vittorio Emanuele of Savoy-Aosta, Count of Turin (1870–1946), Italian nobility

Major General John Upperton (1836–1924), Bengal cavalry

Piers Alexander, Viscount Valletort (1865–1944)

Captain Philip Breda Vanderbyl (1867–1930), naturalist

John Robert William Vesey, 4th Viscount de Vesci (1844–1903)

Prince Christian Victor of Schleswig-Holstein (1867–1900), British nobility

George Herbert Hyde Villiers, 6th Earl of Clarendon (1877–1955)

Count Adolf von Waldstein (1868–1930), German nobleman

Reginald William Henry Warneford (1860–1915), railway engineer in Cooch Behar

Prince Wilhelm of Sweden (1884–1965), Swedish nobleman

Lieutenant Frederick Lord Wolverton (1864–1932)

General Arthur Godolphin Yeatman-Biggs (1843–1898).

CHAPTER 33

Protecting the Rhinoceros in Gorumara National Park, West Bengal

First Record: 1895 – Last Record: current
Species: *Rhinoceros unicornis*
Rhino Region 24

33.1 Rhino Habitat of Gorumara

Gorumara is located about 60 km east of Siliguri and 80 km west of Jaldapara, at the junction of the Jaldhaka and Murti Rivers. Extending over a small area of just under 9 km², it became a Reserve Forest on 2 July 1895, a Wildlife Sanctuary on 2 August 1949 and upgraded to a National Park in 1994. The extent of the park was increased in 1993 to 9.6 km² and again, significantly, in 1995 to 79.45 km². The name Gorumara derives from *goru* (cattle) and *mara* (killed), as local people spending the night in the forest after crossing the Jaldhaka River often lost their livestock to tigers. The park is largely covered by dense forest, with grassland restricted to less than 14 km². The rhinos of Gorumara are known to roam northwards throughout the Chapramari and Tendu forests, typically when their habitat is flooded.

Early estimates of Gorumara's rhino population are few. Shebbeare (1953) thought that there were just three animals in the area, although there was an estimate of 12 animals from the 1920s to early 1950s (Table 33.45). Despite natural breeding in the reserve, numbers have remained relatively static, because rhino calves have been taken by tigers and others have been shot after straying into nearby rice fields (figs. 33.1, 33.3). The best rhino habitat is the grassland, which covers only 10 km² in the Jaldhaka-1A, 1B and Dhupjhora-1B compartments. Additional grasslands of 3.28 km² were created in the Shelkapara and Tondu compartments, and in the Bamandanga and Ramsai extensions adjoining the National Park, along the Murti and Jaldhaka Rivers. In 2012, there were plans to expand the park by adding 25 km² of floodplain along the Jaldhaka River, which presently serves as uninhabited communal grazing land. The proposed extension includes about 15 km² in South Diana and Ramsai and another 10 km² of vested land outside the forest area, which is also used for communal grazing. Except for 5 km² of teak plantation in Barohati, Bichabhanga, Tondu, Shelkapara and Bhogalmardi, the National Park consists of dense forest (N.C. Bahuguna, pers.com.).

FIGURE 33.1   Rhino mother and child crossing the Murti river at Gorumara National Park, West Bengal
PHOTO: KAUSHIK DEUTI, ZOOLOGICAL SURVEY OF INDIA, KOLKATA, 2012

© L.C. (KEES) ROOKMAAKER, 2024 | DOI:10.1163/9789004691544_034
This is an open access chapter distributed under the terms of the CC BY-NC-ND 4.0 license.

•••

Dataset 33.38: Chronological List of Records of Rhinoceros in Gorumara

General sequence: Date – Locality (as in original source) – Event – Sources.

1895-07-02 – Gorumara declared as Reserved Forest, Notification No. 3147-FOR

1949-08-02 – Gorumara declared as Wildlife Sanctuary per Government of West Bengal notification 5181-For of 02-08-1949, area 2129 acres (8.62 km²) – Gupta 1958; West Bengal Forest Dept. 1993: 5; Bist 1994; Ghatak et al. 2016

1954-12 – Heavy floods – Bist 1997

1957 – Hunting prohibited – West Bengal Forest Dept. 1993: 4

1976-06-24 – Gorumara Wildlife Sanctuary ratified under Wildlife (Protection) Act, 1972, Notification 5400-For, dated 24 June 1976, comprising a total area of 8.62 km² of Jalpaiguri Division – Ghatak et al. 2016

1981 – Rhino strayed to Nathua Forest from Gorumara – Bahuguna & Mallick 2010: 287

1989-03 – Female rhino strayed into north Bangladesh and had to be physically brought back. She subsequently died of serious injury – Choudhury 2013: 225; West Bengal Forest Dept. 1993: 7

1992 – Poaching case reported in Kathambari area – Talukdar 2009b; Martin & Vigne 2012

1994-01-31 – Intention issued to declare Gorumara as National Park in Jalpaiguri Division, Notification 319-For, with total forest area of 79.99 km² – Ghatak et al. 2016

1995-10-17 – Male rhino (Madhu) rescued in Kaziranga in 1989 at 7 months was moved to Gorumara, and kept in enclosure for some months before release – Vigne & Martin 1996b; Martin 2006

1995-11-21 – Gorumara NP handed over to the administration of the Wild Life Division II under the Conservator of Forest, Wild Life Circle, Notification 4983-For dated 25 September 1995

2005 – Addition of 474.58 hectares (4.74 km²) of land beyond Jaldhaka River – Ghatak et al. 2016

2018-04-02 – Gorumara – Kees Rookmaaker visits the reserve and takes a drive with a tourist vehicle – fig. 33.2

•••

33.2 Two Rhino Introductions from Assam

In the 1990s, when the area of the reserve was increased, the Forest Department of West Bengal decided to obtain two male rhinos from an unrelated genetic population. Hence, for Rs 650,000 ($19,000), the Bengal Government bought two sub-adult males, Ratul and Madhu, which had been rescued in Kaziranga National Park and were kept in the Guwahati Zoo. They were moved on 17 October 1995, without the use of a tranquillizing agent. After being kept in an enclosure to facilitate acclimatization, Madhu was released in Jaldapara, and Ratul in Gorumara (Vigne & Martin 1996b).

33.3 Poachers Kept at Bay

In 1972, seven rhinos were poached, essentially halving the population from 13 to 7 animals. The trade specialist Esmond Bradley Martin and his colleague Lucy Vigne visited West Bengal in 1993, 1998, 2005 and 2012 to assess the rhino conservation programs (34.6). When Martin visited in 1993, he was most interested in the trade routes that were used to smuggle horn out of the country. He found that, while Kolkata had been the main port of exit from the 1960s through the 1980s, increased vigilance had made

this route more difficult for traders. Many horns were then taken to Phuntsholing on the Bhutan border, open to Indian traders without visa requirements. The Gorumara Sanctuary had one case of rhino poaching in June 1992, while another animal strayed into the nearby Apalchand forest and was killed in 1993–1994 (Martin 1996a).

In 1998, Martin spent time in Siliguri, a major trading hub for illegal wildlife products, including rhino horns which went to stations in nearby Nepal, China and Bhutan. From 1983 to 1997, 18 rhino horns were seized in 22 cases in and around Siliguri. In 1996, middlemen could sell rhino horn at Rs 100,000 (US$ 2,800) per kg, while elsewhere in Asia the price was about US$ 8,500. The Gorumara rhino population remained unaffected by the trade and no rhinos were poached, largely thanks to a sufficient budget (Martin 1999).

When Martin visited again in 2005, the rhino population in Gorumara had increased to 28 animals and no poaching had been reported. Successful protection efforts were attributed to the staff being honest, competent, motivated and hardworking, while the state and central governments allocated high budgets to the park. Patrols in the park were conducted on elephant-back in the grasslands and on foot in other areas (Martin 2006). As settlements on the fringes of the park increase, human-wildlife conflicts (especially with wild elephants) remain a problem. One

FIGURE 33.2
Statue of rhinoceros at an entrance gate of
Gorumara National Park
PHOTO: KEES ROOKMAAKER, 2018

FIGURE 33.3
Rhinoceros taking bath in a river at
Gorumara
PHOTO: N.C. BAHUGUNA, FORMER
PRINCIPAL CHIEF CONSERVATOR OF
FORESTS, WEST BENGAL

woman was killed by a rhino in 2009, the family receiving Rs 100,000 in compensation. Members of the local villages are employed in five ecotourism camps in the park and in transportation and guide services. The Forest Department carries out education programmes on the importance of wildlife conservation, emphasizing elephants and rhinos as flagship species. Some rhino deaths have occurred in Gorumara due to male infighting and aggression by males toward females during mating (Yadav 2000). This may be caused by a sex-ratio of 2 males to 3 females, which has remained almost the same ever since the population started to increase after the number had plummeted to one male, one female and a calf in 1954–55. The population

structure of rhinos in Gorumara over time has been analysed by Amal Bhattacharya of Raiganj following his field work in the 1980s (Bhattacharya 1982, 2020).

33.4   Current Protection of the Park

The grassland area suitable to rhino in Gorumara is limited to 10–12 km², which gives a current density of about 4 rhino per km². When a new Rest House was constructed, some 4–5 hectares of forest were cleared to allow wildlife viewing, which also provided a permanent home for some of the rhinos. In recent years, some rhino have left the confines of Gorumara to settle in other protected forests nearby. Two rhinos have taken shelter in the Baikunthapur Forest and Mahananda Wildlife Sanctuary. Another two animals have moved to the Chapramari Wildlife Sanctuary, where there is no grassland. On 15 September 2008, a male rhino identified as Kan Hela strayed from its usual patch in Dhupjhora-1B block to Amguri village, 8 km south of the park, and then to South Barogila. Harassed by villagers, it ran to Khamoner, where it tried to attack a cow and slipped into a small canal. It was then driven back to the park after traveling nearly 20 km. Another adult rhino was

spotted by staff of the Baikunthapur division (Apalchand Range) on 12 November 2009. After it settled near the Tista River, it was decided to develop a new habitat for rhino in this area. Another smaller rhino moved to roughly the same region on 17 January 2011 and is frequently seen near the railway track (N.C. Bahuguna, pers.com.).

In 2012, Gorumara's rhino population was estimated at 43 and no rhinos had been poached since 1992. The census results were verified from a DNA analysis of 60 fecal samples (Borthakur et al. 2016). The population increased to 52 in 2019 (table 33.45).

TABLE 33.45   Status of *R. unicornis* population in Gorumara area 1920–2022

Year	Number	Natural death	Poached	Source
1920	12			R164, R055
1937	4–5			R055, R156
1940	12			R055, R122, R156
1951	12+	3		R164, R055
1952	3	2		R055, R106, R156, R164
1953	3			R204
1954	3			R055
1955	3		1	R055, R164, R034, R156
1956	5			R055, R191, R034, R156, R198
1957	4; 8			R055, R034, R156; R191, R198, R291
1958	7			R055, R156, R191, R198
1959	4; 8			R191, R198; R055, R156
1963	5; 10–12			R207; R267
1965	14			R055, R156, R191, R198, R291
1966	10			R055, R191, R198
1969	12			R156, R198
1972	13		7	R055, R191, R156
1973	7			R055, R191, R156, R198, R291
1974	7			R055, R113, R156
1978	5–8; 8			R191, R198; R055, R064, R113, R149, R156, R291
1980	10; 11			R286, R291; R064
1981	8	0	1	R164, R156,
1983		1	1	R156, R164
1984			1	R156
1985	8			R191, R198, R286, R291
1986	8			R055, R113, R156. R164
1989	12			R055, R113, R149, R156, R164, R191, R198, R286, R287, R291
1990			1	R156
1992	11; 13		1	R286; R198, R291
1993	7; 12; 13; 15			R144; R113; R098, R274, R275; R055, R149, R156, R164
1994	13; 14; 15; 16; 18			R287; R113; R079, R149; R161, R190, R286; R098, R164
1996	18			R161
1997	14			R113
1998	19			R161, R286
1999	19			R113, R161

TABLE 33.45   Status of *R. unicornis* population in Gorumara area 1920–2022 (*cont.*)

Year	Number	Natural death	Poached	Source
2000	19			R041, R113
2001	22			R113
2002	22			R041, R054, R149, R161, R253
2004	25			R041, R054, R113, R149, R161, R253
2005	27	1		R052, R054. R149, R161, R229, R245
2006	27	1		R041, R052, R113, R153, R253, R290
2007		1		R052
2008	31	1		R041, R052, R113
2009	31; 35	1		R290; R052, R149
2010	35			R041, R113, R153
2011	35	1		R052, R290
2012	43			R041, R113, R153, R249, R290
2013	43			R149
2014	48; 50	1	1	R041, R052, R113, R290
2015	49; 50; 51			R041, R290; R251; R113
2016	50			R072
2019	52			R041
2022	52			R329, R331

References for table: R034 Bahuguna & Mallick 2010: 287; R041 Bengal Chief Conservator of Forests 2019; R052 Bhattacharya 2015; R054 Bhutia 2008; R055 Bist 1994; R059 Borthakur et al. 2016; R064 Choudhury 1985a; R072 Choudhury 2016: 154; R079 Dey 1994a; R098 Foose & van Strien 1997: 9; R106 Gee 1952a: 224; R113 Ghatak et al. 2016; R122 Gupta 1958; R144 Lahan 1993: 18; R149 Mallick 2015: 336; R153 Martin & Vigne 2012; R156 Martin 1996a; R161 Martin 2006; R164 Menon 1995: 15; R190 Pratihar & Chakraborty 1996: 235; R191 Raha 2000; R198 Roy 1993; R204 Shebbeare 1953: 143; R207 Simon 1966; R221 Spillett 1966c; R229 Strien & Talukdar 2007; R245 Talukdar 2009b; R249 Talukdar 2012a; R251 Talukdar 2015; R253 Talukdar et al. 2011; R267 Ullrich 1965a: 101; R274 Vigne & Martin 1994; R286 West Bengal Forest Dept. 1999; R287 West Bengal Forest Dept. 1994: 13; R290 West Bengal Forest Dept. 2016; R291 West Bengal Forest Dept. 1993: 20; R329 Sharma 2022; R331 Talukdar 2022.

CHAPTER 34

Protecting the Rhinoceros in Jaldapara National Park, West Bengal

First record: 1892 – Last record: current
Species: *Rhinoceros unicornis*
Rhino Region 25

34.1 The Early Years of Protection in Jaldapara

The area of Jaldapara in the *duars* of West Bengal is managed by the state's Forest Service. When this region came under British rule in 1865, the forests were administered by the Inspector General of Forest of India, then Dietrich Brandis (1824–1907). A separate forest department for Bengal was created in 1876. The practices and guidelines were published in a series of working plans which at first had no reference to wildlife management or to the need for conservation of rhinos. Although Jaldapara and Buxa were close to Cooch Behar, the Maharaja only went there a few times to hunt, and on his excursion in 1892 (b) one rhino was killed, in 1893 (b) also one, and in 1904

(c) accompanied by the Viceroy Lord Curzon some rhinos were seen (32.09). The Bengal Rhinoceros Preservation Act was passed in 1932. First called Torsa Rhinoceros Sanctuary, Jaldapara was recognised as a game reserve in 1932 and officially gazetted in 1941.

The Conservator of Forests, Edward Shebbeare wrote in 1935 that there had been about 200 rhinos in Jaldapara six years earlier. They inhabited a narrow strip of land spanning the Torsa River, about 60 km from north to south and 6 km at its widest point. A few scattered rhino groups were known outside this core area, but there was no room for them to expand their range. A serious spate of poaching began in 1930 with men from Assam joined by tribesmen from Bengal. At the time, a single rhino horn fetched Rs. 2000 or £150 in the Calcutta market and would be destined for export to China. Although this threat was brought under control within a comparatively short period, the rhino population was greatly reduced, possibly down to 40–50 animals in 1932 (Table 34.47).

•••

Dataset 34.39: Chronological List of Records of Rhinoceros and Admistrative Changes in Jaldapara

General sequence: Date – Locality (as in original source) – Event – Sources.

1892 – Chilapata on Torsa River – Nripendra, Maharaja of Cooch Behar kills 1 rhino during shoot of 25 April to 4 May 1892 – 32.9 (shoot 29)

1893 – Chilapata on Torsa River – Nripendra, Maharaja of Cooch Behar kills 1 rhino during shoot of 10–27 April 1893 – 32.9 (shoot 31)

1920–1930 – Jaldapara – Most rhinos lost due to hunting, habitat loss and flood damage – Menon 1995: 30

1929 – Jaldapara – Charles Kenyon Homfray in the Fourth Working Plan of the Reserved Forests of the Buxa Division (1929–30 to 1948–49) suggested to create Jaldapara as a National Park – Guhathakurta 1985

1930–1931 – Torsa River – Period of rampant poaching by Assamese and Mech people – Fawcus 1943; Avari 1957; Roy 1945: 260; Shebbeare & Roy 1948; Meiklejohn 1937: 2

1931 – Torsa River – Rampant poaching first detected when a rhino was found outside the main grass jungle. Poaching continued at least 6 months, shots were exchanged on both sides and arests made and prosecutions instituted before the evil ceased – Fawcus 1943

1932 – Torsa Rhinoceros Sanctuary – Main portion of rhino country along Torsa River was constituted a Game Reserve for the protection of rhino. It is a narrow strip 50 miles from north to south, and at its widest 4 miles from east to west, total perhaps 50–60 square miles – Fawcus 1943; Tyndale 1947; Martin & Martin 1982: 32

1932 – Jaldapara – A search supervised by Thomas Valentine Dent (1909–1970), Assistant Conservator of Forests, led to the discovery of the remains of 40–50 skeletons of rhinos killed by poachers – Fawcus 1943; Roy 1945: 260; Shebbeare & Roy 1948; Bahuguna & Mallick 2010: 287

1932 – Jaldapara – Census indicated presence of 40–50 animals. Since that time stock has been building up – Fawcus 1943

1932 – Torsa River (Jaldapara) – Bengt Magnus Kristoffer Berg (1885–1967) takes photographs and counts rhino from camp near Torsa River – Berg 1933 – 34.2

1932 – Torsa River (Jaldapara) – Bengt Berg measures rhino tracks and takes photos of 9 individual rhinos – Berg 1933: 110 – 34.2

1932-12 – Torsa River – Heavy floods – Menon 1995: 30

1934 – A.N. Roy was made Honorary Game Warden of Jaldapara and he scarcely ever saw a rhino until 1936 – Roy 1945: 260; Shebbeare & Roy 1948; Bahuguna & Mallick 2010: 287

1935 – Torsa River (Jaldapara) – Last stronghold of rhino. Total area 40 × 60 sq.miles – Shebbeare 1935: 1229

© L.C. (KEES) ROOKMAAKER, 2024 | DOI:10.1163/9789004691544_035
This is an open access chapter distributed under the terms of the CC BY-NC-ND 4.0 license.

1936-09-18 – Male rhino calf captured near the Salkumar Forest of Chilapata Range. Died 1936-11-17 – Meiklejohn 1937: 2

1937 – Rhinos found in riparian forests bordering the Torsa and Malangi rivers and wet forests of Panbari – Meiklejohn 1937: 8

1937 – Jalpaiguri Forests – Rhino common (according to Senior Conservator of Forests, Bengal, 1937) – Harper 1945: 377

1937-12 – Torsa River – Heavy floods – Menon 1995: 30

1941-11-13 – Jaldapara Game Sanctuary – West Bengal Notification No. 10549-For, 93 km². Jaldapara constituted through efforts of Edward Oswald Shebbeare (1884–1964) and Thomas Valentine Dent (1909–1970) – Ray 1961: 90; Spillett 1966e: 534; Das 1964; Martin & Martin 1982: 32; West Bengal Forest Dept. 1993: 5

1943-04-03 – Jaldapara Game Sanctuary – Notification No. 5238 on 3 April 1943, amendation of proclamation – Spillett 1966e: 534

1943 – Torsa Region (area II) – Stronghold was a tract along Torsa River of 50–60 sq.miles – Fawcus 1943; Bist 1994

1946 – Jaldapara – Photos by H.E. Tyndale (a tea planter), rhino in Jaldapara, taken from an elephant at about 20 feet – Tyndale 1946; Shebbeare & Roy 1948

1947-05 – Jaldapara – Shebbeare visits for first time since 1938, sees 7 rhino, finds it easier to see rhino – Shebbeare 1948

1948-12 – Torsa River – Heavy floods – Bist 1994: 48; Menon 1995: 30

1954 – Torsa River – Heavy floods. Three rhinos buried – Bist 1994: 48, 1997

1954 – Inspection report by V.S. Rao. Reports rhino in Bhutri and Khairbari – Bist 1994: 46; Bahuguna & Mallick 2010: 287

1955 – Rhino presence in Barobisha and Balapara areas along the Gholani River – Bahuguna & Mallick 2010: 287 (after Annual Report 1954–55)

1963 – Rhino killed by train – Ullrich 1965a

1964 – First rhino census indicating 72 rhino – Das 1964; Bist 1997

1966-04-26 – James Juan Spillett (1932–2018) executed a short survey when 32 counted but at least 50 rhino present – Spillett 1966e: 534

1968–1972 – Period of heavy poaching attributed to organised poaching groups, rather than local inhabitants – Guhathakurta 1985; Menon 1995: 15

1970 – Baradabari area (incl. old tourist lodge) has been taken over by military and needs compensation – Brahmachary et al. 1970; Waller 1972b

1972 – Opening of new tourist lodge at Holong, in the middle of the sanctuary – Waller 1972b; Martin 2006

1972 – An Inquiry Committee appointed in November 1972 by Govt. of West Bengal into the poaching of rhinos at Jaldapara resulted in report with main author P.K. Basu of 1973 – Bist 1997

1974 – Rhino strayed to Patlakhawa 30 km east of Jaldapara until mid 1970s. Last rhino of Patlakhawa was poached (still one more in 1985) – West Bengal Forest Dept. 1993: 7; Bahuguna & Mallick 2010: 287

1976 – Declaration of Jaldapara Wildlife Sanctuary for the purpose of propagating and developing wildlife and its environment under the Wildlife (Protection) Act, 1972 – Guhathakurta 1985

1980 – Visit by Wildlife Status Evaluation Committee of the Indian Board for Wildlife to evaluate the possibilities of translocation of rhinos from another population – Guhathakurta 1985

1980 – Rhino strayed to Titi Forest 12 km north of Jaldapara until early 1980s–West Bengal Forest Dept. 1993: 7

1982 – Jaldapara – Management transferred to Wildlife Division-II under Cooch Behar Forest Division – West Bengal Forest Dept. 1993: 2

1984-06 – Wildlife Advisory Board for West Bengal resolved to enquire about availability of rhinos in Kaziranga with an aim at eventual translocation – Guhathakurta 1985

1984 – Report on immobilization of rhino to remove bullets shot by poachers in Jaldapara – Chowdhuary & Ghosh 1984

1985 – Last rhino in Patlakhawa killed in 1985. Animal strayed from Jaldapara. It was stranded in a mudpool, where teased by teenager. Rhino attacked the boy. Bahuguna visited him in hospital – Bahuguna & Mallick 2010: 287

1992 – Rhinos only present in 80 (out of 216) km², in forest compartments of Malangi, Chllapata, Jaldapara, Torsa, Mondabari and Dania – Bist 1994

1995-10 – One rhino from Assam (Gauhati Zoo) introduced – West Bengal Forest Dept. 1999; Vigne & Martin 1996b

1995 – Jaldapara includes Patlakhawa Reserve – Menon 1995: 30

1996 – Case of poaching using poison – Mukherjee & Sengupta 1998: 20

1996-04-20 – Census using direct sighting method. Results 42 rhino: adults 9.18.2, subadult 0.1.5, calves 9 – Mukherjee & Sengupta 1999

1998-08-27 – Jaldapara Wild Life Sanctuary – Notification No. 2890-For. 11B-13/98, covering 216.51 km², signed by S.M. Chaki, Deputy Secretary to the Govt. of West Bengal – *Calcutta Gazette* 1998-09-01

2007-06-07 – Sighting indicating presence of twin offspring – Kundu & Menon 2016 – fig. 34.8

2012-04-27 – Jaldapara National Park – Notification no: 973-For /FR/O/IIM-44/11 under section 35 of Wildlife (Protection) act. 1972 – park website

2018-03-31 – Jaldapara – Kees Rookmaaker visits the reserve for two days – fig. 34.5, 34.6

••

34.2 The Photographer Bengt Berg

During troubled time in the reserve at the end of the 1920s, Edward Shebbeare was approached in 1929 by the Swedish zoologist, photographer and cinematographer Bengt Magnus Kristoffer Berg (1885–1967). He asked permission to spend time in the jungle, armed with eight cameras and accessories, hoping to capture the tiger, rhinoceros and other wildlife on film. When he arrived in February 1932, he found that the rhino population had suffered much through poaching in the intervening years: "I came in the final hour. The jungle had not changed in the past four years, but it was said that there were no more than two dozen rhinos left in all of Bengal. Nobody knew this for sure, because who would be brave enough to try to count them. But in the last three years, authorities had found 28 skeletons of these giants of the jungle, killed and left after poachers ran off with the horn." Berg decided to focus on an area of 20 km² around his camp on the banks of the Torsa River. He sketched a map of his chosen locality, and took as many photos and movies as he could. After three weeks, when he compared all the images, he was sure that he could identify nine individual rhinos. He then calculated that there would be 20–25 rhinos in Jaldapara, and maybe 40 in the whole of West Bengal (Berg 1933: 118).

Besides this survey, one of the first of its kind, Bengt Berg is still best known for his photography and the books written about his time pursuing the wildlife of the world (Fahlstedt 2017). His black-and-white photographs taken in the jungle in Bengal are stunning, certainly the best of that age (figs. 34.1 to 34.4). They were not the first photographs of a rhino in its natural habitat, but definitely the naturalistic way in which the animals were immortalized

was a great tribute to Berg. Besides several individual rhinos, Berg was able to photograph the actions of a male rhino chasing a female, and a mother rhino followed by her young calf. In total 22 different photos of rhinos were published as a result of his expedition on the Torsa River (table 34.46). A selection of the images was first included, with an explanatory text, in a book published in Swedish in 1932 (Berg 1932). A translation into German by Ilse Meyer-Lüne (b. 1886) followed in the next year (Berg 1933), as well as a Danish edition of 1933 translated by his wife Ilse Illum (1896–1988). A few photographs of the rhino in Bengal appeared in 1935 in two instalments in *The Field* with English explanations (Berg 1935b,c), and others in his later books of 1935 and 1955.

The original negatives or early prints of Berg's photographs have not survived, according to his granddaughter Natasha Illum Berg (email 2017). She followed his footsteps in 'Skypaths', a four-part mini documentary series directed by filmmaker Kire Godal, taking viewers to Sweden, Africa and India (Godal 2020).

TABLE 34.46   Photographs of rhinoceros in different editions of the works of Bengt Berg[a]

No	1932 Swedish	1933 German	1935 Tiger	1935 Field	1955 Abenteuer	Subject
Photographs of specimens of *R. unicornis* in Jaldapara						
1	2	6	1935: 176	1935: 1291	1955: 193	Front view of male
2	12	16				Side view
3	18	22			1955: 186	Side view in stream
4	27	33				Rhino with long horn
5	34	38				Side view from behind
6	86	94				Gera, female rhino
7	95	absent				Head of long horned
8	107	113				Young rhino
9	111	119				Rhino female
10	112	120				As 111, mother of 107
11	119	127				Front view, long horn
12	126	134				Female going into grass
13	133	143	1935: 180			Side view
14	134	144				Front view
15	137	147				Mother with young
16	151	162				Susanna in stream
17	152	absent				Head of rhino
18	175	187				Male rhino
19	176	188		1935: 1471	1955: 34	Male chasing female
20	178	34				Rhino rear
21	absent	161				Side view
22			1935: 183			Rhino one year later

TABLE 34.46   Photographs of rhinoceros in different editions of the works of Bengt Berg[a] (*cont.*)

No	1932 Swedish	1933 German	1935 Tiger	1935 Field	1955 Abenteuer	Subject
Rhino-related subjects						
	45	45				Dead body of rhino
	46	50				Skulls with horns removed by poachers
	85	93				Rhino path in high elephant grass
	108	114				Rhino dung heap
	17	21				E.O. Shebbeare

a   Berg, B. (1932) *På jakt efter Enhörningen*. Stockholm. (1933) *Meine Jagd nach dem Einhorn*. Frankfurt am Main. (1935a) *Tiger und Mensch*. Berlin. (1935b) Article in the *Field* 1935-11-30, 1935-12-21. (1955) *Meine Abenteuer unter Tieren*. Guttersloh.

FIGURE 34.1
Photograph of mother and baby rhinoceros taken by Bengt Berg in Jaldapara in 1932
BERG, *PÅ JAKT EFTER ENHÖRNINGEN*, 1932, P. 137.
COURTESY: NATASHA ILLUM BERG

FIGURE 34.2
Female rhino in Jaldapara, photographed by Bengt Berg in 1932. Notice the head of her young between the front legs
BERG, *PÅ JAKT EFTER ENHÖRNINGEN*, 1932, P. 133.
COURTESY: NATASHA ILLUM BERG

34.3   Tourist Potential in the 1930s and 1940s

Shebbeare visited the reserve in 1938 and again in May 1948. In 1938, the rhino population had not quite recovered from the poaching episode and the animals were still very

FIGURE 34.3   Male rhino with large horn photographed by Bengt Berg in Jaldapara in 1932
BERG, *PÅ JAKT EFTER ENHÖRNINGEN*, 1932, FRONTISPIECE. COURTESY: NATASHA ILLUM BERG

wary. During the latter visit, he found that it was easy to view rhinos in the park because the Malangi, one of the reserve's main rivers, had turned into a large swampy mud wallow in recent years, providing prime habitat for rhinos and opportunities for people to view them. The number of tourists in the area had greatly increased, which meant that the rhinos had become habituated to the presence of elephants and were less likely to be alarmed, especially in open spaces. Shebbeare warned that the stability of the rhino population was less due to conservation efforts than to a temporary drop in the demand for horns in China. He said to "believe that the best protection they can have is visitors and publicity, and if this could be combined with a scheme which would secure them national protection for all time, the sanctuary would have an educative value which it lacks at present" (Shebbeare 1948). Conservation then was still in its infancy.

Yusuf Salahuddin Ahmad (1902–1984), who worked as a forest official in the Buxa Division in the 1930s, said that the forest department's elephants were trained to enter the reserve to attract tourist traffic in the Haribari area. Elephants and rhinos thus got accustomed to each other, which helped tourists get close enough to view wildlife:

> Every time I went to the sanctuary, I would take at least three elephants. The dung heaps, if fresh, would indicate the presence of rhinoceros in the neighbourhood. The Game Guards also knew about their number and location pretty well. At the beginning, I used to go after a single bull rhinoceros, the mother with a calf would be more dangerous. The rhinoceros would see the elephants from a distance

FIGURE 34.4
Male rhino chasing a female in Jaldapara in 1932 taken by Bengt Berg
BERG, *PÅ JAKT EFTER ENHÖRNINGEN*, 1932, P. 176.
COURTESY: NATASHA ILLUM BERG

and the elephants would also hesitate to get too close. Then the rhinoceros would bolt and stop again after running about two hundred yards. I used to surround it with the elephants, making a detour and blocking its escape. It would snort and run back to the original place. In course or time it realised that the elephants were harmless and the elephants also found it good fun to go near the rhinoceros. Before I left Buxa Division, I found them in groups of four or five who would allow themselves to be photographed from the forest department's elephants (Ahmad 1981: 52).

34.4 Management in the 1960s

In 1951, management of the sanctuary was brought under the Cooch Behar Division. In the first 10-year working plan of the division, prepared in 1962–1963, an area of 24,950 acre (100 km^2) was managed to maintain glades and salt-licks, and these same activities were incorporated in a subsequent working plan prepared in 1972–1973. The park was brought under the control of Wildlife Division-II on 10 February 1982. The 1981–1986 management plan prescribed the creation of fodder plantations to supplement food resources for herbivores, to provide saltlicks and wallow pools, to eradicate *Mikania* and to cut back trees that encroached upon grasslands. Commercial tree-felling was prohibited and the infrastructure was strengthened for protection. The plan also had targets for creating habitat corridors between the two portions of the sanctuary,

as well as the reintroduction of swamp deer (*Rucervus duvaucelii*) and wild buffalo (*Bubalus arnee*), while spotted deer (*Axis axis*) were successfully reintroduced in 1982–1983 (N.C. Bahuguna, pers.com.). In 1957, Jaldapara's rhino population had increased to about 50 animals. It was believed that this was close to the carrying capacity of forests along the Torsa River, which provided ideal rhino habitat of tall grasses and reeds, and low-lying swamps that contained favourite foods such as wild ginger and the tender shoots of marsh reeds.

James Juan Spillett conducted a short survey of rhinos on 26 April 1966, counting only 32 rhinos, but actually "the day was very warm and most rhino were observed in or near wallows or streams. During seven days in the field prior to survey, I observed a total of 46 rhinos: 2 non-sexed, 12 adult males, 21 adult females, 11 young. I feel certain that there are at least 50 rhino" (Spillett 1966e).

34.5 Measures to Reduce Poaching

Poaching continued for the next 10 years, including a serious episode in 1971 when 29 animals were killed. This was likely linked with the aftermath of fighting in East Pakistan which in that year became Bangladesh. The seven-bedroom Hollong Forest Lodge was built in 1972–1973 to attract tourists to the area. It was thought that there were only 14–22 rhino remaining in 1980. Jaldapara has a long border which is very hard to patrol, and many villages surround the entire reserve. Poachers commonly used guns to kill the animals, but there was a case in

FIGURE 34.5
Jaldapara National Park signboard
PHOTO: KEES ROOKMAAKER,
MARCH 2018

FIGURE 34.6   Entrance to Jaldapara National Park, with author
               Kees Rookmaaker
               PHOTO: MANMOHAN SINGH, MARCH 2018

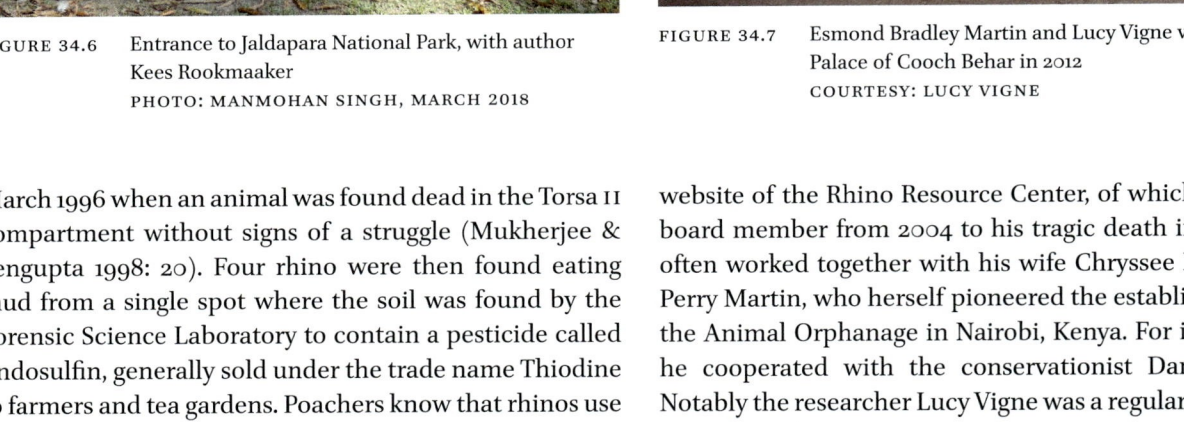

FIGURE 34.7   Esmond Bradley Martin and Lucy Vigne visiting the
               Palace of Cooch Behar in 2012
               COURTESY: LUCY VIGNE

March 1996 when an animal was found dead in the Torsa II compartment without signs of a struggle (Mukherjee & Sengupta 1998: 20). Four rhino were then found eating mud from a single spot where the soil was found by the Forensic Science Laboratory to contain a pesticide called Endosulfin, generally sold under the trade name Thiodine to farmers and tea gardens. Poachers know that rhinos use salt licks and mixed the pesticide with salt and mud.

34.6   Esmond Bradley Martin

Esmond Bradley Martin (1941–2018) is well-known for his meticulous undercover investigations of the trade in elephant ivory and rhino horn, producing a series of baseline reports about the markets in Africa, South and South-East Asia, East Asia, Europe, USA, Vietnam, Hong Kong, Laos, and South Africa (fig. 34.7). His diplomatic efforts to ban the legal sale of rhino horn in Yemen to stop their use for dagger handles in the 1980s and 1990s are legendary (Milliken 2018; Vigne 2018). He published about 260 articles and books, mostly documented on the website of the Rhino Resource Center, of which he was a board member from 2004 to his tragic death in 2018. He often worked together with his wife Chryssee MacCasler Perry Martin, who herself pioneered the establishment of the Animal Orphanage in Nairobi, Kenya. For ivory trade he cooperated with the conservationist Daniel Stiles. Notably the researcher Lucy Vigne was a regular co-author (Vigne 2000).

On the basis of personal visits to Bengal in 1993, 1998, 2005 and 2012, Martin found that low level poaching continued in Jaldapara in the 1990s, mostly perpetrated by West Bengalis, landless Bangladeshi refugees or even Assamese. Gang members would use a muzzle-loader or a more modern rifle to kill a rhino, remove the horn and hooves, and in rare instances (recorded in 1991) also the male reproductive organs, finally burying the carcass to avoid detection. In 1993, a horn fetched $640 to $896 per kg. To improve security, in 1993 the Government of West Bengal employed a total of 105 staff and increased the budget available for conservation activities. Poaching was reduced and the rhino population increased to 34 in 1993 (Martin 1996a, b).

In 1998, Martin and Vigne reported that the grassland areas had been improved and extended, and an electric fence of 80 km (about 75% of the boundary) prevented rhinos from wandering into neighbouring farmland. There was an increase in community eco-development projects and an improvement in the anti-poaching communications network (Martin 1999).

34.7   Noteworthy Developments in Jaldapara

An anthrax infection was detected among elephants in Jaldapara in February 1994 (Pandit & Sinha 2006). To prevent spread of the disease, a rhino vaccination programme was implemented and in one week (10–16 February 1994) 34 rhino were individually identified in the park. Of these, 24 rhinos were vaccinated with live anthrax vaccine. Since then, no incidence of anthrax has been recorded in any of Jaldapara's wildlife species.

The rhino population was believed to stand at 96 animals in 2004. This is a rather large increase from the estimate of 54 animals just five years earlier, which brings into question the reliability of one or both censuses. Poaching at the time was negligible. One animal was killed in 1996 after it wandered out of the sanctuary and in 2000 a horn was stolen from a rhino that had died naturally in the Torsa River. In both cases, the poachers or thieves were arrested and the horns recovered.

On 6 October 2009, sanctuary staff spotted an adult male rhino about 25 years old on an island in the Torsa river, standing still with head hung low for hours. Two days later the animal managed to reach the river bank and attempts were made to drive him away from the water. Blood was observed oozing from a wound on his snout. The people provided unripe paddy (relished by rhinos), water, salt and grass. It took some food but could not chew, and smelled the water but did not drink. Its left lower jaw showed swelling. It neither urinated nor defecated during the period. Next morning, the tranquillization team administered Immobilon, and the veterinarians cleaned and dressed the wound. Even after being treated, the rhino remained restless and refused to eat or drink. Around noon, as it reached the river bank, the ground beneath its hind legs gave way and its hind quarters lurched towards the river. The staff immediately placed ropes around the hind quarters, hoping to help it climb the bank. The staff managed to hold the animal for about 20 minutes, when the ground below it gave way and the rhino was pulled down the fast flowing Torsa. In the water, it managed to keep its head above water for about five minutes, but ultimately drowned. The body was retrieved 300 m downstream. Post-mortem findings revealed a bullet lodged in the left lower mandible. This was the last attempt at rhino poaching documented in West Bengal (N.C. Bahuguna, pers. comm).

The rhino population has been increasing in Jaldapara recently and there are now well over 200 animals in the park, more than have been recorded ever before (Table 34.47). This may well be close to carrying capacity. With improved facilities, it could become a less daunting destination which might be enjoyed by local as well as international tourists.

FIGURE 34.8
Mother *R. unicornis* in Jaldapara accompanied by two young rhinos of similar age, indicating the possibility of a twin birth
PHOTO: SUMANTA KUNDU, 2007

TABLE 34.47   Status of *R. unicornis* population in Jaldapara NP from 1865 to 2022

Year	Number	Natural death	Poached	Source
1865–66	75			R191
1920	12; 200			R079; R055, R164, R172, R184
1929	80; 200			R197, R264; R306, R309
1930	80	4		R055, R164
1931			50	R055, R156, R164, R197, R309
1932	40–50		3	R055, R156, R164, R172, R184, R309; R044
1936	good number	1		R042, R055, R164
1937	56	1	0	R055, R164, R184
1941	increasing	2		R055, R164
1942	25			R305
1945	60			R197
1947	60			R103, R264
1948	60			R055, R184
1949	60+	2	0	R055, R164
1950	60	4	0	R164, R184
1951	increasing	1		R055, R164
1952	30			R055, R106, R164
1953	10; 30			R204; R055
1954	30–56			R034, R055, R156, R164, R184
1955	promising			R055
1956		2	2	R156, R164
1957	50; 57; 65			R055, R156, R184; R041; R191, R198, R291
1958	40; 50	2	0	R108; R164
1959	65			R055, R156
1960	70			R172
1961	72			R192
1963	30; 60; 80			R226; R207; R267
1964	72			R041, R055, R074, R156, R184, R191, R198, R221, R286, R291
1965	75			R221
1966	50; 75			R221, R222; R041, R055, R156, R172, R184, R198
1967	76			R055
1968	76	1	0	R164
1968–72		12	28	R055, R164
1969	75; 80			R055, R156, R184; R191, R041, R198
1971	60		29	R010, R282
1972	74			R281
1973	21	0	6	R156, R164
1973–80	23	4	5	R164
1974	21			R055, R156, R184
1975	23			R041, R055, R064, R149, R156, R172, R184, R191, R198, R286, R291, R332
1978	19		1	R041, R055, R149, R156, R184, R191, R198, R286, R291, R332
1980	22		2	R041, R055, R064, R149, R156, R172, R184, R191, R198, R286, R291, R332
1981	22; 25	1	1	R064, R156, R164, R332
1982			3	R156, R164
1983		3	1	R156, R164

TABLE 34.47   Status of *R. unicornis* population in Jaldapara NP from 1865 to 2022 (*cont.*)

Year	Number	Natural death	Poached	Source
1984		4	2	R156, R164
1985	14		2	R041, R156, R164, R191, R198, R286, R291
1986	14	1		R034, R055, R156, R161, R164, R184
1987		1		R164
1988	24	1		R041, R055, R149, R156, R164, R184, R191, R198
1989	27			R041, R055, R149, R156, R164, R184, R191, R198, R286, R291
1991			1	R156, R164
1992	33	1	1	R041, R055, R149, R156, R164, R184, R191, R198, R286, R291
1993	33 +			R098
1993	29; 34		1	R144, R149; R137, R156, R164, R274, R275
1994	33–35			R041, R079, R097, R183, R184
1995	35			R097, R098, R161, R164
1996	42			R041, R149, R161, R173, R184, R286
1997	44			R161
1998	42; 49; 55			R172; R286; (c) R041, R161, R184
1999	55			R161
2002	74; 84			R149, R184; R041, R054, R161, R253
2004	96			R041, R054, R149, R161, R253
2005	100; 108	3		R052; R054, R161, R229
2006	108	2		R041, R052, R149, R253, R288
2007		4		R052
2008	108	2		R052, R244
2009	125	4	1	R052, R149, R288
2010	155	2	1	R041, R052
2011	149	3		R052, R153, R177, R288
2012	186	7		R041, R052, R177
2013	186	8		R052, R149, R288
2014		6	1	R052
2015	204			R041, R288
2016	200			R072
2019	237			R041
2020				
2021				
2022	287–292			R329, R331 (census 2022–03)

References for table: R010 Anon. 1972; R034 Bahuguna & Mallick 2010: 287; R041 Bengal Chief Conservator of Forests 2019; R042 Meiklejohn 1936: 2; R044 Berg 1933: 118; R052 Bhattacharya 2015; R054 Bhutia 2008; R055 Bist 1994; R064 Choudhury 1985a; R072 Choudhury 2016: 154; R074 Das 1964; R079 Dey 1994a; R097 Foose & van Strien 1995: 10; R098 Foose & van Strien 1997: 9; R103 Gee 1950a: 1732; R106 Gee 1952a: 224; R108 Gee 1958: 353; R137 Khan & Foose 1994: 6; R144 Lahan 1993: 18; R149 Mallick 2015: 336; R153 Martin & Vigne 2012; R156 Martin 1996a; R161 Martin 2006; R164 Menon 1995: 15; R172 Mukherjee & Sengupta 1998: 20; R173 Mukherjee & Sengupta 1999; R177 Nandi 2013; R183 Pandit & Sinha 2006; R184 Pandit 2004: 71; R191 Raha 2000; R192 Ray 1961: 92; R197 Roy 1945: 260; R198 Roy 1993; R204 Shebbeare 1953: 143; R207 Simon 1966; R221 Spillett 1966c; R221 Spillett 1966e: 534; R226 Stracey 1963: 243; R229 Strien & Talukdar 2007; R244 Talukdar 2009c; R253 Talukdar et al 2011; R264 Tyndale 1947; R267 Ullrich 1965a: 101; R274 Vigne & Martin 1994; R281 Waller 1972b: 580; R282 Waller 1972a: 6; R286 West Bengal Conservator of Forests 1999; R288 West Bengal Forest Dept. 2016; R291 West Bengal Forest Dept. 1993: 20; R305 Ahmad 1981: 53; R306 Shebbeare 1935; R309 Fawcus 1943; R329 Sharma 2022; R331 Talukdar 2022; R332 Guhathakurta 1985.

CHAPTER 35

Historical Records of the Rhinoceros in Bhutan

First Record: 1934 – Last Record: 2017 (strays only)
Species: *Rhinoceros unicornis*
Rhino Region 27

35.1 Rhinos in Bhutan

Bhutan is a small country nestled in the Himalayan mountains. With its renowned rugged terrain, there is unlikely to be much habitat suitable for rhinos, especially *R. unicornis*. On the southern border with India, however, there are some low-lying patches which could have attracted rhinos, mainly in the region of the current Royal Manas National Park. When studying the old records, it must be recalled that the "Bhutan Duars" (or Dooars) only became part of British India after the Duar War of 1864–65 (31.1). For instance, when Robert Cecil Beavan in 1865 made an excursion into territory which he called Bhutan, he was still in the area of current Cooch Behar in North Bengal. The larger territory of "Bhotan" (Bootan, Bhootan, Bhutan) is clearly shown on a map of 1851 edited by John Tallis encompassing parts of North Bengal, Assam north of the Brahmaputra as well as most of Arunachal Pradesh (fig. 35.1). This chapter only explores the territory of the current country of Bhutan.

•••

Dataset 35.40: Chronological List of Records of Rhinoceros in Bhutan

General sequence: Date – Locality (as in original source) – Event – Source – § number.

Rhino Region 27: Bhutan

1850 – Territory of Butan [maybe Bhutan Duars] – Rhino in forest – Anon. 1850a: 411

1905 – Chagla, part of Tsangpo oasis, Tibet – Habitat suitable for rhino – Waddell 1905: 315

1930 – Bhutan [but probably North Bengal] – Carl Hagenbeck (1844–1913) heard that rhinos had come to Bengal after fires in Bhutan – Berg 1933: 20

1934 – North of Manas River – Rhinos move north from Manas (India) in dry season. No poaching detected – Bhadran 1934; Milroy 1935: 19

1949 – Manas, Bhutan (?) – Frank Ludlow (1885–1972) photographed Raja Sonam Topgay Dorji (1896–1953), Prime Minister of Bhutan, standing on dead rhino at "Manas / Bhutan" – British Library, Frank Ludlow Collection, Photo 743/6(1–19). Locality not clear, could be Manas NP in Assam – fig. 35.4

1966 – Manas Wildlife Sanctuary gazetted in 1966 – Tshering et al. 2017

1993 – Manas WS merged with Namgyal Wangchuk WS – Tshering et al. 2017

1993 – Royal Manas National Park established, covering 1057 km^2 – Tshering et al. 2017

1985 – Royal Manas NP, Bhutan part – Rhino present – Choudhury 1985a

1992-02 – Royal Manas NP – Last sighting of rhino in the park in February 1992, mentioned in a diary of one of the retired foresters, according to Senior Park Ranger Dorji Wangchuk – *Bhutan News* 2018

1993 – Royal Manas NP – Rhinos move between Manas in India and Bhutan – Lahan 1993: 34; Vigne & Martin 1994 – fig. 35.5

2005 – Royal Manas NP – Rhino in Bhutan adjacent to Manas Tiger Reserve – Sinha 2005

2006 – Royal Manas NP – Reports of 10–20 rhino in safe riparian zone – Bezbarua 2008: 32

2013 – Royal Manas NP – Wanderers from India – Choudhury 2013: 225, 2016: 154

2017 – Royal Manas NP – *R. unicornis* listed in checklist – Tshering et al. 2017

2018-05 – Royal Manas NP – Rhino spotted on camera trap – *Bhutan News* 2018

•••

Dataset 35.41: Localities of Records of Rhinoceros in Bhutan

The numbers and places are found on map 38.20 (West Assam)

Rhino Region 27: Bhutan

A 385   Royal Manas NP – 26.80N; 90.80E-1985ff. – see map 38.20
*   Chagla, Tibet – 29.30N; 89.60E-1905 – unlikely record
* Not on map.

© L.C. (KEES) ROOKMAAKER, 2024 | DOI:10.1163/9789004691544_036
This is an open access chapter distributed under the terms of the CC BY-NC-ND 4.0 license.

FIGURE 35.1   Detail of the map of "Northern India" in an atlas by John Tallis (1817–1876). Published in 1851, it shows the historical
extent of "Bhootan" beyond the current borders of Bhutan
TALLIS, *ILLUSTRATED ATLAS*, 1851, MAP 32

35.2   Rhinoceros in the Culture of Bhutan

The rhinoceros has played only a marginal role in the culture of the Bhutanese people. Recently, a series of postal stamps issued in 1999 show the current link of culture and biodiversity (fig. 35.2).

Shields made from rhino hide originating from Bhutan are occasionally recorded, like one with a diameter of 141 cm from the 18th century in the National Museum of Bhutan (NMB 2001). Another undated example was photographed in 2013 on the walls of a temple in Trashigang

(Leaming 2013). As the raw material could be obtained easily from bordering regions, the significance of these shields is hard to judge. I have not seen a reference to their use in ceremonial or combat settings within Bhutan itself.

Carvings of rhino horn from the Himalayan regions are rare. These were occasional pieces rather than part of an ongoing industry. In her detailed study of rhino horn carvings, the art-historian Jan Chapman (b. 1937) listed one (undated) example from Bhutan which was once presented to the Tango Monastery by the monarchy. This horn (size 4 cm high, 5.5 cm wide) was donated to the Government of Switzerland in 1968 by Ashi Kesang Choden (b. 1930), wife of the 3rd Druk Gyalpo (King) of Bhutan, and is preserved in the Rietberg Museum in Zurich, Switzerland (Chapman 1999, fig. 39.9). An early series of five rhino horn statues possibly carved in Tibet by the tenth Karmapa, Chos Ying Dorje (1604–1674) are now preserved in the Rumtek monastery, Sikkim (Douglas 1976). A carved rhino horn associated with the Kagyu lineage of Tibetan Buddhism, presumed to be from Bhutan, shows figures of holy men in a cosmic mountain setting (Sotheby 2007), and a similar carving from

the Alsdorf Collection is preserved in the Art Institute in Chicago (Pal 1997: 344 and colour plate p. 228) (fig. 35.3).

35.3 Stray Rhinos Crossing the Manas River into Bhutan

Without doubt, *R. unicornis* is sometimes present on Bhutanese soil. Assam's Manas National Park is separated from Bhutan's Royal Manas National Park by the Manas River, which is quite shallow for large parts of the year

FIGURE 35.2
"Animals of the Himalayas", set of stamps issued by the Kingdom of Bhutan in 1999. This includes a stamp with value 10 Ngultrum with head of *R. unicornis*, 25 Ngultrum with *R. unicornis* standing in river, and 100 Ngultrum showing a white rhino (*Ceratotherium simum*)
COLLECTION KEES ROOKMAAKER

FIGURE 35.3
Rhino horn shaped like a cosmic mountain showing several figures, one of which is Shabdrung Ngawang Namgual, who arrived from Tibet in 1616, and 35 years later created the state of Bhutan. Size 20.5 × 9.8 × 21.5 cm. Since 1974 In the collection of James W. Alsdorf (1913–1990) and Marilyn B. Alsdorf (1925–2019)
ART INSTITUTE OF CHICAGO: NO. 2021.253

FIGURE 35.4
Raja Sonam Topgay Dorji (1896–1953), Prime Minister of Bhutan, standing on dead rhino at "Manas / Bhutan." Photograph taken around 1949 by Frank Ludlow (1885–1972)
WITH PERMISSION © BRITISH LIBRARY: FRANK LUDLOW COLLECTION, PHOTO 743/6

FIGURE 35.5
View of the Royal Manas National Park from the Indian side of the river, showing the shallow waters where rhinos can cross
PHOTO: KEES ROOKMAAKER, FEBRUARY 2013

(fig. 35.5). Rhinos cross easily, attracted by a supply of food and by saltlicks. However, the habitat does not support a permanent population. Although exact data could not be retrieved, it is said that historically this area was maintained as an unofficial royal hunting reserve (Green 1993; Riley & Riley 2005), thereby providing some protection to the wildlife. The Manas Wildlife Sanctuary was gazetted in 1966, established in 1993 as the Royal Manas National Park merging with the Namgyal Wangchuk Reserve, covering 1057 km². A submission was made on 8 March 2012 to UNESCO to add Royal Manas to the list of World Heritage Sites, based on its scenic beauty and abundance of rare animals. This has not taken effect as of 2021.

The Chief Conservator of the Indian Forest Service in the 1930s, Chakrapani Ayyangar Rama Bhadran (1907–1977) was one of the first to notice the wandering of rhinos across the Manas River, especially in the dry season (Bhadran 1934). In most cases the animals returned southwards within a day. In 1992–1993, park staff noticed a decline in these movements, but there have never been reports of carcasses found in Bhutan (Vigne &

Martin 1994). When poaching increased in India and the rhino population in Manas National Park declined, some animals might have found refuge in Bhutan, as in 2006 there was a report, yet unverified, of the presence of 10–20 rhinos in Royal Manas (Bezbarua 2008: 32).

35.4 Protection and Illegal Trade

The rhinoceros is protected in Bhutan through the *Forest and Nature Conservation Act of Bhutan* (1995), revised as the *Forest and Nature Conservation Rules of Bhutan* (2006). This provides for fines to be imposed when a rhino is killed or when a trophy is sold. National Park staff are vigilant and no cases of rhino poaching have been officially recorded on the Bhutan side of the border.

Traders in Bhutan have been implicated in the trade of rhino horns from India to the Far East. From the mid-1980s until at least the end of the 1990s, merchants in Phuentsholing bought rhino horns. The town is on the Indian border, close to both Jaldapara and Manas, and was therefore a suitable gateway. It is likely that the horns were taken to Timphu and then on to China. A famous case was the interception on 17 September 1993 of 22 Indian rhino horns in Taiwan's Chiang Kai-Shek airport. The horns were imported by the Bhutanese Princess Deki Choden Wangchuck (b. 1946), the daughter of King Jigme Wangchuck (1905–1952). She claimed that she had obtained them from Indian traders coming to Bhutan, not through direct negotiations with poachers (Kumar 1993).

35.5   Species of Rhinos in Bhutan

While *R. unicornis* is sometimes seen near the southern border with India, the other two rhino species cannot be counted among the nation's fauna. References to the historical occurrence of *R. sondaicus* and *D. sumatrensis* are all recent (table 35.48). These notices in the literature are erroneous, based on confusion with the Bhutan Duars or incorrect interpretation of older sources (Rookmaaker 2016). There is no continuous range of either species through Arunachal Pradesh where no rhino records are known in the west, while the occurrence of these species in North Bengal and North-West Assam remains insufficiently known (52.1, 64.1). It needs no further emphasis that there is no evidence in published hunting records or travel journals, nor any known specimens, to attest to the presence of *R. sondaicus* or *D. sumatrensis* within the current borders of Bhutan.

TABLE 35.48   References to occurrence of different rhinoceros species in Bhutan

	R. unicornis	*R. sondaicus*	*D. sumatrensis*
Sterndale 1886, pl. 9	x		
Lydekker 1894, vol. 2: 470			x
Lydekker 1897, vol. 2: 1057			x
Lydekker 1900: 29			x
Clark 1931, vol. 2: 5	x		
Hobley & Shebbeare 1932: 20	x		
Bhadran 1934: 802	x		
Pocock 1936: 706	x		x
Mukherjee 1966: 29	x		
Singh 1972: 175	x		
Rookmaaker 1980: 255	x	x	
Mukherjee 1982: 56			x
Khan 1989: 2 (map)	x		
Dorji 1989: 9	x		
Corbet & Hill 1992: 243		x	
West Bengal Forest Directorate 1993: 1	x		
Vigne & Martin 1994	x		
Baillie & Groombridge 1996: 19	x		
Foose & van Strien 1997: 13		East	South
Mukherjee & Sengupta 1998: 20	x		
Kemf & van Strien 2002: 6	South		x
Guldenschuh 2002	x		
Bahuguna & Mallick 2004	x		
Amin et al. 2006: 114			x
Srinivasulu & Srinivasulu 2012: 350	East		x
Choudhury 2013: 225	x		
Sharma et al. 2013	x		x
Havmøller et al. 2015: 356 (map)			South-East

TABLE 35.48   References to occurrence of different rhinoceros species in Bhutan (*cont.*)

	R. unicornis	*R. sondaicus*	*D. sumatrensis*
Choudhury 2016: 155	x	x	
Rookmaaker 2016	x		
Tshering et al. 2017: 91	x		
Rookmaaker, Sharma et al. 2017	x		

x means presence of that species recorded in the source.

CHAPTER 36

Historical Records of the Rhinoceros in Western Arunachal Pradesh

First Record: 1952 (strays) – Last Record: 2007 (strays)
Species: *Rhinoceros unicornis*
Rhino Region 28

36.1 *R. unicornis* in Arunachal Pradesh

The current state of Arunachal Pradesh was separated in 1987 from Assam on the southern border and is divided into 15 districts. As part of the Eastern Himalayas, the region is mostly hilly or mountainous. This chapter will deal with the largest section of Arunachal Pradesh on the northern side of the Brahmaputra River. There are no early reports at all for this part of the state, probably partly because the region was inaccessible and less explored.

However, if a resident population of rhinos had been present in historical times, there would at least have been suggestions of this. There is a single possible sighting in the eastern part from the Mishmi Hills (Dibang Valley) in 1952, with debatable identity, of an animal possibly straying from Sadiya about 80 km away (63.6). Rhinos once inhabiting the Sonai Rupai Sanctuary in Assam might have crossed into Arunachal territory before the population went extinct (41.1). The fauna of Arunachal Pradesh has been reviewed recently by the naturalist Anwaruddin Choudhury (b. 1959) in a series of papers and a monograph of 2003. All records from 1985 onwards pertain to *R. unicornis* straying northwards from adjacent parts of Assam.

•••

Dataset 36.42: Chronological List of Records of Rhinoceros in West and Central Arunachal Pradesh

General sequence: Date – Locality (as in original source) – Event (species identifications according to original source) – Sources – Map reference – § no. – map number. The map numbers are explained in dataset 36.43.

Rhino Region 28: Arunachal Pradesh (*Western Part*)

1952 – Mishmi Hills – Sighting of a two-horned rhino, but when shown photos, identified as *R. unicornis* – 63.6 – map B 396

1985 – Kameng Dt. – Rhino often seen – Choudhury 1985a–map B 391

1985 – Arunachal Pradesh – *R. unicornis*. Lower hills. Strays from Assam – Negi 1985a: 15; Alfred et al. 2006: 408

1987 – Panir RF – *R. unicornis* from Kakoi RF – Choudhury 1996, 1997a – map B 394

1990 – Pakhui (Pakke) WS – Stray from Assam via Nameri – Choudhury 1997a, 2003: 65, 2013: 225 – map B 392

1990 – Papum RF – 2 strays from Kaziranga – Choudhury 1997, 2003: 65, 2013: 225 – map B 393

1995 – Drupong RF – *R. unicornis* from Kakoi RF and from Narayanpur, Assam – Choudhury 1996, 1997a, 2003: 65, 2013: 225 – map B 394

2000 – Panir RF – Stray from Assam – Choudhury 2003: 65, 2013: 225 – map B 394

2007 – Doimukh RF, Kimin – Stray from Assam – Choudhury 2013: 225 – map B 395

Species in western Arunachal Pradesh stated as *R. sondaicus*: Mukherjee 1966; Kemf 1994; Foose & van Strien 1997: 11; WWF 2002; Amin et al. 2006; Schenkel et al. 2007; Divyabhanusinh 2018c.

Species in western Arunachal Pradesh stated as *R. unicornis*: Fisher et al. 1969: 111; Negi 1985: 15; Alfred et al. 2006: 408; Choudhury 1997a, 2003: 65, 2013: 225, 2016: 155;

Species in western Arunachal Pradesh stated as *D. sumatrensis*: Corbet & Hill 1992; Kemf & van Strien 2002; Amin et al. 2006; Havmøller et al. 2015.

•••

Dataset 36.43: Localities of Records of Rhinoceros in West and Central Arunachal Pradesh

The numbers and places are shown on maps 38.19 and 38.20.

Rhino Region 28: Arunachal Pradesh

B 391   Kameng Dt. – 27.00N; 92.50E-1985

B 392   Pakke WS, Pakhui WS, East Kameng Dt.– 27.05N; 92.82E-1990, 2009

© L.C. (KEES) ROOKMAAKER, 2024 | DOI:10.1163/9789004691544_037
This is an open access chapter distributed under the terms of the CC BY-NC-ND 4.0 license.

B 393 Papum RF, East Kameng Dt. – 27.09-93.50E-1990

B 394 Panir RF, Drupong RF, Papum Pare Dt. – 27.33N; 93.88E-1987, 1995, 2000

B 395 Doimukh RF, Papum Pare Dt. – 27.31N; 93.96E-2007

B 396 Mishmi Hills – 28.30N; 95.75E-1952

FIGURE 36.1
Rhinoceros in a forest. Oil painting by the German artist Christian Wilhelm Karl Kehrer (1770–1869), who worked most of his life for the Counts of Erbach. Dated circa 1816, size 80 × 114 cm. Sold at Sotheby's New York on 24 January 2002
WITH PERMISSION © WOLFGANG FUHRMANNEK, HESSISCHES LANDESMUSEUM DARMSTADT: GK 1419

FIGURE 36.2
The fight between rhino and elephant was a common theme in the popular press. This example of 1876 was drawn by the French illustrator William Henry Freeman (fl. 1839–1875) and engraved by Alfred Sargent (b. 1828). Published by Adrien Linden (1825–1904) in a work of natural history for children
LINDEN, *LE PETIT BUFFON ILLUSTRÉ DES ENFANTS*, 1876, FRONTISPIECE

36.2 *R. sondaicus* in Arunachal Pradesh

The single-horned *R. sondaicus* is unknown in Arunachal Pradesh. When connecting the records of this species in North Bengal with those elsewhere, the state has mistakenly been included in some maps of historical distribution. There is no evidence to support this.

36.3 *D. sumatrensis* in Arunachal Pradesh

The two-horned *D. sumatrensis* was known from the eastern parts (Tirap and Changlang Districts) formerly known as the Tirap Frontier Tract bordering Myanmar, Nagaland and Assam (63.1). There is no evidence that it ever occurred in the remaining parts of Arunachal Pradesh from Bhutan eastwards to Dihing, even though the hilly habitat may appear suitable. The existence of *D. sumatrensis* here has been inadvertently introduced on some recent maps of the historical distribution of the species and stands in need of correction (Rookmaaker 2016).

Historical Records of the Rhinoceros in Meghalaya

First Record: 1837 – Last Record: 1898
Species: *Rhinoceros unicornis*
Rhino Region 29

37.1   Rhinos in Meghalaya

The hill state of Meghalaya on the southern side of the Brahmaputra valley was in 1972 separated from Assam. Traditionally, there were always three regions, from west to east, known as the Garo, Khasi and Jaintia Hills. The distance between the prime rhino habitat along the Brahmaputra River and the current state border is often less than 30 km, hence some dispersal might be expected.

The absence of regular and persistent reports from the hills shows rather convincingly the dislike of *R. unicornis* for these types of hills and forests. The species was recorded on a few occasions during the 19th century, especially in the west on the edge of the valley of the Brahmaputra (table 37.49).

In his discussion of rhinos in the Garo Hills, Edward Blyth appears to provide proof that both *R. unicornis* and *R. sondaicus* were found (30.2, 37.06). He then extends this range for both these species all across the Khasi and Jaintia Hills. This extrapolation by Blyth was generally followed by subsequent authors, confident therefore of the presence of two species throughout Meghalaya. However, there is no evidence substantiating this at any time.

• • •

Dataset 37.44: Chronological List of Records of Rhinoceros in Meghalaya (Garo, Khasi, Jaintia Hills)

General sequence: Date – Locality (as in original source) – Event – Sources – § number – Map number.

Species identification follows original source (names updated). Persons involved in rhino events are listed in table 37.49. The map numbers are explained in dataset 37.45 and shown on map 37.18. The records are divided in the three traditional regions from west to east: Garo Hills, Khasi Hills and Jaintia Hills.

Rhino Region 29: Meghalaya – Garo, Garrow Hills

1837 – Garrow Hills: Gozgotto – *R. sondaicus*. Without discussion, Blyth 1870: 146, 1872e: 3105 puts this record in the Garrow Hills, while his source (Gallovidian 1837) actually refers to Gozgotto close to Rangpur, Bangladesh. Blyth referred the male rhino killed to *R. sondaicus* on account of its small size, 145 cm height. He made a mistake which has never been corrected – 37.6

1864-03 – Tickri-killa (Tikrikilla) – Charles Richard Cock (1838–1879) in company of Abraham Charles Bunbury (1823–1904) kills total 29 rhinos – Cock 1865 – 37.7 – map A 386

1864-03-17 – Tickri-killa (Tikrikilla) – C.R. Cock and A.C. Bunbury (1823–1904) saw rhino in saltlick – Cock 1865 – 37.7 – map A 386

1864-03-19 – Tickri-killa (Tikrikilla) – C.R. Cock shot female, captured young one – Cock 1865 – 37.7 – map A 386

1864 – Garo Hills – *R. sondaicus*. C.R. Cock saw both *R. sondaicus* and *D. sumatrensis* – Cockburn 1883: 64 – 37.7

1867 – Hills south of Berrampooter river – *R. unicornis* present – Jerdon 1867: 233 – map A 386

1868-05 – Tikri Killah (Tikrikilla) – Fitzwilliam Thomas Pollok (1832–1909) killed 2 rhinos (7, 8) on 1 and 2 May, one claimed to be *R. sondaicus* – Pollok 1879, vol. 2: 111 – 52.3, 39.4 (trip 07) – map A 386

1868-05-01 – Tikri Killah – Augustus Kirkwood Comber (1828–1895), Commissioner of Lakhipur, killed a female rhino – Pollok 1868 (part 5), 1879, vol. 2: 111–112, 1900: 454 – 39.4 – map A 386

1869-03 – Tikrikillah – Pollok kills rhinos 9 and 10 – 39.4 (trip 11) – map A 386

1870 – Garrow Hills – *R. unicornis*. Pair killed – Blyth 1870: 144

1870 – Garrow Hills – *R. unicornis* present – Blyth 1870: 144, 1872a, 1875: 50

1870 – Garrow Hills – *R. sondaicus* present – Blyth 1870: 144, 1872a, 1875: 50

1872 – Valley of Assam to the hills immediately on its southern border – *R. unicornis*, maybe co-existing with *R. sondaicus* and *D. sumatrensis* – Blyth 1872e: 3106

1878 – Garo Hills – George Peress Sanderson (1848–1892) recorded presence of *R. unicornis*. Skull, skeleton in NHM 1884.1.22.1(-3) – Sanderson 1878: 234; Pocock 1944b: 838; Barua & Das 1969: 4 – map A 386

1878 – Marshes at the feet of the hills south of Assam – *R. sondaicus* present – Brehm 1891, vol. 3: 102

1879 – Garo hills – Rhino present – Hunter 1879b, vol. 2: 145, 1909: 501

1879 – Brahmaputra valley, south of Goalpara – *R. sondaicus* is plentiful with its larger congener (*R. unicornis*) on the south bank – Pollok 1879, vol. 1: 94; Mukherjee 1966: 27, 1982: 54

1879 – Garrow Hills – *R. sondaicus*. Lesser rhino found – Pollok 1879, vol. 1: 98; Sterndale 1884: 410; T. Bhattacharya & Chakraborty 1984

1879 – Garrow hills, swamps along the base – *R. unicornis*. Presence of greater one-horned rhino – Pollok 1879, vol. 1: 94

1879 – Garo Hills – *D. sumatrensis* [mistake for *R. sondaicus*], historically (after Pollok 1879) – Rookmaaker 2003; Choudhury 2013: 225, 2016: 155

© L.C. (KEES) ROOKMAAKER, 2024 | DOI:10.1163/9789004691544_038

This is an open access chapter distributed under the terms of the CC BY-NC-ND 4.0 license.

1884 – Foot of Garo Hills – Allan Octavian Hume (1829–1912), specimen with 10¾ in (27.3 cm) horn length – Rowland Ward 1899: 433 – map A 386

1886 – Garo hills – Prince Louis Josef Jérôme Napoleon Bonaparte (1864–1932) and Count Mario Michela killed one rhino during the *Khedda* camp organised by George Peress Sanderson. He was accompanied by the staff officers Thomas Fourness Wilson (1819–1886) and Major Thomas Deane (1841–1897) – *Englishman's Overland Mail* 1885-12-29, 1886-01-12; *Figaro* 1886-11-24, 1886-02-18 – 37.9 – map A 386

1900 – Garo Hills – Rhino found occasionally – Hunter 1909: 501

Rhino Region 29: Meghalaya – Khasi, Cossyah Hills

1870 – Khasya hills – *R. unicornis* and *R. sondaicus* extend from Garo hills – Blyth 1870: 144

1872 – Khasi (Cossyah) – *D. sumatrensis*, historical occurrence. This listing of *D. sumatrensis* for Cossyah Hills in Rookmaaker (2003) is erroneous. It was based on a reference in Anderson (1872) of a rhino at Charyolah, which locality has now been identified as Chargola in Karimganj, lower Assam – Rookmaaker 2003; Choudhury 2013: 225, 2016: 155 – 60.2

1872 – Khasia (or Cossiah) hills – *R. sondaicus* is prevalent species – Blyth 1872b

1879 – Khasi hills – Rhino present – Hunter 1879b, vol. 2: 214

1879 – Cossyah hills, swamps along the base – Presence of *R. unicornis* – Pollok 1879, vol. 1: 94; Sterndale 1884: 410

1883 – Khassia hills – Charles Richard Cock (1838–1879) had seen both *R. sondaicus* and *D. sumatrensis* – Cockburn 1883: 64 – 37.7

1898 – Jyrung (Jirang), about 50 km south of Gauhati – Rhino tolerably abundant in a low-lying belt of swamp – Synteng 1898 – 37.10 – map A 387

1914 – Khasis – In former days shields of rhino hide are said to have been used – Gurdon 1914: 24

Rhino Region 29: Meghalaya – Jaintia, Jhyntea Hills

17th cent. – Tongseng Nongkhlieh – U Moo-Kyndad (Image of the Rhinoceros), carving on big slab of rock – Bareh 1964: 58 – 37.5 – map M 871

1870 – Jhyntea hills – *R. unicornis* and *R. sondaicus* extend from Garo hills – Blyth 1870: 144

1879 – Jaintia hills – Presence of rhino – Hunter 1879b, vol. 2: 214

1879 – Jaintia hills: from Jynteapoor (Jaintiapur) to Nowgong (Nagaon) – Rhino met with – Pollok 1879, vol. 2: 77 – map A 388

1894 – Jaintrapuri jungle "in Sylhet" – Refers to Pollok (1879) record of journey to Nowgong: rhinos of larger species, exterminated 36 years ago – Wood 1930: 68

1970s – Northern Jaintia Hills – Dead rhino "few years ago", stray from Nagaon Dt. – Choudhury 1985a, 2013: 225

1975-03-14 – Border of Mikir Hills and Jaintia Hills – Rhino reportedly died – Report in Khasi weekly *U Naphang* of 14 March 1975 with title 'Ka Rhino Kynthei Bapli' – map A 388

• • •

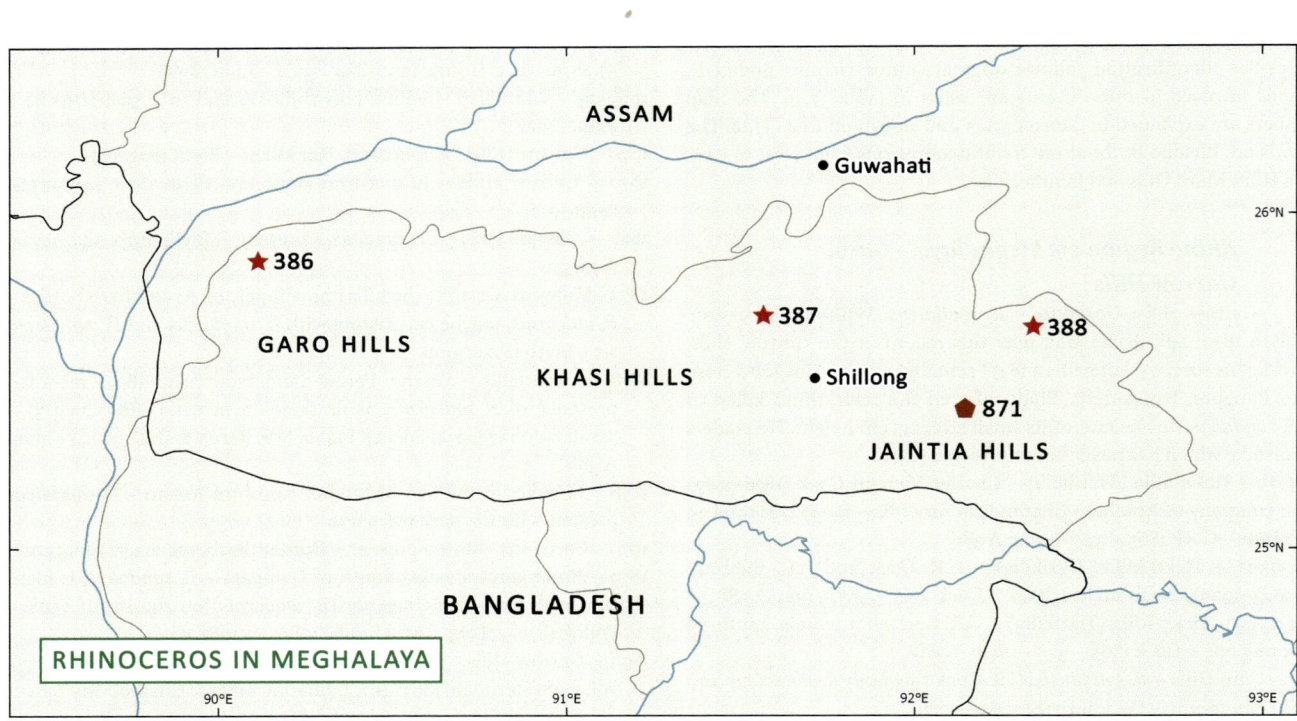

MAP 37.18   Records of rhinoceros in Meghalaya. The numbers and places are explained in Dataset 37.45

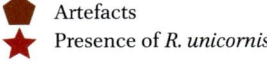   Artefacts

★   Presence of *R. unicornis*

Dataset 37.45: Localities of Records of Rhinoceros in Meghalaya

The numbers and places are shown on map 37.18

Rhino Region 29: Meghalaya
Garo, Garrow Hills

A 386 Tickri-killa, Tikri Killah (Tikrikilla) – 25.91N; 90.16E-1864, 1868, 1869, 1886, Pollok (1860s)

A 386 Hills south of Berrampooter river – 1867

A 386 Valley of Assam to the hills immediately on its southern border – 1872

A 386 Marshes at the feet of the hills south of Assam – 1878, 1879, 1884

Khasi, Cossyah Hills

A 387 Khassia, Cossyah, Khasya, Khaisa, Khasi hills – 1870, 1872, 1879, 1883, 1884, 1914

A 387 Jyrung (Jirang), about 50 km south of Gauhati – 25.90N; 91.60E-1898

Jaintia, Jhyntea Hills

M 871 Tongseng Nongkhlieh – 25.16N; 92.38E-17th century

A 388 Jhyntea, Jaintia hills – 1870, 1879

A 388 Jaintia hills: from Jynteapoor (Jaintiapur) to Nowgong – 25.15N; 92.35E-1879

A 388 Border of Mikir Hills and Jaintia Hills – 25.65N; 92.35E-1970s, 1975

A 388 Jaintia hills towards Nowgong – 25.60N; 92.30E-Pollok (1860s)

37.2 Garo Hills

In the Garo Hills, close to the western Assam border, rhinos were recorded by Blyth for 1837 based on mistaken locality (37.6), Cock for 1864 (37.7) and Pollok for 1868 (37.8). Blyth set a precedent, with rather unfortunate consequences. The only evidence for the existence of a rhino is from the foot hills where the Garo Hills reach the Brahmaputra valley. *R. unicornis* certainly once lived there, while the evidence for *R. sondaicus* remains equivocal (52.1). There is no reference to *D. sumatrensis* in these hills in the 19th century, even though that species is presumed to be better suited to a hilly terrain.

37.3 Khasi Hills

In the Khasi Hills, rhinos were rarely recorded. In fact, only Synteng writing in 1898 knew some rhinos in a hidden swamp (37.10). This locality was close enough to the Brahmaputra valley to assume that the species was *R. unicornis*.

37.4 Jaintia Hills

In the Jaintia Hills, a medieval rock sculpture is said to portray a rhino (37.5). Pollok (1879) also referred to rhinos in this part of the hills, albeit without any detail (37.8). In 1975, a dead rhino was found in the northern Jaintia Hills, which had strayed from the Nagaon plains of Assam.

TABLE 37.49   List of persons who were involved in shooting or capturing rhino in Meghalaya

Person	Date	Type of interaction
Charles Richard Cock (1838–1879)	1864	Garo Hills. Killed 29 rhinos and captured 1. With Bunbury
Abraham Charles Bunbury (1823–1904)	1864	Garo Hills. Killed rhinos. With Cock
Fitzwilliam Thomas Pollok (1832–1909)	1868, 1869	Garo Hills. Killed 4 rhino
George Peress Sanderson (1848–1892)	1878	Garo Hills. Collected 1 skull
Allan Octavian Hume (1829–1912)	1884	Garo Hills. Collected 1 specimen
Prince Louis Josef Jérôme Bonaparte (1864–1932)	1886	Garo Hills. Killed 1 rhino
Synteng	1898	Khasi Hills. Observed rhino in swamp

FIGURE 37.1   Rhinoceros tossing a leopard. Engraving after a watercolour of the German artist Heinrich Leutemann
(1824–1905) in a geographical work by Hermann Wagner (1824–1879)
WAGNER, *DE WARANDE*, 1869, P. 14

37.5   Medieval Rock Sculpture in Jaintia Hills

In South-East Meghalaya, in a place called Tongseng
Nongkhlieh, there is a big slab of rock engraved with a rhinoceros. It is said to be the work of a young local leader
and artist U Sajar Nangli, legendary in medieval times
(Bareh 1964). No photograph was located, but it would be
interesting to see of the animal had one or two horns.

37.6   Blyth Suggesting Presence of *R. sondaicus* in Garo Hills

In 1870, Edward Blyth had found a notice about "a rhinoceros killed on the Garrow Hills described to have a
height of four feet five inches" in the "Bengal Sporting
Magazine, 1837, p. 276" (Blyth 1870, 1872e). The locality
was not discussed, but reference to the original signed
by 'A Gallovidian' on 24 February 1837 shows that in fact
the event happened at Guzgottah, just north of Rangpur,
Bangladesh (30.2). Blyth suggested that the measurements

of this male rhino indicated that it was smaller than usual,
and therefore *R. sondaicus*. Consequently, he concluded
that both single-horned species inhabited the Garo Hills.
His next step was less logical, to say that their ranges
included "probably also the Khasya and Jhyntea hills, if
not still farther eastward" (Blyth 1870). That was an unfortunate and too grand conclusion. As it was also based on a
wrong interpretation of a locality in an old magazine, it is
impossible to agree with Blyth.

37.7   Richard Cock in Garo Hills in 1864

Pollok (1879, vol. 2: 108, 1894: 66) mentioned that the army
officers Cock and Bunbury had shot many rhinos in the
Goalpara region south of the Brahmaputra. Major Charles
Richard Cock (1838–1879), 43rd Regiment Native Infantry,
was apparently famous among sportsmen at the time, but
the only detailed account of his exploits is an extract of
his hunting diary (Cock 1865). During a period of leave in
March–April 1864, Cock and his fellow officer Abraham

FIGURE 37.2
Although unrelated to Captain Richard Cock, he would have cherished this kind of experience. This is an imaginary scene illustrated by Paul Hardy (1862–1942) on the front cover of a weekly boy's newspaper started by Cassell & Company in 1892
CHUMS, LONDON, 21 JUNE 1899, VOL. 7 NO. 354, P. 689

Charles Bunbury (1823–1904) set out from Goalpara and hired five riding elephants in Rakuldooly, a small village 16 miles (25 km) downstream from Goalpara, near Lakhipur. In this area of Assam, close to the river up to 10 March, they shot a male rhino, 6 ft 3 in (190.5 cm) in height at the withers, with horn 13 inches (33 cm), as well as at least two others. Via Lakhipur and the fringes of the Garro hills they moved on 14 March about 30 km further west to Tickri-killa (Tikrikilla), now just across the border into Meghalaya. For few days Cock moved further into the hills where a rhino was seen in a saltlick. They continued shooting in the plains around Tickri-killa until about 10 April 1864. On 19 March they shot a female rhino and captured her young one, which they sold. Their total bag at the time included 29 rhinos. The government in Goalpara refused to pay them the usual Rs.10 bounty per animal because they had already profited from the sale of the horns. There is no indication in Cock's journal that there were different types of rhino in that area.

Cockburn (1883: 64) later claimed that Cock had seen both *R. sondaicus* and *R. sumatrensis* in "very similar localities at the foot of the Garo, Khassia and Naga hills." Cockburn might have known Cock or heard about his hunting experiences. It is quite possible that Cock saw *D. sumatrensis* elsewhere, maybe in the Naga Hills, and the same is true for *R. sondaicus*. There is no evidence that any of these species were obtained or observed by him in the Garo Hills.

37.8 Pollok in Garo Hills in 1868

Fitzwilliam Pollok found and shot rhinos in May 1868 in the Tikrikilla area, close to a small army outpost (39.1). Pollok claimed to have shot both *R. unicornis* and at least one *R. sondaicus* at that time (52.3). As he did not recognize the latter species while in the field, arguably Pollok later became convinced of the difference as the specimen was smaller than others.

Elsewhere, in a general sentence, Pollok (1879, vol. 2: 77) mentioned that in the Jaintia Hills, on a journey from Jaintiapur in the south to Nagaon, rhinos could be found, although maybe this could apply just to the Assam part of his trip.

37.9 Prince Napoleon in Garo Hills in 1886

The son of Napoleon's brother Jérôme Bonaparte, Prince Napoleon Louis Josef Jérôme Bonaparte (1864–1932) visited India in 1885–1886 as part of a world tour. On 22 December 1885 he was expected in Kolkata, where he heard that he could take part in the *kheddah* (elephant catching) operations organised by the military department (*Figaro* 1886-11-24). He left Kolkata on 28 December accompanied by his ADC Count Mario Michela, as well as the staff officers Thomas Fourness Wilson and Major Thomas Deane (*Englishman's Overland Mail* 1885-12-29). From

Dhubri the party joined the team of the Dacca Kheddah Establishment led by George Peress Sanderson and spent 18 days in camp (*Englishman's Overland Mail* 1886-01-12). Prince Bonaparte killed a large male elephant, a tiger and a rhino, dated 13 January 1886 in the French press (*Figaro* 1886-02-18). The date was more probably 3 January because the party returned to Kolkata on 12 January and then proceeded to Madras and Mysore. Sanderson's *kheddah* was an annual event which took place in the Garo Hills. The exact locality has not been found. However, it may have been in the region of Tikrikilla in the far west of the present Meghalaya, where rhinos had previously been found at least to the 1860s.

37.10   Synteng in Khasi Hills Swamps in 1898

In 1898, a writer who called himself Synteng contributed some papers to the *Field* in London, which read like a guide book for the aspiring traveler and hunter. He had reached a place some 50 km south of Guwahati, in the Khasi Hills, advising that "In the outer hills of the Jyrung [Jirang] wilderness, as also on the lower slopes of the mountains proper, the rhino is not found; but there is a low-lying belt of swamp, extending some forty miles, immediately between the two, where the beasts are tolerably abundant. They are exceedingly shy here, though one would not have thought so, the place being so secluded." He recommended a novel way to get to the animals: "To force a rhinoceros from his slushy home is best done by Bengal bombs – small earthenware grenades filled with coarse powder and having a fuse attached. If you can get him in a circumscribed space and can pitch your bomb accurately, thus forcing him out on to comparatively dry land, all well and good; even then your bullet must be well placed or your labour will be all in vain." Synteng himself appears not to have fancied the challenge and he was content to give the advice to his readers.

Historical Records of the Rhinoceros in Assam

First Record: 1755 – Current in National Parks and Reserves. Strays throughout region.
Species: *Rhinoceros unicornis*. Few historical records of *D. sumatrensis*.
Rhino Regions 30 to 37

38.1 Assam: Rhino Country

The rhinoceros being the state animal of Assam symbolizes the special place of this animal in the minds of its inhabitants. Like myself, budding naturalists in the west immediately think of the rhino when the name Assam is mentioned. Rhinos were always widespread throughout the valley of the Brahmaputra River. Assam has important national parks protecting them, the pinnacle of which is Kaziranga where every tourist can be sure to have a close encounter with wildlife, and with very little effort. The Assamese people take great pride to be the hosts of a magnificent animal like the rhinoceros and the administration takes special care to support their protection.

The rhinoceros forms a quintessential part of the life and culture of Assam. The people are proud to share their land with one of the rarest large mammals on earth. During all my visits to this part of the world, I have been struck by the great feeling of ownership and dedication to the preservation of rhinos and their habitats. Take for instance Robin Banerjee (1909–2004), a roving ambassador for all Assamese wildlife who moved to Assam after a visit to Kaziranga and was a co-founder of the Kaziranga Wild Life Society (KWS), which evolved from 'The Rhino Club' at Kohra. Other outstanding naturalists include Arap Kumar Dutta (b. 1946), author of *Unicornis: the Great Indian one-horned rhinoceros* (1991), and Anwaruddin Choudhury (38.5), who has found evidence of rhinos in many unsuspected places. The Director General of Forest C.P. Oberai teamed up with the Director of the Delhi Zoo Bishan Singh Bonal in a book dedicated to Kaziranga (Oberai & Bonal 2002). Many non-governmental organizations such as the World Wide Fund for Nature or Aaranyak (p. xxvii) employ conservationists and scientists who are actively engaged in rhino work. Through their hard work, dedication and love for rhinos and wildlife, visitors today can be sure to encounter rhinos in Assam.

Assam was the center of the Ahom kingdom for some six centuries, a strong dynasty resisting attacks by the Mughals and only surrendering to the British East India Company in 1826 in a greatly changed world order.

FIGURE 38.1
Broadsheet showing a 'true delineation' of the Dutch rhinoceros Clara when she was exhibited by her Dutch owner Douwe Mout at the fair of St. Germain in Paris in January to April 1749. Engraved plate 37 × 49 cm issued by the Paris publisher Jean François Daumont (1717–1768), Rue de la Feronnerie à l'Aigle d'Or
WITH PERMISSION © MUSÉE CARNAVALET-HISTOIRE, PARIS

© L.C. (KEES) ROOKMAAKER, 2024 | DOI:10.1163/9789004691544_039

This is an open access chapter distributed under the terms of the CC BY-NC-ND 4.0 license.

FIGURE 38.2
"Before he can turn, the hunter buries
a spear in his heart." Drawing by
Auguste-André Lançon (1836–1887)
and engraved by Édouard Berveiller
(1843–1875). First published by Louis
Jacolliot in his *Animaux Sauvages* (1884,
p. 409), reproduced by Warren F. Kellogg,
Hunting in the Jungle (1888, p. 213)

The state of Assam was separated from Bengal in 1874. This relative isolation must explain the scarcity of early references to rhinos in western literature. There are no sources indicating that the animal was prominent in the art, cultural expressions, or religion of the people in this region. Curiously, Assam or "the Gross Moguls territories in the province of Asem" was associated in Europe in the 1740s as the birthplace of the female Dutch Rhinoceros 'Clara' touring with her owner Douwe Mout van der Meer (5.4, fig. 38.1).

The rhinoceros of Assam was first recorded in the western parts in Goalpara and Dhubri, then gradually further towards the east along the banks of the Brahmaputra. It must have been more widespread and more common than these observations tell us. Assam was never particularly targeted by the sport hunters in the absence of infrastructure and easier access to hunting grounds in North Bengal (fig. 38.2). For instance, nobody wrote about encountering a rhino in the area of Kaziranga until it was gazetted as a reserved forest early in the 20th century.

• • •

Dataset 38.46: Chronological List of Records of Rhinoceros in Assam

General sequence: Date – Locality (as in original source) – Event – Sources – § number – Map number.

Persons associated with rhino events are listed in Table 38.50. Records in the protected areas after their proclamation are found in the chapters relating to their history: North Kamrup or Manas from 1912 (chapter 40), Sonai Rupai from 1934 (chapter 41), Orang from 1915 (chapter 42), Pabitora from 1959 (chapter 43), Laokhowa from 1907 (chapter 44) and Kaziranga from 1908 (chapter 45). The records without locality are listed in the first section, followed by those located in Rhino Regions 30 to 37 (defined in 38.7). The localities are shown on maps 38.19 (western regions) and 38.20 (eastern regions), and explained in datasets 38.47 and 38.48.

Assam – General

This section lists records without definite locality. Quite a few specimens preserved in museums or once exhibited in captivity are included here.

1738 – Assam – Rhino 'Clara' captured – Rookmaaker 1998: 61–68; van der Ham 2022 – 5.4, figs. 4.6, 4.7, figs. 5.24 to 5.28, fig. 38.1

1765-10 – Assam – Presence of rhino stated in journal of James Rennell (1742–1830), Surveyor of India – Phillimore 1945, vol. 1: 20; La Touche 1910

1830s – Assam – Bruce, a planter of Calcutta, shot a rhino between the eyes on one of his properties – Arago 1840: 238 – Property unidentified, possibly fictional, see 50.7

1833 – Assam – Rhino expected but not seen – Shikarophilos 1833

1835 – Assam – Presence noted by Francis Jenkins (1793–1866), Commissioner of Assam – Saikia 2005: 259 (after Jenkins, *Report on the Judicial and Revenue Administration of Assam for 1835*, para. 200)

1836 – Assam, densest and most retired parts of the country – Rhino present – McCosh 1836b

1839 – Assam – Rhino present. The collection of John Maclelland (1805–1883), Assistant Surgeon, was donated to the Museum of the East India Co. in London – Maclelland 1839; Horsfield 1841

1843 – Assam – *R. unicornis* present – H. Walker 1843: 267

1844 – Assam – Francis Jenkins (see 1835) donates head of rhino to Royal Institution of Cornwall, Truro – Lemon 1845: 17

1855 – Assam – William Jamrach (1843–1891) transported 2 rhinos captured in Assam to UK, sold in Liverpool 1855-06-14. Possibly the subadult female and young male exhibited in Belle Vue Zoo, Manchester from May 1856 – Lucas's Repository 1855; *Guardian* (London) 1856-05-03; Rookmaaker 1998: 92

1862 – Assam – Captain Wilfred Dakin Speer (1835–1867) killed a rhino. His photo might be the first of *R. unicornis* in the wild – Kinloch 1903: 164; Rookmaaker 2021b – 38.18, fig. 38.7

1863 – [Assam] – *R. unicornis* skull and teeth donated on 6 May 1863 to Asiatic Society of Bengal, by the magistrate of Rungpore, Alexander George Macdonald, through J.D. Gordon – Dutt 1863: 288

1863–1864 – Assam – Capture of 2 rhinos exhibited in London Zoo: male 1864-07-25 to 1904-12-12, female 1864-07-25 to 1865-06-20 (to Paris) – Sclater 1876a; Rookmaaker 1998: 86

1869 – *R. unicornis*. Skeleton of male (no locality) prepared and placed in the Indian Museum (Kolkata) in March 1869 – Indian Museum 1870

1870 – Bramaputra – Capture of male rhino, exhibited in Hamburg Zoo 1870-03-14 to 1900-06-12 – Noll 1873a: 51; Rookmaaker 1998: 76

1870s – Assam – Rhinos are captured and transported to Kolkata before shipment to Europe – Blyth 1875: 50

1875 – Assam – Edmund Giles Loder (1849–1920) displayed a horn of 15¼ in (38.7 cm) in his house in Leonardslee, West Sussex. The date 1921 is recorded by Rowland Ward, while Loder might have obtained it in 1874–1875 when he traveled in India – Ward et al. 1922: 456, 1935: 335; Pease 1923

1876 – Assam – Tea planters hunt rhino in cool season – Barree 1876: 139

1879 – *R. unicornis*. Skull (no locality) in Indian Museum, Kolkata, attributed to William Thomas Blanford (1832–1905). Specimen no longer recognized – Sclater 1891, vol. 2: 202, no.m; Lydekker 1880a, pl. 7 (figured)

1882 – Assam – Fictional story of the capture of a young rhino with nets after killing the mother, published by Thomas Mayne Reid (1818–1883) in an American pulp magazine – Reid 1882 – 38.21

1884 – Assam – George Peress Sanderson (1848–1892) obtained a female rhino, preserved NHM 1884.1.22.1/2 – Pocock 1944b: 838

1884 – Assam – Rhino fairly plentiful, sometimes found in tea gardens – Barker 1884: 211

1887 – Assam – Rhino were plentiful and fearless. He was shown a spot behind a bungalow of a tea planter where a rhino prevented the servants to serve dinner – Baker 1887: 257

1889 – Assam – Start of hunting regulations by Dennis Fitzpatrick (1827–1920), Chief Commissioner of Assam from 1887 to 1889, prohibiting all shooting in reserved forests from November to June (the cool season when sports hunting was prevalent) without the permission of a range officer – F.C. Dukes, Secretary to the Chief Commissioner of Assam to the Deputy Commissioners, Shillong, 1889-01-14 in ASP, Revenue and Agriculture, January 1889, after Saikia 2009.

1890s – Assam – William Parker Yates Bainbrigge (1853–1934), physician in London, killed rhino with horn 14¼ in (36.2 cm) – Ward 1896: 282

1890 – Assam – Charles Thomas Buckland (1824–1894) looked for rhino without success – Buckland 1890: 230

1892 – Assam – Herbert Christian Holland (1858–1916) killed rhino with horn of 12½ in (31.8 cm) – Ward 1896: 282 – 38.24

1892 – Assam – Sir Peter Carlaw Walker (1854–1915) killed rhino with horn 8¼ in (21.0 cm) – Ward 1896: 282 – 38.24

1894-03 – Assam – Author with friend A. killed 3 rhino. One charged and wounded a tracker with his teeth – A.H.M.T. 1895

1896 – Assam – Ipswich Museum, exchanged from British Museum in March 1907 (no. IPSMG:R.1906.79): mounted rhino with horn 19½ in (49 cm). Horn stolen in 2011 – Ward 1896: 282, 1910: 464; Markham 1990 – 38.26

1898 – Assam – George Alexander Dolby (1854–1939), tea planter, obtained rhino with horn of 13 in (33 cm) – Ward 1899: 433

1900s – India (no data) – Single horn mounted as trophy and a trophy head with small horn, preserved in Cotehele House, Cornwall, ancestral home of William Henry Edgcumbe, 4th Earl of Mount Edgcumbe (1832–1917) – 38.22

1900 – Assam – *R. sondaicus* (identity uncertain) – Reported by Mr. A.C. Green, some years ago – Burke 1908: 44, 52.9

1902 – Closed season for shooting in reserved forest extended throughout the year for female rhinos with calf – *Amrita Bazar Patrika* 1920-12-02; Milroy 1932, 1934a: 97; Guggisberg 1966: 133; Martin & Martin 1982: 32; Martin 1983: 20; Saikia 2009

1902 – Assam – Maxwell Hannay Logan (1873–1947) killed rhino with trophy horn 18 in (45.7 cm) – Ward 1903: 409

1903 – Assam – Walter Alexis Rene Doxat (1873–1942), Conservator of Forests, killed rhino with trophy horn of 13 in (33.0 cm), circumference at base 21½ in (54.6 cm), weight 3 lbs (1.4 kg), circumference of foot 33 in (83.8 cm) – Ward 1907: 464; Burke 1908: 43

1904 – Assam – Numbers have terribly reduced; but there are enough left to enable a well-equipped sportsman to be pretty sure of obtaining one or two specimens – Kinloch 1904: 61

1908 – Assam – Surya Kanta Acharya Choudhury of Muktagacha, Zamindar of Mymensingh (1849–1908) and another sportsman killed 2 rhino – *Nottingham Journal* 1908-10-24 – 38.16

1909-02-19 – Brahmaputra Valley – Charles Victor Alexander Peel (1869–1931) killed a rhino. Mounted in Peel's Museum in Oxford, transferred to Royal Albert Memorial Museum, Exeter, no. 99/1919.9 – Peel 1908, 1909; Howes & Bamber 1970 – 38.27

1909 – Assam – A. Butcher killed rhino. Head, skull preserved in Bombay Natural History Society – Bombay NHS 1909

1910 – Assam – The army officer Frederick William Gore (1825–1909) killed rhino. Horn in Bombay Natural History Society – Bombay NHS 1914

1913 – Assam – Dennis Herbert Felce (1875–1932), manager of a tea-estate, killed rhino with horn 18 3/8 in (46.7 cm) – Ward et al. 1914: 463

1920s – Brahmaputra River, Assam – R.V. Yates and F.J. Price kill rhino with horn 16¾ in (42.5 cm) – Ward et al. 1928: 438

1921 – Assam – Sales Depot stock: 3 rhino horns valued together at Rs. 300 – Assam Forest Dept. 1921: 40

1921 – Assam – Lionel Walter Rothschild, 2nd Baron Rothschild (1868–1937) owned a rhino horn of 16¼ in (41.3 cm). Not shot by himself – Ward et al. 1922: 456; NHM London 1948.1.28.9

1925 – Assam – Sportsmen are rarely allowed to kill rhino – Willcocks 1925: 40

1932 – Assam – Authors find shooting rhino unexciting, while obtaining a permit is difficult. Illustration of an impression of plastercast of back foot – Alexander & Leake 1932: 80

1935 – Proposal of Rhinoceros in Coat of Arms for the state of Assam. No motto was attached, but one was suggested: 'Arva, Flumina, Montes' – Sword 1935: 16 – 38.34

1937-07 – Assam – On a tour of Assam the Viceroy commented on rhinos straying outside the reserves. Victor Alexander John Hope, 2nd Marquess of Linlithgow (1887–1952), Viceroy 1936–1943 – Glendevon 1971: 71

1940–1945 – Assam – During the war years, wildlife was killed indiscriminately by all military personnel, which declined afterwards – Assam Forest Dept. 1948: 22; Gee 1950b: 82

1943–1944 – Assam – 3 rhinos killed in reserve forests – A. Ghosh 1945

1944–1945 – Assam – 5 rhinos shot (poached?) – Assam Forest Dept. 1947

1949 – Assam – *R. unicornis* skull donated by J. McAlpine-Woods to Royal Albert Memorial Museum, Exeter in 1949. No data where or when acquired – Howes & Bamber 1970

1954 – Assam – Assam Rhinoceros Preservation Act – Assam Government 1954

1954 – Assam – Bishnuram Medhi (1888–1981), Chief Minister of Assam, writes to Prime Minister Jawaharlal Nehru (1889–1964) that rhinoceros is on verge of extinction – Saikia 2005: 277

1963 – Assam: several isolated pockets – Rhino present – Stracey 1963: 66

2003 – Rhinoceros is chosen as the State Animal of Assam, through Notification dated 31 March 2003 by Government of Assam (signed by Anwaruddin Choudhury) – Mazumdar & Mahanta 2016: 126

Rhino Region 30: Assam – Sankosh River, North of Brahmaputra

1864-12 – Jogeegopah (Jogighopa) – Tracks of rhino, seen by Falkland George Edgworth Warren (1834–1908) – Warren 1866: 122 – map A 404

1864-12 – Sidli – Tracks of rhino – Warren 1866: 124, 126 – map A 409

1865-01 – Coochabaree; Dewagaon – Tracks of rhino – Warren 1866: 135, 139, 140 – map A 408

1868 – Putimari (Patamari), west of Dhubri – Rhino scarce, but some seen by another party – Anon. 1869c – map A 401

1869 – Goma Dooar (around Chapar) – W. was informed that 200 rhinos were killed same year – W. 1869: 638 – map A 404

1869-04-18 – Chumpamuttee River, Chapar area – 2 rhino seen not shot. Next day also rhino seen – W. 1869 – map A 404

1869-04 – Chandodingah Hill, Chapar area – Kills 2 female rhinos, one with a really good horn – W. 1869 – map A 404

1872 – Gompore (Gauripur) – Lieutenant-General William Russell, 2nd Baronet (1822–1892) killed rhino with horn of 15 in (38.1 cm), skull in NHM, no. 722k, later 1872.12.30.1 – Gray 1873a: 46; Pocock 1944b: 838 – map A 402

1873 – Dhubri – Fitzwilliam Thomas Pollok (1832–1909) killed rhino 39 – 39.4 (trip 29) – map A 402

1873 – Dhobree (Dhubri) – Lieutenant William John Williamson (1842–1883) killed a young rhino while out hunting with Fitzwilliam Pollok – Pollok & Thom 1900: 488 – 39.4 – map A 402

1875 – Sankosh River – *D. sumatrensis*. Colonel Charles Napier Sturt (1832–1886) shot two-horned rhino – 64.4 – map D 091

1876-03 – Dhubri – Friend F. killed a rhino – Anon. 1877b: 245 – map A 402

1880 – Khagrabari – Maharaja of Cooch Behar and party killed 4 rhino – 32.9 Shoot 10, 1880 (a) – map A 409

1881 – Khagrabari – Maharaja of Cooch Behar and party killed 4 rhino – 32.9 Shoot 11, 1881 (a) – map A 409

1892 – Sapkata (Patgaon) – Maharaja of Cooch Behar and party killed 4 rhino – 32.9 Shoot 28, 1892 (a) – map A 407

1893 – Sapkata (Patgaon) – Maharaja of Cooch Behar and party killed 1 rhino – 32.9 Shoot 30, 1893 (a) – map A 407

1894 – Dudnath – Maharaja of Cooch Behar and party killed 1 rhino – 32.9 Shoot 32, 1894 (a) – map A 403

1900 – Roymona (Raimona) – Maharaja of Cooch Behar and party killed 1 rhino – 32.9 Shoot 42, 1900 (a) – map A 405

1901 – Patgaon – Maharaja of Cooch Behar and party killed 6 rhino – 32.9 Shoot 43, 1901 (a) – map A 407

1905 – Kachugaon – Maharaja of Cooch Behar and party killed 1 rhino – 32.9 Shoot 50, 1905 (a); Choudhury 1996 – map A 406

1916 – Goalpara, on opposite bank – Rhino present – Thackeray 1916: 164 – map A 404

1919 – Goalpara – Male rhino shot during shikar party of Raja of Gauripur – Blunt & Todd 1919; Funk 2014 – §38.16 – map A 404

Rhino Region 31: Assam – Goalpara, South of Brahmaputra

fossil – Goalpara – Fossil molar of Rhino (no species) in Indian Museum, Kolkata, donated by E.N. de Garnier – Lydekker 1886a: 13 – map A 413

1755-07-11 – Goalpara – Jean-Baptiste Chevalier (1729–1789) hears rhino in forest – Chevalier in Deloche 1984: 127, 2008: 133 – 38.8 – map A 413

1755-07-30 – Mountain near Goalpara – J.B. Chevalier shoots at rhino which escapes – Chevalier in Deloche 1984: 142, 2008: 149 – 38.8 – map A 413

1777 – Gwalpara (Goalpara) – Thomas Craigie killed a rhino. Skull, donated by Edw. Smith, in Museum of the East India Company, London; since 1879 in NHM, no. ZD.1879.11.21.487 – Horsfield 1851: 195 – 38.10 – map A 413

1801 – Lukkhipore (Lakhipur) – Maharaja Harendra Narayan (1783–1839) of Cooch Behar hunted rhinos, tigers and buffaloes – Ghose 1874: 172 – 31.7 – map A 412

1818 – Givalpara (Goalpara) – An officer makes an unsuccessful attempt to penetrate the hide of an old rhino carcass with musket ball – Anon. 1818a – map A 413

1827-05-08 – Singamaree (near Goalpara) – David Scott (1786–1831), Agent to the Governor General on the North East Frontier of Bengal, saw 6 rhino in a day, none killed – Watson 1832: 90 – map A 411

1820s – Goalpara – The magistrate Captain Alexander Davidson (1792–1856) wrote on 1830-03-01 that "One of the Zemindars here (as I mentioned to you) had a tame female which produced almost every year" – Burrough Archive (William L. Clements Library, Ann Arbor, Michigan; Extracted from originals) – map A 413

1830-03 – Goalpara – The magistrate Captain Alexander Davidson obtained one male rhino which was 10 weeks old on 1830-03-15. It was sent to the American businessman and consul Marmaduke Burrough (1797–1844) in Kolkata on 1830-03-29. The animal was transported on the ship *Georgian* arriving in Philadelphia, USA on 1830-10-15. Burrough sold it by auction on 1831-01-03 after which it toured with a circus company until its death, probably in 1837 – Burrough Archive (William L. Clements Library, Ann Arbor, Michigan; Extracted from originals); Reynolds 1968; Cox 1997; Thayer 1975, 1998; Rookmaaker 1998: 107 – 38.13, fig. 38.6 – map A 413

1830-12 – Goalpara – Another rhino obtained by Alexander Davidson in Goalpara was sent to George Hough in Kolkata, who wrote to Marmaduke Burrough on 1831-05-09: "I had it here at the Ghaut, gave his servants places to live in, on our premises, and attended to the animal myself. After a lapse of a week I sold her to Mr. Austin, Supercargo of the brig *Neponset* of Boston for 1500 Rs and put the money into Fergusson & Co's house." The *Neponset* arrived in Boston on 1831-05-10 with a rhino on board. It was exhibited in Boston in 1831, possibly later in an American circus, although records are too confused to be certain that it did not die soon – Burrough Archive (William L. Clements Library, Ann Arbor, Michigan; Extracted from originals) – 38.13 – map A 413

1832 – Goalpara – John Leslie (1800–1831), Assistant Surgeon, died of jungle fever brought on by exposure while skinning two rhinos – McCosh 1837: 99; Watson 1832: 106 – map A 413

Undated, maybe 1840s – Without locality, author might have lived in Goalpara – "Old Shekarree" stops with a hunting party at the house of an indigo planter (unnamed). They hear about 2 rhino in a sugarcane field, of which one (male) is killed, the other (female) is wounded, but not found again. No rhino generally known within 80 to 100 miles from this location – Old Shekarree 1860: 104 (author not identified; volume advertised in newspapers in 1858)

Undated, maybe 1840s – Without locality, author might have lived in Goalpara – Hunting party with 17 elephants. They kill 4 rhinos in 3 days (p. 130). The first rhino killed was a male "average height and size, and had a prettily shaped horn on his nose, but decidedly blunt" of which head and horn are preserved (pp. 106–114). Second rhino is a female (p. 114). Third rhino is a male, with thick skin folds and good horn indicating age (p. 123). Fourth rhino is half-grown animal (p. 127). Afterwards more rhino killed but number not specified (p. 130) – Old Shekarree 1860

1860's – Lukheepore (Lakhipur) – Son of Raja killed 34 rhinos in the region. Some trophies were kept in the Raja's Palace – Pollok 1868e: 801, 1879, vol. 2: 108; Pollok & Thom 1900: 453; S. Ward 1884: 149 – 38.14 – map A 412

1860s – Assam – A party of 2 sportsmen killed 30 rhino in a month. Probably refers to Cock and Bunbury in 1864 – Hunting Knife 1868: 216

1860s – Assam – Unnamed sportsman killed 13 rhino. Probably refers to Cock and Bunbury in 1864 – Homeward Mail from India 1864-06-18

1862 – Assam, to the south of the great bend of Brahmaputra – D. sumatrensis. Friend of Edward Blyth saw two horns attached to a skin – 64.2 – map D 090

1864-03 – Rakuldooly, near Lakhipur – Charles Richard Cock (1838–1879), Captain in Bengal Staff Corps, killed several rhinos with Abraham Charles Bunbury (1823–1904) – Cock 1865 – 37.7 – map A 412

1868-05 – Dhubri to Goalpara – Pollok kills rhinos 7 and 8 – 39.4 (trip 7) – map A 412

1869-03 – Lukheepore (Lakhipur) – Pollok kills rhinos 11 to 13 – 39.4 (trip 11) – map A 412

1870s – Lakhipur – George Udny Yule (1813–1886) found game abundant. Saw herd of 7 rhinos – Yule 1903: 47 – map A 412

1870-07 – Luckeepore (Goalpara) – Pollok sees one rhino while in a boat – Pollok & Thom 1900: 474 – 39.4– map A 412

1872 – Valley of Assam to the hills immediately on its southern border (Goalpara) – R. unicornis, maybe co-existing with R. sondaicus and D. sumatrensis – Blyth 1872e: 3106 – map A 412

1879 – Goalpara, Eastern Dwars – Rhino present – Hunter 1879b, vol. 2: 27, 114 – map A 413

1879 – Goalpara Dt. – Rhino present – Pollok 1879, vol. 1: 95; Wood 1930, 1934 – map A 413

1880s – Foot of Garro Hills (Lakhipur area) – Allan Octavian Hume (1829–1912) was a colonial administrator, best known for his ornithological expertise. He killed a rhino with horn of 10¾ in (27.3 cm). Mounted head in NHM London, no. ZD.1912.10.31.105 – Ward 1899: 433, 1903: 409; Lydekker 1913 – map A 412

1883-03-26 – Lower Assam – Henry Meysey-Thompson (1845–1929) killed 3 rhino – Times of India 1883-03-26

Before 1887 – Luckipore (Goalpara) – Edward Biscay Marinatus Baker (1828–1887), Deputy Inspector General of Police, killed 4 rhinos, still plentiful – Baker 1887: 252 – map A 412

1894 – Khalsamari – Maharaja of Cooch Behar and party killed 2 rhino – 32.9 Shoot 32, 1894 (a) – map A 412

Rhino Region 32: Assam – Manas, North of Brahmaputra

(Manas area records to 1933, later in chapter 40)

1815 – Bijni (Bijnee) – Kishen Kant Bose (or Krishnakanta Basu) went to Bhutan on a diplomatic mission in 1815, and on the way northwards records presence of rhino – Bose 1825: 153, 1865: 204; Bray 2009 – map A 415

1822 – Bignee (Bijnee) – Acting Collector of Rungpore hopes to procure a living rhino from the Zamindar of Bignee "as his people are constantly in the habits of taking them" – National Archives of India, Home Department, Public, 30 May 1822, no. 27 – map A 415

1834 – Baksa (from Guwahati to Tawang) – Rhino present – Wilcox 1834: 455 – map A 421

1856 – Assam (Manas) – Skull in Bayerische Staatssammlung, Munich (AM 0415) – Schlagintweit 1869, vol. 1: 433 – §38.17 – map A 417

1864-12 – Bijnee – Rhino in the swamps – Warren 1866: 124, 126 – map A 415

1865-04 – Manas – Lieutenant-General Henry Andrew Sarel (1823–1887) shoots in Manas area from 20 April to 1 June 1865. After borrowing eight elephants from his fellow officer General Henry Tombs (1825–1874), he traveled from the Manas to the Teesta River in Bengal. Rhinos were numerous, and he killed three on the first day and a total 8 during the expedition. One day he saw four rhinos together in a mud hole – Bombay Gazette 1865-07-19; Sarel 1887: 358 – map A 417

1865 – Bhootan Duars south of Dewangiri – Rhino stopped passage of troops in a low pass, observed by Leopold John Herbert Grey (1840–1921) – Grey 1912: 44 – map A 421

1867-06 – Manas – Fitzwilliam Thomas Pollok (1832–1909) kills his first rhino on 11 June 1867 – 39.4 (trip 3) – map A 417

1868-01 – Manas – Pollok kills rhinos 2 to 6 – 39.4 (trip 5) – map A 417

1868-01 – Manas area – Major George Machardy Bowie (1835–1876) and John Barry, tea planter, killed a splendid rhino with horn of 13 in (33 cm) long and weighing 2 seers and 2 chittacks (2.65 kg) – 39.4 (trip 5) – map A 417

1868-01 – Burpettah (Barpeta) – Major G.M. Bowie killed 1 rhino with horn of 13 inch (33 cm), in company of Fitzwilliam Pollok – 39.4 (trip 8) – map A 420

1869-01 – Kumblepore, Monass (Manas) – Pollok, in company of John Barry and Butler, sees many old tracks of rhinos – 39.4 (trip 10) – map A 421

1869-04 – Manas – Pollok kills rhinos 14 and 15 – 39.4 (trip 12) – map A 417

1870-01 – Manas – Pollok kills rhinos 16 and 17 – 39.4 (trip 14) – map A 417

1870-04 – Manas – Pollok kills 12 rhinos (nos. 19 to 30). They were very numerous in the area – 39.4 (trip 16) – map A 417

1870-05 – Kookooriah (Manas area) – G.H. Jackson and John Barry killed a rhino with horn of 13 in (33 cm), and saw 2 others. Pollok was not present – Pollok 1879, vol. 2: 133, 1882b: 59 – map A 416

1871-05 – Manas – Pollok kills rhinos 31 and 32 – 39.4 (trip 23) – map A 417

1871-05 – Kumlabaree, Manas – A. Anley, District Superintendent of Police in Assam, is credited with one rhino while shooting in company of Fitzwilliam Pollok. It had a horn of 12 in (30.5 cm) – Pollok 1879, vol. 2: 137 – 39.4 (trip 23) – map A 419

1871 – Manas – Colonel George Campbell (d. 1876), 52nd Foot, during his service of over 30 years in Assam, probably killed more game in the East than any man now living. He once fired 2 shots rapidly, and 2 rhinos which were standing together fell dead – Pollok 1882a, 1898a: 176 – map A 417

1871-05 – Poho-Marah River, Manas – Colonel George Campbell and John Barry came across 3 rhino, and killed one with a fair horn – Pollok 1879, vol. 2: 148 – 39.4 (trip 23) – map A 419

1871 – Manas area – George Campbell went to the area with friends and killed several rhinos – Pollok 1874b: 310, 1879, vol. 2: 155 – 39.4 (trip 23) – map A 417

1872-04 – Rungiah (Rangia) – Pollok kills rhino 33 – 39.4 (trip 25) – map A 421

1872-04 – Manas – Pollok kills 5 rhinos (nos. 34 to 38) – 39.4 (trip 26) – map A 417

1872-04 – Manas area – Augustus Kirkwood Comber (1828–1895) killed one rhino while shooting with Charles Orchard Cornish (1839–1905) and Fitzwilliam Pollok – 39.4 (trip 26) – map A 417

1872-03 – Barpetta (Barpeta) – Hon. Talbot (unidentified) killed 6 rhino – Talbot 1872 – map A 420

1873-04 – Manas – Pollok kills 5 rhino (nos. 40–44) – 39.4 (trip 30) – map A 417

1873-04 – Manas area – John Barry kills one rhino while hunting with Fitzwilliam Pollok – 39.4 (trip 30)

1875 – Manas area – James Matthie (1806–1865), Magistrate and Collector of Guwahati, hunted rhino north of Brahmaputra – Fayrer 1900: 60 – map A 417

1879 – Kamrup – Rhino common, especially in the north of the district, which swarms with wild animal life of all kinds – Hunter 1879b, vol. 2: 25 – map A 417

1887 – Monass (Manas) river – Rhino abound – Baker 1887: 256 – map A 417

1887-04-14 – Assam (Manas, see next entry) – Henri Charles Joseph Le Tonnelier, Marquis de Breteuil (1847–1916) and his companion Mr. Saulti had returned to Kolkata from hunting in Assam shooting 15 rhinos – Philippe d'Orléans 1892: 234 – Breteuil participated in the Cooch Behar hunt 1887 (a) 1887-02-18 to 1887-03-09 and may have stayed on afterwards – map A 417

1887-04 – North Kamrup (Manas) – Rhino shot by party consisting of "Marquis de Baetmel and his brother Count Goston accompanied by Mr. Ridgeway, an American gentlemen." This refers to Marquis Henri de Breteuil, and his brother Gaston de Breteuil (1864–1937). Ridgeway is not identified – *Englishman's Overland Mail* 1887-05-03, p. 10 – map A 417

1895 – Bhuiapara – Maharaja of Cooch Behar and party killed 6 rhino – 32.9 Shoot 33, 1895 (a) – map A 418

1899 – Bhuiapara – Maharaja of Cooch Behar and party killed 5 rhino – 32.9 Shoot 40, 1899 (a) – map A 418

1899 – Bansbari (Bahbari) – Philip Breda Vanderbyl (1867–1930) killed 2 rhino, in company of Sidney Herbert Buller-Fullerton, 16th Lord Elphinstone (1869–1955) – Vanderbyl 1901, 1915 – 32.9 Shoot 41, 1899 (b), §38.25, fig. 38.12 – map A 418

1899-03 – Jalagaon (Jalahgaon, Manas) – Philip Breda Vanderbyl killed rhino, preserved in NHM, nos. ZD.1901.3.10.1 and ZD.1941.90. A horn of 9 in (22.9 cm) long is among records – Pocock 1944b: 836; Ward 1903: 409 – §38.25 – map A 419

1901 – Bijni RF – Rhino present – Assam Forest Dept. 1901: 7, 1902: 7 – map A 415

1901 – Manas River – Capt. H.R. Beaumont and Arthur killed 1 rhino – 32.9 Shoot 44, 1901 (b) – map A 417

1902 – Lahapara – Maharaja of Cooch Behar and party killed 1 rhino – 32.9 Shoot 45, 1902 (a) – map A 419

1902 – Manaas River – Rhino found on muddy tributaries of the river – O.W. 1902: 952 – map A 417

1903 – Bijni RF (Ripu-Chirang) – Rhino decreasing. Possibility to protect the forest to preserve rhino, but there is no benefit for forest improvement – Assam Forest Dept. 1903: 7; *Globe* 1904-02-25 – map A 415

1905 – North Kamrup (Manas) Reserved Forest, intention published – chapter 40

1911 – North Kamrup (Manas) Reserved Forest, officially notified – chapter 40

1911 – Manas River – Guy Douglas Arthur Fleetwood Wilson (1851–1940) killed a rhino, with trophy preserved in his house in Dees, UK – Wilson 1920: 268, 1921: 123 – 38.30, fig. 38.19 – map A 417

1933 – Ripu RF – Rhino present. Area should be converted to national park – Hanson 1933: 51 – map A 415

1933 – Chirang RF – Rhino present. Area should be converted to national park – Hanson 1933: 51 – map A 415

1934 – Assam near Bhutan (Manas?) – Frank Buck (1884–1950) filmed rhinos for his movie *Fang and Claw* – Buck & Fraser 1942: 201 – §38.31

1934 – Assam near Bhutan (Manas?) – Frank Buck captured a young female, taken to his facility in Amytiville, Long Island, NY, where she lived 1935-09-13 to 1936-02 – Buck & Fraser 1942: 201–204; Rookmaaker 1998: 44 – 38.31

Rhino Region 33: Assam – Darrang, North of Brahmaputra

1836 – Bisnauth (Biswanath) – Woodsman and 'P-re' shoot rhinos, one with horn of 14 in (35.5 cm) – Woodsman 1836 – 38.14 – map A 434

1845-11 – Bahali, near Howrah-Ghat – Lieutenant George Campbell (d. 1876) killed 2 rhino with one shot – Butler 1855: 14 – map A 438

1876 – Tezpore region – A young tea planter came across 2 rhino, which wounded his riding elephant – Baldwin 1876: 145 – map A 432

1879 – Durrung (Darrang) – Rhino present – Pollok 1879, vol. 1: 95; Hunter 1879a, vol. 1: 108; Wood 1930, 1934 – map A 427

1882 – Mungledye (Mangaldoi) – Arthur John Primrose (1853–1888) killed a rhino – Willcocks 1904: 44 – 38.20, figs. 38.8, 38.9 – map A 427

1890 – Tezpore (Tezpur) – Addison Yalden Thomson (1863–1931) killed rhino with horn of 6½ in (16.5 cm) – Ward 1899: 433 – map A 432

1890 – Darrang Dt. – Rhino were plentiful – Wood 1930: 68 – map A 427

1896-04 – Bisnath (Biswanath) – Sidney Herbert Buller-Fullerton, 16th Lord Elphinstone (1869–1955) hunted here in April 1896. Part of this group, Humphrey Brooke Firman (1858–1916) killed a rhino with a horn of 12¾ in (32.4 cm), the trophy head still preserved in the In & Out Club, London – Ward 1896: 282; Elphinstone Game Book 1896 (32.03); Newark 2015: 110 – 38.23, fig. 38.11 – map A 434

1897-06-12 – Darrang Dt. – Great earthquake – Allen 1905, vol. 5: 14 – map A 427

1898 – Belsire (Belsiri) – Walter Stanhope Sherwill (1815–1890) shot rhino with horn of 16¾ in (42.5 cm) – Ward 1899: 433 – map A 431

1909-05 – Tezpur Dt. – Thomas Charles Briscoe (1857–1909) shot a rhino with longest known horn of 24¼ in (61.3 cm) – Shikar 1909; Anon. 1909e; *Morning Post* 1909-11-23; Briscome 1909; Lydekker 1910; Ward 1910: 464; Barclay 1938 – 38.28, figs. 38.13, 38.14 – map A 432

1909 – Bisnath Dt. (Biswanath) – Henry Stotesbury Wood (1865–1956) lost an old shikari following a rhino – Anon. 1909b (see 1930 entry) – map A 434

1911 – Gohpur (former Tezpur Dt.) – Rhino present – Wood 1930: 68 – map A 437

1911 – Tezpur Dt.: right [north] bank of the Brahmaputra between Behali and Boregaon – Rhino present – Wood 1930: 68 – map A 428

1911 – Sonarupa River, Tezpur Dt. – Rhino present – Wood 1930: 68 – map A 429

1911 – Borsola jungles at Singrighat, Tezpur Dt. – Rhino present – Wood 1930: 68 – map A 431

1911 – Orang, Tezpur Dt. – Rhino present – Wood 1930: 68 – map A 430

1915 – Orang Game Reserve notified – chapter 42 – map A 430

1920s – Bargang Tea Estate, Kettela, Biswanath Dt. – Frank Nicholls (b. 1889) kills a rogue rhino – Nicholls 1970: 54 – map A 436

1930 – Booreegaon, Darrang Dt. (north bank) – Henry Stotesbury Wood. Unable to shoot a rhino on foot. Once his hunter Loloong was attacked by rhino and killed – Wood 1930: 60, 1934, 1950: 183 – map A 428

1932 – Darrang Dt. – Only 7 rhino known in district – A. Milroy 1933: 4 – map A 427

1932 – Balipara Frontier Tract – Rhinos present – A. Milroy 1932, 1934a: 100 – see chapter 41 – map A 432

1934 – Sonai Rupai Wildlife Sanctuary notified – chapter 41 – map A 429

1975–1995 – Kuruwa-Mandakata – Stray rhinos from Pabitora – Choudhury 1985a, 1996, 2002a: 16–19; Talukdar 1995: 1 – map A 426

1980 – Kurua area: Kurua Hill, Baman Hill, Kholihoi, Ganesh Hill, Teteliguri Hill, Chaolkhowa, with links to Pabitora – Rhino present – Lahan 1993: 46 – map A 426

1980–1995 – Tezpur – On a few occasions a rhino entered Tezpur town – Choudhury 1996, 2013: 225 – map A 432

1982 – Darrang Dt. – Rhino present – Mukherjee 1982: 49 – map A 427

1985 – Panpur RF – Visited by rhino from Kaziranga south of river. Reserved Forest of 6.1 km² was constituted in 1971 by Gazette Notification dated 9 December 1971 (Ext. 3). In 2020 absorbed in western range of Kaziranga NP – Choudhury 1985a, 1996 – map A 434

1990-04 – Nameri WS – One rhino traveled into reserve – Choudhury 1996 – map A 433

1990 – Gohpur – Stray rhino – Lahan 1993: 46; Choudhury 2002b: 6 – map A 437

1990-09 – Khatonibari (Khatani Bari), Sonitpur Dt. – 2 rhino strayed from Kaziranga along the Borgang River up to Arunachal Pradesh, then chased back to Khatonibari – Choudhury 1996 – map A 435

1994 – Silikaguri – 3 rhinos seen – Choudhury 1996 – map A 437

1995 – Narayanpur to Silikaguri – Lone rhino strayed and was poached – Choudhury 1996, 1997a–map A 437

1995 – Kochmara RF – Presence of rhinos – Talukdar 1995: 1 – map A 431

2021-05 – Konakata Para and Gadhaijhar (west of Orang NP) – two stray rhinos captured and returned to Orang – News release (Rhino Review) 2021-04-05 – A430

Rhino Region 34: Assam – from Guwahati to Golaghat, South of Brahmaputra

Records for region of Laokhowa from 1907 (chapter 44) and Kaziranga from 1908 (chapter 45).

1755-12-22 – Laokhowa area – Jean-Baptiste Chevalier (1729–1789) sees rhino from boat on Brahmaputra – Chevalier in Deloche 1984: 165, 2008: 175 – 38.08 – map A 449

1755 – Assam – J.B. Chevalier noted that 2 rhinos were included in animal sacrifices in the Kamakhya Temple, Guwahati – Deloche 1984: 35, 2008: 36 – 38.09 – map A 446

1845 – Nowgong district (Nagaon) – Rhino found in reed jungles – Butler 1855: 218 – map A 448

1847 – Gowahatty – Tame rhinos can be seen grazing in the fields – Butler 1847: 29; Flex 1873: 4 – 38.12 – map A 446

1848 – Nowgaon (Nagaon) – Reward of Rs 5 paid for killing of rhino – G. Singh 2014: 35 (Bengal Government Papers, File No. 340, 1848; Assam Secretariat Proceedings) – map A 448

1855 – Gowahatti (Guwahati) – Hermann Schlagintweit (1826–1882) observes tame rhino during visit of November to December 1855 – Schlagintweit 1869, vol. 1: 433; Depping 1864; Werner 1884: 253; Brescius et al. 2015; Driver 2018; Brescius 2019 – 38.17 – map A 446

1865-04 – Lowqua Lake (Laokhowa) – An exceedingly fine specimen was shot by fellow officers (68th Native Infantry) of Captain John Henry Baldwin (1841–1908) – Baldwin 1876: 146 – map A 449

1867 – Nowgong – Traveling from Jynteapoor over the hills direct to Nowgong, rhino are met with – Pollok 1879, vol. 2: 77 – map A 448

1867-03 – Loqua Ghat (Laokhowa) – General Charles Reid (1819–1901), Bengal Staff Corps, hunting in the company of Fitzwilliam Pollok, killed 2 rhinos in one day, and lost another the next day – 39.4 (trip 2) – map A 449

1867 – Loqua Ghat – Colonel James Cathorne Cookson (1821–1900), Madras Cavalry, together with Fitzwilliam Pollok shot 1 male and 2 female rhinos – Pollok 1879, vol. 2: 81, 1894: 74 – 39.04 (trip 2) – map A 449

1868 – Gauhati – Skull in Indian Museum, Kolkata (no.p). Specimen no longer recognized – Sclater 1891, vol. 2: 202 – map A 446

1870 – Gauhati – Skull and incomplete skeleton in Indian Museum, Kolkata (no.c), donated by Fitzwilliam Pollok. Specimen no longer recognized – Sclater 1891, vol. 2: 202 – map A 446

1870-03 – Myung (Mayung) – Pollok kills rhino 18 – 39.4 (trip 15) – map A 447

1879 – Nawgong (Nagaon) – Rhino present – Pollok 1879, vol. 1: 95; Wood 1930, 1934 – map A 448

1890s–Nowgong (Nagaon) – Louis Eusebo Fabre-Tonnerre (1854–1901), Assistant Superintendent of Police, shoots rhinos with horns of 14 in (35.6 cm) and 16 in (40.6 cm) – Ward 1896: 282 – map A 448

1892 – Gowhati (Guwahati) – Rhino recorded – Sanyal 1892: 130 – map A 446

1897 – Brahmaputra River – Earthquake affected water levels in the river – Mahanta & Bora 2001

1905 – Laokhowa Reserved Forest notified – chapter 44 – map A 449

1906 – Kaziranga Game Reserve, intention published – chapter 45

1908 – Kaziranga Game Reserve notified – chapter 45 – map A 452

1909 – Mikir Hills – Rhino present – Stack 1908: 3 – map A 453

1959 – Sal Chapori (island near Kaziranga) – Reports of 6 rhinos – Spillett 1966d: 532 – map A 454

1959 – Raja Mayong (Pabitora) Reserved Forest notified – chapter 43 – map A 447

1960 – An area a little way up the Brahmaputra river from the Kaziranga – E.P. Gee estimates presence of 10 rhino – Talbot 1960: 191 – map A 454

1964 – Doboka RF, Nagaon Dt. – 2 rhinos present – Choudhury 2013: 225 – map A 449

1964-04 – Doboka RF – Adult male rhino died – Spillett 1966d: 532 – map A 449

1965-12 – Dihdubi Chapori (Nagaon), 3 miles west of Silghat – Mr. Gohain observed a female with calf – Spillett 1966d: 532 – map A 449

1966 – Ferryghat near Silghat – Presence of solitary rhino – Spillett 1966d: 532 – map A 450

1966 – Doboka RF – Presence of 2 rhino – Spillett 1966d: 532 – map A 449

1966 – Kukurata RF (west of Kaziranga) – Presence of 7 rhino including 2 calves – Spillett 1966d: 531; Simon 1966 – map A 450

1975 – Tatimora Chapori, near Guwahati – Stray rhinos from Pabitora – Choudhury 2002a: 16–19 – map A 446

1980 – Burhachapori RF (44.1 km²). Stray rhinos from Kaziranga – Choudhury 1985a, 1996 – map A 449

1982 – Nowgaon Dt. (Nagaon) – Rhino present – Mukherjee 1982: 49 – map A 448

1984 – Pabitora – One rhino had traveled about 200 km from Pabitora WS to Goalpara. It was captured for translocation to Dudhwa NP in 1984 – Choudhury 1985a, 2013: 225; Lahan 1993: 46 – 19.3 – map A 447

1985 – Kukrakata Hill, west of Kaziranga – Rhino present – Choudhury 1985a; Talukdar 1995: 1 – map A 451

1985 – Northern slopes of the Karbi plateau, adjacent to Kaziranga – Rhino present – Choudhury 1985a–map A 453

1990s – Bagser RF – Stray rhinos from Kaziranga – Choudhury 1996 – map A 451

1990s – Karbi Anglong Dt. – Stray rhinos from Kaziranga – Choudhury 1996 – map A 453

1996 – Kukurakata RF – Rhino from Kaziranga take shelter from floods during monsoon – Choudhury 1996 – map A 451

1996 – Tatimara Chapori – Rhinos stray from Pabitora WS – Choudhury 1996 – map A 446

2008-11 – Sonapur, Kamrup Dt. – Rhino strayed from Pabitora – Choudhury 2010 – map A 446

Rhino Region 35: Assam – North Cachar

1850s – Tularam country, North Cachar – Ruler Senapoti Tularam (d. 1854) extracted tax on rhino horns – Anon. 1855a – 38.11

1856 – North Cachar – Rhino common in the jungles and jheels to the north – Stewart 1856: 596

1880s – Cachar – Rhinos found irregularly – Stuart Baker 1928:175 – 38.22

1888 – North Cachar, 70 miles from headquarters – Edward Charles Stuart Baker (1864–1944) was injured by female rhino when hunting. She had killed 2 women and 1 man. Rhino later found dead – Stuart Baker 1928: 175 – 38.22 – map A 462

1889 – North Cachar – E.C. Stuart Baker shoots 2nd rhino which had killed a man, 6 ft (180 cm) high, horn 18 in (45 cm) – Stuart Baker 1928: 191 – 38.22 – map A 462

1890 – North Cachar – E.C. Stuart Baker shoots 3rd rhino, 6 ft 3 in (190 cm) high, poor horn of 7 in (17.8 cm) – Stuart Baker 1928: 191–198 – 38.22 – map A 462

1897 – North Cachar, 48 miles (77 km) from Haflong – Two rhino shot during a local earthquake – *Derby Mercury* 1897-08-04 – map A 463

1931 – North Cachar –*D. sumatrensis* present – A. Milroy 1931: 4, 1932, 1934a: 102

1970s – North Cachar Hills District: in Sikilangso village, a Karbi (tribal) hamlet near Garampani (Umrongso) – Animal strayed south from Nagaon Dt. along Kopili River valley – Choudhury 1996 – map A 461

Rhino Region 36: Assam – Upper Assam from Golaghat to Tinsukia, South of Brahmaputra

15th cent. – Assam – Medieval Ahom rulers were addicted to rhino hunting – A.K. Dutta 1991: 52; Mazumdar & Mahanta 2016: 25 – 38.8 – map M 875

1756 – Garghaon, former Ahom capital east of Sivasagar – Jean-Baptiste Chevalier (1729–1789) witnesses animal destruction – Chevalier in Deloche 1984: 28, 2008: 29 – 38.8 – map M 875

1856 – Upper Assam (Dibrugarh) – Skull in Bayerische Staatssammlung Munich, no. AM0416 – Schlagintweit 1869, vol. 1: 433 – map A 487

1866-04-25 – Tetellee Gooree (Teteliguri) – W. Millett had young rhino for sale at Rs 350 – F. Buckland 1868: 217 – map A 485

1872 – Dunseree River (going to Dimapur, after leaving Borpathar) – James Johnstone (1841–1895) saw tracks made by elephant and rhino – Johnstone 1896: 7 – map A 481

1875 – Dhansiri valley (below Dimapur) – *R. unicornis*. Major John Butler (1843–1876), Political Agent of Naga Hills, reported the presence of "Rhinoceros Indicus" in the Dhansiri Valley to the east of Golaghat in Assam in 1875 – Butler 1875: 331 – map A 481

1879 – Sibsagar – Rhino present in district – Hunter 1879a, vol. 1: 232 – map A 485

1880 – Komaragaon, near Dhansiri River (Golaghat) – Rhino attacked elephant – *Englishman's Overland Mail* 1880-10-20 – map A 482

1882–1883 – Sadiya – Charles Alfred Elliott (1835–1911), Lieutenant Governor of Bengal, purchased two rhino horns, the largest measuring 19 in (48 cm). They were supposed to be from Singpho (Singphaw) across the border in Myanmar, which cannot be confirmed – Lydekker 1905 – 63.4 – map A 490

1886 – Dhunsiri Mukh, Golaghat – James Willcocks (1857–1926) kills rhino, which were common – Willcocks 1904: 63; Gee 1952: 219, 1964a: 156; Saikia 2009 – 38.20, figs. 38.8, 38.9 – map A 483

1900 – Sibsagar District (Sivasagar) – Former presence – Harper 1945: 380 – map A 485

1916 – Dibrogarh (Dibrugarh) – Edward Talbot Thackeray (1836–1927) finds rhino tracks, animal not seen – Thackeray 1916: 164 – map A 487

1916 – Sudiya (Sadiya) – Major K – hunted rhino, none killed – Thackeray 1916: 164 – map A 490

1946 – Furkating (Golaghat) – Rhino hit by running train – Assam Forest Dept. 1949: 19 – map A 483

1950 – Golaghat – Rhino which had killed a boy was shot by Srijut Protap Chandra Barua, under orders of the Government – Assam Forest Dept. 1950: 28 – map A 482

1966 – Sibsagar Dt. – Isolated rhino present – Mukherjee 1966: 25, 1982: 49; Tiwari 1999, vol. 1: 162 – map A 485

1970 – Kochuoni Pather near Dighaltarang, Tinsukia Dt. – Lone rhino roamed in area – Choudhury 1996 – map A 489

1971 – Nagaon village of Sadiya sub-division, Tinsukia Dt. – Stray rhino – Choudhury 1996 – map A 490

1979-01 – Mathurapur, Sibsagar Dt. – Stray rhino injured a man – Choudhury 1996 – map B 486

1985 – Rangali RF, near Sonari, Sibsagar Dt. – Rhino present – Choudhury 1985a: 27 – map B 488

1985 – Desangmukh RF, Sibsagar Dt. – Rhino present. Forest reserve of 38 km² – Choudhury 1985a–map A 485

1985 – Namdang RF, Dibrugarh Dt. – 2 rhinos observed – Choudhury 1996 – map A 487

1987 – Pani-Dihing WS – Last rhino killed in 1987 – Choudhury 1996 – map A 485

1990s – Ghiladari, Barua Bamungaon, Golaghat Dt. – Stray rhino from Kaziranga – Choudhury 1996 – map A 482

1990 – Saidya, Dibrugarh Dt. – Stray rhino – Lahan 1993: 46 – map A 490

1993 – Pani-Dihing WS – 2 rhino stray from Kaziranga – Choudhury 1996; Talukdar 1995: 1 – map A 485

1993–1994 – Dihingmukh RF – 1 rhino strayed from Pani-Dihing – Choudhury 1996, 2013: 225 – map A 485

1994 – Jorhat town – Stray rhino from Kaziranga on fringe of town – Choudhury 1996 – map B 484

2002 – Panidihing – Team visited. Reserve lost all rhino – Khan et al. 2002 – map A 485

2007-03-15 – Panidihing – One animal from Dikhowmukh straying from Kaziranga – Choudhury 2010 – map A 485

2007-04 – Kolakhowa and Rongdoi – 4 rhino, maybe stray from Kaziranga – Choudhury 2007a: 24 – map A 485

2009 – Dihingmukh RF – Carcass without horn found – Choudhury 2010 – map A 485

2009-03-04 – Sissimukh Haldhibari near Machkhowa, Dhemaji Dt. – Person injured by rhino – Choudhury 2010 – map A 485

2013 – NE Golaghat – Stray rhino – Choudhury 2013: 225 – map A 482

Rhino Region 37: Upper Assam from Lakhimpur to Tinsukia, North of Brahmaputra

1893 – Lakhimpur – Lieutenant-Colonel Lewis killed a large rhino in a flooded area. Head taken. Horn measured 10 in (25.4 cm) high and 13 in (33 cm) round the curve from front base to tip – Lewis 1893 – map A 472

1908 – Lakhimpur Dt. – Rhino scarce, occasionally found – Allen 1908a, vol. 8: 16; Choudhury 2003: 65, 2013: 225 – map A 472

1917 – (Upper?) Assam – The teaplanter Walter M. Nuttall shot a rhino – Playne 1917; Nuttall 1917: 633 – 38.29, fig. 38.16 – map A 472

1935 – North Lakhimpur – Need to check if any rhino is present – Milroy 1935: 21 – map A 472

1958 – Subansiri River, North Lakhimpur – Rhino present – Bertrand 1958: 183 – map A 472

1966-03 – Merbeel in Dhakuakhana Subdivision – Rhino poached – Choudhury 1996 – map B 476

1979 – Lamugaon near Merbeel – Rhino seen – Choudhury 1996 – map B 476

1985 – Matmora – 7–8 rhino in 1980s. Last poached in 1985 – Choudhury 1996 – map B 475

1985–1987 – Bebejia area of Dhakuakhana – 2 rhino seen, traveling east in 1988 – Choudhury 1996 – map B 476

1986 – Dulung RF – Female with full-grown calf traveled to Pabha RF, then Dulung RF, then to Sibsagar – Choudhury 1996 – map B 473

1986 – Kadam RF – Lone rhino seen – Choudhury 1996 – map B 473

1987 – Kakoi RF near Joihing River – Lone rhino. Later moved to Panir RF (Arunachal Pradesh) – Choudhury 1996, 1997a–map B 473

1988 – Thekeraguri – Rhino seen near the confluence of Korha and Charikaria rivers – Choudhury 1996 – map B 475

1990 – Dhakuakhana – Stray rhino – Lahan 1993: 46 – map B 475

1990 – Majuli RF (Molai) – Stray rhino from Kaziranga – Lahan 1993: 46; Choudhury 1996 – map B 474

1992 – Dhunaguri – 2 rhino stray from Kaziranga, chased back – Choudhury 1996 – map B 477

1992–1993 – Borchapori – Villager injured by rhino – Choudhury 1996 – map A 475

1993-01 – Luit Suti of Majuli RF – Carcass found – Choudhury 1996 – map B 474

1994 – Majuli RF (Molai) – 2–3 rhino found (stray from Kaziranga) – Choudhury 1996, 2013: 225 – map B 474

1995 – Sisi-Kalghar area, NE of Matmora – Stray rhino seen – Choudhury 1996 – map B 475

2007-02-24 – Lakhimpur Dt. – Rhino straying from Kaziranga was darted but did not survive – B. Choudhury et al. 2009 – map A 472

2012-08 – Molai Kathoni (Majuli RF) – adult male rhino killed – Times of India (online) 2012-08-04 – map B 474

2013 – Majuli RF – Stray rhino – Choudhury 2013: 225 – map B 474

2013 – Lakhimpur – Stray rhino – Choudhury 2013: 225 – map A 472

2016-07 – Bhuyapara village, Sootea – rhino straying from Kaziranga was tranquilized and taken to the Assam State Zoo, Guwahati to recover from injuries – WWF India press release 2016 – A434

2022 – Majuli RF (Molai) – Indian rhino present in this forest created and planted by Jadav Payeng (b. 1959) – Bhuyan & Gayan 2022 – map B 474

2022-09 – Gotaimari, Makuwa Chapori in the south-west part of Jamugurihat – stray rhino from 6th extension of Kaziranga – *Sentinel Assam* 2022-09-15 – map A434

• • •

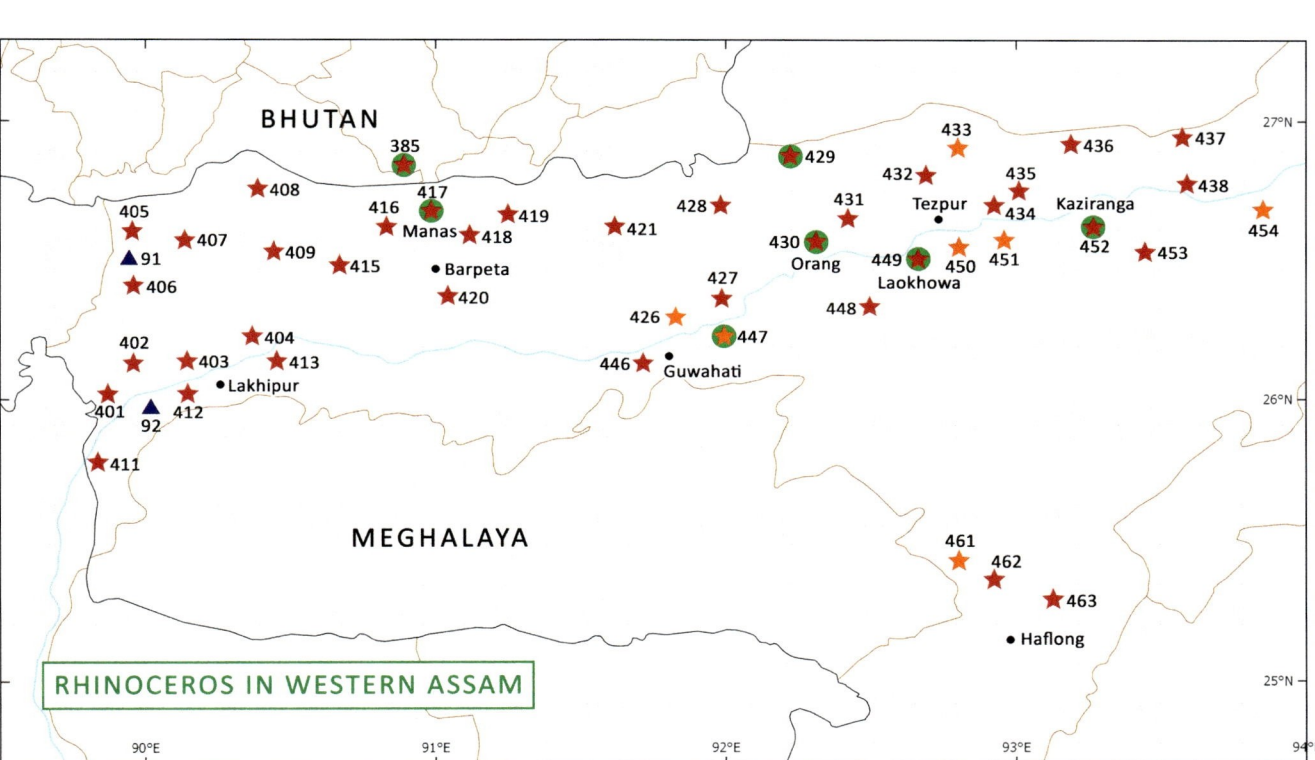

MAP 38.19   Records of rhinoceros in the western and middle parts of the state of Assam. These are in Rhino Regions 30 to 35. The current reserves are circled in green. There are two records of *D. sumatrensis*, but none for *R. sondaicus* in this area. The numbers and places are explained in Dataset 38.47.

★   Presence of *R. unicornis* – dark colour: records before 1950; lighter colour: records only after 1950

⬠   Artefacts

▲   Presence of *D. sumatrensis*

Dataset 38.47: Localities of Records of Rhinoceros in Western and Middle Assam

Localities are arranged roughly from west to east. Numbers starting A combine all dates, those starting B are records after 1950 only (mostly strays, not permanent populations).

The numbers and places for western Assam are shown on Map 38.19.

Rhino Region 30: Assam – Sankosh River, North of Brahmaputra

A 401	Putimari (Patamari), west of Dhubri – 25.95N; 89.85E-1868
A 402	Gompore (Gauripur) – 26.08N; 89.96E-1872
A 402	Dhubri – 26.02N; 90.00E-1873, 1876, (Pollok)
A 402	North of Dhubri, towards Gauripur – 26.05N; 90.00E-1877
A 403	Dudnath, north bank Brahmaputra – 26.20N; 90.31E-1894
A 404	Goma Dooar (Chapar) – 26.30N; 90.30E-1869
A 404	Chumpamuttee River (Chapar) – 26.30N; 90.30E-1869
A 404	Chandodingah Hill (Chapar) – 26.30N; 90.30E-1869
A 404	Jogeegopah (Jogighopa) – 26.22N; 90.56E-1864
A 404	Goalpara, on opposite bank – 26.30N; 90.60E-1916, 1919
A 405	Raimona – 26.65N; 89.86E-1900
D 091	Sankosh River – 26.60N; 90.02E-1875 – Presence of *D. sumatrensis*
A 406	Kachugaon RF – 26.38N; 90.00E-1905
A 407	Patgaon – 26.57N; 90.21E-1901
A 407	Sapkata near Patgaon – 26.59N; 90.21E-1892, 1893
A 408	Coochabaree; Dewagaon – 26.78N; 90.36E-1865
A 409	Khagrabari (north of Sidli) – 26.73N; 90.44E-1880, 1881
A 409	Sidli – 26.54N; 90.46E-1864

Rhino Region 31: Assam – Goalpara, South of Brahmaputra

A 411	Singamaree (Singrimari) – 25.74N; 89.89E-1827
D 090	Lower Assam, probably Goalpara – 26.00N; 90.30E-1862 – Presence of *D. sumatrensis*
A 412	Rakuldooly, near Lakhipur – 26.05N; 90.25E-1864
A 412	Lukheepore, Luckipore (Lakhipur) – 26.00N; 90.30E-1801, 1860s, 1870s, 1887, (Pollok 1860s)
A 412	Foot of Garo Hills (Lakhipur) – 26.00N; 90.30E-1870s, 1872, 1880s
A 412	Khalsamari, south bank of Brahmaputra – 25.68N; 89.94E-1894
A 413	Goalpara, Gwalpara, Givalpara – 26.10N; 90.50E-1755, 1777, 1818, 1820s, 1830, 1832, 1879, 1919, 1985
W 639	Goalpara – 26.10N; 90.50E-fossil
A 413	Valley of Assam; Eastern Dwars (Goalpara) – 26.10N; 90.50E-1872, 1879

Rhino Region 32: Assam – Manas, North of Brahmaputra

A 415	Ripu RF, Chirang RF – 26.54N; 90.65E-1933
A 415	Bijnee, Bijni RF – 26.48N; 90.70E-1815, 1864, 1901, 1903
A 416	Kookooriah (Manas) – 26.70N; 90.90E-1870
A 417	North Kamrup (Manas) – 26.70N; 91.00E-1856, 1865, 1871, 1875, 1879, 1887, 1902, 1911
A 417	Manas NP – 26.77N; 91.00E-passim
A 418	Bansbari, Busbaree (Bahbari) – 26.61N; 91.00E-1899
A 418	Bhuiapara, east of Manas River – 26.70N; 91.00E-1895, 1899

A 419	Lahapara – 26.75N; 91.10E-1902
A 419	Kumlabaree – 26.65N; 91.15E-1871 (Pollok)
A 419	Poho-Marah (Pahumara) River – 26.65N; 91.15E-1871 (Pollok)
A 419	Jalagaon (Jalahgaon, Manas) – 26.65N; 91.22E-1899
A 420	Burpettah (Barpeta) – 26.33N; 91.00E-1868, 1872 (Pollok)
A 421	Bhootan Duars south of Dewangiri – 26.78N; 91.45E-1865
A 421	Baksa (from Guwahati to Tawang) – 26.70N; 91.50E-1834
A 421	Rungiah (Rangia) – 26.45N; 91.65E-1869, 1872 (Pollok)

Rhino Region 33: Assam – Darrang, North of Brahmaputra

B 426	Kurua, Kuruwa – 26.25N; 91.80E-1980, 1975–1995
B 426	Mandakata (west of Kurua) – 26.27N; 91.74E-1975–1995
A 427	Durrung, Darrang Dt. – ca. 26.50N; 92.00E-1879, 1890, 1932, 1982
A 427	Mungledye (Mangaldoi) – 26.45N; 92.02E-1882
A 428	Booreegaon (Borigaon) – 26.80N; 92.03E-1911, 1930
A 429	Sonarupa (Sonai Rupai) – 26.88N; 92.33E-1911, 1934
A 430	Orang (National Park) – 26.55N; 92.30E-1911, 1915
A 431	Borsola jungles at Singrighat (Singari) – 26.62N; 92.50E-1911
A 431	Belsire (Belsiri) – 26.80N; 92.54E-1898
A 431	Kochmara RF – 26.60N; 92.59E-1995
A 432	Tezpur, Tezpore – 26.60N; 92.70E-1890, 1909, 1980–1995
A 432	Balipara – 26.89N; 92.44E-1932
B 433	Nameri WS – 26.90N; 92.87E-1990
A 434	Bisnath (Biswanath) – 26.72N; 93.13E-1836, 1896, 1909
A 434	Panpur RF – 26.67N; 93.16E-1985
A 434	Bhuyapara village (Sootea) – 26.73N; 93.09E-2016
A 434	Gotaimari (Jamugurihat) – 26.67N; 92.91E-2022
A 435	Khatonibari (Khatani Bari) – 26.75N; 93.25E-1990
A 436	Bargang Tea Estate, Ketela – 26.83N; 93.31E-1920s
A 437	Gohpur – 26.87N; 93.60E-1911, 1990
A 437	Silikaguri – 26.93N; 93.85E-1994, 1995
A 438	Bahali, near Howrah-Ghat – 26.87N; 93.80E-1845

Rhino Region 34: Assam – from Guwahati to Golaghat, South of Brahmaputra

A 446	Gowahatty, Gowhati, Gauhati (Guwahati) – 26.15N; 91.80E-1847, 1855, 1868, 1892
A 446	Tatimora Chapori, near Guwahati – 26.22N; 91.89E-1975, 1996
A 446	Sonapur, Kamrup Dt. – 26.11N; 91.98E-2008
B 447	Raja Mayang (Pabitora) – 26.22N; 92.05E-1870 (Pollok), 1967
B 447	Pabitora – 26.22N; 92.05E-1959, 1984
A 448	Nagaon, Nowgong, Nawgaon – 26.50N; 92.70E-1845, 1848, 1867, 1879, 1890s, 1982
A 449	Laokhowa (Lowqua Ghat) – 26.50N; 92.70E-1755, 1865, 1867, 1905
A 449	Burhachapori RF – 26.50N; 92.70E-1980s
A 449	Doboka RF – 26.15N; 92.85E-1964, 1966
A 449	Dihdubi Chapori west of Silghat – 26.60N; 92.90E-1965
B 450	Ferryghat near Silghat – 26.60N; 92.90E-1966
B 450	Kukurata RF (west of Kaziranga) – 26.60N; 92.95E-1966, 1967
B 451	Bagser RF (SW of Kaziranga) – 26.45N; 92.95E-1990s
B 451	Kukrakata Hill (Kukurakata RF) – 26.60N; 93.17E-1966, 1985, 1996
A 452	Kaziranga NP headquarters – 26.60N; 93.44E-passim
A 452	Numalighar Tea Estate – 26.57N; 93.73E-1933

A 453   Mikir Hills – ca. 26.50N; 93.50E-1909
A 453   Karbi Anglong – 26.50N; 93.60E-1985, 1990s
B 454   Sal Chapori (island near Kaziranga) – 26.80N; 93.85E-1959
B 454   Area east of Kaziranga – 26.75N; 93.90E-1960

Rhino Region 35: Assam – North Cachar

B 461   Sikilangso village near Garampani (Umrongso) – 25.50N; 92.67E-1970s
A 462   70 miles from Haflong – 25.40N; 92.90E-1888, 1889, 1890
A 463   48 miles [sic] from Haflong – 25.30N; 93.15E-1897
*   Toolaram country (North Cachar) – 25.30N; 93.30E-1850s

• • •

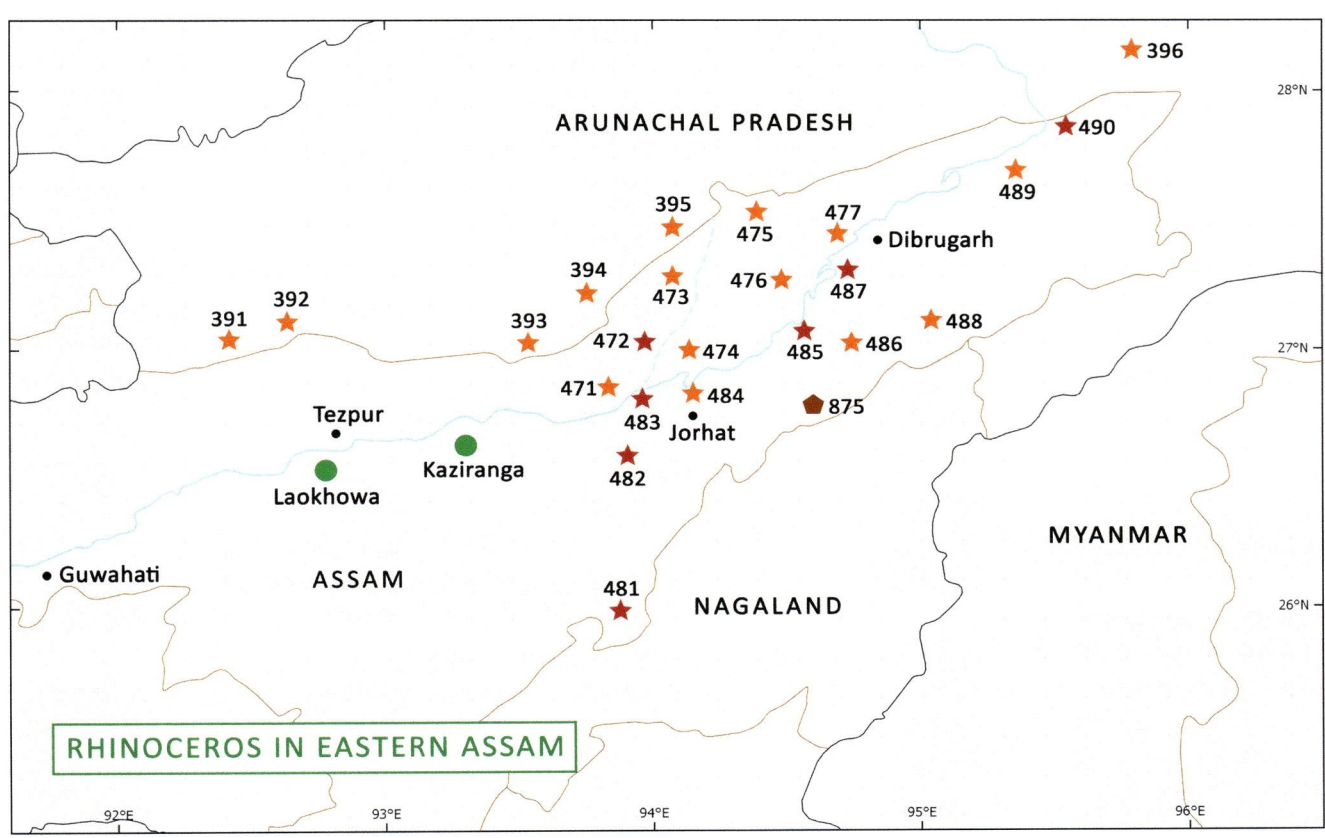

MAP 38.20   Records of rhinoceros in the eastern (upper) part of the state of Assam. These localities are in Rhino Regions 36 and 37. The numbers and places are explained in Dataset 38.48. For symbols, see map 38.19

Dataset 38.48: Localities of Records of Rhinoceros in Eastern Assam

The numbers and places for eastern or upper Assam are shown on Map 38.20.

Rhino Region 36: Assam – Upper Assam from Golaghat to Tinsukia, South of Brahmaputra

A 481   Dhansiri, Dunseree River (Dimapur) – 25.93N; 93.83E-1872, 1875
A 482   Golaghat – 26.46N; 94.00E-1950, 2013
A 482   Komaragaon (Golaghat) – 26.60N; 94.00E-1880
A 482   Ghiladari, Barua Bamungaon (Golaghat) – 26.46N; 94.00E-1990s
A 483   Furkating (Golaghat) – 26.46N; 94.01E-1946
A 483   Dhunsiri Mukh (Golaghat) – 26.79N; 94.06E-1886
B 484   Jorhat town – 26.76N; 94.17E-1994
M 875   Garghaon, former Ahom capital east of Sivasagar – 26.93N; 94.75E-1756

A 485   Desangmukh FR, Sivasagar Dt. – 27.08N; 94.56E-1985
A 485   Sibsagar Dt. (Sivasagar) – 27.10N; 94.60E-1879, 1900, 1966
A 485   Pani-Dihing – 27.09N; 94.62E-1987, 1993, 2002, 2007
A 485   Tetellee Gooree (Teteliguri) – 27.28N; 94.72E-1866
A 485   Dihingmukh RF – 27.28N; 94.72E-1993, 2009
A 485   Kolakhowa and Rongdoi – 27.26N; 94.76E-2007
B 486   Mathurapur, Sibsagar Dt. – 26.98N; 94.90E-1979
A 487   Dibrugarh – 27.20N; 94.90E-1916
A 487   Namdang RF, Dibrugarh Dt. – 27.20N; 94.90E-1985
B 488   Rangali forest near Sonari – ca. 27.02N; 95.05E-1985
B 489   Dighaltarang, Tinsukia Dt. – 27.62N; 95.41E-1970
A 490   Sudiya, Sadiya – 27.86N; 95.63E-1882, 1916, 1990
A 490   Nagaon village of Sadiya – 27.82N; 95.64E-1971

Rhino Region 28: Arunachal Pradesh (East)

B 396   Mishmi Hills – 28.30N; 95.75E-1952 – 63.6

Rhino Region 37: Upper Assam from Lakhimpur to Tinsukia, North of Brahmaputra

B 471   Borchapori – 26.98N; 93.89E-1992

A 472   Lakhimpur Dt. – 27.15N; 94.10E-1893, 1908, 1917, 1935, 2007, 2013

A 472   Subansiri River (North Lakhimpur) – 27.25N; 94.15E-1958

B 473   Kakoi RF near Joihing River – 27.35N; 94.06E-1987

B 473   Kadam RF (Dulung) – 27.40N; 94.16E-1986

B 473   Dulung RF – 27.41N; 94.20E-1986

B 474   Majuli RF (Molai) – 26.86N; 94.17E-1990, 1994, 2012, 2013, 2022

B 474   Luit Suti of Majuli – 26.96N; 94.21E-1993

B 475   Thekeraguri – 27.42N; 94.25E-1988

B 475   Dhakuakhana – 27.30N; 94.40E-1990

B 475   Matmora – 27.20N; 94.50E-1985, 1995

B 476   Bebejia area of Dhakuakhana – 27.25N; 94.52E-1985

B 476   Merbeel in Dhakuakhana Subdivision – 27.27N; 94.52E-1966

B 476   Lamugaon near Merbeel – 27.27N; 94.52E-1979

B 477   Dhunaguri – 27.48N; 94.64E-1992

∴∴

TABLE 38.50   List of persons who were involved in shooting or capturing rhino in Assam

Person	Date	Locality. Type of interaction
Jean-Baptiste Chevalier (1729–1789)	1755	Goalpara. Wounds 1 rhino
Thomas Craigie	1777	Goalpara. Kills 1 rhino
James Rennell (1742–1830)	1765	Assam. Observed rhino
David Scott (1786–1831)	1827	Goalpara. Saw 6 rhino
Bruce (maybe fictional)	1830s	Assam. Killed 1 rhino
Marmaduke Burrough (1797–1844)	1830	Assam. Transport 1 rhino to USA
John Leslie (1800–1831)	1832	Goalpara. Dies while skinning a rhino
Woodsman (pseudonym) with P-r	1836	Biswanath. Killed rhino
P-r (no data) with Woodsman	1836	Biswanath. Killed rhino
John Maclelland (1805–1883)	1839	Assam. Observed rhino
Francis Jenkins (1793–1866)	1855	Assam. Donates rhino head to museum in Cornwall
Lieutenant George Campbell (d. 1876)	1845	Bahali. Killed rhino. With Butler
John Butler (1809–1874)	1845	Bahali. Killed rhino. With Campbell
William Jamrach (1843–1891)	1855	Obtained 2 rhinos captured in Assam, taken to UK
Hermann Schlagintweit (1826–1882)	1856	Guwahati. Saw tame rhino, and collected skulls
Son of Rajah (Zamindar) of Lukheepore	1860s	Goalpara. Killed 34 rhino
Captain Wilfed Dakin Speer (1835–1867)	1862	Killed a rhino, shown in photograph
Charles Richard Cock (1838–1879)	1864	Lakhipur. Killed 29 rhino. With Bunbury
Abraham Charles Bunbury (1823–1904)	1864	Lakhipur. Killed 29 rhino. With Cock
Henry Andrew Sarel (1823–1887)	1865	Manas. Killed 3 or more rhino
Captain John Henry Baldwin (1841–1908)	1865	Laokhowa. Fellow officers killed 1 rhino
W. and friend C. (no details)	1869	Chapur (Goalpara). Killed several rhino
Fitzwilliam Thomas Pollok (1832–1909)	1867–1872	Assam. Killed 44 rhinos (chapter 37)
Col. James Cathorne Cookson (1821–1900)	1867	Laokhowa. Killed 1 rhino
General Charles Reid (1819–1901)	1868	Laokhowa. Killed 2 rhino
Major George Machardy Bowie (1835–1876)	1868	Manas. Kills 1 rhino with Pollok
Augustus Kirkwood Comber (1828–1895)	1868, 1872	Lakhipur, Manas. Killed 2 rhino. With Pollok
G.H. Jackson, 43rd Regiment Native Infantry	1868, 1873	Manas. Killed several rhino. With Pollok
George Udny Yule (1813–1886)	1870s	Lakhipur. Saw 7 rhino
John Barry, tea planter	1870s	Manas. Killed several rhino. With Pollok
Col. J. Macdonald, Survey of India	1871	Manas. Killed several rhino. With Pollok
Col. George Campbell (d. 1876)	1871	Manas. Killed several rhino. With Pollok
A. Anley, Superintendent of Police	1871	Manas. Killed several rhino. With Pollok
William Russell, 2nd Baronet (1822–1892)	1872	Gauripur. Kills rhino with horn of 38.1 cm, in NHM

TABLE 38.50   List of persons who were involved in shooting or capturing rhino in Assam (*cont.*)

Person	Date	Locality. Type of interaction
Hon. Talbot (no details)	1872	Barpetta. Killed 6 rhino
Lieutenant William John Williamson	1873	Dhubri. Kills rhino
Edmund Giles Loder (1849–1920)	1875	Assam. Horn of 38.7 cm in his collection. Perhaps bought in 1874–1875
James Matthie (1806–1865)	1875	Manas. Hunted rhino
F. (no details)	1876	Dhubri. Kills 1 rhino
Allan Octavian Hume (1829–1912)	1880s	Lakhipur. Killed 1 rhino, in NHM London
Arthur John Primrose (1853–1888)	1882	Darrang. Killed 1 rhino
Henry Meysey-Thompson (1845–1929)	1883	Lower Assam. Killed 3 rhino
George Peress Sanderson (1848–1892)	1884	Assam. Obtained female, in NHM London
General Sir James Willcocks (1857–1926)	1886	Golaghat. Killed 1 rhino
Edward Biscay Marinatus Baker (1828–1887)	< 1887	Lakhipur. Killed 4 rhino
Henri Charles Joseph Le Tonnelier, Marquis de Breteuil (1847–1916)	1887	Kamrup. Killed 1 rhino. With Gaston Breteuil
Gaston de Breteuil (1864–1937)	1887	Kamrup. Killed 1 rhino. With Henri Breteuil
Ridgeway (unidentified American)	1887	Kamrup. Killed 1 rhino. With Breteuil
Edward Charles Stuart Baker (1864–1944)	1888–1890	North Cachar. Killed 3 rhino
William Parker Yates Bainbrigge (1853–1934)	1890s	Assam. Horn 36.2 cm recorded in Rowland Ward
Hugh Barclay (1799–1884)	1890s	Assam. Horn 21 cm recorded in Rowland Ward
Louis Eusebo Fabre-Tonnerre (1854–1901)	1890s	Nowgong. Horns of 40.6 cm and 35.6 cm recorded in Rowland Ward
Addison Yalden Thomson (1863–1931)	1890s	Tezpur. Horn 16.5 cm recorded in Rowland Ward
Herbert Christian Holland (1858–1916)	1892	Assam. Horn 31.8 cm recorded in Rowland Ward
Peter Carlaw Walker (1854–1915)	1892	Assam. Horn 21.0 cm recorded in Rowland Ward
Lieutenant-Colonel Lewis (n.d.)	1893	Lakhimpur. Killed rhino with horn 33.0 cm
A.H.M.T. (not identified) and friend	1894	Assam. Killed 3 rhino
Humphrey Brooke Firman (1858–1916)	1896	Biswanath. Horn 32.4 cm recorded in Rowland Ward
Walter Stanhope Sherwill (1815–1890)	1898	Belsiri. Horn 41.5 cm recorded in Rowland Ward
George Alexander Dolby (1854–1939)	1898	Assam. Horn of 33 cm recorded in Rowland Ward
Philip Breda Vanderbyl (1867–1930)	1899	Manas. Killed 2 rhino
Maxwell Hannay Logan (1873–1947)	1902	Assam. Horn 45.7 cm recorded in Rowland Ward
Walter Alexis Rene Doxat (b. 1873)	1906	Assam. Horn of 33 cm recorded in Rowland Ward
Surya Kanta Acharya Choudhury of Muktagacha (1849–1908)	1908	Assam. Killed 2 rhino
Thomas Charles Briscoe (1857–1909)	1909	Tezpur. Record horn of 61.3 cm recorded in Rowland Ward
Charles Victor Alexander Peel (1869–1931)	1909	Assam. Killed 1 rhino, preserved in Exeter
A. Butcher (pseudonym?)	1909	Assam. Killed 1 rhino, preserved in BNHS Mumbai
Frederick William Gore (1825–1909)	1910	Assam. Killed 1 rhino, preserved in BNHS Mumbai
Guy Douglas Arthur Fleetwood Wilson (1851–1940)	1911	Manas. Killed 1 rhino
Dennis Herbert Felce (1875–1932)	1913	Assam. Horn of 46.7 cm recorded in Rowland Ward
Major K.	1916	Sadiya. Hunted rhino
Edward Talbot Thackeray (1836–1927)	1916	Dibrugarh. Found rhino tracks
Walter M. Nuttall	<1917	Upper Assam. Killed 1 rhino
Raja of Gauripur	1919	Goalpara. Party killed 1 rhino
R.V. Yates (n.d.) with F.J. Price	1920s	Assam. Horn of 42.5 cm recorded in Rowland Ward
F.J. Price (n.d.) with R.V. Yates	1920s	Assam. Horn of 42.5 cm recorded in Rowland Ward
Frank Nicholls (b. 1889–)	1920s	Biswanath. Killed a rogue rhino
Jamun (possible pseudonym)	1925	North Assam. Tracks 1 rhino

TABLE 38.50   List of persons who were involved in shooting or capturing rhino in Assam (*cont.*)

Person	Date	Locality. Type of interaction
J. (friend of Jamun)	1925	North Assam. Killed 1 rhino
Henry Stotesbury Wood (1865–1956)	1930s	Darrang. Hunts rhino
Y.D.	1933	Numalighar. Killed 1 stray rhino (see Kaziranga)
Frank Buck (1884–1950)	1934	Manas. Filmed rhinos for movie, and captures one
J. McAlpine-Woods	< 1949	Assam. Donated skull to Exeter Museum
Srijut Protap Chandra Barua	1950	Golaghat. Killed stray rhino under Govt. orders

38.2   The Testimony of Fitzwilliam Pollok

A rare glimpse into the mind of the sportsmen of the mid-19th century in India is provided in the books by Fitzwilliam Thomas Pollok (1832–1909). He published his *Sport in British Burmah, Assam, and the Cossyah and Jyntiah Hills* in 1879, which, in the words of a review published in the *Spectator* of 11 October 1879, finds its place among works on sport in India and Africa "evincing a blood-thirstiness which is repellent to most natures." It was redeemed partly by the fact that it dealt with provinces seldom visited, embodying the results of twenty-six years of wanderings. We must acknowledge that Pollok presented us with an honest and immediate insight into practices which must have been only all too common in his days. If we skip through some of the more vivid descriptions of the hunts, there is much valuable detail in this book which is impossible to find elsewhere, certainly as far as Assam is concerned. His encounters with the rhino are combined in chapter 39.

Pollok is one of the few writers of that period who explicitly comments on the value of rhino horns and other body parts. He found it both amusing and annoying that the Assamese would follow him on his expeditions hoping to obtain some meat of a dead rhino: "Although many castes of Brahmins, Hindoos and Nawarries eschew all flesh, living on grain only, some of them make an exception in favour of the flesh of this pachyderm. I have been asked to dry the tongues for them, and these they pulverise, bottle, and indulge in a pinch or two if unwell. The Assamese prefer its flesh to all other, and used to follow me about like as many vultures. No sooner was the life of one extinct, than they would rush knife in hand and not leave a scrap on the skeleton. Even the hide they roast and cut as we do the crackling of pig." Besides the meat, it was of course the horn which had the real commercial value. Most horns were too small to be displayed as trophies, but there was certainly a market for them. Majors C.R. Cock and Bunbury became famous among hunters for their ability to finance their shooting expeditions by selling rhino horn to traders in Guwahati. In the 1870s, the price was Rs. 45 per 2 lbs. It is thought that these horns were mostly exported through Kolkata to Chinese markets.

38.3   Population Estimates of Rhino in Assam

When the officers of the Assam administration enquired about the presence of rhinos in 1902, they were told that very few were left, which soon led to the notification of Kaziranga, Laokhowa and Manas as reserved forests. Estimates changed from indications like plentiful or common to reduced or scarce (table 38.51). There was no attempt at the time to obtain a more definite figure of the actual rhino population in the state. There was a sense that game in general was less easy to view or obtain, which was put in words by the naturalist-hunter Alexander Kinloch (1904: 61) saying that rhino numbers "have terribly reduced; but there are enough left to enable a well-equipped sportsman to be pretty sure of obtaining one or two specimens." Later in the century, this impression was translated to mean a reduction of the population to even just a dozen, or maybe slightly more. The protection measures were certainly timely because the rhinos had become threatened by a reduction of habitat and an increase in the availability of guns. However, for instance the German naturalist Wolfgang Ullrich (1971: 20) is probably much closer to the mark by calculating a possible 100–150 rhinos in Kaziranga at the start of the century, to which the animals in Laokhowa, Manas and other pockets in the state should be added. Genetic tests have shown that it is very unlikely that any sort of bottleneck occurred in the last hundred years (P. Das 2014; Zschokke 2016).

The pressures on rhinos were real and the populations suffered great reductions over the years, until such time when the problem became apparent and required immediate action. Laws regarding the use of game were in a state of flux, as they had to change from the presentation of rewards for shooting a rhino to a system where the same action was a punishable offence (38.32). Legislation was developed to regulate access to forests and the property rights of forest products. The Indian Forest Act of 1878 stipulated that all products, including rhino horns, were government property if they originated in Reserved Forests. A similar clause was included in the Assam Forest Regulation of 1891. However, for wildlife existing outside designated forest areas, it is unclear how the laws would apply to anyone who killed a rhino to sell its horns or other body parts. Possibly, the number of rhinos residing outside reserve forests had already diminished to such an extent that there was no need to enact general legislation.

The idea to protect wildlife in general and rhinos in particular in designated areas gained momentum at the end of the 19th century. At first the 'reserved forests' were more aimed at harvesting their commercial value, later they were transformed to game reserves in order to protect rhinos and other wildlife. When the state government realized the need of protection, they were able to survey forests which had few villages and had not yet been converted to tea plantations. Famous reserves like Kaziranga and Manas were quickly notified and have continued to serve the purpose, as is seen in the chapters that follow about each of them. Outside the reserves, there have always remained little pockets of rhinos or individuals straying outward from the main centers (table 38.52).

38.4   Threats in the 20th Century

In 1929, reports were received of large-scale poaching activities by local Kacharis and Assamese around the Manas River in north-west Assam. The Government responded by dispatching a detachment of the Assam Rifles under a British officer. They were able to settle the issue, although the same problems reoccurred in adjacent areas of North Bengal. It was estimated that 90 to 100 rhinos were killed in a few years. Poaching has since remained a continuous, although oftentimes opportunistic threat. It is said that in the war years 1943–1944, only three rhinos were killed in reserve forests over the entire Assam area.

TABLE 38.51   Estimates of the status of rhinoceros in Assam

Year	Estimates	Source[a]
1860	extremely numerous	R140
1867	common	R316
1869	becoming rare	R322
1869	common	R318
1876	plentiful	R312
1879	common	R314
1882	very plentiful	R187
1886	numerous	R324
1887	plentiful	R311
1889	numerous	R319
1900	very abundant	R313
1900	less than 20	R234, R317
1904	reduced	R140
1905	very scarce	R104
1906	12	R175
1908	20	R231
1927	increasing	R310
1930	50	R079
1931	small remnant	R315
1933	threatened	R303
1950	53–200	R005
1952	240	R106, R199
1953	240	R107
1954	547	R007
1957	350	R233
1958	350	R108, R119, R120
1963	300	R226
1964	375	R112
1966	486	R196
1972	20	R282
1993	1405	R278
1995	1398/1500	R278, R239
1997	1406	R278

a References in Table. R005 Ali 1950: 472; R007 Amritaghata 1955; R020 Assam Forest Dept. 1948: 100; R079 Dey 1994a; R103 Gee 1950a: 1732; R104 Gee 1950b: 82; R106 Gee 1952a: 224; R107 Gee 1953a: 341; R108 Gee 1958: 353; R112 Gee 1964a: 153; R119 Grzimek 1958: 118; R120 Grzimek 1960: 22; R140 Kinloch 1904: 61; R171 Mountfort 1991; R175 Mukherjee 1966: 25; R187 Pollok 1882b: 31; R196 Rowntree 1981: 67; R199 Ryhiner 1952b: 161; R226 Stracey 1963: 243; R231 Sumthane et al. 2017; R233 Talbot 1957: 390; R239 Talukdar 1995: 1; R278 Vigne & Martin 1998: 25; R282 Waller 1972a: 6; R310 Ali 1927a: 860; R311 Baker 1887: 252; R312 Baldwin 1876: 145; R313 Russell 1900: 339; R314 Handique 2004: 20; R315 Hobley 1931: 21; R316 Jerdon 1867: 233; R317 Laurie et al. 1983; R318 Lewin 1869: 16; R319 Marston 1889: 55; R322 Schlagintweit 1869, vol. 1: 433; R324 Willcocks 1925: 40.

TABLE 38.52   Estimates of status of rhinoceros in pockets outside
               the reserves in Assam

Year	Numbers	Poached	Source[a]
1953	10		R204
1958	10		R108
1963	43		R207, R267
1966	35, 40, 52		R208; R220; R164
1978	25		R145
1979	15		R164
1985	30		R186
1986	25		R164, R273
1988		4	R088
1989	42	7	R273
1990		6	R088
1991		2	R088
1992	60	2	R088, R198
1993	60	3	R164, R274, R275
1994		2	R088, R278
1995	20	2	R088, R098, R164, R278
1996		1	R088, R278
1997	25	2	R088, R278
1998		1	R088

a   References in Table. R088 Doley 2000; R098 Foose & van Strien 1997:
    9; R108 Gee 1958: 353; R137 Khan & Foose 1994: 6; R138 Khan 1989:
    3; R145 Laurie 1978: 12; R164 Menon 1995: 15; R186 Penny 1987: 48;
    R198 Roy 1993; R204 Shebbeare 1953: 143; R207 Simon 1966; R208
    Simon 1967; R211 Sinha & Sawarkar 1993; R220 Spillett 1966d: 529;
    R267 Ullrich 1965a: 101; R273 Vigne & Martin 1991: 215; R274 Vigne &
    Martin 1994; R278 Vigne & Martin 1998: 25.

The devastation of the second World War was followed by India's independence in 1947. The Government of Assam took a serious approach to its responsibility to preserve wild resources. In 1949, the Bombay Natural History Society, being the foremost institution of its kind in India, was invited to send a party of naturalists to survey the province's wildlife and to recommend improvements to the sanctuaries. The team, headed by the eminent naturalist Salim Ali (1896–1987) and the American ornithologist Sidney Dillon Ripley (1913–2001), visited the main reserves in March 1949 with the help of the Conservator of Forests, Patrick Donald Stracey (1906–1988), C.G. Baron in Manas and E.P. Gee in Kaziranga. They submitted valuable suggestions for improvements to the sanctuaries and, at the same time, emphasized the fact that the status of rhinos did not allow for any complacency in attitude or action.

38.5   E.P. Gee, Ambassador of Assam's Wildlife

In 1927, a young man came from England to start a career as a tea planter in the hills of Assam. Edward Pritchard Gee (1904–1968) was to become one of the greatest ambassadors for the rhino in Assam fuelled by his great passion for nature and wildlife photography (fig. 38.3). He believed that the maintenance of reserves was the only way in which large mammals like the rhino would survive the pressures of poaching and habitat reduction. His home was only about 50 km from Kaziranga, and he was a regular visitor to the reserve once it opened to the public in 1938. Through his articles and books he was able to convey his love for India and its wildlife to naturalists abroad (Gee 1948 to 1964). To travel to Assam and be taken on an elephant in the hopes of seeing a rhino or a wild buffalo was still a rare privilege, available to few people. E.P. Gee was an incredible ambassador both for Indian wildlife and for Kaziranga.

As a member of the Indian Board of Wildlife since 1952 and of the Survival Service Commission of IUCN, Gee was influential in shaping the conservation policies of independent India. Born in Durham, educated at Emmanuel College in Cambridge, he arrived in India in 1927 where he became a tea planter, first at Hathikuli Estate on the fringe of Kaziranga, from 1946 to February 1959 at Doyang Tea estate south of Golaghat (A.N. Dutta 2014: 11–20). After retirement he lived in Evergreen Cottage in Shillong surrounded by a garden of Himalayan orchids which were one of Gee's passions. James Spillett stayed there for ten days in the 1960s and came to appreciate Gee as a friend whose knowledge of Indian wildlife was legendary (Spillett 2015: 262). Gee took thousands of pictures with a Hasselblad camera and numerous wildlife movies, which he bequeathed to the Bombay Natural History Society where they are still kept in several special cabinets (Anon. 1969; Barthakur & Sahgal 2005; Mrs. Nirmala Reddy Barure, BNHS, 2018-08-21). He wrote numerous reports and articles in a wide variety of magazines, but he is best remembered for his book *The Wild Life of India*, first published in London in July 1964 and often reprinted (Gee 1964). This provided a window into an exotic world. He told his story simply, authoritatively and passionately. In the chapter on the "rhinos of Kaziranga" in this great popular book, he conveyed the message well – if we do not take urgent measures to preserve these rare spots of nature, it may soon be too late and nothing will remain to show future generations.

FIGURE 38.3
Edward Pritchard Gee (1904–1968), staunch protagonist of
rhino conservation, with elephants and mahouts in Assam
COLLECTION © BOMBAY NATURAL HISTORY SOCIETY

38.6 Survey of Rhino Populations

When E.P. Gee attended the First World Conference on National Parks, held in Seattle in July 1962, he emphasised the need for a trained ecologist to conduct a survey of the rhino throughout northern India. This would not only pinpoint the conservation requirements of this species, but also benefit other wildlife in the reserves. Funding for the survey was secured through the World Wildlife Fund in 1965, and an American Ph.D. student working on lesser bandicoot rats in the grain warehouses of Kolkata was approached for the job. James Juan Spillett (1932–2018) of John Hopkins University embarked on a general ecological survey of all northern India's parks, spending a few days in each (Spillett 2015). He traveled more than 22,000 km in airplanes, on bicycles and in boats, and spent 21 days on elephant back. It was important to obtain reliable numbers of rare animals, as these could be used to inform the government to take appropriate action and allocate sufficient funding to the reserves. The results of Spillett's survey were published in the December 1966 issue of the *Journal* of the Bombay Natural History Society. This Society, through its secretary Salim Ali, had supported Spillett's survey from the start and was anxious to establish a source of authentic knowledge regarding the status of species and habitats across the Indian subcontinent.

Spillett's figures regarding the status of the Indian rhino were authoritative at the time. He had not been able to visit all areas in Assam, and some of the figures were provided by E.P. Gee. It was estimated that there were 525 rhinos in Assam (of which 400 occurred in Kaziranga),

55 in West Bengal and 100 in Nepal, giving a total of 680 animals. Spillett was convinced that any animals living outside the reserves contributed little to the preservation of the species, and that the only hope for increasing rhino numbers was through strictly protected sanctuaries. Problems facing wildlife included overgrazing by domestic livestock, human encroachment and exploitation of the remaining habitats, as well as poaching.

The message was clear. In a growing environmental crisis, the establishment and maintenance of inviolate and extensive national parks constituted a major part of the solution. The Wildlife Protection Act of 1972 was the result of persistent lobbying by international and national conservation institutions and natural history societies. The major reserves in Assam were soon elevated to national park status, providing a framework in which rhinos could be preserved.

The battle to preserve Assam's rhinos continues. The demand for rhino horn has not diminished and prices being paid by traders and consumers continue to rise. Poaching pressure is ever-present, but two peaks in the last decades of the 20th century have coincided with political unrest. In 1982–1986, a wave of poaching hit the Laokhowa, Kaziranga and Orang sanctuaries in central Assam, at a time when the United Liberation Front of Assam (ULFA) was active in the region. In 1989–1993, a second wave hit the Laokhowa and Manas reserves in the midst of the Bodoland dispute. It appears likely that the poachers were opportunists profiting from a breakdown in law and order, rather than gangs organized by political leaders.

38.7 Definition of Eight Rhino Regions in Assam

Given the size of Assam and the wide distribution of rhinos in most districts, the state has here been divided into 8 Rhino Regions. Like elsewhere in this book, these Regions are chosen to combine records with shared historical and geographical characteristics. They do not correspond directly with any administrative entities, nor do the rhinos recognize such borders. The chronological list as well as the localities on the maps are structured accordingly for practical purposes. The Rhino Regions are numbered 30 to 37 for all rhinos in Assam. It is useful to summarize the available information for each Rhino Region. The people involved in rhino encounters are listed in Table 38.50.

Rhino Region 30

North of Brahmaputra, from Sankosh River (border with West Bengal) to Champamati River. Districts Dhubri (north), Kokrajhar, Chirang (west), Bongaigaon.

There are neither any early nor any more recent reports from this region. There were rhinos along the north bank of the Brahmaputra with sightings between 1864 and 1919. In the *terai* further north contiguous with Rhino Region 23 of West Bengal, information only pertains to shoots of the Maharaja of Cooch Behar from 1880 to 1905 (32.9). *R. unicornis* was extinct here after 1920.

Rhino Region 31

South of Brahmaputra. Districts Dhubri (south), Goalpara, Kamrup (south) and Kamrup Metropolitan.

Being closest to settled areas along the Ganges, a few sportsmen penetrated the forest along the bend of the Brahmaputra as early as the 18th century. The relatively narrow strip of land in Lakhipur and Goalpara along the south bank of the river is contiguous southwards with the hill regions of Meghalaya. The reports are often too general to distinguish the two states, and rhinos wandered across the borders. The family of the Rajah of Lakhipur reduced the population in the 1860s. The two well-known rhino hunters C.R. Cock and F.T. Pollok crossed the area and reported numbers of rhinos. *R. unicornis* was largely absent from the region in the mid-1880s, with just a single subsequent record of 1894.

Rhino Region 32

North of Brahmaputra. From Champamati River to Lokhaitora River. Districts Baksa, Barpeta, Nalbari, Kamrup (north). The region contains Manas National Park (since 1911).

Rhinos were always locally abundant in the grass jungles of the northern part of the region. F.T. Pollok and friends regularly hunted here 1867–1873, and Maharaja of Cooch Behar from 1895 to 1909. In the 1900s, there were three hunting parties of the Maharaja, twice to support the Viceroy, 4th Earl of Minto, who took the only rhino obtained during these trips (32.9). The first steps towards the recognition of North Kamrup, later Manas as a wildlife sanctuary were taken in 1905 leading to full protection in 1911. All records pertaining to the area from 1905 onwards are combined in chapter 40. There were a few reports outside the protected area in Ripu and Chirang Reserved Forest until 1962. All rhinos in Manas were exterminated in the early 1990s. The program of translocation and re-introduction started in 2006.

Rhino Region 33

North of Brahmaputra. From Lokhaitora River to border of Lakhimpur Dt. Districts Udaiguri, Darrang, Sonitpur. The region contains the Sonai Rupai (1934), Nameri and Orang (1915) wildlife sanctuaries.

There were always local populations or rhinos in this region from the Brahmaputra northwards to the hills and mountains now in Arunachal Pradesh. Records are relatively infrequent but continuous and point at rhinos straying across the territory from concentrations along the river. There were rhinos in the land notified as Orang NP in 1915 and Sonai Rupai WS in 1934, but none definitely in Nameri NP. After 1932 all reports pertain to strays from the reserves including Kaziranga on the south bank. The data on Sonai Rupai are combined in chapter 41, those for Orang NP in chapter 42.

Rhino Region 34

South of Brahmaputra. Districts Morigaon, Karbi Anglong, Nagaon. The region contains the Pabitora (1959), Laokhowa (1905), Burhachapori (1905), and Kaziranga (1908) sanctuaries.

Despite today's world fame of Kaziranga, there are relatively few reports before 1908 about the presence of rhinos in this region. The reports from the wildlife sanctuaries are combined in chapters 43, 44, 45. There continued to be rhinos outside the reserves after independence in 1947, however they were usually straying animals (§38.33).

Rhino Region 35

South of Brahmaputra. District North Cachar Hills (Dima Hasao), Karbi Anglong (south).

Rhinos have been noticed a few times in North Cachar (former Cachar Hills). There is one sighting of a stray *R. unicornis* from the Brahmaputra valley in the 1970s. Stuart Baker hunted rhinos on different occasions at the end of the 1880s (38.22). The northern part of this district borders on Region 34, where *R. unicornis* is still the common species. Further south in Rhino Region 46 the few available records pertain to *D. sumatrensis*.

Rhino Region 45

Southwards of North Cachar. Districts Karimganj, Hailakandi, and Cachar.

Only *D. sumatrensis* occurred here, shown by the records in chapter 60.

Rhino Region 36

South of Brahmaputra, upper reaches. Districts Golaghat, Jorhat, Sivasagar, Dibrugarh, Tinsukia.

The earlier records are equivocal. Only three rhinos were reportedly killed here. Saidya appears to be the easternmost locality. The animal was largely extinct in the region in the 1930s.

Rhino Region 37

North of Brahmaputra, upper reaches. Districts Lakhimpur, Dhemaji, Tinsukia.

Reports show rhinos present in Lakhimpur. Further east only animals straying from areas like Kaziranga.

38.8   J.B. Chevalier in the Ahom Kingdom in 1755

Jean-Baptiste Chevalier (1729–1789) was sent by the French East India Company to India in 1752, where he was placed in charge of the factory at Dacca and proceeded on a two-year trade mission to Assam in 1755–1757. He wrote two manuscripts about his experiences, a *Journal* and a *Mémoire Historique* (Historical Memoir), which were transcribed and edited by the French scholar Jean Deloche (1929–2019) in French and in English (Deloche 1984, 2008).

Soon after reaching Assam near Goalpara, Chevalier saw rhinos in the wild. On 11 July 1755, he could hear the animals: "Throughout the night we had the charming symphony of buffaloes, tigers, rhinoceros and elephants howling continuously; a frightening noise which echoed through the mountains." Again, on 30 July 1755, Chevalier went hunting in the mountains and "a monstrous rhinoceros of the height of the strongest elephant stopped us. A few gunshots were not enough to make him escape. It looked at us rather carefully and finally came towards us. We did not wait for him; we all escaped separately and left the field open for him."

Chevalier took two months to travel up the Brahmaputra to Garhgaon. On 22 December 1755, just five days after leaving the town of Guwahati, he spotted a rhino on the riverbanks. It would have been nice to suggest that this was the first recorded rhino sighting in Kaziranga, but his narrative appears to indicate that he had only reached the vicinity of Laokhowa.

Garhgaon was the old capital of the Ahom kingdom, now situated 13 km east of Sivasagar in the upper valley of the Brahmaputra, where a palace was built by King Rajeswar Singha (reigned 1751–1769). During Chevalier's stay at the court in early 1756, the king organised a hunting party. Animals were herded together into a fenced area by army units of 1000 to 1200 men armed with stakes. When the royal guest arrived, the edges of the field were set aflame, resulting in quite an horrific scene:

> The rising flames excite the animals which, in their rage and fury, try to escape from every side. As they throw themselves on the fence trying to break it, their dreadful howling and roaring fill the air. But instead of freedom, it is the spears held by the people positioned outside the fence that pierce their bodies. Firing guns and darting arrows thrown by the amused king and his retinue spare no animal that passes back and forth below their scaffold. At last the hunt is over in a day as the flames eat up all the ferocious animals; thousands of wild buffaloes and elephants, quantities of tigers and rhinoceros are destroyed in an instant. It is a superb horror to see and to hear. It is the safety of the country and of the crops that authorise such bloody pleasure and, in fact, it is necessary. The various animals are in such great number and they multiply at such a rate that if they were not destroyed that way, the inhabitants would not be able to step out of their houses without risking their lives and the seeds would be eaten up as soon as they germinate (Chevalier in Deloche 2008: 30).

38.9   Rhinos at the Kamakhya Temple of Guwahati

This Hindu temple dedicated to the mother goddess Kamakhya has been in its present location just outside Guwahati at the top of Nilachal Hill since the 8th century. In the 10th century religious text *Kalika Purana*, allegedly composed in Assam, among the animal sacrifices it is said that "the flesh of antelope and rhinoceros give my beloved (Kali) delight for five hundred years" (Wilkins 1882: 262). The *Yogini Tantra*, a 16th century Tantric text, also mentions the rhino as a suitable item of diet even for the goddesses (G. Singh 2014: 21). When the French officer Jean-Baptiste Chevalier visited the Kamakhya temple in early 1757, he witnessed a tri-annual ceremony in which the offerings included a male and female of every species known in the kingdom, which must have resulted in a remarkable temporary menagerie: "What a surprise it was to see an immense park surrounded with fences containing a couple (male and female) each of all species of quadrupeds and birds known in the entire kingdom. All the governors of the provinces and the chiefs of the villages have to provide their share and send it to the pagoda every three years. When the time arrives for the sacrifices, this entire fauna is immolated at the altar of the goddess" (Chevalier in Deloche 2008: 36). Chevalier specifically mentioned the rhinoceros in his list of offerings, which followed the slaughter of a pair of elephants: "Then came the rhinoceros, tigers, buffaloes and quantities of other animals of all kinds and all were subjected to the same ritual. When the sacrifice of the quadruped, lasting for several days, is over, comes the one of the volatiles and finally the one of the fishes." One of the early kings had vowed to present the goddess with as many golden statues of animals as were offered to her during the festival, a vow to be carried forward by his descendants: "Therefore it is difficult to count the amount of golden animals placed around the temple. I have seen elephants, rhinoceros of medium and even small sizes, and quantities of life sized replicas of other species. The king added two golden stags, as big as the ones we have in our forests in Europe, to this treasure. Although all these statues are hollow and the gold is of a very pale colour, the entire collection as a whole represents a treasure of great value." This golden pair of rhinos at Kamakhya has remained hidden ever since.

38.10 Thomas Craigie at Goalpara in 1777

An early specimen was obtained in 1777 by Thomas Craigie in Goalpara on the borders of Assam. Craigie was an army officer who fought at the battle of Seringapatam in 1799. The rhino horn was first part of the Museum of the East India Company, London, donated by Edw. Smith, and from 1879 in the NHM London. The animal had been found asleep close to the house where Craigie was visiting. He went out on foot accompanied by three gentlemen on horseback: "Mr. Craigie (having knelt down on one knee) levelled his piece, and the ball entered the head just between the eyes; the beast rushed forward, but Mr. Craigie avoided him by springing on one side, and the animal fell dead near the spot where he had knelt" (Horsfield 1851: 195).

38.11 Tax on Rhino Horn in 19th Century Cachar

It is said that Senapoti Tularam Thaosen, the ruler of the North Cachar region until its annexation in 1832, extracted a tax on each elephant killed, amounting to Rs 880 annually, and half the value of every horn of rhinos slain in hunting, which gave him Rs 450 every year (Anon. 1855a). Maybe these amounts are hard to translate into today's money, but it shows that rhinos were relatively common.

38.12 Tame Rhinos Pulling Ploughs near Guwahati

There are regular rumours that rhinos were used to pull ploughs, and these are often set in Assam. Fact or fiction, the evidence is not entirely clear (Rookmaaker 2020a). As early as 1826, the French naturalist Victor Jacquemont (1801–1832) was told that rhinos were used in agriculture in the lands on the other side of the Ganges (Jacquemont 1841: 169). Major John Butler (1847: 29) seems to speak from personal experience when he refers to tame rhinos near Guwahati "grazing on the plains as harmless as cows, guarded by a single man." Half a century later visiting the same town, Money (1893: 618) talks about a dhooby (washerman) who was seen collecting clothes from the wash riding on a rhino.

Earlier than any of these observations, an anonymous author writing in the *Bombay Gazette* of March 1823 takes us to a lecture in Kolkata by Triptolemus Yellowley, the agricultural expert known from *The Pirate* (1822), one of the Waverley Novels by Walter Scott. This speculative genius had brought with him "several engraved copies" of a re-invented machine drawn by three rhinos, with the grand title "The Rhinoceros plough with the Living principle; or self-acting Antediluvian Coulter, respectfully submitted to the agricultural society, as adapted in a peculiar manner to the Light soils of Hindostan" (*Bombay Gazette* 1823-03-12). Although this appears to take the reader well into the realm of fantasy and fabrication, yet this lithographic specimen was mentioned in a letter dated 3 October 1823 written by Edward Gardner (1784–1861), the British Resident in Kathmandu, to Nathaniel Wallich (1786–1854), the Superintendent of the Calcutta Botanic Garden, where he remarked that "The 'Rhinoceros Plough' amused me much – where is it?" (Wallich Correspondence, Central National Herbarium, Howrah). An illustration of this description has remained elusive, if it in fact existed beyond the literary pages.

At the end of the century, in another context, Louis Jacolliot (1837–1890) produced an engraving of a rhino ploughing a field watched by two farmers. His account was set in Egypt, but this did not deter the illustrator Auguste-André Lançon (1836–1887) to draw the animal single-horned and armour-plated (fig. 38.4). Another depiction without more than humorous intention appeared in 1895 (fig. 38.5).

FIGURE 38.4

"The Rhinoceros fulfils the mission of the Ox." Drawing by Auguste-André Lançon (1836–1887) illustrating a scene set in Egypt. First published by Louis Jacolliot in 1884, it was reproduced by W.F. Kellogg (*Hunting in the Jungle*, 1888, p. 215), and again by Jacolliot in *Journal des Voyages*, 22 October 1899, p. 333
JACOLLIOT, *ANIMAUX SAUVAGES*, 1884, P. 417

FIGURE 38.5

The German artist Adolf Oberländer (1845–1923) here depicts an invented scene how rhinos assist in ploughing the field. Although the accompanying story is set in Cameroon (Africa), the animals look decidedly Indian
FLIEGENDE BLÄTTER, MUNICH, 1895,
VOL. 103 NO. 2624, P. 173. DIGITAL COPY BY
UNIVERSITÄTSBIBLIOTHEK HEIDELBERG

FIGURE 38.6   Drawing of rhinoceros taken to USA by Marmaduke Burrough (1797–1844) in 1830
WITH PERMISSION © WILLIAM L. CLEMENTS LIBRARY, ANN ARBOR, MICHIGAN: MARMADUKE BURROUGH PAPERS

38.13   Rhinos from Goalpara in American Circuses in the 1830s

Marmaduke Burrough (1797–1844), a businessman of Philadelphia (Pennsylvania, USA) lived in Kolkata 1828–1830 serving as American Consul. The papers relating to Burrough's endeavours in India are part of an archive which is preserved in the William L. Clements Library, Ann Arbor, Michigan (Cox 1997). Interested in natural history, he must have realized that an animal like the rhino had never been seen alive in his home country. With a surge in popularity of traveling menageries and circus shows, there could well be a lucrative opportunity if he were able to obtain and transport a young rhino. Apparently he first approached Raja Buddynath Roy, who had a young male rhino on his premises in a suburb of Kolkata. As he was traveling in the provinces, he delayed and the animal was sold to another (unnamed) American entrepreneur for Rs 7000. This was the first rhino to arrive in America and it toured from arrival in Boston on 9 May 1830 with the menagerie of June, Titus and Angevin, part of the Flatfoots Association, until it died in 1835 (5.4).

Burrough was touring in lower Assam with his friend George Hough, where he met Captain Alexander Davidson, Magistrate in Goalpara. After his return to Kolkata,

Davidson wrote to him several times about a larger female and a small male rhino obtained in the region (no exact locality). The female never left Goalpara. The male was troublesome but small enough to be transported down to Kolkata and then abroad, as on 15 March 1830 it was said to be 10 weeks old, with a body length (including tail) of 5 ft 6 in (167.6 cm), length of head 1 ft 3 in (38.1 cm) and height at front shoulder of 2 ft 5 in (73.6 cm). Davidson shipped the young male from Goalpara on 29 March 1830, and Burrough then took it with him on his return journey to Philadelphia on the ship *Georgian*, arriving on 15 October 1830 (*New York Evening Post* 1830-10-15). This was the second rhino to arrive in America, shown with his native keeper in a drawing made at the time (fig. 38.6). Burrough wrote on 8 November 1830 that it bore the voyage well and arrived in good health. From December, he exhibited it privately at 48 South Fifth Street, Philadelphia, but soon sold it by auction, as from 1831 the male rhino toured with the circus of Raymond & Ogden and later Purdy, Welch, Macomber & Co. The rhino probably died in 1837 (Rookmaaker 1998: 107).

Davidson was interested in continuing the animal trade. In December 1830 he sent a young male rhino to be sold to an American captain, as stated in a letter by George Hough dated at Howrah 1831-05-09: "I had it here at the Ghaut, gave his servants places to live in, on our premises, and attended to the animal myself. After a lapse of a week I sold her to Mr. Austin, Supercargo of the brig *Neponset* of Boston for 1500 Rs and put the money into Fergusson & Co's house and for all that I have not so much as received a thanks." This third American rhino arrived in Boston on the brig *Neponset* from Kolkata (*New York Evening Post* 1831-05-10). It was shown to the public at the Lion Tavern, Washington St., Boston in June 1830. The trail from here is uncertain due to possible confusion with other rhinos in the country. Maybe it died soon, maybe it lived for few years in the 1830s.

38.14 Rhino Group at Biswanath in 1836

The pseudonymous Woodsman was hunting in the vicinity of Biswanath in July 1836 where he was told about a group of 5 rhinos. He killed one female with a horn of 14 in (35.5 cm), while three others escaped after being shot. The next day he saw a mother rhino with calf, and he killed the mother.

38.15 Rhino Hunts at Lakhipur in the 1860s

Rhinos were common in the region of Lakhipur in western Assam in the mid-19th century. The local Raja or Zamindar had a house filled with trophies (Pollok 1868e). Pollok must have written from personal experience about the pastimes of the Raja's son, who killed 34 rhinos: "His son, a lad of twenty, has been very successful as a sportsman, having cleared the country all round of rhinoceros and buffaloes and tigers; but he is a bit of a poacher for all that. Mounted on Mainah, and armed with a single-barreled cannon, carrying a 6-oz. ball, he goes out on moonlight nights, when the rhinoceros are feeding, and do not suspect danger, and firing into one, he does not follow it up if it be only wounded, but leaves it to his shikarees to retrieve hereafter, which they generally do. In this way, in about six weeks he killed, I believe, thirty-four rhinoceros and ten tigers, besides other game, and has depopulated these jungles as far as game is concerned."

38.16 The Zamindar of Gauripur

The rulers of the small zamindar (tax district) of Gauripur were the hereditary family Barua of Rangamati or Gauripur. Since 1860 they had their seat in the Rajbari in Matiabag on the outskirts of Dhubri. It had been occupied by the last acting Raja, Prabhat Chandra Barua (1878–1940), and despite lack of maintenance it is still in family hands (R. Das 2014). The journalist Mckenzie Funk visited here in recent years and found hunting trophies and a hunting log which had belonged to the Raja, who had been known as a great hunter. The undated log (maybe still present) included "one tiger, two male rhinos, and one female" (Funk 2014). The Zamindar of Gauripur assisted in organizing the 1908 annual hunt of the Maharaja of Cooch Behar.

38.17 Hermann Schlagintweit in Assam in 1855

Hermann Schlagintweit (1826–1882) went with his two brothers to India in 1854–1857 on a journey commissioned by the East India Co. for scientific investigations. Hermann visited the North-East on his own, traveling in Assam from Gowahatti up to Dibrugarh by steamer. He only saw one rhino, tame, in Gowahatti between 16 November and 21 December 1855. He also received a rhino skull with horn which had been killed north of the Brahmaputra (Schlagintweit 1869, vol. 1: 433). On 7 November 1862 he wrote to the Smithsonian Institution, Washington DC, offering zoological specimens including a rhino skull from the Bhutan *terai*, for £8 (Smithsonian 1862: 84). There are 2 skulls in the Bayerische Staatssammlung, Munich, attributed to him, with a 1856 date: AM0415, juv. skull, Assam *terai*; and AM0416, adult skull, Upper Assam. The localities are general, but the Assam *terai* could be the Manas area, while Upper Assam could be around Dibrugarh. Schlagintweit also mentioned the occurrence of rhinos at the Kosi River in East Nepal (Dataset 27.25).

38.18 An Early Photograph of a Rhino Taken in 1862 by Captain Speer

Captain Wilfred Dakin Speer (1835–1867) made two sporting trips to India, from September 1859 to May 1862 and from November 1864 to June 1865 (Anon. 1870). A photograph "Rhinoceros shot by late Captain Speer in Assam, 1862" was published in part 5 of the *Sports of the World* edited by Frederick George Aflalo (1870–1918) in January 1903 illustrating an article by Kinloch (1903). As there is no reference to the event in the text, it is likely that Aflalo sourced the photograph from the Speer family (fig. 38.7). The rhino is shown lying on its side with the head facing the camera, surrounded by over 30 native assistants. Captain Speer is absent from the photograph so he probably was behind the camera. If the dates are

FIGURE 38.7   Rhinoceros shot by Captain Wilfred Dakin Speer (1835–1867) in Assam in 1862. This might be the first photograph of a (dead) rhinoceros taken in the wild

F.G. AFLALO, *THE SPORTS OF THE WORLD*, 1903, P. 164

correct, meaning that the photograph was taken in the first months of 1862, this makes it the first known photograph taken of a rhino in the wild, earlier than the one taken in Botswana by the pioneer James Chapman (1831–1872) on 13 May 1862 (Rookmaaker 2006c, 2021b).

38.19   The Hunt by Henry Andrew Sarel in 1865

Lieutenant-General Henry Andrew Sarel (1823–1887), later Governor of Guernsey, went to shoot in the Manas area of Assam 20 April to 1 June 1865, with an unnamed British friend (Sarel 1887). He borrowed eight elephants from his fellow officer General Henry Tombs (1825–1874) stationed in the Nalbari area, and engaged the assistance of Kurruch Singh, a small land-owner known as Hathi-Raja for his knowledge of the local wildlife. On his first day in the field, he killed 3 rhinos, one of which measured 14 ft 2½ in. (433 cm) from the nose to the tail, with a horn of 11¾ in (29.8 cm) long and 22 in (55.8 cm) round the base. He saw many rhinos in the forests and killed at least 8 of them. At the end of his trip he moved eastwards to the Buxa region. He is one of the few to describe the process of removing the horn, which he called the most objectionable part of

the sport: "these had to be chopped off the bone of the nose with an axe, a process which splashed the operator with blood; on reaching camp the horn and thick skin attached had to be boiled, when the skin became loose, and was easily detached from the base of the horn; the horn itself looks like a conglomeration of coarse hair, with the tip rounded and polished by digging for roots" (Sarel 1887: 360).

38.20   Two Rhinos Killed by Willcocks Drawn by Helen Graham

In 1882, General James Willcocks (1857–1926) traveled by steamer from Dhubri up the Brahmaputra to Mungledye (Mangaldoi) on the north bank (Willcocks 1904: 44). After meeting A.J. Primrose, Deputy Commissioner, he killed a rhino with 14 bullets on the third day of his trip, setting up the horn as a snuff-box. In 1886 Willcocks shot a rhino at Dhunsiri Mukh, a place on the south bank of the Brahmaputra about 24 km from Golaghat (Willcocks 1904: 63). The two events were illustrated by drawings made by Helen Violet Graham, daughter of Douglas Beresford (figs. 38.8, 38.9).

FIGURE 38.8
"My first rhinoceros" illustrating a hunting trip of
General James Willcocks (1857–1926). The drawing is
attributed to Helen Violet Graham (1879–1945)
WILLCOCKS, *FROM KABUL TO KUMASSI*, 1904, P. 44

FIGURE 38.9
"In another instant we were within ten yards of the
turmoil". Hunting scene of James Willcocks in the jungle
drawn by Helen Violet Graham
WILLCOCKS, *FROM KABUL TO KUMASSI*, 1904, P. 66

38.21 Mayne Reid's Pulp Fiction of 1882

Thomas Mayne Reid (1818–1883) wrote fictional romance
and adventure stories published as books and as stories in
a long range of pulp magazines. One of these was *Beadle's
Weekly*, published 1882–1897 by the successful team of
William Adams and Erastus Flavel Beadle (1821–1894) in
New York. The first issues contained an account in seven
parts of a "Chase in Assam" edited by Reid and said to be
the real experiences of a British officer. It is unfortunately
difficult to separate this possible reality from a fictional
element. The story appeared rather late in Reid's life after
he had retired to England, and might have been written or
even published before. Reid (1882) told about a party of
hunters starting out from Rungpoor (Rangpur) and hunting along the Brahmaputra. The unnamed British officer
is accompanied by a friend who is Superintendent of
Police in Assam, his 16-year-old brother Henry B., and two
indigo-planters Mr. Edwards and Mr. James. The second
day after arrival was 25 April, without a year. They capture

FIGURE 38.10   "The mother rhinoceros takes her revenge."
Illustration to a story about a chase in Assam.
The signature of the artist in the lower right corner
is indecipherable
BEADLE'S WEEKLY, 12 DECEMBER 1882, P. 8

a quarter grown rhinoceros in nets after killing the mother with a hog-hunting spear. The young one is tamed and proposed to be sold to an Afghan agent of Jamrach for $110. The account has elements reminiscent of the books by Pollok (1879), but the details are inconsistent. Reid's fiction is illustrated with an engraving of the mother rhino attacking a horse (fig. 38.10).

38.22   Stuart Baker in North Cachar

E.B.M. Baker's son, Edward Charles Stuart Baker (1864–1944) worked in the Indian police service from 1883, stationed in the North Cachar Hills of lower Assam from 1886. He is well-known for his ornithological work editing the volumes on birds of the *Fauna of British India* (N.B.K. 1944). He shot three rhinos, in 1888, 1889 and 1890 (Stuart Baker 1928). His book *Mishi* contains many hunting stories, unfortunately with very few details about his companions or the locations of their camps. He clearly stated that "In the North Cachar Hills rhinoceri were not everyday occurrences and most of those we came across were solitary wanderers from the swamps at the foot of these hills" (Stuart Baker 1928: 175). The first rhino of 1888 "had taken up her abode in a wide stretch of rolling hill country, some 70 miles from my headquarters." There is no certainty, but possibly this points at a place some 100 km north of Haflong, where the hills appear to fade slowly into lower lands, like Lumding or Lanka. The first rhino attacked him and left him injured with six broken ribs. The second rhino measured over 6 ft (180 cm) at the shoulder and had a horn of 18 in (45 cm). The third rhino stood 6 ft 3 in (190 cm) and had a short 7 in (17.8 cm) horn. He did not identify them, except that all shared three characters: they were solitary brutes, bad-tempered and wicked to the core. The measurements point at *R. unicornis*, as does Stuart Baker's silence about their specificity.

38.23   The Trophy of Firman

The *Game Book* kept in India in the 1890s by Sidney Herbert Buller-Fullerton Elphinstone, 16th Lord Elphinstone (1869–1955) is described in 32.3. There are entries of his participation in the shoots organized by the Maharaja of Cooch Behar in 1895, 1896 and 1899. After attending shoot 36 of 1896 (a), he went further east for a few weeks traveling together with E. Fraser Tytler, Sidney Parker (1852–1897), Lieutenant Humphrey Brooke Firman (1858–1916) and Lieutenant-Colonel Dugald

FIGURE 38.11   Trophy of *R. unicornis* killed by Humphrey Brooke Firman in 1896, with a horn of 12¾ inch (32.4 cm). Donated to the In & Out Club, London, where it was photographed in 2018
PHOTO © YASSEN YANKOV, MAY 2018

McTavish Lumsden (1851–1915). From the end of March until 12 April 1896 this party shot at Bisnath (Biswanath), Assam, killing a total of 5 rhinos.

One of these latter rhinos was credited to Firman, a Lieutenant in 16th Queen's Lancers. It had a horn of 12¾ in (32.4 cm) as recorded by Rowland Ward from 1896 onwards. The trophy head was donated in 1903 to the In & Out Club (Naval and Military Club), London, where it still is, without a plaque (fig. 38.11). Lieutenant H.B. Firman was a member of the Club since 1873 (Newark 2015).

38.24   Record Horn Obtained by Holland and
Walker in 1892

Captain Herbert Christian Holland (1858–1916), Chief Constable of Derbyshire, went with his wife (Edith Laurentia Cave) to Assam in 1891–1892 (Randall 1901, vol. 2: 121). Here he went hunting with Peter Carlaw Walker (1854–1915), Sidney Parker and Humphrey Brooke Firman. This trip was not part of a Cooch Behar shoot where Parker

FIGURE 38.12
"Rhinoceros shot in Kuch Behar." Photo by
Philip Breda Vanderbyl (1867–1930). The
same photograph was also published as
"An Assam Rhinoceros"
F.G. AFLALO, *THE SPORTS OF THE WORLD*,
1903, P. 165 ILLUSTRATING ARTICLE BY
KINLOCH

and Firman are recorded in 1895, and Walker in 1907. No data are available about the locality. Holland obtained a trophy with a horn of 12½ in (31.8 cm), and Walker one of 8¼ in (21.0 cm), both said to be from Assam (Ward 1896).

38.25   Vanderbyl and Elphinstone in Assam in 1899

Philip Breda Vanderbyl (1867–1930) was an officer in the Royal Horse Artillery. He participated in the annual shoot of the Maharaja of Cooch Behar in 1899 (shoot 40). He then continued to hunt with Lord Sidney Elphinstone from 24–29 March 1899 securing 2 rhinos (Cooch Behar shoot 41). The Elphinstone Game Book (32.3) puts the locality of the shoot at Jalagaon (Jalahgaon) on the eastern side of Manas. Vanderbyl donated two skulls to the Natural History Museum in London. One had a horn 9 in (22.9 cm) long, recorded as from "Cooch Behar" (Ward 1903: 409). Among photographs taken during this expedition, one shows a dead rhino (fig. 38.12). This was published on three separate occasions either as "Assam rhinoceros" (Vanderbyl 1901: 464; 1915, pl. 32) or as "Rhinoceros shot in Kuch Behar" illustrating Kinloch (1903: 165). The legends make clear that Assam and Cooch Behar were used quite loosely.

38.26   Rosie in the Ipswich Museum

The museum in Ipswich, Suffolk, opened in 1881, received a mounted *R. unicornis* in exchange from the British Museum when Mark Woolnough (1845–1930) was curator

1893–1920 (Markham 1990). This rhino has one of the longest horns known, measuring 19 1/8 in (49 cm), which was listed by Rowland Ward as early as 1892 when the specimen was still in London. The origin is irretrievable. The heavy mounted hide arrived on Friday 15 March 1907 in a horse-drawn van owned by R.D. and J.B. Fraser, a removal firm on Princes Street founded in 1833, and required ten men to help with the offloading (*Ipswich Evening Star* 1907-03-19). The horn was stolen in 2011 during a spate of thefts of rhino horns from collections around the world. The mounted hide is still on display in Ipswich, and was named 'Rosie' by Miss Debbie Brown of Ipswich winning the competition in 1981. Two photographs from the time of arrival in the museum were illustrated by Markham (1990: 48–49).

38.27   The Museum of Charles Peel in Oxford

Charles Victor Alexander Peel (1869–1931) was a naturalist and big-game hunter, who displayed the trophies from around the world in his private museum on 12 Woodstock Road, Oxford. In 1919 he donated the collection to the Royal Albert Memorial Museum, Exeter (Howes & Bamber 1970). A mounted specimen of *R. unicornis* is still present. Peel (1908) enquired where he could best obtain a rhino for his museum. The animal was shot on 20 February 1909 (museum records). The date closely follows the hunt 1909 (a) of the Maharaja of Cooch Behar, which doesn't list Peel among the participants.

38.28 The Longest Horn Obtained by Thomas Briscoe in Tezpur District in 1909

In 1909, there were a series of media statements that a rhino with a record horn had been shot in Assam. The hunter was Thomas Charles Briscoe (1857–1909), who managed the Borelli Tea Estate near Ballupara (Balipara) north of Tezpur (*Bengal Directory* 1884). Briscoe went after the rhino alone and on foot to give the animal a fair chance. His friend who signed his letter of 27 April 1909 to the newspaper as 'Shikar' went to the spot after a few hours and measured the animal's shoulder height between pecks as 6 ft 4 in (193 cm). The horn was 24¼ in (61.3 cm) which shrunk after drying to 24 in (61 cm) (Shikar 1909). The identity of this friend is revealed in a letter written from Tezpur on 8 May 1909 to the Natural History Museum in London urging them to acquire the horn, signed by J.C.H. Mitchell (NHM archives). Mitchell is listed as a member of the Bombay Natural History Society from 1901 with a postal address at Halem P.O., Darrang, Assam. Briscoe had left for the UK in April and died just a few months later in Scotland on 13 November 1909 (*Morning Post* 1909-11-23). The NHM subsequently corresponded with the executor of the estate, Briscoe's son-in-law Claude Fraser De La Fosse (1868–1950), then in Oxford, who on 13 January 1910 confirmed that the family had decided to donate the mounted trophy head and horn to the museum stating that it must be credited to T.C. Briscoe. The specimen set up by the firm of Rowland Ward was soon added to the collection and was registered as NHM ZD.1910.1.23.1 (Lydekker 1910).

This longest known horn entered the record book of Rowland Ward (1910: 464 and later editions), where it was first illustrated separate from the head (fig. 38.13). As shown in a photograph published by Barclay (1938), it was later attached to the trophy (fig. 38.14). This was the true length of the horn, unlike some in the more wishful illustrations (fig. 38.15). From the available documents it is clear that the animal was shot by Thomas Charles Briscoe in March or April 1909. The exact locality remains unknown, except that it was in Tezpur District. This should mean in the region north of the Brahmaputra between Orang NP and the land opposite Kaziranga NP, within easy reach of both Balipara and Halem. Given that there was major concern by the forest administration of dwindling rhino numbers in Assam leading to the proclamation of reserves for their protection between 1907 and 1915, it is curious to note that at least 10 rhinos were killed (presumably with permits) by different hunters like Briscoe in the same period 1900 to 1919 (table 38.50). Like the further historical record, this provides some evidence against the supposed bottleneck of rhino numbers.

FIGURE 38.13 Record horn of Great One-Horned Rhinoceros, measuring 61 cm. Shot by Thomas Charles Briscoe
ROWLAND WARD, *RECORDS OF BIG GAME*, 6TH EDITION, 1910, P. 464

FIGURE 38.14 The record horn of *R. unicornis* mounted on the head of the rhino shot in Assam in 1909 by Thomas Charles Briscoe (1857–1909)
BARCLAY IN *THE FIELD, THE COUNTRY GENTLEMAN'S NEWSPAPER*, 8 JANUARY 1938, P. 65

FIGURE 38.15
"Rhinocéros des Indes (unicorne)." Rhinos with very long anterior horns, illustrating the general reference work on mammals by the French zoologist Louis Figuier (1819–1894). Engraved by Harden Sidney Melville (1824–1894) in London
FIGUIER, *MAMMIFÈRES*, 1869, P. 121, FIG. 37

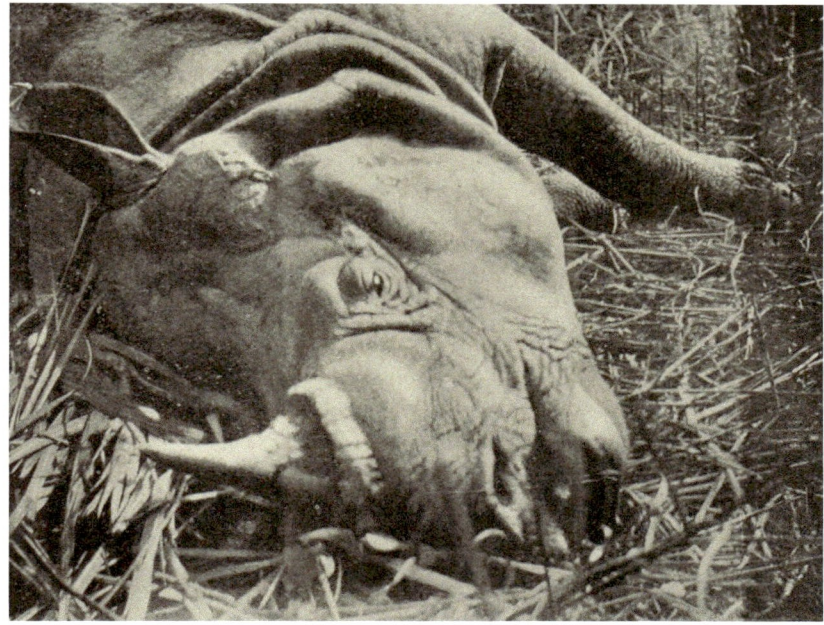

FIGURE 38.16
"Head of rhinoceros, shot by W.M. Nuttall"
PLAYNE, *BENGAL AND ASSAM, BEHAR AND ORISSA*, 1917, P. 633

38.29   Rhino Photograph by W.M. Nuttall

In a chapter on Indian fauna in Playne (1917: 633), there is a photograph of a head of a rhino killed by W.M. Nuttall (fig. 38.16). That is all that remains and further data are elusive. Walter M. Nuttall (or Nuttal) was manager of Digulturrung tea estate in Doomdooma, Upper Assam. A keen naturalist judging from his contribution on Indian fauna, Nuttall (1917) mentions his own shooting exploits of game other than rhino. The photograph was perhaps taken in the 1910s in Upper Assam where he lived.

38.30   Guy Wilson on the Manas River in 1911

Guy Douglas Arthur Fleetwood Wilson (1851–1940) was an army officer working in India for much of his life. He traveled to Assam in 1911, where near the Manas River, he

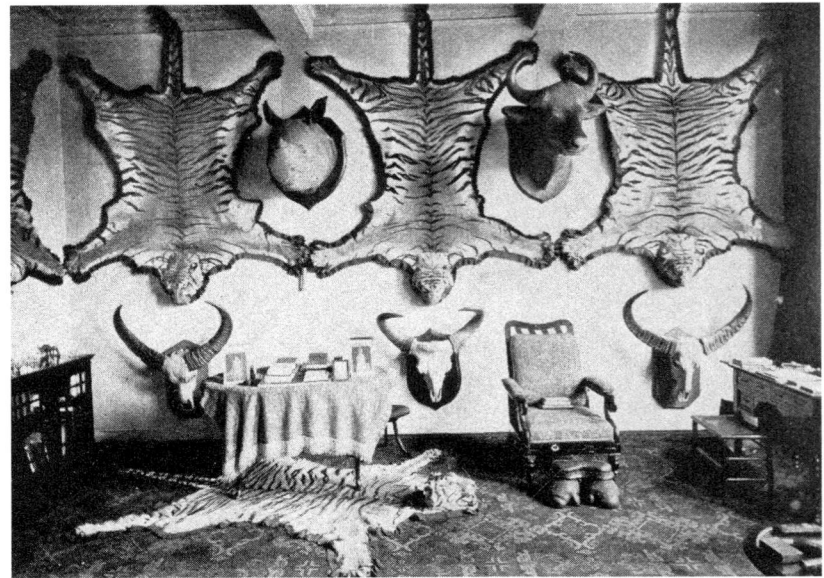

FIGURE 38.17
Skins and heads, including a trophy rhino, in the
smoking-room at Dees, the residence of Sir Guy
Wilson in 1920
THE SPHERE, 6 MARCH 1920, P. 268

killed a rhino on 8 April 1911. The animal had a shoulder height of 5 ft 6½ in (169 cm), with a much worn and very short horn (Wilson 1921: 123). The head was mounted as a trophy and was displayed in Dees, his house near East Grinstead, UK with several tiger skins (Wilson 1920: 268) (fig. 38.17).

38.31   Frank Buck's Movie *Fang and Claw* 1935

The animal dealer Frank Buck (1884–1950) became famous through his role in the successful movies *Bring 'Em Back Alive* (1932) and *Wild Cargo* (1934). The third movie *Fang and Claw* was produced in 1935 for the New York based corporation of Amadee J. Van Beuren (1879–1938). Sections of the movie were shot in Assam "near the Bhutan border", possibly in the region of the current Manas NP (Buck & Fraser 1941). There is a scene where Buck stumbles on a baby rhino just when it was attacked by a large tiger. Buck kills the tiger to save the rhino, and then amputates the animal's mangled ear. The movie appears to have been shot silent, with a soundtrack added later, by the photographers Nicholas Cavaliere (1899–1995) and Harold Edmund Squire (1890–1977). The opening title of the film of 73 minutes reads: "Frank Buck's Fang and Claw" and in a written onscreen statement, the contents are described as an "official and authentic motion picture record of Frank Buck's wild animal collecting expedition in the Asiatic jungles." It was first shown at the Rialto Theatre in New York on 20 December 1935 (*New York Times* 1935-12-28). Despite the similarity in title, Buck's novel *Fang and Claw* co-authored by Ferrin Fraser (1903–1969) makes no reference to a rhino. While filming in Assam, the team

captured a young female rhino, which was transported via Singapore on the freighter *Steel Navigator*, and then kept at Frank Buck's Jungle Camp at Amytiville, Long Island, NY. She only lived there 1935-09-13 to 1936-02 due to a wrong diet (Rookmaaker 1998: 44).

38.32   Hunting Regulations and Protection Measures

Hunting in Assam was regulated in 1889, when Dennis Fitzpatrick (1827–1920), Chief Commissioner from 1887 to 1889, prohibited all shooting in reserved forests from November to June (the cool season when sports hunting was prevalent) without the permission of a range officer. This regulation was tightened in 1902 when the closed season was extended throughout the year for female rhinos accompanied by a calf. There was a further extension of the rules by prohibiting shooting of wildlife across the whole state including unclassed forests. In the absence of a budget for patrolling land outside the reserved forests, the protection was in practice very meagre (Milroy 1932).

The Darrang Game Association was formed in 1910 to aid the government efforts to protect the big game (Saikia 2005: 268). The first meeting was held on 18 August 1910 in Tezpur and was attended by about 20 people (*Englishman's Overland Mail* 1910-09-01). The meeting was largely due to the perseverance of Arthur Milroy, then Assistant Conservator of Forests. The participants felt that two measures would be imperative to achieve the goals, being a strict adherence to the closed season, and a curtailment of the presence of Nepalese herdsmen who used common grazing grounds. Among those present were the Conservator of Forests A.V. Munro and the Deputy

Commissioner of Forests D.P. Copeland. The first officers of the Association were the Deputy Commissioner Major Hugh Maclean Halliday (1866–1920) and Arthur Milroy (1883–1936), with seven members: Felce, Hick, Wilde (Tezpur Dt.), Davidson, Lawes (Bishnath Dt.), Bruce and Bridge (Mangaldai Dt.). They negotiated with the government hoping to obtain a lease of the sporting rights in the reserved forests and thus incidentally assisting the Forest Department in the prevention of poaching (Assam Forest Dept. 1914: 28). These efforts were unsuccessful, which is the reason that the existence of this Game Association is poorly documented. Milroy stated that it disintegrated due to inaction after a few years, and in 1916–17 only six members were reported (Saikia 2005: 268). In a list of sponsors, "Mr. H.O. Allan, Darrang Game Association, Tezpur" is recorded in 1926 (*Journal of the Bombay Natural History Society* 26: 1036). Maybe it was revived in 1930 only to be discontinued soon afterwards (Hanson 1933; Milroy 1934a).

The Wild Birds And Animals Protection Act of 1912 issued by the Government of India prohibited shooting of rhino at any time, with these paragraphs:

Act No. 8 of 1912 [18 September 1912]
para 3. Close time. The State Government may, by notification in the Official Gazette, declare the whole year or any part thereof to be a close time throughout the whole or any part of its territories for any kind of wild bird or animal to which this Act applies, or for female or immature wild birds or animals of such kind; and, subject to the provisions hereinafter contained, during such close time, and within the areas specified in such notification, it shall be unlawful-

(a) to capture any such bird or animal, or to kill any such bird or animal which has not been captured before the commencement of such close time;

(b) to sell or buy, or offer to sell or buy, or to possess, any such bird or animal which has not been captured or killed before the commencement of such close time, or the flesh thereof;

(c) if any plumage has been taken from any such bird captured or killed during such close time, to sell or buy, or to offer to sell or buy, or to possess, such plumage.

The reserves in Assam not only remained closed for shooting from 1926 to 1938, but no permits were issued that would allow any kind of harm to the wildlife. Visitors except forest staff were discouraged and generally prohibited. There must have been hope that the animal populations would increase if the reserves could be managed without outside interference. In 1933 the Assam Legislative Council passed a Bill declaring rhino horns to be forest produce wherever found, inside or outside a reserved forest. Any horn found could thus be declared illegal and the carrier could be booked for this offence (Milroy 1933).

The Assam Rhinoceros Preservation Act was published in the *Assam Gazette* of 9 June 1954, with some amendments in October 1970. It has different clauses that afford protection to the rhino in the whole of Assam (partly paraphrased):

(clause 2) No person shall kill, injure or capture, or attempt to kill, injure or capture any Rhinoceros or be in possession of any limb or part of a Rhinoceros unless so permitted by the State Government by a license granted or an order made under this Act.

Provided that a person will be entitled to kill or injure Rhinoceros in defense of himself or some other person.

(clause 3) Every Rhinoceros captured and the horn or carcass or any part of every Rhinoceros killed in contravention of this Act or any condition of a license or order issued under this Act shall be the property of the State Government.

(clause 4) License can only be granted if (1) any Rhinoceros has become a cause of imminent danger to human life, or (2) such Rhinoceros is required for any zoological, scientific or other special purpose.

(clause 5) The penalty for killing, injuring or capturing a rhino in contravention of this is a fine of 2000 rupees and imprisonment which may extend to three years.

There are six rhino-bearing protected areas with their legal frameworks evolving from reserved forests, to wildlife sanctuaries, game reserves or national parks. Laokhowa was announced in 1907, Kaziranga in 1908, Manas in 1912, Orang in 1915, Sonai Rupai in 1934 and Pabitora in 1959.

38.33 Strays from the Parks

Rhinos in Assam have always been known to wander over large distances. From around the middle of the 20th century, the rhino population has effectively been limited to six reserves or national parks. Animals leave the reserves when the Brahmaputra floods its banks, sometimes for extended periods before returning when their prime habitat is restored with new plant growth. Rhinos have also been encountered in localities at quite a distance from the reserves, especially but not exclusively in Assam north of the Brahmaputra River. In two main papers, Choudhury (1996, 2010) provided documentation of such movements. There is no consistent evidence that rhinos settle in such areas on a permanent basis and then constitute new populations. Outside reserves rhinos are in greater danger of poaching or being shot by farmers defending their land. The records of these stray animals or groups of animals shown on the maps give the impression that the rhino has a much wider range in Assam than is actually the case.

There can be no reason to believe that the numbers of rhinos straying out of the reserves is diminishing. Park authorities will ensure that such movements are kept to a minimum and in some cases take action to return individual rhinos to the protected areas. However, the population in a national park like Kaziranga is increasing which may result in a larger number of rhinos wandering to other places in search of food or companionship. The evidence largely available in local press will need to be continuously monitored and published at intervals.

38.34   The Spreading Exposure in Popular Culture

The rhinoceros has increasingly become part of the cultural fabric of Assam and of India as a whole. There is good reason why such an imposing and iconic species was the one chosen as the State Animal of Assam on the initiative of Anwaruddin Choudhury (Mazumdar & Mahanta 2016). Much earlier in 1935, the rhinoceros was incorporated in the Coat of Arms of the state of Assam (Sword 1935), although this never gained traction. The rhinoceros was chosen as the Emblem of the Indian Air Force 11 Squadron 'Charging Rhinos' when it moved to

Jorhat in February 1961 as 'Vishwambarah Prandah', meaning 'Supporters of the Universe.' There are multiple further occurrences of rhinos in titles, logos and emblems (Mazumdar & Mahanta 2016; Barbora 2017), like those of Assam Oil, the Assam State Transport Corporation, the State Zoological Gardens in Guwahati, and the Rhino Foundation for Nature in North-East India (fig. 38.18).

On a national level, the image of the rhinoceros was used in banknotes, coinage and stamps. It is found on a banknote of 10 rupees since 1996 (fig. 38.19) and on the reverse side of the small 25 paise coin from 1988 to 2011 (fig. 38.20). In 1962 two stamps with the value of 15 new paise were issued (figs. 38.21, 38.22) and another rhino appeared in 2015 (fig. 38.23).

The popularity of the rhinoceros is firmly established and is still spreading.

FIGURE 38.18   Logo of the Rhino Foundation for Nature in North-East India, a local NGO founded in 1985 by Dr Anwaruddin Choudhury in Assam

FIGURE 38.19
In 1996, Rupees Ten notes were issued by the Republic of India in a new model. The reverse of the Mahatma Gandhi portrait series has a motif with a collage of a rhino, an elephant and a tiger in the centre
COLLECTION KEES ROOKMAAKER

FIGURE 38.20
Coin issued by the Republic of India for a value of 25 paise. It is a small coin, 19 mm in diameter, 2.84 g in weight, of stainless steel. The dot below the date represents the Indian Mint at Noida. This was produced from 1988 to 2002 and remained legal tender until 30 June 2011
COLLECTION KEES ROOKMAAKER

FIGURE 38.21
First Day cover (1 October 1962) of the stamp valued at 15 new paise issued by the Indian Posts and Telegraphs, to commemorate wildlife week. First Indian stamp featuring wildlife
COLLECTION KEES ROOKMAAKER

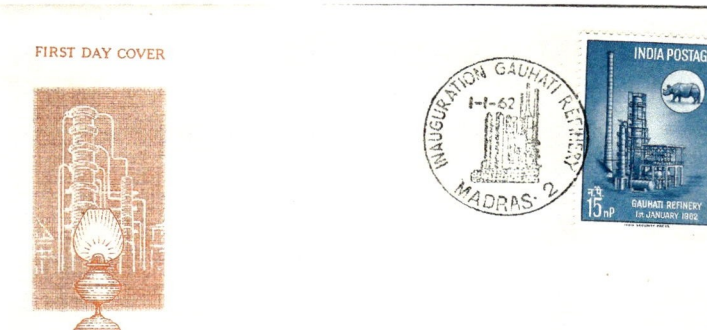

FIGURE 38.22
Stamp issued at the inauguration of the Gauhati Refinery on 1 January 1962 by the Indian Posts and Telegraphs
COLLECTION KEES ROOKMAAKER

FIGURE 38.23   Series of stamps issued for the 3rd India-Africa Forum Summit held in New Delhi in October 2015. Animals from both continents are juxtaposed

COLLECTION KEES ROOKMAAKER

Fitzwilliam Thomas Pollok in Assam

39.1 Biography of Fitzwilliam Thomas Pollok

Fitzwilliam Thomas Pollok (1832–1909) is well-known for his books on game hunting in India and Burma in the middle of the 19th century. Pollok was born on 19 October 1832 in Chennai (Madras). His father William Pollok (1811–1838) was an army officer living together with his mother known as Jumal Bee (1812–1839), who might have originated from Pondicherry. After the early deaths of his parents, Fitzwilliam went to England with his sisters Caroline and Augusta, staying with his paternal grandparents General Thomas Pollok (1772–1848) and Caroline Augusta née Thomas (1773–1848), at Kent Terrace off Park Road, on the fringe of Regent's Park in London. When his grandmother passed away, he was under the care of his guardian Lieutenant-General James Lushington (1779–1859). His last name is also found as Pollock. His first names are spelled either Fitzwilliam or separately Fitz William.

Pollok enlisted as a junior officer in the army on 20 December 1848 and returned to India in 1849 when he reached his first station in Palaveram, on the outskirts of Madras. He received the usual promotions to Lieutenant (1856-11-23), Captain (1861-02-18), Major (1868-12-20), Lieutenant-Colonel (1874-12-20) and Colonel (1878-11-30), at which date he retired (Hart 1875). Working as an engineer in the Public Works Department under the Madras Staff Corps (8th Regiment Native Infantry), he was responsible for surveying and constructing roads. After three years in Central India, he went to Burma in 1853, where he worked mostly in the Pegu district and surrounding areas for 13 years. From 1866 to 1873 he was posted to Assam. It is not quite clear where he was posted afterwards or where he stayed after retirement, but he may have lived in Bangalore. Pollok married twice, first in 1859 to Agnes Paterson Campbell (1839–1864) and second in 1866 to Emilie Anne Rothes Leslie (1844–1931), and he had a total of nine children. Pollok lived for 20 months in Fayal on the Azores (Pollok 1898e) before settling in England in the 1890s, where he was buried in St. James' Cemetery, Dover.

39.2 Pollok's Publications

This short background of Pollok gives an idea about the man who was one of the more prolific hunters of his time.

Even if there were many other officers engaged in sports hunting, he stood out from the crowd merely for the fact that he took time to write down his experiences and publish them in magazines and books. Pollok was responsible for four books and a long series of shorter contributions to both scientific and popular magazines. The contents generally amplify his adventures in the wild areas of Assam and Myanmar. He wrote about the rhinoceros in *Oriental Sporting Magazine* (1868 to 1874), *Illustrated Sporting and Dramatic News* (1882), *Field* (1882), *Navy and Army Illustrated* (1898), *Zoologist* (1898) and *Wide World Magazine* (1898 to 1902), as well as at least two book chapters.

Pollok's works are no longer fashionable being a string of tales how different animals were wounded, captured or killed. Yet they give an insight in practices which were all too normal at a time when the jungle was hard to penetrate, when animals were plentiful, and when only few could afford the time and expense to travel. Pollok must have kept a regular hunting diary which he extracted for his published writings. Such records were probably not uncommon, but very few are known to have survived over the years, and that is a great pity as these could have told us much of a time gone forever.

Pollok started to publish in the *Oriental Sporting Magazine* (Calcutta) [OSM hereafter] with his tales of sport appearing in 15 parts between 1868 and 1874. Although Pollok (1879) stated that he used the penname Poonghee, I have been unable to retrieve those. In fact, his contributions to the OSM were all anonymous, except the final one with the unexplained signature "J.P." There is absolutely no question that Pollok was the actual author, obvious from the style and the coincidence with events found in his books.

His first book with the rather exact title *Sport in British Burmah, Assam, and the Cassyah and Jyntiah Hills* [*Sport* hereafter] was published by Chapman and Hall, of 193 Piccadilly, London, and was available from the first day of April 1879 (advertised in *Pall Mall Gazette*, Friday 1879-04-04 and elsewhere). The two volumes of the usual octavo size (21.5 × 14 cm) cost 24 shillings at the time. These contained a total of 10 lithographed plates, 2 monochrome, 8 coloured, signed by the British sporting artist Alfred Chantrey Corbould (1852–1920). There is one coloured plate showing a rhino hunted by a man in an elephant howdah (1879, vol. 1 facing p. 70) illustrating

© L.C. (KEES) ROOKMAAKER, 2024 | DOI:10.1163/9789004691544_040
This is an open access chapter distributed under the terms of the CC BY-NC-ND 4.0 license.

FIGURE 39.1
"I had just time to turn round and let drive, as rhino's nose was within a few inches of my elephant's posterior." This shows Fitzwilliam Pollok hunting in Assam in 1869. Coloured chromolithograph signed by Alfred Chantrey Corbould (1852–1920). Inscribed (lower right) "P. 70 vol. I." and "Vincent Brooks, Day & Son, Lith." referring to the lithographic firm of William Day (1797–1845), William Day Jr. (b. 1824) after its purchase by Vincent Brooks (1815–1885) in 1867
POLLOK, *SPORT IN BRITISH BURMAH, ASSAM*, 1879, VOL. 1, FACING P. 70

FIGURE 39.2
"A dead rhinoceros." From a photo by Peter Burges (1857–1938). This was taken during a hunt organized by the Maharaja of Cooch Behar in 1896. The same photograph, without the heads of the men, in Moffett (1898a: 508, 1898b: 145)
POLLOK AND THOM, *WILD SPORTS OF BURMA AND ASSAM*, 1900, P. 85

a scene during Pollok's hunt with Macdonald in 1869 in the Manas area (fig. 39.1). We would probably still agree with the reviewer in *The Graphic* (Saturday 1879-05-24) that the book is "roughly written, and at little pains to save the reader by judicious retrenchment from the inevitable repetitions of a hunter's diaries," yet possibly to be ranked among the best books in its class. Pollok had said himself that he was a man who was used to stay in wild places, and could not be expected to write like a novelist. Although Pollok has been widely quoted in later literature, his style cannot have convinced everybody. Privately President Theodore Roosevelt (1858–1919) wrote to the well-known African hunter and traveller Frederick Courteney Selous (1851–1917) that "the Colonel Pollok whom I wrote you about is, I am quite sure, what we should term a fake, although I also have no doubt that he has actually done a good deal of big game hunting; but I am certain that, together with his real experiences, he puts in some that are all nonsense" (Roosevelt 1898-02-15).

The *Wild Sports of Burma and Assam* of 1900 was an updated and revised edition of the 1879 *Sport*. As Pollok himself had no recent information about Burma, he worked together with the younger William Sinclair Thom (b. 1868), Superintendent of Police in the Arakan Hill Tracts. It was published in one volume, the same octavo size, but all pages bound together. Published by Hurst & Blackett of 13 Great Marlborough Street in London, it was available on 7 September 1900. Not only was the text structured differently, the illustrations were drastically altered. There were 57 photographic images, 3 drawings and 3 maps. The photographs with Burmese subjects were contributed by the firm run by Frederick Albert Edward Skeen (b. 1862) and Henry Walker Watts (1862–1936) in Rangoon, as well as William Onslow Hannyngton (b. 1874). The Assamese subjects were credited to P. Burges of Clifton (near Bristol, UK) and Sir B. Simpson. Neither of these men appear elsewhere in Pollok's books, which must indicate that these items were not photographed while Pollok hunted in

FIGURE 39.3
Result of a fortnight's sport (Assam). From a photo by "P. Burges, Esq." There are three skulls of rhinos on the ground in front of the tiger skins. It is likely that this is the display of trophies associated with the shoot of the Maharaja of Cooch Behar in February to March 1896 attended by Peter Burges (1857–1938) of Clifton near Bristol. Moffett (1898b: 136) illustrated the same photo
POLLOK AND THOM, *WILD SPORTS OF BURMA AND ASSAM*, 1900, P. 448

FIGURE 39.4
"Shot in my old hunting-grounds, Dooars, Assam. From a photo by Sir B. Simpson, Bart." Benjamin Simpson participated in 14 hunts of the Maharaja of Cooch Behar, but the actual date is not recorded for this photograph
POLLOK AND THOM, *WILD SPORTS OF BURMA AND ASSAM*, 1900, P. 467

Assam. It cannot be a coincidence that they participated in the annual hunts of the Maharaja of Cooch Behar. Peter Burges (1857–1938), sometimes Burgess, attended in 1893, 1896 and 1897, and Sir Benjamin Simpson (1831–1923) of the Indian Medical Service was a regular visitor recorded for 14 hunts between 1877 and 1902 (32.9). Therefore I suggest that these photos were taken in Cooch Behar. Three plates show a rhinoceros:

(1) Pollok & Thom 1900: 85. Photograph with title "A dead rhinoceros, shot in Assam." Photo by P. Burges (fig. 39.2). Burges is credited with the killing of two rhinos, on 24 February 1896 and in April 1896, and this may be one of them.

(2) Pollok & Thom 1900: 448. Photograph "Result of a fortnight's sport" taken by P. Burges, Esq. Among the trophies there were at least 3 rhino skulls in front of several tiger skins (fig. 39.3).

(3) Pollok & Thom 1900: 467. Photograph with title "Shot in my old hunting-grounds, Dooars, Assam."

From a Photo by Sir B. Simpson, Bart. This is a display of game trophies, including 5 rhino heads (not skulls) with horns (fig. 39.4). The actual occasion is not retrievable.

Besides the two principal treatises of hunting in India and Burma, Pollok was a frequent contributor to a variety of journals and edited volumes about the same subject. He also wrote more popular stories aimed at a younger public full of his adventures in the wilds, like a short series in the *Wide World Magazine* in 1898 to 1902. One of these called "a bout with a rhinoceros" even includes an illustration of an attack of a double-horned rhinoceros in Tenasserim, Myanmar, signed by William Barnes Wollen (1857–1936), where the animal is so obscured in the bushes to be almost invisible (Pollok 1902b: 197).

Pollok's *Fifty Years' Reminiscences of India* published in 1896 presents an overview of his life as an army officer and sportsman. There is one relatively short chapter on the rhino described in more general terms than elsewhere.

FIGURE 39.5
Rhinoceros chasing an elephant. "B.'s elephant did not
wait, but bolted through the long grass." Engraving
signed (lower left) "GWL" and (lower right) "Corbould"
for Alfred Chantrey Corbould (1852–1920)
POLLOK, *FIFTY YEARS' REMINISCENCES OF INDIA*,
1896, P. 197

One plate (1896: 197) is a drawing by Alfred Corbould, again with a generalized picture of the rhino (fig. 39.5).

39.3   Pollok Hunting the Rhinoceros in Assam

Pollok was in Assam from December 1867 to April 1873. His writings may not excite us any longer, but certainly he gave us all the details needed to reconstruct where he saw rhinos. He provides dates, localities and names of his companions, and some details of the guns and bullets to be used. It must be noted that the contents of the excerpts of his diary found in OSM and *Sport in British Burmah* overlap, but they are not identical. Starting from the 15 parts in OSM, parts 1 and 2 are about British Burmah; parts 3, 4, 5, 6 are similar to entries in the second volume of *Sport*; parts 7, 8, 9 with information about trips in 1868–1869 are not in the book except very briefly; while the later trips of 1870–1873 are in the book only. I have collated the available dates about all his trips below (39.4).

When it comes to localities, Pollok was definitely eccentric in his spelling. Some of the larger towns and rivers can be identified, but many of the villages and ponds in the field remain obscure, either because these landmarks disappeared, changed name, or are written in such a way that only a local person might identify them. In most cases we can know the general area where he was traveling, while the exact details are difficult to find. Pollok's maps in 1879 and 1900 are of absolutely no assistance, as only a few of the larger towns are named, probably showing that the maps were actually made in England from existing examples. Places mentioned in Pollok's text are only rarely found on the maps.

The map "Assam" in the *Wild Sports* (1900: 425) includes many red marks + indicating "good sporting areas", which must have been added on advice of Pollok (fig. 39.7). It is unlikely that he himself visited all those places indicated. "Rhinoceros" is printed twice on the map, first in the Manas area north of Barpeta, and second in the Laokhowa area south of Brahmaputra. Most of the red marks have no further information why they are included and are therefore of limited use, except maybe to know where there were still wilderness areas.

Pollok fortunately did not follow the custom of contemporaries to identify their companions by initials or pseudonyms. Sometimes traveling alone, most often he was accompanied by fellow officers, police personnel, or tea planters. Pollok gives us their surnames, but only in few cases tells us their first name or their position. This makes certain identification complicated, especially if the surname is a common one. I have listed his companions in alphabetical order to get an idea who were interested in such sport in remote places (table 39.53). It may be assumed that most of them also killed one or more rhinos, even though the details would be hard to distill from Pollok's writings. Apparently none of these people left any trace of their experiences, which just shows how important Pollok was in a historical analysis.

TABLE 39.53   List of companions of F.T. Pollok in Assam

Pollok mentioned the following companions while hunting in Assam. He used their surnames only, and several have not been further identified.

Arthur Anley, District Superintendent of Police in Assam
John Barry, tea planter – sold Kookooriah garden to Pollok, see Pollok & Thom (1900: 473)
General Henry William Blake (1815–1908), Madras Infantry (*Sport* of 1879 book was dedicated to him, p.iv)

TABLE 39.53   List of companions of F.T. Pollok in Assam (*cont.*)

John Henry Bourne (1844–1873), 44th Sylhet, Bengal Native
Infantry

Major George Machardy Bowie (1835–1876), Deputy Inspector
General

Butler (not identified)

Colonel George Campbell (d. 1876), 52nd Foot Regiment

Campbell's wife

Augustus Kirkwood Comber (1828–1895), Commissioner of
Lakhimpur

Colonel James Cathorne Cookson (1821–1900), Madras Cavalry

Charles Orchard Cornish (1839–1905), Superintending Engineer

J.A. Floyd, Police Detective, Goalpara

G.H. French, Assistant Superintendent of Police

Gordon, tea planter

William John Hicks (1823–1873), Bengal Staff Corps

G.H. Jackson, 43rd Regiment Native Infantry

Colonel J. Macdonald, Survey of India

Masters, Police officer

Ormiston Galloway Reid McWilliam (1841–1905), Deputy
Commissioner of Cachar

Edmund Pipon Ommanney (1841–1910), 44th Sylhet Bengal
Native Infantry

Raja of Luckeepore

General Charles Reid (1819–1901), Bengal Staff Corps

Sookur – local hunter

Sybroodeen – local hunter

Thomas, police officer

Ernest Tye (d. 1919), tea planter in Koliabar, Secretary of the
Indian Tea Association

Lieutenant William John Williamson (1842–1883), Deputy
Commissioner of Garrow Hills

39.4   Excursions in Assam and Shillong

Pollok wrote about 31 excursions in Assam and Meghalaya
from 1867 to 1873. The details found in his publications are
extracted for information on the rhinos killed or encountered. The trips are numbered here for reference only.
His companions are identified in Table 39.53. Pollok published his hunting diary in the *Oriental Sporting Magazine*
(osm) in 15 parts in 1868, 1871, 1873, and 1874. All page reference of the *Sport* of 1879 are included in volume 2, and
Pollok 1900 refers to the *Wild Sports of Burma and Assam*.
The rhinos killed by Pollok are numbered consecutively
as nos. 1–44. The main localities are shown on map 37.20.

(Trip 1) 1867, February. Manas area. No rhino killed.

(Trip 2) 1867, March. Laokhowa area. Pollok 1879: 81, 1882a.
Joined by Reid and Cookson. The locality is given as Logva
Ghat (Loquaghat, now Laokhowa), which was opposite
Tezpur according to Pollok (1896: 197). In this place "rhinoceros" is written on the map of 1900. Although Pollok
was present, he obviously gave chance to his superiors,
and it was General Reid who killed "in one day two full
grown rhinos, each with one bullet of 20 to the pound"
(Pollok 1882a). Cookson shot one, thought it was dead, but
it got up and disappeared. In Pollok (1894: 74), he refers to
an excursion with Colonel Cookson where they traveled
on foot (not on elephant) and his companion shot 1 male
and 2 female rhinos. I have not identified which trip this
could have been. Pollok & Thom (1900: 434) state that 6–7
rhinos were wounded near Dowkagong.

(Trip 3) 1867, 9 to 21 June. Manas area. Pollok 1868 (part
3), 1879: 82–83, 1883, 1894: 72, 1900: 434. After reaching
Burpettah (Barpeta), Pollok kills his first rhino (no. 1)
on 11 June. This was a large animal with a massive horn
weighing 2.2 kg, despite its length of only 8 inch (20.3 cm).
In the following days he saw large numbers of rhino, but
he was unable to shoot any others.

(Trip 4) 1867, July. Shillong. No rhino killed.

(Trip 5) 1868, 8 to 20 January. Manas area. Pollok 1868 (part
4), 1879: 95, 1900: 445. Joined by Bowie and Barry. Pollok kills
5 rhinos (nos. 2–6) on 14 January (3) and 16 January 1867
(2), in the vicinity of Burpettah (Barpeta). Bowie killed 2
others, one of which was huge and had a horn of 13 inch
(33 cm). One of the rhinos was a half-grown one which
they failed to capture.

(Trip 6) 1868, 11 February to 2 March. Tezpur and Laokhowa.
No rhino killed.

(Trip 7) 1868, 27 April to 2 May. Dhubri-Goalpara area.
Pollok 1868 (part 5), 1879: 111–112, 1900: 454. Joined by
Comber. Pollok killed 2 rhinos (nos. 7, 8) on 1 and 2 May,
both within miles of Tikri Killah (Tikrikilla). Comber shot
at least one more. It is important to compare Pollok's various accounts of what happened on 1 May 1868. In his first
write-up, published soon afterwards, he says that first
"Comber came across one and shot it dead, and called
out to me to look out, as there was another. I saw first the
ear of one rushing past me, and guessing for the shoulder,

FIGURE 39.7   Hunting the Rhinoceros. Illustrating an extract of
Trip 12 with Macdonald. The figure was inserted
by the publishers, Frederick Warne & Co. from the
original first found in the plate shown in Figure 39.6
POLLOK, *THE PICTORIAL MUSEUM OF SPORT AND
ADVENTURE*, 1880, P. 21

FIGURE 39.6   "Hunting the Rhinoceros." Engraving of hunters
aiming at a single-horned rhinoceros, although
illustrating the travels of David Livingstone
(1813–1873) in Africa. The signature on lower right is
illegible. Note the rhino escaping up the hill in the
background
J. EWING RITCHIE, *THE LIFE AND DISCOVERIES OF
DAVID LIVINGSTONE, PICTORIAL EDITION*, 1876,
VOL. 1, P. 328

fired and brought it down, but unfortunately it turned out
to be but a half-grown one. We cut off the horn – a very
small one – of the one Comber killed, and went on beating" (Pollok 1868: 803). In his book of 1879, he describes
this rhino as "a three-parts grown rhinoceros, for which
I was very sorry, as it was of a nice size to catch." Definitely
no mention that this animal differed in any way from any
of the others, except for it not being full-grown or small.
In a later book, Pollok (1900: 454) claimed this to be the
only specimen of *R. sondaicus* which he ever killed: "leaving this oasis, we got into heavy grass, and I shot one of
the lesser rhinoceros." When Pollok read in the literature
of his days that not one, but two species of rhino existed
in the region where he spent so much time observing and
chasing game, he must have felt obliged to say that he also

killed it. This exceptionally small rhino would suit the
bill, unless it was indeed just a young animal that had not
reached its full size. It would have been curious, but not
impossible, that both *R. unicornis* and *R. sondaicus* would
live in such close proximity, in fact in a mixed population
(52.3).

(Trip 8) 1868, May. North of Guwahati, East of Manas. No
rhino killed.
(Trip 9) 1868, June. Kookooriah (Manas). No rhino killed.

(Trip 10) 1869, 16 January to 6 February. Manas area.
Pollok 1871 (part 6), 1871 (part 7), 1879: 117; Pollok &
Thom 1900: 457. Joined by Barry and Butler. They find
many old rhino tracks near Kumblepore. No rhino killed.

(Trip 11) 1869, 10 to 17 March. Dhubri-Goalpara area.
Pollok 1871 (part 7), 1879: 118–119. The dates of this trip are
not clearly recorded and may combine two concurrent
events. Pollok first killed 2 rhinos (nos. 9, 10) about six
miles from Tikri Killah (Tikrikilla). This was followed by 3
more (nos. 11, 12, 13) when out near Luckeepore (Lakhipur)
with Floyd.

(Trip 12) 1869, 29 April to 11 May. Manas area. Pollok 1871
(part 7), 1871 (part 8), 1879 I: 67–73, 1880 (fig. 39.6, 39.7),

1882. Joined by Macdonald. Pollok kills 2 rhinos (nos. 14, 15) on 9 and 11 May near Busbaree (Bahbari). Macdonald also kills a female rhino and her calf. A young rhino was captured at Busbaree and brought to the camp at 10 pm: "carried by some 15 coolies; they got him entangled in the meshes of a net, closed upon him and had him securely tied in a very short time. When we saw him next morning, he was far more savage than any tiger would have been. He was tied to four posts by the hind and fore-legs, and also had a stout rope round his neck. Whenever anybody went near him, he tried all he could to break his bondages and to charge open-mouthed, yet in two days he would feed out of Sookur's hand, and in a week follow him about anywhere" (Pollok 1871, part 8: 469; see 1898a: 175). It is likely that the animal was sold to a dealer, as it was worth about Rs 600 or £60 (Pollok 1882). The story of the capture is mirrored with a fictional twist by Thomas Mayne Reid (1818–1883), who might have taken Pollok as his example, while changing several of the personal details (Reid 1882, see 36.19).

(Trip 13) 1869, 26 to 27 October. Fishing trip from Sylhet. No rhino killed.

(Trip 14) 1870, January. Manas area. Pollok 1879: 119, 1900: 458. Joined by Blake and his wife, Ommaney, Masters and Campbell. Pollok kills 2 rhinos (nos. 16, 17).

(Trip 15) 1870, March. Myung. Pollok 1879: 121, 1900: 439. Pollok only devotes one line to this short trip. His locality Myung is irretrievable, as he doesn't provide any context, but maybe it was Mayang, on the south bank of the Brahmaputra east of Guwahati. Pollok kills 1 rhino (no. 18).

(Trip 16) 1870, 15 to 30 April. Manas area. Pollok 1871 (part 8), 1873 (part 9), 1873 (part 10), 1879: 125–129, 1882: 59, 1894: 73–80, 1900: 465–473. Joined by Jackson. This must have been one of Pollok's most exciting excursions as far as rhinos were concerned. Traveling on the eastern side of Manas, he mentions the Phoomarah (Pahumara) River, and camps at Basharee (Bahbari) and Matagoorie (Mathanguri). Rhinos were particularly plentiful: "Going on I found myself in the midst of a whole herd of rhinoceros; there were probably ten or twelve rhinoceros in the grass and five or six immediately round me, all making their diabolical noises, at hearing which elephants generally go mad with fear, become ungovernable and bolt; but the old mucknah I was on never moved. Firing quickly, I wounded at least four, and had just time to reload my battery, when I had one brute charging me on the left, another on the right, and one in the rear, and several others making

feints all round, and I fired as quickly as I could snatch up the rifles. I had four rifles and my smooth-bore with me, and I emptied every one before I was quit of my foes" (1879: 129, see 1898a: 174). Pollok (1894) stated that 13 rhinos were killed in 14 days, some of which must have been credited to Jackson. His personal count was probably 12 rhinos (nos. 19 to 30), secured on 20 April (1), 21 April (4), 22 April (1), 23 April (1), 24 April (4) and 27 April (1). They also tried to capture a young rhino, but it was injured in the process and died.

(Trip 17) 1870, June. Kookooriah (Manas). No rhino killed.
(Trip 18) 1870, July. Luckeepore (Goalpara), along trunk road. No rhino killed.
(Trip 19) 1870, 13 to 19 September. Shillong to Sylhet. No rhino killed.
(Trip 20) 1870, 16 to 22 October. Shillong to Sylhet. No rhino killed.
(Trip 21) 1871, 18 to 27 February. Manas. No rhino killed.
(Trip 22) 1871, 25 March to 1 April. Dhubri. No rhino killed.

(Trip 23) 1871, 8 to 21 May. Manas area. Pollok 1879: 146–149. Joined by Barry, Anley and Campbell. Kills 2 rhinos, on 17 May (no. 31) at Kumlabaree and on 18 May (no. 32) along the course of the Poho-Marah river. His companions also killed several rhinos. The animals were numerous in the area, as one day Pollok recorded seeing at least 8 rhino, and 12 on another afternoon.

(Trip 24) 1872, 13 to 18 March. Singhamaree near Dhubri. No rhino killed.

(Trip 25) 1872, 3 to 8 April. Rungiah (Rangia) area. Pollok also refers to Dewangiri, which is further to the north. Pollok 1879: 152. Pollok kills 1 rhino (no. 33) on 5 April, at Demoo Nuddie.

(Trip 26) 1872, 13 to 23 April. Manas area. Pollok 1874 (part 14), 1879: 154–156. Joined by Comber and Cornish. Pollok kills 5 rhinos, on 18 April (nos. 34–36) and 20 April (nos. 37, 38). Cornish shot a female on 18 April. This mother had a small one, which ran away at first, but on 19 April they went back "to the dead rhinoceros to try and catch its young one. We found the poor little thing lying down by its dead mother's side. We tried to catch it, but the nets had not arrived, and after a while it ran away and I believe joined another rhinoceros, which came up as we were trying to catch it."

(Trip 27) 1872, 12 to 19 September. Fishing to Sylhet. No rhino killed.

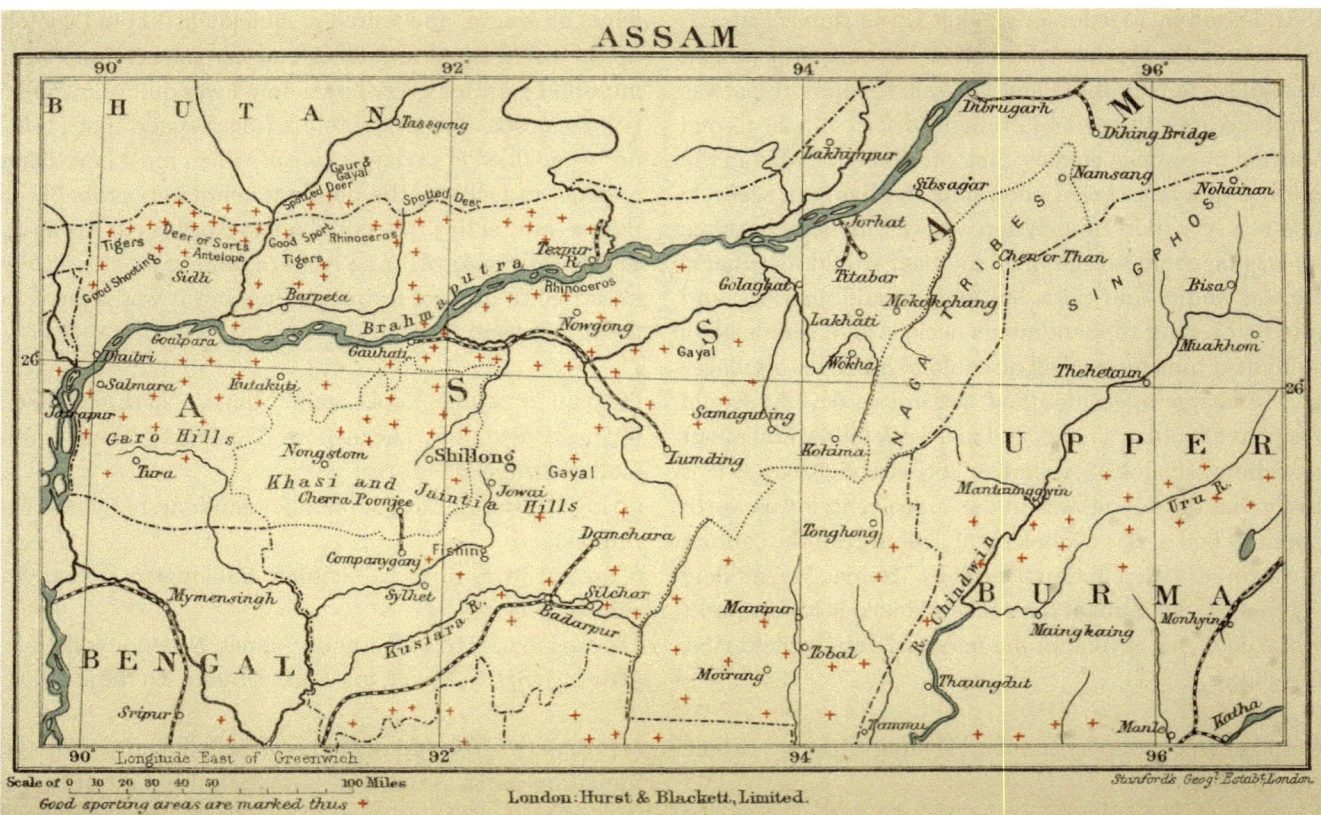

FIGURE 39.8 Map of Assam in the second edition of Pollok's *Wild Sports* printed by "London: Hurst & Blackett, Limited". The red crosses indicate good sporting areas

POLLOK & THOM, *WILD SPORTS OF BURMA AND ASSAM*, 1900, P. 425

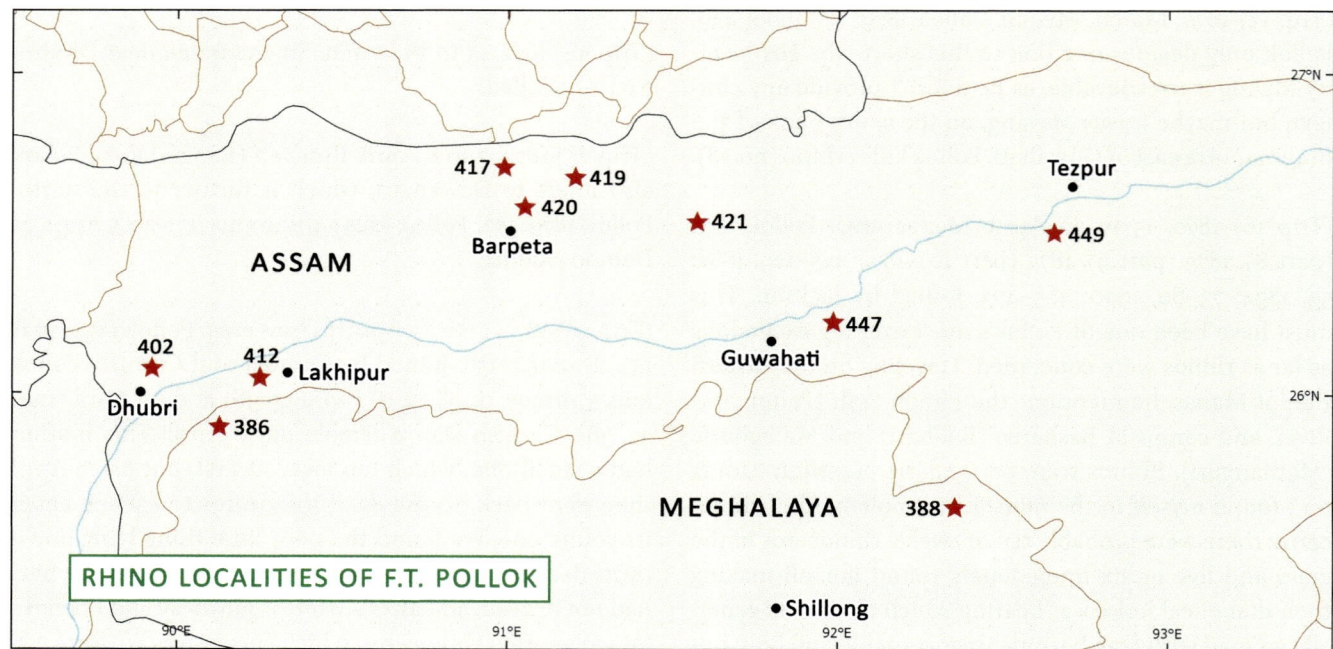

MAP 39.21 Localities associated with rhino events described in the writings of F.T. Pollok 1867-1873. The numbers and places are explained in Dataset 39.49

(Trip 28)1872, 28 October to 6 November. Fishing to Sylhet. No rhino killed.

(Trip 29) 1873, before April (?). Brahmaputra below Dhobree (Dhubri). Pollok & Thom 1900: 488. Joined by Williamson. Pollok went pig-sticking riding on ponies. The event is not dated, but it was in his last year in Assam, as he was transferred to Secunderabad soon after. Williamson died not much later in his late twenties on an unverified date. Pollok shot a female rhino (no. 39), and Williamson killed the calf, about three-quarter grown, with a spear while riding his pony.

(Trip 30) 1873, 15 April to 3 May. Manas area. Pollok 1874 (part 15), 1879: 160–166. Joined by Barry. Pollok kills 5 rhino (nos. 40–44), on 19 April (1), 25 April (1), 27 April (2) and 29 April (1). Barry shot at least one in addition. The animals were plentiful: "The whole place at times must be full of rhinoceros, but they have been so molested that at early morn they betake themselves to the tree-jungle, where they are quite safe from us" (Pollok 1879: 163).

(Trip 31) 1873, 30 May to 9 June. Manas. Pollok 1874 (part 15). Pollok encounters rhino, but he is unable to get close. The country is much flooded by an early monsoon.

• • •

Dataset 39.49: Localities of Records of Rhinoceros Associated with F.T. Pollok

The numbers and places are found on map 39.21 and again included in map 38.19. Numbers in line with chapter 38 (Assam).

*	Singhamaree, near Dhubri – 25.70N; 89.85E-no rhino recorded
A 402	Dhubri-Goalpara area – 26.00N; 89.90E
A 386	Tikri Killah (Tikrikilla) – 25.90N; 90.15E
A 412	Luckeepore (Lakhipur) – 26.00N; 90.30E
A 417	Manas area – 26.70N; 90.90E
A 417	Matagoorie (Mathanguri) – 26.78N; 90.95E
A 417	Kookooriah (Manas) – not located
A 418	Busbaree (Bahbari) – 26.50N; 91.00E
A 419	Phoomarah, Poho-Marah (Pahumara) River – 26.65N; 91.15E
A 419	Kumlabaree – 26.65N; 91.15E
A 420	Burpettah (Barpeta) – 26.33N; 91.00E
A 421	Demoo Nuddie – 25.65N; 91.60E
A 421	Towards Dewangiri – 26.85N; 91.65E
A 421	Rungiah (Rangia) – 26.45N; 91.65E
A 421	Kumblepore, not located (Rangia area)
A 447	Myung (not located, perhaps Mayang) – 26.25N; 92.05E
A 449	Laokhowa, Logva Ghat (Loquaghat) – 26.54N; 92.75E
A 449	Dowkagong – not located (Laokhowa area)
A 388	Jaintia hills towards Nowgong – 25.60N; 92.30E (see Meghalaya)

* Not on map.

•• •

39.5 The Outcome of the Sporting Trips

Pollok (1879, vol. 1: 87, 1896: 195) stated to have killed 44 rhinos in Assam. Later he elaborates that rhinos "are not easy beasts to kill. I was exceptionally lucky, for in 6 years, to my own gun, I killed 44. I helped to kill some 30 others, and saw some 20 others killed by comrades who were out with me, and I lost, and saw lost, fully 50 others. I probably came across some 300 and more during my wanderings in Assam" (Pollok 1882b: 31). Elsewhere Pollok (1894: 69) states that he killed 47 or 48, which number must have been due to faulty memory at the time of writing. I have traced 44 rhinos claimed by Pollok in his writings, although the actual shooter is not always clear from the text.

Pollok apparently did not usually bother to take measurements of the animals killed in Assam. If he had, it would have been a nice dataset which might have slightly vindicated his slaughter. As it is, he only provides the size of two animals, although maybe just one because both were the largest found (table 39.54).

TABLE 39.54 Measurements of two rhinos killed in Assam by F.T. Pollok from his own writings

	Specimen 1 Assam	Specimen 2 Manas
Length of body	381 cm (12 ft 6 in)	406 cm (13 ft 4 in)
Length of tail	61 cm (2 ft)	
Height at shoulder	188 cm (6 ft 2 in)	188 cm (6 ft 2 in)
Horn	35.6 cm (14 in)	33 cm (13 in)
Pollok reference	1879, vol. 1: 95, 1882a, 1898d: 256	1879, vol. 2: 126, 1882b: 58, 1900: 466

In a few cases he recorded the length of the horns of the rhinos killed during his trips. There were two with horns of 8 inch (20.3 cm), one with 12 inch (30.5 cm), two with 13 inch (33 cm), one with 14 inch (35.5 cm) and the longest was shot by Bengal hunters in 1870 with 18 inch (45.7 cm). These figures did not indicate any sexual difference.

Although Pollok claimed to have shot one individual of *R. sondaicus* in May 1868, the evidence is equivocal as there is nothing in his anecdotes written at the time to support this (trip 7). Pollok (1898a: 174) claimed a wide range of *R. sondaicus* throughout North-East India, but avoided to say that he ever met the species. He later changed his mind based on the supposition that *R. sondaicus* was much smaller than *R. unicornis* (52.3)

Pollok sometimes tried to capture young rhinos because they could fetch a good price with animal dealers. He was certainly successful once, in May 1869 when out in the Manas area with Macdonald (trip 12) . However, he stated to have sold two to an Afghan agent of Jamrach "for 1200 rupees, delivered in Gowhatty; but I believe I ought to have got double the amount" (Pollok 1898a: 175).

Sport and entertainment were Pollok's main raison d'être. These excursions were expensive, even if he could use departmental elephants without charge. The absence of a large number of trophies seems to indicate that he usually sold the horns, maybe also parts of the skins. At his time, the authorities still paid a reward of Rs 5 for each rhino, a relatively small amount (Pollok 1882: 31). Any rhino caught and sold of course added to the budget.

There is no indication that Pollok kept any trophies of any of the game species. He never mentioned a private collection or museum. One horn of 13 inch was donated to Martin Andrew Dillon (1826–1913), Secretary to Field Marshal Robert Cornelis Napier (1810–1890) around 1870 (Pollok & Thom 1900: 445). He also made no attempt to immortalize any events in photographs. He stated only once, for an excursion in 1870 at Kookooriah (Manas area), that he preserved some bones of animals which died from cattle disease (Pollok 1879: 133). The Indian Museum in Kolkata reported the accession in July 1870 of one complete and one incomplete skeleton of *Rhinoceros Indicus* from "Gowhatty, Assam – Major Pollok" (Indian Museum 1871). The complete skeleton may have been exchanged, because Sclater (1891: 202, no.c) only records the incomplete one together with its skull. This specimen too, if still present, is no longer recognized.

Protecting the Rhinoceros in Manas National Park, Assam

Established: North Kamrup Reserve in 1907
Species: *Rhinoceros unicornis*
First Record: 1815 – Last Record: current (relocations from 2006)
Rhino Region 32

40.1   Rhinos in Manas

The grasslands and forests in the Himalayan foothills along the Manas River in western Assam provide an excellent wildlife habitat where rhinos used to thrive (Choudhury 2019). Manas National Park, formerly known as North Kamrup Reserve, lies in the Baksa and Chirang districts of Assam, some 40 km north of Barpeta Road. It is contiguous eastwards with the Chirang and Ripu Reserved Forests towards the border of West Bengal. North of the Manas River it borders the Royal Manas National Park in Bhutan, which sometimes harbours some stray rhinos (35.1).

After the Duar War (or Bhutan War) in the 1860s the area opened up to sportsmen and travelers who found an area abounding with wild animals. On the map produced for the 1900 book, *Wild Sports of Burmah and Assam* by Fitzwilliam Pollok, the presence of "rhinoceros" is clearly stated in the *terai* north of Barpeta (fig. 39.7). In 1865 General Henry Andrew Sarel (1823–1887) had what he called sport in the region, at a time when rhinos were still said to be plentiful.

Manas was officially declared a reserve especially to protect the rhinoceros, the first of its kind in South Asia, in 1905. The Maharaja of Cooch Behar came four times from 1905 to 1909, twice hosting the Viceroy, without attempting to hunt rhinos in particular (40.3).

There was a major poaching episode in the 1930s (40.4). The area was lawless in the 1990s as a result of political aspirations (40.5). This devastated the rhino population and none were left at the end of the century. A series of translocations from other areas in Assam followed from 2006 onwards, restoring the population to a decent level in 2022 (40.6).

FIGURE 40.1
Two rhinos in the landscape of Manas National Park
PHOTO: DEBA KUMAR DUTTA

• • •

Dataset 40.50: Chronological List of Records of Rhino and Administrative Changes in Manas National Park

General sequence: Date – Locality (as in original source) – Event – Sources.

   Reports before notification as a reserve in 1902 are included in dataset 38.46: Rhino Region 32

1902-11-04 – Kamrup – John Campbell Arbuthnott (1858–1923) wrote to the Chief Commissioner of Assam, Joseph Bampfylde Fuller (1854–1935) about plight of rhino – Arbuthnott 1902 (Letter No. 75 of No. 2409G); Barthakur & Sahgal 2005: 27; Saikia 2009; Choudhury 2019: 28

© L.C. (KEES) ROOKMAAKER, 2024 | DOI:10.1163/9789004691544_041
This is an open access chapter distributed under the terms of the CC BY-NC-ND 4.0 license.

1902-12-18 – Kamrup – Joseph B. Fuller (No. 2160Misc. – 9628G.) replied to J.C. Arbuthnott that he would consider establishing an asylum for the rhino as a Reserved Forest – Barthakur & Sahgal 2005: 27; Choudhury 2019: 29

1903-08-28 – Kamrup, proposed Reserved Forest – Arbuthnott sends a trace map drawn by Major Philip Richard Thornhagh Gurdon (1863–1942) showing area in North-West Kamrup proposed as a reserve, covering 379.68 square miles (983.36 km²). Gurdon reports: "All along the foot of the hills there is a strip of Khair and Karai forest varying in breadth from about 4 miles to about ½ a mile with some swamp, near Oosla and Lahapara. The whole area does not contain a single village, for people will not live there for fear of the Bhutias. The Kacharis have no rights in the proposed reserve, but they go up sometimes to fish and also to shoot" – Arbuthnott 1903 (Letter No. 77 dated Jowai); Choudhury 2019: 29; Barthakur & Sahgal 2005: 27; Gokhale & Kashyap 2005: 21. Trace map lithographed by Ashrauf Ali reproduced by G. Singh 2014: 185; Choudhury 2019: 26.

1904-03-15 – Kamrup Reserve (proposed) – Letter from Francis John Monahan (1865–1923), Secretary of the Chief Commissioner of Assam (J.B. Fuller) to J.C. Arbuthnott, Deputy Commissioner, Sylhet. The proposals are agreed to in principle subject to three conditions to safeguard existing cultivation – G. Singh 2014: 182 ((ASA) AARP, Revenue A, September 1905)

1904-05-31 – Kamrup Reserve (proposed) – Letter from Frank Ernest Jackson, Deputy Commissioner of Kamrup to Arbuthnott, Commissioner of Assam Valley Districts. He agreed to the urgent need of a game reserve to protect rhino. The area proposed was too large and contained a considerable area under cultivation. Jackson proposed a reduced area of 163 square miles (422.17 km²) – G. Singh 2014: 183 ((ASA) AARP, Revenue A, September 1905)

1904-06-07 – Kamrup Reserve (proposed) – Memo by Arbuthnott, Commissioner of Assam Valley Districts, agreeing with Jackson and asking for further enquiry in this matter – G. Singh 2014: 183 ((ASA) AARP, Revenue A, September 1905); Choudhury 2019: 18

1904-09-20 – North Kamrup Reserve (proposed) – Edward Statter Carr (1857–1925), Conservator of Forests, agrees with Jackson and recommends the reserve with an area of 163 square miles. Carr suggests the name North Kamrup Reserve – Choudhury 2019: 18; G. Singh 2014: 183; Barthakur & Sahgal 2005: 29

1904 – Kamrup Dt. – Rhino live in the swamps that fringe the Brahmaputra and the Manas, and in the sparsely populated tracts in the north-west corner of the district – Allen 1904, vol. 4: 17

1905-02 – Kalabari, Hali River, Kachugaon, Pechadabri – Shooting trip of Maharaja of Cooch Behar around Manas, 18 to 26 February 1905. Hosting the Viceroy, George Curzon (1859–1925). Rhinos seen but not shot – 32.9, shoot 1905 (a); Choudhury 2019: 25

1905-06-01 – North Kamrup Reserve – First declaration of intention to constitute Kamrup as a Reserved Forest of 233.09 km² (57,600 acres). Notification No. 2441R, 1 June 1905, signed by Louis James Kershaw (1869–1947), Officiating Secretary to the Chief Commissioner of Assam (J.B. Fuller) – Assam Forest Dept. 1905: 9; Choudhury 2019: 179 (full text); G. Singh 2014: 188; Gee 1950b: 83; Saikia 2005: 273

1905-06-01 – North Kamrup Reserve – Boundaries of North Kamrup Reserve: (North) Bhutan boundary from the Manos River in an easterly direction to the eastern boundary of the Bijnimauza. (East) The eastern boundary of the Bijnimauza from the Bhutan boundary in a southerly direction to the north-west corner of Kaligaon village lands in Mauzachapguri. (South) A line running due south-west from the north-west corner of kalegaon village lands to the GeateeNadi and then due west to Gohanigaon on the river Manas. (West) Up to east bank of the Manas river from Gohaigaon to the foot of the Bhutan hills – Notification No. 2441R in Choudhury 2019: 179

1905 – North Kamrup Reserve – There are 3 former shooting grounds in the Sanctuary, of which 2 were included. The third was omitted because all rhino had been shot out, but the Local Administrations has been asked to add it, as it is uninhabited and for the most part undrainable swamp – Milroy 1916

1905-06-30 – North Kamrup Reserve – Some 243 square miles of forest, including 90 square miles in Kamrup, for which preliminary notifications under section 5, Assam Forest Regulation, issued during the year, remained under settlement on 30 June 1905 – Assam Forest Dept. 1905: 1

1906–03 – Barpeta – Shooting trip of Maharaja of Cooch Behar around Manas, 22 March to 30 April 1906, no rhino taken – 32.9, shoot 1906 (a); Choudhury 2019: 26

1907–02 – Kalabari – Shooting trip of Maharaja of Cooch Behar in Chiru RF, 18 to 24 February 1907. Hosting the Viceroy, Gilbert John Elliot-Murray-Kynynmound, 4th Earl of Minto (1845–1914). No rhino taken – Nripendra 1908: 411–429; Choudhury 2019: 26 – 32.9, shoot 1907 (a); 40.3

1907-07-29 – Settlement of North Kamrup Forest under notification No. 6784F with 90 square miles (233.09 km²) – Assam Forest Dept. 1908: 30

1909 – Kamrup Dt. – There are three reserves – Anon. 1909b

1909-02-13 – Manas – During the second shooting trip of the Viceroy Minto, his military secretary Victor Reginald Brooke (1873–1914) was attacked by two rhinos. He took a photograph just before the collision, published in the *Illustrated London News* of 1912, probably the second image taken in the wild – Anon. 1909b, 1909c; Brooke 1909, 1912; Minto 1934: 274; Rookmaaker 2021b – 32.9, 40.3, figs. 40.2, 40.3

1909-02-13 – Lohapara, Manas (north of Barpeta) – Shooting trip of Maharaja of Cooch Behar around Manas, 11 to 17 February 1909, hosting for the second time the Viceroy, 4th Earl of Minto (1845–1914). Minto shot a male rhino on 13 February – Anon. 1909c; Milroy 1916; *Times of India* 1909-02-17 – 32.9, shoot 1909 (a); 40.3

1909 – Kamrup Sanctuary – There were considered to be 15–16 rhino in the Sanctuary at the time of Earl of Minto's shoot – Milroy 1916; Hunter 1909: 514

1912-07-23 – North Kamrup RF – Notification No. 1091R for 233.59 km² – Lahan 1993: 33

1912-07-23 – North Kamrup Sanctuary is closed to shooting other than that allowed to authorised permit holders – Stebbing 1912: 37

1916 – North Kamrup Reserve – Rhino shot by poacher. The case was complicated but finally a suitable fine was recovered from the various persons implicated and the gun was confiscated – Blunt 1916; Milroy 1916: 456

1916-02-11 – Koklabari RF, constituent part of Manas – Notification 520R – Lahan 1993: 33

1917 – Manas – Area constituted as a protected region with prohibition of hunting and trapping – Choudhury 2019: 11

1924-08-20 – Kahitama RF, constituent part of Manas – Notification 2051R – Lahan 1993: 33

1925-12-15 – North Kamrup, first addition – Notification 3296M, added 37.04 km² with new total area 270 km² – Lahan 1993: 33

1925-12-19 – Koklabari first addition, constituent part of Manas – Notification 3341R – Lahan 1993: 33

1925 – Northern Assam (Manas?) – Tracks of large male rhino with good horn, not killed – Jamun 1925

1925 – Northern Assam (Manas?) – Friend J. killed a rhino – Jamun 1925

1927-07-11 – Manas Forest Reserve constituted by Notification 1886R, area 770 km² – Lahan 1993: 33

1928 – Manas Wildlife Sanctuary of 270 km² (North Kamrup RF) plus 120 km² (part of Manas RF) constituted in 1928 – Lahan 1993: 33; Talukdar & Sarma 2007: 14

1928 – Monas Game Sanctuary – Arthur John Wallace Milroy (1883–1936), the Conservator of Forests, Assam, reported increased insecurity on the reserve. A British Officer (unnamed) and a Company of Assam Rifles were detailed to spend six weeks touring a district where the inhabitants had got out of hand and were poaching in the Monas Game Sanctuary on a commercial scale, while at the present moment an energetic Assistant Conservator is on special duty at the head of an anti-poaching campaign that is effective – Milroy 1931: 4; 1932, 1934a: 98; G. Singh 2014: 164

1931–06 – North Kamrup Game Sanctuary – About 40 rhino carcasses were removed, all without horns – Saikia 2005: 274, 2009 (Report on Inspection of Manas Reserve no. 368, ASP, No. 286–294, Revenue Department, Forest Branch, Forest B, June 1931 (ASA))

1933 – Manas – Rhino present. Area should be converted to national park – Hanson 1933: 51

1934–1935 – Monas Sanctuary – Reserve of 159 square miles (411 km²), partly in Haltugaon and partly in Kamrup Divisions, in charge of Chakrapani Ayyangar Rama Bhadran (1907–1977), Assistant Conservator of Forests – Milroy 1935: 19

1935–1936 – Monas Sanctuary – In charge of Justman Swer, Extra Assistant Conservator of Forests. Poaching not serious – Milroy 1936: 14

1941 – North Kamrup Sanctuary – Rhino scarce. Poaching by concessionaires, who remove produce in exchange for free labour – Assam Forest Dept. 1941: 26

1949 – Monas Game Sanctuary – Salim Ali (1896–1987) and Sidney Dillon Ripley (1913–2001) visit 6 days. Saw no rhino, only fresh tracks of two – Gee 1950a, 1950b: 83

1950 – North Kamrup Sanctuary – Forest village was instituted, which will provide a threat to one of the best localities for rhino – Assam Forest Dept. 1950: 27

1950 – Kamrup – One rhino migrated from Darrang to Kamrup. Killed by villager for causing injury to a man – Assam Forest Dept. 1950: 28

1950 – Manas – Upgraded to Wildlife Sanctuary – Choudhury 2019: 11

1952-05-21 – Panbari RF, constituent part of Manas – Notification AFR 73/51 – Lahan 1993: 33

1955 – Manas WS – Reserve enlarged to 391 km² – Unesco website

1955–1958 – North Kamrup Wild Life Sanctuary, formerly Manas – Poaching may not be all that prevalent, because staff of park is active and rhino is partial to the rice grown along the southern boundary. Three carcasses without horns found recently – Burnett 1958: 324

1962 – No link from Manas to Chirang, Ripa, Buxa through human settlements – Assam Forest Dept. 1994b: 10

1962 – Ripu RF – Last rhino in Haziraguri area of Ripu RF poached – Lahan 1993: 32

1964 – Manas – Forestry operations continued during 1940s and 1950s, but these were stopped entirely in 1964 – Assam Forest Dept. 1995b

1968 – Manas – Rhino rarely seen due to amount of cover – Wayre 1968

1970-07-17 – Manas came under the Western Assam Wildlife Division. Notification No. FOR 363/66

1972 – Manas – Rhinos increasing. Poaching only on southern border near cultivation. Visitors are few – Waller 1972b: 582

1973 – Manas Tiger Reserve under Project Tiger – Lahan 1993: 38 (date 1974); Choudhury 2019: 20

1984 – Kamrup District – A single rhino strayed from Pabitora WS to Goalpara and Nagarbera (Kamrup). Captured and transferred to Dudhwa NP – Choudhury 1996

1985 – Manas – Inscribed on World Heritage List of UNESCO – Choudhury 2019: 11

1987 – Manas – Start of the Bodoland agitation, demanding greater political rights and powers, causing large-scale damage to wildlife – Vigne

& Martin 1991: 215; Rahmani et al. 1992; Menon 1995: 20; Goswami & Ganesh 2014; Talukdar & Sarma 2007: 14; Choudhury 2019

1989-03 – Manas – The sanctuary had to be officially closed in March 1989 – Vigne & Martin 1991: 215, 1996a

1989 – Manas WS – Declared a core zone of the newly formed national Manas Biosphere Reserve (283,700 ha) – Choudhury 2019

1989 – Manas – One source reported that he counted poaching of 106 rhino since the start of the political turmoil – Bezbarua 2008: 32

1989–1993 – Manas NP – Rhinos wiped out due to poaching – Foose & van Strien 1995: 10; Talukdar & Sarma 2007: 14; Bezbarua 2008: 32; Choudhury 2013: 225; Dutta et al. 2017

1990-09-07 – Manas National Park constituted, notification FRW.55/ 86/64. Original area plus 317 km² portion of Manas Reserved Forests (west block part) covering area westwards to Kamakra River and including entire Koklabari RF, also Panbari RF, Kahitama RF – Lahan 1993: 39; Talukdar & Sarma 2007: 14; Choudhury 2019

1992-03 – Kamrup Dt. – Two people were killed by a rhino, one near Kamalpur and the other near Rangiya – Choudhury 1996

1992 – Manas – Downlisted by UNESCO as Heritage Site in Danger (Article 11, paragraph 4, of the Convention) – Vigne & Martin 1996a; Choudhury 2019

1993 – Manas – Rhino habitat limited to Makhibaha, Semajhora, Rupahi, Kokilabari, Uchila, Bhatgali and Goberkunda areas – Lahan 1993: 38

1993 – Manas – Six personnel on protection duty were killed. Buildings, antipoaching camps and bridges have been burnt – Lahan 1993: 38

1993-03 – Manas – Range officer, Ajoy Brahama, was nearly stabbed to death by a gang who stole rifles and a shotgun, and 9 horns weighing 6 kg from the safe. In the few days that followed, 13 rhinos were poached – Vigne & Martin 1996a

1994 – Manas – Construction commenced on a new road in the adjoining Royal Manas National Park in Bhutan – Vigne & Martin 1996a

1995-10 – Manas NP re-opened to the public – Vigne & Martin 1996a

2000 – Bodoland Liberation Tiger (BLT) surrendered in early 2000. The area became stable afterwards – Rabha 2008

2001-03 – Manas – Rhino poached by villagers. This incident indicated that a few rhinos still survive in the park but complete lack of armed guard in the entire park has made their survival doubtful – Choudhury 2001b; Ghosh 2015

2003 – The Bodo Territorial Council (BTC) was formed in 2003 signaling the return of peace in the Manas area – Goswami & Ganesh 2014

2003 – Chirang-Ripu Elephant Reserve – Notified as part of Buxa-Manas Elephant Reserve (283,700 ha). Name suggested by Anwaruddin Choudhury as Joint Secretary in the Environment and Forest Dept. – Choudhury 2019: 20; Goswami & Ganesh 2014

2006-08 – Rhino sighted in Batabari forest towards east of Manas NP, Baksa Dt., of which a plastercast was obtained – Choudhury 2007c: 24

2006 – Start of translocations to Manas NP (see tables 40.56, 40.57, 40.58) – figs. 40.4, 40.5, 40.6

2011-06 – Manas – UNESCO removed their "in danger" tag – Choudhury 2019

2013-03-23 – Manas NP – First visit of Kees Rookmaaker and his wife Sandy assisted by Deba Dutta and Nilutpal Kashyap, through sponsorship of the WWF Asian Elephant and Rhino Programme directed by Amirtharaj Christy Williams

2018 – Study resulting in Ph.D. Thesis by Deba Kumar Dutta (b. 1976) on the translocation and settlement of rhinos in Manas, supervised by Rita Mahanta, Cotton College, Guwahati – Dutta 2018a–fig. 40.1

TABLE 40.55   Status of *R. unicornis* population in North Kamrup Reserve and later Manas NP, 1916–2022

Year	Numbers	Natural Deaths	Poaching	Sources
1916	12–15		1	R165
1930			90–100	R303
1935	30–50			R018
1939	30–40			R302
1940	scarce			R029
1948	a few			R101
1949	45; 100			R061, R104; R224
1950	45			R103
1952	45			R106
1953	45			R204
1958	20; 45			R061; R108
1962			1	R164, R274, R275
1963	15			R207, R164, R274, R275
1964	15		0	R283, R164, R274, R275
1965			1	R164, R274, R275
1966	15		0	R114, R164, R222, R274, R275
1967			0	R164, R274, R275
1968			0	R164, R274, R275
1969			0	R164, R274, R275
1970			0	R164, R274, R275
1971	40		1	R164, R274, R275, R279
1972	30–40		0	R164, R274, R275
1973		1	0	R144, R164, R274, R275
1974		1	0	R144, R164, R274, R275
1975		1	0	R144, R164, R274, R275
1976	40	2	4	R144, R164, R274, R275
1977	75	1	0	R144, R164, R274, R275
1978	40	0	1	R144, R164, R274, R275
1979		2	5	R144, R274, R275
1980	75	2	0	R064, R144, R164, R274, R275
1981		4	2	R144, R164, R274, R275
1982		2	1	R144, R164, R274, R275
1983		3	3	R144, R164, R274, R275
1984		1	4	R067, R144, R164, R274, R275
1985	75	2	1	R067, R144, R164, R274, R275
1986	75–80	1	1	R067, R144, R164, R273, R274, R275
1987		6	7	R067, R144, R164, R273, R274, R275
1988		5	1	R067, R088, R144, R164, R273, R274, R275
1989	60; 80; 85; 90–100; 100–130	1	6	R211; R237; R067, R164, R274, R275; R067; R046
1990	85–100	3	2	R067, R088, R114, R164, R209, R274, R275, R277
1991		3	3	R067, R088, R144, R164, R209, R274, R275
1992	80	3	11	R067, R088, R144, R164, R198, R209, R274, R275
1993	30–60; 50; 60	1	22	R164; R144, R274, R275, R278
1994	60	0	4	R067, R079, R088, R209, R278
1995	< 10; 16; 12–30; 20; 30; 60	0	1	R067; R164; R277, R278; R114

TABLE 40.55   Status of *R. unicornis* population in North Kamrup Reserve and later Manas NP, 1916–2022 (*cont.*)

Year	Numbers	Natural Deaths	Poaching	Sources
1996	12	1	0	R067, R164, R209, R278
1997	5	0	0	R067, R209, R278
1998		1	0	R067, R209
1999	0	0	0	R032, R067
2000	<8	1	0	R067
2001	0	0	1	R067, R114
2002		0	0	R067
2003		0	0	R067
2004		0	0	R067
2005	< 3	0	1	R067
	Translocations started			
2006	1			R032
2007	3			R237, R253
2008	<5			R067
2009	2, 5			R032, R330
2011	11		1	R114
2012	22		1	R032, R114, R330
2013			5	R114
2014			1	R114
2015	30			R032, R251
2016	31 (29 translocated, 2 original)			R067, R072
2018	38, 41			R067, R330
2019			0	
2020			0	
2021	48		0	R330
2022	40			R331 (incomplete census 2022–04)

References for Manas population estimates: R005 Ali 1950: 472; R018 Milroy 1935: 19; R024 Assam Forest Dept. 1994b: 10; R029 Assam Forest Dept. 1941: 26; R032 Assam Forest website; R046 Bezbarua 2008: 32; R061 Burnett 1958: 324; R064 Choudhury 1985a; R067 Choudhury 2019; R072 Choudhury 2016: 154; R079 Dey 1994a; R088 Doley 2000; R101 Gee 1948a: 64; R103 Gee 1950a: 1732; R104 Gee 1950b: 83; R106 Gee 1952a: 224; R108 Gee 1958: 353; R114 Ghosh 2015; R144 Lahan 1993: 18; R164 Menon 1995: 15; R165 Milroy 1916: 455; R198 Roy 1993; R204 Shebbeare 1953: 143; R207 Simon 1966; R209 Singh 2000; R211 Sinha & Sawarkar 1993; R221 Spillett 1966c; R224 Stracey 1949; R237 Talukdar & Sarma 2007: 10; R251 Talukdar 2015; R253 Talukdar et al. 2011; R273 Vigne & Martin 1991: 215; R274 Vigne & Martin 1994; R277 Vigne & Martin 1996a; R278 Vigne & Martin 1998: 25; R279 Waller 1971: 12; R283 Wayre 1968; R302 Bhadran 1939; R303 Hanson 1933; R330 Choudhury 2022b; R331 Talukdar 2022.

40.2   Protecting Rhino Habitats in Manas

The history of protection in Manas in the early 20th century runs parallel to that of Kaziranga (45.2). On 4 November 1902, the Officiating Commissioner of the Assam Valley Districts, John Campbell Arbuthnott expressed concern about the disappearance of wildlife in a letter to the Chief Commissioner, stating that rhinos had become restricted to three areas including "remote localities at the foot of the Bhutan hills in Kamrup." The Chief Commissioner of Assam at the time, Joseph Bampfylde

Fuller responded favourably. The area near the Manas River was surveyed by Major Philip Richard Thornhagh Gurdon, the Deputy Commissioner of Eastern Bengal and Assam and Superintendent of Ethnography of Assam. He recommended a largely uninhabited area of 379.69 km^2 which he outlined on a map of the proposed reserve in March 1904. The area was subsequently reduced in size because one of Fuller's conditions was that the reserve would not interfere with existing or potential cultivation. Hence on 1 June 1905, the Chief Commissioner's Office issued their declaration of the intention to establish the

North Kamrup Reserve of 233.09 km². Shooting of game was prohibited except under special license. When this was ratified on 29 July 1907, North Kamrup (later Manas) became the first reserve initiated particularly to protect the rhinoceros.

40.3 Earl de Minto in Manas in 1907 and 1909

While these administrative procedures were being finalized, Gilbert, 4th Earl of Minto, Viceroy of India 1905–1910, went hunting in Assam twice, in February 1907 and in February 1909. On both occasions, he was hosted by Nripendra, the Maharaja of Cooch Behar, who could only record the 1907 trip in his book (Nripendra 1908, 32.9, 32.10). In 1907 Minto camped in Kalabari, later included in the southern part of Chiru Reserved Forest (about 120 km west of Bansbari). Then in 1909 he occupied a camp which must have been just north of Bansbari on the eastern shore of the Beki River, inside the recently constituted North Kamrup Reserve. For that reason, Minto did not allow his staff to shoot any rhinos (32.10). Minto was told that the reserve was swarming with rhino, although Milroy (1916) put the number at just 15–16 animals at the time. There are no records of any attempt to establish the actual numbers of rhinos, but in fact the party saw quite a few, including young animals. This shows that it is unlikely that there was a great bottleneck in the *R. unicornis* populations in the early 20th century.

The Viceroy's feat was rather overshadowed, definitely in the world's press, by two rhinos attacking (on 12 February) the elephant carrying his ADC Victor Reginald Brooke in a howdah, resulting in a painful injury. This is probably best told from the recollection of Minto himself: "The same day Victor Brooke had a different and very dangerous adventure. He was always casual as to personal safety, and when he saw a cow rhino and her calf emerge from the jungle he did not shoot but hurriedly grasped his kodak, being anxious to obtain a picture. On sighting the elephant the rhino charged, and these huge beasts met with a tremendous concussion, like two battleships ramming each other; the shock was terrific. Victor was hurled against the iron bar of the howdah, his rifle was flung to the ground, and his right arm severely damaged. Meanwhile one young rhino attacked the elephant from the rear. With his left hand Victor managed to cock his small Winchester rifle and fired three shots. After much trampling round and round, the rhino eventually made off with her calf. Francis [Scott], who had seen the encounter from the next howdah, attempted to come to Victor's rescue, but his elephant turned and careered away into the open" (Minto 1934: 274).

Brooke was taking photographs during his outings in the Manas jungle. He could have used one of the cameras of the Kodak company, like the 3A Pocket Kodak, or the more expensive 1A Speed Kodak (advertised in *Harper's Magazine*). He took a photograph showing the Viceroy in the howdah of an elephant published in *The Graphic* (London) of 24 April 1909 (Anon. 1909c). While concentrating how to take a decent picture of rhinos standing in the grass, the animals started to attack, resulting in a photograph of the animals charging, possibly only the second one of a living rhino in the wild (Brooke 1909; Rookmaaker 2021b). This photographic record of the rhino's attack first appeared in the *Illustrated London News* of 13 January 1912 on a large full page (Brooke 1912), which was repeated in several later newspapers (fig. 40.2). The Paris based *Petit Journal* (Anon. 1909d) seems to have thought that Brooke had been unable to get a shot with his camera, resulting in a large coloured engraving in their Supplement of 14 March 1909 (fig. 40.3).

FIGURE 40.2 Female rhinoceros and calf charging towards the elephant carrying Victor Reginald Brooke on 12 February 1909. Possibly only the second photograph of *R. unicornis* in the wild, this occupied one large folio page in the magazine, size 24 × 30 cm
ILLUSTRATED LONDON NEWS, 13 JANUARY 1912, P. 43

FIGURE 40.3
"Rhinocéros contre Eléphant. Le secretaire militaire
du vice-roi des Indes a failli être tué dans la collision"
(Rhinoceros versus elephant. The Military Secretary of
the Viceroy of India was almost killed in the collision).
Artistic (anonymous) rendering of a rhinoceros attacking
the elephant bearing Victor Reginald Brooke (1873–1914) on
12 February 1909
PETIT JOURNAL, SUPPLÉMENT ILLUSTRÉ, 14 MARCH 1909,
VOL. 20, NO. 956, P. 82

40.4   Problems with Poachers

For some twenty years the rhino population was left
largely undisturbed, or at least there were few reports
to the contrary. One rhino was found dead in 1916, and
the poachers were fined and had their gun confiscated
(Milroy 1916). On 1 October 1928, the Assam Government
combined several reserved forests into the Manas Game
Sanctuary with a total area of 360 km². The security situation deteriorated considerably at the same time, as
reported by Arthur John Wallace Milroy, the Conservator
of Forests. Milroy (1916) confessed that "to be in charge
of a Game Sanctuary is a piece of luck which comes to
few", and through his career he showed considerable
interest and conveyed his enthusiasm to readers around
the world in papers published, among others, by the
Bombay Natural History Society and the Society for the
Preservation of the Fauna of the Empire (now the Fauna
and Flora Preservation Society based in Cambridge, UK).
He was proud of the achievements made in Assam, and in
India in general, being among the forerunners in wildlife
preservation (MacKarness 1937).

Nature conservation is not an easy task. Difficulties
started in 1928, after some poachers who were caught
practically red-handed were given such minor fines by the
Civil Department that the local people realized the limited
power of the Forest Department. The Bodo-Kacharis and
Assamese villagers living on the southern border of the
sanctuary seized their opportunity, and soon were joined
by Nepalese men working on a road on the Bhutan side of
the river. Milroy did not know the number of rhinos which
were poached in the course of a few months, but a rumour
that one man had sold eight horns gave some indication
of the problem (Cavendish 1929: 19). In the annual report
of the Forest Department for 1929–1930, Milroy (1930)
stated that the Manas Sanctuary "is infested by bands of
Cacharis armed with unlicensed guns for the destruction
of rhino and, so they aver, of any human who dares to interfere with them." The Governor of Assam from 1927–1932,
Egbert Laurie Lucas Hammond (1854–1936) read this
with surprise, but promised immediate action if the local
authorities agreed with the assessment. Soon afterwards,
a detachment of Assam Rifles under a British Officer was
sent to spend some six weeks in North Kamrup. Although

any unlicensed guns were of course quickly buried by the villagers, the army action showed that threats of violence would no longer be tolerated (Milroy 1931; C. Milroy 1933: 413). This also disturbed a gang of local shopkeepers who had been offering rewards for rhino horns which were generally quickly disposed of in Kolkata, where prices equaled that of gold (Milroy 1932). The Forest Department personnel were now emboldened to show greater anti-poaching activity, led until January 1932 by Mazhuancharry Chakka Jacob (b. 1887) and continued by Mandayam Mudumbai Srinivasan (Owden 1932; G. Singh 2014: 264). There was a series of arrests which led to serious convictions, acting as a deterrent, even if the poaching continued on a smaller scale. The Assam Forest Regulation (section 24) was amended so that killing a rhino was punishable by a jail term or heavy fine. For the time being, with increased vigilance and action, poaching subsided. There are few figures about the actual number of rhinos poached in these years, only Hanson (1933) estimates this at 90 to 100.

The forest officers in the early 1930s were aware that it would need constant manpower and investment to check the poaching. Funds could be generated through tourism, which at the time was almost impossible for want of roads and camping huts (Milroy 1932). Another option was to open up the reserve to limited shooting, where licenses would be granted for a very large price, attracting local and international hunters: "The Sanctuary is only 10 miles north of the Amingaon branch of the Eastern Bengal Railway, and the interior could very cheaply be made accessible to lorries and cars running up from the station by constructing a few cold weather roads in flat country; the whole journey from Calcutta would take less than twenty-four hours of comfortable travel" (Milroy 1932; Bhadran 1939). A scheme which was never implemented.

40.5   War and Bodo Movement

The rhino population in Manas bounced back adequately, even though actual numbers were never published or verified by census. Two eminent naturalists, Salim Ali and Sidney Dillon Ripley visited Manas in March 1949 for six days, seeing tracks only of two rhinos, but agreed to put the total figure at around 45 rhinos. They were guided around the reserve by C.G. Baron, who in 1933 took a flashlight photograph there (Baron 1933) and himself in 1950 estimated the number of rhinos around 150. When asked by E.P. Gee (1950a), others were less optimistic: the Divisional Forest Officer believed there were 40–50 rhinos, and the Conservator of Forests 100.

There were several changes in the extent and status of the park. It became the core part of the 'Project Tiger' reserve that encompassed a total area of 2,837 km² in western Assam, and it was declared one of UNESCO's World Heritage Sites in December 1985. It was upgraded in 1990 as the Manas National Park with an area of 519 km².

While the importance to protect the biological diversity in the Manas region was acknowledged by these changes in the park's status, Manas was threatened by local politics. This region of Assam was traditionally settled by the Bodo people, who argued for a separate state within the Indian union. The official Bodoland Movement started on 2 March 1987 under the leadership of Upendranath Brahma (1956–1990) of the All Bodo Students' Union. This caused general unrest in the region and a breakdown of the authority of police and army (Vigne & Martin 1991). In February 1989, the Manas Wildlife Sanctuary was invaded by several hundred Bodo fighters. The park headquarters were attacked, causing severe injury to a range officer and the loss of 22 rifles and 9 rhino horns. In fear of their lives, all forest guards left the area, which was followed by illegal encroachment and tree-felling by the Bodo people. At least six rhinos were killed. The park was closed to all Indian and foreign tourists, as their safety could not be guaranteed.

The devastation and subsequent rebuilding of Manas was closely followed by Anwaruddin Choudhury (b. 1959) and carefully documented in his monograph of the park (Choudhury 2019). He was not only stationed in different parts of Assam as Divisional Commissioner, but is one of the foremost naturalists in the region. His extensive knowledge of the fauna of North-East India is exemplified in a series of 908 publications including 27 books (as of August 2021). Mammalogist, ornithologist, explorer, photographer, those are just a few of his accolades. He is Honorary Chief Executive of the Rhino Foundation for Nature in North East India founded in 1995.

Choudhury wrote a letter to the largest English-language daily newspaper in Assam, the *Sentinel*, on 7 May 1989 highlighting the breakdown of protection in Manas and the severe threat to the staff of the Forest Department. In his book on *Manas* (2019), he included a long list of militant incidents threatening the existence of the park from 1989 through the end of the 1990s. The violence and the poaching continued. In March 1993, a Bodo gang leader from Khabsinpara village entered the park and killed 13 rhinos (Vigne & Martin 1996a). The home of the Divisional Forest Officer was bombed in March 1994 and other park buildings were burnt down, while possibly as many as 18–20 rhinos were poached in the park. In 1992, UNESCO included Manas on their List of World Heritage

Sites in Danger. The security situation remained unstable through most of the 1990s and only stabilized in the first years of the new millennium. The park was re-opened to tourists in 1995. Park infrastructure needed to be restored, including improvements to the road system and the construction of staff accommodation. There is a proposal first mentioned in 2008 to extend the area of Manas National Park to 950 km² by adding areas on the western boundary. In June 2011, UNESCO reinstated Manas to the List of World Heritage sites and commended it for its preservation efforts.

The rhino population suffered greatly during this time. Optimistic reports in the early 1990s estimated that half the rhinos had been killed, leaving about 40–45. However, the poaching went on steadily, and in 2006 it was feared that the original Manas rhino population had been extirpated. Possibly some animals had fled to the Bhutan foothills on the northern bank of the Manas River. A case of poaching in 2001 and a sighting in 2006 suggested that maybe two animals of the original population remained (Choudhury 2019).

40.6   New Translocations in the 21st Century

Indian Rhino Vision (IRV) 2020 is a program for the long term conservation of the one-horned rhinoceros in Assam launched in 2005. It was developed and is implemented by the Forest Department, Government of Assam in collaboration with the World Wide Fund for Nature, International Rhino Foundation, US Fish and Wildlife Service, and Bodoland Territorial Council. The program aims to increase rhino populations in new or potential habitats throughout the state of Assam by the year 2020. In June 2005, the Government of Assam created the Rhino Task Force to implement measures advocated in IRV 2020.

At the first meeting, held at Guwahati in November 2005, the tasks were divided between a Security Assessment Group and a Habitat Assessment Group. Manas National Park was recommended for the first phase of a translocation programme aimed to increase rhino numbers and rhino-bearing locations. When Manas celebrated its centennial anniversary in December 2005, the Bodo Territorial Council endorsed IRV 2020 and committed their full support to its implementation. Surveys showed there was still pressure from poaching and deforestation in some parts of the park, especially in the Panbari range, which has remained a concern for many years. It was decided that under the auspices of IRV 2020, a total of 20 rhinos should be translocated to Manas to re-establish the population.

After the main security concerns had been identified and improvements were made, the Task Force met on 23 November 2007 at the Assam State Zoo and decided that it would soon be possible to translocate rhinos to Manas. A Translocation Core Committee was created to plan and execute every detail involved in capturing, transporting, releasing and monitoring of the rhinos. On 11 April 2008, the team captured two male rhinos in Pabitora WS, one 10-years old weighing 1570 kg, the other 7-years old weighing 1540 kg. The animals were transported by road in crates and were released the next morning near Buraburijhar camp. Rhino1 moved toward the south and settled in an area along the southern boundary between Kasimdoha and Kureebeel, while Rhino2 moved north and was seen in the vicinity of Charpoli camp. Both had been fitted with radio-collars to facilitate long-term monitoring of their activities. From mid-May 2008 the rhinos remained together in the southern part of the reserve, often straying outside the boundaries despite efforts to contain them.

FIGURE 40.4
Rhinos arrive in Manas in a translocation exercise
PHOTO: SUBHAMAY BHATTACHARYYA, 2010

The Centre for Wildlife Rehabilitation and Conservation (CWRC), which was established near Kaziranga National Park in 2002, has brought 18 new rhinos to Manas between 2006 and 2020 (tables 40.56, 40.57, 40.58; Dutta 2018a, b). To facilitate the reintroduction of these rehabilitated animals, a large enclosure (*boma*) was constructed in the Kuribeel area of Manas in November 2005. The vegetation was cleared, and an electrified fence was installed around the twin-paddocks, each about 25,000 m^2 in size. The first rehabilitated rhino to arrive in the new *boma* on 21 February 2006 was a female named 'Mainao'. She was followed by two hand-reared young rhinos from CWRC on 28 January 2007. A fourth female had been rescued from the Hatikhuli Tea Estate near Kohora after her

mother was killed by poachers, but at an age of 1½ years did not need to be hand-reared, and arrived in Manas on 23 February 2008.

The rhinos which were free-released into Manas all have shown tendencies to wander outside park boundaries, possibly attracted to crops grown in nearby fields. One of them (R-2) in September 2008 kept moving in an easterly direction, covering a distance of almost 100 km, creating panic among villagers. He was captured on 14 September and returned the *boma* with the other rhinos.

Two new batches of rhinos were translocated from Pabitora in a three week period in 2010–2011 (figs. 40.4, 40.5, 40.6). First, on 29 December 2010 an adult female

FIGURE 40.5
Rhino 17 with calf in Manas National Park
PHOTO: JAMIR ALI, WWF INDIA

FIGURE 40.6
Doimalo was hand-reared in Manas from March 2020
PHOTO: DHIPUL NATH, WWF INDIA

with her female calf, and then on 18 January 2011 one sub-adult male, one adult female and another adult female with her male calf were released. This was the first time that four rhinos were released simultaneously in Manas under IRV2020. All rhinos at first tended to move south from the release area at Buraburijhar and occupied areas close to Kuribeel and Rhino camp. In 2012 a further eight rhinos were translocated to Manas, and all animals appear to have settled well (Dutta & Mahanta 2015; Divyabhanusinh 2018c). There have been 27 births in the park in the period 2006 to 2020 (table 40.57). Some seven animals were lost to poaching, in 2012, 2013, 2014 and 2016. Photographs of all rhinos translocated under the auspices of the Assam Government (R1–R18) were included in the thesis submitted by Deba Dutta (2018a).

The translocation project offers good hope that a new population can establish itself again in Manas NP.

TABLE 40.56   Changes in the population of rhinos translocared to Manas National Park 2006–2020

Year	New rhino translocated	New rhino rehabilitated	New birth	Rhino poached	Rhino died	Total rhino end of year
2006	0	1	0	0	0	1
2007	0	2	0	0	2	1
2008	2	2	0	0	1	4
2009	0	0	0	0	0	4
2010	2	0	0	0	0	6
2011	4	0	0	0	0	10
2012	10	2	1	1	0	22
2013	0	1	11	5	0	29
2014	0	3	2	1	0	33
2015	0	0	0	0	2	31
2016	0	0	0	2	1	28
2017	0	0	3	0	0	31
2018	0	3	3	0	3	34
2019	0	4	3	0	2	39
2020	2	0	4	0	1	44
2021	0	0	1	0	0	45
TOTAL	20	18	27	9	12	

Information from Deba Kumar Dutta, Assam (in 2021).

TABLE 40.57   Individual rhinos released in Manas National Park from 2006 to 2020

No.	Origin	Sex	Age	Name	Date of release	Comment	Source[a]
C-1	CWRC	F	3½ yrs	Mainao	2006-02-21	Birth B-06, B-14	R301, R302, R308
C-2	CWRC	F		Ganga	2007-01-28	Births B-04, B-12, B-17	R301, R302, R308
C-3	CWRC	F		Jamuna	2007-01-28	Births B-07, B-15, B-23	R301, R302, R308
C-4	CWRC	F	2 yrs		2008-02-23	Died 2008-10	R301
R-1	Pabitora	M	10 yrs	Sat Hazar	2008-04-12 Bura Burijhar	Poached 2011-10-14 Bhuyanpara	R306
R-2	Pabitora	M	7 yrs	Iragdao	2008-04-12 Bura Burijhar	Poached 2013-01-13 Bansbari	R306
R-3	Pabitora	F	adult	Laisri	2010-12-28	Birth B-09	R307
R-4	Pabitora	F	calf of R-3	Anida	2010-12-28	Poached 2016-08-28 Panbari	R307

TABLE 40.57 Individual rhinos released in Manas National Park from 2006 to 2020 (*cont.*)

No.	Origin	Sex	Age	Name	Date of release	Comment	Source[a]
R-5	Pabitora	M	12 yrs	Manas	2011-01-18	Poached 2014-11-01 Bhuyanpara	R307
R-6	Pabitora	F	adult	Xavira	2011-01-18	Births B-05, B-13, B-20	R307
R-7	Pabitora	M	calf of R-6	Syra	2011-01-18		R307
R-8	Pabitora	F	12 yrs	Giribala	2011-01-18	Birth B-03. Poached 2013-12-31 Bansbari	R307
R-9	Pabitora	F		Pobitra	2012-01-09	Births B-11, B-19	R307, R310
R-10	Pabitora	F	13 yrs	Udangshri	2012-01-09	Birth B-01 Poached 2013-10-30 Bhuyanpara	R307, R310
R-11	Kaziranga	F	calf of R-12	Maidangshri	2012-02-20		R307, R310
R-12	Kaziranga	F		R-12	2012-02-20	Poached 2012-05-22 Bhuyanpara	R307, R310
R-13	Kaziranga	F		Swimli	2012-02-20	Births B-08, B-18, B-26	R307, R310
R-14	Kaziranga	M	calf of R-13	Adidiga	2012-02-20		R307, R310
C-5	CWRC	M	sub-ad	Maju	2012-03-11		
-6	CWRC	M	sub-ad	Raja	2012-03-11		
R-15	Kaziranga	F		Malati	2012-03-12	Births B-10, B-16, B-22	R307, R310
R-16	Kaziranga	M	calf of R-15	Mann	2012-03-12	Poached 2013-08-06 Bansbari	R307, R310
R-17	Kaziranga	F	12 yrs	Hainari	2012-03-12	Birth B-02. Poached 2013-03-02 Bansbari	R307, R310
R-18	Kaziranga	M	calf of R-17	Khwamajinai	2012-03-12		R307, R310
C-7	CWRC	M		Dwimalu	2013-04-06		
C-8	CWRC	F		Purabi	2013-09-23	Birth B-25	
C-9	CWRC	M		Hari	2014-01-28		
C-10	CWRC	M		Maju-2	2014-01-28		
C-11	CWRC	M		Gopal	2014-01-28		
C-12	CWRC	F		Rhino-74 Sili	2018-01-06		
C-13	CWRC	F		Rhino-68	2018-01-06		
C-14	CWRC	F		Rhino-24	2018-01-06		
C-15	CWRC	M		Bagori	2019-01-22		
C-16	CWRC	F		Bagori-55	2019-01-22		
C-17	CWRC	M		Ghera	2019-02-11		
C-18	CWRC	F		Harmoti-23	2019-02-11		

a   References in table: R301 Barman et al. 2008; R302 Barman et al. 2014; R306 Bonal et al. 2009b; R307 Dutta et al. 2017; R308 Kumar 2010; R310 Sharma et al. 2012.
For details about births see table 40.56.
CWRC = Centre for Wildlife Rehabilitation and Conservation, near Kaziranga, Assam

TABLE 40.58   Births of *R. unicornis* population recorded in Manas National Park, 2012–2020

no.	Sex	Name	Mother	Date of birth and range	Comment
B-01	M	R-10A	R-10 Udangshri	2012-09-26 Bhuyanpara	poached 2016-05-06 Bhuyanpara
B-02	F	Doimalo	R-17 Hainari	2013-03-20 Bansbari	hand-reared
B-03			R-8 Giribala	2013-03-23 Bansbari	
B-04			C-2 Ganga	2013-04-05 Rhino Camp	
B-05			R-6 Xavira	2013-05-14 Bansbari	
B-06	F		C-1 Mainao	2013-06-02 Bhuyanpara	Birth B-21
B-07			C-3 Jamuna	2013-07-04 Bansbari	
B-08			R-13 Swimli	2013-09-27 Bansbari	
B-09	F		R-3 Laisri	2013-09-27 Bansbari	Birth B-24
B-10			R-15 Malati	2013-11-02 Bansbari	
B-11			R-9 Pobitra	2014-03-04 Bansbari	
B-12			C-2 Ganga	2015-06-18 Bansbari	
B-13			R-6 Xavira	2015-05-10 Bhuyanpara	
B-14			C-1 Mainao	2015-12-08 Bhuyanpara	
B-15			C-3 Jamuna	2017-09-03 Bansbari	
B-16			R-15 Malati	2017-09-03 Bansbari	
B-17			C-2 Ganga	2017-09-18 Bhuyanpara	
B-18			R-13 Swinli	2018-04-28 Bansbari	
B-19			R-9 Pobitra	2014-04-28 Bansbari	
B-20			R-6 Xavira	2018-10-15 Bhuyanpara	
B-21			B-06 Mainao A	2019-04-03 Bhuyanpara	
B-22			R-15 Malati	2019-08-26 Bhuyanpara	
B-23			C-3 Jamuna	2019-09-27 Bansbari	
B-24			R3A (B-09)	2020-01-02 Bansbari	
B-25			C-8 Purabi	2020-02-02 Bansbari	
B-26			R-13 Swinli	2020-10-21 Bansbari	
B-27			Unknown	2020-11-14 Bansbari	
B-28			Unknown	2020-11-27 Bansbari	
B-29			Unknown	2021-08-28	

CHAPTER 41

The Rhinoceros in Sonai Rupai Wildlife Sanctuary, Assam

First Record: 1910 – Last Record: 1983
Species: *Rhinoceros unicornis*
Rhino Region 33

41.1 Rhinos in Sonai Rupai

The Sonai Rupai forest in Sonitpur District of Assam was for a long time listed as the home of a small population of rhinos. This forest in the foothills of the Himalayas was declared as a Game Sanctuary in 1934 out of the existing Charduar Reserved Forest (constituted in 1878), on the instigation of Arthur John Wallace Milroy (1883–1936), with an initial area of 67 square miles or 173 km² (Milroy 1935: 20, Gee 1950b: 83, Assam Forest Dept. 2014). Milroy had visited the area around 1910 when he was able to see far more rhinos than in the 1930s. Two English tea planters living in the area, R. Erskine Scott (of Chatwal Tea Estate) and William Milburne, were made Honorary Forest Officers to increase protection (Milroy 1935). In 1935 the Deputy Ranger reported the presence of 12 or 13 rhino, and found footprints in localities which they did not usually visit. He ascribed this to increasing confidence following a lack of human interference (Milroy 1936: 14). In 1937, 5738 acres (23 km²) were added to the reserve.

Sonai Rupai was the home of *Rhinoceros unicornis*. Milroy never commented on the identity of the rhinos in Sonai Rupai, which confirms this. Gee (1948a: 64) hinted at the possibility of the presence of *R. sondaicus* or *D. sumatrensis* "as Manas and this sanctuary are near the Himalayas." This was repeated by Mukherjee (1966: 27). As there is no further evidence for this, possibly they were mistaken. There are no known remains or photographs of any rhino derived from this area.

There were few subsequent reports from Sonai Rupai. The rhino population was generally estimated to be 5 animals, and once 8–10 (table 41.59). There is no explanation for the decrease in numbers in the 1940s, but probably there was a period of poor vigilance. In 1950, two rhino were found dead (Assam Forest Dept. 1950: 28). In 1978, a female and a calf were seen by the forest staff in a location not far from the border of West Kameng district of Arunachal Pradesh (Choudhury 1996, 1997a). Laurie (1978: 397) said that about 15 rhinos were found in a small swampy area beside the Gabru River. Irregular sightings continued until 1983 (Assam Forest Dept. 2014). Lahan (1993: 46) declared the species extinct in the area, which was confirmed during a more recent survey of Assamese rhino areas (Khan et al. 2002). In 1998, the reserve was notified (notification No. 172 dated 22.10.1998) as Sonai-Rupai Wildlife Sanctuary (Assam Forest Dept. 2014).

Sonai Rupai is no longer home to any resident rhinos since the 1980s.

FIGURE 41.1
Postcard of rhinoceros threatening a snake, issued c. 1910 by Kunstverlag Leo Stainer, Innsbruck. Signed by F. Perlbera. (Serie 682. Fauna II. No. 1)
COLLECTION KEES ROOKMAAKER

© L.C. (KEES) ROOKMAAKER, 2024 | DOI:10.1163/9789004691544_042
This is an open access chapter distributed under the terms of the CC BY-NC-ND 4.0 license.

TABLE 41.59   Estimates of the status of rhinoceros in Sonai Rupai

Year	Numbers	Source[a]
1910	more numerous than in 1935	R305
1911	present	R295
1935	few pairs	R018
1936	12–13	R304
1937	17	R019
1948	6	R101
1949	5; a few	R104; R224
1950	1; 5	R005; R103
1952	5	R106
1953	5	R204
1958	5	R108
1963	5	R207
1966	5	R222
1968	8–10	R144
1978	15	R145
1985	rhino exist	R064
1991	strays from Kaziranga	R053
1993	0	R144
1995	rhino present	R239

a   Sources in Table. R005 Ali 1950: 472; R018 Milroy 1935: 19; R019
MacKarness 1938: 10; R101 Gee 1948a: 64; R103 Gee 1950a: 1732;
R104 Gee 1950b: 83; R106 Gee 1952a: 224; R108 Gee 1958: 353; R144
Lahan 1993: 18; R145 Laurie 1978: 12; R204 Shebbeare 1953: 143;
R207 Simon 1966; R222 Spillett 1966c: 575; R224 Stracey 1949; R064
Choudhury 1985a; R053 Bhattacharyya 1991: 84; R239 Talukdar 1995:
1; R295 Wood 1930: 69; R304 Milroy 1936: 14.

CHAPTER 42

Protecting the Rhinoceros in Orang National Park, Assam

Established as Orang Game Reserve in 1915
Species: *Rhinoceros unicornis*
First Record: 1879 – Last Record: current
Rhino Region 33

42.1 Rhinos in Orang

Orang National Park is located in the civil districts of Darrang and Sonitpur, Assam on the north bank of the Brahmaputra River, 65 km from Tezpur town and 120 km from Guwahati. The reserve is surrounded by villages and there are also people living on the islands in the Brahmaputra on the southern side. The Dhansiri and Panchnoi rivers that originate in the Himalayas flow through the area. The annual floods of the Brahmaputra affect the rhino population only slightly as there are enough high water terraces. The land of the park was sparsely populated until about 1900 when most people vacated the area due to prevalence of an unexplained bacterial disease or transmittable malarial fever (Allen 1905: 219).

There is little information available regarding the wildlife of Orang prior to its establishment as a game reserve in 1915. This is not particularly remarkable, given the general patchiness of early data. At most, it shows that Orang was not one of the places where rhinos were traditionally hunted or observed. There also seems to be no record of the reasons why this particular stretch of land was chosen to be gazetted, except perhaps for the simple fact that many people had recently left the area for health reasons.

There were about 10 rhinos in Orang in 1949, and 12 in 1964. Juan Spillett (1966d) did not take time to visit the reserve, but in 1966 he was informed about the presence of 25 animals. As many of these rhinos were not permanent residents, frequently entering or exiting the reserve, he thought that the region was inhabited by a total of 50 rhinos. Only in 1985 a detailed census was undertaken, covering 62 hectares, showing 23 adult males, 23 adult females and 19 subadults, for a total of 65 rhinos (Menon 1995: 23). Figures regarding the number of animals poached are available from 1978 (table 42.60). Serious poaching incidents occurred from 1982 to 1985, when 20 animals, about a third of the existing population, were killed during the Assam Agitation. Rhinos were then caught within park boundaries in pit traps near which the poachers would camp. The organizer usually provided a gang of four or five men with firearms, and paid US$171–514 per person for a single horn (Vigne & Martin 1994).

Some respite to the situation occurred in 1985 when Orang was gazetted as a wildlife sanctuary, with increased interest in the region. The Range Officer Bhupen Talukdar actively combatted illegal entry into the park and reduced the number of poaching incidents, even to zero in 1990. The number of staff was increased, a jeep was purchased, equipment was updated, the road system was improved, and funds were allocated to support an intelligence network. Another deterrent to poaching was the presence of five rogue elephants in the area, one of which killed 16 women and 2 men outside the park in 1987–1992 (Vigne & Martin 1994).

In 1994 there was a general breakdown of law enforcement in this part of Assam due to local agitation by the Bodo tribe. Rhino killings increased and the population was halved to about 45 animals in 1997. While poachers mainly used guns, in 1995 two rhinos were killed by poisoning their drinking water with insecticide. Most poachers entered the park from the southern side, where a forest camp was constructed in 1997 by the Range Officer Probin K. Deka to combat another upsurge of poaching (Vigne & Martin 1998). In November 1997 one camp was looted, negatively impacting staff morale in a situation where local threats were increasing. The park was closed to visitors for the 1997–98 winter season due to damaged roads and bridges. Hence 25 rhinos were killed in Orang in 1998–2000, leaving just 46 rhinos counted in the 1999 census.

The situation improved when Orang was declared a National Park in 1999. With the help of International Rhino Foundation, US Fish and Wildlife Service, Care for the Wild, David Shepherd Conservation Foundation, Early Birds and Aaranyak, new equipment was made available, including walkie-talkies, solar panels, a Gypsy jeep and speed boat. This led to a decrease in poaching, and in 1999 forest guards killed a poacher. In 2001 only a single rhino was killed, while four poachers were arrested and two were killed. Thanks to increased vigilance and protection measures, the rhinos in Orang recovered slightly to a total of 68 animals in 2006 (Martin et al. 2009b). Orang is geographically isolated and therefore less visited by

© L.C. (KEES) ROOKMAAKER, 2024 | DOI:10.1163/9789004691544_043

This is an open access chapter distributed under the terms of the CC BY-NC-ND 4.0 license.

tourists. After a census in March 2012, Sushil Kumar Daila, Mangaldoi Wildlife Divisional Officer stated that there were again about 100 rhinos in the park, which therefore places it among the more significant rhino areas in Assam.

• • •

Dataset 42.51: Chronological List of Records of Rhino and Administrative Changes in Orang National Park

General sequence: Date – Locality (as in original source) – Event (species identifications according to original source) – Sources.

1905 – Darrang – Rhino now extremely scarce – Allen 1905, vol. 5: 14

1905 – This part of Darrang west of Tezpur sparsely populated. The area is prone to diseases sometimes thought to be an infectious malarial fever, or an infectious worm-transmitted bacterial disease, affecting many people in area – Allen 1905, vol. 5: 78, 219

1905 – Darrang – Area was once heavily settled but villagers left the area because of chronic black-water fever – Talukdar 2008a

1914 – Darrang – Shooting of rhino has been prohibited throughout the province – Assam Forest Dept. 1914: 28

1914 – Orang Reserve – Submitted to government, proposed Orang Reserve, 31 square miles, being reserved in the interests of game – Assam Forest Dept. 1914: 3

1915-05-31 – Orang Game Reserve – Notification 2276/R for Orang Game Reserve with area 19,936 acres (80.54 km²). Formerly an abandoned village where 26 manmade ponds still exist with an old temple. Part of Mazbat Forest Range under Darrang Forest Division – Blunt 1915; Gee 1950b: 84; Lahan 1993: 24; Assam Forest Dept 1993b; Sarma 2010; Hazarika 2015: 41

1931-11-30 – Orang GR – Notification 3778 R, dereserved 17.29 km² from northern border, to allow settlement of immigrants from Bangladesh as part of a campaign to encourage Muslim immigration from Mymensingh District at the time of Syed Muhammad Saadulla (1885–1955), Chief Minister of Assam – Lahan 1993: 24; Sarma 2010; Hazarika 2015: 41

1933-34 – A male rhino calf which was found wandering and captured by some villagers in Darrang, was subsequently sold by the Department for Rs. 3,000 to Paris Zoo – Milroy 1934a: 19

1969-06-18 – Orang GR – Notification of addition of 8.73 km². In mid-1960s a program of afforestation started – Lahan 1993: 24; Hazarika 2015

1972 – Orang GR – Transferred to Western Assam Wildlife Division. No forestry operations permitted since then – Lahan 1993: 25

1985-09-20 – Orang WS – Notification FRS.133/85/5 dated 1985-09-20, intention to declare Wildlife Sanctuary covering 75.60 km² – Talukdar 1995: 2; Talukdar & Sarma 2007: 6; Sarma 2010; Hazarika 2015: 41

1991 – Orang WS – Area of 3.21 km² was added on western boundary – Hazarika 2015: 41

1992-10-01 – Rajiv Gandhi Wildlife Sanctuary – Renamed – Assam Forest Dept. 1995b

1994 – Rajiv Gandhi Orang WS – Breakdown of law enforcement due to local agitation – Martin 1996b: 12

1998-03-17 – Rajiv Gandhi Orang WS – Notification no. FRW. 28-98-116 of 1998-03-17, officially declared Wildlife Sanctuary – Vashishth 2002

1999-04-08 – Rajiv Gandhi Orang NP – Notification FRW 28-90-154 of 1999-04-08, declared National Park – Vashishth 2002; Sarma 2010; Hazarika 2015: 41

2003 – Orang NP – Study of rhino behaviour and ecology conducted from 1999–11 to 2000-03 and 2000-04 to 2003-03 by Buddhin Ch. Hazarika – Hazarika 2007, 2015; Hazarika & Saikia 2010.

2016-02 – Rajiv Gandhi Orang NP – Status changed to become a Tiger Reserve

2018-12 – Orang Tiger Reserve – Change of name

2021-04 – Two rhinos strayed from the reserve and were immobilized to release them again in Magurmari Grassland area – Ahmed et al. 2021

2022-01-03 – Orang NP – Notification of second expansion of the park towards Laokhowa measuring 200.13 km² – Choudhury 2022b

•••

TABLE 42.60 Status of *R. unicornis* population in Orang NP, from 1940 to 2022

Year	Number	Natural deaths	Poached	Sources
1940			2	R029
1947	a few			R030
1949	6, 10			R224; R104
1950	8, 10			R005; R103
1952	10			R106
1953	12			R204
1958	15			R108
1963	12			R207

TABLE 42.60   Status of *R. unicornis* population in Orang NP, from 1940 to 2022 (*cont.*)

Year	Number	Natural deaths	Poached	Sources
1966	12, 12–25			R208; R220
1972	35			R064, R170, R201, R238
1978	25–30			R145
1979			2	R274
1980	25–30	0, 2	2, 3	R164, R016, R022, R144, R239, R274
1981	60	0, 3	2	R164, R016, R022, R144, R239, R274
1982		4, 8	5	R164, R016, R022, R144, R239, R274
1983		4, 6, 9	4, 8, 10	R164, R016, R022, R144, R239, R274, R201
1984		1, 4, 7	3, 6, 7	R164, R016, R022, R144, R239, R274, R201
1985	65	1	8	R016, R022, R088, R126, R128, R144, R164, R170, R201, R238, R239, R272, R274
1986	65	1	3	R016, R022, R144, R164, R201, R239, R274
1987		2, 3	1, 4	R016, R144, R164, R201, R239, R274
1988		0, 1, 2	4, 5, 8	R016, R144, R164, R201, R239, R274
1989	65	1, 2, 3	2, 3	R211, R201, R016, R022, R144, R164, R239; R274
1990		1	0	R016, R022, R144, R164, R239
1991	97	2	1	R016, R022, R088, R126, R128, R144, R164, R170, R201, R238, R239, R272, R274
1992	100	3	2	R016, R022, R088, R144, R164, R201, R239, R272, R274
1993	90, 100	2	1, 2	R016, R137, R088, R144, R201, R274, R278
1994	90	4	6	R016, R201, R272, R278
1995	90	6, 8	9, 10, 11	R016, R088, R164, R201, R272, R278
1996		4	9, 10	R088, R201, R272, R016, R278
1997	45	2	11	R088, R106, R201, R272, R278
1998		3, 4	11, 12	R088, R106, R201, R241, R272
1999	46	0	7	R032, R088, R126, R128, R170, R201, R238, R241, R272, R330
2000		5	8	R201, R242, R272
2001	50	2	1	R201, R241, R242, R272
2002		1	0	R201, R241, R242
2003		1	1	R201, R242
2004		2	0	R201, R242
2005		1	3	R201, R242
2006	68	6	3	R032, R126, R170, R201, R238, R242, R330
2007		4	3	R201
2008		3	6	R201
2009	64	3	6	R032, R126, R201, R330
2010			2	R246
2012	100			R032, R330
2013	no census			R032
2014	no census			R032
2015	100			R251
2016	100			R072
2018	101			R252, R330
2019				
2020				
2021				
2022	125			R329, R330, R331 (census 2022-03)

References in table: R005 Ali 1950: 472; R016 Assam Forest Dept. 2000a; R022 Assam Forest Dept 1993b; R029 Assam Forest Dept. 1941: 26; R030 Assam Forest Dept. 1949: 19; R032 Assam Forest website; R053 Bhattacharya 1991: 84; R064 Choudhury 1985a; R072 Choudhury 2016: 154; R079 Dey 1994a: 2; R088 Doley 2000; R103 Gee 1950a: 1732; R104 Gee 1950b: 83; R106 Gee 1952a: 224; R108 Gee 1958: 353; R126 Hazarika 2015: 45; R128 Hussain 2001; R137 Khan & Foose 1994: 6; R144 Lahan 1993: 18; R145 Laurie 1978: 12; R164 Menon 1995: 15; R170 Momin 2008; R201 Sarma 2010; R204 Shebbeare 1953: 143; R207 Simon 1966; R208 Simon 1967; R211 Sinha & Sawarkar 1993; R220 Spillett 1966d: 529; R224 Stracey 1949; R238 Talukdar & Sarma 2007: 5; R239 Talukdar 1995: 1; R241 Talukdar 2000; R242 Talukdar 2006; R246 Talukdar 2010b; R251 Talukdar 2015; R252 Talukdar 2018; R272 Vashishth 2002; R274 Vigne & Martin 1994; R278 Vigne & Martin 1998: 25; R329 Sharma 2022; R330 Choudhury 2022b; R331 Talukdar 2022.

FIGURE 42.1
Leopards attacking a rhino in a popular book with plates drawn by
the German artist Gustav Jacob Canton (1813–1885)
CANTON, *DER THIERGARTEN*, 1854, PL. 16

FIGURE 42.2
Rhinoceros female and calf in the landscape of
Orang National Park
PHOTO: NEJIB AHMED. COURTESY:
JAHAN AHMED

CHAPTER 43

Protecting the Rhinoceros in Pabitora Wildlife Sanctuary, Assam

Established: Pabitora Reserve Forest in 1971, Wildlife Sanctuary 1987
Species: *Rhinoceros unicornis*
First Record: 1925 – Last Record: current
Rhino Region 34

43.1 Rhinos in Pabitora

Pabitora is a small sanctuary of 38 km² in the Morigaon District, located 50 km east of Guwahati on the southern bank of the Brahmaputra River. It lies in a basin surrounded by the low hills of the Mayong, Kamarpur and Monoha ranges. Villages surround the park on all sides. Pabitora, also spelled Pobitora, was first designated as a Reserve Forest in November 1971. According to Juan Spillett, who visited the area in 1966, the first proposals for the conservation of the Raja Mayang Forest were then made by the Forest Department. It became a Wildlife Sanctuary on 16 July 1987. About three-quarters of the reserve is covered by alluvial grassland, the rest by forests, rivers and wetlands. During the winter, most of the landscape is dry and access to water is limited (Borthakur 2017).

As elsewhere on the banks of the Brahmaputra, there are no definite early records of rhinos in the precise area of Pabitora. Local people claim that rhinos were present there at least as early 1925, and it is likely that rhinos may have wandered through the area for a long time. In 1966, Forest Department officials estimated that there were 12 rhinos in Pabitora. One adult male was killed in an intraspecific fight in February 1966. A case of poaching was uncovered in 1964, from which the 1.75 kg horn of

an adult female was recovered. Reportedly 8 rhinos were present in 1971 when Pabitora was proclaimed a Reserve Forest. The population has increased to 54 in 1987, 77 in 2011 and 107 in 2022.

The main challenges of the sanctuary are related to its small size. People from the surrounding villages graze their livestock on sanctuary grassland. This causes competition with the rhinos during winter months, forcing them to leave the reserve, mainly at night, to seek food. During the flooding season rhinos stray outside the sanctuary's boundaries. Rhinos occasionally wander quite far, as was seen in December 1983 and January 1984 when a lone individual moved from Pabitora to Nalbari, across the Brahmaputra to Nagarbera and then to Goalpara, covering a distance of at least 150 km. It was captured in mid-March 1984, and released near five other rhinos that resided closer to Pabitora. As these animals could not be protected properly, they were selected for the first national translocation of rhinos to Dudhwa National Park at the request of the Government of Uttar Pradesh. They were kept in a stockade for veterinary care for a fortnight before being flown to Delhi and transported overland to Dudhwa on 31 March 1984 (19.3).

In September 1989, poachers devised a new method for killing rhinos in the park and avoiding detection. Using the high-voltage power lines that traverse the sanctuary, poachers would suspend them a meter above the ground across known rhino trails. During the night, a rhino using that trail would contact the wire and electrocute itself. Burn marks on the carcasses indicated that the animals might live for as long as five minutes before

FIGURE 43.1
In Pabitora Wildlife Sanctuary, rhinos graze in herds like domestic cows
PHOTO: KEES ROOKMAAKER, MARCH 2013

© L.C. (KEES) ROOKMAAKER, 2024 | DOI:10.1163/9789004691544_044
This is an open access chapter distributed under the terms of the CC BY-NC-ND 4.0 license.

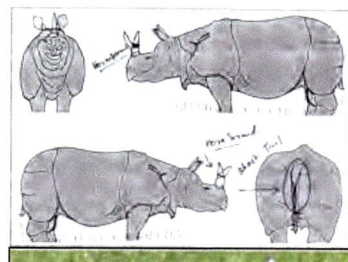

Reference No. M10

Sex-M   Age- Adult (>10 years)
Calf Position-NA
Remarks- Dominant Male
Last updated- 06.11.2008

Key Identity-	*Tail is very short, the base is thick around 2 inch rest 6inch is slender, Horn Broken.*
Horn-	*Horn broke, was well developed and around 8-12 inch in length.*
Ear-	*Right ear loss one part, left ear is clean with short hair.*
Skin Folds-	*Prominent and clean.*
Tail-	*is extreamly short; the base is thick around 2 inch rest 6inch is slender.*
Remarks-	*Dominant male.*

FIGURE 43.2   Sample identification sheet of a rhino in Pabitora prepared by Ramesh Bhatta during his investigations in 2007–2011
COURTESY: RAMESH BHATTA

dying. Between 1989 and 1993, one out of every five rhinos poached in Pabitora were killed by that method (Vigne & Martin 1994). During 1997–2000 range officer Mrigen Barua was effective in arresting 16 poachers in at least six separate incidents, recovering rifles, ammunition and some of the horns. The rhino population in Pabitora continues to grow despite the pressures of poaching and overgrazing, and its rhino density (2.3 animals per km²) is the highest in India (figs. 43.1, 43.3). Poaching declined in the 1990s, thanks to intensive patrolling and local public awareness programs. Tourism has also increased, with Guwahati serving as a center for foreign visitors (Vigne & Martin 1998).

A number of studies of rhino ecology and demographics have been conducted in Pabitora. It was found that the rhinos use three principal routes when they stray outside the reserve, which actually helped authorities in locating anti-poaching camps. All individual rhinos were identified between 2007–2011, using a marking system developed by Ramesh Bhatta (b. 1975; later with WWF-India) for a project by Guwahati University (fig. 43.2). This confirmed the presence of 77 rhinos (Bhatta 2011).

Several animals have been translocated from Pabitora to Manas National Park under the auspices of the Indian Rhino Vision (IRV) 2020 program. Translocations took place on 12 April 2008 (two rhinos: Sat Hazaar and Iragdao), on 27 December 2010 (two rhinos: Lasiri and Anida), on 17 January 2011 (four rhinos: Manas, Xavira, Syra and Giribala), and on 8 January 2012 (two rhinos: Pabitora and Udangshri). Three of the translocated females gave birth in Manas (40.6).

• • •

Dataset 43.52: Chronological List of Records of Rhino and Administrative Changes in Pabitora Reserve

General sequence: Date – Locality (as in original source) – Event – Sources.

1923 (summer) – Pabitora – First rhinos visited from adjacent areas and gradually settled – Talukdar 1999

1925 – Pabitora and Mayong – Presence of rhino first recorded – Talukdar & Barua 2006; Talukdar et al. 2007

1925 – Mayong – Rhino calf kept as pet by Mayong King (probably Bakat Singha Rahan) for few months until it died of diarrhea – Kalita 1992; Talukdar 2000; Talukdar & Barua 2006 – 5.4 (Mayong)

1950s – Raja Mayong – According to past official information, 3 rhinos were found dead, one each at Raja Mayong, Sildubi and Barhampur. The villagers handed over the rhino horns to Nagaon Police – Talukdar & Barua 2006

1950s – Raja Mayong, Lunmati and Burha Mayong – Villagers demanded that the area be declared a reserve forest to prevent migrants from encroaching, and to protect rhinos – Choudhury 2005b

1959-10-21 – Notification AFR.472/54/40 of 21-10-1959, declaration of Raja Mayan Reserved Forest, 11.98 km² – Lahan 1993: 43

1961–1962 – Mayong – Nagaon Forest Division confirmed the presence of a few rhinos – Choudhury 2005b

1965 – Raya Mayang RF – Forest Dept. proposed to constitute 4464 acre as Pabitora RF – Spillett 1966d: 532

1971-11-18 – Notification FOR/Sett/542/65/54 of 18-11-1971, declaration of Pabitora Reserved Forest, 15.86 km² – Lahan 1993: 43; Assam Forest Dept. 1993c, 1994d: 12; Choudhury 2005b

1984 – Pabitora RF – Rhino present – Choudhury 1984

1987 – Pabitora Wildlife Sanctuary – Notification FNP/19/87/39, Wildlife Sanctuary constituted by combining Pabitora RF and Raja Mayang RF – Choudhury 2005b; Bhatta & Saikia 2011: 15

1989 – Pabitora WS – First rhino found electrocuted by poachers – Vigne & Martin 1994: 42

1998 – Pabitora WS – Notification FRS/19/87/152 extension of boundary by adding Dubaritali and Kamarpur – Bhatta & Saikia 2011: 15

2007 – Pabitora ws – Total 38.81 km² but only 2.82 km² under Forest Department – Talukdar & Sarma 2007: 9

2009 – Pabitora ws – Study of food plants and eating habits by Pradip Konwar, Malabika Kakati Saikia and Prasanta Kumar Saikia, Department of Zoology, Gauhati University – Konwar et al. 2009

2011-01-17 – Pabitora – Capture of 4 animals: 1 adult male, 1 adult female, a mother-calf pair. Loaded in crates and moved by truck to Manas – Rakshit 2011

2012 – Pabitora – Photo of 2 rhinos electrocuted in Pabitora – Choudhury 2012: 29

2013-03-21 – Pabitora – First visit of Kees and Sandy Rookmaaker hosted by Pobitora Conservation Society (Nripen Das)

2014 – Pabitora – one case of poaching reported – Ashraf et al. 2022

FIGURE 43.3
Rhinos among the trees of the Pabitora Wildlife Sanctuary
PHOTO: KEES ROOKMAAKER, 2013

TABLE 43.61   Status of *R. unicornis* population in Pabitora, 1962–2022

Year	Number	Death	Poached	Sources
1962	a few, 14	1		R069, R236, R254
1966	6–12			R220
1970	20			R069
1971	8			R014, R036, R236
1979	40			R164
1980			0	R274
1981			0	R274
1982			0	R274
1983			0	R274
1984			4	R164, R274
1985	67	2	2	R144, R164, R274
1986	40		0	R273
1987	54	3	2	R014, R088, R093, R144, R164, R236, R237, R254, R273, R274
1988		5	3	R144, R164, R236, R237
1989	40	1	4	R014, R036, R088, R144, R164, R236, R237
1990		2	2	R014, R036, R088, R144, R164, R236, R237, R274
1991		1	1	R036, R088, R144, R164, R236, R237, R274
1992	70	2	3	R014, R036, R088, R144, R164, R236, R237, R274
1993	56	1	4	R014, R036, R069, R088, R093, R144, R164, R236, R237, R254, R274, R278
1994	56	2	4	R014, R036, R088, R164, R236, R237, R278

TABLE 43.61   Status of *R. unicornis* population in Pabitora, 1962–2022 (*cont.*)

Year	Number	Death	Poached	Sources
1995	68	3	2	R014, R036, R088, R093, R164, R236, R237, R254, R278
1996			5	R014, R236, R237, R278
1997	76	1	3	R014, R036, R236, R237, R278
1998		2	4	R014, R088, R236, R237
1999	74	3	6	R032, R069, R093, R236, R237, R241, R253, R254
2000		1	2	R236, R237, R241
2001	70		2	R241
2002		1	1	R236, R237
2003		2	2	R236, R237
2004	79			R069, R093, R236, R237, R253
2005		4	1	R236, R237
2006	81	2	1	R032, R093, R236, R237, R253, R254
2007	74 (38/29 + 10)			R048, R050
2009	84			R032, R248
2012	93			R032
2014			1	R328
2015	92		0	R251
2016	90		0	R072
2018	102		0	R252
2019			0	
2020			0	
2021			0	
2022	107			R328, R329, R331

References in table: R014 Assam Forest Dept. 2000b; R032 Assam Forest website; R036 Bairagee 2004; R048 Bhatta & Saikia 2011; R050 Bhatta 2011; R069 Choudhury 2005b; R072 Choudhury 2016: 154; R088 Doley 2000; R093 Dutta 2008; R144 Lahan 1993: 18; R164 Menon 1995: 15; R220 Spillett 1966d: 529; R236 Talukdar & Barua 2006; R237 Talukdar & Sarma 2007: 10; R241 Talukdar 2000; R248 Talukdar 2011b; R251 Talukdar 2015; R252 Talukdar 2018; R253 Talukdar et al. 2011; R254 Talukdar et al. 2007; R273 Vigne & Martin 1991: 215; R274 Vigne & Martin 1994; R278 Vigne & Martin 1998: 25; R328 Ashraf et al. 2022; R329 Sharma 2022; R331 Talukdar 2022.

FIGURE 43.4
Rhino sculpture made of grass at the Forest Office of Pabitora
PHOTO: KEES ROOKMAAKER, 2013

Protecting the Rhinoceros in Laokhowa Wildlife Sanctuary, Assam

Established: Laokhowa Reserved Forest in 1905, Wildlife Sanctuary in 1979
First Record: 1867 – Last Record: 1991; translocated 2016, none current
Species: *Rhinoceros unicornis*
Rhino Region 34

44.1 Rhinos in Laokhowa

Rhinos once inhabited the Laokhowa Wildlife Sanctuary as well as the adjacent reserved forests of Burachapori or Burhachapori (44.06 km²) and Kochmara (21.55 km²) closer to the Brahmaputra.

Fitzwilliam Pollock hunted rhinos near Lowqua Lake in 1867 and probably again in 1868 (39.4). It is impossible to tell how numerous the animals were in this area. In 1902 John Campbell Arbuthnott included Laokhowa (besides Kaziranga) in his proposal to reserve parts of Assam to regulate sport. Laokhowa was declared as a Reserved Forest in 1907 and a Game Sanctuary in 1916.

The status of rhinos was never recorded until around 1950, when it was thought that there were maybe 20 in the park. During his survey of wildlife areas, Juan Spillett visited in March 1966. He was told of a mother and calf seen in the reserved forest in December 1965 and a lone male near Ferryghat. The forest of Laokhowa, with an area of 70.14 km², then consisted of about 40% woodland, 30% grassland and 30% wetland. The reserve often flooded during the monsoon season from May to July. Spillett found that at least 1000 acres were under cultivation by local villagers, 5000 heads of livestock were grazing in the reserve, while the Ruphiahi Co-operative for Fishery and Farming had leased 400 acres for agricultural cultivation since 1958.

These economic activities probably continued even after Laokhowa was declared a Wildlife Sanctuary in 1979. However, the rhino population was reasonably healthy and a high number of 73 was estimated in 1981. During the ethnic tensions at the time of the Assam Agitation in 1983, a temporary breakdown of law and order resulted in a quick and fatal massacre during which at least 41 (unofficially 70) rhinos were killed. The last remaining rhino was spotted in October 1991. Vivek Menon, director of the Wildlife Trust of India, writing of the sanctuary's fate in 1995, found no forest, all was just one large rice field.

While there have been occasional reports of rhinos which strayed to Laokhowa from Orang across the Brahmaputra or from Kaziranga further east, apparently the animals never settled again to build a new population.

A survey in 2015 by Smarajit Ojah among villagers and forest staff established that rhinos were once largely concentrated in Laokhowa and the adjoining southern part of Burhachapori (see map in Ojah et al. 2015). A translocation of 2 rhinos from Kaziranga in 2016 was unsuccessful. Any further plans to reintroduce rhinos to Laokhowa require support, and this initiative probably should include the Burhachapori forest.

...

Dataset 44.53: Chronological List of Records of Rhino and Administrative Changes in Laokhowa and Adjoining Burhachapori Reserves

General sequence: Date – Event – Sources. All events refer to Laokhowa unless Burhachapori is specified.

1902-11-04 – John Campbell Arbuthnott (1858–1923) wrote letter (no. 75) regarding proposed reservation of Laokhowa – R. Das 2005: 17

1905-06-01 – Notification No. 2440R. Intention of Reserved Forest. Area undergoing settlement, 40 square miles (103.6 km²) – Assam Forest Dept. 1905: 9; G. Singh 2014: 191 (map of 1905); Sivakumar 2015

1907-01-29 – Notification 227/71/5 declaring Laokhowa as Reserved Forest, with contiguous Kochmara RF and Burhachapori RF.

Forest activities continued until late 1960s – Lahan 1993: 40 (date uncertain)

1907-03-01 – Notification No. 11 declaring Laokhowa as Reserved Forest, and final notification 1907-10-01, area 25,760 acres (104.2 km²) – Talukdar & Sarma 2007: 17 (date uncertain)

1907 – Formation of Laokhowa game sanctuary with an area of 25,760 acres. Closed for shooting – Anon. 1909b; Stebbing 1912: 37; Friel 1951: 2; Saikia 2005: 273

1916-11-10 – Notification declaring Laokhowa Game Sanctuary – Talukdar & Sarma 2007: 17

1929 – Laokhowa established as a forest reserve – Spillett 1966d; Martin & Martin 1982: 32

© L.C. (KEES) ROOKMAAKER, 2024 | DOI:10.1163/9789004691544_045
This is an open access chapter distributed under the terms of the CC BY-NC-ND 4.0 license.

1936 – 2 rhino shot, forest guard dismissed – Milroy 1936: 15

1941 – The Hon. Minister-in-charge-Forests visited Laokhowa and saw 3 rhino – Assam Forest Dept. 1941: 26

1947 – One old rhino had died of old age, horn lost – Assam Forest Dept. 1949: 19

1949 – Visit of Salim Ali (1896–1987) and Sidney Dillon Ripley (1913–2001) – Gee 1950b: 83

1950 – Laokhowa is seldom visited and received little attention – Gee 1950b: 84

1958 – Laokhowa recognized as a 'multiple use' area since 1958 – Spillett 1966d

1965 – Laokhowa proposed as a Wildlife Reserve in 1965 – Spillett 1966d

1966-03-12 – Visit by James Juan Spillett (1932–2018) – Spillett 1966d: 529

1966-03-28 – Rhino census observed 41 rhino: 1 unsexed, 25 adult male, 12 adult female, 3 young – Spillett 1966d: 529

1979 – Laokhowa declared as a wildlife sanctuary – Lahan 1993: 40

1983 – Ethnic unrest (Assam Agitation) resulted in widescale poaching wiping out much of the rhino population – Choudhury 1985a; Lahan 1993: 40; S. Roy 1993: 19; Menon 1995: 26; Talukdar & Sarma 2007: 18; Baura & Talukdar 2008; Choudhury 2013: 225; Ojah et al. 2015

1991 – Infrequent strays from Kaziranga – Dutta 1991: 54

1991 – Population in Laokhowa extinct – Khan & Foose 1994: 6; Sivakumar 2015; Choudhury 2016: 154

1991-10-05 – Last recorded rhino died on 5 October 1991, none seen afterwards – Menon 1995: 26

1995 – Burachapori RF – Rhino present – Talukdar 1995: 1

2002 – Rhino area visited. Population extinct – Khan et al. 2002

2007 – Infrequent strays from Kaziranga – Talukdar & Sarma 2007: 18

2015 – Survey by Smarajit Ojah (b. 1980), Anup Saikia and P. Sivakumar among villagers and forest staff to establish habitat suitability – Ojah et al. 2015

2016-03-29 – Burachapori – Female rhino (35 yrs) and calf were captured in Kaziranga (Bagori Range) and released in Burachapori the next morning. Mother died 2016-05-22 from natural causes. The calf 'Ramani' died on 2016-10-26 – Dasgupta 2017; Talukdar 2017; Dutta et al. 2021

2018-05 – A sub-adult female rhino strayed out of Orang NP to islands (chuars) in the Brahmaputra. The animal apparently was poached in 2019 and her remains were found buried on 2019-06-24 – B. Bhattacharya 2019.

∵

INDIAN RHINOCEROS FIGHTING AN ELEPHANT.

FIGURE 44.1

"Indian Rhinoceros fighting an Elephant"

A. LUNDEBERG AND F. SEYMOUR, *BIG GAME HUNTING IN AFRICA AND OTHER LANDS*, 1910, P. 274

A GREAT HUNTER IN A PERILOUS POSITION.

FIGURE 44.2

A great hunter in a perilous position. Engraving by the American illustrator Armand Welcker for the adventure book by James William Buel (1849–1920) published in 1889

BUEL, *SEA AND LAND*, 1889, P. 615

TABLE 44.62   Status of *R. unicornis* population in Laokhowa, 1937–2022

Year	Total number	Natural Deaths	Poached	Sources[a]
1937	12			R124
1941	frequent			R301
1949	20; 20–30			R104; R224
1950	18; 20			R005; R103
1951	12			R300
1952	20			R106
1953	25			R204
1955	41			R216
1958	25			R108
1963	25			R207
1966	40; 41			R220; R144
1978	40			R145
1979		6	6	R144; R274
1980	70; 40	2; 3	1	R182; R216; R237; R164; R144; R274; R078
1981	73	2; 4	5; 6	R064; R144; R164; R237; R274
1982	40	12; 15	5; 6; 9	R198; R144; R164; R237; R274
1983	0	6; 7	31; 40; 41; 43; 51	R064; R144; R164; R182; R198; R237; R274
1984		0	0; 5	R144; R164; R274
1985	2; 40	0	0; 2	R144; R164; R274
1986	5	1	0; 1	R144; R164; R273; R274
1987		3	0	R144; R164; R274
1988		1	0; 1	R144; R164; R274
1989	5	1	0; 1; 3	R144; R164; R273; R274
1990	0	2	0; 3	R164; R274; R078
1991	0	0	0; 1	R164; R216; R274
1992	20		0	R198; R274
1993	0; 5; 5–6; 6–10; 10		0	R098; R144; R198; R274
1994	0		0	R079; R278
1995	0		0	R098, R278
1996			0	R278
1997	5		0	R278
2007	0			R237
2016	2 (translocated)	2		R078, R333
2022	0			R333

a Sources on population of Laokhowa: R005 Ali 1950: 472; R064 Choudhury 1985a; R078 Dasgupta 2017; R079 Dey 1994a: 2; R098 Foose & van Strien 1997: 9; R103 Gee 1950a: 1732; R104 Gee 1950b: 83; R106 Gee 1952a: 224; R108 Gee 1958: 353; R124 Harper 1945: 378; R144 Lahan 1993: 18; R145 Laurie 1978: 12; R164 Menon 1995: 15; R182 Ojah et al. 2015; R198 Roy 1993; R204 Shebbeare 1953: 143; R207 Simon 1966; R216 Sivakumar 2015; R220 Spillett 1966d: 529; R224 Stracey 1949; R237 Talukdar & Sarma 2007: 10; R273 Vigne & Martin 1991: 215; R274 Vigne & Martin 1994; R278 Vigne & Martin 1998: 25; R300 Friel 1951; R301 Assam Forest Dept. 1941: 26; R333 Dutta et al. 2021.

Protecting the Rhinoceros in Kaziranga National Park, Assam

Established: Kaziranga Reserved Forest in 1908, as Wildlife Sanctuary 1916
Species: *Rhinoceros unicornis*
First Record: 1902 – Last Record: current
Rhino Region 34

45.1 The Name of Kaziranga

The first time that Kaziranga is mentioned in print, as far as this can be ascertained, was by John Peter Wade, who went to the valley of the Brahmaputra in 1792 as Assistant Surgeon. He compiled a geographical sketch of Assam, published posthumously in 1807, which included Casirunga, with all names of locations rather curiously spelled:

Casirunga lies to the east and south-east of Rungulighur; and Namdoyungh to the eastward above Khonarmook or Sonarmook. The country here is low, and subject to inundation. It extends about six miles in length, from the causeway to Bassa, and four in breadth to the foot of the mountains from Namdoyungh. Namdoyungh is forty miles long, and ten broad; it has Colarphaut on the west, Ouperdoyungh on the east, Casirunga on the south, and the Berhampooter flows on the north. Toquharrurgown, Khoololgown, Atooneagown, and Dehinghiagown, are the principal towns of this flourishing province (Wade 1807: 121).

FIGURE 45.1   Detail of map of Assam and Eastern Provinces compiled by the London based cartographer Edward Weller (1819–1884) in 1859. He names the area "Kuzuranga"
MAP ENTITLED INDIA: THE EASTERN PROVINCES. LONDON: DAY AND SON, 1859

© L.C. (KEES) ROOKMAAKER, 2024 | DOI:10.1163/9789004691544_046
This is an open access chapter distributed under the terms of the CC BY-NC-ND 4.0 license.

FIGURE 45.2
Stamp issued on 31 May 2007 by the Indian Posts and Telegraphs, in a series of wildlife highlighting the rhinoceros of Kaziranga
COLLECTION KEES ROOKMAAKER

Almost the same text, again with the spelling "Casirunga", is found in a general history of India by Robert Montgomery Martin of 1838. The spelling changes in the gazetteer by Edward Thornton: "Kazuranga. – A town of Assam, in the British district of Nowgong, 42 miles E.N.E. of Nowgong. Lat. 26°37′, long. 93°24′′′" (Thornton 1854). "Kuzuranga" is found on a map prepared by Edward Weller in 1859 (fig. 45.1). It is spelled "Kazeeranga" on the map produced by John Butler (1847) of the Bengal Native Infantry.

Kaziranga is never associated with wild animals in these early works. Apparently F.T. Pollok never shot in this particular area in the 1860s or 1870s (39.1). When he said that rhinos are met with on a journey between Jynteapoor and Nowgong (Sylhet to Nagaon), this doesn't leave the impression that such a route would pass close to the Brahmaputra (Pollok 1879, vol. 2: 77). The sportsman visiting the area in 1886 mentioned anonymously by Gee (1952a) was General James Willcocks, who shot rhino near Dhansirimukh after setting out from Golaghat, therefore at the eastern extremity of Kaziranga (38.20).

• • •

Dataset 45.54: Chronological List of Records of Rhino and Administrative Changes in Kaziranga National Park

General sequence: Date – Locality (other than Kaziranga) – Event – Sources – § number.

Records for Rhino Region 34 previous to 1900 are in chapter 38. The localities are included in maps 38.19 and 38.20.

1807 – First printed mention of "Casirunga" by John Peter Wade (1762–1802) of the Bengal Medical Service – Wade 1807: 121; Gokhale & Kashyap 2005: 18

1838 – Casirunga mentioned by Robert Montgomery Martin (1803–1869) – Martin 1838c: 633; Gokhale & Kashyap 2005: 18

1847 – Kazeeranga, shown on "General Map of Assam" with title: By an officer in the Hon. East India Company's Bengal Native Infantry in civil employ. With illustrations from sketches by the author. Scale 16 miles to 1 inch. Smith, Elder and Co. Litho: 65, Cornhill, London. Dimensions 58.4 × 35.1 cm. John Butler (1809–1874) served in the Bengal Native Infantry until 1859 – Butler 1847

1854 – Kazuranga listed in the gazetteer by Edward Thornton (1799–1875) – Thornton 1854, vol. 3: 47

1859 – Kuzuranga, shown on map by cartographer Edward Weller (1819–1884) of Bloomsbury – Weller 1854 – fig. 45.1

1875 – Kajiranga listed in table of Mauzas in Sibsagar District in 1875 – Hunter 1879a, vol. 1: 280

1876 – Tea estates around Kaziranga with opening years, chronological: Jagdamba 1876 – Bokakhat 1877 – Borchapori 1879 – Amgoorie 1880 – Seconee 1880 – Diffloo 1890 – Nahorjan 1895 – Hatikhuli 1904 – Behora 1905 – Methoni 1914 – Anandpur 1930 – Sagmootea 1935 – Burrapahar 1938 – Yadava 2014: 192

1900 – Rhinos number barely a dozen, maybe 20 maximum – Gee 1948a: 64, 1948b: 106, 1964a: 153; Talbot 1957: 390; Barua & Das 1969: 3; Lahan 1993: 4; Roy 1993: 5; Menon 1995: 40; Sarma 2023: 6

1902-11-04 – John Campbell Arbuthnott (1858–1923), Commissioner of Assam valley, reported the scarcity of rhinos – Gokhale & Kashyap 2005: 11; Saikia 2009, 2020; G. Singh 2014: 179

1902-07-16 – General Secretary of the Alipore Zoological Gardens in Calcutta writes to the Assam Government seeking assistance to procure a young rhino – Gokhale & Kashyap 2005: 20; Sarma 2023: 7 – 45.2

1902-11-04 – The letter from the Alipore Zoo was actioned by the Commissioner of the Assam Valley Districts, John Campbell Arbuthnott (1858–1923), Indian Civil Service 1879–1914. He wrote about the status of rhinos in Assam to Joseph Bampfylde Fuller (1854–1935), Chief Commissioner of Assam from 1902 to 1906 – Gokhale & Kashyap 2005: 19; Barthakur & Sahgal 2005: 27; Das 2005: 17 (facsimile) – 45.2

1902-12-18 – J.B. Fuller writes to J.C. Arbuthnot that he would consider establishing an asylum for the rhinoceros – Gokhale & Kashyap 2005: 20; Das 2005: 17 (facsimile); Sarma 2023: 7 – 45.2

1903 – Naharjan tea estate in Golaghat, owned by Messrs. Pringle and Forbes, situated in Namdayang, on 13 December 1903 comprising 1514 acres of which 505 were planted, with a labour force of 480 workers – Allen 1906, vol. 7: 14

1903 – Anxiety expressed in *Times of Assam* about indiscriminate killing of wildlife by Mikir tribe – Saikia 2009

1903-08-28 – J.C. Arbuthnot writes to J.B. Fuller with recommendations for protected areas – Gokhale & Kashyap 2005: 21; Barthakur & Sahgal 2005: 27; Sarma 2023: 8 – 45.2

1904-03-15 – J.B. Fuller, in a letter signed by his Secretary Francis John Monahan (1865–1923), comments on Arbuthnot's proposals – Barthakur & Sahgal 2005: 29; Saikia 2005: 270; Singh, G. 2014: 182; Saikia 2020 – 45.2

1904-06-18 – Major Hugh Maclean Halliday (1866–1920), Deputy Commissioner of Nowgaon, suggests to extend the southern boundary – Halliday 1904 – 45.2

1904-09-20 – Edward Statter Carr (1857–1925), Conservator of Forests from 1902, suggests to revise the regulations for sport hunting and would only employ two forest guards – Gokhale & Kashyap 2005: 23; Barthakur & Sahgal 2005: 29 – 45.2

1904-12-22 – J.B. Fuller replied to E.S. Carr that villages should be excluded from the proposed reserve – Gokhale & Kashyap 2005: 23; Barthakur & Sahgal 2005: 29 – 45.2

1905-06-01 – Preliminary notification 2442R, intention to constitute Kaziranga as reserved forest, area of 57,273.6 acres (231.77 km²) – Allen 1906, vol. 7: 14; Choudhury 2004: 1; Gokhale & Kashyap 2005: 22; Barthakur & Sahgal 2005: 29 (text quoted); Sarma 2023: 9 – 45.2

1906 – No progress has been made with the two proposed game reserves in the Nowgong and Sibsagar Divisions, but it is hoped that the game reserves will be taken in hand by the Forest Settlement Officer during 1906–1907 – Assam Forest Dept. 1906: 1, 23

1907 – Visit by German traveller Oscar Kauffmann (1874–1924) – Kauffmann 1911 (2nd edn 1923: 152–158); Ullrich 1971: 19 – 45.5

1908-01-03 – Notification No. 37 F for Kaziranga Reserved Forest, area of 56,544 acres (226.17 km²), signed by Herbert Carter (b. 1867), Conservator of Forests of the Eastern Circle (until 1912) – Assam Forest Dept. 1908: 30; Choudhury 1987a; Lahan 1993: 5; Das 2005: 16; Gokhale & Kashyap 2005: 24; G. Singh 2014: 193; Yadava 2014: 145; Sarma 2023: 10

1909 – Kasaranga is reserve in Nowgong District – Anon. 1909b

1910-09-01 – Inaugural meeting of Darrang Game Association in Tezpur – *Englishman's Overland Mail* 1910-09-01; Saikia 2005: 268 – §38.33

1911-04-18 – Notification No. 2069 F. Dereserved 1441 acres (5.84 km²) opposite Kuthori and Baguri villages to allow free access to Mora Diflu River, and despatch of tea by boat – Assam Forest Dept. 1911: 1; Lahan 1993: 5

1911-10-17 – Notification No. 5311 F. Addition of 16,347 acres (66.15 km²) to protect rhino because hunting was still allowed outside the Reserved Forest – Lahan 1993: 5

1912 – Kaziranga Sanctuary of 57,273.6 acres (231.77 km²) is closed to all shooting except permit-holders – Stebbing 1912: 37

1913-01-28 – Notification No. 295 R. Addition of 13,506 acres (54.65 km²) – Allen 1915, vol. 7: 2; Gokhale & Kashyap 2005: 24

1914-04-23 – Notification No. 1813 R. Addition of some land – Allen 1915, vol. 7: 2

1915 – Assam Rhino Preservation Act 1915 – Barthakur & Sahgal 2005: 32 – No other trace of this Act was found

1916-11-10 – Upgrade as Game Sanctuary (68,608 acres or 277.65 km²) by William Frederick Loftus Tottenham (1866–1936), Conservator of Forests, Eastern Circle – Wood 1930: 68; Choudhury 2004: 1; Das 2005: 16; Gokhale & Kashyap 2005: 27; Tirkey 2005; Barthakur & Sahgal 2005: 31; Saikia 2005: 271; Yadava 2014: 145; Saikia 2020

1917-07-26 – Notification No. 3560 R. Addition of 37,529 acres (151 km²) on northern side – Yadava 2014: 145

1918 – First mention of Kaziranga in annual reports: Division Sibsagar, Civil district Sibsagar and Naga Hills, Reserved Forest: Kaziranga. Area on 20 June 1918: 106,138 acres (429.52 km²) – Blunt 1918

1922–1923 – Poachers caught, accused and later acquitted. Guns confiscated and horn sold in auction for Rs. 1230. Appeal for retrial was not allowed – Milroy 1935: 19; Handique 2004: 140; G. Singh 2014: 160

1926 – Kaziranga as Game Sanctuary. No hunting licenses issued. Gee (1950b: 84) has 1928. In fact, it may refer to the 1916 change in status – Cavendish 1929; Gee 1964a: 153; Ullrich 1971: 20; Choudhury 1985b; G. Singh 2014: 193

1930 – Kaziranga fairly free from poachers – Milroy 1930: 4

1933-01-14 – Numalighar Tea Estate, 10 km east of Kaziranga. Stray rhino settled near tea garden, made nightly visit to labourer's village. Attempt to drive back failed, then shot – Y.D. 1933: 755 – fig. 45.4 – map no. A452

1933 – Arthur John Wallace Milroy (1883–1936) has no definite information on Kaziranga as it is never visited – Milroy 1933: 4; Gee 1964a: 154; Saikia 2005: 271

1934 – Systematic poaching by shooting and pitting. Suppression taken on by Divisional Forest Officer – Milroy 1934b: 4

1934 – E.P. Gee applied for permission to visit. District Officer refused, "No one can enter the place. It is all swamps and leeches and even elephants cannot go there" – Gee 1964a: 154

1934 – Sanctuary unknown. Milroy was warned that the swamps made marching through it with elephants quite impossible. This proved to be a myth, and Sanctuary has now been thoroughly explored – Milroy 1935: 19

1934 – Charles Edward Simmons (1902–1971), Deputy Conservator of Forests, investigated numerous and very complicated cases of rhino poaching, which began to come to light as soon as a definite policy of Wild Life Preservation was inaugurated – Milroy 1935: 19

1934 – Milroy appointed Mahi Chandra Miri (1903–1939), Extra Assistant Commissioner (EAC) Forests. First based in Baguri to stop poaching, moved to Golaghat in 1937, died 29 July 1939. Area adjacent to Karbi Anglong district was named after him – Milroy 1935: 19; Barthakur & Sahgal 2005: 35; Deka 2005, 2012

1937 – Kaziranga opened to visitors, with permission from the Divisional forest officer. There is an Inspection Bungalow, and two inspection elephants named Akbar and Sher Khan. On an average visit one would see 10 rhino – Gee 1948a: 64, 1948b, 1949a, 1950b; Barua & Das 1969: 4; Martin 1983: 20; Saikia 2005: 271; Gokhale & Kashyap 2005: 28; G. Singh 2014: 166; Yadava 2014: 145 – 45.7

1937 – A rhino died in pit after capture – India Forest Dept. 1939: 578

1938-05-25 – First comment in Kaziranga visitor book by A.R. Tomas: "Six rhinos seen in Daflong Beel; bathing with 10 buffaloes" – Gokhale & Kashyap 2005: 86, Barthakur & Sahgal 2005: 146–147 – table 45.63

1938–07 – Male rhino swam on 1938-07-25 to Mikir Hills, then caught near Sildubi in a pit on 1938-10-19. Animal wounded by the fall. Transported on a train to Kolkata on 1938-11-02. Sold for Rs 4,000 to National Zoo in Washington, DC, where 1939-07-07 to 1959-01-09 – Miri 1939; Assam Forest Dept. 1939: 13; Gee 1948a: 65

1939-04-06 – Edward Pritchard Gee (1904–1968) visited with M.C. Miri and two friends. The party carried two rifles for self-defense, one on each elephant, but this practice of taking defensive weapons into a sanctuary was soon discontinued. Gee also visited on 1949-02-17, 1952-11-01, 1957-03-21, 1959-07-19 – Gee 1954, 1964a: 154; Gokhale & Kashyap 2005: 86

1939 – 'Boorra Goonda' in Kohora, seen by Gee and Miri. First photographed on 6 April 1939 with old wound on the hindquarters, again 8 January 1950 in close company with a cow. Mentioned by visitors

J.B. Rowntree (DFO) 1945-05-12; Lt Col Chinwan of ARC 1950-04-28; J.S. Mehta with daughter G. Mehta 1951-02-12; Brigadier Bhagwati Singh 1951-09-17. Rhino died 8 June 1953 when height at shoulder 5 ft 9 in (175.4 cm) – Gee 1953c, 1954, 1964a: 159; Burton 1955a, b; Vaidya 1967: 154; Gokhale & Kashyap 2005: 86 – fig. 45.5

1941 – Rhino are reported to be increasing and a number of young calves have been observed – Assam Forest Dept. 1941: 26

1945 – During the Second World War the reserve was visited by military personnel from across the world. There were frequent reports of killings – Assam Forest Dept. 1947; Waller 1971, 1972a: 6; Saikia 2005: 272

1946–03 – A visiting group saw 30 rhino, 80 buffaloes, 1 elephant and numerous deer – Assam Forest Dept. 1947; Saikia 2005: 272

1946 – There were 192 visitors in 1945–1946 – Assam Forest Dept. 1948: 22

1947 – Reduction in rhino numbers due to disease, probably anthrax – Assam Forest Dept. 1949: 19; Gee 1952a: 225; Stracey 1963: 66; Street 1963: 102; Ullrich & Ullrich 1962: 186

1947 – Efforts are being made to improve the conditions for visitors and to increase control over trespass and poaching, and to remedy the rather neglected state of affairs – Assam Forest Dept. 1949: 19

1947-02 – Male Mohan captured 1947-02 in pit 5 ft deep, 5 ft wide, for Whipsnade Zoo. Transported by L.M. Flewin. Whipsnade #13, alive 1947-08-07 to 1961-03-07. Skull in NHM 1961.5.10.1 – Vevers 1947; Gee 1948b: 106, 1950a: 1731, 1952b

1947-02 – Female rhino captured in February 1947 died of septicaemia – Gee 1948a: 65, 1948b; Street 1963: 106

1948-02 – Capture of 2 rhino for Chicago Zoological Society (Brookfield Zoo) as witnessed by Ralph Graham (1901–1980) from 20 February to 25 April 1948 – Gee 1948a: 65, 1950a: 1731; Assam Forest Dept. 1953: 19; Graham 1949, 1949b, 1954; Gokhale & Kashyap 2005: 86 – fig. 45.6

1948 – Rhino killed by a running train near Furkating Railway Station – Assam Forest Dept. 1953: 19

1949 – Rhinos captured for Cairo Zoo, Rs 10,000 each – Gee 1950a: 1731, 1953c; Assam Forest Dept. 1950: 27

1949–03 – Aerial survey by Rustom Phirozsha. Only 14 rhino seen – Gee 1950a: 1732; Gokhale & Kashyap 2005: 86

1949 – Rhino survey sets number at 150, rather than 500 predicted – Gee 1950a: 1732

1950 – Fishery Mahal was terminated and efforts are being made to minimize, or exclude, grazing – Assam Forest Dept. 1950: 27; Waller 1971, 1972a: 6, 1972b, 1999

1950 – Earthquakes in 1950 affected water levels of the Brahmaputra – Mahanta & Bora 2001

1950 – Name was officially altered to Wild Life Sanctuary. R.C. Das in charge. Patrick Donald Stracey (1906–1977) is Chief Conservator – Gee 1964a: 153; Waller 1971, 1972a; Gokhale & Kashyap 2005: 28; Yadava 2014: 145

1951 – Male Gadadhar caught by animal dealer Peter Ryhiner (1920–1975) for Basel Zoo. He also caught female in 1953. Gadadhar mounted after death for Zoological Museum Bern – Assam Forest Dept. 1951: 29; Anon. 1952; Huber 1964

1951 – Fair-weather road from trunk road to edge of reserve was constructed. Forest Bungalow at Baguri was improved with electricity and sanitation. Elephants are now five: Akbar and Sherkhan, plus Mohan, Shivasingha and Mohanprosad – Gee 1952a: 226

1951 – Proposal for reservation of a corridor connecting Kaziranga with the Mikir Hills on the southern side near Hatikhuli Tea Estate to help animals during flood time – Assam Forest Dept. 1951

1951-11 – E.P. Gee visit. Saw 19 rhinos with 3 calves. Mating was observed in the park on 1938-04-17, 1940-02-24, 1944-04-25 – Gee 1952a: 226

1951-12 – Team of 31 staff and students from the Veterinary College of Gauhati inoculated 50,000 domestic cattle around Kaziranga against rinderpest – Street 1963: 103

1953 – Famous rhino Kan Kota (torn ear), mentioned between 1939 and 1953 in visitors book, first on 1939-04-28 by M.C. Miri – Gee 1964a: 139; Vaidya 1967: 154; Gokhale & Kashyap 2005: 86

1953 – Ylla, or Camilla Koffler (1911–1955), photographer, recorded capture of animals for Philadelphia Zoo – Ylla 1958 – figs. 45.7, 45.8, 45.9

1953-02-07 – Movies taken by Ellis Roderick Dungan (1909–2001), American cinematographer, for "Big Hunt" released in 1959 – Gee 1953c; Vaidya 1967: 154 – fig. 45.10

1954 – Assam Rhinoceros Preservation Act passed – Assam Government 1954; Yadava 2014: 145

1958 – Capture of male rhino, first to be shown in Japan. Tokyo #22 1958-11-10 to 1995-07-16 – Hara 1961

1959 – Visit by Wolfgang Ullrich (1923–1973), director of Dresden Zoo and his wife Ursula (b. 1932) – Two books: Ullrich & Ullrich 1962, Ullrich 1971: 17 and detailed zoological observations: Ullrich 1964, 1965a, 1965b, 1967, 1970

1961 – Robin Banerjee (1908–2003) released first documentary movie "Kaziranga" (50 minutes). Banerjee was the co-founder of the Kaziranga Wild Life Society (KWS), which evolved from 'The Rhino Club' at Kohra – Banerjee 1972, 2011

1967-04-07 – Notification FOF/WL/512/66/17, addition of 151 acres (0.61 km²) to the south of National Highway 37, providing a corridor to the Karbi Anglong Hills during the flood – Yadava 2014

1968-10-22 – Death of E.P. Gee, in United Kingdom – Vaidya 1967: 154; Anon. 1968, 1969; Dutta 2014 – 38.5

1969-04-29 – Assam National Parks Act of 1968 notified, approved by India's President Zakir Husain (1897–1969) on 29 April 1969 – Yadava 2014: 145

1969-09-23 – Notification no. FOR/WL/722/45 dt. 23-9-1969, proposal to elevate Kaziranga to National Park – Lahan & Sonowal 1973: 247; Gokhale & Kashyap 2005: 29; Yadava 2014: 145

1970 – Assam Rhinoceros Preservation (Amendment) Act – Assam Government 1970

1972-01 – Study of rhino food habits by Ratan Lal Brahmachary (1932–2018), B. Rakshit and B. Mallik of Indian Statistical Institute, Calcutta – Brahmachary et al. 1970, 1974

1972-03 – Rhino census methods – Lahan & Sonowal 1973: 253–259

1974-02-11 – Notification No. FOR/WL/722/68, final for Kaziranga National Park covering an area of 429.9 km² – Gokhale & Kashyap 2005: 29; Yadava 2014: 145

1975 – Kamal Chandra Patar, Indian Forest Service, studies food preferences and behavioural patterns of rhino – Patar 1977

1980–1983 – Poachers reigned supreme during the Assam disturbances on the foreigners issue 1980–1983, when most guards were withdrawn for law and order duty – Choudhury 1985a

1984-09-20 – 1st Addition, preliminary notification issued – Yadava 2014: 145

1985 – Declared World Heritage Site by UNESCO – Yadava 2014: 145

1985-05-31 – Preliminary notification of 3rd addition at Panbari Reserve Forest, 0.69 km² – Yadava 2014

1985-06-13 – Preliminary notification of 4th addition at Kanchanjuri, 0.89 km² – Yadava 2014

1985-06-13 – Preliminary notification of 5th addition at Haldibari, 1.15 km² – Yadava 2014

1985-07-10 – Preliminary notification of 2nd addition at Sildubi, 6.47 km² – Yadava 2014

1988 – Flood accounted for death of 105 rhinos – Menon 1995: 16; Martin & Vigne 1989; Choudhury 2004: 34

1997-05-20 – Final notification of the 1st Addition, Burapahar, 43.7 km² – Yadava 2014: 145

1998 – Floods killed 39 rhinos – Choudhury 1998b: 83, 2004: 34

1998 – P.O. Mary, Observations on feeding and territorial behaviour of Indian rhino – Mary et al. 1998

1999-08-07 – Notification of 6th addition, Panpur Reserve Forest and stretch of Brahmaputra on north side, 376.50 km² – Yadava 2014

2000 – Rhino habitat suitability analysis – Kushwaha et al. 2000

2002 – Kukrakata RF was added under the Burapahar Range – Yadava 2014: 145

2002 – The Director General of Forest C.P. Oberai and the Director of the Delhi Zoo Bishan Singh Bonal dedicated a book to Kaziranga – Oberai & Bonal 2002

2005–02 – Kaziranga celebrated centenary 14–17 February 2005 – Khan et al. 2005; Agarwalla et al. 2005; Choudhury 2005c; Gokhale & Kashyap 2005

2005 – Declared Elephant Reserve – Yadava 2014: 145

2006 – Launch of "Rhino Vision 2020" aiming to increase the total number to 3000 in 2020 – D. Sharma 2006

2007 – Increase in poaching incidents – Martin et al. 2009b; Choudhury 2007b: 24

2007 – Declared Tiger Reserve – Yadava 2014: 145

2017 – Appreciation of the existence of the rhino in Kaziranga by Pranab Borah and Dileep Chandan – Borah 2017; Chandan 2018

2021-09-22 – The Assam Forest Department ordered the burning of 2479 rhino horns from the Assam State Treasury – Yadava 2022 – 45.10, figs. 3.9, 3.10, 45.16, 45.17

2022-03-28 – The census undertaken in Kaziranga NP counted 2613 rhinos, of which 866 males, 1049 females, 273 adult unsexed, and 279 juveniles and 146 calves – Information released by Jatindra Sarma, Director of the park.

TABLE 45.63   Population estimates of *R. unicornis* in Kaziranga National Park, 1900 to 2022

Year	Number	Natural Deaths	Poached	References[a]
1900	12			R144
1904	12; 100–150			R108, R227; R270
1905	12; 10–20			R110
1906	12; 20; 20–40			R100
1907	12			R198
1908	12			R051, R103, R106, R112, R154, R164, R233
1911	40			R144
1912	< 100			R081
1913	40			R116
1918	100			R307
1927	32–33			R295
1930	150			R108, R227
1935	60–100			R018
1936	100			R304
1937	100; 500			R124; R103
1938	70–80			R099
1939	100			R166
1940	300			R104, R106, R227
1946	1000			R021, R223
1947		14		R104
1949	150; 500			R102, R104, R224
1950	24; 150			R005; R103, R106
1953	250			R204
1956	250			R232
1957	250			R233
1958	250			R108
1959	250; 350			R265
1960	260			R227
1962			32	R267, R270

TABLE 45.63   Population estimates of *R. unicornis* in Kaziranga National Park, 1900 to 2022 (*cont.*)

Year	Number	Natural Deaths	Poached	References[a]
1963	250; 275		30	R267, R270; R207
1964	350			R266
1965	150–200	19	18	R143, R154
1966	366	9	6	R057, R068, R076, R088, R143, R144, R154, R198, R296
1967		23	12	R143, R154
1968	400	12	10	R143, R154
1969		12	8	R143, R154
1970		20	2	R143, R154
1971	400	13	8	R143, R154
1972	658; 670	16	5	R057, R068, R076, R088, R144, R147, R198, R296; R143
1973		39	3	R051, R144, R147, R154
1974	700	13	3	R051, R068, R144, R147, R154
1975		29	5	R051, R068, R144, R147, R154
1976		15	1	R051, R068, R144, R147, R154
1977		25	3	R051, R068, R144, R147
1978	900; 939	16	5	R057, R068, R076, R088, R144, R147, R154, R198, R296
1979		15	2	R068, R144, R147, R154
1980	939	58	11	R057, R068, R144, R147, R154, R164, R198
1981		39	24	R057, R068, R144, R147, R164, R198
1982	960	48	25	R057, R068, R144, R147, R164, R198, R296
1983		46	37	R057, R068, R144, R147, R164, R198, R296
1984	946; 1080	50	28	R057, R147, R296; R068, R076, R088, R144, R164, R198
1985	1195	37	44	R057, R068, R144, R147, R164, R198, R296
1986	1080	38	45	R057, R068, R144, R147, R164, R198, R296
1987		41	23	R057, R068, R144, R147, R164, R198, R296
1988		105	24	R051, R057, R068, R088, R144, R147, R164, R198, R296
1989	1080	54	44	R051, R057, R068, R088, R144, R147, R164, R198, R296
1990	1500	57	35	R051, R057, R068, R088, R144, R147, R164, R198, R296
1991	1129	79	23	R051, R057, R068, R076, R088, R144, R147, R164, R198, R296
1992		67	48	R057, R068, R076, R088, R144, R147, R164, R198, R296
1993	1164 ± 136	58	40	R028, R057, R068, R076, R088, R144, R147, R164, R198, R278, R296
1994		37	14	R057, R068, R076, R088, R147, R278, R296
1995	1200	53	27	R057, R068, R076, R088, R147, R164, R278, R296
1996		52	26	R057, R068, R076, R088, R147, R278, R296
1997	1250	48	12	R057, R068, R076, R088, R147, R278, R296
1998		87	8	R057, R068, R076, R088, R147, R296
1999	1552	46	4	R076, R088, R147, R296
2000		44	4	R068, R076, R147, R242, R296
2001		35	8	R068, R076, R147, R242, R296
2002	1649	62	4	R068, R076, R147, R242, R296
2003		63	3	R147, R242, R296
2004		100	4	R147, R242, R296
2005		49	7	R147, R242, R296
2006	1855	52	7	R147, R242, R296
2007		79	16	R147, R296,
2008		105	6	R147, R296
2009	2048	59	6	R147, R296

TABLE 45.63   Population estimates of *R. unicornis* in Kaziranga National Park, 1900 to 2022 (*cont.*)

Year	Number	Natural Deaths	Poached	References[a]
2010		68	5	R147, R296
2011		67	3	R147, R296
2012	2290	109	11	R147, R296
2013	2329	74	27	R296
2014		36	17	R296
2015	2401		17	R251
2016	2400		12	R072
2017			7	
2018	2413		6	R252
2019			3	
2020			2	
2021			2	
2022	2613			R329, R331 (census 2022-03)

a   References in table: R005 Ali 1950: 472; R018 Milroy 1935: 19; R021 Assam Forest Dept. 1948: 22; R028 Assam Forest Dept 1995a; R051 Bhattacharya 1993; R057 Bonal 2000; R068 Choudhury 2004: 10; R072 Choudhury 2016: 154; R076 Das 2005: 114; R088 Doley 2000; R099 Gass 1971; R100 Gee 1948b: 106; R102 Gee 1949a; R103 Gee 1950a: 1732; R104 Gee 1950b: 83; R106 Gee 1952a: 224; R108 Gee 1958: 353; R110 Gee 1963a; R112 Gee 1964a: 153; R116 Gokhale & Kashyap 2005: 25; R124 Harper 1945: 378; R143 Lahan & Sonowal 1973: 250; R144 Lahan 1993: 18; R147 Lopes 2014: 21; R154 Martin 1983: 19; R164 Menon 1995: 15; R166 Miri 1939: 207; R198 Roy 1993; R204 Shebbeare 1953: 143; R207 Simon 1966; R223 Stonor 1946: 47; R224 Stracey 1949; R227 Street 1963: 100; R232 Talbot 1956a; R233 Talbot 1957: 390; R242 Talukdar 2006; R251 Talukdar 2015; R252 Talukdar 2018; R265 Ullrich & Ullrich 1962: 187; R266 Ullrich 1964: 226; R267 Ullrich 1965a: 101; R270 Ullrich 1971: 20; R278 Vigne & Martin 1998: 25; R295 Wood 1930: 69; R296 Yadava 2014: 108; R304 Milroy 1936: 14; R307 Blunt 1918; R329 Sharma 2022; R331 Talukdar 2022.

45.2   The Declaration of Kaziranga Reserved Forest in 1908

The need to protect the wildlife of Assam, and rhinos in particular, was suddenly realized by the officers in charge of the forest department and the state administration at the start of the 20th century. The recent claim about the involvement of Lady Curzon is likely to be apocryphal (45.3). The start of the proceedings came from unexpected quarters. On 16 July 1902, the Assam Government received a letter from the General Secretary of the Alipore Zoological Gardens in Calcutta, probably instigated by its Superintendent Ram Brahma Sanyal (1858–1908). The Zoo was seeking assistance to procure a young rhino for a sum not to exceed Rs. 1,000. Reviewing the feasibility of the request brought a stark sense of urgency to the Commissioner of the Assam Valley Districts John Campbell Arbuthnott. He wrote an important letter to the Chief Commissioner of Assam, dated 4 November 1902 (no. 75):

the animal [rhinoceros] which was formerly common in Assam has been exterminated except in remote localities at the foot of the Bhutan hills in Kamrup and Goalpara and in a very narrow tract of country between the Brahmaputra and Mikir Hills in Nowgong and Golaghat where a few individuals still exist. – There is, I think, still time to preserve the very few that are left. I understand that the shooting of rhinoceros has been prohibited in Bengal. I would therefore suggest that the destruction of rhinoceros in Assam by shooting or by pitfalls to be prohibited until further orders. I am convinced that, unless an order of the kind is issued, the complete extinction of a comparatively harmless and most interesting creature is only a question of a very short space of time (facsimile in Das 2005: 17).

Arbuthnott's letter was received by the highest administrative authority, Joseph Bampfylde Fuller, who was Chief Commissioner of Assam from 1902 to 1906. He replied to Arbuthnott on 18 December 1902 (no. 76):

The Chief Commissioner agrees with you in thinking that it would be most regrettable if the rhinoceros became extinct in Assam, but that it would be impossible without special legislation to penalise the unlicensed shooting of this animal, and that such legislation is not very likely to be undertaken. ... However, gladly consider the possibility of establishing an asylum for the rhinoceros by taking up as Reserved Forest a sufficient area of suitable land (facsimile in Das 2005: 17).

Among the areas recommended by Arbuthnott in a letter of 28 August 1903 for this special purpose was an uninhabited stretch of land between the Leterijan and Brahmaputra Rivers "in the vicinity of Kaziranga." There were no villages in the area, it was forested and partly swampy, although sometimes people went there to fish or shoot. Other possible localities were Laokhowa and North Kamrup (Manas), whose administrative histories run parallel to that in Kaziranga.

Fuller agreed to pursue the matter, as can be seen in the letter signed by his Secretary Francis John Monahan on 15 March 1904. There were only three conditions that had to be met to preserve some forests, namely that these tracts did not injure existing cultivation, that those tracts which were suitable for cultivation were not selected for game preservation, and that not much public money would be spent on these undertakings.

Next Arbuthnott asked further guidance from the local Deputy Commissioners on the feasibility of the plans. The Deputy Commissioner of Nowgaon, Major Hugh Maclean Halliday noted on 18 June 1904 that the area included two non-cadastral villages, suggesting to extend the southern boundary of the Kaziranga tract to the government road (Halliday 1904). Edward Statter Carr, Conservator of Forests from 1902, suggested on 20 September 1904 that a reserved forest in the area only needed two foresters to be employed at Rs. 15 per month and two forest guards at Rs. 8 per month, while suggesting to revise the regulations for sport hunting. The Chief Commissioner's office replied to Carr on 22 December 1904 that it would be best to exclude the villages from the proposed reserve. The good news was that he "would be glad if you will take early action towards procuring their notification. Meanwhile they should be closed to shooting by executive orders."

On 16 January 1905 Carr duly submitted a draft notification to the Chief Commissioner's Office regarding the proposed game reserve of Kaziranga. He followed this up on 16 March 1905 with a set of amended rules for the regulation of sport hunting.

The investigations completed, the government issued a preliminary Notification 2442R on 1 June 1905:

> In exercise of the powers conferred by Section 5 of the Assam Forest Regulation VII of 1891, the Chief Commissioner hereby declares that it is proposed to constitute as Reserved Forest in the interest of the preservation of game the lands described in the schedule hereto annexed, and appoints the Deputy Commissioner of the Sibsagar district to be the Forest Settlement Officer to enquire into and determine the existence, nature and extent of any rights claimed by or alleged to exist in favour of, any person in or over any land comprised within the limits as described in the schedule hereto annexed, and to deal with the same as provided in Chapter II of the Regulation.

> Under the provision of Section 5 sub-section (2) of the Assam Forest Regulation, VII of 1891, the Chief Commissioner appoints the Divisional Forest Officer, Sibsagar, to assist the Forest Settlement Officer in enquiries prescribed by Chapter II of the Regulation in connection with the proposed reservation. Under the provision of Section 15 of the Assam Forest Regulation, the Chief Commissioner appoints the Commissioner of Assam valley Districts to hear appeals from the orders of the Forest Settlement Officer.

The area proposed was 57,274 acres (231.77 km²) in extent, divided between the Golaghat and Nagaon Sub-Divisions.

On 3 January 1908, all formalities concluded, the Kaziranga Reserved Forest was established with an extent of 56,544 acres (226.17 km²) as per Notification No. 37F. As Fuller had resigned from his post as Lieutenant Governor of East Bengal and Assam, the order was signed by Herbert Carter. Regulations dating back to 1889 allowed Carter to ban all hunting, shooting, trapping and fishing in the newly-created reserve.

45.3 The Apocryphal Involvement of Lady Curzon

The investigations and recommendations which led to the notification of Kaziranga as a Reserved Forest were completed entirely on state or district level, while the agreement of the higher authorities is implied. There is no doubt of the genuine concern about the diminition of wildlife in general and the rhino in particular by administrators like Fuller, Arbuthnott, Carr, Halliday and Carter. There are two quite similar anecdotes that the proceedings were accelerated or even initiated by the wife of a high official who noticed the absence of rhinos personally on a visit to the Kaziranga area. According to Richard Waller (1971, 1972a), this happened in 1906 and involved the wife of the Chief Commissioner, hence probably Laura Maud Nation (1861–1916), married to Commissioner

FIGURE 45.3   Lord and Lady Curzon with First Day's Bag in Camp
near Nekonda, Hyderabad, 1902. Photograph by Lala
Deen Dayal (1844–1905), court photographer
WITH PERMISSION © NATIONAL ARMY MUSEUM,
LONDON: 1963-04-2-2-46

of Assam Lancelot Hare (1851–1922). The more common variant since the centenary of 2005 is that this happened in 1905 to the American-born Lady Curzon, Mary Victoria Leiter (1870–1906), married to George Curzon, 1st Marquess Curzon of Kedleston (1859–1925), who was Viceroy from 1899 to 1905 (fig. 45.3). As can be seen from the historical sequence, the dates 1905 or 1906 are unlikely because by that time the wheels had already been set in motion. The story involving Lady Curzon was told in 1969 to Sanjay Debroy, Chief Conservator of Forests (Wildlife), by a villager living near Kaziranga who was then over 80 years old. He said that Lady Curzon had expressed an interest in visiting the Naharjan tea estate, as a planter named Forbes had told her that wildlife was abundant in that area. Forbes arranged three elephants to facilitate game viewing, and had asked a local *shikari* called Nigona to accompany her. Nigona, later identified as Bapiram Hazarika, emphasized the plight of the rhino, as the party could only find a footprint and never saw the animal itself (Gokhale 2004; Gohkale and Kashyap 2005; Saikia 2005; Sarma 2023).

It is almost certain that a 1905 date cannot be correct, as Lady Curzon went to England in January 1904 with her husband, delivered a baby on 20 March 1904, after which she fell ill, and only arrived back in Kolkata in March 1905 – and not long after, on 18 July 1906, she died (Nicolson 1977; Bradley 1985). The Lady Curzon papers preserved in the British Library have no information about her trip to Assam (Thomas 2004). There is of course a chance that the visit to the tea estate was part of the Viceroy's official visit to Assam from 6 to 16 March 1900, when they traveled from Guwahati to Dibrugrah (Deka 2005, 2012; Rookmaaker 2019b). This trip was closely followed by the press and there were regular updates in the *Times of India*. The Chief Commissioner of Assam from 1896 to 1902, Sir Henry John Stedman Cotton (1845–1915) attended the whole trip (*Times of India* 1900-03-10). Traveling by steamer and train, they reached Dibrugarh in Upper Assam on 7 March (*Times of India* 1900-03-09). On the return journey, they arrived in Tezpur on Friday 9 March: "A most enthusiastic welcome was accorded to the Viceroy and Lady Curzon on their arrival at Tezpur on Friday. A large number of guests and officials from all parts of Assam were present to welcome their Excellencies. A number of ladies were presented to Lady Curzon by Colonel Buckingham. In the evening the Viceregal party were entertained at dinner by the Planters" (*Times of India* 1900-03-13). The next morning, they "left Tezpur for the rubber forests at Balipara [north of Tezpur]. A mounted escort of the Assam Valley Light Horse Volunteers met and conducted his Excellency to Adabarrie, the residence of Mr. M. Chamney, where chhota-hazri [early breakfast] was served. After visiting and inspecting the rubber forests and planting a tree in honour of the occasion, the Viceroy and party left for Borjula Garden, the residence of Mr. Moore, where a large company of ladies and gentlemen had been invited to breakfast to meet his Excellency" (*Times of India* 1900-03-15). That afternoon they traveled by steamship from Tezpur back to Guwahati, reaching on 12 March.

There is no mention of a visit to a tea plantation near Nowgong in these proceedings. In the 13 March 1900 newspaper report of the Saturday excursion, Lady Curzon is not mentioned specifically, but it appears unlikely that she would have traveled independently in an opposite direction without this having been noticed. However, to tie in with the possible history of the start of Kaziranga, it is quite possible that she talked to the planter Forbes of the Naharjan Tea Estate located on the edge of the future reserve, maybe during the evening function in Tezpur. The Naharjan estate was first planted in 1895 (Yadava 2014) and is mentioned in the Assam District Gazetteer of 1906:

"Naharjan, owned by Messrs. Pringle and Forbes, situated in Namdayang, on 13 December 1903 comprising 1514 acres of which 505 were planted, with a labour force of 480 workers" (Allen 1906: 254). The owner might have been Herbert Russell Forbes (1863–1920), who was Honorary Magistrate of Naharjan in Sibsagar District.

45.4   Lord Curzon on Game Preservation

Lord Curzon showed himself well aware of the current trends and appeared sensitive to attempts to halt the disappearance of all wild animals. He had first-hand knowledge of wild sports, as it was customary for Viceroys to be invited by Indian Maharajas for a week of sport in the cold season. He had shot at least one rhino in 1901 at Morang in Nepal, yielding what was then the longest rhino horn ever obtained in that country (27.5). On a visit to Rangoon on 10 December 1901, he addressed the Burma Game Preservation Society, emphasizing the conservation needs:

> There are some persons who doubt or dispute the progressive diminution of wild life in India. I think that they are wrong. The facts seem to me to point entirely in the opposite direction. Up to the time of the Mutiny lion were shot in Central India. They are now confined to an ever-narrowing patch of forest in Kathiawar. I was on the verge of contributing to their still further reduction a year ago myself; but fortunately I found out my mistake in time, and was able to adopt a restraint which I hope that others will follow. Except in Native States, the Terai, and forest reserves, tigers are undoubtedly diminishing. This is perhaps not an unmixed evil. *The rhinoceros is all but exterminated save in Assam* (Curzon 1906).

Curzon was one of the founding members of the Society for the Preservation of the Fauna of the Empire in 1903, which is still active as the Fauna & Flora Preservation Society based in Cambridge, UK (Anon. 1905b). At first mainly focused on conservation measures in Africa, Curzon was able to report on the situation in India (Prendergast & Adams 2003). On a national level, efforts were focused on a strengthening of game laws, as he discussed in a speech of 15 June 1906:

> Now, in regard to India, your familiarity with that country, my Lord, will remind you that we have existing in India, owing to natural causes, perhaps, the greatest extent of reserves in the world. In the first place we have our forest reserves, which, no doubt, were created in the first place for the growth and preservation of timber, but which constitute indirectly a sort of reserve for game; then the Native States in India, particularly when you have a sporting Rajah at the head, are in themselves a sort of reserve; and, finally, all along the north of India you have under the mountains the long strip of Nepaul, which, as at present administered, is perhaps the finest natural game preserve in the world. Therefore in India we have not to look at the question from the same point of view as in Africa; we have not got to create reserves, because they exist; and in India we were devoting ourselves, when I left the country, to an alteration and a strengthening of the Game Laws (Curzon 1908).

45.5   Oscar Kauffmann in Kaziranga in 1907

At first the officials and local residents were none too certain about the game laws applicable in Assam or about the actual meaning of a reserved forest. This shows from the remarks by the German traveller and hunter Oscar Kauffmann (1874–1924), who visited in 1907 at a time when the notification of the reserved forest was being finalized. He mentions the "Kasironga Reserve" in his book published in German in 1911, with a second edition in 1923. He had traveled from Kolkata by train and steamer to Kokilamukh on the Brahmaputra near Jorhat and then on to Golaghat. Shooting in the area had been recommended to him by Percy Comyn Lyon (1862–1952), Chief Secretary of the Governor of Assam, but on arrival in Assam the Deputy Commissioner Arthur William Botham (1874–1963) advised him that the reserve was in fact closed. Although this was contradicted by the tea planters and by the Assistant Deputy Commissioner of Golaghat Louis James Kershaw (1869–1947), he decided not to venture into the reserved forest (Kauffmann 1911: 157). Rather pointedly, he heard that Botham himself was camping in the forest with his wife. Kauffmann himself saw just two rhinos for a short moment and found that the local Assamese had put up huts in the swamps near Bokakhat where they shot at every rhino in sight, because the horn could fetch up to 300 rupees.

45.6   Kaziranga as a Closed Reserve

Once Kaziranga was notified as a reserved forest in 1908, there were some initial difficulties. It was found that the

FIGURE 45.4
"The end of a raider". Stray rhino in 1933 shot at
Numalighar, 10 km east of Kaziranga
Y.D. IN *THE FIELD, THE COUNTRY
GENTLEMAN'S NEWSPAPER*, 8 APRIL 1933,
P. 755

people of Kuthori and Baguri villages had been cut off from the Moridiphlu River, to which access was needed for water and grazing. The managers of the Hatikhuli and Kuthori tea estates also needed an approach to the river in order to ship tea by boat. Hence, on 18 April 1911, an area of 1442 acres near these villages was dereserved, excluded from the protected forest. At the same time, the authorities planned an extension on the eastern side towards the Bokakhat Dhansirimukh road, as many rhinos were known to stray outside the gazetted area in search of food. There were objections both from villagers and the European tea planters. The former were concerned about reduced grazing ground, fishing facilities and the availability of thatch and firewood. The latter felt that hunting grounds for deer and pigs would be significantly reduced. The Deputy Commissioner of Sibsagar District Colonel Alan Playfair (1868–1952) overruled the objections, stating:

> There are several keen sportsmen among them and it is an undoubted fact that the areas of their operations have been greatly restricted. Our object is to preserve the rhinoceros and the herds of buffaloes in these parts, and were it only the real sportsman whom we had to deal with, the newly made game laws might be sufficient. There are however, certain persons who have few scruples against whom further restrictions have to be aimed. An alternative to constituting an addition to the Kaziranga reserve would be to prohibit entirely the shooting of rhinoceros for a certain number of years and then to issue only a limited number of permits per annum. I have consulted persons who are well acquainted with the game in the part of the district and the highest estimate made of the number of rhinoceros in the Kaziranga reserve is twenty pairs (Lahan 1993: 6).

Therefore, the reserved forest was extended on the eastern side by 16,347 acres on 17 October 1911. There were further extensions in the same area to the extent of 13,506 acres on 28 January 1913 and on 23 April 1914. The extensions were made at comparatively higher altitudes than the original reserved land, thereby providing additional shelter to the rhinos during the annual floods, as well as a buffer against diseases that might be transmitted by livestock on the surrounding lands. Further land on the northern side close to the Brahmaputra, 37,529 acres in extent, was added on 26 July 1917.

Kaziranga was declared a Game Sanctuary on 10 November 1916 by William Frederick Loftus Tottenham (1866–1936), Conservator of Forests, Eastern Assam. Browsing through the early progress reports of Assam's Forest Administration, I found that Kaziranga was listed there for the first time in 1918 after it was declared as a game sanctuary with a total area of 68,609 acres (277 km^2) including an extension granted the previous year (Blunt 1918).

There is remarkably little information about any developments in Kaziranga from 1916 to 1938. The sanctuary was essentially closed to all visitors including hunters, officials, villagers and tourists. Any application for a permit was denied on the grounds that there was nothing there but swamps full of leeches, and there were no facilities or domesticated elephants. Left undisturbed, as intended by the Forest Administration, it would be natural to expect that the rhino population would stabilize and grow. Although no census was undertaken, estimates found in the literature for 1916 range from 50 to 100 animals, and for 1938 from 100 to 300 animals. An initial population of 50 animals, with a growth rate of 5% per annum, would yield a population of 132 rhinos. An initial population of 100 animals, with the same annual growth rate, would yield a population of 265 rhinos.

Arthur John Wallace Milroy (1883–1936), nearing the end of his long career in the Indian Forest Service as the Conservator of Forests Assam (1933–1936), took an interest in Kaziranga. He did not like what he found. Although the park was off the map for most Europeans, he found that there were many illegal settlers, cattle grazers and poachers in the sanctuary. At the same time, there was no budget to pay staff and no apparent revenue-producing opportunities. Working with Mahi Chandra Miri (1903–1939), the first Indian officer in Kaziranga employed as the Extra Assistant Conservator of Forests in 1934, Milroy attempted to deal with the poaching gangs (Milroy 1933, 1934b, 1935). One of his measures stemmed from his frustration in getting sufficient funding and his fear that soon the whole reserve would be settled and cultivated. Knowing that most poachers entered the reserve from the Brahmaputra River, he made an agreement with the professional grazers to report anyone seen entering the reserve. Allowing livestock to be grazed within the park boundaries had been a mixed blessing through the years. A second approach was to allow visitors into the park, as the increased human presence would also deter poachers (Milroy 1936). He set the wheels in motion which would give the legal background to open up Kaziranga to all lovers of wildlife, which materialized in 1937, almost two years after Milroy's death.

45.7   Opening Kaziranga for Visitors in 1937

This was the signal that many expatriate residents had been waiting for (table 45.65). Among the first visitors was E.P. Gee, who in the following years would be the person who helped to give Kaziranga international exposure through his writings which with his simple love for the nature of India awakened a lasting fascination in readers across the world (38.5). The visitor's book of the Tourist Bungalow at Kohora shows an entry of one of E.P. Gee's first visits to Kaziranga with two friends on 6 April 1939:

"First rhino charged us flat out. Last rhino (which had an old wound) we followed on foot and cine-photographed." Gee had been accompanied by M.C. Miri, and the second rhino with the old wound was soon to become a favourite with visitors:

> Near the boundary we encountered this large bull. He dwelled, we were told, in solitary exile and had come to be called the *Boorra Goonda*. Miri told us the animal's story and pointed out to us an old wound on his plated hindquarters. 'It might even be safe to dismount and approach on foot,' said Miri, and we lost no time in following his suggestion. I was covered on one side by Miri and on the other by one of my friends. Both men carried rifles for self-protection, though the practice has long since been discontinued in Kaziranga. I walked cautiously toward the *Boorra Goonda* and was able to take a motion picture of him at a range of about ten yards. He stopped grazing, glared with resentment in my direction, and then resignedly turned away. We felt elated at our bravery, and – more important – we had saved face for both our elephants and ourselves (Gee 1954).

For many years, *Boorra Goonda* would be the focal point for all park visitors (fig. 45.5). It was known that the rhino wandered towards the Kohora grasslands every morning to graze and wallow, and the mahouts, mounted on elephant-back, were usually able to find him. He became accustomed to the presence of elephants, which is important when tourists want to take pictures at close range. *Boorra Goonda* survived the Second World War and all other changes in the area, and died on 8 June 1954. He had been the one to introduce his kind to all those who were privileged to visit this distant corner of north-eastern India. Gee continued to visit Kaziranga and to learn more about its elusive rhinos. On 24 July 1955, riding on an elephant near the Mora Difloo stream, Gee observed a

FIGURE 45.5
"Boorra Goonda" photographed by E.P. Gee in Kaziranga National Park in the early 1950s
NATURAL HISTORY, NEW YORK, OCTOBER 1954, VOL. 63, P. 366

mother with two calves, and presumed these to be twins (Gee 1955), which was ascertained again in Jaldapara (fig. 34.6).

45.8   Rhino Captures for Exhibition

Miri and Milroy were instrumental in facilitating the capture of the first rhinos within Kaziranga that were destined for zoos abroad. Indian rhinos were rare exhibits in zoos in the early 20th century and were regarded as a great prize by many zoo directors. During heavy floods in July 1938, a rhino managed to swim to safety on the Mikir Hills in the southern part of Kaziranga. Plans were made to capture the rhino in a pit, which was done successfully on 19 October 1939. After the rhino, a male that was named 'Gunda', was treated for a wound sustained during his fall, he was transported by train to Kolkata on

FIGURE 45.6
Three details of Ralph Graham's account and sketches of the capture and transportation of Kashi Ram destined for the Chicago Zoo
GRAHAM, *RHINO! RHINO!*, 1949

FIGURE 45.7
A rhino observed from elephant back, seen in February 1955 by
the wildlife photographer Ylla in Kaziranga National Park
WITH PERMISSION © UNIVERSITY OF ARIZONA, TUCSON
AZ, CENTER FOR CREATIVE PHOTOGRAPHY: ARCHIVE
NUMBER AG138.8

FIGURE 45.8
Rhino photographed by Ylla in Kaziranga NP during a
capture operation in 1955
WITH PERMISSION © UNIVERSITY OF ARIZONA

FIGURE 45.9
Another view of a rhino capture operation seen by Ylla
WITH PERMISSION © UNIVERSITY OF ARIZONA

FIGURE 45.10
Lobby card for the movie *The Big Hunt* of 1959 starring Ellis Roderick Dungan (1909–2001), including scenes of rhino capture and behaviour taken in Kaziranga in 1953
COLLECTION KEES ROOKMAAKER

2 November 1938 (Miri 1939). Gunda's long journey ended at the National Zoo in Washington, DC, USA, where he lived from 7 July 1939 to 9 January 1959.

After the Second World War, Kaziranga continued to allow the capture of rhinos for exhibits in zoos in America and Europe (table 45.64). Ralph Graham (1901–1980), Associate Director of the Chicago Zoological Society, vividly described his experiences of 1948 in letters which were later combined in one of the rarest booklets on rhinos, written in the form of letters home and illustrated with his own pencil sketches (Graham 1954). When Graham arrived in Kolkata, he had difficulty even finding Kaziranga on a map. He traveled there by train and found that people had already begun preparations to capture rhinos. A male rhino was captured in a pit on 29 February 1948, much later followed by a female (fig. 45.6).

The rhinos at first were captured using pit traps measuring about 3 m long, 1.5 m wide and 1.8 m deep. These are dug in the middle of a rhino path, and then covered with sticks and grass for camouflage. Any rhino which falls in the trap is quickly removed to a cage on wheels, which is then dragged by an elephant to a nearby stockade (Gee 1964a: 159). Typically a large number of people would be involved and the operation was of course not without danger.

One of the capture operations in 1955 was followed by the wildlife photographer Camilla Koffler, known as Ylla (1911–1955). She was in Kaziranga in February when they tried to catch a young male rhino for the Philadelphia Zoo, this time unfortunately unsuccessful (Ylla 1958). She met E.P. Gee and Patrick D. Stracey, who allowed her to spend some days on elephant back and in an observation hut in the reserve (Vaidya 1967: 151). As Ylla died in an accident

in March, only a fragmentary record of her observations was published. Only three rhino photographs taken in Kaziranga are still present in her archives (figs. 45.7, 45.8, 45.9).

The American film director Ellis Roderick Dungan (1909–2001) had also been attracted to the possibilities offered in Kaziranga to take moving pictures of a capture operation and observations in the field. On 7 February 1953 he was taking shots of a couple of rhinos known as Romeo and Juliet while they were playing, chasing and courting each other. Suddenly one of them ran towards the Assistant Conservator of Forests, who unfortunately was pushed by the animal and broke his collar bone (Gee 1953c). Dungan used the sequence of the charging rhino in his movie released as *The Big Hunt* in 1959 (fig. 45.10), directed by George Sherwood (1892–1983) and narrated by Sidney Hertzberg (1910–1983).

The pair of rhinos taken by Graham to Chicago never bred, which was not unusual for Indian rhinos in captivity. This situation changed at Switzerland's Basel Zoo through the efforts of Ernst Michael Lang (1913–2014), who served as Director from 1953 to 1978. Lang was a brave and innovative man. The first pair of rhinos obtained by the Basel Zoo was captured by the Swiss dealer Peter Ryhiner (1920–1975) in Kaziranga, the male Gadadhar on 30 May 1951 and the female Joymothi on 8 July 1952. The animals came to live in a large enclosure, where Lang and his team kept them together, even during the female's oestrous period, when serious fights between the sexes sometimes occur. As a result Lang was able to announce the birth of the first Indian rhino in captivity, that of a male, Rudra, on 14 September 1956. Joymothi became the matriarch of the captive rhino population. Until her death on 10 November 1983, she produced 10 calves (six male, four female), and those between them produced an additional 17 male and 19 female offspring. Although there is a danger of inbreeding in the captive rhino population, most of these effects have been contained thanks to the management of a successful studbook of the species by Basel Zoo, with details published regularly.

TABLE 45.64   Specimens of *R. unicornis* captured in Kaziranga 1939–1991

Sex	Age	Date of arrival	Date of death	Zoological garden
M		1939-07-07	1959-01-09	Washington
M		1947-08-07	1961-03-07	Whipsnade
F		1948-06-24	1968-05-06	Chicago Brookfield
M		1948-06-24	1970-11-13	Chicago Brookfield
M		1949	1955	Cairo
F		1949	1955	Cairo
M		1951-05-30	1964-11-25	Basel
M		1951-09-05	1983-02-28	Rome
F	2 yrs	1952-07-06	1985-04-25	Whipsnade – London – Amsterdam
F	5 yrs	1952-07-08	1983-11-10	Basel
F	3 yrs	1953-06-17	1977-05-12	Philadelphia – San Diego
M		1955-09-14	1996-01-06	Philadelphia
F		1957-06-11	1971-01-22	Hamburg Stellingen – Basel – Los Angeles
F	10 yrs	1958-10-05	1964-10-28	Gauhati
M		1958-11-10	1995-07-16	Tokyo Tama
M	7 yrs	1959-09-22	1983-04-15	Berlin Zoo – Basel
M		1960-05-26	1989-09-25	Washington – New York – Oklahoma
F	2 days	1960-06-29	1988-07-05	Gauhati – Delhi
F	9 yrs	1960-06-29	1967-04-03	Gauhati – Paris Vincennes
M	5 yrs	1960-09-24	1986-09-23	Gauhati
F		1961-06-06	1991-12-03	Calcutta – Tokyo Tama
F	9 yrs	1962-10-29	1963-12-28	Gauhati – Washington
M	4 mo	1964-10-28	1964-11-13	Gauhati
F	5 yrs	1965-09-29	1982-10-19	Gauhati
F	6 yrs	1968-03-28	1984-11-10	Delhi
F		1968-06-26	1983-08-19	Hyderabad
F	2 weeks	1968-07-29	1970-01-31	Gauhati – Omaha
M	1.5 yrs	1971-06-28	1973-12-03	Gauhati – Brownsville
F	2.5 mo	1974-01-22	2004-08-09	Gauhati – Calcutta
M	4 yrs	1974-02-12	2003-01-23	Gauhati – Calcutta
F	1.5 mo	1974-08-10	1977-08-20	Gauhati
F	3 mo	1974-09-16	2007-05-28	Gauhati – Bhubaneswar
M	6 mo	1976-02-02	1976-08-07	Gauhati
M	1 mo	1976-04-06	1976-08-07	Gauhati
M	20 mo	1977-10-17	1993-07-03	Gauhati – Chandigarh
M	old	1978-03-03	1987-01-18	Gauhati
F	7 yrs	1978-06-15	1986-05-26	Gauhati – Chandigarh
M	2 mo	1980-02-27	1986-02-23	Gauhati
M	4 mo	1980-09-03	1989-07-07	Gauhati – Madras
F	1 mo	1981-07-03	1981-10-11	Gauhati
M	12 yrs	1982-05-05	2001-04-24	Gauhati – Delhi
F	1.5 mo	1982-06-04	2001-03-25	Gauhati – Delhi
M	10 yrs	1982-07-15	2006-11-19	Gauhati

TABLE 45.64   Specimens of *R. unicornis* captured in Kaziranga 1939–1991 (*cont.*)

Sex	Age	Date of arrival	Date of death	Zoological garden
M	1 yr	1989-07-26		Gauhati – Trivandrum
M	1 mo	1989-07-26	1989-09-04	Gauhati
M	1 mo	1989-07-26	1995-10-17 to Jaldapara	Gauhati
M	3 mo	1990-08-20	2014-09	Gauhati – Tripura
F	4 mo	1991-08-06	1991-08-17	Gauhati
F	4 mo	1991-08-10	1991-09-19	Gauhati
F	1 yr	1991-08-10		Gauhati

45.9   Studies and Visitors in the 1950s and 1960s

In 1950, the then Conservator of Forests, Patrick Donald Stracey (1906–1988), changed the designation of Kaziranga to that of Wildlife Sanctuary. Through the years there were many visitors, ranging from thousands of daily tourists to some high-ranking officials and eminent scientists (table 45.65, figs. 45.11, 45.12).

There was an attempt on 23 March 1949 to census Kaziranga's wildlife from the air. A six-seater airplane was flown over the reserve but, due to its size, it could not descend below 400 feet. This was too high to identify individual rhinos or other wildlife in the high grass. However, Rustom Phirozsha was able to take the first aerial photographs of the reserve. In 1961, noted wildlife expert Robin Banerjee (1908–2003) released the first documentary movie on the park "Kaziranga" (50 minutes).

FIGURE 45.12   Visit in April 1962 to Kaziranga of John Kenneth Galbraith (1908–2006), US Ambassador to India 1961–1963, with his wife Catherine Merriam Atwater (1913–2008)
COURTESY: PETER GALBRAITH AND JAMES K. GALBRAITH

FIGURE 45.11   Visit in 1956 to Kaziranga of Jawaharlal Nehru (1889–1964), first Prime Minister of India 1947–1964
PHOTO: E.P. GEE

James Juan Spillett (1932–2018) was an American ecologist who was in India in the 1960s to research his dissertation on the Lesser Bandicoot Rat (*Bandicota bengalensis*). He received a grant from the World Wildlife Fund to conduct a survey of the status of the rhinoceros in Assam and Bengal (Spillett 2015). He visited most of the rhino-bearing reserves and published his reports in the renowned *Journal of the Bombay Natural History Society*. In March 1966 he took part in the first scientific census of Kaziranga's rhinos (Spillett 1966b). The park was divided into blocks and people went out on elephant-back for several consecutive days. The census was undertaken mainly to gauge the effect of poaching activities, which reportedly had been very high during 1964 and 1965. Poaching was most prevalent in the north-eastern part of the sanctuary, farthest away from the Kohora headquarters, and rhinos fared best in areas frequented by visitors. The census yielded 366 rhinos that could be individually counted, but

they were unevenly distributed throughout the park. The total number of rhinos was obviously higher, so a conservative estimate of 400 animals was generally quoted at the time. Spillett recommended prohibiting the exploitation of natural resources, phasing out the grazing of livestock, and the designation of a part of the Mikir Hills south of the reserve as Reserved Forest to provide refuge during the flooding season. Spillett's final suggestion was that Kaziranga be declared a national park to facilitate implementation of his other proposals.

The German biologist Wolfgang Ullrich made several visits to Kaziranga in the 1950s and 1960s (figs. 45.13, 45.14, 45.15). He had been appointed Director of the Dresden Zoo, besides which he took part together with his wife Ursula in a series of television documentaries *Der gefilmte Brehm* (Brehm on film) shown in Germany from 1960. They wrote two books on their experiences in India and Kaziranga, only available in German, well-written and

FIGURE 45.13   Photo of a rhino in the high grass of Kaziranga NP taken by Wolfgang Ullrich in the 1960s
ULLRICH, *IM DSCHUNGEL DER PANZERNASHÖRNER*, 1962, P. 163

FIGURE 45.14   Two rhinos in Kaziranga NP photographed by Wolfgang Ullrich
ULLRICH, *IM DSCHUNGEL DER PANZERNASHÖRNER*, 1962, P. 163

FIGURE 45.15
Transport of cage to the pit where the rhino is trapped
ULLRICH, *IM DSCHUNGEL DER PANZERNASHÖRNER*,
1962, P. 238

extensively illustrated with their own photographs, 'In the Jungle of the Rhinoceros' (1962) and 'Kaziranga, wildlife paradise on the Brahmaputra' (1971). Ullrich published several papers on his zoological surveys of the park, and brought the plight of nature conservation to the attention of the German public. Such efforts were important to change attitudes in Europe, as ideas of nature conservation were still developing and were then in the process of becoming more generally accepted and more scientific in approach.

TABLE 45.65   List of selected visitors to Kaziranga reported in the literature since it opened for tourism in 1938

Date	Visitor	Reference
1938-05-25	A.R. Tomas – first entry in visitors book	Gokhale & Kashyap 2005: 86; Barthakur & Sahgal 2005: 146–147
1938	F.T. Gass, Gurkha Rifles	Gass 1971
1939-04-06	Edward Pritchard Gee (1904–1968) as Honorary Forest Officer	Gee 1954
1944-11-27	Botha Van Ingen (1904–1996), taxidermist in Bangalore	Gokhale & Kashyap 2005: 86
1947	M.L. Flemin of Whipsnade Zoo	Vevers 1947
1947-01-02	Salim Ali (1896–1987), Indian naturalist	Gee 1950: 83; Gokhale & Kashyap 2005: 86
1947-01	Sidney Dillon Ripley (1913–2001)	Gee 1950: 83; Gokhale & Kashyap 2005: 86
1948-02	Ralph Graham (1901–1980), Curator of Chicago Zoological Society	Graham 1949, 1954
1948-04-20	Sir Akbar Hydari (1894–1948), Governor of Assam	Oberai & Bonal 2002: 34; Tirkey 2005; Barthakur & Sahgal 2005: 146 (photo of letter)
1949	Gerald B. Eastmure	Eastmure 1949
1949-03	Rustom Phirozsha (b. 1902), Master of the Films Division, Mumbai	Gokhale & Kashyap 2005: 86
1951	Peter Ryhiner (1920–1975), animal dealer	Assam Forest Dept. 1951: 29; Huber 1964; Ryhiner & Mannix 1964
1951	Jairam Das Daulatram (1891–1979), Governor of Assam 1949–1956	Assam Forest Dept. 1951

TABLE 45.65   List of selected visitors to Kaziranga reported in the literature since it opened for tourism in 1938 (*cont.*)

Date	Visitor	Reference
1951	Ronald Ivelaw-Chapman (1899–1987), Air Marshall, India	Assam Forest Dept. 1951
1953-02-07	Ellis Roderick Dungan (1909–2001), American cinematographist	Gee 1953c, Vaidya 1967: 154
1953-03-07	Chester Bliss Bowles (1901–1986), American Consul General	Gee 1953c
1953	Jagjivan Ram (1908–1986), Minister for Agriculture, India	Tirkey 2005, Oberai & Bonal 2002: 33
1953	Ylla, or Camilla Koffler (1911–1955), photographer	Ylla 1958
1953	Kenn Reed, American photographer	Reed 1956; Talbot 1957
1955	Lee Merriam Talbot (b. 1930), American zoologist	Talbot 1956a, 1956b
1956	Eugen Schuhmacher (1906–1973), photographer	Schuhmacher 1968: 180–181
1956-10-20	Jawaharlal Nehru (1889–1964), Prime Minister of India	Barua & Das 1969: 19, 25 (photos); Tirkey 2005; Gee 1964, pl. 63 – fig. 45.11
1956	Indira Gandhi (1917–1984), daughter of J. Nehru	Barua & Das 1969; Tirkey 2005
1956	Feroze Gandhi (1912–1960), husband of Indira Gandhi	Barua & Das 1969; Tirkey 2005
1956	Bimala Prasad Chaliha (1912–1971), Chief Minister Assam	Oberai & Bonal 2002: 34
1959	Wolfgang Ullrich (1923–1973), director of Dresden Zoo, with his wife Ursula Ullrich (b. 1932)	Ullrich & Ullrich 1962; Ullrich 1971
1961	Robin Banerjee (1908–2003)	documentary movie
1962-04	Prof John Kenneth Galbraith (1908–2006), US Ambassador and his wife Catherine Merriam Atwater (1913–2008)	Galbraith 1969: 369, Barua and Das 1969: 23; fig. 45.12
1965	Heinz-Georg Klös (1926–2014), director of Berlin Zoo from 1954	Klös 1966a, 1966b
1965-05	George Beals Schaller (b. 1933), American ecologist	Spillett 1966b
1966-03	James Juan Spillett (1932–2018), American conservationist	Spillett 1966b
1966-11	Zdenek Veselovsky (1928–2006), director of Prague Zoo, and his wife Alena Veselovsky	Veselovsky & Veselovsky 1968
1968	Clyde A. Hill, Assistant Curator of San Diego Zoo	Hill 1968
1970	B. Cash, A. Cash	Cash & Cash 1971
1975-11	Peter Lüps, Museum of Natural History of Bern, Switzerland	Lüps 1978
1979	Dennis O'Connor, American travel-writer	O'Connor 1980
1980s	Esmond Bradley Martin (1941–2018), trade investigator	34.6
1980s	Lucy Vigne, trade investigator	34.6
1982	Manmohan Singh (b. 1932), Prime Minister of India	Tirkey 2005
1988	Rajiv Gandhi (1944–1991) and his wife Sonia Gandhi (b. 1946)	Tirkey 2005
1991	Janis Burger, Amsterdam Zoo	Burger 1991
1994	Constance Alderlieste, Rotterdam Zoo	Alderlieste 1995
2001	Robert Dean Blackwill (b. 1939), US Ambassador and his wife Wera Hildebrand	Gokhale & Kashyap 2005: 88; Tirkey 2005
2004	Anwaruddin Choudhury (b. 1959), Indian naturalist	Choudhury 2004

TABLE 45.65   List of selected visitors to Kaziranga reported in the literature since it opened for tourism in 1938 (*cont.*)

Date	Visitor	Reference
2004	David Campbell Mulford (b. 1937), US Ambassador, and his wife Jeannie	Tirkey 2005; Das 2005: 131
2005	Gustasp and Jeroo Irani, Indian travel writers	Irani & Irani 2005
2005	Mark Roland Shand (1951–2014), British travel writer	Gokhale & Kashyap 2005: 89
2005	Maharani Gayatri Devi (1919–2009)	Gokhale & Kashyap 2005: 89
2013-03	Kees Rookmaaker (b. 1953) visits with his wife Sandy (1947–2016) in preparation of a WWF report, assisted by Christy Williams, Amit Sharma and Pranab Bora. First visit was 1980-03	Rookmaaker et al. 2017
2016-04	Prince William (b. 1982), Duke of Cambridge, and Catherine (b. 1982), Duchess of Cambridge	45.11, fig. 45.18
2017	Gautam Prasad Baroowah (b. 1942)	Baroowah 2017
2022-02	Ram Nath Kovind (b. 1945), 14th President of India	Media reports

45.10   Kaziranga as National Park and World Heritage Site

The suggestion by Spillett to designate Kaziranga as a national park was championed by Prabhakar Barua (b. 1918), Chief Conservator of Forests. After all objections by local people had been addressed, Kaziranga National Park came into being on 11 February 1974. In December 1985, Kaziranga was declared a UNESCO World Heritage Site.

Kaziranga celebrated its centennial anniversary in February 2005. The park had seen many challenges, as well as many successes. The number of visitors increased greatly over the years, and so has the availability of accommodation along the road. Traffic on the National Highway continues to increase, which brings its own set of problems, especially during the floods when animals migrate to the hills. The ecology of the park is also forever in flux, with siltation in and around the Brahmaputra, and erratic changes in flood levels. Poaching remains a problem that is costly to combat, and the park staff must be constantly vigilant. Much has been done to reduce poaching, both by the government and through the help of numerous international and national conservation agencies who provide additional equipment, expertise and education.

An important event showing the strength of rhino conservation in Assam was the destruction of all rhino horns accumulated in the Assam State Treasury on World Rhino Day, 22 September 2021 (Yadava 2022). With the full support of Himanta Biswa Sarma (b. 1969), Chief Minister, Government of Assam, a total of 2479 horns

FIGURE 45.16
Scenes during the first Horn Burning Exercise of all horns preserved in the Assam State Treasury on 22 September 2021. The furnaces with burning pyres were especially constructed for the purpose
COURTESY © ASSAM FOREST DEPARTMENT

FIGURE 45.17
"Abode of the Rhinoceros" constructed from
the ashes of horns burnt in 2021 and unveiled
in September 2022 in Kaziranga National Park
PHOTO © CHANDANA SARMA, OCTOBER 2022

FIGURE 45.18
Prince William and Catherine visiting the
Centre for Wildlife Rehabilitation and
Conservation in April 2016
PHOTO © CWRC KAZIRANGA, COURTESY:
RATHIN BARMAN

were completely burned (fig. 45.16). The ashes were collected and sculpted into a monumental Abode of the Rhinoceros, unveiled on 25 September 2022 at Mihimukh, Kaziranga Range, representing the supreme sacrifice of Forest staff in the protection of Kaziranga National Park & Tiger Reserve (fig. 45.17).

45.11 Centre for Wildlife Rehabilitation and Conservation

The Centre for Wildlife Rehabilitation and Conservation (CWRC) was established in 28 August 2002 in the Panbari reserve forest near Kaziranga. Rathin Barman was the first Director, followed in October 2020 by Bhaskar Choudhury. The Centre's primary objective is the care of animals displaced by the floods, injured during predation or orphaned by poaching (Ashraf et al. 2005; Barman et al. 2014). Young rhinos attacked by tigers are often seriously mauled and some don't survive despite medical care. The center is run in collaboration with the Wildlife Trust of India (WTI) and the International Fund for Animal Welfare (IFAW).

The first rhino calf, rescued in July 2002, was less than a month old when it was separated from its mother during a period of heavy floods. Two additional calves, both only a few months old, were rescued in July 2004. Several rescued rhinos were translocated to Manas National Park (40.6). In April 2016, Prince William, Duke of Cambridge (b. 1982), and Catherine, Duchess of Cambridge (b. 1982) visited the CRWC to enable them to get first-hand knowledge of the efforts of rhino conservation (fig. 45.18).

45.12 Indian Rhino Vision 2020

Beginning in February 2012, under the auspices of Indian Rhino Vision (IRV) 2020, Assam government authorities began translocating rhinos from Kaziranga National Park to Manas National Park, in collaboration with local coalitions and international wildlife conservation institutions. On 19 February 2012, four rhinos (Maidangshri, Swimli, Adidiga and one unnamed animal) were released in Manas, followed by another four (Malati, Mann, Hainari and Kwajinai) on 11 March 2012. Four of the females

FIGURE 45.19
Rhinoceros mother with her baby happily frolicking in the jungle. Drawing by Sarah Hewitt in 2016
WITH PERMISSION © SARAH HEWITT 2016 'BOOKS BY SARAH'

translocated from Kaziranga have subsequently produced calves in Manas, but four of the translocated animals were also killed by poachers in 2012–2013.

The number of rhinos in Kaziranga National Park is rising steadily, from 366 recorded during the 1966 census to 2,613 in 2022 (table 45.63). This represents an increase of 4% annually, which is within the limits of expectation. Rising rhino numbers, of course, bring their own set of challenges. Today, Kaziranga is an international park iconic for rhino conservation, as well as a premier destination for wildlife enthusiasts and nature photographers.

PART 2

The Javan Rhinoceros Rhinoceros sondaicus *in South Asia*

∵

CHAPTER 46

The Javan Rhinoceros *Rhinoceros sondaicus*

The Javan Rhinoceros is one of just two single-horned recent species of rhinoceros, both endemic to Asia. It looks quite similar to *R. unicornis*, with which it is classified in the same genus. The two species started to diverge in the middle to late Pleistocene around 4.32 million years ago (Liu et al. 2021). Among all the useful works written about the Javan Rhinoceros in the past years, the review by Henri Jacob Victor Sody (1892–1959) is invaluable for an insight in older sources. First published in Dutch in 1941, most copies fell victim to the Japanese invasion of Batavia (Jakarta), but it has been rescued through the German translation by Erna Mohr (1894–1968) of 1959. The large monograph of 1970 about Ujung Kulon by the Dutch naturalist Andries Hoogerwerf (1906–1977) contains many personal observations as well as rare early photographs. The condensed paper by Colin Groves and David Leslie (2011) summarizes all biological aspects in the series of *Mammalian Species* published by the American Society of Mammalogists to provide encyclopaedic accounts of all living mammals.[1]

In English, the species is generally called the Javan Rhinoceros. Alternative names include Lesser one-horned and Smaller one-horned Rhinoceros in juxtaposition to the larger one-horned species, as well as Sundarbans Rhinoceros. A list of indigenous names is found in Sody (1941: 18) and some additional references are in Rookmaaker (1983a: 110). As with the other species, the geographical part of the name "Javan" looks odd when describing animals seen in South Asia. For that reason only, in this book the Javan Rhinoceros is identified using the scientific name *Rhinoceros sondaicus*, in the abbreviated form *R. sondaicus*.

R. sondaicus was widely distributed across South-East and South Asia. It occurred in Java and Sumatra, through Malaysia westwards in southern parts of Thailand and Myanmar to Bangladesh and India. In an eastern direction its range extended to Cambodia, Laos and Vietnam. The records from the Indochinese region are quite sparse, and after my first review in Rookmaaker (1980, 1999a, 2002),

the literature was extensively investigated through the efforts of the French mining engineer Henri Carpentier (1923–2017), all made public through his unique "Carino" database on the Rhino Resource Center (Carpentier 2004, 2011). The presence of the rhinoceros in parts of China in historical times remains poorly documented in western works (Rookmaaker 2006b). The reports from the island of Hainan are spurious (Rookmaaker & Carpentier 2007). The species has been found in Borneo only in fossil deposits (Cranbrook & Piper 2007). There is no record from Singapore. The last population of the Javan Rhinoceros on the Asian continent in Cat Tien National Park, Vietnam was exterminated in 2011 (Brook et al. 2011). At this time, the species only survives in Ujung Kulon National Park on the western tip of Java, Indonesia. *R. sondaicus* is listed as Critically Endangered in the IUCN's Red List in 2021 (Strien et al. 2011).

Rhinoceros sondaicus was partially sympatric with both the other Asian species *D. sumatrensis* and *R. unicornis*. Morphologically, *R. sondaicus* and *D. sumatrensis* are too different in appearance to be confused. The two species differ in size, in hairiness, in the arrangement of the folds and the number of horns. However, in the field, where observations often last mere fractions of seconds in dense jungle, misidentifications always remain a possibility, hence greatly complicating any study of literature. The single-horned species *R. sondaicus* and *R. unicornis* are much more similar, still probably they can rather easily be told apart when a direct comparison can be made. The word "probably" needs to be added because in practise the two species have rarely been seen together in one place, except maybe in a few museums.

The habitat of *R. sondaicus* is confined to lowland forests and fertile floodplains, where water is plentiful. The animals are rarely seen in the wild, not only because few populations are now left, but also because they have always been able to hide well in the dense undergrowth of the forests. Even photographs of *R. sondaicus* are a real prize, although nowadays the use of camera-traps has produced some useful images. The rhinos have been able to avoid most hunters, therefore specimens in museums are few. They have been rarely captured, ensuring that the species has remained almost unknown in captivity.

R. sondaicus differs from *R. unicornis* in four main aspects: morphology of the skin, the arrangement of the skin folds, the size, and the horn length in females. The

---

1 The study of *R. sondaicus* is complicated because the animals have become extremely rare in the wild. The standard reviews are those by Sody (1941) with the German translation of 1959, and the chapters about the rhinoceros by Hoogerwerf (1970). The ecological study by Rudolf and Lotte Schenkel (1969) was followed recently by several of Indonesian researchers. Groves & Leslie (2011) summarized all aspects of biology.

© L.C. (KEES) ROOKMAAKER, 2024 | DOI:10.1163/9789004691544_047

This is an open access chapter distributed under the terms of the CC BY-NC-ND 4.0 license.

70

FIGURE 46.1   Watercolour by Joseph Wolf (1820–1899) of the Javan rhinoceros (*Rhinoceros sondaicus*). Drawn after the male specimen from Java exhibited in London Zoo 7 March 1874 to 23 January 1885. Signed lower right "J. Wolf 1874". This was later engraved for Sclater, *Transactions*, 1876, pl. 96
WITH PERMISSION © ZOOLOGICAL SOCIETY OF LONDON, ARCHIVES: DRAWINGS BY WOLF, VOL. III. MAMMALIA III. UNGULATA I, NO. 70

discussions in some older literature about the differences between these two species in the mosaic-like polygons of the epidermis are generally too hard to understand to be particularly useful. To some degree, divisions in the upper skin layers occur in both species, and it may well depend on the light when the animal is seen, or on the skills of the taxidermist after its death. Any consistent characteristics which may exist are almost impossible to be used to identify animals in written accounts or on artistic drawings. Having said that, in some cases it can definitely add to the certainty that a specimen is actually *R. sondaicus*. For instance, the animal in the photograph taken in Junagadh in 1900 carrying Lady Curzon (Rookmaaker 2018), especially when seen in close-up, gives a strikingly obvious example of these epidermal polygons all over the body (fig. 48.9). They do exist, they can help, but not consistently in all cases.

The skin folds of *R. sondaicus* and *R. unicornis* are similar, with one exception. There is a significant difference in the arrangement and structure of the folds in the neck region. In *R. sondaicus*, the skin forms a clear shield shaped like a triangular saddle, which is immediately diagnostic. This never occurs in *R. unicornis*. All specimens of *R. sondaicus* have this saddle, in all localities, in all age groups, in both sexes. Unfortunately, it doesn't mean that the saddle is always visible in all photographs when the position of the animal obscures the feature. Besides, it is not always properly understood and displayed by artists, and it only rarely features in accounts of encounters with the animals in nature. There may be some special circumstances, but if no saddle is visible in specimens of South Asia, the likelihood is that the animal is actually *R. unicornis*.

One of the photographs taken in January 1941 by Hoogerwerf (1970, pl. 20) shows two males taking a bath in Ujung Kulon, of which the one on the left shows this saddle in a particularly obvious arrangement (figs. 46.2, 46.3). The photographs taken by Hoogerwerf were for a long time the best available, and there was even a movie

FIGURE 46.2
Two adult male (horned) *R. sondaicus*
in a wallow. Photograph by Andries
Hoogerwerf (1906–1977) in Ujung Kulon,
Java on 2 January 1941
HOOGERWERF, *UDJUNG KULON*, 1970,
PL. 20

FIGURE 46.3
Detail of the triangular saddle formed
by skin folds in the neck region,
characteristic of *R. sondaicus*
PHOTO: NICO J. VAN STRIEN

showing rhinos in the wild (Rookmaaker 1983a: 125). Hoogerwerf's notes and correspondence are preserved at the National Archives of The Netherlands in The Hague (no. 2.21.281.27).

The similarity of *R. unicornis* and *R. sondaicus* explains the slow recognition of the Javan animal as a separate species. Ever since Bontius (1658), the presence of a rhinoceros in Java was well-known. Linnaeus (1758) was aware of this, yet did not include the South-East Asian islands in the known range. The first to study the morphology of the skull and skeleton of animals from Java was Petrus Camper (1722–1789), an influential comparative anatomist working at the University of Groningen, The Netherlands (Rookmaaker & Visser 1982). In 1786, he received a skeleton, four heads, a tongue and a penis of rhinoceros collected by Jacob van der Steege (1746–1811), who had been

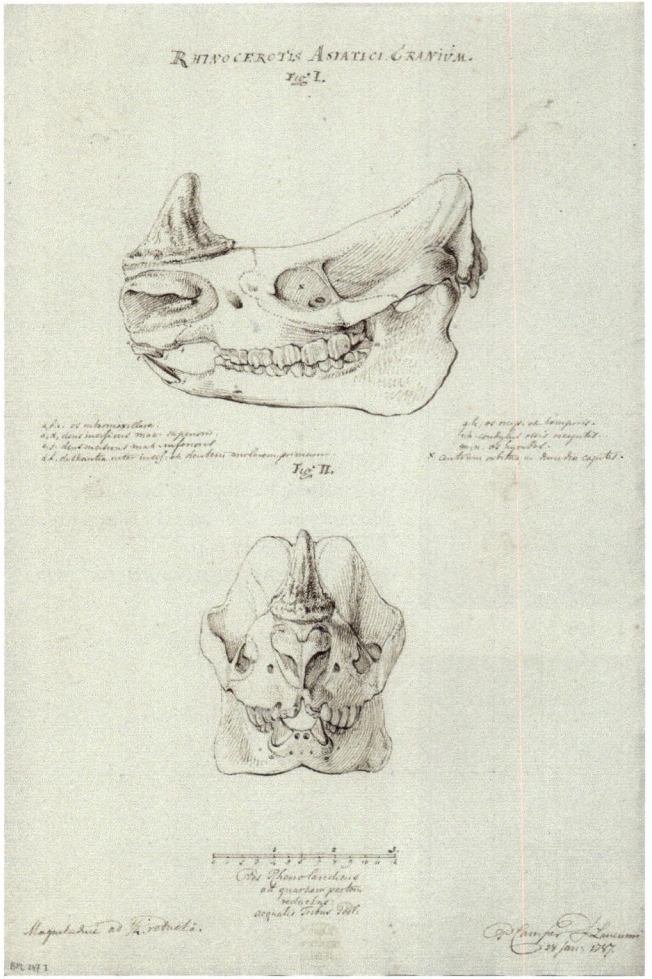

FIGURE 46.4　　"Rhinocerotis Asiatici Cranium", Skull of the Asian Rhinoceros. Signed (lower right) "P. Camper, K. Lancoum, 28 Jan: 1787", therefore drawn by Petrus Camper (1722–1789), at his home in Klein Lankum, Friesland, The Netherlands. The scale represents one foot, and the drawings are sketched at a quarter of the original size
WITH PERMISSION © LIBRARY OF THE UNIVERSITY OF LEIDEN: BPL 247-I (DRAWINGS) OMSLAG 20

his student and worked as a physician in Batavia (Jakarta) from 1744 to 1788 (fig. 46.4).

There is a recurring perception in works on Indian wildlife that *R. sondaicus* is smaller than *R. unicornis*. In a time when *shikar* was still a regular pursuit in the regions of North Bengal and Assam, this was an important distinction, because hunters and their helpers thought to identify animals as *R. sondaicus* if they looked smaller than usual. Jerdon (1867: 233) was quite clear in his comparison of *R. sondaicus* to *R. unicornis*, differing in greatest length of 243 cm compared to 305 or even 365 cm, and in shoulder height of 114 cm compared to 152 cm. Those are major and presumably recognizable differences, but in the case of

R. sondaicus, the specimen(s) on which the size was based may not have been adult. Blyth (1875: 50) also maintained that *R. sondaicus* was about a third smaller than *R. unicornis*. On the contrary, the sportsman Pollok (1879, vol. 1: 98) was certain to "have killed them at least a foot [30 cm] higher" than the published figures.

In a rare animal like *R. sondaicus*, exact data are almost impossible to obtain. Hoogerwerf (1970: 74), for instance, who saw many of them in Java, was never able to measure even a single specimen. As animals obviously continue to grow until full adulthood and as some observers may not have judged the age (or sex) of an animal correctly, really only the maximum sizes can be used for comparisons. There is no certainty that the species had the same size across its wide range, which is now impossible to verify. Nevertheless, some limited data exist, both from Java and from South Asia (tables 46.66, 46.67), in summary as follows:

Size of *R. sondaicus*
Length of body	maximum 373 cm
Height at shoulders	maximum 178 cm
Horn length	maximum 27 cm, males only

Therefore the maximum dimensions of *R. sondaicus* reach those of average specimens of *R. unicornis* which however can grow considerably larger. Size is not a straightforward characteristic which in itself would be completely diagnostic.

The horn in *R. sondaicus* never grows to any great length. The record in Rowland Ward (1928) is a horn from Java reaching 27.3 cm (10¾ inch). Additionally, the adult male which lived in Adelaide Zoo from 1886 to 1907 had a

TABLE 46.66　　Measurements of the greatest body length of specimens of *R. sondaicus*

Locality	Sex	Length in cm	Source
India		213 to 243 = 7 to 8 ft	Jerdon 1867
General		305–344	Groves & Leslie 2010: 193
Java		305	Sody 1959
Java		315 and 316	Sody 1959
Java		320 and 325	Sody 1959
Java		335	Sody 1959
Sundarbans		365.7 = 12 ft	Shekarea 1832
Sundarbans	F ad	373.3 = 12 ft 3 in	Rainey 1875c

Figures in cm without the tail. The data from Sody (1959) and Groves & Leslie (2010) are based on literature surveys.

TABLE 46.67   Measurements of the shoulder height of specimens of *R. sondaicus*

Locality	Sex	Height in cm	Source
India		106.6 to 114.3 = 3 ½ to 3 ¾ ft	Jerdon 1867
Sundarbans		132.6 = 13 hands	Rainey 1872c: 302
Garro Hills	M	134.6 = 4 ft 5 in	Blyth 1870, 1872e; Gallovidian 1837
Java		135, 138	Sody 1959
Assam		137.2 = 4 ½ ft or more	Pollok 1879, vol. 1: 98, 1882
Java		160, 166	Sody 1959
Sundarbans	F ad	167.6 = 5 ft 6 in	Rainey 1875c; Fraser 1875: 10
General		120–170	Groves & Leslie 2010: 193
Java		178	Sody 1959
Sundarbans		213.4 = 7 ft	Shekarea 1832

Figures in cm arranged from least to greatest.

horn measuring 36.8 cm (14½ inch) along the dorsal contour (Groves 1971; Brooks 2017).

Horns are absent in females of *R. sondaicus*, or if present at all, they do not grow beyond a small hump of a few centimetres. This topic continues to be a matter of discussion, mainly because it is almost impossible to find straightforward evidence. The animals are rarely seen in the wild and generally cannot be observed for a longer period. Specimens that are captured or killed are often not described in enough detail (fig. 46.5). There is a distinct possibility of geographical variation across the large (former) range of the species in South and South-East Asia. There is no recognizable sexual difference of horn length in any of the other living species of rhinoceros, and therefore none is expected in *R. sondaicus*. However, the continuous rumours might well have a strong basis in fact. The issue has been carefully reviewed by Sody (1941, 1959), Hoogerwerf (1970) and Groves (1971).

The evidence on horn length in females is limited. The records of *R. sondaicus* in captivity specify the sex only in 4 males and 6 females (Rookmaaker 1998: 120). There are no drawings or photographs of any of the females. A female imported from the Sundarbans by Jamrach to London in 1877 was still young and was hornless. For specimens preserved in museums or observed in the wild, there is no question that some of them are hornless. This is seen quite clearly in the two animals collected by Lamare-Picquot in the Sundarbans and described as the types of *R. sondaicus inermis* (51.5). Baldwin (1910) stated that all rhinos of the Sundarbans are hornless. Sody concluded that horns in males were always said to be larger than in females. The question remains therefore if all females are hornless, or if only females in specific populations are always or predominantly hornless. Hoogerwerf spent much time in the small reserve of Ujung Kulon, now still the last stronghold

LA CHASSE AUX RHINOCÉROS.

Publié par Furne à Paris

FIGURE 46.5
"Chasse aux Rhinocéros" (Hunting the rhinoceros), set in Java but the species is undefined by the artist, and the horn size seems imaginary. Drawing by Adolphe Rouargue (1810–1870), engraved by Charles Beyer (1808–1873) and Charles Lalaisse (1811–1863), and printed in Paris by Charles Furne (1794–1859). Size 20 × 14 cm
JACQUES DUMONT D'URVILLE, *VOYAGE AUTOUR DU MONDE*, NOUVELLE ÉDITION, 1848, VOL. 2, FACING P. 322

of the species. From his own observations, Hoogerwerf (1970) came to the conclusion that females in Ujung Kulon are always without horn. Of course, not all animals could be sexed in the field, but there was enough information to state this as a fact. However, the Swiss ecologist Rudolf Schenkel (1914–2003) had a different opinion after a field study in the park conducted during April to November 1967 and March to October 1968, together with his wife Lotte (Hulliger) Schenkel. He had seen only animals with a horn, assumed an equal sex ratio, hence suggested that females carried horns just like the males (Schenkel & Schenkel 1969: 130). He may have changed his mind, because in his posthumous rhino book it is mentioned that the horn is reduced to a horny knob (Schenkel et al. 2007: 67).

A survey of museum specimens and the available literature by Groves (1971) found only limited evidence of the presence of horns in females. There are reports of females carrying a small horny protuberance where males have horns. Maybe these can grow to a height up to 4 cm, like the knob found in the animal shot by Arthur Stannard Vernay (1877–1960) on the Sungai Lampan in Perak, Malaysia, which was 3.7 cm high (NHM 1932.10.21.1, figured in Loch 1937, pl. 3; Groves 1971, fig. 13). Another female specimen in the NHM London (1921.5.15.1) was shot by Theodore Rathbone Hubback (1872–1942) in Tenasserim, South Myanmar (Groves 1982: 13). Although the original horn was replaced for a while to provide a better exhibit, it was suggested to Groves that the horn now attached of 19.2 cm was the original one. However, when the specimen arrived, Harmer (1921) exhibited the specimen and drew attention to the lack of horns in females. As this is the only documented female *R. sondaicus* with a decent-sized horn, it is most likely that in fact there was some confusion about the actual horn in the museum. The conclusion, in agreement with the statements by Hoogerwerf and Groves, is that females of *R. sondaicus* are always hornless, or carry a horny knob at most 4 cm in height.

Taxonomy and Nomenclature of *Rhinoceros sondaicus* in South Asia

47.1 Taxonomy of the Javan Rhinoceros

The Javan Rhinoceros is one of two living single-horned species in the genus *Rhinoceros*. At first this genus was used for all types of rhinoceros both living and extinct. Now distinguished as *Rhinoceros sondaicus*, morphological and genetic investigations indicate the existence of a number of subspecies mainly distinguished by overall size. The modern subspecific classification is based on the meticulous research of the British mammalian taxonomist and anthropologist Colin Peter Groves (1942–2017), who has made great discoveries and large-scale surveys across all orders of mammals. Educated in Britain, he worked since 1973 at the Australian National University in Canberra, Australia as Professor of Biological Anthropology (Behie & Oxenham 2015). He wrote 791 publications including 16 books about a wide variety of his interests (Rookmaaker & Robovsky 2019, 2020). Ever since his student days in London and Cambridge, he was passionate about his research on rhinos, he was able to identify taxa when the remains were fragmentary, he was generous in sharing his results and helping other scientists, and he was a great collector of rhino objects and books.

Groves (1967) at first recognized three subspecies *R. s. sondaicus*, *R. s. floweri* and *R. s. inermis*, acknowledging that material from some parts of the range was very limited. Further research showed that this arrangement needed adjustments. The latest classification is summarized in Groves & Grubb (2011: 22) and Groves & Leslie (2011), with the suggested historical range.

1. *Rhinoceros sondaicus sondaicus* Desmarest, 1822. Found in Java, Sumatra, Malaysia, South Myanmar and adjoining regions of Thailand.
2. *Rhinoceros sondaicus inermis* Lesson, 1836. Found in India, Bangladesh, North and East Myanmar.
3. *Rhinoceros sondaicus annamiticus* Heude, 1892. Found in Vietnam, Cambodia, Laos, maybe southern China.

47.2 Name of the Genus *Rhinoceros*

Genus *Rhinoceros* Linnaeus, 1758
Carl Linnaeus. 1758. *Systema Naturae* [10th ed]. Holmiae, vol. 1, p. 56.
Details see 5.2.

Since its first description, the specific name *sondaicus* has mainly been used in combination with the generic name *Rhinoceros* (Rookmaaker 1983a: 110; Groves & Leslie 2011). A number of variations were proposed:
Rhinoceros (Eurhinoceros) javanicus by Gray 1868a: 1009
Ceratorhinus sondaicus by Steinmann & Döderlein 1890: 774
Rhinoceros (Rhinoceros) sondaicus by Lydekker 1916: 48
Eurhinoceros sondaicus by Heissig 1972: 29.

47.3 Names of the Living Species *Rhinoceros sondaicus*

List of synonyms of *Rhinoceros sondaicus* in chronological order used for South Asian populations. There are 12 names used for the recent species *R. sondaicus* and its subspecies, of which seven are part of the South Asian fauna (below), and five are extralimital (47.4). The summary is followed by historical and nomenclatorial annotations for names relating to the recent taxon in South Asia.

Rhinoceros sondaicus Desmarest, 1822
Anselme Gaëtan Desmarest. 1822. *Mammalogie*. Paris, vol. 2, p. 399.
Type-specimen: Mounted skin, articulated skeleton and skull of a young male rhinoceros collected by Pierre-Médard Diard (1794–1863) in 1820, preserved in the Muséum national d'Histoire Naturelle, Paris, number MNHN-ZM-MO-1981–561. Refer 47.5.
Type-locality: Java. Here restricted to the western part of Java.
Derivation of name *sondaicus*: latinized adjective form of Sunda or Sonda (region encompassing most of South-East Asia).

Rhinoceros javanicus F. Cuvier, 1824
Fréderic Cuvier. 1824. *Histoire Naturelle des Mammifères*. Paris, part 45–46, p. 2.
Type-specimen: Young rhino seen alive by Pierre-Médard Diard, probably kept in captivity in or near Batavia (Jakarta) in 1820–1821. Refer 47.6.
Type-locality: Java.
Derivation of name *javanicus*: latinized adjective form of Java.

© L.C. (KEES) ROOKMAAKER, 2024 | DOI:10.1163/9789004691544_048
This is an open access chapter distributed under the terms of the CC BY-NC-ND 4.0 license.

Rhinoceros camperis Griffith, 1827

Edward Griffith. 1827. *The Animal Kingdom*. London, vol. 5, p. 291.

Type-specimen: Skull figured by Petrus Camper on a broadsheet privately published in 1787 (fig. 46.5). The skull was donated after his death to the Muséum National d'Histoire Naturelle, Paris, where it was examined by Georges Cuvier (1769–1832). Refer 47.7.

Type-locality: Java.

Derivation of name *camperis:* commemorating Petrus Camper (1722–1789), Dutch physician and anatomist.

Rhinoceros javanus Cuvier, 1829

Georges Cuvier. 1829. *Le Règne Animal*. Nouvelle édition. Paris, vol. 1, p. 247.

Type-specimen: Young rhino seen alive by Pierre-Médard Diard, probably kept in captivity in or near Batavia (Jakarta) in 1820–1821. Refer 47.8.

Type-locality: Java.

Derivation of name *javanus*: latinized adjective form of Java.

Rhinoceros camperii Jardine, 1836

William Jardine. 1836. *The Naturalist's Library: Pachydermes*. Edinburgh, p. 181.

Type-specimen: Skull figured by Petrus Camper on a broadsheet privately published in 1787 (as *R. camperis* in 1827). Refer 47.9.

Type-locality: Java.

Derivation of name *camperii*: commemorating Petrus Camper (1722–1789), Dutch physician and anatomist.

Rhinoceros inermis Lesson, 1836

Réné-Primevère Lesson. 1836. *Complément des Oeuvres de Buffon*. Paris, vol. 10, p. 399.

Type-specimen: Lectotype: Skin and skull of an adult female collected in November 1828 by Christoph-Augustin Lamare-Picquot (1785–1873) in the Museum für Naturkunde of Berlin, no. ZMB_Mam_1957, previous An. 10603 (see Peters 1877; Rookmaaker 2019d). Paralectotype is the young female collected at the same time, no. ZMB_Mam_1958, previous An. 10602. Refer 47.10, 51.5.

Type-locality: "Les Sundries" = Sundarbans, restricted to the area around Mongla (22.50N; 90.60E), Bangladesh.

Derivation of name *inermis*: Latin adjective *inermis* (unarmed, without defenses, *i.e.* hornless).

Rhinoceros jamrachi Jamrach, 1875

William Jamrach. 1875. *On a new species of Indian rhinoceros*. London.

Type-specimen: Female rhino imported by Jamrach in 1874 to London, and exhibited in Berlin Zoo from 30 June 1874 (still living in 1884). No remains survive. Refer 47.10, 47.11.

Type-locality: "Mooneypoor" (Jamrach 1875) or "Munipoor" (Jamrach 1876), i.e. Manipur, India. As this locality is hearsay, it is here changed to Sundarbans, Bangladesh.

Derivation of name *jamrachi*: commemorating William Jamrach (1842–1923), British animal dealer.

47.4   Names for Extralimital Taxa

There are five names used for populations of *R. sondaicus* outside South Asia:

Rhinoceros nasalis Gray, 1868 (p. 1012) – Borneo, probably actually Java.

Rhinoceros floweri Gray, 1868 (p. 1015) – Sumatra.

Rhinoceros frontalis von Martens, 1876 (p. 257) – lapsus for *R. nasalis*.

Rhinoceros annamiticus Heude, 1892 (p. 113) – Vietnam.

Rhinoceros unicornis [= *sondaicus*] var. *sinensis* Laufer, 1914 (p. 159) – China. Preoccupied by the fossil *Rhinoceros sinensis* Owen, 1870.

The names for fossil taxa associated with the genus *Rhinoceros* in South Asia are listed in 8.6. In some cases it remains uncertain if these are referable to *R. unicornis* or *R. sondaicus* or to another species. There are three extralimital extinct populations very closely allied to *R. sondaicus*:

Rhinoceros sivasondaicus Dubois, 1908 (p. 1245) – Java; fossil (probably Upper Pleistocene).

Rhinoceros boschi von Koenigswald, 1933 (p. 121) –Java; fossil (late Pliocene).

Rhinoceros sondaicus guthi Guérin, 1973 (p. 19) – Cambodia; fossil (Pleistocene).

47.5   History of the Specific Name *sondaicus* in Genus *Rhinoceros*

As part of a larger *Encyclopédie Méthodique*, the French naturalist Anselme Gaëtan Desmarest (1784–1838) published a list of all known mammals in two volumes dated 1820 and 1822, accompanied by an Atlas with plates. In the genus *Rhinoceros*, he distinguished 4 recent species named *R. indicus*, *R. sondaicus*, *R. africanus* and *R. sumatrensis*. The second species was listed as follows:

617ᵉ. Esp. Rhinocéros des Îles de la Sonde, *rhinoceros sondaïcus*.

(Non figuré.) Espèce nouvelle, découverte par MM. Diard et Duvaucel, envoyée au Muséum d'histoire naturelle en 1821. – *Rhinoceros sondaicus*, Georg. Cuv.

This was a new species, discovered by Diard and Duvaucel, and sent to the Museum of Natural History in Paris in 1821, with locality Sumatra (Desmarest 1822, vol. 2: 399–400, 1827: 362). Although Desmarest spelled *sondaïcus* with a diacritic mark, these are no longer used in zoological nomenclature (ICZN 1999, art. 27), hence the current correct spelling *sondaicus*. Desmarest added a Supplement at the end of volume 2, where he corrected his previous description of *R. sondaicus*, which was actually "found in Java, and not in Sumatra, as we have said by mistake, by Diard and Duvaucel" (Desmarest 1822, vol. 2: 547).

Alfred Duvaucel (1793–1824) and Pierre-Médard Diard (1794–1863) had both traveled out to collect animals for the Museum of Natural History in Paris (Weiler 2019; Dorai & Low 2021; Rookmaaker 2021a). They worked together in India in the vicinity of Kolkata and Chandernagor from May to December 1818. They were engaged by Thomas Stamford Raffles (1781–1826) to assist in his zoological researches in Malaysia, Singapore and Sumatra from December 1818 to March 1820. When the partnership broke up, Diard proceeded to Java, and Duvaucel returned to India, where he visited Sylhet, traveled up the Ganges and was mauled by a rhinoceros in the Rajmahal Hills on 24 January 1823 causing his death the next year (29.6). Among the written sources, Cuvier mentions a manuscript memoir concerning the rhinoceros from Java as well as the double-horned rhino of Sumatra, written by Diard and Duvaucel. This memoir is still available in the library of the Paris Museum (MNHN, MS 625 no. 13). Desmarest's confused original attribution of the type specimen was probably just a miscommunication. A careful examination of the sources shows that Desmarest was mistaken to include Duvaucel in the history of the skeleton from Java.

Desmarest attributed the name *sondaicus* to "Georg. Cuvier", the influential anatomist and taxonomist Georges Cuvier (1769–1832). Although regularly repeated in later literature, Cuvier never actually used this name in his own publications as he preferred *R. javanus* (Hooijer 1946: 34; Sody 1959: 133). In his osteological description of the rhinoceros, Cuvier (1822) listed six rhino specimens available to him (belonging to all recent species) including the skeleton of the animal which lived in Versailles (1770–1793), a skull and a jaw from the collection of Petrus Camper, the skull and a skeleton of a two-horned (African)

species, and as no. 4: "The skeleton of a one-horned rhinoceros of that species [= the species first made known by Camper], from Java, adult, which Mr. Diard, correspondent of our Museum, has recently sent us from that island" (Cuvier 1822, vol. 2: 4). Note that Duvaucel here is not mentioned in connection with the skeleton. Cuvier (1822, vol. 2: 29) is clear about the origin of the skeleton, stating that he received "from Java, a skeleton of a one-horned rhinoceros, collected in the forests by Mr. Diard." An accurate drawing of this skeleton was prepared by Nicolas Huet (fig. 47.1). This was later engraved by Jean-Louis-Denis Coutant (1776–1831) and published by Georges Cuvier in the second revised edition of his authoritative *Recherches sur les Ossemens Fossiles* as "Squelette du Rhinocéros unicorne de Java" (Cuvier 1822, vol. 2, pl. XVII) and repeated in the third edition (Cuvier 1825) and fourth edition (1836) without alterations.

Desmarest (1822) clearly described the external characters of the skin of a young rhinoceros, with a small and rounded horn. He included a set of measurements including the total length (without tail) of 1.76 m and height at shoulder 97 cm. G. Cuvier (1821: 481), in a lecture of 14 May 1821, mentioned that the Museum had just, probably within the previous week, received from Diard a shipment of animals collected in Java. Among 24 skins, 6 specimen in alcohol and 9 skeletons of mammals, there was a new species of single-horned rhinoceros from Java, of which previously the existence had only been rumoured.

The specimen, a young male, of *R. sondaicus* received in May 1821 is still preserved in the Muséum national d'Histoire Naturelle in Paris, correctly as the type specimen or holotype (Paris MNHN 2023). It was listed by De Beaufort (1963: 555) as a skeleton ("animal monté") with number 43–698. There have been various numbers and catalogues over the years. The mounted skin displayed in the exhibits is known in the catalogue of naturalizations no. 42, and in the catalogue of the Galleries no. 698. The skeleton and skull are listed in the Catalogue of Comparative Anatomy no. A7966, and the Catalogue of Ancient Galleries no. BVI-196. The new overall number of this specimen is MNHN-ZM-MO-1981–561 (Cécile Callou, Paris, email of 2023).

Diard had sent the rhino parts from Java, according to Cuvier (1822, vol. 2: 29) "receueilli dans les bois" (collected in the forests). As the place where the animal was obtained would be the type-locality, it is useful to explore if there is any evidence where Diard traveled. On leaving the company of Raffles on 1 April 1820, Duvaucel traveled to Padang and then back to India, while Diard proceeded to Batavia (Jakarta) on Java. Diard embarked on the

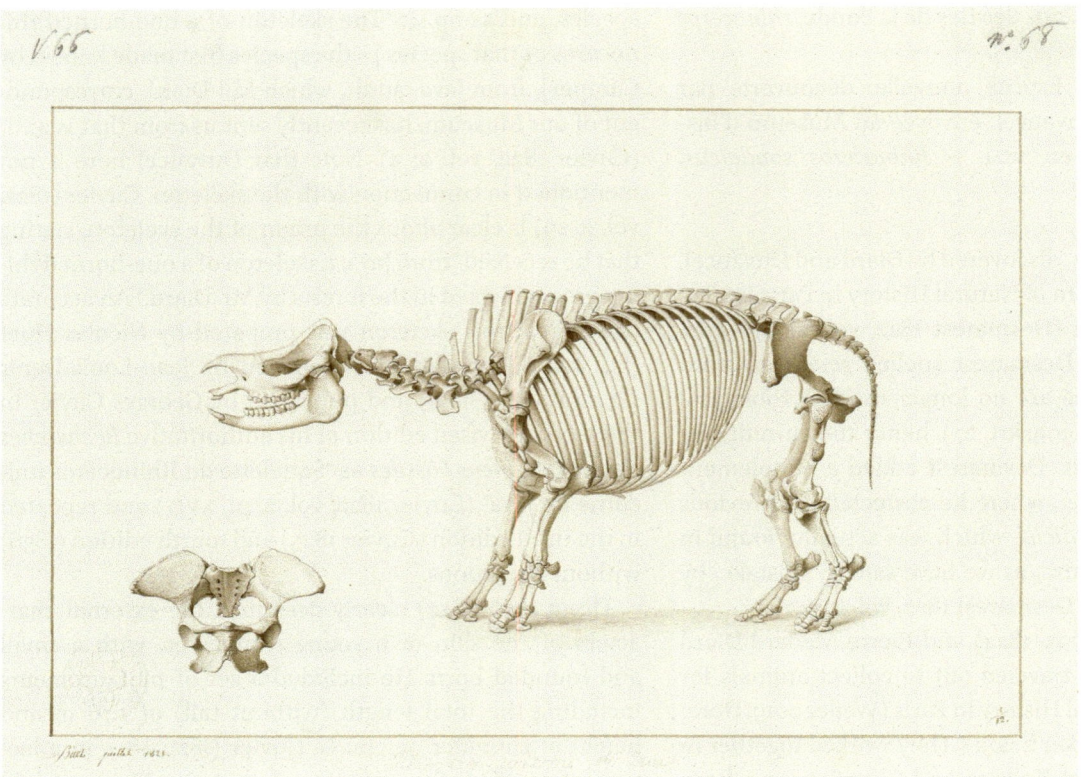

FIGURE 47.1   Skeleton of the one-horned rhinoceros from Java. Type specimen of *R. sondaicus* Desmarest, 1822, sent from Java by Pierre-Médard Diard in 1821. Original drawing by Nicolas Huet (1770–1830), signed lower left "Juillet 1822" (July 1822)

WITH PERMISSION © MUSÉUM NATIONAL D'HISTOIRE NATURELLE, DIST. RMN-GRAND PALAIS / IMAGE DU MNHN, BIBLIOTHÈQUE CENTRALE, PARIS: VÉLINS, ANATOMIE COMPARÉE, PORTEFEUILLE 66 FOLIO 68

Indiana (captain J. Pearl) in Padang on 21 March (a little earlier than other sources suggest) and arrived in Batavia on 7 April 1820 (*Bataviasche Courant* 1820-04-15). He left on 29 April 1821 on the *La Larose* destined for Cochin China and Manilla (*Bataviasche Courant* 1820-05-05). Diard would spend time in Indochina and again in the East Indies studying their natural productions for the rest of his life (Brébion 1914; Peyssonaux 1935). There appear to be no indications where he went in 1820–1821, while it is likely that he did not stray too far from Batavia (Jakarta) and Buitenzorg (Bogor), generally in the western part of Java. Therefore, in line with Rookmaaker (1982), this should be the restricted type-locality of *Rhinoceros sondaicus*.

47.6   History of the Specific Name *javanicus* in Genus *Rhinoceros*

The *Histoire Naturelle des Mammifères* jointly edited by Étienne Geoffroy-Saint Hilaire (1772–1844) and Frédéric Cuvier (1773–1838) was issued in parts 1819–1842. Each part had one or more coloured plates accompanied by descriptive text, which according to the introduction was always written by F. Cuvier. The text on the "Rhinocéros de Java" is dated December 1824, and in the table of contents for volume III the plate is listed as appearing in part 45 and the text in part 46, both published in the same month. At the end of the description, F. Cuvier (1824: 2) stated that "my brother has given to this species the name of *Javanicus*." I believe that despite previous practice, this name should be attributed to F. Cuvier alone.

The engraved plate is entitled "Rhinocéros de Java" and states that it is 1/10th of the natural size (fig. 47.2). F. Cuvier (1824) stated that Pierre-Médard Diard had sent a drawing (now unknown) of the animal made in Java from a young living animal. This is clearly the type specimen. No information is available where it was kept when alive, and it has hitherto been overlooked. The animal recorded and drawn by Thomas Horsfield (1773–1859) in Surakarta 1816–1821 must be a different specimen given the distance from Jakarta (Rookmaaker 1998: 125). *Rhinoceros javanicus* F. Cuvier, 1824 is a junior subjective synonym of *Rhinoceros sondaicus* Desmarest, 1822.

Rhinocéros de Java

FIGURE 47.2
"Rhinocéros de Java." Young specimen observed by Pierre-Médard Diard in Java in 1820–1821. Type-specimen of *Rhinoceros javanicus* F. Cuvier, 1824. Drawn by Jean Charles Werner (1798–1856), engraved by Charles-Philibert de Lasteyrie (1759–1849). Size 50.6 × 33.6 cm FRÉDÉRIC CUVIER, *HISTOIRE NATURELLE DES MAMMIFÈRES*, PART 45, 1824, UNNUMBERED PLATE

47.7 History of the Specific Name *camperis* in Genus *Rhinoceros*

The Animal Kingdom was a translation of G. Cuvier's *Règne Animal*, of which the first four volumes were edited by Edward Griffith (1790–1858), Charles Hamilton Smith (1776–1859) and Edward Pidgeon (1790–1834). This attribution is changed on the title-page of vol. 5 (the Synopsis), which has "Griffith and others." In the description of *R. unicornis*, Griffith (1827: 291) observes that "Camper has described a rhinoceros with two incisors in each jaw, as distinct from this. M. Cuvier thinks it the same species, but M. de Blainville otherwise. He has called it *R. Camperis*." This refers to the classification proposed by Henri Marie Ducrotay de Blainville (1777–1850) in 1817, where he discusses the "Rhinocéros de Camper" without giving a formal scientific name (Blainville 1817; Rookmaaker 1983b, 2021c). Although the reference is indisputably to De Blainville, the text in Griffith is not a direct quote, and for that reason I believe that authorship of *R. camperis* must be attributed to Griffith. *Rhinoceros camperis* Griffith, 1827 is a junior subjective synonym of *Rhinoceros sondaicus* Desmarest, 1822.

47.8 History of the Specific Name *javanus* in Genus *Rhinoceros*

According to Groves & Leslie (2011), this might be an incorrect subsequent spelling of *Rhinoceros javanicus* F. Cuvier,

1824. It is likely that G. Cuvier (1822: 4) based his distinction of this species on all the material which he had listed in his *Ossemens Fossiles* (47.5). This is the first time the Javan Rhino is separated as a distinct species in Cuvier's works, being based on specimens and drawings sent home by Diard in 1821. As Cuvier refers to the plate published by his brother F. Cuvier (1824), the animal depicted (fig. 47.2) can be taken as the type specimen of this name. *Rhinoceros javanus* G. Cuvier, 1829 is a junior subjective synonym of *Rhinoceros sondaicus* Desmarest, 1822.

47.9 History of the Specific Name *camperii* in Genus *Rhinoceros*

William Jardine (1784–1843) used the name *R. camperii* referring to Blainville (1817), but he was not convinced that it was actually a different species (Rookmaaker 1983b, 2021c). *Rhinoceros camperii* Jardine, 1836 is a junior subjective synonym of *Rhinoceros sondaicus* Desmarest, 1822.

47.10 History of the Specific Name *inermis* in Genus *Rhinoceros*

The history of the two type-specimens was elucidated by Rookmaaker (2019d). The French pharmacist and explorer Christoph-Augustin Lamare-Picquot (1785–1873) traveled in the Sundarbans from 2 November to 13 December 1828 (Lamare-Picquot 1835). His party shot two rhinos, mother

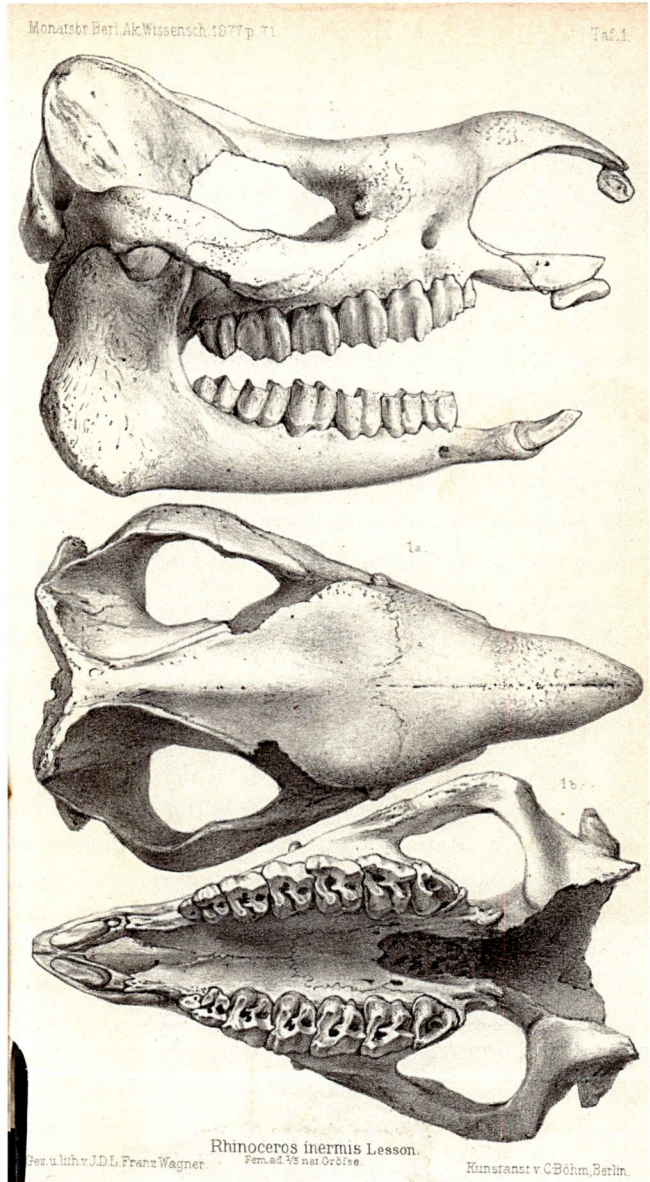

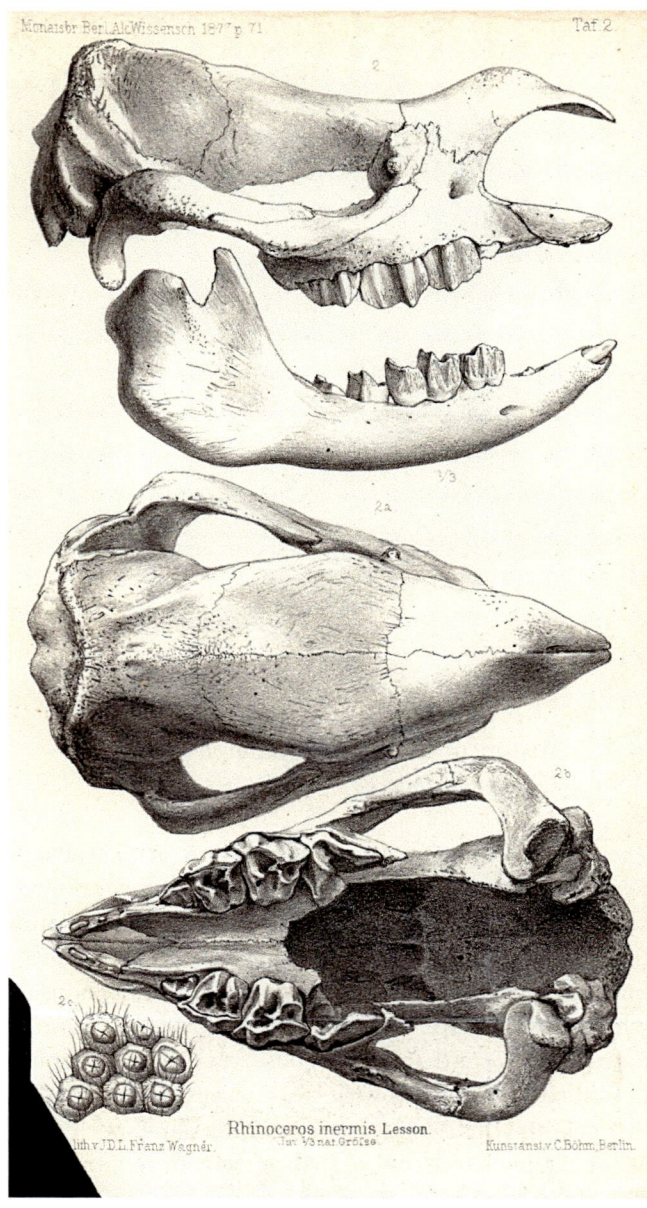

FIGURE 47.3   "Rhinoceros inermis Lesson. Fem. ad. 1/5 nat. Grösse."
Skull from different aspects of the adult female
Rhinoceros inermis collected by Lamare-Picquot in
the Sundarbans in November 1828 and preserved
in the Museum für Naturkunde in Berlin (no.
ZMB_Mam_1957). Plate drawn and engraved by
Johann Daniel Leberecht Franz Wagner (1819–1883),
published by C. Böhm, Berlin
PETERS, *MONATSBERICHTE DER KÖNIGLICHEN
PREUSSISCHEN AKADEMIE DER WISSENSCHAFTEN
ZU BERLIN*, 1877, PL. 1. REPRODUCED BY KIND
PERMISSION OF CAMBRIDGE UNIVERSITY LIBRARY

FIGURE 47.4   "Rhinoceros inermis Lesson. Juv. 1/3 nat." Skull of
juvenile *Rhinoceros inermis* from the Sundarbans
PETERS, *MONATSBERICHTE DER KÖNIGLICHEN
PREUSSISCHEN AKADEMIE DER WISSENSCHAFTEN
ZU BERLIN*, 1877, PL. 2

and calf, on Monday 17 November 1828. Lamare-Picquot
reached Paris with a large ethnographical and zoological collection in 1830. As the two rhino specimens were
both entirely hornless, they were thought to be a new
species or a variety (Anon. 1833). After the publication

of the account of the Indian journey in 1835, the species
was described with a new name *Rhinoceros inermis* by the
French zoologist Réné-Primevère Lesson (1794–1847). As
part of Lamare-Picquot's collection, the rhino specimens
were bought by the Prussian King Friedrich Wilhelm III in
1836 and deposited in the local museums. They were again
examined by Wilhelm Karl Hartwich Peters (1815–1883),
curator of the Zoologische Museum in Berlin (now
Museum für Naturkunde). Peters (1877) upheld the distinction as a separate species *R. inermis* and illustrated the
skulls on two plates (figs. 47.3, 47.4). The mounted hides

FIGURE 47.5
The mother and calf *Rhinoceros inermis*
obtained by Christoph-Augustin
Lamare-Picquot (1785–1873) in 1828 in
the Sundarbans. The specimens were
mounted and displayed in the Museum
für Naturkunde in Berlin
COURTESY: RENATE ANGERMANN, 1997

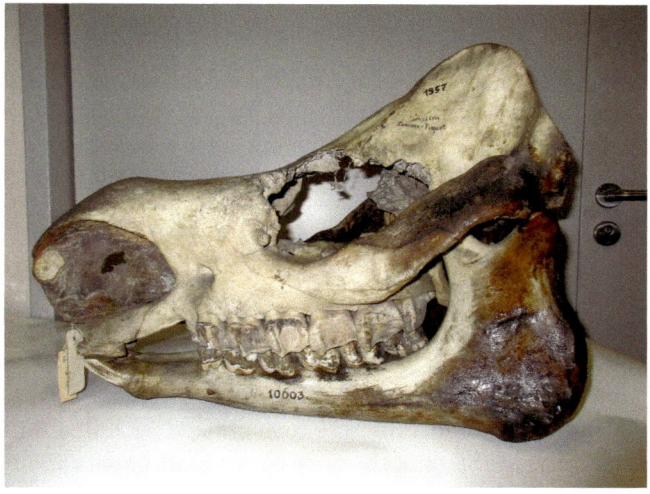

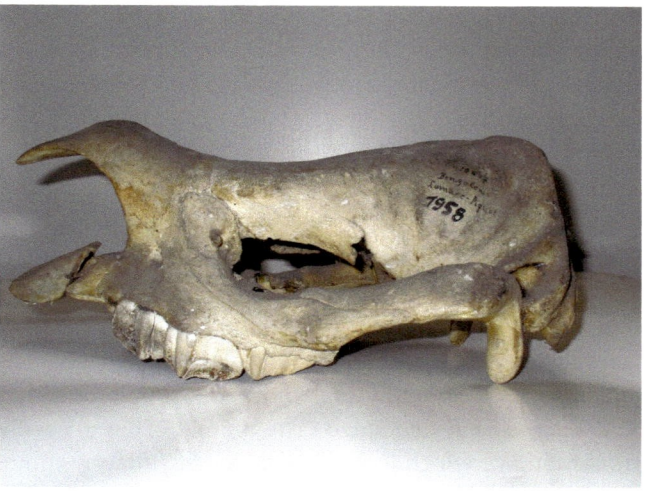

FIGURE 47.6   Skull of the adult female collected by Lamare-Picquot
in the Sundarbans, the type of *Rhinoceros inermis*
Lesson, 1836. Preserved in the Museum für
Naturkunde in Berlin, old number An 10603, new
number ZMB_Mam_1957
PHOTO: JAN ROBOVSKỲ, OCTOBER 2018

FIGURE 47.7   Skull of the young female collected by Lamare
Picquot in the Sundarbans, paralectotype of
Rhinoceros inermis Lesson, 1836. Preserved in the
Museum für Naturkunde in Berlin, old number
An 10602, new number ZMB_Mam_1958
PHOTO: JAN ROBOVSKỲ, 2018

and skulls of both female specimens are still in the Berlin Museum (Rookmaaker 1997a, 2019d) (figs. 47.5, 47.6, 47.7 and paragraph 51.5, figs. 51.7, 51.8, 51.9).

The name *R. inermis* has had a checkered history and there has been confusion about the date of first publication. Gray (1843) listed it as a synonym of *R. unicornis* stating that the name was published by Lesson (1842: 159) in his *Nouveau Tableau du Règne Animal*. Peters (1877) found the name in a Supplement to Buffon's *Histoire Naturelle* edited by Lesson with title *Races Humaines et Mammifères* (2nd edition, vol. 1), dated 1838 on the title-page but probably also reprinted in 1840 and 1848 (Rookmaaker 2019d).

However, there also was an earlier series of *Complément des Oeuvres de Buffon*, equally edited by Lesson, with species added since the death of Buffon, of which volume 10 was definitely available in 1836 (Lesson 1836). This volume had a separate title *Histoire naturelle générale et particuliere des mammifères et oiseaux découverts depuis la mort de Buffon: Oiseaux et mammifères*, and was published by Pourrat Frères, Editeurs, Rue des Petits-Augustins, 5 and Roret, Libraire, Rue Hautefeuille, Paris. This set of the *Complément* consisted of 10 volumes published from 1828 to 1837. The sequence of appearance was irregular. Fortunately, a book review of volumes 5 and 10 in

Germany of October 1836 ensures their date of publication. Anon. (1836) states that there was a gap between volumes 1, 2, 3, 4, and 6 dating 1828–1830, probably due to the political unrest in France, and these volumes 5 and 10 were issued together in 1836. The reviewer mentions text about the rhinoceros, which again strengthens the evidence (Rookmaaker 2019d).

When treating the rhinoceros family in the body of the text, Lesson (1836: 238) mentioned that he had previously forgotten to include an unpublished species discovered by Lamare-Picquot, but he did not add a scientific name. In an appendix ("Additions") in the same volume, Lesson included a full four-page description of the Hornless Rhinoceros or Gaindar ("Rhinocéros sans cornes ou Gaindar"), named *Rhinoceros inermis* (Lesson 1836: 399–402). The French text of this new description is reproduced in Rookmaaker (2019d: 6).

Since Groves (1967), the name is used in the combination *Rhinoceros sondaicus inermis* Lesson, 1836 for the subspecies of the Javan Rhinoceros on the South-East Asian mainland from Malaysia to Bangladesh and India.

47.11   The Life of Jamrach's Rhinoceros of 1874

Rhinoceros jamrachi or Jamrach's Rhinoceros is based on the characteristics of a unique type specimen. The history and taxonomy of this particular animal have proved frustratingly confused and inconclusive. Unless it really was a distinct species, being single-horned it should be either *R. unicornis* or *R. sondaicus*. The animal came allegedly from Manipur which is an unusual locality for any rhino; it lived many years in the Berlin Zoo where arrival, departure or death remained unrecorded; the remains were not preserved; and a small number of drawings were executed at such angles that the specific characteristics cannot be seen. The animal dealer William Jamrach imported the animal from India in 1874 and himself convinced that it was a new species, he named it *Rhinoceros Jamrachi*, even that in such a way that there is discussion if the name is available. Rather hoping to solve this mystery, I have spent time researching the animal's history, related art work and proposed classification.

William Jamrach (1842–1923) was part of a family concern trading in all types of wildlife from their base in London (H. 1894; Larsson 2021; Reichenbach 2022). William was often mentioned in relation to the business in India, centered in the port of Kolkata, working with agents as well as paying personal visits. Jamrach collaborated with other firms, like that of Charles Rice (1841–1879)

FIGURE 47.8   Photograph taken in London in the 1870s, showing from left to right: Charles Rice (1841–1879), Carl Hagenbeck (1844–1913), Clarence Bartlett (1848–1903) and William Jamrach (1842–1923)
WITH PERMISSION © ARCHIV CARL HAGENBECK, HAMBURG. COURTESY: HERMAN REICHENBACH

in London and Carl Hagenbeck (1844–1913) in Hamburg (Rookmaaker 2014b: 231). These important animal dealers were photographed together in the 1870s (fig. 47.8).

Jamrach's Rhinoceros was first noticed in Kolkata in early 1874. At the time there was a series of papers about the rhinoceros in the *Oriental Sporting Magazine* written by Young Nimrod, who contributed regularly about Indian wildlife. This pseudonym definitely resolves to H. James Rainey, a planter in Khulna: "H. James Rainey, manages with his brother J. Rudd Rainey, the Khulna Estate south of Calcutta. H.J. Rainey, in his character as Shikari and naturalist is well and popularly known throughout India under his nom de plume of 'Young Nimrod' and has amply justified his sporting fame by slaying many tigers on foot, instead of from the comparatively safe retreat of a howdah" (Anon. 1895). Rainey knew the rhino living in the Sundarbans from personal experience in the field, and first established that this was the same species as the one described from Java, hence *R. sondaicus* (Rainey 1872a). He continued in a second paper to explain an easy method to distinguish *R. sondaicus* from *R. unicornis*, based on the "projecting and conspicuous scales on the skin" found only in *R. sondaicus* (Rainey 1874b).

The presence of two rhinos captive in Kolkata was first mentioned in a footnote added by the editors of the *Oriental Sporting Magazine* (OSM) to a paper by Rainey (Young Nimrod) in May 1874, stating that they had seen "two of the tubercled rhinoceros – as we suppose – at Mr. Routledge's to the East of the Circular Road, Calcutta, not long ago" (Rainey 1874a: 240). The two animals had arrived a few weeks before May 1874, hence in late March or early April 1874. William Rutledge (1839–1905) was a well-known animal dealer in Kolkata, based then at 32 Jaun Bazar Street and Entally (5.4). He obtained several rhinos from the Sundarbans in the 1870s, and those were certainly indicated by the name "tubercled rhinoceros." There was a note in the OSM of July 1874 that one of these rhinos had been shipped to England (Rainey 1874c). However, this was a mistake, corrected by Rainey (1874d) in September 1874, because in fact both rhinos were taken on board. One of them, larger (Jamrach 1875) and showing more prominent tubercles died during the journey and was cast overboard (Rainey 1874c).

Jamrach (1875) stated that he received a telegram from Kolkata informing that two rhinos had been shipped in early April 1874, but this date probably written from memory was slightly off. The two rhinos were shipped on board the *Sultan*, a single-screw steamer built for Messrs. Green and Company in 1873. It left Kolkata on 30 April 1874 (*Homeward Mail* 1874-06-01) and arrived at Gravesend near London on 16 June 1874 (*London Daily News* 1874-06-16). Several English newspapers noted the arrival of the *Sultan* with a "perfect cargo of wild animals including a rhinoceros" because a leopard had escaped and had to be killed on board (e.g. *Shipping and Mercantile Gazette* 1874-06-20).

When the remaining young female rhino arrived in London, she was "very seedy, but was, after a deal of trouble, restored to health" (Rainey 1874d). There is no actual record where she was kept in London, but Jamrach probably used premises of his partner Charles Rice or his brother Anton Jamrach Jr. (1841–1885). She might have stayed only a few days before being shipped again to their partner Hagenbeck in Hamburg, where a drawing (unknown) was made (Sclater 1876a: 650).

The rhino was then transported to Berlin, where she entered the Zoological Gardens on the morning of Tuesday 30 June 1874. In a notice repeated on the same date in Berlin Zoo (1874), the local newspaper *National Zeitung* (1874-07-02), was for once rather explicit: "The Elephant House received on Tuesday morning a new resident, which is a Javan Rhinoceros which is a different species than the Indian Rhinoceros. The animal is about 1½ years old and was sold by the London animal dealer Jamrach for such a good price that the opportunity could not be missed even though there were already other rhinos in the park" (original: "Das Elephantenhaus hat am Dienstag [30. Juni] Morgens einen neuen Bewohner bekommen, nämlich das von dem indischen Nashorn spezifisch verschiedene javanische Nashorn. Es its ungefähr anderthalb Jahre alt und wurde von dem Londoner Thierhändler Jamrach zu so günstigen Bedingungen offerirt, daß der Ankauf trotz der vorhandenen Vertretung dieser Gattung nicht aus der Hand gelassen werden durfte.") The same newspaper reported on 7 July 1874 that the acquisition of a young Javan Rhino had been completed.

The date of arrival on 30 June 1874 is earlier than has hitherto been expected. It had been said that the rhino had been bought cheaply during the regular animal auction organized by the Antwerp Zoo in September 1874 (Anon. 1878b: 201, followed by Schlawe 1969: 28). Details about these auctions are hard to find, which is disconcerting as they were the foremost events in the zoo world of the time to exchange and procure stock for the European zoos. A poster associated with the auction of 1898 was reproduced by Rookmaaker (1998a: 136). A feature article was edited by Du Bosch (1893) with illustrations by F. de Harnen (fig. 47.9). It is unlikely that Jamrach was still soliciting buyers for this rhino at the sale of 1874.

The Berlin Zoo was expanding quickly from 1869 under the directorship of the zoologist and physician Heinrich Bodinus (1814–1884). Already the Zoo was exhibiting a pair of *R. unicornis* from 19 September 1872, as well as a female *Diceros bicornis* from 6 July 1870, all transferred to the new Elephant House when that was completed on 12 November 1873 (Berlin Zoo 1873; Klös 1969: 68). It appears that the female obtained from Jamrach was always kept in a separate stall. Certainly at first, she was displayed as a Javan Rhino. In the Guidebook of 1874, Bodinus (1874) listed both species, *R. unicornis* being represented by a male and female which were still growing, and *R. sondaicus* (with locality Java) said to be smaller.

Jamrach (1875) saw the female rhino in Berlin at the end of September 1875. Sclater (1880) discussed her after a visit in 1879. In his detailed description of the inhabitants of Berlin's Elephant House, August Woldt (1882) recalled a fight in June 1881 between the male *R. unicornis* and Jamrach's female (well identified and definitely not the female of 1872), also mentioned in a newspaper (Berlin Zoo 1881): "the male suddenly got angry, attacked the female and harassed her so much, that they could only just be separated" ("Das Männchen wurde plötzlich böse, überfiel das Weibchen und setzte ihm derart zu, daß

FIGURE 47.9   An animal auction at the Zoological Gardens of Antwerp. Engraving signed by F. de Harnen
L'ILLUSTRATION, JOURNAL UNIVERSEL, 10 SEPTEMBER 1893, P. 232

es nur mit genauer Noth seiner Wuth entzogen werden konnte"). One wonders, were these animals allowed together despite being different species? Jamrach's rhino was sketched at that time by Mützel (below). Bruno Gaebler (1883: 243) saw her in December 1882. Philipp Leopold Martin (1882, vol. 1: 574) mentions her presence. A Guidebook of the zoo, published after 1881, has the same text as in 1874 (Bodinus 1884). Schlawe (1969: 28) refers to a report of 1884 by a Mr. Bau (unidentified) about the presence of this rhino, which might refer to the Guidebook. The Director of the Berlin Zoo from 1888, Ludwig Heck (1860–1951) stated in his recollections that the zoo had a Javan rhino in the time of Bodinus (Heck 1940: 231), maybe implying that the animal was no longer there in his time. The records of the Berlin Zoo for the 1880s are no longer present. The death of the Javan rhino cannot be traced, despite the fact that very few Javan rhinos have ever been in captivity.

It has generally been suggested that Jamrach's Rhinoceros died around 1884 (Reynolds 1961: 23; Schlawe 1969: 28), because there are no later reports

about her presence in the Zoo. There is another possibility which has been suggested by my friend, the meticulous zoo-historian Lothar Schlawe. Because Berlin had two female one-horned rhinos at the time, there might have been some confusion. The Hagenbeck firm has an archival record that on 5 April 1883 the Berlin director Bodinus exchanged with Hagenbeck "1 w. ind. Rhinoceros" [one female Indian Rhino]. Although Hagenbeck himself was well aware of the difference between the species, this transfer could have been Jamrach's Rhino, but then identified as *R. unicornis*. Unfortunately, there is no trace of that animal's destination, therefore more likely to be in the USA than in Europe.

The remains of Jamrach's Rhinoceros are considered lost. Around 1930 the mammalogist Hermann Pohle (1892–1982) was unable to find them in the Zoological Museum of Berlin. There is a mounted specimen of *R. son-daicus* in the Museum of Comparative Zoology of Harvard University, Cambridge, Mass., which according to Barbour & Allen (1932: 147) might have come from Berlin via the Natural Science Establishment of Henry Augustus Ward

FIGURE 47.10   Drawing by Paul Meyerheim (1842–1915) of a pair of
              R. unicornis in Berlin Zoo around 1872. Collection
              Martin Sperlich (1919–2003), Berlin
              WITH PERMISSION © SCHLOSS
              CHARLOTTENBURG, STIFTUNG PREUSSISCHE
              SCHLÖSSER UND GÄRTEN, BERLIN-BRANDENBURG

(1830–1906). However, this specimen with a very small horn looks too young to be identified with the Berlin animal.

In view of the extended presence of the three one-horned rhinos in Berlin – the male *R. unicornis* for 37 years, the female *R. unicornis* for 24 years, and Jamrach's female for at least ten years – the small number of known depictions or photographs is perhaps remarkable. The pair of *R. unicornis* was drawn in pencil by the illustrator Paul Meyerheim (1842–1915) soon after their arrival in 1872 (Heikamp 1980, fig. 23) (fig. 47.10). The same artist depicted a male and female engaged in a fight in an engraving of 1881 (Meyerheim 1881; Schlawe 1994) (fig. 47.11). An impression of the interior of the new Elephant House in Berlin by Meyerheim of 1874 shows a single adult *R. unicornis* in one of the stalls (E.L. & Meyerheim 1874: 421) (figs. 47.12, 47.13). A notice by Friedrich Lichterfeld (1803–1878) is illustrated with a vignette of the head of a *R. unicornis* (Lichterfeld 1874) (fig. 47.14). Two photographs of the male were published by Heck (1899), one showing him with a massive horn, the other after the horn was shed, which occurred in 1882 and again in 1891 (Wunderlich 1892). Even rarer are depictions of Jamrach's rhino. In fact, sadly, there is only one, where she is drawn next to the black rhino 'Molly' as part of scenes in the Elephant House by the Berlin artist Gustav Mützel (1839–1893) illustrating the description by Woldt in the *Gartenlaube* of December 1882 (fig. 47.15). Mützel was definitely an experienced animal painter contributing extensively to the 1877 edition of Brehm's animal encyclopedia (Wilke 2018: 85). However, his depiction of Jamrach's rhino is too artistic to show the animal's characteristics fully, seen in an unnatural position from the back. The typical saddle formed by the neck

FIGURE 47.11
Fight between a male and female
R. unicornis in Berlin Zoo engraved by Paul
Meyerheim in 1881. The same depiction was
included in the Barcelona magazine *El Mundo
Ilustrado*, 1881, No. 122
ILLUSTRIRTE ZEITUNG, LEIPZIG,
6 AUGUST 1881, NO. 1988, P. 119

FIGURE 47.12   "Die Dickhäuter im zoologischen Garten zu Berlin."
Interior of the Pachyderm House in Berlin Zoo, drawn
by Paul Meyerheim in 1873 soon after it was opened
to the public. There is an adult single-horned rhino in
the stall on the left. The building was destroyed on
22 November 1943. Engraving 27 × 35.5 cm
*ÜBER LAND UND MEER, ALLGEMEINE ILLUSTRIRTE
ZEITUNG*, 1874, VOL. 31, NO. 22, P. 421

FIGURE 47.13   The Pachyderm House in the Zoo of Berlin in a
slightly different setting. This lithographed plate
was part of a series of images designed for use in the
popular Guckkasten (Peep-box) issued by Johann
Christian Winckelmann (1766–1845) under the name
Winckelmann und Söhne in Berlin from the 1860s.
Size 36 × 43 cm
WITH PERMISSION © SAMMLUNG STIFTUNG
STADTMUSEUM BERLIN: NO. IV 61/1524 S A

FIGURE 47.14
"Die Dickhäuter des Berliner Zoologischen
Gartens und ihr neuer Palast." Front view
of *R. unicornis* and head of *Diceros bicornis*,
drawn as part of a full-page illustration by
Paul Meyerheim in the Berlin Zoo in 1874,
illustrating a notice by Friedrich Lichterfeld
(1803–1878). The Zoo then exhibited a pair
of *R. unicornis* since 19 September 1872, as
well as a female black rhino *Diceros bicornis*
from 6 July 1870, all transferred to the new
Elephant House when that was completed
on 12 November 1873
ILLUSTRIRTE ZEITUNG, 15 AUGUST 1874,
NO. 1642, P. 137

folds in *R. sondaicus* doesn't show, but that rather begs the question if it was lost in artistic translation.

47.12   History of the Specific Name *jamrachi* in Genus *Rhinoceros*

The type of *Rhinoceros jamrachi* was bought by the Berlin Zoo in 1874. It doesn't appear that the Berlin authorities were particularly interested in the taxonomy of the animal which they bought from Jamrach in 1874. Locally there was no debate, maybe indicating that the difference

from *R. unicornis* exhibited next to her was just obvious. However, when she was young, soon after capture, she caused quite a bit of debate in her home country, still lingering in some ways today. It is best to follow the arguments chronologically.

March–April 1874 (Kolkata). The two rhinos arrived in the quarters of the dealer Rutledge a few weeks before May 1874 (note to Rainey 1874a), therefore in late March or early April 1874. In that period, the naturalists in Kolkata were better acquainted with *R. sondaicus* living not too far away in the Sundarbans than with *R. unicornis* living further north.

Das Elephantenhaus in dem zoologischen Garten zu Berlin.
Originalzeichnung von G. Mützel.

FIGURE 47.15   "Das Elephantenhaus in dem zoologischen Garten zu Berlin." Engraving by Gustav Mützel (1839–1893) with the black rhino Molly on the left pointing her horn at a female one-horned rhinoceros in the Berlin Zoo. Produced in the workshop of Wilhelm Aarland (1822–1906) by Max Bach (1841–1914). Size 29 × 20 cm
GARTENLAUBE, DECEMBER 1882, NO. 52, P. 461. COLLECTION KEES ROOKMAAKER

April 1874 (Kolkata). Jamrach's agent in Kolkata (probably Rutledge) and Fraser believed that two rhinos about to be shipped to London were species hitherto undescribed (Jamrach 1875). Oscar Louis Fraser (1848–1894) worked as a naturalist at the Indian Museum in Kolkata from 1871 (51.12). Commenting on rhino specimens from the Sundarbans (Fraser 1875), he wanted to differentiate between rhino species using the presence or arrangement of tubercles on the skins (Rainey 1874a). Rainey (1874b, 1877: 263) believed that the two Kolkata specimens were definitely different from the Sundarbans rhino (*R. sondaicus*), because in the latter the tubercles were always fully formed, even in the foetal stage.

May 1874 (Kolkata). Rainey (1874b) must have discussed the characteristics of the one-horned rhinos with Fraser, because he added that there were good indications that besides *R. unicornis* and *R. sondaicus* there might well be one, maybe two, additional species of rhino, "but do not consider myself at present warranted in making public." Doubtless this remark was based on his examination of the two rhinos waiting for transport at Rutledge's premises.

July 1874 (Kolkata). The editors of the OSM mentioned that one of Rutledge's rhinos had been shipped to England, which might well allow the naturalists in London to verify that this tubercled rhino is an "altogether new and distinct species" (Rainey 1874d: 340). Later sources show that in fact both rhinos were shipped.

June 1874 (London). Being alerted by his Indian contacts, Jamrach (1875) examined the rhino carefully on arrival (16 June 1874), and found "the surface of the skin being tuberculated, the head rather long, and behind the usual single fold, as with *Rh. Indicus*, I noticed the appearance of an additional one." Hoping for expert confirmation, he first asked a second opinion from Abraham Dee Bartlett (1812–1897), the Superintendent at the London Zoo, who thought that it was a young *R. sondaicus*. At the time, London Zoo exhibited a male *R. unicornis* (25 July 1864 to 12 December 1904) as well as a young male *R. sondaicus* from Java received on 7 March 1874, which lived until 1885.

June–July 1874 (London). Not content with Bartlett's identification, Jamrach persevered, and the next day the rhino was examined by other naturalists (Jamrach 1875), identified as Sclater and Garrod (editorial comment in Rainey 1875; Sclater 1876a: 650). Philip Lutley Sclater (1829–1913) was the influential Secretary of the Zoological Society of London (ZSL) from 1859, a keen taxonomist and able administrator, who submitted a definitive monograph on the Zoo's rhinos on 15 June 1875 (published in the *Transactions* in December 1876). Alfred Henry Garrod (1846–1879) was the Zoo's Prosector from 1872 to

1880. At first Sclater identified the animal as *R. unicornis* (Rainey 1875), but on second thought "on being shown the differences" he agreed with Garrod and Bartlett that it was "a young *R. sondaicus*, although it appeared to have a rather squarer, shorter upper lip than is usual in that species" (Sclater 1876a: 650). No word about tubercles, and the shape of a lip seems a rather minute difference between species.

July 1874 (London). After listening to Sclater and Garrod, Jamrach again told Bartlett that the animal was a new species, but the latter did not agree (Jamrach 1875). Jamrach next wrote a letter to *The Field*, which was then the foremost journal for naturalists with contributions from all over the world. In this letter he proposed to name the animal "Rhinoceros Jamrachi", because he was the one who had made it known to science (Jamrach 1875). Although the letter was rejected and never published, Jamrach also mentioned his discovery in letters to correspondents. This was the reason why this new name appeared twice in print before it was actually published by Jamrach in the pamphlet of October 1875. First, in the OSM issue of September 1874, Rainey (1874d) stated that "afterwards Mr. Jamrach described it, I hear, in the columns of the *Field*, as *R. Jamrachii*, and sold it to some one in Berlin." Also, the editors of *The Field* added a footnote to Rainey's (1875) contribution published on 6 March 1875 that Jamrach had named the animal himself as "*Rhinoceros Jamrachii.*" Note that at this time, the specific name was spelled with double -ii at the end. As a mention of a vernacular name, locality or specimen only does not constitute an indication in the sense of the ICZN (1999, art. 12.3), these two uses of "Rhinoceros jamrachii" are invalid and have no standing in zoological nomenclature.

June 1874 (Hamburg). While in transit in Carl Hagenbeck's premises in Hamburg, a drawing was made, which Sclater (1876a) was to "exhibit" in the ZSL meeting of June 1875 when his paper was read. He did not reproduce it, and this evidence is now lost (no record was found during my search in the ZSL library). Apparently, the audience at the time did not greatly object to it being *R. sondaicus*.

June 1874 (Berlin). The female rhino arrived in Berlin on 30 June 1874. Bodinus (1874) wrote in the Guidebook of the Berlin Zoo that Jamrach's rhino differed from *R. unicornis* in having a lower skull, longer upper lip, and especially by the presence of small tubercles arranged like a mosaic on all the skin, from which small hairs are growing.

September 1875 (Berlin). Jamrach (1875) visited Berlin at the end of September and found "the animal in excellent condition with the posterior fold fully developed. It was placed in juxta-position to the two *Rh. Indicus*, and the most thick-headed schoolboy would, at a glance, now be able to discern the difference between the different species."

8 October 1875. Jamrach was still convinced that the female in Berlin was a new species, despite the advice of authorities like Sclater, Garrod and Buckland. He published his opinion himself as a pamphlet of 3 pages on green-coloured paper, dated 8 October 1875. In his description of *R. jamrachi*, Jamrach (1875) gave two features in which it differed from *R. unicornis*: (1) folds on knee joints very distinct, other folds similar with an additional one at the nape of the neck; (2) head much longer, and laterally more compressed. In comparison with *R. sondaicus* (from memory), this female had (3) a square lower lip; (4) skin covered with bosses, varying in size from a sixpenny piece to a shilling, and are all detached, not joining as in *R. sondaicus*; (5) and the space between the anterior and posterior folds quite smooth. The animal lived in "Mooneypoor and probably Upper Burma."

January 1876 (Calcutta). The text of Jamrach's pamphlet of 1875 was reprinted without alteration in the January 1876 issue of the OSM (Jamrach 1876).

1875–1877 (Berlin). Possibly occasioned by the receipt of Jamrach's pamphlet in October 1875, Sclater tried to find a solution. He pursued this in the course of a series of letters to Wilhelm Karl Hartwich Peters (1815–1883), curator of the Zoological Museum in Berlin. Peters (1877) was also working on a paper describing two hornless female specimens obtained in the Sundarbans by Lamare-Picquot (49.9), which when published on 19 February 1877 confirmed their separation as the species (not subspecies) *Rhinoceros inermis*. While no letters to Sclater are preserved in the ZSL library, it is fortunate that Sclater's own correspondence is preserved in the archives of the Zoological Museum of Berlin (Z.M. Archiv, Abteilung II). I have extracted the entries relevant to Jamrach's rhino, as follows.

Sclater to Peters, 12 November 1875 – Is the type of "Rhinoceros jamrachii, Jamrach" still alive at Berlin and what do you think of it?

Sclater to Peters, 9 December 1875 – I have written to Dr. Bodinus about the Rhinoceros jamrachii as Mr. Jamrach calls it. I wish to have a small photograph of this beast if possible.

Sclater to Peters, 14 December 1875 – Is the Rhinoceros living in the Berlin Zoological Gardens (which Jamrach has named after himself, R. jamrachi) = R. unicornis, = or *R. sondaicus*, or distinct from both, that is the question.

Sclater to Peters, 30 March 1876 – I am still not quite convinced that Rhinoceros jamrachii (in Berlin Zool. Gard.) is really = *R. sondaicus*. According to my recollection (and a drawing now before me) *R. jamrachii* has a short square upper lip – not pointed as in *R. sondaicus*. How is this? Perhaps you would see when you next go to the Gardens.

Sclater to Peters, 12 April 1876 – Thanks many for your sketches of the snout of *Rhinoceros jamrachii*. I will compare them with the beast, but I am still a little doubtful. If I could see my way to a spare week I would run over to Berlin and see it myself.

Sclater to Peters, 14 June 1876 – I have never replied to your letter of April about *Rhinoceros sondaicus*, but now thank you for it and the sketches.

Sclater to Peters, 2 January 1877 – Your note about the Rhinoceros [= Peters 1877] is of great interest to me, although my little memoir to accompany the figures of Wolf is already printed off [= Sclater 1876]. I have recently obtained information that in the Rhinoceros of the Sunderbans the female has no horn. There can be little doubt therefore that *R. jamrachii* which you have in your Gardens is the male of *R. inermis*. I had always thought it singular that the *R. sondaicus* should extend into the Sunderbans of Bengal. Now it appears from your observations that the species [*sondaicus* of Java, and *inermis* of Sundarbans] are really distinguishable.

Sclater to Peters, 11 May 1877 – I have received with great interest your article on *Rh. inermis*, but cannot say that I am satisfied as to the specific difference of it and *Rh. Sondaicus*. I have no doubt our *Rh. sondaicus* was really obtained in Java, but whether the so-called *jamrachii* is really from Munipore is quite doubtful – that is what Jamrach told me. A short time ago Jamrach had another *Rh. Sondaicus* from Calcutta for sale, but I could not examine it well inside its den.

The drawings, even possibly the photograph, sent from Berlin to Sclater are now unknown (not found in the ZSL archives). Of course, Peters's use of *R. inermis* for the Sundarbans rhino confuses the arguments somewhat, since today we think of it at most as a subspecies of *R. sondaicus*. But note that on 2 January 1877 Sclater seemed to conclude that Jamrach's rhino was a male, not a female, probably because the drawings might have shown a horn on the nose. Peters had confirmed reports from India that the females of the Sundarbans rhinos were hornless. On the basis of this correspondence starting in late 1875, it is likely that the paragraph and footnote (with the wrong date 1874 for Jamrach's pamphlet) was inserted just before Sclater's paper was published in December 1876.

1877 (Berlin). In his comparisons of the two specimens of *R. inermis* from the Sundarbans, Peters (1877: 70) mentioned Jamrach's rhino (as a female) with a well developed horn ("mit wohl entwickeltem Horn").

1879 (Berlin). When Sclater (1880) visited the Berlin Zoo again in 1879, he saw Jamrach's rhino "now quite adult". He changed his mind and now thought that "the animal was much too large for *R. sondaicus* and did not show the peculiar shoulder-fold that characterizes that species. He believed it to be merely *R. unicornis*."

1882 (Berlin). August Woldt (1840–1890), although not a zoologist, said that the female of 1874 differed from *R. unicornis* and should maybe be regarded as a variety of that species – not of *R. sondaicus* (Woldt 1882). The skin had far more tubercles, while on the shoulder the folds form a saddle ("besitzt es auf der Schulter eine Faltenbildung, welche einen Sattel darstellt"), and it also had a shorter and somewhat thicker skull. The engraving illustrating Woldt's text in the *Gartenlaube* of December 1882 shows the female black rhino Molly (in Berlin from 6 July 1870) in an aggressive position, and the female of 1874 in a submissive attitude which she took when she was attacked by the male. Although Woldt's story definitely identifies the animal as this female, it is strange that she would have been in one enclosure with the male of another species, hence maybe the engraving in fact shows the female *R. unicornis* which had been imported together with the male.

William Jamrach disregarded the taxonomic conventions of his day. He named the new species after himself, and he publicized his view merely in a leaflet rather than a solid paper in a journal. This has led to the recent suggestion that this leaflet does not constitute a 'publication' in the sense of the Code of Zoological Nomenclature, which consequently would make the name invalid (Groves 1967: 234, 2003; Corbet & Hill 1992: 242). As one of the few persons who has ever handled a copy of the leaflet (the only one known, in the library of the Museum of Natural History in Paris), I disagree. In my view, it was printed in multiple copies, it was obtainable probably free of charge from the author, and it was meant to be a permanent scientific record, thereby satisfying the criteria of publication set by the Code (ICZN 1999, art. 8). Sclater (1876a: 650), the only one to refer to the existence of the pamphlet, said that Jamrach "in October 1874 printed an account of the supposed new species on a sheet of green paper, and proposed to call it R. jamrachii!" Sclater's date 1874 is incorrect, his use of *R. jamrachii* (with double i) is an incorrect

subsequent spelling, but he did accept it as published. It may be added that Jamrach's entire text was reprinted in the January 1876 issue of the *Oriental Sporting Magazine*, which invalidates the arguments against the 1875 pamphlet. *R. jamrachi* is unlikely to be a name which will ever be resurrected, still in my view it should be dated to 1875 as proposed by William Jamrach in his pamphlet.

The type locality was stated to be Mooneypoor, which must be a rather awkward spelling of the state of Manipur. As such it is an unexpected locality, because rhinos of any species have rarely been reported from that remote and unexplored region (61.1). Jamrach gave no details about who sold the animal to his agents in India. I am unaware if there is a contemporary source stating that this rhino was captured in the Sundarbans, as was mentioned by Sody (1941: 84), Schlawe (1969: 28) and Anhalt (2008: 274). The early reports written about the tubercled rhinoceros in Kolkata before shipment (in Rainey 1874b,d, 1875a), even though contorted through discussions about possible new species, always compare the animal to *R. sondaicus*, the only species known in the Sundarbans.

A few careful taxonomists have listed *R. jamrachi* or *R. jamrachii* in lists of synonyms, with the name credited to Sclater's paper of 1876 rather than Jamrach's pamphlet (Laurie et al. 1983; Corbet & Hill 1992; Groves 2003). Even though Sclater did not believe in the name, his use of the name may be available, but of course it postdates the original by Jamrach (1875 or 1876). Jamrach's name was treated both as a synonym of *R. sondaicus* or of *R. unicornis*.[1]

It appears to be impossible to solve the identity of Jamrach's rhinoceros exhibited in Berlin from 1874 with full regard of all descriptions and opinions (Rookmaaker 2021c). As it definitely wasn't a separate species, the animal was either *R. unicornis* or *R. sondaicus*. If it really came from Mooneypoor (Manipur) as asserted by Jamrach, it was *R. unicornis*. If it actually came from the Sundarbans as the early history in Kolkata suggests, it was *R. sondaicus*. The presence of a horn in this female points at *R. unicornis*, the tubercled skin and neck folds forming a triangle are characteristic of *R. sondaicus*. The 1882 engraving by Mützel shows a small horn but no neck

---

1 Jamrach's name was treated as a synonym of *R. sondaicus* by Sclater (1876a: 650); Sody (1941: 84); Mukherjee (1966: 27, 1982: 54). Most taxonomists and some historians favour it as a synonym of *R. unicornis*: Reynolds (1961: 23); Klös (1969: 72); Rookmaaker (1977, 1983b: 45, 1983c); Laurie et al. (1983); Corbet & Hill (1992: 242); Choudhury (2013: 225, 2016: 154); Groves (1967: 23, 2003: 345); Groves & Guérin (1980: 206); Groves & Grubb (2011); Srinivasulu & Srinivasulu (2012: 351); Heard (2020: 92).

shield, which could be artistic license. In my own previous work (Rookmaaker 1977, 1983c, 1998a), I have advocated it to be *R. unicornis*, based on Mützel's print as well as the stated origin. However, a careful analysis of the reports by Rainey (Young Nimrod) and others while the animal was still in Kolkata to me now unequivocally point at an origin in the Sundarbans and hence *R. sondaicus*. This was

the identification when she was in transit in London and probably while she was exhibited in Berlin.

For these reasons, Jamrach's Rhinoceros – the female rhino imported in 1874 and exhibited in Berlin Zoo – was a specimen of *R. sondaicus*, and *Rhinoceros jamrachi* Jamrach, 1875 is a junior subjective synonym of *R. sondaicus*.

The Javan Rhinoceros (*Rhinoceros sondaicus*) in Captivity in South Asia

48.1   Javan Rhinoceros in Captivity

Only a small number of *R. sondaicus* was definitely exhibited in captivity (Rookmaaker 1998). Although my previous historical assessment listing 22 specimens needs review, the total will remain surprisingly small. The species has been exhibited in four zoological gardens: in Adelaide (1886–1907), Berlin (1874–>1884), Kolkata (twice: 1877–1880, 1887–>1892) and London (1874–1885). Most were captured in South-East Asia, while out of the list of 22 only six were captured or exhibited in South Asia.

Given that *R. sondaicus* and *R. unicornis* are both one-horned and morphologically similar in many respects, it is expected that some animals of *R. sondaicus* were wrongly identified as *R. unicornis*. My own working default has been to require some irrefutable proof that an animal must have been *R. sondaicus* (Rookmaaker 2019b). This has the benefit that at least the real *R. sondaicus* have been properly identified. It also leads to some careful and wide-ranging historical enquiries, as in the cases of the Liverpool Rhinoceros of 1834–1841 and Huguet's Rhinoceros of 1845–1862, both suggested in the past to be *R. sondaicus*, but definitely identified as *R. unicornis* (Rookmaaker 1993, 2015b).

The list of *R. sondaicus* captured in South-East Asia will need further analysis of the additional information.

One was taken from Java to Bali in 1839 to be killed in a cremation ceremony of the King of Klungklung in 1842 (Rookmaaker 2005). During the current investigation, three specimens captive in Indonesia were encountered. First is the young animal sketched while alive in the vicinity of Batavia (Jakarta) around 1820 by Pierre-Médard Diard (1794–1863), which is the type specimen of *Rhinoceros javanicus* F. Cuvier, 1824 (47.5). The second is the calf captured by the Hungarian Count Emanuel Andrássy (1821–1891) in Java in the 1850s (fig. 48.1). A third is found in a list of various observations about rhinos collected by the Dutch naturalist Ernst Brinck (1582–1649), who is introduced in 12.5. He heard from the Dutch minister Constantinus Citharaeus (b. 1596), who worked on Java until 1629, that there was a living rhinoceros and several tigers in the Castle of Batavia during his time.

48.2   Addition of 9 *Rhinoceros sondaicus* in Captivity in South Asia

According to my assessment of 1998, there were four specimens of *R. sondaicus* shown in India, and a fifth exported from Kolkata to Europe. A sixth specimen from Malaysia was supposedly transported to India where the trail disappeared. These figures relating to South Asia need to be

FIGURE 48.1
Hunting a rhinoceros in Java, and capturing the baby. Scene depicted by Count Emanuel Andrássy de Mano Csikszentkiraly Et Krasznahorka (1821–1891). Plate absent in the Hungarian edition of 1853. Attributed to the "Comte Andrasi" as draughtsman, and to Victor Adam (1801–1866) and Eugene Ciceri (1813–1890) as engravers. Set in Java, the actual animals should represent *R. sondaicus*
ANDRASSY, *REISE IN OSTINDIEN*, 1859. COURTESY: S.P. LOHIA FOUNDATION

© L.C. (KEES) ROOKMAAKER, 2024 | DOI:10.1163/9789004691544_049
This is an open access chapter distributed under the terms of the CC BY-NC-ND 4.0 license.

TABLE 48.68   Number of specimens of *R. sondaicus* in captivity in comparison with Rookmaaker (1998)

	1998	Present	Origin in South Asia
South Asia			
List (1998) captured in South Asia	4	4	4
Specimens previously *R. unicornis*	(1)	1	1
Imported from Malaysia	1	1	0
Additions in South Asia	–	9	9
Exported from South Asia to Europe	1	2	2
Exported (previous *R. unicornis*)	(2)	2	2
Subtotal (South Asia)		19	18
South-East Asia			
Specimens in SE Asia only	12	12	0
Additions in Bali and Java (48.1)	–	4	0
Exported from SE Asia to Europe	1	1	0
Exported from SE Asia to Australia	1	1	0
Unknown origin			
Exported to Europe	2	2	0
Subtotal (others)		20	0
Total	22	39	18

Numbers in brackets refer to specimens previously listed as *R. unicornis*.

adjusted according to the historical records found during the current investigation.

At this time three animals formerly identified as *R. unicornis* can be firmly added to the list of *R. sondaicus*, including the type of *R. jamrachi* in Berlin from 1874 (47.10).

The evidence in this chapter points to an additional 9 animals kept in South Asia and a 10th one exported to Europe.

The animal from the Dindings in Malaysia is more definitely documented in Chennai.

This gives a new total of 19 specimens of *R. sondaicus* exhibited in South Asia (table 48.68). Four of these were exported, of which only one survived more than a year in a zoological garden in Europe (in Berlin). Specimens are often known by just one or two events, which give us a snapshot without providing full details of the entire stay in a captive facility (dataset 48.55).

In summary, there have been at least 19 specimens of *R. sondaicus* in captivity in South Asia, and in total 39 worldwide (an increase of almost 60% over previous assessments).

• • •

Dataset 48.55: Chronological Summary of Captive Specimens of *R. sondaicus* Exhibited in or Exported from South Asia

* means recorded in Rookmaaker (1998) as *R. sondaicus*
** means recorded in Rookmaaker (1998) as *R. unicornis*
(nd) no data about transfer or death

    1700 India: animal in Mughal painting, no data (nd)
** 1770s Dhaka: Matthew Day (nd)
    1799 Kolkata: Wellesley (nd)
    1867 India: Traveling show (nd)
** 1874-04 Kolkata: Rutledge 1874 To London, died in transit 1874

** 1874-04 Kolkata: Rutledge 1874-06-16 London: Jamrach 1874 Hamburg 1874-06-30 Berlin Zoo (Jamrach's Rhinoceros) (nd, around 1884, 47.10)
* 1875-11 Kolkata: Rutledge 1875-12-02
    1875 Possibly South Asia 1875 To London or Hamburg for Charles Rice (nd)
* 1876-05 Capture Sundarbans 1876 Kolkata: Jamrach 1876
* 1877 Capture Sundarbans 1877 Kolkata: Jamrach 1877-03 London: Jamrach 1877-09

*   1876 Capture Sundarbans 1876-0-09 Kolkata: Rutledge 1876-09
Kolkata: Garden Reach (King of Oudh) 1887–12 Kolkata: Alipore
Zoo (alive 1892)
*   1877-11-17 Kolkata: Alipore Zoo 1880
1879 Kolkata: Rutledge 1879
1880 Kolkata: Rutledge 1880

1880 Kolkata: Alipore Zoo 1881-04-10
1881 Kolkata: Rutledge 1881
1900 Junagadh: Nawab (one of two) (nd)
1900 Junagadh: Nawab (one of two) (nd)
*   1905-09 Capture Malaysia: Dindings 1905 Singapore 1905
Chennai Zoo (nd)

48.3   List of Specimens of *R. sondaicus* in South Asia up to 1905

Listed alphabetically according to the locality of the collection, there are details for 19 specimens of *R. sondaicus* in captivity in India and Bangladesh. The last one was recorded in 1905.

Chennai (Madras), Tamil Nadu – People's Park

1. M 1905

The Indian forest officer Granville Maurice O'Hara was present during the capture of a rhino in the Dindings, near Sitiawan, Perak, Malaysia in September 1905. This male rhinoceros was almost full grown, stood an estimated 4½–5 ft (137–152 cm) at the shoulder and had a horn of 2½–3 inches (6–8 cm). When he enquired a month later, he heard that the animal had first been bought by a local trader for $200, sold on to a trader in Singapore for $500,

who sent it to the People's Park in Chennai for Rs. 1500 (O'Hara 1907; Rookmaaker 1998: 126). The arrival or stay in Chennai (Madras) has not been substantiated.

There are two photographs of *R. sondaicus* without date or locality taken in a zoological garden. Both photographs are likely to date from the early 20th century. As *R. sondaicus* was extinct in South Asia by that time, probably these photos show an animal obtained in South-East Asia. Potentially this could be the male caught in the Dindings in 1905.

The first item is a stereo view photograph of a "Rhinoceros captured by natives" issued by the Keystone View Company of Meadville, Pennsylvania, USA on 12 December 1919 (Library of Congress 1919). On the back is a text describing the "Rhinoceros of Asia" without detail about the animal depicted. The same image was also sold as a glass positive slide, with the animal in reverse compared to the stereo view (fig. 48.2). It is an adult

FIGURE 48.2
Glass positive slide of a specimen of *R. sondaicus*, possibly the male captured in the Dindings in 1905 and then sent to Chennai. The view is identical (in reverse) to a stereo view photograph issued by Keystone View Co., Meadville, PA. in December 1919
COLLECTION DAN ZIEGLER

FIGURE 48.3
Postcard of a "Rhinoceros" chained by its left hind foot to a post in a fence. The facility is not identified
COLLECTION JOHN EDWARDS, LONDON

R. sondaicus with a decent horn, chained on its right hind foot and tethered to an invisible post, with a man sitting on the fence behind the animal.

The second photograph (fig. 48.3) is found on a postcard with title "Rhinoceros." It shows a rhinoceros trying to reach a branch held out by a man outside the picture. Again it is chained on its left hind foot. The fence has the same characteristics as that in the Keystone stereo view, but the trees and scene in the background are different (Rookmaaker 1998, fig. 68).

Dhaka, Bangladesh – Matthew Day

1. 1770s (?)

In a list of gentlemen owning a rhinoceros, Thomas Williamson (1807: 45) included Mr. Matthew Day of Dacca (Dhaka). Day was a collector or revenue agent in Dacca from the 1760s to around 1787. He kept the animal "in a park into which it was not very safe to venture." There are no further data. It was previously listed as *R. unicornis* (Rookmaaker 1998: 61), but is here included as *R. sondaicus* because the town Dacca had good access to the Sundarbans where only this species could have been obtained.

India – Unknown location

1. Period between 1650 and 1750

There is a painting in the style of Mughal miniatures depicting the meeting of Layla and Majnun in the presence of a single *R. sondaicus*, elephants and other animals (fig. 48.4). This romantic tale was a popular theme in Asian art for a long time. Indian examples often included a variety of animals, in a few cases including one or more rhinos (Ettinghausen 1950: 49; Bautze 1985: 420). These rhinos always show a horn, while the actual characteristics of

FIGURE 48.4 Illustration of the ancient romantic story of Layla and Majnun. Ink, opaque watercolour and gold on paper, 25.7 × 15.5 cm. Undated, around 1650–1750. The rhino shows the characteristics of *R. sondaicus*
WITH PERMISSION © THE TRUSTEES OF THE BRITISH MUSEUM: 1974,0617,0.21.25

the animal often are more poorly defined. However, the specimen in the painting in the British Museum has all the features of *R. sondaicus*, including the shield in the neck region. The date is tentatively estimated between 1650 and 1750. While this doesn't help to identify where or when this animal might have been observed, it seems likely that it was drawn after a specimen in captivity, or at a stretch after a mounted hide in a collection of natural history.

India – Traveling show

1. 1867

Given the rarity of *R. sondaicus*, it is unexpected that Jerdon (1867: 234) rather explicitly said that individuals of that particular species in India "are not unfrequently taken about the country as a show" (repeated by Nott 1886: 334). This could mean that Jerdon had at least seen one of these animals himself. However, the reference is too vague to elucidate. In fact, no rhinos of any kind in traveling shows have been recorded with any detail in South Asia.

Junagadh, Gujarat – Nawab of Junagadh

The Nawabs (rulers) of the princely state of Junagadh (Kathiawar) in Gujarat had an interest in wildlife. They were keepers of the lions (*Panthera leo*) of the Gir Forest with a dwindling population at the end of the 19th century (Divyabhanusinh 2006). In 1863 the Sakkarbaug Zoological Park was initiated on the outskirts of Junagadh, with four cages to keep injured Gir lions during their rehabilitation. It is unknown if the rhinos owned by the Nawabs were kept there or elsewhere. There is still a Genda Agad Road in Junagadh named after a rhinoceros (*genda*) although the circumstances are unknown. Rhinos have been reported in Junagadh in 1876, 1889 and 1900. The 1900 report is documented with a photograph of one animal which is clearly *R. sondaicus* and the mention of another. The earlier ones are included as *R. unicornis* in 3.4.

1–2. 1900

The Viceroy George Curzon (1859–1925) included Junagadh in a state visit arriving on 3 November 1900 at

FIGURE 48.5   Lady Mary Victoria Curzon riding on a *R. sondaicus* owned by Nawab Mohammad Rasul Khanji of Junagadh at the time of her visit on 3 November 1900

WITH PERMISSION © BRITISH LIBRARY: CURZON COLLECTION, PHOTO 430/76. PRINT 39 FROM AUTUMN TOUR OF 1900

Veraval, traveling to Junagadh on a special train to stay for a day and departing on 4 November afternoon to Rajkot (*Homeward Mail from India* 1900-11-12). He was received by Nawab Mohammad Rasul Khanji (1858–1911), who included "two harnessed rhinoceroses, ridden by postilions, among the processional steeds with which Lord Curzon was greeted" (*Edinburgh Evening News* 1900-12-26).

One of the two rhinos seen in the procession welcoming Curzon to Junagadh on 3 November 1900 took the fancy of Curzon's wife Mary Victoria (1870–1906). These animals must have looked quite tame, which gave her the idea that she could sit on one while posing for a snapshot. One print of this event is preserved in the British Library (fig. 48.5) and might well be the same as the image illustrated by Bradley (1985) from the collection of Curzon's third daughter, Lady Alexandra Metcalfe (1904–1995). It is not known if the occasion was witnessed by the American orientalist Edmund Russell (1905: 360), who commented to have seen that "the royal rhinoceroses wander as freely through the bazaars as elephants, Lady Curzon was snapshotted when she rode on one of them, but claimed the negative."

The photograph of 1900 clearly shows an adult *R. sondaicus*, with the knobbly hide and the triangular shoulder-shield, and just a short horn (Rookmaaker 2018). The second animal is not in the photographs, but it is fair to assume that both were acquired together in the course of the 1890s.

Kolkata (Calcutta), West Bengal: Fort William – Governor-General Wellesley

1. 1799

A 9-months old animal was owned by the Governor-General Richard Colley Wellesley, 1st Marquess Wellesley (1760–1842), either in Fort William or maybe in Barrackpore. This young *R. sondaicus* is depicted on a coloured drawing, known from two copies: one in the Wellesley Albums in the British Library (fig. 48.6), the other in the Victoria Memorial Hall, Kolkata (fig. 48.7).

The first has a written text: "Rhinoceros. From the life, about nine months old in the possession of the Earl of Mornington. [scale] one foot – Rhinoceros sondaicus (fide Blyth)" (Archer 1962: 6). Mornington was an alternative title of Marquess Wellesley.

The second includes a text with additional information: "From the Life by Mr. Home [deleted: Singhy Hey]. 1799. The animal about 9 months old. – [scale] one foot." This second copy came from a collection made by John Fleming (1747–1829), a surgeon in the Indian Medical Service in Bengal from 1768 to 1813 (Vaughan 1997: 94).

The drawing was not done by the local artist Singhy Hey (his name was deleted), but by the painter Robert Home (1752–1834), who had been in India since 1791.

Kolkata (Calcutta) – Zoological Gardens, Alipore

The Zoo opened in 1876, still operating in 2021. State-owned.

Besides the three specimens listed here, there is another reference which is hard to allocate to these. Walter Samuel Burke (b. 1861) in his *Indian field shikar book* first published in 1904 (consulted 4th edition of 1908) writes under the heading of *R. sondaicus* that "the horn of the specimen in the Calcutta Zoological Gardens is under 8 inches [20 cm]." There is no date, and Burke was aware that females of the species lack the horn. As there was no specimen of *R. unicornis* in the zoo around 1900 either, this sentence remains mysterious.

1. 1877-11-17 to 1880 †

This juvenile was listed as a "Malayan one-horned rhinoceros, Rhinoceros sondaicus" purchased in 1877 (Calcutta Zoo 1878: 31). The animal could have come from Malaysia as the naming implied, or even from the Sundarbans. It died in the first months of 1880 (Sanyal 1892: 132).

2. M 1880 to 1881-04-10 †

When both the *R. unicornis* of April 1877 and the *R. sondaicus* of November 1877 died in the first months of 1880, the first president of the zoo Charles Thomas Buckland (1824–1894) tried to obtain a replacement. He wrote to Tyjumal Ali, a magistrate in the Sundarbans, and after several months he received a reply: "Honored Sir, Herewith I send you a rhinoceros, which my shikaris have caught after much labour. They shot the mother and then secured the young one. Please forgive me for sending such a small one, but it will soon get bigger. I am your obedient servant, Tyjumal Ali" (Buckland 1890: 230). This may have been in the second half of 1880 (Rookmaaker 2011a). It was reported in the *Times of India* (1881-03-21, p. 3) that the animal was tame enough to carry people: it "is ridden by his keeper round the enclosure of his abode. We understand that anyone willing to pay a rupee may also have a ride, and the ride is certainly worth the money." Buckland said that the animal grew fast, but got fever when the large teeth began to erupt resulting in his untimely death. This was attributed to chronic pleurisy, when "he" was reported to have died on Sunday 10 April 1881 (*Times of India* 1881-04-16). Possibly this specimen contributed the juvenile male skull recorded by Sclater (1891: 203, no. r) in the Indian Museum.

FIGURE 48.6
Drawing of a nine months old rhinoceros in
the possession of the Governor-General Lord
Mornington or Wellesley, dated 1799
WITH PERMISSION © BRITISH LIBRARY:
NHD32/46, WELLESLEY ALBUMS

FIGURE 48.7
Drawing of a nine months old rhinoceros
by Robert Home (1752–1834). From a
folio of watercolours assembled by John
Fleming (1747–1829). Size 30.2 × 47.2 cm.
Provenance: Maharaja Prodyot Coomar Tagore,
15 March 1934
WITH PERMISSION © VICTORIA MEMORIAL
HALL, KOLKATA

3. F 1887–12 and alive in 1892 (from King of Oudh, Kolkata)

After King Wajid Ali Shah of Oudh (1822–1889) was exiled to Kolkata in 1856, he soon established a new menagerie in his property Garden Reach in Metiabruz (5.4). The *R. sondaicus* obtained in September 1876 from the Sundarbans through the dealer William Rutledge was one of the animals sold by auction in December 1887 after the death of the King (Calcutta Zoo 1888). This female was then obtained by the Alipore Zoo and apparently she was still alive when Sanyal (1892) wrote his textbook.

Kolkata (Calcutta): Metiabruz – Wajid Ali Shah, King of Oudh

1. F 1876–09 (from Kolkata: Rutledge) to 1887–12 (to Kolkata: Alipore Zoo)

Rutledge obtained a female rhino from the Sundarbans and sold her to Wajid Ali Shah, the exiled King of Oudh, for his establishment in Metiabruz. At the auction of the King's menagerie in December 1887, she was obtained by the Alipore Zoo, Kolkata.

Kolkata (Calcutta) – William Rutledge (dealer)

William Rutledge (1832–1905) was a dealer in birds and wild animals in Kolkata. There is no information about his personal life. His business is referred to by Rainey (1874b) and listed in 1875 as a bird shop and menagerie at 32 Jaun Bazar Street and Entally (*Bengal Directory* 1875: 410, 679). He supplied large numbers of animals to King Wajid Shah of Oudh who lived in Metiabruz from 1856 to 1887. The ornithologist Frank Finn (1868–1932), who was working at the Indian Museum 1894–1903 often took a short walk to "Mr. W. Rutledge's establishment in South Road, Entally. There business has been carried on for nearly half a century, Mr. Rutledge dealing in living animals of all kinds; and many very choice birds pass through his hands, though he naturally does not trouble himself greatly about the commoner species. To him I have long been indebted for much information concerning birds and the methods pursued in keeping them" (Finn 1901). In 1899 Finn bought a specimen of the Yellow Weaver in the shop and named it *Ploceus rutledgii* (*Proceedings of the Asiatic Society*, July 1899) but soon found that it was the summer plumage of *Ploceus megarhynchus*.

The references to rhinos traded by Rutledge largely date from the 1870s. In 1874 he was located "east of the Circular Road", in 1875 at 32 Jaun Bazar Street and Entally. The *Bengal Directory* for 1884 (p. 310) lists W. Rutledge as ornithologist at 31 Mott's Lane and 3 Wellesley Street, and the menagerie (p. 459) still located at 35 South Road, Entally. In 1885 a baby *D. sumatrensis* was born in his premises (55.2). It is likely that he worked together with William Jamrach providing premises as well as local expertise. This cooperation continued regularly, shown by a notice in the *Englishman's Overland Mail* (1890-10-29) of a consignment of ostriches and zebras sent by William Jamrach to Rutledge at Entally.

1 1874-04 to 1874-04 †
2 F 1874–04 (exported to London)

In the *Oriental Sporting Magazine* (OSM) of May 1874, the editors commented that they saw "two of the tubercled rhinoceros (as we suppose) at Mr. Routledge's to the east of the Circular Road, Calcutta, not long ago" (footnote to Rainey 1874b: 240). The name "tubercled rhino" points to *R. sondaicus*, most likely procured from the Sundarbans. When Rainey (1874c: 340), suggested that in fact they belonged to a new (unnamed) species, he also mentioned that one of them had been shipped to England.

Almost certainly, these were the two animals mentioned in the notorious description of *Rhinoceros jamrachi* by William Jamrach (1875) when he said that in April 1874 he received a telegram from Kolkata about the shipment of two rhinos (47.10). That telegram must have been sent off before the animals were actually meant to be loaded on board. However, in the OSM of September 1874, Rainey (1874d: 431) rather confusingly declared that both these animals were shipped, although one of them died in transit. In short, one of the rhinos seen at Rutledge in March–April 1874 by the OSM editors either died while still in Kolkata, or otherwise during the voyage to London. The second of these rhinos is the specimen which arrived in London on 16 June 1874 and was the subject of Jamrach's description of *Rhinoceros jamrachi* (47.10).

3. M 1875–11 to 1875-12-02

Members of the Asiatic Society of Bengal were expecting to see a living specimen of *R. sondaicus* at their meeting on Wednesday 1 December 1875. However, only photographs were shown by the Secretary James Wood-Mason (1846–1893), because the animal was sick and died in Rutledge's quarters on the following day (Wood-Mason 1875). The skull (later badly damaged) of a juvenile male *R. sondaicus* was donated by Rutledge to the Indian Museum in 1875 (Sclater 1891, vol. 2: 203, no.j) and is still present (Groves & Chakraborty 1983: 254, no. 19368).

The Arader Gallery in New York offered for sale a copy of the *Mammals of India* by Thomas Caverhill Jerdon (1811–1872) with additional watercolours and photographs of related interest (Petretti 2018: 210). These include two photographs (undated, without annotations) each showing the same young *R. sondaicus* petted by a man with a topi (figs. 48.8, 48.9). One of them has an imprint in lower right corner "Marion & Co. London" which may mean that it was printed on photographic paper manufactured by the firm of Augustin Marion, which was an established retailer in London since the 1840s. The other photograph has a signature: 27 / 8 Stretton. While the significance of the numbers is unknown, it was taken by Captain William George Stretton (1836–1899), who had a photographic studio at 5 Chowringhee, Kolkata listed in the *Bengal Directory* (1875: 679). Stretton worked at least five years in Kolkata as a photographer from 1875 onwards, after which the records are patchy (Bautze 2018).

As these two photographs are not dated, it will remain difficult to be certain about the circumstances. However, it would be quite feasible to suggest that when Wood-Mason heard that Rutledge could not manage to show the young *R. sondaicus* due to sickness, he walked to Stretton's shop and asked for some photographs to be taken. Rutledge's premises in Entally were just a short distance away. These were the photographs exhibited at the Asiatic Society. The man shown with the rhino could well be William Rutledge, although no comparative material has been found. Unfortunately it remains unclear how these

FIGURE 48.8
Young *R. sondaicus* photographed by Captain William George Stretton, probably in the premises of William Rutledge at Entally, Calcutta on 1 December 1875. From a miscellaneous collection associated with T.C. Jerdon offered for sale by the Arader Galleries in 2018
WITH PERMISSION © ARADER GALLERIES, PHILADELPHIA

FIGURE 48.9
A second photograph of *R. sondaicus* taken in Calcutta in 1875 by W.G. Stretton
WITH PERMISSION © ARADER GALLERIES, PHILADELPHIA

items would have found their way into a collection associated with Jerdon who died in England three years earlier.

4. F 1876-09-09 to 1876-09 (to Wajid Ali Shah, Kolkata)
The *Times of India* (1876-09-09) reported that Rutledge had obtained a "splendid specimen of a rhinoceros standing only 2½ feet (76 cm)" from the Sundarbans. This should be the animal sold to King Wajid Ali Shah of Oudh for his menagerie.

5. M 1879
In the catalogue of the Indian Museum published by William Lutley Sclater (1863–1944) in 1891, three young specimens of *R. sondaicus* each represented by a skin and skeleton are attributed to W. Rutledge, with dates 1879, 1880 and 1881. As Rutledge was actively dealing in these animals at the time, it is likely that these are animals obtained in the Sundarbans and dying soon after capture before they could be sold (49.11). The 1879 donation was

a juvenile male, skin and skeleton (Sclater 1891, vol. 2: 203, no. g). Apparently it is no longer present in the collection of the Zoological Survey of India (Groves & Chakraborty 1983).

6. F 1880

The 1880 donation was a juvenile female, skin and skeleton (Sclater 1891, vol. 2: 203, no. h). Apparently no longer present in the collection of the Zoological Survey of India (Groves & Chakraborty 1983).

7. F 1881

The 1881 donation was a juvenile female, skin and skeleton (Sclater 1891, vol. 2: 203, no. f). Apparently no longer present in the collection of the Zoological Survey of India (Groves & Chakraborty 1983).

Kolkata (Calcutta) – William Jamrach (dealer)

1. F 1876-05 to 1876-05 †

In May 1876, William Jamrach procured a young female rhinoceros from Ray Mangal (Raimangal) River in the Sundarbans. It lived only 24 hours after arrival in Kolkata. Jamrach sent the skin to England, where it was exhibited on his behalf by P.L. Sclater at the meeting of the Zoological Society of London on 21 November 1876 (Sclater 1876b).

2. F 1877 – 1877 (to London: Jamrach)

A female rhino from the Sundarbans was obtained by Jamrach and transported to London.

London – William Jamrach (dealer)

1. F 1877-02 (from Kolkata: Jamrach) to 1877-09 †

P.L. Sclater (1877) reported in the meeting of the Zoological Society of London on 6 March 1877 that he had been to examine a young female *R. sondaicus* from the Sundarbans, measuring 3 feet (91 cm) in height, imported by William Jamrach (51.8). The animal died after it had been in London for just over six months, when it was dissected by Alfred Henry Garrod (1846–1879), the ZSL Prosector (Garrod 1877a). There is no explanation why this rare animal wasn't sold and its health might have been precarious. The Zoological Society of London already exhibited an example of *R. sondaicus* and might therefore not have been interested (fig. 48.10).

London or Hamburg – Charles Rice (dealer)

1. 1875

There is an advertisement in *The Era* (London) of 5 December 1875 where the dealer Charles Rice offers a female *R. sondaicus* (curiously misspelled "Loudaicus"), about 4 feet (122 cm) high (fig. 48.11). Charles William

THE NEW RHINOCEROS IN THE GARDENS OF THE ZOOLOGICAL SOCIETY.

FIGURE 48.10
Male *R. sondaicus* from Java, in London Zoo 7 March 1874 to 23 January 1885
ILLUSTRATED LONDON NEWS,
18 APRIL 1874, P. 377

RHINOCEROS.—FOR SALE, Male One-horn Indian RHINOCEROS, about 4½ Feet High; also Female Loudaicus Rhinoceros, about 4 Feet High; One Full-grown Indian Tapir, Two Black Leopards, Pair Black Jackals, Pair Full-grown Cheetahs. For price and particulars address, Mr C. RICE, at Umlauff's, No. 8, Spielbudenplatz, Hamburg.

FIGURE 48.11 "Female Loudaicus" (Sondaicus) offered for sale by Charles Rice
ADVERTISEMENT IN *THE ERA*, 5 DECEMBER 1875

Rice (1841–1879) was trading in animals often in collaboration with Jamrach and with his brother-in-law Carl Hagenbeck (1844–1913) in Hamburg, Germany (Rookmaaker 2014b). The address given for a reply "Umlauff's, No.8, Spielbudenplatz, Hamburg" names the Italian animal dealer Johann Friedrich Gustav Umlauff (1833–1889), who worked in partnership with Hagenbeck. This female *R. sondaicus* may have come together with the *R. unicornis* imported from the Bhutan Duars (North Bengal) in 1875 which died in Rice's stables in London, also mentioned in the advertisement (Rookmaaker 1998: 88). Although it is unknown if this female was actually captured in the Sundarbans, it is listed here to point at the possibility. There is no information about a further sale of the specimen.[1]

---

1   This chapter has benefited from reviews of earlier drafts by Helmut and Gertrud Denzau.

CHAPTER 49

Historical Records of the Rhinoceros in Odisha

No definite records
Species: *Rhinoceros sondaicus*
Rhino Region 38

49.1 Rhinos in Odisha (Orissa)

Nobody has ever claimed to have seen themselves a rhinoceros in the wild in the state of Odisha (Orissa) or neighbouring Chhattisgarh. Any rhino in this territory should have been *R. sondaicus* which were found nearby in the Sundarbans of West Bengal. Thomas Caverhill Jerdon, in his authoritative *Mammals of India* (1867 and 1874), asserted that a few individuals of *R. sondaicus* occurred in the forest tract along the Mahanuddy (Mahanadi) River, extending northwards towards Midnapore (West Bengal). Jerdon had no personal experience in the area, and his source remains obscure. Very soon after, the geologist Valentine Ball (1843–1895) traveled widely around the state and queried the veracity of Jerdon's statement (Ball 1877). He was supported by William Thomas Blanford in the highly-regarded *Fauna of British India* of 1891.

The records in the chronological list (dataset 49.56, shown on map 49.22) mostly refer to artefacts, rock art and a toponym. Combined with the uncertainty and some potential errors, this gives a false impression that rhinos were once widespread in this part of India. The testimony by Schouten in 1664 was based only on hearsay, and the rhinos seen wandering through the city of Bengalla in 1638 need further explanation. The occurrence of the rhinoceros in Odisha is therefore unsubstantiated.

•••

Dataset 49.56: Chronological List of Records of Rhinoceros in Odisha (Orissa)

General sequence: Date – Locality (as in original source) – Event – Source – § number – Map number.

The map numbers are explained in dataset 49.57 and shown on map 49.22.

Rhino Region 38: Odisha (Orissa) and Chhattisgarh

Neolithic – Vikramkhol – Rock painting – Manuel 2008: 34 – 49.6 – map no. P 717

5th cent. BCE – Tel River – Grammarian Panini mentions trade in rhino hide – Mishra 2003: 199 – 49.6 – map no. M 882

2nd cent. BCE – Khandagiri Caves [in Bhubaneswar] – Satghara cave: sculpture of Shreyansanath, the 11th Tirthankar (28.6), with symbol rhinoceros – Ganguly 1912: 54 – 49.6 – map no. M 884

1st cent. BCE – Asurgarh – Crystal pendant of rhino, excavated in 1973, now Sambalpur University Museum – Mishra 2003: 199 – 49.6 – map no. M 881

1st cent BCE – Asurgarh – Punch marked coins embossed with the rhino motif – Mishra 2003: 199 – 49.6 – map no. M 881

12th cent. – Meghesvara Temple, Bhubaneswar – Sculptures include rhino – Ganguly 1912: 327 – 49.6 – map no. M 885

1638 – Bengalla. Ancient city, locality debated, here taken to be near Puri (Odisha) – The seaman William Bruton reported that the City of Bengalla "is likewise famous for its multitude of Rhinoceroes" – Bruton 1638: 32, annotated in Bruton 1924 – 49.3 – map no. Z 923

1664-01 – River Jillisar (Jaleswar), Pipley (Pipli) River – Wouter Schouten (1638–1704) reports the presence of "Renocerots" – Schouten 1676a (pt. 2): 125, 1676a (pt. 3): 59; Schouten 1676b (translation of 1676a) – 49.2, figs. 49.1, 49.2 – map no. Z 924

1800 – Keonjhar Dt – Place name Gandograma. Possibly derived from *ganda* meaning rhino – Mahajan 2003 – 49.5 – map no. M 883

1800 – Mahanadi delta – *R. sondaicus*, former presence – Harper 1945; Groves & Leslie 2011: 195 – map no. Z 922

1850s – Bailadila range, Bastar Dt., Chhattisgarh – George Whitty Gayer (d. 1945), administrator of Bastar, heard from Murias in Indravati hills about the presence of rhinos in their father's time – De Brett 1909: 32 – map no. Z 921

1867 – Mahanuddy (Mahanadi) River – *R. sondaicus*, very few individuals – Jerdon 1867: 234, 1974: 234; Blyth 1872a; Mukherjee 1980: 124 – map no. Z 922

1867 – From Mahanuddy north to Midnapore – *R. sondaicus* present – Jerdon 1867: 234

1867 – Mahanadi River – Jerdon's statement ascertained to be a mistake – Ball 1877: 171; Blanford 1891: 475; Thompson 1932: 154; Loch 1937: 132

1877 – Jeypore – Captive rhino was marched down from Kolkata to be sold to the Raja of Jeypore – Ball 1877: 171, 1880: 570 – Identity uncertain, supposed *R. unicornis*, see 5.4

1877 – Sambalpur – The Raja of Sambalpur had received a captive rhino as part of a dowry – Ball 1877: 171 – Identity uncertain, supposed *R. unicornis*, see 5.4

1980 – Peace Stupa (Shanti stupa) in Dhauligiri, Bhubaneswar – Modern rhino sculpture, made by Japanese artists (no illustration retrieved) – van der Geer 2008: 386

© L.C. (KEES) ROOKMAAKER, 2024 | DOI:10.1163/9789004691544_050

This is an open access chapter distributed under the terms of the CC BY-NC-ND 4.0 license.

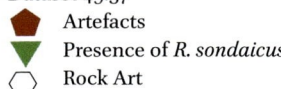

MAP 49.22   Records of rhinoceros in the state of Odisha (former Orissa). The numbers and places mentioned in the text are explained in
                    Dataset 49.57

🔷 Artefacts

🔻 Presence of *R. sondaicus*

⬡ Rock Art

• • •

Dataset 49.57: Localities of Records of Rhinoceros in Odisha and Chhattisgarh

The numbers and places are shown on map 49.22. The numbers are for
reference only due to the uncertainty of the records.

Rhino Region 38: Odisha (Orissa) and Chhattisgarh

Z 911   Bailadila range, Bastar Dt., Chhattisgarh – 18.75N;
             81.25E – 1850s

M 881   Asurgarh – 20.12N; 83.36E – 1st cent. BCE – artefact

P 717   Vikramkhol – 21.85N; 83.75 – rock art

M 882   Tel River – 20.80N; 83.85E – 5th cent. BCE – trade

M 883   Keonjhar Dt. – 21.65N; 85.60E – 1800 – toponym

M 884   Khandagiri Caves [in Bhubaneswar] – 20.27N; 85.78E – 2nd
             cent. BCE – artefact

M 885   Meghesvara Temple [Bhubaneswar] – 20.27N; 85.78E – 12th
             cent. – artefact

Z 912   Mahanadi delta – 20.35N; 86.65 E – 1800, 1867 – considered
             mistaken

Z 913   Bengalla (perhaps Puri, 19.82N; 85.85E) – 1638 – uncertain
             locality

Z 914   Jillisar (Jaleswar) or Pipley River – 22.00N; 87.35E – 1664

• •
•

49.2 The Dutch Factory at Pipli

The early factory of the Dutch East India Company at Pipley (Pipli) at the mouth of the Subarnarekha River near Jaleswar was abandoned in the 1670s. The Dutch physician Wouter Schouten (1638–1704) visited here during his stay in the region from 1658 to 1664. For the Jillisar River (Jaleswar) on 16 January 1664, he recorded that the forests around him were full of snakes, wild buffaloes, tigers and the occasional rhino ("Renocerot"). As Schouten did not travel far inland, his detailed description of the rhino of the Hooghly and Pipley (Pipli) rivers must be partly second-hand, in fact he confesses that he only saw a stuffed example at the Cape of Good Hope on his return journey (Schouten 1676a: 125). Written originally in Dutch,

Schouten's book appeared almost simultaneously in 1676 in a German translation by Johannes Deusing of Bremen, with the author's name changed to Walter Schultz. A French translation of 1707 changes his name to Gautier Schouten. Both the Dutch and German editions have an engraved title-page, with different letterings, which prominently includes a rhino, with a single horn on the nose and two additional hornlets on the stern. An oriental prince sits on the animal, in a scene which may represent an event in Indonesia where Schouten spent most of his time abroad (figs. 49.1, 49.2). The plate itself was one of many prepared by artists working for the Amsterdam publisher Jacob van Meurs (1619–1680), possibly Romeyn de Hooghe (1645–1708) or one of his colleagues (Schmidt 2015).

FIGURE 49.1   Frontispiece of the first Dutch edition of Wouter Schouten, *Oost-Indische voyagie*, 1676. The plate was produced in the workshop of the Dutch publisher Jacob van Meurs and probably shows an imaginary scene. In later editions, the rhino is replaced by an elephant
SCHOUTEN, *OOST-INDISCHE VOYAGIE*, 1676. REPRODUCED BY KIND PERMISSION OF CAMBRIDGE UNIVERSITY LIBRARY

FIGURE 49.2   Frontispiece of the German translation of Schouten's *Voyagie*
WOUTER SCHOUTEN OR WALTER SCHULTZ, *OST-INDISCHE REYSE*, 1676. REPRODUCED BY KIND PERMISSION OF CAMBRIDGE UNIVERSITY LIBRARY

FIGURE 49.3 "Waltz with a Rhinoceros." Portrayal of a hunt of *R. sondaicus* (despite the fictitious long horn) set in Burma. Drawn by the English illustrator John Swain (1820–1909), and engraved by Alfred Thomas Elwes (1841–1917)
ILLUSTRATED SPORTING AND DRAMATIC NEWS, 31 MARCH 1883, P. 56

49.3 William Bruton in Bengalla in 1638

The seaman William Bruton reported that the city of Bengalla "is likewise famous for its multitude of Rhinoceroes; it hath a Beast much like unto a Unicorne, and because it hath but one Horne, some doe beleeve and take it for the Unicornes Horne for the vertue it hath in it" (Bruton 1638). There is still much debate about the actual position of this large city called Bengalla. Some scholars believe it to be Gaur in West Bengal, others (followed here) take it to have been near Puri in Odisha. The claim that the city was infested by large numbers of rhinos cannot add much to the debate, but presents a rather unusual twist.

49.4 Travels of Valentine Ball

Valentine Ball (1843–1895) had another explanation for the references to the presence of rhinos in Odisha by Jerdon (1867). During tours in the Barpali and Sambalpur area, some 350 km inland from Bhubaneswar, he heard about two rhinos which had been kept in captivity. He met a man in Sambalpur who claimed that he had marched a rhino down from Kolkata to sell it to the Raja of Jeypore

(Ball 1877, 1880). Another Raja of Sambalpur had at an earlier date received a rhino as part of a dowry and kept the animal for several years (Ball 1877). These animals are listed here as specimens of *R. unicornis* given the limited data available (5.4).

49.5 Rhinoceros in the Toponym Gandograma

In northern Odisha there once was a place called Gandograma, which seems to indicate a big region with plenty of rhinos. The name may still exist as Gandabaraie in Baria, Keonjhar district (Mahajan 2003). As the origin of similar place names including animals like a rhino is often impossible to ascertain, they are generally poor indications of historical distribution (Boshoff & Kerley 2013: 25). The same applies to the rock painting found at Vikramkhol.

49.6 Old and New Representations

The Sanskrit scholar Panini (ca. 5th century BCE) mentioned the Tel River in southern Odisha in connection with trade in rhino hide (Mishra 2003). Archaeological excavations in the same area (Asurgarh) have uncovered a rhinoceros pendant made of crystal and a punch-marked coin embossing the rhino motif. In the Khandagiri Caves (2nd cent BCE) in Bhubaneswar, one sculpture represents Śreyāṅsanātha, the 11th Tirthankar of the Jain tradition, whose chosen symbol was a rhino (28.6). A rhino sculpture is also found in the 12th century Temple of Meghesvara (Ganguly 1912).

Historical Records of the Rhinoceros in the Western Sundarbans

First Record: 1737 – Last Record: 1910
Species: *Rhinoceros sondaicus*
Rhino Region 39

50.1 Rhinos in the Western Sundarbans

Travel at any time is defined by the sense of excitement to reach a new destination. A passage from Europe to India used to take months before the invention of airplanes. Many ships and steamers were bound for Kolkata (Calcutta), the vibrant metropolis on the eastern side of the Indian peninsula. Imagine reaching the mouth of the River Hooghly, just a few hours to the landing stage, when the pilot boats help the vessels navigate the river's treacherous waters. As one of the journey guides put it, "here we begin to feel more powerfully than we have yet done that we have reached the famous land of India; and every object that meets our wondering gaze tends to deepen and intensify this feeling" (Anon. 1859: 41). The crew and seasoned travelers would point out that the jungle seen from the deck was full of man-eating tigers, rhinoceroses, monkeys and crocodiles, definitely out of bounds for the faint-hearted.

True enough, however long the new arrivals would spend in the fascinating diversity of the South Asian continent, few would venture to the marshy and forested lands south of Kolkata and Dacca, an area called the Sundarbans (or Sunderbunds, Soonderbuns). This was the delta of the Ganges and Brahmaputra, stretching from the Hooghly River in the west to the Meghna River in the east. The region consisted of overgrown waterways surrounding heavily forested islands, not only hiding tigers well-known for their man-eating tendencies, but also notorious for the dangers of attacks by pirates and bandits, all that added to the threat of jungle fever and other unusual diseases. An elephant or a horse, used elsewhere for relatively exhilarating conveyance, were of no use here. Provisions were almost impossible to obtain locally. A trip to the Sundarbans was therefore nothing short of an expedition, undertaken by a brave handful. The Sundarbans began to be surveyed in the first quarter of the 19th century, and soon after the government sold parcels of land to those willing to start cultivating the land. Canals linking the two main centers of Kolkata and Dacca were built and maintained, all slowly eroding the space for wildlife in the patches of forest where sweet water might be available.

The Sundarbans are spread from the Indian state of West Bengal in the west to the Bangladesh provinces of Khulna and Barisal in the east. Dacca and especially Kolkata are within reach, just kept at bay by the dangers of the journey. When early travelers mentioned in their travel journals that they had heard about the existence of rhinos in India, how many might have inadvertently referred to the animals living in the Sundarbans close to Kolkata rather than to *R. unicornis* living further north. We will surely never know. On the other hand, many of the rhinos, both dead and alive, which reached the museums, zoos and private collections in Kolkata in the course of the 18th and 19th centuries, were more likely to be *R. sondaicus* than other species. Even though records are vague or unobtainable, this is exemplified by the natural history collection of the Asiatic Society of Bengal, the precursor of the Indian Museum in Kolkata. When they received their first *R. unicornis* in 1871, they already had a dozen *R. sondaicus*. The famous curator Edward Blyth, who wrote important taxonomic papers on the rhinoceros, had never examined a skeleton or skull of *R. unicornis*. He was the first to identify the specimens in the Sundarbans with the species *R. sondaicus* described from Java and Sumatra, as late as 1862.

The Sundarbans are indeed the realm of *R. sondaicus*. This needs to be stressed to overcome a long-standing confusion. No other species of rhino has been seen or reported from the Sundarbans in the last millennium (or ever), neither the one-horned *R. unicornis* nor the two-horned *D. sumatrensis*. While this fact is pivotal to an understanding of rhino distribution in South Asia, it continues to surprise, because we are so used to hear how rare *R. sondaicus* is, for decades limited to Ujung Kulon, a small national park on the western tip of Java in Indonesia. The Sundarbans were hiding a true treasure, overlooked by hunters and naturalists alike, and yet reduced in numbers to extinction, just over a century ago. Nowadays we classify these animals occurring in mainland Asia as a subspecies, using the name combination *Rhinoceros sondaicus*

© L.C. (KEES) ROOKMAAKER, 2024 | DOI:10.1163/9789004691544_051
This is an open access chapter distributed under the terms of the CC BY-NC-ND 4.0 license.

inermis, first described in 1836 after specimens from the Sundarbans (47.9).

The first president of the Calcutta Zoological Gardens, Charles Thomas Buckland (1824–1894) rather pointedly stated that "when recent editions of popular Indian hand-books solemnly inform the reader that rhinoceros hunting is an ordinary amusement in the suburbs of Calcutta, it is much easier to acquiesce in that information than to urge respectfully that alligators may sometimes be found in Battersea Park" (Buckland 1884: 3). His remark might have developed from an observation in Jerdon's *Mammals of India* under *R. sondaicus* that several rhinos had been killed recently within a few miles of Kolkata (Jerdon 1867: 234). Thomas Caverhill Jerdon (1811–1872) arrived in Madras on 21 February 1836 as Assistant Surgeon in the service of the East India Company. He made the usual advances in rank, to Surgeon Major on 1 October 1858, and retired in December 1868 as Deputy Inspector General of Hospitals. Jerdon returned to the United Kingdom in June 1870 (Elliot 1874; Knox 2014). He is well-known for his catalogues of Indian birds which were partly based on personal collections. His venture to compile information about all the mammals of India in a single volume was a true pioneer effort. The *Mammals of India* was first privately printed in Roorkee in 1866 but not commercially available until 1867. As this first edition was difficult to obtain outside India, it was reprinted by the London publisher John Wheldon in June 1874, not as a new edition but merely to correct some typographical errors found in the first "Calcutta" printing. Jerdon's work was well-known to all naturalists in India, but of course as a first attempt invited some additions. Although Jerdon did not elucidate, rhinos were indeed killed close

to Kolkata up to the 1860s, on the far side of the salt lakes towards the east (50.9) and by the Tent Club towards the south (50.10). Jerdon was correct to insert his comment in the section about *R. sondaicus*, because this is the only species known here and in the Sundarbans.

Perhaps due to the position of the Sundarbans close to the first point of entry of many visitors to the Indian subcontinent, the region was chosen for a number of imaginary adventure stories. The mixture of dangers of the jungle, exotic living conditions, threats of pirates and other wild folks, tigers intent to kill and eat people, what better background is there for adventures entertaining the young (and older) people in Europe. A bulky rhino ready to charge and gore fits perfectly well in the general dramatic events, and in fact were not out of place. Set in the Sundarbans, adventures involving a rhino were written in Germany by Körber in the 1850s (50.11) and in Italy by Salgari in the 1890s (50.12). Another type of fiction were the illusions of Arago located near the French enclave at Chandernagore to the north of Kolkata (50.7). Illustrations enlivened these stories, which through their wide popular appeal have introduced the rhino linked to the Sundarbans and Bengal. Naturalists might have known little about the wildlife of this remote region, yet the reading public in many European countries were all well aware that there were dangerous rhinos in the tropical forests of the Sundarbans.

Publications about the rhino in the Sundarbans in the 19th century, when the species was still found there, give few clues about the zoological identity of the animals. There were few contemporary illustrations, and by a quirk of fate all those reproduced in this chapter, largely produced remotely by artists working for the popular

FIGURE 50.1
Budgerow used to navigate the Sundarbans, drawn by Frans Balthazar Solvyns (1760–1824) SOLVYNS, *LES HINDOÛS, OU DESCRIPTION DE LEURS MOEURS, COUTUMES ET CÉRÉMONIES*, 1811, VOL. 3 NO. 2: BADJÉRA. COURTESY © BIBLIOTHÈQUE NATIONALE, PARIS

press, show a "rhinoceros" but not one with the characteristics (triangular neck shield, short horn) of the actual *R. sondaicus*.

The majority of records of *R. sondaicus* for the Sundarbans up to Kolkata date from the 19th century. A total of 26 specimens were killed (table 50.69) and 18 were captured (48.3). The discussion is divided between the western part of the Sundarbans in the state of West Bengal in this chapter, and the eastern part of Bangladesh in chapter 51. This division is not always clear when authors write about wildlife in the Sundarbans in general without further definition of the locality. However, the rhinoceros had largely disappeared from the west by the 1850s, and most later records are referable to the eastern section.

TABLE 50.69   List of persons who were involved in shooting or capturing rhino in the Sundarbans region

Person	Date	Locality and interaction	Museum[b]
Matthew Day	1770s	No place. Captive rhino in Dacca	
Charles Pigot (1710–1791)	1780	Sundarbans. Attacked by rhino	
Anonymous	1799	No place. Captive rhino in Barrackpore	
Hugh Morrieson (1788–1859)	1812[a]	Raimangal River. Kills 1	
Christoph-Augustin Lamare-Picquot (1785–1873)	1828[a]	Mongla. Kills female and young	Berlin
Lewis	1830	Salt water lake east of Calcutta. Kills 1	
Shekarea	1832	Sagar. Kills 1	
John Henry Barlow (1795–1841)	1834[a]	Bagundee. Kills 1	
Lieutenant Souter and Mr. Lewis	1838	Sagar. Kills 1	
Anonymous	1841	Sagar. Tracks observed	
Anonymous	1849	Sundarbans. Kills juvenile	Munich
Armenian Superintendent of a Zemindary	1850	Sundarbans. Attacked by rhino	
Francis Bruce Simson (1827–1899)	1855[a]	Phuljhuri. Kills 1	
Captain of steamer	1857	Sundarbans. Rhino observed	
Arthur Grote (1814–1886)	1859	Sundarbans. Female skull donated	RCS London
Calcutta Tent Club	1860	Piyali River, Calcutta. Rhino pursued	
W.W. Sheppard	1867[a]	Sundarbans. Kills 1	I.M.
Anonymous	1868[a]	Sundarbans. Kills male	I.M.
Prince Alexander Philipp Maximilian zu Wied-Neuwied (1782–1867)	1869	Sundarbans. Rhino skull donated	AMNH
Edward Biscay Marinatus Baker (1828–1887)	1870s[a]	Eastern Sundarbans. Kills male	
Friend of H. James Rainey	1872[a]	Khulna. Rhino observed	
J.F. Barkley	1872[a]	Sundarbans. Kills young female	I.M.
Oscar Louis Fraser (1848–1894)	1874[a]	Sundarbans. Kills female	I.M.
Anonymous	1874[a]	Mooneypoor. Captured female, to Europe	Berlin
Louis Jacolliot (1837–1890)	1875	Sundarbans. Kills 1	
Party of 4 hunters	1875[a]	Matabangah River. Kills male	I.M.
William Rutledge (1830–1905)	1875[a]	Sundarbans. Juvenile male lived few days	I.M.
Anonymous	1876	Sundarbans. Rhino skull obtained	NHM
William Rutledge (1830–1905)	1876[a]	Sundarbans. Rhino captured	Metiabruz
William Jamrach (1843–1891)	1876[a]	Raymangal. Obtained alive	Jamrach
William Jamrach (1843–1891)	1877[a]	Sundarbans. Rhino transported to London	Jamrach
W.	1879[a]	Sundarbans. Kills adult female	
William Rutledge (1830–1905)	1879[a]	Sundarbans. Juvenile male died	I.M.
Captain Charles Sharling (b. 1839)	1879[a]	Chilipangpi. Kills female	I.M.
William Rutledge (1830–1905)	1880[a]	Sundarbans. Juvenile female died	I.M.
Alipore Zoo	1880[a]	Sundarbans. Rhino exhibited	Calcutta Zoo
Anonymous	1880[a]	Sundarbans. Juvenile male purchased	I.M.
William Rutledge (1830–1905)	1881[a]	Sundarbans. Juvenile male died	I.M.

TABLE 50.69   List of persons who were involved in shooting or capturing rhino in the Sundarbans region (*cont.*)

Person	Date	Locality and interaction	Museum[b]
Raisaheb Nalini Bushan	1885[a]	Ishwaripur. Rhino observed	
Commissioner of Fisheries	1887	Sundarbans. Rhino observed	
Anonymous	1888[a]	Khulna. Kills 1	
Prince Henri d'Orleans (1867–1901)	1888[a]	Ishwaripur. Rhinos observed	
Ram Bramha Sanyal (1850–1908)	1890s	Sundarbans. Rhino observed	
Sportsman	1892	Matla. Kills 2 rhino	
Edmond de Montaigne, Viscount of Poncins (1866–1913)	1892[a]	Ishwaripur. Observes 4 rhino	
Nawab of Junagadh	1900	No place. Two captive rhinos in Junagadh	
Officer of Survey Department	1900	Sundarbans. Rhino observed	

a Record pertaining to eastern Sundarbans of Bangladesh (chapter 51).

b Abbreviations: AMNH (American Museum of Natural History, New York); Berlin (Zoological Museum); Calcutta Zoo (in Alipore); IM (Indian Museum, Kolkata); Jamrach (animal dealer based in London); Metiabruz (King of Oudh menagerie near Kolkata); Munich (Bayerische Staatssammlung); NHM (Natural History Museum, London); RCS London (Royal College of Surgeons of England).

• • •

Dataset 50.58: Chronological List of Records of Rhinoceros in Western Sundarbans and Adjoining West Bengal

General sequence: Date – Locality (as in original source) – Event – Source – § number – Map number

The map numbers are explained in dataset 50.59 and shown on map 50.23.

The records are divided in two sections for Rhino Region 39:

1. General records for Sundarbans: without further locality
2. Western Sundarbans: records with locality

Rhino Region 39: Sundarbans (Sunderbunds, Soonderbuns) – General, without Indication of Locality

1737-10 – Rhinos drowned in cyclone – Anon. 1739; Martineau 1920: 339 – 50.5

1772 – The orientalist Alexander Dow (1735–1779) recorded "marshy islands, near the mouth of the Ganges, which are, at present, the wild haunts of the rhenoceros and tiger" – Dow 1772: cxxvi

1780s (?) – Charles Pigot (1710–1791) of Peplow attacked by rhino (not fatal) – Pennant 1793, vol. 1: 155; 1798, vol. 2: 153; Rookmaaker 1997a – 50.6

1800 – Rhino present – Carey 1800: 139

1826 – Rhino present – Ranking 1826: 288

1841 – Rhino present – Anon. 1841c: 306

1844 – Rhino abound – Weitbrecht 1844: 326

1845 – *R. sondaicus*, specimen attributed to Tucker, no further data. Transferred from the museum of the Zoological Society of London to the Natural History Museum, London, NHM 1845.12.29.5

1848 – Rhino present – Buyers 1848: 37

1849 – *R. sondaicus*, juvenile skull. No data – Bayerische Staatssammlung, Munich, Germany, no. AM0417

1857 – Captain of steamer has seen rhinos regularly – Polehampton 1858: 373

1858 – Single-horned rhino present – Reilly 1858: 408

1859 – *R. sondaicus*. Arthur Grote (1814–1886) obtained a female skull. He worked for the Civil Service in Calcutta 1834–1868, serving as President of the Asiatic Society of Bengal 1859–1862 and 1865. He took the skull to England and in 1882 donated it to Royal College of Surgeons of England, London, no. 2132. Destroyed in war bombing – Blyth 1872e: 3106; Flower 1884, vol. 2: 419; Rookmaaker 1997a

1862 – Rhino present – *Church Missionary Intelligencer* 1862: 92

1867 – Rhino far from uncommon, but rarely seen – R.H.D. 1867: 292

1867 – Charles Thomas Buckland (1824–1894) visits rhino area; none seen – Buckland 1890a: 230, 1890b: 17

1867 – Great number of rhinos destroyed in the cyclone of 1867 – Baldwin 1910: 367

1869 – *R. sondaicus* skull donated by Prince Alexander Philipp Maximilian zu Wied-Neuwied (1782–1867) to American Museum of Natural History, New York – Carter & Hill 1942

1872 – Numerous in some parts – Blyth 1872a, 1875: 50

1875 – Rhino in groups of 2–4 gather at dungheaps in dried-up streams – William Jamrach in Leutemann 1875

1875 – Louis Jacolliot (1837–1890) writes about killing a rhino, 2.25 m high – Jacolliot 1875: 114–121

1876 – *R. sondaicus* mounted skin and skull in Natural History Museum, London, no. ZD.1876.3.30.1 and ZD.1972.734, purchased from the taxidermist Edward Gerrard Jr. (1832–1927). Greatest length of skull 525 mm – Loch 1937: 147; Pocock 1944b: 835; Rookmaaker 1997a; Groves & Leslie 2011, figs. 4, 5

1877 – Last tracks seen here about 1877 – Loch 1937: 132

1887 – Sancar Island – Fictional story of Major Newbridge hunting on one of the islands in the Ganges delta, where he saw a young rhino attacked by a tiger and then rescued by the rhino mother. Scene shown in a large engraved plate signed by Gustav Adolf Closs (1864–1938) and Oskar Frenzel (1855–1915). The same plate

Katholieke Illustratie of 1891 – Decinthel 1887; Closs & Frenzel 1891 – fig. 33.2 (showing *R. unicornis*, and probably published elsewhere previously)

1887 – *R. sondaicus*. Commissioner of Fisheries reported seeing rhinos in the marshy flats – Thompson 1932: 154

1887 – *R. sondaicus*. Last tracks seen in 1887 – Mukherjee 1982: 54

1888 – Last killed in 1888 – Mallick 2011: 59

1889 – Rhino present – Pargiter 1889

1890 – Felice Scheibler (1856–1921) hunts in Sundarbans, but no rhino mentioned – Scheibler 1900

1890s – Rhino seen when Ram Bramha Sanyal (1850–1908) visited the area – Sanyal 1896: 5

1892 – Rhinos exterminated by poachers after 1892 – Guggisberg 1966: 119

1900 – Officer of Survey Department saw rhino tracks – Shebbeare 1953: 145

1904 – Only met one gentleman who had shot a rhino – Kinloch 1904: 65

1910 – Extinct during the last 100 years – Rahman & Asudazzaman 2010: 45

1910 – Thomas Goldsmidt Baldwin (1857–1925) states rhinos very scarce, only found with difficulty – Baldwin 1910: 367

1925 – No definite news could be obtained – Jamun 1925

Rhino Region 39: West Bengal – Western Sundarbans

sub-fossil – Ramnager village, Sonarpur (Kolkata) – *R. unicornis* fossilized remains, in 1990 found 4 m below surface, dated 3030–2500 BP – Ghosh et al. 1992a, 1992b: 361 – 50.4, fig. 50.4 – map S 121

sub-fossil (undated) – Ramnager village, Sonarpur – Specimens identified as *R. unicornis* reported by Ghosh et al. (1992a) preserved in the Paleozoology Division, Zoological Survey of India, Kolkata: right astragalus (WB01), 7th thoracic vertebra (WB02), broken mandible with teeth (WB03), three fragments of ribs (WB04), distal end of right radius (WB05), cuneiform bone of right manus (WB06) and sternal bone (WB07). Collected in May 1990 at Sonarpur, South 24 Parganas, West Bengal by M. Ghosh and two others. Determination by M. Ghosh and U. Sahar– Information from Dr. Supriyo Nandi, Scientist B, Paleozoology Division, Zoological Survey of India, Kolkata, in September 2021 – 50.4, fig. 50.4 – map S 121

sub-fossil (undated) – Meenaklan, Gobra (Kolkata) – One-horned skeleton excavated in 1977 – Biswas 1992: 11; C. Das 2008 – map S 121

sub-fossil (undated) – Chhota Mollakhali, Gosaba – Fossilized remains of rhino, 2.7 m below surface, excavated April 2000 – Ananda Bazar 2000-04-24; Bahuguna & Mallick 2010: 288; Pandit 2013 – 50.4 – map S 130

sub-fossil (undated) – Bakkhali – Large bone dug up in 1984 – Bahuguna & Mallick 2010: 288; Pandit 2013: 7 – map S 126

sub-fossil (undated) – Kashinagar – Rhino mandible and long bones possibly of rhinoceros excavated by Deenbandhu Naskar (d. 2021) and preserved in Khari Chatrabhog Sangrahashala, Kashinagar – 50.4, fig. 50.3 – map S 120

1st cent. (Kushan) – Chandraketugarh – Excavation in 1956–1957 brought fragment of a terracotta plaque with rhino – Das Gupta 1959: 50; Anon. 1965; Biswas 1981: 101, pl. 47a; Bautze 1985: 415, 1995: 28; Christie's 1989: 141; Mukherjee 1990; Haque 2001, no. 786; C. Das 2002; Sengupta et al. 2007: 237; Manuel 2008: 36; Bose 2018b: 55; Rookmaaker & Edwards 2022: 78 – 50.2, fig. 50.2 – map M 886

1st cent. – Dum-dum (Kolkata) near Clive House on Rashtraguru Avenue – Excavation in 2001–2003 yielded a terracotta plaque

made out of a single mould showing rhino – C. Das 2008 – 50.3 – map M 887

1669 – Creeks and rivelets at or about the entrance into the Ganges – Thomas Bowrey (c.1650–1713) reported "Rhinocerots" – Bowrey 1905: 199

1727 – Sagor to Chittagoun (Sagar to Chittagong) – Captain Alexander Hamilton (d. 1732) found rhinos on those islands – A. Hamilton 1727, vol. 2: 24

1800 – Gandargadi, located near Raidhigi – Place name meaning 'plenty of rhinos' – Gertrud Denzau (email 2021–08) – map M 888

1830s – Salt water lake within reach of Calcutta – Rhino shot by Lewis from Calcutta on a parcel owned by James H. Lewis, Frederick Broadhead (d. 1846), Thomas Pitkin (retired in 1844) and Colmer Symes (d. 1839). Rhino head of this "bicornuous brute" was preserved as a trophy – *Sydney Morning Herald* 1854-05-12; Auceps 1867 – 50.9 – Land grant parcels closest to eastern Kolkata on 1873 map are near Canning – map S 123

1831 – Saugor (Sagar) Island – Rhino is rare – Gogerly 1871: 252 – map S 125

1832 – Saugor (Sagar) Island, Middleton Point – Rhino killed by the pseudonymous author and the resident of Middleton Point. The horn was preserved. Animal was 365 cm long and 213 cm high – Shekarea 1832, 1835; Rainey 1872a – 50.8, figs. 50.6, 50.7 – map S 125

1832 – Chandernagore (Chandannagar) – Fictional story that a rhino was wounded after injuring a man – Arago 1840: 221 – 50.7

1838 – Saugor (Sagar) Island near the western shore – Rhino killed Lieutenant Souter and Mr. Lewis of the pilot service – Anon. 1838, 1854a – 50.8 – map S 125

1841 – Saugor (Sagar) Island, near the western shore – Rhino tracks seen by three tiger hunters – Anon. 1841a – 50.8 – map S 125

1849 – Chandernagore (Chandannagar; fictional) – "Black rhino" was first blinded and then killed – Anon. 1867, a work of fiction – 50.7

1850 – Calcutta to Khulna – A hunter who wounded a rhino was gored open and died. Profit from a rhino is Rs.25–35 – Asmodeus 1850: 71

1850 – Area north of Sunderbunds – Stray rhino, wounded Armenian Superintendent of a Zemindary – Asmodeus 1850: 71

1853 – Saugor (Sagar) Island – Rhino very rarely seen – Everett 1853: 63 – map S 125

c.1860– Pealee [Piyali] River east of Barrapoor (Baruipur) – Rhino pursued by members of Calcutta Tent Club, including Edward Biscay Marinatus Baker (1828–1887) as well as G.G.M., C.B.S., D.R.S. and D. – Baker 1887: 279 – 50.10, fig. 50.8 – map S 122

1867 – Calcutta vicinity – Rhino once seen within few miles of Calcutta. Maybe refers to the 1860 record by the Calcutta Tent Club – R.H.D. 1867: 292; Grève 1898: 320 – map S 122

1867 – Within few miles of Calcutta – *R. sondaicus*. Several killed recently – Jerdon 1867: 234, 1874: 234; Blyth 1872a, 1875: 50; Brandt 1878: 36; Craig 1897, vol. 2: 474; Kinloch 1904: 65; Pocock 1946b– map S 122

1891 – Halliday Island – Rhinos numerous – *Englishman's Overland Mail* 1891-11-25 – map S 129

1891 – Bulcherry Island – Rhinos numerous – *Englishman's Overland Mail* 1891-11-25 – map S 128

1892 – Bulcherry Island – Hunting party wounded a rhino but it got away in impenetrable jungle – *Englishman's Overland Mail* 1892-02-17 – map S 128

1892 – Sea face of the Mutlah (Matla) River – Sportsman succeeded in bagging two rhinos – *Englishman's Overland Mail* 1892-02-24, p. 14 – map S 127

1902 – Diamond Harbour – Rhino found dead; villagers had never seen such an animal – *Homeward Mail* 1902-01-13– map S 124

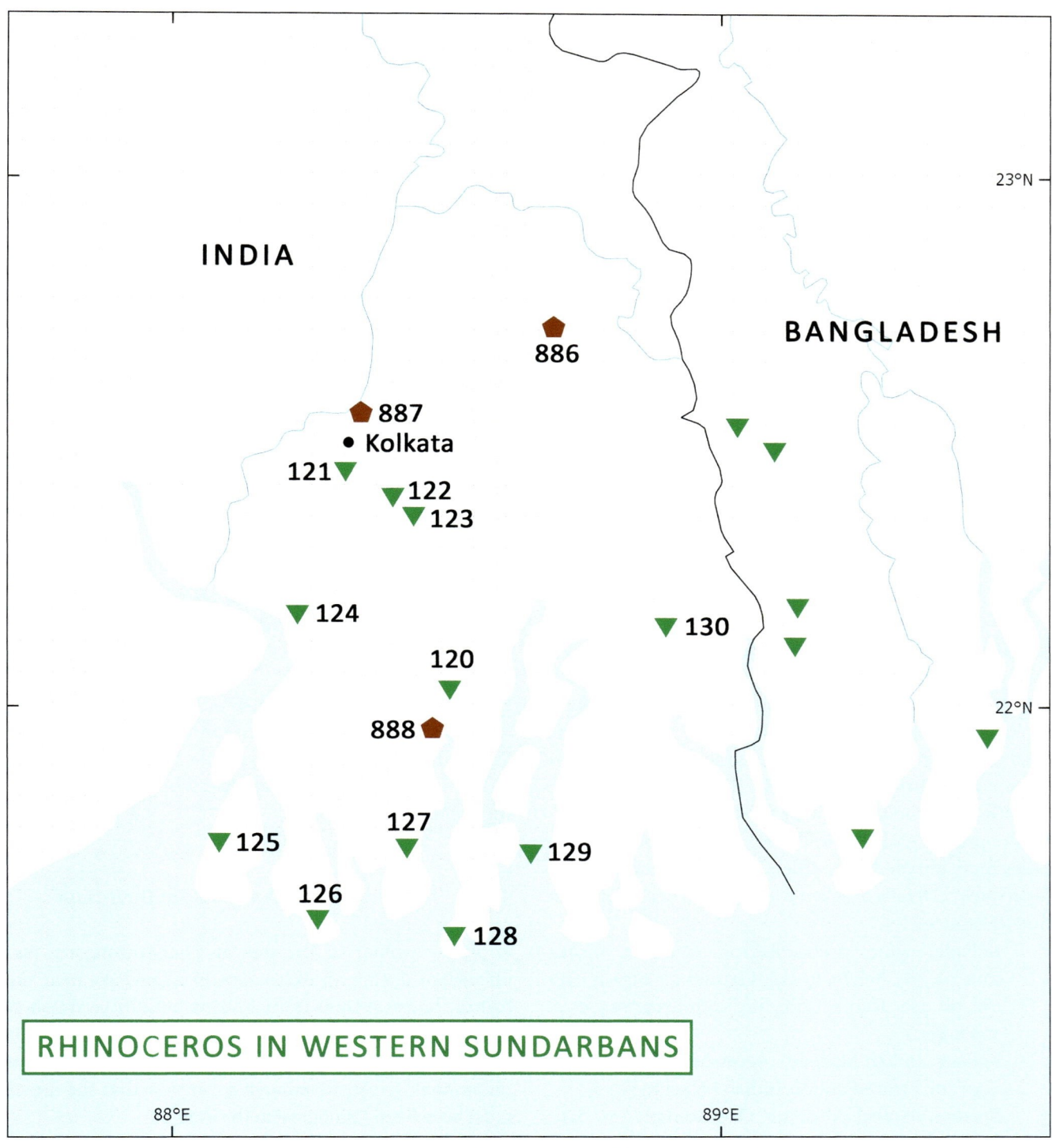

MAP 50.23   Records of rhinoceros in the western Sundarbans (West Bengal, India). The numbers and places are explained in Dataset 50.59

▼   Presence of *R. sondaicus*

⬠   Artefacts

...

Dataset 50.59: Localities of Records of Rhinoceros in the Western Sundarbans

The numbers and places are shown on map 50.23.

Rhino Region 39: West Bengal – Western Sundarbans

M 886   Chandraketugarh – 22.69N; 88.69E – 1st cent. – artefact
M 887   Dum-dum (Clive House) – 22.62N; 88.41E – 1st cent. – artefact
S 121   Ramnager village, Sonarpur (Kolkata) – 22.41N; 88.40E – fossilized (unconfirmed *R. unicornis*)
S 121   Meenaklan, Gobra (Kolkata) – 22.54N; 88.38E – fossilized
S 122   Pealee [Piyali] River SE of Barrapoor [Baruipur] – 22.35N; 88.52E – 1860, 1867, 1872, 1878
S 123   Salt water lake (Canning area, ca. 22.30N; 88.65E – 1830s

S 124   Diamond Harbour – 22.20N; 88.19E – 1902
S 125   Saugor (Sagar) Island – 21.65N; 88.10E – 1831, 1832, 1838, 1841, 1853
S 126   Bakkhali – 21.57N; 88.29E – fossilized
S 127   Bulcherry Island – 21.58N; 88.51E – 1891, 1892
S 127   Mutlah (Matla) River – 21.75N; 88.45E
S 128   Halliday Islands – 21.66N; 88.63E – 1891, 1892
S 120   Kashinagar – 22.08N; 88.43E – subfossil
M 888   Gandargadi – 22.01N; 88.40E – toponym
S 129   Mollakhali, Gosaba – 22.20N; 88.88E – fossilized
*   Chandernagor (Chandannagar) – 22.86N; 88.36E – fictional (50.7)

* Not on map.

..

50.2 Terracotta Plaques at Chandraketugarh

During excavations at Chandraketugarh in 1956–1957, a terracotta plaque with a depiction of a standing rhinoceros was found. Located near Berachampa, about 35 km north-east of Kolkata, objects were dated to the first century CE on stratigraphic grounds. Several similar small plaques with some lateral damage were subsequently unearthed. The rhino faces left and is identified by the presence of skin folds and a single horn, in some cases more abraded than others. The use of these objects is unknown. Details of seven plaques have been published and illustrated:

1. Kolkata, State Archaeological Museum. Accession 04.376, height 17 cm. Figured by Biswas 1981: 101, pl. 47a; Haque 2001, no. 786; Sengupta et al. 2007: 237.
2. Kolkata, Indian Museum. Accession 90/391, size 6.3 × 5.7 cm. Figured online: Indian Museum 2020.
3. Kolkata, Asutosh Museum. One example (no data) figured by Das Gupta 1959: 50, pl. 16 fig. 19.
4. Kolkata, Archaeological Survey of India. One example (no data) figured by Bose 2018b: 55.
5. Kolkata, Archaeological Survey of India. Another example (no data) figured by Anon. 1965: 46, pl. 95B; C. Das 2002.
6. London, Julian Sherrier Collection. Height 12 cm. Figured by Bautze 1985: 39, pl. XLVb; Christie's 1989: 141, Rookmaaker & Edwards 2022, pl. 4.8. See fig. 50.2.

7. Allahabad, Museum. Accession AM-TC-91.1871. Size 5.5 × 6.5 cm. Figured online: http://museum sofindia.gov.in/repository/record/alh_ald-AM-TC -91-1871-31479

Biswas (1981) and Das (2002) connect these plaques from Chandraketugarh with the rhino population in the Sundarbans. While this is acceptable, there are no characteristics on the plaques themselves which clearly identify the animals as *R. sondaicus*.

50.3 Terracotta Plaque Excavated at Dum-Dum

A plaque similar to the one at Chandraketugarh was uncovered during an excavation in 2001–2003 near the historic house at Dum-Dum, Kolkata built by Lord Robert Clive (1725–1774) when he was Governor General of the East India Company (C. Das 2008). The representation of the animal is realistic enough to suggest that the artists must have been familiar with the animal.

50.4 Archaeological Excavations of Fossilized Bones

Several sets of fossilized bones of a rhino have been excavated in Gobra in 1977, in Bakkhali in 1984 and in Gosaba in 2000, of which the published evidence is very scant. A small local museum known as Khari Chatrabhog Sangrahashala in Kashinagar, West Bengal, founded by

FIGURE 50.2
Plaque with rhinoceros excavated at
Chandraketugarh. Formerly in the collection of
Julian Sherrier
PHOTO: JOACHIM K. BAUTZE

FIGURE 50.3
Mandible and long bones possibly of rhinoceros
excavated from the soil near Kashinagar, West
Bengal and preserved in a local museum aimed
at education. Kashinagar is about 70 km south
of Kolkata
WITH PERMISSION © KHARI CHATRABHOG
SANGRAHASHALA, KASHINAGAR

the amateur archaeologist Deenbandhu Naskar (d. 2021), has at least a broken rhino mandible, which together with bones from a variety of mammals, requires further investigation (fig. 50.3).

In 1990 the Zoological Survey of India recovered a rhino mandible from a pit in Ramnager village, Sonarpur, just to the south of Kolkata (Ghosh et al. 1992a). Six skeletal elements and a mandible are preserved in the Paleozoology Division, Zoological Survey of India, Kolkata (fig. 50.4). Bones from this site are estimated to date from 2500–3030

years BP. The ramus of the mandible was 532 mm long and the bone had a vertical height of 265 mm. In comparison, adult *R. unicornis* have a mandibular length of 526–600 mm and condylar height of 277–309 mm; and adult *R. sondaicus* have a length 467–518 mm and height 208–247 mm (Guérin 1980: 53). The shape and size of the specimen suggest that it belonged to *R. unicornis*. This is difficult to reconcile with the known ranges of these species. The conclusion that *R. unicornis* should be added to the fauna of the Sundarbans needs further evidence before it can be firmly asserted.

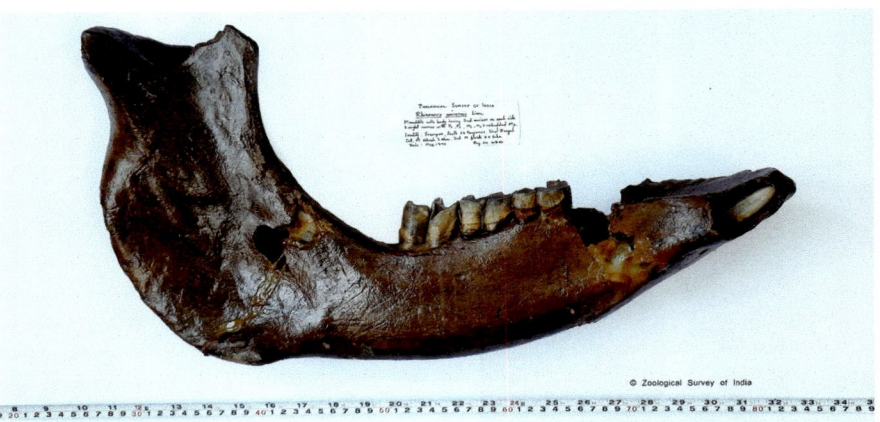

FIGURE 50.4
Rhinoceros unicornis excavated at Ramnager village, West Bengal. Mandible having 2nd incisor on each side and right ramus with P2, P3, M1, M2 and embedded M3
WITH PERMISSION © PALEOZOOLOGY DIVISION, ZOOLOGICAL SURVEY OF INDIA, KOLKATA: REG.NO. WB03

50.5 Effects of the Cyclone of 1737

There is no doubt that the Sundarbans experienced many changes in the habitat. Being a tidal landscape bordering on the ocean, watercourses and islands were often unstable. In addition, when the land started to be claimed for agricultural purposes from around 1775 onwards, the areas of land suitable for wildlife changed dynamically (Mallick 2019). However, for a long time it remained a challenging environment for human settlement, with the threats of malaria, pirates and tigers. On 12 October 1737, the area including Kolkata was hit by a violent cyclone, maybe accompanied by a minor earthquake (Bilham 1994). When water levels rose by half a meter in the violent deluge, it was said that many "cattle, tygers and even rhinoceroses" were drowned (Anon. 1739: 321).

50.6 Pennant's View of Hindoostan of 1798

The *View of Hindoostan* by the English naturalist Thomas Pennant (1726–1798) is a series of impressions from earlier authors woven together. When he places the rhino

correctly in the Sundarbans, he quotes a gentleman who once "roused a rhinoceros, which rushed on him, flung him down, and ripped open his belly; the rhinoceros proceeded without doing him any further injury; the gentleman survived the wound, and lived to a very advanced age" (Pennant 1798: 153). There is no time frame for the event, but elsewhere Pennant (1793: 155) identified the gentleman as Charles Pigot (1710–1791) of Peplow, Shropshire.

50.7 The Fabrications of Jacques Arago

Jacques Étienne Victor Arago (1790–1855) was an artist and an author popular with his French public. He had the good fortune to accompany Louis de Freycinet (1779–1841) for the circumnavigation on the *Uranie* in 1817–1820 working as the official draughtsman. Before the official grand report, Arago published his personal recollections in 1820, expanded into *Souvenirs d'un Aveugle* (Memoirs of a Blind Man) of 1840. Arago wrote plays for the theater and all types of popular stories, besides editing a number of magazines (Liesen 2001). The contents of some of his work were of such nature that he could inspire an author like

Chasse au Rhinocéros.

FIGURE 50.5
Chasse au Rhinocéros (Hunt of a rhino). Engraved by Nicolas Eustache Maurin (1799–1850), after a drawing by Jacques Arago (1790–1855). Printed by the firm of Rose-Joseph Lemercier (1803–1887) and Jean François Bénard (b. 1796) in Paris. While set in Chandernagore on the Hooghly, rhinos are unknown in that locality, and the animal depicted is unlike the species found in the nearby Sundarbans
ARAGO, *SOUVENIRS D'UN AVEUGLE*, 1840, FACING P. 221

Jules Verne (1828–1905) to publish his books of imaginary travels to fantastic new worlds. What Arago writes about the rhinoceros and other wildlife of India must be judged in this context where reality mixes with fiction. Arago never came to India, despite impressions to the contrary in his accounts. He did his homework well, because what he writes could have happened, but for the purpose of this book I have decided that there are too many discrepancies in the details. He wrote about the encounter of Alfred Duvaucel with the rhinoceros in the Rajmahal Hills in 1824 (29.5) in a way which differs from actual happenings, while somehow he must have heard about the events. Arago (1840) writes that in 1832 near Chandernagore a rhino was wounded after it carried a man suspended on its horn into the fields. Then he quotes a pamphlet by Mr. Stephen of Calcutta about a domesticated rhino carrying a family along the banks of a river, which "suddenly changed its disposition, and despite being beaten on its skin, it wanted to go and swim in the river, and dived in. After having followed the stream for an hour, it went on the opposite bank, with all what he carried being wet." In the southern part of India, Mr. Huskisson of Pondicherry had to contend with a pair of rhinos causing havoc in his plantation. None of this can be linked to other sources, hence all must be relegated to the realms of fiction. Arago was an accomplished artist in his own way and provides a good drawing of a rhino being hunted by two men armed with gun and spear (fig. 50.5). The claims of Arago were repeated in later editions and translations of his work, as well as providing inspiration for many stories of adventures.

Sounding quite Arago-esque in its fictitious reality, an anonymous hunter writing in an English *Boys' Journal* recalls an incident of meeting a rhino near Chandernagore in 1849 (Anon. 1867). He was walking along a small stream when he saw a "black rhinoceros" charging him at full speed. He hid behind a tree, the rhino followed him, was blinded by two shots from the gun and was killed afterwards.

Chandernagore was the French factory on the Hooghly where French travellers might have found refuge, as did Alfred Duvaucel and his colleague Pierre-Médard Diard while they were in India (Rookmaaker 2021a). With detailed dates of rhino encounters supposedly in 1832 and 1849, one wonders if Arago actually based his stories on local information. They must, however, be treated as fictitious.

50.8   Saugor Island in the 1830s

Saugor (Sagar) Island is situated where the Hooghly River meets the Bay of Bengal, and was therefore passed by travellers arriving in Kolkata after their sea journey from Europe. The last stretch from the Bay to the harbours needed a pilot to guide the larger ships through the sands. Few had reason to disembark on the little island where tigers were killed in 1786 and 1792, and remained at least until the 1830s (Burton 1931: 124). Rhinos were rarely seen, yet became well-known through a series of anecdotal accounts in popular magazines and in newspapers. These indicate three different events.

FIGURE 50.6
Destruction of a rhinoceros in India, illustrating a hunt on Saugor Island in 1832. The rhinoceros image was invented by the anonymous illustrator and doesn't show the correct species or the actual specimen
PARLEY'S MAGAZINE, 1838, VOL. 6, P. 377

FIGURE 50.7
A second illustration accompanying the account of the destruction of the rhinoceros on Saugor Island in 1832. The artist CLB is not identified
MERRY'S MUSEUM AND PARLEY'S MAGAZINE, 1 DECEMBER 1854, P. 368

The first was written in Kolkata on 16 March 1832 to the editor of the *Bengal Hurkara and Chronicle* by A. Shekarea, which as a corruption of 'a shikari' is undoubtedly a pseudonym. The original publication was reprinted in an early issue of the *Oriental Sporting Magazine*. The story clearly piqued the curiosity of readers as it was reprinted at least 8 times in contemporary English and American magazines, translated into German 3 times, and still incorporated in anthologies in 1867 and 1872 (Shekarea 1832, 1835; Rainey 1872a). On a visit to Edmonton Island, the author heard about a rhino seen close to the residence at Middleton Point near the lighthouse on Saugor Island. A resident European joined Shekarea sitting up overnight at a tank and together they shot the animal as soon as it

appeared. The animal could escape when one of the guns exploded and injured the resident. The hunters returned 1½ month later and then killed the rhino on the third night. The horn was preserved together with strips of the hide. The dimensions, possibly exaggerated, were stated to be 12 ft (365 cm) body length, 2 ft (61 cm) length of tail, 7 ft (213 cm) high, and 13 ft (396 cm) circumference. The flesh was eaten by the Burmese crew of a passing vessel. The identity of the two hunters is unknown, the locality near Middleton Point is clear, and the date might be either 1831 or early 1832.

The second event was again widely circulated as "Destruction of a rhinoceros" in magazines and newspapers in 1838, with one reprint in 1854 (Anon. 1838, 1854a).

The simple illustrations accompanying two of these publications in the American *Parley's Magazines* of 1838 and 1854 bear no relationship with the actual events and give a general impression of a *R. unicornis*, not of the species or specimen actually seen (figs. 50.6, 50.7). In this case the hunters are named as Lieutenant Souter and Mr. Lewis of the pilot service, neither of whom can be identified. They heard that a rhino had killed several ryots (land tenants). Entering Saugor Island from the vessel, they killed the animal on the first night near a large tank. The body of the animal was retrieved on the following day, when some strips of hide were recovered to be taken to Kolkata. The two hunters are named, the locality is known only generally as Saugor Island, and the date is early 1838.

The third event was reported in the London *Times* in 1841 and then widely copied in the British press (Anon. 1841a). Three people entered the forest of Saugor Island one evening from a vessel moored on the shore. They killed a male tiger measuring 9 ft 1 in (277 cm), and saw clear footprints of a rhino. The people remain anonymous, the locality is on Saugor, and the date may be in the beginning of 1841.

50.9   Slaughter in the Salt Lakes

Entitled "Slaughter of a male rhinoceros in the Soonderbunds," there is a curious account known from the *Sydney Morning Herald* of 12 May 1854 and from an article authored by Auceps in the *New Sporting Magazine* of 1867. There is a wide discrepancy in the publication dates, and both must have had a precursor in an Indian newspaper or magazine which so far has remained elusive. Four residents of Kolkata working with the Pilot Service clubbed together to apply for a piece of land in the Sundarbans to be developed for agricultural purposes. When they arrived there, they found a dangerous rhino in the area, which soon after was killed by a friend who had come from Kolkata.

While Auceps only identifies the four persons with initials, their surnames are found in the Sydney newspaper. Using the listings in the *East India Register and Directory*, they can be identified as Second or First Mates working with the Marine Board (pilot service), named James H. Lewis, Frederick Broadhead, Thomas Pitkin and Colmer Symes. With the death of Symes in 1839, the events must have occurred in the 1830s, or even 1820s. A second scheme to develop the Sundarbans through grants of land started in 1816. The locality of the potential farm is unnamed, but clearly was within a few hours travel from town. Auceps states that the men initially "left Calcutta

in three beaulieughs (covered boats), and passing into the salt-water lake, arrived at their new settlement." The next day they were able to paddle their way back to Kolkata. The only salt lake in the region is on the eastern fringes of Kolkata, now known as a new suburbs. This might give a general, if imprecise idea where this rhino was found.

The animal was shot by Lewis from Calcutta, with the same surname as one of the pilots but he could have been the person who killed another rhino on Saugor Island in 1838 (50.8). Lewis preserved the head of this male rhino as a trophy, and cut some strips from the hide. Auceps states that "he was a bicornuous brute, which are to be met with only occasionally in the dense forests of the Indian Peninsula." There is no reason for this curious statement which is not found in the Sydney newspaper. Therefore it must have been a misinterpretation or faulty transfer of information. Nevertheless, the account is important as evidence of a rhino in the vicinity of Kolkata in the 1830s.

50.10   An Excursion of the Calcutta Tent Club

The Calcutta Tent Club was a loose association of men interested in the sport of pig-sticking. They would go out on irregular excursions within reach of the town usually just for a day (fig. 50.8). On one of these trips, the party heard that a rhino had been spotted, as reported in *Sport in Bengal* (1887). The author of this remarkable volume full of interesting facts was Edward Biscay Marinatus Baker (1828–1887), Deputy Inspector-General of the Bengal Police Force for 35 years, and the father of the famous ornithologist Edward Charles Stuart Baker (1864–1944). According to Baker, one of the gentlemen stayed behind at the end of the day to verify the villager's report, and was conducted the following morning to a patch of jungle on the banks of the Pealee river, a mile or two from camp, where he found fresh footprints of a large rhino. The news was conveyed to town, and next evening five members returned to the spot. They drove in their dog-carts to within a few miles of Barrapoor, where they mounted horses, and on reaching the Pealee River, they hired boats and traveled against the tide until an hour after daybreak. Here they found four elephants under command of G.G.M. The rhino was in a patch of jungle 8 to 10 acres in extent. The hunters waited, C.B.S. and Baker on one side, D.R.S. fifty paces west, and D. at another angle. When a huge bull rhino came out, all fired, but it ran away, and was not seen again. Local villagers had heard about these animals occurring somewhat further south, which could mean that rhinos had wandered north from the edges of the Sundarbans.

Percy Carpenter del. E. Walker lith.   London, Published Dec.r 2.nd 1861 by Day & Son, Lithographers to the Queen, W. Thacker & Co. London & Thacker, Spink & Co. Calcutta.   Day & Son, Lithog.rs to the Queen

N.o 8.— TENT CLUB AT TIFFIN.

FIGURE 50.8   The Tent Club at Tiffin. Plate painted by Percy Carpenter (1820–1895) depicting a scene on the Sowerra Burrea Plains in March 1860
CARPENTER, *HOG HUNTING IN LOWER BENGAL*, 1861, PL. 8

Baker's account is expansive, yet lacks details. The Pealee is the Piyali River, which runs to the east of Baruipur. He never gave a date, nor does he identify his companions except by initials. Baker (1887: 44–52) had introduced his involvement with the Tent Club in a previous chapter of his book. It had been in existence for about 25 years before it was disbanded in 1863. He mentioned judge James Patton and the merchant William Ferguson as successive presidents of the club which had no more than 12 members. Therefore the encounter with the rhino must have taken place in the early 1860s. A new Tent Club was formed in Calcutta in 1870 with up to 20 members (Shrestha 2009: 154–163). The manuscript records (British Library, Mss Eur C335) show that the men would travel on horses or elephants and would usually shoot less than ten boars. It was a popular pursuit among some classes of society, but a rhino was definitely a rare prize.

50.11   Adventures Invented by Philipp Körber

As an insight in the perception of popular western culture, the rhino appeared in adventure stories set explicitly in the jungles of the Sundarbans. These introduced the animal to the European readers through word and illustration.

The German teacher Philipp Wolfgang Körber (1811–1873) was a prolific writer of adventure stories for children with a moralistic twist. His hero Julius Bath and friends traveled to the Sundarbans, where they were chased into a tree by a rhino (Körber 1853). The unsigned illustration of this fictitious scene is far removed from reality and shows a rhino closest to *R. unicornis* (fig. 50.9). The artist can be excused because *R. sondaicus* was not yet included in the Indian fauna by 1853.

FIGURE 50.9   "Noch schlug es sterbend mit dem Horn." (Even close to death, it continued to wield its horn and jerk its feet). Unsigned plate illustrating an adventure story by Philipp Wolfgang Körber (1811–1873) set in the Sundarbans
KÖRBER, *JULIUS BATH'S MALERISCH-ROMANTISCHE REISE NACH CALCUTTA*, 1853, FRONTISPIECE

50.12   Sundarbans Fiction of Emilio Salgari

The adventure stories by the Italian writer Emilio Salgari (1862–1911) were immensely popular in his home country ever since the first one appeared in 1895. They have continued to be in print in a wide variety of formats, they have been translated into many languages, and some even made it to the big screen (Torri 2012). Ten novels in the Sandokan series are all set in the Indian environment and are filled with fast-paced action and unexpected thrills. The first novel of this series *I Misteri della Jungla Nera* (Mystery of the Black Jungle) of 1895 is set in the Sundarbans of 1851. The tiger hunter Tremal-Naik sees the fleeting apparition of a beautiful young woman. Together with his servant Kammamuri he sets off to find her but she is held hostage by thugs and worshippers of the goddess Kali. One terrible night, they are attacked by a tiger, close to death, when the situation is saved by a rhino changing the scene. In the next novel, Salgari introduces his hero Sandokan, who comes face to face with a rhino in the Sundarbans in 1857 as told in *Le Duo Tigre* (The two tigers) of 1904.

Places, dates, characters, they were all fictional. Why Salgari set these adventures in the Sundarbans is a question beyond my remit. He was right of course that there were plenty of tigers, pirates, thugs, dangers, and even the occasional rhino in the Sundarbans. Through him this fact became well-known in Italian popular culture, and through translations elsewhere. A rhino was illustrated in several editions of Salgari's novels throughout the 20th century. The artists imposed their own interpretations on what they believed a Sundarbans rhino could

FIGURE 50.10
Tiger attacked by a single-horned rhinoceros in the Sundarbans, following the fictional story of Tremail-Naik by Emilio Salgari (1862–1911). The scene is imaginary, the rhino shows a horn never seen in the wild. Illustration by the Italian artist Giuseppe Pipein Gamba (1868–1954) in the first edition of 1895
SALGARI, *I MISTERI DELLA JUNGLA NERA*, GENOVA, A. DONATH, 1895. COURTESY: FELICE POZZO

FIGURE 50.11 Rhinoceros and tiger in the Sundarbans. From the fiction of Emilio Salgari, this illustration by the French artist Edouard-Auguste Carrier shows a rhino with an impossibly long horn
SALGARI, *LES MYSTÈRES DE LA JUNGLE NOIRE*, PARIS, MONTGREDIEN, 1899. COURTESY: FELICE POZZO

FIGURE 50.12 An imaginary scene in the Sundarbans after the novel by Emilio Salgari. Drawing unsigned
SALGARI, *I MISTERI DELLA JUNGLA NERA*, MILANO, CARROCCIO, 1950. COURTESY: FELICE POZZO

have looked like, which in reality was far removed from what occurred. None of these illustrated rhinos showed the characteristics of *R. sondaicus*, and it is likely that neither Salgari nor the artists were aware of the zoological complexity. The first edition of *Jungla Nera* of 1895 was illustrated by Giuseppe Gamba, who was followed in portraying the encounters with the rhino by other anonymous artists (figs. 50.10, 50.11, 50.12, 50.13). Far removed from the world of scholars, but exerting great influence in the perception of the rhino in popular culture.

50.13 Last Vestiges and Extinction

R. sondaicus became extinct in the Sundarbans of West Bengal and Bangladesh no later than 1910. Maybe the Sundarbans weren't really remote, but they remained inaccessible and underdeveloped. In such an inhospitable landscape, it is hard to find a good reason for the rhino's decline and extinction. It would be unfair to blame the handful of foreign hunters who have left accounts of their experiences. Habitat destruction would be mainly small scale, although definitely with an expanding population, a search for new lands would have entailed an eradication of all possible dangers. Local hunters were known to have killed scores of rhinos and this must have had enough of an impact to lead to the disappearance of rhinos in the region.

An unexpected type of recommendation written by a journalist, not a tour operator, appeared in the *Englishman's Overland Mail* (and probably in other Calcutta based newspapers) on 25 November 1891:

FIGURE 50.13
Rhinoceros in the Sundarbans, drawn by an anonymous artist to illustrate the story by Emilio Salgari. Again this is not like the animal in the wild
SALGARI, *I MISTERI DELLA JUNGLA NERA*. BOLOGNA, CARROCCIO-ALDEBARAN (COLLANA NORD-OUEST NO. 39), 1958. COURTESY: FELICE POZZO

"sportsmen with leisure and the command of a steam launch, can within a day's run from Calcutta secure with ease rhino, tiger, and deer. Bulcherry and Halliday Islands, off the mouth of the Mutlah river, swarm with rhinoceros, and though the jungle is heavy, it can be penetrated in many places." Edmond De Poncins (§51.10) might have read that and arranged to do just as the newspaper said, even though he did not believe that there were still ten rhinos left. Rhinos may not have been swarming in large groups, but they still existed.

References to the presence of rhinos in the Sundarbans vanish at the start of the 20th century. The last well-documented sightings were those near Ishwaripur by Prince Henri d'Orléans in 1888 and Edmond de Poncins in 1892. When a rhino was found dead near Diamond Harbour in 1902, even the villagers had never seen such an animal. Hopeful references to the continued existence in later years became increasingly indeterminate.[1]

---

1   This chapter has benefited from reviews of earlier drafts by Helmut and Gertrud Denzau.

Historical Records of the Rhinoceros in the Eastern Sundarbans

First Record: 1599 – Last Record: 1909
Species: *Rhinoceros sondaicus*
Rhino Region 40

51.1   Rhinos in the Eastern Sundarbans of Bangladesh

Rhinoceros sondaicus was the only species of rhinoceros ever found in the Sundarbans (Rookmaaker 1997a). This region in the delta of the Ganges and Brahmaputra Rivers extended from the Hooghly in the southern part of West Bengal eastwards to the Meghna River, now administratively divided between India and Bangladesh (50.1).

Sightings of rhinos have been most frequent between the Raimangal River in the west and the Baleshwar River in the east. These were concentrated in the surroundings of Mongla and Ishwaripur, where rhinos continued to be observed until the end of the 19th century. In the eastern part, there are just two records east of the Baleshwar River dated 1642 and 1855.

The rhinoceros had largely disappeared from the western parts of the Sundarbans by the 1850s (50.1). Hence all the general references to rhinos in the Sundarbans after 1850 are most likely to refer to the eastern part, although the early authors probably would not have differentiated these areas. The Indian Museum had a remarkable collection of *R. sondaicus* specimens consisting of skins mounted for exhibition, several complete skeletons, as well as skulls, horns and loose hides. As many of these items were added in the course of the 1870s and 1880s, the animals must have been obtained east of the Raimangal River (51.11). The history of this collection, originating in the Asiatic Society of Bengal and more recently transferred to the care of the Zoological Survey of India, is explored in this chapter. This information provides valuable insight to scientists using these specimens up to the present (51.12, 51.13).

The rhinoceros was still seen along the Kadamtala River south of Ishwaripur in 1888 and 1892. Although consisting of muddy and marshy terrain, it was within relatively easy reach of Kolkata or Dacca. This must have been the area where the remaining rhino population was concentrated. The rhino probably didn't disappear overnight after Viscount de Poncins still observed a few in this specific locality in 1892. Records become very sparse afterwards and it is unlikely that *R. sondaicus* continued to exist after 1910 at the latest (50.13).

• • •

Dataset 51.60: Chronological List of Records of Rhinoceros in Eastern Sundarbans (Bangladesh)

General sequence: Date – Locality (as in original source) – Event – Source – § number – Map number.

The map numbers are explained in dataset 51.61 and shown on map 51.24.

The records are divided in two sections for Rhino Region 40:

1. General records for eastern part of Sundarbans: without further locality
2. Eastern Sundarbans between Raimangal and Meghna Rivers, Bangladesh

Rhino Region 40: Bangladesh – Eastern Sundarbans, Records without Exact Locality

This includes most *R. sondaicus* specimens in the Indian Museum.

1850 – The Indian Museum obtained a rhino specimen from Charles Huffnagle (1808–1860), listed in Blyth (1863a) as *R. sondaicus* "stuffed specimen, under 3½ feet [106 cm] high." It was probably a specimen of *R. unicornis*, now no longer recognized in a museum collection – Blyth 1850: 88, 1863a: 137; Cockburn 1883: 60; Anderson 1916: 8 – 51.11

1862 – Rhino still common in eastern Sundarbans – Blyth 1862a: 151

1864 – *R. sondaicus* skull in Indian Museum, transferred from Asiatic Society of Bengal. No history – Sclater 1891, no.o, now ZSI 17689

undated (after 1864) – *R. sondaicus* skin of juvenile male in Indian Museum. No history – Sclater 1891, no.k – 51.13

undated (after 1864) – *R. sondaicus* lower jaw (mandible) in Indian Museum. No history – Sclater 1891, no.v, now ZSI 20386

undated (after 1864) – *R. sondaicus* lower jaw (mandible) in Indian Museum. No history – Sclater 1891, no.w, now unknown

1867 – *R. sondaicus* skull donated by W.W. Sheppard to Indian Museum, still in collection ZSI 19241 – Indian Museum 1868; Sclater 1891, no.q; Groves & Chakraborty 1983: 254, skull measurements in table Ia – 51.12, fig. 51.13.

1868 – *R. sondaicus* male skeleton purchased for Rs 10 by Indian Museum – Indian Museum 1869; Sclater 1891, no.u

1870 – Last rhino collected in Sundarbans. Displayed in Indian Museum – Barbour & Allen 1932, pl. 11; Mukherjee 1980: 124; Das 2002; Mallick 2011: 59 (found in excavated pond) – This refers to the specimen in Sclater 1891 no.b collected in 1872, see next entry (1872).

© L.C. (KEES) ROOKMAAKER, 2024 | DOI:10.1163/9789004691544_052
This is an open access chapter distributed under the terms of the CC BY-NC-ND 4.0 license.

1872 – *R. sondaicus* remains of young female donated to Indian Museum by J.F. Barkley, Inspector of Tolls. The hide was mounted and the skeleton set up. Hide still in display gallery, GRM 11 – Indian Museum 1873: 68,76; Sclater 1891, no.b; Barbour & Allen 1932, pl. 11; Chakraborty 2004: 40 – 51.12, 51.13, figs. 51.15, 51.16

1874 – *R. sondaicus* female hide, skeleton, skull obtained by Oscar Louis Fraser (1848–1894) and J.F. Barkley for Indian Museum (no longer present) – Fraser 1875; Sclater 1891, no.a; Rainey 1874c – 51.12, fig. 51.14

1875 – *R. sondaicus*. The animal dealer William Rutledge (1830–1905) donated juvenile male skull to Indian Museum of an animal which lived only few days in Kolkata – Wood-Mason 1875; Sclater 1891, no.j – 48.2

1875 – In Indian Museum, H. James Rainey saw "a very young one of that species, about three months old." Possibly the mounted specimen still on exhibit listed as GRM 273 – Rainey 1875c; Chakraborty 2004: 40 – 51.13, fig. 51.18

1876–09 – Rhino captured alive for William Rutledge, who sold it to Wajid Ali Shah, the exiled King of Oudh, for his establishment at Metiabruz, Kolkata – *Times of India* 1876-09-09 – 48.2

1877 – *R. sondaicus*, young female taken alive to London by William Jamrach; died after 6 months – Sclater 1877b; Garrod 1877a,b – 48.2, 51.8

1879 – Author (initialed W.) on a hunting expedition in unknown locality in Sundarbans first finds female together with a male ¾ grown, which had "only a small horn." He shoots the full-grown adult female. Finds more rhinos in the same area – W. 1879

1879 – Local shikaris (hunters) regularly shoot rhinos in the region – W. 1879

1879 – *R. sondaicus* juvenile male skeleton and skin donated by William Rutledge to Indian Museum – Sclater 1891, no.g – 48.2, 51.12

1880 – *R. sondaicus* juvenile female skeleton and skin donated by William Rutledge to Indian Museum – Sclater 1891, no.h – 48.2, 51.12

1880 – *R. sondaicus* exhibited in Alipore Zoo, 1880 to 10 April 1881 – Buckland 1890a; Rookmaaker 2011a – 48.2

1880 – *R. sondaicus* juvenile male skin and skeleton purchased by Indian Museum – Sclater 1891, no.e – 51.12

1881 – *R. sondaicus* juvenile male skin and skeleton and skin donated by William Rutledge to Indian Museum – Indian Museum 1881; Sclater 1891, no.f – 48.2, 51.12

1886 – Rhino present – Simson 1886: 188

1893 – Last rhino of Sundarbans displayed as mounted specimen in Indian Museum – Mandal & Nandi 1989: 62 – 51.13

1950 – Rhino still present in 1950 – Sundarbans Project 2003: 14

Rhino Region 40: Bangladesh – Eastern Sundarbans between Raimangal and Meghna Rivers

sub-fossil (undated) – Dhumkhola, 3 km south of Ishwaripur – Gazi Mohammad Sekandar Hossain (d. 2014), village doctor in Dhumkhola village, found a rhino skull. Shown by his son Assadure Rahman Shamim Hossain, a retired army officer, to Helmut and Gertrud Denzau in 2018; photos taken – 51.3, fig. 51.3 – map S 133

sub-fossil (undated) – Harinagar, near Shyamnagar – Skeleton of large rhino was found at house of Affaz Tulla Sarkar during pond excavation. Skeleton was preserved – Jalil 2000: 90 – map S 131

sub-fossil (undated) – Sreefal Kathi village (near Ishwaripur) – Rhino skeleton recovered – Jalil 2000: 90 – map S 133

sub-fossil (undated) – Dhumghat village (near Ishwaripur) – 6 rhino skeletons recovered – Jalil 2000: 90 – map S 133

sub-fossil (undated) – Ishwaripur – There are 6 skeletons preserved at house of Makhanlal Adhikari – Jalil 2000: 90 – map S 133

sub-fossil (undated) – Shyamnagar, Sathkira – Rhino skull excavated by local archaeologist, photographed by Isma Azam Rezu in 2021 – 51.3, fig. 51.4 – map S 131

sub-fossil (undated) – Shibsha River near Adachai FD camp – Rhino bone excavated by local archaeologist, photographed by Isma Azam Rezu in 2021 – 51.3, fig. 51.5 – map S 137

sub-fossil (undated) – Angrakona Khal, on Arpangasia River – Rhino molar excavated by local archaeologist, photographed by Isma Azam Rezu in 2021 – 51.3, fig. 51.6 – map S 136

1599 – Towards Chandecan (Jessore region) – The Jesuit priest Melchior Fonseca heard that rhinos were roaming the forest, but none were seen – Jarric 1614, vol. 3: 833; Hosten 1925: 64; Beveridge 1876: 175 – 51.2

1642 – Xavaspur Island in the estuary of the Meghna River – Sebastien Manrique (1585–1669), a Portuguese Catholic priest of the Augustinian Order "came across many Rhinos ("Rinocerontes"), whose horns, offensive in life, are after their death used in a defensive drug" – Manrique 1649: 222, 1927: 395; Rookmaaker 1997a: 38 – 51.2 – map S 141

1800 – Srifalkati and Khagraghat (Ishwaripur area) – Rhino present early 19th century – Jalil 2000: 90 – map S 133

1800 – Ganra Khal, Gandarkhalib (means: rhinoceros creek) – Rhino present early 19th century (possibly based on name) – A.Gupta 1964: 236; Jalil 2000: 90 – (toponym) map M 891

1812 – Roymungol (Raimangal) River – Hugh Morrieson (1788–1859) and his brother William Elliot Morrieson (1791–1815) killed a rhino during survey of Sundarbans, and found tank with many traces of rhino – Morrieson 1859: 19 – 51.4 – map S 135

1812 – Mouths of Roymungul (Raimangal) and Mollinchew (Malancha) Rivers – Hugh Morrieson found the land infested by rhinos – Morrieson 1859: 19 – 51.4 – map S 135

1828 – Mongla area – Christoph-Augustin Lamare-Picquot (1785–1873) shoots a female and calf, preserved in Museum für Naturkunde, Berlin – Lamare-Picquot 1835, 1835b; Jomard et al. 1832; Rookmaaker 1997a, 2019c – 47.9, 51.5, figs. 47.5, 47.6, 47.7; figs. 51.7, 51.8, 51.9 – map S 137

1834 – Bagundee (Bagundi, Basirhat) – John Henry Barlow (1795–1841) kills female rhino. Remains donated to Asiatic Society of Bengal, later Indian Museum – Prinsep 1834: 142; Pearson 1840: 518; Blyth 1841a: 841, 1841b: 928, 1862a: 156, 1863a: 137; Sclater 1891: 203, no.c; Barbour & Allen 1932: 147 – §51.12, fig. 51.12 – map S 131

1842 – No locality – *R. sondaicus* mounted specimen (with horn) dating from 1842 preserved in Natural History Museum, Mainz from the Rheinische Naturforschende Gesellschaft. During restoration in 1980s, it was found that the head belonged to a male, the back part to a female – Renker et al. 2018

1855 – "Isla Foolzurree" (Phuljhuri) – Francis (Frank) Bruce Simson (1827–1899) kills one rhino, and saw at least 6 others – Simson 1886: 190; Rookmaaker 1997a: 40 – 51.6 – map S 140

1855 – Burrisawel (Barishal) – Horn offered for sale – Simson 1886: 192

1857 – Boyrah Bheel – Rhino said to exist; never seen their footprints – Smyth 1857: 38 – map S 132

1857 – Pergunnah Dhooleapoor (Dhuliapur) – Surveyor Ralph Smyth (1812–1886) stated that rhino visit the low lands – Smyth 1857: 38; Hunter 1875a, vol. 1: 37 – map S 132

1860 – [Khulna] – Despite searches, fails to find rhino – Rainey 1860 – map S 137

1867 – Morrellgunge (Morrelganj) – Tracks were found on spot where Henry Morrell has his house – R.H.D. 1867: 292 – map S 138

1867 – Mouths of Roymungul and Mollinchew Rivers (see 1812) – Rhinos very numerous before 1867 cyclone – Baldwin 1910: 367; Jalil 2000: 90 – map S 135

1870s – Eastern Soonderbuns, near coast – Edward Biscay Marinatus Baker (1828–1887) kills male with horn, and a female. Head and horn preserved – Baker 1887: 258, 274; Rookmaaker 1997a: 40 – 51.7 – map S 140

1870s – Eastern Soonderbuns – Baker kills a second male – Baker 1887: 258, 274 – 51.7 – map S 140

1872 – [Khulna] – Friend saw rhino swimming across a wide river, not killed – Rainey 1872b: 139, 1872c: 302 – map S 137

1873 – Chandesar (Chandeshar), where Baleswar and Haringhata Rivers flow into Bay of Bengal – The cartographer James Ellison (1838–1891) inserted the word "Rhino:" on a new map – Ellison 1873 – fig. 51.2 – map S 139

1874 – "Mooneypoor" corrupted for a place in the Sundarbans – *R. sondaicus* female captured, seen in Kolkata in May 1874, transported by William Jamrach to London arriving 16 June 1874 and then to Berlin Zoo arriving on 30 June 1874; date of death unknown. Type of *Rhinoceros jamrachi* Jamrach, 1875. There is a locality Manipur or Monipur in the Sundarbans near Atkira (22.45N; 89.24E), however the name mentioned by Jamrach is more likely to have been transmitted incorrectly – Jamrach 1875, 1876 – §47.13

1875 – Matabangah River (Mathabanga Khal, Maithbhanga Khal) – *R. sondaicus* male skull and feet purchased by Indian Museum, still present as zsi no. 17688 – Sclater 1891 no.s; Groves & Chakraborty 1983: 254, skull measurements in table Ia – 51.12, fig. 51.16 – map S 134

1875 – South of Issuripore (Ishwaripur) – Hunting party of 4 people sighted 2 male and 1 female rhino. One male was killed. Skin forwarded to Asiatic Museum, Calcutta – *Sheffield Daily Telegraph* 1875-11-18. This was possibly Sclater 1891 no. s, which refers to a skull, not a skin – 51.12 – map S 134

1875 – Khulna – Rhino present. Rainey saw innumerable *R. sondaicus* in all stages of growth – Rainey 1875a, 1878 – map S 137

1876 – Ray Mangal (Raimangal) River – *R. sondaicus*, young female obtained alive by William Jamrach (1842–1923), died in Kolkata within 24 hours – Sclater 1876b; Rookmaaker 1997a – 48.2 – map S 135

1879 – Chillipangpi (Chillichang) Creek, corruption of Chili and Chandpi (Chandpai) near Mongla – *R. sondaicus* female skull from "Capt. Charling" registered in Indian Museum, still in collection zsi 3521. Charling is an error for Charles Sharling (b. 1839), captain of steamers of the Indian General Steam Navigation Co. (*Bengal Directory* 1878, p. 800) – Sclater 1891, no.t; Groves 1967; Groves & Chakraborty 1983: 254, skull measurements in table Ia; Ghosh et al. 1992a – 51.12 – map S 137

1885 – Ishwaripur region – Raisaheb Nalini Bushan, brother of Acharya P.C. Roy, reported the sighting of the last rhino in 1885 – Jalil 2000; Bahuguna & Mallick 2010: 282 – map S 133

1886 – Backergunge (Bakergonj) – No rhino ever known near Backergunge – Simson 1886: 189

1888 – Khulna division – Records in Sundarbans Divisional Office at Khulna mention killing of rhino – Gupta 1964: 236 – map S 137

1888-02 – Ishwaripur region – Prince Henri d'Orléans (1867–1901) observes a family of 6 rhinos. He was accompanied by Paul Marie Joseph Mercier de Boissy (1850–1897) and Antoine-Amédée-Marie-Vincent Manca Amat de Vallombrosa, Marquis de Morès (1858–1896) – Henri d'Orléans 1888; Cardane 1888; Droulers 1932: 59; Jacques d'Orléans 1999 – 51.9 – map S 134

1888-03 – Kudumtolle (Kuddum Tullee River, Pizon Khalee River, Kadamtala River) near Ishwaripur – Prince Henri d'Orléans and party heard rhinos in jungle – Henri d'Orléans 1889: 97,145; Jacques d'Orléans 1999: 60 – 51.9 – map S 134

1892-01 – Pizon Khalee River (Kadamtala River) – Edmond de Montaigne, Viscount of Poncins (1866–1913) sees 3 or 4 rhino – Poncins 1935; Rookmaaker 1997a: 41; Groves & Leslie 2011: 195 – 51.9 – map S 134

1896 – Baleshwar River – Tracks of rhino – Clay 1896: 110 – map S 137

1908 – Khulna region – Rhino rare, only found within the southern portion of the reserved forests – O'Malley 1908: 20,197 – map S 137

1909 – Backergunge (Bakergonj) – Rhino now nearly extinct – Hunter 1909: 363

1918 – Baleshwar River – Rhino not seen for many years, as for Sundarbans generally – Jack 1918: 11 – map S 137

• • •

Dataset 51.61: Localities of Records of Rhinoceros in the Eastern Sundarbans

The numbers and places are shown on map 51.24.

The Sundarbans are an area of mangrove forests, cultivated lands crisscrossed by large rivers and a multitude of smaller creeks. The courses of these waterways change with times. The team of Helmut Denzau, Gertrud Neumann-Denzau and Peter Gerngross have compiled a gazetteer with an atlas of geographical names found in this administratively complicated region, available as the Sundarbans Atlas of 2015 (Denzau et al. 2015). This resource has been of great assistance in clarifying some of the older place names attached to rhinos found here.

Rhino Region 40: Bangladesh – Eastern Sundarbans between Raimangal and Meghna Rivers

S 131	Bagundee (Bagundi, Basirhat) – 22.65N; 88.85E – 1834
S 131	Shyamnagar, Sathkira – 22.60N; 88.58E – subfossil
S 132	Boyrah Bheel – ca. 22.60N; 89.05E – 1857
S 132	Pergunnah Dhooleapoor (Dhuliapur) – 22.45N; 89.00E – 1857
S 133	Ishwaripur – 22.30N; 89.10E – subfossil, 1800, 1875, 1888
S 133	Dhumkhola village south of Ishwaripur, 22.25N; 89.10E – subfossil
S 134	Matabangah River (Mathabanga Khal) – 22.18N; 89.10E – 1875
S 134	Kudumtolle (Kuddum Tullee, Kadamtala River) – 22.15N; 89.17E – 1888
S 134	Pizon Khalee River – 22.15N; 89.17E – 1892
S 135	Mouth of Roymungul (Raimangal) River – 21.75N; 89.11E – 1800, 1812, 1867, 1876
S 135	Mouth of Mollinchew (Malancha) River – 21.75N; 89.25E – 1800, 1812, 1867, 1876
S 136	Angrakona Khal, on Arpangasia River – 22.02N; 89.26E – subfossil
M 891	Gandarkhal, Ganra Khal, Gander Khal – 22.43N; 89.34E – toponym
S 137	Adachai – 22.26N; 89.49E – subfossil
S 137	Mongla – 22.48N; 89.60E – 1828

S 137   Chillipangpi (Chillichang) Creek – 22.43N; 89.62E – 1879
S 137   Baleshwar River – 22.40N; 89.60E – 1896, 1918
S 137   Khulna division – ca. 22.50N; 89.65E – 1875, 1888, 1908
S 138   Morrellgunge (Morrelganj) – 22.46N; 89.53E – 1867
S 139   Chandesar – 21.90N; 89.85E – 1873

S 140   Isla Foolzurree (Phuljhuri) – 22.20N; 90.05E – 1855
*   Backergunge (Bakergonj), 20 km south of Barishal – 22.54N;
         90.33E – absent in 1886, 1909
S 141   Xavaspur Island (Bhola Island) – 22.30N; 90.75E – 1642

51.2 Portuguese Clergy in the 16th Century

The first definite record of a rhino in historical times in the Sundarbans region relates to a mission of Portuguese Jesuit priests traveling to the kingdom of Maharaja Pratapaditya (1561–1611), who had his capital in the Jessore region. Just after reaching Chandecan (in Jessore) on 20 November 1599, Melchior Fonseca reported that even then deer and monkeys were retreating away from civilized areas, and while the forests still had some rhino, he had no chance to see them.

51.3 Excavated Skeletal Remains

Skeletal material of *R. sondaicus* has been excavated in the Sundarbans in recent years (Jalil 2000). Most likely, such bones would have become buried in the course of the 19th century. One skull was collected from his neighbourhood by Dr. Gazi Mohammad Sekandar Hossain in Dhumkhola village in Shyamnagar Upazila. It is still preserved by his son Assadure Rahman Shamim Hossain in the ancestral house, where photographs were taken in 2018 by Helmut and Gertrud Denzau (fig. 51.3).

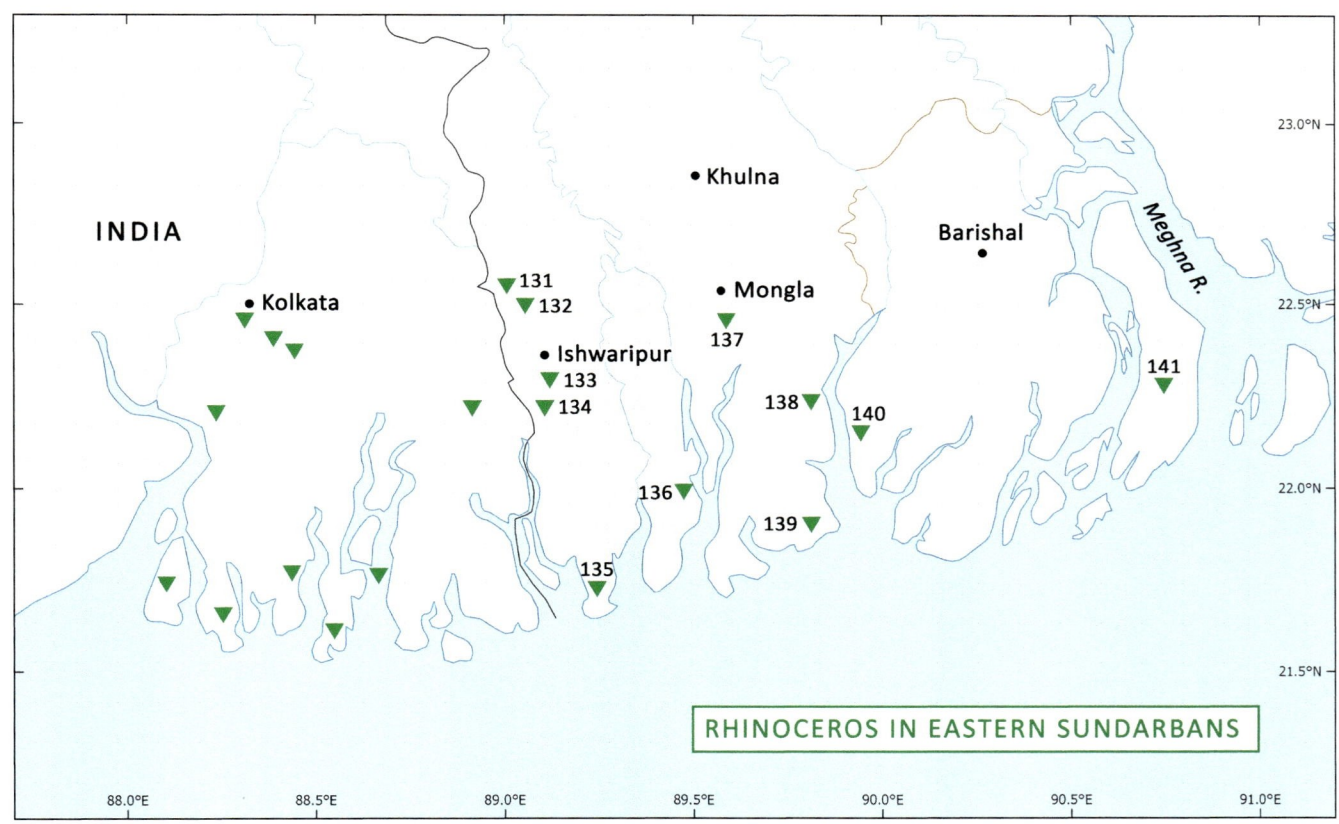

MAP 51.24   Records of rhinoceros in the eastern Sundarbans (Bangladesh). The numbers and places are explained in Dataset 51.61
             Presence of *R. sondaicus*

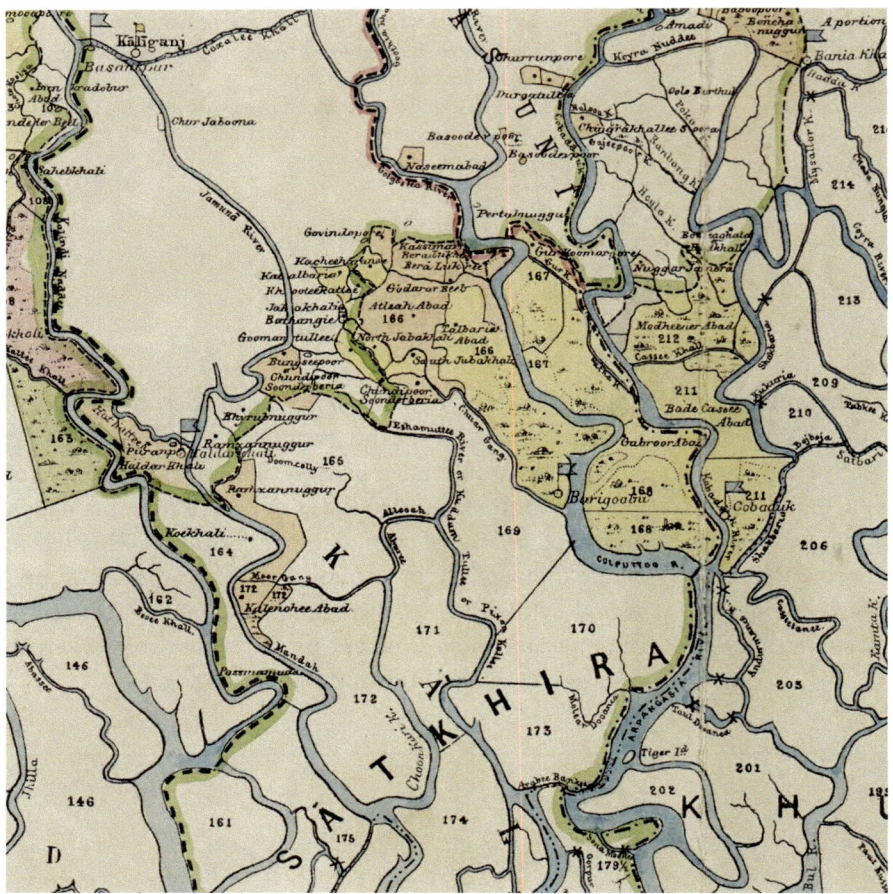

FIGURE 51.1
Section of the Map of the Sundarbans
prepared by James Ellison (1838–1891) in
1873, showing the region where many rhinos
were killed in the 1870s and 1880s south of
Ishwaripur, especially in lot numbers 165,
169, 170, 171, 172. Size 90 × 125 cm
UNIVERSITY OF MINNESOTA LIBRARIES,
JOHN R. BORCHERT MAP LIBRARY. MAP
DIGITIZED HTTP://PURL.UMN.EDU/246294

FIGURE 51.2
Detail of the Map of the Sundarbans
prepared by James Ellison in 1873. He wrote
the word "Rhino:" near Chandesar and
Tiger-Point on the Baleswar River
UNIVERSITY OF MINNESOTA LIBRARIES,
JOHN R. BORCHERT MAP LIBRARY

FIGURE 51.3
Skull of *R. sondaicus* in the private collection of G.M. Sekandar Hossain (d. 2014), Dhumkhola village in Shyamnagar Upazila, Sundarbans
PHOTO: HELMUT AND GERTRUD DENZAU, 2018

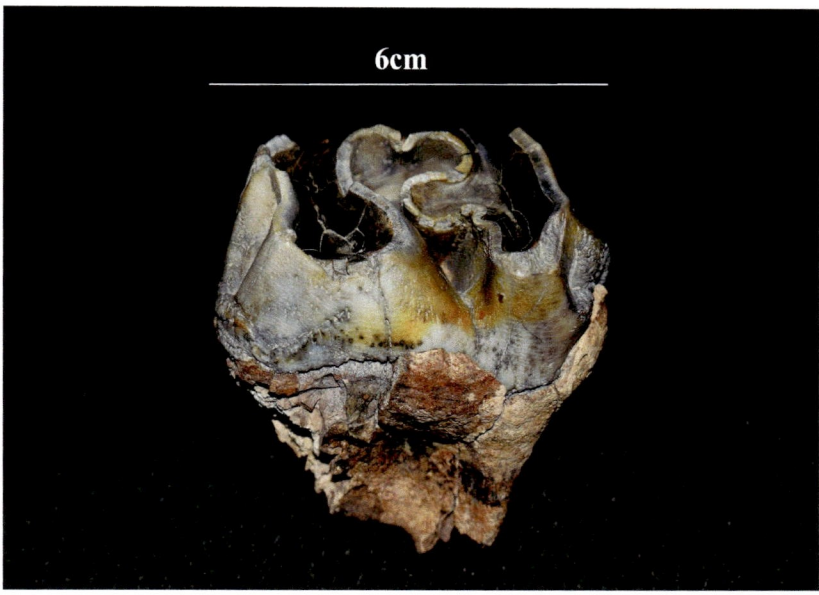

FIGURE 51.4   Rhinoceros skull and molar excavated by a local collector at Shyamnagar, Satkhira
PHOTO: ISMA AZAM REZU, 2021

Another skull was found around 2020 by a local collector at Shyamnagar, Satkhira, Bangladesh (fig. 51.4) and a molar and skeletal part were retrieved in the same general area figs. 51.5, 51.6.

51.4   The Sundarbans Survey of Hugh Morrieson in 1812

As part of the general Survey of India, the Sundarbans were mapped and investigated by Lieutenant Hugh Morrieson (1788–1859) of the 4th Regiment Bengal Native Infantry, and his brother Lieutenant William Elliot Morrieson (1791–1815) of the Bengal Engineers. Their notebook records how they saw a rhino drinking on the banks of a river, but they were unable to shoot it. The rest of their party had gone on shore in search of water and found that the only good well was surrounded by long grass, "the haunts of many rhinoceroses, they had made a regular bed in it." They were at the mouths of the Mollinchew (Malancha) and Roymungul (Raimangal) Rivers, where the land was reportedly "infested by rhinoceroses and deer, the whole ground being cut up by their feet" (Morrieson 1859).

51.5 The Hornless Female Rhinos of Lamare-Picquot 1828

The French pharmacist and explorer Christoph-Augustin Lamare-Picquot (1785–1873) shot two rhinos in the Sundarbans south of Khulna on Monday 17 November 1828. This was part of his extensive travels around India in 1821–1823 and 1826–1829 to make collections relating to ethnography, anthropology and natural history (Rookmaaker 2019c). From his base in the French factory of Chandernagor, he made one excursion to the

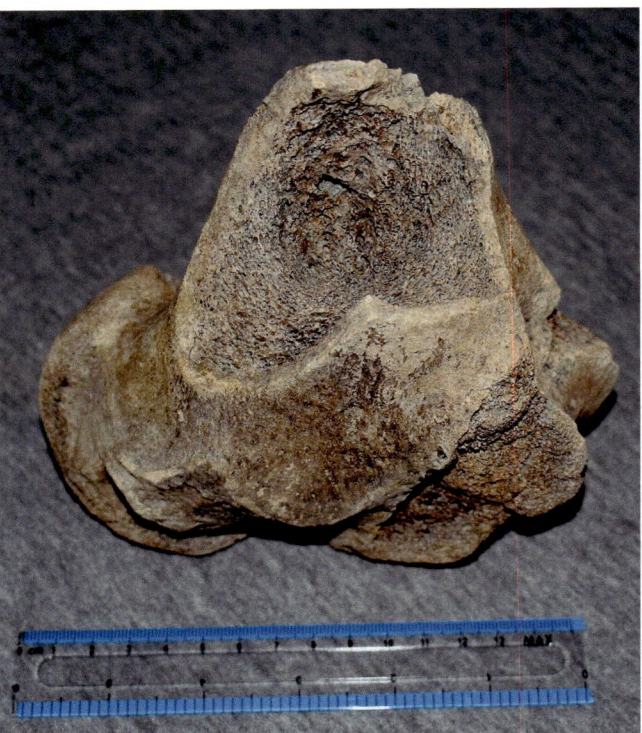

FIGURE 51.5 Rhino bone excavated on the bank of Shibsha River near Adachai Forest camp
PHOTO: ISMA AZAM REZU, 2021

Sundarbans, the account of which he published after his return home in a rather obscure appendix to a contribution about serpent venom, just 11 pages without illustrations (Lamare-Picquot 1835). He left on 2 November 1828 and was away for 42 days, to 13 December 1828. He had hired two large boats each with five local sailors to accommodate his two domestic servants and nine hunters said to be Portuguese, Indian and Muslim. From the Hooghly River just south of Kolkata at Keedrepoor (Khidirpur), he traveled along the 'canals' or waterways connecting to Dacca. His crew was unhappy when they reached the forested areas full of tigers and other wildlife after five days (6 November 1828), so he made his way to Kulna (Khulna). He engaged an additional six hunters experienced with the local situation, added a third boat, and on 15 November set out again southward towards the islands of the Sundarbans. Lamare-Picquot is vague about the exact locality of the rhino hunt, stating that it was over 60 miles ("plus de soixante lieues") from Kolkata. Going such a distance south of Khulna along the Passur River would have brought him south of Mongla. Here he found a group of woodcutters from Jessore, who told him that they had seen a rhino. He made his way to the location, and around 1 pm on Monday 17 November 1828 he was told that two rhinos had been spotted. He saw a large female rhino with her young, which he was afraid to scare away. Later in the day, his headhunter named Sobol shot the mother, gaining a reward of 30 rupees (75 francs). On examination, he found that this female rhino was absolutely hornless. He was careful to preserve the head, skin and bones of the animal. Then he tried to capture the young female animal, but when this proved impossible, she was shot. The experience allowed him to taste the milk of a rhino, sweeter than cow's milk and to try the meat of the young animal prepared as a steak and the liver of the mother, all said to be tasty.

FIGURE 51.6
Molar of rhino found at Angrakona Khal, on Arpangasia River, Bangladesh
PHOTO: ISMA AZAM REZU, 2021

FIGURE 51.7
Adult female *R. sondaicus* collected in 1828 by Lamare-Picquot in the region of Mongla, Sundarbans. Type specimen of *Rhinoceros inermis* Lesson, 1836. Museum für Naturkunde in Berlin, Germany, old number An 10603, new number ZMB_Mam_1957
PHOTO: JAN ROBOVSKÝ, OCTOBER 2018

FIGURE 51.8
The nasal region of the hornless adult female from the Sundarbans. Museum für Naturkunde in Berlin, Germany, new number ZMB_Mam_1957
PHOTO: JAN ROBOVSKÝ, 2018

The remains of these two rhinos from the Mongla area were shipped to Paris, where they were examined and described as a new species, *Rhinoceros inermis* Lesson, 1836 (45.9). In 1836, they were sold to Berlin (47.10), where they were mounted and are still preserved in the Museum für Naturkunde in Berlin, Germany (figs. 47.5, 47.6, 47.7 and 51.7, 51.8, 51.9).

51.6 Frank Simson at Phuljhuri (Barisal)

Francis Bruce Simson (1827–1899), Commissioner of Dacca, published his *Letters on Sport in Eastern Bengal* in 1886, combining incidents from 1847 to his retirement in 1872. While stationed in the 1850s at Backergunge (now Bakergonj), south of Barisal, he was told that no rhinos were known to reside near the town. The animal should be plentiful further south towards the shores of the Bay of Bengal. It took Simson two nights and a day by boat to reach the area, near a village established by some pirates of Arakan which he called 'Isla Foolzurree', a name still known as Phuljhuri on the Biskhali River. It did not take him long to find a rhino, which was killed, and the head removed. Another rhino managed to escape, but Simson thought that maybe the horn of this animal was soon afterwards offered for sale in Burrisawel (Barishal).

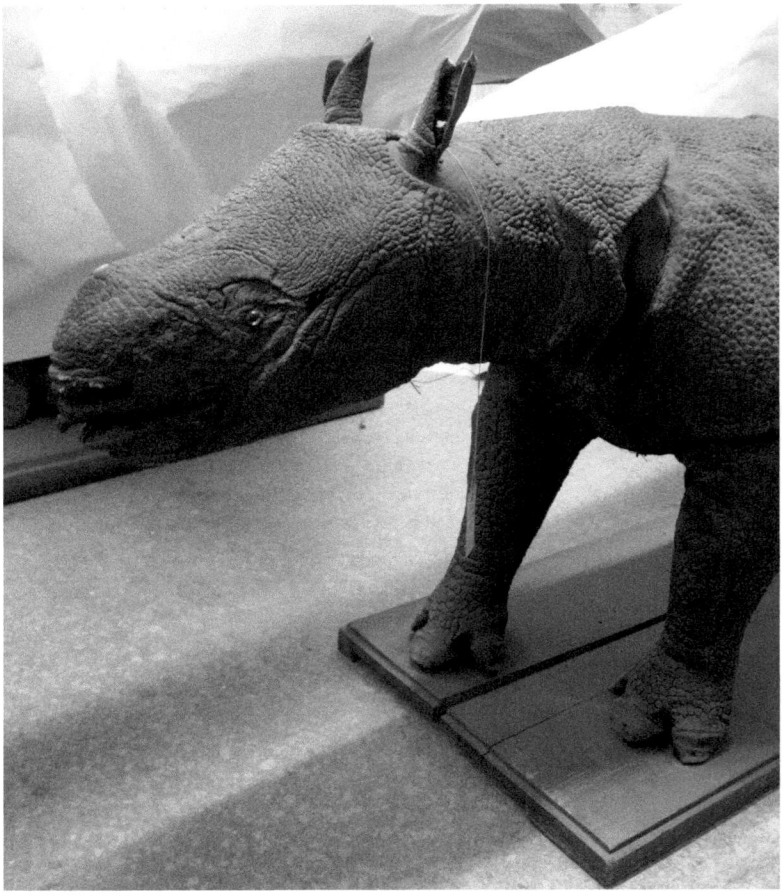

FIGURE 51.9
The young female from the Sundarbans. Possible
paralectotype of *Rhinoceros inermis* Lesson, 1836.
Preserved in the Museum für Naturkunde in Berlin, old
number An 10602, new number ZMB_Mam_1958
PHOTO: JAN ROBOVSKÝ, 2018

There were definitely more rhinos in the area, and Simson thought to have seen at least six, which were moving too fast to be targeted.

51.7   Edward Baker in the Eastern Sundarbans

Edward B. Baker, in full Edward Biscay Marinatus Baker (1828–1887), Deputy Inspector of Police in Calcutta, spent some weeks annually in pursuit of sports like pig-sticking and occasionally looking for larger game. On one occasion he had been one of a party organized by the Calcutta Tent Club which killed a rhino where normally they were not seen (50.10). His book *Sport in Bengal* published in 1887 just after his death in April of that year is a series of anecdotes of 40 years, with few indications of date or exact place. So one January, Baker went by boat through the eastern Sundarbans, and made his camp "within reach of an ebb tide of the mouth of one of the many rivers". This is what he saw: "On the margin of a mud-hole twenty or thirty feet in diameter stood a huge rhinoceros in deep contemplation of two shapeless slate-coloured lumps just showing above the muddy water; in other words, two companions enjoying a mud-bath, while he, having had his, as his

well-plastered hide testified, was basking in the sun half asleep, working his ears and stamping with a foot now and then as flies pestered him" (Baker 1887: 258). What a sight that must have been, one that would be worth its weight in gold to any tour company, yet Baker was not surprised to find three rhinos together taking a mud-bath. In fact, he summarily proceeded to shoot two of them, one a male "of the largest size, carrying a well-worn horn of moderate size", the other a full-grown cow. The heads and some parts of the hide were taken. Later, he shot a third rhino, "a large male, with a better horn than the other two had." Baker is one of the few people to observe male rhinos and state with confidence that they were carrying horns of reasonable size. He did not actually comment on the horn of the female, except in comparison, and his silence allows for it to have been very short.

51.8   Young Female Alive in London Imported by Jamrach

In the Zoological Society of London meeting of 6 March 1877, the Secretary Philip Sclater reported that the animal dealer Charles Jamrach had imported a young

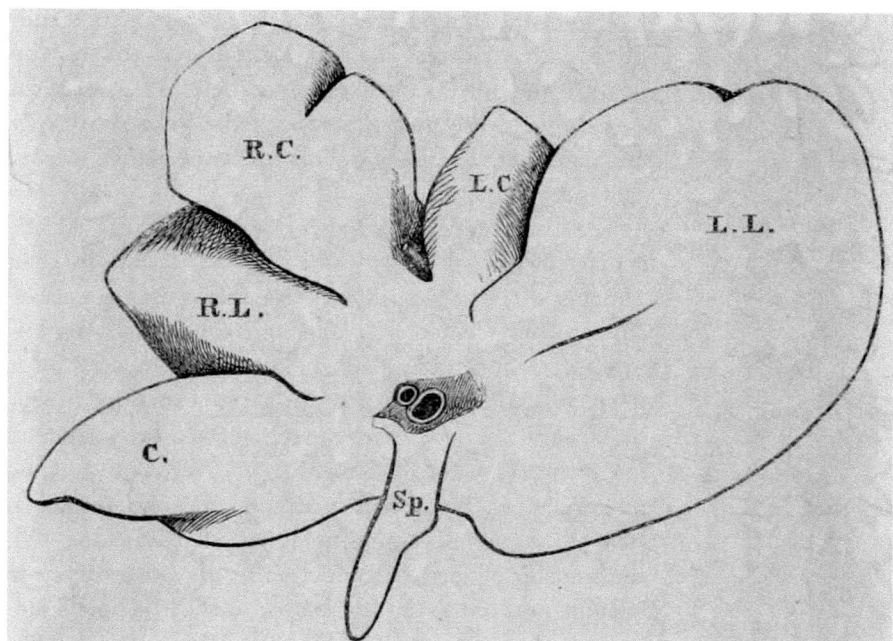

FIGURE 51.10
Liver of a female *R. sondaicus* which died in
the premises of Jamrach in London in 1877.
Dissection by Alfred Henry Garrod
*PROCEEDINGS OF THE ZOOLOGICAL
SOCIETY OF LONDON*, 6 NOVEMBER 1877,
P. 710, FIG. 3

female rhino from the Sunderbunds. It stood 3 feet (91.5 cm) high, and showed no sign of a horn (Sclater 1877). Apparently the animal was kept in Jamrach's premises on Ratcliffe Highway, London. There are no records why it remained unsold, but maybe she was not well, as she died some 6½ months later, probably early in September 1877. That date is suggested because on 1 October 1877 the zoo's prosector Alfred Henry Garrod (1846–1879) submitted two papers, one on a species of *Taenia* found on the animal (Garrod 1877a), the other on the visceral anatomy accompanied by figures showing the mucous surface of duodenum, the shape of the liver (fig. 51.10), and the surface of ileum (Garrod 1877b). The stuffed hide measured 6 ft 2 in (188 cm) from the tip of the nose to the base of the tail, while the tail added another 1 ft (30 cm). Garrod states that he borrowed the skull from the firm of taxidermists run by Edward Gerrard Jr. (1832–1927). There is no record of any of the remains being preserved.

51.9   Prince Henri d'Orléans at Ishwaripur in 1888

The French aristocratic explorer Prince Henri d'Orléans (1867–1901) arrived on 20 January 1888 in Kolkata. He was accompanied by two French officers, Paul Mercier de Boissy and the Marquis de Morès. Morès had brought his wife Medora (1856–1921), who came with the ambition to be the first woman to kill a rhino (*Pall Mall Gazette* 1888-02-03). They soon made their way south for a hunting expedition in the Sundarbans, where they engaged a local guide called Diptchai and two local

hunters, the muscular and handsome Chaiem and the small-statured Haim known to have slain six rhinos in his time. On 17 February 1888, Morès wrote excitedly in a letter from "Issaripore" that in company of the Prince he had approached a rhino within ten steps, but the jungle had been too dense to allow a shot. On the next day pushing through the mud up to their stomachs to follow the tracks, they observed six rhinos, which they followed closely all day long (Henri d'Orléans 1889: 97). On 10 March 1888, near the village Kudumtolle, just south of Ishwaripur, Chaiem found a good locality for rhino. They set out and could hear the animals foraging in the distance, with time proving insufficient to follow them in this difficult terrain (Henri d'Orléans 1889: 145). They managed to shoot a tiger, which remained particularly common and dangerous to the locals. Photographs taken during their 3 weeks hunting trip in the Sundarbans do not include the rhinoceros (Jacques d'Orléans 1999).

51.10   Viscount de Poncins on the Pizon Khalee in 1892

One of the last persons to pursue the Sundarbans rhinos, certainly the last to tell his story, was the French aristocrat Edmond de Montaigne, Viscount of Poncins. He traveled by boat through the Sundarbans in January–February 1892. Aware that rhinos had become rare, he suggested that there could be no more than six specimens surviving in a maze of forests with access to sweet water. He might have heard from the party of Henri d'Orléans that they

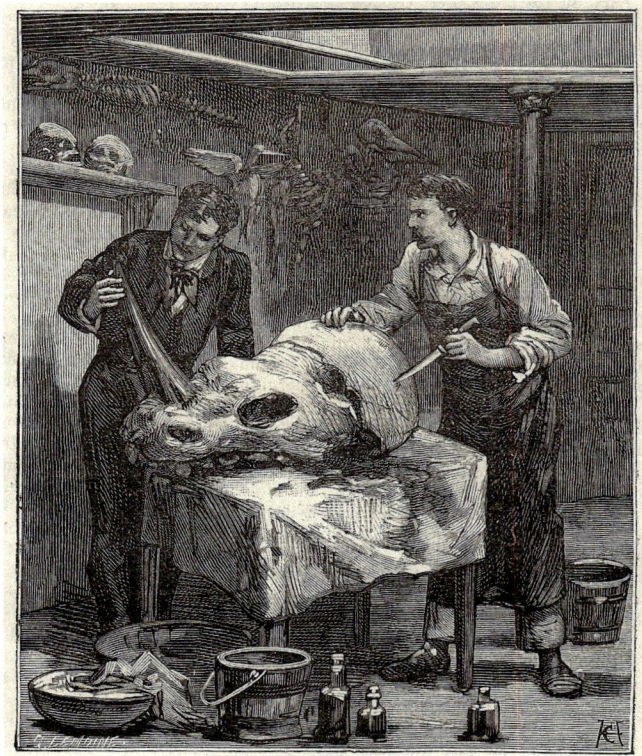

FIGURE 51.11   "Préparation de la tête du rhinocéros" (Preparing a
rhino head). A scene showing taxidermists working
on the head of a single-horned rhino illustrating a
fictional story by Louis Henri Boussenard (1847–1910)
set in Burma. Engraved by Georges Lemoine
(1859–1922) in France. The horn is too long for a true
R. sondaicus
BOUSSENARD, *JOURNAL DES VOYAGES,*
11 OCTOBER 1885, NO. 431, P. 228

encountered rhinos in exactly this area in 1888. Poncins
referred to islands numbered 165, 172, 171, 170, 169, which
are identified on the map of land grants published by
Ellison (1873) and are located along the Pizon Khalee
River, some 15 miles (24 km) south of Ishwaripur (fig. 51.1).
Only three rhinos were observed after tracking them for
many days: "For the first and, I am sorry to say, the last
time in my life I saw that long, grey, hornless head and
everything was explained: these rhinos were *R. sondaicus*, they had no trophy worth having and shooting them
was without excuse" (Poncins 1935). His restraint was
exemplary.

51.11   Specimens of *R. sondaicus* in the Museum in Kolkata

The Indian Museum in Kolkata had an incredible collection of *R. sondaicus* at the end of the 19th century, comprising 22 specimens. Probably 18 of these had come from

Jessore or the Sundarbans, even though a locality is only
provided in six cases. The documentation is not straightforward and needs careful consideration. All *R. sondaicus*
specimens in the museum supposedly from South Asia
are listed with available details in the chronological list
(dataset 51.60).

The first catalogue of the mammals in the Asiatic
Society of Bengal (ASB) was published in 1840 by John
Thomas Pearson (1801–1851). At this time *R. sondaicus* was represented by a skeleton collected by Barlow
in 1834 and a skull by Wallich, both listed incorrectly as
R. unicornis. Pearson's early efforts were expanded in 1863
by Edward Blyth (1810–1873), who curated the collection
located in Park Street from 1841 until his return home
in 1863 (4.10). Blyth was succeeded by John Anderson
(1833–1900), whose appointment in 1864 coincided with
the transfer of the ASB collection to the Government to
become part of the Indian Museum (IM). The new building on Chowringhee designed by Walter Long Bozzi
Granville (1819–1874) was completed in 1875, and opened
to the public on 1 April 1878 with two galleries for archaeology and birds. Anderson completed the first volume of
the new mammal catalogue (excluding ungulates) in 1881
before retiring in 1886. From 1869, Anderson was assisted
by James Wood-Mason (1846–1893), who became superintendent from 1887 to 1893.

William Lutley Sclater (1863–1944) arrived in Kolkata in
1887 engaged as deputy superintendent, and during his six
year tenure he completed the second volume of the mammal catalogue published in 1891. He noted that 3,542 specimens had been added to the collection since the time of
Blyth. If the section on the genus *Rhinoceros* is represent-
ative, Sclater completely overhauled the work, identifying
each specimen with a letter (following the contemporary
British Museum catalogues). As these letters differed from
the ones used by Blyth, apparently there was no accession
register which could be used as template. The material
available of the three Asian species of rhinoceros was
quite diverse. In 1891 a high proportion of specimens was
without known locality (table 51.70).

The collection was legally transferred to the Zoological
Survey of India (ZSI) in 1916 while maintained in the
building of the Indian Museum. In 1942 a large part was
moved to Lucknow due to safety concerns, but the osteological specimens remained in Kolkata. All was moved
to premises in Bhowanipore, Kolkata in 1964 and to the
ZSI Headquarters at 8 Lindsay Street in 1966 (Groves &
Chakraborty 1983). These moves resulted in losses both of
actual specimens as well as documentation. The humid
climate of Kolkata is detrimental to animal skins generally
and will have taken its toll (fig. 51.11).

TABLE 51.70   Specimens of the species of rhinoceros in the Indian Museum listed by W.L. Sclater (1891)

	Total		India		Burma		Elsewhere		No locality	
	1863	1891	1863	1891	1863	1891	1863	1891	1863	1891
R. unicornis	[1]	17	0	3	0	0	0	4	[1]	10
R. sondaicus	8	22	1	6	4	2 [3]	1	1	2	13 [12]
D. sumatrensis	8	18	0	0	6	4	1	7	1	7
	17	57								

The current Bangladesh is included in India.
The square brackets refer to uncertainties in the listings.

TABLE 51.71   Specimens of *R. sondaicus* from South Asia in the collection of the Asiatic Society of Bengal

Date	Donor	Sex, age	Skin (7)	Skeleton (9)	Skull (6)	Sclater 1891	Current	Locality after Sclater 1891. Additional references. Figure
1834	J.H. Barlow (from ASB)	F ad	–	skeleton	–	no. c	–	Jessore dist. Pearson 1840: 518 as *R. indicus*; Blyth 1862: 159, 175; Blyth 1863 no.b. See 51.12
< 1864	ASB	–	–	–	skull	no. o	ZSI 17689	–
1867 Aug	W.W.Sheppard	–	–	–	skull	no. q	ZSI 19241	Sunderbunds. IM Annual 1868: placed in museum Sep 1867. Lacks mandible. See 51.12
1868 Dec	Purchased Rs10	M	–	skeleton	–	no. u [1869]	–	Anderson 1869: skeleton set up.
1870s	Capt. Charling (= C. Sharling)	F	–	–	skull	no. t	ZSI 3521	Chillichang Creek, Sunderbunds. See 51.12
1872 June	J.F. Barkley	F juv	mount	skeleton	–	no. b	GRM 11 (mount)	Sunderbunds. IM Annual 1873. Skin displayed in gallery IM. See 51.12, 51.13.
1874	O.L.Fraser, J.F. Barkley	F	mount	skeleton	–	no. a	–	Sunderbunds. See 51.12
[1874]	O.L.Fraser, J.F. Barkley	–	foetus	–	–	[no. a]	–	Fraser 1875, Rainey 1875a. From female 1874 (no.a)
1875	W. Rutledge	M juv	–	–	skull	no. j	ZSI 19378	Badly damaged.
1875	Purchased	M ad	–	feet bones	skull	no. s	ZSI 17688	Matabangah R., Sunderbunds. See 51.12.
1879	W. Rutledge	F juv	skin	skeleton	–	no. g	–	–
1880	W. Rutledge	F juv	skin	skeleton	–	no. h	–	–
1880	Purchased	M juv	skin	skeleton	–	no. e	–	–
1881	W. Rutledge	F juv	skin	skeleton	–	no. f	–	–
[1881]	Alipore Zoo	M juv	–	–	skull	no. r	–	Exhibited Calcutta Zoo 1880–1881
1884	Babu Hurry Mohun Roy	M	–	skeleton	–	no. d	–	IM Annual 1885

TABLE 51.71   Specimens of *R. sondaicus* from South Asia in the collection of the Asiatic Society of Bengal (*cont.*)

Date	Donor	Sex, age	Skin (7)	Skeleton (9)	Skull (6)	Sclater 1891	Current	Locality after Sclater 1891. Additional references. Figure
n.d.	No history	M juv	skin	–	–	no. k	GRM 273, ZSI 76	On display in the Indian Museum, a 3-months old baby acquired before 1875 (Rainey 1875c). See 51.13.
n.d.	No history	–	–	lower jaw	–	no. v	ZSI 20386	–
n.d.	No history	–	–	lower jaw	–	no. w	–	–

Specimens are arranged in chronological order of accession. The main list follows the catalogue by W.L. Sclater of 1891, the current status follows Groves & Chakraborty (1983) and Chakraborty (2014). The elements belonged to 18 different individuals, plus one foetus extracted from one of these.*

* Abbreviations in table: ASB = Asiatic Society of Bengal. IM = Indian Museum. ad = adult. juv. = juvenile. M = Male. F = Female. Mount = mounted hide. n.d. = no data

When Sclater published his catalogue in 1891, the collection of *R. sondaicus* was definitely impressive, boasting no less than 22 specimens (table 51.71). As he combined all parts of the same specimen under one letter, the actual inventory is somewhat disguised. Unraveling this, he listed 7 skins of which 2 were mounted; 9 skeletons presumably all including skulls and horns; 6 single skulls; one set of foot bones; and 2 lower jaws. Three skulls were obtained outside South Asia, in Tenasserim (no.m), Tavoy Point (no.n) and Java (no.o). A fourth skull without locality (no.l, ZSI 17685) was donated by the surgeon Nathaniel Wolff Wallich (1786–1854) and is supposed to originate from lower Burma (Groves & Chakraborty 1983: 254). Without these four extralimital specimens (excluded from the discussions below), there were therefore 18 presumably from the Sundarbans or Jessore District.

Given the relative scarcity of *R. sondaicus* in world zoos, it may seem counter-intuitive to suggest that five of these specimens had been in captivity (nos. f, g, h, j, r), even if, presumably, only for a very short time (48.2). The indication that these were juveniles would support that possibility, as it is likely that young or small rhinos were captured in the Sundarbans and brought to Kolkata for sale. William Rutledge (1830–1905) was a well-known animal dealer in Kolkata in that period. I have no details about Babu Hurry Mohun Roy (no.d), but the skeleton could well have come from the Sundarbans.

There was a further specimen listed by Blyth (1863a) as *R. sondaicus*, "a. Stuffed specimen, under 3½ feet high [106 cm]", the size indicating a skin of a juvenile animal. Blyth (1850: 88) recorded it as a donation of a carcass of a rhinoceros, which was being mounted for the museum.

It was obtained from Charles Huffnagle (1808–1860) of Philadelphia, who was in Kolkata from 1826 and acted for a while as US Consul to India. Cockburn (1883: 60) seemed to recognize a mistake and said that it was "a characteristic stuffed specimen of *R. indicus* [*unicornis*]." There is no trace of a stuffed skin of similar description in Sclater (1891) or other later sources.

The fate of the 18 specimens available at Sclater's time is incredibly hard to determine because several of them appear to have lost all documentation. The annual reports of the Indian Museum available to 1910 record the exchange of several of their rhino specimens with other museums, unfortunately without details which could help to identify the actual individual. Enquiries sent to the museums concerned have not provided further data:

> 1881 – To Queensland Museum, Brisbane: 1 *Rhinoceros sondaicus*. Female (skin, skull and bones).
> 1895 – To Lucknow Museum: rhinoceros skeleton in return for the skeleton of a female elephant.
> 1899 – To Stavanger Museum: skeleton of rhinoceros (*D. sumatrensis*)
> 1902 – To University of Aberdeen: skull of a rhinoceros.

Barbour & Allen (1931) published a succinct list, almost certainly supplied at the time by the curatorial staff in Kolkata. This included 3 mounted skins, 7 skeletons, 4 skulls and 1 mandible, belonging to 13 different specimens. In the period from 1891 to 1931 therefore, the collection had lost 5 loose skins, 2 skeletons, 2 skulls and 1 mandible. However 1 mounted skin was added (51.13).

A more recent survey of the osteological material by Groves & Chakraborty (1983) provides a list of 5 skulls and 1 mandible (besides 3 extralimital specimens), but the status of "a nearly complete skeleton, one stuffed skin" (p. 252) are not discussed further. Chakraborty (2004: 40) tabulated 1 skull (from Java) and 2 mounted specimens in the mammal gallery in the Indian Museum, separated from the reference collection (51.13). This points at some further loss between 1931 and 1983, while the number of skins mounted or otherwise is quite confusing.

51.12 Six Rhino Specimens in the Kolkata Collection from the Sundarbans

The best fit for the current inventory of the collection of *R. sondaicus* in the Zoological Survey of India and the Indian Museum shows the presence of one mounted skin, one foetus (or baby), 6 skulls and 1 mandible (table 51.71). Fortunately these include 3 skulls with good locality data (Sclater's nos. q, s, t). The six specimens with stated provenance from the Sundarbans (Sclater's nos. c, q, b, a, s, t) will add to our knowledge of the range of rhinos from this region.

No. c in Sclater 1891 (Barlow 1834). At the 20 March 1834 meeting of the Asiatic Society of Bengal (ASB) it was minuted that "the skin and skeleton of a large Rhinoceros were presented for the museum, by J.H. Barlow" (Prinsep 1834). John Henry Barlow (1795–1841), the son of the Governor-General George Hilaro Barlow (1763–1846), worked for the last 27 years of his life as Magistrate and Superintendent of Salt Works at Bagundee or Baugundee (Bagundi, Basirhat), at the time part of Jessore District. The animal might have been killed in the vicinity of the station or just further south. The ASB proceedings at first referred to Barlow's specimen to include a skin and skeleton (Prinsep 1834). When Edward Blyth joined the Society as Curator in 1841, he found the skeleton but never mentioned the skin. The remains had first arrived in Kolkata at the Mint building on Strand Road, where they were prepared by W.E. Templeton and then donated to the Society by the orientalist James Prinsep (1799–1840) on behalf of Barlow (Blyth 1862a: 198). Because the rhino skeleton was much soiled and badly set up, Blyth (1841b: 928) gave orders to have it remounted, which was completed in October 1841. He noted, for the first time, that it lacked the digital bones of all four extremities, the sternal pieces, the penultimate pair of ribs, and one of the diminutive last pairs. In his revision of Asian rhinos, Blyth (1862a, pl. 1 fig. 2) referred to it as the broad type skull and it was included in a plate (fig. 51.12). The skeleton appeared in the catalogues by Blyth (1863a) "b. Skeleton of nearly full-grown female" and Sclater (1891) "c. skeleton." Although listed by Barbour & Allen (1932: 147), Loch (1937) and Choudhury (2016: 155), it was absent during the survey of Groves & Chakraborty (1983) and in fact might have been missing for a long time.

No. q in Sclater 1891 (Sheppard 1867). In August 1867, a skull of *R. sondaicus* from the Sundarbans was donated to the museum by W.W. Sheppard (fig. 51.13). It was among items prepared and placed in the museum in September 1867 (Indian Museum 1868). The skull was catalogued by Sclater (1891, no.q) and is now ZSI 19241, lacking premaxillae and lower jaw (Groves & Chakraborty 1983: 254). There is no information about the donor, whose name Sheppard is incorrectly spelled Shepperd in some later documents.

No. b in Sclater 1891 (Barkley 1872). In June 1872, the hide and skeleton of a young female *R. sondaicus* from the Sundarbans were donated by J.F. Barkley. He was

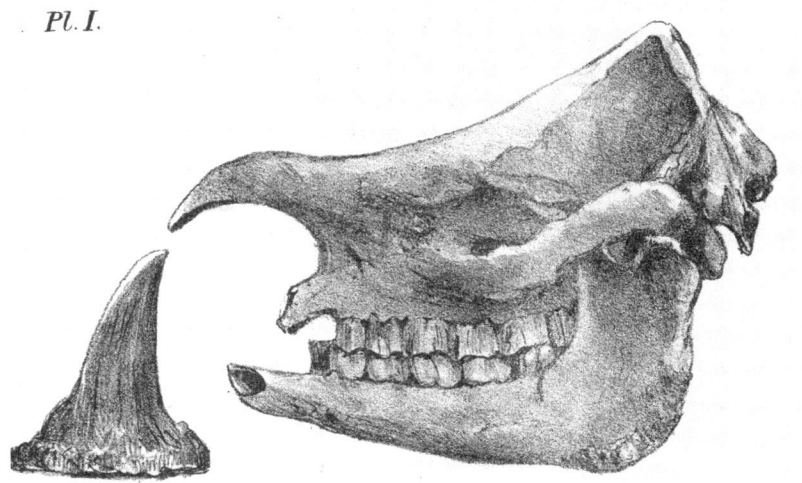

Pl. I.

f. 2.

FIGURE 51.12
Skull of *R. sondaicus* presented by John Henry Barlow in 1834 to the Indian Museum. The horn in the plate belongs to another specimen
BLYTH, *JOURNAL OF THE ASIATIC SOCIETY OF BENGAL*, 1862, VOL. 31, PL. 1 FIG. 2

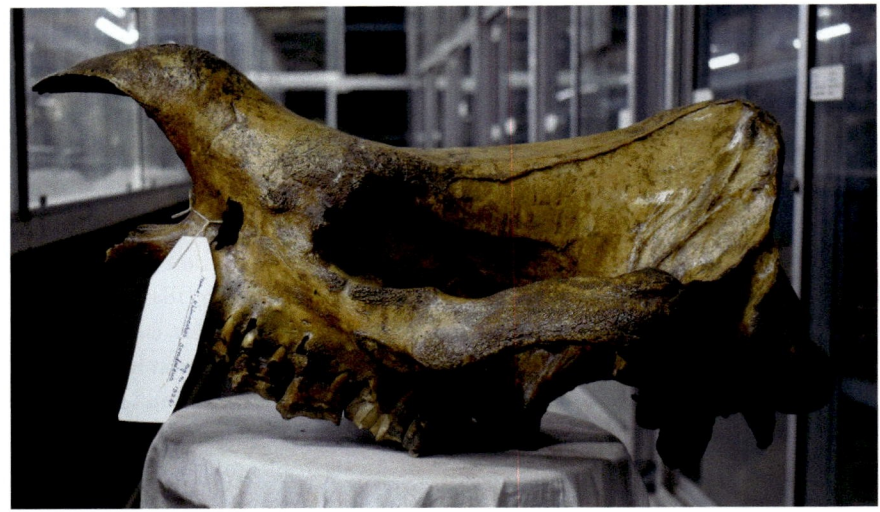

FIGURE 51.13
Skull of *R. sondaicus* in the Zoological Survey
of India, donated by W.W. Sheppard in 1867.
Sclater no. q, now ZSI 19241
PHOTO: © TANOY MUKHERJEE, 2018

Inspector of Tolls at Tolly's Nullah, which was the start of the main waterway towards the Sundarbans commencing at the southern end of Calcutta (*Calcutta Directory* of 1874: 36). The skin was mounted and the skeleton set up (Indian Museum 1873: 68, 76), both listed by Sclater (1891) as no.b. The skin is discussed in 51.13, while the skeleton is no longer in the collection.

No. a in Sclater 1891 (Fraser 1874). Oscar Louis Fraser (1848–1894) wrote a letter on 21 November 1870 from London to apply for the position of Osteologist at the Calcutta Museum (Indian Museum 1871). He was the son of Louis Fraser (1810–1866), who was a naturalist working at times for the Zoological Society of London and for Stanley, 13th Earl of Derby (1775–1851) at Knowsley near Liverpool (Moore 2004). Oscar Fraser arrived in Kolkata on 10 May 1871 and was promised accommodation in Kyd Street (Indian Museum 1871). He is listed as an assistant in the museum in 1888, and at his death on 1 November 1894 he is still attached to the Indian Museum (*Englishman* 1894-11-07).

In 1874 Fraser added a full specimen of *R. sondaicus* to the collection "lately obtained by me in the Sunderbuns." Sclater (1891) added J.F. Barkley, Inspector of Tolls, as a second donor. This female was 12 ft 3 in (373 cm) long and 5 ft 6 in (167 cm) high (Fraser 1875; Rainey 1875c), which are on the upper scale of *R. sondaicus* sizes. Her age was confirmed by the fact that she was found to be carrying a foetus (Fraser 1875; Rainey 1875a). Fraser (1875) wrote a paper discussing the ossified nasal septum observed in this skull, which surprised him (and Rainey 1875b) because the animal had been hornless. Different views of the skull were illustrated in a plate, now the only evidence of its existence, as it disappears from the records after 1891 (fig. 51.14). Fraser compared this skull with the

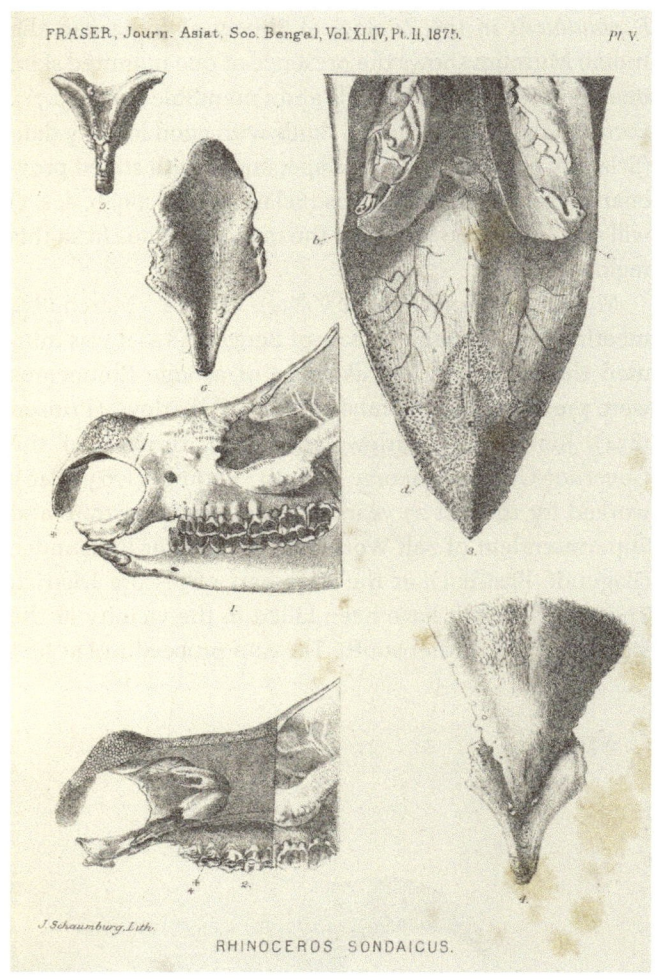

FIGURE 51.14   Skull of *R. sondaicus* illustrating the ossified nasal septum, described by Oscar Louis Fraser in 1875. Plate lithographed by Jules Henri Jean Schaumburg (1839–1886)
BLYTH, *JOURNAL OF THE ASIATIC SOCIETY OF BENGAL*, 1875, VOL. 44, PL. 5

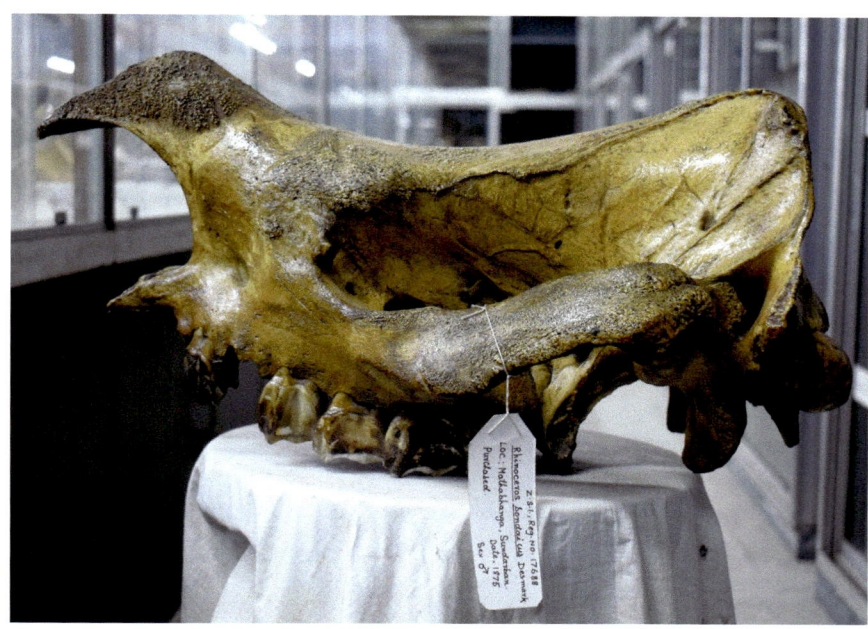

FIGURE 51.15
Skull of *R. sondaicus* in the Zoological Survey of India, no. ZSI 17688
PHOTO: © TANOY MUKHERJEE, 2018

adult female of Sheppard in 1868 (Sclater's q), and then added (possibly after his first draft) that he had examined two other specimens "shot at the same place", probably the adult male no.s (purchased in 1875) and the juvenile male no.j (Rutledge, 1875). While Sclater (1891, no.a) listed both a stuffed skin and skeleton (with skull), none of these specimens are currently recognized.

No. s in Sclater 1891 (Purchase 1875). In 1875, the museum purchased a male skull and foot bones of a *R. sondaicus* from the Matabangah River, Sunderbunds. The skull is still present, listed by Groves & Chakraborty (1983: 254) as ZSI 17688 (fig. 51.15). They locate the Mathabhanga River in Barisal District on the eastern end of the Sundarbans, where I cannot find it on the maps available to me. Denzau et al. (2015) in the *Sundarbans Atlas* include a Mathabanga Khal and Maithbhanga Khal close to each other in the area south of Ishwaripur, Sathkira District, where rhinos were still reported in the 1880s.

There was a report in newspapers in October 1875 about a party of four (anonymous) hunters who penetrated the jungle about 6 miles from Ishwaripur (*Sheffield Daily Telegraph* 1875-11-18). They sighted 2 males and 1 female rhinos in difficult terrain full of marshes and boggy soil. One male was killed, "so fine a specimen that his skin was forwarded to the Asiatic Museum." There is no record of this skin, while the circumstances are similar to the skull no.s.

No. t in Sclater 1891 (Sharling 1879). Sclater (1891) lists this female skull, obtained at Chillichang Creek, Sunderbunds by "Capt. Charling" without date. Groves (1967: 235, 1971: 245) added from museum information that there was

a distinct indication of a horn, occipital height 171 mm, breadth of upper M^3 51 mm, and gave a slightly changed locality as the lower mouth of the Chillichangpi Creek, obtained in 1879 or earlier. The skull is still in the collection, number ZSI 3521 (Groves & Chakraborty 1983: 254; Ghosh et al. 1992). The locality is a corruption of Chila and Chandpi or Chandpai, two streams close to Mongla south of Khulna. Charling was actually Captain Charles Sharling (b. 1839), listed as commanding the steamer *Calcutta* of the Indian General Steam Navigation Co. (*Bengal Directory* 1878, p. 800).

51.13 Mounted Hides in the Indian Museum

Sclater listed two "stuffed" hides of *R. sondaicus* in the collection in 1891, the adult female of Fraser (no.a) and the juvenile female of Barkley (no.b). Writing on 23 April 1874, Rainey (1874c) mentions a skin of a large *R. sondaicus* being set up in the museum "when in town a few days ago." The next year Rainey (1875a) saw a foetus "preserved in spirits" from a large female shot in the past year. Both these references are compatible with the information about Fraser's specimen of 1874 (no.a), confirming it was adult and full-grown. The foetus in spirits is lost from the records.

Barbour & Allen (1932, pl.11) published a photograph of a hide on display, with the caption "Specimen from the Sundarbans of Bengal. In the Indian Museum, Calcutta; the last example known to have been taken in that region" (fig. 51.16). It has a plaque identifying it as *R. sondaicus*.

FIGURE 51.16
Mounted skin of *R. sondaicus* in the exhibition gallery of the Indian Museum in Kolkata. The plaque reads: "Rhinoceros sondaicus Desm. Distr.: Bengal, Burma, Malay Peninsula, to Java. This specimen is from the Sunderbunds, where the species has become extinct"
BARBOUR & ALLEN, *JOURNAL OF MAMMALOGY*, 1932, PL. 11. COURTESY: DAVID M. LESLIE

FIGURE 51.17
Mounted skin of a young specimen of *R. sondaicus* in the Indian Museum. The current label identifying it as *R. unicornis* is a mistake. Listed as GRM 11 by Chakraborty in 2004
PHOTO: © TANOY MUKHERJEE, 2018

FIGURE 51.18
Very young specimen of *R. sondaicus* displayed in the mammal gallery of the Indian Museum. Possibly the 3-months old baby seen by J.H. Rainey in the museum around 1875. Listed as GRM 273 by Chakraborty in 2004
PHOTO: © TANOY MUKHERJEE, 2018

Although there is no scale, this appears to be a young specimen showing the characteristic triangular folds in the shoulder region. The American zoologists Thomas Barbour (1884–1946) and Glover Morrill Allen (1879–1942) compiled a list of *R. sondaicus* in world musea. From information provided by the curators of the Kolkata collection, this included 3 mounted skins: those of specimens a and b (without ages), preceded by "one adult female, mounted skin, from the Sundarbans, mouth of the Ganges, about 1870." This is likely to be the source of the statement in more recent literature that the last specimen was killed in the Sundarbans in 1870, which is overtaken by evidence of later observations. This 1870 specimen in the list should indicate the animal in Barbour & Allen's photograph, but it is not the skin of an adult animal. Groves (1971: 244) suggested that the photograph depicted Fraser's 1874 adult female (no.a), meaning that there were only two mounted specimens in the collection.

The Fraser 1874 female (cat.a) was a full-grown adult considering that a foetus was obtained from her when she was with the taxidermists. Therefore she was not the specimen shown in the photograph of Barbour & Allen (1932), clearly also indicated by the difference in size. Therefore I suggest that the photograph shows Barkley's juvenile female (no.b), now GRM 11. At the time of entry this was the last rhino obtained in the Sundarbans, a notice which remained with the records ever since, despite its obvious error.

In an inventory of specimens displayed in the mammal gallery of the Indian Museum, Chakraborty (2004: 40) included two skins of *R. sondaicus*: no. GRM 11 being a mounted skin (Mt.) of a young *R. sondaicus* "last collected specimen from Sunderbans"; and GRM 273 (or ZSI 76) a "ME Baby", where ME means a mounted exhibit. The GRM numbers are part of the inventory of the "Museum and Taxidermy Section", and ZSI 76 is not recorded elsewhere. There are still (in 2018) two mounted specimens of *R. sondaicus* in the galleries, one a juvenile or subadult specimen currently mislabelled as *R. unicornis* (fig. 51.17), and another very young one correctly labelled (fig. 51.18). Both these animals have lost their original documentation which could help to match them to entries in Sclater's catalogue.

The skin of the very young *R. sondaicus* GRM 273 is without documentation. It could be the specimen seen by Rainey (1875c) in the Indian Museum being "a very young one of that species, about three months old." With this restriction in the date, there is no corresponding entry in Sclater (1891). The only possibility is no.k, the skin of a juvenile male without history, but this must remain speculative.[1]

---

1 This chapter has benefited from reviews of earlier drafts by Helmut and Gertrud Denzau.

Records of *Rhinoceros sondaicus* in North Bengal and Assam

First Record: 1877 – Last Record: 1900
Species: *Rhinoceros sondaicus*
Rhino Regions: 24, 25, 26

52.1 *Rhinoceros sondaicus* in North Bengal and Assam

The *terai* belt where Bengal approaches the foothills of the Himalayas was a favourite habitat for *R. unicornis*. Few authors cared to comment on the characteristics of the animals which they encountered, rightly so, because the significant majority belonged to the same species which was by then generally well-known. However, a few anomalies were spotted over the years, introducing persistent comments in the literature about the possible presence of *R. sondaicus* in this same region. There is some hesitation to admit these records largely because there have been so few instances, because it means that the two single-horned species were virtually sympatric, and because North Bengal and Assam are far from any other known populations (in the Sundarbans) of *R. sondaicus*.

Ever since Kinloch (1885) wrote about his friend shooting an animal like *R. sondaicus* in the Sikkim *terai*, the presence of this species in North Bengal has been incorporated in texts about the rhino in India. Pollok (1879) had already claimed its existence in Assam south of the Brahmaputra (Goalpara). In their review of fauna in Jaldapara, Inglis et al. (1919) mentioned that it was often hard to tell the two single-horned species apart, because they had identical tracks and habitats. Years later, Milroy (1932, 1934a) suggested that both species might be present in the Manas Sanctuary, because "pairs of smaller, less truculent, and definitely less armoured rhino can be put up in the Sanctuary and these, if not cases of *R. unicornis* pairing while still far from mature, must be specimens of *R. sondaicus*."

The evidence of instances where *R. sondaicus* was found in this part of India was reviewed by Rookmaaker (1980, 2002b, 2006a) and Choudhury (2013). Only three possible instances were identified, to which five more are added here. At least four of these prove to be spurious. The evidence is scant, yet with the existence of two museum specimens, the conclusion is inevitable. Nevertheless, Milroy's statement shows that the authorities presented odd reasons for their identification. Generally these included the smaller size of the animal, or the small size or absence of a horn, while nobody clearly commented on the conspicuous saddle formed by the skin in the shoulder region. More unusually, the specimens attributed to *R. sondaicus* were always found sympatric with *R. unicornis*. In fact, hunting parties might shoot a *R. unicornis* in the morning, and a *R. sondaicus* in the evening in exactly the same locality. This sympatry is unexpected among two large mammals which are very similar in many aspects of behaviour and habitat requirements.

There are two records of *R. sondaicus* for Assam (1868 Pollok, 1890 Green), one for the Garo Hills or Rangpur (1870 Blyth), and five in a belt of *terai* land in North Bengal (1877 Money; 1878 Kinloch; 1881 Möller; 1895 Sunder; 1900 Grieve). There are no reports later than 1900, when it is suggested that the species went locally extinct.

∙ ∙ ∙

Dataset 52.62: Chronological List of Records of *R. sondaicus* in Assam, West Bengal, Bhutan and Meghalaya

General sequence: Date – Locality (as in original source) – Event (species identifications according to original source) – Sources – § number – Map number.

The map references refer to map 52.25 (*R. sondaicus* only), explained in Dataset 52.63.

Rhino Regions 24, 25, 26: North Bengal

1872 – Bhutan [duars] – According to Thomas Andrew Donogh (b. 1818) (§31.11), rhino of Bhutan is a distinct species, "different from all others, even that of the Sunderbuns [*R. sondaicus*], and a very powerful beast" – Donogh 1872: 248 – map S 147

1877 – Ghorabhandah (? Gorubathan) – Robert Cotton Money (1835–1903) shot smaller species of rhino – Money 1893 – 52.4 – map S 146

1878-05 – Left bank of the Ti'sta (Teesta) River – A friend (known by initial S.) of Alexander Angus Airlie Kinloch (1838–1919) shot a lesser rhino – Kinloch 1885: 201, 1892: 200–201, 1903: 162, 1904: 65 – 52.4 – map S 147

© L.C. (KEES) ROOKMAAKER, 2024 | DOI:10.1163/9789004691544_053
This is an open access chapter distributed under the terms of the CC BY-NC-ND 4.0 license.

1878 – Sikhim terai – Kinloch shot a *R. sondaicus* – Blanford 1891: 475; Brehm 1891: 102; Watt 1892, vol. 6: 489; Shebbeare 1953: 145. These authors mistake Kinloch for his unnamed friend – 52.4 – map S 147

1881 – Moraghat, Bhutan Duars – Frederik August Möller or Møller (1852–1915) shot a female *R. sondaicus* – Zoological Museum, Copenhagen, female skull, accession CN524 – Groves 1967: 234, 2003: 345; Groves & Leslie 2011; Rookmaaker 1980: 254. Details about the life of F.A. Möller in Darjeeling and his interests in natural history (52.6) are found in Junk 1906; Cassino 1914: 187; *Kungliga Svenska Vetenskapsakademiens Årsbok* 1903: 43 – 52.2, figs. 52.1, 52.2, 52.3 – map S 148

1882 – Islands in the Brahmapootra River above Dubri – Presence of lesser rhino – Pollok 1882a, 1896: 195 – 52.3

1882 – Brahmaputra valley – *R. sondaicus*: same locality as *R. unicornis* – Hunter 1882: 520, 1893: 759

1895 – Jalpaiguri – Donald Herbert Edmund Sunder (1857–1935), in a Settlement Report, states the presence of a rhino species named 'Kuku gera'. Body rough and tuberculated. It has a very bad temper – Sunder 1895: 103; Gruning 1911; Mitra 1951: xxvii – 52.7 – map S 147

1900 – Chilapata, Buxa division – James Wyndham Alleyn Grieve (1872–1939) shot one *R. sondaicus* – Milroy 1932; Meiklejohn 1937: 2; Shebbeare 1953. Specimen in NHM London ZE.1963.7.25.1 (with date *c.*1903) – 52.8 – map S 149

1919 – Jalpaiguri – *R. sondaicus* has been shot in past 20 years – Inglis et al. 1919: 825 – map S 147

Rhino Regions 30, 31, 32: Assam

1837 – Guzgottah, near Rangpur (mistakenly located in Garrow Hills by Blyth) – Animal killed was 4 ft 5 in high (145 cm), male, pretty large horn – Gallovidian 1837. Specimen due to size identified as *R. sondaicus* by Blyth 1870: 146, 1872e: 3105. Actually *R. unicornis* – 52.2 – map A 343

1865 – Manas – Henry Andrew Sarel (1823–1887) refers to Kurruch Singh (Hathi Rajah), who "described three kinds of rhino, of which the smallest has the longest horn, but it seemed to us that the largest and oldest animals had the shortest horns only from more wear in digging for roots" – Sarel 1887: 363. Actually *R. unicornis* – map A 417

1868 – Left (south) bank of Brahmapootra River – Fitzwilliam Thomas Pollok (1832–1909) stated to have shot the "lesser rhinoceros" [*R. sondaicus*] when in the company of Augustus Kirkwood Comber (1828–1895), Commissioner of Lakhimpur – Pollok 1898d: 255; Pollok & Thom 1900: 454 – 52.3 – map A 386

1868-05-01 – Tikrikillah (Tikrikilla) – F.T. Pollok killed a small rhino – Pollok 1868e: 803, 1879, vol.2: 111. Pollok & Thom (1900: 454) stated that "leaving this oasis, we got into heavy grass, and I shot one of the lesser rhinoceros," where lesser rhino means *R. sondaicus* – 52.3 – map A 386

1879 – South bank of Brahmaputra (Goalpara to Lakhipur) – *R. sondaicus* plentiful on the south bank – Pollok 1879, vol. 1: 94. Actually *R. unicornis* – 52.3 – map A 386

1880 – Assam – *R. sondaicus* present until about 1880 – Roy 1993: 4

1883 – Khassia hills (Meghalaya) – Charles Richard Cock (1838–1879) had seen both *R. sondaicus* and *D. sumatrensis* – Cockburn 1883: 64 – The locality of this general remark possibly based on communication between Cockburn and Cock is uncertain, and may well be garbled – 52.7

*c.*1900 – Assam – *R. sondaicus*. A.C. Green killed one some years ago with a horn of 10 ¾ in (27.3 cm) – Burke 1908: 44 – 52.9

1906 – Assam – *R. sondaicus* occasionally in same locality as *R. unicornis* – Assam Government 1906: 7

1932 – Monas Sanctuary (Manas) – *R. sondaicus* might exist – Milroy 1932 – map A 417

1934 – Monas Sanctuary (Manas) – Two types of footprints indicate two different types: "One indicating a heavy-bodied animal is more or less rounded and the three nails are set out almost equally from the edge. Second indicating a lighter build animal is distinctly oval, and the median nail is set well projected in front compared to the side nails." The second may be *R. sondaicus*, but uncertain – Bhadran 1934 – map A 417

R. sondaicus recorded in Sikkim, without further details (in chronological order): Brehm 1891: 102; Watt 1892, vol. 6: 489; Lydekker 1894, vol. 2: 470, 1897, vol. 2: 1057, 1900: 26; Ward 1896: 283 and later editions; Pocock 1915: 709; Kauffmann 1911: 158; Dollman 1931: 844; Barbour & Allen 1932: 145; Pocock 1946; Shebbeare 1953: 145; Talbot 1960: 205; Walker 1964: 1351; Guggisberg 1966: 118; Mukherjee 1966: 27; Fisher et al. 1969: 113.

R. sondaicus recorded in North Bengal, without further details (in chronological order): Briggs 1931: 279; Berg 1933: 121; Prater 1934: 90; Thom 1943: 257; Pocock 1946; Prater 1948: 193, 1965: 232; Tikader 1983: 91; Corbet & Hill 1992: 243; Amin et al. 2006: 113; Choudhury 2013: 225; Mallick 2019: 73.

R. sondaicus recorded in Bhutan, without further details (in chronological order): Corbet & Hill 1992: 243; Srinivasulu & Srinivasulu 2012: 351; Choudhury 2013: 225.

R. sondaicus recorded in Meghalaya (Garo Hills), without further details (in chronological order): Blyth 1870: 144, 1872a,b,e, 1875: 50; Pollok 1879, vol. 1: 94, vol.2: 111; Cockburn 1883: 64; Sterndale 1884: 410; Brehm 1891: 102; Mukherjee 1966: 27, 1982: 54; Bhattacharya & Chakraborty 1984.

R. sondaicus recorded in Assam, without further details (in chronological order): Hunter 1882: 520; Lydekker 1892: 903; Sanyal 1892: 131; Trouessart 1898, vol. 2: 753; Lydekker 1900: 26; Ward 1903: 440; Hunter 1908a, vol. 6: 20, 1909: 22; Nuttall 1917: 632; Moulton 1922: 10; Woodyatt 1922a: 148; Ward 1928: 438; Dollman 1931: 844; Prater 1934: 90; Loch 1937: 133; Thom 1943: 257; Harper 1945: 381; Prater 1948: 193, 1965: 232; Walker 1964: 1351; Mukherjee 1966: 27, 1982: 54; Guggisberg 1966: 118; Tikader 1983: 91; Vigne & Martin 1994; Amin et al. 2006: 113; Schenkel et al. 2007: 64.

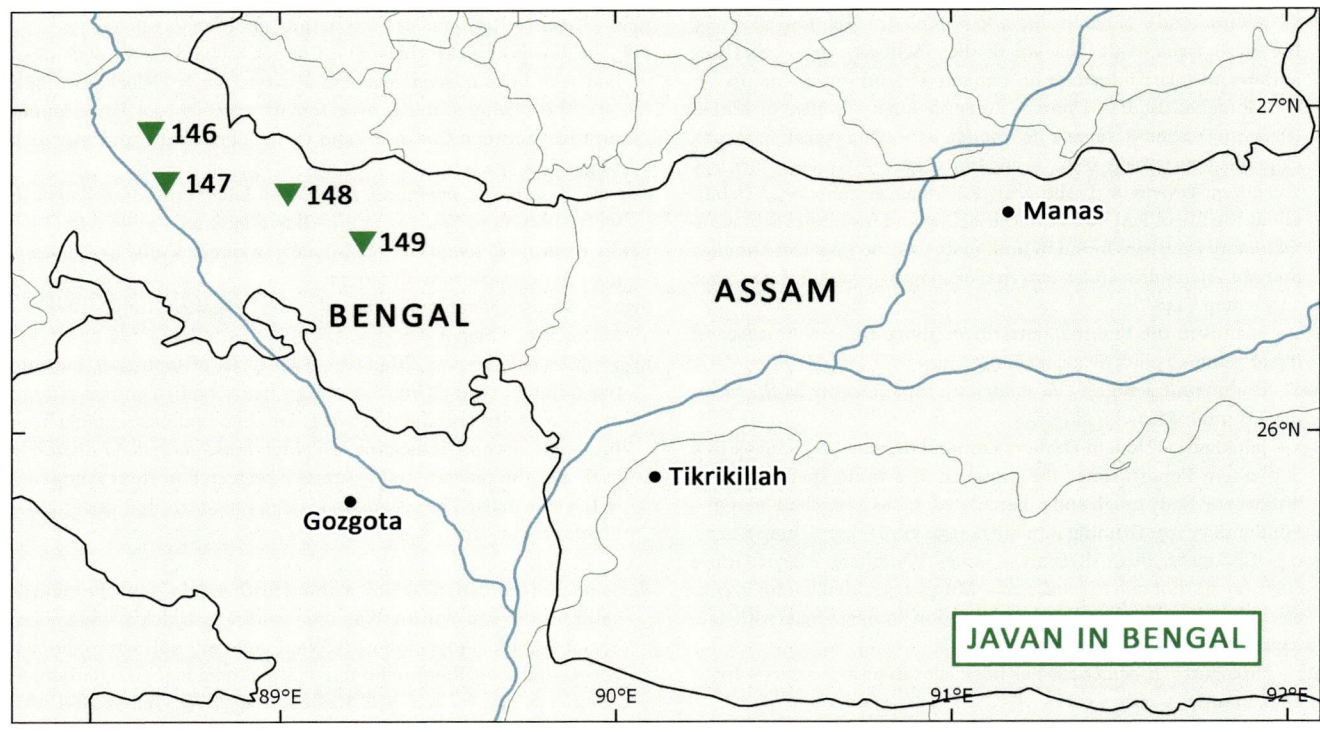

MAP 52.25   Records of *Rhinoceros sondaicus* in North Bengal. The numbers and places are explained in Dataset 52.63
▼   Presence of *R. sondaicus*

• • •

Dataset 52.63: Localities of Records of Possible *R. sondaicus* in North Bengal and Assam

The numbers and places are shown on map 52.25.

North Bengal (Rhino Regions 24, 25, 26)

S 146	Ghorabhandah (Gorubathan) – 26.95N; 88.70E – 1877
S 147	Teesta River left bank – 26.70N; 88.85E – 1878
* (S 147)	Jalpaiguri – 26.70N; 88.80E – 1895, 1919 – both uncertain records
S 148	Moraghat – 26.77N; 89.02E – 1881
S 149	Chilapata, Buxa Dt. – 26.55N; 89.40E – 1900

Assam (Rhino Regions 30, 31, 32). Questionable records (actually *R. unicornis*):

*A 343	Gozgota (Guzgottah), Rangpur – 25.83N; 89.25E – 1837 – refers *R. unicornis*
*A 386	Brahmaputra River south bank (Tikrikillah), Assam – 25.92N; 90.14E – 1879 – refers *R. unicornis*
*A 417	Monas (Manas) Sanctuary, Assam – 26.70N; 90.80E – 1865, 1932, 1934 – refers *R. unicornis*

* Not on map.

• •
•

52.2   The Small Rhino of Rangpur in 1837

Hunting (shikar) as a sporting pursuit was highlighted by the participants in popular accounts in journals and magazines through most of the 19th century. The authors usually disguised themselves under pseudonyms, and their companions were identified using initials. The *Bengal Sporting Magazine* published such a short account signed by "A Gallovidian" and dated "R – pore 24th Feb. 1837." The author had heard about a rhino terrorizing people "in a patch of bamboos near the peaceful village of Gozgotto, six miles from this station." The town in the signature must be Rungpore (now Rangpur) and the village is found about 8 km due northwards as Guzgottah on contemporary maps. A rhino was unusual in the area even then, because there was no large jungle within 40 or 50 miles. Gallovidian

(1837) pursued the animal and after it was killed, he found that it was "a male, with a pretty large horn, and he was a very powerful animal." He measured the length 12 ft 2 in (370 cm) and the height 4 ft 5 in (134.7 cm).

At first sight such a height could indicate a juvenile rhino. Growth charts maintained in Basel Zoo show a height of around 135 cm combined with a length of 275 cm after 18 months for a male born in 1956, and with a length of 293 cm after 20 months for a female born in 1958 (Lang 1961: 408). However, Gallovidian's terminology of the pretty large horn and powerful animal led Edward Blyth (1870, 1872e) to the conclusion that it was full-grown and presumably adult. Blyth calculated correctly that a full-grown *R. unicornis* should be larger (at least taller) than the measurements provided by Gallovidian, hence this account indicated the smaller *R. sondaicus*.

Blyth curiously placed this episode in the Garrow (Garo) Hills, without further comment. All evidence points to the fact that this cannot be correct. It had the unfortunate consequence that he could state that both *R. unicornis* and *R. sondaicus* were known in those hills and made him jump to the less logical next step to say that their ranges included "probably also the Khasya and Jhyntea hills, if not still farther eastward" (Blyth 1870), or throughout the current state of Meghalaya (37.6).

Gallovidian was careless in his writing, which crumbled under the rigorous scientific approach applied by Blyth. On the one hand, the rhino shot by Gallovidian near Rangpur could have been an adult *R. sondaicus*, on the other hand maybe it was a juvenile *R. unicornis*. In any direction it is at least a distance of 100 km from Rangpur to areas where any type of rhino population has historically been found, or it concerned an isolated stray animal. Further evidence will not be forthcoming. My cautious conclusion is that Gallovidian killed a juvenile male *R. unicornis* near Rangpur in 1837.

52.3 Pollok and the Lesser Rhinoceros at Tikrikilla in 1868

Fitzwilliam Thomas Pollok (1832–1909) was one of the more active hunters in North Bengal and Assam from the 1860s (39.1). In most of his anecdotal writings, the species of rhino is left open, suggesting *R. unicornis* rather than another. In his first book summarizing his experiences, Pollok (1879, vol. 1: 94) mentioned that *R. sondaicus*, which he called the "lesser rhinoceros", was never found on the north bank of the Brahmaputra, "but it is plentiful with its larger congener [*R. unicornis*] on the south

bank. In appearance it very much resembles the larger rhinoceros, and often it has a larger horn." He diverted from this first impression in later writings when he stated that they "are met with in many of the churs or islands in the Brahmapootra river above Dubri, and on both banks, where the ground is suitable" (Pollok 1882a, 1896). He assumed that north of the river "doubtless it exists there too, as all these beasts wander about a good deal in search of food" (Pollok 1894: 65, 1898d: 255). Pollok was generally silent about his own encounters with *R. sondaicus*, for instance never even mentioned the possibility in Pollok (1898a: 174). In fact only once, in a rather late article, Pollok (1898d: 256) explicitly said that "I have shot [*R. sondaicus*] on the left bank of the Brahmapootra river." This in his terminology meant the area around and south of Lakhipur.

Pollok saw three differences between *R. sondaicus* and *R. unicornis*. Regarding the hide, "the folds are less prominent and fewer; the hide is covered with square angular tubercles; the fold at the setting on of the head is at the most only indicated" (Pollok 1882). Regarding size, he was not so sure that *R. sondaicus* was much smaller, stating once that it "is perhaps a foot less in height than the larger" (Pollok 1896: 195), elsewhere that it is at least a foot higher than the 3 to 3½ feet mentioned by Jerdon (Pollok 1882), which at 135 cm shoulder height would indeed be smaller than an adult *R. unicornis*. Regarding the horn, it is "often longer than in the other variety" (Pollok 1896: 195). Pollok, therefore, would look for a relatively small or short animal, with a normal or longer horn, and less prominent skin folds. Maybe significantly, he has no word about the triangular neck saddle characteristic for *R. sondaicus*.

While reading the early accounts of Pollok's hunting excursions, I have searched in vain for the "lesser rhinoceros" which he suggested to have shot. However, in his later summary, he mentioned "leaving this oasis, we got into heavy grass, and I shot one of the lesser rhinoceros" (Pollok & Thom 1900: 454). This referred to an excursion (trip 07 in 39.3) on 1 May 1868 from Tikrikilla, on the south bank of the Brahmaputra. His companion Augustus K. Comber killed one rhino, and then Pollok "saw first the ear of one rushing past me, and guessing for the shoulder, fired and brought it down, but unfortunately it turned out to be but a half-grown one. We cut off the horn – a very small one – of the one Comber killed, and went on beating" (Pollok 1868e: 803). Elsewhere he mentioned the same animal as a "a three-parts grown rhinoceros" (Pollok 1879, vol. 2: 111). Neither in 1868 nor in 1879 he had stated that it belonged to the "lesser rhino", but of course it was definitely on the small side, which made him first think that it was not yet full-grown.

Pollok's suggestions do not stand under scrutiny. In my view, he had read in the papers by Blyth (52.2) about the existence of *R. sondaicus* in the Garo Hills, now in Meghalaya, but encompassing Tikrikilla. How could a great naturalist and hunter like himself have missed such an animal? Surely he was convinced that he didn't miss it, even though he did not realize it at the time. Blyth was mistaken, and Pollok was led astray. There is no evidence for the veracity of his late claim. Yet the fact remains that his books were influential and read by many hunters and naturalists in his days. He therefore established the spurious possibility that *R. sondaicus* was found in the Garo Hills to his contemporary public.

52.4 The Pugnacious Rhino of the Teesta River in 1877

Robert Cotton Money (1835–1903) managed the estates of Lakshmeshwar Singh Bahadur (1858–1898), the Maharaja of Darbhanga (1860 to 1898) in Bihar from 1876 to 1878. During this time, in March 1877, Money organized a hunt in Jalpaigori District of Bengal, which he described in a story submitted to the *Boy's Own Paper* of June 1893 (31.14). They were camped at Ghorabhandah in Rangbah Bagnah, "one of the best bits of rhino country I have ever seen." The locality might have been Gorubathan on the upper Teesta River. Money knew about two species of rhino in India, one being *R. indicus* [= *unicornis*], the other "smaller, and its skin is covered with square angular tubercles; it resembles the *Rhin. Javanus* of Cuvier [= *sondaicus*]. This smaller animal is the least common of the two. Both are very stolid, plucky beasts, and greatly feared by elephants. These folds are not so fully developed in the smaller sort. Both kinds have one horn on the nose, varying in size with age, but seldom exceeding a foot in height" (Money 1893).

The party had shot several rhinos, nothing out of the ordinary, when Money writes: "Went to find my friends. I had not gone far when one of the smaller species of rhinoceros, at all times the most pugnacious, with a sudden rush charged the line and scattered the elephants. Indescribable confusion followed. All the elephants seemed to go mad with excitement as the rhino, with a peculiar whistling, squalling grunt, charged one after the other. Four times it charged, but fortunately did no special damage to any of the elephants. Presently I got my chance, and a couple of shots bowled him over."

There is no further information, about the sex of the animal, the size of the horn, or the preservation of a trophy.

52.5 Kinloch's Encounters on the Teesta River in 1878

When William Thomas Blanford (1832–1905) compiled the chapter on *Rhinoceros sondaicus* in his *Fauna of British India* (1891: 475), he stated its presence in the Sundarbans and added that "Kinloch shot an undoubted specimen in the Sikhim Terai." As the current Sikkim state is mountainous, without the undulating *terai*, he obviously meant the more southern region around Siliguri. Blanford's statement has been repeated regularly ever since, although some doubts surfaced as the locality was far from other known occurrences of the species and because the event was apparently absent from Kinloch's own works (Loch 1937; Guggisberg 1966).

Alexander Angus Airlie Kinloch (1838–1919) was in India from 1870 to 1895 and took every opportunity to try to achieve his goal to shoot every species of large game (31.15). In his book of 1885, he included the details of his first attempt in April 1878 to find the rhino in North Bengal, adding new details of excursions in May 1878 and 1886 to the edition of 1892. Easily overlooked, there is a miscellaneous chapter at the end of both editions of the book, almost an appendix, with details of six species which he had not been able to shoot personally. One of these was the "Javan or Sunderbun rhinoceros, Rhinoceros Sondaicus" (Kinloch 1885: 200, 1892: 259). A few weeks after shooting his first *R. unicornis*, he went on to hunt in May 1878, "not far from the left bank of the Ti'sta river, with two friends, S. and L." This person S., maybe Thomas Smith, was the one to bring down a female rhino, accompanied by a calf which ran away. Kinloch took the credit of recognizing it "as *R. sondaicus*; and on telling the natives who were with us that this was not the ordinary Rhinoceros, they informed me that they were aware that there were two kinds."

Kinloch (1903: 162) knew the general characteristics of *R. sondaicus* in comparison with *R. unicornis*: "not much difference in size, but females of *sondaicus* have no vestige of a horn, and the skin appears to be formed of mosaic instead of being studded with protuberant knobs. The arrangement of the folds of the skin is moreover quite different." In the case of the animal shot on the Teesta River, probably the absence of a horn brought him to this identification. On the map showing Kinloch's routes over the years, there are several excursions north of Jalpaiguri on the eastern bank of the Teesta River (fig. 31.9). This can be taken as the locality where S. killed the *R. sondaicus*. Although S. had some shields and trays made from the hide, these trophies have been lost.

Kinloch never had the chance to kill a *R. sondaicus* himself. He may have found traces of the species later in his travels in this part of Bengal, because he recorded that "*sondaicus* is more of a forest-loving animal, and I have found its tracks among low hills, where entanglements of the thorny cane render its pursuit on elephants almost impossible, and on foot nearly equally difficult" (Kinloch 1903: 162, 1904: 65).

52.6 The Rhino Skull Collected from Moraghat in 1881

In June 1887, the Zoological Museum of Copenhagen, Denmark, received a donation of 20 skulls from Engineer J.A. Möller, which included two rhino skulls (figs. 52.1, 52.2). There is no mention of their horns (Lütken 1888: 462). The entries in the register, on different sheets according to species, are as follows:

> Rhinoceros sondaicus [skull, lower mandible]
> 6.1887. Nr. 1. – F juv. – Moraghat, Bhootan Dooars, 24.2.1881, J.A. Möller
> Rhinoceros unicornis [skull, lower mandible]
> 6.1887. Nr. 2. – M juv. – Moraghat, Bhootan Dooars, 6.6.1881. J.A. Möller

Moraghat near Gairkata and Binnaguri is still part of the name of a tea garden as well as a small forest between the Torsha and Kaljani rivers (Patra et al. 2015). Moraghat Tea Garden was established in 1919 by Maraglet Tea Co. (fig. 52.3). In 1873–1874 two blocks in the Moraghat forest having a total area of 13.77 square miles (35.6 km²) were constituted Reserved Forest. The forest extended quite widely from Oodlabari in the east to Haldibari in the west (N.C. Bahuguna, email March 2018). Although there are no wild large mammals now in Moraghat, it would have been different in the mid-19th century.

J.A. Møller, the donor of the specimens to the Danish museum, remains unidentified. He must have been related to members of the Møller or Möller family who had settled as teaplanters in Darjeeling. One of those was Frederik August Möller (1852–1915), who was born in Helsingor, Finland, and went to manage the tea estates at Mohurgong in the 1870s and later at Tukvar in Darjeeling. He married Harriet Maria Rehling in 1892, and died in Kolkata on 23 June 1915. F.A. Möller lived in the house "Rockwood" in Darjeeling. He was interested in natural history, as he is known to have collected butterflies and in 1903 he donated bird specimens to the Museum of Natural History in Stockholm. An account of the fauna of Darjeeling in the district gazetteer was credited to him (O'Malley 1907: 12). There are no records that any of the Möller family were involved in excursions to hunt big game.

The two skulls in the museum were both obtained at Moraghat in 1881, some three months apart. They were identified, probably by the museum staff on arrival in Copenhagen, as a male *R. unicornis* and female *R. sondaicus* as shown in the register and on the labels (31.16). The identification of the female skull was supported by the rhino taxonomist Colin Peter Groves (1967: 234, 2003: 345; Groves & Leslie 2011). This female specimen with number CN-524 on the label is a stage 4 skull (not yet full-grown), remarkable for its very large teeth, but with skull measurements comparable with Sundarbans examples of similar age.

52.7 Sunder's Survey of Jalpaiguri in 1895

In his Settlement Report of 1895, Donald Herbert Edmund Sunder (1857–1935) included a list of the fauna of Jalpaiguri District (more information in 64.5). He listed three species of rhinoceros known as Hati gera (*Rhinoceros Indicus*), Sheng Shengi gera (*Rhinoceros Malayan*) and Kuku gera (*Rhinoceros Sondaicus*). This last species had the additional information of "Body rough and tuberculated. It has a very bad temper." There is no further information why he included the species, but the temperament is reminiscent of that in the story of Money (1893) which Sunder might have heard.

52.8 The Small Rhinoceros of Chilapata Killed by Grieve in 1900

Arthur John Wallace Milroy (1883–1936), Conservator of Forests in Assam, wrote that "it is on record that Rowland Ward identified the head and shield from a rhino shot by a Forest Officer in the Bengal Dooars as belonging" to *R. sondaicus* (Milroy 1932). The lack of detail may indicate that Milroy was writing from memory without access to relevant documentation. Milroy was friendly with Edward Oswald Shebbeare (1884–1964), who later added that the rhino was killed in the Buxa Forest Division by Grieve, and that at the time it was thought to be a small *R. unicornis* (Shebbeare 1953). The sex was not recorded.

James Wyndham Alleyn Grieve (1872–1939) joined the Indian Forest Service in 1894 and his first posting was in Bengal. In 1903 he was elected a member of the Asiatic Society of Bengal, when his address was given

FIGURE 52.1
View of the tooth row of the skull of *R. sondaicus* obtained in Moraghat in 1881. It was donated to the Zoological Museum of Copenhagen in 1887. Accession number CN-524
PHOTO © HANS J. BAAGØE, 2018

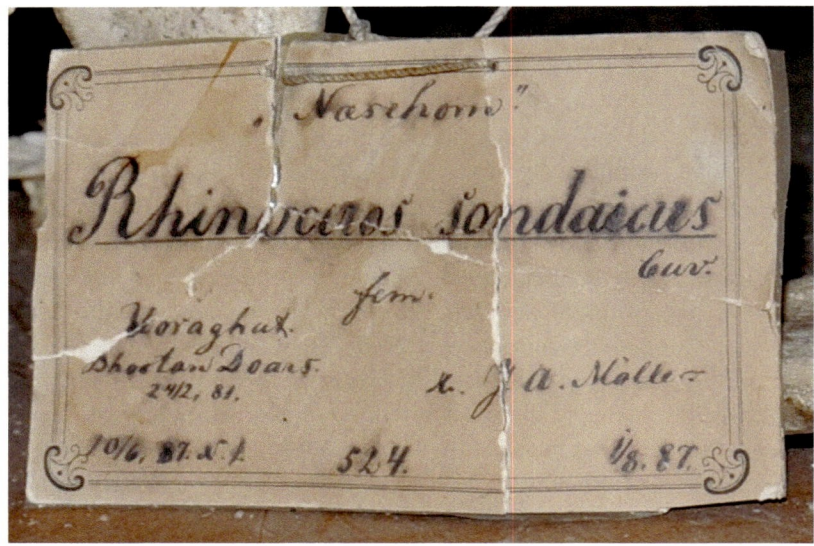

FIGURE 52.2
Label attached to the skull of *R. sondaicus*. Zoological Museum of Copenhagen, no. CN-524
PHOTO © HANS J. BAAGØE, 2018

FIGURE 52.3
Signboard of the Moraghat Tea Estate taken in 2019 on location
PHOTO: KEES ROOKMAAKER

as Kalimpong. He was Conservator of Forests in Burma in 1916, and in 1921 became Chief Conservator in Punjab (Anon. 1939). After retirement, he settled in northern England with his second wife Marjorie Patmore, moved south to Beech in Hampshire and died in a nursing home in London (*Scotsman* 1939-04-12). Grieve never wrote about killing a rhino, but this probably took place in North Bengal where he was stationed.

At the time most trophies were sent to the famous taxidermist Rowland Ward (1835–1912) in Piccadilly, London, who was certainly well versed with all large mammals and would have separated the two single-horned species of rhino instantly. On the other hand, it is strange that Rowland Ward did not refer to Grieve's specimen in his *Records of Big Game*. In the second edition (Ward 1896), there is a listing under *R. unicornis* of a horn 12½ in (32 cm) long, circumference 21 in (53 cm), attributed to J.W. Grieve, with locality unknown. This remained unchanged in all editions published during Ward's lifetime. The range of *R. sondaicus* was given as extending from the Sundarbans "to the Tarai, Sikhim, Assam", where Sikhim means the duars where Grieve might have been.

Inglis et al. (1919), with Shebbeare as co-author, stated that the smaller species was shot in Jalpaiguri within the last twenty years, around the turn of the century, and that the two species can only be separated if a good view is obtained. Meiklejohn (1937) combined these data to say that Grieve shot the rhino in 1900 in Chilapata, part of Buxa, in the eastern part of Jaldapara (Bist 1994; Pandit 2004: 64; Bahuguna & Mallick 2010: 281). This is the first time that the locality Chilapata is specified, as well as the date 1900. Quite possibly these were found in manuscript records of the Forest Department, and should be reliable.

Grieve's trophy (head and horn, maybe a skull) remained with the Grieve family and was donated by J.W.A. Grieve and M.R. Grieve in the 1960s to the Natural History Museum, London. No correspondence could be found. The specimen identified as *R. sondaicus* is registered as NHM ZE.1963.7.25.1, collected c.1903 at "Chilapata, Buxa Forest Division, Bengal (Jaldapara Sanctuary)." It is not entirely clear if the specimen in the museum collection is a skull or a mounted head with horn. Besides the Moraghat skull, this is the second specimen of *R. sondaicus* from North Bengal in a public collection.

52.9 A.C. Green's Long-Horned Rhinoceros in Assam

The notice about Green's rhinoceros is a good example of the challenges surrounding the reconstruction of historical rhino distribution. Like others, this account is second-hand, without date, without locality, without description or reason for the identity of the species, without context about the hunter or observer. Some similar ones can be resolved, others remain elusive. In this case, Burke (1908: 44) recorded under the heading of *R. sondaicus* that "Mr. A.C. Green, some years ago, shot one in Assam which carried a horn of 10 ¾ inches [27.3 cm]." The author Walter Samuel Burke (b. 1861) had lived in Kolkata since 1877, where he contributed to several newspapers, edited the *Indian Field* (from 1901), and published the large and popular (but unexpectedly rare) *Indian field shikar book*. This oblong book containing hints for hunters and all the game laws appeared first in March 1904 and then at least five reprints or revised editions to 1928. Burke had obtained measurements of big game from scores of sportsmen. One of these correspondents was A.C. Green, who only appears once in the book, and has proved impossible to identify. Assam as a locality is too undefined to be helpful, especially because the term was often used rather loosely and could include Bengal. The period "some years ago" might be the end of the 19th century. The horn of 27.3 cm equals the longest ever recorded by Rowland Ward from Java, and remains the longest record from the northern part of the range. This great horn length combined with the unusual locality leaves many potential questions unanswered. Information can get distorted. If more was known, we could have added a record of *R. sondaicus* in Assam confidently and accurately. Now really the evidence is too circumstantial to be allowed.

52.10 The Existence of *R. sondaicus* in North Bengal and Absence in Assam

From the eight instances referring to *R. sondaicus* in North Bengal and Assam, four are most likely spurious. The rhino shot in Rangpur in 1837 was a juvenile *R. unicornis* (52.2), the lesser rhino killed by Pollok at Tikrikilla in 1868 was an author's boast (52.3), the listing in the survey by Sunder in 1895 is without evidence (52.7), and the second-hand information about Green in Assam around 1900 is unbelievable (52.9). Two encounters, by Money in 1877 (52.4) and by Kinloch in 1878 (52.5) are a little short on detail, yet as reliable as many others. The most significant evidence is provided by the skulls obtained by Möller in 1881 (52.6) and by Grieve in 1900 (52.8). These latter four reliable records all pertain to North Bengal.

In my own previous reviews, I have been skeptical about the reality of the existence of *R. sondaicus* in either North Bengal or Assam. First, the species has only been positively recorded very rarely. Second, all occurrences are so distant from other populations that it is hard to imagine how they were once joined. Third, the external characters are never clearly mentioned. Fourth, *R. sondaicus* was found in localities where *R. unicornis* was also recorded, often at close range. Like in the case of Moraghat, both species were found in one locality, one of which shot at the start of the monsoon season when most hunters would not venture out. Statements or histories could have been mixed up. I have stated that during all the hunts of the Maharaja of Cooch Behar (chapter 32), over many years in the same general area, he never even hinted at the possibility that *R. sondaicus* might be found (Rookmaaker 2002b). Bahuguna & Mallick (2010: 283) may be right that he just never had the opportunity to hunt it. If he had, I believe his staff would have known the difference between the species.

Given the small numbers of indications that *R. sondaicus* existed in North Bengal, the way this is interpreted on range maps for the species in the western part of its range is generous. Although prepared carefully based on the literature, Foose & van Strien (1997: 11) in the IUCN Action Plan extend the range without interruption from Myanmar through eastern Bangladesh and most parts of NE India, Assam, into eastern Bhutan and even a portion of China, however leaving out the more western localities in North Bengal. Published in an authoritative and widely circulated work, this map is often repeated. Only regarding Indian territory, Choudhury (2016: 253) extends an unbroken range of *R. sondaicus* from the Sundarbans northwards to North Bengal, and then eastwards including most of Assam and Meghalaya and the Indian border states. The merits of these interpretations are explored in the epilogue (chapter 66), but the few records discussed here in North Bengal may not warrant such an extensive distribution area in historical times.

R. sondaicus has never been reliably identified anywhere in Assam. There are four acceptable records for *R. sondaicus* in North Bengal from the Teesta River in the west to Buxa in the east. *R. sondaicus* in these places was sympatric with *R. unicornis*, which deserves further analysis. The species went extinct in the first decade of the 20th century.

PART 3

The Sumatran Rhinoceros Dicerorhinus sumatrensis *in South Asia*

∵

CHAPTER 53

The Sumatran Rhinoceros *Dicerorhinus sumatrensis*

The Sumatran Rhinoceros is the smallest living species of rhinoceros. It is among the most endangered animals. It is easily distinguished from other rhinos by a smooth and often hairy skin, by size, and by the presence of two horns (figs. 53.1, 53.2). Although this latter characteristic is shared with the rhino species in Africa, the Sumatran rhino is phylogenetically and genetically much closer to the Asian species (Liu et al. 2021).

The English name Sumatran Rhinoceros has a geographical element which becomes an anomaly in some sentences. For that reason, in this book the species is identified using the scientific name *Dicerorhinus sumatrensis*, in the abbreviated form *D. sumatrensis*.

The species once had a wide distribution across South-East Asia, from Borneo and Sumatra, through Malaysia and Myanmar into India and Bangladesh. The former range is not known in great detail due to a lack of data and possible confusion with sightings of *R. sondaicus*. There has been no record from Singapore. In Thailand it was found only in regions bordering southern Myanmar. The evidence from the Indochinese countries Cambodia, Laos and Vietnam was reviewed by Rookmaaker (1980), without a firm conclusion due to the lack of specific data. It is most likely, however, that *D. sumatrensis* did not in fact extend into this region (Groves & Grubb 2011). It is listed as Critically Endangered in the IUCN's Red List in 2021 (Khan 2014; Payne 2022).

Adult Sumatran rhinos are up to 283 cm long excluding a tail of about 50 cm, and up to 144 cm high at the shoulder. The skin is dark grey-brown in colour (figs. 53.3). There are two major skin folds, one encircling the body just behind the front-legs, and another over the belly and flanks. There is no indication of difference between males and females in general appearance. The animals have two horns, of which the posterior one above the eyes is rarely more than a small knob. The anterior horn can grow up to a record of 81 cm, which is a very rare occurrence, while an average horn is usually no more than 40 cm (Hubback 1939; van Strien 1974; Groves & Kurt 1972).

Size of *D. sumatrensis*	
Length of body	maximum 283 cm
Height at shoulders	maximum 144 cm
Maximum horn length	anterior maximum 81 cm

Historically the Sumatran rhino occurs in all kinds of habitat, from swampy places at sea-level to forests on high mountains, as long as there is access to food, water and shade. However, for about a century it has been rare to find it anywhere near the coast or near human settlements. It prefers to stay in dense forests where it is incredibly difficult to encounter. The engineer and conservationist Theodore Rathbone Hubback (1872–1944) working in Malaysia had much experience with rhinos in the wild (fig. 53.3). He found the animals almost invariably in the densest jungle where visibility was limited. When he tried to follow their tracks, the animals would disappear at great speed regardless of steep slopes or any possible obstacles (Hubback 1939). He found that they often wandered over large distances.

D. sumatrensis is a cryptic species in the sense that it knows how to avoid encounters with people in the wild. Actual sightings have always been rare, and even fewer have made their way into the literature. Our knowledge of their behaviour and ecology therefore needs to be pieced together from a study of their tracks and feeding signs. The Dutch ecologist Nicolaas Jan van Strien (1946–2008) collected this type of evidence in the mountains of Northern Sumatra resulting in one of the few detailed descriptions of their habits in the forest (van Strien 1985, 1986). During his long-term study he only saw a glimpse of a rhinoceros on very few occasions. Van Strien made a remarkable contribution to the conservation of rhinos in South-East Asia with his in-depth knowledge of the Indonesian fauna and his ability to encourage scientists on all levels. He was advisor to the International Rhino Foundation, Chair of the IUCN/SSC Asian Rhino Specialist Group, and the first Chairman of the Rhino Resource Center. His personal engagement has made a lasting inpact and his publications were always full of new research and best conservation practices (van Strien 1974 to 2011; Foose & van Strien 1995a, b, 1998, 2000; Amin et al. 2006). His work together with Thomas John Foose (1945–2006) stood at the basis of the first Sumatran Rhino Sanctuary at the edge of Way Kambas National Park in southern Sumatra, where several babies have been born and reared successfully providing a glimpse of hope for the survival of the Sumatran Rhinoceros into the future.

Given that encounters with a Sumatran Rhinoceros in the wild are extremely rare and generally last only a few

© L.C. (KEES) ROOKMAAKER, 2024 | DOI:10.1163/9789004691544_054

This is an open access chapter distributed under the terms of the CC BY-NC-ND 4.0 license.

FIGURE 53.1   Watercolour by Joseph Wolf (1820–1899) of the Sumatran rhinoceros ("Rhinoceros sumatrensis"). Signed lower
left "J. Wolf", without a date. Size 24 × 34.8 cm. Two views of the female exhibited in London 14 April 1872 to
21 September 1872. This was later engraved (not reversed) for Sclater, *Transactions*, 1876, pl. 97
WITH PERMISSION © ZOOLOGICAL SOCIETY OF LONDON, ARCHIVES: DRAWINGS BY WOLF, VOL. IIL.
MAMMALIA III. UNGULATA I, NO. 69

FIGURE 53.2   Watercolour by Joseph Wolf (1820–1899) of the Hairy-eared rhinoceros ("Rhinoceros lasiotis"). Signed
lower right "J. Wolf 1872". Size 24 × 34.8 cm. These are two views of the female Begum imported from
Chittagong in 1872. This was later engraved (not reversed) for Sclater, *Transactions*, 1876, pl. 98
WITH PERMISSION © ZOOLOGICAL SOCIETY OF LONDON, ARCHIVES: DRAWINGS BY WOLF, VOL. IIL.
MAMMALIA III. UNGULATA I, NO. 68

FIGURE 53.3 A rare photograph of a wild *D. sumatrensis* taken in the forest of Pahang in the 1930s by Theodore Hubback (1872–1942). Note the decent sized anterior horn, while the posterior horn is hardly visible
HUBBACK, *JOURNAL OF THE BOMBAY NATURAL HISTORY SOCIETY*, 1939, VOL. 40, PL. 4

seconds, there is always a possibility that this species is confused with another species of rhinoceros or even another large mammal. The second horn is too small to be noticed during such fleeting experiences. Rhinos are of course mentioned in the literature in many places, but every instance needs to be investigated and the evidence evaluated (van Strien 1974; Rookmaaker 1983a). This is a task that rarely leads to certainty. The Sumatran rhino is easily distinguished from the other species when a comparison can be made. In practice, this is really never going to be possible in the wild. For instance, the rhinos living in Way Kambas in South Sumatra were thought to belong to *R. sondaicus*, until a study of four footprints led to the realization that these were actually *D. sumatrensis* (Reilly et al. 1997).

The Sumatran rhinoceros was never considered part of the Indian fauna until Edward Blyth in a paper read at the Asiatic Society of Bengal in August 1862 mentioned the "unexpected fact" that a specimen had been killed in Assam. This he heard from "a friend, whose informant (when in the province) had seen the two horns attached to the skin" (Blyth 1862c: 367, also 1863a: 137, 1863b: 157). There was no immediate response from the scientific community, partly because the same species was known from Burma (Myanmar), at that time under the same administration. This revelation was soon overtaken when

a rhinoceros with two horns was captured in Chittagong in November 1867 (54.9). The first announcement in a Kolkata newspaper in early 1868 was followed by a longer account by Frederick Henry Hood (1837–1875) who had captured the animal and cared for her (Hood 1869). This single instance of a double-horned rhinoceros in South Asia became well-known after Philip Lutley Sclater in London decided that her features required the description of a new species in 1872 (54.8).

In this survey of the references to the rhinoceros in the eastern states of India and Bangladesh, it is found that the animals definitely occurred in the hills and mountains close to the borders shared with Myanmar. This is rugged terrain which for large parts was rarely entered by those responsible for the colonial administration. Reports have been few but persistent. Rhinos may actually have continued to exist in parts of the region, away from human interference, until relatively recently. This is easier said than actually proven, and in fact the less we investigate, the safer any remaining individuals might be.

Over the years, authors have identified the rhinoceros in these eastern regions of South Asia not only as *D. sumatrensis*, but also as *R. sondaicus* and *R. unicornis*. During the course of my investigation, I have come to the conclusion that there is really no evidence to accept the presence of either of the one-horned species in those regions. This is not in any way meant as a criticism of past authors. Maybe some could have been more careful, while over the years the perception of the different species has seen many changes. It is clear that even some of the more indistinct early reports have come to lead their own life once the historical detail was lost, and hence mushroomed in subsequent literature. When you only see an animal for few seconds in dense jungle, it is natural to turn to books written by the best authorities. Those authors again tried to make sense of conflicting reports from a poorly researched region.

All sightings of rhinoceros in this eastern border region are here referred to *D. sumatrensis*. This too is the only species known in adjoining parts of Myanmar. Although there are quite a few references to *R. sondaicus*, these are not convincing enough and actually would be difficult to explain. Only *D. sumatrensis* therefore occurred in Chittagong, Mizoram, Tripura, Sylhet, Hailakundi, Karimganj, Cachar, Manipur, Nagaland and the eastern part of Arunachal Pradesh. Where the northern states meet the valley of the Brahmaputra some specimens of *R. unicornis* might have wandered into the lower lands on the state borders of this territory.

Taxonomy and Nomenclature of *Dicerorhinus sumatrensis* in South Asia

54.1 Taxonomy of the Sumatran Rhinoceros

The Sumatran Rhinoceros is the only living two-horned species in Asia. It belongs to the genus *Dicerorhinus*, which also includes a large number of forms which have become extinct. There is some variation in the different populations, mainly expressed in the sizes of the skulls and teeth. The study of the geographical variation in this species is constrained by the lack of specimens from most parts of the range. Colin Peter Groves (1942–2017) is the main authority for the taxonomic analysis of the species in recent times (Rookmaaker 2015). The latest classification by Groves in cooperation with the mammalian taxonomist Peter Grubb (1942–2006) recognizes three subspecies (Groves 1967; Groves & Kurt 1972; Groves & Grubb 2011: 23).

1. *Dicerorhinus sumatrensis sumatrensis* (Fischer, 1814). Large size. Teeth medium to small. Occiput low and narrow. Found in Sumatra, Malaysia and southern Myanmar up to Pegu, possibly locally crossing the border into Thailand. Assertions of its presence in Indo-China (Cambodia, Vietnam, Laos) remain tenuous. The author's name and date of this and other names are enclosed in brackets because the name is classified in a different genus than used in its first description.

2. *Dicerorhinus sumatrensis harrissoni* (Groves, 1965). Smaller in size. Teeth small. Occiput narrow but proportionally high. Found in Borneo (Sabah, Sarawak, Brunei, Kalimantan). Named after the Borneo specialist Tom Harnett Harrisson (1911–1976).

3. *Dicerorhinus sumatrensis lasiotis* (Sclater, 1872). Largest size. Teeth large. Occiput broad and high. Found in Myanmar north of Pegu, in Bangladesh and North-East India. The only subspecies known in South Asia.

The genus name has three alternatives as discussed in 54.2. There are 11 names on a specific level of which 4 refer to the populations in South Asia (54.3, 54.4). In English, this species is generally referred to as the Sumatran Rhinoceros, with alternatives Two-horned Asiatic, Two-horned Asian, Asiatic two-horned, Hairy, Hairy-eared and Lesser two-horned Rhinoceros. The same names in translation are used in other modern western languages.

A list of names in the languages of the range states of this species was collected by Van Strien (1974: 68). Rookmaaker (1983a: 91) has some additional references.

54.2 The genus *Dicerorhinus* and Its Synonyms

The Sumatran rhinoceros is now classified in the genus *Dicerorhinus* after it was separated from the genus *Rhinoceros* (4.2). Although *Dicerorhinus* is not the earliest name proposed for the genus, it is the oldest correct valid name after the suppression of *Didermocerus* by the International Commission of Zoological Nomenclature (Rookmaaker 1983a: 28, 1983b). There are three synonyms for the genus: *Didermocerus*, *Dicerorhinus* and *Ceratorhinus*.

Didermocerus Brookes, 1828
Joshua Brookes. 1828. *A Catalogue of the Anatomical & Zoological Museum*. London, p. 75.
Type species *Didermocerus sumatrensis* (now *Dicerorhinus sumatrensis*).

Joshua Brookes (1761–1833), who conducted lectures on anatomy and surgery in his home on Great Marlborough Street, London, had a collection of over 7000 anatomical and osteological specimens (HWD 1833). When his Museum was auctioned in July 1828 by George Henry Robins (1777–1847), a catalogue was prepared with each lot identified by a genus name, species name and common name. On the 12th day of the sale, lots 16–22 were assigned to "*Didermocerus sumatrensis*", selling one "stuffed, junior", five horns and the skin of feet. The fate of these items is not known, and they may have been widely dispersed. The name *Didermocerus* was forgotten, largely because taxonomists at the time saw no need to separate *sumatrensis* in a separate genus. After it was resurrected by Ellerman & Morrison-Scott (1951: 339), palaeontologists still continued to use the junior name *Dicerorhinus* for fossil species. To ensure stability, the International Commission of Zoological Nomenclature suppressed *Didermocerus* in Opinion 1080 (Boylan 1967; Boylan & Green 1974; Melville 1977), hence this generic name has no current status in zoological nomenclature and cannot be used.

© L.C. (KEES) ROOKMAAKER, 2024 | DOI:10.1163/9789004691544_055
This is an open access chapter distributed under the terms of the CC BY-NC-ND 4.0 license.

Derivation of *Didermocerus*: Greek δίς dis (twice, doubly), Greek δέρμα derma (skin, hide) and Greek κέρας keras (horn).

Dicerorhinus Gloger, 1841

Constantin Wilhelm Lambert Gloger. 1841. *Gemeinnütziges Hand- und Hilfsbuch der Naturgeschichte*. Breslau, vol. 1, p. 125.

Type species *"Rhinoceros sumatrensis*, Cuvier" (now *Dicerorhinus sumatrensis*).

Name proposed by Constantin Wilhelm Lambert Gloger (1803–1863), a German ornithologist working in Breslau (Glaubrecht & Haffer 2010), as part of a general review of the animal kingdom.

Derivation of *Dicerorhinus*: Greek δίς dis (twice, doubly), Greek κέρας keras (horn) and Greek ῥίς rhis (genitive ῥινός rhinos) (nose, nostril).

Ceratorhinus Gray, 1868

John Edward Gray. [1867] 1868. Observations on the preserved specimens and skeletons of the Rhinocerotidae. *Proceedings of the Zoological Society of London*, p. 1021.

Type species *Ceratorhinus sumatranus* (now *Dicerorhinus sumatrensis*).

Proposed by the curator of mammals at the British Museum, John Edward Gray (1800–1875) in a paper read at a meeting of the Zoological Society of London on 12 December 1867, and published in the third part of the *Proceedings* in April 1868 (as 4.3).

Derivation of *Ceratorhinus*: Greek κέρας keras (genitive κέρατος, keratos) (horn) and Greek ῥίς rhis (genitive ῥινός, rhinos) (nose, nostril).

54.3   Names of the Species *Dicerorhinus sumatrensis*

List of synonyms of *Dicerorhinus sumatrensis* used for South Asian populations in chronological order. The summary is followed by historical and nomenclatorial annotations. There are 11 names used for the recent species *D. sumatrensis* and its subspecies, and one for a fossil subspecies (54.4). Four of these relate to the populations in South Asia.

Rhinoceros sumatrensis Fischer, 1814

Gotthelf Fischer. 1814. *Zoognosia*. Mosquae, vol. 3, p. 301.

Type specimen: Male rhino described and figured by William Bell (1793). The skull was recorded in the Royal College of Surgeons of England, London, but it is now lost. Refer 54.5, 54.6.

Type locality: Sumatra, 10 miles (16 km) from Fort Marlborough (Bintuhan) on the south-west coast.

Derivation of name *sumatrensis*: latinized adjective form of Sumatra.

Rhinoceros sumatranus Raffles, 1822

Thomas Stamford Raffles. 1822. Descriptive catalogue of a zoological collection. *Transactions of the Linnean Society of London* 13, p. 268.

Type specimen: not designated. There is no need to select a new type from specimens sent home by Raffles. Refer 54.7.

Type locality: Sumatra.

Derivation of name *sumatranus*: latinized adjective form of Sumatra.

Rhinoceros lasiotis Buckland, 1872

Frank Trevelyan Buckland. 1872. A new rhinoceros. *Land and Water* 10 August 1872, p. 89.

Type specimen: Female animal 'Begum' living in the London Zoo 15 February 1872 to 31 August 1900 (record longevity of 28 years 6½ months). Skull, skin of head and horns preserved in the Natural History Museum, London, no. 1901.1.22.1. Refer 54.8, 54.9.

Type locality: Sungoo (Sangu) River, Chittagong, SE Bangladesh (place of capture in November 1867).

Derivation of name *lasiotis*: Greek λάσιος (shaggy, woolly) and Greek οὖς ous, (genetive ὠτός otos) (ear).

Rhinoceros malayan Sunder, 1895

Donald Herbert Edmund Sunder. 1895. *Survey and Settlement of the Western Duars in the District of Jalpaiguri*. Calcutta, p. 103.

Type specimen: not designated.

Type locality: Dalgaon Forest, North Bengal, India.

Listed by Donald Herbert Edmund Sunder (1857–1935) with few remarks. Unclear if the author meant this to be a binomial or english name (64.5). This is probably a *nomen nudum* but definitely a *nomen oblitum* (forgotten name), in either case without any standing in zoological nomenclature.

Derivation of name *malayan*: from Malaya.

54.4   Names for Extralimital Taxa

There are seven names used for recent populations of *D. sumatrensis* outside South Asia:

Rhinoceros crossii Gray, 1854 (p. 251) – continental South-East Asia, locality unknown.

Ceratorhinus niger Gray, 1873b (p. 357) – Malaysia.

Ceratorhinus blythii Gray, 1873b (p. 360) – Myanmar. Proposed conditionally (hence unavailable).

Rhinoceros malayanus Newman, 1874 (p. 3950) – probably Malaysia.

Rhinoceros borniensis Hose and McDougall, 1912 (vol. 1, p. 3) – Borneo. *Nomen nudum* (hence unavailable).

Rhinoceros bicornis [= *sumatrensis*] var. *sinensis* Laufer, 1914 (p. 159) – China. Preoccupied by the fossil *Rhinoceros sinensis* Owen, 1870.

Didermocerus sumatrensis harrissoni Groves, 1965 – Borneo.

There is one name used for an extinct population of *D. sumatrensis* outside South Asia:

Dicerorhinus sumatrensis eugenei Sody, 1946 – Sumatra; fossil (early Holocene).

54.5 History of the Specific Name *sumatrensis* in Genus *Rhinoceros* (later *Dicerorhinus*)

Gotthelf Fischer von Waldheim (1771–1853) was a zoologist and palaeontologist working in Moscow, Russia. His multi-volume *Zoognosia* of 1814 was meant to present an overview of the animal kingdom (Bessudnova 2013). He distinguished three recent species of rhinoceros (*R. asiaticus, R. sumatrensis, R. africanus*). The new name *sumatrensis* was based on the first clear description of a two-horned rhinoceros from Sumatra by William Bell published in 1793, which therefore is designated the type specimen (54.6).

Fischer also referred to another reproduction of Bell's sketch of the skull published by Cuvier (1806: 30, pl. 2 fig. 8). Working in the Museum of Natural History in Paris (MNHN), Georges Cuvier (1769–1832) in his paper of 1806 referred to Bell (1793) and included the side-view of the skull in his plate of living rhino species, using the French name "rhinocéros de Sumatra" for this probably new species (fig. 3.3). When Cuvier compiled the first edition of his comprehensive *Règne Animal* (dated 1817, but available in 1816, see 4.9), he included "Le *Rhinocéros de Sumatra*. (*Rh. Sumatrensis*. Cuv.)" with reference to Bell (Cuvier 1816, vol. 1: 240). While this name clearly is a homonym and junior synonym of the same epithet used by Fischer in 1814, *Rhinoceros sumatrensis* was credited to G. Cuvier consistently by all authors from publication in 1816 to the middle of the 20th century. The reference to Fischer was included in the unsurpassed *Index animalium* compiled by Charles Davies Sherborn (1861–1942) without

attracting further attention (Sherborn 1922: 6320). As far as I can ascertain, the first to combine *sumatrensis* with the author G. Fischer was Chasen (1940).

Due to this collective oversight, it is understandable that the combination "*R. sumatrensis* Cuvier, 1816" has led its own life, and that it has been conferred its own type specimen. De Beaufort (1963: 555) included a "♀; animal monté entier; Cat. Galerie des Mammifères", or a complete articulated skeleton of a female animal in the collections of the Muséum national d'Histoire naturelle of Paris (MNHN). Groves (1967: 235) suspected that the "type" examined by Cuvier was a mounted adult male in the Muséum d'Histoire Naturelle in La Rochelle, France (no. M.275). The museum in Paris has details about both these specimens (Paris MNHN 2023; Cécile Callou, email 2023):

1. *Rhinoceros sumatrensis.* Holotype. Collection number: MNHN-ZM-MO-1981-562. Sent by Diard and Duvaucel in 1821 from Sumatra. Female animal, mounted skin (old no. 43, catalogue of Galleries 700a), and complete skeleton with skull (Catalogue of Comparative Anatomy A7965; Catalogue of Ancient Galleries – Comparative Anatomy: BVI-192).

2. *Rhinoceros sumatrensis.* Paratype. Collection number: MNHN-ZM-2011-160. Sent by Diard and Duvaucel in 1821 from Sumatra. Male animal, mounted skin (old no. 699). This specimen was deposited in the Musée de la Rochelle from 1931 to 1995.

Both these specimens are attributed to the French collectors Alfred Duvaucel and Pierre-Médard Diard (54.7), who stayed in Bencoolen on the west coast of Sumatra from August 1819 to March 1820 while engaged by Thomas Stamford Raffles (Weiler 2019; Oury 2021). Here they must have been able to shoot several specimens of rhinoceros, because one was among the collections sent home by Raffles and then preserved in the Museum of the Zoological Society of London (S. Raffles 1830: 644), while at least two arrived in Paris. The date of this shipment is unclear, but it is most likely that the specimens reached the Museum of Natural History in Paris in 1820 or 1821 (Rookmaaker 1983b: 46). An engraved plate after a drawing by Duvaucel was published by Frédéric Cuvier in 1825 (fig. 54.1).

If this chronology is accurate, then Georges Cuvier could not have examined the collections sent by Duvaucel and Diard from Sumatra before he compiled the *Règne Animal* which was printed in 1816. Hence, it appears to me that the specimen described by William Bell in 1793 remains the type of *Rhinoceros sumatrensis* regardless who first named it.

FIGURE 54.1   "Rhinocéros de Sumatra". Engraving of a two-horned rhino, probably after an original sent from Sumatra by Alfred Duvaucel
(1793–1824). Drawn by Jean-Charles Werner (1798–1856), lithographed by Charles-Philibert de Lasteyrie (1759–1849). Size 50.6 ×
33.6 cm
FRÉDÉRIC CUVIER, *HISTOIRE NATURELLE DES MAMMIFÈRES*, 1825, NO. 47, UNNUMBERED PLATE

54.6   William Bell's Rhinoceros from Fort Marlborough, Sumatra

The type of *Rhinoceros sumatrensis* is a specimen sent from Sumatra by William Bell (1759–1792). Following his training as a zoological amanuensis and draughtsman by the great surgeon and collector John Hunter (1728–1793) in London from 1778 to 1791, he was appointed as surgeon at Fort Marlborough (Bintuhan), situated to the south of the East India Company's station at Bencoolen (Bengkulu, Sumatra), through the instrumentality of the famous naturalist Joseph Banks (1743–1820), President of the Royal Society (Rookmaaker 2015a). Arriving early in 1792, he succumbed to putrid fever on 3 July 1792 (Harfield 1995: 554). During these few months in Sumatra, Bell had the opportunity to examine a rhino killed within 10 miles (16 km) of Fort Marlborough (Bintuhan): it was a male, double-horned, 4 feet 4 inches (132.2 cm) high, 8 feet 5 inches (256.5 cm) long, and coloured brownish-ash. Soon afterwards he could study a female, of a lead colour,

younger than the male, and showing less conspicuous folds in the skin (Bell 1793).

Bell compiled his observations and sent them to London where they were received by the surgeon Everard Home (1756–1852). His shipment included a careful description of the double-horned rhinoceros, a drawing of the adult male, and two (or maybe three) drawings of the skull. Home mentioned to Banks on 21 September 1792 that Bell "has sent a letter to you with a description & drawings of the double-horned Rhinoceros" (Dawson 1958: 419). The contents of the shipment were examined by Jonas Dryander (1748–1810), who in turn wrote about this to Banks on 30 September 1792, stating that Bell sent a description, with a drawing of the Rhinoceros of Sumatra and 3 drawings of the cranium. Dryander had compared this material with information of the African (black) rhino, concluding that "the Sumatra beast is a new species, intermediate between the Asiatic one-horned and African two-horned" – diagnosed by dentition and skin folds (Chambers 2007: 155–156).

FIGURE 54.2

Lateral view of the male Sumatran Rhinoceros described and drawn by William Bell (1759–1792), found near Fort Marlborough in eastern Sumatra. Signed "Basire sc." for James Basire (1769–1835). This represents the type-specimen of *Rhinoceros sumatrensis* G. Fischer, 1814

BELL, *PHILOSOPHICAL TRANSACTIONS OF THE ROYAL SOCIETY*, 1793, PL. 2. COURTESY: TINEKE VAN STRIEN

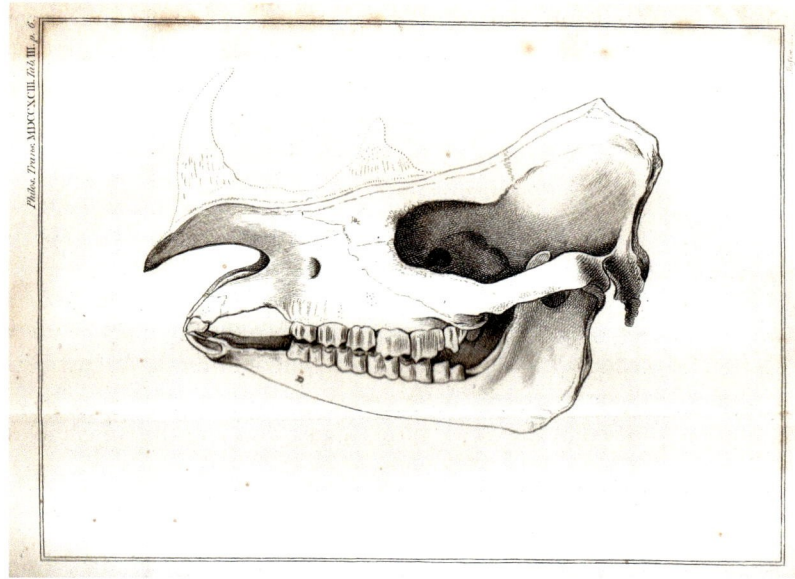

FIGURE 54.3

"The cranium" of the male Sumatran Rhinoceros, drawn by William Bell in Sumatra in 1792

BELL, *PHILOSOPHICAL TRANSACTIONS OF THE ROYAL SOCIETY*, 1793, PL. 3. COURTESY: TINEKE VAN STRIEN

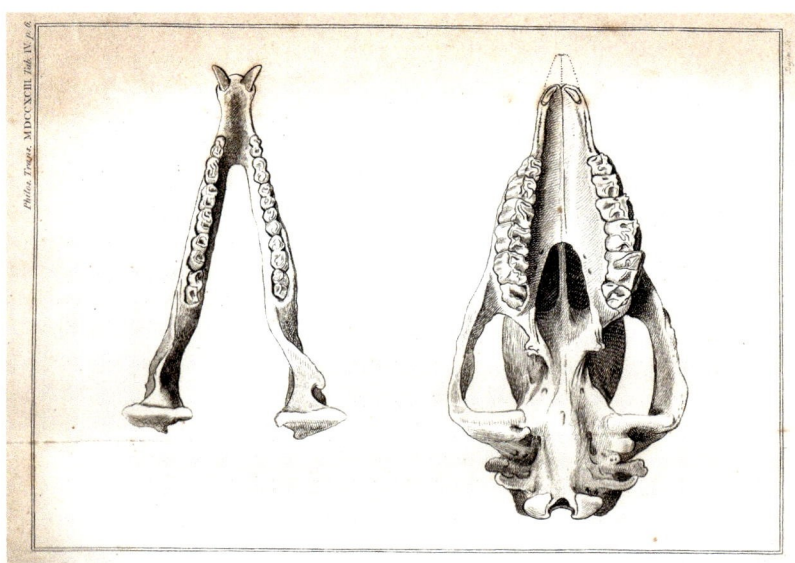

FIGURE 54.4

"The upper and under jaw, separated from each other" of the male Sumatran Rhinoceros examined near Fort Marlborough by William Bell

BELL, *PHILOSOPHICAL TRANSACTIONS OF THE ROYAL SOCIETY*, 1793, PL. 4. COURTESY: TINEKE VAN STRIEN

Banks duly read Bell's report in a meeting of the Fellows of the Royal Society of London on 10 January 1793, and ensured its publication in the *Philosophical Transactions* (Bell 1793). This morphological and anatomical description was then accompanied by three plates based on Bell's (lost) drawings, ostensibly all depicting the male specimen (figs. 54.2, 54.3, 54.4).

Dryander (1796: 66) actually took the next logical step even in print, stating clearly that this rhino found in Sumatra was a "nova species" with reference to Bell's paper. However, nobody at the time used the opportunity to name this new species.

Together with the manuscript material, Bell also shipped the skull of this male rhino, which was donated by Banks to the Royal College of Surgeons (RCS) in London, according to the catalogue by William Clift (1775–1849), once the anatomical assistant of John Hunter. He thought that no. 816 was the actual skull depicted on Bell's plates (Clift 1831: 118). In a later catalogue of the RCS, William Henry Flower (1876: 451; 1884, no. 2144) only mentioned a young skull in connection with Bell's plates, doubting that it was actually the specimen sent in 1792. This skull was subsequently destroyed and is no longer available.

54.7 History of the Specific Name *sumatranus* in Genus *Rhinoceros* (later *Dicerorhinus*)

The famous British statesman Thomas Stamford Raffles (1781–1826) made a catalogue of the animal life in Sumatra, based on material and papers submitted to him by the French collectors Alfred Duvaucel (1793–1824) and Pierre-Médard Diard (1794–1863) during their tenure from December 1818 to March 1820 (47.4). Besides referring to the description by William Bell, Raffles commented on the number of teeth, and therefore might have examined one or more specimens himself. In a large shipment of zoological specimens of 29 March 1820, Raffles forwarded the skeletons and intestines of both a male and a female rhinoceros (Rookmaaker 1983b: 13). The adult female was examined by Everard Home and depicted both in his report of 1821 (pl. 22) and again to illustrate one of his lectures in 1823 (fig. 54.5). The anatomical preparations were placed in the East India House (since lost), while the skeletons and skulls were added to the museum in the Royal College of Surgeons where they were still present in 1884, but destroyed in the London Blitz of 1941 (Clift 1831: 118, nos. 813–814; Flower 1876: 451; Flower 1884, nos. 2141–2142). *Rhinoceros sumatranus* Raffles, 1822 is a junior subjective synonym of *Rhinoceros sumatrensis* G. Fischer, 1814.

54.8 History of the Specific Name *lasiotis* in Genus *Rhinoceros* (later *Dicerorhinus*)

Philip Lutley Sclater (1829–1913) proposed the new specific name *Rhinoceros lasiotis* at the annual meeting of the British Association for the Advancement of Science held in Brighton, in the afternoon of Friday 16 August 1872 (Sclater 1873a). A summary of the meeting was published in the *Times* of 19 August 1872 and the *Athenaeum* of 24 August 1872 (Sclater 1872c,d). The announcement

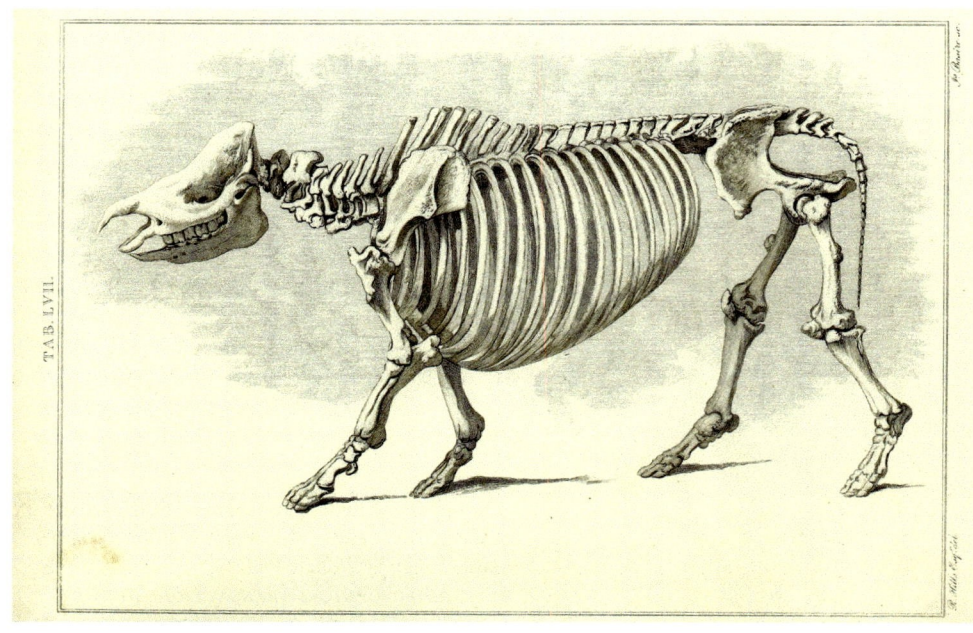

FIGURE 54.5
Skeleton of an adult female of the double-horned rhinoceros, sent from Sumatra by Thomas Stamford Raffles in 1820. Plate published by Everard Home (1756–1852), signed by Robert Hills (1769–1844) and James Basire (1769–1835). It was first published in the *Philosophical Transactions* for 1821, pl. XXII
HOME, *LECTURES ON COMPARATIVE ANATOMY*, 1823, VOL. 4, PL. LVII

was widely repeated in the contemporary press showing how much this meant to him (Sclater 1872e,f,g,h, 1873b; Newman 1872b). Sclater was the Secretary of the Zoological Society of London (ZSL) 1860–1902 during an important time for zoological discoveries. The type of his new name was a female double-horned rhinoceros captured in Chittagong in 1867 and purchased by the London Zoo on 15 February 1872 (54.9). Called Begum, she lived in the zoo until 1900.

When Begum was still in India, observers such as John Anderson (1872) already commented on the long hairs on the ear fringes. After arrival in London, Sclater as a keen naturalist certainly contemplated the importance of this feature, but took no action, probably because comparative material was lacking. Then on 2 August 1872 a second female double-horned female rhinoceros was imported by Jamrach from Malacca (Malaysia), and purchased by the Zoo for £600. This new animal was generally smaller, had a longer tail, and most importantly, showed no signs of the long hairs on the ears. Sclater decided that these were two separate species, of which the Malacca specimen represented the previously known *Rhinoceros sumatrensis*. Noteworthy perhaps that there was only a fortnight between the arrival of the second rhino in London and the Brighton meeting, and apparently it took Sclater only a day or two to reach his conclusions. This was a period in taxonomic history when new species were described in large numbers, often based on very limited material. Sclater himself described at least 1067 new mammals and birds (Goode 1896: ix), a number which must have been surpassed by his contemporary John Edward Gray, the curator at the British Museum.

Begum from Chittagong belonged to a new species, which Sclater decided to name *Rhinoceros lasiotis*. Even the short interval between the arrival of the Malacca rhino on 2 August and the Brighton meeting on 16 August was no guarantee to give him priority. He must have mentioned his discovery in private discussions with the zoologist and author Francis Trevelyan Buckland (1826–1880). Rather naively Buckland introduced both rhinos in his regular column in the weekly magazine *Land and Water* in the issue dated Saturday 10 August 1872: "The other beast, which comes from Chittagong, therefore is a new species, hitherto unknown and undescribed. It has been named by Dr. Sclater *Rhinoceros lasiotis*, or the hairy-eared rhinoceros. It has a tuft of hair like a curtain fringe round the edge of each ear" (Buckland 1872b).

Sclater reported on new additions to the menagerie in most ZSL meetings, which were then made public in the *Proceedings* some months afterwards, appearing in three parts per annum. Sclater (1872b) highlighted Begum's arrival in the meeting of 19 March 1872, which appeared in print in the second part in November 1872 (Duncan 1937). The delay gave him time to add a footnote (dated 28 August 1872) highlighting his description of *Rhinoceros lasiotis*, as well as the first coloured plate labelled with this new name signed by Keulemans (fig. 54.6). The colour of the animal is quite dark, and again both horns are of almost equal small size. In his report on the arrival of the Malacca rhino in the meeting of 5 November 1872, he compared this animal with Begum, both in text and in figures (Sclater 1872k). Begum was still said to be 4 ft 4 in (132 cm) high at the shoulder, the same figure quoted when she was in Kolkata. In three separate

FIGURE 54.6
"Rhinoceros lasiotis, F." Coloured engraving with printed signatures: (left) "J.G. Keuleman lith." for Johannes Gerardus Keulemans (1842–1912), and (right) "M. & N. Hanhart imp.", for Michael Hanhart (1810–1884) and Nicholas Hanhart (1815–1902). Note the spelling Keuleman instead of Keulemans
SCLATER, *PROCEEDINGS OF THE ZOOLOGICAL SOCIETY OF LONDON* FOR 19 MARCH 1872, PL. 23

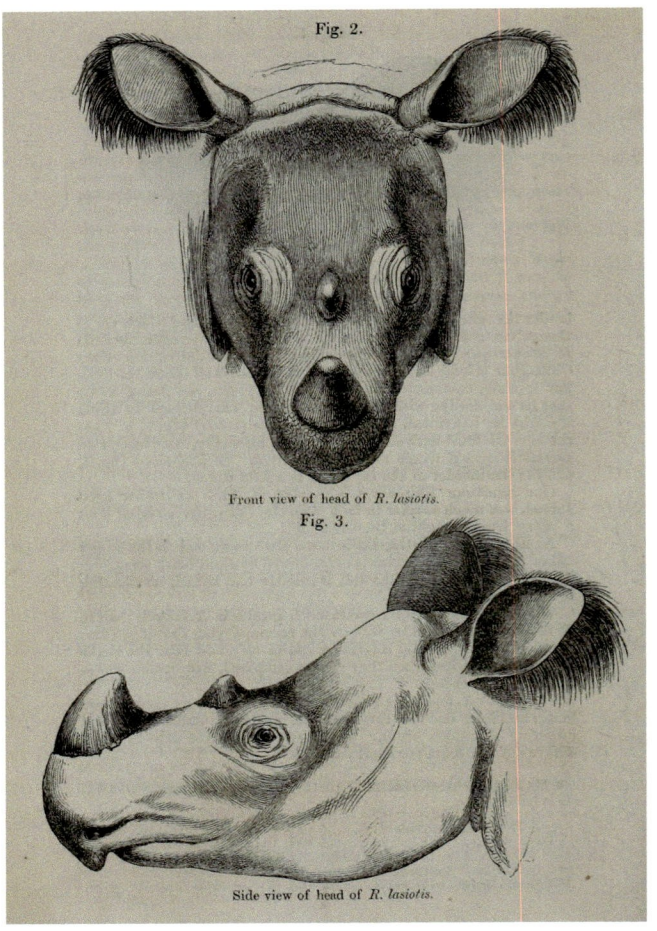

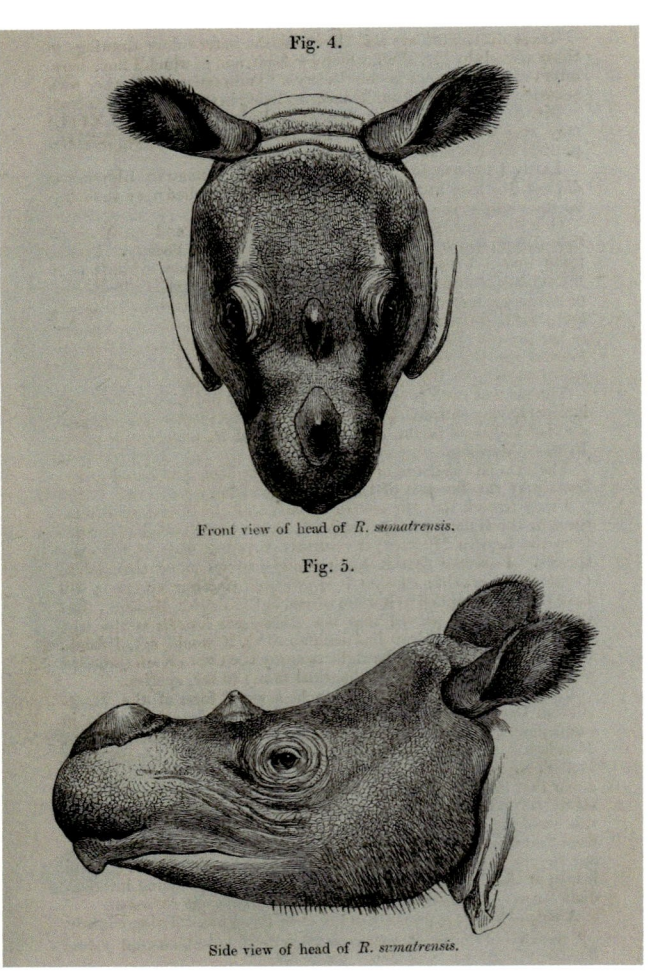

FIGURE 54.7   Head and side-view of *Rhinoceros lasiotis* after Begum
in the London Zoo
SCLATER, *PROCEEDINGS OF THE ZOOLOGICAL
SOCIETY OF LONDON* FOR 5 NOVEMBER 1872, P. 792

FIGURE 54.8   Head and side-view of *Rhinoceros sumatrensis* after
the female from Malacca in the London Zoo
SCLATER, *PROCEEDINGS OF THE ZOOLOGICAL
SOCIETY OF LONDON* FOR 5 NOVEMBER 1872, P. 793

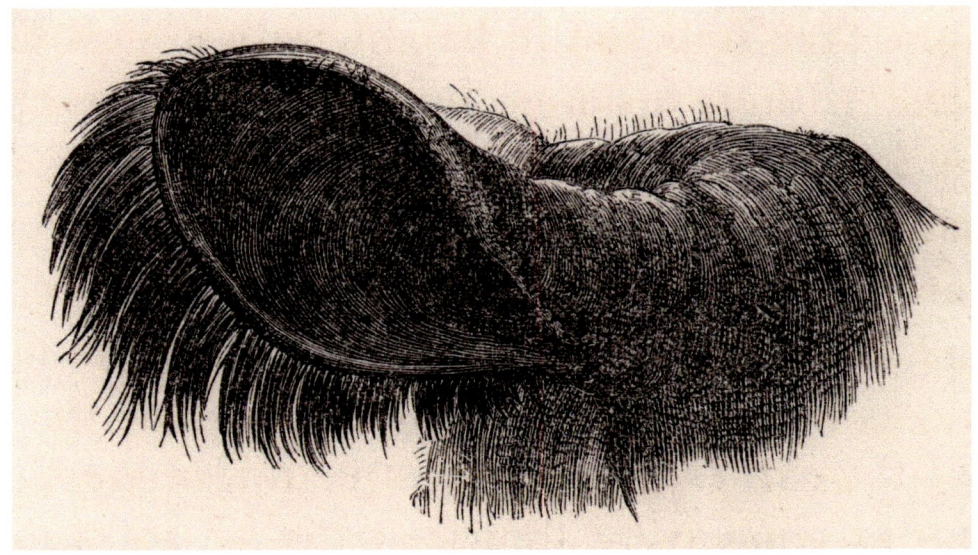

FIGURE 54.9
Right ear of *Rhinoceros lasiotis*.
Figure to show long hairs
on ear-conch, characteristic
for the species described by
Philip L. Sclater in 1872. This
unsigned drawing was published
in Sclater (*Proceedings of the
Zoological Society of London* for
5 November 1872, p. 791, Figure 1)
and again in Sclater (*Transactions
of the Zoological Society of London*,
1876, vol. 9, p. 653, Figure 4)

FIGURE 54.10
The meeting of the Zoological Society, Hanover Square, London. Drawn by the illustrator Harry Furniss (1854–1925). An imaginary scene showing Philip Sclater in the middle. The head of the female *Rhinoceros lasiotis* oversees the proceedings from a vantage point on the right wall
PUNCH, 23 MAY 1885, P. 251

FIGURE 54.11
Detail of the head of *Rhinoceros lasiotis* in an imaginary setting
PUNCH, 23 MAY 1885, P. 251

unsigned figures, Sclater illustrated the right ear, the front view of head, and the side view of head of *Rhinoceros lasiotis* (figs. 54.7, 54.8, 54.9). These were designed to show the difference in the length of hairs on the ear fringes. The first two figures were again used in Sclater's grand paper in the *Transactions of the Zoological Society* for 1876.

As an example of the taxonomic complexities discussed at the time, it may be noted that John Edward Gray objected to Sclater's assessment (Rookmaaker 1984b). Based presumably almost entirely on Buckland's short report in *Land and Water*, Gray (1872) agreed that the Chittagong and Malacca specimens were two separate species. As he decided that the long hairs on the ears were an individual peculiarity, he advanced that the Chittagong animal was in fact in all aspects identical to the nominate *R. sumatrensis*, adding rather pointedly that "I cannot conceive how the idea originated of giving another name to this species." He suggested that the Malacca animal with its long black hair was *Rhinoceros crossii*, using a name which he had proposed in 1854 for a strangely formed horn of unknown locality. When Gray's paper appeared on 1 September 1872, the Malacca female was still alive in the zoo. Sclater (1872e) defended his position in the next issue of the *Annals and Magazine of Natural History* (1 October 1872), criticizing Gray for advancing an opinion without seeing the animals, stating that "I believe that any naturalist who has an opportunity of examining the two animals in the Zoological Society's Gardens will come to the same conclusion." Obviously Gray had to take one further step. When the Malacca animal died on 21 September 1872, the skin was mounted and the skeleton and skull were set up by the taxidermist Edward Gerrard Jr. (1832–1927) for the British Museum (no. 1872.12.31.1). Gray (1873b) examined the evidence in January 1873, and found that the black hairs of the animal were sufficient to separate it from the Sumatra (and Chittagong) species, providing the new name *Ceratorhinus niger*. Sclater's (1872k) paper comparing the Chittagong and Malacca animals with illustrations of their heads and a plate of the latter appeared in a part of the *Proceedings* available in April 1873. Gray (1873c) in September 1873 continued to maintain that the Malacca female was too black to be similar to the animals from Sumatra according to their descriptions, therefore substantiating his *Ceratorhinus niger*. There probably was no clear winner at the time, and Gray's death on 7 March 1875 prevented a continuation of the discussions. When Begum died, Thomas (1901) upheld Sclater's opinion and quickly relegated Gray's arguments on geographical grounds. Neither *R. lasiotis* nor *C. niger* were used much in the 20th century, due to a general lack of interest in the classification of the recent species of rhinoceros (Rookmaaker 2015a).

Sclater always maintained that he was the author of the new species *Rhinoceros lasiotis*. He was proud to have named a new megamammal. An (imaginary?) bust of the animal took pride of place on the walls of the meeting room of the Zoological Society, at least according to the humorous magazine *Punch* with a drawing by Harry Furniss (figs. 54.10, 54.11).

Harper (1940) was the first to notice Buckland's priority. His correction of author's name has generally been accepted. Following Groves (1967), the name is used for the continental subspecies, in the combination *Dicerorhinus sumatrensis lasiotis* (Buckland, 1872).

54.9 The Life of Begum, the Chittagong Rhinoceros

In November 1867, unexpected news reached the town of Chittagong that a rhinoceros had been captured in the region. In charge of the Kheddah (elephant capture) department, Captain Frederick Henry Hood (1837–1875) was one of the English officers stationed in the small town, later described as "a pleasant fellow, fond of the jungle, a keen sportsman with rod and rifle, and had seen service in the mutiny" (Clay 1896: 203). Hood served in the 30th regiment of the Bengal Native Infantry, from 20 August 1855 as 2nd Lieutenant, promoted on 17 November 1857 to Lieutenant and on 20 August 1867 to Captain (Hart 1875).[1]

Captain Hood was told by some headmen of a village near the Sungoo River that they had found an unknown large animal stuck in quicksand or bog. Some 200 villagers had rescued it by throwing a noose over the head and hauling it to dry land, where they had tied it securely to a tree. They did not know what to do with this animal which they had never seen before. Hood lost no time, he gathered all eight elephants available in camp, and set out accompanied by the engineer Thomas Haines Wickes (1840–1899). After covering about 30 miles (50 km) through difficult terrain in 16 hours, they reached the village on the evening of the following day. They tied the rhino so that she could walk between two elephants. The journey back was even slower and not without challenges, because the rhino at least once struggled free and rushed into a muddy patch. "When Capt Hood was wondering by what means he could succeed in extricating the rhinoceros from the mass of mud, in which she had embedded herself, a female elephant of more than usual intelligence, that was one of the guides, quietly walked into the morass and placed

---

1 The story of the rescue and capture of the young rhinoceros was told by Hood (1869), as well as M.D. (1872), Tegetmeier (1872b), Sclater (1872a, 1876a: 652), summarized recently by Rookmaaker & Edwards (2022).

herself behind the rhinoceros, applied her gigantic forehead to the animal, and forcibly pushed her clean out of the morass" (M.D. 1872). The caravan also had to cross two large rivers (Sclater 1876a): the Sungoo (Sangu) River was crossed by towing the rhino between two elephants as she could not swim, while the Kurnafoolie (Karnaphuli) River was crossed on the cattle ferry. Often during the journey the curious onlookers were so numerous that the queue was up to a mile long (Tegetmeier 1872). They reached Chittagong six days after leaving the capture site.

Close to Hood's house near Tempest Hill (Clay 1896), a stockade was quickly erected, a small pool or mud bath was dug out, and a covered shed was built. At first, the rhinoceros was extremely restless, and Hood decided that he needed to investigate. He told his Indian helpers to feed her pieces of sugarcane, while he entered the stockade. The investigation revealed two large ulcers beneath each shoulder as a result of the ropes used to tie her during the transport. Hood cleaned the wounds, and continued to dress them daily. Apparently from that time she became completely tame and even allowed herself to be ridden about (M.D. 1872). Although the rhino was most particular to sugarcane and bananas, she ate such great quantities costing some Rs. 2 per diem, that she was slowly accustomed to the fodder and browse given to the elephants (Hood 1869: 168).

The rhinoceros was a female, and Hood called her Begum. On arrival, probably still in November 1867, Hood thought that she was full-grown but only just adult "cutting her incisor teeth", standing 4 ft 4 in (132 cm) at the shoulder, with a length of 7 ft (210 cm) to the root of the tail. The anterior horn was 3 inches (7.5 cm) high, the posterior one some 2 inches (5 cm). Already the front horn was starting to be malformed: "the anterior horn she has worn much away from rubbing it against the palings of her stockade, but it has developed considerably at the base" (Hood 1869).

There are two drawings of the rhinoceros in her enclosure. The first probably made by Frederick Hood and illustrating his paper in the *Oriental Sporting Magazine* (1869) gives a close-up of the animal in front of her shed (fig. 54.12). She is depicted with two small horns of equal size, and noticeably no hairs are shown on the ear fringes. The second drawing was made by Arthur Lloyd Clay of the Bengal Civil Service, who stayed with Hood in Chittagong until June 1868 (fig. 54.13). This is signed "A.Ll.C. 23.7.95" being the initials of the artist and presumably a date. If he actually prepared this drawing in 1895 rather than on the spot in 1867, this might explain the shape of the horns, which both are large and of a peculiar shape, also seen in photographs of Begum taken much later in her life. I suggest that Clay went to see Begum in London Zoo to refresh his memory and drew the animal accordingly.

Begum was to stay in Chittagong for about four years. For a short time, she was cared for by the writer's brother, Charles F. Manson, Deputy Magistrate (Manson 1876: 177). The editor of Alexander Manson's paper also saw her there, commenting that "she was neither savage nor tame; quiet, but there was risk in going into her enclosure." When a Kolkata newspaper wrote about the rhino in early 1868, the news reached London, where Philip Lutley Sclater, the Secretary of the Zoological Society of London, realized an opportunity to add the species to the collection for the first time in history (Sclater 1876a). He immediately wrote to Hood, who was willing to part with Begum, but "as they steadily decline to bear any of the cost of passage, or risk, and merely offer a certain sum for her on delivery in good condition in London, we have not, up to the present, been able to come to terms" (Hood 1869: 169). Sclater (1872h) claimed that it wasn't entirely his fault, as "there seemed to be some question of the true ownership." Sclater may have asked for help from the well-known animal dealer William Jamrach (1843–1891), who maintained a base in Kolkata and who sold some special animals to the

FIGURE 54.12
The two-horned rhinoceros Begum when she was in Chittagong, drawn by Captain Frederick Henry Hood (1837–1875) in 1867 or 1868
ORIENTAL SPORTING MAGAZINE,
MARCH 1869, FACING P. 167

FIGURE 54.13
Drawing of Begum in her enclosure in
Chittagong, signed "A.Ll.C. 23.7.95", for Arthur
Lloyd Clay (1842–1903). This was probably
sketched on 23 July 1895, hence the shape of
the horns which Begum developed in later
life in London Zoo
CLAY, *LEAVES FROM A DIARY IN LOWER
BENGAL*, 1896, PL. FACING P. 209

London Zoo in the 1870s. According to William Bernhardt Tegetmeier (1872) who interviewed Jamrach, the dealer made three journeys to Chittagong to make all necessary arrangements. The handover was attended by Clay (1872). From Tempest Hill to the harbour was a few miles distance, but the authorities had refused permission to transport the animal on the main road through the villages. The rhino was made to walk through back roads at night, following her keeper who carried a lantern and sang all the way. On arrival in Chittagong, the rhino was embarked in a small vessel, chained to the deck, to make the journey to Kolkata in a few days.

When Begum arrived in Kolkata, "she was so exhausted by her efforts to escape that she lay down after being landed, and had to be dragged by main force into the bazaar" (Tegetmeier 1872: 3059). Here she was spotted on 5 December 1871 by a correspondent of the English naturalist Henry Lee (1826–1888), who signed his letter "O.L.", therefore possibly Oscar Louis Fraser (b. 1849), an assistant at the Indian Museum: "The other morning when riding through Jawn Bazaar I was stopped by an immense crowd who were apparently very much interested in something in the middle of the road. Working my way through I found the obstruction was caused by a huge animal which had taken a fit into its head to lie down there, and no effort of its attendants could persuade it to get up again. As a last resort they threw about 50 buckets of water over it" (Lee 1872). Jawn (Jaun, Jan) Bazar was where the animal dealer William Rutledge (1832–1905) had premises, sometimes called a menagerie, at Jaun Bazar Street 32, just off Chowringhee, in the centre of Kolkata (48.3). It is likely that in fact Rutledge was Jamrach's main agent and contact in Kolkata in the 1870s.

Lee's correspondent "a good zoologist" noticed the animal's large ears with a fringe of long red hair, the whole of the body being also covered with reddish hair about 4 cm long (Lee 1872). Soon afterwards the rhino was examined by John Anderson (1833–1900), Superintendent of the Indian Museum 1864–1886. He sent his notes together with the animal to London, where they were then read and published with Sclater's help. Anderson (1872) noticed as the most striking feature of this animal the long drooping hair of the margins of the ears, although he suggested that these might wear off later in life. He also enclosed some sketches made by a native artist (Sclater 1872f), which remained unpublished and are now presumed lost. The animal was then 4 ft 6 in (137 cm) at the shoulder, about 8 feet (243 cm) from snout to root of tail, and weighed nearly 2000 lbs (900 kg).

While in Kolkata, Jamrach arranged for transport and had a new crate made of teak, 12 ft × 9 ft (365 × 275 cm) and 8 ft (243 cm) high (Tegetmeier 1872; Buckland 1872a: 191). The rhino was shipped on the Screw Steamer *Petersburg* (Captain Blake) through the Suez Canal (Buckland 1872a), departing on 27 December 1871, reaching London Gravesend on 13 February 1872 (*Englishman's Overland Mail* 1871-12-27; *Homeward Mail from India* 1872-02-19). Two days later, on Thursday 15 February 1872, the rhino was taken from Millwall Docks to Regent's Park. On arrival, it was found that the crate could not fit through the entrance, and the only option was to let the animal walk to her new home (Tegetmeier 1872). This scene was depicted by Ernest Henri Griset (fig. 54.14) showing the procession directed by Abraham Dee Bartlett (1812–1897), Superintendent of London Zoo 1859–1897 (Buckland 1872a; E.G. 1872). After walking like this for some 60 or 80 yards, Begum arrived

TRANSFERRING THE HAIRY RHINOCEROS FROM HER TRAVELLING DEN TO HER CAGE

FIGURE 54.14

"Transferring the hairy rhinoceros from her traveling den to her cage." Wood engraving by Ernest Henri Griset (1844–1907). Size 15 × 22.7 cm. The man wearing a top hat behind the animal's head must be Abraham Dee Bartlett

GRAPHIC, 2 MARCH 1872, P. 208

FIGURE 54.15   The new Elephant-House in the Zoological Society's Gardens, Regent's Park, London. Note the sculptured heads of rhinoceros above the first and third entrances

ILLUSTRATED LONDON NEWS, 26 JUNE 1869, P. 640

THE SUMATRAN OR HAIRY RHINOCEROS.

FIGURE 54.16
"The Sumatran or Hairy Rhinoceros." Drawn by
Thomas W. Wood (1839–1910), engraved by the
firm of Butterworth & Heath
*FIELD, THE COUNTRY GENTLEMAN'S
NEWSPAPER*, 16 MARCH 1872, P. 233

HAIRY-EARED RHINOCEROS.

FIGURE 54.17
"The Hairy-eared Rhinoceros." Engraving in
Sclater's Zoo Guide for 1872 and subsequent
editions
SCLATER, *GUIDE TO THE GARDENS OF THE
ZOOLOGICAL SOCIETY OF LONDON*, 26TH
EDITION, 1872, P. 52

at her new home in the Elephant and Rhinoceros House which had been constructed by the architect Anthony Salvin (1779–1881) in the northern section of the Zoo and opened in May 1869 (Anon. 1869a; Edwards 2012: 80). The house had sculptured heads of rhinos by the sculptor William Plows (1836–1885) above the first and third entrances, with an elephant in the middle (fig. 54.15). The building was demolished in 1939.

London Zoo paid £1250 to Jamrach, and was justly proud of their new acquisition (Sclater 1872a, 1876a). Begum was only the second *D. sumatrensis* ever exhibited in any zoological garden (Rookmaaker 1998: 129). Begum was also the type specimen of a new species of rhinoceros (54.8). Once Begum entered her new accommodation in

February 1872, she would not move again for 28 years until she died on 31 August 1900. The Elephant and Rhinoceros House had eight interior stalls and two (east and west) paddocks with a pool. The inhabitants of the house would be allowed outside on rotation. During her lifetime the London Zoo exhibited another eight Sumatran rhinos, most of them short-lived and possibly not all kept in the same house.

Begum was often illustrated, especially in publications associated with the Zoological Society of London. She figured prominently in the important monograph by Sclater on the rhinoceros now or lately living in the London Zoo, issued in the *Transactions* (volume 9 part 11) in March 1877 with an 1876 imprint (ZSL 1878: 12). Sclater (1876a) was

FIGURE 54.18
"Figure 2. Hairy-eared Rhinoceros"
illustrating Sclater's contribution in *Nature*,
24 October 1872, p. 519. The same engraving
had appeared with caption "Sumatran
Rhinoceros" in *Nature*, 28 March 1872, p. 427.
It is again found in the *List of Vertebrated
Animals*, 6th edition (1877), p. 101; 7th edition
(1879), p. 111; and 8th edition (1883), p. 126

illustrated with five coloured lithographs (Palmer 1895: 103–105), which were drawn by the artist Joseph Wolf (1820–1899) and engraved by Joseph Smit (1836–1929). On plate XCVIII of *Rhinoceros lasiotis*, two animals are shown, meaning that both should be representations of Begum in two different positions. Wolf first produced this as a watercolour painting, which is still in the Archives of the Zoological Society of London (Vol. III. Mammalia III. Ungulata I, no. 68), and reproduced in print here for the first time (fig. 53.2). The Malacca female figured in plate XCVII and was equally preceded by a watercolour (fig. 53.1). According to the Society's *Annual Report* for 1878, these five original watercolours were part of an exhibition arranged "in the Picture Gallery, and will shortly be open for inspection by the Fellows and Visitors to the Gardens." This 'Picture Gallery' seems to have been a section of the Reptile House, at the eastern end of the Middle Garden. How long the exhibition lasted is not recorded.

The presence of a totally new type of rhinoceros combined with the description of a new species created understandable excitement in the press. Several artists were sent to the Zoo to sketch the new arrival. The plate in the *Illustrated London News* of 23 March 1872 with two rhinos was drawn by the German artist Johann Baptist Zwecker and engraved by "J.G.", possibly John Gilbert, who was known to have worked for the magazine elsewhere (Anon. 1872). The second rhino in the back appears quite young, and must be the artist's imagination. The rhinos show the hairs on the ear fringes and two normal-sized horns (fig. 56.2).

Illustrating the text by William Bernhardt Tegetmeier (1816–1912), an engraving after a drawing by Thomas W. Wood (1839–1910) appeared in the *Field* for 16 March 1872 (fig. 54.16). The rhino faces left, has long drooping

hairs on the ear fringes, and has two rather short horns. A very similar unsigned engraving is found in the *Guide to the Gardens*, first in the 25th edition (Sclater 1873c) and repeated in subsequent annual editions, at least until the 54th edition of 1900. The animal faces to the right, and has two short horns (fig. 54.17). Two papers by Sclater (1872g,h) about the types of rhinoceros published in *Nature* for 28 March 1872 and 24 October 1872 are illustrated with an unsigned figure with the animal facing left, and with two short horns (fig. 54.18). In March, the animal is called "Sumatran Rhinoceros", but in October this is changed to "Hairy-Eared Rhinoceros." Exactly the same engraving is found in the *List of Vertebrated Animals* edited by Sclater from 1877 to 1883.

The anterior horn of Begum developed a peculiar shape after her arrival in the zoo. It was bending backwards at the tip and reached close to the skin between the two horns. The posterior horn was a bulky mass standing in an upright position. Parts of the anterior horn were removed by the staff of the zoo, maybe once in 1880s, definitely on 24 February 1898 when it was said to be the second procedure. As the tip of the horn had penetrated the skin, Begum was secured with ropes held by 18 keepers under the supervision of Arthur Thomson. The horn was sawed off by the superintendent Clarence Bartlett (1849–1903) while she was lying on her side close to the fence (Anon. 1898).

The shape of Begum's horns can be seen in the photographs and can assist in dating some of them. The first known photograph was taken by Major John Fortuné Nott and published in 1886 (fig. 54.19). Two examples by unknown photographers were taken around 1890 and show Begum on the edge of a pool (Edwards 1996: 124, 125) (figs. 54.20, 54.21). Charles Knight (b. 1853), a professional

FIGURE 54.19
Photograph of Begum taken by Major John Fortuné Nott
(1847–1930) before 1886
NOTT, *WILD ANIMALS PHOTOGRAPHED AND DESCRIBED*,
1886, FOLLOWING P. 338

FIGURE 54.20
Begum, the hairy-eared rhinoceros, at her pool in London
Zoo. Anonymous photograph, ca. 1890
COLLECTION JOHN EDWARDS, LONDON; SEE
EDWARDS 1996, P. 124

FIGURE 54.21
Begum emerging from the pool in London Zoo. Anonymous
photograph, ca. 1890
COLLECTION JOHN EDWARDS, LONDON; SEE
EDWARDS 1996, P. 125

FIGURE 54.22
"Rhinoceros", Begum in her paddock, photographed by
Charles Knight, Newport, Isle of Wight
THE SKETCH, 4 NOVEMBER 1896, P. 64

FIGURE 54.23
Photograph by Henry Sandland showing Begum in the
outdoor paddock in London Zoo, ca. 1899
COLLECTION JOHN EDWARDS, LONDON; SEE
EDWARDS 2012, P. 141

photographer listed both at 6 Royal Victoria Arcade, Ryde
and in High Street, Newport, Isle of Wight, published a picture of Begum (Knight 1896) (fig. 54.22). Henry Sandland
of Birling House, Thurnham, Maidstone, was a photographer adept at taking photographs in the Zoo, as seen in
a small childen's book *Half Holidays at the Zoo* (Morley
& Friederichs 1895). His three photographs of Begum in
her outside paddock probably date from 1899, because
the front horn is small after the tip was removed in 1898,
and because in one of them Begum shares the enclosure

with another Sumatran rhino (figs. 54.23, 54.24, 54.25).
The male Jackson (living 1886–1910) had a large anterior
horn, hence this second animal should be the female
exhibited from 26 September 1898 to 13 February 1900
(Edwards 2012: 141, 146). Another photograph from the
same period (fig. 54.26) was taken by a different photographer (Edwards 1996: 125). The animal photographer
Gambier Bolton (1903, pl. 44 fig. 1) was often seen in the
London Zoo, where he photographed Begum probably not
long before she died (fig. 54.27).

FIGURE 54.24
Two Sumatran rhinos sharing the outside
paddock in London Zoo. Begum is in
front, the other animal is probably the
female exhibited 1898–1900. Anonymous
photograph
COLLECTION JOHN EDWARDS,
LONDON; SEE EDWARDS 2012, PP.
146–147

FIGURE 54.25
Begum on the side of the pool.
Photograph by Henry Sandland, ca. 1899
COLLECTION JOHN EDWARDS,
LONDON; SEE EDWARDS 2012, P. 141

FIGURE 54.26
Begum in London Zoo, ca. 1899.
Anonymous photograph
COLLECTION JOHN EDWARDS, LONDON;
SEE EDWARDS 1996, P. 125

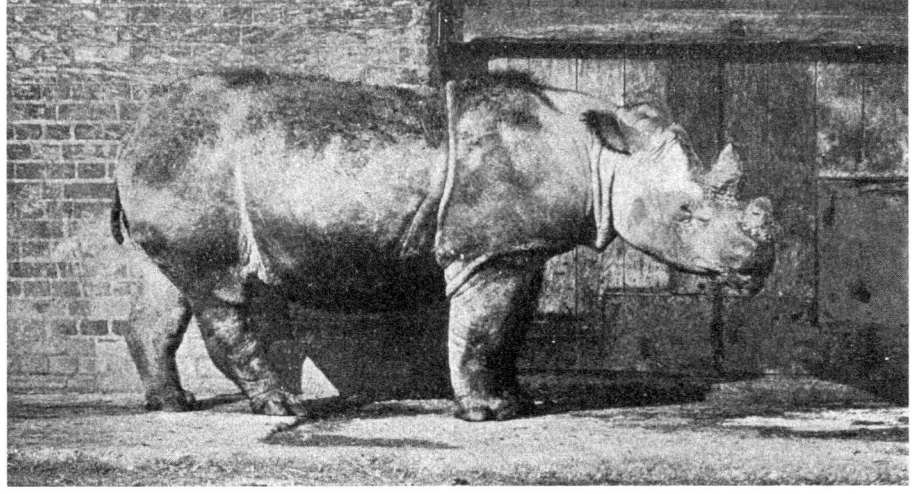

FIGURE 54.27
The Hairy-ear'd Rhinoceros by
Gambier Bolton (1854–1928), probably
photographed 1899–1900
BOLTON, *BOOK OF BEASTS AND BIRDS*,
1903, PL. 44 FIG. 1 FOLLOWING P. 80

Begum died on Friday 31 August 1900, after 28 years 6 ½ months (10,425 days) in London Zoo. She must have been at least 32 years old, but more likely 35 years old at the time. Her death was attributed to old age, adding to which Thomas (1901: 155) remarked that "the animal had become very much diseased" so that only skull and head-skin were worth preservation. He did not say anything about the horns. The remains were catalogued in the Natural History Museum of London (1901.1.22.1) and are preserved as the type-specimen of *Rhinoceros lasiotis* Buckland, 1872.

Captain John Kaye Kendall (1869–1952), writing under the pseudonym Dum-Dum, wrote an elegy to Begum (fig. 54.28). First published in the *Homeward Mail from India, China and the East* for 21 January 1901, and reprinted in Kendall (1915), it is a poem reflecting the tastes of the time, shedding a last tear for a famous rhinoceros.

FIGURE 54.28
A sketch of Begum in the water, by George
Morrow (1869–1955)
JOHN KAYE KENDALL, *ODD CREATURES*, 1915,
P. 79

Dum-Dum's elegy for Begum, from Kendall (1901 and
1915).
R.I.P.
Elegy on a Rhinoceros, Lately Deceased.

Come, let us weep for Begum; he is dead.
Dead ; and afar, where Thamis' waters lave
The busy marge, he lies unvisited,
Unsung; above, no Cyprus branches wave,
Nor flowers, fertilise around his grave.
But ours it is to mourn, with welling eyes,
Th' anachronistic pachyderm's demise.

Blithesome was he and beautiful; the Zoo
Hath nought to match with Begum; he was one
Of infinite humour ; well indeed he knew
To catch with mobile lips the jocund bun
Cast him-ward, by some sire-encouraged son
Half-fearful, yet of pride fulfilled note
The dough, swift-homing down th' exultant throat.

Whilome in pensive-wise he stood, ornate
With comfortable mud, and idly stirred
His rearward caudal, disproportionate,
But not ungraceful, while wanton herd
Of revellers the mystic lens preferred;
Whereof the focus rightly they addressed,
And, Phoebus being kind, the button pressed.
Then, being frolic, he, with mien distraught,
Would, blindly groping, seek the watery verge

And sink, nor rise again ; but when, untaught
In craft, the mourners raised th' untimely dirge,
Lo ! otherwhere himself would swift emerge
Incontinent, and shake his tasselled ears;
And, all-vivacious, own the sounding cheers.

Nothing of base suspicion nor of guile
Was limned on Begum. His the mirthful glance,
The genial port, the comprehensive smile;
The very sunbeams shimmering loved to dance
Adown that honest, open countenance;
And, far as eye could pierce, his roomy grin
Was pink, as 'twere Aurora dwelt therein.

Yet he is dead. Whether the wheaten feast
Some lawless lodgement made, nor found escape;
Or if, perchance, the wild and ravening East
Had howled adown that hospitable gape,
And, ill requiting, knocked him out of shape,
We nothing know: only the fact is spread.
Not how died: simply that he is dead.

Still, the callous bards neglect to hymn
Thy praises, Begum: tho', on dross intent,
The hireling sculptor pauseth not to limn
Thy spacious visage, kindly hands are bent
E'en now, to stuff thy frail integument.
Then sleep in peace, Beloved; blest Sultan
Of some Rhinokeraunian Devachan.
[Dum-Dum.]

CHAPTER 55

Sumatran Rhinoceros (*Dicerorhinus sumatrensis*) in Captivity in South Asia

55.1 Sumatran Rhinoceros in Captivity

There have been nine specimens of *D. sumatrensis* in captivity in South Asia, all in India (dataset 55.64). One was born in the Calcutta Zoo, another was born in transit in the harbour. Five of the adult animals were captured in South-East Asia. Only two animals of this species were captured in the region (in Chittagong), one of which was exported to London, the other was exhibited in Kolkata. The reference to a specimen in Aurangzeb's court is not included in the total count due to lack of evidence. There are no additions to the list provided by Rookmaaker (1998).

• • •

Dataset 55.64: History of the Specimens of *D. sumatrensis* in Captivity

Chronological summary of captive specimens of *D. sumatrensis* exhibited in or exported from South Asia. Total 9 animals. Each line shows: arrival date – facility – date of transfer or death.

(nd) means no date known.

*1867 Chittagong 1871 Transport via Calcutta Harbour 1872 London Zoo 1900

*1882 Chittagong 1882 Calcutta Zoo (nd, died 1892–1905)

*1883 Malaysia 1883 Calcutta Zoo 1888

*1883 Malaysia 1883 Calcutta Zoo 1883

*1884 Burma 1884 Calcutta Zoo 1886 London Zoo 1910

*1885 Malaysia 1885 Calcutta: Rutledge 1885 London transit 1885 Hamburg Hagenbeck (nd)

*1885 Born in Calcutta: Rutledge 1885 London transit 1885 Hamburg Hagenbeck (nd)

*1889 Born in Calcutta Zoo (nd, died 1892–1905)

*1889 Chennai Zoo 1913

* recorded in Rookmaaker (1998: 129–153)

∴

55.2 List of Specimens of *D. sumatrensis* in South Asia

Listed alphabetically according to locality of the collection, there are details of nine specimens of *D. sumatrensis* in captivity in India and Bangladesh.

Agra – Court of Aurangzeb

1. 1690s
Danish merchants procured a double horn of a rhino said to have been killed at the court of the Mughal Emperor Aurangzeb (see §11.16). The origin is too circumstantial to add this specimen to the list.

Chennai (Madras), Tamil Nadu – Zoological Gardens

1. F 1899 to 1913
Captured by a team working for the dealer John Hagenbeck (1866–1940) in Sumatra in 1899. It was still in the zoo in 1913, after which there are no data (Flower 1914; Rookmaaker et al. 2002).

Chittagong, Bangladesh – Frederick Henry Hood

1. F 1867–11 to 1871-12-27 (transferred via Kolkata to London)
Begum, sold to London Zoo by Frederick Henry Hood (1837–1875). Her story is found in 54.9.

Kolkata (Calcutta), West Bengal – Zoological Gardens, Alipore

1. F 1882-06-02 to [1892–1905] †
Female called Muni Begum, captured in Chittagong (Anderson 1883; Sclater 1884; Knox 1891: 256; Renshaw 1907: 186). More information in 56.4.

2. M 1883–06 to 1888 (to Afghanistan)
Imported from Malaysia with next female. Father of 1889 calf (Anderson 1883; Sanyal 1892: 132). In 1888 to collection of Amir Abdul Rahmann of Kabul, Afghanistan (Buckland 1889). Previously, the arrival of the pair from Malaysia (nos. 2, 3) was taken to be June 1882 (Rookmaaker 1998), but as it was first mentioned in the *Friends of India* (1883-07-07), a 1883 date is more likely. At that time the rhino enclosure was subdivided.

© L.C. (KEES) ROOKMAAKER, 2024 | DOI:10.1163/9789004691544_056

This is an open access chapter distributed under the terms of the CC BY-NC-ND 4.0 license.

3. F 1883–06 to 1883 †

Imported from Malaysia with previous male (Anderson 1883; Sanyal 1892). She seems to have been sickly, and probably did not survive very long.

4. M 1884-04-21 to 1886–01 (to London)

Captured in Burma near Rangoon 1884-03-27. To London Zoo, where he lived (called Jackson) from 1886-04-27 to 1910-11-22 (Flower 1931: 203).

5. M Birth 1889-01-30 to [1892–1905] †

Birth to the Malaysian male of 1883 and the Chittagong female of 1882 (Buckland 1889; Sanyal 1892; Flower 1914; Noack 1886). Although Muni Begum from Chittagong and the Malaysian male were then supposed to be separate species (*Rhinoceros lasiotis* and *R. sumatrensis*), they were the parents of the first rhino baby of any species bred in a zoo (figs. 55.1, 55.2). This was, rather comically, announced in the Personal Pages of the *Englishman* (Calcutta) for Tuesday 5 February 1889 in the section of Births:

"Rhinoceros. At the Zoological Gardens, Alipore, on the 30th January, Rhinoceros Lasiotis, the wife of Rhinoceros Sumatrensis, of Cabool, of a son" (fig. 55.3). At the time, the father had already been transferred. The birth proceeded well, albeit that the baby took almost a day to drink milk from the mother. The birth was described by Ram Bramha Sanyal (1850–1908), the Zoo's Superintendent (Mittra 1992). Sanyal (1892: 134) stated that the baby was alive and well in August 1891, then 2 years 7 months old. Earlier I had misread Flower (1914), who visited Calcutta in June 1913, to say that the young one of 1889 was still alive in 1913 (Rookmaaker 1989: 138). However, in fact there is total silence about these zoo rhinos, no photographs, no recollections of zoo visitors, nothing after 1892. The Calcutta Zoo only acquired a next rhino in February 1905, hence maybe both mother and baby had died in or before 1892.

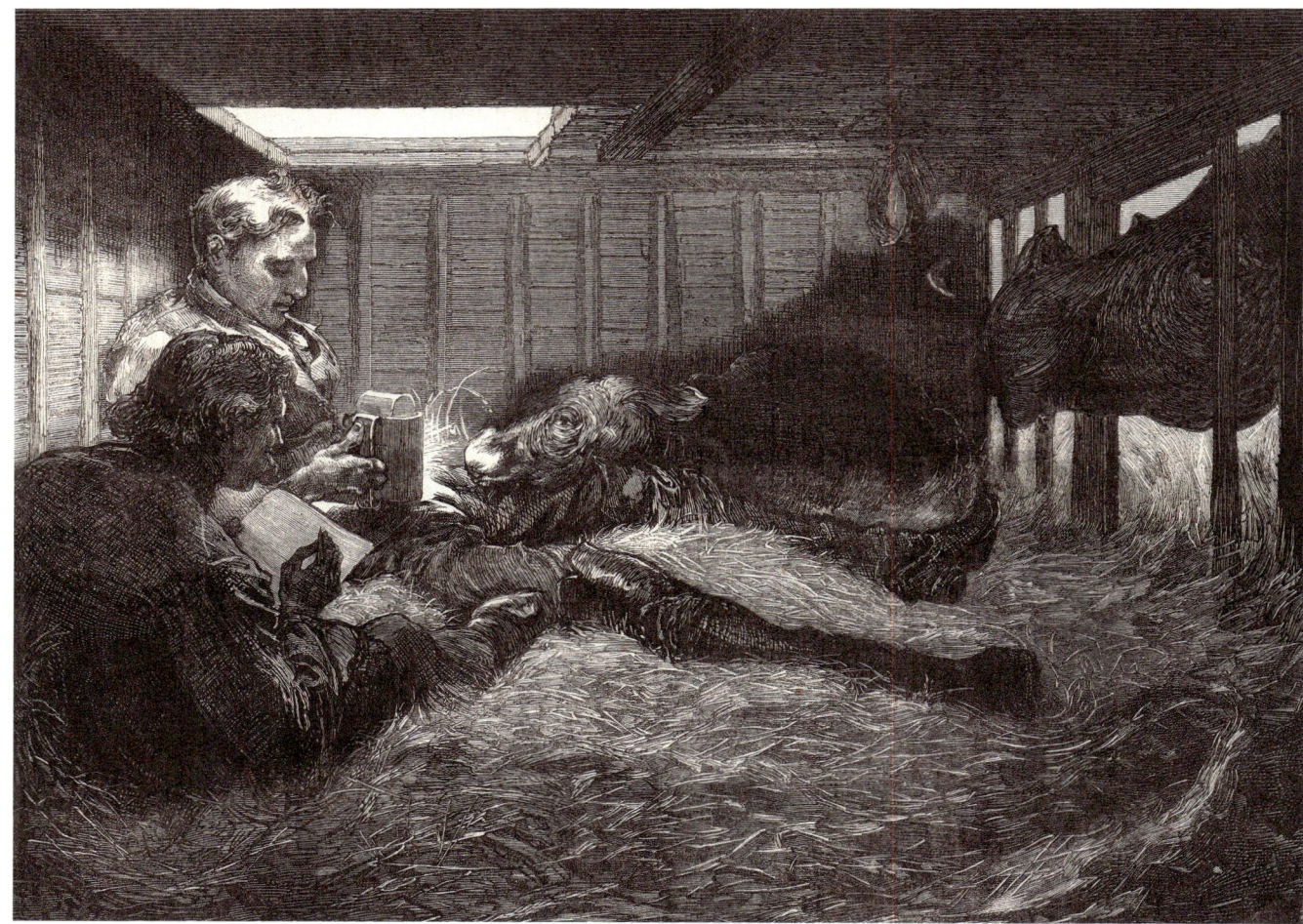

FIGURE 55.1   "The late baby rhinoceros and his mother. A sketch from life." Scene drawn in the stables of Charles Rice in London, showing the mother rhino imported from Malaysia on the right. The baby was born in the evening of Friday 6 December 1872 on board the *Orchis* on arrival at the Victoria Docks in London. The two men are unidentified. Size 30 × 22 cm
GRAPHIC, AN ILLUSTRATED WEEKLY NEWSPAPER, SATURDAY 11 JANUARY 1873, P. 41

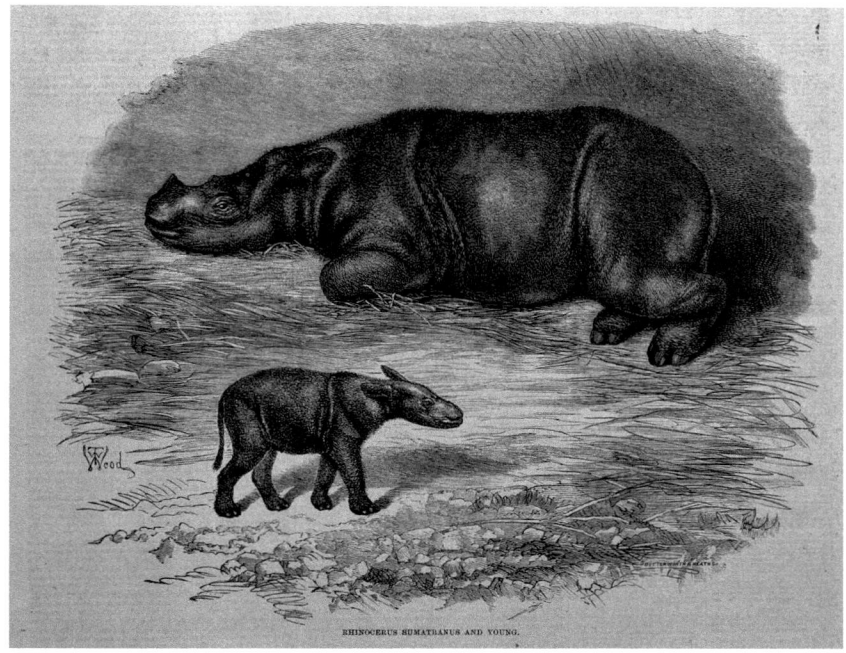

FIGURE 55.2
Mother *D. sumatrensis* and baby delivered on 6 December 1872 in the London Docks. Drawn by Thomas W. Wood (1839–1910), engraved by the firm of Butterworth & Heath. Plate discovered after publication of Rookmaaker (2014)
FIELD, THE COUNTRY GENTLEMAN'S NEWSPAPER, 4 JANUARY 1873, P. 2

Domestic Occurrences.

BIRTHS.

BAHR.—At Ebenezer Mission Station, on January 30th, 1889, the wife of Halfdan Bahr, of a son.
BUSKIN.—On the 1st February, at Kehunia Factory, Gorakhpur, the wife of A. Buskin, of a son.
DRION.—At Ranchi, on the 25th instant, the wife of W. H. P. Drion, of a daughter.
RHINOCEROS.—At the Zoological Gardens, Alipore, on the 30th January, Rhinoceros Lasiotis, the wife of Rhinoceros Sumatrensis, of Cabool, of a son.
MAXWELL.—At Jalapahar, on the 27th January, the wife of Major Maxwell, Derbyshire Regt., of a son.

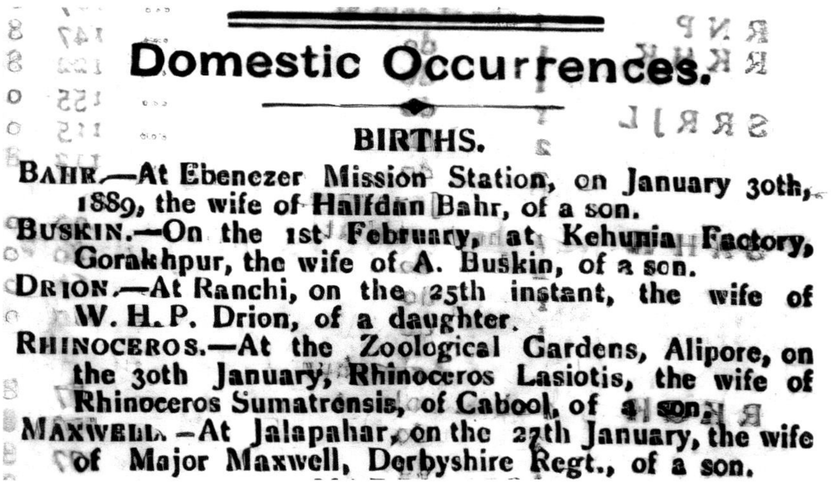

FIGURE 55.3
Announcement of the birth of a rhinoceros in Calcutta Zoo in 1889
ENGLISHMAN'S OVERLAND MAIL, CALCUTTA, TUESDAY 5 FEBRUARY 1889, P. 18

Kolkata (Calcutta) – Harbour

1. F 1871-12-03 (from Chittagong) to 1871-12-27 (exported to London)

Begum, the female captured in Chittagong 1867, was in Kolkata in December 1871 waiting for transport to England. She lived in the London Zoo for a record-breaking 28½ years from 15 February 1872 to 31 August 1900 (54.9).

Kolkata (Calcutta) – William Rutledge (Dealer)

The animal dealer William Rutledge had premises in Entally (48.2). He often worked together with Jamrach and other dealers shipping to Europe.

1. F 1885–02 to 1885-11-05 (exported via London to Hamburg)

On arrival in London, the correspondent of *Land and Water* heard that this female was one of four rhinos sent from Singapore after presumed capture in Malaysia. The animal was found to be pregnant and delivered a baby "under the charge of Mr. Rutledge, the animal agent in Calcutta" (Anon. 1885b). The two animals were shipped on the City Line steamer *City of Edinburgh* (captain W.B. Berham) leaving Kolkata on 1885-11-05 and arriving in the Royal Albert Docks in London on 1885-12-11. She was accompanied by a native attendant Mooram, and received special care from the chief steward R. Wood. In London it was said that the birth took place 9–10 months ago. The baby stood 2 ft 6 in high (76 cm) and had one horn "well advanced." The mother was without front horn "knocked to pieces in Calcutta against a tree and other objects." Owned by the German dealer Carl Hagenbeck, mother and baby were transported to Hamburg where they were examined and described by the zoologist

Theophil Johann Noack (1840–1918). Noack (1886) stated that mother and baby were bought by the circus owner Barnum in the USA for 25,000 Mark. They were to be sent to Hagenbeck's American agent William A. Conklin, then also superintendent of the Central Park Menagerie in New York. Apparently both mother and young (said to be a male) soon died, maybe still in Germany, or during the sea voyage, or soon after arrival (Dittrich & Rieke-Müller 1998: 134). No mention was found in the New York newspapers in early 1886 of any event regarding these animals.

2. M Birth 1885–02 to 1885-11-05 (exported via London to Hamburg)

Born to the female in transit in February 1885, the first ever of this species in captivity. If the date of birth was calculated after arrival in London in December 1885, with a gestation period of around 16 months, the birth may have taken place around August 1884 (Schwarzenberger & Hermes 2023). He was transported with his mother to London and onwards to Hamburg, where both were drawn by Theophil Johann Noack (fig. 55.4). The young animal died in 1886 (see no. 1).

Rhinoceros lasiotis mit Jungem.
N. d. L. gez. v. Th. Noack.

FIGURE 55.4
Mother Sumatran rhinoceros with her baby born in Kolkata in February 1885. They were drawn by Theophil Johann Noack (1840–1918) after their arrival in Germany in December 1885
NOACK, *ZOOLOGISCHE GARTEN*, 1886, P. 14

CHAPTER 56

Historical Records of the Rhinoceros in Chittagong

First Record: 1867 – Last Record: 1967
Species: *Dicerorhinus sumatrensis*
Rhino Region 41

56.1 Rhinos in Chittagong

The Chittagong (or Chattogram) Division of Bangladesh is part of the range of the two-horned Sumatran rhino *D. sumatrensis*. Neither *R. unicornis* nor *R. sondaicus* are known in the region. The rhinos were restricted to the mountainous areas running along the eastern border with Mizoram and the Rakhine and Chin states of Myanmar. None were recorded north of the Muhuri River in the Feni District.

Two rhinos were captured in Chittagong. A female called Begum was found near the Sangu River in 1867, and afterwards exhibited in London Zoo until 1900 (54.9). A second female was captured near Cox's Bazar in 1880 and then shown in Calcutta Zoo until 1892 (56.4). These two captures were unexpected, unusual, surprising events. There was only one known sighting in 1872 (56.3). The information about a rhino killed in 1891 is scant, yet not impossible (56.5). Since that time there have been rumours, but no significant encounters. Probably the areas where they lived were too inaccessible or remote, or maybe the animals in Chittagong were strays from a larger population across the borders.

When the rhino was first seen on the Sangu River in 1867, this was only the second instance, and probably the first known internationally, that a rhino with two horns was reported to exist in South Asia. From 1867 onwards, there have been many listings of *D. sumatrensis* for Chittagong in the general literature. Although the presence of *R. sondaicus* has also been asserted in a few cases, it can safely be assumed that none of these were based on direct evidence. There is no reason to believe that the one-horned species ever existed in Chittagong.

Given the scarcity of reports, it is best to conclude that rhinos occasionally strayed into Chittagong by chance. The animal killed in 1967 in the eastern part of Chittagong near the Burmese border was the last known report. Hence the species should now be considered extinct in this division.

...

Dataset 56.65: Chronological List of Records of Rhinoceros in Chittagong

General sequence: Date – Locality (as in original source) – Event – Source – § number, figures – Map number.

The list includes occasional reports from the regions of Myanmar across the eastern border.

The map numbers are explained in dataset 56.66 and shown on map 56.26.

Rhino Region 41: Bangladesh – Chittagong

*c.*1855 – Chittagong – Francis Bruce Simson (1828–1899), Commissioner in Dacca went hunting with the Commissioner of Chittagong who had shot many rhino – Simson 1886: 46 – 56.3

1860s – Chittagong Hill Tracts – Rhino present, possibly *D. sumatrensis* – Hood 1869

1864 – Chittagong – *R. sondaicus*. Last one shot – Mukherjee 1966: 27, 1982: 54. This date not found elsewhere

1867-11 – Sungoo (Sangu) River – Capture of female 'Begum', later kept in London Zoo 1872-02-15 to 1900-08-31. Type of *R. lasiotis* – Hood 1869; Lewin 1869: 16, 1870: 45; Clay 1896: 209 – 54.9, 56.2, fig. 56.2 – map D 23

1870s – Chittagong Hill Tracts – Rhino present – Lewin 1875: xv

1872-12 – Southern border – Thomas Herbert Lewin (1839–1916) saw rhino on a path – Lewin 1885: 441, 1912: 296 – 56.3 – map D 20

1876 – Chittagong – Assam rhino [*R. unicornis*] is common – Hunter 1876a, vol. 6: 33

1879 – Chittagong – *R. sondaicus* present – Pollok 1879, vol. 1: 98

1880 – Chittagong – Rhino has two horns – Oldfield 1880, vol. 1: 237

1881 – Ekserree (Ekshari), tributary of the Myanee – Tracks of mother and young rhino, animals not seen – R.P. 1884 – map D 24

1881-12 – Chittagong, estate of Begum Latifa Khatum of Ramu – Two-horned rhino, donated to Calcutta Zoo (female 'Muni Begum', June 1882 – *c.*1892). Mother of calf born 1889-01-30 in the Zoo – *Englishman's Overland Mail* 1882-06-17; Anderson 1883: 74; Sanyal 1892: 131 – 56.4 – map D 20

1881-12 – Chittagong, estate of Begum of Ramu – Second rhino seen, but not caught – *Englishman's Overland Mail* 1882-06-17 – map D 20

1890 – Chittagong – *R. sondaicus* extinct since late 1880s or early 1890s – Stebbing 1920: 98

1891 – Ropley [unknown locality, probably just east of Chittagong] – Frederick William Higgins, teaplanter near Chittagong town, killed a large rhino – Higgins 1891 – 56.5 – map D 23

1906 – Valleys of the Mynee River (Kassalong) – Rhino present – Hutchinson 1906: 196 – map D 25

1909 – Kasalong River – *D. sumatrensis* present – Hutchinson 1909: 5 – map D 25

© L.C. (KEES) ROOKMAAKER, 2024 | DOI:10.1163/9789004691544_057
This is an open access chapter distributed under the terms of the CC BY-NC-ND 4.0 license.

1909 – Kassalong RF – Rhino present up to early 20th century – Bahuguna & Mallick 2010: 279 – map D 25

1912 – Lakher Land (Myanmar, border with Chittagong) – Rhino in this dense jungle – Lorrain 1912: 17 – map D 3

1920s – Chittagong – Rhino said to exist but never seen when he visited 1923–1930 – Shebbeare 1953: 141

1943 – Lemro River (Myanmar) – Rhino present – Thom 1943: 258 – map D 2

1945 – Arakan Mountains (Myanmar) – 3 sightings, all said be 1-horned but actually 2-horned – Christison 1945; Ansell 1947 – map D 1

1950 – Sangu-Matamuhiri Valley – 90-year old Mro man mentioned that he encountered a rhino 70 years ago – Creative Common Alliance 2016: 24 – map D 22

1965 – Chittagong Hill Tracts – Fauna survey, no rhino found – Husain 1965: v

1967 – Hills east of Cox's Bazar – Rhino straying from Myanmar was killed and parts sold in Chittagong town – Mountfort 1969a: 230, 1991; Choudhury 1997b: 151, 1998: 7; Asmat 2009: 160 – 56.6 – map D 21

1970s – Kassalong RF – Rhino encounter reported by bamboo collectors – Chakma 2016: 48 – map D 25

1990 – Cox's Bazar Forest Division – No rhino listed – Rashid et al. 1990 – map D 21

Species in Chittagong stated as *D. sumatrensis*: listed by over 50 authors since 1872.

Species in Chittagong stated as *R. sondaicus*: Pollok 1879, vol. 1: 98; Meiklejohn 1937: 8; Stebbing 1920: 98; Talbot 1960: 205; Guggisberg 1966: 118; Mukherjee 1966: 27, 1982: 54.

Species in Chittagong stated as *R. unicornis*: Hunter 1876a, vol. 6: 33.

•••

Dataset 56.66: Localities of Records of Rhinoceros in Chittagong (Bangladesh)

The numbers and places are shown from south to north on map 56.26.

Rhino Region 41: Bangladesh – Chittagong

D 20   Estate of Begum Latifa Khatum of Ramu – 21.42N; 92.10E – 1881

D 20   Southern border – 1872

D 21   Hills east of Cox's Bazar – 21.50N; 92.50E – 1967

D 22   Sangu-Matamuhiri Valley – 22.00N; 92.30E – 1950

D 23   Sungoo (Sangu) River – 22.20N; 92.15E – 1867

D 23   Ropley, tea estate of Higgins – 1891

D 24   Ekserree (Ekshari), tributary of the Myanee – 23.50N; 92.00E – 1881

D 25   Kassalong RF (NE Chittagong) – 23.50N; 92.20E – 1909, 1970s

D 25   Mynee River (Kasalong) – 23.55N; 92.10E – 1906, 1909

Myanmar

D 1   Arakan Mountains, Myanmar – 21.30N; 93.00E – 1945

D 2   Lemro River Myanmar – 21.40N; 92.90E – 1943

D 3   Lakher Land Myanmar – 21.50N; 92.75E – 1912

• •

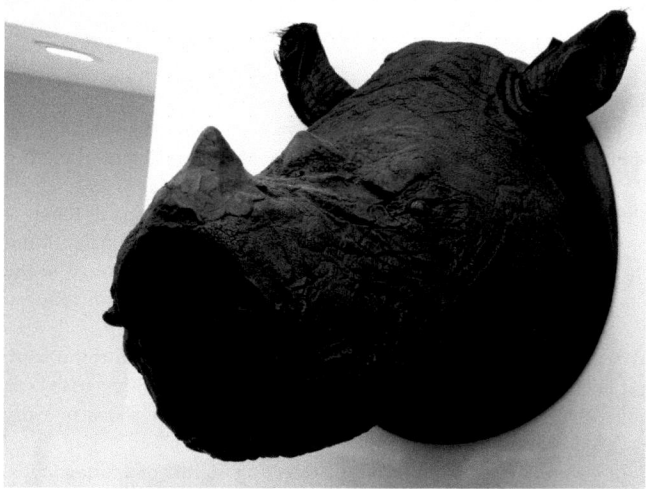

FIGURE 56.1   Mounted head of *D. sumatrensis* in the collection of the Zoological Survey in Kolkata. The origin of the specimen is not established

PHOTO: © TANOY MUKHERJEE, 2018

56.2   Begum, the First Captive Rhino from Chittagong

The first known report of a rhino reached the town of Chittagong (now Chattogram) in November 1867. A live rhino had been captured by villagers near the Sungoo (Sangu) River, about 50 km from town (fig. 56.2). Captain Frederick Henry Hood (1837–1875) managed to walk her to his compound in Chittagong, where she stayed until December 1871. This female rhino called 'Begum' was eventually purchased by the Zoological Society of London and exhibited in Regent's Park from 15 February 1872 to 31 August 1900, setting the longevity record for the species at 10,425 days (54.9). Soon after arrival in London, she was considered to belong to a new species, *Rhinoceros lasiotis*, a name still used in the combination *Dicerorhinus sumatrensis lasiotis* (Buckland, 1872) for the northern subspecies.

MAP 56.26 Records of rhinoceros in Chittagong. The numbers and places are explained in dataset 56.66

▲ Presence of *D. sumatrensis*

⬠ Artefacts

THE SUMATRA RHINOCEROS AT THE ZOOLOGICAL SOCIETY'S GARDENS.

FIGURE 56.2
"The Sumatra Rhinoceros at the Zoological Society's Gardens" depicting Begum in London Zoo after arrival in 1872. Engraving signed (lower left) by Johann Baptist Zwecker (1814–1876), engraved (lower right) by J.G., maybe John Gilbert (1817–1897). Size 17 × 23 cm
ILLUSTRATED LONDON NEWS, 23 MARCH 1872, P. 284

56.3   Recollections 1850–1872

In the 1850s, Francis Bruce Simson (1828–1899), Commissioner in Dacca, went out hog hunting with the Commissioner of Chittagong, possibly Horace Abel Cockerell (1832–1908). The latter was a good companion because he was known to have killed many tigers and rhinos (Simson 1886). Simson did not elaborate where these events took place, hence this does not provide proof that a rhino was killed in Chittagong.

Thomas Herbert Lewin (1839–1916), Deputy Commissioner and Political Agent for the Chittagong Hill Tracts, was one of the few to have seen a rhino in the mountains near the border between Chittagong and Burma (Whitehead 1992). In December 1872, while traveling from his home in Demagree (Tlabung, S. Mizoram) to the Arakan border, "one morning, as we marched along, a large rhinoceros trotted playfully in front of us for some distance" (Lewin 1885).

56.4   Muni Begum, the Second Captive Rhino from Chittagong

A second rhino was captured alive near Cox's Bazar in 1881. She was donated to the Calcutta Zoo by the Begum of Ramoo, a small town near Cox's Bazar. This was probably Begum Latifa Khatum of Ramu on behalf of her brother Nawab Moulvi Ali Amjad Khan (1871–1905). The rhino was discovered in December 1881 and captured in a novel fashion: "They then tied a number of ropes to the branches of the trees letting them hang down as nooses, in the course the animal was following. In a short time, their labour was rewarded, as it ran its head first into one noose and then into another, tearing them away, however, from the trees and in its excitement rushing out onto the open slope leading to the village, dragging the ropes after it. By this time it was somewhat exhausted, and fell in a muddy hollow where it was immediately surrounded, secured by ropes and ultimately dragged into the village" (*Englishman's Overland Mail* 1882-06-17). 'Muni Begum', as she was called, soon became tame and was a favourite with the children in the Begum's house allowing them to ride on her back (Renshaw 1907: 186). Once the Begum had decided to send her to the Zoo, the animal was walked in easy daily marches to Chittagong with the help of the Commissioner Edmond Elliot Lowis (1836–1918). From there she was conveyed on the British India Steam Navigation Co's steamer *Arabia* and arrived at Tollah's Nullah in the afternoon of Friday 2 June 1882. The rhino was housed in a new enclosure situated between the hornbill house and the superintendent's bungalow, at the south-west corner of the gardens, described by Anderson (1883) measuring 229 feet in length and 116 feet in breadth (70 × 35 m), with a pond of 165 by 40 feet (50 × 12 m).

56.5   Teaplanter Higgins in 1891

A newspaper report of June 1891 tells about a surprising hunt of a rhino: "On the ultimo Mr. F.W. Higgins added to his bag a much coveted rhinoceros which he had been

FIGURE 56.3
Impression of a Hairy-eared Rhinoceros
FRANK EVERS BEDDARD, *CAMBRIDGE NATURAL HISTORY*, 1902, P. 257, FIG. 132

after for years past. He shot it at Ropley, close to the station. Mr. Higgins was on foot; his dogs brought the beast out, and he got a shot at it at about 15 paces, smashing its shoulder. The rhino fell, recovered and then ran about 200 yards, when a well-directed bullet in the centre of the forehead prevented his escape. I have not heard the measurements, but those who saw the head, which Mr. Higgins sent round to the Club the same evening, say it was enormous" (Higgins 1891). The locality of Ropley is unknown. Frederick William Higgins was a teaplanter and hunter, who later in life owned an estate near Patiya, east of Chittagong. There is no confirmation elsewhere.

56.6   Rhino Trade in Cox's Bazar in 1967

In 1967, Guy Mountfort (1905–2003), co-founder of the World Wildlife Fund (WWF), heard rumours that a rhino had been shot in the mountains on the Indo-Burmese border east of Cox's Bazar. The animal had strayed from Myanmar, and was shot as soon as it crossed the border. The carcass was taken to Chittagong, where small pieces were sold for a total of Rs. 15,000, then about £1,150 (Mountfort 1969a). No rhinos have been seen since.

Historical Records of the Rhinoceros in Mizoram

First Record: 1872 – Last Record: 1930s, then one in 2009
Species: *Dicerorhinus sumatrensis*
Rhino Region 42

57.1   Rhinos in Mizoram

The two-horned rhino *Dicerorhinus sumatrensis* was once abundant in parts of the current state of Mizoram, formerly the Lushai Hills, a region of steep terrain inhabited mainly by the Mizo and Kuki tribes. Reports from the 1870s onwards are concentrated in the southern part of the province. These are relatively few, because foreign sportsmen only rarely ventured into this inaccessible landscape of sharp mountain ranges and dense forests. This range was contiguous with rhino sightings in eastern Chittagong and with the Arakan Mountains of Myanmar.

There is little evidence to decide if rhinos ever occurred in the central and northern parts of Mizoram. There are naturally quite a few general records of Mizoram in the literature, most of which are hardly based on personal knowledge. When Harper (1945) lists *R. sondaicus* in "North Lushai", he may not have realized the implications of the locality, and therefore this must be discounted.

Few outsiders were aware of the presence of rhinos, and even fewer saw one or tried to kill one. There are really only four specific records: Tanner in 1872, Stewart in 1888, Hutchinson in 1906 and Gordon in 1911 (57.2 to 57.6). The general report by H.S. Wood needs to be discussed due to the uncertain classification (57.7).

To most of the ethnic groups living in Mizoram it was unethical to kill a rhino, because the folds in the rhino's skin were derived from the clothes of ancestors who were transmogrified (Shakespear 1912). However, to have a chance to reach 'Peira', the best heaven, after death, one had to kill a human, as well as a long list of animal species including rhino (Parry 1932: 396). If a rhino was killed for any reason, the spirit of the dead rhino would follow the hunter to harm him and his family, so to deceive the spirit, on their way home, the hunter would split a bamboo and say that the animal "was struck by a thunderbolt moving horizontally" (Lalthangliana 2017). In similar subterfuge, the hunter might not return home directly, but first go to the forge in the village to stay there for one day, after which he was safe, provided that he left his gun and haversack behind and had changed all his clothes (Shakespear 1912).

This respect for the rhino changed over the years, when the Lusheis became envious of the Lakhers (Mara) who were earning large sums of money from the sale of rhino horns and other products (Pachuau & Schendel 2015). This switch in ecological perception had a major impact on rhino populations, although naturally it was still a feat to kill one. There is a memorial stone near Siatlal in Saiha district for Mr. H. Hmokha of Zyhno village, who had killed one hundred rhinos in his life (Zohra 2010). This must be one example out of many others. If he would have kept and displayed the skulls around his house, which he might have done, the collection would have wildly surpassed that of any of the major museums of the world.

Although all three species of rhinoceros have been listed for Mizoram, there is absolutely no reason to believe that either *R. unicornis* or *R. sondaicus* were ever found in the province. Lalthanzara (2017: 115) listed the name 'samak' applying to all rhino species, with a designation of 'samak ki hnih nei' for the two-horned animals. All contemporary reports write about the rhino being two-horned, or *D. sumatrensis*. Although there is no evidence in photographs or specimens, it is indisputable that this was the only species ever found in Mizoram.

The Conservator of Forests, Arthur John Wallace Milroy (1883–1936) wrote in 1932 that the rhino was almost exterminated in Mizoram. Although not personally acquainted with the interior, his information might well be correct. Mizoram remained a region with difficult terrain, where the secretive rhinos could still have their hiding places. R.C. Hanson (1933) assumed that a few rhinos inhabited forest reserves near the Lushai foothills, which had been recommended for conversion to national park status to provide much needed protection for wildlife. The IUCN Red List for Mammals first published by Simon (1966) leaves the possibility open that there may still be some rhinos in those forests. Choudhury (2013) quotes an event in 2009 when a rhino crossed the border from Bangladesh into the Dampa area of western Mizoram.

It cannot be denied that the local hunters, like the one at Siatlal claiming to have killed a hundred rhinos, must have driven the animals to extinction, or nearly so. Although the state has not been on the radar to be surveyed by the conservation community, it is likely that

© L.C. (KEES) ROOKMAAKER, 2024 | DOI:10.1163/9789004691544_058

This is an open access chapter distributed under the terms of the CC BY-NC-ND 4.0 license.

human disturbance and other developments have indeed meant the end of the rhinos. They could have gone extinct in the 1930s, or in any of the following decades, something which we will never know with certainty.

• • •

Dataset 57.67: Chronological List of Records of Rhinoceros in Mizoram

General sequence: Date – Locality (as in original source) – Event – Source – § number – Map number.

The map numbers are explained in dataset 57.68 and shown on map 57.27.

Rhino Region 42: Mizoram

1872-02 – Uiphum range south of Demagiri – Henry Charles Baskerville Tanner (1835–1898) finds that rhino are abundant – Macdonald 1872: 48 – 57.2 – map D 31

1888 – Valley below Lung Leh (Lunglei) – Leslie Waterfield Shakespear (1860–1933) says a two-horned rhino was shot by others – L. Shakespear 1929: 92; Chatterjee 1995: 55; Choudhury 2013: 228, 2016: 155 – map D 32

1888 – Lushai Hills, east of Rangamati – John Fraser Stewart (1863–1888), Patrick McCormik (d. 1888) and John Owens (d. 1888) received permission to shoot as many rhino as they liked; no trophy recorded – Shakespear 1928: 11 – 57.3 – map D 32

1890 – Lushai – Presence of two-horned rhino – Woodthorpe 1890: 20

1890 – Camp Teriot, 15 miles (24 km) from Fort Lungleh – "Many signs of rhinoceros have been met with" during Chin-Lushai expedition – *Civil & Military Gazette* 1890-01-28: 4–5 – map D 31

1893 – Lushai – Rhino present – Reid 1893: 1; Hunter 1909: 457

1896 – Letha Range; Imbukklang; Tuimong and Tuinan country (Myanmar, east of Mizoram border) – Presence of both *D. sumatrensis* and *R. sondaicus* – Carey 1896: 10, 209 – map D 4

1899 – Lushai – "Assam Rhino" is common – Chambers 1899:57

1903 – Mizoram – Contemporary reports about great value of rhino horn, fetching 100 rupees per seer – Jackson 2016: 51

1903 – Samang Tlang (Myanmar, south of Mizoram border) – Rhino present – Pachuau & Schendel 2015: 185 (after newspaper *Mizo Leh Vai*, April 1903) – map D 3

1903 – Southern Lushai – 9 rhino hunted in competition with Mara people – Pachuau & Schendel 2015: 185 (after *Mizo Leh Vai*, April 1903)

1906 – Valleys of the Thega and Tuichong (Demagiri area) – Two-horned rhino present. One shot by a Gurkha of the Police Batallion. The animal obtained by Robert Henry Sneyd Hutchinson (1866–1930) for a friend – Hutchinson 1906: 196 – 57.4 – map D 33

1906 – Thega River – *D. sumatrensis* present – Hutchinson 1909: 5 – 57.4 – map D 33

1906–1907 – Lungleh area – Government prohibits export of rhino horns – Document in Mizoram State Archives (retrieved by Kyle Jackson, email 2018) – map D 32

1908 – Lushai Hills – *D. sumatrensis* present – Allen 1908b: 5

1908 – Mizoram – Theodor Krummel Sap advertises for purchase of living rhino, adult or calf – Krummel 1908, translated in Pachuau & Schendel 2015: 184 – 57.5

1909–1910 – Lungleh area – Government issues orders to prohibit shooting of rhino – Document in Mizoram State Archives (retrieved by Kyle Jackson, email 2018) – map D 32

1911–1912 – Lushai Hills – Archibald Douglas Gordon (1888–1966), Inspector General, Bengal Police, once followed rhino tracks for 1½ days – Gordon 2015 – 57.6

1929 – [Mizoram or Nagaland] – Specimen: Body armour of rhino hide, with straps for fastening, 34.5 × 23.5 cm. Acquired in March 1930 by Pitt Rivers Museum, Oxford, no. 1930.72.1 (online catalogue) – fig. 57.1.

1930 – Lushai – Smaller rhino found in the foothills in considerable numbers - Wood 1930:60,69, 1934:171. Probably refers to *D. sumatrensis* – 57.7, fig. 57.2

1932 – Lushai – *D. sumatrensis*. Almost hunted to extinction – Milroy 1932, 1934: 102; Choudhury 2013: 225

1932 – Lushai – *D. sumatrensis*. Survived until recently – Hobley & Shebbeare 1932: 20

1933 – Lushai – Rhino present – R.C. Hanson 1933: 51

1935 – Mizo hills – *D. sumatrensis*. Presence up to about 1935 – Gee 1964a: 157; Seshadri 1969: 106; Sinha et al. 2005; Bahuguna & Mallick 2010: 278

1936 – North Lushai Hills – *R. sondaicus*. Few may survive – Harper 1945: 381 (information from Bombay Natural History Society, December 1936). This needs confirmation (Mukherjee 1982: 54)

1943 – Lemro River (Myanmar, south of Mizoram border) – William Sinclair Thom (b. 1868) of the Burma Police killed a double-horned rhino – Thom 1943: 274 – map D 2

1945 – Arakan Mountains (Myanmar, south of Mizoram border) – Alexander Frank Philip Christison (1893–1993) listed 3 sightings, all said to be one-horned, but in fact two-horned – Christison 1945; Ansell 1947 – map D 2

1960s – Lushai Hills – *D. sumatrensis*. Few may survive – Simon 1966; Mukherjee 1966: 29; Fisher et al. 1969: 115; Goodwin & Holloway 1972

1999 – Lushai Hills – *D. sumatrensis*. Possibility of isolated population – Tiwari 1999: 162

2009 – Dampa Tiger Reserve, across border from Chittagong – Chittagong hunters chased one, but animal escaped by crossing the Indian border into Mizoram – Choudhury 2013: 225 – map D 34

2010 [for earlier events] – Siatlal village, Saiha district – Memorial stone commemorates H. Hmokha of Zyhno village who killed 100 rhino – Zohra 2010; Choudhury 2013: 225, 2016: 155 – map D 30

2014 – Chin Hills (South Mizoram) – Rhino now inhabit only very few remote regions – Strait 2014: 27.

Species in Mizoram stated as *D. sumatrensis*: Milroy 1932, 1934: 102; Mukherjee 1982: 54; Choudhury 2013: 225; Groves 1967: 225; Rookmaaker 2003; Schenkel et al. 2007: 72; Strait 2014: 27; Havmøller et al. 2015: 356; Lalthanzara 2017: 115.

Species in Mizoram stated as *R. sondaicus*: Wood 1930: 60, 69, 1934; Guggisberg 1966: 118; Groves 1967: 224; Talbot 1960: 205.

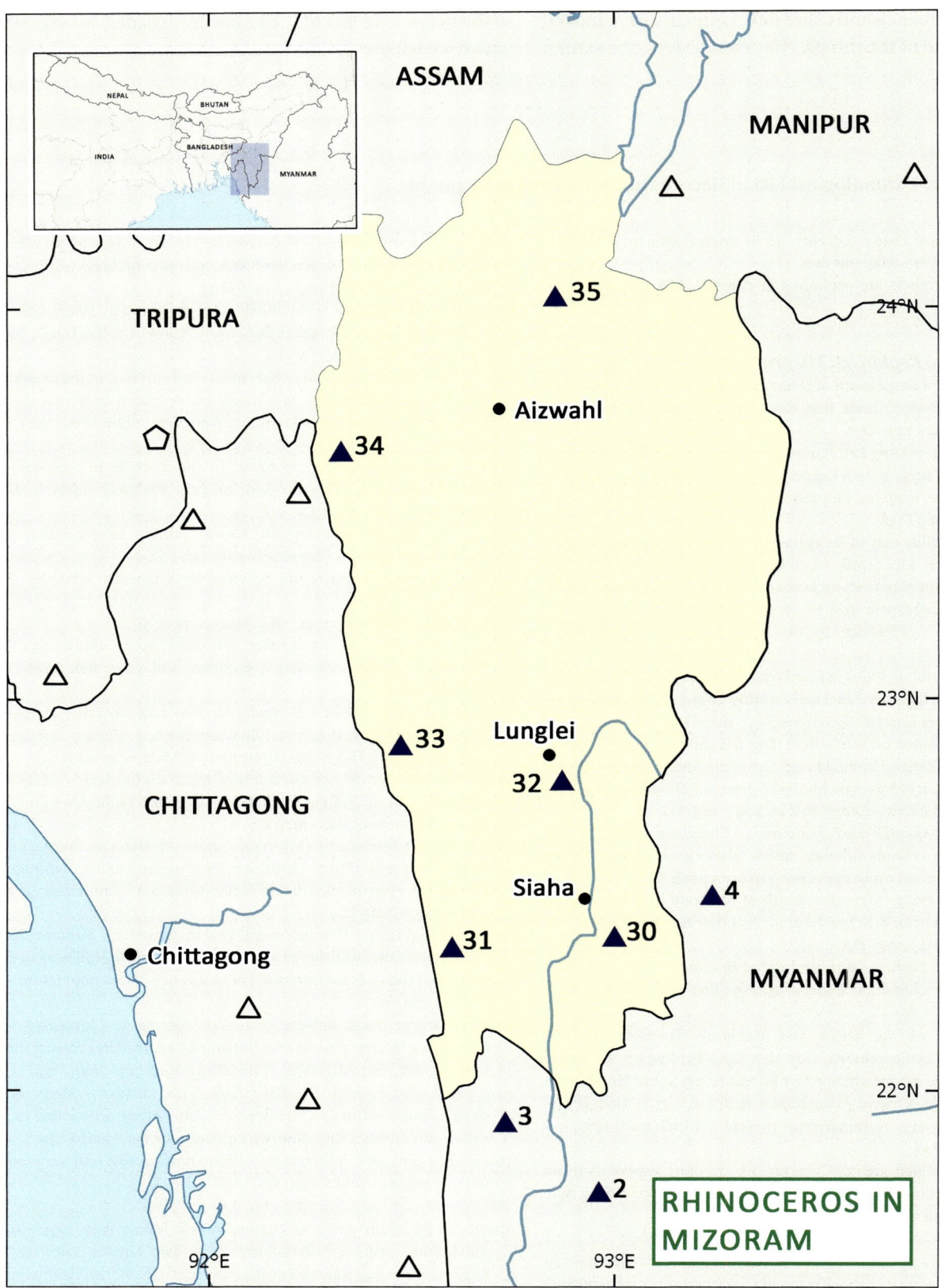

MAP 57.27   Records of rhinoceros in Mizoram. The numbers and places are explained in Dataset 57.68

▲ Presence of *D. sumatrensis*

⬠ Artefacts

•••

Dataset 57.68: Localities of Records of Rhinoceros in Mizoram

The numbers and places are shown from south to north on map 57.27.

Rhino Region 42: Mizoram

D 30   Siatlal village, Saiha district – 22.26N; 93.03E – 2010
D 31   Uiphum range south of Demagiri – 22.50N; 92.60E – 1872
D 31   Camp Teriot, 24 km from Lunglei – 22.50N; 92.60E – 1890
D 32   Valley below Lung Leh (Lunglei) – 22.75N; 92.75E – 1888
D 32   Lushai Hills east of Rangamati – 22.75N; 92.75E – 1888
D 32   Lungleh area – 22.75N; 92.75E – 1906–1907, 1909–1910
D 33   Valleys of the Thega and Tuichong (Demagiri area) – 22.80–92.50 – 1906
D 33   Thega River – 22.80N; 92.50E – 1906

D 34   Dampa TR across border from Chittagong – 23.50N; 92.30E – 2009
D 35   North Lushai Hills – 24.00N; 93.00E – 1936

Myanmar

D 2   Lemro River, Arakan Mountains, Myanmar – 21.40N; 92.90E – 1943
D 2   Arakan Mountains, Myanmar – 21.30N; 93.00E – 1945
D 3   Samang Tlang, Myanmar – 21.96N; 92.82E – 1903
D 4   Letha Range; Imbukklang; Tuimong and Tuinan country, Myanmar – 23.00N; 94.50E – 1896

∴

57.2   Tanner Follows Rhino Trails in 1872

Captain Henry Charles Baskerville Tanner (1835–1898), Deputy Superintendent of the Revenue Survey who participated in the Lushai Expedition, undertook a journey in February 1872 down the Uiphum Range (south of Lunglei), where his progress through the mountains was facilitated by following animal trails: "Wild elephants and rhinoceros appear to abound in great numbers in this tract, and between them they had managed to keep portions of the old deeply-worn Kookie path open, and in such places we found no difficulty" (Macdonald 1872).

57.3   Stewart Received Permit in 1888

Setting out on a punitive expedition in the interior in 1888, Lieutenant John Fraser Stewart chose Lance-Corporal Patrick McCormick and Private John Owens to accompany him because they were active sportsmen. His application to shoot elephants was denied, but he "cheered up on receiving permission to shoot as many 'rhinos' as they liked" (Shakespear 1928). This never happened, as the whole party was killed when their camp was attacked.

57.4   Gurkhas Kill a Rhino in South Lushai in 1906

The first to actually refer to the shooting of a rhino in Mizoram was Robert Henry Sneyd Hutchinson, Superintendent in the Indian Police. A former colleague

FIGURE 57.1   Body armour of rhinoceros hide, with straps for fastening, from Nagaland or Mizoram, July 1929
WITH PERMISSION © PITT RIVERS MUSEUM, UNIVERSITY OF OXFORD: NO. 1930.72.1

from Bengal had requested some rhino flesh or powdered horn, believed to be a certain cure for barrenness in women. This was put to the test and proved effective, after he heard that the Gurkhas of the Police Battalion, of the South Lushai hills, had recently killed a rhino (Hutchinson 1906).

57.5 Theodor Krummel Advertising for Live Rhinos in 1908

In 1908 there appeared a surprising advertisement in one of the local Mizo newspapers, the *Mizo Leh Vai* (Pachuau & Schendel 2015: 184). For several months, the readers were asked to go out and catch a rhino (called sazukte): "A very profitable business. For a live rhinoceros, and an adult one, I am willing to pay Rs. 3000/– or Rs. 4000/–. There is also a price for Rhino calves. My name is Theodor Krummel Sap. My address is 21 Park Lane, Calcutta." That was a large amount to pay, but the possibilities were few, and there is no sign that there were any lucky takers. There were some members of a German family Krummel in Kolkata at the time, one of which imported pigs from Carl Hagenbeck in 1903 for an unsuccessful home-farm in Ballygunge (*Englishman's Overland Mail* 1903-05-07).

57.6 Gordon Follows Rhino Tracks in 1911

Archibald Douglas Gordon (1888–1966) of the Bengal Police, found tracks of a rhino in the Lushai hills in 1911. He started to follow them, but abandoned his chase when he heard from a woodcutter that the animal was at least a day's walk ahead (Gordon 2015).

57.7 Henry Stotesbury Wood and His Rhino Classification

The works by Henry Stotesbury Wood (1865–1956) contain some paragraphs on the rhino of Mizoram, Tripura and Sylhet which need careful examination. For the most part of his army career, he worked as a Civil Surgeon in the Indian Medical Service, first in Manipur 1891–1898, then in Assam 1898–1913. After his retirement in July 1922, he settled in North Cachar with his wife Florence Laura Emily (Skinner) Wood (d. 1955), who accompanied him in many of his shooting and walking expeditions. After their return to Europe in 1927, they traveled widely in Europe. Wood wrote *Shikar Memories* (1934) and *Milestones of Memory* (1950) with rhino anecdotes, and *Glimpses of the Wild* (1936) without any text on the rhino. Wood is known to have participated in the annual hunt of the Maharaja of Cooch Behar in 1903 (32.9). He was an active hunter of big game, and his trophies were presented in a photograph in *The Sphere* of 1910 (fig. 57.2). That photo was taken in Tezpur, where in 1910 he joined the Darrang Game Association (*Englishman's Overland Mail* 1910-09-01). Wood went to hunt rhinos on foot in Booreegaon, north of the Brahmaputra opposite Kaziranga. Together with an Assamese tracker called Laloong, he found rhinos but only wounded some and when the rhinos charged, Laloong was fatally injured (Wood 1930, 1934). Apparently the tragedy can be dated to May 1909 from a remark in Anon. (1909b).

Wood had seen two species of rhino, one *R. unicornis* and another: "The smaller rhino is found at the base of hilly country, where the hills are undulating and the forest dense and mixed with streams and cane brakes. I have seen them in considerable numbers in the foot hills of Tipperah, Sylhet and Lushai. In the small ravines there are swamps and streams" (Wood 1930: 60, 1934: 171). He lived in that general area now distributed across the Bangladesh Division of Sylhet and the Indian states of Tripura and Mizoram for some years from 1898 and again after retirement in 1922, but of course might well have visited at other times. Maybe we should note that Cachar or Assam are not included in the list. Rhinos have been recorded by others from Sylhet and nearby lower Assam, but for Tripura and northern Mizoram his testimony is all that remains – and he would have seen these animals there in "considerable numbers." Wood (1930) wrote that he had been "after both the Great Indian rhino and the smaller, the Sumatran one." This "Sumatran one" is an unfortunate ambiguous term. In almost all literature, including the works about Indian fauna accessible to Wood, it denotes the two-horned *D. sumatrensis*. However, in a sentence in his paper of 1930, but removed from the book of 1934, he continues to say that "I have never seen the third variety of rhino, the Asiatic two-horned one (*R. sumatrensis*) the greater part of whose body is clad with hair of some length." Therefore, did he mean to say that his lesser "Sumatran one" was *R. sondaicus*? According to him, this lesser rhino was a great wanderer, which should help, but not a great deal in fact. *R. sondaicus* definitely wanders when there is a need according to older sources reviewed by Sody (1959), while *D. sumatrensis* enjoys his long walks (van Strien 1974: 38), maybe best put in words by Theodore Rathbone Hubback (1872–1944), a naturalist in Malaysia: "They seem impervious to any physical feeling of discomfort. Their walking powers in bad country are phenomenal. I have often followed up a rhino which has laboriously climbed – I should say we have laboriously

THE WEALTH OF SPORT: In Far Assam and in the United States.

SOME FINE TROPHIES OF BIG GAME SHOT IN ASSAM

FIGURE 57.2
Henry Stotesbury Wood (1865–1956)
with some trophies obtained in Assam
THE SPHERE, SATURDAY
24 SEPTEMBER 1910, P. 281

climbed – a high mountain merely to go down the other side" (Hubback 1939: 602).

Besides the unusual name and the unusual localities, Wood also added that he saw these animals in considerable numbers. As the available accounts testify, there have been people who have seen *R. sondaicus* or *D. sumatrensis* in South Asia, but if they did, in the majority of cases those must have been encounters lasting minutes with single specimens rushing through the undergrowth of the jungle. Wood provided no further detail where he went after this lesser rhino, or if he ever actually killed one. Our knowledge of his shooting excursions is not detailed enough to help. His books read like straightforward

unadulterated accounts of some of his experiences, but definitely not all of them. It would not surprise me if Wood had followed the example of Pollok (1879, vol. 1: 98), who also referred to *R. sondaicus*, ranging "throughout Assam down through Sylhet, the Garrow hills, Tipperah, Chittagong into Arrakan, Burmah." Pollok had no personal experiences with rhinos in most of these tracts.

This presents a dilemma, and rather an important one, because Wood was one of very few with data on rhinos in that part of the continent. Maybe I am unable to fully justify it, but my feeling is that Wood must have meant *D. sumatrensis* when adding his sentence on the "Sumatran one."

CHAPTER 58

Historical Records of the Rhinoceros in Tripura

First Record: 1750 – Last Record: 1920s
Species: *Dicerorhinus sumatrensis*
Rhino Region 43

58.1 Possible Absence of Rhinos in Tripura

Tripura, formerly known as Tipperah, is not typical rhino country. Rhinos were equally unknown in regions towards the west, including the northern part of Chittagong, Dacca and Mymensingh divisions. However, towards the east in the Mizoram hills the animal has been reported. Towards the north, Sylhet division (Bangladesh), Karimganj and Hailakandi (Assam, India) had occasional reports about rhinos. Before the partition of 1947, Tipperah extended on both sides of the Indo-Bangladesh border.

The *Rajmala*, a chronicle of Tripuri kings, records that during the 18th century, rhinos were common throughout the state. The only definite record is a rhino which was thought to have strayed from Tripura before it was killed south of Comillah in Chittagong territory in 1876 (58.2). In his taxonomically confused remarks, Henry Stotesbury Wood (1865–1956) claimed to have seen the smaller type of rhino (meaning *R. sondaicus* or *D. sumatrensis*) "in considerable numbers" in the foothills of Tipperah, Sylhet and Lushai (57.7). The combination of these three areas is unfortunate, just somewhat too general to know where in Tipperah territory he encountered these rhinos. Despite the absence of definite records, it may well be that the rhinos existed in Tripura, especially in the hills in the southern part of the state. The changes in administrative responsibilities and the vagueness of existing records probably distort the real picture.

When Fitzwilliam Thomas Pollok (1879, vol. 1: 98) included Tipperah in a longer list of localities where rhinos were known, he referred to the animals as *R. sondaicus*, even though arguably he had no personal knowledge of the area. The same type of confusion is present in later literature. A short list of authors were aware of the existence of a rhinoceros in Tripura, adding a specific identity without further evidence. Clearly, *R. unicornis* was never found in Tripura, and the former presence of *R. sondaicus* is in fact highly unlikely. Only one rhino from Tripura has ever been collected, a two-horned *D. sumatrensis,* which leads to the conclusion that this species was the only one ever found in the state.

•••

Dataset 58.69: Chronological List of Records of Rhinoceros in Tripura

General sequence: Date – Locality (as in original source) – Event – Source – § number – Map number.

The map numbers are explained in dataset 58.70 and shown on map 58.28.

Rhino Region 43: Tripura

Undated – Gandacherra (Gondachara) – Place named after rhino. Indicates former presence of *R. unicornis* (Darlong & Alfred 2002: 22) or *D. sumatrensis* (Choudhury 2013: 225) – map M 42

9–10th century – Shyam Sundar Tilla, Jolaibari – Rhino on terracotta plaque of Buddhist stupa. There is no pictorial evidence and this may represent a boar. Record needs confirmation – Chauley & Lourdusamy 2000; Manuel 2006, 2008: 36; Dey 2019 – map M 41

1750 – Tripura – Rhino common, according to the *Rajmala*, an 18th-century chronicle of Tripuri kings – Long 1978; Darlong & Alfred 2002: 21 – 58.1

1876-02 – 20 miles south of Comillah (Cumilla), from Tippera hills – Two-horned rhino killed. Specimen obtained, once preserved in Museum of Royal College of Surgeons of England, London, no. 2146 (destroyed) – Manson 1876: 176; Sclater 1877a: 269; Flower 1878: 634; Flower 1884: 422; Blanford 1891: 477; Pollok & Thom 1900: 166; Lydekker 1900: 29; Prater 1939: 619; Harper 1945: 394; Mukherjee 1982: 56 – 58.2 – map D 40

1886 – Tippera – *R. sondaicus* occasionally killed – Simson 1886: 188

1910 – Tripura – Rhino in large numbers during 1900s – Menon 1975: 42

1922 – Foot hills of Tipperah – *R. sondaicus* (=*D. sumatrensis*, 57.7). Seen in considerable numbers. Disappearing due to cultivation – Wood 1930: 60, 69 – 57.7

1932 – Tippera – *D. sumatrensis* surviving until short time back – Hobley & Shebbeare 1932: 20

1996 – Tripura – Rhino virtually extinct – Debbarma 1996: 17.

Species in Tripura stated as *R. sondaicus*: Pollok 1879, vol. 1: 98; Sterndale 1884: 410; Simson 1886: 188; Wood 1930: 60,69.

Species in Tripura stated as *R. unicornis*: A.K. Gupta 1998; Darlong & Alfred 2002: 22.

© L.C. (KEES) ROOKMAAKER, 2024 | DOI:10.1163/9789004691544_059
This is an open access chapter distributed under the terms of the CC BY-NC-ND 4.0 license.

Species in Tripura stated as *D. sumatrensis*: Manson 1876: 176; Sclater 1877a: 269; Flower 1878: 634; Blanford 1891: 477; Sclater 1891: 204; Sanyal 1892: 131; Trouessart 1898, vol. 2: 756; Pollok & Thom 1900: 166; Lydekker 1900: 29; Finn 1929: 189; Hobley & Shebbeare 1932: 20; Prater 1939: 619; Harper 1945: 394; Shebbeare 1953: 147; Talbot 1960: 174; Groves 1967: 225; Mukherjee 1982: 56; Bahuguna & Mallick 2010: 275; Choudhury 2013: 225.

MAP 58.28   Records of rhinoceros in Tripura. The numbers and places are explained in Dataset 58.70

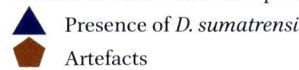 Presence of *D. sumatrensis*

Artefacts

Dataset 58.70: Localities of Records of Rhinoceros in Tripura

The numbers and places are shown on map 58.28.

Rhino Region 43: Tripura

D 40   20 miles south of Comilla, from Tippera hills – 22.26N; 91.36E – 1876

M 41   Shyam Sundar Tilla, Jolaibari – 23.30N; 91.60E – 10th century – uncertain identification

M 42   Gandacherra (Gondachara) – 23.61N; 91.81E – toponym

58.2   Rhino Killed in Comillah in 1876

There is really only one definite record of a rhino in Tripura, close to the Chittagong border. Alexander Manson (1833–1902) of the Bengal Civil Service, stationed in Chittagong from 1864 to 1889, wrote a letter to the *Oriental Sporting Magazine* of 1876, reporting that a two-horned rhino had been killed in February 1876 about 20 miles south of the station of Comillah (Cumilla). It was suggested that the animal had strayed down from the Tipperah hills into a highly cultivated and thickly populated country, "16 miles from the jungle." When Manson (1876) obtained the head, the skin was already too much rotten to be preserved, but the skull was intact, with a front horn of 8½ inches (21.6 cm), and a small stud for the second horn. The skin had very little hair and was cut in pieces by the villagers. Manson was told that the animal was a male, full-grown and rather old, standing 4½ feet (137 cm) in height. In London, Philip Sclater (1877a) read this letter and endeavoured to obtain the evidence for scientific examination. With the help of William Day Stewart (1840–1892), a civil surgeon based in Cuttack, Orissa, the skull was added to the collection of the Royal College of Surgeons of England, London. Here it was studied by the comparative anatomist William Henry Flower (1831–1899), who found that the animal must have been a young adult. The skull stood out by the superior breadth proportional to the length and the greater size of teeth. Sclater had hoped that the characteristics might support his designation in 1872 of *Rhinoceros lasiotis* as a separate species, based only on an examination of the living animal in the zoo at that point. Although Flower (1878) recognized it as *D. sumatrensis*, he did not want to decide if the differences in skulls from various regions should be considered of specific value.

Rauhohr-Nashorn, Dicerorhinus sumatrensis lasiotis *Scl.* 1/20 natürlicher Größe.

FIGURE 58.1
"Rauhohr-Nashorn, Dicerorhinus sumatrensis lasiotis Scl. 1/20 natürlicher Große". An impression of the Hairy-eared rhinoceros drawn and engraved by Wilhelm Kuhnert (1865–1926), published by Lutz Heck (1892–1987). There is no information which specimen was depicted
HECK IN *BREHMS TIERLEBEN*, 1915, VOL. 3, P. 602

Historical Records of the Rhinoceros in Sylhet Division

First Record: 1910 – Last Record: 1910
Species: *Dicerorhinus sumatrensis*
Rhino Region 44

59.1 Absence of Rhinos in Sylhet, Bangladesh

Rhinos were mostly absent in Sylhet, formerly part of Assam but since 1947 a division in north-east Bangladesh. The animals were equally rare or absent westwards in the bordering divisions (Dacca, Mymensingh) of northern Bangladesh, southwards in Tripura and Chittagong, or northwards in Meghalaya. Towards the east there were infrequent sightings in the southern part of Assam (Karimganj and Cachar).

There is only one definite record of a rhino in Sylhet, shot in the first part of the 20th century, known from a report by the army officer Henry S. Wood (59.3). His identification as *R. sondaicus* is questionable (55.7). Any rhino seen in Sylhet should have been *D. sumatrensis*.

∙∙∙

Dataset 59.71: Chronological List of Records of Rhinoceros in Sylhet

General sequence: Date – Locality (as in original source) – Event – Sources – § number – Map number.

Species identification follows original source (names updated). The map numbers are explained in dataset 59.72 and shown on map 59.29.

Rhino Region 44: Bangladesh – Sylhet

7th century – Jaintiapur (ancient capital) – The son of the 7th century King Joymalla tried to fight a freshly-caught rhino but met with accidental death – Nath 1948: 99 – map M 46

1800 – Bangladesh – *D. sumatrensis*: former occurrence – Simon 1966

1800 – Sylhet – Rhino, former presence, not seen lately – Hunter 1879b, vol. 2: 269

1822 – Sylhet District – George Theophilus Collins (1790–1833), Commissioner, found that only one rhino had ever been seen many years ago – Collins 1822 – 59.2

1850 – Sylhet – Sylhet is market for rhino hides to make shields – Baker 1887: 249

1879 – Sylhet – *R. sondaicus*. Lesser rhino present – Pollok 1879, vol. 1: 98

1900 – Sylhet – *R. sondaicus* (= *D. sumatrensis*, 57.7). Seen in considerable numbers. Disappearing due to cultivation – Wood 1930: 60, 69, 1934 – 57.7, 59.1

1898–1914 – Shamshernagar – *R. sondaicus* (= *D. sumatrensis*). The teaplanter Gordon Fraser shot one in a tank supplying the teahouse – Wood 1930: 61, 1934; Fraser 1932; Rookmaaker 2002b; Choudhury 2013: 225, 2016: 155; Watt 2017 – 59.3 – map D 45

1975 – Sylhet – Rhino is found – Nyrop 1975: 70

∙∙∙

FIGURE 59.1
Female *D. sumatrensis* caught in Pahang, Malaysia on 11 July 1988. Named Seputih, she was kept in the Sumatran Rhino Conservation Centre, Sungai Dusun, Selangor, until her death on 28 October 2003. Photograph taken in the 10-acre paddock in Sungai Dusun in 2002. Note the hairiness of her limbs and ear conches, as well as the relatively small size of both horns
PHOTO: ZAINAL ZAHARA ZAINUDDIN, 2002

© L.C. (KEES) ROOKMAAKER, 2024 | DOI:10.1163/9789004691544_060
This is an open access chapter distributed under the terms of the CC BY-NC-ND 4.0 license.

MAP 59.29   Records of rhinoceros in Sylhet. The numbers and places are explained in Dataset 59.72

▲   Presence of *D. sumatrensis*

⬠   Artefacts

• • •

Dataset 59.72: Localities of Records of Rhinoceros in Sylhet

The numbers and places are shown on map 59.29.

Rhino Region 44: Bangladesh – Sylhet

D 45   Shamshernagar – 24.39N; 91.89E – 1898–1914
M 46   Jaintiapur (ancient capital) – 25.14N; 92.12E – 7th century

• •
•

59.2   Captures Impossible in 1822

When the Bombay Government was trying to obtain a rhino for the Gaekwar in Baroda, the reply dated 21 May 1822 was quite clear. The Commissioner George Theophilus Collins (1790–1833) wrote that "I shall not succeed in capturing one, as but one animal of that description has ever been seen in this District, and that many years ago; consequently they must be exceedingly scarce, if indeed there are any now to be found in the District of Sylhet" (Collins 1822).

59.3   Gordon Fraser at Shamshernagar

When Henry S. Wood was discussing the range of his "Sumatran" rhino (57.7), he noted that these animals are great wanderers, shown by the fact that "it is recorded that one was shot in a tank, that supplied the engine of a tea house with water, by the late Gordon Fraser, a tea planter in Sylhet" (Wood 1930). Wood supplied neither date nor exact locality. The record is repeated by Wood (1934) omitting Fraser's name. Gordon Fraser arrived in India in 1876 and managed the Shamshernagar Tea Estate of the Lungla Company from 1880 to 1914 (*Englishman's Overland Mail* 1914-03-05; Fraser 1932). As Wood himself was in Cachar from 1898 to 1922, he might well have known Fraser. There is no further record of the shooting of the rhino in Shamshernagar, which must have occurred between 1898 and 1914.

CHAPTER 60

Historical Records of the Rhinoceros in Lower Assam

First Record: 1840 – Last Record: 1967
Species: *Dicerorhinus sumatrensis*
Rhino Region 45

60.1 Rhinos in Lower Assam

Rhinos were regularly observed in the southern districts of Assam comprising Karimganj, Hailakandi, Cachar and North Cachar, wedged between Meghalaya and Bangladesh in the west, Tripura and Mizoram in the south, and Manipur in the east. The animals were seen on the hills in the south, as well as in the valleys of the Barak, Surma and Longai Rivers.

In the south of Cachar, which until the 1980s included the current southern districts Karimganj and Hailakandi, there are a handful of records, especially near the western border with Sylhet and the southern border with Mizoram. This includes the rhino reported south of Chargola by John Anderson in 1872 (60.2) and the hunting account by Synteng in 1885 (60.3). Edward Baker had also heard about rhinos in this general region (60.4).

There is no consensus about the species of rhino living in this region. Anderson (1872) spoke about *D. sumatrensis*. Synteng (1885) never gave a clue. Baker (1887) referred the rhinos in Cachar and in neighbouring Sylhet to

R. unicornis, apparently without ever observing one or providing further background. Milroy (1932, 1934a) thought the animals in the south were *D. sumatrensis*, adding that the local people "recognize three varieties of rhino, and though their classification is not made on scientific lines, it does not follow therefrom that it is all moonshine." Hunter (1908a) and other authors of gazetteers preferred *D. sumatrensis*. Wood (1930, 1934) seems to be a less reliable witness (57.7), but could have meant *D. sumatrensis* when he listed the rhino in Cachar. In line with other listings, I would therefore suggest that all rhinos in Karimganj, Hailakandi and Cachar south of the Barak River belonged to the two-horned *D. sumatrensis*.

From the late 1920s, the spread of cultivation threatened the existence of rhinos in Cachar. Then in 1929 there was a major flood in the Barak valley (Misra 2011), which drove the few remaining rhinos to the hills, where they were killed by the Lushais (Milroy 1932). Villagers in South Cachar had opened up large patches of marshy land, which reduced the habitat available to rhino, but "some special steps have already been taken to try and look after the few rhino still left alive in this difficult country where little control can be exercised over the shikaries" (Milroy 1931; Owden 1932). Maybe this had success, as Choudhury (1997b) states that a rhinoceros was seen in 1967 by local people in the Punikhal area of Sonai Reserved Forest.

...

Dataset 60.73: Chronological List of Records of Rhinoceros in Cachar and South Assam

General sequence: Date – Locality (as in original source) – Event – Source – § number – Map number

The map numbers are explained in dataset 60.74 and shown on map 60.30. This includes records from Lower Assam: districts Karimganj, Hailakandi and Cachar

Rhino Region 45: Lower Assam

1840s – Cachar – Jungles and swamps are particularly suited to its habits. Never saw even tracks. Disappeared *c.*50 years ago – Baker 1887: 249 – 60.4 – map D 51

1856 – North Cachar – Rhino common in the jungles and jheels to the north – Stewart 1856: 596

1868 – Cachar – Rhino present – Reynolds 1868: 324

1870s – Langai River (Sunai River, in Karimganj) – Rhino seen by elephant catchers, assumed *R. unicornis* – Baker in Woodthorpe 1873: 29; Baker 1887: 249 (as Sylhet) – 60.4 – map D 51

1872 – Charyolah (actually Chargola) – John Henry Bourne (1844–1873) heard about presence of smooth-skinned rhino. – Anderson 1872: 132; Grève 1898: 320. Probably *D. sumatrensis* – 60.2 – map D 54

1879 – Cachar – Presence noted – Hunter 1879b, vol. 2: 376

1880s – Cachar – Rhinos found irregularly – Stuart Baker 1928: 175

1885 – Langai River – Rhino seen, shot missed – Synteng 1885 – 60.3 – map D 51

1885 – Swampy land immediately beneath the Chatachura peak to the south-west – Rhinos killed by Kuki tribesmen in pitfalls – Synteng 1885 – 60.3 – map D 50

1887 – Gaindamara ('rhino killed') in Cachar [not located] – Place name commemorates rhino – Baker 1887: 249 – 60.4

1890 – Katakhal RF and Innerline RF of Hailakandi Dt. – Rhino present (after A. Majid Choudhury) – Choudhury 1997b: 151, 1998a: 7, 2013: 225 – map D 53

1906 – Surma valley – *D. sumatrensis*. Occasionally seen – Assam Government 1906: 7; Hunter 1908a, vol. 6: 20 – map D 54

© L.C. (KEES) ROOKMAAKER, 2024 | DOI:10.1163/9789004691544_061
This is an open access chapter distributed under the terms of the CC BY-NC-ND 4.0 license.

to 1922 – Sylhet (= Cachar) – *R. sondaicus* (= *D. sumatrensis*, 57.7). Seen in considerable numbers. Disappearing due to cultivation – Wood 1930: 60

1929-07 – Cachar (Barak River) – Flood drove rhinos into hills, where killed by Lushais – Milroy 1932, 1934a: 102

1931 – Surma valley – *D. sumatrensis*. Occasionally seen – Milroy 1931: 4 – map D 54

1931 – Hati-Thal – Steps taken to protect few remaining rhino – Milroy 1931: 4 – map D 53

1932 – Cachar – Steps have been taken to patrol the country where these animals are supposed to live. Rhino are fond of low-lying country – Owden 1932: 4

1967 – Sonai RF, Cachar – Stray record. (Forest Dept. thought *R. sondaicus*) – Choudhury 1997: 151; 2016: 155 – map D 52

2017 – Patharia Hills RF – All 3 species vanished from area – Talukdar & Choudhury 2017: 133 – map D 51

2018 – Barak Valley (comprising districts Cachar, Hailakandi, Karimganj) – All 3 species vanished – Talukdar et al. 2018 – map D 54.

Species in Cachar stated as *R. unicornis*: Reynolds 1868: 324; Hunter 1879b, vol. 2: 376; Baker 1887: 249.

Species in Cachar stated as *R. sondaicus*: Pollok 1879, vol. 1: 98; Sterndale 1884: 410; Simson 1886: 188; Wood 1930: 60, 69; Milroy 1932, 1934a.

Species in Cachar stated as *D. sumatrensis*: Milroy 1931: 4, 1932, 1934a: 102; Talbot 1960: 174; Groves 2003: 345; Choudhury 1997a: 151, 1998a: 7, 2016: 155.

•••

Dataset 60.74: Localities of Records of Rhinoceros in Karimganj, Hailakandi, Cachar

The numbers and places from south to north are shown on map 60.30.

Rhino Region 45: Lower Assam

D 50   Chatachura peak to the south-west – 24.35N; 92.45E – 1885

D 51   Langai River – 24.50N; 92.30E – 1870s, 1885

D 51   Patharia Hills RF – 24.60N; 92.35E – 2017

D 52   Sonai RF – 24.73N; 92.89E – 1967

D 53   Katakhal RF and Innerline RF, Hailakandi Dt – 24.80N; 92.60E – 1890

D 53   Hati-Thal – 24.80N; 92.60E – 1931

D 54   Chargola (= Charyolah) – 24.84N; 94.42E – 1872

D 54   Surma valley – 24.90N; 92.50E – 1906, 1931

D 54   Cachar (Barak River) – 1929-07, 2018

*   Gaindamara, Cachar [not found] – 1887 – toponym

•• •

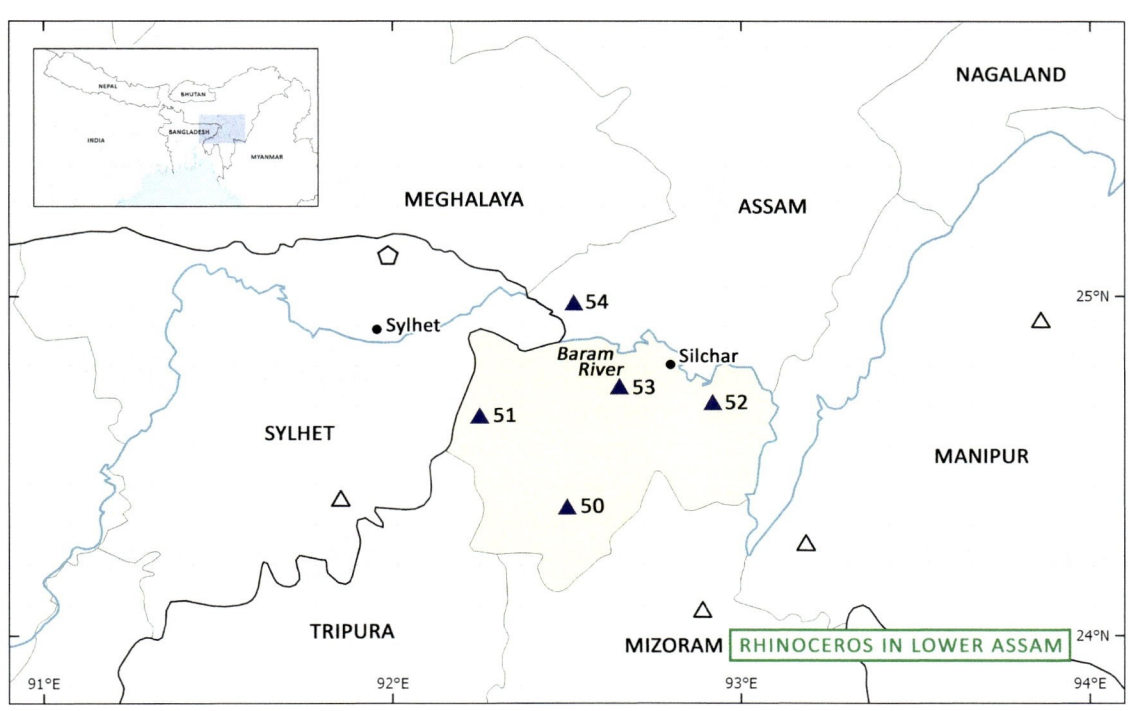

MAP 60.30   Records of rhinoceros in Lower Assam. The numbers and places are explained in Dataset 60.74

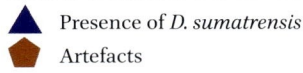   Presence of *D. sumatrensis*

    Artefacts

60.2 Double-Horned Rhino near Chargola in 1870s

John Anderson (1833–1900) of the Indian Museum in Kolkata heard in 1872 from his friend Lieut. Bourne, "that a smooth-skinned Rhinoceros is said by the Cossyahs to occur in their hills, two days' journey to the south of Charyolah. These men know *Rhinoceros sondaicus*, so that it seems very probable that *R. sumatrensis* extends into the heart of the Cossyah hills" (Anderson 1872: 132). Cossyah is another spelling of Khasi, and Charyolah (never explained in rhino literature) is Chargola in Karimganj District, Assam, just south of the Barak River. John Henry Bourne (1844–1873) was an officer of the Bengal Native Infantry,

FIGURE 60.1 "Charged by a Rhinoceros." A two-horned rhinoceros, with deep skin folds and a slightly hairy skin attacking one of his assailants. Drawn by Stanley Berkeley (1855–1909). This is probably intended to show an African scene with a rhino reminiscent of the Asian species
CHATTERBOX, BOSTON, 1890, NO. 52, P. 410

44th regiment stationed in Sylhet (now Bangladesh), who is known to have hunted with Fitzwilliam Pollok in the early 1870s (Pollok 1879, vol. 2: 38). They never went after rhino together, and if they did discuss the animal, Pollok was not aware of Bourne's claim because Pollok (1882a) clearly stated the absence of a two-horned rhino in Assam. Chargola is not in the heart of the Khasi hills, but that was only Anderson's interpretation unaware of the actual geography. It must also be noted that Bourne himself never saw a rhino on his travels there.

60.3 Synteng in South Assam

Although Synteng did not reveal his identity, he wrote some papers in the *Field* with tips for sportsmen. He must have hunted in the southern parts of Karimganj and Hailakandi. When he missed a shot at a rhino, his Kuki tracker remained totally calm. Rhinos were found in swamps, especially beneath the Chatachura peak to the south-west. Synteng (1885) is the only one to tell that rhinos were usually captured alive by the local Kukis in pits, not to be encouraged because "the spearing of them to death when so entrapped is a tedious and cruel piece of business."

60.4 Reports Submitted by Edward Baker in 1887

Edward Biscay Marinatus Baker (1828–1887) wrote in his *Sport in Bengal* (1887) that he heard from elephant catchers that rhino were known from the valley of the Langai (Longai) River, south-east of Sylhet in Assam's Karimganj District. He states that they were *R. unicornis*, but provides no reason for this. He added that the animal also lived in Cachar, because the jungles and swamps there are particularly suited to its habits and tastes. He mentions a place called 'Gaindamara' (rhinoceros-killed), without providing an exact location. Until perhaps the 1870s, this would have been an ideal spot for rhinos, but when he went there no spoor or trace was found. In fact, Baker concluded that rhino had left the area some fifty years earlier. Baker had also participated in the Lushai Expedition of 1871, and his report in Woodthorpe (1873) mentions the presence of rhinos near the Longai River, this time without saying more about the species.

CHAPTER 61

Historical Records of the Rhinoceros in Manipur

First Record: 1720 – Last Record: 1990s
Species: *Dicerorhinus sumatrensis*
Rhino Region 46

61.1 Rhinos in Manipur

There is no doubt that the eastern and southern parts of Manipur were once inhabited by rhinos. This is certain despite the fact that there is actually not a single person who claimed (in print) to have seen a rhino alive in the state. The forested hills towards the Burmese border would be the type of habitat where *D. sumatrensis* would feel at home.

The only record of a living animal was the one donated to Emperor Meidingu Pamheiba in 1720, of which little evidence remains. Grant saw tracks in 1832 (61.2), Godwin-Austen mentioned their former existence (61.3),

Higgins in 1913 saw a skull preserved in a village (61.4) and Choudhury in 1996 saw an old skin and found that people in the interior still knew the animals (61.5). A rhino supposedly captured in Manipur and sold to the Berlin Zoo, the type of *Rhinoceros jamrachi*, is now believed to have been obtained in the Sundarbans, and therefore was *R. sondaicus* (47.10). Rhinos probably continued to live in the north-eastern parts of Manipur, in the hill ranges that border on Myanmar. Reports referred to them as late as the 1970s and 1990s (61.5).

Higgins (1934) discussed the rhinos in the south-west as *R. sondaicus*, even though he never saw one. The two known specimens preserved locally were much damaged and were kept in places where detailed investigation was impossible. All available evidence combined with the probabilities of the terrain point to the suggestion that the rhinos found in Manipur belong to *D. sumatrensis*.

...

Dataset 61.75: Chronological List of Records of Rhinoceros in Manipur

General sequence: Date – Locality (as in original source) – Event – Source – § number, figures – Map number.

This list includes records from Myanmar just east of Manipur. The map numbers are explained in dataset 61.76 and shown on map 61.31.

Rhino Region 46: Manipur

1720 – Chandel Dt. – Tarao Palli Nagas caught a rhino, and presented it to Emperor Meidingu Pamheiba (1690–1751) – Hodson 1912 – map D 62

1800 – Manipur – *D. sumatrensis* absent as habitat lacking – Bahuguna & Mallick 2010: 279

1832 – Angoching hills – Francis John Grant (1787–1843) of Manipur Levy saw tracks of rhino – Grant 1834: 124 – 61.2 – map D 65

1859 – Munnipoor – In dense forests. Probably not seen by William McCulloch (1816–1885), Political Agent. Name for rhino in Munnipooree: Samoogunda; in Undro and Sengmai: Keegunda; in Chairel: Gunda – MacCulloch 1859: 72

1873-03-04 – Leimakhong, NW Manipur – Rhino disappeared from valley due to fires – Godwin-Austen 1873; Moorehead 2013 – 61.3 – map D 63

1874 – "Mooneypoor", but now thought to be mistaken for Sundarbans – Rhino captured and transported to Europe by William Jamrach (1843–1891), where exhibited in Berlin Zoo 1874–1884. Type of *R. jamrachi* – §47.10, 61.1

1874 – Manipur, hills to the east and south – Rhino present – Brown 1874: 11; Far East 1875: 225

1884 – Munipur – *R. sondaicus* present – Sterndale 1884: 410

1891 – Manipur – Rhino present – Anon. 1891b: 292

1894 – Manipur – *R. sondaicus* from Manipur eastwards – Heber Percy 1894: 234

1895 – Manipur – Rhino common – Ehlers 1894, vol. 2: 84

1913 – Khuga River – John Comyn Higgins (1882–1952) saw a skull in house of Kuki chief – Higgins 1934: 304 – 61.4 – map D 61

1915 – Barak River, near Tipaimukh – 2 or 3 killed by Kukis in last 20 years. Assumed *R. sondaicus* but not verified – Higgins 1934: 304; Groves & Leslie 2011: 195; Choudhury 2013: 225, 2016: 15 – map D 60

1919 – Fort Keary, Myanmar – Tracks of rhino near wallows in a swamp at 8000 ft – E.M.W. 1923 – map D 5

1929 – Thadou villages – John Henry Hutton (1885–1968) states that Thadou people destroyed all the rhino – Hutton 1929; J.H. Hutton in Shaw 1929: 23

1930s – Barak River – *D. sumatrensis*. Part of skin seen at Kaikao village, Tamenglong Dt. (24.51N; 93.27E) in 2001 by A.U. Choudhury – Choudhury 2013: 225 – map D 60

1934 – Manipur – *R. sondaicus*. Occasional straggler, formerly not uncommon in the hills – Higgins 1934: 304; Choudhury 2013: 225, 2016: 15 – 61.4

1936 – Manipur – *R. sondaicus*. Few may survive – Harper 1945: 381 (info from Bombay NHS 1936)

1940s – Uyu River, Myanmar – Rhino present – Peacock 1933; Ansell 1947 – map D 6

© L.C. (KEES) ROOKMAAKER, 2024 | DOI:10.1163/9789004691544_062
This is an open access chapter distributed under the terms of the CC BY-NC-ND 4.0 license.

1970s – Khamsong, NE of Ukhrul – *D. sumatrensis*. One shot by Tangkul Naga tribe – Choudhury 1997b: 151, 1998a: 7, 2013: 225 – 61.5 – map D 66

1982 – Manipur – *R. sondaicus*. Presence needs confirmation – Mukherjee 1982: 54

1990s – Konkan village (88 km SE Ukhrul) – *D. sumatrensis*. Local tribesmen told Anwaruddin Choudhury (b. 1959) that they encountered a stray rhino – Choudhury 1997b: 151, 1998a: 7, 2013: 225 – 61.5 – map D 64

2000s – Manipur – *D. sumatrensis*. Apparently still occurs – Groves 2003: 345; Alfred et al. 2006: 411; Srinivasulu & Srinivasulu 2012: 350.

Species in Manipur stated as *R. sondaicus*: Sterndale 1884: 410; Heber Percy 1894: 234; Higgins 1934: 304; Harper 1945: 381; Talbot 1960: 205; Guggisberg 1966: 118; Mukherjee 1982: 54; Groves & Leslie 2011: 195; Choudhury 2013: 225, 2016: 155.

Species in Manipur stated as *D. sumatrensis*: Milroy 1932; Talbot 1960: 174; Choudhury 1997b: 151, 1998a; Rookmaaker 2003; Groves 2003: 345; Alfred et al. 2006: 411; Bahuguna & Mallick 2010: 279; Srinivasulu & Srinivasulu 2012: 350; Choudhury 2013: 225; Havmøller et al. 2015: 356; Choudhury 2016: 155.

MAP 61.31   Records of rhinoceros in Manipur. The numbers and places are explained in Dataset 61.76

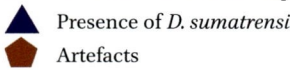

Presence of *D. sumatrensis*

Artefacts

•••

Dataset 61.76: Localities of Records of Rhinoceros in Manipur

The numbers and places from south to north are shown on map 61.31.

Rhino Region 46: Manipur

D 60   Barak River, near Tipaimukh – 24.25N; 93.00E – 1915, 1930s
D 61   Khuga River – 24.25N; 93.74E – 1913
D 62   Chandel Dt. – 24.30N; 94.00E – 1720
D 63   Leimakhong – 24.95N; 93.84E – 1873

D 64   Konkan village (88 km SE Ukhrul) – 24.80N; 94.50E – 1990s
D 65   Angoching hills – 25.06N; 94.66E – 1832
D 66   Khamsong, NE of Ukhrul – 25.20N; 94.48E – 1970s

Myanmar

D 5   Fort Keary, Myanmar – 25.23N; 94.73E – 1919
D 6   Uyu River, Myanmar – 24.90N; 94.85E – 1940s

•••

CCCXVII. G.

Rhinoceros Sumatranus Raffl.

Werner del.

FIGURE 61.1
"Rhinoceros Sumatranus Raffl."
Plate published in the zoological
work of Johann Christian Daniel
von Schreber (1739–1810),
revised by Andreas Johann
Wagner (1797–1867). Drawn
by Carl Werner (1808–1894),
engraved by Andreas
Fleischmann (1811–1878). This
plate of 1835 is a reversed copy
of that of Frédéric Cuvier in
Figure 54.1
SCHREBER AND WAGNER,
*DIE SÄUGTHIERE IN
ABBILDUNGEN*, 1835, VOL. 6, P.
325, PL. CCCXVII.G

61.2   Grant in the Angoching Hills in 1832

In January 1832, Captain Francis John Grant (1787–1843) followed a route across the Angoching hills in eastern Manipur. Upon reaching a small stream named Maylung Nala, the road passed through a forest in which they found "innumerable and recent tracks of the wild elephant, tiger, rhinoceros, bear, boar, cattle, and deer." The party saw six elephants, reportedly of quite a large size, that had come to drink, while rhinos stayed out of sight.

61.3   Godwin-Austen Believed Rhinos Were Extinct in 1873

Rhinos were reported in central Manipur west of Imphal in 1873. Lieutenant-Colonel Henry Haversham Godwin-Austen (1834–1923) visited the valley of the Laimakhong River, similar in its vegetation and landscape to the Imphalturel and Eeril Rivers. Some of the dry jungle was burning, and these fires, he believed, were responsible for the rhino's extirpation.

61.4   Higgins Finds a Rhino Skull in 1913

John Comyn Higgins (1882–1952) entered the Indian Civil Service in 1905 and served as Political Agent for Manipur State from 1917 to 1933. His work provided invaluable opportunities to observe native wildlife, yet he never saw a rhino. He did state, however, that formerly they were "not uncommon" in the hill regions. As early as 1913, he examined a rhino skull in the house of a Kuki chief shot at the Khuga (Tuitha) River, which flows through Manipur south of Churachandpur. Higgins had also heard that two or three rhinos had been killed by Kuki tribesmen during that same decade, somewhere in the lower valley of the Barak River near Tipaimukh, in the far south-western corner of Manipur close to Mizoram and Cachar. He believed that these animals would have wandered in from areas farther to the south and referred to them to *Rhinoceros sondaicus*, however without any supporting evidence. In the local Tangkhul language, the animal is known as *selho*. In Manipuri, it is called 'black animal rhinoceros' *samu ganda* (Higgins 1934).

61.5   Choudhury Records Recent Traces in Ukhrul

When the naturalist and author Anwaruddin Choudhury visited the Ukhrul District in January 1996, he found that the villagers knew the two-horned rhino from the Anko range, an area of inaccessible tropical forests. In Konkan village (88 km SE of Ukhrul town), he was told that a stray rhino had been spotted in the early 1990s. Earlier, in the 1970s, a rhino was shot by tribesmen of Khamsong village (NE of Ukhrul town).

CHAPTER 62

Historical Records of the Rhinoceros in Nagaland

First Record: 1879 – Last Record: 1999
Species: *Dicerorhinus sumatrensis*
Rhino Region 47

62.1 Rhinos in Nagaland

Although rhinos occurred in Nagaland, records from the region are few and personal sightings are all but absent. Most of the records came from the higher hills in the eastern part, especially from Noklak District (62.5) and from the extended slopes of Saramati Peak across the border into Myanmar, in a range continuous towards Htamanthi Wildlife Sanctuary further eastwards (62.3). If rhinos were ever found in the regions along the border between the western part of Nagaland and Assam, these would have been strays of *R. unicornis* (dataset 38.48, Rhino Region 36).

Several skulls of rhinos killed by local villagers and preserved in their compounds have been found over the years. The evidence of these trophies points to the presence of *D. sumatrensis* in the Noklak-Saramati region. This is the species found in the mountains all along the Indo-Burmese border from Chittagong northwards. In Nagaland, this is a heavily forested and largely inaccessible region. Rumours of rhinos are few yet persistent, and as noted by Choudhury (2019b) there is just a slight possibility that the species still exists in some undisclosed localities. The rhinos on the eastern border and on into Myanmar were *D. sumatrensis*.

● ● ●

Dataset 62.77: Chronological List of Records of Rhinoceros in Nagaland

General sequence: Date – Locality (as in original source) – Event – Source – § number, figures – Map number.

The map numbers are explained in dataset 62.78 and shown on map 62.32.

The records are divided in two sections: 1. Nagaland, and 2. Areas in Myanmar bordering Nagaland.

Rhino Region 47: Nagaland

1879 – Naga Hills – *R. unicornis* present – Hunter 1879b, vol. 2: 177

1879 – Foot of Naga Hills – Major Charles Richard Cock (1838–1879) had seen both *R. sondaicus* and *D. sumatrensis* – Cockburn 1883: 64

1884 – Naga Hills – *R. sondaicus*, incorrectly after "Pollock" – Sterndale 1884: 410

1899 – Noklak – Rhino trophy, maybe obtained late 19th century, in Khiamniungan house seen in 1967–1968 – Ganguli 1984: 236; Stirn & Van Ham 2003: 87 – 62.4 – map D 70

1899 – Khiamniungan house in Noklak – Head of dead rhino attached to hut – Sardeshpande 1987, pl. 6; Stirn & Van Ham 2003: 87 – 62.4, fig. 62.1 – map D 70

1900 – Noklak, Tuensang Dt. – One rhino killed. Photo of skull in 2004 – Choudhury 1997b, 2005a, 2013: 237, 2016: 155 – 62.3, fig. 62.2 – map D 70

1900 – Nantaleik River (Nantalet) – Former presence – Saul 2005: 16 – map D 71

1900 – Upper Konyak (north of Saramati) – Former presence – Saul 2005: 16 – map D 71

1905 – Nagaland – Shield of rhino skin, 95.5 × 43 cm, obtained by the Commissioner Keith Cantlie (1888–1977) possibly in the 1920s, donated to the Museum of the University of Aberdeen, Scotland, in 1905 (no. Abdua: 56680) – Aberdeen 2021

1921-11-21 – Near Hukpang, peak known as 'The Rhino Horn' or 'Piyongkung' – John Henry Hutton (1885–1968) climbed mountain of 7000 ft (2100 m) – Hutton 1997: 21 – map D 70

1922 – Naga hills – Lhota Nagas use rhino bone charm – Mills 1922: 169; Crooke 1926: 257

1937 – Noklak – Charles Pawsey (1894–1972) took photo of animal skulls on side of house, stated to be rhino but not visible on image – Pawsey 1937 – map D 70

1953 – Naga east of Kohima – *R. unicornis* present – Ali & Santapau 1953

1959 – Langnawk on slopes of Saramati – Expedition of Oliver Milton (b. 1927) and Richard Despard Estes (1927–2021) in 1959 reported rhino – Milton & Estes 1963: 42; Sayers 2017 – map D 71

1967–68 – Tuensang: lower slopes of Saramati – Rhino reported – Choudhury 1997b: 151 (after S. Hukiye), 1998a, 2013: 225, 2016: 155 – map D 71

1994 – Fakim area south of Saramati – Yimchunger Nagas reported seeing a 'dwarf' rhino – Choudhury 1997b: 151 – map D 71

1999-11 – Area below Saramati – One rhino came down to plains. Same seen by villagers at Thanamir in Kiphire Dt. – Choudhury 2013: 225, 2016: 155 – map D 71

Records from Areas in Myanmar bordering Nagaland

1900 – Payaw near Saramati – Rhino shot. Skull tied to post in community house seen in 1959 – Milton & Estes 1963: 42; Sayers 2017 – 62.3 – map D 7

1959 – Lay Shi, a locality east of Mizoram – Rumours of rhino, which could not be verified – Oliver Milton in Sayers 2017 – 62.3 – map D 9

1965 – Tamanthi WS – 4 rhino reported – Tun Yin 1967: 162 – map D 8

© L.C. (KEES) ROOKMAAKER, 2024 | DOI:10.1163/9789004691544_063

This is an open access chapter distributed under the terms of the CC BY-NC-ND 4.0 license.

1967 – Tamanthi ws – *D. sumatrensis* present – Tun Yin 1956, 1967: 162; Rabinowitz et al. 1995; Rabinowitz 2002: 32, 40 – map D-8

1971–80 – Saramati, Burma side – 1 rhino observed – Rabinowitz et al. 1995 – 62.3 – map D 7

1980s – Tamanthi ws – 9 rhino kills reported (6 male, 3 female) – Rabinowitz et al. 1995 – map D 8

1981–90 – Saramati, Burma side – 2 rhino observed – Rabinowitz et al. 1995 – 62.3 – map D 7

1991 – Tamanthi ws – Tracks seen – Rabinowitz et al. 1995 – map D 8

1991–93 – Saramati, Burma side – 1 rhino observed – Rabinowitz et al. 1995 – map D 7

1995 – Saramati – *D. sumatrensis* present – Rabinowitz et al. 1995 – map D 7

Species in Nagaland stated as *D. sumatrensis*: Cockburn 1883: 64; Tun Yin 1967: 162; Ganguli 1984; Rabinowitz et al. 1995; Choudhury 1997a: 151, 1998b: 7, 2005a, 2013: 237, 2016: 155; Groves 2003: 345; Alfred et al. 2006: 411; Srinivasulu & Srinivasulu 2012: 350.

Species in Nagaland stated as *R. sondaicus*: Cockburn 1883: 64; Sterndale 1884: 410; Rookmaaker 2002b.

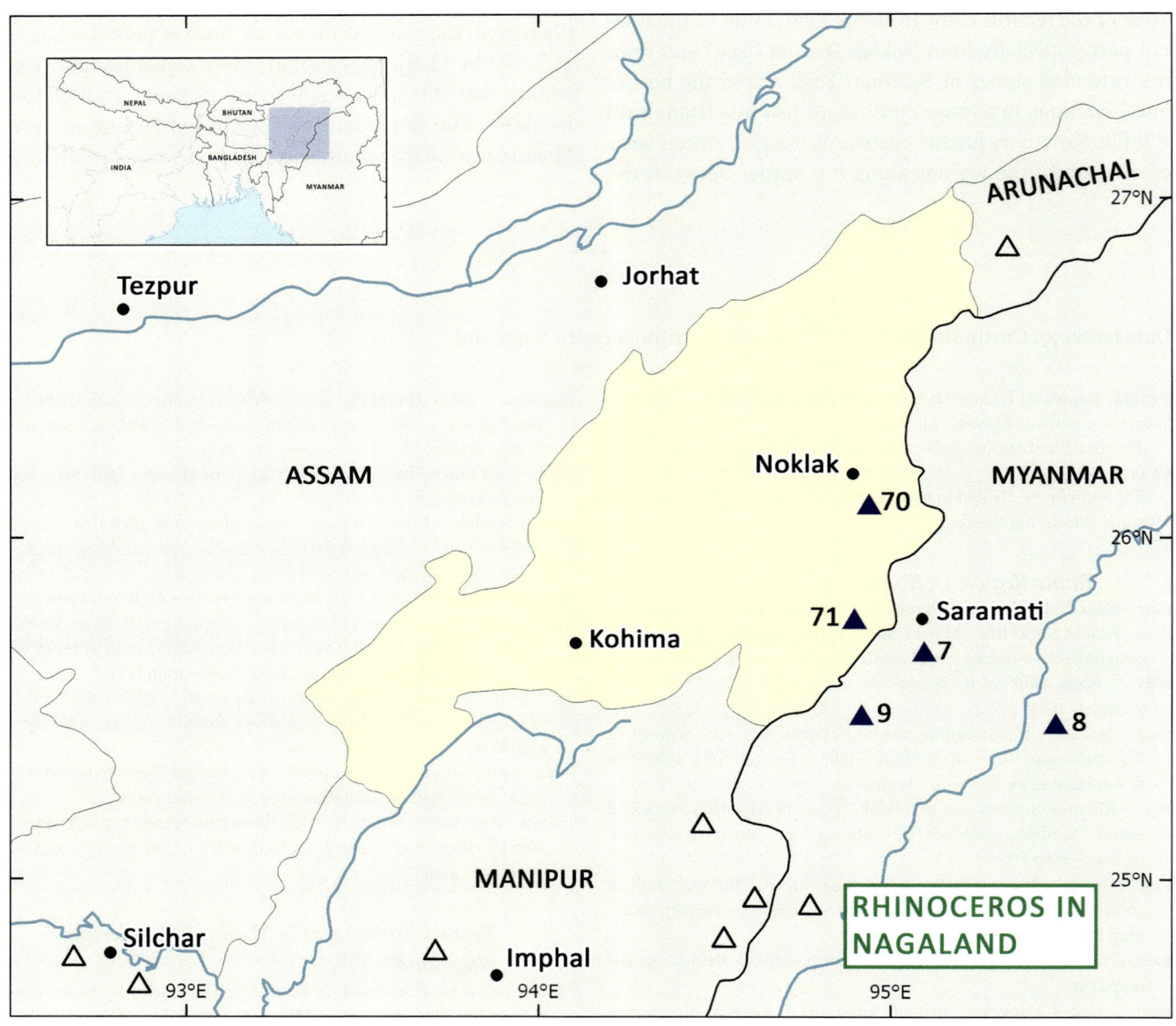

MAP 62.32   Records of rhinoceros in Nagaland. The numbers and places are explained in Dataset 62.78

▲ Presence of *D. sumatrensis*

⬠ Artefacts

•••

Dataset 62.78: Localities of Records of Rhinoceros in Nagaland

The numbers and places are shown on map 62.32.

Rhino Region 47: Nagaland

D 70 Noklak – 26.20N; 95.00E – 1899, 1900, 1937
D 70 Peak known as 'The Rhino Horn' or 'Piyongkung' – 26.45N; 94.80E – 1921
D 71 Nantaleik River (Nantalet) – 25.65N; 94.85E – 1900
D 71 Langnawk on slopes of Saramati – 25.75N; 95.05E – 1959
D 71 Tuensang: lower slopes of Saramati – 25.78N; 95.00E – 1967

D 71 Fakim area near Saramati – 25.78N; 95.00E – 1994, 1999
D 71 Upper Konyak (west of Saramati) – 1900

Myanmar

D 7 Saramati, Myanmar – 25.75N; 95.05E – 1971–80, 1981–90, 1991–93. 1995
D 7 Payaw near Saramati, Myanmar – 25.75N; 95.05E – 1900
D 8 Tamanthi ws, Myanmar – 25.40N; 95.75E – 1965, 1967, 1980, 1991
D 9 Lay Shi, Myanmar – 25.45N; 94.95E – 1959

∵

62.2   The Slopes of Mt. Saramati on the Indo-Burmese Border

In 1959, the American zoologist Oliver Milton explored the slopes of Mount Saramati (3840 m) on the border of Nagaland and Myanmar. Although unable to find direct evidence of rhino, he saw a rhino skull without jawbone in a dilapidated condition among human and animal skulls tied to posts around a community house in Payaw, on the Burmese side of the border. One of the village elders told him that he shot it when he was a young boy, probably at the start of the 20th century. Milton also heard rumours about rhinos in Laisai (Lay Shi) a little further south, which he could not confirm. Others, he said, could be coming over from the Shilloi reserve in Nagaland (Sayers 2017). Probably he meant Shilloi Lake, about 100 km SW of Saramati, but the report remains obscure.

According to information collected by Anwaruddin Choudhury, a rhino was seen on the lower slopes of Saramati Peak in 1967–1968. When the zoologist Alan Rabinowitz (1953–2018) surveyed wildlife on the Burmese side of the border, he also retrieved reports of sightings of rhino in the Saramati area: 1 observed 1971–1980, 2 in 1981–1990, 1 in 1991–1993. South of the mountain in 1994, the Yimchunger Nagas of the Fakim area claimed they had seen a 'dwarf' rhinoceros. This, like all other evidence from eastern Nagaland, points to the (former) presence of the two-horned rhinoceros.

62.3   Imaginary Sighting by Travis on the Burmese Border

In an earlier paper I referred to a rhino seen alive in a Kachin village in north-central Burma by Colonel Ian John Travis (b. 1953) in 1996 (Rookmaaker & Klee 2010). There was some scepticism about the reality of the event, which on further investigation has proved to be the case. Maybe this referred to a visit by Travis to Tenasserim in the 1980s, but actually it is more likely to have been a flight of imagination.

62.4   Skulls Preserved by Villagers in Noklak

The skull found in Payaw by Oliver Milton was not the only rhino trophy near Saramati Peak. A photograph taken by the geographer Charles Pawsey, at Noklak at the end of 1937, shows a hut with animal heads, labelled "rhino head" by the author (Pawsey 1937), but the skulls shown belong to gayals (*Bos frontalis*). The Czech ethnologist Milada Ganguli (1913–2000) was in Noklak in December 1967, when in the highest morung (community dormitory) she saw "the stuffed head of a rhino, captured by the grandfathers of the present generation in the valley below Noklak" (Ganguli 1984: 236). The capture probably occurred in the late 19th century (Peter van Ham, pers.com. 2020). The same age can be ascribed to a rhino skull photographed in 1985 by Lt. Colonel Shrikrishna Sardeshpande (1934–2023) attached to a pole in the house of Makum, a Khiamniungan tribesman in Noklak (fig. 62.1).

FIGURE 62.1
Rhino and buffalo skulls attached to a pole in the house
of Makun in Kamnoi Noklak in 1985
FROM SARDESHPANDE, *THE PATKOI NAGAS*, 1987,
PL. 6. COURTESY: PETER VAN HAM

FIGURE 62.2
Skull in village Noklak, Nagaland, in 2004
PHOTO: ANWARUDDIN CHOUDHURY

An exciting find was made in February 2004 when Anwaruddin Choudhury received reports of a rhino skull that had been preserved in a Noklak village near the northern part of the Saramati. He was told that two brothers, Musanj and Shanji, had speared the animal to death around 1900. The skull was hung on one of the pillars in their house (fig. 62.2). The nasal bones are absent because they were damaged when the two tiny horns were removed. This skull is an important piece of evidence and should be preserved in a good local museum, but as the local Keimnugan Naga tribesmen revere the skull and splash water on it when there has been no rain for many days, maybe this will be a difficult task (Choudhury 2005).

CHAPTER 63

Historical Records of the Rhinoceros in Eastern Arunachal Pradesh

First Record: 1826 – Last Record: 1966
Species: *Dicerorhinus sumatrensis* (south-east), *Rhinoceros unicornis* (north)
Rhino Region 28

63.1   Rhinos in Eastern Arunachal Pradesh

In eastern Arunachal Pradesh, there have been regular reports of rhinos in the Changlang and Tirap districts, along the headwaters of the Dihing River and in the area of the current Namdapha National Park. Here the hills and mountain ranges run in a north-south direction separating India from the adjoining parts of Myanmar.

The various reports of rhinos in the eastern part of Arunachal Pradesh indicate their presence, even if there are no specimens and just a few actual sightings. Prince Henri's guide said that the animals were plentiful, and the existence of well-maintained tracks would support that (63.3). The few available records seem to indicate that the rhinos in this hilly, even mountainous region, were double-horned, hence *D. sumatrensis*. Of course, the lower parts of Arunachal Pradesh border the Brahmaputra valley, and strays of *R. unicornis* might not be impossible at lower elevations. However, in view of the type of terrain and the absence of clear records, clearly *R. unicornis* never extended eastwards towards or across the border with Myanmar (63.5).

There is little information about the status of the rhinoceros in this part of Arunachal Pradesh. The authors of the Red Data Book of the 1960s still suspected a few specimens (Simon 1966). At the time of the 1997 status survey a question mark appeared on the map (Foose & van Strien 1997). It is certainly to be expected that the rhinos which once lived in or traveled through the Upper Dihing and the mountains of the Burmese border have disappeared, albeit the mechanism of this process is unclear. This is probably one of the few places on earth where a question mark is not totally out of place.

• • •

Dataset 63.79: Chronological List of Records of Rhinoceros in Eastern Arunachal Pradesh

General sequence: Date – Locality (as in original source) – Event – Sources – § number – Map number

Species identification follows original source (names updated). The map numbers are explained in dataset 63.80 and shown on map 63.33.

Rhino Region 28: Arunachal Pradesh – Eastern Regions

The list includes some records across the border eastwards into Myanmar.

1826 – Dihing River – Richard Wilcox (1802–1848) of the 46th regiment expected rhino in this district – Wilcox 1834: 419 – 63.2 – map D 82

1837 – Namtuseek, Myanmar – William Griffith (1810–1845) saw tracks – Griffith 1847: 70 – map D 11

1845 – Upper Assam – Double horn donated by Francis Jenkins (1793–1866) to Royal Institution of Cornwall, Truro – Lemon 1845: 18 – 63.2 – map D 83

1869 – Dehing River – Thomas Thornville Cooper (1839–1878) said rhino occurred – Cooper 1870, 1873: 91 – 63.2 – map D 82

1882–83 – Singpho area, Myanmar – *R. unicornis*. Charles Alfred Elliott (1835–1911) obtained in Saidya (Assam) from a trader a long and a short horn, presumably from same specimen. The longer horn was 19 inch (48 cm). The date 1921 listed in Ward 1986: 564 is probably incorrect – Lydekker 1905, 1907; Ward 1899: 433, 1986: 564 – 63.4 – map D 13

1885-02 – Kumki area – Tracks seen by Charles Reginald MacGregor (1847–1902) and Robert Gosset Woodthorpe (1844–1898) – MacGregor 1887; Blanford 1891: 475; Shebbeare 1953 – 63.2 – map D 84

1895-12-01 – Nam-Tsaï River, Burma (Myanmar) – Guide of Prince Henri d'Orléans (1867–1901) stated that paths were maintained by rhinos "with two horns and that their meat is quite good"; their tracks were smaller than those in the Sundarbans – Henri d'Orléans 1898a: 320, 1898b: 333; Ward 1921: 241; Hubback 1939: 596; Choudhury 2003: 66, 2013: 225 – 63.3, fig. 63.1 – map D 13

1898 – Upper Burma, border with India – Bor Khamti hunters kill rhinos. Horns sold to traders in "Borua pothar" and Chongkham, for Rs 80–100 per seer [kg] – Errol Gray 1893: 16; Allen 1908a: 16, 86; Hunter 1909: 152 – map D 13

c.1900 – Burma – *R. unicornis* skull said to be from Burma – Groves 1982: 253, but no collection given and apparently a mistake

1905 – Upper Burma – Information of existence of a type of rhino larger than the known species – Lydekker 1905

1928 – Eastern Assam – *D. sumatrensis* – Stockley 1928: 101 – map D 83

1942 – Namlang Valley, Myanmar – Single-horned rhino was shot by a Lisu man – Tun Yin 1956, 1967: 153; Talbot 1960: 175, 188 – 63.5 – map D 12

1944 – Tirap Frontier Tract – James Philip Mills (1890–1960), District Commissioner in the Naga Hills, proposes a game sanctuary – Mills 1944 – map D 83

© L.C. (KEES) ROOKMAAKER, 2024 | DOI:10.1163/9789004691544_064

This is an open access chapter distributed under the terms of the CC BY-NC-ND 4.0 license.

1944 – Tirap Frontier Tract NP – *D. sumatrensis*, few suspected – Gee 1950b: 84 – map D 83

1950 – Tirap Frontier Tract NP – *D. sumatrensis*, existence doubtful – Gee 1950b: 84 – map D 83

1950s – Tirap Frontier Tract NP – Species uncertain. Only tracks seen in recent years – Ali & Santapau 1953 – map D 83

1950 – Tirap Frontier Tract, hills 2–3 days march from Marghuerita – *R. unicornis* has been seen – Ali & Santapau 1953; Groves & Grubb 2011: 22

1950s – Kachin State, Myanmar – *R. unicornis*. Lowlands near Putao in the winter – Talbot 1960: 175; Tun Yin 1956 – 63.5 – map D 12

1952 – Mishmi Hills – Sighting of a two-horned rhino, but when shown photos, identified as *R. unicornis* – Shebbeare 1953: 147 (after *Daily Telegraph* 1952-09-01); Ali & Santapau 1953; *Times of India* 1955-07-03; Choudhury 2013: 225 – 63.6 – map A 396

1956 – Kachin State, Myanmar – *D. sumatrensis*. Sighting – Tun Yin 1967: 153 – 63.5 – map D 12

1962–02 – Bumhpabum, Myanmar – Presence of small group of rhinos, about 6–8 animals – Tun Yin 1967: 153 – map D 10

1963 – Noa-Dihing valley – Rhino suspected to be *D. sumatrensis* – Stracey 1963: 239 – 63.5 – map D 82

1964 – Upper Dihing – Footprints have been observed – Gee 1964a: 121; Choudhury 1997a – map D 82

1964 – Southern Tirap – Few isolated rhinos, supposed to be *R. unicornis* – Mukherjee 1963: 46; Gee 1964a: 121; Choudhury 1997a; Groves & Grubb 2011: 22 – 63.7 – map D 80

1966 – Tirap Frontier Tract – *D. sumatrensis* may have survived – Simon 1966; Fisher et al. 1969: 115; Goodwin & Holloway 1972 – 63.7 – map D 83

1990 – Hukawn Valley, Myanmar – Rhino present – Choudhury 1997a, 2013: 225 – map D 10

Species in Arunachal Pradesh (east) stated as *D. sumatrensis*: Gee 1950b: 84; Shebbeare 1953: 147; *Times of India* 1955-07-03; Talbot 1960: 174; Stracey 1963: 239; Simon 1966; Fisher et al. 1969: 115; Goodwin & Holloway 1972; Choudhury 1997a, 2003: 66, 2013: 225, 2016: 153; Havmøller et al. 2015: 356

Species in Arunachal Pradesh (east) stated as *R. unicornis*: Ali & Santapau 1953; Talbot 1960: 191; Fisher et al. 1969: 111; Chowdhury 1983: 55; Choudhury 1985a, 1997, 2003: 65, 2013: 225; Alfred et al. 2006: 408; Groves & Grubb 2011: 22.

Maps where distribution of *R. unicornis* is shown to extend to Myanmar: Talbot 1960; Mukherjee 1966; Laurie 1978: 10; Trense 1989: 306; Foose & van Strien 1997: 9; Kemf & van Strien 2002: 20; Schenkel et al. 2007: 54; Amin et al. 2006: 110; Dinerstein 2015: 98; Rookmaaker et al. 2017: 12–13.

MAP 63.33
Records of rhinoceros in the eastern part of Arunachal Pradesh. The numbers and places are explained in Dataset 63.80

▲ Presence of *D. sumatrensis*
 Artefacts
★ Presence of *R. unicornis*

• • •

Dataset 63.80: Localities of Records of Rhinoceros in Eastern Arunachal Pradesh

The numbers and places are shown on map 63.33.

Rhino Region 28: Arunachal Pradesh – Eastern Regions

D 80   Southern Tirap – 26.95N; 95.37E – 1964
D 81   Tirap Frontier Tract, hills 2–3 days march from Marghuerita – 27.20N; 95.95E – 1950
D 82   Dihing River – 27.45N; 96.40E – 1826, 1869, 1963, 1964
D 83   Tirap Frontier Tract NP, Namdapha NP – 27.50N; 96.70E – 1944, 1950, 1950s, 1966

D 84   Kumki area – 27.37N; 96.87E – 1885
A-396   Mishmi Hills – 28.25N; 96.00E – 1952

Myanmar

D 10   Hukawn Valley, Myanmar – 26.50N; 96.57E – 1990
D 10   Bumhpabum, Myanmar – 26.66N; 97.05E – 1962
D 11   Namtuseek, Myanmar – 27.00N; 96.80E – 1837
D 12   Namlang Valley, Myanmar – 27.35N; 97.40E – 1942
D 12   Kachin State, Myanmar – 27.35N; 97.30E – 1950s, 1956
D 13   Singpho area, Myanmar – 27.42N; 97.00E – 1882–83
D 13   Nam-Tsaï River, Myanmar – 27.42N; 97.00E – 1895, 1898, 1905

∴

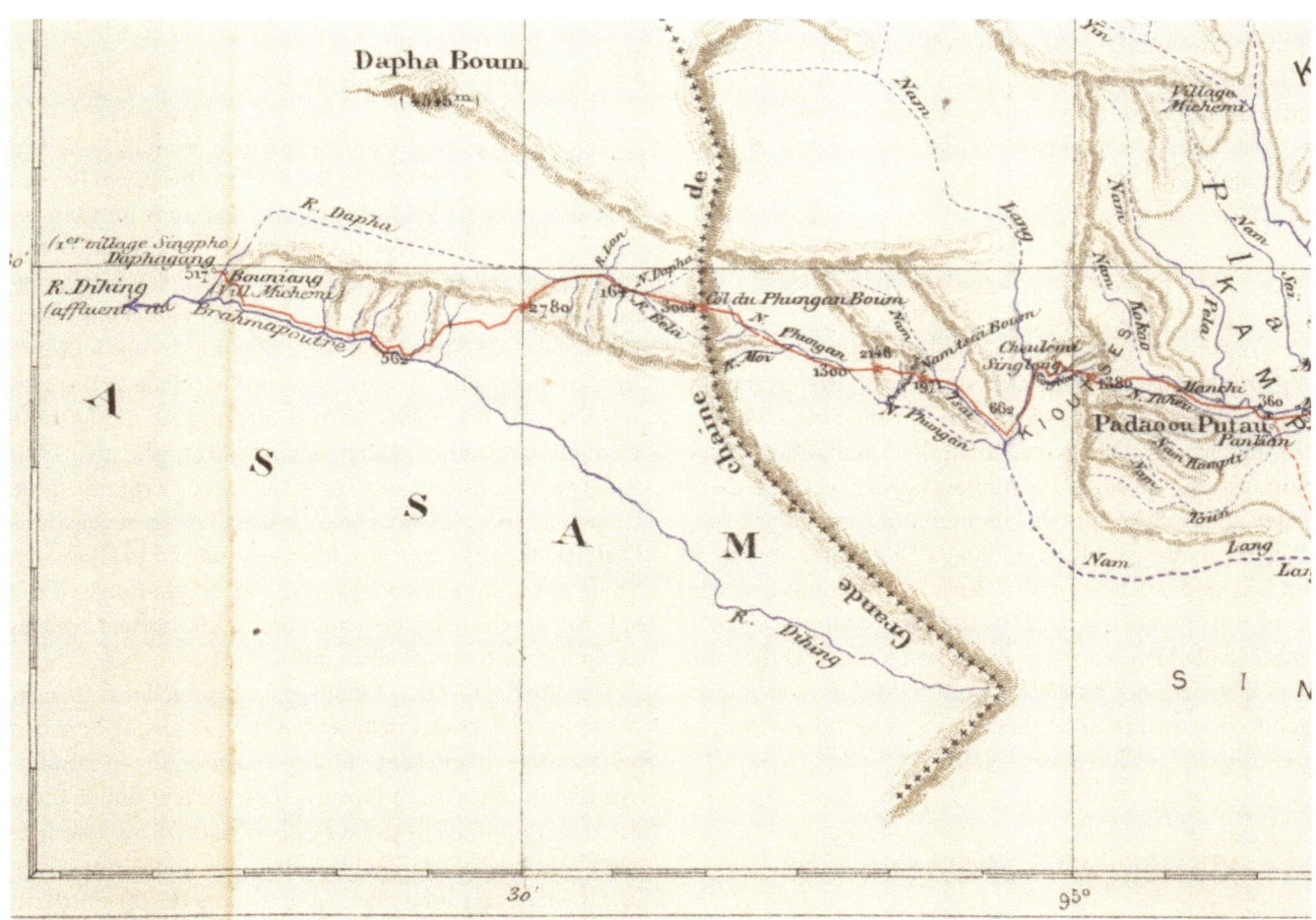

FIGURE 63.1   The south-western part of the itinerary followed by Henri d'Orléans from the River Dihing in Myanmar to Sadiya in Assam. Map drawn by the cartographer Èmile Roux (1853–1933), entitled "Itinéraire de Fong Chouan à Daphagang (Assam)". The position of the valley of the Nam Tsaï is shown west of 'Padao ou Putau' towards the Assamese border
DETAIL FROM MAP IN HENRI D'ORLÉANS, *DU TONKIN AUX INDES*, 1898, P. 294

63.2   Reports from the Dihing River

An early survey of the Dihing River was made by Richard Wilcox of the 46th regiment in 1826, and he reported the existence of rhinoceros in the district. Similar reports were heard by Thomas Thornville Cooper in 1860. This was not an area which outsiders could visit easily. In 1845, Francis Jenkins, Commissioner of Assam, donated a "double horn of rhinoceros" from "Upper Assam" to the Royal Institution of Cornwall in Truro. The locality is vague, and the specimen is no longer recognized, but might well have come from this eastern part of Arunachal Pradesh, then part of Assam.

In February 1884, Charles Reginald MacGregor and Robert Gosset Woodthorpe surveyed mountains around Kumki beyond the source of the Dihing River. Here they "noticed on some of the less precipitous ridges where the stunted oak and the gorgeous rhododendrons abounded, that rhinoceros had traveled over them, presumably when making their way to the salt-licks in the valley of the Turong, the source of the Khyendwen [Chindwin] River. I have noticed the marks of wild elephant at even higher altitudes than 7000 feet, but never before that of rhinoceros so high." They personally did not see any of these animals.

63.3   The Expedition of Prince Henri d'Orléans

In a daring expedition at the end of the century resulting in the discovery of the sources of the Irrawaddy, Prince Henri d'Orléans (1867–1901) traveled from Vietnam through Myanmar into Assam. In the uninhabited mountainous area close to the Indian border, he records on 1 December 1895 that "we climbed into the valley of Nam Tsai on a reasonably good road. The public works on bridges and roads are done here by the rhinoceros who level and enlarge the paths by their frequent passage. The guide (called Poulanghing) explained to me that these are rhinoceroses with two horns and that their meat is quite good" (Henri d'Orléans 1898b: 333). The presence of *D. sumatrensis* was thus established.

63.4   Two Horns of the Singpho Rhinoceros

The naturalist Richard Lydekker (1849–1915) reviewed the evidence of rhino in the Singpho region in 1905, after he had heard from a gentleman that a rhino larger in size than the known species would exist in parts of upper Burma inhabited by the Singpho (Singphaw). He found

that Rowland Ward in the *Records of Big Game* under *R. unicornis* listed a horn from "Singpo" (later Singpho) measuring 19 inch (48 cm) owned by Sir Charles Alfred Elliott (1835–1911), Lieutenant Governor of Bengal. Elliott confirmed to Lydekker that in Sadiya (Assam) in 1882–1883 he had bought two horns brought from Myanmar, a long one which was set up as a trophy, and a smaller one made into an inkstand (neither of these are now known to exist). Two horns, argued Lydekker, meant that the animal was not *R. unicornis*, but it could quite possibly be a related species endowed with this additional horn. Such an approach was indicative for that age, and let's not forget that just few years later, in 1908, Lydekker could describe a new subspecies of rhino in Central Africa, now known as the Nile Rhinoceros, *Ceratotherium simum cottoni* (Lydekker, 1908). There is no indication in Lydekker's Singpho paper that the horns might have belonged to *D. sumatrensis*. No new species of rhinoceros was ever found in Singpho.

63.5   Kachin State in Myanmar

U Tun Yin, the expert on Burmese mammals, reviewed rhino records in the Kachin State of northwestern Burma. In a survey of December 1955, the Assistant Resident of Putao Sub-division reported sightings of tracks of two smaller rhinos in the Namlang River, and a large solitary animal on the border of Putao and Hukawng Valley. Previously a one-horned rhino had been shot by a Lisu in the Namlang valley in 1942, and another killed in the area was also one-horned. Tun Yin (1956) suggested that these were all *R. unicornis* which wandered into the area from the Tirap Frontier Tract National Park. Certainly these thoughts were the rationale behind recent suggestions that the range of *R. unicornis* once extended into Myanmar. There are quite a few maps of the historical range of the species which definitely show the single-horned species extending into Myanmar (dataset 63.79).

Despite the fact that I inadvertently joined the ranks of these authors, I don't believe that there is any evidence to substantiate the presence of *R. unicornis* in the mountainous region of the Indo-Burmese border. The rhinos there are much more likely to be *D. sumatrensis*. This was definitely the view of Alan Rabinowitz (1953–2018), one of the few western scientists allowed to explore the far-northern part of Myanmar, who in his observations only talks about the Sumatran rhinos, which were probably hunted to extinction there in the 1980s (Rabinowitz 2002). The idea that one-horned rhinos were regularly straying from Assam to Myanmar across the hill and mountain ranges,

FIGURE 63.2
Watercolour by Charles Hamilton Smith (1776–1859), named "R. bicornis Sumatrensis – Two Horned Sumatran Rhinoceros." He sketched animals in many zoos and museums in Europe in the early 19th century. Size 40.6 × 33.6 cm
WITH PERMISSION © ARADER GALLERIES, PHILADELPHIA

even if the distance is only about 50 km, seems far-fetched, because *R. unicornis* is never seen in such terrain elsewhere (Choudhury 2022). New evidence is unlikely to be forthcoming. My conclusion is that the range of *R. unicornis* never extended eastwards as far as Singpho and anywhere into Myanmar.

63.6 A Rhino Sighting in the Mishmi Hills in 1952

The Mishmi Hills are located north of the upper Brahmaputra River. Edward Oswald Shebbeare (1884–1964), in his rhino status survey of 1953, refers to a report in the *Daily Telegraph* (London) of 1 September 1952 that a two-horned rhino was seen there. No further particulars were available from the newspaper office and Shebbeare leaves the conclusion open. The same instance may have been referred to in the *Times of India*, 3 July 1955: "About two years back, reports reached the N.-E.F.A. [North-East Frontier Agency] that a rare two-horned rhinoceros was sighted in the jungles of the Mishmi Hills but no planned search could be made for it, and the existence of the animal has not yet been established. However, a directive

has been issued to local authorities that if the animal is seen, it must not be destroyed." This may or may not be the event mentioned by Ali & Santapau (1953), where the villager identified the rhino as *R. unicornis* rather than the two-horned species from photographs sent by E.P. Gee. This sighting was a rare occurrence of which the details have never become clear.

63.7 Conservation in Tirap

When the Tirap Frontier Division was surveyed in the 1940s, James Philip Mills, District Commissioner in the Naga Hills, discussed the possibility to establish a game sanctuary or national park in the region. It was still 'proposed' when Stracey (1963) wrote, while both Stracey and Gee (1964a) suspected that the area could harbour a few rhinos, probably *D. sumatrensis*, or maybe these were strays from the upper Brahmaputra basin and therefore *R. unicornis*. Without doubt, if there ever was a sighting of a rhinoceros, even a track, the evidence has now been lost. The Namdapha National Park in the same area was finally established in 1972.

CHAPTER 64

Records for *Dicerorhinus sumatrensis* in North Bengal and Assam

First Record: 1862 – Last Record: 1895
Species: *Dicerorhinus sumatrensis*
Rhino Regions: 26, 30, 31

64.1   *Dicerorhinus sumatrensis* in Bengal and Assam

There are only four records when a two-horned rhinoceros was killed, or believed to be killed, in West Bengal and Assam (Rookmaaker 1980; van Strien 1974). Blyth had heard about a skin with two horns attached in 1862 (64.2), Cutter obtained a specimen in an undefined locality (64.3), Napier Sturt sent a horn from Assam to London in 1875 (64.4) and the surveyor Sunder stated that a rhino of such description was killed once in Dalgaon Forest of Bengal (64.5). Despite the small number of reports, many zoologists nowadays believe that *D. sumatrensis* was "previously", until the 19th century, part of the fauna of those territories. The list for Assam has 50 references from 1862 to present, that for North Bengal has 23 references from 1939 and that for Bhutan 12 references from 1936 (dataset 64.81). However, because these four historical records are far from other known instances of the species in South Asia, a further examination is necessary (64.6).

Rhinoceros lasiotis.
(*R. Indicus* and *R. Sondaicus* in the distance.)

FIGURE 64.1
"Rhinoceros lasiotis". An artist's impression, comparing the two-horned species with specimens of *R. unicornis* and *R. sondaicus*
ROBERT A. STERNDALE, *NATURAL HISTORY OF THE MAMMALIA*, 1884, P. 410

• • •

Dataset 64.81: Chronological List of Records of Two-Horned *D. sumatrensis* in West Bengal and Assam

General sequence: Date – Locality (as in original source) – Event – Source – § number, figures – Map number

The map numbers are explained in dataset 64.82 and shown on map 64.34.

Records Combining Rhino Regions 26, 30 and 31

1862 – Assam, to the south of the great bend of Brahmaputra – A friend saw two horns attached to a skin – Blyth 1862c: 367, 1863b: 157, 1870: 150, 1872a, 1872d – 64.2 – map D 92

1862 – Lower Assam – Indigo-farmer saw dried head of two-horned rhino – Blyth 1872e: 3106 – 64.2 – map D 92

1867 – India (locality uncertain) – Specimen obtained from the dealer William George Cutter (1807–1888) in University Museum of Zoology, Cambridge, UK (parts H.6385 to H.6403) – 64.3, fig. 64.3

1875 – Sankosh River – Colonel Charles Napier Sturt (1832–1886) shot two-horned rhino. He was hunting in the company of Archibald Colin Campbell (1836–1890), Deputy-Commissioner of Dohbree (Dhubri), and Captain William John Williamson (1842–1883),

© L.C. (KEES) ROOKMAAKER, 2024 | DOI:10.1163/9789004691544_065
This is an open access chapter distributed under the terms of the CC BY-NC-ND 4.0 license.

Commissioner of the Garo Hills stationed in Tura (or Towra) – Sclater 1875b; Harper 1945: 394; Shebbeare 1953: 147 – 64.4 – map D 91

1875 – Bhootan terai – *R. lasiotis* has been shot once – Lydekker 1894, vol. 2: 470, 1897, vol. 2: 1057, 1900: 29 – 64.4

1875 (1864) – Sankosh River – Rhino shot (refers to Sturt's specimen, year 1864 must be mistaken) – Inglis et al. 1919: 825; Choudhury 2013: 225, 2016: 155; Mallick 2019: 73 – 64.4 – map D 91

1895 – Dalgaon Forest – "Rhinoceros Malayan" shot, but very rare – Sunder 1895: 103; Gruning 1911: 13; Mitra 1951: xxvii; Choudhury 2016: 155; Mallick 2019: 73 – 54.3, 64.5, fig. 64.5 – map D 90

1920 – Assam – *D. sumatrensis* present until first decades of 20th century – Choudhury 2013: 225

1930 – Assam foothills – *D. sumatrensis* lost in late 1920s or early 1930s – Roy 1993: 3

1936 – Assam – *D. sumatrensis,* on verge of extinction – Talbot 1960: 174

1966 – Assam – *D. sumatrensis,* some may survive – Mukherjee 1966: 28

1999 – Assam – *D. sumatrensis,* survival needs confirmation – Tiwari 1999: 162

Historic occurrence of *D. sumatrensis* recorded in North Bengal (in chronological order): Prater 1939: 620; Sanderson 1956: 269; Rohr 1959: 955; Street 1963: 110; Mukherjee 1966: 29, 1982: 56;

Guggisberg 1966: 102; A.K. Dutta 1991: 48; Kemf & van Strien 2002: 18; Bahuguna & Mallick 2010: 275. Note species not listed by Bist 1994.

Historic occurrence of *D. sumatrensis* recorded in Bhutan (in chronological order): Pocock 1936: 706; Mukherjee 1966: 29, 1982: 56; Kemf & van Strien 2002: 6; Amin et al. 2006: 114; Srinivasulu & Srinivasulu 2012: 350; Sharma et al. 2013; Havmøller et al. 2015: 356. – These records probably refer to the former Bhutan Duars.

Historic occurrence of *D. sumatrensis* recorded in Assam (in chronological order): Blyth 1862: 367, 1863a: 137 (rare), 1863b: 157 (great rarity), 1872d, 1872c (sparingly); Schlegel 1872: 133; Blyth 1875: 51 (rare); Blanford 1891: 477 (rare); Watt 1892, vol. 6: 489 (rare); Lydekker 1894, vol. 2: 470; Lydekker 1897, vol.2: 1057 (rare); Lydekker 1900: 29; Beddard 1902: 257; Woodyatt 1922a: 148; Lang 1924: 527; Ward et al. 1928: 440 (very rare); Finn 1929: 189 (rare); Dollman 1931: 844; Briggs 1931: 279; Prater 1934: 91 (practically exterminated); Pocock 1936: 706; Harper 1945: 394; Prater 1948: 194 (practically exterminated); Ellerman & Morrison-Scott 1951: 340; Shebbeare 1953: 147; Rohr 1959: 955; Maliepaard & de Vos 1961: 96; Stracey 1963: 256 (practically exterminated); Street 1963: 110; Walker 1964: 1352; Prater 1965: 233 (until end 19th century); Mukherjee 1966: 28; Guggisberg 1966: 102; Simon 1966; IUCN 1969: 26; Honacki et al. 1982: 311; Corbet & Hill 1986: 130, 1992: 244; Mountfort 1991 (once); Roy 1993: 3; Vigne & Martin 1994; Tiwari 1999: 162; Grubb 2005: 635; Bahuguna & Mallick 2010: 275; Choudhury 2013: 225 (until first decades 20th century).

.·.

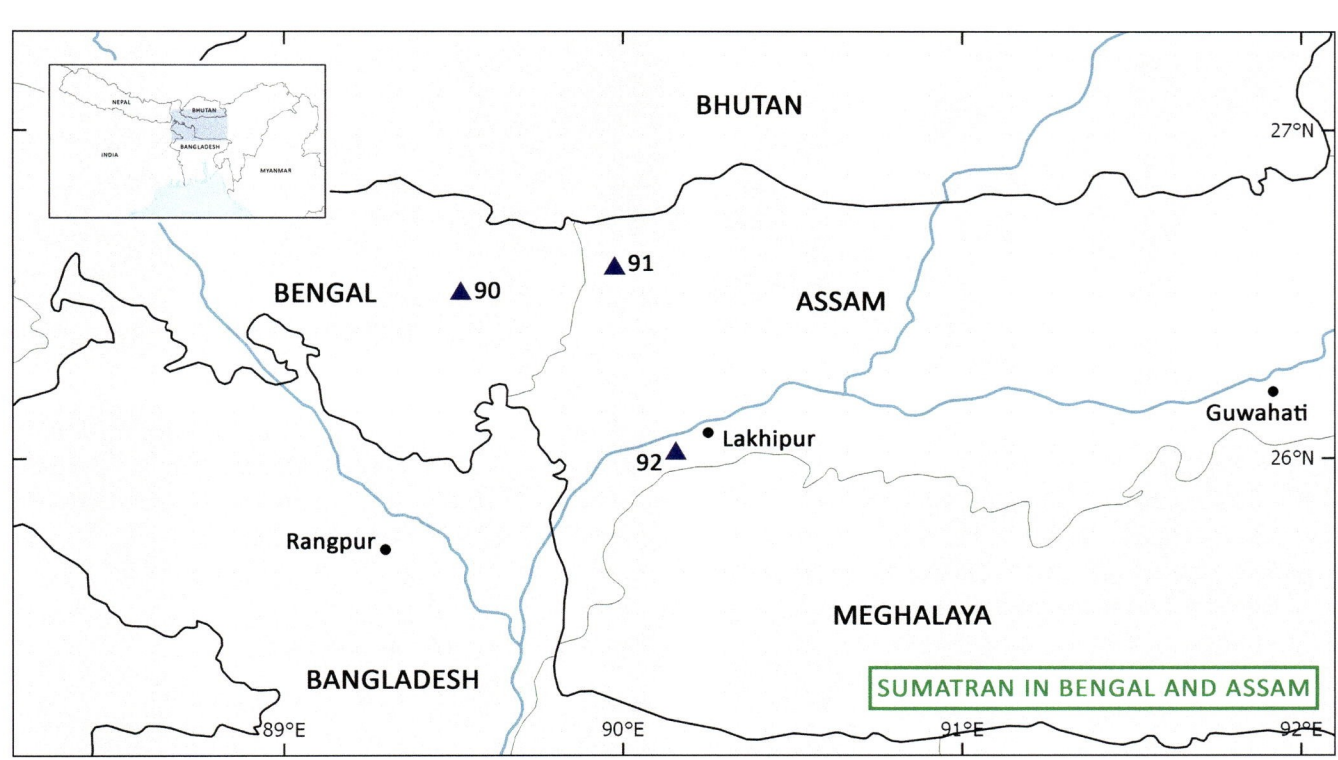

MAP 64.34   Records of *Dicerorhinus sumatrensis* in North Bengal and Assam. The numbers and places are explained in Dataset 64.82

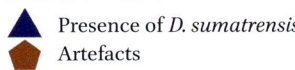   Presence of *D. sumatrensis*
       Artefacts

...

Dataset 64.82: Localities of Records of Two-Horned *D. sumatrensis* in West Bengal and Assam

The numbers and places are shown on map 64.34.

Rhino Region 26: North Bengal – Cooch Behar

D 90   Dalgaon Forest – 26.65N; 89.15E – 1895

Rhino Region 30: Assam (North-West) – Sankosh River, North of Brahmaputra

D 91   Sankosh River – 26.60N; 90.00E – 1875

Rhino Region 31: Assam (South-West)– Goalpara

D 92   Lower Assam, probably Goalpara – 26.00N; 29.30E – 1862

∴

FIGURE 64.2
Fictional scene of a double-horned rhinoceros killing his hunter. The skin folds are distinctly Indian, although this probably illustrates an African adventure. From an anonymous tale "Clara: Novelle" of 1859
DIE ILLUSTRIRTE WELT, 1859, VOL. 7, P. 145

64.2   Blyth on a Two-Horned Rhino in Lower Assam

Edward Blyth, the Curator of the Asiatic Society Museum in Kolkata, wrote in several of his papers about a friend who had reliable information that a two-horned rhino was killed in Assam. He expanded on this twice, saying that a European planter (Blyth 1872a) or an Indigo planter (Blyth 1872b) had seen the two horns attached to the skin of the head. As to the locality Assam, Blyth depended solely on his source, saying that it was in "Lower Assam" (Blyth 1872b), elsewhere qualifying this as "hill ranges bordering the valley of the Brahmaputra, to the south of the great bend of that river" (Blyth 1870: 150, 1872a). Blyth never divulged the name of his friend, who of course was not even the actual shooter. His locality seems to indicate some part of the western Meghalaya hills (then part of Assam). Despite the uncertainties, *D. sumatrensis* had entered the fauna of Assam in print.

64.3   Cutter's Specimen of 1867

There is a specimen of *D. sumatrensis* (skull, lower jaw, loose bones of skeleton) in the University Museum of Zoology, Cambridge. This was obtained in 1868 from "W. Cutter, London", who is probably the dealer in natural history and ethnography William George Cutter (1807–1888) of 36 Great Russell Street, Bloomsbury,

FIGURE 64.3
Skull and lower jaw of *D. sumatrensis* said to be from India, obtained in 1868 from the dealer William George Cutter by the Museum of Zoology, University of Cambridge
PHOTO © MATHEW LOWE, 2022

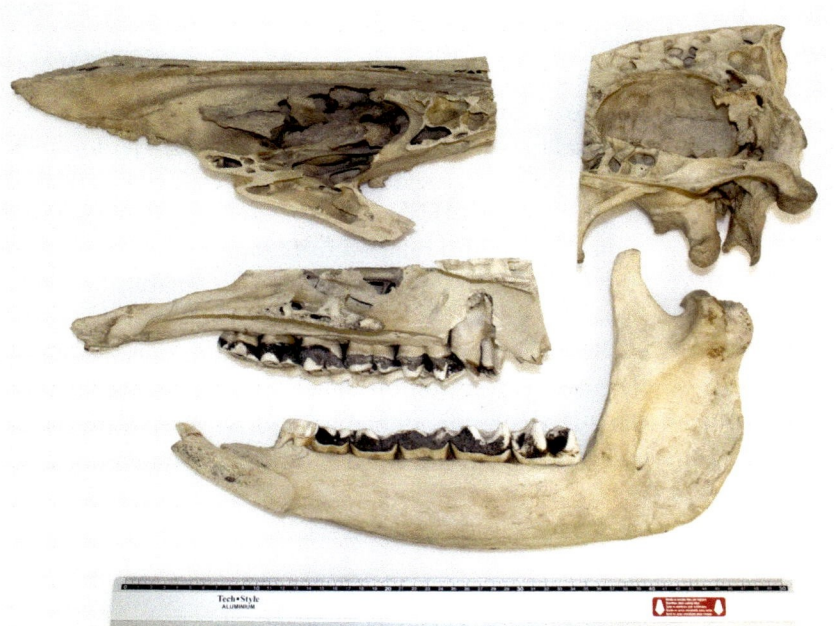

FIGURE 64.4
Lower jaw and bone fragments of the specimen in the Museum of Zoology, University of Cambridge
PHOTO © MATHEW LOWE, 2022

London. Cutter had obtained it from a naturalist exhibiting at the Exposition Universelle in Paris in 1867, who had stated that it had come from India (figs. 64.3, 64.4). "India" as a locality for a two-horned rhino could mean any of the eastern or north-eastern states, or even Burmese territory. It is added here as a specimen possibly obtained in North Bengal or Assam for the sole reason that this type of rhino at that time had never been reported even in Chittagong, and therefore potentially might have attracted more attention among the specialists.

64.4 Napier Sturt on Sankosh River in 1875

Charles Napier Sturt (1832–1886), Colonel of the Grenadier Guards, went to India visiting his brother-in-law the Viceroy, Thomas George Baring, 1st Earl of Northbrook (1826–1904). He was shooting tigers in Nepal for ten days with General Henry Ramsay (1816–1893) and Lord Hardwick in April 1875 (*Aberdeen Press Journal* 1875-05-12 after Calcutta *Englishman*). Before that, in March 1875, Sturt planned a trip to the north together with Archibald

Colin Campbell and Captain William John Williamson. On his return, Napier Sturt must have written to Philip L. Sclater, the Secretary of the Zoological Society of London, because in the ZSL meeting of 16 November 1875, Sclater (1875b) "exhibited the upper horn of a two-horned rhinoceros from the Valley of the Brahmapootra," about 40–50 miles (70–90 km) north-east of Dhubri. He had been informed that "the rhino was found near the gorge where the Sunkos River issues from the Bhotan range, and is actually within the old boundary of Bhotan." The Sankosh or Sunkos River is now the division between Bengal and Assam, both bordered in the north by the country Bhutan. A locality north-east of Dhubri is on the east bank of the river in Assam. There is no description of the horn, no measurements, no explanation why only one of the two horns was on show, nor any indication what might have happened to the trophy. The Zoological Society no longer maintained a museum (Wheeler 1997), so possibly the horn was returned to Napier Sturt, without leaving further trace.

It would not have been unusual to find a rhino in this locality on the Sankosh River. The Maharaja of Cooch Behar often hunted here from 1871 onwards (32.10). Inglis et al. (1919) refer to the same incident, but gave an arguably mistaken date 1864. This is the "single example" of *D. sumatrensis* "obtained from Bhutan" referred to by Lydekker (1894, 1897). Sclater's notice (1875b) was widely available, and apparently authoritative, so this instance found its way into much of the later literature.

64.5 Sunder on Dalgaon Forest

An unexpected enumeration of the rhino species in North Bengal is found in the Settlement Report submitted in 1895 by Donald Herbert Edmund Sunder (1857–1935), the magistrate or settlement officer of Jalpaiguri. He had visited many parts of the district in 1889–1894 to collect data for his wide-ranging survey incorporating information on a multitude of subjects. This included lists of the mammals, birds, reptiles and fishes, produced without introduction or statement about sources. Sunder had a special interest in natural history, because in November 1909 he was elected as a Fellow of the Linnean Society of London. In his zoological list, he provided for each species the English name, the Native name, the Scientific name, and short remarks. There were three species of rhinoceros (Sunder 1895: 103).

The native names all include 'gera', a local variation of 'genda' in Hindi (fig. 64.5). The other parts of the native names are quite new, and as far as I know have never been used in the literature before or after his publication. The scientific names *R. indicus* (= *R. unicornis*) and *R. sondaicus* were in common use, but his "Rhinoceros Malayan" is a new combination, maybe meant as a binomen, again unique in the literature. This last species should be the two-horned Sumatran rhino, definitely "Malayan." It was unusual to distinguish rhino species by temperament, while Sunder fails to mention the number of horns. Why did he invent a new name for this type of rhino, while any general book on the mammals of India like Jerdon (1874) would have suggested *R. sumatranus* or *R. sumatrensis* for a third species? Despite these anomalies, the conclusion must be that this referred to *D. sumatrensis*. Sunder was the first to claim all three species to inhabit the Bengal Duars.

Sunder stated that this two-horned species was shot in Dalgaon Forests, unfortunately without further particulars. Dalgaon, mentioned as a forest reserve by Sunder (1895: 100), is located just south of Birpara and north of

English name	Native name	Scientific name	Remarks
Rhinoceros	Hati gera	Rhinoceros Indicus	Found in the forests; also in swampy khas lands. Are becoming scarce. Meches, Garos and Rajhansis eat the flesh.
Rhinoceros	Kuku gera	Rhinoceros Sondaicus	Body rough and tuberculated. It has a very bad temper.
Rhinoceros	Sheng Shengi gera	Rihnoceros Malayan	Is small and ill-tempered. Shot in Dalgaon forest, but very rare.

FIGURE 64.5 Transcript of the page with information on rhinoceros in the Settlement Report submitted in 1895 by Donald Herbert Edmund Sunder
SUNDER, *SETTLEMENT REPORT OF THE WESTERN DOARS*, 1895, P. 103

Falakata in North Bengal. Although the presence of a rhino there would not be remarkable, I have found no earlier records stating that any type of rhino was shot in Dalgaon using the same place name. Therefore Sunder's record is hard to verify. Possibly, at a stretch, this may refer back to the animal shot by Sturt in 1875, even though the locality is not exactly the same. The Maharaja of Cooch Behar hunted in this area near the Torsa River west of his capital in his shoots of 1884 (a), 1892 (a), 1893 (b), 1904 (c), without discussing the possibility that there were any rhino species other than *R. unicornis*.

The information in the Gazetteers was often repeated from one edition to the next. In this case, Sunder's list of mammals was still reproduced without change by Gruning (1911) and Mitra (1951). Based on Sunder (1895), the existence of *D. sumatrensis* in this part of North Bengal was ascertained by Inglis et al. (1919), followed again by Bahuguna & Mallick (2010), Choudhury (2016) and with an incorrect 1915 date by Mallick (2019).

64.6 The Existence of *D. sumatrensis* in North Bengal and Assam

None of the reports of two-horned rhinos in Bengal and Assam are totally straightforward. The single surviving museum specimen is not definitely from either Bengal or Assam. There is a lack of historical context. This is of course a problem repeated throughout the potential former range of *D. sumatrensis*. The most positive record is that of Napier Sturt on the Sankosh River, who must have realized that he had shot something unusual, because it would have been a futile effort to write to Sclater about a rhino type of which scores were found every year. There was a horn, which was presumably examined by zoologists with an understanding of the various species, but why only a single horn of a two-horned animal was displayed remains enigmatic.

Surveying the published maps of the historical distribution of *D. sumatrensis*, most extend the range from the Indo-Burmese border westward to include these records, sometimes as separate dots (van Strien 1974; Rookmaaker 1980; Choudhury 2013), otherwise as a continuous population through Assam (Talbot 1960; Guggisberg 1966; Mukherjee 1966; Penny 1987) or through Arunachal Pradesh (Kemf & van Strien 2002; Havmøller et al. 2015; Amin et al. 2016). The best way forward may be to accept these few records at face value, in the localities as stated, even if there is only little hope to discover how they fit in a larger range.

Epilogue

∵

The Historical Range of the Greater One-Horned Rhinoceros *Rhinoceros unicornis* in South Asia

65.1 The Historical Distribution of *Rhinoceros unicornis*

The Greater One-Horned Rhinoceros is only found in South Asia. Their distribution in the past is different from their occurrence in the wild at present. In the available maps of former range, their presence is usually depicted in a wide belt along the southern edge of the Himalayas from the northern part of Pakistan eastwards, along the valley of the Ganges to the eastern extremity of the valley of the Brahmaputra in Assam. At present, rhinos have disappeared from large parts of these regions, being confined to seven national parks in India and three in Nepal (fig. 1.4).

The monumental difference in the former and present ranges of the rhinoceros in South Asia may be well-known in this general format, but has remained tentative in many details. Clearly the maps represent a generous and perhaps optimistic view that rhinos occurred in each and every locality in this wide region. It does not need to be explained that this is not the case. There were no wild rhinos in the streets of Delhi or Patna in the 19th century, farmers had excluded most wildlife from their farms in Bihar or Uttar Pradesh, the rhino only occurred where the habitat provided them with shelter, food and water. Beyond these clear exclusions, the borders of the range, even if dynamic, should be established accurately in every direction.

The limits of the historical range are shown either too generous or too restricted in most available maps. The historical detail is either obscure or is lost in the wider context. The northern limit may be too far into the higher mountains of Nepal, Darjeeling and Bhutan. The western limit fails to account for some of the records which it tries to incorporate. The eastern limit across the border into Myanmar based on horns obtained in Singpho is

FIGURE 65.1   A group of *R. unicornis* stranded on small islands during annual floods of the Brahmaputra River in Assam, India
PHOTO: DEBABRATA PHUKON, 2017

© L.C. (KEES) ROOKMAAKER, 2024 | DOI:10.1163/9789004691544_066
This is an open access chapter distributed under the terms of the CC BY-NC-ND 4.0 license.

unlikely to be correct. The southern limit remains to be established almost everywhere across continental India. Although this may sound like a condemnation of the decisions taken by past researchers, it is actually the opposite. Records are hidden in many elusive corners, their interpretation is rarely straightforward, the time-scale is cumbersome.

The current project was designed to address these questions by examining as many relevant sources about the occurrence of the rhinoceros as possible. There are thousands of data listed in the chapters of this book. Some were obvious, others brought surprising insights. As the sources are diverse, in time, in reliability, in kind, in almost every aspect possible, it is necessary to look at their potential relevance to this enquiry. The use of the various types of sources is explored in 1.2.

65.2   Definition of Historical Distribution

Although the term 'historical distribution' is widely used in conservation literature, historical may mean any period of time before the present. The actual timeframe clearly depends on the availability of sources, which for instance might go back to the 17th century in southern Africa and to the 19th century in eastern Africa. The general principle is that if a species is observed in a certain place once, it also occurred there in earlier times. It is logical to assume that most observations are limited to the age of printing, from the 16th century onwards. In practice, most descriptions of historical distribution show the status in the 19th or early 20th century, because in almost all cases earlier reports are very patchy, and to zoologists hard to access or to interpret. In their detailed and careful survey of mammalian distribution in South Africa, Boshoff & Kerley (2013: 23) have chosen the period 1820s–1920s to be represented on their maps.

In the South Asian context, this is too restrictive. There are definitely earlier sources which cannot be excluded arbitrarily, like the writings of the Mughal Emperors from the 16th century onwards or the wall paintings of Rajasthan of the 17th and 18th centuries. The definition of 'historical' needs reasonable flexibility. Rhinos were always found in suitable habitats of Assam and North Bengal, but the earliest definite records date from 1755 and 1768. This is a combination of poor accessibility of the regions and absence of written or other sources.

For these reasons, the historical range of *R. unicornis* in South Asia is defined as a set of localities where it occurred from 1500 onwards and slowly merging with the current range by the middle of the 20th century.

65.3   Protohistorical Distribution of *R. unicornis*

The definition of historical distribution to include records from 1500 onwards eliminates several large and defined sets of earlier evidence. The rhinoceros is clearly depicted in the petroglyphs of central India. There are archaeological finds representing rhinos, like those of the Harappan civilisation in Pakistan and western India. This evidence needs to be incorporated into our interpretations, but would extend the usual understanding of the historical distribution in zoological terms too far back in time.

Hence these earliest records are added as the "protohistorical distribution", defined as the possible range of the rhinoceros during the Holocene, which started 11,700 years before present. This excludes the fossil occurrences of the Pleistocene and earlier, which remains the realm of palaeontologists.

65.4   Mapping the Distribution of *R. unicornis*

The distribution of *R. unicornis* is reconstructed in two separate maps, one for the protohistorical distribution (map 65.35), the other for the historical period (map 65.36). These show two complementary sets of records, in the west including Afghanistan and Pakistan, towards the east in India, Nepal, Bangladesh with strays in Bhutan. The maps follow the evidence set out in detail in Section 1 of this book as accurately as possible. The following justification for each region draws attention to some special features.

Afghanistan
The mountainous landscape and dry habitat of Afghanistan appear unsuitable to a rhino. Yet there are three records: one of a possible sighting in the 16th century and two artefacts. The large carved rhino sculpture at Rag-i-Bibi dated to the 3rd century is especially powerful. None allow to actual identification of the species. The two early artefacts belong to the protohistorical period, while the sighting by Babur in the 16th century is in a rather unusual locality.

Southern Pakistan
Artefacts with rhinos are known from Harappa, Mohenjo-Daro and other settlements of the Harappan civilisation, which belong to the protohistorical period. There is a single sighting of 1333 in an isolated locality. These all may point at the existence of an early rhino population in the valley of the Indus River.

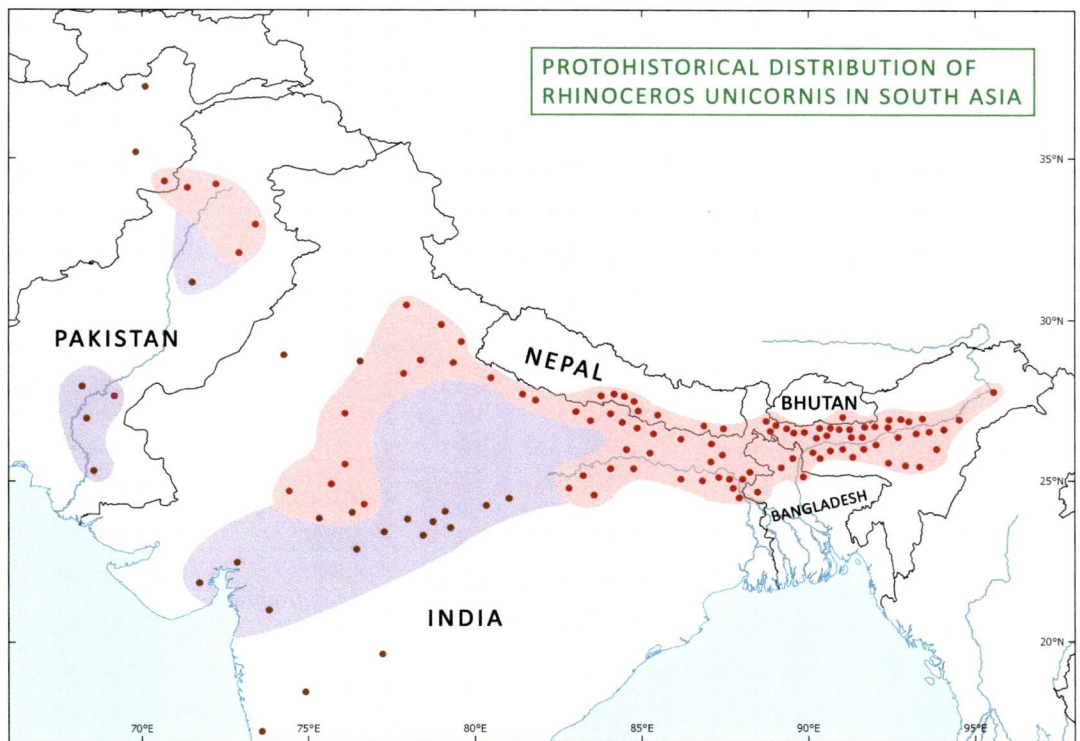

MAP 65.35   Map showing the Protohistorical Distribution of *Rhinoceros unicornis*. The three regions in purple represent known records of the early period from the start of the Holocene (11,700 years before present). The two regions in light red represent historical records from 1500 CE, identical to map 65-36. This excludes the fossil occurrences of the Pleistocene and earlier epochs

MAP DESIGN BY AJAY KARTHICK AND RICHARD KEES. © KEES ROOKMAAKER

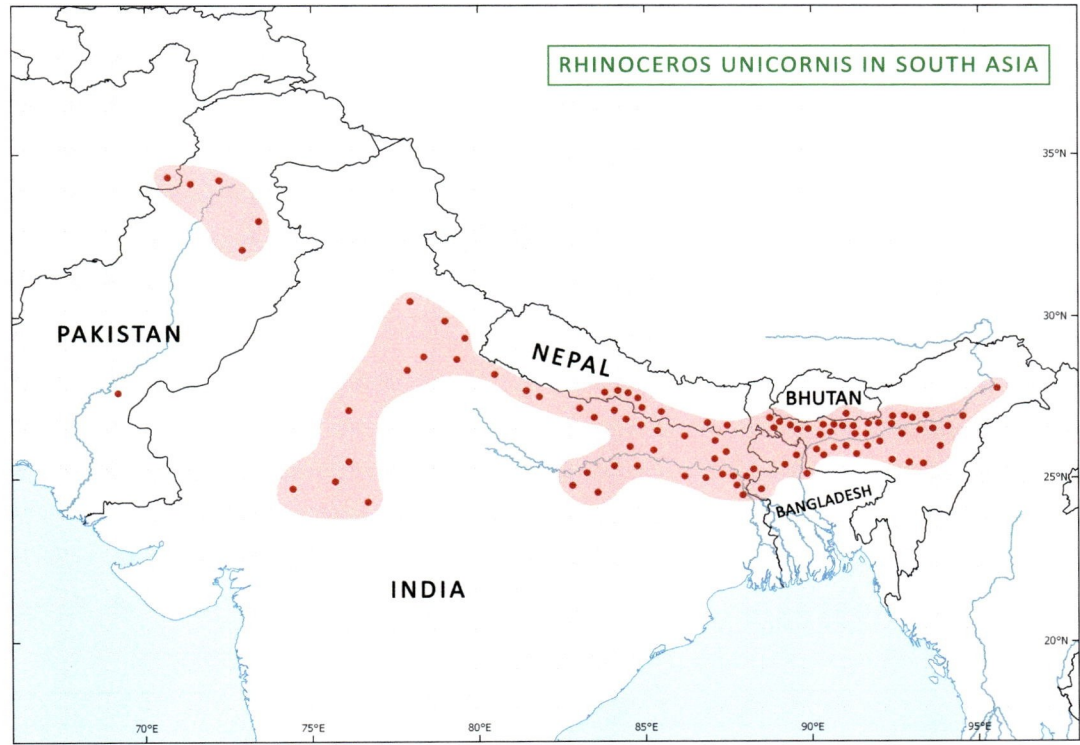

MAP 65.36   Map showing the Historical Distribution of *Rhinoceros unicornis*. This incorporates records from around 1500 CE. The single record in South Pakistan on the Indus dates from 1333

MAP DESIGN BY AJAY KARTHICK AND RICHARD KEES. © KEES ROOKMAAKER

TABLE 65.72   First and last records of *R. unicornis* in the Rhino regions of South Asia

					R. unicornis		
Rhino Region	Area	Protected area	Neolithic rock art	Earliest artefact	Earliest record	Last record	Species[a]
RR 01	Afghanistan			200	1556	1556	RU
RR 02	Pakistan			< 2000	1333	1525	RU
RR 03	Rajasthan		Present		1600	1850	RU
RR 04	Gujarat		Present				
RR 04	Maharashtra		Present				
RR 07	Madhya Pradesh		Present	< 700	1695	1695	RU
RR 08	NW India				1387	1387	RU
RR 09	Uttarakhand			1200	1755	1878	RU
RR 10	Uttar Pradesh S	Dudhwa		< 400	1017	reintroduction	(RU)
RR 11	Uttar Pradesh NW				1341	1876	RU
RR 12	Uttar Pradesh NE				1525	1933	RU
RR 13	Nepal West	Suklaphanta Bardia			nil	reintroduction	(RU)
RR 14	Nepal Centre	Chitwan			1750	current	RU
RR 15	Nepal East				1797	1995	RU
RR 16	Bihar 1 SW			100	1766	1766	RU
RR 17	Bihar 2 S			300 BCE	1632	1893	RU
RR 18	Bihar 3 NW				1665	current	RU
RR 19	Bihar 4 NE				1632	1959	RU
RR 20	Jharkhand			1650	1788	1860	RU
RR 21	Bengal Central			1700	1806	1876	RU
RR 22	Bangladesh NE			1000	1609	1886	RU
RR 23	Bangladesh NW				1640	1886	RU
RR 24	North Bengal 1	Gorumara			1869	current	RU, SON
RR 25	North Bengal 2	Jaldapara			1800	current	RU, SON
RR 26	North Bengal 3				1785	1904	RU, SON, SUM
RR 27	Bhutan	Royal Manas			(1934)	(2007)	(RU)
RR 28	Arunachal W				(1952)	(2007)	(RU)
RR 28	Arunachal E				1955	1955	RU, SUM
RR 29	Meghalaya			1600	1837	1898	RU
RR 30	Assam NW				1864	1919	RU, SUM
RR 31	Assam SW				1755	1894	RU, SUM
RR 32	Assam N	Manas			1815	reintroduction	RU
RR 33	Assam N	Sonai Rupai Orang			1836	current	RU
RR 34	Assam S	Pabitora Laokhowa Kaziranga			1755	current	RU
RR 35	Assam S (Lower)			1850s	1856	1897	RU
RR 36	Assam NE			1600	1756	1886	RU
RR 37	Assam SE				1893	1917, strays	RU
Rhino Region	Area	Protected area	Rock Art	Earliest Artefact	Earliest Record	Last Record	

a   RU = *Rhinoceros unicornis*. SON = *Rhinoceros sondaicus*. SUM = *Dicerorhinus sumatrensis*. < = previous to date. Abbreviations enclosed in brackets refer to potentially introduced or stray populations.

Northern Pakistan

The records for this region date from the times of the Mughal Emperors in the 16th century. Some are far closer to the Afghan border than shown on existing maps of rhino distribution. The reports about rhino hunts of Babur are generally accepted, even though they are in places where no rhinos have been seen before or after. When rhinos were depicted in the manuscripts relating to these accounts, these did not represent contemporary knowledge which could have been corrected as they were made at least half a century later. The possibility of mistranslation finds little resonance with experts of the history and language of the period. This section of the range is kept as a separate unit due to the absence of records which could link it with occurrences further to the east.

North-West India

There are no reports of rhinos from Northern Pakistan eastwards to Himachal Pradesh (13.2). While this might be a consequence of patchy historical writings, it could also be a true state of affairs. The gap shown in the map will focus attention which will hopefully lead to a solution to this phenomenon.

Himachal Pradesh and Uttarakhand

Nestled at the foot of the western Himalayas, rhinos existed here in the foothills. The sighting by Thomas and William Daniell at Kotdwara in 1789 is an iconic early record accompanied by a sketch prepared on the spot. The western extremity is located in the Sirmaur Hills, close to the localities where many of the Siwaliks fossils were unearthed in the 1830s.

Gujarat and Maharashtra

On the basis of three artefacts associated with Harappan settlements in Gujarat and Maharashtra, the protohistorical distribution is extended to the Arabian Sea. This is largely the valley of the Narmada River which also sustained rhino populations further upstream.

Rajasthan and Madhya Pradesh

The extension of the range to include the eastern part of Rajasthan and Raghogarh in Madhya Pradesh is entirely based on the art work of local artists from the 16th to early 19th centuries. Although these must all be classed as artefacts, their sheer abundance and accuracy of portrayal must lead to the conclusion that rhinos existed in the region. This seems counterintuitive given the relative aridity of the western part of India. The evidence is strong that the rhino was indeed hunted by the rulers of Rajasthan in whose palaces the murals are located. In

theory the hunting parties could have traveled further northwards, in reality there is no evidence which supports this. It would be better if the artistic representations had been reflected in hunting diaries or some other kind of written testimony. On the other hand, real rhinos easily identified in drawings or paintings cannot be ignored. The area is traversed by the Chambal River which still supports forests along its banks. Even today the Ranthambore and Keoladeo (Bharatpur) reserves have the right habitat for wild mammals and birds. Therefore the possibility of the presence of rhinos in southern and eastern Rajasthan must be approached with an open mind. The existence of the wall paintings in the palaces of Kota and other towns is discussed here by Joachim Bautze (chapter 15) for the first time in a zoological work.

South-West Nepal

There are no records for this western part of Nepal. Wedged between Uttar Pradesh and the higher Himalayas, this absence might merely reflect a lack of travellers when the country remained off-limits. After the translocations of rhinos to Suklaphanta and Bardia in this region, the dynamic has forever changed.

Central and South-East Nepal

The rhino populations remained restricted to the *terai* region of low hills and river valleys south of the higher mountains. The central area of Chitwan has always been a famous stronghold. The Kings and Maharajas of Nepal hunted here during the 19th and 20th centuries. They hosted the British royal family in 1876, 1911, 1921 and 1961.

Uttar Pradesh and Bihar

These two states had significant rhino populations between the borders of Nepal and the Ganges River. South of the river In Uttar Pradesh, there is only one record for 1529. Further eastwards in Bihar these become more common. The track found at Rohtas on the Sone River in 1769 points at a rhino population in the southern part of the region.

Rajmahal Hills

This region, recently included in the state of Jharkhand, was known as a hunting location from early times. Traffic passing on the Ganges could see the hills. Rhinos were regularly mentioned and might have been restricted to some lower areas along the river banks.

From Bihar to North Bengal

Rhinos were rarely recorded in this administratively divided region, which is wedged between more attractive wilderness areas in Purnia and North Bengal. The reliefs depicting rhinos in the terracotta temples could resonate the presence of rhinos until the 19th century.

North Bengal

This northern part of West Bengal is prime rhino habitat. This is reflected in the large numbers of records in the *terai* towards the north. The Maharaja of Cooch Behar organized annual shooting camps from 1871 to 1911 hosting scores of international aristocrats and distinguished army officers. Although wildlife is now restricted to national parks, this remains an important stronghold of the rhino.

Bhutan

The only records of rhinos in Bhutan pertain to strays from the south.

Arunachal Pradesh

There is no indication that there was a permanent rhino population in Arunachal Pradesh except *D. sumatrensis* at the eastern extremity towards Nagaland and Myanmar.

Meghalaya

Rhinos were relatively rare in this hilly region and may have been restricted to northern parts bordering on the valley of the Brahmaputra River.

West and Central Assam

The rhino was regularly encountered in many wilderness areas of Assam both north and south of the Brahmaputra River. The expansion of settlement and agriculture from west to east is reflected in the dates of encounters with rhinos. Wildlife has become limited to national parks of which today Kaziranga is the most important rhino habitat. Assam remains prime Rhino Country.

East Assam

Records become sparse at the upper reaches of the Brahmaputra River. Rhinos were only found in the lower lying valleys. In my view, it is highly unlikely that *R. unicornis* ever penetrated in the higher hills towards Myanmar, which were elsewhere locally inhabited by *D. sumatrensis*.

65.5 The Road to Extinction and Successful Conservation

For a long time in the historical period, the strongholds of rhinos were in south-central Nepal, in North Bengal and in Assam. Elsewhere *R. unicornis* was largely exterminated by the 1930s (table 65.72). The creation of national parks has been an indispensable lifeline without which few might have survived in the wild. The extinction elsewhere is often blamed on sports hunting by foreign personnel and the wealthy Maharajas and their guests. While this was a contributing factor, the success of the population in Chitwan shows that the effects of hunting can be exaggerated. Yet rhinos disappeared from large parts of their range, through a combination of poaching, urban expansion, increased agriculture, development of indigo and tea plantations, political unrest, and changes in habitat which left no room for a mighty herbivore like the rhino.

The establishment of game sanctuaries and national parks made a careful start early in the 20th century and has gained momentum ever since. Sports hunting has declined with changing attitudes about the value of wildlife, while poaching of rhinos for their horns remains a constant threat. The engagement of the respective governments as well as a series of environmental agencies is exemplary. The rhino is indeed a symbol of nature conservation and of love for the wilderness across South Asia.

65.6 New Maps of the Historical Distribution of *Rhinoceros unicornis* in South Asia

The evidence collated in previous chapters leads to a new representation of the protohistorical range of *R. unicornis* (previous to 1500 CE) shown in map 65.35.

The more recent (since 1500 CE) historical distribution of *R. unicornis* in South Asia is represented in the new synthesis in map 65.36.

CHAPTER 66

The Historical Range of the Javan Rhinoceros *Rhinoceros sondaicus* in South Asia

66.1 The Historical Distribution of *Rhinoceros sondaicus*

The Javan rhinoceros *R. sondaicus* of South Asia had its stronghold in the Sundarbans. It has always been the only species of rhino known in that part of the region. There was an isolated group in North Bengal sympatric with other rhino species. If there ever were rhinos in Odisha and further to the south-west, these would have been *R. sondaicus* as well.

There is definite certainty that *R. sondaicus* was found in the Sundarbans, a region of islands and creeks formed by the delta of the Ganges and Brahmaputra Rivers in the southern parts of West Bengal and Bangladesh. This is an important point of reference which helps to understand the zoogeography of all rhinos in South Asia. The Sundarbans were a challenging destination for explorers, sportsmen and settlers, partly due to the riverine landscape, largely due to diseases, bandits and an abundance of man-eating tigers. Being within reach of the large towns of Kolkata and Dacca, it was inevitable that over time many rhinos were killed, observed and captured, and all evidence shows that these belonged to *R. sondaicus* only. All passengers arriving in the harbours of the two cities had to pass the extreme ends of this region, and if ever a rhino was noticed, it was this species – and not the other single-horned *R. unicornis* which was so much better known in Europe.

There is a great similarity between *R. sondaicus* and *R. unicornis*. Both are single-horned, both show heavy skin folds, and any difference in size is difficult to interpret in the field. Examined side by side, dead or alive, the two species are distinguished with relative ease. As these are rare occasions, every record of every event involving a rhino in South Asia must be weighed carefully to be allocated to either of these species. My analysis of the records in the preceding chapters bears testimony to this constant battle.

66.2 Mapping the Distribution of *R. sondaicus*

The new map of the historical distribution of *R. sondaicus* (map 66.37) is my interpretation based on a combination of all the records listed and explained in Section 2 (chapter 46 to 52). *R. sondaicus* was found in two widely separated regions, the Sundarbans in the south and North Bengal in the north. The species might have existed further to the west in past centuries. There is no substantiated evidence of a single-horned rhino living in Chittagong or even in any of the states in North-East India close to the border with Myanmar. The Javan Rhinoceros *R. sondaicus* is definitely an integral element in the history of the rhinoceros in South Asia.

Odisha (Orissa)

The records from Odisha are too few and too circumstantial to extend the range confidently in this direction (chapter 50). Most pertain to artefacts which introduce uncertainty in any type of locality. The single 19th century report based on oral tradition of local people is isolated in time and place, hence best treated with care until further evidence emerges. On the other hand, if *R. sondaicus* in fact extended westwards from the Sundarbans, it means that the species must be considered as a possible contender for any findings in the more southern parts of India. For instance, the Pleistocene species *Rhinoceros karnuliensis* found in Andhra Pradesh has been said to be most closely associated with *R. sondaicus* (8.7).

North-East India

In the existing literature, mainly dating from the 19th century but with regular repetitions later, there are references to the existence of *R. sondaicus* in North-East India. As explored in Section 3, these relate to Mizoram, Tripura, Sylhet, Cachar, Manipur, and Nagaland. Most, and probably all, of these instances were introduced by authors without personal acquaintance of a rhino in these states. For that reason, the evidence positively suggests that only *D. sumatrensis* was known in these hilly regions along the border with Myanmar. Hence the range of *R. sondaicus* did not include North-East India.

Chittagong

Despite allusions in the literature to the contrary, only *D. sumatrensis* has been found in this eastern part of Bangladesh, and even those only infrequently. The range of *R. sondaicus* did not include Chittagong. The

© L.C. (KEES) ROOKMAAKER, 2024 | DOI:10.1163/9789004691544_067
This is an open access chapter distributed under the terms of the CC BY-NC-ND 4.0 license.

single-horned species was found further east in southern Myanmar and beyond towards Indonesia.

North Bengal

Towards the north, no definite sightings of *R. sondaicus* are known in the large tract between the vicinity of Kolkata and the *terai* of North Bengal. The existence of this isolated population in North Bengal remains difficult to interpret. There are just four substantiated records between 1877 and 1900. In the map I have combined these as a separate region, left unconnected to other populations. The species has not been found in Assam.

In the course of my current investigation, I have tried to test the possibility that *R. sondaicus* ranged from the Sundarbans past Kolkata northwards along the Ganges River, then through the Rajmahal Hills of Jharkhand to the Malda and Rangpur regions ending in North Bengal, either allopatric or sympatric locally with *R. unicornis*. The evidence does not support this theory with any kind of confidence and finally had to be firmly rejected.

FIGURE 66.1   A rare photograph of *R. sondaicus* taken on 13 July 2009 on the Cigenter River in Ujung Kulon, Java by the French wildlife photographer Alain Compost
WITH PERMISSION © ALAIN COMPOST

TABLE 66.73   First and last records of *R. sondaicus* in the Rhino regions of South Asia

Rhino region	Area	Rock art	Earliest artefact	Earliest record	Last record	Species in region[a]
RR 24	North Bengal 1			1877	1877	SON, RU
RR 25	North Bengal 2			1878	1878	SON, RU
RR 26	North Bengal 3			1895	1900	SON, RU, SUM
RR 38	Odisha	Yes	before 200	[1664]	[1850s]	SON
RR 39	West Bengal Sundarbans		100	1737	1910	SON
RR 40	Bangladesh Sundarbans			1599	1909	SON

a   SON: *Rhinoceros sondaicus*; RU: *Rhinoceros unicornis*; SUM: *Dicerorhinus sumatrensis*. Dates enclosed in square brackets refer to indefinite and probably spurious records.

While the distribution of *R. sondaicus* in Myanmar needs further evaluation, the records seem sparse, and there is the continuous possibility of confusion with *D. sumatrensis*. Their former occurrence in eastern and central parts of Myanmar is insufficiently known, yet maybe never included regions close to the border with India or Bangladesh.

66.3 Extinction of *R. sondaicus* in South Asia

The dates of all accepted records of *R. sondaicus* in South Asia show that the species was extinct in North Bengal before 1900 and in the Sundarbans before 1910 (table 66.73). These final dates may not be entirely exact, but they definitely give an indication when the rhinos disappeared. The route to extinction will need further analysis. The Sundarbans with its limitations of accessibility at first glance cannot rank high as a region where sports hunting by foreign personnel or even extension of agricultural practices could have been the deciding factor. Tigers and other wildlife are still found in the Sundarbans. While other factors including changes in climate or availability of good water should be considered, the processes leading to the extinction of *R. sondaicus* are far from clear and evident.

66.4 New Map of the Historical Distribution of *Rhinoceros sondaicus* in South Asia

A new representation of the historical distribution of *R. sondaicus* (from around 1500 CE to 1920) in the South Asian part of its range, based on the discussions in this book, is presented in map 66.37.

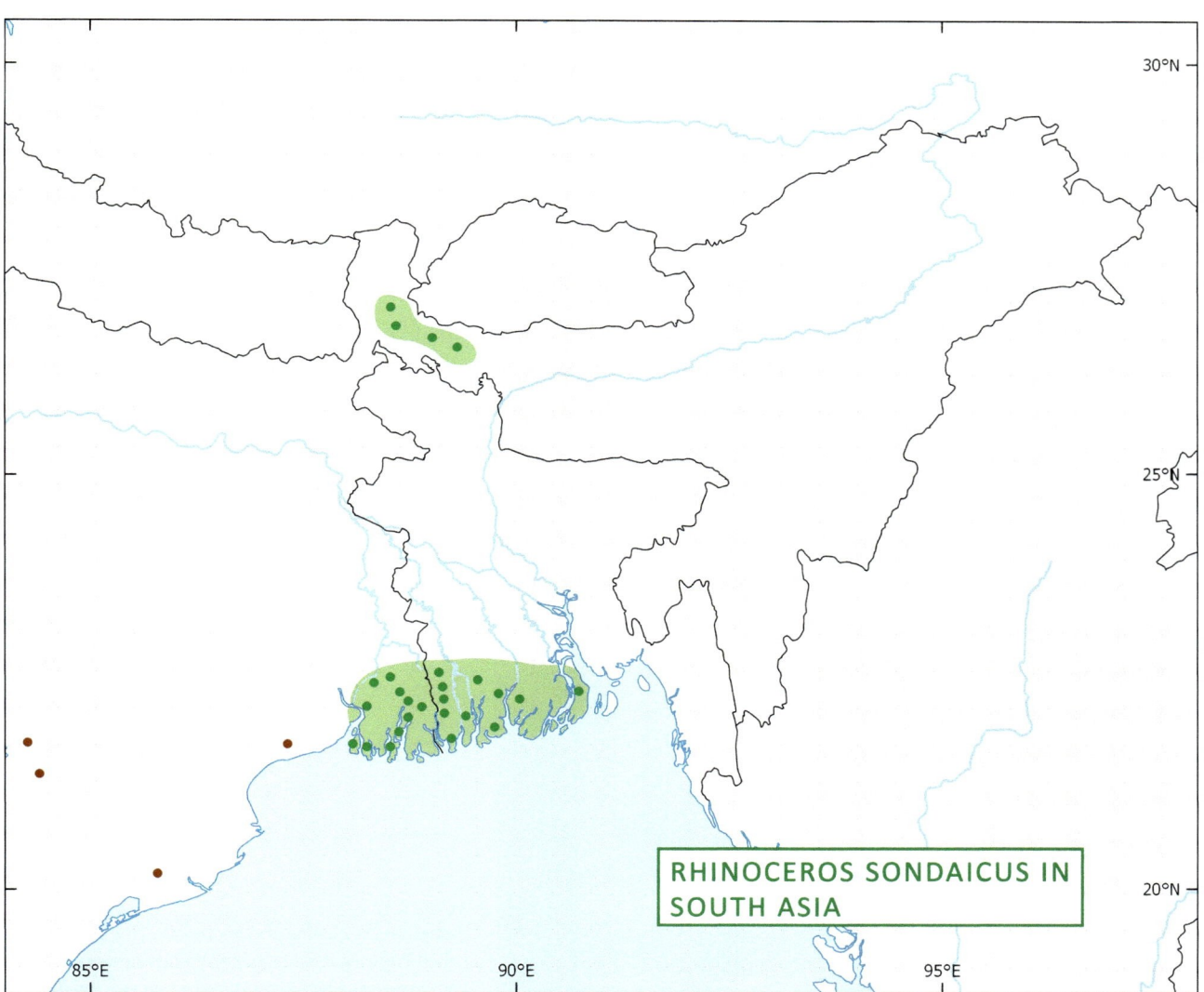

MAP 66.37   Map showing the Historical Distribution of *Rhinoceros sondaicus inermis* in South Asia. This shows the extent of their range from records up to 1920. The dots in the region of Odisha and Chhattisgarh represent unverifiable reports
MAP DESIGN BY AJAY KARTHICK AND RICHARD KEES. © KEES ROOKMAAKER

The Historical Range of the Sumatran Rhinoceros *Dicerorhinus sumatrensis* in South Asia

67.1 The Historical Distribution of *Dicerorhinus sumatrensis*

The two-horned Sumatran rhinoceros *D. sumatrensis* is a marginal element in the South Asian fauna. It was the only species of rhino in the eastern parts of both Bangladesh and India along the border with Myanmar, from Chittagong in the south to Arunachal Pradesh in the north. The species was sympatric with others in an isolated group in North Bengal and Assam. The presence of the species in India was first suspected by Edward Blyth in 1862.

This sketch of historical distribution is correct with a fair degree of certainty, yet based on limited evidence (Section 3, chapter 54 to 63). This was to be expected when surveying a cryptic species which favours mountain ranges and forests rarely penetrated by humans and then with great difficulty. If a rhino is seen or heard, inevitably the animal will have disappeared from view within a few seconds.

67.2 Rhino Species in North-East India and Chittagong

This all leads to continued uncertainty about the identity of the species in this region. The existing literature is confused because the records are rarely equivocal. There are no photographs taken in the field, only one complete specimen (of a captive animal) preserved in any scientific institution, and the written accounts have lost their impact over time. There is no doubt, however, that *D. sumatrensis* was found in Chittagong, in view of the two rhinos captured in 1867 and 1881. This certainty, combined with the occasional mention of two horns and the type of terrain, allows my suggestion that *D. sumatrensis* was the

FIGURE 67.1
Sumatran rhinoceros, *D. sumatrensis*. This female called Puntung lived in Tabin Wildlife Reserve, Sabah, East Malaysia. She was captured when 12 years old on 2011-12-18 and had to be euthanized on 2017-06-15
PHOTO: JOHN PAYNE, JANUARY 2012

© L.C. (KEES) ROOKMAAKER, 2024 | DOI:10.1163/9789004691544_068
This is an open access chapter distributed under the terms of the CC BY-NC-ND 4.0 license.

TABLE 67.74   Instances of direct evidence of the occurrence of *D. sumatrensis* in South Asia

Rhino region	Capture	Sighting (T = tracks)	Killed	Specimen
RR 26 – North Bengal			Sunder 1895	
RR 30, 31 – Assam			Blyth 1870	Cutter 1867
			Sturt 1875	
RR 41 – Chittagong	London Zoo 1867[a]	Lewin 1872	Higgins 1891	ZSL 'Begum' died 1900[a]
	Kolkata Zoo 1882		Poached 1967	
RR 42 – Mizoram		Gordon 1911 (T)	Local 1903	
			Hutchinson 1906	
			Siatlal 20th cent.	
RR 43 – Tripura			Manson 1876	1 destroyed
RR 44 – Sylhet			Fraser 1910	
RR 45 – Lower Assam		Anderson 1872		
		Synteng 1885		
RR 46 – Manipur				Local 2 skulls
RR 47 – Nagaland		Local hunters		Local 2 skulls
RR 28 – East Arunachal Pradesh		Griffith 1837 (T)		Elliot 1883 (Singpho)
		McGregor 1885 (T)		
		Henri 1895 (T)		
TOTAL RECORDS	2	8	10	7

a   The specimen called 'Begum' is recorded twice, in captivity and as specimen.

only species of rhino ever known along the eastern border of the South Asian region.

67.3   Paucity of Records

The records are truly sparse. There are no more than 26 instances of captures, sightings or killings of a rhinoceros since 1862 (table 67.74). This is clearly an insignificant number for a period of 150 years and an area of 500,000 km². Nevertheless, there is certainty that rhinos lived or visited the area and must be counted as part of the fauna of Bangladesh and India.

67.4   Mapping the Distribution of *D. sumatrensis*

The instances of the occurrence of *D. sumatrensis* are plotted in detail on the maps in chapters 56 to 64. These include a few records from Myanmar across the border from India and Bangladesh to help to elucidate the historical distribution. The species has never been known in Bhutan.

All records are combined in one map showing the historical distribution of *D. sumatrensis* in South Asia (map 67.38). This map best reflects the situation 1860–1930, incorporating the earlier reports and still allowing the possibility of continued survival. No sightings of *D. sumatrensis* in South Asia are known in the 21st century.

Chittagong

Two rhinos were captured in the second half of the 19th century in Chittagong. The first of these was described as a new species, which was later shown to be a geographic subspecies *Dicerorhinus sumatrensis lasiotis*. These specimens are important as they are clear and undeniable evidence of the existence of the two-horned species. It is likely that the main population inhabited the hills or mountains further east in adjoining Myanmar.

North Bengal and West Assam

The four records from the eastern part of North Bengal and the western part of Assam are isolated from any known occurrence of the species elsewhere in the region. These may constitute a separate population, or a misrepresentation of the historical record, all requiring further research if ever further data become available. This is best shown as an area of distribution detached from other known records. There is no evidence of sightings of *D. sumatrensis* in the large area connecting western Assam with the range in Nagaland or Manipur.

67.5 Extinction of *D. sumatrensis* in South Asia

The dates of reports combined in table 67.75 show that rhinos might still have been present at the end of the 20th century. It is best not to speculate further to give these animals a chance for survival if the evidence is actually current or reflected in reality.

67.6 New Map of the Historical Distribution of *Dicerorhinus sumatrensis* in South Asia

A new representation of the historical distribution of *D. sumatrensis* (from 1860–1930) in the South Asian part of its range, based on the discussions in this book, is presented in map 67.38.

TABLE 67.75 First and last records of *D. sumatrensis* in the Rhino Regions of South Asia

		D. sumatrensis		
Rhino Region	Area	Earliest Record	Last Record	Species in region[a]
RR 26	North Bengal	1895	1895	SUM, SON, RU
RR 28	Arunachal Pradesh East	1826	1966	SUM, RU
RR 30	Assam – North-West	1875	1875	SUM, RU
RR 31	Assam – South-West	1862	1862	SUM, RU
RR 41	Chittagong	1867	1967	SUM
RR 42	Mizoram	1872	1930s, 2009 ?	SUM
RR 43	Tripura	1750	1920s	SUM
RR 44	Sylhet	1910	1910	SUM
RR 45	Lower Assam	1840	1967	SUM
RR 46	Manipur	1720	1990s	SUM
RR 47	Nagaland	1879	1999	SUM

a SUM: *Dicerorhinus sumatrensis*; SON: *Rhinoceros sondaicus*; RU: *Rhinoceros unicornis*

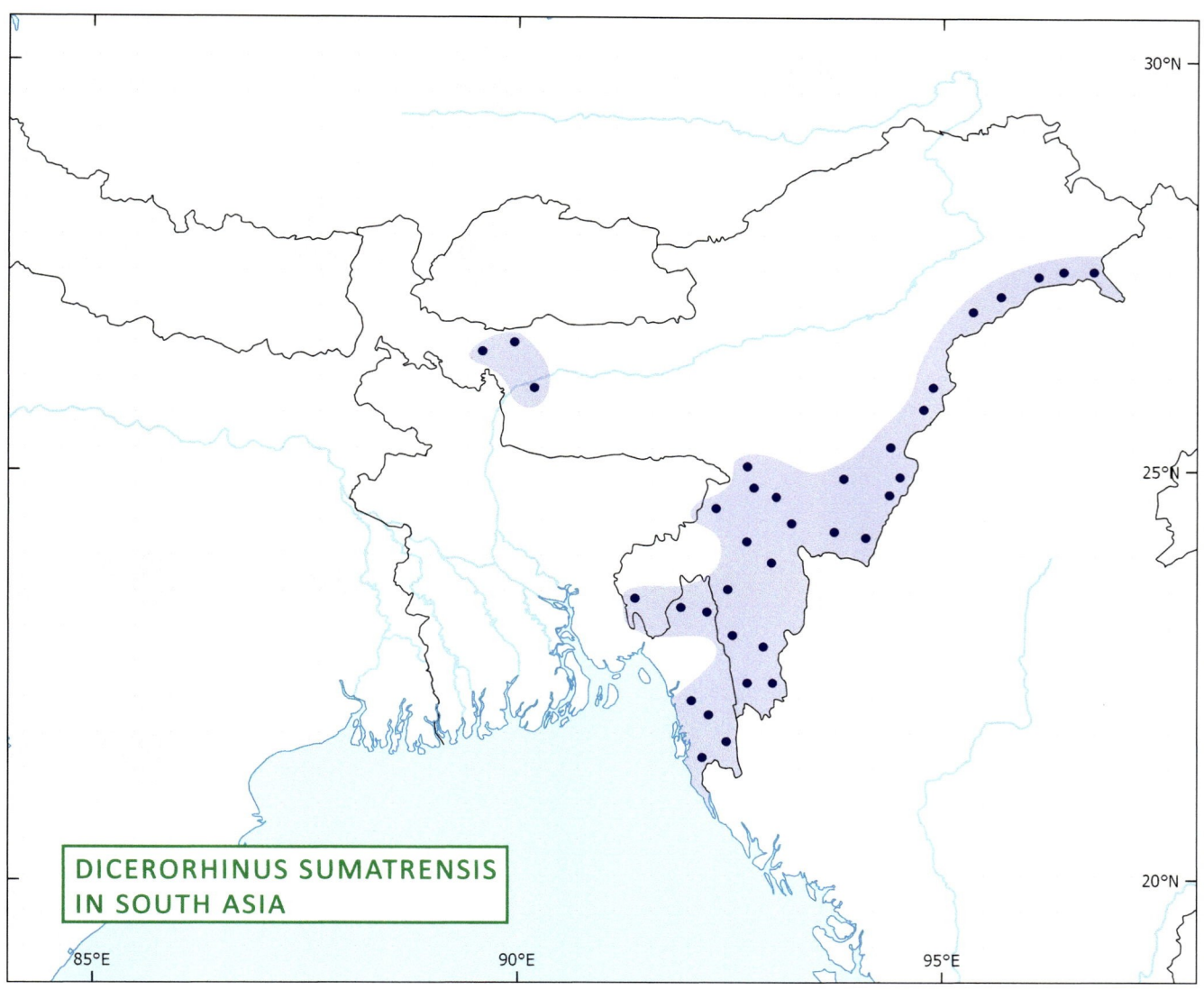

DICERORHINUS SUMATRENSIS
IN SOUTH ASIA

MAP 67.38   Map showing the Historical Distribution of *Dicerorhinus sumatrensis lasiotis* in South Asia. This incorporates data from 1860 to 1930
© MAP DESIGN BY AJAY KARTHICK AND RICHARD KEES. © KEES ROOKMAAKER

Bibliography

This bibliography combines all sources listed in all chapters of this book. It includes 3081 items, all of which were verified either in hard copy or as digitized versions provided by online libraries. In the Introduction of this book, there is a full explanation of the different kinds of data used. The bibliography is complicated due to the date range of the entries (1515 to present), the 16 languages of the originals, and the wide range of disciplines represented in the titles, as well as the large number of items. It is my aim that users can find each item in a library or in their home with minimal effort.

Literature retrieval will always remain challenging. My colleagues will know that the journal *Bijdragen tot the Dierkunde* was renamed *Contribution to Zoology*, that *Der Zoologische Garten* was the journal of the German zoo directors, that *Bongo* was published by the Berlin Zoo and *Artis* by Amsterdam Zoo, that *Oryx* is the journal of the Fauna & Flora Preservation Society (formerly Fauna Preservation Society). However, none of that is obvious to people from other disciplines or language areas. There is no easy solution to this, unless the listings are greatly expanded, which again might be too cumbersome for the average user.

The digital age has changed the way that bibliographies can be composed. There is nothing wrong with full and complete references listing all elements *in extenso*, and in some works that is clearly necessary. There is definitely something wrong when elements are deleted or abbreviated to such an extent that users are left wondering how to obtain the item. References must be clear and within one contribution all must follow the same consistent format. I have tried to find a happy medium: to keep the references relatively short, yet to allow quick retrieval. If further information is required, the title or author can be looked up in an internet search engine or using online library catalogues, many of which are accessible through the Karlsruhe Virtual Katalog (kvk.bibliothek.kit.edu). In the case of the rhinoceros, the Rhino Resource Center (www.rhinoresourcecenter.com) includes most references listed here, often with extensive information or digital access to the pages on rhinos.

The references have the following elements:

(1) Author's surname followed without punctuation by the initials of the given names. A title without author is listed as Anonymous (abbreviated Anon.).

(2) The date of publication of the book, chapter, or paper. A date is given in the YYYY-MM-DD format.

(3) In all cases, there are no details about plates, figures or tables.

Books:

(4) The title, mostly without subtitles, decapitalized where possible.

(5) If needed, the edition number or the volume number.

(6) Place of publication, without publisher.

Papers:

(7) The title of the paper, decapitalized, unabbreviated.

(8) The title of the journal, unabbreviated, without leading article. The aim is to allow unproblematic retrieval of the journal, hence the title is sometimes followed by place of publication.

(9) Other elements in this sequence: (series number) volume number (issue date or number): page range.

Newspapers:

(10) All newspaper entries were retrieved using digital platforms.

(11) Signed contributions are listed under the name of the author.

(12) Unsigned notices are recorded using the newspaper's title, with place of issue where needed, and the date.

Chapters:

(13) The title of the chapter.

(14) The edited volume in which it is published, treated in the format of books.

(15) The page range for the chapter.

Web resources:

(16) The URL is provided together with the accession date, for items only available on the internet.

The 3081 references listed here were mostly written in English: 2658, others in Danish 2, Dutch 25, French 141, German 173, Italian 21, Latin 42, Polish 3, Portuguese 3, Spanish 5, Swedish 2, and five Asian languages: Bengali 2, Hindi 1, Gujarati 1, Mizo 1, Nepali 1.

In summary: 136 references were published in 1799 or earlier. There were 974 of the 19th century, 1273 of the 20th century, and 694 since the start of the 21st century (see fig. 68.1).

Digital platforms. Research depends to a large extent on the availability of sources. These are found in libraries and archives, both public and private. I have spent many happy hours chasing rhinos on the pages of the great books of the past and present in many shapes and forms. In this century there has been a significant shift to digital media, of enormous importance to anybody who is not fortunate enough to live near a great institution. This has revolutionised all research of written and figurative sources, with enormous savings of time and effort. I have used a great variety of websites with sources related to my field of interest, and can only thank their creators and all those involved in the daily additions. Here are some which have helped me over the past years:

Archive.org – many Indian scans – https://archive.org/

Asiatic Society of Bombay – https://www.granthsanjeevani.com/jspui/

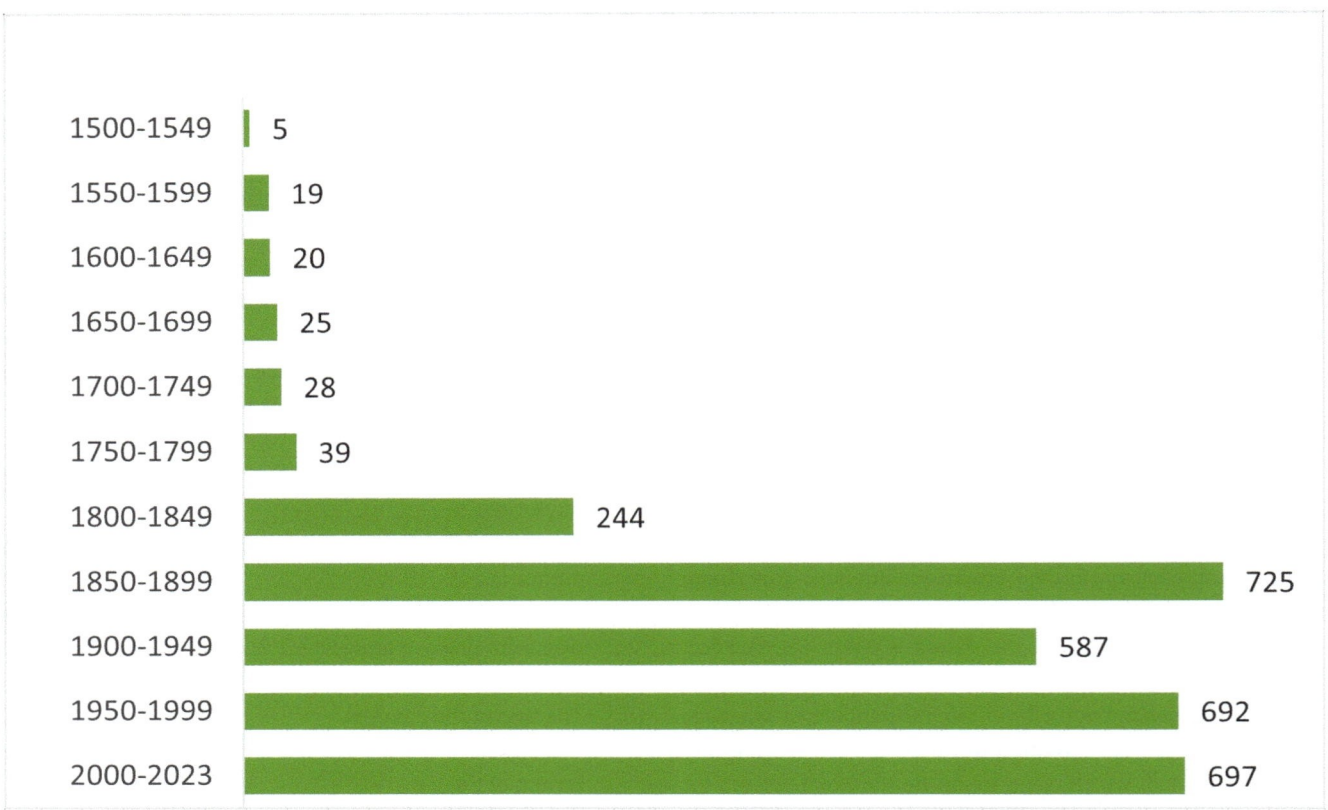

1500-1549	5
1550-1599	19
1600-1649	20
1650-1699	25
1700-1749	28
1750-1799	39
1800-1849	244
1850-1899	725
1900-1949	587
1950-1999	692
2000-2023	697

FIGURE 68.1   Graph showing the dates of the publications used in this work

Bibliothèque Nationale de France – https://gallica.bnf.fr/

Biodiversity Heritage Library – https://www.biodiversitylibrary.org/

British Newspaper Archive – https://www.britishnewspaperarchive.co.uk/

Delpher for Dutch literature – https://www.delpher.nl/

Europeana.eu – https://www.europeana.eu/en

French newspaper archive – https://www.retronews.fr/

Google Books – https://books.google.com/

Hathi Trust – https://www.hathitrust.org/

Heidelberg Historic Literature – https://digi.ub.uni-heidelberg.de/

Indian Museums – http://museumsofindia.gov.in/repository

Karlsruhe Virtual Katalog – https://kvk.bibliothek.kit.edu/

Old maps – https://www.oldmapsonline.org/

Qatar Digital Library – https://www.qdl.qa/

Swiss institutions – https://www.e-rara.ch/

UK Library Hub – https://discover.libraryhub.jisc.ac.uk/

Rhino-related material is frequently added to the RRC (edited by Kees Rookmaaker since 2004):

Rhino Resource Center – www.rhinoresourcecenter.com.

Titles of Works Consulted for This Research

Aa P. van der. 1729. *La galerie agréable du monde*, vol. 2 *des Perses et des Mogol*. Leiden.

Abdul A.M.; Cerdeño E.; Akhtar M.; Khan M.A.; Siddiq M.Kh. 2014. An account of the Upper Siwalik rhinocerotids of Pakistan. In: *Proceedings 4th International Palaeontological Congress, Mendoza, Argentina*, p. 549.

Aberdeen Free Press 1885-07-08: Death of photographer Peter Arthur Johnston.

Aberdeen Press Journal 1875-05-12: Hunting party of Napier Sturt in Nepal (after *Englishman*).

Aberdeen University Museum. Shield from Nagaland, 1905. https://calm.abdn.ac.uk/museums/Record.aspx?src=CalmView.Catalog&id=ABDUA%3a56680 [accessed 2023-04].

Acharya D. 2008. *Records of rhinos poached in Chitwan National Park 1998–2004*. Unpublished report.

Adams W.H.D. 1888. *India, pictorial and descriptive*. London.

Adhikari K. 2015. *Ecology, demography, conservation and management of greater one horned rhinoceros* (*Rhinoceros unicornis*) *in Chitwan National Park, Nepal*. Thesis presented to Saurashtra University, Rajkot.

Adhikari P. 2003. *Status, dispersal and habitat use of greater one-horned rhinoceros (Rhinoceros unicornis) in Royal Suklaphata Wildlife Reserve*. Dissertation presented to Tribhuvan University, Kathmandu.

Africa Hunt Lodge. 2022. https://africahuntlodge.com/hunting -packages/white-rhinoceros-hunt. [accessed 2022-12-02].

Agarwalla R.P.; Vasu N.K.; Bhargava A.; Das J.; Mahanta N.; Hazarika S.; Saikia A.; Gokhale N.; Tirkey A.; Dey M.C. 2005. *Kaziranga centenary 1905–2005*. Gauhati.

Agra Messenger 1851-03-08: Escape of rhinoceros.

Agrawal V.C.; Das P.K.; Chakraborty S.; Ghose R.K.; Mandai A.K.; Chakraborty L.K.; Poddar A.K.; Lal J.P.; Bhattacharyya T.P.; Ghosh M.K. 1992. Mammalia. In: *Fauna of West Bengal*, part 1. Calcutta, pp. 27–170.

Ahmad Y.S. 1981. *With the wild animals of Bengal*. Dacca.

Ahmed J.; Sarma S.; Ahmed N. 2021. Rescue and rehabilitation of two stray greater one-horned rhinoceros (Rhinoceros unicornis) in Orang National Park, Assam. *Indian Journal of Animal Health* 60 (2): 278–280.

A.H.M.T. 1895. Tracking rhinoceros in India. *Field, the Country Gentleman's Magazine* 86 (1895-12-28): 1045–1046.

Akhandananda Swami. 1979. *From holy wanderings to service of God In man*. Chennai.

Alberti T. 1889. *Viaggio a Costantinopoli di Tommaso Alberti (1609–1621)*, edited by Alberto Bacchi della Lega. Bologna.

Albinus B.S. 1747. *Tabulae sceleti et musculorum corporis humani*. Lugduni Batavorum. [English edition: London, 1749].

Albuquerque A. de. 1884. *The commentaries of the Great Afonso Dalboquerque, second viceroy of India*. Translated from the Portuguese edition of 1774, with notes and an introduction, by Walter de Gray Birch, vol. 4. London: Hakluyt Society, vol. 69.

Alderlieste C. 1995. Neushoorns in de mist. *Dieren*, Amsterdam 11: 152–157.

Aldrovandi U. 1616. *De quadrupedibus solidipedibus volumen integrum*. Bononiae.

Aldrovandi U. 1621. *Quadrupedum omnium bisulcorum historia*. Bononiae.

Aldrovandi U. 1642. *Quadrupedum omnium bisulcorum historia*. Bononiae.

Alexander R.D.T.; Leake A.M. 1932. *Some signposts to shikar*. Calcutta.

Alfred Duke of Saxe-Coburg. 1870. *His Royal Highness the Duke of Edinburgh in the Oudh and Nepal Forests*. Edinburgh, London.

Alfred J.R.B.; Das A.K.; Sanyal A.K. 2006. *Animals of India*. Kolkata.

Alfred J.R.B.; De J.K. 2006. *Checklist of Indian ungulates*. Kolkata.

Alfred J.R.B.; Ramakrishna; Pradhan M.S. 2006. *Validation of threatened mammals of India*. Kolkata.

Ali S.A. 1927a. The Moghul Emperors of India as naturalists and sportsmen, part 1. *Journal of the Bombay Natural History Society* 31: 833–861.

Ali S.A. 1927b. The breeding of the Indian rhinoceros (Rhinoceros unicornis) in captivity. *Journal of the Bombay Natural History Society* 31: 1031.

Ali S.A. 1950. The great Indian one-horned rhinoceros (Rhinoceros unicornis Linn) in Assam Province, India. *Proceedings and Papers, International Technical Conference on the Protection of Nature* 1949: 470–472.

Ali S.A.; Santapau H. 1953. [Rhino on Indo-Burmese border]. *Journal of the Bombay Natural History Society* 53: 694.

Allen B.C. 1904. *Assam district gazetteers*, vol. 4: *Kamrup*. Allahabad.

Allen B.C. 1905. *Assam district gazetteers*, vol. 5: *Darrang*. Allahabad.

Allen B.C. 1906. *Assam district gazetteers*, vol. 7: *Sibsagar*. Allahabad.

Allen B.C. 1908a. *Assam district gazetteers*, vol. 8: *Lakhimpur*. Calcutta.

Allen B.C. 1908b. *Assam district gazetteers*, vol. 10: *The Khasi and Jaintia Hills, the Garo Hills and the Lushai Hills*. Calcutta.

Allen B.C. 1915. *Assam district gazetteers*, vol. 7: *Sibsagar*. New edition. Calcutta.

Allen W.O.B. 1885. *A parson's holiday: being an account of a tour in India, Burma, and Ceylon, in the winter of 1882–83*. Tenby.

Allen's Indian Mail 1848-02-22, p. 121: Presentation of plate to Lord Ellenborough.

Allen's Indian Mail 1862-10-27: Death of Bedee Bikaman Singh (Lahore).

Alli D.U. 1874. *The Lucknow album: containing a series of fifty photographic views of Lucknow and its environs together with a large sized plan of the city*. Calcutta.

Altekar A.S. 1954. *Catalogue of the Gupta gold coins in the Bayana hoard*. Bombay.

Altekar A.S. 1955. Buried 1500 years ago and discovered by chance in 1946: a rich Indian's hoard of the Gupta dynasty. *Illustrated London News* 227 (1955-09-24): 524.

Altekar A.S. 1957. *The coinage of the Gupta Empire and its imitations*. Banaras (Corpus of Indian Coins, vol. 4).

Ambrose C.T. 2010. Darwin's historical sketch – an American predecessor: C.S. Rafinesque. *Archives of Natural History* 37: 191–202.

Amin R.; Jnawali S.R.; Chapagain N.R.; Subedi N.; Updhayay G.; Pradhan N.; Nepal R.C.; Pandey M.; Kharel P.; Paudel B.N.; Thapa K.; Murphy S.; Kock R. 2009. *The status and distribution of the greater one-horned rhino in Nepal*. Kathmandu.

Amin R.; Thomas K.; Emslie R.H.; Foose T.J.; Strien N.J. van. 2006. An overview of the conservation status and threats to rhinoceros species in the wild. *International Zoo Yearbook* 40: 96–117.

Amin R. et al. (13 others). 2018. The status of Nepal's mammals. *Journal of Threatened Taxa* 10: 11361–11378.

AMNH 1930. *An album of the groups in the Vernay-Faunthorpe Hall of South Asiatic Mammals of the American Museum of Natural History.* New York.

Amrita Bazar Patrika 1887-12-01: The late King of Oudh's birds and animals (from *The Englishman*).

Amrita Bazar Patrika 1920-12-02: Shooting of rhinoceros.

Amritaghata. 1955. The preservation of wild life in India. *Journal of the United Service Institution of India* 85 (360): 249–254.

Ananda Bazar 2000-04-24: Bones of rhino found in Sundarbans [in Bengali].

Anderson J. 1872. Notes on Rhinoceros sumatrensis, Cuvier. *Proceedings of the Zoological Society of London* 1872-02-06: 129–132.

Anderson J. 1883. *Guide to the Calcutta Zoological Gardens.* Calcutta.

Anderson J.A. 1916. *Springdale, the Huffnagle home.* Doylestown, PA.

Anderson V.D. 2006. *Creatures of empire: how domestic animals transformed early America.* Oxford.

Andrassy E. 1853. *Utazás Kelet Indiákon: Ceylon, Java, Khina, Bengal.* Pesten.

Andrassy E. 1859. *Reise in Ostindien: Ceylon, Java, China und Bengalen.* Pest.

Anhalt U. 2008. *Tiere und Menschen als Exoten.* Ph.D. thesis presented to University of Hanover.

Anon. 1739. Historical chronicle: hurricaine in the Bay of Bengal on 30 September 1738. *Gentleman's Magazine* 8: 321.

Anon. 1814. *La Ménagerie Royale ou collection de 300 figures représentant tous les quadrupèdes ou les animaux les plus curieux des quatre parties du monde,* vol. 1. Paris.

Anon. 1817. An account of a rhinoceros hunt in India. *Asiatic Journal and Monthly Register* 3: 112–113.

Anon. 1818a. The rhinoceros. *Asiatic Journal and Monthly Register* 6: 295.

Anon. 1818b. Rhinoceros-hunting. *Literary Journal and General Miscellany of Science, Art, History, Politics, Morals, Manners, Fashion, and Amusements* 1 (35): 535–536.

Anon. 1818c. Rhinoceros hunting. *Weekly Entertainer or, Agreeable and Instructive Repository* 58 (1818-11-30): 953–954.

Anon. 1830. Rhinoceros in Boston. *American Turf Register and Sporting Magazine* 1 (12): 620.

Anon. 1833. Histoire naturelle: le rhinocéros du Boulevart St-Denis. *Le Temps, Journal des Progrès* 4 (1448), 1833-10-05: 1–3.

Anon. 1834a. The rhinoceros. *Penny Magazine of the Society for the Diffusion of Useful Knowledge* 3 (1834-04-26): 153–155.

Anon. 1834b. Oude. *Asiatic Journal and Monthly Register* (n.s.) 15 (59): 215–225.

Anon. 1834c. Indian sports. *Asiatic Journal and Monthly Register* (n.s.) 15 (60): 304–312.

Anon. 1835. The rhinoceros. *Dublin Penny Journal* 4 (1835-08-29): 68–69.

Anon. 1836. [Review of the Complément de Buffon, vols. 5 and 10]. *Gelehrte Anzeigen,* München 3 (207), 1836-10-15: 614–616.

Anon. 1837. *Short statement relative to the presents transmitted to England in 1835, by the King of Oude, under the charge of P. Friell, Esq., one of His Majesty's Aides-du-Camp, to be laid before their majesties the King and Queen of England, as a mark of attachment and fidelity, and in acknowledgement for the horses presented to his father by His late Majesty George the Fourth, when recognising him as King of Oude, together with the correspondence relating thereto.* London, pp. 1–45.

Anon. 1838. Destruction of a rhinoceros in India. *Parley's Magazine,* New York and Boston 6 (December): 377–378. Also in newspapers: (1) *Morning Herald* (London) 1838-08-11; (2) *London Evening Standard* 1838-08-11; (3) *Sun* (London) 1838-08-14; (4) *Birmingham Journal* 1838-08-18; (5) *Leeds Times* 1838-08-18; (6) *Hampshire Chronicle* 1838-08-20; (7) *Caledonian Mercury* 1838-08-20; (8) *Belfast Commercial Chronicle* 1838-08-20; (9) *Inverness Courier* 1838-08-22; (10) *Warwick and Warwickshire Advertiser* 1838-09-01.

Anon. 1841a. Tiger shooting by moonlight in Bengal. *Times,* London 1841-10-21. Also in: (1) *London Evening Standard* 1841-10-21; (2) *Evening Mail* 1841-10-22; (3) *New Court Gazette* 1841-10-23; (4) *Northampton Mercury* 1841-10-23; (5) *Newry Telegraph* 1841-10-28; (6) *Morning Herald,* London 1841-10-28; (7) *Coventry Standard* 1841-10-29; (8) *Exeter and Plymouth Gazette* 1841-10-30; (9) *Kerry Examiner and Munster General Observer* 1841-11-02; (10) *Londonderry Sentinel* 1841-11-20; (11) *Liverpool Standard* 1842-01-11.

Anon. 1841b. Statistical memoranda: District Rungpore. *Bengal and Agra Annual Guide and Gazetteer* 1841 (2): 267–272.

Anon. 1841c. Statistical memoranda: District Soonderbons. *Bengal and Agra Annual Guide and Gazetteer* 1841 (2): 304–307.

Anon. 1842. Guide for parties proceeding by the Honorable Company's inland flats and steamers. *Bengal and Agra Annual Guide and Gazetteer* 1842 (1): 30–40.

Anon. 1848. *Alfred in India, or, scenes in Hindoostan.* Edinburgh, London.

Anon. 1850a. Annexation of the Indian hill states. *Colburn's United Service Magazine* 1850 (March): 401–412.

Anon. 1850b. Das Nashorn oder Rhinozeros und seine Jagd. *Archiv für Natur, Kunst, Wissenschaft und Leben,* Braunschweig 8: 92–94.

Anon. 1851a. *Bengal almanac for the year 1851, with a companion and appendix.* Calcutta: Bengal Hurkaru Office.

Anon. 1851b. The Ellenborough plate: Hunt and Roskell. *Illustrated London News* 18 (494), Exhibition Supplement of 1851-06-14: 571.

Anon. 1852a. Capture of the outlaw Rundheer Singh near Allahabad. *Illustrated London News* 20 (1852-05-15): 387–388.

Anon. 1852b. Mr. R.H. Dunlop. *Friend of India and Statesman* 1852-03-18: 180.

Anon. 1852c. *The Crystal Palace and its contents: an illustrated cyclopaedia of the great exhibition of 1851.* London, Part 19, 1852-02-17, pp. 296–297.

Anon. 1854. Destruction of a rhinoceros in India. *Merry's Museum and Parley's Magazine* 28 (1854-12-01): 368.

Anon. 1855a. Annexation of Cachar. *Friend of India and Statesman* 1855-09-06: 564–565.

Anon. 1855b. Tschindabra: eine Tragödie in 3 Akten. *Fliegende Blätter*, München 21 (493): 98–99.

Anon. 1857. The wild huntsman. *Harper's Weekly* 1857-03-07: 156–158.

Anon. 1859. *Overland route to India and China.* London.

Anon. 1862. Jerdon in Golcong district. *Ibis* 4: 303.

Anon. 1867. The wild huntsman. *Boy's Journal: a Magazine of Literature, Science, Adventure, and Amusement*, London 1867: 264–271. Also in *Harper's Weekly* 1857-03-07.

Anon. 1869a. The Zoological Society's Garden [opening of Elephant House]. *Illustrated London News* 54 (1869-06-26): 641–642, pl.on 640.

Anon. 1869b. Sport on the Berhampooter Churs [Goalpara District]. *Oriental Sporting Magazine* (n.s.) 2 (17, May): 328–333.

Anon. 1870. Obituary notices: Captain Wilfred Dakin Speer. *Journal of the Linnean Society of London* 10: cviii–cviv.

Anon. 1871a. A lady's reminiscences of tiger-hunting. *Oriental Sporting Magazine* (n.s.) 4 (no. 44, August): 374–388.

Anon. 1871b. Baroda – how we got there, and what we saw. *Once a Week* (n.s.) no. 195 (1871-09-23): 280–284.

Anon. 1872a. Asiatic rhinoceroses. *Leisure Hour* no. 1063 (1872-05-11): 295–298; no. 1064 (1872-05-18): 309–311.

Anon. 1872b. The Sumatra rhinoceros. *Illustrated London News* 60 (1872-03-23): 284.

Anon. 1873. The young rhinoceros. *Field, the Country Gentleman's Magazine* 41 (1873-01-04): 2–3.

Anon. 1874. Addition to Zoological Society. *Illustrated London News* 64 (1874-04-18): 377–378.

Anon. 1875a. The Prince of Wales in India. *Illustrated London News* 67 (1875-12-25): 615, 634.

Anon. 1875b. The Prince of Wales in India. *Graphic, an Illustrated Weekly Newspaper* 12 (317), 1875-12-25: 619–620.

Anon. 1877a. Mitten unter Elefanten: eine interessante Jagdgeschichte. *Neuigkeits Welt-Blatt*, Wien, no. 253, 1877-11-03: 8.

Anon. 1877b. Tiger-shooting in the Rungpore District. *Oriental Sporting Magazine* (n.s.) 10 (115, August): 241–245.

Anon. 1878a. *Gazetteer of the Province of Oudh*, vol. 3: *N to Z*. Allahabad.

Anon. 1878b. Berlin: Aus dem Zoologischen Garten. *Isis, Zeitschrift für alle naturwissenschaftlichen Liebhabereien* 3 (25): 200–201.

Anon. 1880. Marriage of the Gaikwar of Baroda. *Graphic, an Illustrated Weekly Newspaper* 21 (543), 1880-04-24: 411, 420.

Anon. 1882. *The history of the year: a narrative of the chief events and topics of interest from October 1, 1881 to September 30, 1882.* London.

Anon. 1884a. The Earl of Mayo. *County Gentleman Sporting Gazette and Agricultural Journal* 1884-11-29: 1520.

Anon. 1884b. A hunting tour in Nepaul. *Graphic, an Illustrated Weekly Newspaper* 30 (767), 1884-08-09: 126, 128–129.

Anon. 1885a. *Administration report of Cooch Behar State 1884–1885.* Cooch Behar.

Anon. 1885b. A baby rhinoceros. *Land and Water* 1885-12-12 (as reported in *Morning Post* 1885-12-12 and other newspapers).

Anon. 1888. An Indian prince at home [Maharajah of Darbhanga]. *Graphic, an Illustrated Weekly Newspaper* 38 (991), 1888-12-01: 566–567, 584–585.

Anon. 1889. The Duc d'Orleans and the tigress. *Graphic, an Illustrated Weekly Newspaper* 39 (1889-06-01): 590, 591.

Anon. 1890. Breeding of animals in captivity. *National Magazine* (n.s.) no. 6: 228–240.

Anon. 1891a. The late Mr Charles Jamrach. *Field, the Country Gentleman's Magazine* 78 (1891-09-12): 431.

Anon. 1891b. Manipur. *Proceedings of the Royal Geographical Society* 13 (5): 291–293.

Anon. 1895. Two English landowners in Bengal: Messrs. J.R. and H.J. Rainey. *Colonies and India*, London 1895-04-06: 19.

Anon. 1898. Horn removed from the female Rhinoceros lasiotis in London Zoo. *Zoologist, a popular miscellany of natural history* (ser.4) 2: 142.

Anon. 1899a. A record bag (in Cooch Behar). *Field, the Country Gentleman's Magazine* 93 (1899-04-15): 536.

Anon. 1899b. El Conde de Turin en la India. *Caras y Caretas*, Buenos Aires 11 (49), 1899-09-09: 12.

Anon. 1899c. Die vier Temperamente: oder, das Nashorn und sein Horn. *Lustige Blätter* 14 (36): 2.

Anon. 1899d. Shooting party of the Maharajah of Kuch Behar. *Country Life Illustrated* 5 (1899-04-22): 484.

Anon. 1899e. Shoot by Lord Elphinstone and Van der Byl. *Country Life Illustrated* 5 (1899-06-10): 708.

Anon. 1903a. Indian rhinoceros at South Kensington. *Field, the Country Gentleman's Magazine* 101 (1903-02-14): 265.

Anon. 1904a. Lord Curzon of Keddleston in Kooch Behar. *Illustrated Sporting and Dramatic News* 1904-06-04: 518.

Anon. 1904b. Big game shooting in Nepaul – Native skinners with rhinoceros heads. *Navy and Army Illustrated* 17 (1904-01-30): 421.

Anon. 1905a. Colonel E.W. Baird. *Bailey's Magazine* 84: 1–5.

Anon. 1905b. List of members. *Journal of the Society for the Preservation of the Wild Fauna of the Empire* 2: 2–4.

Anon. 1906a. Indian animals for the Prince of Wales. *Field, the Country Gentleman's Magazine* 107 (1906-06-02): 900.

Anon. 1906b. London Zoo receives collection of Nepalese animals donated by Prince of Wales. *Illustrated Sporting and Dramatic News* 1906-06-23: 647.

Anon. 1906c. Curiosities of big game hunting in India, and a distinguished sportsman. *Illustrated London News* 128 (1906-02-10): 200.

Anon. 1907. A shooting camp in Cooch Behar. *Highland Light Infantry Chronicle* 7 (3): 97–100.

Anon. 1908. *Administration report of Cooch Behar State 1907–08*. Cooch Behar.

Anon. 1909a. Big game shooting in Cooch Behar. *Illustrated Sporting and Dramatic News* 1909-11-06: 396.

Anon. 1909b. The game preserves of Assam. *Forest and Stream* 73 (26): 1036–1037.

Anon. 1909c. A Viceroy's holiday: Lord Minto shooting in Assam. *Graphic, an Illustrated Weekly Newspaper* 79 (1909-04-24): 521.

Anon. 1909d. Rhinocéros contre éléphant. *Petit Journal, Supplément Illustré* 20 (956), 1909-03-14: 82, 88.

Anon. 1909e. Record horn of Indian rhinoceros. *Illustrated Sporting and Dramatic News* 1909-07-31: 922.

Anon. 1909f. The shooting party of H.H. the Maharaja of Cooch Behar. *Baily's Magazine of Sports and Pastimes* 86: 79.

Anon. 1911a. Aux Indes Anglaises: une grande chasse en l'honneur de S.M. George V. *Petit Journal, Supplément Illustré* 22 (1102), 1911-12-31: 424.

Anon. 1911b. *A catalogue of Purshotam Vishram Mawjee Museum (Walkeshwar Malbar Hill Bombay)*. Bombay.

Anon. 1912a. The King's Indian animals. *Field, the Country Gentleman's Magazine* 119 (1912-02-17): 304.

Anon. 1912b. The imperial coronation tour in the East – His Majesty the King shooting in the Nepal jungle. *Illustrated Sporting and Dramatic News* 1912-01-20: 920–921.

Anon. 1922a. The Prince of Wales in a tiger hunt. *Graphic, an Illustrated Weekly Newspaper* 105 (1922-01-14): 30, 37.

Anon. 1922b. The Prince of Wales in India: sport in the Nepal terai. *Field, the Country Gentleman's Magazine* 140 (1922-12-25, Christmas issue): 27–29.

Anon. 1923a. The Faunthorpe-Vernay Indian expedition of 1923. *Natural History*, New York 24 (1): 113.

Anon. 1939. Obituary: Mr. J.W.A. Grieve, I.F.S. *Indian Forester* 60: 580–581.

Anon. 1952. Catching a rare rhino: Assam hunters easily capture one-horned specimen in trap but have tough time getting it out again. *Life Magazine* 1952-04-28: 131–132–135.

Anon. 1956. The Maharajkumar of Vizianagram. *Illustrated Weekly of India* 77 (4): 57.

Anon. 1965. Excavation at Chandraketugarh, District 24-Parganas. *Indian Archaeology: a Review* 1962–63: 46–47.

Anon. 1966. Gods, beasts and lovers: the alluring art of India. *Life International* 40 (6), 1966-03-21: 46–56.

Anon. 1968. Death of E.P. Gee. *Oryx* 9: 396.

Anon. 1969. Obituary: E.P. Gee. *Journal of the Bombay Natural History Society* 69: 361–364.

Anon. 1971. End sought on rhino captures in Nepal. *IUCN Bulletin* 2 (18): 157.

Anon. 1972. Rhinos slaughtered. *Animals* 14 (12): 568.

Anon. 1975. Evelyn Arthur Smythies. *Commonwealth Forestry Review* 54: 101–103.

Anon. 2011. Rhino count 2011 (Nepal). *Wildlife Times (Nepal)* 5 (32): 3–5.

Anon. 2016. Siliguri Bengal Safari. *Zoo News, the Bengal Safari Newsletter*, January–March 2016: 5.

Anquetil-Duperron A.H. 1771. Voyage aux Indes Orientales. In: *Zend-Avesta, ouvrage de Zoroastre*, vol. 1 part 1. Paris.

Ansell W.F.H. 1947. A note on the position of rhinoceros in Burma. *Journal of the Bombay Natural History Society* 47: 249–276.

Anthony H.E. 1930. Southern Asia in the American Museum: the new Vernay-Faunthorpe Hall of South Asiatic mammals. *Natural History*, New York 30: 577–592.

Antoine P.O. 2012. Pleistocene and holocene rhinocerotids (Mammalia, Perissodactyla) from the Indochinese Peninsula. *Comptes Rendus Palevol* 11: 159–168.

Antoine P.O. 2023 (in press). Rhinocerotids from the Siwalik faunal sequence. In: Badgley C.; Pilbeam D.; Morgan M., *At the foot of the Himalayas: paleontology and ecosystem dynamics of the Siwalik record of Pakistan*. Baltimore.

Antoine P.O.; Downing K.F.; Crochet J.Y.; Duranthon F.; Flynn L.J.; Marivaux L.; Métais G.; Rajpar A.R.; Roohi G. 2010. A revision of Aceratherium blanfordi Lydekker, 1884 (Mammalia: Rhinocerotidae) from the Early Miocene of Pakistan: postcranials as a key. *Zoological Journal of the Linnean Society* 160: 139–194.

Antoine P.O.; Duranthon F.; Welcomme J.L. 2003. Alicornops (Mammalia, Rhinocerotidae) dans le Miocène supérieur des collines Bugti (Balouchistan, Pakistan): implications phylogénétiques. *Geodiversitas* 25: 575–603.

Antoine P.O.; Métais G.; Orliac M.J.; Crochet J.-Y.; Flynn L.J.; Marivaux L.; Rajpar A.R.; Roohi G.; Welcomme J.L. 2013. Mammalian neogene biostratigraphy of the Sulaiman Province, Pakistan. In: Fortelius M.; Xiaoming Wang; Flynn R., *Fossil mammals of Asia: neogene biostratigraphy and chronology*. Columbia, pp. 400–422.

ANU. 2021. Australian National University. Album: His Imperial Majesty's shoot in Nepalese Terai, December 1911. Mhow, India: Herzog & Higgins, https://openresearch-repository.anu.edu.au/handle/1885/131971 [accessed November 2022].

Appel L.C. 2004. *Dancing with GIs: a Red Cross club worker in India, World War II*. London.

Ara J. 1948. Wildlife reserves in India: Bihar Province. *Journal of the Bombay Natural History Society* 48: 283–289.

Ara J. 1954. Conservation problems in Bihar. *Journal of the Bengal Natural History Society* 27: 66–73.

Arago J. 1840. *Souvenirs d'un aveugle. Voyage autour du monde: chasses, drame.* Paris.

Arago J. 1856. *Oeuvres illustrées de Jacques Arago.* Paris.

Arbuthnott J.C. 1902. Letter No. 75 (No.2409G) dated Gauhati, 4 November 1902, to the Secretary to the Chief Commissioner of Assam, on destruction of the rhinoceros. Assam Secretariat Proceedings, September 1905 (reproduced in Choudhury 2019:28).

Arbuthnott J.C. 1903. Letter No. 77 dated Jowai, 28 August 1903, to the Secretary to the Chief Commissioner of Assam, on establishing an asylum for the rhinoceros. Assam Secretariat Proceedings, September 1905, p. 2 (reproduced in Choudhury 2019: 29).

Archer E. 1833. *Tours in upper India, and in parts of the Himalaya mountains; with accounts of the courts of the native Princes,* vol. 1. London.

Archer M. 1962. *Natural history drawings in the India Office Library.* London.

Archer M. 1980. *Early views of India: the picturesque journeys of Thomas and William Daniell 1786–1794.* London.

Argos. 1912. El Rey Jorge V de Inglaterra caza en la India. *Alrededor del Mundo* no. 662 (1912-02-07): 101–104.

Arluke A.; Sanders C. 1996. *Regarding animals.* Philadelphia.

Arriëns P. 1853. *Dagboek eener reis door Bengalen in 1837 en 1838.* 's Gravenhage.

Aryal A.; Acharya K.P.; Shrestha T.K.; Dhakal M.; Raubenheimer D.; Wright W. 2017. Global lessons from successful rhinoceros conservation in Nepal. *Conservation Biology* 31: 1494–1497.

Ashraf B.; Barman R.; Mainkar K.; Choudhury B. 2005. The principles for rehabilitation of large mammals. In: Menon V.; Ashraf N.V.K.; Panda P.; Mainkar K., *Back to the wild: studies in wildlife rehabilitation.* New Delhi, pp. 91–102.

Ashraf M.; Saikia A.; Sharma S. 2022. Legacy of co-existence between rhino and people in a protected area in India. *Environmental Challenges* 9 (100539): 1–10.

Asmat G.S.M. 2009. Rhinoceroses – Rhinocerotidae. In: Ahmed Z.U., *Encyclopedia of flora and fauna of Bangladesh,* vol. 27: *Mammals.* Dhaka, pp. 157–165.

Asmodeus. 1850. Eastward Ho! *India Sporting Review* 12 (July–December): 68–80.

Assam Forest Dept. 1901. *Progress report of forest administration in Assam 1900–1901.* Shillong.

Assam Forest Dept. 1902. *Progress report of forest administration in Assam 1901–1902.* Shillong.

Assam Forest Dept. 1903. *Progress report of forest administration in Assam 1902–1903.* Shillong.

Assam Forest Dept. 1905. *Progress report of forest administration in Assam 1904–1905.* Shillong.

Assam Forest Dept. 1906. *Progress report of forest administration in Assam 1905–1906.* Shillong.

Assam Forest Dept. 1908. *Progress report of forest administration in Assam 1907–1908.* Shillong.

Assam Forest Dept. 1911. *Progress report of forest administration in Assam 1910–1911.* Shillong.

Assam Forest Dept. 1914. *Progress report of forest administration in Assam 1913–1914.* Shillong.

Assam Forest Dept. 1914–1919. Progress reports: see Blunt A.W.

Assam Forest Dept. 1921. *Progress report of forest administration in Assam 1920–1921.* Shillong.

Assam Forest Dept. 1924–1929. Progress reports: see Cavendish F.H.

Assam Forest Dept. 1929–1931. Progress reports: see Milroy A.J.W.

Assam Forest Dept. 1931–1932. Progress reports: see Owden J.S.

Assam Forest Dept. 1932–1936. Progress reports: see Milroy A.J.W.

Assam Forest Dept. 1936–1938. Progress reports: see MacKarness C.G.M.

Assam Forest Dept. 1939. *Progress report of forest administration in Assam 1938–1939.* Shillong.

Assam Forest Dept. 1941. *Progress report of forest administration in Assam 1940–1941.* Shillong.

Assam Forest Dept. 1947. *Progress report of forest administration in Assam 1944–45.* Shillong.

Assam Forest Dept. 1948. *Progress report of forest administration in Assam 1945–46.* Shillong.

Assam Forest Dept. 1949. *Progress report of forest administration in Assam 1946–47.* Shillong.

Assam Forest Dept. 1950. *Progress report of forest administration in Assam 1949–1950.* Shillong.

Assam Forest Dept. 1951. *Progress report of forest administration in Assam 1950–1951.* Shillong.

Assam Forest Dept. 1953. *Progress report of forest administration in Assam 1952–1953.* Shillong.

Assam Forest Dept. 1993a. *Status report on Kaziranga National Park* (compiled by S.K. Sen). Gauhati.

Assam Forest Dept. 1993b. *Status report on Orang (Rajiv Gandhi) Wildlife Sanctuary.* Gauhati.

Assam Forest Dept. 1993c. *Status report on Pabitora Wildlife Sanctuary.* Gauhati.

Assam Forest Dept. 1994a. Kaziranga National Park. *Zoos' Print (Magazine of Zoo Outreach Organisation)* 9 (3–4): 7–9.

Assam Forest Dept. 1994b. Manas Wildlife Sanctuary. *Zoos' Print* 9 (3–4): 10.

Assam Forest Dept. 1994d. Pabitora Wildlife Sanctuary. *Zoos' Print* 9 (3–4): 12.

Assam Forest Dept. 1995a. *Status report on Kaziranga National Park.* Gauhati.

Assam Forest Dept. 1995b. Manas Wildlife Sanctuary. In Molur S. et al. PHVA *workshop: great Indian one-horned rhinoceros, Jaldapara, 1993.* Coimbatore.

Assam Forest Dept. 2000a. Status report Orang. In: *Report on the Regional Meeting for India and Nepal of IUCN/SSC Asian Rhino Specialist Group*. Doorn, pp. 123–129.

Assam Forest Dept. 2000b. Status report Pabitora. In: *Report on the Regional Meeting for India and Nepal of IUCN/SSC Asian Rhino Specialist Group*. Doorn, pp. 130–134.

Assam Forest Dept. 2014. *A draft proposal for declaring eco-sensitive zone around Sonai-Rupai Wildlife Sanctuary*. Tezpur.

Assam Government. 1906. *Provincial gazetteer of Assam*. Shillong.

Assam Government. 1954. *Assam Rhinoceros Preservation Act*. Gauhati.

Assam Government. 1970. *Assam Rhinoceros Preservation (Amendment) Act*. Guwahati.

Atkins T. 1834. *List of the animals in the Liverpool Zoological Gardens, with notices respecting them*. Liverpool.

Atlas, London 1847-01-02: Advertisement of future parts of Fauna Antiqua Sivalensis.

Atre S. 1985. The Harappan riddle of unicorn. *Bulletin of the Deccan College Research Institute* 44: 1–10.

Atre S. 1990. Harappan seal motifs and the animal retinue. *Bulletin of the Deccan College Research Institute* 49: 43–51.

Aubertin J.J. 1892. *Wanderings and wonderings*. London.

Auceps. 1867. Slaughter of a male rhinoceros in the Soonderbunds. *New Sporting Magazine* 49 (March): 233–237.

Avari E.D. 1957. The Jaldapara Game Sanctuary, West Bengal. *Journal of the Bengal Natural History Society* 29 (3): 65–68.

Awasthi A.B.L. 1978. *Studies in Skanda Purana, part 2: Education, economic life, religion and philosophy*. Lucknow.

Aziz T. 1990. *Some studies on the behaviour of the Indian one-horned rhinoceros (Rhinoceros unicornis) in the Dudhwa national Park (U.P.)*. Thesis presented to Aligarh Muslim University.

B. 1876. The zoological gardens, Calcutta. *Oriental Sporting Magazine* (n.s.) 9 (108, December): 451–454.

Babur. 1826. *Memoirs of Zehir-ed-din Muhammed Baber, Emperor of Hindustan*. Translated by J. Leyden and W. Erskine. London.

Babur. 1871. *Mémoires de Baber Zahireddin Mohammed, fondateur de la dynastie mongole dans l'Hindoustân*. Traduits pour la première fois sur le texte djagataï par A. Pavet de Courteille, vol. 2. Paris.

Babur. 1922. *Babur-Nama (Memoirs of Babur)*. Translated by A.S. Beveridge. London.

Badam G.L. 1979. *Pleistocene fauna of India, with special reference to the Siwaliks*. Pune.

Badam G.L. 1985. The late quaternary fauna of Imamgaon. In: Misra V.N.; Bellwood P.S., *Recent advances in Indo-Pacific prehistory*. Leiden, pp. 413–416.

Badam G.L. 2013. An integrated approach to the quaternary fauna of South and South East Asia, a summary. *Journal of the Palaeontological Society of India* 58: 93–114.

Badam G.L.; Jayakaran S.C. 1993. Pleistocene vertebrate fossils from Tamil Nadu, India. In: Jablonski N.G., *Evolving landscapes and evolving biotas of East Asia since the Mid-Tertiary*. Hong Kong, pp. 241–264.

Badam G.L.; Sankhyan A.R. 2009. Evolutionary trends in Narmada fossil fauna. In: Sankhyan A.R., *Asian perspectives on human evolution*. Delhi, pp. 89–99.

Bahuguna N.C.; Mallick J.K. 2004. Ungulates of West Bengal and its adjoining areas including Sikkim, Bhutan and Bangladesh. *Envis Bulletin*, Dehradun 7 (1): [1–15].

Bahuguna N.C.; Mallick J.K. 2010. *Handbook of the mammals of South Asia*. Dehradun.

Baidya K.N. 1982. Alarm call for great Indian rhino (Rhinoceros unicornis). *Tiger Paper* (FAO) 9 (2): 6–7.

Baidya K.N. 1983. Alarming status of the great Indian rhinoceros (Rhinoceros unicornis). *Environmental Conservation* 9 (4): 346–347.

Baidya K.N. 1984. Alarm call by great Indian rhino (Rhinoceros unicornis). *Cheetal (Journal of the Wild Life Preservation Society of India)* 23 (4): 8–9.

Baillie J.; Groombridge B. 1996. *1996 IUCN Red List of threatened animals*. Gland.

Bairagee A. 2004. A study on the population status and conservation approach for Rhinoceros unicornis in Pabitora Wildlife Sanctuary, Assam, India. *Tiger Paper* (FAO) 31 (1): 11–14.

Baker E.B. 1887. *Sport in Bengal: and how, when, and where to seek it*. London.

Baker W.E. 1850. *Memoir on the fossil remains, presented by himself and Colonel Colvin, to the Museum of Natural History, at Ludlow*. Ludlow.

Baker W.E.; Durand H.M. 1836. Sub-Himalayan fossil remains of the Dadapur collection. *Journal of the Asiatic Society of Bengal* 5: 486–504.

Baker E.C. Stuart: see Stuart Baker.

Baldwin J.H. 1876. *The large and small game of Bengal and the North-Western provinces of India*. London.

Baldwin J.H. 1877. *The large and small game of Bengal and the North-Western provinces of India*, 2nd edition. London.

Baldwin T.G. 1910. Animals in the Sundarbans. *Zoologist, a popular miscellany of natural history* (ser.4) 14: 365–367.

Ball V. 1877. Notes on certain mammals occurring in the Basin of Mahanadi. *Proceedings of the Asiatic Society of Bengal* 1877 (July): 168–172.

Ball V. 1880. *Jungle life in India; or, the journeys and journals of an Indian geologist*. London.

Ball V. 1888. On the identification of the animals and plants of India which were known to early Greek authors. *Proceedings of the Royal Irish Academy* 2: 302–346.

Ballantine H. 1895. *On India's frontier: or, Nepal: the Gurkhas' mysterious land*. New York.

Banerjee A. 1964. Rhinoceros: a historical treatment. In: *West Bengal forests: centenary commemoration volume*. Calcutta, pp. 241–243.

Banerjee R. 1972. Where flying vultures reveal secrets: Kaziranga. *Cheetal (Journal of the Wild Life Preservation Society of India)* 15 (1): 48–50.

Banerjee R. 2011. Treasures from the past: our vanishing wildlife. *The Rhino, Journal of the Kaziranga Wildlife Society* 17: 38–42.

Banerjee S.; Chakraborty S. 1973. Remains of the great one-horned Rhinoceros, Rhinoceros unicornis Linneus, from Rajasthan. *Science and Culture* 39: 430–431.

Barbora S. 2017. Riding the rhino: conservation, conflicts, and militarisation of Kaziranga National Park in Assam. *Antipode* 49 (5): 1145–1163.

Barbosa D. 1518. The book. In: Dames M.L. 1918. *The book of Duarte Barbosa: an account of the countries bordering on the Indian Ocean and their inhabitants*. Translated by Mansel Longworth Dames. London: Hakluyt Society, Second Series, vol. 44.

Barbour T.; Allen G.M. 1932. The lesser one-horned rhinoceros. *Journal of Mammalogy* 13: 144–149.

Barclay E.N. 1932. *Big game shooting records, together with biographical notes and anecdotes on the most prominent big game hunters of ancient and modern times*. London.

Barclay E.N. 1938. Big game trophies for the Field Exhibition. *Field, the Country Gentleman's Magazine* 171 (1938-01-08): 64–65.

Bareh H. 1964. *The history and culture of the Khasi people*. Calcutta.

Barker G.M. 1884. *A tea planter's life in Assam*. Calcutta.

Barman R.; Choudhury B.; Ashraf N.V.K.; Menon V. 2014. Rehabilitation of greater one-horned rhinoceros calves in Manas National Park, a World Heritage Site in India. *Pachyderm*, Nairobi 55: 78–88.

Barman R.; Talukdar A.; Pal M.; Ashraf N.V.K.; Choudhury B.; Menon V. 2008. Bringing back rhinos (Rhinoceros unicornis) to Manas National Park. In: Menon V.; Kaul R.; Dutta R.; Ashraf N.V.K.; Sarkar P. (eds.), *Bringing back Manas*. Delhi, pp. 102–113.

Baron C.G. 1933. Jungle snapshots. *Illustrated London News* 183 (1933-10-21): 645.

Baroowah G.P. 2017. *Where wise owls dare: the life of the rhinoceros in Kaziranga*. Chennai.

Barree C. 1876. Life of a tree planter in India. *Field, the Country Gentleman's Magazine* 47 (1876-02-05): 139.

Barras J. 1885. *India and tiger-hunting, series 1*. London.

Barros J. de. 1553. *Segunda decada da Asia*. Lisboa.

Barthakur R.; Sahgal B. 2005. *The Kaziranga inheritance*. Mumbai.

Barua P.; Das B.N. 1969. *Kaziranga: the rhinoland in Assam*. Gauhati.

Basel Zoo. 2017. *International studbook for the greater one-horned or Indian rhinoceros, Rhinoceros unicornis, 31 December 2017*. Basel (studbook keeper F. von Houwald).

Basel Zoo. 2019. *International studbook for the greater one-horned or Indian rhinoceros, Rhinoceros unicornis, 31 December 2019*. Basel (studbook keeper F. von Houwald).

Basel Zoo. 2021. *International studbook for the greater one-horned or Indian rhinoceros, Rhinoceros unicornis, 31 December 2021*. Basel (studbook keeper B. Steck).

Basu B. 1910. *Report of the Honorary Committee for the Management of the Zoological Garden 1909–10*. Calcutta.

Basu B. 1912. *Report of the Honorary Committee for the Management of the Zoological Garden 1911–12*. Calcutta.

Basu B. 1913. *Report of the Honorary Committee for the Management of the Zoological Garden 1912–13*. Calcutta.

Basu K.K. 1935. Some old accounts of Bhagalpur. *Journal of the Bihar and Orissa Research Society* 21: 140–179.

Bataviasche Courant 1820-04-15 and 1820-05-05: Arrival and departure of Pierre-Médard Diard in Java.

Bates and Hindmarch. 2019. Sale (online catalogue) of an album of photographs by Ernest Brooks. Cheltenham.

Bath Chronicle and Weekly Gazette 1789-08-13, p. 2: Shooting party on Buckarah Lake.

Bauer A.M.; Lavilla E.O. 2022. *J.G. Schneider's Historiae Amphibiorum. Herpetology at the dawn of the 19th century*. Topeka (Society for the Study of Amphibians and Reptiles).

Bauer E.A. 1972. *Treasury of big game animals*. New York.

Bauer J.J. 1988. A preliminary assessment of the reintroduction success of the Asian one-horned rhinoceros (Rhinoceros unicornis) in Bardia Wildlife Reserve, Nepal. *Tiger Paper* (FAO) 15 (4): 26–32.

Baura M.; Talukdar B.K. 2008. Status and rhino poaching in Assam and trade on rhino horn. In: Syangden B. et al., *Report on the regional meeting for India and Nepal IUCN/SSC Asian Rhino Species Group, March 5–7, 2007*, p. 14.

Bautze J.K. 1985. The problem of the Khaḍga (Rhinoceros unicornis) in the light of archaeological finds and art. In: Schotsmans J.; Taddei M., *South Asian archaeology 1983: papers from the seventh International conference of the Association of South Asian archaeologists in Western Europe, held in the Musées Royaux d'Art et d'Histoire, Brussels*. Naples: Istituto universitario orientale (Series Minor 23), vol. 1, pp. 405–433.

Bautze J.K. 1985b. Portraits of Bhao Singh Hara. *Berliner Indologische Studien* 1: 107–122.

Bautze J.K. 1986a. Portraits of Rao Ratan and Madho Singh Hara. *Berliner Indologische Studien* 2: 87–106.

Bautze J.K. 1986b. Sporting pastimes of the Hara Kings: murals of Bundi and Kota. *The India Magazine of her People and Culture* 6: 64–71.

Bautze J.K. 1986c. A contemporary and inscribed equestrian portrait of Jagat Singh of Kota. In: Bhattacharya G. (ed.), *Deyadharma: studies in memory of Dr. D.C. Sircar*. Delhi, pp. 47–64.

Bautze J.K. 1987a. *Drei "Bundi"-Rāgamālās: ein Beitrag zur Geschichte der rajputischen Wandmalerei*. Stuttgart (Monographien zur indischen Archäologie, Kunst und Philologie, Band 6).

Bautze J.K. 1987b. Une représentation des bārahmāsā dans la Baḍe Devtājī kī Havelī de Kota, exemple du rapport entre les peintures murales et les miniatures indiennes. *Berliner Indologische Studien* 3: 195–251.

Bautze J.K. 1989. Shri Brijnathji and the murals in the Chattar Mahal, Kota. In: Bhattacharya D.C.; Handa D. (ed.), *Prācī-Prabhā: perspectives in Indology (Essays in honour of Professor B.N. Mukherjee)*. New Delhi, pp. 319–326.

Bautze J.K. 1990. Ikonographie und Datierung der späteren Jagdszenen und Portraits Bundis. *Tribus: Jahrbuch des Linden-Museums* 39: 87–112.

Bautze J.K. 1991. *Lotosmond und Löwenritt: Indische Miniaturmalerei*. Stuttgart.

Bautze J.K. 1993. German private collections of Indo-Islamic paintings. In: Haase C.P; Kröger J.; Lienert U., *Oriental splendour: Islamic art from German private collections*. Hamburg, pp. 247–294.

Bautze J.K. 1994. Jhala Zalim Singh of Kotah, a great patron of Rajput Painting. In: Balbir N.; Bautze J.K. (eds), *Festschrift Klaus Bruhn zur Vollendung des 65. Lebensjahres dargebracht von Schülern, Freunden und Kollegen*. Reinbek, pp. 105–127.

Bautze J.K. 1995. *Early Indian terracottas*. Leiden (Iconography of Religions XIII, 17).

Bautze J.K. 1996. The royal murals of Rajasthan: art in peril. In: *Silk & Stone – The Art of Asia. The third Hali Annual*. London, pp. 24–31.

Bautze J.K. 1997. The history of Kotah in an art-historical context. In: Welch S.C. (ed.), *Gods, kings, and tigers: the art of Kotah*. New York, pp. 39–60.

Bautze J.K. 2000. Early painting at Bundi. In: Topsfield A. (ed.), *Court painting in Rajasthan*. Mumbai: Mārg Publications (vol. 51, no. 3, March 2000), pp. 12–25.

Bautze J.K. 2005. The early royal murals of Dungarpur, Rajasthan. In: Franke-Vogt U.; Weisshaar H.J. (eds), *South Asian Archaeology 2003*. Aachen, pp. 509–514.

Bautze J.K. 2006. Examples of unlicensed copies and versions of views from Benares: their authorship and identification. In: Gänszle M.; Gengnagel J., *Visualizing space in Banaras: images, maps, and the practice of representation*. Wiesbaden, pp. 213–232.

Bautze J.K. 2007. Uncredited photographs of India and Burma by F. Beato. *Indo-Asiatische Zeitschrift* 9: 68–78.

Bautze J.K. 2013. [Review] A. van der Geer, Animals in stone. *Orientalistische Literaturzeitung* 108 (2): 132–135.

Bautze J.K. 2018. The Calcutta cyclone of 1864: the pictorial evidence. *Journal of Bengal Art* 23: 29–41.

Bayley J.A. 1875. *Reminiscences of school and army life, 1839 to 1859*. London.

Beach M.L. 1974. *Rajput painting at Bundi and Kota*. Ascona (Artibus Asiae Supplementum XXXII).

Beach M.L. 1981. *The imperial image: paintings for the Mughal court*. Washington, DC.

Beach M.L. 1992. *Mughal and Rajput painting*. Cambridge.

Beach M.L.; Galbraith J.K.; Welch S.C. 1965. *Indian painting, 15th–19th centuries. From the collection of Mrs. John F. Kennedy, John Kenneth Galbraith, Stuart Cary Welch, Fogg Art Museum*. Cambridge (Mass).

Beach M.L.; Nahar Singh, R. 2005. *Rajasthani painters Bagta and Chokha, master artists at Devgarh*. Zürich (Artibus Asiae Publishers, Supplementum XLVI).

Beaufort F. de. 1963. Catalogue des types d'ongulés du Muséum National d'Histoire Naturelle, Paris et recherches sur ces types. *Bulletin du Muséum d'Histoire Naturelle, Paris* (2) 35: 551–579.

Beauvau H. de. 1615. *Rélation journalière du voyage du Levant*. Nancy.

Beavan R.C. 1865. The rhinoceros in Bhotan (Rhinoceros indicus, Cuv). *Intellectual Observer* 6: 170–174.

Beavan R.C. 1868. Notes on various Indian birds. *Ibis* (ser. 2) 4: 370–406.

Bechhold J.H. 1930. Das indische Panzernashorn. *Umschau* 34: 887.

Beddard F.E. 1902. *The Cambridge natural history*, vol. 10: *Mammalia*. London.

Bedini S.A. 1997. *The Pope's elephant*. Manchester.

Beebe W.C. 1910. The Calcutta Zoological Garden. *New York Zoological Society Bulletin* 11 (September): 692–695.

Beglar J.D. 1876. Photograph 'View of gateway to the Caravanserai, Nurmahal.' http://www.bl.uk/onlinegallery /onlineex/apac/photocoll/v/019pho000001003u00925000.html [accessed April 2023].

Behie A.M.; Oxenham M.F. (eds.). 2015. *Taxonomic tapestries: the threads of evolutionary, behavioural and conservation research*. Canberra.

Beijing Palace Museum. 2002. *Bamboo, wood, ivory and rhinoceros horn carvings: the complete collection of treasures of the Palace Museum*. Hong Kong.

Bell W. 1793. Description of the double horned rhinoceros of Sumatra. *Philosophical Transactions of the Royal Society of London* 83 (1): 3–6.

Bellew F.J. 1843. *Memoirs of a griffin, or, a cadet's first year in India*, vol. 2. London.

Bending Z.J. 2018. Improving conservation outcomes: understanding scientific, historical and cultural dimensions of the illicit trade in rhinoceros horn. *Environment and History* 24 (2): 149–186.

Bengal Chief Conservator of Forests. 2019. *Estimation of Indian rhinoceros (Rhinoceros unicornis) 2019 West Bengal.* Jalpaiguri.

Bengal Directory 1875, comprising, amongst other information, official directory, military directory, mofussil directory, commercial directory, trades directory, street directory Calcutta, alphabetical list of residents &c. 13th annual publication. Calcutta: Thacker, Spink and Co.

Bengal Directory. 1878. 16th annual publication. Calcutta.

Bengal Directory. 1884. 22nd annual publication. Calcutta.

Bengal Forest Dept. 1921. *Annual progress report on forest administration in the Presidency of Bengal 1920–21.* Calcutta.

Bengal Government. 1932. *The Bengal Rhinoceros Preservation Act, 1932 – Act 8 of 1932.* Calcutta.

Benn R.A.E. [1916]. *Notes on Jaipur.* Jaipur.

Beresford C. 1914. *The memoirs of Admiral Lord Charles Beresford,* vol. 1. Boston.

Berg B. 1932. *På jakt efter Enhörningen.* Stockholm.

Berg B. 1933. *Meine Jagd nach dem Einhorn.* Frankfurt am Main.

Berg B. 1935. *Tiger und Mensch.* Berlin.

Berg B. 1935. A price on his nose. *Field, the Country Gentleman's Magazine* 166 (1935-11-30): 1291.

Berg B. 1935. What mad pursuit (in Bengal's jungle). *Field, the Country Gentleman's Magazine* 166 (1935-12-21): 1471.

Berg B. 1955. *Meine Abenteuer unter Tieren.* Gütersloh.

Bergen C.A. von. 1746. *Oratio de rhinocerote.* Francofurti ad Viadrum.

Berkenmeyer P.L. 1729. *Le curieux antiquaire, ou recueil géographique et historique des choses les plus remarquables qu'on trouve dans les quatre parties de l'univers,* vol. 3. Leiden.

Berlin Zoo. 1873. Neue Behausung für Elephanten und Rinocerosse. *Berliner Börsen-Zeitung* No. 530 (1873-11-13): 7.

Berlin Zoo. 1874. Anlagen des Zoologischen Gartens. *Berliner Börsen-Zeitung* No. 301 (1874-07-02): 7.

Berlin Zoo. 1881. Zweikampf der Nashörner. *Berliner Börsen-Zeitung* No. 317 (1881-07-02): 4.

Bernier F. 1671. *Suite des mémoires sur l'Empire du Grand Mogol.* Paris.

Bernier F. 1699. *Voyages, contenant la description des états du Grand Mogol, de l'Hindoustan, du Royaume de Kachimire,* vol. 1. Amsterdam.

Bernier F. 1891. *Travels in the Mogul Empire A.D. 1656–1668.* Translated, on the basis of Irving Brock's version and annotated by Archibald Constable. London.

Bertrand G. 1959. *The jungle people: men, beasts and legends of the Moi country.* London.

Bertuch F.J.J. 1816. *Bilderbuch für Kinder,* vol. 9, nos. 1–100. Weimar.

Bessudnova Z. 2013. Grigory (Gotthelf) Fischer von Waldheim (1771–1853): author of the first scientific works on Russian geology and palaeontology. *Earth Sciences History* 32 (1): 102–120.

Beusterien J. 2019. Comedy and environmental cultural studies: an image of a Spanish rhinoceros and Sancho with his donkey. *Environmental Cultural Studies Through Time: The Luso-Hispanic World* 24: 251–268.

Beusterien J. 2020. *Transoceanic animals as spectacle in early modern Spain.* Amsterdam.

Beveridge H. 1876. *The district of Bakarganj; its history and statistics.* London.

Beveridge H. 1909. *The Tuzuk-i-Jahangiri, or memoirs of Jahangir,* vol. 1. *From the first to the twelfth year of his reign.* Translated by A. Rogers. London.

Beveridge H. 1914. *The Tuzuk-i-Jahangiri, or memoirs of Jahangir,* vol. 2. *From the thirteenth to the nineteenth year of his reign.* Translated by A. Rogers. London.

Bezbarua P. 2008. *One horned rhinoceros conservation in Manas Tiger Reserve. Final technical report.* Guwahati.

Bezbarua P.; Hazarika B. 2007. *One horned rhinoceros conservation in Manas Tiger Reserve.* Guwahati.

Bhadran C.A.R. 1939. The Monas Game Sanctuary, Assam. *Indian Forester* 60: 802–811.

Bhatt D.B. 1977. *Natural history and economic botany of Nepal.* Revised edition. New Delhi.

Bhatt N. 2003. Kings as wardens and wardens as king: post-Rana ties between Nepali royalty and national park staff. *Conservation and Society* 1 (2): 247–268.

Bhatta R. 2011. *Ecology and conservation of great Indian one horned rhino Rhinoceros unicornis in Pobitora Wildlife Sanctuary, Assam, India.* Thesis presented to Guwahati University.

Bhatta R.; Saikia P.K. 2011. Determining population size and demography of great Indian one-horned rhino, Rhinoceros unicornis in Pobitora Wildlife Sanctuary, Assam, India. *NeBio* 2 (4): 14–19.

Bhatta S.R. 2008. Status of rhinos and horn stockpiles in Nepal. In: Syangden B. et al., *Report on the regional meeting for India and Nepal IUCN/SSC Asian Rhino Species Group, March 5–7, 2007,* p. 11.

Bhattacharya A. 1982. Daily activity cycle of great Indian one horned rhinoceros at Gorumara and Jaldapara Wildlife Sanctuaries, West Bengal, India. *Journal of the Bengal Society of Natural History* (n.s.) 1 (2): 30–32.

Bhattacharya A. 1993. The status of the Kaziranga rhino population. *Tiger Paper* (FAO) 20 (4): 1–6.

Bhattacharya A. 2020. *Ecology, behaviour and management practices of great Indian one horned Rhinoceros at Gorumara, Jaldapara and Kaziranga national parks, India.* Thesis presented to Raiganj University.

Bhattacharya B. 2019. Death on the Brahmaputra: the rhino, the rangers, and the usual suspects. www.mongabay.com 2019-08-28.

Bhattacharya B.C. 1974. *The Jaina iconography*. Delhi.

Bhattacharya T.; Chakraborty D. 1984. *Wild mammalian resources of Tripura: Proceedings of Seminar on Planning for Forestry Development in N.E. Region*. Agartala.

Bhattacharya U. 2015. West Bengal. In: Ellis S. et al., *Indian Rhino Vision 2020 Population Modeling Workshop final report*. Fort Worth, pp. 26–30.

Bhattacharyya B.K. 1991. *Studies on certain aspects of the biology of the one horned rhino, Rhinoceros unicornis*. Thesis presented to Gauhati University.

BHBAMBT. 1822. Death of a large rhinoceros. *Calcutta Journal of Politics and General Literature* 2 (1822-03-28): 295.

Bhikkhu T. 2013. Khaggavisana Sutta: a rhinoceros (Sn 1.3), translated from the Pali. Access to Insight (BCBS Edition), 2013-11-30. https://www.accesstoinsight.org/tipitaka/kn/snp /snp.1.03.than.html. [accessed 2022-10-18].

Bhuju U.R.; Aryal R.S.; Aryal P. 2009. *Report on the facts and issues on poaching of mega species and illegal trade in their parts in Nepal*. Kathmandu.

Bhutan Government. 1995. *Forest and nature conservation act of Bhutan*. Timphu.

Bhutan Government. 2006. *Forest and nature conservation rules of Bhutan*. Timphu.

Bhutan News 2018. Camera trap photograph in Bhutan. http://www.kuenselonline.com/one-horned-rhinoceros -caught-on-camera-in-royal-manas-national-park/ [accessed April 2023].

Bhutia P.T. 2008. Status of rhino and horn stockpiles in Jaldapara and Gorumara National Parks. In: Syangden B. et al., *Report on the regional meeting for India and Nepal IUCN/SSC Asian Rhino Species Group, March 5–7, 2007*, pp. 7–8.

Bhuyan R.; Gayan P.J. 2022. Conservation of forest: a case study report on a single man-made forest in Assam, Molai Forest located in Majuli district near Kokilamukh, Jorhat, Assam. *International Journal of Research in Engineering and Science* 10 (4): 15–17.

Bilder aus Nepal. 1941. Postcard from Sawari Hunting Camp, 1941. Online at https://www.bilder-aus-nepal.de/Pages/All gemein/index2.html [accessed April 2023].

Bilham R. 1994. The 1737 Calcutta earthquake and cyclone evaluated. *Bulletin of the Seismological Society of America* 84: 1650–1657.

Bingley W. 1804. *Animal biography*, 2nd edition, vol. 1. London.

Bist S.S. 1994. Population history of great Indian rhinoceros in North Bengal and major factors influencing the same. *Zoos' Print (Magazine of Zoo Outreach Organisation)* 9 (3–4): 42–51.

Bist S.S. 1997. Wildlife management in West Bengal (1947–1997): the first fifty years. *Himalayan Paryavaran* 5: 4–15.

Bist S.S. 2008. Some reminders from the past. *Banabithi (West Bengal)* 2008 (November): 3–9.

Bist S.S. 2009. Beginning of forest legislation in West Bengal. *Banabithi (West Bengal)* 2009 (July): 4–10.

Bista A. 2011. *Proximate determinants of ungulate distribution and abundance in Pilibhit Forest Division, Uttar Pradesh, India*. Dissertation presented to Saurashtra University, Rajkot.

Biswas O. 1992. *Calcutta and Calcuttans: from Dihi to Megalopolis*. Calcutta.

Biswas S.S. 1981. *Terracotta art of Bengal*. Delhi.

Blainville H.M.D. de. 1817. Lettre de M.W.J. Burchell sur une nouvelle espèce de rhinocéros, et observations sur les différentes espèces de ce genre. *Journal de Physique, de Chimie et d'Histoire Naturelle* 85: 163–168.

Blanc V. Le. 1648. *Les voyages fameux du Sieur Vincent Le Blanc Marseillois, qu'il a fait depuis l'aage [âge] de douze ans iusques [jusques] à soixante, au quatre parties du monde*. Paris.

Blanchard P. 1849. *Le Buffon de la jeunesse: zoologie, botanique, minéralogie*. Revu, corrigé et augmenté par M. [Jean Charles] Chenu. Paris.

Blanford W.T. 1891. *The fauna of British India, including Ceylon and Burma: Mammalia* [part 2]. London.

Blanford W.T. 1892. The Mammalia of India. *Journal of the Bombay Natural History Society* 7: 533–535.

Blaszkiewitz B. 2012. Zoologische Notizen einer Nepalreise. *Milu, Mitteilungen aus dem Tierpark Berlin-Friedrichsfelde* 13: 748–773.

Blathwayt R. 1898. The best of all lives: a talk with H.W. Seton-Karr. *Cassell's Magazine* 26: 339–344.

BL-EAP. 2021a. Album preserved at Madan Puraskar Pustakalaya, Kathmandu. Digitized under Endangered Archives Program of the British Library (EAP166/2/1/8): https://eap.bl.uk /archive-file/EAP166-2-1-8 [accessed March 2021].

BL-EAP. 2021b. Album preserved at Madan Puraskar Pustakalaya, Kathmandu. Digitized under Endangered Archives Program of the British Library (EAP166/2/1/12): https://eap.bl.uk /archive-file/EAP166-2-1-12 [accessed March 2021].

BL-EAP. 2021c. Album preserved at Madan Puraskar Pustakalaya, Kathmandu. Digitized under Endangered Archives Program of the British Library (EAP166/2/1/17): https://eap.bl.uk /archive-file/EAP166-2-1-17 [accessed March 2021].

BL-EAP. 2021d. Album preserved at Madan Puraskar Pustakalaya, Kathmandu. Digitized under Endangered Archives Program of the British Library (EAP166/2/1/27): https://eap.bl.uk/archive-file/EAP166-2-1-27 [accessed March 2021].

BL-EAP. 2021e. Dirgha Man collection of photographs. Digitized under Endangered Archives Program of the British Library (EAP838/1): https://eap.bl.uk/collection/EAP838-1/search [accessed March 2021].

Blower J. 1973. Rhinos – and other problems – in Nepal. *Oryx* 12: 272–280.

Blumenbach J.F. 1779. *Handbuch der Naturgeschichte* [1st edition]. Göttingen.

Blumenbach J.F. 1782. *Handbuch der Naturgeschichte* [2nd edition]. Göttingen.

Blumenbach J.F. 1788. *Handbuch der Naturgeschichte* [3rd edition]. Göttingen.

Blumenbach J.F. 1791. *Handbuch der Naturgeschichte* [4th edition]. Göttingen.

Blumenbach J.F. 1796. *Abbildungen naturhistorischer Gegenstände*. Göttingen.

Blumenbach J.F. 1797. *Handbuch der Naturgeschichte* [5th edition]. Göttingen.

Blumenbach J.F. 1810. *Abbildungen naturhistorischer Gegenstände* Nos. 1–100. Göttingen.

Blumenbach J.F. 1814. *Handbuch der Naturgeschichte* [9th edition]. Göttingen.

Blumenbach J.F. 1821. *Handbuch der Naturgeschichte* [10th edition]. Göttingen.

Blunt A.W. 1915. *Progress report of forest administration in Assam 1914–1915*. Shillong.

Blunt A.W. 1916. *Progress report of forest administration in Assam 1915–1916*. Shillong.

Blunt A.W. 1918. *Progress report of forest administration in Assam 1917–1918*. Shillong.

Blunt A.W.; Todd F.H. 1919. *Progress report of forest administration in Assam 1918–1919*. Shillong.

Blyth E. 1841a. Report for the month of September 1841. *Journal of the Asiatic Society of Bengal* 10: 836–842.

Blyth E. 1841b. Report for the month of October 1841. *Journal of the Asiatic Society of Bengal* 10: 917–929.

Blyth E. 1850. Report of Curator, Zoological Department. *Journal of the Asiatic Society of Bengal* 19: 83–88.

Blyth E. 1862a. A memoir on the living Asiatic species of rhinoceros. *Journal of the Asiatic Society of Bengal* 31: 151–175.

Blyth E. 1862b. A further note on elephants and rhinoceroses. *Journal of the Asiatic Society of Bengal* 31: 196–200 [Reprinted *Zoologist* 21 (1863): 8518–8520].

Blyth E. 1862c. A further note on wild asses, and alleged wild horses. *Journal of the Asiatic Society of Bengal* 31: 363–366.

Blyth E. 1862d. Letter dated 8 April 1862 on species of rhinoceros. In: Owen R., *On the extent and aims of a national museum of natural history*. London, pp. 32–33.

Blyth E. 1863a. *Catalogue of the Mammalia in the Museum Asiatic Society*. Calcutta.

Blyth E. 1863b. On some horns of ruminants. *Proceedings of the Zoological Society of London* 1863-04-21: 155–158.

Blyth E. 1870. [Additional notes.] In: Figuier L., *Mammalia, their various forms and habits*. London.

Blyth E. (signed Z). 1872a. The new rhinoceros at the Zoological Gardens. *Field, the Country Gentleman's Magazine* 40 (1872-08-24): 184.

Blyth E. (signed Z). 1872b. Concerning rhinoceroses. *Field, the Country Gentleman's Magazine* 40 (1872-08-31): 214.

Blyth E. (signed Z). 1872c. More about rhinoceroses. *Field, the Country Gentleman's Magazine* 40 (1872-10-05): 328.

Blyth E. 1872d. On the species of Asiatic two-horned rhinoceros. *Annals and Magazine of Natural History* (ser.4) 10: 399–405.

Blyth E. 1872e. Asiatic rhinoceroses. *Zoologist, a popular miscellany of natural history* (ser.2) 7: 3104–3108.

Blyth E. 1875. Catalogue of the mammals and birds of Burma. *Journal of the Asiatic Society of Bengal* 44 (Extra no.): 1–167.

Boch J. 1595. *Descriptio publicae gratulationis in adventu Sereniss. Principis Ernesti Archiducis Austriae*. Anyverpiae.

Bochart S. 1663. *Hierozoicon sive bipertitum opus de animalibus Sacrae Scripturae*. Londini.

Bodinus A. 1874. *Die Thierwelt im Zoologischen Garten von Berlin*. Berlin.

Bodinus A. [1884]. *Die Thierwelt im Zoologischen Garten von Berlin*, 2nd edition. Berlin.

Boillot I. 1592. *Nouveaux pourtraitz et figures de termes pour user en l'architecture*. Lègres.

Bol L.J. 1949. Philips Angel van Middelburg en Philips Angel van Leiden. *Oud Holland* 64: 2–19.

Bolling F.A. 1678. *Oost-Indiske reise-bog hvor udi befattis hans reise til Oost-Indien*. København.

Bolton G. 1903. *A book of beasts and birds*. London.

Bolton M. 1976. *Royal Karnali wildlife reserve: management plan 1976–1981*. FAO, National Parks and Wildlife Conservation Project.

Bolton M. 1977. New parks and reserves in Nepal. *Oryx* 13: 473–479.

Bombay Gazette 1821-01-17, p. 30: Governor General's party in Rajmahal.

Bombay Gazette 1823-03-12, pp. 7–9: Mrs. Casement's at home.

Bombay Gazette 1865-07-19, p. 5: H.A. Sarel's shooting expedition in Bhootan Dooars.

Bombay Gazette 1866-02-14, p. 3: A visit to the Guicowar's Menagerie.

Bombay Gazette 1866-03-17: A visit to the Guicowar's Menagerie (reprint of 1866-02-14).

Bombay Gazette 1866-03-20, p. 3: Elephant and rhino fights in Baroda.

Bombay NHS. 1886. Catalogue of the Mammalia in the collection of the Bombay Natural History Society. *Journal of the Bombay Natural History Society* 1: 9–15.

Bombay NHS. 1909. Contributions to the Museum. *Journal of the Bombay Natural History Society* 19: 770.

Bombay NHS. 1914. Contributions to the Museum. *Journal of the Bombay Natural History Society* 23: 383–384.

Bombay Times and Journal of Commerce 1853-04-22, p. 762: A hunting party from Rampore Bauleah.

Bombay Times and Journal of Commerce 1856-03-01, p. 138: Rhinoceros fight in Baroda.

Bonal B.S. 2000. Status report Kaziranga. In: Strien N.J. van et al., *Report on the Regional Meeting for India and Nepal of IUCN/SSC Asian Rhino Specialist Group*. Doorn, Asian Rhino Specialist Group, pp. 58–83.

Bonal B.S.; Sharma A. 2009. *The first translocation operations within Assam (Pobitora WLS to Manas NP): a photographic presentation*. Presentation prepared for Translocation Core Committee, Assam.

Bonal B.S.; Sharma A.; Dutta D.K.; Swargowari A.; Bhobora C.R. 2008. *An account of the released rhinos in Manas National Park, Assam (April–September 2008)*. Report to WWF India (Indian Rhino Vision 2020).

Bonal B.S.; Sharma A.; Dutta D.K.; Swargowari A.; Bhobora C.R. 2009a. *An account of the released rhinos in Manas National Park, Assam (October 2008–March 2009)*. Report to WWF India (Indian Rhino Vision 2020).

Bonal B.S.; Talukdar B.K.; Sharma A. 2009b. Translocation of rhino in Assam. *Tiger Paper* (FAO) 36 (1): 7–12.

Bonhams Auctioneers. 2018. [Auction catalogue] *Islamic and Indian art*. London: Bonhams, 4 October 2011.

Bonhams Auctioneers. 2015. [Auction catalogue] *Indian, Himalayan & Southeast Asian Art*. New York: Bonhams, 14 September 2015.

Bonhams Auctioneers. 2015. [Auction catalogue] *Travel and exploration*. London: Bonhams, 3 November 2015.

Bonhams Auctioneers. 2018. [Auction catalogue] *Fine Chinese ceramics and works of art*. Hong Kong: Bonhams, 27 November 2018.

Bontius J. 1658. Historia naturalis et medicae Indiae Orientalis libri sex. In: Piso G., *De Indiae utriusque re naturali et medica, libri quatuordecim*, part 3. Amstelaedami.

Borah M.I. 1936. *Baharistan-i-Ghaybi: a history of the Mughal Wars in Assam, Cooch Behar, Bengal, Bihar and Orissa during the reigns of Jahangir and Shahjahan*. Gauhati.

Borah P. 2017. *Kaziranga: a horny affair*. Delhi.

Borthakur U. 2017. Pabitora Wildlife Sanctuary. *Biolink* 14: 11–14.

Borthakur U.; Das P.K.; Talukdar A.; Talukdar B.K. 2016. Non-invasive genetic census of greater one-horned rhinoceros Rhinoceros unicornis in Gorumara National Park, India: a pilot study for population estimation. *Oryx* 50: 489–494.

Bose B.K.K. 1825. Some account of the country of Bhutan. *Asiatic Researches* 15: 128–156.

Bose B.K.K. 1865. *Political missions to Bootan*. Calcutta.

Bose S. 2014. From eminence to near extinction: the journey of the greater one-horned rhino. In: Rangarajan M.; Sivaramakrishnan K., *Shifting ground*. Delhi, pp. 65–87.

Bose S. 2018a. Before the written word. In: Divyabhanusinh; Das A.K.; Bose S., *The story of India's unicorns*. Mumbai, pp. 33–42.

Bose S. 2018b. A search through antiquity. In: Divyabhanusinh; Das A.K.; Bose S., *The story of India's unicorns*. Mumbai, pp. 45–60.

Bose S. 2020. *Mega mammals in ancient India: rhinos, tigers, and elephants*. New Delhi.

Boshoff A.F.; Kerley G.I.H. 2013. *Historical incidence of the larger mammals in the Free State Province (South Africa) and Lesotho*. Port Elizabeth.

Bostock S. 2021. Clara in Qatar: a new life for a Meissen porcelain rhinoceros. *Luxury* 8 (1): 59–75.

Bourne & Shepherd. 1883. Scenes of a hunting expedition in Nepal, 1883. Photographic album in J. Paul Getty Museum, number 84.XA.1170. Online: http://www.getty.edu/art/collection/objects/33147/ [accessed November 2022].

Boussenard L. 1885. Aventures d'un gamin de Paris au pays des tigres. *Journal des Voyages* no. 425: 432–435; no. 426: 147–151; no. 427: 164–167; no. 428: 180–184.

Boussingault A. 1677. *Le nouveau theatre du monde, ou l'abrégé des états et empires de l'univers*. Part 3: *L'Asie*. Nouvelle édition. Paris.

Bowe P. 2012. Lal Bagh: the botanical garden of Bangalore and its Kew-trained gardeners. *Garden History* 40: 228–238.

Bowrey T. 1905. *A geographical account of countries round the Bay of Bengal, 1669 to 1679*. Edited by R.C. Temple. London: Hakluyt Society, Second Series, vol. 12.

Boylan P.J. 1967. Didermocerus Brookes, 1828 vs. Dicerorhinus Gloger, 1841 (Mammalia, Rhinocerotidae). *Bulletin of Zoological Nomenclature* 24 (1): 55–56.

Boylan P.J.; Green M. 1974. Request for the suppression of Didermocerus Brookes, 1828 (Mammalia). *Bulletin of Zoological Nomenclature* 31 (3): 135–139.

Boyle C.L. 1961. A rhinoceros shot. *Times*, London 1961-03-03: 13.

Braddon E. 1893. Thirty years of shikar. *Blackwood's Edinburgh Magazine* 154: 498–517.

Braddon E. 1895. *Thirty years of shikar*. Edinburgh and London.

Bradley J. 1985. *Lady Curzon's India: letters of a Vicereine*. London.

Bradshaw G. 1878. *Bradshaw's through route overland guide to India, and colonial handbook*. London.

Brahmachary R.L.; Mallik B.; Rakshit B. 1970. An attempt to determine the food habits of the Indian rhinoceros. *Journal of the Bombay Natural History Society* 67: 558–560.

Brahmachary R.L.; Rakshit B.; Mallik B. 1974. Further attempts to determine the food habits of the Indian rhinoceros at Kaziranga. *Journal of the Bombay Natural History Society* 71: 295–299.

Brandon-Jones C. 1997. Edward Blyth, Charles Darwin, and the animal trade in nineteenth-century India and Britain. *Journal of the History of Biology* 30: 145–178.

Brandon-Jones C. 2006. A clever, odd, wild fellow: the life and work of Edward Blyth, zoologist, 1810–1873. *Hamadryad, Journal of the Centre for Herpetology* 31: 1–175.

Brandt J.F. 1878. Tentamen synopseos rhinocerotidum viventium, et fossilium. *Mémoires de l'Académie Impériale des Sciences de St. Petersbourg* (ser.7) 26 (5): 1–66.

Braun G. 1599. *Urbium praecipuarum mundi theatrum quintum.* Coloniae Agrippinae.

Bray J. 2009. Krishnakanta Basu, Rammohan Ray and early-nineteenth century contacts with Bhutan and Tibet. *Tibet Journal* 34 (Special Issue): 1–28.

Brébion A. 1914. Diard, naturaliste français dans l'Extrême-Orient. *T'oung Pao* (ser.2) 15 (2): 203–213.

Brehm A.E. 1891. *Brehms Tierleben* (edited by E. Pechuel-Lösche), vol. 3: *Rüsseltiere, Unpaarzeher, Paarzeher, Sirenen, Waltiere, Beuteltiere, Gabeltiere.* Leipzig und Wien.

Brentjes B. 1969. Eine Darstellung des bengalischen Javanashorns Rhinoceros sondaicus inermis Lesson, 1840. *Säugetierkundliche Mitteilungen* 17: 209–211.

Brentjes B. 1978. Die Nashörner in den alten orientalischen und afrikanischen Kulturen. *Säugetierkundliche Mitteilungen* 26: 150–160.

Brescius M. 2019. *German science in the Age of Empire: enterprise, opportunity and the Schlagintweit brothers.* Cambridge.

Brescius M.; Kleidt S.; Kaiser F. 2015. *Über den Himalaya: die Expedition der Bruder Schlagintweit nach Indien und Zentralasien 1854 bis 1857.* Köln, Weimar, Wien.

Breteuil H. Marquis de. 2007. *Journal secret: 1886–1889.* Paris.

Briggs H.G. 1849. *The cities of Gujarashtra: their topography and history illustrated.* Bombay.

Briggs G.W. 1931. The Indian rhinoceros as a sacred animal. *Journal of the American Oriental Society* 51: 276–282.

Briggs G.W. 1938. *Gorakhnāth and the Kānphaṭa yogīs.* Calcutta.

Brijraj Singh, Mahaharaj Kumar. 1985. *The kingdom that was Kotah: paintings from Kotah.* New Delhi.

Brijraj Singh. 2004. A prince slays a rhinoceros from an elephant. In: Welch S.C.; Masteller K. (eds.), *From mind, heart, and hand: Persian, Turkish, and Indian drawings from the Stuart Cary Welch Collection.* New Haven and London, pp. 46–47.

Brinck E. [undated]. [Manuscript notes] Adversaria. Harderwijk, Streekarchivariaat Noordwest-Veluwe, Oud-Archief Harderwijk, toegang 5299, inv. 2058 – online: https://snwv.nl [accessed January 2023].

Brion de la Tour L. 1781. [map sheet] Presqu'îsle des Indes Orientales en deçà du Gange, comprenant l'Indostan ou Empire du Mogol, différens royaumes ou états: les vastes possessions des Anglais (d'après leurs propres cartes) et les autres établissemens européens avec les grandes routes. Paris: chés Esnauts et Rapilly, rue St. Jacques à la ville de Coutances. (University Library of Bern, Ryhiner Collection, MUE Ryh 7102: 6).

Briscoe T. 1884. Shooting in Cooch Behar. *Indian Forester* 10: 424.

Briscome T. 1909. A record great Indian rhinoceros (R. unicornis). *Journal of the Bombay Natural History Society* 19: 746.

British Press 1806-04-18, 1806-06-03, 1806-12-09: Advertisement for Oriental Field Sports.

Broberg G. 2006. Review of Faust, Zoologische und Botanische Einblattdrücke. *Svenska Linnesallskapets Årsskrift* 2006: 167–171.

Brod W.M. 1958. Jungfer Clara. *Altfränkische Bilder* 58: 7–10.

Brook S.M.; Dudley N.; Mahood S.P.; Polet G.; Williams A.C.; Duckworth J.W.; Thinh Van Ngoc; Long B. 2014. Lessons learned from the loss of a flagship: the extinction of the Javan rhinoceros Rhinoceros sondaicus annamiticus from Vietnam. *Biological Conservation* 174: 21–29.

Brook Northey W. 1937. *The land of the Gurkhas or the Himalayan kingdom of Nepal.* Cambridge.

Brooke V. 1909. Sport abroad. *Baily's Magazine of Sports and Pastimes* 91: 507.

Brooke V. 1912. Sport akin to that arranged for the King-Emperor. *Illustrated London News* 140 (1912-01-13): 41–43.

Brookes J. 1828. *A catalogue of the anatomical and zoological museum.* London.

Brooks E. 1922. The King and the Prince of Wales: some intimate and amusing anecdotes of the royal family. *Strand Magazine* 62: 204–211.

Brooks G. 2017. *Game in transit: a history of the rhino in South Australia.* Mile End.

Brown B. 1925. Glimpses of India, including an account of the field experiences of the Siwalik Hills Indian Expedition of the American Museum. *Natural History*, New York 25: 108–125.

Brown F.H. 1920. The big bag of the Maharajah of Bikanir [Bikaner]. *Graphic, an Illustrated Weekly Newspaper* 102 (1920-07-24): 136.

Brown J. 1980. A memoir of Colonel Sir Proby Cautley, F.R.S., 1802–1871, engineer and palaeontologist. *Notes and Records of the Royal Society of London* 34: 185–225.

Brown M. 2001. A complimentary mission from Nawab Nasir-ud-din Haider to King William IV. *Asian Affairs* 32: 279–286.

Brown R. 1874. *Statistical account of the native state of Manipur and the hill territory under its rule, 1873.* Calcutta.

Brown T. 2014. Zoo proliferation: the first British zoos from 1831–1840. *Zoologische Garten* 83: 17–27.

Bruiser. 1870. Sport in Purneah [1870]. *Oriental Sporting Magazine* (n.s.) 3 (July): 1213–1223.

Bruiser. 1871. Sport in Purneah, 1871. *Oriental Sporting Magazine* (n.s.) 4 (July): 299–309.

Bruton W. 1638. *Newes from the East-Indies; or, a voyage to Bengalla, one of the greatest kingdomes under the high and mighty prince Pedesha Shassallem, usually called the Great Mogull.* London.

Bruton W. 1924. The first coming of the English to Bengal [annotated text]. *Bengal Past and Present* 27: 127–147.

Bryden H.A. 1914. Big-game notes: two remarkable rhinoceros heads. *Sphere, an illustrated newspaper for the home* 56 (1914-03-21): 358.

Buchan J. 1920. *Francis and Riversdale Grenfell, a memoir.* London.

Buchanan-Hamilton F. 1833. *A geographical, statistical, and historical description of the district, or zila, of Dinajpur, in the province, or soubag, of Bengal.* Calcutta.

Buchanan-Hamilton F. 1926. *Journal of Francis Buchanan kept during the survey of the district of Shahabad in 1812–1813.* Patna.

Buchanan-Hamilton F. 1928. *An account of the District of Purnea in 1809–10.* Patna.

Buchanan-Hamilton F. 1930. *Journal of Francis Buchanan kept during the Survey of the District of Bhagalpur in 1810–1811.* Patna.

Buchanan-Hamilton F. 1939. *An account of the district of Bhagalpur in 1810–11.* Patna.

Buck F. 1931. From jungle to zoo on the wild animal trail. *Modern Mechanics and Inventions* 6 (2): 86–90.

Buck F. 1934. My most thrilling adventure. *Official Handbook for members of Frank Buck's Adventurers Club* 1934: 6–12.

Buck F. 1936. Live tigers preferred: an interview with Frank Buck, the famous game hunter. *American Girl* 1936 (August): 5–7, 42–43.

Buck F. 1953. *Bring 'em back alive* (cartoon version). Classics Illustrated no. 104.

Buck F.; Anthony E. 1930. *Bring 'em back alive.* New York.

Buck F.; Fraser F. 1935. *Fang and claw.* London.

Buck F.; Fraser F. 1942. *All in a lifetime.* New York.

Buckland C.T. 1884. *Sketches of social life in India.* London.

Buckland C.T. 1889. Birth of a rhinoceros at Calcutta. *Field, the Country Gentleman's Magazine* 73 (1889-03-02): 308.

Buckland C.T. 1890a. Some Indian wild beasts. *Longman's Magazine* 16: 220–234.

Buckland C.T. [C.T.B.] 1890b. In search of tigers in the Bengal Sunderbunds. *Field, the Country Gentleman's Magazine* 75 (1890-01-04): 17.

Buckland F.T. 1861. *Curiosities of natural history.* Second series. New edition. London.

Buckland F.T. 1868. *Curiosities of natural history.* Third series. London.

Buckland F.T. 1872a. The hairy rhinoceros (from Land and Water). *Graphic, an Illustrated Weekly Newspaper* 5 (118), 1872-03-02: 191, 208.

Buckland F.T. 1872b. A new rhinoceros at the Zoological Gardens. *Land and Water* 1872-08-10: 89.

Buckland W. 1837. Lecture on a large and valuable collection of fossil bones from the sub-Himalayan mountains. *Proceedings of the Ashmolean Society* 14 (1837-06-05): 1–4.

Buffon G.L. Leclerc de. 1764. *Histoire naturelle, générale et particulière*, vol. 11. Paris.

Burbach O. 1873. *Gemeinnützige Naturgeschichte von H.O. Lenz*, vol. 1: *Die Säugethiere.* 5th edition. Gotha.

Burger J. 1991. A passage to India. *Wildlife Conservation* 94: 72–77.

Burges B. 1790. *A series of Indostan letters.* New York.

Burke W.S. 1908. *The Indian field shikar book*, 4th edition. Calcutta.

Burl P. 1998. *Bexhill's Maharajah, HH Nripendra Narayan, Maharajah of Cooch Behar.* Bexhill.

Burnett J.H. 1958. The Manas: Assam's unknown wild life sanctuary. *Oryx* 4: 322–325.

Burt N. 1791. *Delineation of curious foreign beasts and birds in their natural colours: which are to be seen alive at the great room over Exeter Change and at the Lyceum, in the Strand.* London.

Burton M. 1955a. On the trail of Asiatic rhinoceros. *Illustrated London News* 226 (1955-05-07): 840.

Burton M. 1955b. Further notes on rhinoceros. *Illustrated London News* 227 (1955-08-27): 344.

Burton R.G. 1931. *A book of man-eaters.* London.

Burton W.E. 1839. Curious facts in the natural history of the rhinoceros. *Gentleman's Magazine and American Monthly Review* 4 (6): 313–314.

Butler J. 1847. *A sketch of Assam, with some account of the hill tribes.* London.

Butler J. 1855. *Travels and adventures in Assam, during a residence of fourteen years.* London.

Butler J. 1875. Rough notes on the Angami Nagas and their language. *Journal of the Asiatic Society of Bengal* 44: 307–350.

Buttrey T.V. 2007. Domitian, the rhinoceros, and the date of Martial's Liber De Spectaculis. *Journal of Roman Studies* 97: 101–112.

Buyers C. 2020. Genealogy of Nripendra Narayan, Maharajah of Cooch Behar, https://www.royalark.net/India/cooch6.htm [accessed June 2021].

Buyers W. 1848. *Recollections of northern India.* London.

Byrne P. 2008. *The White Grass Plains Wildlife Reserve, the Sukila Phanta: the jewel in the crown.* Varanasi.

Caine W.S. 1890. *Picturesque India: a handbook for European travellers.* London.

Calcutta Directory. 1874. Calcutta: Printed for Cones and Co.

Calcutta Gazette 1808-04-07, Supplement p. 1: Six rhinoceroses killed in Rajmahal.

Calcutta Gazette 1814-06-09, pp. 8–9: Killing of rhinos near Rajmahal and Kolasse.

Calcutta Gazette 1814-08-11: Earl of Moira hunt at Terriagully.

Calcutta Gazette 1998-09-01: Notification of Jaldapara Wild Life Sanctuary.

Calcutta Zoo. 1878. *List of animals in the Zoological Gardens, Calcutta.*

Calcutta Zoo. 1888. Annual report: Zoological gardens. In: *Report on the administration of Bengal, 1887–1888*. Calcutta, p. 310.

Cambridge Independent Press 1877-02-10, p. 3: The Prince's Indian hunting trophies.

Campbell A.C. 1907. *Glimpses of Bengal*, vol. 2. Calcutta.

Camper P. 1782. *Natuurkundige verhandelingen over den orang outang; en eenige andere aapsoorten. Over den rhinoceros met den dubbelen horen, en over het rendier*. Amsterdam.

Camper P. 1787. [Broadsheet] Rhinocerotis Africae Catagraphum. Rhinoc: Asiae Catagr: Cranium a fronte. Viro Doctissimo Expertissimo D. Jacobo van der Steege. (No place). One leaf. Large folio. ca. 54 × 36.5 cm. Library of the University of Leiden, BPL 247-I (drawings) omslag 20. Reproduced in Rookmaaker & Visser 1982, fig. 1.

Canton G. 1854. *Der Thiergarten: eine Aufforderung an die Jugend zu heiterer Betrachtung der lebendigen Natur*. Stuttgart.

Cardane J. 1888. Le voyage du Prince Henri d'Orléans. *Le Soleil*, Paris 1888-09-26: 1–2.

Carey B.S. 1896. *The Chin Hills*. Rangoon.

Carey W. 1800. Letter to R.B. on Sunderbunds. *Periodical Accounts relative to the Baptist Missionary Society* 1: 136–139.

Carlisle Journal 1836-04-02: Exhibition of rhinoceros in Carlisle, Penrith and Kendal.

Carlleyle A.C.L. 1879. *Report of tours in the Central Doab and Gorakhpur in 1874–75 and 1875–76*. Calcutta.

Carpenter P. 1862. *Hog hunting in lower Bengal*. London.

Carpentier H. 2004. Des rhinocéros à la Société de Géographie. *La Géographie* 175 (1515): 82–85.

Carpentier H. 2011. Comment je suis devenu fan des rhinocéros. *Pachyderm*, Nairobi 50: 19.

Carter T.D.; Hill J.E. 1942. Notes on the lesser one-horned rhinoceros, Rhinoceros sondaicus, 1. A skull of Rhinoceros sondaicus in the American Museum of Natural History. *American Museum Novitates* 1206: 1–3.

Carwitham J. 1739. *A natural history of four-footed animals: of the rhinoceros*. London.

Cash B.; Cash A. 1971. The Indian Rhinoceros at home. *Field, the Country Gentleman's Magazine* 238 (1971-07-08): 89.

Cassar B.J. 2004. *The rock relief discovered in the village of Shamarq, Baghlan province*. Report to Society for the Preservation of Afghanistan's Cultural Heritage.

Casserly G. 1914. *Life in an Indian outpost*. London.

Cassiano da Macerata P. 1767. *Memorie istoriche delle virtù, viaggj e fatiche del P. Giuseppe Maria de' Bernini da Gargnano, delle missioni del Thibet*. Verona.

Cassino S.E. 1914. *Naturalists' universal directory*. Salem, Mass.

Catto T.S. 1944. The new Governor of the Bank of England. *Tatler and Bystander*, London 172 (1944-04-26): 15.

Caughley G. 1970. Nepal's rhinos are disappearing. *Oryx* 10: 212–213.

Cautley P.T. 1835. Discovery of further fossils in the Sewalik range. *Journal of the Asiatic Society of Bengal* 4: 585–587.

Cautley P.T.; Falconer H. 1835. Synopsis of fossil genera and species from the upper deposits of the tertiary strata of the Sivalik hills, in the collection of the authors. *Journal of the Asiatic Society of Bengal* 4: 705–706.

Cavendish F.H. 1929. *Quinquennial review of forest administration in Assam for the period 1924–25 to 1928–29*. Shillong.

Cecil G. 1911. The sportsman's paradise. *Illustrated Sporting and Dramatic News* 1911-02-25: 1080.

Central Zoo Authority. 2016. *National studbook of one horned rhinoceros (Rhinoceros unicornis)*, 3rd edition. Dehradun.

Chaigneau M. 1982. Christophe-Augustin Lamare-Picquot, pharmacien, naturaliste, explorateur. *Revue d'Histoire de la Pharmacie* 70: 5–26.

Chakma S. 2016. *Assessment of large mammals of the Chittagong Hill Tracts of Bangladesh with emphasis on tiger (Panthera tigris)*. Thesis presented to University of Dhaka.

Chakraborty R. 2004. A catalogue of mammalian exhibits of the Zoological Galleries of the Indian Museum. *Records of the Zoological Survey of India, Occasional Paper* 219: 1–99.

Chambers N. 2007. *Scientific correspondence of Sir Joseph Banks, 1765–1820*, vol. 4: *The Middle Period, 1785–1799. Letters 1790–1799*. London.

Chambers O.A. 1899. *Hand-book of the Lushai country*. Calcutta [copy in LSE Library, London].

Champion A. 1765. *Journal*. London: India Office Library, Home Miscellaneous Series, vol. 198, pp. 293–294.

Champion F.W. 1933. *The jungle in sunlight and shadow*. London.

Chandan D. 2018. *Ballad of Kaziranga*. New Delhi.

Chandra M. 1973. Studies in the cult of the mother goddess in ancient India. *Prince of Wales Museum Bulletin* 12: 1–47.

Chandra P. 1970. *Stone sculpture in the Allahabad Museum: a descriptive catalogue*. Bombay.

Channa S.M. 2005. The descent of the Pandavas: ritual and cosmology of the Jads of Garhwal. *European Bulletin of Himalayan Research* 28: 67–87.

Chapman J. 1999. *The art of rhinoceros horn carving in China*. London.

Chardin J. 1711. *Voyages en Perse, et autres lieux de l'Orient*, vol. 3. Amsterdam.

Charles. 1840. Recollections of Rungpore. *Bengal Sporting Magazine* 15: 109–116.

Chasen F.N. 1940. A handlist of Malaysian mammals. *Bulletin of the Raffles Museum* 15: i–xx, 1–209.

Chatterbox, Boston 1890-11-08, pp. 409–410: Charged by a rhinoceros.

Chatterjee S. 1995. *Mizo chiefs and the chiefdom*. New Delhi.

Chatterji B. 1929. *Short sketch of Maharaja Sukhmoy Roy Bahadur and his family*. Calcutta.

Chaudhuri H.N. 1903. *The Cooch Behar state and its land revenue settlements*. Cooch Behar.

Chauhan P.R. 2008. Large mammal fossil occurrences and associated archaeological evidence in Pleistocene contexts of peninsular India and Sri Lanka. *Quaternary International* 192: 20–42.

Chauley G.C.; Lourdusamy K. 2000. Excavation at Shyam Sundar Tilla, District South Tripura. *Indian Archaeology: a Review* 1999–2000: 154–155.

Chesemore D.L. 1970. Notes on the mammals of southern Nepal. *Journal of Mammalogy* 51: 162–166.

Chhetry D.T.; Pal J. 2010. Diversity of mammals in and around of Koshi Tappu Wildlife Reserve. *Our Nature* 8: 254–257.

Chicago FMNH (Field Museum of Natural History). 1926. *Annual report of the Director to the Board of Trustees 1926*. Chicago.

Chicago FMNH (Field Museum of Natural History). 1929. *Annual report of the Director to the Board of Trustees 1929*. Chicago.

Chinmulgund P.J. 1955. Another rhinoceros slayer type coin of Kumāragupta I. *Journal of the Numismatic Society of India* 17: 104–106.

Chitalwala Y.M. 1990. The disappearance of rhino from Saurashtra. *Bulletin of the Deccan College Research Institute* 49: 79–82.

Chodynski A.R. 2020. Dzielo M.B. Valentiniego Museum museorum. *Opuscula Musealia* 27: 131–167.

Choegyal L. 2017. Returning to Chitwan. *Nepali Times* 2017-12-01.

Choudhury A. 1985a. Distribution of Indian one-horned rhinoceros (Rhinoceros unicornis). *Tiger Paper* (FAO) 12 (2): 25–30.

Choudhury A. 1985b. Kaziranga cannot be allowed to die. *Sunday Telegraph*, Guwahati 1985-03-17: 1.

Choudhury A. 1987a. Railway threat to Kaziranga. *Oryx* 21: 160–163.

Choudhury A. 1987b. A day in Pabitara. *Sunday Sentinel*, Guwahati 1987-04-26: 1–2.

Choudhury A. 1989. S.O.S. Manas. *Sunday Sentinel*, Guwahati 1989-5-07 (reprinted in Choudhury 2019: 173).

Choudhury A. 1996. The greater one-horned rhino outside protected areas in Assam, India. *Pachyderm*, Nairobi 22: 7–9.

Choudhury A. 1997a. Indian one-horned rhinoceros Rhinoceros unicornis Linnaeus 1758, in Arunachal Pradesh. *Journal of the Bombay Natural History Society* 94: 152–153.

Choudhury A. 1997b. The status of the Sumatran rhinoceros in north-eastern India. *Oryx* 31: 151–152.

Choudhury A. 1998a. Sumatran rhinoceros rediscovered in India. *Newsletter of the Rhino Foundation of Nature in North-East India* 2 (1): 7.

Choudhury A. 1998b. Flood havoc in Kaziranga. *Pachyderm*, Nairobi 26: 83–87.

Choudhury A. 2001. Rhino poaching in Manas. *Newsletter of the Rhino Foundation of Nature in North-East India* no. 3: 7.

Choudhury A. 2002a. Big cats, elephant, rhino and gaur in Guwahati. *Newsletter of the Rhino Foundation of Nature in North-East India* no. 4: 16–19.

Choudhury A. 2002b. Rhino poaching in Gohpur. *Newsletter of the Rhino Foundation of Nature in North-East India* no. 4: 6.

Choudhury A. 2003. *The mammals of Arunachal Pradesh*. New Delhi.

Choudhury A. 2004. *Kaziranga: wildlife in Assam*. Delhi.

Choudhury A. 2005a. A new historic record of the Sumatran rhinoceros in north-eastern India. *Pachyderm*, Nairobi 39: 91–92.

Choudhury A. 2005b. Threats to the greater one-horned rhino and its habitat, Pabitora Wildlife Sanctuary, Assam, India. *Pachyderm*, Nairobi 38: 82–88.

Choudhury A. 2005c. Kaziranga completes a hundred years. *Ishani* 1 (5): 1–6.

Choudhury A. 2007a. Rhino population increased. *Newsletter of the Rhino Foundation of Nature in North-East India* no. 7: 24.

Choudhury A. 2007b. Rhino poaching (in Kaziranga). *Newsletter of the Rhino Foundation of Nature in North-East India* no. 7: 24.

Choudhury A. 2007c. Rhino sightings (in Batabari and Kolakhowa). *Newsletter of the Rhino Foundation of Nature in North-East India* no. 7: 24.

Choudhury A. 2010. Recent records of rhinoceros from outside traditional areas. *Newsletter of the Rhino Foundation of Nature in North-East India* no. 8: 12–13.

Choudhury A. 2012. *The secrets of wild Assam*. Guwahati.

Choudhury A. 2013. *The mammals of North East India*. Guwahati.

Choudhury A. 2016. *The mammals of India: a systematic & cartographic review*. Guwahati.

Choudhury A. 2019a. *Manas: India's threatened World Heritage*. Guwahati.

Choudhury A. 2019b. There still could be a couple of Sumatran rhinos in Myanmar. Online: www.downtoearth.org 2019-05-30 [accessed 2019-05].

Choudhury A. 2022a. Records of greater one-horned rhinoceros Rhinoceros unicornis (family Rhinocerotidae) in hilly terrain. *Journal of the Bombay Natural History Society* 119: 1.

Choudhury A. 2022b. Towards a new rhino conservation landscape in Assam for the increasing population of the greater one-horned rhino (Rhinoceros unicornis). *Pachyderm*, Nairobi 63: 179–182.

Choudhury B.; Dutta B.; Talukdar A.; Shukla U.; Smith M.L. 2009. *Formulating a working protocol for rescue and translocation of greater one horned rhinoceros based on experiences of three cases in India*. Report (only available on the Rhino Resource Center).

Choudhury P.C.R. 1960. *Bihar district gazetteers: Champaran*. Patna.

Choudhury P.C.R. 1963. *Bihar district gazetteers: Purnea*. Patna.

Chowdhuary M.K.; Ghosh S. 1984. Operation Rhino – Jaldapara Sanctuary. *Indian Forester* 110: 1098–1108.

Chowdhury J.N. 1983. *Arunachal Pradesh: from frontier tracts to union territory*. Delhi.

Christie's. 1971. [Auction catalogue] *Oriental books printed and manuscript, miniatures and one Hebrew manuscript*. London: Christie's, 1971-07-21.

Christie's. 1989. [Auction catalogue] *South-, Southeast Asian and Indonesian sculpture and works of art and antiquities*. Amsterdam: Christie's, 1989-12-05.

Christie's. 1992. [Auction catalogue] *Cameras, photographs and related material*. London: Christie's, 1992-0-10.

Christison A.F.P. 1945. A note on the present distribution of the Sumatran rhinoceros (Dicerorhinus sumatrensis) in the Arakan district of Burma. *Journal of the Bombay Natural History Society* 45: 604–605.

Church Missionary Intelligencer 1862 (vol. 13), pp. 88–93: The valley of the Ganges and its spiritual destitution.

Cimino R.M. 2001. *Wall paintings of Rajasthan (Amber and Jaipur)*. New Delhi.

Cimino R.M. 2011. *Leggende e fasti della corte dei Grandi Re. Dipinti murali di Udaipur, Rajasthan*. Torino.

Civil & Military Gazette, Lahore 1890-01-28, pp. 4–5: The Chin Lushai Expedition.

Civil & Military Gazette, Lahore 1901-03-26: Camp of Lord Curzon in Nepal.

Clark A.H. 1931. *Nature narratives*, vol. 2. Baltimore.

Clarke C.P. 1898. *Catalogue of the collection of Indian arms and objects of art presented by the princes and nobles of India to H.R.H. the Prince of Wales on the occasion of his visit to India in 1875–1876*. London.

Clarke C.P. 1910. *Arms and armour at Sandringham*. London.

Clarke T.H. 1973. The iconography of the rhinoceros from Dürer to Stubbs, part I: Dürer's Ganda. *Connoisseur* 184 (no. 739, September): 2–13.

Clarke T.H. 1974. The iconography of the rhinoceros, part II. The Leyden Rhinoceros. *Connoisseur* 185 (no. 744, February): 113–122.

Clarke T.H. 1976. The rhinoceros in European ceramics. *Keramik Freunde der Schweiz* 89: 3–20.

Clarke T.H. 1986. *The rhinoceros from Dürer to Stubbs 1515–1799*. London.

Clarke T.H. 1987. 'I am the horn of a rhinoceros'. *Apollo* 1987 May: 344–349.

Clason A.T. 1977. Wild and domestic animals in prehistoric and early historic India. *Eastern Anthropologist* 30: 241–289 [also as book, Lucknow, 1979].

Clay A.L. 1872. Hairy rhinoceros in Chittagong. *Graphic, an Illustrated Weekly Newspaper* 5 (133), 1872-06-15: 555.

Clay A.L. (signed C.S.) 1896. *Leaves from a diary in lower Bengal*. London and New York.

Clift W. 1831. *Catalogue of the Hunterian collection in the museum of the Royal College of Surgeons in London*, part 3. London.

Closs G.A.; Frenzel O. 1891. Neushoorn en tijger. *Katholieke Illustratie: Zondags-lektuur voor het Katholieke Nederlandsche Volk* 24 (5): 36, 38.

Clubman. 1911. Shooting in Nepal. *Sketch*, London 1911-12-20: 325.

Clutterbuck C. 1983. The Lydekker family, 1650–1983. Harpenden History website: http://www.harpenden-history .org.uk/page/the_lydekker_family_1650-1983 [accessed November 2021].

Clutton-Brock J. 1965. *Excavations at Langhnaj: 1944–63*, part 2: *The fauna*. Poona.

C.N. 1913. Rhinoceros head shot by King George V in Nepal. *Sketch*, London 1913-11-26: 227.

Cock C.R. 1865. Sport in Assam. *Field, the Country Gentleman's Magazine* 25 (1865-02-25): 137–138.

Cockburn J. 1879. Notes on stone Implements from the Khasi Hills, and the Banda and Vellore Districts. (By John Cockburn, Late Curator of the Allahábád Museum, Officiating Assistant Osteologist, Indian Museum, Calcutta). *Journal of the Asiatic Society of Bengal* 48 (2): 133–143.

Cockburn J. 1882a. On an abnormality in the horns of the Hog-deer, Axis porcinus, with an amplification of the theory of the evolution of antlers in ruminants. (By John Cockburn, Offg. 2nd Asst. to Supdt. Indian Museum Calcutta). *Journal of the Asiatic Society of Bengal* 51 (2): 44–49.

Cockburn J. 1882b. On the habits of a little known Lizard, Brachysaura ornata. (By John Cockburn, 2nd Assistant to Superintendent Indian Museum). *Journal of the Asiatic Society of Bengal* 51 (2): 50–54.

Cockburn J. 1883. On the recent existence of Rhinoceros indicus in the North Western Provinces, and a description of a tracing of an archaic rock painting from Mirzapore representing the hunt of this animal. *Journal of the Asiatic Society of Bengal* 52: 56–64.

Cockburn J. 1884a. On the recent extinction of a species of rhinoceros in the Rajmahal Hills, and Bos gaurus in the Mirzapur District. *Proceedings of the Asiatic Society of Bengal* 1884 September: 140–141.

Cockburn J. 1884b. On the durability of haematite drawings on sandstone rocks. *Proceedings of the Asiatic Society of Bengal* 1884: 141–145.

Cockburn J. 1884c. Note on Rhinoceros indicus. In: Sterndale R.A., *Natural history of the Mammalia of India and Ceylon*. London, p. 407.

Cockburn J. 1888. On palaeolithic implements from the drift-gravels of the Singrauli basin, south Mirzapur. *Journal of the Royal Anthropological Institute* 17: 57–65.

Cockburn J. 1893. On the durability of haematite drawings on sandstone rocks. *North Indian Notes and Queries*, 1893 (September): 98–99.

Cockburn J. 1894. On flint implements from the Kon ravines of South Mirzapore. *Journal of the Asiatic Society of Bengal* 63 (3): 21–27.

Cockburn J. 1899. Cave drawings in the Kaimur range, North-West provinces. *Journal of the Royal Asiatic Society* 31 (1): 89–97.

Cocker M., Inskipp C. 1988. *A Himalayan ornithologist: the life and work of Brian Houghton Hodgson*. Oxford.

Colbert E.H. 1934. A new rhinoceros from the Siwalik beds of India. *American Museum Novitates* 749: 1–13.

Colbert E.H. 1935. Siwalik mammals in the American Museum of Natural History. *Transactions of the American Philosophical Society* (n.s.) 26: 1–401.

Cole F.J. 1953. The history of Albrecht Dürer's rhinoceros in zoological literature. In: Underwood E.A., *Science, medicine and history: essays of the evolution of scientific thought and medical practice*. London, vol. 1, pp. 337–356.

Coleridge N. 1997. Dungarpur. *World of Interiors* 17 (12): 134–145.

Colin A. 1602. *Histoire des drogues, episceries, et de certains médicamens simples, qui naissent és Indes & en l'Amerique*. Lyon.

Collins G.T. 1822. Correspondence dated 21 May 1822. National Archives of India, Home Department, Public.

Comte A. 1854. *Musée d'histoire naturelle*. Paris.

Conrad R. 1968. Die Haustiere in den frühen Kulturen Indiens. *Säugetierkundliche Mitteilungen* 16: 189–258.

Coomaraswamy A.K. 1910. *Indian drawings*. London.

Coomaraswamy A.K. 1912. *Indian drawings: second series, chiefly Rajput*. London.

Cooper T.T. 1870. Travels in northern Assam. *Field, the Country Gentleman's Magazine* 36 (1870-09-10): 242.

Cooper T.T. 1873. *The Mishmee Hills*. London.

Corbet G.B.; Hill J.E. 1986. *A world list of mammalian species*, 2nd edition. London.

Corbet G.B.; Hill J.E. 1992. *The mammals of the Indomalayan region: a systematic review*. Oxford.

Cork Advertiser 1842-11-09, 1842-11-21: Exhibition of rhinoceros in Cork, Ireland.

Coronation Durbar. 1911. *Loan exhibition of antiquities. An illustrated selection of the principal exhibits*. Delhi.

Correia G. 1862. *Lendas da India*, vol. 3 part 1. Lisboa.

Coryat T. 1616. *Travailler for the English Wits, and the good of his Kingdom*. London.

Costa A. Fontoura da. 1937. *Deambulations of the rhinoceros (ganda) of Muzafar, King of Cambaia, from 1514 to 1516*. Lisboa [also in Portuguese and French].

Coste C. 1946. Ancienne figurations du rhinocéros de l'Inde: à propos du frontispiece de l'ouvrage de Wilhelm Pison 'De Indiae utriusque re naturali et medica' edité a Amsterdam en 1658. *Acta Tropica* 3: 116–129.

Cotton H.A.E. 1923. The Daniells in India: an unpublished account of their journey from Calcutta to Garhwal in 1788–1789. *Bengal Past and Present* 25: 1–70.

Coulmann J.J. 1862. *Réminiscences* (XIX. Alfred Duvaucel). Paris.

Cox R. 1997. The Jacksonian unicorn. *The Quarto, Clements Library* 1 (7): 2–3.

Craig H. 1897. *Johnson's household book of nature*, vol. 2. New York.

Cranbrook, Earl of; Piper P.J. 2007. The Javan rhinoceros Rhinoceros sondaicus in Borneo. *Raffles Bulletin of Zoology* 55: 217–220.

Creative Conservation Alliance. 2016. *A preliminary wildlife survey in Sangu-Matamuhuri Reserve Forest, Chittagong Hill Tracts, Bangladesh*. Unpublished (online) report to Bangladesh Forest Department, Dhaka.

C.R.H. 1883. Hunting from boats during inundation in Burmah. *Illustrated Sporting and Dramatic News* 1883-03-31: 55–56.

Cronon W. 1995. The trouble with wilderness; or, getting back to the wrong nature. In: Cronon W., *Uncommon ground: rethinking the human place in nature*. New York, pp. 69–90.

Crooke W. 1899. The hill tribes of the central Indian hills. *Journal of the Anthropological Institute of Great Britain and Ireland* 28: 220–248.

Crooke W. 1926. *Religion and folklore of Northern India*. London.

Cummerbund C. 1860. *From Southampton to Calcutta*. London.

Cummins J.M. 1994. Indian drawings in the Brooklyn Museum. In: Poster A.G.; Canby S.R.; Chandra P.; Cummins J.M., *Realms of heroism: Indian paintings at The Brooklyn Museum*. New York, pp. 307–323.

Cunningham A. 1879. *The stupa of Bharhut*. London.

Cunningham A. 1882a. *Report of a tour in Bihar and Bengal in 1879–80 from Patna to Sunargaon*. Calcutta.

Cunningham A. 1882b. Report of a tour in the Punjab in 1878–79. *Reports of the Archaeological Survey of India* 14: 1–155.

Curtis B.R. 1878. *Dottings round the circle*. Boston.

Curzon G.N. 1906. *Lord Curzon in India, being a selection from his speeches as Viceroy & Governor-General of India 1898–1905*. London.

Curzon G.N. 1908. Address on game preservation in India. In: Minutes of proceedings at a deputation of the Society for the Preservation of the Wild Fauna of the Empire to the Right Hon. The Earl of Elgin, His Majesty's Secretary of State for the Colonies. *Journal of the Society for the Preservation of the Wild Fauna of the Empire* 3: 22–24.

Curzon G.N. 1925. *British government in India: the story of the viceroys and government houses*. London.

Curzon M.V. 1900. British Library, Curzon Collection, Photo 430/76. Print 39 from Autumn Tour of 1900: Lady Mary Victoria Curzon seated on a rhinoceros.

Cuvier F. 1824. Rhinocéros de Java. In: Geoffroy Saint-Hilaire É.; Cuvier F. 1819–1842. *Histoire naturelle des mammifères*, vol. 6, no. 45. Paris.

Cuvier F. 1825a. Rhinocéros de Sumatra. In: Geoffroy Saint-Hilaire É.; Cuvier F. 1819–1842. *Histoire naturelle des mammifères*, vol. 6, no. 47. Paris.

Cuvier F. 1825b. Croo Gibbon. In: Geoffroy St. Hilaire É.; Cuvier F. 1819–1842. *Histoire naturelle des mammifères*, vol. 6, no. 49. Paris.

Cuvier G. 1806. Sur les rhinocéros fossiles. *Annales du Muséum d'Histoire Naturelle* 7: 19–52.

Cuvier G. 1816. *Le règne animal*, vol. 1. Paris [dated 1817, available late 1816].

Cuvier G. 1821. Notice sur les voyages de MM. Diard et Duvaucel, naturalistes français, dans les Indes orientales et dans les îles de la Sonde. *Revue encyclopédique* 10 (June): 473–482.

Cuvier G. 1822. *Recherches sur les ossemens fossiles, où l'on rétablit les caractères de plusieurs animaux dont les révolutions du globe ont détruit les espèces*, vol. 2 part 1: *contenant l'histoire des rhinocéros, de l'élasmothérium, des chevaux, des sangliers et cochons, du daman, des tapirs et des animaux fossiles voisins des tapirs, et le résumé général de la première partie*. New [second] edition. Paris and Amsterdam.

Cuvier G. 1825. *Recherches sur les ossemens fossiles*, vol. 2 part 1. Third edition. Paris and Amsterdam.

Cuvier G. 1829. *Le règne animal*, 2nd edition, vol. 1. Paris.

Cuvier G. 1836. *Recherches sur les ossemens fossiles. Atlas*, vol. 1. Paris and Amsterdam.

Dadlani C. 2015. Transporting India: the Gentil Album and Mughal manuscript culture. *Art History* 38: 748–761.

Daftary F. 1990. *The Isma'ilis: their history and doctrines*. Cambridge.

Dahal A. 2019. Wildlife parts, trophies stolen from Singha Durbar. *MyRepublica* (online) 2019-08-25.

Dahmen-Dallapiccola A.L. 1976. *Indische Miniaturen: Malerei der Rajput-Staaten*. Baden-Baden.

Daily Graphic, London 1902-10-18, p. 316: Sport in Indian art: a rhinoceros hunt of two hundred years ago.

Daily Mirror, London 1922-02-15, p. 16: The Prince's Indian tour.

Daily Mirror, London 1961-03-03, pp. 1, 8–9: Jungle death trap.

Daily Pioneer, Patna 2013-03-20: Train crushes 10 year old rhino in Champaran, Uttar Pradesh, India.

Daisy, Princess of Pless. 1931. *From my private diary*. London.

Dales G.F. 1993. Figurines. In: Possehl G.L., *Harappan civilization, a recent perspective*. New Delhi, pp. 502–510.

Dando W.P. 1911. *Wild animals and the camera*. London.

Daniell W.; Caunter H. 1834a [1835a]. *The oriental annual, or scenes in India*, vol. 2 (for 1834, dated 1835 on title-page). London.

Daniell W.; Caunter H. 1834b. Wanderungen und Scenen in Indien. *Bilder-Magazin für Allgemeine Weltkunde* 1834: 321–331.

Daniell W.; Caunter H. 1835b. *Tableaux pittoresques de l'Inde*. Paris [8vo].

Darlong V.T.; Alfred J.R.B. 2002. *The state fauna of Tripura: an overview*. Calcutta: Zoological Survey of India (State Fauna Series 7).

Darwin C.R. 1844. *Geological observations on the volcanic islands visited during the voyage of H.M.S. Beagle*. London.

Darwin C.R. 1868. *The variation of animals and plants under domestication*, vol. 2. London.

Das A.K. 1995. Notes on an Aurangzeb period album of birds and animals. In: Goswamy B.N.; Bhatia U., *Indian painting. Essays in honour of Karl J. Khandalavala*. New Delhi, pp. 71–85.

Das A.K. 2012. *Wonders of nature: Ustad Mansur at the Mughal court*. Mumbai.

Das A.K. 2018a. The unicorn and the great Mughals. In: Divyabhanusinh; Das A.K.; Bose S., *The story of India's unicorns*. Mumbai, pp. 63–78.

Das A.K. 2018b. At large in art and culture. In: Divyabhanusinh; Das A.K.; Bose S., *The story of India's unicorns*. Mumbai, pp. 81–96.

Das C. 2002. A note on plaques depicting rhinoceros from Chandraketugarh, West Bengal. *Journal of the Department of Museology, Calcutta University* 1: 21–23.

Das C. 2008. Further discovery of a plaque representing Rhinoceros unicornis from Dum Dum Excavation, West Bengal: an observation. In: Sahai B. (ed.), *Recent researches in Indian art and iconography*. Delhi, pp. 93–95.

Das P. 2014. *Genetic diversity of Indian rhinoceros, Rhinoceros unicornis (Lin, 1758)*. Thesis presented to Gauhati University.

Das P.K. 1964. Jaldapara wild life sanctuary. In: *West Bengal forests: centenary commemoration volume*. Calcutta, pp. 251–258.

Das R. 2014. Contribution of Gauripur zamindar Raja Prabhat Chandra Barua. *IOSR Journal of Humanities and Social Science* 19: 56–60.

Das R.K. 2005. *Kaziranga: leaves from a forester's notebook*. Guwahati.

Das Gupta P.C. 1959. Early terracottas from Chandraketugarh. *Lalit Kalā* 6: 45–52.

Das Gupta P.C. 1961. *Paharpur and its monuments*. Calcutta.

Dasgupta S. 2017. Saving orphaned baby rhinos in India. www.mongabay.com (Mongabay Series: Asian Rhinos) 2017-03-17.

Datta A. 2004. Brian Hodgson and the mammals and other animals of Nepal. In: Waterhouse D.M., *The origins of Himalayan studies: Brian Houghton Hodgson in Nepal and Darjeeling 1820–1858*. London, pp. 154–171.

Daukes C. 1933. Capturing a Nepal rhino. *Field, the Country Gentleman's Magazine* 162 (1933-09-23): 769–770.

Daukes C. 1934. Capturing a Nepal rhino (price of rhino horn). *Field, the Country Gentleman's Magazine* 163 (1934-04-14): 802.

Davidson C.J.C. 1843. *Diary of travels and adventures in upper India: with a tour in Bundelcund, a sporting excursion in the kingdom of Oude, and a voyage down the Ganges*, 2 vols. London.

Davies B. 2005. *Black market: inside the endangered species trade in Asia*. San Rafael.

Davis H. 1940. *Land of the eye: a narrative of the labors, adventures, alarums and excursions of the Denis-Roosevelt Asiatic Expedition to Burma, China, India and the lost Kingdom of Nepal*. New York.

Dawson W.R. 1958. *The Banks letters*. London.

Day J.W. 1935. *King George V as a sportsman: an informal study of the first country gentleman in Europe*. London.

Debans C. 1881. L'enjeu du Guicowar. *Figaro, Supplément litteraire du Dimanche* 1881-06-18: 90.

Debbarma S. 1996. *Origin and growth of Christianity in Tripura*. New Delhi.

Debnath T.; Sarkar T. 2021. Bag size, record specimen and masculinity: an analysis of the hunting diary of Maharaja Nripendra Narayan of the princely state of Cooch Behar. *Sidho Kanho Birsha University Journal of History* 1 (1): 13–47.

De Brett E.A. 1909. *Central Provinces district gazetteers*, vol. 9: *Chhattisgarh feudatory states*. Bombay.

Debrett J. 1801. Oriental literature. *Asiatic Annual Register*, for the year 1800: 85–89.

Decinthel. 1887. Un beau coup … manqué (un drame dans la jungle). *Chasse Illustrée, journal des chasseurs et la vie à la campagne* 20 (40), 1887-10-01: 316–318, 1 plate (signed by G.A. Closs and G. Frenzel).

Deka R.K. 2005. Curzons, Miri and Kaziranga. In: Agarwalla R.P. et al., *Kaziranga centenary 1905–2005*. Guwahati, pp. 85–86.

Deka R.K. 2012. Curzons, Miri and Kaziranga. *The Rhino, Journal of the Kaziranga Wildlife Society* 18: 51–53.

Deka R.J.; Sarma N.K. 2015. Studies on feeding behaviour and daily activities of Rhinoceros unicornis in natural and captive condition of Assam. *Indian Journal of Animal Research* 49: 542–545.

Delessert A. 1843. *Souvenirs d'un voyage dans l'Inde executé de 1834 à 1839*. Paris.

Delhi Gazette. 1849. Capture of Oonah, the seat of Gooroo Beekram Singh. *Indian News and Chronicle of Eastern Affairs* 1849-02-02: 56.

Deloche J. 1984. *Les aventures de Jean-Baptiste Chevalier dans l'Inde Orientale (1752–1765)*. Paris.

Deloche J. 2008. *The adventures of Jean-Baptiste Chevalier in Eastern India, 1752–1765*. Guwahati.

DeMello M. 2021. *Animals and society: an introduction to human-animal studies*, 2nd edition New York.

Denis A. 1963. *On safari: the story of my life*. London.

Dennell R.; Coard R.; Turner A. 2006. The biostratigraphy and magnetic polarity zonation of the Pabbi Hills, northern Pakistan: an Upper Siwalik (Pinjor Stage) Upper Pliocene – Lower Pleistocene fluvial sequence. *Palaeogeography Palaeoclimatology Palaeoecology* 234: 168–185.

Denzau H.; Neumann-Denzau G.; Gerngross P. 2015. *Sundarbans atlas: Bangladesh forest compartment maps and gazetteer*. Dhaka.

Depping G. 1864. Les frères Schlagintweit et leur voyage dans l'Inde. *L'Illustration, Journal Universel* 43: 91–94, 151–154, 187–190.

Deraniyagala P.E.P. 1938. Some fossils from Ceylon, part II. *Journal of the Ceylon Branch of the Royal Asiatic Society of Great Britain and Ireland* 34: 231–239.

Deraniyagala P.E.P. 1946. Some mammals of the extinct Ratnapura fauna of Ceylon (part II). *Spolia Zeylanica* 24: 161–171.

Deraniyagala P.E.P. 1958. *The pleistocene of Ceylon*. Colombo.

Derby Daily Telegraph 1887-06-03: Manufacture of curiosities from rhino hide by Rowland Ward.

Derby Mercury 1897-08-04 (from *Daily Mail*, London): A weird jungle story [rhino in Ha-Flong, Assam].

Desai R.B.G.H. 1916. *Visitor's guide to Baroda*. Baroda.

Desmarest A.G. 1822. *Mammalogie, ou description des espèces des mammifères*. Paris.

Desmarest A.G. 1827. Rhinocéros. In: *Dictionnaire des sciences naturelles*. Strasbourg, vol. 45, pp. 351–365.

Devee S. [Suniti Devi]. 1916. *Bengal dacoits and tigers*. Calcutta.

Devkar V.L. 1957. Some recently acquired miniatures in the Baroda Museum. *Bulletin of the Baroda Museum and Picture Gallery* 12: 19–24.

Dey S. 2019. Terracotta plaques of Pilak in Tripura: a hidden cultural heritage. *Heritage: Journal of Multidisciplinary Studies in Archaeology* 7: 945–953.

Dey S.C. 1994a. Overview of rhinoceros. *Zoos' Print (Magazine of Zoo Outreach Organisation)* 9 (3–4): 2–5.

Dhavalikar M.K. 1982. Daimabad bronzes. In: Possehl G.L., *Harappan civilization: a contemporary perspective*. Warminster, pp. 362–366.

Dhavalikar M.K. 1993. Daimabad bronzes. In: Possehl G.L., *Harappan civilization, a recent perspective*, 2nd edition. New Delhi, pp. 422–426.

Dias N. 2012. Les trophées de chasse au Musée du Duc d'Orléans. In: Cros M.; Bondaz J.; Michaud M., *L'Animal cannibalisé: festins d'Afrique*. Paris, pp. 99–110.

Dickinson E.C.; Overstreet L.K.; Dowsett R.J.; Bruce M.D. 2011. *Priority! The dating of scientific names in ornithology: a directory to the literature and its reviewers*. Northampton.

Dikshit R.B.K.N. 1938. Excavations at Paharpur, Bengal. *Memoirs of the Archaeological Survey of India* 55: 56–72.

Dinerstein E. 2003. *The return of the unicorns: natural history and conservation of the greater one-horned rhinoceros*. Columbia.

Dinerstein E. 2006. A jungle in mourning: Nepal loses its leading defenders of large mammals. *Pachyderm*, Nairobi 41: 107–109.

Dinerstein E. 2013. *The kingdom of rarities*. Washington DC.

Dinerstein E. 2015. Greater one-horned rhinoceros, Rhinoceros unicornis. In: Johnsingh A.J.T. et al., *Mammals of South Asia*, vol. 2. Hyderabad, pp. 95–111.

Dinerstein E.; Jnawali S.R. 1993. Greater one-horned rhinoceros populations in Nepal. In: Ryder O.A., *Rhinoceros biology and conservation: Proceedings of an international conference, San Diego, U.S.A.*, pp. 196–207.

Dinerstein E.; MacCracken G.F. 1990. Endangered greater one-horned rhinoceros carry high levels of genetic variation. *Conservation Biology* 4: 417–422.

Dinerstein E.; Price L. 1991. Demography and habitat use by Indian rhinoceros in Nepal. *Journal of Wildlife Management* 55: 401–411.

Dinerstein E.; Shrestha S.; Mishra H. 1990. Capture, chemical immobilization, and radio-collar life for greater one-horned rhinoceros. *Wildlife Society Bulletin* 18: 36–41.

Dittrich L. 1997. The first painting of a Javan rhinoceros in Europe. *Contributions to Zoology* 67 (2): 151–154.

Dittrich L.; Rieke-Müller A. 1998. *Carl Hagenbeck (1844–1913): Tierhandel und Schaustellungen im Deutschen Kaiserreich*. Frankfurt am Main.

Divyabhanusinh. 1999. Hunting in Mughal painting. In: Verma S.P., *Flora and fauna in Mughal art*. Mumbai, pp. 94–108.

Divyabhanusinh. 2006. Junagadh State and its lions: conservation in Princely India, 1879–1947. *Conservation and Society* 4: 522–540.

Divyabhanusinh. 2018a. India's unicorns. In: Divyabhanusinh; Das A.K.; Bose S., *The story of India's unicorns*. Mumbai, pp. 13–30.

Divyabhanusinh. 2018b. Under the British and beyond. In: Divyabhanusinh; Das A.K.; Bose S., *The story of India's unicorns*. Mumbai, pp. 98–114.

Divyabhanusinh. 2018c. Resurrection of the unicorn. In: Divyabhanusinh; Das A.K.; Bose S., *The story of India's unicorns*. Mumbai, pp. 116–129.

Divyabhanusinh; Das A.K.; Bose S. 2018. *The story of India's unicorns*. Mumbai.

Dodwell E. 1838. Alphabetical list of the officers of the Indian Army. London.

Doley S. 2000. Action plan in Assam. In: *Report on the Regional Meeting for India and Nepal of* IUCN/SSC *Asian Rhino Specialist Group*. Doorn, pp. 44–47.

Dollman J.G. 1931. Mammalia. In: Pycraft W.P., *The standard natural history from amoeba to man*. London and New York.

Dominic Winter. 2020. Auction at Dominic Winter Auctioneers, Cirencester, 16 December 2020, lot 399: photographs of Nepali aristocracy. https://www.dominicwinter.co.uk/ [accessed 2021-01].

Donogh T.A. (signed T.A.D.). 1869. Extract from my journal in the Dooars. *Oriental Sporting Magazine* (n.s.) 2 (August): 557–564.

Donogh T.A. (signed T.A.D.). 1870. Extract from my Bhutan journal. *Oriental Sporting Magazine* (n.s.) 3 (December): 1567–1572.

Donogh T.A. (signed T.A.D.). 1872. My Bhutan journal of tiger-shooting &c. in the western dooars of Bhutan. *Oriental Sporting Magazine* (n.s.) 5 (June): 248–253.

Donogh T.A. (signed T.A.D.). 1873a. My Bhootan journal of tiger-shooting &c. in the western dooars of Bhootan. *Oriental Sporting Magazine* (n.s.) 6 (February): 63–67.

Donogh T.A. (signed T.A.D.). 1873b. My Bhootan journal of tiger-shooting &c. in the western dooars of Bhootan. *Oriental Sporting Magazine* (n.s.) 6 (May): 217–223.

Donogh T.A. (signed T.A.D.). 1873c. My Bhootan journal of tiger-shooting &c. in the western dooars of Bhootan. *Oriental Sporting Magazine* (n.s.) 6 (December): 575–582.

Donogh T.A. (signed T.A.D.). 1875. My Bhutan journal of tiger-shooting &c. in the western dooars of Bhutan. *Oriental Sporting Magazine* (n.s.) 8 (June): 215–222.

Doornbos W.G. 2006. Clara de Rinoceros. *Stad & Lande*, Groningen 15: 32–35.

Dorai F.; Low M.E.Y. 2021. *Diard & Duvaucel: French natural history drawings of Singapore and Southeast Asia*. Singapore.

Dorji D.R. 1989. *A brief religious, cultural and secular history of Bhutan*. New York.

Douglas N. 1976. *Karmapa: the black hat Lama of Tibet*. London.

Dow A. 1772. Plan for restoring Bengal to its former prosperity. In: *The history of Hindostan, from the death of Akbar, to the complete settlement of the Empire under Aurungzebe*. London, pp. cxv–cliv.

Downing J. 1844. The rhinoceros in his native wilds. *Rover, a weekly magazine of tales, poetry and engravings*, New York 2 (25): 385.

D'Oyly C. 1829. Rhinoceros hunting. Plate in *Indian Sports No.2*. London [copy in British Library, London X294. f.37].

Dozey E.C. 1922. *A concise history of the Darjeeling district since 1835 with itinerary of tours in Sikkim and the district*. Calcutta.

Drake-Brockman D.L. 1911. *District gazetteers of the United Provinces of Agra and Oudh*, vol. 27: *Mirzapur*. Allahabad.

Driver F. 2018. Face to face to Nain Singh: the Schlagintweit collections and their uses. In: MacGregor A., *Naturalists in the field* (Emergence of Natural History, vol. 2). Leiden, pp. 441–469.

Droulers C. 1932. *Le Marquis de Morès, 1858–1896*. Paris.

Drouot. 2019. [Auction catalogue] *Vendredi 8 Mars 2019: Bestiaire*. Paris.

Dryander J. 1796. *Catalogus bibliothecae historico-naturalis Josephi Banks*, vol. 2: *Zoologi*. Londini.

Du Bosch G. 1893. Une vente au Jardin Zoologique d'Anvers. *L'Illustration, Journal Universel* 102 (1893-09-10): 230–233.

Dubey M. 1992. Rock paintings of Pachmarhi hills. In: Lorblanchet M., *Rock art in the old world*. Darwin, pp. 131–135.

Dubois E. 1908. Das geologische Alter der Kendeng- oder Trinilfauna. *Tijdschrift van het Koninklijk Nederlandsch Aardrijkskundig Genootschap* (ser.2) 25: 1235–1270.

Dudley A. 2017. In Nepal, the rhino evokes national pride. www.mongabay.com 2017-08-01.

Duncan F.M. 1922. The Prince of Wales' collection at the 'zoo'. *Country Life* 51 (1922-07-01): 886–887.

Duncan F.M. 1937. On the dates of publication of the Society's Proceedings, 1859–1926. *Proceedings of the Zoological Society of London* 107: 71–84.

Dundee Evening News 1900-12-29: Rhino in Gwalior.

Dunlap T. 1988. *Saving America's wildlife*. Princeton.

Dunlop R.H.W. 1860. *Hunting in the Himalaya*. London.

Durand E. 1911. *Rifle, rod and spear in the East, being sporting reminiscences*. London.

Durand G. 1893. A panic at an Indian fair: a runaway elephant in the sweetmeat bazaar at Sonepore. *Graphic, an Illustrated Weekly Newspaper* 48 (1893-09-09): 320.

Durand H.M. 1883. *The life of Major-General Sir Henry Marion Durand of the Royal Engineers*, vol. 1. London.

Dutt L. 1863. Proceedings of the Asiatic Society, May 1863. *Journal of the Asiatic Society of Bengal* 32: 288–290.

Dutta A.K. 1991. *Unicornis: the great Indian one-horned rhinoceros*. Delhi.

Dutta A.N. 2017. *Monograph on golden langur Trachypithecus geei*. Assam.

Dutta D.K. 2018a. Studies on greater one horned rhinoceros behaviour and ecology, with special references to wild to wild translocated rhinoceros: a review. *Indian Forester* 144: 922–928.

Dutta D.K. 2018b. *A study on behaviour and habitat preferences of translocated rhinos Rhinoceros unicornis*. Thesis presented to Gauhati University, Assam. [See *Pachyderm* 64: 92-106 (2023)].

Dutta D.K.; Bora P.J.; Sharma R.; Sarmah A.; Sharma A.; Kumar J.; Dekaraja K.S.; Saikia R.B. 2021. The greater one-horned rhino (Rhinoceros unicornis) behaviour during high floods at Kaziranga National Park and the Burhachapori Wildlife Sanctuary, Assam. *Pachyderm*, Nairobi 62: 63–73.

Dutta D.K.; Mahanta R. 2015. A study on the behavior and colonization of translocated greater one-horned rhinos Rhinoceros unicornis during 90 days from their release at Manas National Park, Assam India. *Journal of Threatened Taxa* 7: 6864–6877.

Dutta D.K.; Sharma A.; Mahanta R.; Swargowari A. 2017. Behavior of post released translocated greater one-horned rhinoceros (Rhinoceros unicornis) at Manas National Park, Assam, India. *Pachyderm*, Nairobi 58: 58–66.

Dutta S. 2008. Status of rhino and horn stockpiles in Pabitora Wildlife Sanctuary. In: Syangden B. et al., *Report on the regional meeting for India and Nepal IUCN/SSC Asian Rhino Species Group, March 5–7, 2007*, pp. 9–10.

Dutt-Mazumdar S. 1954. The vanishing fauna of North Bengal. *Journal of the Bengal Natural History Society* 26: 156–161.

Duvaucel A. 1824. Les voyages dans l'Inde, ayant pour objet plus particulier, l'histoire naturelle (troisième notice). *Journal Asiatique* 5: 277–285.

Dwivedi S. 2002. *Die Maharadschas und die indischen Fürsten-staaten*. Frechen: Komet (originally published as: *The Maharaja and the Princely States of India*. New Delhi, 1999).

Dyott G.M. 1923a. After royal game in Nepal: rhinoceros-shooting. *Illustrated London News* 163 (1923-09-08): 432–435.

Dyott G.M. 1923b. Shooting swamp deer in India (Gond). *Illustrated London News* 163 (1923-08-25): 360–364.

Dyott G.M. 1924. Los rinocerontes des Nepal [Vernay-Faunthorpe expedition]. *Iberica, El Progeso de las Ciencias y de sus Aplicaciones* 21 (1924-04-12): 230–231.

E.L.; Meyerheim P. 1874. Aus dem Zoologischen Garten zu Berlin [Elephant Pagoda]. *Über Land und Meer, Allgemeine Illustrirte Zeitung* 31 (22): 426, figure p. 421.

Eardley-Wilmot S. 1910. *Forest life and sport in India*. London.

Earls Court. 1895. *Official catalogue of the Empire of India exhibition: Earl's Court, London, S.W., 1895*. London.

Eastmure G.B. 1949. Indian one-horned rhinoceros. *Field, the Country Gentleman's Magazine* 194 (1949-12-17): 891–892.

Eastwick E.B. 1857. *Autobiography of Lutfullah, a Mohamedan gentleman*, 2nd edition. London.

Ebeling K. 1973. *Ragamala painting*. Basel, Paris, New Delhi.

Eden A. 1865. Report on a visit to Bootan (1863–1864). *Parliamentary Papers* 1865 (47), Papers relating to Bootan: 173–255.

Eden E. 1866. *Up the country: letters written to her sister from the upper provinces of India*, vol. 1. London.

Eden E. 1872. *Letters from India, by the Hon. Emily Eden*, edited by her niece [Eleanor Eden], vol. 1. London.

Edinburgh Evening News 1900-12-26: Arrival of Lord Curzon in Junagadh.

Edwardes M. 1960. *The Orchid House: splendours and miseries of the kingdom of Oudh, 1827–1857*. London.

Edwards G. 1758. *Gleanings of natural history*, vol. 5. London.

Edwards J. 1996. *London Zoo from old photographs 1852–1914*. London.

Edwards J. 2012. *London Zoo from old photographs 1852–1914*, 2nd edition. London.

Edwards O. 1980. Captain Thomas Williamson of India. *Modern Asian Studies* 14: 673–682.

E.G. 1872. Wild beasts for sale (by Jamrach). *Notes and Queries* (ser.4) 9: 207.

Egen E. 2011. Memories of big game hunting of her father Victor Egen. https://solitaryw.blogspot.com/search/label/Victor [accessed April 2023].

Egerton F. 1852. *Journal of a winter's tour in India: with a visit to the court of Nepaul*, vol. 1. London.

Egerton W. 1880. *An illustrated handbook of Indian arms, exhibited at the India Museum*. London.

Egerton W. 1896. *A description of Indian and Oriental armour*. London.

Ehlers O.E. 1894. *An indischen Fürstenhöfen*, 2 vols. Berlin.

Ehnbom D.J. 1985. *Indian miniatures: the Ehrenfeld Collection*. New York.

Ehnbom D.J. 1988. *Indische Miniaturen: die Sammlung Ehrenfeld*. Stuttgart, Zürich.

Ellerman J.R.; Morrison-Scott T.C.S. 1951. *Checklist of palaearctic and Indian mammals 1758 to 1946*. London.

Ellerman J.R.; Morrison-Scott T.C.S. 1966. *Checklist of palaearctic and Indian mammals 1758 to 1946*, 2nd edition. London.

Elliot H.M. 1872. *The history of India as told by its own historians*, vol. 4: *The Muhammadan period*. London.

Elliot W. 1874. Memoir of Dr. T.C. Jerdon. *History of the Berwickshire Naturalists' Club* 7: 143–151.

Ellis S.; Talukdar B.K. 2019. Rhinoceros unicornis, greater one-horned rhino. In: *IUCN Red list of threatened species*. <www.iucnredlist.org>.

Ellison B.C. 1922. H.R.H. The Prince of Wales' shoots in India in 1921 and 1922, part 1. *Journal of the Bombay Natural History Society* 28: 675–697.

Ellison B.C. 1925. *H.R.H. The Prince of Wales's sport in India*. London.

Ellison J. 1873. Map of the Sundarbans showing the extent of available land, of land granted under the rules of the 24th September 1853, or held in fee simple, permanently settled estates. Corrected in 1917. digital: http://purl.umn.edu/246294 [accessed April 2023].

Elphinstone Game Book. 1896. Album arranged for Sidney Herbert Buller-Fullerton-Elphinstone. In private ownership.

Elwes H.J. 1930. *Memoirs of sports, travel and natural history*. London.

Emslie R.H.; Milledge S.; Brooks M.; Strien N.J. van; Dublin H.T. 2007. African and Asian rhinoceroses: status, conservation and trade. In: *CITES Secretariat CoP 14 Doc.54, Annex 1*, pp. 6–22.

E.M.W. 1923. The Singpho rhinoceros. *Field, the Country Gentleman's Magazine* 142 (1923-12-06): 805.

Englishman's Overland Mail 1871-01-18: Governor of Bombay in Baroda.

Englishman's Overland Mail 1871-12-27: Steamer Petersburg from Calcutta to London.

Englishman's Overland Mail 1880-10-20, p. 9: Komaragaon, Assam (adventure with rhinoceros).

Englishman's Overland Mail 1881-10-31: Assistant to the Superintendent of the Indian Museum.

Englishman's Overland Mail 1882-03-07, p. 10: The Lieutenant Governor in Kuch Behar.

Englishman's Overland Mail 1882-06-17, p. 3: The Zoological Gardens [Calcutta]: A new arrival.

Englishman's Overland Mail 1885-03-24: Cockburn employed in Opium Department.

Englishman's Overland Mail 1887-05-03, p. 10: A shooting party in North Kamrup.

Englishman's Overland Mail 1889-02-05, p. 18: Domestic occurrences – Births.

Englishman's Overland Mail 1889-02-05, p. 20: Birth of a rhinoceros.

Englishman's Overland Mail 1890-02-25: Shoot of Maharaja of Cooch Behar.

Englishman's Overland Mail 1890-03-18: Rhinos shot in Cooch Behar.

Englishman's Overland Mail 1890-10-29: Shipment of animals to Rutledge at Entally.

Englishman's Overland Mail 1891-02-18: Shoot of Maharaja of Cooch Behar.

Englishman's Overland Mail 1891-03-18: Camp life in Cooch Behar.

Englishman's Overland Mail 1891-11-25: Visit to Halliday and Bulcherry Islands.

Englishman's Overland Mail 1892-02-17: Hunting on Bulcherry Island.

Englishman's Overland Mail 1892-02-24, p. 14: Sport on Mutlah River.

Englishman's Overland Mail 1892-03-02, p. 5: Sport in Cooch Behar with the Viceroy.

Englishman's Overland Mail 1893-03-08: Death of rhinoceros in Lal Bagh Park.

Englishman's Overland Mail 1893-05-03: Lieutenant-General Sir Charles Elliott in Cooch Behar.

Englishman's Overland Mail 1894-01-31: Return home of Lord Beresford.

Englishman's Overland Mail 1894-03-14: Shoot of the Maharaja of Cooch Behar.

Englishman's Overland Mail 1894-04-04: Kheddah operation in Julpaiguri.

Englishman's Overland Mail 1894-11-07: Death of Oscar Fraser.

Englishman's Overland Mail 1895-03-20: Hunting party of Charles Alfred Elliott.

Englishman's Overland Mail 1895-04-10: Big shoot in Assam.

Englishman's Overland Mail 1895-12-18: Sport in Cooch Behar.

Englishman's Overland Mail 1898-10-06: The Archaeological Survey of India.

Englishman's Overland Mail 1901-04-25: Shoot of Maharaja of Cooch Behar.

Englishman's Overland Mail 1901-04-25: P.A. Herzog in Nepal.

Englishman's Overland Mail 1906-04-05: Arrival of animals from Nepal in Calcutta.

Englishman's Overland Mail 1909-12-09: Arrival of pair of rhinos in Calcutta Zoo.

Englishman's Overland Mail 1910-09-01: Meeting of the Darrang Game Association.

Englishman's Overland Mail 1912-01-25: Trophies from Nepal in Mysore [also: *Civil & Military Gazette*, Lahore, 1912-01-25].

Englishman's Overland Mail 1914-03-05: Gordon Fraser of Shamshernagar.

Englishman's Overland Mail 1916-03-10, p. 3: Hunting shoot by the Maharaja of Cooch Behar.

Enright K. 2008. *Rhinoceros*. London: Reaktion Books.

Enright K. 2010. Why the rhinoceros doesn't talk: the cultural life of a wild animal in America. In: Brantz D., *Beastly natures: animals, humans, and the study of history*. Charlottesville.

Enright K. 2012. None tougher. *Antennae: the Journal of Nature in Visual Culture* 23: 95–97.

Era, London 1875-12-05: Advertisement offering a rhinoceros for sale.

Errol Gray J. 1893. *Diary of a journey to the Bor Khamti Country, 1892–3*. Simla.

Ettinghausen R. 1950. *The unicorn*. Washington DC.

Everett I. 1853. *Observations on India, by a resident there many years*. London.

Examiner 1875-12-18, pp. 1408–1410: Englishmen in India.

Eyries M. 1824. Duvaucel: voyage dans le Silhet. *Abrégé des Voyages Modernes depuis 1780 jusqu'à nos jours* 14: 35–47.

Ezra A. 1917. Big-game shooting in Cooch Behar, Assam, and the Bhutan Duars, India. *Proceedings of the Zoological Society of London* 1917: 210.

Fabb J. 1986. *The British Empire from photographs: India*. London.

Fahlstedt K.K. 2017. The cinematic fauna of Bengt Berg. *Journal of Scandinavian Cinema* 7 (3): 243–251.

Fairservis W.A. Jr. 1982. Allahdino: an excavation of a small Harappan site. In: Possehl G.L., *Harappan civilization: a contemporary perspective*. Warminster, pp. 107–112.

Falconer H. 1837. Note on the occurrence of fossil bones in the Sewalik Range, eastward of Hardwar. *Journal of the Asiatic Society of Bengal* 6 (3): 233–234.

Falconer H. 1845. Description of some fossil remains of Dinotherium, Giraffe, and other Mammalia, from the Gulf of Cambay, western coast of India, chiefly from the collection presented by Captain Fulljames, of the Bombay Engineers, to the Museum of the Geological Society. *Quarterly Journal of the Geological Society* 1: 356–372.

Falconer H. 1868a. Introduction. In: Murchison C., *Palaeontological memoirs and notes of the late Hugh Falconer*. London, pp. 1–29.

Falconer H. 1868b. On the species of fossil Rhinoceros found in the Sewalik Hills. In: Murchison C., *Palaeontological memoirs and notes of the late Hugh Falconer*. London, pp. 157–185.

Falconer H.; Cautley P.T. 1847. *Fauna antiqua sivalensis, being the fossil zoology of the Sewalik Hills, in the North of India*. Illustrations, part VIII: Suidae and Rhinocerotidae. London.

Falconer H.; Walker H. 1859. *Descriptive catalogue of the fossil remains of vertebrata from the Sewalik Hills, the Nerbudda, Perim Island, etc. in the museum of the Asiatic Society of Bengal*. Calcutta.

Fane H.E. 1842. *Five years in India*, vol. 1. London.

Far East: a Monthly Illustrated Journal 1875 (vol. 6 no. 10), pp. 221–228: Account of the hill country and tribes under the rule of Munnipore.

Faunthorpe J.C. 1923. After big game in India, Anglo-American expedition: stalking rhinoceros on foot. *Times*, London 1923-09-07: 11–12, 14.

Faunthorpe J.C. 1924a. Jungle life in India, Burma, and Nepal: some notes on the Faunthorpe-Vernay Expedition of 1923. *Natural History*, New York 24: 174–198.

Faunthorpe J.C. 1924b. The disappearance of wildlife in India. *Natural History*, New York 24: 204–207.

Faunthorpe J.C. 1924c. Hunting big game in India. *Wide World Magazine* 53: 212–221.

Faunthorpe J.C. 1930. Shikar in India: hunting big and small game. *Times*, London 1930-02-18: 55.

Faust I. 1976. Jungfer Clara im Ballhof. *Zoofreund* (Zoo Hannover) 18: 2–3.

Faust I. 2003. *Zoologische Einblattdrucke und Flugschriften vor 1800*, vol. 5: *Unpaarhufer*. Unter Mitarbeitung von Klaus Barthelmess und Klaus Stopp. Stuttgart.

Fawcus L.R. 1943. *Report of the Game and Game Fishes Preservation Committee on the existing species of game in Bengal*. Alipore.

Fayrer J. 1870. *H.R.H. the Duke of Edinburgh in India*. Calcutta.

Fayrer J. (signed M.D.). 1872. Arrival of a two-horned rhinoceros. *Land and Water* 13 (1872-01-20): 47.

Fayrer J. (signed M.D.). 1873. Journal of a tiger-shooting expedition in Oude in 18-. *Oriental Sporting Magazine* (n.s.) 6 (63, March): 111–117.

Fayrer J. (signed M.D.). 1875. Journal of a tiger-shooting expedition in Oude in 18-. *Oriental Sporting Magazine* (n.s.) 8 (91, July): 274–281.

Fayrer J. 1875. *The royal tiger of Bengal, his life and death*. London.

Fayrer J. 1879. *Notes of the visits to India of their Royal Highnesses the Prince of Wales and the Duke of Edinburgh 1870–1875-6*. London.

Fayrer J. 1900. *Recollections of my life*. Edinburgh, London.

Feith J.A. 1915. De Bengaalse Sichterman. *Groningse Volksalmanak* 1914: 14–74.

Feudge F.R. 1880. *India: the gorgeous East with richest hand showers on her kings barbaric pearl and gold*. Boston.

Figaro, Paris 1886-11-24: Josef Jérôme Bonaparte in Jaipur.

Figuier L. 1869. *Les mammifères*. Paris.

Figuier L. 1870. *Mammalia, their various forms and habits popularly illustrated by typical species*. London.

Figuier L.; Wright E.P. 1883. *Mammalia, their various forms and habits popularly illustrated by typical species*, 3rd edition. London.

Filoux A.; Suteethorn V. 2018. A late Pleistocene skeleton of Rhinoceros unicornis (Mammalia, Rhinocerotidae) from western part of Thailand (Kanchanaburi). *Geobios* 51: 31–49.

Finch W. 1625. Travels in India. In: Purchas S., *His pilgrimage, or relations of the world*. London, vol. 1, pp. 414–440.

Finn F. 1901. The cage-birds of Calcutta. *Ibis* 43: 423–444.

Finn F. 1906a. *Guide to the Prince of Wales collection of Indian animals, Zoological Gardens, Regents Park*. London.

Finn F. 1906b. Latest notes from the Zoo – a baby rhinoceros. *Country-Side* 3 (1906-06-23): 103.

Finn F. 1929. *Sterndale's Mammalia of India*: a new and abridged edition, thoroughly revised. Calcutta, Simla.

Finotello P.; Agnelli P. 2015. La collezione dei trofei di caccia di Vittorio Emanuele Duca di Savoia-Aosta, Conte di Torino, al Museo di Storia Naturale dell'Universita degli Studi di Firenze. *Museologia Scientifica* (n.s.) 9: 13–29.

Fischer von Waldheim G. 1814. *Zoognosia tabulis synopticis illustrata*, vol. 3. Mosquae.

Fisher A.H. 1912. *Through India and Burmah with pen and brush*. London.

Fisher J.; Simon N.; Vincent J. 1969. *The red book: wildlife in danger*. London.

Fitter R. 1968. *Vanishing wild animals of the world*. London.

Fleming R.L. 1969. Nepal fauna and flora: comments on present status. *IUCN Bulletin* 2 (13): 108.

Flex O. 1873. *Pflanzerleben in Indien: kulturgeschichtliche Bilder aus Assam*. Berlin.

Flower S.S. 1908. *Notes on zoological collections visited in Europe, 1907*. Cairo (Zoological Gardens, Giza, Special report no. 2).

Flower S.S. 1914. *Report on a zoological mission to India in 1913*. Cairo (Zoological Service, Publication no 26).

Flower S.S. 1931. Contributions to our knowledge of the duration of life in vertebrate animals, v. Mammals. *Proceedings of the Zoological Society of London* 1931: 145–234.

Flower W.H. 1876. On some cranial and dental characters of the existing species of rhinoceroses. *Proceedings of the Zoological Society of London* 1876-05-16: 443–457.

Flower W.H. 1878. On the skull of a rhinoceros (R. lasiotis, Scl.?) from India. *Proceedings of the Zoological Society of London* 1878-06-04: 634–636.

Flower W.H. 1884. *Catalogue of specimens illustrating the osteology and dentition of vertebrated animals, recent and extinct, contained in the Museum of the Royal College of Surgeons of England* (assisted by John George Garson), vol. 2. London.

Förschler S. 2017. Das Horn des Nashorns: Objekt der Parzellierung, Ästhetisierung und Wissensgenese. In: Cremer A.C.; Mulsow M., *Objekte als Quellen der historischen Kulturwissenschaften: Stand und Perspektiven der Forschung*. Köln, pp. 195–207.

Foose T.J.; Strien N.J. van. 1995a. Population and distribution figures for all rhino species. *Asian Rhinos* (newsletter of the Asian Rhino Specialist group) 1: 24.

Foose T.J.; Strien N.J. van. 1995b. Population & distribution figures for all rhino species. *Asian Rhinos* 2: 16.

Foose T.J.; Strien N.J. van. 1997. *Asian rhinos: status survey and conservation action plan*, new edition. Gland.

Foose T.J.; Strien N.J. van. 2000. Asian rhino areas and estimates. *Asian Rhinos* 3: 19–21.

Foote R.B. 1874. Rhinoceros deccanensis, a new species discovered near Gokak, Belgaum district. *Memoirs of the Geological Survey of India, Palaeontologia Indica*, Series 10, vol. 1 part 1: 1–17.

Foote R.B. 1876. The geological features of the South Mahratta country and adjacent districts. *Memoirs of the Geological Survey of India* 12 (1): 1–176.

Foote R.B. 1884. Work at the Billa Surgam Caves. *Records of the Geological Survey of India* 17: 200–208.

Forberg C. 2017. What does the Emperor of India look like? European representations of Indian rulers (1650–1740). In: Malekandathil P., *The Indian Ocean in the making of early modern India*. London, pp. 175–212.

Forge and Lynch. 2023. [auction catalogue] *Court painting from India*. New York: Oliver Forge and Brendan Lynch, auction 16–24 March 2023.

Forrest C.R. 1824. *A picturesque tour along the river Ganges and Jumna, in India*. London.

Fortescue J.W. 1912. *Narrative of the visit to India of their Majesties King George V and Queen Mary and of the Coronation Durbar held at Delhi 12th December 1911*. London.

Foster W. 1921. *Early travels in India 1583–1619*. London.

Foster W. 1951. *British artists in India, 1760–1820*. London.

Francfort H.P. 1983. Excavations at Shortughai in Northeast Afghanistan. *American Journal of Archaeology* 87: 518–519.

Franzius W. 1612. *Historia animalium sacra*. Witebergae.

[Franz Ferdinand, Archduke of Austria]. 1895. *Tagebuch meiner Reise um die Erde, 1892–1893*, vol. 1. Wien.

Fraser O.L. 1875. Note on a partially ossified nasal septum in Rhinoceros sondaicus. *Journal of the Asiatic Society of Bengal* 44: 10–12.

Fraser W.M. 1935. *The recollections of a tea planter*. London.

Freedman A.M. 2003. *Mala ke manke, 108 Indian drawings: from private collection of Subhash Kapoor*. New York.

Freytag F.G. 1747. *Rhinoceros e veterum scriptorum monimentis descriptus*. Lipsiae.

Friel R. 1951. *Assam district gazetteers*, Supplement to volume VI: *Nowgong*. Shillong.

Friend 1835-03-14, pp. 177–178: Splendours of the East: Oudh.

Frisch J.L. 1775. *Das Natur-System der vierfüssigen Thiere*. Glogau.

Frost J. 1856. *Grand illustrated encyclopedia of animated nature*. New York and Auburn.

Fryer J. 1698. *A new account of East-India and Persia, in eight letters being nine years travels, begun 1672 and finished 1682*. London.

Fudge E. 2004. *Animal*. London.

Funk M. 2014. *Windfall: the booming business of global warming*. London.

Furniss H. 1885. The meeting of the Zoological Society, Hanover Square. *Punch*, London 88 (1885-05-23): 251.

Gabriel V.; Luard C.E. 1914. *The historical record of the imperial visit to India, 1911, compiled from the official records under the orders of the Viceroy and Governor-General of India*. London.

Gaebler B. 1883. Der Zoologische Garten zu Berlin. *Zoologische Garten* 24: 240–248.

Gaekwad F. 1980. *The palaces of India*. London.

Gailer J.E. 1839. *Wunderbuch für die reifere Jugend: eine Gallerie der merkwürdigsten und interessantesten Werke der Natur und Kunst in Erzählungen und Bildern zur Belehrung und Unterhaltung* [first edition]. Stuttgart.

Gairola H.K. 1987. Study of parturition in the great Indian rhinoceros. *Indian Veterinary Medicine Journal* 11: 185–190.

Galbraith J.K. 1969. *Ambassador's journal: a personal account of the Kennedy years*. London.

Gallovidian. 1837. Rhinoceros hunting. *Bengal Sporting and General Magazine* 9: 275–277.

Gangoly O.C. 1961. *Critical catalogue of miniature paintings in the Baroda Museum*. Baroda.

Ganguli M. 1984. *A pilgrimage to the Nagas*. Oxford.

Ganguly M.M. 1912. *Orissa and her remains: ancient and mediaeval (District Puri)*. Calcutta.

Gardner N. 2016. *Illustrated pursuits: W.S. Sherwill in India*. Calcutta.

Garfagnini M.D. 1992. *Sigismondo Tizio, Historiae Senenses*. Roma.

Garrod A.H. 1877a. On some points in the visceral anatomy of the rhinoceros of the Sunderbunds (Rhinoceros sondaicus). *Proceedings of the Zoological Society of London* 1877-11-06: 707–711.

Garrod A.H. 1877b. On the Taenia of the rhinoceros of the Sunderbunds. *Proceedings of the Zoological Society of London* 1877-11-20: 788–789.

Gass F.T. 1971. Ill met, the Indian rhino. *Field, the Country Gentleman's Magazine* 238 (1971-08-05): 326.

Gay J.D. 1876. *The Prince of Wales in India; or, from Pall Mall to the Punjaub*. London.

Gee E.P. 1948a. The great one-horned rhinoceros in Kaziranga Sanctuary. *Journal of the Bengal Natural History Society* 23: 63–65.

Gee E.P. 1948b. The great Indian one-horned rhinoceros: how Mohan came from Kaziranga, Assam, to Whipsnade. *Zoo Life* 3 (4): 106–107.

Gee E.P. 1949a. Home of India's rhino. *Journal of the Society for the Preservation of the Fauna of the Empire* 60: 25–27.

Gee E.P. 1949b. The Indian rhinoceros. *Loris* 5 (2): 73–78.

Gee E.P. 1950a. The Indian rhino faces extinction. *Country Life* 107 (1950-06-09): 1731–1735.

Gee E.P. 1950b. Wild life reserves in India: Assam. *Journal of the Bombay Natural History Society* 49: 81–89.

Gee E.P. 1952a. The great Indian one-horned rhinoceros. *Oryx* 1: 224–227.

Gee E.P. 1952b. Catching a bride for 'Mohan': how the rhinoceros is trapped and transported in Assam. *Zoo Life* 7: 102–105.

Gee E.P. 1953a. The life history of the great Indian one-horned Rhinoceros (R unicornis Linn.). *Journal of the Bombay Natural History Society* 51: 341–348.

Gee E.P. 1953b. Information wanted on the Indian rhinoceros. *Journal of the Bengal Natural History Society* 26: 123–126.

Gee E.P. 1953c. Further observations on the great Indian one-horned rhinoceros (R. unicornis Linn.). *Journal of the Bombay Natural History Society* 51: 765–772.

Gee E.P. 1954. The most famous rhino. *Natural History*, New York 63: 366–369.

Gee E.P. 1955. Great Indian one-horned rhinoceros (R. unicornis Linn) cow with (presumptive) twin calves. *Journal of the Bombay Natural History Society* 53: 256–257.

Gee E.P. 1958. Four rare Indian animals. *Oryx* 4: 353–358.

Gee E.P. 1959a. The great Indian rhinoceros (R. unicornis) in Nepal: Report of a fact-finding survey, April–May 1959. *Journal of the Bombay Natural History Society* 56: 484–510.

Gee E.P. 1959b. Report on a survey of the rhinoceros areas of Nepal, March and April 1959. *Oryx* 5: 53–85.

Gee E.P. 1959c. The Great Indian rhinoceros. *Wildlife Observer* no. 16 (1959-01): 10–12.

Gee E.P. 1962. The management of India's national Parks and wildlife sanctuaries, part IV. *Journal of the Bombay Natural History Society* 59: 453–485.

Gee E.P. 1963a. Armour-plated rhino. *Animals* 2: 510–512.

Gee E.P. 1963b. Wildlife in India. *Field, the Country Gentleman's Magazine* 222 (1963-11-21): 976.

Gee E.P. 1963c. Report on a brief survey of the wild life resources of Nepal, including the rhinoceros, March 1963. *Oryx* 7: 67–76.

Gee E.P. 1964a. *The wild life of India*. London.

Gee E.P. 1964b. Chased by a rhino! *Animals* 3: 626–631.

Gee E.P. 1964c. Obituary: E.O. Shebbeare. *Oryx* 7: 274.

Geer A. van der. 2008. *Animals in stone: Indian mammals sculptured through time.* Leiden (Handbook of Oriental Studies. Section two, vol. 21).

Gentil J.B.J. 1770. *Empire Mogol divisé en 21 soubas ou Gouvernements tiré de differents écrivains du pais à Faisabad MDCCLXX.* Single copy preserved in the British Library, India Office Library, London (Add.MS. Or. 4039).

Gentil J.B.J. 1774. Album with scenes of India. Copy in Victoria & Albert Museum, London (Atlas IS25).

Gentil J.B.J. 1822. *Mémoires sur l'Hindoustan ou Empire Mogol.* Paris.

Gentlewoman 1901-09-21: Travel of Robert Baker-Carr.

Geoffroy St. Hilaire I. 1842. *Description des collections de Victor Jacquemont: mammifères et oiseaux.* Paris.

George W. 1969. *Animals and maps.* London.

Georges É. 1996. *Les petits palais du Rajasthan.* Paris.

Geraads D.; Cerdeño E.; Fernandez D.G.; Pandolfi L.; Billia E.; Athanassiou A.; Albayrak E.; Codrea V.; Obada T.; Deng T.; Tong H.; Lu X.; Pícha S.; Marciszak A.; Jovanovic G.; Becker D.; ZervanovaJ.;ChaïdSaoudiY.;BaconA.M.;SévêqueN.;PatnaikR.; Brezina J.; Spassov N.; Uzunidis A. 2021. A database of Old World Neogene and Quaternary rhino-bearing localities. http://www.rhinoresourcecenter.com/about/fossil-rhino -database.php [accessed April 2023].

Gessner C. 1551. *Historia animalium lib. I de quadrupedibus viviparis.* Tiguri.

Gessner C. 1563. *Thierbüch: das ist ein kurtze Beschreybung aller vierfüssigen Thieren, so auff der Erden und in Wassern wonend, sampt jrer waren Conterfactur.* Zürych.

Gessner C. 1620. *Historia animalium lib. I de quadrupedibus viviparis*, 2nd edition. Francofurti.

Ghatak S.; Bhutia P.T.; Mitra A.; Raha A.K. 2016. Time series study of rhino habitat and its impact on rhino population in Gorumara National Park through remote sensing technology. *International Journal of Environment, Agriculture and Biotechnology* 1: 328–335.

Ghose J. 1874. *The Rájopákhyán: or, history of Kooch Behar.* Calcutta.

Ghose M. 2015. Nathdwara: a personal journey. In: Ghose M. et al., *Gates of the Lord: the tradition of Krishna paintings.* Chicago, pp. 14–25.

Ghosh A. 1975. *Jaina art and architecture*, vol. 3. New Delhi.

Ghosh A.K. 1945. The Indian fauna during 1943–44. *Current Science* 14: 240.

Ghosh D.; Siddiqui S.S. 2017. Garden Reach: The forgotten kingdom of Nawab Wajid Ali Shah. Internet blog: The Concrete Paparazzi.

Ghosh M.; Saha U.; Roy S.; Talukdar B.K. 1992. Subrecent remains of great one-horned rhinoceros from southern West Bengal, India. *Current Science* 62: 577–580.

Ghosh M.; Saha K.D.; Saha U.; Roy S.; Talukder B. 1992. Archaeological remains from West Bengal, India. In: *Fauna of West Bengal*, part 2. Calcutta, pp. 349–381.

Ghosh S. 2015. Manas National Park. In: Ellis S. et al., *Indian Rhino Vision 2020 Population Modeling Workshop final report.* Fort Worth, pp. 12–15.

Ghosh S.C. 1970. *The social condition of the British Community in Bengal: 1757–1800.* Leiden.

Gilbey W. 1900. *Animal painters of England from the year 1650*, vol. 2. London.

Giorgi A.A. 1762. *Alphabetum Tibetanum missionum apostolicarum commodo editum.* Romae.

Giovio P. 1559. *Dialogo dell'imprese militari et amorose.* Lione.

Girdlestone H. 1864. *Genesis: its authenticity and authority discussed, the first eleven chapters.* London.

Gissibl B.; Hohler S.; Kupper P. 2012. *Civilizing nature: national parks in global historical perspective.* New York.

Gladstone H.S. 1930. *Record bags and shooting records.* New and enlarged edition. London.

Glasenapp H. von. 1925. *Der Jainismus: Eine indische Erlösungsreligion. Nach den Quellen dargestellt.* Berlin.

Glaubrecht M.; Haffer J. 2010. Classifying nature: Constantin W.L. Gloger's (1803–1863) tapestry of a "Natural System of the Animal Kingdom." *Zoosystematics and Evolution* 86: 81–115.

Glendevon J. 1971. *The Viceroy at bay: Lord Linlithgow in India 1936–1943.* London.

Globe, London 1887-06-02: The Jungle display of Rowland Ward in London.

Globe, London 1895-04-27: A Maharajah's tiger hunt.

Globe, London 1903-01-30: Indian rhinoceros at South Kensington.

Globe, London 1904-02-25, p. 8: Big game in Assam: a proposed reserve.

Gloger C.W.L. 1841. *Gemeinnütziges Hand- und Hilfsbuch der Naturgeschichte*, vol. 1. Breslau.

Glynn C. 2011. The "Stipple Master". In: Beach M.C.; Fischer E.; Goswamy B.N. (eds.), *Masters of Indian painting*, vol. 2: 1650–1900. Zürich (Artibus Asiae Publishers, Supplementum 48 I/II), pp. 515–530.

Goblet d'Alviella E. 1877. *Inde et Himalaya: souvenirs de voyage.* Paris.

Goblet d'Alviella E. 1892. Combats d'animaux et chasse à Baroda. In: Lanier M.L. 1892. *Choix de lectures de géographie*, vol. 2: *L'Asie.* Paris, pp. 128–132.

Godal K. 2020. Skypaths, documentary in four parts, https://sky paths.wordpress.com [accessed February 2021].

Gode P.K. 1945. The menagerie of the Peshwa, its site and description in contemporary records, between AD 1778 and 1794. *Proceedings of the Deccan history conference* 1945: 403–407.

Godfrey R. 2012. *Olly: the life and times of Frederick Oliver Robinson, 2nd Marquis of Ripon*. London.

Godwin-Austen H.H. 1873. Journal of a tour in Assam, 26th November 1872 to 15th April 1873. Manuscript in Royal Geographical Society, London: http://himalaya.socanth.cam.ac.uk/collections/naga/record/r87078.html [accessed April 2023].

Gogerly G. 1871. *The pioneers: a narrative of facts connected with early christian missions*. London.

Gois D. de. 1567. *Chronica do felicissimo Rei D Manuel*, part 4. Lisboa.

Gokhale N.A. 2004. The birth of Kaziranga. *Focus Northeast* 2004-12-28: 1–2.

Gokhale N.A.; Kashyap S.G. 2005. *Kaziranga: the rhino century*. Guwahati.

Gole S. 1988. *Maps of Mughal India: drawn by Colonel Jean-Baptiste-Joseph Gentil, agent for the French Government to the Court of Shuja-ud-Daula at Faizabad, in 1770*. Delhi.

Gommans J.J.L.; Hond J. de. 2015. Willem Schellinks en India: tussen werkelijkheid en illusie. In: Bange P.; Geurts J., *Onbegrensd perspectief: cultuurhistorische verkenningen*. Amersfoort, pp. 69–96.

Goode G.D. 1896. *The published writings of Philip Lutley Sclater, 1844–1896*. Washington.

Goodrich S.G. 1859. *Illustrated natural history of the animal kingdom*, vol. 1. New York.

Goodwin H.A.; Holloway C.W. 1972. *Red Data Book*, vol. 1: *Mammalia*. Morges.

Gorakshkar S. 1979. *Animal in Indian art: catalogue of the exhibition held at the Prince of Wales Museum of Western India, Bombay from 30 September to 21 October 1977*. Bombay.

Gordon D. 2015. *Memoirs of life as a police officer in India: 1907 to 1946*. London.

Gordon-Alexander W. 1898. *Recollections of a Highland subaltern, during the campaigns of the 93rd Highlanders in India*. London.

Gore's Liverpool General Advertiser 1834-06-19, 1834-06-26: Arrival of rhinoceros owned by Atkins.

Goswami R.; Ganesh T. 2014. Carnivore and herbivore densities in the immediate aftermath of ethno-political conflict: the case of Manas National Park, India. *Tropical Conservation Science* 7: 475–487.

Gouldsbury C.E. 1909. *Dulall the forest guard: a tale of sport and adventure in the forests of Bengal*. London.

Gouldsbury C.E. 1913. *Tigerland: reminiscences of forty years' sport and adventure in Bengal*. London.

Govindrajan R. 2018. *Animal intimacies: interspecies relatedness in India's Central Himalayas*. Chicago.

Gowers W.F. 1950. The classical rhinoceros. *Antiquity* 24: 61–71.

Graaff N. de. 1701. *Reisen na de vier gedeeltens des Werelds, als Asia, Africa, America en Europa*. Hoorn.

Graham G. 1878. *Life in the Mofussil; or, the civilian in Lower Bengal*. London.

Graham J.A. 1897. *On the threshold of three closed lands: the Guild outpost in the eastern Himalayas*. Edinburgh.

Graham R. 1949a. *Rhino! Rhino!* Chicago.

Graham R. 1949b. Great Indian Rhinoceros (for the Brookfield Zoo). *Parks and Recreation* 32 (3): 172–176.

Graham R. 1954. Rhino! Rhino! A postwar adventure to capture two rhinos for the Brookfield Zoo. *Ex-C-B-I Round Up* 8 (8): 10–18.

Grant C. 1860. *Rural life in Bengal, illustrative of Anglo-Indian suburban life: letters from an artist in India*. London.

Grant F.J. 1834. Extracts from a journal kept during a tour of inspection on the Manipur Frontier along the course of the Ningthee River, &c. in January 1832. *Journal of the Asiatic Society of Bengal* 3: 124–134.

Graphic, an Illustrated Weekly Newspaper 19 (495), 1879-05-24, pp. 514–515: Review of Pollok, Sport in British Burmah.

Gratz A. 1898. Auch ein Röslein. *Fliegende Blätter*, München 108 (2747): 117.

Gray J.E. 1843. *List of the specimens of Mammalia in the collection of the British Museum*. London.

Gray J.E. 1846. *Catalogue of the specimens and drawings of Mammalia and Birds of Nepal and Thibet, presented by B.H. Hodgson to the British Museum*. London.

Gray J.E. 1847. *List of the osteological specimens in the collection of the British Museum*. London.

Gray J.E. 1854. On a new species of rhinoceros. *Proceedings of the Zoological Society of London* 1854-11-28: 250–251.

Gray J.E. 1862. *Catalogue of the bones of Mammalia in the collection of the British Museum*. London.

Gray J.E. 1868a. Observations on the preserved specimens and skeletons of the Rhinocerotidae in the collection of the British Museum and Royal College of Surgeons, including the description of three new species. *Proceedings of the Zoological Society of London* 1867-12-12: 1003–1032.

Gray J.E. 1868b. Letter to Charles Darwin, 6 February 1868. In: Burkhardt F. et al., *The correspondence of Charles Darwin: January–June 1868*, vol. 16. Cambridge, pp. 77–78.

Gray J.E. 1872. On the double-horned Asiatic rhinoceros (Ceratorhinus). *Annals and Magazine of Natural History* (ser.4) 10: 207–209.

Gray J.E. 1873a. *Hand-list of the edentate, thick-skinned and ruminant mammals in the British Museum*. London.

Gray J.E. 1873b. On the dentition of rhinoceroses (Rhinocerotes), and on the characters afforded by their skulls. *Annals and Magazine of Natural History* (ser.4) 11: 356–361.

Gray J.E. 1873c. On the black and ash-grey double-horned Asiatic rhinoceroses (Ceratorhinus sumatrensis, C. niger and C. lasiotis). *Annals and Magazine of Natural History* (ser.4) 12: 252–253.

Green M.J.B. 1993. *Nature reserves of the Himalaya and the mountains of Central Asia.* Gland.

Gregson J. 2002. *Massacre at the palace: the doomed royal dynasty of Nepal.* New York.

Grenet F. 2005. Découverte d'un relief sassanide dans le Nord de l'Afghanistan. *Comptes Rendus des Séances de l'Académie des Inscriptions et Belles-Lettres* 149: 115–134.

Grenet F.; Lee J.; Martinez P.; Ory F. 2007. The Sasanian relief at Rag-i Bibi (Northern Afghanistan). *Proceedings of the British Academy* 133: 243–267.

Grenfell F.W. 1925. *Memoirs of Field-Marshal Lord Grenfell.* London.

Grève C. 1898. Die geographische Verbreitung der jetzt lebenden Perissodactyla, Lamnunguia und Artiodactyla non ruminantia. *Nova Acta Academiae Caesareae Leopoldino-Carolinae Germanicae Naturae Curiosorum* 70: 290–370.

Greville F.R.C.G. 1877a. Shikar by Lord Brooke. *Indian Mirror*, Sunday edition, 1877-01-07: 3.

Greville F.R.C.G. 1877a. Shooting in Bengal. *Indian Mirror* 1877-04-08: 8.

Greville F.R.C.G. 1878. Notes of a sporting tour in India. *Journal of the Household Brigade* 1878: 244–249.

Greville F.R.C.G. (Earl of Warwick and Brooke). 1917. *Memories of sixty years.* London.

Grew N. 1681. *Musaeum Regalis Societatis.* London.

Grewal R. 2007. *In the shadow of the Taj: a portrait of Agra.* New Delhi.

Grey L.J.H. 1912. *Tales of our grandfather: or, India since 1865.* London.

Griffith E. 1827. *The animal kingdom, arranged in conformity with its organization by the Baron Cuvier, with additional descriptions of all the species hitherto named, and of many not before noticed*, vol. 5. London.

Griffith W. 1847. *Journals of travels in Assam, Burma, Bootan, Afghanistan and the neighbouring countries.* Calcutta.

Grigson C. 2015. New information on Indian rhinoceroses (Rhinoceros unicornis) in Britain in the mid-eighteenth century. *Archives of Natural History* 42: 76–84.

Grigson C. 2016. *Menagerie: the history of exotic animals in England 1100–1837.* Oxford.

Grimmer C.F. 1834. *Neues Dresdener Bilder-Cabinet zur belehrenden Unterhaltung für die Jugend.* Dresden.

Griset E. 1868. A hunting story, (not to say, fib), being the adventures of Cornelius Cracker, Esq. In: Hood T., *The 5 alls: a collection of stories, charades, acrostics, and puzzles.* London (Warne's Christmas Annual), pp. 38–39.

Gronovius L.T. 1778. *Museum Gronovianum sive index rerum naturalium* [belonging to] Laur. Theod. Gronovius [auctioned in Leiden 7 October 1778 and following days]. Lugduni Batavorum.

Grote A. 1875. Memoir of Edward Blyth. *Journal of the Asiatic Society of Bengal* 44 (Extra no.): i–xxiv.

Groves C.P. 1967. On the rhinoceroses of South-East Asia. *Säugetierkundliche Mitteilungen* 15: 221–237.

Groves C.P. 1971. Species characters in rhinoceros horns. *Zeitschrift für Säugetierkunde* 36: 238–252.

Groves C.P. 1982. The skulls of Asian rhinoceroses, wild and captive. *Zoo Biology* 1: 251–261.

Groves C.P. 1983. Phylogeny of the living species of rhinoceros. *Zeitschrift für Zoologische Systematik und Evolutionsforschung* 21: 293–313.

Groves C.P. 1993. Testing rhinoceros subspecies by multivariate analysis. In: Ryder O.A., *Rhinoceros biology and conservation: Proceedings of an international conference, San Diego, U.S.A.*, pp. 92–100.

Groves C.P. 2003. Taxonomy of ungulates of the Indian subcontinent. *Journal of the Bombay Natural History Society* 100: 341–362.

Groves C.P.; Chakraborty S. 1983. The Calcutta collection of Asian rhinoceros. *Records of the Zoological Survey of India* 80: 251–263.

Groves C.P.; Grubb P. 2011. *Ungulate taxonomy.* Baltimore.

Groves C.P.; Guérin C. 1980. Le Rhinoceros sondaicus annamiticus d'Indochine. *Geobios* 13: 199–208.

Groves C.P.; Kurt F. 1972. Dicerorhinus sumatrensis. *Mammalian Species* no. 21: 1–6.

Groves C.P.; Leslie Jr. D.M. 2011. Rhinoceros sondaicus (Perissodactyla: Rhinocerotidae). *Mammalian Species* 43 (887): 190–208.

Grubb P. 1993. Order Perissodactyla. In: Wilson D.E. et al., *Mammal species of the world*, 2nd edition. Washington, pp. 369–372.

Grubb P. 2005. Order Perissodactyla. In: Wilson D.E. et al., *Mammal species of the world*, 3rd edition, vol. 1. Baltimore, pp. 634–636.

Gruning J.F. 1911. *Eastern Bengal and Assam district gazetteers*, vol. 11: *Jalpaiguri.* Allahabad.

Grzimek B. 1958. Die gegenwärtige Zahl der Nashörner auf der Erde. *Säugetierkundliche Mitteilungen* 6: 117–120.

Grzimek B. 1960. Die gegenwärtige Zahl der Nashörner auf der Erde (Teil 2). *Säugetierkundliche Mitteilungen* 8: 21–25.

Guardian, London 1856-05-03: Arrival of rhinos from Assam in Liverpool.

Guardian, London 1911-12-18: King's trip to Nepal.

Gubernatis A. de. 1878. *Gli scritti del Padre Marco della Tomba: missionario nelle Indie Orientali.* Firenze.

Guérin C. 1973. Rhinocerotidae. In: Beden M.; Guérin C. 1973. Le gisement des vertèbres du Phnom Loang (Province de Kampot, Cambodge). *Travaux et Documents de l'ORSTOM* 27: 3–97 (pp. 9–50).

Guérin C. 1980. Les rhinocéros (Mammalia, Perissodactyla) du Miocène terminal au Pleistocene supérieur en Europe occidentale. *Documents du Laboratoire de Géologie de la Faculté des Sciences de Lyon* 79: 3–1183.

Guggisberg C.A.W. 1966. *S.O.S. Rhino*. London.

Guhathakurta P. 1985. The rhinoceros in Jaldapara Wildlife Sanctuary. *Indian and Foreign Review* 22 (6): 24–26.

Guhathakurta M.; Schendel W. van. 2013. *The Bangladesh reader: history, culture, politics*. Durham.

Guldenschuh G. 2002. Conservation status and taxonomy. In: Guldenschuh G.; Houwald F.F. von, *Husbandry manual for the greater one-horned or Indian rhinoceros Rhinoceros unicornis Linné, 1758*. Basel.

Gundevia Y.D. 1984. *Outside the archives*. Hyderabad.

Gupta A.C. 1958. Gorumara Game Sanctuary. *Journal of the Bengal Natural History Society* 29: 132–139.

Gupta A.C. 1964. Wild life of Lower Bengal with particular reference to the Sundarbans. In: *West Bengal forests: centenary commemoration volume*. Calcutta, pp. 233–238.

Gupta A.K. 1998. Status and management of wildlife in Tripura. *Indian Forester* 124: 787–793.

Gupta P.L.; Hardaker T. 2013. *Punchmarked coinage of the Indian subcontinent: Magadha-Mauryan series*. Mumbai.

Gupta S.P. 1980. *The roots of Indian art*. Delhi.

Gurdon P.R.T. 1914. *The Khasis*. London.

Gurung B.B. 2014. *Historical introduction of Narayanhiti Palace Museum*. Kathmandu.

Gurung M.K. 2004. *Human dimensions in one-horned rhinoceros conservation in Royal Chitwan National Park Nepal*. Thesis presented to Uni Boku Vienna.

H. 1894. A stroll through Jamrach's. *Sketch*, London 1894-09-05: 310.

Hackin J. 1954. Nouvelles recherches archéologiques à Begram (Ancienne Kâpici): 1939–1940. *Mémoires de la Délégation Archéologique Française en Afghanistan* 11: 1–354.

Hajra P.K.; Shukla U. 1984. Dudhwa National Park: some botanical aspects of the proposed new habitat for rhino. In: Singh S.; Rao K., *India's rhino-reintroduction programme*. New Delhi, pp. 52–62.

Halliday H.M. 1904. Letter to the Chief Commissioner of Assam, dated 18 June 1904. Assam State Archive, Assam Agricultural and Revenue Proceeding, Revenue-A, September 1905.

Ham G. van der. 2022. *Clara de neushoorn*. Amsterdam: Rijksmuseum (also in English: *Clara, the rhinoceros*).

Hamilton A. 1727. *A new account of the East Indies*, 2 vols. Edinburgh.

Hamilton F. 1819. *Account of the Kingdom of Nepal and of the territories annexed to this dominion by the House of Gorkha*. Edinburgh.

Hamilton F.S. 1919. *The vanished pomps of yesterday, being some random reminiscences of a British diplomat*. London.

Hamilton F.S. 1921. *Here, there and everywhere*. London.

Hamilton W. 1815. *The East-India gazetteer*. London.

Hamilton W. 1820. *Geographical, statistical, and historical description of Hindostan, and the adjacent countries*. London.

Hamilton W. 1828. Purneah. In: *East India Gazetteer*, vol. 2, pp. 429–435.

Handique R. 2004. *British forest policy in Assam*. New Delhi.

Hanson C.A. 2010. Representing the rhinoceros: the Royal Society between art and science in the eighteenth century. *Journal for Eighteenth-Century Studies* 33 (4): 545–566.

Hanson R.C. 1933. The fauna of Assam. *Journal of the Society for the Preservation of the Fauna of the Empire* 20: 50–53.

Haque E. 2001. *Chandraketugarh: a treasure house of Bengal terracottas*. Dhaka.

Haque Z. 2014. *Terracottas of Bengal: an analytical study*. Dhaka.

Hara K. 1961. How I brought Indian rhinoceros from India. *Animals and Zoo* 13: 11–13.

Haraway D. 1990. *Primate visions: gender, race, and nature in the world of modern science*. New York.

Hardy P. 1899. The ungainly brute shot past him. *Chums*, London 7 (no. 351), 1899-06-21: 689.

Hare S. 1929. *The annals of the King's Royal Rifle Corps*, vol. 4: *The 60th, the K.R.R.C. 1872–1913*. London.

Harfield A.G. 1995. *Bencoolen: a history of the Honourable East India Company's garrison on the West Coast of Sumatra (1685–1825)*. London.

Harmer S.F. 1921. Exhibition of a specimen of Rhinoceros sondaicus. *Proceedings of the Zoological Society of London* 1921-05-10: 643.

Harnath Singh. 1970. *Jaipur and its environs*. Dundlod.

Harper F. 1940. Nomenclature and type localities of certain old world mammals. *Journal of Mammalogy* 21: 191–203.

Harper F. 1945. *Extinct and vanishing mammals of the old world*. New York.

Harris J. 1744. *Navigantium atque itinerantium bibliotheca, or, a complete collection of voyages and travels*, vol. 1. London.

Harrison J.J. 1892. *A sporting trip through India*. Beverley [copy in Scarborough Museums Trust].

Harrison J.J. 1910. The marriage of a famous sportsman (at Brandesburton Hall). *Tatler*, London 38 (1910-11-16): xvi.

Hart H.G. 1875. *The new annual army list, and Indian civil service list, for 1875*. London.

Hart-Davis D. 1986. Chitwan's rescue of the rhino. *Field, the Country Gentleman's Magazine* 266 (1986-05-22): 60–61.

Hart-Davis D. 2005. *Honorary tiger: the life of Billy Arjan Singh*. Delhi.

Hartwig G.F. 1875. *The polar and tropical worlds: a description of man and nature in the polar and equatorial regions of the globe*. Springfield, Mass. and Chicago.

Hastings F. 1858. *The private journal of the Marquess of Hastings, Governor-General and Commander-in-Chief in India*, vol. 1. London.

Hastings F.E. 1821. Hunting excursion (in Rajmahal Hills). *Asiatic Journal and Monthly Register* 1821-12: 581.

Haughton J.C. 1879. *Report of the police of the Cooch Behar Division for the year 1869*. Calcutta.

Havmøller R.G.; Payne J.; Ramono W.; Ellis S.; Yogana K.; Long B.; Dinerstein E.; Christy Williams C.; Putra R.; Gawi J.; Talukdar B.K.; Burgess N. 2015. Will current conservation responses save the Critically Endangered Sumatran rhinoceros Dicerorhinus sumatrensis? *Oryx* 50: 355–359.

Hays D.L. 2002. Francesco Bettini and the pedagogy of garden design in eighteenth-century France. In Hunt J.D.; Conan M., *Tradition and innovation in French garden art: chapters of a new history*. Philadelphia, pp. 93–120.

Hazarika B.C. 2007. *Studies on the eco-behavioural aspects of Great Indian one-horned rhinoceros (Rhinoceros unicornis Linn.) in the Orang National Park, Assam, India*. Thesis presented to Gauhati University.

Hazarika B.C. 2015. *Ecology and behaviour of Indian rhino in the Brahmaputra floodplain habitat of Assam, India*. Guwahati.

Hazarika B.C.; Saikia P.K. 2010. A study on the behaviour of great Indian one-horned rhinoceros (Rhinoceros unicornis) in the Rajiv Gandhi National Park, Assam, India. *NeBio* 1 (2): 62–74.

H.C. 1836. The rhinoceros' horn. *Bengal Sporting Magazine* (n.s.) 7: 158–159.

Heard S.B. 2020. *Charles Darwin's barnacle and David Bowie's spider: how scientific names celebrate adventurers, heroes, and even a few scoundrels*. Yale University Press.

Heber R. 1828a. *Narrative of a journey through the Upper Provinces of India, from Calcutta to Bombay, 1824–1825*. 1st edition (in 2 volumes). London.

Heber R. 1828b. *Narrative of a journey through the Upper Provinces of India, from Calcutta to Bombay, 1824–1825*. 2nd edition (in 3 volumes). London.

Heber Percy R. 1894. Indian shooting. In: Phillipps-Wolley C., *Big game shooting*. London, vol. 2, pp. 182–362.

Heck L. 1899. *Lebende Bilder aus dem Reich der Tiere*. Berlin.

Heck L. 1915. Perissodactyla. In: *Brehms Tierleben: Säugetiere*. Leipzig, vol. 3, pp. 599–625.

Heck L. 1940. Heiter-ernste Erinnerungen an Tiergärtner. *Zoologische Garten* 12: 228–238.

Hediger H. 1970. Ein Nashorn mit Dürer-Hornlein. *Zoologische Garten* 39: 101–106.

Heidegger E.M.; Houwald F. von; Steck B.; Clauss M. 2016. Body condition scoring system for greater one-horned rhino (Rhinoceros unicornis). *Zoo Biology* 35: 432–443.

Heikamp D. 1980. Seltene Nashörner in Martin Sperlichs Nashorngalerie und anderswo. In: Heikamp D., *Schlösser, Gärten, Berlin (Festschrift für Martin Sperlich)*. Tübingen, pp. 301–325.

Heinen J.T.; Kattel B. 1992. A review of conservation legislation in Nepal: past progress and future needs. *Environmental Management* 16: 723–733.

Heise U.K. 2016. *Imagining extinction: the cultural meanings of endangered species*. Chicago.

Heissig K. 1972. Rhinocerotidae (Mammalia) aus den unteren und mittleren Siwalik-Schichten. *Abhandlungen der Bayerischen Akademie der Wissenschaften, Mathematisch-Naturwissenschaftliche Klasse*, München (N.F.) 152: 1–112.

Henderson G. 1834. Review of the Oriental Annual for 1834. *Monthly Review* 1834 (November): 423.

Hendley T.H. 1876. *The Jeypore Guide*. Jeypore.

Hendley T.H. 1895. *Handbook to the Jeypore Museum*. Calcutta.

Henri d'Orléans. 1888. La chasse au tigre. *Le Soleil*, Paris 1888-04-27: Supplement p. 1; also *Gazette Nationale ou le Moniteur Universel*, Paris 1888-04-29: 472–473.

Henri d'Orléans. 1889. *Six mois aux Indes, chasses aux tigres*, 4th edition (editions 3–7 all dated 1889). Paris.

Henri d'Orléans. 1898a. *Du Tonkin aux Indes, Janvier 1895–Janvier 1896*. Paris.

Henri d'Orléans. 1898b. *From Tonkin to India by the sources of the Irawadi*. London.

Herbaut E. 1876. Voyage du Prince de Galles aux Indes. *Univers Illustré (Journal hebdomadaire)* 19 (1876-03-11): 167, 171.

Herne P. (pseudonym) 1858. *Perils and pleasures of a hunter's life, or, the romance of hunting*. Philadelphia.

Herzog H. 2010. *Some we love, some we hate, some we eat: why it's so hard to think straight about animals*. New York.

Heude P.M. 1892. Ètudes odontologiques, 1. Herbivores trizygodontes et dizygodontes. *Mémoires concernant l'Histoire Naturelle de l'Empire Chinois* 2 (2): 65–84.

Hewett J.P. 1938. *Jungle trails in northern India, reminiscences of hunting in India*. London.

Hewitt S. 2016. *Clara, the true story of the rhino who became a superstar*. London.

Hewson E. 2014. *Burials in Assam & N.E. India 1793–1974*. Wem (The Kabristan Archives).

Heywood T. 1638. *Porta pietatis, or, the port of harbour of piety*. London.

Hibbert C. 2007. *Edward VII: the last victorian King*. London.

Hickmann R.; Enderlein V. 1993. *Meisterwerke der Moghul-Zeit: Indische Miniaturen des 17. und 18. Jahrhunderts aus dem Museum für Islamische Kunst der Staatlichen Museen zu Berlin, Stiftung Preussischer Kulturbesitz*. Lachen am Zürichsee.

Higgins F.W. 1891. Shooting a rhinoceros in Bengal. *Englishman's Overland Mail* 1891-06-09: 3.

Higgins J.C. 1934. The game birds and animals of the Manipur state with notes of their numbers, migration and habits. *Journal of the Bombay Natural History Society* 37: 298–309.

Hill C.A. 1968. Kaziranga. *Zoonooz*, San Diego 41 (9): 4–8.

Hill D. 1937. Bengal rhino. *Field, the Country Gentleman's Magazine* 169 (1937-05-22): 1383.

Hilton E.H. 1891. *The tourist's guide to Lucknow*. Lucknow.

Hinchingbrook E.G.H.M. 1879. *Diary in Ceylon and India 1878–9.* London.

Hinton M.A.C.; Fry T.B. 1923. Bombay Natural History Society's mammal survey of India, Burma and Ceylon. Report no. 37, Nepal. *Journal of the Bombay Natural History Society* 29: 399–428.

Hirst F.C. 1917. *The surveys of Bengal by Major James Rennell, 1764–1777.* Calcutta.

Hobbes R.G. 1893. *Reminiscences of seventy years' life, travel, and adventure,* vol. 1. London.

Hobhouse. 1986. *Indian painting during the British period.* London.

Hobhouse C.P. 1916. *Some account of the family of Hobhouse and reminiscences.* Leicester.

Hobley C.W. 1931. The rhinoceros. *Journal of the Society for the Preservation of the Fauna of the Empire* 14: 18–23.

Hobley C.W.; Shebbeare E.O. 1932. The rhinoceros. *Journal of the Society for the Preservation of the Fauna of the Empire* 17: 20–21.

Hodgkin H.; Topsfield A.; Filippi G.G. 1997. *Indian miniatures and paintings from the 16th to the 19th century. The collection of Howard Hodgkin.* Verona.

Hodgson B.H. 1825. Remarks on the procreation of the rhinoceros. *Quarterly Oriental Magazine* 3: 155–156.

Hodgson B.H. 1826a. Remarks on the rate of growth and habits of the Rhinoceros indicus. *Asiatic Journal and Monthly Register* 22: 193–197.

Hodgson B.H. 1826b. De dragt van den Rhinoceros. *Bijdragen tot de Natuurkundige Wetenschappen* 1 (1): 154.

Hodgson B.H. 1827. On the growth and habits of a young rhinoceros. *Edinburgh Journal of Science* 7: 165–166.

Hodgson B.H. 1832. On the mammalia of Nepal. *Journal of the Asiatic Society of Bengal* 1: 335–349.

Hodgson B.H. 1834. On the mammalia of Nepal. *Proceedings of the Zoological Society of London* 1834-08-26: 95–104.

Hodgson B.H. 1844. Classified catalogue of mammals of Nepal. *Calcutta Journal of Natural History* 4: 284–294.

Hodson T.C. 1912. Meithei literature. *Folk-lore, a Quarterly Review* 23: 175–184.

Höfer R. 2010. *Imperial sightseeing: die Indienreise von Erzherzog Franz Ferdinand von Österreich-Este.* Wien.

Hoffmeister W. 1847. *Briefe aus Indien. Nach dessen nachgelassenen Briefen und Tagebüchern.* Braunschweig.

Hoffmeister W. 1848. *Travels in Ceylon and continental India: including Nepal and other parts of the Himalayas to the borders of Thibet, with some notices of the overland route.* Edinburgh, London.

Hogendorp D.C.A. van. 1887. *Mémoires du General Dirk van Hogendorp, Comte de l'Empire.* La Haye.

Holeckova D. 2009. *Breeding of endangered species in Dvur Kralove Zoo,* vol. 3: *Rhinos.* Dvur Kralove.

Home E. 1821. An account of the skeletons of the dugong, two-horned rhinoceros, and tapir of Sumatra, sent to England by Sir Thomas Stamford Raffles, Governor of Bencoolen. *Philosophical Transactions of the Royal Society of London* 111: 268–275.

Home E. 1823. *Lectures on comparative anatomy,* vol. 4. London.

Homem L.; Reinel P.; Reinel J. 1519. The Atlas Miller (Lopo Homem-Reineis Atlas). Manuscript cartographic atlas preserved in Bibliothèque Nationale, Paris, Registre C; 28836 (rhino on folio 3). Available online: https://gallica.bnf.fr /ark:/12148/btv1b55002605w/f1.

Homeward Mail from India 1864-06-18, pp. 533: Assam as shikar paradise.

Homeward Mail from India 1872-02-19, p. 202: Steamer Petersburg from Calcutta to London.

Homeward Mail from India 1873-04-07, p. 352: Baring's shikar expedition.

Homeward Mail from India 1874-06-01, p. 572: Departure of steamer Sultan from Calcutta.

Homeward Mail from India 1876-02-14, p. 171: The Prince of Wales in India.

Homeward Mail from India 1880-03-25, p. 334: Nripendra receives title of Maharaja Bahadur.

Homeward Mail from India 1881-04-25, p. 391: The Durbhangah shooting party.

Homeward Mail from India 1884-06-30, p. 632: Destruction of game in Jalpaiguri.

Homeward Mail from India 1892-03-07, p. 323: The Cooch Behar shooting camp.

Homeward Mail from India 1892-03-14, p. 326: The Viceroy in Cooch Behar.

Homeward Mail from India 1892-03-21, p. 357: The Viceroy in Cooch Behar.

Homeward Mail from India 1900-03-12, p. 970: Shoot of Maharaja of Cooch Behar.

Homeward Mail from India 1900-11-12, p. 1481: Curzon's visit to Junagadh.

Homeward Mail from India 1901-04-27, p. 529: The Viceroy's shooting tour.

Homeward Mail from India 1901-05-04, p. 566: The Viceroy's shooting tour (end of the tour).

Homeward Mail from India 1902-01-13, p. 3: Rhinoceros found in the Sundarbans.

Homeward Mail from India 1902-04-19, p. 519: Bag of Maharaja of Cooch Behar.

Homeward Mail from India 1902-12-29, p. 3: Twenty epsoms combined – Durbar in Delhi.

Homeward Mail from India 1909-02-27, p. 257: The Viceroy at Lohapara, Assam.

Homeward Mail from India 1911-12-30, p. 1604: The royal visit to India.

Homeward Mail from India 1912-01-06, p. 2: The King's bag in Nepaul.

Honacki J.H.; Kinman K.E.; Koeppl J.W. 1982. *Mammal species of the world: a taxonomic and geographic reference*. Lawrence.

Hood F.H. 1869. The Sumatran rhinoceros. *Oriental Sporting Magazine* (n.s.) 2: 167–169.

Hood J. 1912. Luxury and comfort marked King's hunt. *New York Times* 1912-01-20: 3.

Hoogerwerf A. 1970. *Udjung Kulon, the land of the last Javan rhinoceros*. Leiden.

Hooijer D.A. 1946. Prehistoric and fossil rhinoceroses from the Malay Archipelago and India. *Zoologische Mededelingen* 26: 1–138. (Also as thesis presented to the University of Leiden, 1946).

Hornaday W.T. 1919. Citizens of the jungle: the Indian rhinoceros. *Mentor* 7 (16): 1.

Horsfield T. 1841. A list of mammalia and birds collected in Assam by John McClelland. *Annals and Magazine of Natural History* 6: 366–374.

Horsfield T. 1851. *A catalogue of the Mammalia in the Museum of the Hon. East India Company*. London.

Hose C.; MacDougall W. 1912. *The pagan tribes of Borneo*, vol. 1. London.

Hosten H. 1912. Father A. Monserrate's account of Akbar (26th Nov. 1582). *Journal of the Asiatic Society of Bengal* (n.s.) 8: 185–221.

Hosten H. 1914. Mongolicae Legationis commentarius, or the first Jesuit mission to Akbar by Fr. Anthony Monserrate. *Memoirs of the Asiatic Society of Bengal* 3 (9): 513–704.

Hosten H. 1925. Jesuit letters from Bengal, Arakan & Burma (1599–1600). *Bengal Past and Present* 30: 52–76.

Houlton J.W. 1949. *Bihar, the heart of India*. Bombay.

Houwald F.F. von. 2001. *Foot problems in Indian rhinoceroses (Rhinoceros unicornis) in zoological gardens: macroscopic and microscopic anatomy, pathology, and evaluation of the causes*. Thesis presented to Zürich University.

Howes C.A.; Bamber M. 1970. Rarities in a museum. *Oryx* 10: 326–328.

Howitt S. 1799. Hunting the rhinoceros. *Sporting Magazine, or Monthly Calendar* 13: 160–163.

Howitt S. 1801. Announcement of publication of The Indian Sportsman. *Sporting Magazine, or Monthly Calendar* 17 (February): 217–218.

Howitt S. 1814. *Foreign field sports, fisheries, sporting anecdotes*. London: Edward Orme.

Howitt S. 1819. *Foreign field sports, fisheries, sporting anecdotes*. London: H.R. Young.

Howitt S. 1820. Shooting the rhinoceros: an etching. *Sporting Magazine, or Monthly Calendar* (n.s.) 6: 89–90.

Howitt S. 1834. Attaque d'un rhinocéros par des éléphants. *Musée des Familles: Lectures du Soir* 1: 297–298.

Howitt S. 1835. Combattimento tra il rinoceronte e gli elefanti. *Album, Giornale Illustrata* 2 (9), 1835-05-09: 68–69.

Howitt S. 1837. O rhinoceronte da Asia, ou abada (Rhinoceros unicornis, Lin.). *O Panorama (Jornal litterario e instructivo da Sociedade Propagadora dos Conhecimentos Uteis)* 1 (1837-09-16): 154–155.

Howitt S. 1840. Das Nashorn im Kampf mit Elephanten. *Heller-Blatt oder Magazin zur Verbreitung gemeinnütziger Kenntnisse*, Breslau 7: [no page].

Howitt S. 1845. Ein Rhinoceros von Elefanten angegriffen. *Pfennig-Magazin für Belehrung und Unterhaltung* (N.F.) 3 (no. 139), 1845-08-30: 273–274.

Hubback T. 1939. The two-horned Asiatic rhinoceros (Dicerorhinus sumatrensis). *Journal of the Bombay Natural History Society* 40: 594–617.

Huber W. 1964. Der Indische Panzernashornbulle Gadadhar. *Bericht des Vereins des Naturhistorischen Museums Bern* 1964: lv–lviii.

Hügel C. von. 1836. Recent discovery of fossil bones in Perim Island, in the Cambay Gulph. *Journal of the Asiatic Society of Bengal* 5: 288–289.

Hughes J.E. 2009. *Indo-Islamic kingdoms and cultures*. Thesis presented to University of Texas.

Hughes J.E. 2013. *Animal kingdoms: hunting, the environment, and power in the Indian Princely States*. Harvard.

Hunter W.W. 1875a. *A statistical account of Bengal*, vol. 1: *Districts of the 24 Parganas and Sundarbans*. London.

Hunter W.W. 1875b. *A statistical account of Bengal*, vol. 5: *Districts of Dacca, Bakarganj, Faridpur, and Maimansinh*. London.

Hunter W.W. 1876a. *A statistical account of Bengal*, vol. 6: *Chittagong Hill Tracts, Chittagong, Noakhall, Tipperah, Hill Tipperah*. London.

Hunter W.W. 1876b. *A statistical account of Bengal*, vol. 9: *Districts of Murshidabad and Pabna*. London.

Hunter W.W. 1876c. *A statistical account of Bengal*, vol. 10: *Darjiling, Jalpaiguri and State of Kuch Behar*. London.

Hunter W.W. 1877. *A statistical account of Bengal*, vol. 14: *Districts of Bhagalpur and the Santal Parganas*. London.

Hunter W.W. 1879a. *A statistical account of Assam*, vol. 1: *Districts of Kamrup, Darrang, Nowgong, Sibsagar, and Lakhimpur*. London.

Hunter W.W. 1879b. *A statistical account of Assam*, vol. 2: *Districts of Goalpara, the Garo Hills, the Naga Hills, the Khasi and Jaintia Hills, Sylhet, and Cachar*. London.

Hunter W.W. 1882. *The Indian empire: its history, people and products*. London.

Hunter W.W. 1893. *The Indian empire: its history, people and products*. 3rd edition. London.

Hunter W.W. 1908a. *The Imperial gazetteer of India*, vol. 6: *Argaon to Bardwan*. New edition. Oxford.

Hunter W.W. 1908b. *The Imperial gazetteer of India*, vol. 12: *Einme to Gwalior*. New edition. Oxford.

Hunter W.W. 1909. *The Imperial gazetteer of India, provincial series: Eastern Bengal and Assam*. Calcutta.

Hunting knife. 1868. Sport in Assam. *Field, the Country Gentleman's Magazine* 32 (1868-09-12): 216–217.

Husain M. 1976. *The Rehla of Ibn Battuta (India, Maldive Islands and Ceylon)*. Translation and commentary. Baroda.

Husain S.S. 1965. Report of meeting 27 September 1964. *Journal of the Asiatic Society of Pakistan*, Dacca 10 (2): iv–v.

Hussain B. 2001. Status of Rhinoceros unicornis in Orang National Park, Assam. *Tiger Paper* (FAO) 28 (1): 25–27.

Hussain K.Z. 1985. Last live captive rhino of Bangladesh [in Bengali]. *Bichitra* 1985 January: 1–3.

Hutchins F.G. 1980. *Young Krishna. Translated from the Sanskrit Harivamsa. Illustrated with paintings from historic manuscripts*. West Franklin, New Hampshire.

Hutchinson R.H.S. 1906. *An account of the Chittagong Hill Tracts*. Calcutta.

Hutchinson R.H.S. 1909. *Eastern Bengal and Assam district gazetteers: Chittagong Hill Tracts*. Allahabad.

Hutchisson W.H.F. 1883. *Pen and pencil sketches*. London.

Hutton J.H. 1997. Tour diary for the months of November and December 1921. *Journal of the North-East India Council for Social Science Research* 21: 24–32.

HWD. 1833. Joshua Brookes. *Gentleman's Magazine* 1833 February: 184–185.

Ibn Battuta. 1855. *Voyages d'Ibn Batoutah. Texte arabe, accompagné d'une traduction de C. Defrémery et B.R. Sanguinetti*, vol. 3. Paris.

Ibn Battuta. 1971. *The travels of Ibn Battuta AD 1325–1354*, edited by H.A.R. Gibb, vol. 3. London: Hakluyt Society, vol. 141.

ICZN – International Commission on Zoological Nomenclature. 1999. *International Code of Zoological Nomenclature*, 4th edition. London.

I.J.W. 1909. Record Indian rhinoceros. *London Evening Standard* 1909-06-04.

Illustrated London News 68 (1876-05-13), pp. 1–10: Welcome home number, The Prince of Wales in India.

Illustrated Sporting and Dramatic News 1877-04-07, p. 70: Indian hunting trophies and zoological specimens, collected by H.R.H. The Prince of Wales. From the Picture Gallery of the Zoological Gardens, Regent's Park.

Illustrated Sporting and Dramatic News 1909-11-06: Review of Nripendra, Thirty-seven years of big game shooting.

Imig K.; Purohit M. 2006. *Juna Mahal Dungarpur. Ein Rajputen-Palast in Rajasthan, Indien*. Zürich.

India Forest Dept. 1939. Statement of wild animals shot in some of the Indian Provinces, Indian States and Burma during 1937–38. *Indian Forester* 65 (9): 574–575.

Indian Daily News 1901-03-28: Lord Curzon in Nepal.

Indian Daily News 1901-04-04: Lord Curzon in Nepal.

Indian Express 2016-12-01 (online): Oldest rhino at Dudhwa National Park dies.

Indian Museum. 1868. Annual report, and list of accessions. April 1867 to March 1868. Calcutta.

Indian Museum. 1869. Annual report, and list of accessions: April 1868 to March 1869. Calcutta.

Indian Museum. 1870. Minutes of the Trustees. April 1869 to March 1870. Calcutta.

Indian Museum. 1871. Annual report, and list of accessions, April 1870 to March 1871. Calcutta.

Indian Museum. 1872. Annual report, and list of accessions, April 1871 to March 1872. Calcutta.

Indian Museum. 1873. Minutes of the Trustees. April 1872 to March 1873. Calcutta.

Indian Museum. 1881. Annual report and list of accessions, April 1880 to March 1881. Calcutta.

Indian Museum. 1883. Annual report and list of accessions, April 1882 to March 1883. Calcutta.

Indian Museum. 1885. Annual report and list of accessions, April 1884 to March 1885. Calcutta.

Indian Museum. 1911. Annual report 1910–1911. Calcutta.

Indian Museum, Kolkata, AT/95/1586: Shield made of rhino skin used by Rana Pratap in the battle of Haldighat in the year 1576. Online at https://museumsofindia.gov.in/repository/record/im_kol-AT-95-1586-15544.

Indian Times 2018-05-27 (online): Repopulation effort in Patlakhawa: after 60 years, rhinos to walk grasslands.

Inglis C.M.; Travers W.L.; O'Donel H.V.; Shebbeare E.O. 1919. A tentative list of the vertebrates of the Jalpaiguri District, Bengal. *Journal of the Bombay Natural History Society* 26: 819–825.

Inglis J. (Maori). 1874. What a sell – or shooting near the Koosee. *Oriental Sporting Magazine* (n.s.) 7 (January): 21–25.

Inglis J. (Maori). 1878. *Sport and work on the Nepaul frontier or twelve years sporting reminiscences of an indigo planter*. London.

Inglis J. 1888. *Tent life in tigerland, being sporting reminiscences of a pioneer planter in an Indian frontier district*. London.

Inglis J. 1892. *Tent life in tigerland*, 2nd edition. London.

International Cultural Exhibition. 1945. *Catalogue of the paintings, bronzes, sculptures, etc. shown at the Exhibition in the Convocation Hall (University of Bombay) by the India Section of the International Cultural Exhibition from 19th January 1945 to 3rd February 1945*. Bombay.

International Rhino Foundation. 2021. 2021 State of the Rhino report (September 21, 2021). https://rhinos.org/blog/news-room/2021-state-of-the-rhino-report. [accessed 2022-12-02].

International Rhino Foundation. 2022a. https://rhinos.org. [accessed 2022-10-18].

International Rhino Foundation. 2022b. About IRF: mission and values. https://rhinos.org/about-irf/mission-and-values. [accessed 2022-10-18].

Ipswich Evening Star 1907-03-19: Arrival of rhinoceros in the Museum.

Irani G.; Irani J. 2005. Kaziranga: wild and beautiful. *Jetwings* 5 (2): 16–32.

Irvine A.A. 1938. *Land of no regrets*. London.

Islam S.U.; Islam Z. 2004. *Hunting dangerous game with the Maharajas in the Indian sub-continent*. New Delhi.

IUCN. 1969. *Conservation in India: Proceedings of the special meeting of the Standing Committee of the Indian Board for Wild Life, New Delhi, 24 November 1965*. Morges.

Jack J.C. 1918. *Bengal district gazetteers*, vol. 36: *Bakarganj*. Calcutta.

Jackson C.E. 2018. The Ward family of taxidermists. *Archives of Natural History* 45: 1–13.

Jackson K. 2016. Globalizing an Indian borderland environment: Aijal, Mizoram, 1890–1919. *Studies in History* 32: 39–71.

Jacobaeus O. 1696. *Museum Regium, seu catalogus rerum tam naturalium quam artificialium, quae in basilica bibliothecae Monarchae Christiani Quinti Hafniae asservantur, descriptus*. Hafniae.

Jacobaeus O.; Laürentzen J. 1710. *Museum Regium*, 2nd edition. Havniae.

Jacolliot L. 1875. *Trois mois sur le Gange et le Brahmapoutre*. Paris.

Jacolliot L. 1884. *Les animaux sauvages*. Paris.

Jacolliot L. 1899. La chasse au rhinocéros. *Journal des Voyages* (ser. 2) 151: 332–333.

Jacomb-Hood G.P. 1925. *With brush and pencil*. London.

Jacquemont V. 1841. *Voyage dans l'Inde pendant les années 1828 à 1832: Journal*, vol. 1. Paris.

Jacques, Duc d'Orléans. 1999. *Chasses des Princes d'Orléans*. Paris.

Jaipur School of Art. 1920. *Illustrated catalogue, School of Art, Jaipur, Rajputana*. Gwalior.

Jalil A.F.M.A. 2000. *Sundarbaner Itihas* [History of the Sundarbans] (in Bengali). Kolkata.

Jamrach C. 1858. Arrival of an Indian monster. *Morning Advertiser* 1858-04-10: 5.

Jamrach W. 1875. *On a new species of Indian rhinoceros*. London [pamphlet; copy in Central Library of Muséum national d'Histoire Naturelle, Paris].

Jamrach W. 1876. On a new species of Indian rhinoceros. *Oriental Sporting Magazine* (n.s.) 9 (January): 25–26.

Jamun. 1925. The rhinoceros in India. *Field, the Country Gentleman's Magazine* 146 (1925-07-30): 181.

Jansen M. 1985. Mohenjo-Daro, city of the Indus valley. *Endeavour* (n.s.) 9: 161–169.

Jardine W. 1836. *The naturalist's library*, vol. 9, *Mammalia part 5: Pachydermes*. Edinburgh.

Jardine W. 1840. *Leaves from the book of nature*. Edinburgh.

Jarrett H.S. 1891. *The Ain I Akbari by Abul Fazl Allami*, vol. 2. Calcutta.

Jarrett H.S.; Sarkar J.N. 1949. *Ain-i-Akbari of Abul Fazl-i-Allami*, vol. 2: *A gazetteer and administrative manual of Akbar's empire and past history of India*. Calcutta.

Jarric P. du. 1614. *Histoire des choses plus mémorables advenues tant ez Indes orientales*, vol. 3. Bourdeaus.

Jefferys T. 1768. (Map-sheet) The East Indies, with the roads. By Thomas Jefferys, Geographer to the King. MDCCLXVIII. The second edition. London, published according to Act of Parliament, 30th Apr. 1768 by Robt. Sayer, no. 53 in Fleet Street. Scale 1: 2,600,000. Published in: *A general atlas, describing the whole universe: being a complete collection of the most approved maps extant; corrected with the greatest care, and augmented from the latest discoveries. The whole being an improvement of the maps of D'Anville and Robert. Engraved in the best manner on sixty-two copper-plates, by Thomas Kitchin, Senior, and others*. London: printed for Robert Sayer, no. 53, Fleet-Street.

Jeffrey R. 1863. Friends traveling in the ministry. *British Friend, a monthly journal chiefly devoted to the interests of the Society of Friends* 21: 114–115.

Jenkins A.E. 1893. Sixteen days shooting in Cooch Behar. *Rifle Brigade Chronicle* 1893: 206–208.

Jennison G. 1937. *Animals for show and pleasure in ancient Rome*. Manchester.

Jentink F.A. 1887. Catalogue ostéologique des mammifères. *Muséum d'Histoire Naturelle des Pays Bas* 9: 1–360.

Jentink F.A. 1892. Catalogue systématique des mammifères (singes, carnivores, ruminants, pachydermes, sirènes et cétacés). *Muséum d'Histoire Naturelle des Pays Bas* 11: 1–219.

Jenyns S. 1954. The Chinese rhinoceros and Chinese carvings in rhinoceros horns. *Transactions of the Oriental Ceramic Society* 29: 31–62.

Jerdon T.C. 1867. *The mammals of India; a natural history of all the animals known to inhabit continental India*. Roorkee.

Jerdon T.C. 1874. *The mammals of India*, 2nd printing. London.

Jesse W. 1899. A day's egging on the sandbanks of the Ganges. *Ibis* (ser. 7) 5: 4–9.

Jeyes S.H. 1896. *The life and times of the Right Honorable the Marquis of Salisbury*, vol. 2. London.

Jha D.K.; Kshetry N.T.; Pokharel B.R.; Panday R.; Aryal N.K. 2015. Comparative study of some morphological and microscopic identifying features of genuine rhino (Rhinoceros unicornis) horns and fake horns. *Journal of Forensic Research* 6 (6): 1–5.

J.H.C. 1836. A rhinoceros attacked by elephants. *Saturday Magazine* 8 (225), 1836-01-02: 1–3.

Jnawali S.R.; Wegge P. 1993. Space and habitat use by a small re-introduced population of greater one-horned rhinoceros

(Rhinoceros unicornis) in Royal Bardia national park in Nepal – a preliminary report. In: Ryder O.A., *Rhinoceros biology and conservation: Proceedings of an international conference, San Diego, U.S.A.*, pp. 208–217.

Jnawali S.R. 1995. *Population ecology of greater one-horned rhinoceros (Rhinoceros unicornis)*. Thesis presented to Agricultural University, Ås.

Jnawali S.R.; Baral H.S.; Lee S.; Acharya K.P.; Upadhyay G.P.; Pandey M.; Shrestha R.; Joshi D.; Laminchhane B.R.; Griffiths J.; Khatiwada A.P.; Subedi N.; Amin R. 2011. *The status of Nepal mammals*. Kathmandu.

Jörg C.J.A.; Knol E.; Campbell D.A. 2014. *Jan Albert Sichterman 1692–1764: een imponerende Groninger liefhebber van kunst*. Groningen.

Joglekar P.P.; Sharada C.V. 2016. Faunal remains from Madina, Rohtak District, Haryana. In: *Excavations at Madina, District Rohtak, Haryana, India*, pp. 209–247.

Johnsingh A.J.T.; Ramesh K.; Qureshi Q. 2007. Dhikala grasslands in Corbett Tiger Reserve, a potential site for reintroduction of the one-horned rhinoceros in India. *Pachyderm*, Nairobi 43: 108–110.

Johnson R.F. 2003. *Reverie and reality: nineteenth-century photographs of India from the Ehrenfeld Collection*. San Francisco.

Johnstone J. 1896. *My experiences in Manipur and the Naga Hills*. London.

Jomard E.F.; Bianchi T.X.; Eyriès J.B. 1832. Rapport sur la collection ethnographique de M. Lamare-Picquot, par une commission spéciale. *Bulletin de la Société de Géographie* 17: 86–96.

Jombert C.A. 1755. *Méthode pour apprendre le dessein*. Paris.

Jones O.J. 1859. *Recollections of a winter campaign in India, in 1857–58*. London.

Jonston J. 1653. *Historiae naturalis de quadrupedibus libri*. Francofurti ad Moenum.

Jonston J. 1657a. *Historiae naturalis de quadrupedibus libri*. Amstelodami.

Jonston J. 1657b. *An history of the wonderful things of nature, set forth in ten severall classes*. London.

Jonston J. 1660. *Beschrijving van de natuur der viervoetige dieren neffens haer beeldenissen in koper gesneden*. Uyt 'et Latyn vertaelt door M. Grausius. Amsterdam.

Jonston J. 1678. *A description of the nature of four-footed beasts*. Translated into English by John Rowland. London.

Jordan Gschwend A. 2000. A masterpiece of Indo-Portuguese art: the mounted rhinoceros cup of Maria of Portugal, Princess of Parma. *Oriental Art* 46: 48–58.

Jordan Gschwend A. 2015a. "[...] underlasse auch nit mich in Portugal vnnd ander orten umb frömbde sachen zu bewerben": Hans Khevenhüller and Habsburg menageries in Vienna and Prague. In: Haag S., *Echt Tierisch: die Menagerie des Fürsten*. Innsbruck, pp. 31–41.

Jordan Gschwend A. 2015b. Two portraits of an Indian rhinoceros, called the "Wonder of Lisbon" (Catalogue 2.3). In: Haag S., *Echt Tierisch: die Menagerie des Fürsten*. Innsbruck, pp. 134–135.

Jordan Gschwend A. 2018. The Emperor's exotic and new world animals: Hans Khevenhüller and Habsburg menageries in Vienna and Prague. In: MacGregor A., *Naturalists in the field* (Emergence of Natural History, vol. 2). Leiden, pp. 76–103.

Jørgenson D. 2019. *Recovering lost species in the modern age: histories of longing and belonging*. Boston.

Joshi J.P.; Parpola A. 1987. *Corpus of Indus seals and inscriptions*, vol. 1: *Collections in India*. Helsinki (Memoirs of the Archaeological Survey of India, no. 86).

Junk W. 1906. *The entomologist's directory*. Berlin.

J.W.G. 1900. Rhinoceros tracking in Upper Bengal. *Field, the Country Gentleman's Magazine* 96 (1900-07-14): 49.

Kala S.C. 1961. *Indian miniatures in the Allahabad Museum*. Allahabad.

Kalita D.K. 1992. A study of the magical beliefs and practices in Assam, with special reference to the magical lore of Mayong. Thesis presented to Gauhati University.

Kalof L. 2017. *The Oxford handbook of animal studies*. Oxford.

Kalof L.; Montgomery G. 2011. *Making animal meaning*. East Lansing.

Kalra M.; Sinha S.P. 2004. Reintroduction and rehabilitation of rhinoceros in the sub Himalayan Terai grasslands, India. *Proceedings of the 5th International Symposium on Physiology, Behaviour and Conservation of Wildlife*, Berlin 1: 135.

Kalus L. 1974. Boucliers circulaires de l'orient musulman. *Gladius* 12: 59–133.

Kandel R.C. 2012. Wildlife use of Bharandabhar forest corridor: between Chitwan National Park and Mahabharat foothills, Central Tarai, Nepal. *Journal of Ecology and the Natural Environment* 4: 119–125.

Kangle R.P. 1963. *The Kautiliya Arthasastra*, part II. An English translation with critical and explanatory notes. Bombay.

Kanpur Zoo. 1994. First Indian national studbook for the great Indian (one-horned) rhinoceros. *Zoos' Print (Magazine of Zoo Outreach Organisation)* 9: 55–63.

Karlsson K. 1999. Face to face with the absent Buddha: the formation of Buddhist aniconic art. *Acta Universitatis Upsaliensis: Historia Religiorum* 15: 1–200.

Karttunen K. 1989. India in early Greek literature. *Studia Orientalia (Finnish Oriental Society)* 65: 1–269.

Kauffmann O. 1911. *Aus Indiens Dschungeln: Erlebnisse und Forschungen*. Leipzig.

Kauffmann O. 1923 *Aus Indiens Dschungeln: Erlebnisse und Forschungen*, 2nd edition. Bonn.

Kaufmann W. 1968. *The Ragas of North India*. Bloomington, London.

Kawata K. 2014. Injuries by rhinos. *International Zoo News* 61 (2): 147–149.

Keating J.; Markey L. 2011. 'Indian' objects in Medici and Austrian-Habsburg inventories: a case-study of the sixteenth century term. *Journal of the History of Collections* 23 (2): 283–300.

Keeling C.H. 1984. *Where the lion trod.* Guildford.

Keeling C.H. 1991. *Where the elephant walked.* Guildford.

Kellogg W.F. 1888. *Hunting in the jungle with gun and guide after large game.* Boston.

Kemf E.; Strien N.J. van. 2002. *Asian rhinos in the wild: a WWF species status report.* Gland.

Kendall J.K. (Dum-Dum). 1901. R.I.P. Elegy on a rhinoceros, lately deceased. *Homeward Mail from India, China and the East* 1901-01-21: 67.

Kendall J.K. (Dum-Dum). 1915. *Odd creatures: a selection.* London.

Kennard A.C.H. 1910. Letter to his mother-in-law from a shooting camp in Assam. National Archives, London – https://discovery.nationalarchives.gov.uk/details/r/2c7b4c7c-6a14-41ad-a07c-df97c4220a18 [accessed February 2021].

Kennion R.L. 1922. A royal shooting ground. *Field, the Country Gentleman's Magazine* 139 (1922-04-22): 540; and 139 (1922-05-03): 603–604.

Kennion R.L. 1928. The rhinoceros in Nepal. *Field, the Country Gentleman's Magazine* 151 (1928-02-23): 333.

Kennion R.L. 1932. *Diversions of an Indian political.* Edinburgh, London.

Kessel O. van. 1862. *Colani de gemsenjager en andere reis- en jagtavonturen voor de jeugd.* Leiden.

Kevin Standage. 2020. The Konkan petroglyphs – Chave Dewood. https://kevinstandagephotography.wordpress.com/2019/03/14/the-konkan-petroglyphs-chave-dewood/ [accessed October 2020].

Khan A.M. 2009. *Taxonomy and distribution of rhinoceroses from the Siwalik Hills of Pakistan.* Thesis presented to University of the Punjab, Lahore.

Khan A.M.; Akhtar M.; Khan M.A.; Shaheen A. 2012. New fossil remains of Brachypotherium perimense from the Chinji and Nagri formations of Pakistan. *Journal of Animal and Plant Sciences* 22: 347–351.

Khan E. 1971. Punjabitherium, gen.nov., an extinct rhinocerotid of the Siwaliks, Punjab, India. *Proceedings of the Indian National Science Academy* 37 (2) A: 105–109.

Khan K.S.M.A.A.; Stapleton H.E. 1931. *Memoirs of Gaur and Pandua.* Calcutta.

Khan M. 1989. *Asian Rhinos: an action plan for their conservation.* Gland.

Khan M. 2014. *The lesser two-horned rhinoceros.* Kuala Lumpur.

Khan M.; Foose T.J. 1993. Asian Rhino Specialist Group. *Pachyderm*, Nairobi 18: 3–8.

Khan M.; Foose T.J.; Strien N.J. van. 2002. Asian Rhino Specialist Group report. *Pachyderm*, Nairobi 32: 9–11.

Khan M.; Foose T.J.; Strien N.J. van. 2003. Asian Rhino Specialist Group report. *Pachyderm*, Nairobi 34: 11–12.

Khan M.; Foose T.J.; Strien N.J. van. 2005. Asian Rhino Specialist Group report. *Pachyderm*, Nairobi 38: 16–18.

Khandalavala K.; Chandra M. 1965. *Miniatures and sculptures from the collection of the late Sir Cowasji Jehangir, Bart.* Bombay.

Khandalavala K.; Chandra M. 1969. *New documents of Indian painting – a reappraisal.* Bombay.

Khandalavala K.; Mittal J. 1974. The Bhāgavata MSS from Palam and Isarda: a consideration in style. *Lalit Kalā* 16: 28–31.

Kheiri S. 1921. *Indische Miniaturen der islamischen Zeit.* Berlin (Orbis Pictus, vol. 6).

Kincaid C.A. 1908. *The tale of the Tulsi plant.* Bombay.

King P. 1985. *The shooting field: one hundred and fifty years with Holland & Holland.* London.

King Pippin. 1870. Game in the western Dooars. *Oriental Sporting Magazine* (n.s.) 3 (June): 1192–1194.

Kinloch A.A.A. 1885. *Large game shooting in Thibet, the Himalayas, and Northern India.* Calcutta.

Kinloch A.A.A. 1892. *Large game shooting in Thibet, the Himalayas, Northern and Central India.* London.

Kinloch A.A.A. 1903. The great Indian rhinoceros. In: Aflalo F.G., *The sports of the world*, part 5. London, pp. 161–165 (reprinted in 1905).

Kinloch A.A.A. 1904. Indian rhinoceros shooting. In: Aflalo F.G., *The sportsman's book for India.* London, pp. 59–66.

Kinzelbach R. 2011a. Eine bunte Gesellschaft: die Tiere des Artemidor-Papyrus. *Antike Welt* 3 (11): 8–13.

Kinzelbach R. 2011b. Der Artemidor-Papyrus: Tierbilder aus dem ersten Jahrhundert. *Zoologie (Mitteilungen der Deutschen Zoologischen Gesellschaft)* 2011: 13–26.

Kinzelbach R. 2012a. *Augusta, das erste Panzernashorn in Europa: eine Natur- und Kulturgeschichte.* Hohenwarsleben.

Kinzelbach R. 2012b. An Indian Rhino, Rhinoceros unicornis, (V34) on the Artemidorus Papyrus: its position in the antique cultural tradition. In: Althoff J.; Föllinger S.; Wöhrle G., *Antike Naturwissenschaft und ihre Rezeption.* Trier, pp. 93–131.

Kipling R. 1898. 'Just so' stories: how the rhinoceros got his wrinkly skin. *St. Nicholas* 25 (4): 272–275.

Kipling R. 1902a. *Just so stories for little children.* Illustrated by the author [first edition]. London.

Kipling R. 1902b. *Just so stories.* New York: Doubleday & Co.

Kirkpatrick W. 1811. *Account of the Kingdom of Nepaul*, vol. 1. London.

Klenov V. 2007. From Odessa to Kathmandu. *Journal of the Britain-Nepal Society* no. 24: 11–17.

Klös H.G. 1966a. Kaziranga, ein Besuch des Panzernashorn-Reservats. *Kosmos* 62: 153–156.

Klös H.G. 1966b. Im Elefantengrasdschungel der Panzernashörner. *Zoo Berlin Jahresbericht* 1966: 1–15.

Klös H.G. 1969. *Von der Menagerie zum Tierparadies: 125 Jahre Zoo Berlin*. Berlin.

Klös H.G.; Frädrich H.; Klös U. 1994. *Die Arche Noah an der Spree: 150 Jahre Zoologischer Garten Berlin, eine tiergartnerische Kulturgeschichte von 1844–1994*. Berlin.

Klös H.G.; Klös U. 1980. Der Royal Chitwan Nationalpark in Nepal. *Bongo (Beiträge zur Tiergärtnerei und Jahresberichte aus dem Zoo Berlin)* 4: 43–48.

Klös H.G.; Klös U. 1990. *Der Berliner Zoo im Spiegel seiner Bauten 1841–1989: eine baugeschichtliche und denkmalpflegerische Dokumentation über den Zoologischen Garten Berlin*. Berlin.

Kloska T. 2014. King George V's second visit to India: The Durbar of 1911 and the royal hunting expedition in Nepal. *Ot Kontinens* 2: 33–44.

Knight C. 1844. *Natural history: the elephant as he exists in a wild state, and as he has been made subservient, in peace and in war, to the purposes of man*. New York.

Knight C. 1896. Two old-world beasts (female Sumatran rhinoceros in London Zoo). *Sketch*, London 16 (no. 197), 1896-11-04: 64–65.

Knighton W. 1855. *The private life of an eastern king*. By a member of the household of His late Majesty, Nussir-u-Deen, king of Oude. [also 2nd and 3rd printing.] London: Hope & Co.

Knighton W. 1855. *The private life of an eastern king*. New York: Redfield.

Knighton W. 1856. *The private life of an eastern king*. Compiled for a member of the household of His late Majesty, Nussir-u-Deen, king of Oude. New edition, revised. London and New York: G. Routledge & Co.

Knighton W. 1857. *The private life of an eastern king*. New edition, revised. London and New York.

Knighton W.; Révoil B.H. 1858. *Le Roi d'Oude, moeurs de l'Inde*. Paris: Gustave Havard.

Knighton W.; Révoil B.H. 1865. *La cour d'un roi d'Orient, ou les distractions de Nussir-u-Deen, souverain de Luknow*. Traduit de l'Anglais. Paris: J. Vermot.

Knighton W.; Thiele L. 1856. *Ein Indischer Königshof, von einem Mitgliede des Hofstaates zu Audh*. Nach dem Englischen. Leipzig: Carl B. Lorck.

Knowles S. 1889. *The gospel in Gonda*. Lucknow.

Knox M.V.B. 1891. *A winter in India and Malaysia among the Methodist missions*. New York.

Knox A.G. 2014. The first egg of Jerdon's courser Rhinoptilus bitorquatus and a review of the early records of this species. *Archives of Natural History* 41 (1): 75–93.

Kock R.; Amin R.; Subedi N. 2009. Postscript: Rogue army staff involved in poaching in Bardia National Park, Nepal, 2007–2008. *Pachyderm*, Nairobi 45: 119–120.

Koenigswald G.H.R. von. 1933. Beiträge zur Kenntnis der fossilen Säugetierfauna Javas. *Wetenschappelijke Mededelingen Dienst van den Mijnbouw in Nederlandsch-Indie* 23: 1–185.

Körber P. 1853. *Julius Bath's malerisch-romantische Reise nach Calcutta und seinen Umgebungen*. Nürnberg.

Koffler C.: see Ylla.

Koliyal A. 2003. *National studbook for the greater one horned rhinoceros (Rhinoceros unicornis)*. Dehradun.

Konwar P.; Saikia M.K.; Saikia P.K. 2009. Abundance of food plant species and food habits of Rhinoceros unicornis Linn. in Pobitora Wildlife Sanctuary, Assam, India. *Journal of Threatened Taxa* 1 (9): 457–460.

Kourist W. 1970. Die ersten einhörnigen Nashörner (Rhinoceros sondaicus Desmarest, 1822 und Rhinoceros unicornis L., 1758) der grossen Europäischen Zoologischen Gärten in der Malerei des 19. Jahrhunderts. *Zoologische Beiträge* 16: 141–154.

Kourist W. 1974. Nachtrag zu Teil I und Teil II von Frühe Haltung von Großsäugetieren. *Zoologische Beiträge* 20: 543–546.

Kourist W. 1976. *400 Jahre Zoo: im Spiegel der Sammlung*. Köln.

Krishne Gowda C.D. 1975. Plans for breeding colonies of large mammals in India. In: Martin R.D., *Breeding endangered species in captivity*. London, pp. 309–313.

Krumbiegel I. 1960. Die asiatischen Nashörner (Dicerorhinus Gloger und Rhinoceros Linné). *Säugetierkundliche Mitteilungen* 8: 12–20.

Krummel T. 1908. Sum-dawn-na Hlawk Tak (in Mizo). *Mizo leh Vai Chanchin Bu*, for months of April, May, July, October and December 1908.

Kühne-van Diggelen W. 1995. *Jan Albert Sichterman, VOC-dienaar en koning van Groningen*. Groningen.

Kühnel E. 1937. *Indische Miniaturen aus dem Besitz der Staatlichen Museen zu Berlin*. Berlin.

Kühnel E. 1941. Jagdbilder aus Indien. *Atlantis* 13: 417–424.

Kühnel E. 1962. *Die Kunst des Islam*. Stuttgart (Kröners Taschenausgabe).

Kühnel E.; Ettinghausen R. 1933. *Indische Miniaturen*. Berlin (Bilderhefte der Islamischen Kunstabteilung, Heft I).

Kumar A. 2013. Poachers kill rhino in tiger reserve (Valmikinagar). *Deccan Herald* 2013-04-13.

Kumar S. 1993. Taiwan accuses princess of smuggling rhino horn. *New Scientist* 140 (1993-10-16): 11.

Kumar U. 2010. Ecological aspects of reintroduced hand-raised Indian rhinoceros in Manas National Park. *Young Ecologists Talk and Interact*, Bangalore 2010: 135–136.

Kumar V. 2019. Catalogue of antiquities of State Archaeological Museum, Lucknow, U.P. India. Part I: Jain (from Kankali Tila, Mathura & other places of Mathura), Bauddha (from Mathura) & Gandhar antiquities. *Indian Journal of Archaeology* 4 (4): 62–1167.

Kumar V. 2020. Catalogue of terracottas kept at State Museum, Lucknow: Part II: Terracottas dateable from Gupta to Modern Period. *Indian Journal of Archaeology* 5 (4): 1–2942.

Kundu S.; Menon V. 2016. Possible sighting of a twin greater one-horned rhinoceros (Rhinoceros unicornis) in Jaldapara National Park, India. *Pachyderm*, Nairobi 57: 115–116.

Kunwar K.J. 2012. *Four years for the rhino: an experience of anti-poaching operations*. Kathmandu.

Kushwaha S.P.S. 2008. *Mapping of Kaziranga conservation area, Assam: project report*. Dehradun: Forestry and Ecology Division of Indian Institute of Remote Sensing.

Kuster. 1880. Hunting pass issued to Mr. Kuster. Horniman Museum, London: Manuscript 15460.

Kuster T. 2015. Hern Adam Hochreitters Schiffart und Rayss (Catalogue 2.2). In: Haag S., *Echt Tierisch: die Menagerie des Fürsten*. Innsbruck, pp. 132–133.

Kutzner J.G. 1857. *Die Reise seiner Königlichen Hoheit des Prinzen Waldemar von Preußen nach Indien in den Jahren 1844 bis 1846. Aus dem darüber erschienenen Prachtwerke im Auszuge mitgetheilt*. Berlin.

Kvaerne P. 1979. The visit of Prince Waldemar of Prussia to Nepal in February and March 1845. *Kailash* 7 (1): 35–50.

Lach D.F. 1965. *Asia in the making of Europe*, vol. 1: *The century of discovery, Book 1: The visual arts*. Chicago.

Ladvocat J.B. 1749. *Lettre sur le rhinocéros*. Paris.

Lahan P. 1993. *Present status and distribution of the Indian rhinos (Rhinoceros unicornis) in the wild in Assam and its habitat*. Report to the Assam Forest Department.

Lahan P.; Sonowal R.N. 1973. Kaziranga Wild Life Sanctuary, Assam: a brief description and report on the census of large animals (March 1972). *Journal of the Bombay Natural History Society* 70: 245–278.

Lahiri M. 2012. *Mapping India*. Delhi.

Lahiri R.M. 1954. *The annexation of Assam (1824–1854)*. New Delhi.

Lal Baloo. 1902. A shoot with H.H. the Maharajah of Cooch Behar. *Navy and Army Illustrated* 15 (1902-10-25): iv, 145.

Lally T. 1875. *Le Jardin des Plantes, quatrième serie: la girafe, l'hippopotame, l'élan, le rhinocéros, le renne, le sanglier*. Paris.

Lalthangliana B. 2017. *Culture and folklore of Mizoram*. Aizwahl.

Lalthanzara H. 2017. A systematic list of mammals of Mizoram. *Science Vision* 17: 104–121.

Lamare-Picquot C.A. 1835. Relation d'une chasse de rhinocéros sans corne, lequel constitue, sinon une espèce nouvelle, au moins une variété. In: Lamare-Picquot C.A., *Réponse pour servir de refutation aux opinions et à la critique du rapport de M. Constant Duméril, sur mon mémoir concernant les Ophidiens; suivie d'une relation de chasse dans les îles des bouches du Gange*. Paris, pp. 54–64.

Lamare-Picquot C.A. 1838. Rhinocerosjagd des Herrn M. Lamarepicquot auf den Inseln in den Mündungen des Gangesflusses. *Wanderer im Gebiete der Kunst und Wissenschaft, Industrie und Gewerbe, Theater und Gesellligkeit* 25 (no. 114) 1838-05-12: 453–455, and (no. 115) 1838-05-14: 458–459.

Lançon A.A. 1876. Rhinocéros renversant un cavalier. *La Mosaique: revue pittoresque illustrée de tous les temps et de tous les pays* 4: 393–394.

Landon P. 1928. *Nepal*, vol. 1. London.

Lang E.M. 1956. Breeding of the Indian rhinoceros at the Basel Zoo. *Zoo Life* 11: 126–128.

Lang E.M. 1961. Beobachtungen am indischen Panzernashorn (Rhinoceros unicornis). *Zoologische Garten* 25: 369–409.

Lang E.M. 1967. Einige biologische Daten vom Panzernashorn (Rhinoceros unicornis). *Revue Suisse de Zoologie* 74: 603–607.

Lang E.M. 1968. Asiatische Nashörner. In: Grzimek B., *Grzimeks Tierleben: Enzyklopädie des Tierreiches*, vol. 13: *Säugetiere 4*. Zürich, pp. 45–53.

Lang E.M. 1994. *Mit Tieren unterwegs: aus dem Reisebuch eines Zoodirektors*. Basel.

Lang H. 1924. Asiatic rhinoceroses secured by the Faunthorpe-Vernay expedition. *Natural History*, New York 24 (4): 527–528.

Lang H.; Vernay A.S. 1924. The Faunthorpe-Vernay expedition. *Natural History*, New York 24 (4): 525–526.

Larese A.; Sgreva D. 1996. *Le lucerne fittili del museo archeologico di Verona*, vol. 1. Verona.

Larive; Fleury ((pseudonyms). 1889. *Dictionnaire français illustré des mots et des choses, ou Dictionnaire encyclopédique des écoles, des métiers et de la vie pratique*, vol. 3. Paris.

Larsson E. 2021. "On Deposit": animal acquisition at the Zoological Society of London, 1870–1910. *Archives of Natural History* 48: 1–21.

La Touche T.H.D. 1910. The journals of Major James Rennell first surveyor-general of India, during his surveys of the Ganges and Brahmaputra rivers 1764 to 1767. *Memoirs of the Asiatic Society of Bengal* 3 (3): 95–248.

Laufer B. 1914. Chinese clay figures, part 1: Prolegonema on the history of defensive armor, chapter 1: History of the rhinoceros. *Publications of the Field Museum of Natural History* 13 (2): 73–173.

Laurie W.A. 1977. A most preposterous beast. Rhino fact and folklore – a potpourri. *International Wildlife* 7 (4): 4–16.

Laurie W.A. 1978a. *The ecology and behaviour of the greater one-horned rhinoceros*. Thesis presented to University of Cambridge.

Laurie W.A. 1978b. Hoffnung für die Dicken? *Geo Magazin* 1978 (6): 88–102.

Laurie W.A. 1982. Behavioural ecology of the greater one-horned rhinoceros (Rhinoceros unicornis). *Journal of Zoology*, London 196: 307–341.

Laurie W.A. 1983. Nashörner in Asien. *Bongo (Beiträge zur Tiergärtnerei und Jahresberichte aus dem Zoo Berlin)* 7: 1–16.

Laurie W.A. 1984. The rhinoceros in Asia. *International Zoo News* 31: 4–12.

Laurie W.A. 1986. Rhinoceroses. In: Hawkins R.E., *Encyclopedia of Indian natural history*. Bombay Natural History Society, pp. 469–471.

Laurie W.A. 1997. Das indische Panzernashorn. In: [Gansloßer U.], *Die Nashörner: Begegnung mit urzeitlichen Kolossen*. Fürth, pp. 94–113.

Laurie W.A.; Lang E.M.; Groves C.P. 1983. Rhinoceros unicornis. *Mammalian Species* no. 211: 1–6.

Lavers C. 2009. *The natural history of unicorns*. London.

Lawley A.G. 1933. *Algy Lawley*. Plymouth [copy in Bishopsgate Institute, London].

Lawrence H. 1980. *The journals of Honoria Lawrence: India observed 1837–1854*. Edition edited by John Lawrence and Audrey Woodiwiss. London.

Leach L.Y. 1995. *Mughal and other Indian paintings from the Chester Beatty Library*, vol. 2. London.

Leaming L. 2013. Blog about Bhutan. https://www.goodreads.com/author/show/4397507.Linda_Leaming/blog?page=3 [accessed June 2021].

Lee H. 1872. Hairy rhinoceros at Calcutta. *Land and Water* 13 (1872-01-06): 10.

Leeds Times 1856-08-02: Sale of King's menagerie in Oudh.

Leger Galleries. 1976. *Exhibition of watercolours, November 15th–December 24th 1976*. London.

Leicester Daily Post 1891-09-14: The game-book of the Maharaja of Cooch Behar.

Lemon C. 1845. Presents made to the Institution, from 28 October 1843 to 10 November 1844. *Annual Report of the Royal Institution of Cornwall* 26: 17–19.

Lereboullet A. 1837. Notice sur le Musée d'Histoire Naturelle de Strasbourg: Partie historique. *Revue d'Alsace* 2: 131–165.

Lesson R.P. 1836. *Complément des oeuvres de Buffon ou histoire naturelle des animaux rares découverts par les naturalistes et les voyageurs depuis la mort de Buffon*, vol. 10: *Oiseaux et mammifères*. [second title] *Histoire naturelle générale et particulière des mammifères et oiseaux découverts depuis la mort de Buffon: Oiseaux et mammifères*. Paris.

Lesson R.P. 1838. *Compléments de Buffon: Races humaines et mammifères*, 2nd edition (revue, corrigée et augmentée par l'auteur), vol. 1. Paris.

Lesson R.P. 1842. *Nouveau tableau du règne animal: mammifères*. Paris.

Lethbridge R. 1893. *The golden book of India, a genealogical and biographical dictionary*. London.

Leutemann H. 1872. Aus dem Tigerleben. *Gartenlaube: illustrirtes Familienblatt* no. 48: 785–788.

Leutemann H. 1875. Von einem Dickhäuter. *Gartenlaube: illustrirtes Familienblatt* no. 37: 626–628.

Levesque D. 1888. Le Duc d'Orléans aux Indes. *Univers Illustré* (*Journal hebdomadaire*) 31: 618–619, 631–632.

Lévi S. 1905. *Le Népal: étude historique d'un royaume Hindou*, vol. 1. Paris.

Lévi S. 1925. Le Maharaja du Népal, Grand-Officier de la Légion d'Honneur. *L'Illustration, Journal Universel* 83 (1925-01-17): 56–59.

Levine E.I.; Plekhov D. 2019. Reconsidering Rag-i Bibi: authority and audience in the Sasanian east. *Afghanistan* 2: 233–260.

Lewin T.H. 1869. *The hill tracts of Chittagong and the dwellers therein*. Calcutta.

Lewin T.H. 1870. *Wild races of south-eastern India*. London.

Lewin T.H. 1875. [Notes.] In: Grote A., Memoir of Edward Blyth. *Journal of the Asiatic Society of Bengal* 44 (Extra no.): i–xxiv.

Lewin T.H. 1885. *A fly on the wheel: or, how I helped to govern India*. London.

Lewin T.H. 1912. *A fly on the wheel: or, how I helped to govern India*, 2nd edition. London.

Lewis Lt-Col. 1893. My first and only rhinoceros hunt. *Boy's Own Paper* 15 (1893-09-16): 811.

Leyrat J.C. 1982. *Contribution à l'étude des rhinocéros asiatiques*. Thesis presented to National Veterinary College, Alfort.

Library of Congress Copyright Office. 1919. *Catalogue of copyright entries*, part 4. New series, vol. 14, no. 1, p. 17786. Keystone View Co.: Set of 48 photographs, with captured rhinoceros, 7825. Date Dec. 12, 1919. Washington.

Lichterfeld F. 1872. Der Zoologische Garten zu Berlin. *Illustrirte Zeitung* no. 1505 (1872-05-04): 327–328.

Lichterfeld F. 1874. Die Dickhäuter des Berliner Zoologischen Gartens und ihr Palast. *Illustrirte Zeitung* no. 1642 (1874-08-15): 136–138.

Liechty M. 2017. *Far out: countercultural seekers and the tourist encounter in Nepal*. Chicago.

Liechty M. 2017. Jung Bahadur Coapsingha: John Coapman, hunting, and the origins of adventure tourism in Nepal. In: *Far out: countercultural seekers and the tourist encounter in Nepal*. Chicago, pp. 94–125.

Lienhard S. 1974. *Nevarigitimanjari: religious and secular poetry of the Nevars of the Kathmandu Valley*. Stockholm.

Liesen B. 2001. *Jacques Arago: littérateur, auteur dramatique et voyageur français (1790–1855)*. Private report.

Linden A. 1876. *Le petit Buffon illustré des enfants: histoire récréative des animaux d'après les meilleurs auteurs*. Paris.

Linlithgow (Hope V.A.J.). 1938a. Lord Linlithgow Indian Films, 1938–1941. http://collections-search.bfi.org.uk/web/Details/ChoiceFilmWorks/150001125 [accessed March 2021].

Linlithgow (Hope V.A.J.). 1938b. *Speech by His Excellency the Viceroy at Bhikna Thori on 10 December 1938*. [Published in volume with other speeches, without title page].

Linnaeus C. 1735. *Systema naturae, sive regna tria naturae systematice proposita per classes, ordines, genera, & species* [edition 1]. Lugduni Batavorum.

Linnaeus C. 1740a. *Systema naturae, in quo naturae regna tria, secundum classes, ordines, genera, species systematice proponuntur*, edition 2. Stockholmiae.

Linnaeus C. 1740b. *Systema naturae*, edition 3. Halle.

Linnaeus C. 1744. *Systema naturae*, edition 4. Parisiis.

Linnaeus C. 1747. *Systema naturae*, edition 5. Halae Magdeburgicae.

Linnaeus C. 1748a. *Systema naturae*, edition 6. Stockholmiae.

Linnaeus C. 1748b. *Systema naturae*, edition 7. Lipsiae.

Linnaeus C. 1756. *Systema naturae*, edition 9, vol. 1. Lugduni Batavorum.

Linnaeus C. 1758. *Systema naturae per regna tria naturae, secundum classes, ordines, genera, species, cum characteribus, differentiis, synonymis, locis*. Editio decima, reformata [edition 10], vol. 1. Holmiae.

Linnaeus C. 1760. *Systema naturae*, edition 11, vol. 1. Halae Magdeburgicae.

Linnaeus C. 1766. *Systema naturae*, edition 12, vol. 1. Holmiae.

Linschoten J.H. van. 1596. *Itinerario. Voyage ofter schipvaert naer Oost ofte Portugaels Indien*. Amstelredam.

Linschoten J.H. van. 1598. *His discours of voyages into the Easte & West Indies*. London.

Liu S.; Dalen L.; Gilbert T.; Rookmaaker L.C.; and 35 others. 2021. Ancient and modern genomes unravel the evolutionary history of the rhinoceros family. *Cell* 184: 4874–4885. https://doi.org/10.1016/j.cell.2021.07.032.

Liverpool Mail 1842-04-30: Rhinoceros in Liverpool Zoological Gardens.

Liverpool Mercury 1833-05-31: Opening of Liverpool Zoological Gardens.

Liverpool Mercury 1834-06-27: Arrival of rhinoceros owned by Atkins.

Liverpool Standard and General Commercial Advertiser 1842-05-06: Rhinoceros in Liverpool Zoological Gardens.

Liverpool Standard and General Commercial Advertiser 1848-06-13: Thomas Atkins and family.

Liverpool Zoo. 1834. *List of the animals in the Liverpool Zoological Gardens, with notices respecting them*. Liverpool.

Liverpool Zoo. 1841. *The visitor's handbook to the Liverpool Zoological Gardens with notices respecting the animals*. Liverpool.

Llewellyn-Jones R. 2000. *Engaging scoundrels: true tales of old Lucknow*. Delhi.

Llewellyn-Jones R. 2016. *The last king in India: Wajid Ali Shah, 1822–1887*. Delhi.

Loch C.W. 1937. Rhinoceros sondaicus: the Javan or lesser one-horned rhinoceros and its geographical distribution. *Journal of the Malayan Branch of the Royal Asiatic Society* 15: 130–149.

Lohani U. 2011a. Traditional uses of animals among Jirels of Central Nepal. *Ethno Med* 8: 115–124.

Lohani U. 2011b. Eroding ethnozoological knowledge among Magars in Central Nepal. *Indian Journal of Traditional Knowledge* 10: 466–473.

Loisel G. 1912. *Histoire des ménageries de l'antiquité à nos jours*, 3 vols. Paris.

London Daily News 1874-06-16: Shipping arrivals.

London Evening Standard 1835-08-08: Donation by the Nawab of Oudh.

Long J. 1978. *Rajmala: or, an analysis of the chronicles of the kings of Tripura*. Calcutta.

Longair S.; Jones C.S. 2018. Prize possession: the 'silver coffer' of Tipu Sultan and the Fraser family. In: Finn M.; Smith K., *East India Company at home, 1757–1857*. London, pp. 25–38.

Longley P.R.H. 1969. *Tea planter sahib: the life and adventures of a tea planter in North East India*. Auckland.

Longstaff T.G. 1950. *This my voyage*. London.

Lonsdale, Lord (Lowther H.C.). 1899. Attending a hunt at Cooch Behar. *Sketch*, London 1899-04-19: 541.

Lopes A.A. 2014. Civil unrest and the poaching of rhinos in the Kaziranga National Park, India. *Ecological Economics* 103: 20–28.

Lorrain R.A. 1912. *Five years in unknown jungles for God and empire*. London.

Losty J.P. 2002. Towards a new naturalism: portraiture in Murshidabad and Avadh, 1750–80. *Mārg Magazine* 54 (4): 38–55.

Losty J.P. 2015. *Sita Ram's painted views of India: Lord Hastings's journey from Calcutta to the Punjab, 1814–15*. London.

Lott D.F.; MacCoy M. 1995. Asian rhinos Rhinoceros unicornis on the run? Impact of tourist visits on one population. *Biological Conservation* 73: 23–26.

Low M.E.Y. 2019. *Voyageurs, explorateurs et scientifiques: the French and natural history in Singapore*. Singapore.

Lowther D.A. 2019. The art of classification: Brian Houghton Hodgson and the 'Zoology of Nipal.' *Archives of Natural History* 46: 1–23.

Lubbock B. 1934. *Barlow's Journal of his life at sea in King's ships, East and West Indiamen and other merchantmen from 1659 to 1703*, vol. 2. London.

Lucas F.A. 1920. The unicorn and his horn. *Natural History*, New York 20: 532–535.

Lucas's Repository. 1855. Auction of two rhinoceroses from Assam. *Illustrated London News* 26 (1855-06-02): 15.

Lüps P. 1978. Eindrücke von einer Sammelreise zu den Panzernashörnern. *Jahreshefte des Vereins für vaterländische Naturkunde in Württemberg* 6: 109–123.

Lütken C.F. 1888. Det Zoologiske Museum 1887. *Aarbog for Kjøbenhavns Universitet* 1887–1888: 460–466.

Lumsden T. 1822. *A journey from Merut in India, to London, during the years 1819 and 1820*. London.

Lundeberg A.; Seamour F. 1910. *Big game hunting in Africa and other lands*. London.

Lunden S. 1994. A Nepalese labyrinth. *Caerdroia* 26: 117–134.

Lundgren E. 1872. *En malares anteckningar. Utdrag ur dagbocker och bref: Indien*. Stockholm.

Lunsingh Scheurleer P. 1996. Het Witsenalbum: zeventiende-eeuwse Indiase portretten op bestelling. *Bulletin van het Rijksmuseum* 44: 167–254.

Lunsingh Scheurleer P.C.M. 1978. *Miniaturen uit India: de verzameling van Dr. P. Formijne*. Amsterdam.

Luttman-Johnson H. 1988. The earthquake in Assam. *Journal of the Society of the Arts* 46: 473–493.

Lydekker R. 1876. Indian tertiary and post-tertiary vertebrata: Molar teeth and other remains of Mammalia. *Memoirs of the Geological Survey of India, Palaeontologia Indica*, Series 10, vol. 1 part 2: 1–69, pls. 4–6.

Lydekker R. 1878. Notices of Siwalik mammals. *Records of the Geological Survey of India* 11: 64–104.

Lydekker R. 1879. Further notices of Siwalik Mammalia. *Records of the Geological Survey of India* 12: 33–52.

Lydekker R. 1880a. Notes on the dentition of rhinoceros. *Journal of the Asiatic Society of Bengal* 49 (2): 135–142.

Lydekker R. 1880b. A sketch of the history of the fossil vertebrates of India. *Journal of the Asiatic Society of Bengal* 49 (2): 8–40.

Lydekker R. 1880c. Indian tertiary and post-tertiary vertebrata: Preface (to volume 1). *Memoirs of the Geological Survey of India, Palaeontologia Indica*, Series 10, vol. 1: vi–xix.

Lydekker R. 1881a. Indian tertiary and post-tertiary vertebrata: Siwalik Rhinocerotidae. *Memoirs of the Geological Survey of India, Palaeontologia Indica*, Series 10, vol. 2 part 1: 1–62, pls. 1–10.

Lydekker R. 1881b. Indian tertiary and post-tertiary vertebrata: Introductory observations (to volume 2). *Memoirs of the Geological Survey of India, Palaeontologia Indica*, Series 10, vol. 2: ix–xv.

Lydekker R. 1884. Indian tertiary and post-tertiary vertebrata: Additional Siwalik perissodactyla and proboscidea. *Memoirs of the Geological Survey of India, Palaeontologia Indica*, Series 10, vol. 3 part 1: 1–11, pls. 1–5.

Lydekker R. 1885. *Catalogue of the remains of Siwalik vertebrata contained in the Geological Department of the Indian Museum, Calcutta*, part 1: *Mammalia*. Calcutta.

Lydekker R. 1886a. *Catalogue of the remains of pleistocene and pre-historic vertebrata, contained in the Geological Department of the Indian Museum, Calcutta*. Calcutta.

Lydekker R. 1886b. Indian tertiary and post-tertiary vertebrata: The fauna of the Karnul caves. *Memoirs of the Geological Survey of India, Palaeontologia Indica*, Series 10, vol. 4 part 2: 40–58.

Lydekker R. 1886c. Preliminary note on the Mammalia of the Karnul Caves. *Records of the Geological Survey of India* 19: 120–122.

Lydekker R. 1886d. *Catalogue of the fossil Mammalia in the British Museum (Natural History), part III containing the order Ungulata, suborders Perissodactyla, Toxodontia, Condularthia, and Amblypoda*. London.

Lydekker R. 1892. Notes on rhinoceroses, ancient and modern. *Field, the Country Gentleman's Magazine* 79 (1892-06-18): 903.

Lydekker R. 1894. *The royal natural history*, vol. 2. London.

Lydekker R. 1897. *Library of natural history*, vol. 2. New York.

Lydekker R. 1900. *The great and small game of India, Burma and Tibet*. London.

Lydekker R. 1903. Indian rhinoceros at South Kensington. *Field, the Country Gentleman's Magazine* 101 (1903-02-14): 265.

Lydekker R. 1905. The Singpho rhinoceros. *Field, the Country Gentleman's Magazine* 106 (1905-07-22): 152.

Lydekker R. 1907. *The game animals of India, Burma, and Tibet*. London.

Lydekker R. 1908. The great Indian rhinoceros. *Field, the Country Gentleman's Magazine* 111 (1908-03-21): 486.

Lydekker R. 1909. Oriental big game notes. *Field, the Country Gentleman's Magazine* 113 (1909-05-29): 923.

Lydekker R. 1910. The Indian rhinoceros. *Field, the Country Gentleman's Magazine* 115 (1910-01-15): 53.

Lydekker R. 1913. *Catalogue of the heads and horns of Indian big game bequeathed by A.O. Hume to the British Museum (Natural History)*. London.

Lydekker R. 1916. *Catalogue of the ungulate mammals in the British Museum (Natural History), vol. 5: Perissodactyla*. London, pp. 46–58.

Lydekker R. 1918. *Wild life of the world: a descriptive survey of the geographical distribution of animals*, vol. 2. London and New York.

Lynam R. 1961. Tiger hunt and ring around a rhino. *Life Magazine* 1961-03-24: 45–54.

Lyons T. 2004. *The artists of Nathadwara: the practice of painting in Rajasthan*. Indiana, Ahmedabad.

Lytton V.A.G.R. 1942. *Pundits and elephants: being the experiences of five years as governor of an Indian province*. London.

Maarel F.H. van der. 1932. Contribution to the knowledge of the fossil mammalian fauna of Java. *Wetenschappelijke Mededelingen van den Dienst Mijnbouw* 15: 1–208.

MacCracken G.F.; Brennan E.J. 1993. Genetic variation in the greater one-horned rhino and implications for population structure. In: Ryder O.A., *Rhinoceros biology and conservation: Proceedings of an international conference, San Diego, U.S.A.*, pp. 228–237.

MacCulloch W. 1859. *Account of the valley of Munnipore and of the hill tribes*. Calcutta.

MacDonald J. 1872. The Lushai expedition. *Proceedings of the Royal Geographical Society* 17: 42–55.

MacDonell J.C. 1929. Early days of forestry in India. *Empire Forestry Journal* 8: 85–97.

MacDougal C. 1978. Royal Chitawan Nepal's tiger land. *Wildlife* 20: 20–25.

MacFarlane H. 1900. Sportsmen in purple. *Badminton Magazine of Sports and Pastimes* 11: 522–528.

MacGregor A. 2018. *Company curiosities: nature, culture and the East India Company, 1600–1874*. Chicago.

MacGregor C.R. 1887. Journey of the expedition under Col. Woodthorpe from upper Assam to the Irawadi, and return over the Patkoi range. *Proceedings of the Royal Geographical Society* 9: 19–42.

MacKarness C.G.M. 1937. *Progress report of forest administration in Assam 1936–1937*. Shillong.

MacKarness C.G.M. 1938. *Progress report of forest administration in Assam 1937–1938*. Shillong.

Mackay E. 1938. *Die Induskultur: Ausgrabungen in Mohenjo-Daro und Harappa*. Leipzig.

Mackay E.J.H. 1943. *Chanhu-Daro excavations 1935–36*. New Haven.

MacLelland J. 1839. List of mammals and birds collected in Assam. *Proceedings of the Zoological Society of London* 1839-10-22: 146–167.

Madras Weekly Mail 1902-10-30, p. 26: Death of John Cockburn.

Magee J. 2013. *Images of nature: the art of India*. London.

Maggs. 1986. Oriental miniatures & illumination. *Maggs Bulletin* no. 40.

Magon de Clos Doré M. 1822. *Souvenirs d'un voyageur en Asie, depuis 1802 jusqu'en 1815 inclusivement*. Paris.

Mahajan M. 2003. *Orissa, from place names in inscriptions*. Delhi.

Mahanta P.; Bora A.K. 2001. Flood and erosion hazards in the Kaziranga National Park of Assam, India. *Revista Sociedade e Natureza* 13 (25): 37–45.

Mahato G.P.S.; Dahal A. 2021. Boris Lissanevitch and Nepali tourism: history revisited. *Journal of Tourism and Himalayan Adventures* 3: 11–22.

Mahlmann H. 1853. Die Reise durch Hindostan. Von Kalkutta über Patna, Katmandu, Benares und Delhi nach Naini Tal, 3. Januar bis 27. Mai 1845. In Mahlmann H. (ed.), *Zur Erinnerung an die Reise des Prinzen Waldemar von Preußen nach Indien in den Jahren 1844–1846*. Berlin, vol. 1, part 2, pp. 1–45.

Mahul A. 1825. Duvaucel (Alfred). *Annuaire Nécrologique, ou Complément annuel et Continuation de toutes les Biographies* 1824: 109–110.

Majupuria T.C. 1977. *Sacred and symbolic animals of Nepal*. Kathmandu.

Malcolm J. 1836. *The life of Robert, Lord Clive: collected from the family papers communicated by the Earl of Powis*, vol. 2. London.

Maliepaard C.H.J.; Vos A. de. 1961. *Dieren sterven uit: de fossielen van morgen in vijf werelddelen*. Amsterdam.

Mallick J.K. 2011. Status of the mammal fauna in Sundarban Tiger Reserve, West Bengal, India. *Taprobanica* 3: 52–68.

Mallick J.K. 2015. Ecological crisis vis-à-vis intraspecific conflict: a case study with rhinos in Jaldapara and Gorumara National Parks, West Bengal, India. In: Gupta V.K.; Verma A.K., *Animal diversity, natural history and conservation*, vol. 5. Delhi, pp. 335–366.

Mallick J.K. 2019. An updated checklist of the mammals of West Bengal. *Journal on New Biological Reports* 8: 37–123.

Man E.G. 1867. *Sonthalia and the Sonthals*. Calcutta.

Manchester Courier and Lancashire General Advertiser 1836-12-03: Arrival of rhinoceros.

Mancini E. 1899. I viaggi e le caccie del Conte di Torino in India. *Illustrazione Italiana* 26 (1899-07-30): 68–73.

Mandal A.K.; Nandi N.C. 1989. *Fauna of Sundarban mangrove ecosystem, West Bengal, India*. Kolkata.

Mandala V.R. 2019. *Shooting a tiger: big-game hunting and conservation in colonial India*. New Delhi.

Mandelslo J.A. de. 1727. *Voyages célèbres et remarquables, faits de Perse aux Indes Orientales*, vol. 1. Amsterdam.

Mandelslo J.A. de 1669. Travels into the Indies. In: Olearius A., *The voyages and travells of the Ambassadors sent by Frederick Duke of Holstein; whereto are added the travels of John Albert de Mandelslo*. London, pp. 71–125.

Mani V. 1975. *Purāṇic encyclopaedia: a comprehensive dictionary with special reference to the Epic and Purāṇic literature*. Delhi, Patna, Varanasi.

Manners Smith J. 1909. Haunts of the Indian rhinoceros. *Field, the Country Gentleman's Magazine* 114 (1909-07-24): 177 (Reprinted in *Journal of the Bombay Natural History Society* 19: 746–747).

Manrique S. 1643. Itinerario de las Missiones Orientales. In: Luard C.E. et al. 1927. *Travels of Fray Sebastien Manrique 1629–1643*. London: Hakluyt Society, Second Series, vols. 59, 61.

Manrique S. 1649. *Itinerario de las Missiones del India Oriental*. Roma.

Manson A. 1876. Unusual visit by a wild rhinoceros. *Oriental Sporting Magazine* 9 (May): 176–178.

Manucci N. 1907. *Storia do Mogor or Mogul India 1653–1708*. London.

Manuel J. 2005. Harappan environment as one variable in the preponderance of rhinoceros and paucity of horse. *Purātattva* 35: 21–27.

Manuel J. 2006. Cohabitation of rhinoceros and man in the Narmada Valley. *Journal of Academy of Indian Numismatics and Sigillography* 21–22: 629–636.

Manuel J. 2007. Portrayal of rhinoceros in art: some questions. In: Reddy P.C., *Exploring the mind of ancient man*. Delhi, pp. 232–238.

Manuel J. 2008. Depiction of rhinoceros: transition from popular art to state sponsored art. In: Mani B.R.; Tripathi A., *Expressions in Indian art*. Delhi, vol. 1, pp. 33–38.

Maori: see Inglis J.

Markel S.; Gude T.B. 2010. *Une cour royale en Inde: Lucknow, XVIIIᵉ–XIXᵉ siècle*. Paris. [Also: *India's fabled city: the art of courtly Lucknow*. Los Angeles].

Markham R.A.D. 1990. *A rhino in High Street: Ipswich Museum, the early years*. Ipswich.

Marshall J. 1931. *Mohenjo-Daro and the Indus civilisation*, vol. 2. London.

Marshall J.; Foucher A. 1902. *The monuments of Sanchi*, vol. 1. London.

Marston M. 1889. *Korno Siga, the mountain chief; or, life in Assam*. Philadelphia.

Martens E. von. 1876. *Die Preussische Expedition nach Ost-Asien, nach amtlichen Quellen. Zoologischer Theil*, vol. 1: *Allgemeines und Wirbelthiere*. Berlin.

Martin E.B. (Bradley) 1983. *Rhino exploitation*. Hong Kong.

Martin E.B. 1984. They're killing off the rhino. *National Geographic* 165: 404–422.

Martin E.B. 1985. Religion, royalty and rhino conservation in Nepal. *Oryx* 19: 11–16.

Martin E.B. 1992. The poisoning of rhinos and tigers in Nepal. *Oryx* 26: 82–86.

Martin E.B. 1996a. Smuggling routes for West Bengal's rhino horn and recent successes in curbing poaching. *Pachyderm*, Nairobi 21: 28–34.

Martin E.B. 1996b. The importance of park budgets, intelligence networks and competent management for successful conservation of the greater one-horned rhinoceros. *Pachyderm*, Nairobi 22: 10–17.

Martin E.B. 1998. Nepal destroys large stocks of wildlife products. *Pachyderm*, Nairobi 25: 107–108.

Martin E.B. 1999. West Bengal – committed to rhino conservation yet a major entrepot for endangered wildlife products. *Pachyderm*, Nairobi 27: 105–112.

Martin E.B. 2001. What strategies are effective for Nepal's rhino conservation: a recent case study. *Pachyderm*, Nairobi 31: 42–51.

Martin E.B. 2004. Rhino poaching in Nepal during an insurgency. *Pachyderm*, Nairobi 36: 87–98.

Martin E.B. 2006. Policies that work for rhino conservation in West Bengal. *Pachyderm*, Nairobi 41: 74–84.

Martin E.B. 2010. *From the jungle to Kathmandu: horn and tusk trade*. Lalitpur.

Martin E.B.; Martin C.P. 1982. *Run rhino run*. London.

Martin E.B.; Martin C.P. 2006. Insurgency and poverty: recipe for rhino poaching in Nepal. *Pachyderm*, Nairobi 41: 61–73.

Martin E.B.; Martin C.P. 2010. Enhanced community support reduces rhino poaching in Nepal. *Pachyderm*, Nairobi 48: 48–56.

Martin E.B.; Martin C.P.; Vigne L. 2009a. Recent political disturbances in Nepal threaten rhinos. *Pachyderm*, Nairobi 45: 98–107.

Martin E.B.; Talukdar B.K.; Vigne L. 2009b. Rhino poaching in Assam: challenges and opportunities. *Pachyderm*, Nairobi 46: 25–34.

Martin E.B.; Vigne L. 1989. Kaziranga's calamity: a new threat to the Indian rhino. *Oryx* 23: 124–125.

Martin E.B.; Vigne L. 1995. Nepal's rhino – one of the greatest conservation success stories. *Pachyderm*, Nairobi 20: 10–26.

Martin E.B.; Vigne L. 2012. Successful rhino conservation continues in West Bengal, India. *Pachyderm*, Nairobi 51: 27–37.

Martin M. 1949. *Out in the mid-day sun*. Boston.

Martin P.L. 1882. *Illustrirte Naturgeschichte der Thiere*, vol. 1: *Säugethiere*. Leipzig.

Martin R.M. 1838. *The history, antiquities, topography, and statistics of Eastern India*, vol. 1: *Behar (Patna City) and Shahabad* (1838a); vol. 2: *Bhagulpoor, Goruckpoor, and Dinajepoor* (1838b); vol. 3: *Puraniya, Ronggopoor, and Assam* (1838c). London.

Martin R.M. 1838d. *Historical documents of eastern India*, vol. 7: *Ronggopur*. London.

Martineau A. 1920. *Dupleix et l'Inde Française 1722–1741*. Paris.

Martinelli A.; Michell G. 2004. *Palaces of Rajasthan*. Mumbai.

Marvin G.; McHugh S. 2014. *Routledge handbook of human animal studies*. New York.

Marx E.; Koch A. 1910. Neues aus der Schausammlung: das indische Nashorn. *Bericht des Senckenbergischen Naturforschenden Gesellschaft* 41: 161–171.

Mary P.O.; Solanki G.S.; Limboo D.; Upadhaya K. 1998. Observations on feeding and territorial behaviour of Indian rhino (Rhinoceros unicornis) in Kaziranga National Park, Assam, India. *Tiger Paper* (FAO) 25 (4): 25–27.

Mathew J. 2015. Edward Blyth, John M'Clelland, the curatorship of the Asiatic Society's collections and the origins of the Calcutta journal of natural history. *Archives of Natural History* 42: 265–278.

Mathpal Y. 1987. *Prehistoric painting of Bhimbetka*. Delhi.

Mathur P.; Midha N. 2008. *Mapping of national parks and wildlife sanctuaries: Dudhwa Tiger Reserve*. Dehradun.

Matthew W.D. 1929. Critical observations upon Siwalik mammals. *Bulletin of the American Museum of Natural History* 56: 437–560.

Mazumdar J.; Mahanta P. 2016. *Rhino: pride of Assam*. Guwahati.

MCB: see Nripendra 1908.

McCosh J. 1836. Asiatic intelligence: zoology of Assam. *Asiatic Journal and Monthly Register* 20: 30–31.

McCosh J. 1837. *Topography of Assam*. Calcutta.

McInerney T. 2011. Mir Kalan Khan. In: Beach M.; Fischer E.; Goswamy B.N., *Masters of Indian painting, 1650–1900*. Ascona, vol. 2, pp. 607–622.

M.D.: see Fayrer J.

Meghani K. 2017. *Splendours of the subcontinent: a Prince's tour of India, 1875–6*. London.

Mehra S. 1993. *Rajasthan: une longue romance*. New Delhi, Paris.

Meiklejohn W. 1936. *Annual report on game preservation in Bengal 1935–1936*. Calcutta.

Meiklejohn W. 1937. *Annual report on game preservation in Bengal 1936–1937*. Calcutta.

Meisner D. 1638. *Sciographia Cosmica: das ist Newes Emblematisches Büchlein*, part 7 (Littera G). Nürnberg.

Melladew H. 1909. *Sport and travel papers*. London.

Melville R.V. 1977. Opinion 1080: Didermocerus Brookes, 1828 (Mammalia) suppressed under the plenary powers. *Bulletin of Zoological Nomenclature* 34 (1): 21–24.

Menon G. 2014. *Evergreen leaves: recollections of my journeys into wild India*. Delhi.

Menon J.; Varma S.; Dayal S.; Sidhu P.B. 2008. Indor Khera revisited: excavating a site in the upper Ganga plains. *Man and Environment* 33: 88–98.

Menon K.D. 1975. *Tripura district gazetteers: Tripura*. Agartala.

Menon V. 1995. *Under siege: poaching and protection of greater one-horned rhinoceroses in India*. Delhi (Traffic India).

Menzies A.C. 1917. *Lord William Beresford: some memories of a famous sportsman, soldier and wit*. London.

Metzger H.; Brijraj Singh of Kotah. 2000. *Festivals and ceremonies observed by the Royal Family of Kotah*. Zürich and Kotah.

Meuschen F.C. 1787. *Museum Geversianum*. Rotterdam.

Meyerheim P. 1881. Ein Rhinoceroskampf im Zoologischen Garten zu Berlin. *Illustrirte Zeitung*, Leipzig 1881-08-06: 119–120.

Michel E. 1882. Aux Indes. *La Croix* 1882-06-01: 68–75.

Michel E. 1893. *Le tour du monde en 240 jours*, vol. 3: *L'Hindoustan*. Limoges.

Michell G. 1983. *Brick temples of Bengal: from the archives of David McCutchion*. Princeton, NJ.

Michell G. 1994. *The royal palaces of India*. Photographs by Antonio Martinelli. London.

Mighetto L. 1991. *Wild animals and American environmental ethics*. Tucson.

Milligan J.A. 1919. *Final report on the survey and settlement operations in the Jalpaiguri District, 1906–1916*. Calcutta.

Milliken T. 2018. Obituary: Esmond Martin. *Pachyderm*, Nairobi 59: 132–133.

Mills J.P. 1922. *The Lhota Nagas*. London.

Mills J.P. 1944. Tour note on Tirap Frontier Tract, December 1944. http://himalaya.socanth.cam.ac.uk/collections/naga/record/r73734.html.

Milroy A.J.W. 1916. The North Kamrup Game Sanctuary, Assam. *Indian Forester* 42: 452–464.

Milroy A.J.W. 1930. *Progress report of forest administration in Assam 1929–1930*. Shillong.

Milroy A.J.W. 1931. *Progress report of forest administration in Assam 1930–1931*. Shillong.

Milroy A.J.W. 1932. Game preservation in Assam. *Journal of the Society for the Preservation of the Fauna of the Empire* 16: 28–38.

Milroy A.J.W. 1933. *Progress report of forest administration in Assam 1932–1933*. Shillong.

Milroy A.J.W. 1934a. Assam fauna preservation. *Journal of the Society for the Preservation of the Fauna of the Empire* 22: 55–61.

Milroy A.J.W. 1934b. *Quinquennial review 1929–1934, and progress report of forest administration in Assam 1933–1934*. Shillong.

Milroy A.J.W. 1935. *Progress report of forest administration in Assam 1934–1935*. Shillong.

Milroy A.J.W. 1936. *Progress report of forest administration in Assam 1935–1936*. Shillong.

Milroy C.W. 1933. Big game's chances in Assam. *Field, the Country Gentleman's Magazine* 161 (1933-03-04): 413.

Milton O.; Estes R.D. 1963. Burma wildlife survey 1959–1960. *Publications of the American Committee of International Wildlife Protection* 15: 1–72.

Minto, Mary Caroline. 1934. *India: Minto and Morley 1905–1910*. London.

Miri M.C. 1939. Note on the rhinoceros captured during last October for the American zoo. *Indian Forester* 65: 207–209.

Mishra B. 2003. Early history of the Tel River valley, Orissa. *Proceedings of the Indian History Congress* 64: 197–206.

Mishra H.R. 2008. *The soul of the rhino*. (With Jim Ottoway Jr.). Guildford, Conn.

Mishra H.R. 2010. *Bones of the tiger: protecting the man-eaters of Nepal*. London.

Mishra H.R.; Dinerstein E. 1987. New ZIP codes for resident rhinos in Nepal. *Smithsonian* 18: 66–73.

Mishra H.R.; Mierow D. 1976. *Wild animals of Nepal*. Kathmandu.

Mishra K.D.; Mishra U. 1982. One-horned rhinoceros (Rhinoceros unicornis). In: Majupuria T.C., *Wild is beautiful*. Lashkar, pp. 417–422.

Misra B.B. 1984. Exploration in the Mid-Son valley, Mirzapur, Uttar Pradesh, and Sidhl, Madhya Pradesh. *Indian Archaeology: a Review* 1981–82: 44–45.

Misra M. 2011. *Deluging the colonial space: a history of flood in Assam, 1897–1941*. Thesis presented to the University of Assam, Gauhati.

Mitchell R.B. 1977. *Accounts of Nepalese mammals and analysis of the host-ectoparasite data by computer techniques*. Thesis presented to Iowa State University.

Mitra A. 1951. *District handbooks: Jalpaiguri*. Calcutta.

Mitsubishi. 2009. Rhino. https://www.adforum.com/creative-work/ad/player/34444477/rhino/mitsubishi. [accessed 2022-12-01].

Mittal J. 2007. *Sublime delight through works of art: from Jagdish and Kamla Mittal Museum of Indian Art.* Hyderabad.

Mittra D.K. 1992. Role of Ram Bramha Sanyal in initiating zoological researches on the animals in captivity. *Indian Journal of History of Science* 27: 279–289.

Moffett C. 1898a. After big game on elephants: scenes from the experience of a famous hunter of big game. *Pearson's Magazine* 6: 500–508.

Moffett C. 1898b. Hunting on elephants. *McClure's Magazine* 12 (2): 136–146.

Mohapatra R.P. 1975. *Udayagiri and Khandagiri caves in Orissa.* Thesis presented to Utkal University.

Moienuddin M. 2000. *Sunset at Srirangapatam: after the death of Tipu Sultan.* London.

Moller F. 1904. Die Entschuldigung. *Fliegende Blätter*, München 121 (3087): 148.

Molur S.; Sukumar R.; Seal U.S.; Walker S. 1994. Working group reports of Population and Habitat Viability Assessment (PHVA) workshop, great Indian one-horned rhinoceros, Jaldapara, 1993. *Zoos' Print (Magazine of Zoo Outreach Organisation)* 9: 19–39.

Momin S. 2008. Status of rhino and horn stockpiles in Orang National Park. In: Syangden B. et al., *Report on the regional meeting for India and Nepal IUCN/SSC Asian Rhino Species Group, March 5–7, 2007*, p. 10.

Momin K.N.; Shah D.R.; Oza G.M. 1973. Great Indian rhinoceros inhabited Gujarat. *Current Science* 42: 801–802.

Mondal T.N.; Chakraborty S. 2021. Riddle of the rhino: tracing early human migration in India through the cave paintings of Bhimbetka. In: Choudhary M.; Dwivedi A.; Nanda A.Z.G., *Reimagining South Asian art, culture and archaeology.* Delhi, pp. 59–81.

Monde Illustré, Journal Hebdomadaire 20 (979), 1876-01-15, pp. 39, 44: Le voyage du Prince de Galles.

Money R.C. 1891. A fighting tiger. *Baily's Magazine of Sports and Pastimes* 48: 161–168.

Money R.C. 1893. A day after rhinoceros. *Boy's Own Paper* 15 (1893-06-24): 616–619.

Monserrate A. 1922. *The commentary of Father Monserrate on his journey to the court of Akbar*, translated by J.S. Hoyland. London.

Montfort P.D. 1810. *Conchyliologie systématique, et classification méthodique des coquilles*, vol. 2. Paris.

Montgomery, George; Lee S.E. 1960. *Rajput painting.* New York.

Montgomery, Georgina. 2015. *Primates in the real world: escaping primate folklore and creating primate science.* Charlottesville.

Moor E. 1794. *A narrative of the operations of Captain Little's detachment, and of the Mahratta army, commanded by Purseram Bhow.* London.

Moore A. 2004. "Your lordship's most obliged servant": letters from Louis Fraser to the thirteenth Earl of Derby, 1840 to 1851. *Archives of Natural History* 31: 102–122.

Moore L. 2004. *Maharanis: the lives and times of three generations of Indian princesses.* London.

Moorehead C. 2013. *The K2 man (and his molluscs): the extraordinary life of Haversham Godwin-Austen.* Glasgow.

Moray Brown J. 1887. *Shikar sketches with notes on Indian field-sports.* London.

Moris M. 1826. Miroir du pays ou rélation des voyages de Sidi Aly fils d'Housain. *Journal Asiatique* 9: 193–217.

Morley C.; Friederichs H. 1895. *Half-holidays at the Zoo, with photographs by Henry Sandland and other illustrations.* London.

Morning Chronicle, London 1805-11-06: Advertisement for Oriental Field Sports.

Morning Post 1909-11-23: Death of Thomas Briscoe.

Morrieson H. 1859. Mss. field books of Lieut. Hugh Morrieson, of the 4th Regiment Bengal Native Infantry, and Lieut. W.E. Morrieson, Bengal Engineers, surveyors of the Soonderbunds, 1812–1818. *Calcutta Review* 32: 1–25.

Morris P.A. 2003. *Rowland Ward: taxidermist to the world.* Ascot.

Morris P.A. 2004. *Edward Gerrard & Sons: a taxidermy memoir.* Ascot.

Morris P.A. 2006. *Van Ingen & Van Ingen: artists in taxidermy.* Ascot.

Moscardo L. 1656. *Note overo memorie del Museo di Ludovico Moscardo, nobile Veronese.* Padoa.

Moss M.S. 2003. *The magnificent castle of Culzean and the Kennedy family.* Edinburgh.

Motilal S. 1970. *My life: law and other things.* Lucknow.

Moulton J.C. 1922. *Mammals of Malaysia*, part 1: *Malaysian ungulates.* Singapore.

Mountbatten L. 1987. *The diaries of Lord Louis Mountbatten 1920–1922: tours with the Prince of Wales.* London.

Mountfort G. 1969a. *The vanishing jungle: the story of the World Wildlife Fund expedition to Pakistan.* London.

Mountfort G. 1969b. Pakistan's progress. *Oryx* 10: 39–43.

Mountfort G. 1991. *Wild India: the wildlife and scenery of India and Nepal.* London.

Müllenmeister K.J. 1978. *Meer und Land im Licht des 17 Jahrhunderts*, vol. 2: *Tierdarstellungen in Werken niederländischer Künstler A-M.* Bremen.

Münster S. 1552. *Cosmographiae universalis libri VI.* Basileae.

Mukherjee A.K. 1963. The extinct, rare and threatened game of the Himalayas and the Siwalik ranges. *Journal of the Bengal Natural History Society* 32: 36–67.

Mukherjee A.K. 1966. The extinct and vanishing birds and mammals of India. *Indian Museum Bulletin* 1 (2): 7–41.

Mukherjee A.K. 1980. Wild life in the Sundarban, West Bengal. In: *Proceedings of the Workshop on Wildlife Ecology*, Dehradun, January 1978, pp. 123–127.

Mukherjee A.K. 1982. *Endangered animals of India.* Calcutta.

Mukherjee B.N. 1955. Rhinoceros-slayer-type of coins of Kumāragupta I. *Indian Historical Quarterly* 31: 175–181.

Mukherjee S.; Sengupta S. 1998. Rhino poaching in Jaldapara Wildlife Sanctuary, North Bengal, India. *Tiger Paper* (FAO) 25 (1): 20–21.

Mukherjee S.; Sengupta S. 1999. Census of great Indian one horned rhinoceros (Rhinoceros unicornis Linn) at Jaldapara Wildlife Sanctuary, Cooch Behar Forest Division, West Bengal, India. *Tiger Paper* (FAO) 26 (4): 18–21.

Mukherjee S.K. 1990. Terracotta figurines of the Kushana period in the Gangetic delta. In: *Historical Archaeology of India*, pp. 271–280.

Mundy G.C. 1833. *Pen and pencil sketches, being the journal of a tour in India*, vol. 2. London.

Mundy P. 1914. *The travels of Peter Mundy, in Europe and Asia, 1608–1667*, vol. 2. *Travels in Asia, 1628–1634*. London: Hakluyt Society, vol. 35.

Muriel C.E. 1916. *The Bengal Presidency forest manual*, part 1. *Notifications under the Indian Forest Act and other orders affecting the public*. Calcutta.

Murray A. 1866. *The geographical distribution of mammals*. London.

Murray J. 1901. *A handbook for travellers in India, Burma and Ceylon*. London.

Murty M.L.K. 1979. Recent research on the Upper Palaeolithic Phase in India. *Journal of Field Archaeology* 6 (3): 301–320.

Nagar M.M. 1949. A rhinoceros slayer type coin of Kumāra-gupta I. *Journal of the Numismatic Society of India* 11: 7–8.

Nagar S. 1998. *Indian gods and goddesses*, vol. 1: *The early deities from chalcolithic to beginning of historical period*. Delhi.

Nair P.T. 1989. *Calcutta in the 19th century: Company's days*. Calcutta.

Nandi D. 2013. Challenges and prospects for community participation in Community Based Tourism in Jaldapara National Park, Jalpaiguri. In: Huong Ha; Ram H.S.G., *Trends and challenges in global business management*. Coimbatore, pp. 156–161.

Nandi N.C.; Venkataraman K.; Bhuinya S; Das S.R.; Das S.K. 2005. Wetland faunal resources of West Bengal, 4. Darjiling and Jalpaiguri districts. *Records of the Zoological Survey of India* 104: 1–25.

Nash R. 1967. *Wilderness and the American mind*. New Haven.

Nath A. 1998. Rakhigarh: a Harappan metropolis in the Sarasvati-Drishadvati divide. *Puratāttva* no. 28: 39–45.

Nath B. 1961. Animals of prehistoric India and their affinities with those of the Western Asiatic countries. *Records of the Zoological Survey of India* 59: 335–367.

Nath B. 1968. Advances in the study of prehistoric and ancient animal remains in India: a review. *Records of the Zoological Survey of India* 61: 1–63.

Nath B. 1976. On the occurrence of great Indian rhinoceros, Rhinoceros unicornis Linn, from the prehistoric site at Chirand, San District, Bihar. *Newsletter of the Zoological Survey of India* 2: 86–87.

Nath B.; Biswas N.K. 1980. Animal remains from Chirand, Saran District (Bihar). *Records of the Zoological Survey of India* 76: 115–124.

Nath B.; Rao G.V.S. 1985. Animal remains from Lothal excavations. In: Rao S.R., *Lothal, a Harappan port town, 1955–62*. Delhi, pp. 636–675.

Nath R.M. 1948. *The back-ground of Assamese culture*. Shillong.

National Archives of India, Home Department, Public, 16 May 1822, no. 5. Bombay Government to Commissionery in Cooch Behar.

National Archives of India, Home Department, Public, 16 May 1822, no. 7 (No. 480 of 1822). Public Department to C. Lushington Esquire, Acting Chief Secretary to the Supreme Government, Fort William.

National Archives of India, Home Department, Public, 30 May 1822, no. 27. Acting Collector of Rungpore to C. Lushington Esquire, Acting Chief Secretary to Government, Fort William.

National Archives of India, Home Department, Public, 7 August 1823, no. 47. Commissioner Cooch Behar to C. Lushington Esquire, Acting Chief Secretary to Government, Fort William.

National Archives of India, Home Department, Public, 7 August 1823, no. 47, part 2. C. Lushington to W. Newnham Esq., Chief Secretary to the Government of Bombay.

National Archives of India, Home Department, Public, 7 August 1823, no. 47, part 3. Zillah Rungpore, Commissioners Office to C. Lushington Esquire, Secretary to the Government in the Public Department, Fort William.

National Archives of India. 1949. *Calendar of Persian correspondence, being letters which passed between some of the Company's servants and Indian rulers and notables*, vol. 9 (1790–1791), vol. 10 (1792–1793). Delhi.

National-Zeitung, Berlin No. 301, 1874-07-02, p. 4: Aus dem Zoologischen Garten [Berlin].

National-Zeitung, Berlin No. 309, 1874-07-07, pp. 7–8: Aus dem Zoologischen Garten [Berlin].

Naval & Military Gazette 1834-06-07: Arrival of rhinoceros in London from Bengal.

Nawab S.M. 1959. *The oldest Rajasthani paintings from Jain Bhandars*. Ahmedabad.

Nawab S.M. 1980. *Jain paintings*, vol. 1: *Paintings on palm leaves and wooden book-covers only*. Ahmedabad.

Nawaz M. 1982. Re-introduction of wild fauna in Pakistan. *Tiger Paper* (FAO) 9 (2): 5.

N.B.K. 1944. Obituary: Edward Charles Stuart Baker. *Journal of the Bombay Natural History Society* 45: 211–213.

Negi S.S. 1985. *Himalayan wildlife: an introduction*. Dehradun.

Nelson M.P.; Callicott J.B. 1998. *The Great New Wilderness debate*. Athens.

Nelson M.P.; Callicott J.B. 2008. *The Wilderness Debate rages on*. Athens.

Nepal Government. 1995. *Red data book of the fauna of Nepal.* Arnhem.

Nepal Government. 2006. *The greater one-horned rhinoceros conservation action plan for Nepal (2006–2011).* Kathmandu.

Nepal Government. 2008. *Rhino count, Nepal.* Kathmandu.

Nepal Government. 2012. *Chitwan National Park, Annual report 2011–2012.* Kathmandu.

Neubert M. 2007. Jungfer Clara: das Holländische Rhinozeros, portraitiert vom Hofmaler Anton Clemens Lünenschloß, auf dem Marktplatz zu Würzburg. In: Mettenleiter A., *Tempora mutantur et nos: Festschrift Walter M. Brod.* Pfaffenhofen, pp. 328–331.

Neumayer E. 1993. *Lines on stone: the prehistoric rock art of India.* New Delhi.

Neumayer E. 2011. *Rock art of India: the prehistoric cave-art of India.* Oxford.

Neuville H. 1927. Remarques et comparaisons rélatives aux phanères des rhinocéros. *Archives du Muséum d'Histoire Naturelle*, Paris (ser.6) 2: 179–208.

New York Times 1935-12-28: Frank Buck's Fang and Claw.

Newark T. 2015. *The In & Out: a history of the Naval and Military Club.* Oxford.

Newman E. 1872. Another rhinoceros at the Zoological Gardens. *Zoologist, a popular miscellany of natural history* (ser.2) 7: 3232–3233.

Newman E. 1874. Rhinoceros sondaicus at the Zoological Gardens. *Zoologist, a popular miscellany of natural history* (ser.2) 9: 3949–3952.

Nicholls F. 1970. *Assam shikari: a tea planter's story of hunting and high adventure in the jungles of North East India.* Auckland.

Nichols J. 1812. *Literary anecdotes of the eighteenth century*, vol. 5. London.

Nicolson N. 1977. *Mary Curzon.* London.

NMB – National Museum of Bhutan. 2001. *The living religious and cultural traditions of Bhutan: catalogue.* Paro.

Noack T. 1886. Über das zottelohrige Nashorn (Rhinoceros lasiotis). *Zoologische Garten* 27: 138–144.

Noll F.C. 1873a. Die Rhinoceros-Arten, part 1. *Zoologische Garten* 14: 47–55.

Noll F.C. 1873b. Die Rhinoceros-Arten, part 2. *Zoologische Garten* 14: 81–87.

Noll F.C. 1889. Die Zoologischen Gärten zu Kalkutta und Bombay. *Zoologische Garten* 30: 62–63.

Nordlunde J. 2016. *A passion for Indian arms. A private collection.* Denmark.

Norfolk L. 1996. *The Pope's rhinoceros.* London.

Norfolk Chronicle 1809-02-11: Advertisement for Oriental Field Sports.

Nott J.F. 1886. *Wild animals photographed and described.* London.

Nottingham Journal 1908-10-24: Big game shooting record.

Nripendra Narayan (MCB). 1908. *Thirty-seven years of big game shooting in Cooch Behar, the Duars and Assam: a rough diary.* Bombay: Printed at The Times Press [identical to the London-Bombay edition, except initial pages differ].

Nripendra Narayan (MCB). 1908 (second printing). *Thirty-seven years of big game shooting in Cooch Behar, the Duars and Assam: a rough diary.* London: Rowland Ward, Limited, 167, Piccadilly [and] Bombay: Bennett, Coleman & Co. The Times Press [edition available for sale in October 1909].

Nripendra Narayan. 1909. Advertisement for Thirty-Seven Years of Big Game Shooting. *Field*, London 114 (1909-10-23): xix.

NTNC – National Trust for Nature Conservation (Nepal). 2017. *Annual report.* Kathmandu.

NTNC – National Trust for Nature Conservation (Nepal). 2021. Nepal rhino count 2021. NTNC website, 11 April 2021.

Nuttall W.M. 1917. Fauna. In: Playne S., *Bengal and Assam, Behar and Orissa.* London, pp. 631–640.

Nyrop R.F. 1975. *Area handbook for Bangladesh.* Washington.

Oberai C.P.; Bonal B.S. 2002. *Kaziranga: the rhino land.* Delhi.

Oberländer A. 1895. Landwirthschaftliches aus Afrika. *Fliegende Blätter*, München 103 (2624): 173.

O'Connor D. 1980. India's armored giants. *Wildlife* 22 (3): 40–42.

O'Connor F. 1931. *On the frontier and beyond: a record of thirty years' service.* London.

O'Connor F. 1935. A trip to Nepal. *Country Life* 78 (1935-06-15): 613–615.

O'Connor F. 1940. *Things mortal.* London.

Oelschlaeger M. 1991. *The idea of wilderness: from prehistory to the age of ecology.* New Haven.

Ogden L. 2011. *Swamplife.* Minneapolis.

O'Hanlon R. 1984. *Into the heart of Borneo.* New York: Penguin.

O'Hara G.M. 1907. Trapping of rhinoceros in the Dindings, Straits Settlements. *Indian Forester* 53: 383–388.

Ojah S.; Saikia A.; Sivakumar P. 2015. Habitat suitability of Laokhowa Burhachapori wildlife sanctuary complex of Assam, India for Rhinoceros unicornis Linn. *Clarion (Multidisciplinary International Journal)* 4: 39–47.

Okada A. 1992. Le Prince Salim à la chasse: une miniature inédite peinte à Allahabad. *Artibus Asiae* 52: 319–327.

Olagnier P. 1931. Les Jésuites à Pondichéry de 1703 à 1721 et l'affaire Naniapa. *Revue de l'Histoire des Colonies Françaises* 19: 517–550.

Old Shekarree. 1860. *The spear and the rifle; or recollections of sport in India.* London [advertised in newspapers of January 1858].

Oldfield H.A. 1880. *Sketches from Nipal, historical and descriptive*, vol. 1. London.

Oliphant L. 1852. *Journey to Katmandu (the capital of Nepaul), with the camp of Jung Bahadoor.* London.

O'Malley L.S.S. 1907. *Bengal district gazetteers*, vol. 5: *Darjeeling*. Patna.

O'Malley L.S.S. 1908. *Bengal district gazetteers*, vol. 15: *Khulna*. Calcutta.

O'Malley L.S.S. 1911. *Bengal district gazetteers*, vol. 25: *Purnea*. Patna.

Orlich L. von. 1845a. *Reise in Ostindien in Briefen an Alexander von Humboldt und Carl Ritter* [1st illustrated edition]. Leipzig: Verlag von Mayer und Wigand.

Orlich L. von. 1845b. *Reise in Ostindien in Briefen an Alexander von Humboldt und Carl Ritter*. Zweite durchgesehene und vermehrte Auflage, vol. 2. Leipzig: Verlag von Gustav Mayer.

Orlich L. von. 1845c. *Travels in India, including Sinde and the Punjab*, vol. 2. London.

Orme E. 1806. *Prospectus of the Indian Sportsman; being a complete, detailed, and accurate description of the Wild Sports of the East and exhibiting in a novel and interesting manner, the natural history of the elephant, the rhinoceros, the tiger, the leopard, the bear, the deer, the buffalo, the wold, the wild hog, the jackal, the wild dog, the civet, and other undomesticated animals*. London: Orme, pp. 1–22.

Orta G. Da 1567. *Aromatum, et simplicium aliquot medicamentorum*. Antverpiae.

Osborn H.F. 1922. *Annual report of the American Museum of Natural History*, no. 54 for 1922. New York.

Osborn H.F. 1923. *Annual report of the American Museum of Natural History*, no. 55 for 1923, with information on Vernay-Faunthorpe expedition. New York.

Osborn H.F. 1928. Building the American Museum 1869–1927. *Annual Report of the American Museum of Natural History*, New York no. 59 for 1927: 1–30.

Osborn H.F. 1930. *An album of the groups in the Vernay-Faunthorpe Hall of South Asiatic Mammals of the American Museum of Natural History*. New York.

Osborn H.F. 1931. *Annual Report of the American Museum of Natural History*, no. 62 for 1930. New York.

Oury C. 2021. Diard, Duvaucel and the Museum National d'Histoire Naturelle. In: Dorai F.; Low M.E.Y., *Diard & Duvaucel: French natural history drawings of Singapore and Southeast Asia, 1818–1820*. Singapore: Embassy of France in Singapore and National Library Board, pp. 20–33.

O.W. 1902. North Assam as a shooting ground. *Illustrated Sporting and Dramatic News* 1902-02-22: 952.

Owden J.S. 1932. *Progress report of forest administration in Assam 1931–1932*. Shillong.

Owen R. 1870. On the fossil remains of mammals found in China. *Quarterly Journal of the Geological Society of London* 26: 417–434.

Oxford University and City Herald 1808-12-24: Advertisement for Oriental Field Sports.

Pachuau J.L.K.; Schendel W. van. 2015. *The camera as witness: a social history of Mizoram, Northeast India*. Delhi.

Padilla M. 2019. Trophy hunter seeks to import parts of rare rhino he paid $400,000 to kill. *New York Times* 2019-09-08.

Pal J. 2015. *Development of Cooch Behar state under the Maharajas (1847–1949)*. Thesis presented to University of North Bengal, Siliguri.

Pal P. 1997. *A collecting odyssey: Indian, Himalayan, and Southeast Asian art from the James and Marilynn Alsdorf Collection*. Chicago.

Pall Mall Gazette 1879-04-04: Advertisement of Pollok, Sport in British Burmah.

Pall Mall Gazette 1888-02-03: Henri d'Orléans in India.

Palmer A.H. 1895. *The life of Joseph Wolf: animal painter*. London.

Panchamukhi R.S. 1951. *Gandharvas and Kinnaras in Indian iconography*. Dharwar.

Pandey J.N. 1990. Mesolithic in the middle Ganga valley. *Bulletin of the Deccan College Research Institute* 49: 311–316.

Pandit P.K. 2004. Conservation of the great Indian one horned rhino (Rhinoceros unicornis) in Jaldapara Wild Life Sanctuary, West Bengal, India. *Vidyasagar University Journal of Biological Sciences* 10: 60–69.

Pandit P.K. 2013. Biodiversity of mangrove forests of Indian Sundarbans and its conservation. *Tiger Paper* (FAO) 40 (3): 1–11.

Pandit P.K.; Sinha S.P. 2006. Anthrax incidence and its control by vaccinating greater one horned rhino (Rhinoceros unicornis) against anthrax in Jaldapara Wildlife Sanctuary, West Bengal, India. *Zoos' Print (Magazine of Zoo Outreach Organisation)* 21: 54–65.

Pandolfi L.; Maiorino L. 2016. Reassessment of the largest Pleistocene rhinocerotine Rhinoceros platyrhinus (Mammalia, Rhinocerotidae) from the Upper Siwaliks (Siwalik Hills, India). *Journal of Vertebrate Paleontology* 36 (2): 1–12.

Panemanglor K.N. 1926. *The viceregal visit to Baroda 1926*. Baroda.

Pant G.N. 1982. *Indian shield*. New Delhi.

Pappu S. 2008. Prehistoric antiquities and personal lives: the untold story of Robert Bruce Foote. *Man and Environment* 33: 30–50.

Parasnis Rao B.D.B. 1921. *Poona in bygone days*. Bombay.

Pargiter F.E. 1889. Cameos of Indian districts: the Sundarbans. *Calcutta Review* 89: 280–301.

Parika N. 2000. *Jaipur that was: royal court and the seraglio*. Jaipur.

Paris MNHN – Muséum national d'Histoire naturelle. 2023. Database of specimens in the collection: https://science.mnhn.fr/institution/mnhn/search.

Parkinson J. 1640. *Theatrum botanicum: the theatre of plants*. London.

Parks F. 1850. *Wanderings of a pilgrim in search of the picturesque*, 2 vols. London.

Parks F. 1851. *Grand moving diorama of Hindostan*. London.

Parry N.E. 1932. *The Lakhers*. London.

Parsons J. 1743. A letter containing the natural history of the rhinoceros. *Philosophical Transactions of the Royal Society of London* 42: 523–541.

Parsons J. 1747. *Die Natürliche Historie des Nashorns, welche von Doctor Parsons in einem Schreiben an Martin Folkes, Rittern und Präsidenten der Koniglich-Englischen Societät, abgefasset, mit zuverlässigen Abbildungen versehen*. Nürnberg.

Parsons J. 1760. Lettre à M. Martin Folkes, Ecuyer, President de la Société Royale, contenant l'histoire du rhinocéros [translated by Pierre DeMours]. *Transactions Philosophiques* 42 (470) [for 1743]: 237–254.

Patar K.C. 1977. *Food preferences of the one horned Indian rhinoceros Rhinoceros unicornis, in Kaziranga National Park, India*. Thesis presented to Michigan State University.

Pathak A.; Hsu Y.; Sadaula A.; Joshi J.; Gairhe K.P.; Kandel R.C.; Gilbert M.; Paudel P. 2022. Reduced genetic diversity in greater one-horned rhinoceros in Chitwan National Park: a new challenge and opportunity for rhino conservation in Nepal. *Journal of Biological Studies* 5 (4): 693–704.

Patnaik R.; Badam G.L.; Murty M.L.K. 2008. Additional vertebrate remains from one of the Late Pleistocene-Holocene Kurnool Caves (Muchchatla Chintamanu Gavi) of South India. *Quaternary International* 192: 43–51.

Patra A.K.; Das V.; Datta T.; Dastidar S.G. 2015. Zooplankton fauna of Moraghat forest, a territorial forest of Jalpaiguri district, West Bengal, India. *Pelagia Research Library* 5: 39–47.

Pawar K.A. 2015. An archaeological perspective on rock art from Vidarbha region of Maharashtra. *Bulletin of the Deccan College Research Institute* 75: 63–84.

Pawsey C. 1937. Photo of animal heads on house in Noklak. http://himalaya.socanth.cam.ac.uk/collections/naga/record/r57250.html [accessed March 2023].

Payne J. 2022. *The hairy rhinoceros: history, ecology and some lessons for management of the last Asian megafauna*. Kota Kinabalu.

Peacock E.H. 1933. *A game book for Burma and adjoining territories*. London.

Pearson J. 1986. *The ultimate family: the making of the Royal House of Windsor*. London.

Pearson J.T. 1840. Zoological catalogue of the Museum of the Asiatic Society. *Journal of the Asiatic Society of Bengal* 9: 514–530.

Pease A.E. 1923. *Edmund Loder: naturalist, horticulturist, traveller and sportsman, a memoir*. London.

Peel C.V.A. 1908. The great Indian rhinoceros. *Field, the Country Gentleman's Magazine* 111 (1908-03-14): 430.

Peel C.V.A. 1909. Mr Peel's big-game museum. *Oxford Journal Illustrated* 1909-10-27: 6.

Peissel M. 1966. *Tiger for breakfast: the story of Boris of Kathmandu*. New York.

Pelinck E.; Upreti B.N. 1972. A census of rhinoceros in Chitwan National Park and Tamaspur Forest. Typewritten report. Kathmandu.

Pemberton J.J. 1854. *Geographical and statistical report of the district of Maldah*. Calcutta.

Penderel-Brodhurst J. 1911. *The life of His Most Gracious Majesty King Edward VII: from his birth to his coronation*, vol. 3. London.

Pennant T. 1793. *History of quadrupeds*, 3rd edition, vol. 1. London.

Pennant T. 1798. *The view of Hindoostan*, vol. 2: *Eastern Hindoostan*. London.

Penny M. 1987. *Rhinos: endangered species*. London.

Péquignot A. 2013. The rhinoceros (fl. 1770–1793) of King Louis XV and its horns. *Archives of Natural History* 40: 213–227.

Percy S.; Percy R. 1821. *The Percy anecdotes, original and select*, vol. 19. London.

Perrée W.F. 1906. Great Indian rhinoceros in Assam. *Badminton Magazine of Sports and Pastimes* 23: 232.

Peters W. 1877. Über Rhinoceros inermis Lesson. *Monatsberichte der Königlichen Preussischen Akademie der Wissenschaften zu Berlin* 1877: 68–71.

Peters W.; Sclater P.L. 1875–1877. Correspondence about the rhinoceros. Manuscript letters preserved in the Archives of the Museum für Naturkunde of Berlin: Z.M. Archiv, Abteilung II [information in 2021, see chapter 47].

Petis de la Croix F. 1723. *The history of Timur-Bec*, vol. 2. London.

Petite Gironde 1873-11-04: Hunt of Baring family in India.

Petretti A. 2018. *The Arader Galleries natural history watercolor collection*, vol. 3: *Fauna*. Philadelphia.

Pettigrew S.T. 1882. *Episodes in the life of an Indian chaplain*. London.

Peyssonnaux J.H. 1935. Vie, voyages et travaux de Pierre Médard Diard, naturaliste français aux Indes-Orientales (1794–1863), voyage dans l'Indochine (1821–1824). *Bulletin des Amis du Vieux Hué* 22: 1–120.

Phenix Gazette, Alexandria, DC 1830-12-28: Rhinoceros in Boston.

Philippe de la Très Sainte Trinité R.P. 1649. *Itinerarium orientale*. Lugduni.

Philippe d'Orléans. 1892. *Une expédition de chasse au Nepaul*. Paris.

Phillimore R.H. 1945. *Historical records of the Survey of India*, vol. 1: *18th century*. Dehra Dun.

Phillips P. 1922. *The Prince of Wales' eastern book: a pictorial record of the voyage of H.M.S. Renown, 1921–1922*. London.

Pichot P.A. 1908. Les trophées de chasse du Duc d'Orléans à Wood Norton. *Sports Modernes* 10 (35): 5–9.

Pioneer, Allahabad 1872-03-18: Mr. Talbot at Barpetta.

Pioneer, Allahabad 1875-02-18: Advertisement to purchase animals.

Pioneer, Allahabad 1876-01-08: Jamrach advertising to purchase animals.

Pioneer, Allahabad 1901-04-19: The Viceroy's shooting party.

Pittsburgh Weekly Gazette 1830-05-28: Exhibition of a rhinoceros in Boston.

P.L. 1876. Le voyage du Prince de Galles. *L'Illustration, Journal Universel* 67 (1876-01-01): 4, 7.

Plancius P. 1594. Orbis terrarum typus de integro multis in locis emendatus. Map, 23 × 16 inches.

Playne S. 1917. *Bengal and Assam, Behar and Orissa: their history, people, commerce, and industrial resources*. London.

Plumb C. 2010. Strange and Wonderful: encountering the elephant in Britain, 1675–1830. *Journal for Eighteenth-Century Studies* 33: 525–543.

Pocock R.I. 1912. The Zoological Society. *Field, the Country Gentleman's Magazine* 119 (1912-01-20): 143.

Pocock R.I. 1915. The two-horned Asiatic rhinoceros. *Field, the Country Gentleman's Magazine* 126 (1915-10-23): 709.

Pocock R.I. 1936. Mammalia. In: Regan C.T., *Natural history*. London, pp. 705–709.

Pocock R.I. 1944a. The identity of the genotype of Rhinoceros Linn. *Annals and Magazine of Natural History* (ser.11) 11: 616–618.

Pocock R.I. 1944b. The premaxillae in the Asiatic rhinoceroses. *Annals and Magazine of Natural History* (ser.11) 11: 834–842.

Pocock R.I. 1946. Some structural variations in the second upper molar of the lesser one-horned rhinoceros (Rhinoceros sondaicus). *Proceedings of the Zoological Society of London* 115: 306–309.

Poins. 1868. Bravo rhino. *Oriental Sporting Magazine* (n.s.) 1 (August): 597–600.

Pol B. van der. 2016. *Holland aan de Ganges: Prins Willem Frederik Hendrik in India (1837–1838)*. Zutphen.

Polehampton E. 1858. *A memoir, letters and diary of the Rev. Henry S. Polehampton, Chaplain of Lucknow*. London.

Pollen S.H. 2009. A hunting-party with the Maharaja of Cooch Behar, 1895–1896. Photographic album auctioned at Bonhams (London) on 9 April 2009 – https://www.bonhams.com/auctions/16200/lot/20/.

Pollok F.T. 1868a. Records of sport in British Burmah and Assam (part 1). *Oriental Sporting Magazine* (n.s.) 1 (April): 264–272.

Pollok F.T. 1868b. Records of sport in British Burmah and Assam (part 2). *Oriental Sporting Magazine* (n.s.) 1 (May): 367–374.

Pollok F.T. 1868c. Records of sport in British Burmah and Assam (part 3). *Oriental Sporting Magazine* (n.s.) 1 (June): 442–446.

Pollok F.T. 1868d. Records of sport in British Burmah and Assam (part 4). *Oriental Sporting Magazine* (n.s.) 1 (July): 508–514.

Pollok F.T. 1868e. Records of sport in Assam (part 5). *Oriental Sporting Magazine* (n.s.) 1 (November): 796–804.

Pollok F.T. 1871a. Records of sport in Assam (part 6). *Oriental Sporting Magazine* (n.s.) 4 (August): 354–357.

Pollok F.T. 1871b. Records of sport in Assam (part 7). *Oriental Sporting Magazine* (n.s.) 4 (September): 408–412.

Pollok F.T. 1871c. Records of sport in Assam (part 8). *Oriental Sporting Magazine* (n.s.) 4 (October): 467–470.

Pollok F.T. 1873a. Records of sport in Assam (part 9). *Oriental Sporting Magazine* (n.s.) 6 (June): 269–276.

Pollok F.T. 1873b. Records of sport in Assam (part 10). *Oriental Sporting Magazine* (n.s.) 6 (August): 388–393.

Pollok F.T. 1873c. Records of sport in Assam (part 11). *Oriental Sporting Magazine* (n.s.) 6 (September): 444–450.

Pollok F.T. 1873d. Records of sport in Assam (part 12). *Oriental Sporting Magazine* (n.s.) 6 (October): 499–503.

Pollok F.T. 1874a. Records of sport in Assam (part 13). *Oriental Sporting Magazine* (n.s.) 7 (April): 155–162.

Pollok F.T. 1874b. Records of sport in Assam (part 14). *Oriental Sporting Magazine* (n.s.) 7 (July): 308–313.

Pollok F.T. 1874c. Records of sport in Assam (part 15). *Oriental Sporting Magazine* (n.s.) 7 (September): 399–406.

Pollok F.T. 1879. *Sport in British Burmah, Assam, and the Cassyah and Jyntiah Hills*, 2 vols. London.

Pollok F.T. 1880. Hunting the rhinoceros. In: *The pictorial museum of sport and adventure*. London, pp. 20–22.

Pollok F.T. 1882a. Indian rhinoceroses. *Field, the Country Gentleman's Magazine* 59 (1882-05-06): 589.

Pollok F.T. 1882b. A bout with rhinoceros. *Illustrated Sporting and Dramatic News* 1882-09-21: 31; 1882-09-30: 58–59.

Pollok F.T. 1894. *Incidents of foreign sport and travel*. London.

Pollok F.T. 1896. *Fifty years' reminiscences of India: a retrospect of travel, adventure and shikar*. London.

Pollok F.T. 1898a. A chat about Indian wild beasts. *Zoologist, a popular miscellany of natural history* (ser.4) 2: 154–177.

Pollok F.T. 1898b. Buried alive by a dead elephant. *Wide World Magazine* 1: 347–351.

Pollok F.T. 1898c. Sport in the Army. *Navy and Army Illustrated* 5 (1898-02-04): 240–241.

Pollok F.T. 1898d. Rhinoceros – India. In: Howard H.C. Earl of Suffolk; Peek H.; Aflalo F.G., *The encyclopaedia of sport*. London, vol. 2, pp. 255–256 [reprinted in Pollok 1911].

Pollok F.T. 1898e. Rats and rabbits in the Azores. *Field*, London 92 (1898-04-09): 57.

Pollok F.T. 1899. A doctor in the wilds. *Wide World Magazine* 4: 100–105.

Pollok F.T. 1900. Faithful unto death: the story of a pet tiger. *Wide World Magazine* 4: 297–300.

Pollok F.T. 1902a. Fallen among thieves. *Wide World Magazine* 8: 34–37.

Pollok F.T. 1902b. A disastrous trip. *Wide World Magazine* 8: 192–197.

Pollok F.T. 1911. Rhinoceros – India. In: Howard H.C. Earl of Suffolk, *The encyclopaedia of sport & games*. London, vol. 4, pp. 24–25.

Pollok F.T.; Thom W.S. 1900. *Wild sports of Burma and Assam*. London.

Poncins E. de 1935. A hunting trip in the Sunderbunds in 1892. *Journal of the Bombay Natural History Society* 37: 844–858.

Pontevès de Sabran J.B.E.M.C. 1886. *L'Inde à fond de train*. Paris.

Portland W.J.A.C.J. 1937. *Men, women and things: memories of the Duke of Portland*. London.

Potocki J. 1896. *Notatki mysliwskie z Dalekiego Wschodu*, vol. 1: *Indye*. Warszawa.

Powlett P.W. 1878. *Gazetteer of Ulwur*. London.

Pradhan N.M.B. 2001. Rhino translocation [in Nepal]. *Fonarem Newsletter* 1 (1): 4.

Prakash J. 1962. Observations on the Mrigaya coin types of the Guptas. *Journal of the Numismatic Society of India* 23: 152–163.

Prakash J. 1963. Observations on the so-called Khadga-Trātā coin-type of Kumāragupta I. *Journal of the Numismatic Society of India* 25: 29–35.

Prasad I. 1996. *The life and times of Maharaja Juddha Shumsher Jung Bahadur Rana of Nepal*. New Delhi.

Prasad K.N. 1998. Pleistocene cave fauna from Peninsular India. *Journal of Caves and Karst Studies* 58: 30–34.

Prasad S.; Kumar S. 2011. Successful re-introduction and re-habilitation of great one-horned rhinocerous, Dudhwa National Park, UP. *Stripes (bi-monthly outreach journal of National Tiger Conservation Authority)* 2 (4): 10–11.

Prashad B. 1936. Animal remains from Harappa. *Memoirs of the Archaeological Survey of India* 51: 1–76.

Pratap R. 1996. *The panorama of Jaipur paintings*. New Delhi (Contours of Indian Art & Architecture, no. 1).

Prater S.H. 1934. The wild animals of the Indian Empire and the problem of their preservation, part 2. *Journal of the Bombay Natural History Society* 37 (1) Supplement: 57–96.

Prater S.H. 1939. Additional notes on the Asiatic two-horned rhinoceros. *Journal of the Bombay Natural History Society* 40: 618–627.

Prater S.H. 1948. *The book of Indian animals*. Bombay.

Prater S.H. 1965. *The book of Indian animals*, 2nd edition. Bombay.

Pratihar S.; Chakraborty S. 1996. An account of the mammalian fauna of Gorumara National Park, Jalpaiguri, West Bengal. *Records of the Zoological Survey of India* 95: 229–241.

Prendergast D.K.; Adams W.M. 2003. Colonial wildlife conservation and the origins of the Society for the Preservation of the Wild Fauna of the Empire (1903–1914). *Oryx* 37: 251–260.

Preston Chronicle 1871-03-18: Government Gardens, Madras.

Price D. 1836. *Memoirs of the early life and service of a field officer, on the retired list of the Indian Army*. London.

Prinsep H.T. 1844. *A general register of the Hon'ble East India Company's civil servants of the Bengal Establishment from 1790 to 1842*. Calcutta.

Prinsep J. 1834. Proceedings of the Asiatic Society (Thursday evening, the 20th March, 1834). *Journal of the Asiatic Society of Bengal* 3: 141–142.

Proceedings of the Asiatic Society, July 1899: Description of Ploceus rutledgii.

Pudma Jung Bahadur Rana. 1909. *Life of Maharaja Sir Jung Bahadur of Nepal*. Allahabad.

Puṇyavijayajī, munijī. 1951. *jaisalmeranī citra samṛddhi. ahmadāvād: sārābhāī maṇilāl navāb, 2007* (1951 CE) [in Gujarati].

Purohit D.N. 1938. *Mewar history. Guide to Udaipur*. Bombay.

Purohit M. [undated]. *History of Juna Mahal, Dungarpur*. Typescript.

Pye-Smith P.H. 1874. *Catalogue of the preparations of comparative anatomy in the Museum of Guy's Hospital*. London.

Pyrard F. de Laval. 1619. *Voyage, contenant sa navigation aux Indes Orientales, Maldives, Moluques, Brésil*, 3rd edition, vol. 1. Paris.

Pyrard F. de Laval. 1887. *The voyage of Francois Pyrard of Laval to the East Indies, the Maldives, the Moluccas and Brazil*, vol. 1. London: Hakluyt Society, vol. 76.

Qazwini Z. 1547. Manuscript of Aja'ib al-makhluqat translated into Persian at Bijapur in 1548. Preserved in British Library Or. 1621, in Qatar Digital Library https://www.qdl.qa/en/archive/81055/vdc_100035587342.0x000001 [accessed 22 April 2023].

Rabha A. 2008. Status of rhino and horn stockpiles in Manas National Park. In: Syangden B. et al., *Report on the regional meeting for India and Nepal IUCN/SSC Asian Rhino Species Group, March 5–7, 2007*, p. 9.

Rabinowitz A. 2002. *Beyond the last village: a journey of discovery in Asia's forbidden wilderness*. London.

Rabinowitz A.; Schaller G.B.; Uga U. 1995. A survey to assess the status of the Sumatran rhinoceros and other large mammal species in Tamanthi wildlife sanctuary, Myanmar. *Oryx* 29: 123–128.

Raffles Lady S. 1830. *Memoir of the life and public services of Sir Thomas Stamford Raffles*. London.

Raffles T.S. 1822. Descriptive catalogue of a zoological collection, made on account of the Honourable East India Company, in the island of Sumatra and its vicinity, with additional notices illustrative of the natural history of these countries. *Transactions of the Linnean Society of London* 13: 239–274.

Rafinesque C.S. 1815. *Analyse de la nature, ou tableau de l'univers et des corps organisés*. Palerme.

Raha A.K. 2000. Status report West Bengal. In: Strien N.J. van et al., *Report on the Regional Meeting for India and Nepal of IUCN/SSC Asian Rhino Specialist Group*. Doorn, pp. 102–122.

Rani P.N. 2017. Spectacles of empire: the coronation durbar and imperial hunt conjoined. *KAAV International Journal of Arts, Humanities and Social Sciences* 4: 163–187.

Rahman M.R.; Asudazzaman M. 2010. Ecology of Sundarban, Bangladesh. *Journal of the Science Foundation* 8: 35–47.

Rahmani A.R.; Narayan G.; Rosalind L. 1992. A threat to India's Manas Tiger Reserve. *Tigerpaper* (FAO) 19 (2): 22–28.

Rai S. 2016. After 250 years, rhinos set to make comeback in West Uttar Pradesh. *Times of India* 2016-11-17.

Rainey H.J. (signed Young Nimrod). 1860. Trips to the Soonder-bunds. *Bedfordshire Times and Independent* 1860-08-07: 6.

Rainey H.J. (Young Nimrod). 1869. Strictures on the (revised) scale of government rewards for killing dangerous and destructive game. *Oriental Sporting Magazine* (n.s.) 2 (September): 593–598.

Rainey H.J. (Young Nimrod). 1872a. A strange mode of shooting a rhinoceros. *Oriental Sporting Magazine* (n.s.) 5 (February): 86–87.

Rainey H.J. (Young Nimrod). 1872b. Information anent the Soonderbuns. *Oriental Sporting Magazine* (n.s.) 5 (March): 139–140.

Rainey H.J. (Young Nimrod). 1872c. The rhinoceros of the Soonderbuns. *Oriental Sporting Magazine* (n.s.) 5 (July): 300–302.

Rainey H.J. (Young Nimrod). 1874b. An easy method for sportsmen to distinguish the common India rhinoceros, (R. Indicus), from the Soondurbun rhinoceros (R. Sondaicus). *Oriental Sporting Magazine* (n.s.) 7 (May): 239–240.

Rainey H.J. (Young Nimrod). 1874c. Shipment of rhinoceros from Calcutta to London. *Oriental Sporting Magazine* (n.s.) 7 (July): 340.

Rainey H.J. (Young Nimrod). 1874d. The two (supposed new species of) Indian rhinoceroses. *Oriental Sporting Magazine* (n.s.) 7 (September): 431–432.

Rainey H.J. (Young Nimrod). 1875a. New species of the one-horned Indian rhinoceros [with remarks by the editor]. *Field, the Country Gentleman's Magazine* 45 (1875-03-06): 242.

Rainey H.J. (Young Nimrod). 1875b. Curiosity in (Indian) natural history. *Oriental Sporting Magazine* (n.s.) 8 (November): 438.

Rainey H.J. (Young Nimrod). 1875c. Note on the comparative size of the so-called common Indian Rhinoceros (R. indicus, Cuvier) and the Sunderbun Rhinoceros (R. sondaicus, Muller). *Oriental Sporting Magazine* (n.s.) 8 (April): 157.

Rainey H.J. (Young Nimrod). 1877. New species of rhinoceros. *Oriental Sporting Magazine* (n.s.) 10 (September): 262–264.

Rainey H.J. 1878. Note on the absence of a horn in the female of the Sundarban rhinoceros and Javanese rhinoceros (Rh. javanicus, Cuv). *Proceedings of the Asiatic Society of Bengal* 1878 June: 139–141.

Rajak S.; Deo S.G.; Saha S. 2020. Rock art at Isko in Hazaribagh District, Jharkhand: anthropological perspective. *Heritage: Journal of multidisciplinary studies in Archaeology* 8: 1041–1058.

Rakshit A. 2011. Experience of rhino translocation in Assam. *Banabithi* (West Bengal) 2011 (November): 6–11.

Rana, Subodh. 2011. In the belly of the beast. http://historyl essonsnepal.blogspot.com/2011/02/in-belly-of-beast.html [accessed April 2023].

Randall J.L. 1901. *A history of the Meynell Hounds and Country 1780 to 1901*, vol. 2. London.

Ranking J. 1826. *Historical researches on the wars and sports of the Mongols and Romans*. London.

Rao G.V.S. 1985. *Lothal, a Harappan port town, 1955–62*, vol. 2. New Delhi.

Rao S.R. 1962. Further excavations at Lothal. *Lalit Kalā* 11: 14–30.

Rao S.R. 1968. Excavation at Paiyampalli, district North Arcot. *Indian Archaeology: a Review* 1967–1968: 26–30.

Rao S.R. 1978. Bronzes from the Indus valley. *Illustrated London News* 266 (March): 62–63.

Rao S.S.; Sharma A.; Talukdar B.K.; Sarma K.K. 2019. *Protocol for wild to wild translocation of rhinos in Assam*. Guwahati.

Rapier. 1901. Notes [game shot by Earl de Grey]. *Badminton Magazine of Sports and Pastimes* 12: 232–236.

Rashid S.M.A.; Khan A.; Khan M.A. 1990. Mammals of Cox's Bazar Forest Division (South) Bangladesh, with notes on their status and distribution. *Journal of the Bombay Natural History Society* 87: 62–67.

Rauwolf L. 1583. *Aigentliche Beschreibung der Raiß, so er vor diser zeit gegen Auffgang inn die Morgenländer, fürnemlich Syriam, Iudaeam, Arabiam, Mesopotamiam, Babyloniam, Assyriam, Armeniam etc. nicht ohne geringe mühe vnnd große gefahr selbs volbracht*. Laugingen.

Raven E.M. 2012. From Candragupta II to Kumāragupta I: styles in the patterning of Gupta coin designs. *Journal of Bengal Art* 17: 51–62.

Raven E.M. 2019. From third grade to top rate: the discovery of Gupta coin styles, and a mint group study for Kumāragupta I. In: Wessels-Mevissen C.; Mevissen G.J.R., *Indology's pulse: arts in context*. Delhi, pp. 195–222.

Ravenshaw J.H. 1878. *Gaur, its ruins and inscriptions*. London.

Ravi R.; Gupta M.; Pathak S.; Sunakar M.K.; Sharma A. 2022. *Status of rhinoceroses in Re-Introduction Area (RRA)-2, Dudhwa National Park*. New Delhi, WWF-India (Technical report).

Ray B. 1961. *Census 1961: Jalpaiguri*. Calcutta.

Ray J. 1693. *Synopsis methodica animalium quadrupedum et serpentinis generis*. Londini.

Ray N.R. 2015. The princely hunt and Kshatriyahood. *Global Environment* 8: 446–472.

Reading Mercury 1842-01-15: Rhinoceros [broken loose in Barrackpore, Calcutta, killing a man].

Reed K. 1956. Capture d'un rhinocéros en Assam. *Sciences et Voyages*, Paris, no. 130: 15–20.

Reed S. 1906. *The royal tour in India: a record of the tour of H.R.H. The Prince and Princess of Wales in India and Burma, from November 1905 to March 1906.* Bombay.

Reed S. 1912. *The King and Queen in India: a record of the visit of Their Imperial Majesties the King Emperor and Queen Empress to India, from December 2, 1911, to January 10, 1912.* Bombay.

Regmi D.R. 1961. *Modern Nepal*, vol. 1. Delhi.

Regmi M.C. 1971. Ban on killing of rhinoceros, 1798. *Regmi Research Series* 3 (5): 121.

Regmi M.C. 1995. *Kings and political leaders of the Gorkhali Empire 1768–1814.* Hyderabad.

Reichenbach H. 2002. Lost menageries – why and how zoos disappear (part 1). *International Zoo News* 49: 151–163.

Reichenbach H. 2022. 19. Jahrhundert: die Jamrachs, Partner und Konkurrenten Carl Hagenbecks. *Hagenbeck* 2022 (4): 46–47.

Reid A.S. 1893. *Chin-Lushai land.* Calcutta.

Reid M. 1882. The chase in Assam; being real experiences of a British Officer, III: a rhinoceros chase. *Beadle's Weekly* 1 (1882-12-02): 8.

Reilly J.; Hills Spedding G.; Apriawan; Setiawan V.I.; Ahmad M.; Sadjudin H.R.; Ramono W.S. 1997. Recent records of the Sumateran rhino, Dicerorhinus sumatrensis, in Way Kambas National Park, Sumatera. *Tropical Biodiversity* 4: 215–217.

Reilly J.H. 1858. The Soonderbunds – their economical importance. *Calcutta Review* 31: 385–411.

Renié P.L. 1992. Louis Rousselet: le voyageur, le photographe & l'écrivain. In: *L'Inde: photographies de Louis Rousselet 1865–1868.* Bordeaux, pp. 39–49.

Renker C.; Henrich B.; Hildebrand U. 2018. Mainz: the zoological collections of the Mainz Natural History Museum. In: Beck L.A., *Zoological collections of Germany.* Berlin, pp. 519–528.

Rennefort U. Souchu de. 1688. *Histoire des Indes Orientales.* Paris.

Renshaw G. 1907. *Final natural history essays.* London, Manchester.

Reynolds H.J. 1868. *Principal heads of the history and statistics of the Dacca division.* Calcutta.

Reynolds R.J. 1961. Asian rhinos in captivity. *International Zoo Yearbook* 2: 17–42.

Reynolds R.J. 1967. Some photographs of rhinos exhibited by American circuses between 1855 and 1926. *Zoologische Garten* 34: 279–292.

Reynolds R.J. 1968. Circus rhinos. *Bandwagon* 12 (6): 4–13.

Reyssat S. 2011. Auction at Chevau-Legers Enchères. Versailles 19 Feb. 2012, lot 34. *Gazette Drouot International* no. 8: 26–27.

R.H.D. 1866. The wild sports of the terai. *Field, the Country Gentleman's Magazine* 28 (1866-12-22): 485.

R.H.D. 1867. The Bengal Soonderbuns. *Field, the Country Gentleman's Magazine* 30 (1867-10-12): 291–292.

Rhodes M. 2012. *The final curtsey: a royal memoir by the Queen's cousin.* Edinburgh.

Rice C.W. 1873. The late baby rhinoceros and his mother. *Graphic, an Illustrated Weekly Newspaper* 7 (163) 1873-01-11: 41.

Rice C.W. 1875. Advertisement of sale of two rhinoceroses. *Era* 1875-12-05.

Richard F. 1996. Jean-Baptiste Gentil, collectionneur de manuscrits persans. *Dix-Huitième Siècle* 28: 91–110.

Richardson G. 1779. *Iconology; or, a collection of emblematical figures*, vol. 1. London.

Ridinger J.E. 1748. Anno 1748 im Monath Maij und Junio ist dises Nashorn Rhinoceros. In: Die wundersamsten Hirsche und andere Thiere. Print series, Augsburg.

Ridley G. 2004. *Clara's grand tour: travels with a rhinoceros in eighteenth-century Europe.* New York: Atlantic Monthly Press (also London: Atlantic).

Ridley G. 2022. Art and merchandise, followers and fragility: creating the blueprint for animal celebrity. In: Nachumi N.; Straub K., *Making stars: biography and celebrity in eighteenth-century Britain.* Newark, pp. 148–176.

Ridpath C.E. 1912. *With the world's people: an account of the ethnic origin, primitive estate, early migrations, social evolution, and present conditions and promise of the principal families of men*, vol. 2. Washington.

Riley L.; Riley W. 2005. *Nature's strongholds: the world's great wildlife reserves.* Princeton.

Rimal S.; Adhikari H.; Tripathi S. 2018. Habitat suitability and threat analysis of greater one-horned rhinoceros Rhinoceros unicornis Linnaeus, 1758 in Rautahat District, Nepal. *Journal of Threatened Taxa* 10: 11999–12007.

Rissman P. 1985. *Migratory pastoralism in western India in 2nd millenium BC: the evidences from Oriyo Timbo (Chiroda).* Ph.D. thesis presented to University of Pennsylvania.

Ritchie J.E. 1876. *The life and discoveries of David Livingstone. The pictorial edition*, vol. 1. London.

Ritvo H. 1989. *The animal estate: the English and other creatures in Victorian England.* Cambridge.

Ritvo H. 2002. History and animal studies. *Society and Animals* 10 (4): 403–406.

Roberts E. 1835. *Scenes and characteristics of Hindostan, with sketches of Anglo-Indian society*, vol. 1. London.

Roberts P.; Delson E.; Miracle P.; and 12 others. 2014. Continuity of mammalian fauna over the last 200,000 y in the Indian subcontinent. *Proceedings of the National Academy of Sciences of the United States of America* 111: 5848–5853.

Roberts T.J. 1977. *The mammals of Pakistan.* London.

Robin Hood. 1836. A rhinoceros shot dead with one ball. *Bengal Sporting Magazine* (n.s.) 6 (23): 326–327.

Robinson E.K. 1907. Young Indian rhinoceros – Zoo series no. 23. *Country-Side* 4 (1907-02-09): 184.

Robinson H.P. 1910. *Of distinguished animals*. London.

Rode P. 1934. Le Musée du Duc d'Orléans. *Terre et la Vie* 4: 67–75.

Roe D. 1878. Les combats d'animaux dans les Indes. *Journal des Voyages* 9 (1878-05-12): 283–286.

Rohilla. 1848. Sporting rambles. *India Sporting Review* 7 (March–June): 207–233.

Rohr W. 1959. Sorgen um Nepals Einhörner. *Orion* 1959-12-12: 955–959.

Rolfe-Smith B. 2013. *Colonel John Colvin, 1794–1871: a good and faithful servant, to India, Ludlow and Leintwardine*. Ludlow.

Ronaldshay G.E. 1928. *The life of Lord Curzon: being the authorised biography of George Nathaniel, Viceroy of India*, vol. 2. London.

Rookmaaker, Leendert Cornelis (Kees) 1973. Captive rhinoceroses in Europe from 1500 until 1810. *Bijdragen tot de Dierkunde* 43: 39–63.

Rookmaaker L.C. 1976. An early engraving of the Black Rhinoceros (Diceros bicornis [L.]) made by Jan Wandelaar. *Biological Journal of the Linnean Society* 8: 87–90.

Rookmaaker L.C. 1977. The identity of the one-horned rhinoceros in Berlin 1874–1884. *International Zoo News* 24 (2): 15.

Rookmaaker L.C. 1978a. De neushoorn van 1741. *Ons Amsterdam* 30: 16–17.

Rookmaaker L.C. 1978b. Two collections of rhinoceros plates compiled by James Douglas and James Parsons in the eighteenth century. *Journal of the Society for the Bibliography of Natural History* 9: 17–38.

Rookmaaker L.C. 1979. The first birth in captivity of an Indian rhinoceros (Rhinoceros unicornis): Kathmandu, May 1824. *Zoologische Garten* 49: 75–77.

Rookmaaker L.C. 1980. The distribution of the rhinoceros in Eastern India, Bangladesh, China and the Indo-Chinese region. *Zoologische Anzeiger* 205: 253–268.

Rookmaaker L.C. 1981. Early rhinoceros systematics. In: Wheeler A.; Price J.H., *History in the service of systematics*. London, pp. 111–118.

Rookmaaker L.C. 1982a. A story of horns: early views on rhinoceros classification. *Zoonooz*, San Diego 55 (4): 4–10.

Rookmaaker L.C. 1982b. The type locality of the Javan Rhinoceros (Rhinoceros sondaicus Desmarest, 1822). *Zeitschrift für Säugetierkunde* 47: 381–382.

Rookmaaker L.C. 1983a. *Bibliography of the rhinoceros: an analysis of the literature on the recent rhinoceroses in culture, history and biology*. Rotterdam, A.A. Balkema.

Rookmaaker L.C. 1983b. Historical notes on the taxonomy and nomenclature of the recent Rhinocerotidae (Mammalia, Perissodactyla). *Beaufortia* 33: 37–51.

Rookmaaker L.C. 1983c. Jamrachs Rhinozeros. *Bongo (Beiträge zur Tiergärtnerei und Jahresberichte aus dem Zoo Berlin)* 7: 43–50.

Rookmaaker L.C. 1983d. Histoire du rhinocéros de Versailles (1770–1793). *Revue d'Histoire des Sciences* 36: 307–318.

Rookmaaker L.C. 1984a. The former distribution of the Indian Rhinoceros (Rhinoceros unicornis) in India and Pakistan. *Journal of the Bombay Natural History Society* 80: 555–563.

Rookmaaker L.C. 1984b. The taxonomic history of the recent forms of Sumatran Rhinoceros (Dicerorhinus sumatrensis). *Journal of the Malayan Branch of the Royal Asiatic Society* 57: 12–25.

Rookmaaker L.C. 1989. *The zoological exploration of Southern Africa 1650–1790*. Rotterdam: A.A. Balkema [the author's Ph.D. dissertation presented to the University of Utrecht on 22 May 1989 at 4.15 pm].

Rookmaaker L.C. 1992. J.N.S. Allamand's additions (1769–1781) to the Nouvelle Edition of Buffon's Histoire Naturelle published in Holland. *Bijdragen tot de Dierkunde* 61: 131–162.

Rookmaaker L.C. 1993. The mysterious Liverpool rhinoceros. *Zoologische Garten* 63: 246–258.

Rookmaaker L.C. 1997a. Records of the Sundarbans rhinoceros (Rhinoceros sondaicus inermis) in India and Bangladesh. *Pachyderm*, Nairobi 24: 37–45.

Rookmaaker L.C. 1997b. Records of an animal collection in Poona at the end of the eighteenth century. *Back When and Then (Zoo Outreach Organisation)* 2 (1): 2.

Rookmaaker L.C. 1997c. The Royal Menagerie of the King of Oudh. *Back When and Then (Zoo Outreach Organisation)* 2 (1): 10.

Rookmaaker L.C. 1997d. Portrait of a rhinoceros in captivity. *Journal of the Bartlett Society* 8: 2.

Rookmaaker L.C. 1997e. A gift from Nepal to London Zoo, 1921. *Journal of the Bartlett Society* 8: 28–29.

Rookmaaker L.C. 1997f. Nashörner und Menschen. In: [Gansloßer U.], *Die Nashörner: Begegnung mit urzeitlichen Kolossen*. Fürth, pp. 7–13.

Rookmaaker L.C. 1998 [1998a]. *The rhinoceros in captivity*. The Hague, SPB Academic Publishing.

Rookmaaker L.C. 1998b. The sources of Linnaeus on the rhinoceros. *Svenska Linnesallskapets Årsskrift* 1996/97: 61–80.

Rookmaaker L.C. 1999a. William Daniell's depictions of the rhinoceros in India. *Archives of Natural History* 26: 205–210.

Rookmaaker L.C. 1999b. The rhinoceros of Kotdwara. *Hornbill (Bombay Natural History Society)* 1999 July–September: 9.

Rookmaaker L.C. 1999c. Records of the rhinoceros in Northern India. *Säugetierkundliche Mitteilungen* 44: 51–78.

Rookmaaker L.C. 1999d. Specimens of rhinoceros in European collections before 1778. *Svenska Linnesallskapets Årsskrift* 1998–1999: 59–80.

Rookmaaker L.C. 2000. Records of the rhinoceros in Pakistan and Afghanistan. *Pakistan Journal of Zoology* 32: 65–74.

Rookmaaker L.C. 2002a. Historical records of the rhinoceros (Rhinoceros unicornis) in northern India and Pakistan. *Zoos' Print Journal* 17: 923–929.

Rookmaaker L.C. 2002b. Historical records of the Javan rhinoceros in North-East India. *Newsletter of the Rhino Foundation of Nature in North-East India* no. 4: 11–12.

Rookmaaker L.C. 2002c. The quest for Roualeyn Gordon Cumming's rhinoceros horns. *Archives of Natural History* 29: 89–97.

Rookmaaker L.C. 2003a. The Rhino Resource Centre. *Ratel* 30 (4): 106–107.

Rookmaaker L.C. 2003b. Rhino Resource Center. *Pachyderm*, Nairobi 34: 101–102.

Rookmaaker L.C. 2003c. Historical records of the Sumatran rhinoceros in North-East India. *Newsletter of the Rhino Foundation of Nature in North-East India* no. 5: 11–12.

Rookmaaker L.C. 2004a. Rhinoceros rugosus – a name for the Indian rhinoceros. *Journal of the Bombay Natural History Society* 101: 308–310.

Rookmaaker L.C. 2004b. Fragments on the history of the rhinoceros in Nepal. *Pachyderm*, Nairobi 37: 73–79.

Rookmaaker L.C. 2004c. Historical distribution of the black rhinoceros (Diceros bicornis) in West Africa. *African Zoology* 39 (1): 63–70.

Rookmaaker L.C. 2005a. A Javan rhinoceros, Rhinoceros sondaicus, in Bali in 1839. *Zoologische Garten* 75: 129–131.

Rookmaaker L.C. 2005b. Review of the European perception of the African rhinoceros. Journal of Zoology, London 265: 365–376.

Rookmaaker L.C. 2006a. The demise of the Lesser Indian Rhinoceros. In: Das J. et al., *Souvenir of Kaziranga Elephant Festival 2006: English section.* Airawat, pp. 27–28.

Rookmaaker L.C. 2006b. Distribution and extinction of the rhinoceros in China: review of recent Chinese publications. *Pachyderm*, Nairobi 40: 102–106.

Rookmaaker L.C. 2006c. The first photographs of a rhinoceros. *Swara (East African Wild Life Society)* 29 (3): 54–55.

Rookmaaker L.C. 2007a. The first rhino in Britain. *The Horn Newsletter (Save the Rhino International, London)* 2007 Spring: 19.

Rookmaaker L.C. 2007b. July 1846: Strickland's invitation to the first meeting of the Cotswold Club. *Proceedings of the Cotteswold Naturalists Club* 44 (1): 108–111.

Rookmaaker L.C. 2008. *Encounters with the African rhinoceros: a chronological survey of the bibliographical and iconographical sources on rhinoceroses in southern Africa from 1795 to 1875, reconstructing views on classification and changes in distribution.* Münster: Schüling Verlag [also in: *Transactions of the Royal Society of South Africa* 62 (2): 55–198 (2007)].

Rookmaaker L.C. 2010a. *Calendar of the scientific correspondence of Hugh Edwin Strickland in the University Museum of Zoology, Cambridge.* Cambridge.

Rookmaaker L.C. 2010b. The Rhino Resource Center. *Pachyderm*, Nairobi 47: 100–101.

Rookmaaker L.C. 2011a. Calcutta Zoo 1880s: evidence of a hitherto unrecorded Javan rhinoceros. *International Zoo News* 58: 24–25.

Rookmaaker L.C. 2011b. How I met Clara, the Dutch rhinoceros. *Pachyderm*, Nairobi 50 (special issue): 10–11.

Rookmaaker L.C. 2011c. The early endeavours by Hugh Edwin Strickland to establish a code for zoological nomenclature in 1842–1843. *Bulletin of Zoological Nomenclature* 68 (1): 29–40.

Rookmaaker L.C. 2011d. Medicinal use of rhino horn. *Newsletter of the Rhino Resource Center* no.25: 4–5.

Rookmaaker L.C. 2014a. Three rhinos on maps of India drawn in Faizabad in the 18th century. *Pachyderm*, Nairobi 55: 95–96.

Rookmaaker L.C. 2014b. The birth of the first Sumatran Rhinoceros Dicerorhinus sumatrensis (Fischer, 1814) – London Docks 1872. *Zoologische Garten* 83: 1–16.

Rookmaaker L.C. 2015a. Rhino systematics in the times of Linnaeus, Cuvier, Gray and Groves. In: Behie A.M.; Oxenham M.F. 2015. *Taxonomic tapestries* (Festschrift to Colin P. Groves). Canberra, pp. 299–319.

Rookmaaker L.C. 2015b. The history and identity of a rhinoceros exhibited through Europe by Huguet of Massillia and sold to the Zoological Garden of Marseille, 1845–1862. *Mésogée (Muséum d'Histoire Naturelle de Marseille)* 69: 41–57, 75–80.

Rookmaaker L.C. 2016a. On the alleged presence of the two-horned Sumatran rhinoceros and the one-horned Javan rhinoceros in the himalayan kingdom of Bhutan. *Pachyderm*, Nairobi 57: 116–117.

Rookmaaker L.C. 2016b. The zoological contributions of Andrew Smith (1797–1872) with an annotated bibliography and a numerical analysis of newly described animal species. *Transactions of the Royal Society of South Africa* 72: 105–173.

Rookmaaker L.C. 2018. The Javan rhino in Junagadh. *Hornbill (Bombay Natural History Society)* 2018 January–March: 23.

Rookmaaker L.C. 2019a. Edward Barlow's depiction of a living rhinoceros in transit to London in 1683. *Archives of Natural History* 46: 156–158.

Rookmaaker L.C. 2019b. Lady Curzon and the establishment of Kaziranga National Park. *Pachyderm*, Nairobi 60: 110–111.

Rookmaaker L.C. 2019c. The hornless rhinoceros (Rhinoceros sondaicus inermis Lesson, 1836) discovered by Lamare-Picquot in the Sundarbans of Bangladesh in 1828, with notes on the history of his Asian collections. *Mammalia* 84: 74–89.

Rookmaaker L.C. 2019d. Mauled by a rhinoceros: the final years of Alfred Duvaucel (1793–1824) in India. *Zoosystema* 41: 259–267.

Rookmaaker L.C. 2019e. Evidence of a rhinoceros in Cork, Ireland, in November 1842. *Journal of the Bartlett Society* 27: 20–22.

Rookmaaker L.C. 2020a. Rhinoceros pulling a plough: fact or fiction? *Pachyderm*, Nairobi 61: 176–178.

Rookmaaker L.C. 2020b. The maximum size of the greater one-horned rhinoceros (Rhinoceros unicornis). *Pachyderm*, Nairobi 61: 184–190.

Rookmaaker L.C. 2020c. *Twenty years of literature on the rhinoceros 2000–2019, extracted from the Rhino Resource Center (RRC)* – www.rhinoresourcecenter.com. Doorn, RRC, pp. 1–346.

Rookmaaker L.C. 2021a. The Diard and Duvaucel collection of drawings. In: Dorai F.; Low M.E.Y., *Diard & Duvaucel: French natural history drawings of Singapore and Southeast Asia, 1818–1820*. Singapore: Embassy of France in Singapore and National Library Board, pp. 46–173.

Rookmaaker L.C. 2021b. Early photographs of the greater one-horned rhinoceros (Rhinoceros unicornis) in the wild. *Pachyderm*, Nairobi 62: 98–104.

Rookmaaker L.C. 2021c. Eponyms associated with the nomenclature of the recent species of rhinoceros. *Pachyderm*, Nairobi 62: 87–97.

Rookmaaker L.C.; Antoine P.O. 2012. New maps representing the historical and recent distribution of the African species of rhinoceros: Diceros bicornis, Ceratotherium simum and Ceratotherium cottoni. *Pachyderm*, Nairobi 52: 91–96.

Rookmaaker L.C.; Carpentier H. 2007. Early references to the rhinoceros on the Chinese island of Hainan. *Journal of the Royal Asiatic Society Hong Kong Branch* 45: 235–236.

Rookmaaker L.C.; Carpentier H.; Reichenbach H. 2002. John Hagenbeck's rhinoceros. *International Zoo News* 49 (5): 274–275.

Rookmaaker L.C.; Edwards J. 2022. The rhinoceros on terracotta temples and in jungles of West Bengal and Bangladesh. *Journal of Bengal Art* 26: 71–88.

Rookmaaker L.C.; Gannon J.; Monson J. 2015. The lives of three rhinoceroses exhibited in London 1790 to 1814. *Archives of Natural History* 42: 279–300.

Rookmaaker L.C.; Klee E.L. 2010. Sighting of a rhinoceros in Upper Myanmar in 1996. *Pachyderm*, Nairobi 48: 73.

Rookmaaker L.C.; Monson J. 2000. Woodcuts and engravings illustrating the journey of Clara, the most popular rhinoceros of the eighteenth century. *Zoologische Garten* 70: 313–335.

Rookmaaker L.C.; Mundy P.J.; Glenn I.; Spary E.C. 2004. *François Levaillant and the Birds of Africa*. Johannesburg: Brenthurst Press (third series, vol. 5).

Rookmaaker L.C.; Nelson B.; Dorrington D. 2005. The royal hunt of tiger and rhinoceros in the Nepalese terai in 1911. *Pachyderm*, Nairobi 38: 89–97.

Rookmaaker L.C.; Reynolds R.J. 1985. Additional data on rhinoceroses in captivity. *Zoologische Garten* 55: 129–158.

Rookmaaker L.C.; Robovsky J. 2019. Bibliography of Colin Peter Groves (1942–2017), an anthropologist and mammalian taxonomist. *Lynx*, Praha 49: 255–294.

Rookmaaker L.C.; Robovsky J. 2020. Addenda and corrigenda: bibliography of Colin Peter Groves (1942–2017). *Lynx*, Praha 51: 227–229.

Rookmaaker L.C.; Sharma A.; Bose J.; Thapa K.; Dutta D.; Jeffries B.; Williams A.C.; Ghose D.; Gupta M.; Tornikoski S. 2017. *The Greater One-Horned Rhino: past, present and future*. Gland: WWF.

Rookmaaker L.C.; Vigne L.; Martin E.B. 1998. The rhinoceros fight in India. *Pachyderm*, Nairobi 25: 28–31.

Rookmaaker L.C.; Visser R.P.W. 1982. Petrus Camper's study of the Javan rhinoceros (Rhinoceros sondaicus) and its influence on Georges Cuvier. *Bijdragen tot de Dierkunde* 52: 121–136.

Roosevelt B. 1926. Rhinoceros shot in Nepal. *Sunday Star*, Washington DC. 1926-05-02 (gravure section).

Roosevelt B.W. 1927. From the land where the elephants are, 1926. In: *Cleared for strange ports*. New York and London, pp. 217–254.

Roosevelt E.B. 1959. *Day before yesterday: the reminiscences of Mrs. Theodore Roosevelt, Jr.* Garden City.

Roosevelt T. 1898. Letter to Frederick Courteney Selous, 15 February 1898. Theodore Roosevelt Papers. Library of Congress Manuscript Division. https://www.theodorerooseveltcenter.org/Research/Digital-Library/Record?libID=0161328 [accessed April 2023].

Roosevelt T. 1910. *African game trails*. New York.

Roscam Abbing M. 2022. Some notes by Ernst Brinck (1582–1649) on painters, collectors and exceptional art. *Oud Holland* 135: 204–224.

Rose D.B.; van Dooren T.; Chrulew M. 2017. *Extinction studies: stories of time, death, and generations*. New York.

Rothfels N. 2008. *Savages and beasts: the birth of the modern zoo*. Baltimore.

Rothfels N. 2021. *Elephant trails: a history of animals and cultures*. Baltimore.

Rousselet L. 1871. L'Inde des Rajahs, part 6. Les fêtes et les chasses du Guicowar. *Le Tour du Monde* 22: 242–248.

Rousselet L. 1872a. Am Hofe des Guikowar zu Baroda in Indien. *Globus* 20: 193–199.

Rousselet L. 1872b. Een bezoek aan het hof van den Guikowar. *De Aarde en haar Volken* 1872: 281–304.

Rousselet L. 1875a. *L'Inde des Rajahs: Voyage dans l'Inde Centrale et dans les Présidences de Bombay et du Bengale*. Paris.

Rousselet L. 1875b. *India and its native princes: travels in Central India and in the Presidencies of Bombay and Bengal*, new edition. London: Chapman and Hall [first English edition].

Rousselet L. 1875c. Le Guicowar, roi de Baroda. *Journal de la jeunesse: nouveau recueil hebdomadaire illustré* 1875 (1): 360–362, 375–378.

Rousselet L. 1875d. India and its native princes. *Scribner's Monthly* 11: 65–79.

Rousselet L. 1876a. *India and its native princes: travels in Central India and in the Presidencies of Bombay and Bengal*, new edition. London: Chapman and Hall [second English edition].

Rousselet L. 1876b. *India and its native princes. Travels in Central India and in the Presidencies of Bombay and Bengal*. Carefully revised and edited by Lieut.-Col. Buckle. New York: Scribner, Armstrong, and Co. [American edition].

Rousselet L. 1876c. Der Rhinozeroskampf in der Arena von Baroda. *Über Land und Meer* 35: 377, 383.

Rousselet L. 1877a. *L'Inde des Rajahs: voyage dans l'Inde centrale et dans les présidences de Bombay et de Bengale*. Paris.

Rousselet L. 1877b. *L'India: viaggio nell'India centrale e nel Bengala*. Milano.

Rousselet L. 1878a. *India and its native princes: travels in Central India and in the Presidencies of Bombay and Bengal*, new edition. London: Bickers and Son [third English edition].

Rousselet L. 1878b. La India de los Rajas. In: *El mundo en la mano: viaje pintoresco a las cinco partes del mundo por los más célebres viajeros*, vol. 3. Barcelona.

Rousselet L. 1882. *India and its native princes: travels in Central India and in the Presidencies of Bombay and Bengal*, new edition. London: Bickers and Son [fourth English edition].

Roux C. 1976. On the dating of the first edition of Cuvier's Règne Animal. *Journal of the Society for the Bibliography of Natural History* 8: 31–49.

Rowntree J. 1981. *A chota sahib: memoirs of a forest officer*. Padstow.

Roy A. 2012. *Jagjivanpur: 1996–2005 excavation report*. Kolkata.

Roy A. 2013. *Buddhist vestiges at Jagjivanpur in North Bengal: a case study of terracotta plaques*. Thesis presented to Utkal University, Bhubaneswar.

Roy A.N. 1945. The great Indian rhinoceros. *Indian Forester* 71: 125–126.

Roy B. 2003. *Fallen cicada: unwritten history of Darjeeling Hills*. Darjeeling.

Roy S.D. 1993. India: rhino conservation action plan. In: Foose T.J., IUCN/SSC *Asian Rhino Specialist Group (AsRSG) meeting, Jaldapara Sanctuary: Briefing book*. Columbus, pp. 1–74.

Royal Academy. 1799. *The exhibition of the Royal Academy of Arts 1799*, no. 31. London.

Royal Tiger. 1836. A few days sport in the Rajmahal Hills, or my first interview with a rhinoceros. *Bengal Sporting Magazine* (n.s.) 7 (25): 3–7.

Royle J.F. 1839. *Illustrations of the botany and other branches of the natural history of the Himalaya mountains, and of the flora of Cashmere*, vol. 1. London.

Rozian. 1912. Ein Opfer der Eitelkeit. *Fliegende Blätter*, München 136 (3479): 152.

R.P. 1884. Fight between a tiger and a rhinoceros. *Madras Mail* 1884-09-04.

Ruge A. (Stein A.W.) 1854. *Jagden und Thiergeschichten für unsere Knaben*. Stuttgart.

Ruggles J. 1906. *Recollections of a Lucknow veteran, 1845–1876*. London.

Russell C.E.M. 1900. *Bullet and shot in Indian forest, plain and hill*. London.

Russell E. 1905. The sacred animals of India. *Everybody's Magazine* 13: 625–634.

Russell W.H. 1876. Our sketches from India. *Illustrated London News* 68 (1876-03-18): 282.

Russell W.H. 1877. *The Prince of Wales' tour: a diary in India*. London.

Ruth M. 2007. Ein Leben mit Katastrophen: der Naturwissenschaftler, Reisende, Sammler und Philanthrop Christophe-Augustin Lamarepicquot (1785–1873). In: Müller C.; Stein W., *Exotische Welten*. Dettelbach, pp. 95–105.

Ryhiner M. 1952. Zur Kulturgeschichte des Rhinozeros. *Du*, Zürich 1952: 66–69, 94.

Ryhiner P.R. 1952b. Das Panzernashorn: eine aussterbende Tierart Indiens. *Prisma (Illustrierte Monatsschrift für Natur, Forschung und Technik)* 7: 161–163.

Ryhiner P.R.; Mannix D. 1964. *The wildest game*. London.

Saar J.J. 1662. *Ost-Indianische fünfzehen-jährige Kriegs-Dienst, und wahrhaftige Beschreibung*. Nürnberg.

Saban R. 1983. La dissection au Muséum du rhinocéros de la Ménagerie de Versailles, en 1793, et les 'Velins du Muséum'. *Compte Rendu du Congres des Sociétés Savantes de Paris* 108 (4): 33–40.

Saban R.; Dupin M.P. 1989. Objets en corne et peau du rhinocéros. *Objets et Mondes: la revue du Musée de l'Homme* 26: 25–33.

Sachau E.C. 1910. *Alberuni's India*, 2 vols. London.

Sachse F.A. 1917. *Bengal district gazetteers*, vol. 34: *Mymensingh*. Calcutta.

Sagreiya K.P. 1969. Wild life of Madhya Pradesh through ages. *Indian Forester* 95: 715–718.

Saha S.; Rajak S. 2019. Digital documentation of rock art site of Isko, Hazaribagh, Jharkhand. *Heritage: Journal of multidisciplinary studies in Archaeology* 7: 493–506.

Sahlström A.; Purohit M. 1985. *The "Library Project" in Dungarpur, Rajasthan or the establishment of "The Maharawal Bijay Singh Research Archive" at Udai Vilas Palace, Dungarpur*. Final Report of a S[wedish] I[nternational] D[evelopment Corporation] A[gency] supported Project. Dungarpur.

Sahni D.R. 1914. *Catalogue of the Museum of Archaeology at Sārnāth*. Calcutta.

Saikia A. 2005. *Jungles, reserves, wildlife: a history of forests in Assam*. Guwahati.

Saikia A. 2009. The Kaziranga National Park: dynamics of social and political history. *Conservation and Society* 7: 113–129.

Saikia A. 2011. *Forests and ecological history of Assam 1826–2000*. Delhi.

Saikia A. 2020. Rhinoceros in Kaziranga National Park: nature and politics in modern Assam. In: Rao M., *Reframing the environment: resources, risk and resistance in neoliberal India*. Delhi, pp. 159–188.

Sale J.B. 1986. Rhinos re-established in Uttar Pradesh. *Indian Forester* 112: 945–948.

Sale J.B.; Singh S. 1987. Reintroduction of greater Indian rhinoceros into Dudhwa National Park. *Oryx* 21: 81–84.

Sale J.B.; Woodford M.H. 1984. Preliminary report on drug immobilisation and transport of the great Indian rhinoceros. In: Singh S.; Rao K., *India's rhino-reintroduction programme*. New Delhi, pp. 25–51.

Salgari E. 1895. *I misteri della jungla nera: racconto*. Genova: A.Donath.

Salgari E. 1899. *Les mystères de la jungle noire*. Paris: Montgredien.

Salgari E. 1904. *Le duo tigre*. Genova.

Salgari E. 1950. *I misteri della jungla nera*. Milano: Carroccio.

Salgari E. 1958. *I misteri della jungla nera*. Bologna: Carroccio-Aldebaran (Collana Nord-Ouest no. 39).

Sali S.A. 1986. *Daimabad, 1976–79*. New Delhi.

Salomon R. 2000. *A Gāndhārī version of the Rhinoceros Sūtra: British Library Kharoṣṭhī fragment 5B*. Seattle.

Sandeman H.D. 1868. *Selections from Calcutta Gazettes of the years 1806 to 1815*, vol. 4. Calcutta.

Sanderson G.P. 1878. *Thirteen years among the wild beasts of India*. London.

Sanderson I.T. 1956. *Knaurs Tierbuch in Farben: Säugetiere*. München.

Sangram Singh, Kumar of Nawalgarh. 1965. *Catalogue of Indian miniature paintings (collection of Kumar Sangram Singh of Nawalgarh)*. Exhibited in the Rajasthan University Library, Jaipur, from 12th to 17th December, 1965. Jaipur.

Sanyal R.B. 1892. *A hand-book of the management of animals in captivity in Lower Bengal*. Calcutta.

Sanyal R.B. 1896. *Hours with nature*. Calcutta.

Sanyal R.B. 1905. *Report of the Honorary Committee for the Management of the Zoological Garden, 1904–1905*. Calcutta.

Sardesai G.S. 1936. *Poona Residency correspondence*, vol. 2: *Poona affairs 1786–1797*. Bombay.

Sardeshpande S.C. 1987. *The Patkoi Nagas*. Delhi.

Sarel H.A. 1887. Note on rhinsoceros shooting in Assam. In: Newall D.J.F., *The highlands of India*, vol. 2: *being a chronicle of field sports and travel in India*. London, pp. 358–368.

Sarkar J. 1928. A description of North Bengal in 1609 A.D. *Bengal Past and Present* 35: 143–146.

Sarker S.U.; Sarker N.J. 1984. Mammals of Bangladesh. *Tiger Paper* (FAO) 11 (1): 8–13.

Sarma B. 2023. Empire, nature and agrarian world: a history of rhino preservation in the Kaziranga Game Reserve, India (1902–1938). *Environment and History* 2023: 1–24.

Sarma P.K. 2010. *Habitat suitability for rhino (Rhinoceros unicornis) and utilization pattern in Rajiv Gandhi Orang National Park of Assam*. Thesis presented to the North Eastern Hill University, Shillong.

Sarwar M. 1971. A Javanese rhinoceros recorded from the Upper Siwaliks of Azad Kashmir, Pakistan. *Geological Bulletin, University of Peshawar* 6: 49–53.

Sass M. 2021. Liebelei und Fehde: Frühneuzeitliche Pokale aus Rhinozeros-Horn als Wissensobjekte. In: Bauernfeind R.; Rudolph P., *Bilder exotischer Tiere. Zwischen wissenschaftlicher Erfassung und gesellschaftlicher Normierung 1500–1800*. Augsburg, pp. 96–124.

Sathe V. 2010. The archaeology of great one-horned Indian rhinoceros (Rhinoceros unicornis Linnaeus 1758). In: Swarup A.; Agrawal S.C., *Indian civilization through the millennia*. Delhi, pp. 22–30.

Sathe V. 2015. Discovery of a fossil stone bed in the Manjra valley, district Latur, Maharashtra. *Bulletin of the Deccan College Research Institute* 75: 1–16.

Saul J. 2005. *The Naga of Burma: their festivals, customs, and way of life*. Bangkok.

Saunders's News-Letter 1835-06-30: Arrival of rhinoceros in Dublin.

Saunders's News-Letter 1835-07-25: Departure of rhinoceros from Dublin.

Sax W.S. 1997. Fathers, sons, and rhinoceroses: masculinity and violence in the Pandav Lila. *Journal of the American Oriental Society* 117: 278–293.

Sax W.S. 2002. *Dancing the Self: personhood and performance in the Pāṇḍav Līlā of Garhwal*. Oxford.

Sayers D. 2017. *Climbing Mt. Saramati: from Myanmar and India with travel on the Chindwin*. Kibworth.

Scarisbrick D.; Wagner C.; Boardman J. 2016. *The Guy Ladrière collection of gems and rings*. London.

Schaffer H. 2011. *Adapting the eye: an archive of the British in India, 1770–1830*. New Haven.

Scheibler F. 1899a. Alla caccia della tigre in India. *Nuova Antologia di Scienze, Lettere ed Arti* 163: 103–118.

Scheibler F. 1899b. Tiger hunting in India. *Chautauquan: organ of the Chautauqua Literary and Scientific Circle* 29: 50–54.

Scheibler F. 1900. *Sette anni di caccia grossa e note di viaggio in America, Asia, Africa, Europa*. Milano.

Schenkel R. 1984. Report on the suitability of Dudhwa National Park U.P. as potential site for re-introduction of the Indian rhinoceros. In: Singh S.; Rao K., *India's rhino-reintroduction programme*. New Delhi, pp. 21–24.

Schenkel R.; Lang E.M. 1969. Das Verhalten der Nashörner. *Handbuch für Zoologie* 8 (46): 1–56.

Schenkel R.; Nievergelt B.; Bucher F. 2007. *8 Hörner auf 5 Nasen: ein Nashornbuch*. Zürich.

Schenkel R.; Schenkel L. 1969. The Javan rhinoceros (Rh. sondaicus Desm.) in Udjung Kulon Nature Reserve: its ecology and behaviour. *Acta Tropica* 26: 97–135.

Scherpner C. 1983. *Von Bürgern für Bürger: 125 Jahre Zoologischer Garten Frankfurt am Main*. Frankfurt.

Schlagintweit-Sakünlünski H. von. 1869. *Reisen in Indien und Hochasien*, vol. 1. Jena.

Schlagintweit E. 1880. *Indien in Wort und Bild: eine Schilderung des Indischen Kaiserreiches*, vol. 1. Leipzig.

Schlagintweit E. 1890. *Indien in Wort und Bild: eine Schilderung des Indischen Kaiserreiches*, 2nd edition, vol. 1. Leipzig.

Schlawe L. 1969. *Die für die Zeit vom 1 August 1844 bis 31 Mai 1888 nachweisbaren Thiere im Zoologischen Garten zu Berlin*. Berlin.

Schlawe L. 1994. Illustrationen nach dem Leben (ndL) aus dem Zoologischen Garten zu Berlin. *Bongo (Beiträge zur Tiergärtnerei und Jahresberichte aus dem Zoo Berlin)* 23: 35–62.

Schlegel H. 1872. *De dierentuin van het Koninklijk Zoologisch Genootschap Natura Artis Magistra te Amsterdam*. Amsterdam.

Schmidt B. 2015. *Inventing exoticism: geography, globalism, and Europe's early modern world*. Philadelphia.

Schönberg E. von. 1852. *Patmakhanda: Lebens- und Charakterbilder aus Indien und Persien*, vol. 1. Leipzig.

Schönberg E. von. 1853. *Travels in India and Kashmir*, vol. 1. London.

Schouten W. 1676a. *Oost-Indische voyagie*. Amsterdam.

Schouten W. [Schultz W.] 1676b. *Ost-Indische Reyse*. Amsterdam.

Schouten W. 1707. *Voiage aux Indes Orientales commencé l'an 1658 & fini l'an 1665*. Traduit du hollandois, vol. 2. Paris.

Schrader S. 2018. Rembrandt and the Mughal line: artistic inspiration in the global city of Amsterdam. In: Schrader S., *Rembrandt and the inspiration of India*. Los Angeles, pp. 5–28.

Schreber J.C.D. von; Wagner J.A. 1835. *Die Säugthiere in Abbildungen nach der Natur mit Beschreibungen*, vol. 6. Erlangen.

Schröder J. 1748. *Pharmacopoeia universalis*, part 3. Nürnberg.

Schuhmacher E. 1968. *The last of the wild: on the track of rare animals*. London [German 1956].

Schwarzenberger F.; Hermes R. 2023. Comparative analysis of gestation in three rhinoceros species (Diceros bicornis; Ceratotherium simum; Rhinoceros unicornis). *General and Comparative Endocrinology* 334 (114214): 1–11.

Scientific American 1884-09-13, p. 167: Khedda operation of Shillingford.

Sclater P.L. 1872a. Announcement of the addition to the Society's collection of a female Sumatran Rhinoceros. *Proceedings of the Zoological Society of London* 1872-02-20: 185.

Sclater P.L. 1872b. Additions to the Society's Menagerie in February 1872. *Proceedings of the Zoological Society of London* 1872-03-19: 493–496.

Sclater P.L. 1872c. British Association: the new rhinoceros. *Times*, London 1872-08-19: 5.

Sclater P.L. 1872d. On Rhinoceros lasiotis. *Athenaeum* 1872-08-24: 243.

Sclater P.L. 1872e. Notes on Propithecus bicornis and Rhinoceros lasiotis. *Annals and Magazine of Natural History* (ser.4) 10: 298–299.

Sclater P.L. 1872f. A new Asiatic rhinoceros. *Popular Science Review* 11: 432.

Sclater P.L. 1872g. Rhinoceroses. *Nature* 5 (1872-03-28): 426–428.

Sclater P.L. 1872h. The new rhinoceros. *Nature* 6 (1872-10-24): 518–519.

Sclater P.L. 1872k. Additions to the Society's Menagerie during the months of June, July, August and September 1872. *Proceedings of the Zoological Society of London* 1872-11-05: 790–794.

Sclater P.L. 1872m. *Guide to the gardens of the Zoological Society of London*, 25th edition. London.

Sclater P.L. 1873a. On a new rhinoceros, with remarks on the recent species of this genus and their distribution. *Report of the British Association for the Advancement of Science* 42nd Meeting 1872 (Sections): 140.

Sclater P.L. 1873b. New Asiatic rhinoceros. *Year-book of Facts in Science and Art* 1873: 224.

Sclater P.L. 1873c. *Guide to the gardens of the Zoological Society of London*, 26th edition. London.

Sclater P.L. 1875a. Exhibition of a drawing of a supposed new rhinoceros from the terai of Bhootan. *Proceedings of the Zoological Society of London* 1875-02-16: 82.

Sclater P.L. 1875b. Exhibition of, and remarks upon, the upper horn of a two-horned rhinoceros from the valley of the Brahmapootra. *Proceedings of the Zoological Society of London* 1875-11-16: 566.

Sclater P.L. 1876a. On the rhinoceroses now or lately living in the Society's Menagerie. *Transactions of the Zoological Society of London* 9: 645–660.

Sclater P.L. 1876b. Exhibition and remarks upon a skin of a young rhinoceros from the Sunderbunds. *Proceedings of the Zoological Society of London* 1876-11-07: 751.

Sclater P.L. 1877a. Remarks upon a two-horned rhinoceros killed in 1876 near Comillah in Tipperah, and on a living specimen of Rhinoceros sondaicus from the Sunderbans. *Proceedings of the Zoological Society of London* 1877-03-20: 269–270.

Sclater P.L. 1877b. *List of the vertebrated animals now or lately living in the Gardens of the Zoological Society of London*, 6th edition. London.

Sclater P.L. 1880. Remarks on animals observed in the zoological gardens of Berlin, Hamburg, Amsterdam, The Hague and Antwerp. *Proceedings of the Zoological Society of London* 1880-06-01: 420.

Sclater P.L. 1884. Remarks upon a copy of the lately issued 'Guide to the Zoological Gardens' and on Rhinoceros lasiotis. *Proceedings of the Zoological Society of London* 1884-02-05: 55–56.

Sclater P.L. 1887. Additions to the Society's Menagerie in December 1886. *Proceedings of the Zoological Society of London* 1887-01-18: 1–2.

Sclater W.L. 1891. *Catalogue of Mammalia in the Indian Museum, Calcutta*, vol. 2. Calcutta.

Scotsman 1836-01-16, 1836-01-23, 1836-01-30: Arrival of rhinoceros in Edinburgh.

Scotsman 1836-02-06, 1836-02-17, 1836-02-20: Departure of rhinoceros from Edinburgh.

Scotsman 1886-01-30: Rhino fight in Baroda.

Scotsman 1938-12-13: Rhinoceros shot by Viceroy.

Scotsman 1939-04-12: Death of J.W.A. Grieve.

Scott R.S. 1971. Rhino roundup in Nepal. *Karatasi Yenye Habari* 19 (2): 30–33.

Seiferheld, Helene C. Gallery. 1962. *Animal drawings from the XV to XX centuries*, no 8. New York.

Sengupta G.; Chowdhury S.R.; Chakraborty S. 2007. *Eloquent earth: early terracottas in the State Archaeological Museum, West Bengal*. Kolkata.

Sengupta J.C. 1969. *West Bengal district gazetteers: Malda*. Calcutta.

Sentinel Assam 2022-09-15: Stray rhino creates panic in Jamugurihat.

Seow A. 2010. Shebbeare: on the shoulders of giants. *Malaysian Naturalist* 2010 (September): 34–37.

Sergeant P.W. 1928. *The ruler of Baroda: an account of the life and work of the Maharaja Gaekwar*. London.

Seshadri B. 1969. *The twilight of India's wild life*. London.

Sethi S. 1999. *Intérieurs de l'Inde; Indian Interiors; Indien Interieurs*. Köln.

Seton-Karr H.W. 1897. After big game in Africa and India. *Century Magazine* 54: 370–381.

Seyller J.; Mittal J. 2019. *Central Indian paintings in the Jagdish and Kamla Mittal Museum of Indian Art*. Hyderabad.

Shaffer H. 2011. *Adapting the eye: an archive of the British in India, 1770–1830*. New Haven, CT.

Shah S.G.M.; Parpola A. 1991. *Corpus of Indus seals and inscriptions*, vol. 2. *Collections in Pakistan*. Helsinki (Memoirs of the Department of Archaeology and Museums, Pakistan, vol. 5).

Shah U.P. 1987. *Jaina-Rūpa-Maṇḍana*, vol. 1: *Jaina iconography*. New Delhi.

Shaha R.; Mitchell R.M. 2001. *Wildlife in Nepal*. New Delhi.

Shahi S.P. 1977. *Backs to the wall: saga of wildlife in Bihar, India*. New Delhi.

Shakespear J. 1912. *The Lushei Kuki clans*. London.

Shakespear J. 1928. Lushai reminisces. *Assam Review* 1: 7–16.

Shakespear L.W. 1912. *History of the 2nd King Edward's own Goorkha Rifles*. Aldershot.

Shakespear L.W. 1929. *History of the Assam rifles*. London.

Sharar A.H. 1975. *Lucknow: the last phase of an oriental culture*. London.

Sharma A. 2022. The population status of the greater one-horned rhino in India and Nepal, and the importance of regular monitoring census intervals. *Pachyderm*, Nairobi 63: 176–178.

Sharma A.; Dutta D.K.; Swargowari A.; Singh S.P. 2012. *The released rhinos in Manas National Park, the fourth year (April 2011–March 2012)*. Report of IRV2020.

Sharma A.; Gupta M.; Kakati P.K.; Maheswari R.; Bonal B.S.; Chowdhary S.; Kaujalagi M.; Ravi R.; Sharma K.K.; Chowdhary B. 2018. *Rhino translocation to RRA 2, Dudhwa Tiger Reserve (April 2018)*. New Delhi: WWF.

Sharma B.N. 1978. *Festivals of India*. New Delhi.

Sharma D.C. 2006. Relocation planned for one-horned rhino. *Frontiers in Ecology and the Environment* 4: 120.

Sharma G.; Kamalakannan M.; Venkataraman K. 2013. *A checklist of mammals of India with their distribution and conservation status*. Kolkata.

Sharma G.R. 1969. Excavations at Kausambi 1949–50. *Memoirs of the Archaeological Survey of India* 74: 1–90.

Sharma G.R. 1975. Excavation at Sarai Nahar Rai, district Pratapgarh. *Indian Archaeology: a Review* 1971–1972: 48–49.

Sharma R.; Gupta M. 2015. *Status and monitoring of the greater one-horned rhinoceros in Dudhwa National Park*. New Delhi.

Sharma R.S. 2005. Was the Harappan culture vedic? *Journal of Interdisciplinary Studies in History and Archaeology* 1: 135–144.

Sharma S.K. 2002. Camundi of Harsacarita of Bana: a critical review. *Sambodhi*, Ahmedabad 25: 196–208.

Sharpe R.B. 1906. Birds. In Lankester E.R., *The history of the collections contained in the natural history departments of the British Museum*. London, pp. 79–515.

Shastri H. 1910. *Syainika sastra: or a book on hawking by Raja Rudradeva of Kumaon*. Calcutta.

Shaw W. 1929. Notes on the Thadou Kukis. *Journal of the Asiatic Society of Bengal* (n.s.) 24: 1–168.

Shebbeare E.O. 1935. Protecting the great Indian rhinoceros. *Field, the Country Gentleman's Magazine* 165 (1935-05-18): 1229–1231.

Shebbeare E.O. 1948. The Bengal rhinoceros sanctuary. *Journal of the Society for the Preservation of the Fauna of the Empire* 56: 33–35.

Shebbeare E.O. 1953. Status of the three Asiatic rhinoceros. *Oryx* 2: 141–149.

Shebbeare E.O.; Roy A.N. 1948. The great one-horned rhinoceros (Rhinoceros unicornis L). *Journal of the Bengal Natural History Society* 22: 88–91.

Sheffield Daily Telegraph 1875-11-18, p. 2 (from *Indian Daily News* 1875-10-22): Hunting in Soonderbunds.

Shekarea A. 1832. The Saugor island rhinoceros. *Oriental Sporting Magazine* 2: 313–314. Reprinted: (1) *Asiatic Journal and Monthly Register for British and Foreign India* 9: 167 (1832); (2) *New-York Traveller and Spirit of the Times* 1, no. 41: 1 (1832-12-08); (3) *Friend. Religious and Literary Journal* 6 no. 19: 149 (1833-02-07); (4) *Chambers's Edinburgh Journal* 3, no. 142: 304 (1834-10-18); (5) *London and Paris Observer* 10, no. 502: 830 (1834-12-28); (6) *Albion*, Stuttgart, no. 23: 177 (1835-06-07); (7) *Odd Fellows' Magazine* 4: 86 (1835-12); (8) In: Shelton E., *The book of battles; or, daring deeds by land and sea*. London, p. 515.

Shekarea A. 1835. Jagd auf ein Rhinozeros. *Magazin für die Literatur des Auslandes*, Berlin 7, no. 4: 16 (1835). Also in: (1) *Allgemeine Forst- und Jagdzeitung* 4, no. 45: 180 (1835-04-13); (2) *Münchener politische Zeitung*, no. 143 (1835-06-19): 949–950.

Shepard O. 1967. *The lore of the unicorn*, 2nd impression. London [first edition 1930].

Sherborn C.D. 1922. *Index animalium, sive index nominum quae ab AD MDCCLVIII generibus et speciebus impositae sunt. Sectio Secunda (1801–1850)*. London.

Shere S.E. 1951. Stone discs found at Murtaziganj. *Journal of the Bihar Research Society* 37 (3/4): 178–190.

Sherwill J.L. 1865. *A geographical and statistical report of the Dinagepore District for 1863*. Calcutta.

Sherwill W.S. 1854. *Geographical and statistical report of the district of Bhaugulpoor*. Calcutta.

Shikar. 1909. A record rhino. *Englishman's Overland Mail* 1909-05-04: 3.

Shikarophilos. 1833. The buffaloe tails elongated and brought to an end. *Bengal Monthly Sporting Magazine* 1: 481–482.

Shikarree. 1830. Sketches of Bengal. *Sporting Magazine* 25 (149): 251–259.

Shillingford J. 1871. Sport in lower Bengal. *Allen's Indian Mail* 1871-05-30: 513.

Shinde V.; Willis R.J. 2014. A new type of inscribed copper plate from Indus Valley (Harappan) civilisation. *Ancient Asia* 5 (1): 1–10.

Shipping and Mercantile Gazette 1874-06-20: Escape of leopard.

Shouldham T.H. 1826. Visit of the Governor General to Lucknow. *Quarterly Oriental Magazine, Review and Register* 6 (12): cxlvii–cli. [Also in *Asiatic Journal* 23: 845–847 (June 1827)].

Showers H.L. 1909. *Notes on Jaipur*. Ajmer.

Shrestha S. 2009. *Sahibs and shikar: colonial hunting and wildlife in British India, 1800–1935*. Thesis presented to Duke University, Durham NC.

Shrestha T.J.; Shrestha S.; Shrestha A.K. 2017. *Wilderness and diversity of life in Nepal's Chitwan National Park*. Kathmandu.

Shrestha T.K. 1997. *Mammals of Nepal, with reference to those of India, Bangladesh, Bhutan and Pakistan*. Kathmandu.

Siddiq M.Kh.; Akhtar M.; Khan M.A.; Ghaffar A.; Sarwar K.; Khan A.M. 2016. New fossils of rhinoceros (Rhinocerotidae) from the Soan Formation (Plio-Pleistocene) of Northern Pakistan. *Pakistan Journal of Zoology* 48: 657–664.

Simon N. 1966. *Red Data Book*, vol 1: *Mammalia*. Morges [and updated in 1967].

Simpson W. 1876. *Shikāre and Tōmasha: a souvenir of the visit of H.R.H. the Prince of Wales to India*. London.

Simson F.B. 1886. *Letters on sport in Eastern Bengal*. London.

Singh A.N. 1983. The rhinoceros-slayer type of Kumāragupta I and its significance. *Journal of the Numismatic Society of India* 45: 150–151.

Singh D.M.; Sharma A.; Dutta D.K.; Swargowari A.; Bhobora C.R.; Bonal B.S. 2010. *The released rhinos in Manas National Park, Assam – the second year (April 2009-March 2010)*. Report to IRV2020.

Singh D.M.; Sharma A.; Talukdar B.K. 2011a. *Translocations of rhinos within Assam: a successful first round of the second phase of translocations under Indian Rhino Vision 2020*. Report of IRV2020.

Singh D.M.; Sharma A.; Talukdar B.K. 2011b. *Translocations of rhinos within Assam: a successful second round of the second phase of translocations under Indian Rhino Vision 2020*. Report of IRV2020.

Singh G. 2014. *Hunting to conservation: a study of British policies towards wildlife in Assam 1826–1947*. Thesis presented to Assam University, Silchar.

Singh K. 1970. *Pocket encyclopaedia of shikar; dealing with game birds and animals*. New Delhi.

Singh N. 1972. *Bhutan, a kingdom in the Himalayas*. New Delhi.

Singh S.; Rao K. 1984. *India's rhino re-introduction programme*. New Delhi.

Singh S.P. 2000. Status report Manas. In: *Report on the Regional Meeting for India and Nepal of IUCN/SSC Asian Rhino Specialist Group*. Doorn, pp. 135–138.

Singh S.P.; Sharma A.; Dutta D.K.; Swargowari A.; Bhobora C.R.; Singh D.M. 2011c. *An account of the released rhinos in Manas National Park – the third year (April 2010–March 2011)*. Report of IRV2020.

Singh S.P.; Sharma A.; Talukdar B.K. 2012. *Translocation of rhinos within Assam: a successful third round of the second phase of translocations under Indian Rhino Vision 2020*. Report of IRV2020.

Singh V.K. 2015. Dudhwa National Park. In: Ellis S. et al., *Indian Rhino Vision 2020 Population Modeling Workshop final report*. Fort Worth, pp. 30–32.

Singh Y.B. 1981. A note on the Rhinoceros Type coin of Kumāragupta I. *Journal of the Numismatic Society of India* 43: 67–70.

Singh, Harnath: see Harnath Singh.

Singh, Sangram: see Sangram Singh.

Sinha K. 1916. *Purnea: a shikar land.* Calcutta.

Sinha K. 1917. Purnea: a shikar land. *Field, the Country Gentleman's Magazine* 129 (1917-05-26): 746.

Sinha K. 1934. *Shikar in hills and jungles.* Calcutta.

Sinha S.K. 2011. Nature-assisted re-establishment of greater one-horned rhinoceros, Rhinoceros unicornis in its historical distribution range. *Current Science* 100: 1765–1766.

Sinha S.P. 2011. Experiences with reintroduced rhinos in Dudhwa National Park, Uttar Pradesh India. *Pachyderm*, Nairobi 50: 16.

Sinha S.P. 1991. Rhino reintroduction programme in Dudhwa National Park, India. *Re-Introduction News* 3: 7.

Sinha S.P. 2003. *Assessment of corridor viability and habitat restoration between Dudhwa N.P. and Katerniaghat WLS.* Unpublished report.

Sinha S.P. 2004. The great indian one-horned rhinoceros (Rhinoceros unicornis Linnaeus, 1758). *Envis Bulletin*, Dehradun 7 (1): 1–8.

Sinha S.P. 2005. Twenty years of rhino re-introduction in Dudhwa National Park, Uttar Pradesh, India. *Re-Introduction News* 24: 19–21.

Sinha S.P.; Sawarkar V.B. 1993. Management of the reintroduced great one horned rhinoceros (Rhinoceros unicornis) in Dudhwa National Park, Uttar Pradesh, India. In: Ryder O.A., *Rhinoceros biology and conservation: Proceedings of an international conference, San Diego, U.S.A.*, pp. 218–227.

Sinha S.P.; Sawarkar V.B. 1994. Ten years (1984–1994) of the Asian Rhino reintroduction programme in Dudhwa National Park, India. *Re-Introduction News* 9: 13–14.

Sinha S.P.; Sawarkar V.B. 2000. Reintroduced rhino in Dudhwa. In: Strien, N.J. van et al., *Report on the Regional Meeting for India and Nepal of IUCN/SSC Asian Rhino Specialist Group.* Doorn, pp. 139–145.

Sinha S.P.; Sawarkar V.B.; Singh P.P. 2005. Twenty years of rhino re-introduction programme in Dudhwa National Park. *Tiger Paper* (FAO) 32 (3): 14–19.

Sinha S.P.; Sinha B.C. 2007. *The great Indian one horned rhinoceros (Rhinoceros unicornis) in India and Nepal: a review.* Saarbrücken.

Sinha S.P.; Sinha B.C. 2010. *Creation of a viable breeding satellite population of great Indian one-horned rhinoceros (Rhinoceros Unicornis) in Bhadhi-Churella Taal sector of Dudhwa National Park.* Report.

Sinha S.P.; Sinha B.C.; Sawarkar V.B. 2011a. *Twenty five years of Rhino Reintroduction Programme in Dudhwa National Park and Tiger Reserve, Uttar Pradesh, India (1984–2010) with management implications.* Saarbrücken.

Sinha S.P.; Sinha B.C.; Qureshi Q. 2011b. *The Asiatic one-horned rhinoceros (Rhinoceros unicornis) in India and Nepal: ecology, management and conservation strategies.* Saarbrücken.

Sinha Y.P. 2004. Mammals. In: *Fauna of Bihar (including Jharkhand).* Kolkata, pp. 15–72.

Sircar D.C. 1966. Rhinoceros-slayer type coins of Kumāragupta I. *Journal of the Numismatic Society of India* 28: 211–213.

Sireng I. 2015. Rhino numbers in Shuklaphanta wildlife reserve, Nepal. *Headlines Himalaya* no. 361: 3.

Siudmak J. 2016. *John Siudmak Asian Art. Indian and Himalayan sculpture, including property from the collection of the late Simon Digby and private collection, Singapore.* Exhibition in New York.

Sivakumar P. 2015. Laokhowa-Burachapori Wildlife Sanctuary complex. In: Ellis S. et al., *Indian Rhino Vision 2020 Population Modeling Workshop final report.* Fort Worth, pp. 15–26.

Skelton J. 1854. *Engraved illustrations of ancient arms and armour from the collection of Llewelyn Meyrick, at Goodrich Court, Herefordshire, after the drawings, and with the descriptions of Samuel Rush Meyrick,* vol. 2. London.

Smart E.S. 1977. *Paintings from the Baburnama: a study of sixteenth century Mughal historical manuscript illustration.* Thesis presented to the University of London.

Smith, Andrew. 1838. *Illustrations of the Zoology of South Africa,* part 1. London.

Smith D.B. 1868. *Annual report of the Sanitary Commissioner for Bengal for 1868.* Calcutta.

Smith L.F. 1806. A letter to a friend, giving an account of a hunting party of the late Nawab, Asuf-ud-Dowlah. *Asiatic Annual Register* 6 (Miscellaneous tracts): 12–16.

Smith L.F. 1816. An account of a hunting party of the late Nawab Usuf-ad-Dowlah. *Asiatic Journal and Monthly Register* 1: 539–542.

Smith L.O.; Porsch L. 2015. *The costs of illegal wildlife trade: elephant and rhino.* Berlin.

Smith T. 1849. Sporting scenes in Nepaul. *Colburn's United Service Magazine* 1849: 338–348 [also *Littell's Living Age* 23: 537–542 (1849)].

Smith T. 1852. *Narrative of a five years' residence at Nepaul.* London.

Smith V.A. 1901. *Jain Stūpa and other antiquities of Mathurā.* Allahabad.

Smith V.A. 1906a. *Catalogue of the coins in the Indian Museum, Calcutta.* Oxford.

Smith V.A. 1906b. *History of India,* vol. 2: *From the sixth century B.C. to the mohammedan conquest.* London.

Smithsonian. 1862. *Annual Report of the Board of Regents of the Smithsonian Institution for 1862.* Washington.

Smyth R. 1857. *Statistical and geographical report of the 24-Pergunnahs District.* Calcutta.

Smythies E.A. 1942. *Big game shooting in Nepal: with leaves from the Maharaja's sporting diary.* Calcutta.

Smythies E.A. 1974. Memories of old Nepal. *Commonwealth Forestry Review* 53: 195–198.

Smythies O. 1953. *Tiger lady: adventures in the Indian jungle.* London.

Smythies O. 1961. *Ten thousand miles on elephants.* London.

Sodikoff G.M. 2012. *The anthropology of extinction: essays on culture and species death.* Bloomington.

Sody H.J.V. 1941. *De Javaansche neushoorn, Rhinoceros sondaicus, historisch en biologisch.* Buitenzorg.

Sody H.J.V. 1946. [Review] Prehistoric and fossil rhinoceros, by D.A. Hooijer. *Natuurwetenschappelijk Tijdschrift voor Nederlandsch-Indië* 102 (7): 151.

Sody H.J.V. 1959. Das Javanische Nashorn, Rhinoceros sondaicus, historisch und biologisch. *Zeitschrift für Säugetierkunde* 24: 109–240 (translation of Sody 1941 by Erna Mohr).

Sohoni S.V. 1955. Khadga-trātā coins of Kumāragupta I. *Journal of the Bihar Research Society*, Patna 41: 378–386.

Sohoni S.V. 1956. Khadga-Trata coins of Kumāragupta I. *Journal of the Numismatic Society of India* 18: 178–186.

Soltykoff, Prince Alexis. 1848. *Lettres sur l'Inde.* Paris.

Soltykoff, Prince Alexis. 1858. *Voyages dans l'Inde*, 3rd edition. Paris.

Somaratne G.A. 2018. The horn of rhinoceros: a text that speaks unorthodoxy. *Journal of Nanasamvara Centre for Buddhist Studies* 1: 235–262.

Sotheby's. 1977. [Auction catalogue] *Indian miniatures.* London: Sotheby's, 7 December 1977.

Sotheby's. 1978. [Auction catalogue] *Important oriental manuscripts and miniatures. The property of Hagop Kevorkian Fund.* London: Sotheby's, 3 April 1978.

Sotheby's. 1981. [Auction catalogue] *Catalogue of important old master paintings.* London: Sotheby's, 8 July 1981.

Sotheby's. 1983. [Auction catalogue] *Fine oriental miniatures and manuscripts including drawings from the Pan Asian Collection.* London: Sotheby's, 20 June 1983.

Sotheby's. 2007. [Auction catalogue] *Old master paintings.* London: Sotheby's, 5 December 2007.

Sotheby's. 2011. [Auction catalogue] *The Stuart Cary Welch Collection. Part Two. Arts of India.* London: Sotheby's, 31 May 2011.

Sotheby's. 2014. [Auction catalogue] *Art of imperial India.* London: Sotheby's, 8 October 2014.

Sotheby's. 2015. [Auction catalogue] *The Sven Gahlin collection.* London: Sotheby's, 6 October 2015.

Sotheby's New York. 1990. [Auction catalogue] *Indian, Himalayan and Southeast Asian Art.* New York: Sotheby's, 21 and 22 March 1990.

Sotheby's New York. 2007. Tashi Tsering Ma and her long-life sisters, polychrome rhinoceros horn, Bhutan. New York: Sotheby's, 23 March 2007, Lot 67.

Sotheby's New York. 2017. [Auction catalogue] *Fine books and manuscripts.* New York: Sotheby's, 11 December 2017.

Spectator 1879-10-11: Review of Pollok, Sport in British Burma.

Spens A.B. 2014. *A winter in India.* New Delhi.

Sphere, an illustrated newspaper for the home 48 (1912-02-17), p. 200: After the Durbar: cleaning off the paint.

Sphere, an illustrated newspaper for the home 88 (1922-01-14), p. 1: The Prince with the Maharajah in the Gurkha Land of Nepal.

Sphere, the Empire's illustrated weekly 158 (1939-07-22), p. 137: When an elephant does battle with a tigress.

Spillett J.J. 1966a. A report on wild life surveys in North India and Southern Nepal. *Journal of the Bombay Natural History Society* 63: 492–493.

Spillett J.J. 1966b. The Kaziranga Wild Life Sanctuary, Assam. *Journal of the Bombay Natural History Society* 63: 494–528.

Spillett J.J. 1966c. Brief summary of the status of the great Indian one-horned rhinoceros. *Journal of the Bombay Natural History Society* 63: 573–575.

Spillett J.J. 1966d. Laokhowa and other rhino areas in Assam. *Journal of the Bombay Natural History Society* 63: 529–534.

Spillett J.J. 1966e. The Jaldapara Wild Life Sanctuary, West Bengal. *Journal of the Bombay Natural History Society* 63: 534–544.

Spillett J.J. 2015. *Happenings: brief accounts about people and events that have affected my life.* New York.

Spillett J.J.; Tamang K.M. 1966. Wild life conservation in Nepal. *Journal of the Bombay Natural History Society* 63: 556–572.

Sports Illustrated 1956-02-27: Safari in Cooch Behar.

Srinivasulu C.; Srinivasulu B. 2012. *South Asian mammals: their diversity, distribution, and status.* Berlin.

Srivastava J.P.; Verma B.C. 1972. New fossil forms of Rhinocerotidae and Suidae from Pinjor Beds Lower Pleistocene near Chandigarh. *Journal of the Palaeontological Society of India* 15: 76–82.

Stables G. 1881. *Wild adventures in wild places.* London.

Stack E. 1908. *The Mikirs.* London.

Stacy L.R. 1836. Fossil remains in the Himalayan mountains. *Proceedings of the Ashmolean Society* 11 (1836-05-20): 9.

Staudinger, M. 1996. Naturstudien Kaiser Rudolfs II (1576–1612): Zur Kunstkammer auf der Prager Burg. Nr. 47/13 Panzernashorn. In: *Thesaurus Austriacus: Europas Glanz im Spiegel der Buchkunst, Handschriften und Kunstalben von 800 bis 1600.* Wien, pp. 261–268.

Stchoukine I. 1966. *La peinture turque d'après les manuscrits illustré, part 1: De Sulayman I à Osman II 1520–1622.* Paris.

Stead R.B.A. 1907. *Adventures on the High Mountains.* Philadelphia.

Stearn W.T. 1981. *The Natural History Museum at South Kensington.* London.

Stebbing E.P. 1912. Game sanctuaries and game protection in India. *Proceedings of the Zoological Society of London* 82: 23–55.

Stebbing E.P. 1920. *The diary of a sportsman naturalist in India.* London.

Stebbing E.P. 1926. *The forests of India*, vol. 3. London.

Stebbing E.P. 1930. H.J. Elwes, as naturalist, explorer and sportsman. *Nature* 126 (1930-08-02): 158–160.

Steinmann G.; Döderlein L. 1890. *Elemente der Palaeontologie.* Leipzig.

Sterndale R.A. 1884. *Natural history of the Mammalia of India and Ceylon.* Calcutta.

Sterndale R.A. 1886. *Denizens of the jungles.* Calcutta.

Stewart R. 1856. Notes on northern Cachar. *Journal of the Asiatic Society of Bengal* 24: 582–610.

Stiles C. 2017. *Memoirs of a Maharani: the true story of an American princess.* Portland (ebook).

Stimpson C.M.; Jukar A.M.; Bonea A.; Newell S.; Howlett E. 2022. A 'large and valuable' Siwalik fossil collection in the archives of the Oxford University Museum of Natural History. *Historical Biology* 2022: 1–14.

Stirn A.; Ham P van. 2003. *The hidden world of the Naga.* Munich.

Stockley C.H. 1928. *Big game shooting in the Indian Empire.* London.

Störk L. 1977. *Die Nashörner: Verbreitungs- und kulturgeschichtliche Materialien unter besonderer Berücksichtigung der Afrikanischen Arten und des Altägyptischen Kulturbereiches.* Hamburg.

Störk L. 1992. Selims rhinoceros. In: Gamer-Wallert I.; Helck W., *Gegengabe: Festschrift für Emma Brunner-Traut.* Tübingen, pp. 331–334.

Stonor C.R. 1946. A visit to the great Indian rhino. *Journal of the Society for the Preservation of the Fauna of the Empire* 53: 46–48.

Storrs W.T. 1877. Taljhari Mission church. *Church Missionary Gleaner* 4: 88–89.

Storrs W.T. 1895. The Santal mission. *Church Missionary Gleaner* 22: 22–23.

Stracey P.D. 1949. The vanishing rhinoceros and Assam's wild life sanctuaries. *Indian Forester* 75: 470–473 (also: *Journal of the Bengal Natural History Society* 25: 92–97).

Stracey P.D. 1951. The vanishing rhinoceros and Assam's wild life sanctuaries. *Journal of the Bengal Natural History Society* 25: 92–97.

Stracey P.D. 1957. On the status of the great Indian rhinoceros (R unicornis) in Nepal. *Journal of the Bombay Natural History Society* 54: 763–766.

Stracey P.D. 1963. *Wildlife in India: its conservation and control.* Delhi.

Strait C.U. 2014. *The Chin people: a selective history and anthropology of the Chin people.* New York.

Street P. 1963. *Vanishing animals: preserving nature's rarities.* New York.

Strien N.J. van. 1974. Dicerorhinus sumatrensis (Fischer), the Sumatran or two-horned rhinoceros: a study of literature. *Mededelingen Landbouwhogeschool Wageningen* 74 (16): 1–82.

Strien N.J. van. 1979. Neushoorns in Zuidoost-Azië. *Panda* (WNF) 15 (11): 149–152.

Strien N.J. van. 1985. *The Sumatran rhinoceros in the Gunung Leuser National Park; its distribution, ecology and conservation.* Ph.D. thesis presented to the Agricultural College at Wageningen.

Strien N.J. van. 1986. *The Sumatran rhinoceros, Dicerorhinus sumatrensis, in the Gunung Leuser National Park, Sumatra, Indonesia: its distribution, ecology and conservation.* Hamburg (Mammalia Depicta 12).

Strien N.J. van. 1997. Das Sumatra-Nashorn. In: [Gansloßer U.], *Die Nashörner: Begegnung mit urzeitlichen Kolossen.* Fürth, pp. 57–74.

Strien N.J. van. 2001. Conservation program for Sumatran and Javan rhino in Indonesia, Malaysia and Vietnam. In: Schwammer H.M. et al., *Recent research on elephants and rhinos: Abstracts of the International Elephant and Rhino Research Symposium, June 7–11, 2001.* Vienna, p. 50.

Strien N.J. van; Manullang B.; Sectionov; Isnan W.; Khan M.K.M.; Sumardja E.; Ellis S.; Han K.H.; Boeadi; Payne J.; Martin E.B. 2011. Dicerorhinus sumatrensis. In: IUCN 2011. *IUCN Red List of Threatened Species.* Version 2011.1. <www.iucnredlist.org>.

Strien N.J. van; Maskey T. 2006. Asian Rhino Specialist Group report. *Pachyderm*, Nairobi 40: 15–23.

Strien N.J. van; Rookmaaker L.C. 2010. The impact of the Krakatoa eruption in 1883 on the population of Rhinoceros sondaicus in Ujung Kulon, with details of rhino observations from 1857 to 1949. *Journal of Threatened Taxa* 2 (1): 633–638.

Strien N.J. van; Steinmetz R.; Manullang B.; Sectionov; Han K.H.; Isnan W.; Rookmaaker L.C.; Sumardja E.; Khan M.K.M.; Ellis S. 2011. Rhinoceros sondaicus. In: IUCN 2011. *IUCN Red List of Threatened Species.* Version 2011.1. <www.iucnredlist.org>.

Strien N.J. van; Talukdar B.K. 2007. Asian Rhino Specialist Group report. *Pachyderm*, Nairobi 42: 17–21.

Stuart Baker E.C. 1928. *Mishi the man-eater and other tales of big game.* London.

Stuckius J.W. 1577. *Arriani Maris Erythri periplus.* Genevae.

Subedi N.; Jnawali S.R.; Dhakala M.; Pradhan N.M.B.; Lamichhanea B.R.; Malla S.; Amin R.; Jhala Y.V. 2013. Population status, structure and distribution of the greater one-horned rhinoceros Rhinoceros unicornis in Nepal. *Oryx* 47: 352–360.

Subedi N.; Thapa R.K.; Kadariya R.; Thapa S.K.; Lamichhane B.R.; Yadav H.K.; Pokheral C.P.; Khanal N.; Yadav S. 2014. *Profiles of the greater one-horned rhinoceros (Rhinoceros unicornis) of Bardia National Park and Shuklaphanta Wildlife Reserve, Nepal.* Kathmandu.

Sugich M. 1992. *Palaces of India. A traveller's companion featuring the Palace Hotels.* London.

Suman D.N.S. 1994. Rhino reintroduction project in Dudhwa National Park. *Zoos' Print (Magazine of Zoo Outreach Organisation)* 9: 16.

Sumthane Y.Y.; Sharma K.R.; Bhutia K.G.; Dhiman B. 2017. A review on current status of one horned rhino's in subcontinent of India. In: *17th International Wildlife Law Conference 6–7 January 2017, Pune*, pp. 1–2.

Sun Gang; Jin Kun; Wang Zhentang. 1998. A spatio-temporal model of rhinoceros extinction in China. *Journal of Forestry Research* 9: 129–130.

Sundarbans Biodiversity Conservation Project. 2003. *Study on the current regional herbivore status and potential stock supply of extirpated herbivore species.* Dacca.

Sunday Star, Washington DC 1926-05-02, gravure section: Rhinoceros in Nepal.

Sunder D.H.E. 1895. *Survey and settlement of the Western Duars in the District of Jalpaiguri, 1889–1895.* Calcutta.

Surius L. 1566. *Commentarius brevis rerum in orbe gestarum.* Coloniae.

Sutton T. 1955. *The Daniells: artists and travellers.* London.

Swan C. 2021. *Rarities of these lands: art, trade, and diplomacy in the Dutch Republic.* Princeton.

Sword V.H. 1935. *Baptists in Assam: a century of missionary service 1836–1936.* New York.

Sydney Morning Herald 1854-05-12, p. 3: Destruction of a rhinoceros in the Soonderbunds.

Symons N.V.H. 1935. *The story of Government House.* Alipore.

Synteng. 1885. Among the Kukis. *Field, the Country Gentleman's Magazine* 66 (1885-11-14): 709.

Synteng. 1898. East Indian sporting ground. *Field, the Country Gentleman's Magazine* 92 (1898-12-10): 952–953.

T.A.D.: see Donogh T.A.

Tagare G.V. 1978. *The Bhāgavata Purāṇa.* Translated and annotated, part IV. Delhi.

Tait W. 1857. The ex-king of Oude and his chroniclers. *Tait's Edinburgh Magazine* 24: 162–163.

Talbot. 1872. Hunting at Barpetta, Assam. *Times of India* 1872-03-18: 3.

Talbot L.M. 1956a. The Indian rhino, a living fossil. *Field, the Country Gentleman's Magazine* 208 (1956-10-04): 576–577.

Talbot L.M. 1956b. An animal whose value is proving its death warrant: the great Indian rhinoceros. *Illustrated London News* 229 (1956-09-01): 346.

Talbot L.M. 1957. Stalking the great Indian Rhino. *National Geographic Magazine* 111: 389–398.

Talbot L.M. 1959. Marco Polo's unicorn. *Natural History*, New York 58: 558–565.

Talbot L.M. 1960. A look at the threatened species: a report of some animals in the Middle East and Southern Asia which are threatened with extermination. *Oryx* 5: 169–215.

Tallis J. 1851. *Illustrated atlas, and modern history of the world* [map 32]. London and New York.

Talukdar A.; Choudhury B. 2008. Rehabilitation of rhinos in Assam. In: Syangden B. et al., *Report on the regional meeting for India and Nepal IUCN/SSC Asian Rhino Species Group, March 5–7, 2007*, p. 12.

Talukdar B.K. 1995. Rhino poaching in Orang Wildlife Sanctuary, Assam (India). *Journal of Nature Conservation* 7: 1–6.

Talukdar B.K. 1999. Status of Rhinoceros unicornis in Pabitora Wildlife Sanctuary, Assam. *Tiger Paper* (FAO) 26 (1): 8–10.

Talukdar B.K. 2000. The current state of rhino in Assam and threats in the 21st century. *Pachyderm*, Nairobi 29: 39–47.

Talukdar B.K. 2006. Assam leads in conserving the greater one-horned rhinoceros in the new millennium. *Pachyderm*, Nairobi 41: 85–89.

Talukdar B.N. 2008a. Rhino census methodology. In: Syangden B. et al., *Report on the regional meeting for India and Nepal IUCN/SSC Asian Rhino Species Group, March 5–7, 2007*, p. 12.

Talukdar B.K. 2008b. Asian Rhino Specialist Group report. *Pachyderm*, Nairobi 44: 14–16.

Talukdar B.K. 2009a. Report on the current status of Rhinoceros unicornis. In: *International studbook for the greater one-horned or Indian rhinoceros, Rhinoceros unicornis, 31 December 2008*. Basel, pp. 5–14.

Talukdar B.K. 2009b. Asian Rhino Specialist Group report. *Pachyderm*, Nairobi 46: 14–17.

Talukdar B.K. 2010. Asian Rhino Specialist Group report. *Pachyderm*, Nairobi 48: 16–17.

Talukdar B.K. 2011. India: restoration by translocation. *The Horn* 2011 Spring: 4.

Talukdar B.K. 2012. Asian Rhino Specialist Group report. *Pachyderm*, Nairobi 51: 22–26.

Talukdar B.K. 2013. Asian Rhino Specialist Group report. *Pachyderm*, Nairobi 53: 25–27.

Talukdar B.K. 2015. Asian Rhino Specialist Group report. *Pachyderm*, Nairobi 56: 40–43.

Talukdar B.K. 2017. Asian Rhino Specialist Group report. *Pachyderm*, Nairobi 58: 36–39.

Talukdar B.K. 2018. Asian Rhino Specialist Group report. *Pachyderm*, Nairobi 59: 27–30.

Talukdar B.K. 2022. Asian Rhino Specialist Group report. *Pachyderm*, Nairobi 63: 33–37.

Talukdar B.K.; Barua M. 2006. *Wonders of Pabitora.* Guwahati.

Talukdar B.K.; Barua M.; Sarma P.K. 2007. Tracing straying routes of rhinoceros in Pabitora Wildlife Sanctuary, Assam. *Current Science* 92: 1303–1305.

Talukdar B.K.; Emslie R.; Bist S.S.; Choudhury A.; Ellis S.; Bonal B.S.; Malakar M.C.; Talukdar B.N.; Barua M. 2011. Rhinoceros unicornis. In: IUCN 2011. *IUCN Red List of Threatened Species.* Version 2011.1. <www.iucnredlist.org>.

Talukdar B.K.; Sarma P.K. 2007. *Indian rhinos in protected areas of Assam: a geo-spatial documentation of habitat changes and threats.* Guwahati.

Talukdar B.K.; Sectionov; Whetham L.B. 2010. *Proceeding of the Asian Rhino Specialist Group meeting held at Kaziranga National Park, India 10–12 February 2010.* Guwahati.

Talukdar B.K.; Sharma A.; Guleria H.; Gupta M. 2012. *Dudhwa's rhinos – a plan for their growth and secured future.* New Delhi: WWF-India (Technical report).

Talukdar B.K.; Sinha S.P 2013. Challenges and opportunities of transboundary rhino conservation in India and Nepal. *Pachyderm*, Nairobi 54: 45–51.

Talukdar N.R.; Singh B.; Choudhury P. 2018. Conservation status of some endangered mammals in Barak Valley, Northeast India. *Journal of Asia-Pacific Biodiversity* 11: 167–172.

Tandon P. 2006. A gold coin of the Pāla king Dharmapāla. *Numismatic Chronicle* no. 166: 327–333.

Tanghe P.F. 2017. *When rhinos are sacred: why some countries control poaching.* Dissertation presented to the University of Denver.

Tavernier J.B. 1676. *Les six voyages qu'il a fait en Turquie, en Perse, et aux Indes.* Seconde partie. Paris.

Tavernier J.B. 1889. *Travels in India.* Translated from the original French edition of 1676 by V. Ball, 2 vols. London.

Taylor E.G.R. 1932. *A brief summe of geographie by Roger Barlow.* London: Hakluyt Society, Second Series, vol. 69.

Taylor J.C. 1980. *Wildlife in India's tiger kingdom.* New York.

Taylor N. 2013. *Humans, animals, and society: an introduction to human-animal studies.* New York.

Tegetmeier W.B. 1872. Arrival of a Sumatran rhinoceros in the Zoological Gardens. *Zoologist, a popular miscellany of natural history* (ser.2) 7: 3057–3060.

Tennant W. 1803. *Indian recreations*, vol. 2. Edinburgh.

Terry E. 1625. Travels in India. In: Purchas S., *His pilgrimage, or relations of the world.* London, vol. 2, pp. 1464–1482.

Terry E. 1655. *A voyage to East India.* London.

Tewari R. 1987. Rhino-hunt scene of the Ghora Mangara Rock-Shelter, Mirzapur – a reappraisal. *Bulletin of Museums and Archaeology (Lucknow Museum)* 39: 25–29.

Tewari R. 1992. Rock paintings of Mirzapur in Uttar Pradesh. In: Lorblanchet M., *Rock art in the old world.* Darwin, pp. 285–301.

Thackeray E. 1916. *Reminiscences of the Indian mutiny (1857–58) and Afghanistan (1879).* London.

Thackray J.C. 1995. *A catalogue of portraits, paintings and sculpture at the Natural History Museum, London.* London.

Thapa K.; Nepal S.; Thapa G.; Bhatta S.R.; Wikramanayake E. 2013. Past, present and future conservation of the greater one-horned rhinoceros Rhinoceros unicornis in Nepal. *Oryx* 47: 345–351.

Thapa M. 2013. *A boy from Siklis: the life and times of Chandra Gurung.* New Delhi.

Thapa R. 2016. Poaching statistics of Rhinoceros unicornis in Chitwan National Park, Nepal: a review. *International Journal of Applied and Natural Sciences* 5: 29–34.

Thapa V. 2005. *Analysis of the one-horned rhinoceros (Rhinoceros unicornis) habitat in the Royal Chitwan National Park, Nepal.* Thesis presented to University of North Texas, Denton TX.

Thapa V.; Acevedo M.F.; Limbu K.P. 2014. An analysis of the habitat of the greater one-horned rhinoceros Rhinoceros unicornis at the Chitwan National Park, Nepal. *Journal of Threatened Taxa* 6: 6313–6325.

Thayer S. 1975. One sheet. *Bandwagon* 19 (3): 31.

Thayer S. 1998. America's second live rhinoceros. *Bandwagon* 42 (5): 30–31.

Theobald W. 1890. Notes on some of the symbols found on the punch-marked coins of Hindustan. *Journal of the Asiatic Society of Bengal* 1890 (3/4): 181–268.

Thevenot J. de 1684. *Voyages, contenant la relation de l'Indostan, des nouveaux Mogols, et des autres peuples et pays des Indes.* Paris.

Thevet A. 1558. *Les singularitez de la France Antarctique.* Paris.

Thevet A. 1575. *La cosmographie universelle.* Paris.

Thom W.S. 1943. A few notes about the five rhinoceros of the world. *Journal of the Bombay Natural History Society* 44: 257–274.

Thomas O. 1901. Notes on the type specimen of Rhinoceros lasiotis Sclater, with remarks on the generic position of the living species of rhinoceros. *Proceedings of the Zoological Society of London* 1901-06-04: 154–158.

Thomas O. 1906. Mammals. In: *The history of the collections contained in the natural history departments of the British Museum.* London, vol. 2, pp. 1–110.

Thomas O. 1911. The mammals of the tenth edition of Linnaeus; an attempt to fix the types of the genera and the exact bases and localities of the species. *Proceedings of the Zoological Society of London* 1911: 120–145.

Thomas N. 2004. Exploring the boundaries of biography: the family and friendship networks of Lady Curzon, Vicereine of India 1898–1905. *Journal of Historical Geography* 30: 496–519.

Thomas P.K.; Joglekar P.P.; Mishra V.D.; Pandey J.N.; Pal J.N. 1995. A preliminary report of the faunal remains from Damdama. *Man and Environment* 20: 29–36.

Thomas P.K.; Joglekar P.P.; Matsushima Y.; Pawankar S.J.; Deshpande A. 1997. Subsistence based on animals in the Harappan culture of Gujarat, India. *Anthropozoologica* 25–26: 767–776.

Thompson E. 1932. *A letter from India.* London.

Thornhill M. 1899. *Haunts and hobbies of an Indian official.* London.

Thornton E. 1854. *A gazetteer of the territories under the government of the East-India company, and of the native states on the continent of India*, vol. 3. London.

Thornton J.H. 1895. *Memories of seven campaigns*. Westminster.

Tiger. 1832. Tiger shooting in the Goruckpore terai. *Calcutta Magazine* no. 31: 373–392.

Tikader B.K. 1983. *Threatened animals of India*. Calcutta.

Tillotson G.H.R. 2006. *Jaipur Nama: tales from the Pink City*. London.

Times, London 1874-10-27, p. 8: Indian contrasts.

Times, London 1961-03-03, p. 24: The Queen at a rhinoceros hunt [in Nepal].

Times of India 1871-01-13: H.E. the Governor on tour [in Baroda].

Times of India 1871-12-12: The rhinoceros hunt at Madras.

Times of India 1872-07-12: The Museum of Allahabad.

Times of India 1875-06-14: Sports at Baroda.

Times of India 1876-09-09: Rhinoceros captured in Sundarbans.

Times of India 1876-10-31: The rhinoceros (Rhinoceros indias and Rhinoceros sondaicus).

Times of India 1877-02-09: Gift of a rhinoceros to the Zoological Garden (Calcutta).

Times of India 1877-04-12: Arrival of a rhinoceros at the Zoological Garden (Calcutta).

Times of India 1881-03-21: The zoological gardens [in Calcutta].

Times of India 1881-04-02: The Durbhungah shooting party.

Times of India 1881-04-16: Death of the rhinoceros at the Calcutta Zoo.

Times of India 1881-12-29: The Baroda festivities, 27 December 1881.

Times of India 1883-02-06: Frederick Oliver Robinson, 4th Earl de Grey shooting in Nepal.

Times of India 1883-02-19: Frederick Oliver Robinson, 4th Earl de Grey shooting in Nepal.

Times of India 1883-03-26: Sir Henry Meysey-Thompson shooting a rhinoceros in Lower Assam.

Times of India 1883-04-20: Shikar of Maharajah of Dhurbanga.

Times of India 1884-02-28: Earl of Mayo on the Jaldhaka River.

Times of India 1884-03-20: Earl of Mayo at Nagrakata.

Times of India 1885-03-09: General Roberts shooting in Cooch Behar.

Times of India 1885-03-17: Curious accident in the harbour.

Times of India 1887-06-21: The Jungle display of Rowland Ward in London.

Times of India 1900-03-06: The Viceroy's movements.

Times of India 1900-03-09: Viceroy Lord Curzon visit to Assam.

Times of India 1900-03-10: Viceroy Lord Curzon visit to Assam.

Times of India 1900-03-13: Viceroy Lord Curzon welcome in Tezpur.

Times of India 1900-03-13: The Viceroy's tour: visit to Tezpur.

Times of India 1900-03-13: Kuch Behar's big shoot.

Times of India 1900-03-15: Viceroy Lord Curzon visits Balipara.

Times of India 1900-03-21: Maharajah of Kuch Behar: successful shooting party.

Times of India 1901-04-18: The Viceroy's shooting party.

Times of India 1904-03-21: Sport in Nepal.

Times of India 1909-02-17: Viceroy's shooting trip.

Times of India 1909-05-04: A record rhino [shot by T. Briscome in Assam].

Times of India. 1922. *The Prince in India 1921–1922*. Souvenir number. Delhi.

Times of India 1938-12-14: Rhinoceros hunt in Nepal: last day of Viceroy's visit.

Times of India 1955-07-03: Preservation of fauna: wild life board set up (rhinoceros in Mishmi Hills).

Times of India 2012-08-04 (online): Poachers kill only rhino in 'forest man's' woods.

Timler A.K. 1888. Colonel Timler's diary in India. *Times of India* 1888-09-24: 6 and 1888-10-04: 5.

Tirkey A. 2005 When Monomoha Gorh smiles. In: Agarwalla R.P. et al., *Kaziranga centenary 1905–2005*. Guwahati, pp. 78–82.

Tiwari S.K. 1999. *Animal kingdom of the world*, vol. 1. New Delhi.

Tiwari S.K. 2000. *Riddles of Indian rockshelter paintings*. New Delhi.

Tiwary S.K. 2014. Newly discovered rock art heritage from Bhagwanpur Block of Kaimur District, Bihar. *Heritage: Journal of Multidisciplinary Studies in Archaeology* 2: 810–829.

TMWTEG. 1874. Opening of Rajpootana railway. *Times of India* 1874-09-21.

Tod J. 1920. *Annals and antiquities of Rajasthan or the Central and Western Rajput States of India*. Edited with an introduction and notes by William Crooke, C.I.E., vols. 1–3. London.

Tomba M. della. 1878. *Gli scritti del Padre Marco della Tomba: missionario nelle Indie Orientali*. Edited by Angelo de Gubernatis. Firenze.

Topsfield A. 1980. *Paintings from Rajasthan in the National Gallery of Victoria. A collection acquired through the Felton Bequests' Committee*. Melbourne.

Topsfield A. 1990. *The City Palace Museum Udaipur. Paintings of Mewar Court Life*. Ahmedabad.

Topsfield A. 2001. *Court painting at Udaipur. Art under the patronage of the Maharanas of Mewar*. Zürich.

Topsfield A. 2008. *Paintings from Mughal India*. Oxford.

Topsfield A.; Beach M.C. 1991. *Indian paintings and drawings from the collection of Howard Hodgkin*. New York, London.

Torri M. 2012. L'India e gli indiani nell'opera di Emilio Salgari. In: Mastrodonato P.I.G.; Dionisi M.G., *Riletture salgariane*. Pesaro, pp. 31–66.

Townsend G. 1890. *Friend and foe from field and forest: a natural history of the Mammalia*. Chicago.

Townsend M. 1855. The age of conquest, is it past. Review of The Private life of an Eastern King. *Calcutta Review* 25: 117–137.

Trafford F. 1911. *The Bengal forest manual*, 2nd edition. Calcutta.

Trense W. 1989. *The big game of the world*. Hamburg.

Trevor-Battye A. 1916. The home of the Gurkha. *Field, the Country Gentleman's Magazine* 128 (1916-11-04): 683.

Tripathi A.K. 2002. *Ecology and behaviour of rehabilitated rhino Rhinoceros unicornis in Kakraha Block of Dudhwa tiger reserve*. Thesis presented to the University of Kanpur.

Tripathi A.K. 2013. Social and reproductive behaviour of great Indian one-horned rhino, Rhinoceros unicornis in Dudhwa National Park, U.P., India. *International Journal of Pharmacy and Life Sciences* 4: 3116–3121.

Tripathi A.K. 2018. Status of Indian one-horned rhino, Rhinoceros unicornis Linn. 1858 in Dudhwa National Park, Uttar Pradesh, India. IJRDO *Journal of Biological Science* 4 (10): 1–4.

Trouessart E.L. 1898. *Catalogus mammalium tam viventium quam fossilium*, nova editio, vol. 2. Berolini.

Tshering U.; Katel O.; Nidup T. 2017. Determining ungulate distribution and habitat utilization in Royal Manas national park, Bhutan. *International Journal of Fauna and Biological Studies* 4: 91–96.

Tuan Yi-Fu. 2004. *Dominance and affection: the making of pets*. New Haven.

Tun Yin U. 1956. Rhinoceros in the Kachin State. *Journal of the Bombay Natural History Society* 53: 692–694.

Tun Yin U. 1967. *Wild animals of Burma*. Rangoon.

Turner S. 1800. *An account of an embassy to the court of the Teshoo Lama, in Tibet, containing a narrative of a journey through Bootan, and part of Tibet*. London.

Turvey S.T.; Sathe V.; Crees J.J.; Jukar A.M.; Chakraborty P.; Lister A.M. 2020. Late Quaternary megafaunal extinctions in India: how much do we know? *Quaternary Science Reviews* 252 (106740): 1–14.

Tyndale H.E. 1946. Rhino hunting with a camera in the Jaldapara Sanctuary, Bengal. *Journal of the Bengal Natural History Society* 21: 50–53.

Tyndale H.E. 1947. The great Indian rhinoceros photographed in its native jungle (Jaldapara, Bengal). *Illustrated London News* 210 (1947-03-22): 300–301.

Uddin M.S.; Rezowana S. 2012. Animal (mammals) representation in Somapura Mahavihara in situ terracotta plaques. *Journal of Bengal Art* 17: 189–210.

Uddin M.S.; Rezowana S. 2015. Terracotta ornamentation. In: Ahmed B., *Buddhist heritage of Bangladesh*. Dhaka, pp. 124–141.

Ullrich U.; Ullrich W. 1962. *Im Dschungel der Panzernashörner*. Radebeul (3rd edition, 1968).

Ullrich W. 1964. Zur Biologie der Panzernashörner (Rhinoceros unicornis) in Assam. *Zoologische Garten* 28: 225–249.

Ullrich W. 1965a. Neue Feststellungen über den Schutz des Panzernashorns (Rhinoceros unicornis) in Bengalen und Assam. *Zoologische Garten* 31: 101–103.

Ullrich W. 1965b. Die Einhörner von Kaziranga. *Freunde des Kölner Zoo* 8 (3): 98–100.

Ullrich W. 1967. Nashornstrassen in Assam. In: Hediger H., *Die Strassen der Tiere*. Braunschweig, pp. 56–67.

Ullrich W. 1970. Die Bedeutung der Gras- und Waldbrände für die Ökologie des Kaziranga-Reservates in Assam. *Zoologische Garten* 38: 97–107.

Ullrich W. 1971. *Kaziranga: Tierparadies am Brahmaputra*. Neudamm.

Unbegaun B.O. 1956. Wie hieß das Rhinoceros im Altrussischen? *Veröffentlichungen der Abteilung für Slavische Sprachen und Literaturen des Osteuropa-Instituts Slavisches Seminar an der Freien Universität (Festschrift für Max Vosmer zum 70. Geburtstag)* 9: 546–551.

Unesco. 1984. *Royal Chitwan National Park: nomination to the World Heritage list*. Report.

Urbain P.J.A.; Daniell W. 1840. *L'Inde pittoresque: Calcutta. Vingt-deux gravures d'après les dessins originaux de Daniell*. Paris.

Vaidya S. 1967. *Ahead lies the jungle*. Bombay.

Valentia, Viscount G.A. 1809. *Voyages and travels to India, Ceylon, the Red Sea, Abyssinia, and Egypt*, vol. 1. London.

Valentijn F. 1726. *Oud en Nieuw Oostindiën*, vol. 5 part 2: *Beschrijving van 't Nederlandsch Comptoir op de kust van Malabar*. Dordrecht, Amsterdam.

Valentim Fernandes de Moravia. 1515. Letter to a correspondent in Germany. [Original manuscript in Italian]. Biblioteca Nazionale Centrale di Firenze, Banco Rari 233: Miscellanea di viaggi assemblata da Alessandro Zorzi, folio CXXb–CXXVIIIa.

Vallard N. 1547. Portalan atlas (or Vallard atlas). Manuscript containing 15 charts preserved in The Huntington Library, San Marino, CA, HM27. Available online at https://catalog.huntington.org/record=b1842785.

Vambéry A. 1899. *The travels and adventures of the Turkish Admiral Sidi Ali Reis in India, Afghanistan, Central Asia, and Persia, during the years 1553–1556*. London.

Vanderbyl P.B. 1901. Photograph of rhinoceros shot in Assam. *Badminton Magazine of Sports and Pastimes* 12: 464.

Vanderbyl P.B. 1915. The Indian Empire. In: *The gun at home and abroad: The big game of Asia and North America*. London, pp. 71–125.

Vannini F. 1977. *Christian settlements in Nepal during the eighteenth century*. New Delhi.

Varmā, badrī nārāyaṇ. 1989. *koṭā bhitti citraṅkan paramparā (hāḍotī bhitti citra kalā kī paṣṭhabhūmi). naī dillī: rādhā pablikeśans* [The tradition of wall-painting in Kota, in Hindi].

Vashishth S.P. 2002. Orang National Park, from losers to winners. *Sentinel*, Darrang, Assam 2002: 23–29.

Vats M.S. 1940. *Excavations at Harappa: being an account of archaeological excavations at Harappā carried out between the years 1920–21 and 1933–34*, 2 vols. Delhi.

Vaughan P. 1997. *The Victoria memorial hall, Calcutta*. Calcutta.

VCPH. 1935. Captain Thomas Williamson, author of Oriental Field Sports. *Bengal Past and Present* 50 (1): 47–48.

Velmurugan M. 2017. Historical development of wildlife protection in India. *International Journal of Current Research and Modern Education* 2: 386–390.

Vendries C. 2016. Les Romains et l'image du rhinocéros: les limites de la ressemblance. *Archeologia Classica* 67: 279–340.

Verhey I. 1992. *Op reis met Clara: de geschiedenis van een bezienswaardige neushoorn*. Rotterdam.

Veritas. 1856. Record of bags for four years. *India Sporting Review* (n.s.) 1 (3): 302–304.

Vernay A.S. 1923a. Big game shooting with an object: the Vernay-Faunthorpe expedition to India. *Spur* 1923-11-15: 29–32, 78–79.

Vernay A.S. 1923b. The Vernay-Faunthorpe film. *Times*, London 1923-11-03.

Vernay A.S. 1923c. An Indian big game film. *Field, the Country Gentleman's Magazine* 142 (1923-11-08): 667.

Vernay A.S. 1931. John Champion Faunthorpe: sportsman, civil servant, soldier, conservationist, and friend. *Natural History*, New York 31 (1): 75–80.

Verney E.H. 1862. *The Shannon's brigade in India*. London.

Veselovsky Z. 1968. Beobachtungen aus dem Kaziranga Naturschutzgebiet. *Freunde des Kölner Zoo* 11: 77–81.

Vevers G.M. 1947. Recent additions to the zoo. *Zoo Life* 2 (3): 86–87.

Vignau S. de. 1992. "Soyez le bienvenu, Sirdar Rousselet Sahib Chamchar Bahadour!": Louis Rousselet & l'Inde 1863–1868. In: *L'Inde: photographies de Louis Rousselet 1865–1868*. Bordeaux, pp. 31–38.

Vigne L. 2018. Esmond Bradley Martin, 1941–2018. *Geographical Journal* 185: 125–126.

Vigne L. 2020. *The rhino horn and ivory trade: 1980–2020*. Ph.D. thesis presented to Oxford Brookes University.

Vigne L.; Martin E.B. 1991. Assam's rhinos face new poaching threat. *Oryx* 25: 215–221.

Vigne L.; Martin E.B. 1994. The greater one-horned rhino of Assam is threatened by poachers. *Pachyderm*, Nairobi 18: 28–43.

Vigne L.; Martin E.B. 1995. Nepal's rhino success story – lessons for Africa? *Swara* (East African Wild Life Society), Nairobi 18 (3): 15–17.

Vigne L.; Martin E.B. 1996a. Assam's Manas National Park re-opens. *International Zoo News* 43: 174–175.

Vigne L.; Martin E.B. 1996b. Assam State Zoo supplies rhinos to West Bengal. *International Zoo News* 43: 513–514.

Vigne L.; Martin E.B. 1998. Dedicated field staff continue to combat rhino poaching in Assam. *Pachyderm*, Nairobi 26: 25–39.

Vigne L.; Martin E.B. 2012. A rhino success story in West Bengal, India. *Oryx* 46: 327–328.

Visser P.C. 1932. Album of photographs taken in Nepal in 1932. National Archives of The Netherlands, 2.21.284 –

https://www.nationaalarchief.nl/onderzoeken/archief/2.21.284/invnr/37/file/NL-HaNA_2.21.284_37_001 [accessed April 2023].

Visser R.P.W. 1985. *The zoological work of Petrus Camper*. Amsterdam.

Vyas N.; Saran S.C. 1982. Exploration at Kishan Bilas and Kapildhara, district Kota. *Indian Archaeology: a Review* 1981–82: 56.

W. 1869. A week's shooting in Assam. *Oriental Sporting Magazine* (n.s.) 2 (October): 631–638.

W. 1879. A lucky shot at rhinoceros in the Soonderbunds. *Oriental Sporting Magazine* (n.s.) 12 (139, July): 150–153.

Waddell L.A. 1905. *Lhasa and its mysteries: with a record of the expedition of 1903–1904*. London.

Wade J.P. 1807. Geographical sketch of Assam, part 1. *Asiatic Annual Register* 5 (for 1805): 116–127.

Wagner J.C. 1686. *Interiora orientis detecta, oder grundrichtige und eigentliche Beschreibung aller heut zu Tag bekandten grossen und herrlichen Reiche des Orients*. Augsburg.

Wagner H. 1869. *De warande: landschappen en dierengroepen uit alle werelddeelen*. Utrecht.

Wakanker V.S. 1985. Bhimbetka: the stone tool industries and rock paintings. In: Misra V.N.; Bellwood P.S., *Recent advances in Indo-Pacific prehistory*. Leiden, pp. 175–176.

Waldau P. 2013. *Animal studies: an introduction*. Oxford.

Walker D. 1922. *The Prince in India: a record of the Indian tour of His Royal Highness the Prince of Wales – Nov. 1921 to March 1922*. Calcutta, Bombay, London.

Walker E.P. 1964. *Mammals of the world*, vol. 2. Baltimore.

Walker H. 1843. A catalogue of the Mammalia of Assam. *Calcutta Journal of Natural History* 3: 265–268.

Walker J. 1957. Oriental coins. *British Museum Quarterly* 21: 46–48.

Walker S. 2003. The great one-horned Rhinoceros (Rhinoceros unicornis) in captivity (zoos) in South Asia (India, Pakistan, Bangladesh and Nepal). *Zoos' Print* (*Magazine of Zoo Outreach Organisation*) 18 (12): 1–5.

Walker W. (Tom Cringle). 1865. *Jottings of an invalid in search of health*. Bombay.

Wallace F. 1929. The Orleans natural history trophies: a world-wide record of big game and an unrivalled work of British taxidermy. *Field, the Country Gentleman's Magazine* 1929-07-27: 133–134.

Waller R. 1971. The great Indian rhino: a struggle for survival. *Animals* 13: 830–833.

Waller R. 1972a. The great Indian one-horned rhinoceros (Rhinoceros unicornis) with special reference to its main sanctuary, Kaziranga, in Assam. *Cheetal* (*Journal of the Wild Life Preservation Society of India*) 15: 5–10.

Waller R. 1972b. Observations on the wildlife sanctuaries of India. *Journal of the Bombay Natural History Society* 69: 574–590.

Waller R. 1999. The great Indian one-horned rhinoceros (Rhinoceros unicornis) with special reference to its main sanctuary, Kaziranga, in Assam. *Cheetal* (*Journal of the Wild Life Preservation Society of India*) 38: 13–20.

Wanderer. 1871. A trip through the Sikhim Himalaya to the Kongra Lama Pass. *Field, the Country Gentleman's Magazine* 37 (1871-01-14): 26–27.

Wanderer. 1873. Sport in the Bhootan Duars. *Oriental Sporting Magazine* (n.s.) 6 (November): 533–544.

Ward A.E. 1887. *The sportsman's guide to Kashmir and Ladak*, 3rd edition. Calcutta.

Ward F.K. 1921. *In farthest Burma*. Philadelphia.

Ward R. 1886. The Jungle trophy at the Colonies. *Pall Mall Gazette* 1886-06-16: 6.

Ward R. 1887. *The sportsman's handbook to practical collecting, preserving, and artistic setting-up of trophies and specimens*, 4th edition. London.

Ward R. 1888. Publication of The sportsman's handbook by Rowland Ward. *Field*, London 71 (1888-02-25).

Ward R. 1892. *Horn measurements and weights of the great game of the world: being a record for the use of sportsmen and naturalists*. London.

Ward R. 1896. *Records of big game containing an account of their distribution, descriptions of species, lengths and weights, measurements of horns and field notes for the use of sportsmen and naturalists*, 2nd edition. London.

Ward R. 1899. *Records of big game*, 3rd edition. London.

Ward R. 1903. *Records of big game*, 4th edition. London.

Ward R. 1907. *Records of big game*, 5th edition. London.

Ward R. 1910. *Records of big game*, 6th edition. London.

Ward R. 1913. *A naturalist's life study in the art of taxidermy*. London.

Ward R. 1986. *Records of big game*, centenary edition. London.

Ward R.; Dollman J.G.; Burlace J.B. 1922. *Records of big game*, 8th edition. London.

Ward R.; Dollman J.G.; Burlace J.B. 1928. *Records of big game*, 9th edition. London.

Ward R.; Dollman J.G.; Burlace J.B. 1935. *Records of big game, African and Asiatic sections*, 10th edition. London.

Ward R.; Lydekker R.; Burlace J.B. 1914. *Records of big game*, 7th edition. London.

Ward S.R. 1884. *A glimpse of Assam*. Calcutta.

Warren F.G.E. 1866. My journal during the Bhootan campaign, 1864–5. *Minutes of Proceedings of the Royal Artillery Institution* 5: 116–160.

Warren T.H. 1903. *Christian Victor: the story of a young soldier*. London.

Watson A. 1832. *Memoir of the late David Scott, agent to the Governor General, on the North-East frontier of Bengal, and Commissioner of Revenue and Circuit in Assam*. Calcutta.

Watt G. 1892. *A dictionary of the economic products of India*, vol. 6 part 1. London.

Watt S. 2017. *Tea, tennis, and turbulent times*. London.

Wayre P. 1968. The golden langur and the Manas sanctuaries. *Oryx* 9: 337–339.

Weeks E.L. 1896. *From the Black Sea through Persia and India*. New York.

Wegge P.; Shrestha A.K.; Moe S.R. 2006. Dry season diets of sympatric ungulates in lowland Nepal: competition and facilitation in alluvial tall grasslands. *Ecological Research* 21: 698–706.

Weil K. 2012. *Thinking animals: why animals studies now?* New York.

Weiler D. 2019. Pierre Médard Diard and Alfred Duvaucel: two French naturalists in the service of Sir Stamford Raffles (December 7, 1818–March 27, 1820). In: Low M.E.Y.; Pocklington K.; Jusoh W.F.A., *Voyageurs, explorateurs et scientifiques: the French and natural history in Singapore*. Singapore, pp. 4–45.

Weitbrecht J.J. 1844. *Protestant missions in Bengal illustrated*. London.

Welch S.C. 1983. Return to Kotah. In: Harper P.O.; Pittman H. (eds.), *Essays on Near Eastern art and archaeology in honour of Charles Kyrle Wilkinson*. New York, pp. 78–93.

Welch S.C. (ed.). 1997. *Gods, kings, and tigers: the art of Kotah*. New York.

Welch S.C. 1997a. Kotah's lively patrons and artists. In: Welch S.C. (ed.), *Gods, kings, and tigers: the art of Kotah*. New York, pp. 15–38.

Welch S.C. 2004. Fresh from mind, heart and hand. In: Welch S.C.; Masteller K. (eds.) *From mind, heart, and hand: Persian, Turkish, and Indian drawings from the Stuart Cary Welch Collection*. New Haven and London, pp. 1–14.

Welch S.C.; Beach M.C. 1965. *Gods, thrones and peacocks. Northern Indian painting from two traditions: fifteenth to nineteenth centuries*. New York.

Weller E. 1859. [Map] India: The Eastern Provinces. London: Day and Son.

Wenley R. 2021. Miss Clara and the celebrity rhinoceros in art, 1500–1800. In: Avery C.; Cowie H.; Shaw S.; Wenley R., *Miss Clara and the celebrity beast in art 1500–1860*. Birmingham, pp. 9–24.

Werner W. 1884. *Das Kaiserreich Ostindien und die angrenzenden Gebirgsländer. Nach den Reisen der Brüder Schlagintweit und anderer neuerer Forscher dargestellt*. Jena.

West Bengal Forest Dept. 1993. *Status report on conservation and management of Indian rhino population in West Bengal*. Presented to meeting of IUCN-SSC-ARSG at Jaldapara Wildlife Sanctuary, West Bengal, India. Calcutta, pp. 1–20.

West Bengal Forest Dept. 1994. N. Bengal wildlife Sanctuaries. *Zoos' Print* (*Magazine of Zoo Outreach Organisation*) 9: 13–15.

West Bengal Forest Dept. 1995. N. Bengal wildlife sanctuaries. In: Molur S. et al., *PHVA workshop: great Indian one-horned rhinoceros, Jaldapara, 1993*. Coimbatore, pp. 1–3.

West Bengal Forest Dept. 1999. *State report on West Bengal forest 1998–99*. Calcutta.

West Bengal Forest Dept. 2016. *Annual administrative report 2015–2016*. Calcutta.

West P. 2016. *Dispossession and the environment: rhetoric and inequality in Papua New Guinea*. New York.

Westaway J. 2018. Thinking like a mountain: the life and career of E.O. Shebbeare. *Alpine Journal* 122: 205–218.

Whalley S.R.L. 1981. Wildlife of Nepal: the Royal Chitwan National Park. *Britain-Nepal Society Journal* no. 5: 13–17.

Wheeler A. 1997. Zoological collections in the early British Museum: the Zoological Society's Museum. *Archives of Natural History* 24: 89–126.

Wheeler G. 1876. *India in 1875–76: the visit of the Prince of Wales*. London.

White J.J. 1909. Hunting ahead of Roosevelt in Africa. *Harper's Weekly* 1909-03-27: 16–17.

Whitehead J. 1992. *Thangliena: the life of T.H. Lewin*. Kiscadale.

Whyte-Melville G.J. 1901. *The gladiators: a tale of Rome and Judaea*. London.

Wicki J. 1970. Letter of Petrus Tavares, Praefectus Major Lusitanorum et Imperator Akbar to Padre Roderico Vicente, 1578. *Documenta Indica* 11: 428–429.

Wielandt K. 2002. Nepal: political turmoil spells trouble for rare Indian rhinos. *The Horn Newsletter (Save the Rhino International, London)* 2002 Autumn: 4.

Wilcox R. 1832. Memoir of a survey of Asam and the neighbouring countries, executed in 1825-6-7-8. *Asiatic Researches* 17: 314–469.

Wild A.E. 1900. *Progress report of forest administration in the lower provinces of Bengal 1899–1900*. Calcutta.

Wilke H.J. 2018. *Die Geschichte der Tierillustration in Deutschland 1850–1950*. Rangsdorf.

Wilkins W.J. 1882. *Hindu mythology, vedic and puranic*. Calcutta.

Willan R.S.M. 1965. Rhinos increase in Nepal. *Oryx* 8: 159–160.

Willcocks J. 1904. *From Kabul to Kumassi: twenty-four years of soldiering and sport*. London.

Willcocks J. 1925. *The romance of soldiering and sport*. London.

William of Sweden (Prince). 1915. *In the lands of the sun: notes and memories of a tour in the East*. London (original 1913).

Williamson T. 1807a. *Oriental field sports; being a complete, detailed, and accurate description of the wild sports of the East; and exhibiting, in a novel and interesting manner, the natural history of the elephant, the rhinoceros, the tiger, the leopard, the bear, the deer, the buffalo, the wolf, the wild hog, the jackall, the wild dog, the civet, and other undomesticated animals: as likewise the different species of feathered game, fishes, and serpents. The whole interspersed with a variety of original, authentic, and curious anecdotes, which render the work replete with information and amusement. The scenery gives a faithful representation of that picturesque country, together with the manners and customs of both the native and european inhabitants. The narrative is divided into forty heads, forming collectively a complete work but so arranged that each part is a detail of one of the forty coloured engravings with which the publication is embellished. The whole taken from the manuscript and designs of Captain Thomas Williamson, who served upwards of twenty years in Bengal; the drawings by Samuel Howett, made uniform in size, and engraved by the first artists, under the direction of Edward Orme*. London: Printed by William Bulmer and Co. Shakespeare printing office, for Edward Orme, Printseller to His Majesty, Engraver and Publisher, Bond Street, the corner of Brook-Street. [Plates and text watermarked 1804]. Oblong folio, (18 3/8 × 23 inches; 46.7 × 58.4 cm), pp. i–ii, 150 pp., plus list of plates.

Williamson T. 1807b. *Oriental field sports*, vol. 1. London: Edward Orme (4to edition).

Williamson T. 1808. *Oriental field sports*, vol. 1. London: Edward Orme and B. Crosby and Co. (4to, 8vo editions).

Williamson T. 1811. Description of the rhinoceros. *Select Reviews of Literature and Spirit of Foreign Magazines*, Philadelphia 6 (34): 287–291.

Williamson T. 1819a. *Oriental field sports*, 2nd edition. London: Edward Orme (folio).

Williamson T. 1819b. *Oriental field sports*, 2nd edition, vol. 1. London: H.R. Young (4to).

Williamson T. 1819c. *Oriental field sports*. London: Thomas McLean (folio).

Williamson T. 1828. *Oriental field sports*, 2nd edition (reprint). London: Edward Orme (folio).

Wilson A. 1876. A run through Kathiawar: Junaghar. *Blackwood's Edinburgh Magazine* 120 (730): 191–210.

Wilson G.D.A.F. 1920. Some of Sir Guy Fleetwood Wilson's trophies. *Sphere, an illustrated newspaper for the home* 80 (1920-03-06): 268.

Wilson G.D.A.F. 1921. *Letters to nobody, 1908–1913*. London.

Wilson O.E.; Pashkevich M.D.; Rookmaaker L.C.; Turner E.C. 2022. Image-based analyses from an online repository provide rich information on long-term changes in morphology and human perceptions of rhinos. *People and Nature* 4 (6): 1560–1574.

Wilson O.E.; Pashkevich M.D.; Turner E.C.; Rookmaaker L.C. 2022. The Rhino Resource Center: accessing and utilizing a unique digital database. *Pachyderm*, Nairobi 63: 190–197.

Windsor, Edward Duke of. 1922. *The Prince of Wales eastern book: a pictorial record of the voyages of HMS Renown 1921–1922*. London.

Woldt A. 1882. Das Elephantenhaus im Berliner Zoologischen Garten. *Gartenlaube* 30 (52): 860–862.

Wood H.S. 1910. The wealth of sport. *Sphere, an illustrated newspaper for the home* 42 no. 557 (1910-09-24): 281.

Wood H.S. 1930. Observations on Indian rhino and their shikar on foot. *Journal of the Darjeeling Natural History Society* 4: 59–69.

Wood H.S. 1934. *Shikar memories: a record of sport and observation in India and Burma*. London.

Wood H.S. 1936. *Glimpses of the wild: an observer's notes and anecdotes on the wild life of Assam*. London.

Wood H.S. 1950. *Milestones of memory: a plain tale of service*. London.

Wood-Mason J. 1875. Meeting of 1st December 1875. *Proceedings of the Asiatic Society of Bengal* 1875: 229–230.

Woodsman. 1836. Rhinoceros hunting. *Bengal Sporting Magazine* (n.s.) 7 (27): 139–141.

Woodthorpe R.G. 1873. *The Lushai expedition 1871–1872*. London.

Woodthorpe R.G. 1890. The Lushai country. *Journal of the United Service Institution of India* 19: 14–48.

Woodyatt N. 1922a. *My sporting memories: forty years with note-book and gun*. London.

Woodyatt N. 1922b. Indian sport. *Illustrated Sporting and Dramatic News* 1922-10-28: 330.

Wright J.C. 2001. The Gandhari Prakrit version of the Rhinoceros Sūtra. *Anusamdhan* 18: 1–15.

Wroughton R.C. 1920. Summary of the results from the Indian Mammal Survey of the Bombay Natural History Society. *Journal of the Bombay Natural History Society* 27: 301–322.

Wüst E. 1922. Beiträge zur Kenntnis der diluvianen Nashörner Europas. *Centralblatt für Mineralogie, Geologie und Paläontologie* 20–21: 641–656, 680–688.

Wunderlich L. 1892. Der Hornwechsel beim indischen Nashorn. *Zoologische Garten* 33: 373–374.

WWF – World Wildlife Fund. 2002. *Securing a future for Asia's wild rhinos and elephants: WWF's Asian rhino and elephant action strategy*. Gland.

Yadav V.K. 2000. Male-male aggression in Rhinoceros unicornis – case study from North Bengal, India. *Indian Forester* 126: 1030–1034.

Yadava M.K. 2014. *Kaziranga National Park: detailed report on issues and possible solutions for long term protection of the greater one-horned rhinoceros in Kaziranga National Park*. Guwahati.

Yadava M.K. 2022. In Assam, a horn has value only on a living rhino. *Pachyderm*, Nairobi 63: 201–204.

Y.D. 1933. Shooting of a proclaimed rhino. *Field, the Country Gentleman's Magazine* 161 (1933-04-08): 755.

Ylla (Camilla Koffler). 1958. *Animals in India*. London.

Ylla (Camilla Koffler). 1955. Photographic items preserved in University of Arizona, Tucson AZ, Center for Creative Photography, archive number AG138.8.

Young Nimrod: see Rainey H.J.

Yule A.F. 1903. *A memoir of Colonel Sir Henry Yule with a bibliography of his writings*. London.

Yule H. 1863. *The wonder of the East, by Friar Jordanus (circa 1330)*. London: Hakluyt Society, vol. 31.

Yule H.; Maclagan R. 1882. *Memoir of General Sir William Erskine Baker*. London.

Yule J.; Burnell A.C. 1886. *Hobson-Jobson: a glossary of colloquial Anglo-Indian words and phrases, and of kindred terms, etymological, historical, geographical and discursive*. London.

Yule P. 1985. Metalwork of the Bronze Age in India. *Prähistorische Bronzefunde*, Abteilung XX Band 8. München.

Yule P. 1997. The copper hoards of Northern India. *Expedition* 39: 22–25.

Zecchini A. 1998. *Le rhinocéros: au nom de la corne*. Paris.

Zeuner F.E. 1952. The microlithic industry of Langhnaj, Gujarat. *Man* 52: 129–131.

Ziegler P. 1990. *King Edward VIII: the official biography*. London.

Zimmermann E.A.W. 1780. *Geographische Geschichte des Menschen und der allgemein verbreiteten vierfüssigen Thiere*, vol. 2. Leipzig.

Zittel K.A. 1891. *Palaeozoologie*, vol. 4. *Vertebrata* (*Mammalia*). München.

Zohra K. 2010. Rhino in Maraland. https://maralandwebs.wordpress.com/maraland [accessed April 2023].

Zschokke S. 2016. Genetic structure of the wild populations of the Indian rhinoceros (Rhinoceros unicornis). *Indian Journal of History of Science* 51: 380–389.

Zschokke S.; Armbruster G.F.J.; Ursenbacher S.; Baur B. 2011. Genetic differences between the two remaining wild populations of the endangered Indian rhinoceros (Rhinoceros unicornis). *Biological Conservation* 144: 2702–2709.

Zschokke S.; Baur B. 2002. Inbreeding, outbreeding, infant growth, and size dimorphism in captive Indian rhinoceros (Rhinoceros unicornis). *Canadian Journal of Zoology* 80: 2014–2023.

ZSL – Zoological Society of London. 1878. *Report of the Council of the Zoological Society of London, read at the Annual General Meeting, April 29, 1878*. London.

Index

This index combines all names, localities and subjects. It includes entries in figures (f.), maps (m.), tables (t.) and datasets (d.), unless the contents are repeated from the text. The states of India and the USA in the localities are abbreviated according to the usual conventions. Other abbreviations are found on p. liii.